PHYSICS
for
DEGREE STUDENTS

B.Sc. Third Year

As per UGC Model Curriculum (For all Indian Universities)

C.L. ARORA
M.Sc.
Formerly Principal
D.A.V. College, Jalandhar, *and*
Dean of Colleges
Guru Nanak Dev University, Amritsar

Dr. P.S. HEMNE
M.Sc. (Physics), Ph.D.
Principal and Head
Postgraduate Department of Physics
Nevjabai Hitkarini College, Bramhapuri
(RTM Nagpur University)

S. CHAND
PUBLISHING

S Chand And Company Limited
(ISO 9001 Certified Company)

S Chand And Company Limited
(ISO 9001 Certified Company)
Head Office: Block B-1, House No. D-1, Ground Floor, Mohan Co-operative Industrial Estate, New Delhi – 110 044 | Phone: 011-66672000
Registered Office: A-27, 2nd Floor, Mohan Co-operative Industrial Estate, New Delhi – 110 044
Phone: 011-49731800
www.**schandpublishing.com**; e-mail: **info@schandpublishing.com**

Branches

Chennai : Ph: 23632120; chennai@schandpublishing.com
Guwahati : Ph: 2738811, 2735640; guwahati@schandpublishing.com
Hyderabad : Ph: 40186018; hyderabad@schandpublishing.com
Jalandhar : Ph: 4645630; jalandhar@schandpublishing.com
Kolkata : Ph: 23357458, 23353914; kolkata@schandpublishing.com
Lucknow : Ph: 4003633; lucknow@schandpublishing.com
Mumbai : Ph: 25000297; mumbai@schandpublishing.com
Patna : Ph: 2260011; patna@schandpublishing.com

First Edition 2014
Reprints 2014, 2015, 2016, 2019, 2020 (Twice)
Reprint 2021

ISBN : 978-81-219-4287-4 **Product Code :** H6PDY68PHYS10ENAA14O

PRINTED IN INDIA

By Vikas Publishing House Private Limited, Plot 20/4, Site-IV, Industrial Area Sahibabad, Ghaziabad – 201 010 and Published by S Chand And Company Limited, A-27, 2nd Floor, Mohan Co-operative Industrial Estate, New Delhi – 110 044.

Preface

We feel great pleasure in presenting this book to the students of B. Sc. Third Year as per the revised syllabus of Physics framed by UGC. This Textbook is prepared keeping in view the syllabi framed by different Indian universities on the basis of Model UGC Curriculum. The present book has eight sections. Sections I, II, III, IV and V cover Course 7 which includes chapters on Relativity, Quantum Mechanics, Atomic Physics, Molecular Physics and Nuclear Physics. Sections VI, VII and VIII cover Course 8 which includes chapters on Solid State Physics, Solid State Devices and Electronics. The complete syllabus of B. Sc. Third Year is covered in 30 chapters. Some salient features of the book are:

- The material is presented in a comprehensive way using a simple language.
- The sequence of articles in each chapter enables the students to understand the gradual development of the subject.
- A large number of illustrations, pictures and interesting examples have been given to illustrate the basic principles involved so as to make reading interesting and understandable.
- Solved examples as well as numerical questions frequently asked in different universities are incorporated, including summer 2011 examination at the end of each chapter.
- Exercises with large number of questions and Multiple Choice Questions are given at the end of each chapter to facilitate the students to prepare for their examination as per the latest pattern of setting question papers in most of the Indian universities.
- And many more is supplemented even for a last student who feels Physics as a killer subject.

We believe that every student and teacher will appreciate the method of presentation of the subject in this book. Any correction of errors and suggestions for further improvement of this book will be highly appreciated.

C. L. ARORA
Dr. P. S. HEMNE

Acknowledgements

While preparing the manuscript, I have referred a large number of books of different authors. I record here my indebtedness to the authors of those original works from where some of the information is liberally drawn.

I (PSH) am greatly indebted to all those who constantly inspired me while writing this book. My special thanks are due for my wife Vaishali, daughter Dr. Vishakha, Ashvini, son Vishal-Vaibhav for their constant encouragement.

I am also thankful to the management and the editorial team of S. Chand & Company Pvt. Ltd. for all help and support in the publication of this book.

Dr. P. S. HEMNE

UGC MODEL CURRICULUM
PHYSICS, B.Sc. THIRD YEAR

COURSE 7: RELATIVITY QUANTUM MECHANICS, ATOMIC, MOLECULAR AND NUCLEAR PHYSICS

7.1 RELATIVITY — 15

Reference systems, inertial frames, Galilean invariance and conservation laws, propagation of light, Michelson-Morley experiment; search for ether. (5)

Postulates for the special theory of relativity, Lorentz transformations, length contraction, time dilation, velocity addition theorem, variation of mass with velocity, mass-energy equivalence, particle with a zero rest mass. (10)

7.2 QUANTUM MECHANICS — 30

Origin of the quantum theory : Failure of classical physics to explain the phenomena such as black-body spectrum, photoelectric effect, Ritz combination principle in spectra, stability of an atom. Plank's radiation law, Einstein's explanation of photoelectric effect, Bohr's quantization of angular momentum and its applications to hydrogen atom, limitations of Bohr's theory. (5)

Wave-particle duality and uncertainty principle: de Broglie's hypothesis for matter waves; the concept of wave and group velocities, evidence for diffraction and interference of 'particles', experimental demonstration of matter waves. (4)

Consequence of de Broglie's concepts; quantisation in hydrogen atom; energies of a particle in a box, wave packets, Heisenberg's uncertainty relation of p and x, its extension to energy and time. (3)

Consequence of the uncertainty relation: gamma ray microscope, diffraction at a slit, particle in a box, position of electron in a Bohr orbit. (3)

Quantum Mechanics: Schrodinger's equation. Postulatory basis of quantum mechanics; operators, expectation values, transition probabilities, applications to particle in a one- and three-dimensional boxes, harmonic oscillator, reflection at a step potential, transmission across a potential barrier. (7)

Hydrogen atom; natural occurrence of n, l and m quantum numbers, the related physical quantities, comparison with Bohr's theory. (4)

7.3 ATOMIC PHYSICS — 15

Spectra of hydrogen, deuteron and alkali atoms spectral terms, doublet fine structure, screening constants for alkali spectra for s, p, d, and f states, selection rules. (6)

Singlet and triplet fine structure in alkaline earth spectra, L-S and J-J couplings. (3)

Week spectra: continuous X-ray spectrum and its dependence on voltage, Duane and Hunt's law. Characteristics X-rays. Moseley's law, doublet structure of X-ray spectra, X-ray absorption spectra. (3)

7.4 MOLECULAR PHYSICS — 15

Discrete set of electronic energies of molecules, quantisation of vibrational and rotational energies, determination of internuclear distance, pure rotational and rotation vibration spectra. Dissociation limit for the ground and other electronic states, transition rules for pure vibration and electronic vibration spectra. (7)

Raman effect, Stokes and anti-Stokes lines, complimentary character of Raman and infrared spectra, experimental arrangements for Raman spectroscopy. (3)

Spectroscopic techniques: Sources of excitation, prism and grating spectrographs for visible, UV and IR, absorption spectroscopy, double beam instruments, different recording systems. (5)

7.5 NUCLEAR PHYSICS **15**

Interaction of charged particles and neutrons with matter, working of nuclear detectors, G-M counter, proportional counter and scintillation counter, cloud cambers spark chamber, emulsions. (7)

Structure of nuclei, basic properties (I μ, Q and binding energy), deuteron binding energy, p-p and n-p scattering and general concepts of nuclear forces. Beta decay, range of alpha particle. Geiger-Nuttal law. Gamow's explanation of beta decay, alpha decay and continuous and discrete spectra. (5)

Nuclear reactions, channels, compound nucleus, direct reaction (concepts). (3)

Shell model; liquid drop model, fission and fusion (concepts), energy production in stars by p-p and carbon cycles (concepts). (2)

COURSE 8: SOLID STATE PHYSICS, SOLID STATE DEVICES AND ELECTRONICS

8.1 SOLID STATE PHYSICS **30**

Overview : Crystalline and glassy forms, liquid crystals, glass transition. (2)

Structure : Crystal structure; periodicity, lattices and bases, fundamental translation vectors, unit cell, Wigner-Seitz cell, allowed rotations, lattice types, lattice planes, common crystal structures. (5)

Laue's theory of X-ray diffraction, Bragg's law, Laue patterns. (2)

Bonding : Potential between a pair of atoms; Lennard-Jones potential, concept of cohesive energy, covalent, Van der Walls, ionic and metallic crystals. (3)

Magnetism : Atomic magnetic moment, magnetic susceptibility, Dia- Para-, and Ferromagnetics, ferromagnetic domains, Hysteresis. (3)

Thermal properties: Lattice vibrations, simple harmonic oscillator, second order expansion of Lennard-Jones potential about the minimum, vibrations of one dimensional monatomic chain under harmonic and nearest neighbour interaction approximation, concept of phonons, density of modes (1-D). Debye model; lattice specific heat, low temperature limit, extension (conceptual to 3-D). (8)

Band structure: Electrons in periodic potential; nearly free electron model (qualitative), energy bands, energy gap, metals, insulators, semiconductors. (3)

Motion of electron: Free electrons, conduction electrons, electron collisions, mean free path, conductivity and Ohm's law. Density of states, Fermi energy Fermi velocity, Fermi-Dirac distribution. (4)

8.2 SOLID STATE DEVICES **15**

Semiconductors : Instrinsic semiconductors, electrons and holes, Fermi level.

Temperature dependence of electron and hole concentrations. Doping; impurity states, n and p type semiconductors, conductivity, mobility, Hall effect, Hall coefficient. (7)

Semiconductors devices: Metal-semiconductor junction, p-n junction, majority and minority carriers, diode, Zener and tunnel diodes, light emitting diode, transistor, solar cell. (8)

8.3 ELECTRONICS **45**

Power supply : Diode as a circuit element, load line concept, rectification, ripple factor, zener diode, voltage stabilization, IC voltage regulation, characteristics of a transistor in CB, CE and CC mode, graphical analysis of the CE configuration, low frequency equivalent circuits, h-parameters, bias stability, thermal runaway. (15)

Field effect transistors: JFET volt-ampere curves, biasing JEET, ac operation of JFET, source follower, Depletion and enhancement mode, MOSFET, biasing MOSFET, FET as variable voltage resister, digital MOSFET circuits. (15)

Small signal amplifiers: General principles of operation, classification, distortion, RC coupled amplifier, gain frequency response, input and output impedance, multistage amplifiers, transformer coupled amplifiers, Equivalent circuits at low, medium and high frequencies; emitter follower, low frequency common-source and common-drain amplifier, Noise in electronic circuits. (15)

Contents

SECTION I

RELATIVITY

1. REFERENCE SYSTEMS 3–28

1.1 Newton's notion of space 3
1.2 Frame of reference, event, observer 4
1.3 Inertial and non-inertial frames of reference 4
1.4 Newtonian principle of relativity 5
1.5 Translational invariance of space 5
1.6 Conservation of momentum from homogeneity of space 6
1.7 Rotational invariance of space 7
1.8 Conservation of angular momentum from rotational invariance of space 7
1.9 Homogeneity of time 8
1.10 Galelian transformations 10
1.11 Galelian transformation of position vector 12
1.12 Velocity is variant to Galelian transformations 13
1.13 Acceleration is invariant to Galelian transformations 13
1.14 Invariance of Newton's laws of motion in Galelian transformations 15
1.15 Inertial frame in uniform translatory motion 16
1.16 Invariance of conservation laws in Galelian transformations 16
1.17 Propagation of light 19
1.18 Galelian transformations not valid for velocity of light 20
1.19 Michelson-Morley experiment 20
1.20 Reason for oblique transverse path 24

2. SPECIAL THEORY OF RELATIVITY 29–88

Introduction 29
2.1 Postulates of special theory of relativity 29
2.2 Lorentz transformations 29
2.3 Velocity of light is absolute 35
2.4 Consequences of special theory of relativity 35
2.5 Length contraction 36
2.6 Time dilation 43
2.7 Experimental verification of time dilation 44
2.8 Simultaneity of events 45
2.9 Four dimensional volume element invariant under Lorentz transformations 47
2.10 Twin paradox 53
2.11 Relativistic addition of velocities 54
2.12 Constancy of speed of light 56
2.13 Relativistic Doppler effect 61
2.14 Relativity of mass 65
2.15 Einstein's mass- energy relationship 69

2.16 Energy-momentum relation 78
2.17 Momentum- kinetic energy relation 79
2.18 Minkowski space 81
2.19 Tachyons 82

SECTION II

QUANTUM MECHANICS

3. ORIGIN OF QUANTUM THEORY **91–131**
Introduction 91
3.1 Failure of classical mechanics 91
3.2 Black body radiation 92
3.3 Planck's quantum theory 94
3.4 Planck's radiation law 96
3.5 Photoelectric effect 98
3.6 Einstein's equation: Quantum theory of photoelectric effect 103
3.7 Stability of atom 108
3.8 Bohr's theory of hydrogen atom 109
3.9 Emission spectrum of hydrogen 112
3.10 Binding energy of electron 114
3.11 Merits of Bohr's theory 114
3.12 Limitations of Bohr's theory 114
3.13 Ritz combination principle 117
3.14 Compton effect 118
3.15 Old quantum theory and its limitations 126

4. QUANTUM THEORY: WAVE-PARTICLE DUALITY **132–163**
Introduction 132
4.1 de-Broglie hypothesis for matter waves 132
4.2 Wave Nature of microscopic particles 141
4.3 Wavelength of electron moving with relativistic velocity 141
4.4 de-Broglie phase velocity and particle velocity 143
4.5 Group velocity 145
4.6 Rest mass of photon 148
4.7 Evidences of matter waves: diffraction and interference of particles 149
4.8 Free and localized particle 155
4.9 Wave packet: amplitude and its velocity 156
4.10 Quantization in hydrogen atom (Bohr quantum condition) 159

5. HEISENBERG'S UNCERTAINTY PRINCIPLE **164–183**
5.1 Heisenberg's uncertainty principle 164
5.2 Time-energy uncertainty relation 165
5.3 Angular momentum- angular position uncertainty 166
5.4 Heisenberg's uncertainty principle from de-Broglie wave concept 166
5.5 Illustration of uncertainty principle 168

5.6 Significant for only microscopic particles 169
5.7 Heisenberg's uncertainty principle by diffraction of electrons 169
5.8 Applications of uncertainty principle 174
5.9 Minimum energy of harmonic oscillator using uncertainty principle 179
5.10 Superiority of an electron microscope 180
5.11 Complementarity principle 181
5.12 Difference between quantum mechanics and classical mechanics 181

6. SCHRÖDINGER'S THEORY OF QUANTUM MECHANICS 184–266

6.1 Wave function (Ψ) of a moving particle 184
6.2 Conditions for the wave function to be well behaved 184
6.3 Need for differential wave equation 185
6.4 Time-dependent Schrödinger's wave equation 185
6.5 Schrödinger's wave equation in terms of Hamiltonian operator 188
6.6 Eigen function and eigen value 188
6.7 Time independent Schrödinger's wave equation 189
6.8 Schrödinger's wave equation in momentum representation 193
6.9 Born's interpretation 195
6.10 Normalisation 196
6.11 Conservation of probability 197
6.12 Steady state 197
6.13 Probability density 197
6.14 Normalisation factor 198
6.15 Importance of normalisation 199
6.16 Superposition of states 201
6.17 Operator 202
6.18 Expectation value 204
6.19 Postulates of quantum mechanics 206
6.20 Schrödinger's equation from postulates of quantum mechanics 209
6.21 Free particle in one dimensional box 210
6.22 Energy levels 214
6.23 Probability density in one dimensional box 219
6.24 Particle with zero energy 222
6.25 Orthogonal wave functions 223
6.26 Parity 223
6.27 Parity of one dimensional harmonic oscillator wave function 224
6.28 Particle in three dimensional rectangular box 225
6.29 Potential step and potential barrier 232
6.30 One dimensional potential step (or barrier) 232
6.31 Transmission across potential barrier 242
6.32 Tunnel effect 247
6.33 Condition for perfect transmission 247
6.34 Harmonic oscillator 249
6.35 Wave function of a harmonic oscillator 253
6.36 Three dimensional harmonic oscillator 256

7. SCHRÖDINGER'S THEORY OF HYDROGEN ATOM — **267–304**

Introduction 267
7.1 Schrödinger's three dimensional time-independent wave equation in spherical polar co-ordinates 267
7.2 Schrödinger's wave equation for hydrogen atom 271
7.3 Solution of Θ (θ) for hydrogen atom 273
7.4 Azimuthal component 275
7.5 Orbital quantum number 276
7.6 Spherical harmonics 277
7.7 Principal quantum number (n) and radial wave function $R(r)$ 279
7.8 Complete wave function for the hydrogen atom 282
7.9 Significance of quantum numbers n, l and m 286
7.10 Degeneracy 288
7.11 Space quantization of $\vec{L}$ and precision 289
7.12 Probability densities of various states of H-atom 291
7.13 Frequency of radiation (quantum mechanical treatment) 294
7.14 Comparison of quantum model and Bohr model of hydrogen atom 300

SECTION III

ATOMIC PHYSICS

8. ATOMS WITH ONE ELECTRON — **307–357**

Introduction 307
8.1 Critical, ionization and excitation potentials 307
8.2 Frank-Hertz experiment 307
8.3 Quantisation of magnetic moment 310
8.4 Larmour precession (atomic dipole in a strong magnetic field $\vec{B}$) 311
8.5 Factors leading to discovery of spin 312
8.6 Spin angular momentum 313
8.7 Space quantisation 313
8.8 Gyromagnetic ratio 314
8.9 Spin-orbit coupling 316
8.10 Vector model of atom 319
8.11 Spin-orbit interaction 321
8.12 Fine structure of hydrogen atomic spectrum 324
8.13 Total magnetic moment and Lande's factor 329
8.14 Hyperfine structure of spectral lines 333
8.15 Relativistic Somerfield theory of hydrogen spectrum 334
8.16 Correction term in fine structure of Balmer lines of hydrogen atom 338
8.17 Stern-Gerlach experiment 339
8.18 Zeeman effect: normal and anomalous 343
8.19 Normal Zeeman effect 345
8.20 Normal Zeeman effect in H, He, Li, Mg, Na atoms 349
8.21 Anomalous Zeeman effect: quantum theory 349

9. ATOMS WITH MANY ELECTRONS 358–385

Introduction 358
9.1 Total wave function of two identical particles 358
9.2 Symmetric, anti-symmetric wave functions lead to Pauli's exclusion principle 360
9.3 Wave functions for bosons and fermions 362
9.4 Singlet and triplet states 362
9.5 Coupling schemes: LS and JJ couplings 366
9.6 Comparison of LS and JJ coupling 371
9.7 Distribution of electrons in shell and sub-shells 373
9.8 Electronic configuration of some elements with spins 375
9.9 Spectral notation representation 376
9.10 Spectra of alkalie atoms 379
9.11 Spectra of alkaline earth elements 381

10. X-RAYS 386–416

Introduction 386
10.1 Production of X-rays 386
10.2 Need of high atomic number material as target 388
10.3 Uses of X-rays 388
10.4 Origin and mechanism of X-rays (Duane-Hunt law) 388
10.5 Soft and hard X-rays 391
10.6 Characteristic X-ray spectrum 391
10.7 Photons of continuous and characteristic X-rays 394
10.8 Bremsstrahlung 395
10.9 Difference in origin of X-rays and visible light 395
10.10 Moseley's work 400
10.11 Moseley's law 400
10.12 Derivation of Moseley's law according to Bohr's theory 401
10.13 Doublet structure of X-ray spectra (fine structure) 407
10.14 Auger effect 409
10.15 Absorption of X-rays 410
10.16 Characteristic absorption spectrum 412

SECTION IV

MOLECULAR PHYSICS

11. VIBRATIONAL AND ROTATIONAL SPECTRA 419–439

11.1 Diatomic molecule 419
11.2 Molecular spectra 419
11.3 Kinds of molecular spectra 420
11.4 Rotational spectra 421
11.5 Vibrational spectra 427
11.6 Vibrational rotational spectra 431
11.7 Potential energy curve 433

11.8 Spectra of homonuclear diatomic molecule 434
11.9 Electronic spectra of di- and polyatomic molecules 434
11.10 Luminescence 435

12. RAMAN EFFECT 440–451
Introduction 440
12.1 Raman effect 440
12.2 Characteristics of Raman lines 441
12.3 Experimental arrangement 441
12.4 Explanation of Raman effect on quantum theory 442
12.5 Intensity of Raman lines 443
12.6 Raman effect– a two step process 443
12.7 Pure rotational Raman spectra 444
12.8 Raman shift 444
12.9 Complimentary character of Raman and infrared spectra 446
12.10 Importance of Raman effect 446
12.11 Application of Raman effect 446
12.12 Raman effect and fluorescence 448
12.13 Raman scattering and Rayleigh scattering 448
12.14 Recent techniques for Raman spectroscopy 449

13. SPECTROSCOPIC TECHNIQUES 452–458
Introduction 452
13.1 Types of spectroscopy 452
13.2 UV- visible spectrophotometry 453
13.3 IR spectrophotometry 453
13.4 Spectrophotometers 454
13.5 Single and double beam spectrophotometers 455

SECTION V

NUCLEAR PHYSICS

14. NUCLEAR DETECTORS 461–494
Introduction 461
14.1 Interaction of charged particles with matter 461
14.2 Neil-Bohr's formula 461
14.3 Limitations to Bethe-Bloch formula 464
14.4 Range and straggling of a charged particle 465
14.5 Energy loss of electrons and heavy ions during interaction 466
14.6 Passage of a fast electron through matter 466
14.7 Interactions of neutrons with matter 467
14.8 Nuclear radiation and detectors 469
14.9 Ionisation chamber 470
14.10 Uses of ionisation chamber 472
14.11 Specific ionisation 473

14.12 Proportional counter 473
14.13 Geiger-Mueller counter 476
14.14 Advantages and disadvantages of G.M. counter 480
14.15 Difference between ionisation chamber and G.M. counter 480
14.16 Comparison between proportional counter and G.M. counter 480
14.17 Scintillation counter 482
14.18 Radiation detectors based on image formation 486
14.19 Wilson cloud chamber 488
14.20 Bubble chamber 490
14.21 Spark chamber 491

15. STRUCTURE OF NUCLEI 495–551

15.1 Structure of the nucleus 495
15.2 Proton-electron model 495
15.3 Proton-neutron hypothesis 496
15.4 Properties of nuclei 498
15.5 Wave mechanical properties 502
15.6 Relation between mass number and radius of nucleus 503
15.7 Experimental determination of nuclear size 505
15.8 Magnetic moment of an atom 506
15.9 Nuclear magnetic dipole moment (μ) 507
15.10 Nuclear electric quadrupole moment (Q) 509
15.11 Atomic mass unit (a.m.u.) 512
15.12 Mass defect 512
15.13 Packing fraction 513
15.14 Binding energy 514
15.15 Difference between mass defect and packing fraction 514
15.16 Binding energy per nucleon or average binding energy 515
15.17 Binding energy and nuclear stability 515
15.18 Deutron binding energy 517
15.19 Non-existence of electrons in nucleus 518
15.20 Nuclear forces 521
15.21 Nuclear forces are exchange forces 523
15.22 Nuclear, electrostatic and gravitational forces 523
15.23 Nuclear stability 526
15.24 α, β and γ radiations 527
15.25 Emission of alpha (α) particles 528
15.26 Range of α-particles 529
15.27 Geiger-Nuttal law 530
15.28 Potential barrier of a nucleus 531
15.29 Tunnel effect 533
15.30 Gamow's theory of α- emission 534
15.31 Fine structure in α- ray spectrum 542
15.32 Long range α- particles 542
15.33 Radioactive emission of β-rays: β-decay 543

15.34 Difference between α-spectra and β-spectra 546
15.35 Evidence for existence of neutrinos 547
15.36 Significance of antineutrino in β-decay 548

16. NUCLEAR MODELS 552–565
Introduction 552
16.1 Liquid-drop model 552
16.2 Merits of liquid-drop model 555
16.3 Failures (limitations) of liquid-drop model 556
16.4 Shell model 556
16.5 Achievements (predictions) 559
16.6 Nuclear magic numbers 563

17. NUCLEAR REACTIONS 566–596
17.1 Nuclear reaction 566
17.2 Types of reactions 566
17.3 Laws governing nuclear reactions 568
17.4 *Q*-value of nuclear reaction 569
17.5 Exo-ergic and endo-ergic reactions 571
17.6 Nuclear reaction cross-section 574
17.7 Direct nuclear reaction 576
17.8 Nuclear reaction channels 577
17.9 Compound nucleus 577
17.10 Nuclear transmutation reactions 578
17.11 Neutron and its properties 581
17.12 Basic properties of a neutron 584
17.13 Role of neutron in nuclear reactions 584
17.14 Nuclear fission 586
17.15 Chain reaction 589
17.16 Nuclear reactors 590
17.17 Breeder reactor 592
17.18 Atomic energy 592
17.19 Nuclear fusion 593
17.20 Source of energy of sun and stars 594
17.21 Radioactive hazards 594
17.22 Radioactive traces 594
17.23 Distinction between fission and fusion 595
17.24 Carbon cycle 595

SECTION VI

SOLID STATE PHYSICS

18. CRYSTAL STRUCTURE 599–631
Introduction 599
18.1 Crystal 599
18.2 Some definitions 600

18.3 Crystal structure 602
18.4 Symmetry operation 603
18.5 Point groups and space groups 606
18.6 Bravais lattice 607
18.7 Characteristics of a unit cell 611
18.8 Coordination number 611
18.9 Atomic radius 612
18.10 Atomic packing fraction (or factor) 612
18.11 Number of lattice points 613
18.12 Hexagonal close packed structure 615
18.13 Some crystal structures 617
18.14 Miller indices 619
18.15 Some important planes 621
18.16 Spacing of planes in crystal lattices 624
18.17 Liquid crystals 627

19. X- RAY DIFFRACTION 632–679

Introduction 632
19.1 X-rays for crystal structure 632
19.2 Bragg's law 633
19.3 Verification of Bragg's law for NaCl crystal 635
19.4 Laue's equations 639
19.5 Experimental X-ray diffraction methods 642
19.6 Atomic scattering factor 645
19.7 Reciprocal lattice 647
19.8 Ewald's construction 655
19.9 Cohesive energy 656
19.10 Cohesive energy of a crystal 658
19.11 Bonding 659
19.12 Ionic crystals 659
19.13 Potential energy of ionic crystal 661
19.14 Lattice energy of ionic crystals 663
19.15 Madelung constant 664
19.16 Cohesive energy of ionic solids 666
19.17 Covalent bonding 668
19.18 Van der Waal bonding 669
19.19 Interaction in inert gas crystals 671
19.20 Lenard-Jones potential energy 672
19.21 Cohesive energy of an inert gas crystal 673
19.22 Hydrogen bonding 675
19.23 Metallic bonding 675

20. MAGNETIC PROPERTIES OF SOLIDS 680–717

Introduction 680
20.1 Intensity of magnetisation 680
20.2 Classification of magnetic materials 681

20.3 Repulsion of a diamagnetic material 681
20.4 Attraction of a paramagnetic material 682
20.5 Diamagnetic, paramagnetic and ferromagnetic substances 682
20.6 Atomic magnetic moment 685
20.7 Langevin's theory of diamagnetism 687
20.8 Spin angular momentum 690
20.9 Langevin's theory of paramagnetism 691
20.10 Ferromagnetism 694
20.11 Variation of $\vec{H}$ with P_m (Hysteresis loop) 696
20.12 Dissipation of energy due to hysteresis 697
20.13 Permanent magnets 700
20.14 Use of hysteresis curve 701
20.15 Magnetism in matter 702
20.16 Weiss theory of ferromagnetim 705
20.17 Anti-ferromagnetism 709
20.18 Ferrimagnetism 712
20.19 Ferrite 713
20.20 Difference between ferrites and ferromagnetic materials 714
20.21 Uses of ferrite material 714

21. THERMAL PROPERTIES OF SOLIDS (Lattice Vibrations) 718–748

21.1 Lattice vibrations 718
21.2 Phonon 718
21.3 Umklapp process 719
21.4 Inelastic scattering of photons by phonons 720
21.5 Comparison between photons and phonons 721
21.6 Lattice waves in a monoatomic linear lattice 722
21.7 Dispersion relation in terms of wave velocity 724
21.8 Vibrations of one dimensional linear diatomic lattice 726
21.9 Specific heat 732
21.10 Dulong and Pettit's law 733
21.11 Agreement of results with experiments 734
21.12 Einstein's theory of lattice heat capacity (specific heat) 734
21.13 Einstein temperature 737
21.14 Failure of Dulong and Pettit's law at low temperature 738
21.15 Debye model for lattice specific heat 738
21.16 Success of Debye model 743
21.17 Limitations (drawbacks or failures) of Debye model 744
21.18 Comparison with Einstein model 745
21.19 Debye temperature 745
21.20 Lattice heat capacity at low temperature 745

22. FREE ELECTRON THEORY OF METALS 749–769

Introduction 749
22.1 Free electron Fermi gas 749
22.2 Static and transport properties 750

22.3 Mean free path (λ) 750
22.4 Electrical conductivity 751
22.5 Thermal conductivity 752
22.6 Widemann Franz law 753
22.7 Sommerfeld theory of metals 754
22.8 Average kinetic energy at 0K 763
22.9 Fermi velocity 764
22.10 Significance of Fermi energy level 765
22.11 Fermi temperature 766
22.12 Fermi-Dirac statistics 766
22.13 Fermi-Dirac energy distribution 766
22.14 Failure of free electron theory 767

23. BAND THEORY OF SOLIDS 770–791

Introduction 770
23.1 Splitting of atomic energy levels into bands 770
23.2 Origin of band structure 771
23.3 Periodic potential in a crystal 774
23.4 Bloch theorem 774
23.5 Kronig Penney model 777
23.6 Origin of energy gap 784
23.7 Valence band, conduction band and forbidden band 784
23.8 Behaviour of conductors, insulators and semiconductors (on the basis of band theory) 785
23.9 Effect of impurity on conductivity of semiconductor 788
23.10 Resistivity range of good conductor and insulator 789

SECTION VII

SOLID STATE DEVICES

24. SEMICONDUCTORS 795–814

Introduction 795
24.1 Semiconducting materials 795
24.2 Intrinsic semiconductor 796
24.3 Recombination 797
24.4 Energy band structure of an intrinsic semiconductor 797
24.5 Electron and hole concentration 799
24.6 Fermi level in intrinsic semiconductor 801
24.7 Effect of temperature on Fermi level 802
24.8 Intrinsic density (n_i) 803
24.9 Intrinsic conductivity (σ) 804
24.10 Doping 805
24.11 Donar or *n*-type semiconductor 805
24.12 Acceptor or *p*-type semiconductor 807
24.13 Temperature dependence of carrier concentration in extrinsic semiconductor 808
24.14 Hall effect 809

24.15 Hall coefficient 811
24.16 Mobility 812
24.17 Extrinsic conductivity 812

25. SEMICONDUCTORS DEVICES 815–831

Introduction 815
25.1 pn junction diode 815
25.2 Depletion region and potential barrier 816
25.3 Drift of minority carriers 817
25.4 Energy band structure of pn junction at equilibrium 818
25.5 Forward biased junction 820
25.6 Reverse breakdown 822
25.7 Zener diode 823
25.8 Light emitting diode (LED) 823
25.9 Metal-semiconductor junction: Schottky diode 825
25.10 Tunnel diode 825
25.11 Solar cell 826
25.12 Transistor 828
25.13 Working of p-n-p transistor 829
25.14 Working of n-p-n transistor 830
25.15 Different techniques for fabrication of BJT 830

SECTION VIII

ELECTRONICS

26. POWER SUPPLY 835–854

Introduction 854
26.1 Diode as a circuit element 835
26.2 Load-line concept 836
26.3 Half-wave diode rectifier 837
26.4 Analysis of half-wave rectifier 837
26.5 Full-wave rectifier 840
26.6 Analysis of full-wave rectifier 841
26.7 Full-wave bridge rectifier 844
26.8 Comparison of a half-wave and full-wave rectifier 845
26.9 Filters 845
26.10 Electronic voltage regulators (stabilizers) 849
26.11 Zener diode as a voltage regulator 849
26.12 Voltage regulation using IC317 852

27. CHARACTERISTICS OF TRANSISTORS 855–880

Introduction 855
27.1 Basic transistor circuits 855
27.2 Characteristics in common base (CB) mode 855

27.3 Characteristics in common emitter (CE) mode 859
27.4 Common collector (CC) configuration 862
27.5 Relation between current gains α and β 862
27.6 Accurate collector current in CB and CE mode 864
27.7 Transistor as an amplifier 865
27.8 Common emitter amplifier 866
27.9 Comparison of CB, CE and CC configuration 868
27.10 Load line and Q point 868
27.11 Graphical analysis of CE voltage amplifier 869
27.12 A.C. load line 870
27.13 Hybrid or *h*-parameters 871
27.14 Low frequency *h*-parameter equivalent circuit 874
27.15 Faithful amplification 874
27.16 Stabilisation 876
27.17 Thermal runaway 877
27.18 Stability factor 877
27.19 Bias stabilising circuits 877

28. FIELD EFFECT TRANSISTORS 881–904

Introduction 881
28.1 Junction field effect transistor (JFET) 881
28.2 Static characteristics of JFET 883
28.3 Transfer characteristic 884
28.4 Parameters of JFET 885
28.5 D.C. biasing of JFET 886
28.6 JFET as an amplifier 887
28.7 Source-follower JFET: a common-drain circuit 889
28.8 Difference between JFET and BJT 890
28.9 Applications of JFET 891
28.10 Application as a voltage variable resistor 891
28.11 MOSFET 892
28.12 Types of MOSFET 893
28.13 Enhancement MOSFET 893
28.14 Depletion MOSFET 896
28.15 Special features of MOSFET 899
28.16 Digital MOSFET circuits 900

29. SMALL SIGNAL AMPLIFIERS 905–918

Introduction 905
29.1 Classification of amplifiers 905
29.2 Multistage amplifier 908
29.3 Cascaded amplifier 909
29.4 Coupling network 910

29.5 RC coupled amplifier 911
29.6 Band width (B.W.) 913
29.7 Advantages of RC coupling 913
29.8 Distortion in amplifier 914

30. HYBRID EQUIVALENT CIRCUITS AND NOISE 919–942
Introduction 919
30.1 Hybrid equivalent circuit of CE amplifier 919
30.2 Analysis of single-stage common emitter amplifier 920
30.3 A.C. equivalent circuit of two stage RC coupled amplifier 924
30.4 Mid frequency response of the amplifier 924
30.5 Low frequency response of the amplifier 926
30.6 High frequency range 929
30.7 Emitter follower 934
30.8 JFET as amplifier 937
30.9 Noise in electronic circuits 939

INDEX 943–946

SECTION I

RELATIVITY

1. REFERNECE SYSTEMS
2. SPECIAL THEORY OF RELATIVITY

THEORY OF RELATIVITY

Albert Einstein

Einstein's theory of special relativity includes electricity and magnetism in a simple logical extension of the relativity of Galileo and Newton. Its conclusions, including time dilation, length contraction, and $E = mc^2$ have changed profoundly our ideas of time and space, matter and energy. General theory of relativity is the *geometric theory of gravitation* published by **Albert Einstein** in 1916. It is the current description of gravitation in modern physics and has a number of physical consequences. General relativity generalises special relativity and Newton's law of universal gravitation, providing a unified description of gravity as a geometric property of space and time or space-time. In particular, the curvature of space-time is directly related to the four-momentum (mass-energy and linear momentum) of whatever matter and radiation are present. The relation is specified by the Einstein field equations, a system of partial differential equations.

Some predictions of general relativity differ significantly from those of classical physics especially concerning the passage of time, the geometry of space, the motion of bodies in free fall, and the propagation of light. Examples of such differences include gravitational time dilation, gravitational lensing, the gravitational red shift of light, and the gravitational time delay. General relativity's prodictions have been confirmed in all observations and experiments to date. Although general relativity is not the only relativistic theory of gravity it is the simplest theory that is consistent with experimental data. However, unanswered questions remain, the most fundamental being how general relativity can be reconciled with the laws of quantum physics to produce a complete and self-consistent theory of quantum gravity.

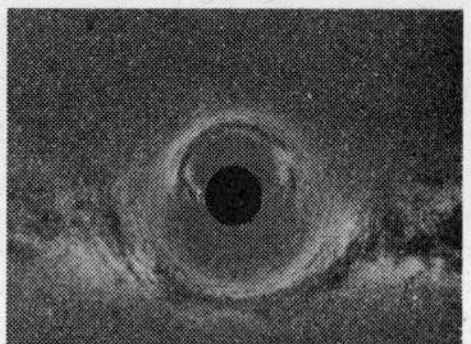

Black-hole

Einstein's theory has important astrophysical implicatons. For example, it implies the existence of **black holes**—regions of space in which space and time are distorted in such way that nothing, not even light, can escape—as an end-state for massive stars.

Whenever the ratio of an object's mass to its radius becomes sufficiently large, general relativity predicts the formation of a black hole. Usually a galaxy has one super massive black hole with a few million to a few billion solar masses in its center and it presence is throught to have played an important role in the formation of the the galaxy and larger cosmic structures.

There is ample evidence that such stellar black holes as well as more massive varieties of black hole are responsible for the intense radiation emitted by certain types of astronomical objects such as active galactic nuclei or micro-quasars. The bending of light by gravity can lead to the phenomenon of gravitatonal lensing, where multiple images of the same distant astronomical object are visible in the sky. General relativity also predicts the existence of gravitational waves, which have since been measured indirectly; a direct measurement is the aim of projects such as LIGO and NASA/ESA Laser Interferometer Space Antenna. In addition, general relativity is the basis of current cosmologicai models of a consistently expanding universe.

REFERENCE SYSTEMS

1 Chapter

1.1 NEWTON'S NOTION OF SPACE

According to Newton, '*Absolute space in its own nature without relation to anything external remains always similar and immovable.*' According to Newton, main classical properties of space are:

(*i*) *Space is three dimensional.* The concept of space and time is fundamental to the study of mechanics. All objects occupy space and have a length, a breadth and a height. Space is, therefore, three dimensional. This is why the position of a point can be specified completely by the three co-ordinates (x, y, z) or (r, θ, ϕ); depending upon the co-ordinate system.

(*ii*) *Space is flat. i.e.*, it possesses *Eucledian flatness*. This means that the shortest distance between any *two* points in space is a sraight line. If we take *three* points in space to form a triangle, the sum of angles is equal to π. If it is a right angled triangle, then the three sides are related by Pythogoras theorem i.e., Hypotenuse^2 = Base^2 + Altitude^2.

However, according to latest theory, *space is not exactly flat but somewhat curved*. The departure from flatness is very small and can be ignored in the study of classical mechanics.

(*iii*) *Space is homogeneous.* In other words, *space is everywhere* alike. Homogeneity of free space (a space in which fields and forces are absent) means *translational invariance* of its properties *i.e.*, the result of an experiment is not altered due to linear displacement of coordinate systems.

(*iv*) *Space is isotropic.* It means that if we consider a point O in space and we move from this point in any direction, the properties are the same *i.e.*, there is nothing to distinguish one direction from the other. In other words there is no preferred direction in space or one direction is as good as any other direction.

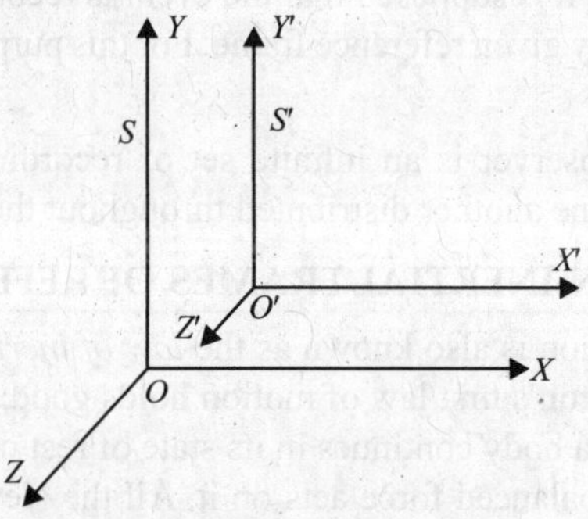

Fig. 1.1

Distinction between homogeneity and isotropy

The homogeneity of space means *translational invariance* of the properties of space.

If we consider two coordinate systems S and S' displaced to one another as shown in fig. 1.1, then an experiment performed in system S will give exactly the same results in the system S'.

On the other hand, isotropy mean *rotational invariance* of free space.

Thus, if we have two coordinate systems S and S' rotating with respect to one another as shown in Fig.1.2, then again an experiement performed in system S' will give exactly the same result in system S.

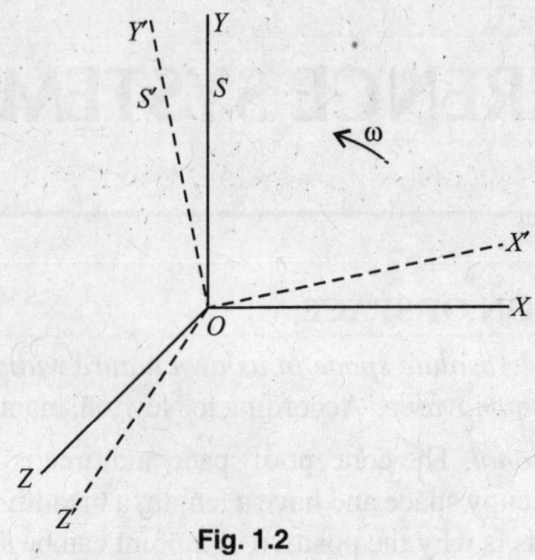

Fig. 1.2

1.2 FRAME OF REFERENCE, EVENT, OBSERVER

Frame of Reference. In order to describe the motion of a particle in space, a 'co-ordinate system' or a 'reference frame' is required. A frame of reference is a coordinate system (designated to measure three space coordinates) plus a clock (designated to register 'time-coordinate')

Event. An event is a physical concept. An event may be collision between two particles or the striking of building by a thunderbolt. An event happens at a point in space at a particular time and is thus assigned four coordinates—the three position coordinates (x,y,z) that measure the distance from the origin of the coordinate system where the observer is located and fourth is the time coordinate (t) that the observer records with his clock.

Observer. An observer is normally considered as a person or an instrument or robot which can make measurements. Such an observer occupies a particular position in a reference frame and records an event only when a light signal from the event reaches it.

But in theory of relativity, it is supposed that the event is recorded instantaneously at the place and time of its occurrence in any given reference frame. For this purpose the observer must be present at all places, and at all times.

Hence in relativity, an observer is an infinite set of recording clocks and measuring scales synchronised with respect to one another distributed throughout the space of the inertial frame.

1.3 INERTIAL AND NON-INERTIAL FRAMES OF REFERENCE

Newton's first law of motion is also known as the *law of inertia*. A reference frame *e.g.*, a co-ordinate system in which Newton's first law of motion holds good, is known as an *inertial frame of reference*. In an inertial frame a body continues in its state of rest or of uniform motion in a straight line as long as no external unbalanced force acts on it. All the Newton's law of motion hold good in an inertial frame. All frames of reference moving with a *constant velocity* with respect to an inertial frame are also inertial frames of reference. So inertial frame of reference can be called an unaccelerated frame of reference.

Example. The best approximation to an inertial frame is the frame of reference in the intergalactic space.

Non-inertial frame. The basic laws of Physics are not changed in form in inertial frames of reference. But when a frame of reference is accelerated relative to an inertial frame the form of basic physical laws such as Newton's second law of motion becomes completely different. Such frames of reference having an accelerated motion relative to an inertial frame are called *non-inertial frames of reference*. Since a uniformly rotating frame has a centripetal acceleration it is also a non-inertial frame.

Example. A non-inertial frame is either a frame having *uniform linear acceleration* or a frame which is *uniformly rotating.*

Earth a non-inertial frame. Earth is not an inertial frame. It is a non-inertial frame. The earth is a rotating sphere as it is rotating about its own axis. It has a centripetal acceleration due to rotation.

In addition, earth also revolves round the sun in a period of one year in an orbit of radius 1.49×10^8 km.

$\therefore$ Its orbital velocity,
$$v = \frac{2\pi r}{T} = \frac{2\pi \times 1.49 \times 10^8 \times 10^3}{365 \times 24 \times 60 \times 60} = 3 \times 10^4 \text{ ms}^{-1}$$

and centripetal acceleration of the earth due to revolution

$$= \frac{v^2}{r} = \frac{3 \times 10^4 \times 3 \times 10^4}{1.49 \times 10^8 \times 10^3} = 6 \times 10^{-3} \text{ ms}^{-2}$$

Thus, the reference frame attached to the earth is an *accelerated frame* and is therefore, non-inertial.

For a rotating particle, its velocity is continuously changing (being a vector quantity and acts along tangent to its radius vector in a circular orbit), it posses acceleration. Hence it is a non-inertial frame.

1.4 NEWTONIAN PRINCIPLE OF RELATIVITY

It states '*By performing physical experiments entirely is one inertial frame, we cannot determine its motion with respect to some other inertial frame and the phenomenon occurring in a closed system are independent of any unaccelerated motion of the system as a whole.*'

In other words, the laws of physics are the same in all inertial frames of reference.

As an example, consider an observer *A* in a train at rest and another observer *B* standing outside on the platform. The observer *A* throws a ball vertically upward and finds that the ball returns vertically downward to him. The same thing is observed by the observer *B* standing on the platform.

Now, suppose the train starts moving with a uniform velocity along a straight track. If the observer *A* repeats his experiment *i.e.*, throws the ball vertically upward he again finds that the ball falls vertically downward to him. But the observer *B* standing on the platform finds that the ball thrown upward by the observer *A* in the moving train follows a parabolic path. Thus we find that:

(*i*) The observer in the train cannot distinguish whether the train is moving or not because he observes exactly the same thing whether the train is at rest or in uniform motion in a straight line. Therefore, we conclude that:

'*It is not possible to detect the state of rest or of uniform rectilinear motion of a system by performing mechanical experiments within the system itself.*'

(*ii*) The vertically upward and vertically downward motion of the ball for observer *A* and the motion of the ball along the parabolic path for the observer *B* can be explained by the same laws of Mechanics *i.e.*, the laws of Mechanics remains unchanged due to uniform rectilinear motion of the frames of reference. Thus

'*All inertial frames are equivalent and the laws of Physics* (*nature*) *are the same in all inertial frames of reference.*'

1.5 TRANSLATIONAL INVARIANCE OF SPACE

Translational invariance of the properties of free space is due to the homogeneity of free space.

To understant it, consider two inertial frames of reference *S* and *S'* with their *X*-axes along the same line and *Y* and *Z* axes parallel to each other. The origin *O'* of *S'* is displaced by a distance *b* along the positive direction of X-axis. If x and x' are the co-ordinates of a point in the two systems respectively , then $x' = x - b$, as shown in Fig. 1.3.

Let there be two particle P_1 and P_2 lying on the X-axis of inertial systems, the x-co-ordinates of P_1 being x_1 and x_1' and those of P_2 being x_2 and x_2' in the systems S and S' respectively. Suppose the two particles have masses m_1 and m_2 respectively and exert gravitational forces on each other, then the force between two particles being only due to their mutual interaction is a conservative force and is given by $F_x = -\frac{\partial U}{\partial x}$ where U is the potential energy of the particle and x its co-ordinate on the X- axis.

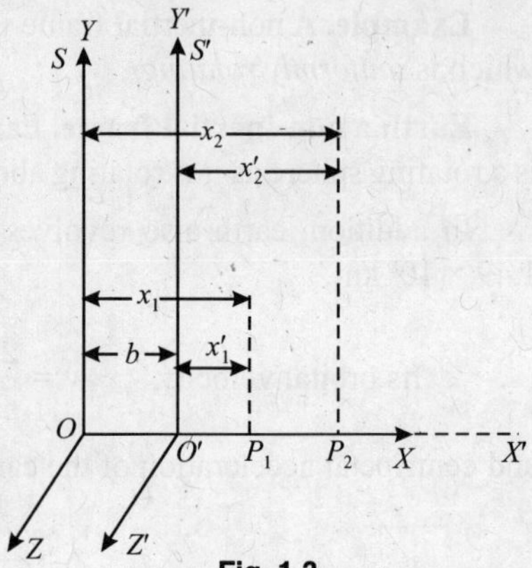

Fig. 1.3

The form of the function U should be such that it is a *scalar function and force derived from it is independent of the inertial system, i.e.*

$$F_x = -\frac{\partial U}{dx} = \frac{\partial U'}{\partial x'}$$

where U and U' are the potential energies and x and x' the co-ordinates of the particle in the S and S' systems respectively. Since U is a function of x_1 and x_2 the co-ordinates of the two particles, the various simple forms which satisfy the above two conditions are

$$U = a\,|(x_1 - x_2|,\ U = \frac{a}{|(x_1 - x_2)|},\ U = (x_1 - x_2)^2$$

We shall select the last form because even if $x_1 - x_2$ is a vector $(x_1 - x_2)^2$ is a scalar quantity.

$\therefore$ $$U(x_1, x_2) = (x_1 - x_2)^2$$

and $$U(x_1', x_2') = (x_1' - x_2')^2$$

$$= [(x_1 - b) - (x_2 - b)]^2 = (x_1 - x_2)^2 = U(x_1 - x_2)$$

This shows that the above form of potential energy function U is independent of the co-ordinate system or the frame of reference.

The physical meaning of the above relation is that U is indpendent of the absolute value of x_1 and x_2 as long as $(x_1 - x_2')^2$ is the same. In other words, so long as the magnitude of $|x_1 - x_2|$ remains constant, it is immaterial where two particles are situated along the x-axis. This means that the properties of space are independent of the positions of x_1 and x_2 or the *free space possesses linear uniformity or has translational invariance.*

1.6 CONSERVATION OF MOMENTUM FROM HOMOGENEITY OF SPACE

We shall make use of the property of linear uniformity of space to prove the law of conservation of linear momentum with the help fo Newton's third law of motion. The linear unifomity of space demands that the force between the two particles P_1 and P_2 be independent of displacement between two frmaes of reference, *i.e.*,

$$F_x = -\frac{\partial U}{\partial x} = -\frac{\partial U'}{\partial x'}$$

The magnitude of the force on the particle P_1 acting along the X-axis

$$= F_{21} = -\frac{\partial U}{\partial x_1} = -\frac{\partial U'}{\partial x_1'}$$

The magnitude of the force on the particle P_2 acting along the X-axis

$$= F_{12} = -\frac{\partial U}{\partial x_2} = \frac{-\partial U'}{\partial x'_2}$$

$$\because \quad U(x_1, x_2) = (x_1 - x_2)^2 \text{ and } U(x_1' - x_2') = (x_1' - x_2')^2$$

the forces are independent of the frames of reference.

Put $\quad x_1 - x_2 = s \; \therefore \; \frac{\partial s}{\partial x_1} = 1$ and $\frac{\partial s}{\partial x_2} = -1$

Hence
$$F_{21} = -\frac{\partial U}{\partial x_1} = -\left(\frac{\partial U}{\partial s}\right)\left(\frac{\partial s}{\partial x_1}\right) = \frac{-\partial U}{\partial s}$$

and
$$F_{12} = -\frac{\partial U}{\partial x_2} = \left(\frac{\partial U}{\partial s}\right)\left(\frac{\partial s}{\partial x_2}\right) = \frac{\partial U}{\partial s}$$

$$\therefore \quad F_{12} = -F_{21}$$

which is Newton's third law of motion.

Therefore, if free space possesses the property of linear uniformity, the forces and potentials are independendent of the linear displacement of the frame of reference which as proved above leads to the third law of motion. We have already proved the law of conservation of momentum with the help of Newton's third law of motion. Hence the linear uniformity (or homogeneity) of space proves the law conservation of momentum with the help of Newton’s third law of motion.

1.7 ROTATIONAL INVARIANCE OF SPACE

Space is *isotropic i.e.* if we move in any direction from a point, there is nothing to distinguish one direction from the another. For example, if we keep a particle p_1 fixed and move another particle p_2 on the surface of a sphere with p_1 as centre, the interaction energy between the two will be the same for all positions of p_2 on the surface of the sphere. *Isotropy* thus means *rotational invariance* of the properties of free space.

The term rotational invariance of space further implies that the potential energy of interaction between two particles is an *invariant,* i.e., does not change if position co-ordinates of both the particles are rotated through the same angle about an arbitrary axis.

1.8 CONSERVATION OF ANGULAR MOMENTUM FROM ROTATIONAL INVARIANCE OF SPACE

The space simultaneously posses the property of

(*i*) Linear unformity and (*ii*) Rotation invariance

Therefore the form of potential energy function $U(r_1, r_2) = U(r)$ where

$r = |\vec{r}| = |\vec{r}_1 - \vec{r}_2|$, $\vec{r}_1$ and $\vec{r}_2$ being the space co-ordinates of the two interacting particles P_1 and P_2 respectively, satisfies both the conditions stated above. The potential U depends upon the magnitude of the separation of the two point $|\vec{r}_1 - \vec{r}_2|$ but is independent of the actual value of the space co-ordinates and their direction.

The force on the particle $P_1 = \vec{F}_{21} = -\left(\frac{\partial U}{\partial r_1}\right)\hat{r} = -\left(\frac{\partial U}{\partial r}\right)\left(\frac{\partial r}{\partial r_1}\right)\hat{r}$

The force on the particle $P_2 = \vec{F}_{12} = -\left(\frac{\partial U}{\partial r_2}\right)\hat{r} = -\left(\frac{\partial U}{\partial r}\right)\left(\frac{\partial r}{\partial r_2}\right)\hat{r}$

where $\hat{r}$ is a unit vector along $\vec{r}$, *i.e.* the direction along which either force of interaction acts.

Two cases arise:

(*i*) $r_1 > r_2$

In such a case $r = r_1 - r_2 \quad \therefore \quad \frac{\partial r}{\partial r_1} = 1$ and $\frac{\partial r}{\partial r_2} = -1$

$$\vec{F}_{21} = -\left(\frac{\partial U}{\partial r}\right)\hat{r} \text{ and } \vec{F}_{12} = +\left(\frac{\partial U}{\partial r}\right)\hat{r}$$

Hence $\qquad \vec{F}_{12} = -\vec{F}_{21}$

(*ii*) $r_2 > r_1$

In such a case $r = r_2 - r_1 \quad \therefore \quad \frac{\partial r}{\partial r_1} = -1$ and $\frac{\partial r}{\partial r_2} = +1$

$$\therefore \qquad \vec{F}_{21} = +\left(\frac{\partial U}{\partial r}\right)\hat{r} \text{ and } \vec{F}_{12} = -\left(\frac{\partial U}{\partial r}\right)\hat{r}$$

Hence again $\qquad \vec{F}_{12} = -\vec{F}_{21}$

Thus the potential function $U(r)$ leads to Newton's third law of the motion whether

$$r_1 > r_2 \text{ or } r_2 > r_1$$

As proved above $\qquad \vec{F}_{21} = \pm\left(-\frac{\partial U}{\partial r}\right)\hat{r}$ and $\vec{F}_{12} = \pm\left(-\frac{\partial U}{\partial r}\right)\hat{r}$

If $-\frac{\partial U}{\partial r} = F(r)$, $\vec{F}_{12}$ as well as $\vec{F}_{21}$ can be expressed as $\pm F(r)\hat{r}$. In other words, the force will depend only on r the distance between the centres of the two particles. It is, therefore, a **central force.** Thus we find that rotational invariance of space requires motion under a central force.

Now the torque $\vec{\tau}$ for a force about the centre of the force is given by

$$\vec{\tau} = \vec{r} \times \vec{F} = \vec{r} \times F(r)\hat{r} = F(r)\, \vec{r} \times \hat{r} = F(r)\, r\left[\hat{r} \times \hat{r}\right] = 0$$

$\therefore$ For such a force, torque $\vec{\tau} = 0$

But torque is the rate of change of angular-momentum $\vec{J}$.

$$\therefore \qquad \vec{\tau} = \frac{d\vec{J}}{dt} = 0$$

Integrating we get, $\vec{J} = a$ constant. this proves the law of conservation of angular momentum. Thus, the rotational invariance or isotropy of space shows that:

(*i*) There is a central force between two particles interacting with each other.

(*ii*) $\vec{F}_{21} = -\vec{F}_{12}$, i.e. Newton's third law of motion holds good.

These two properties, on combination, show that the angular momentum is a constant. Hence, *the rotational invariance of space leads to the principle of conservation of angular momentum.*

1.9 HOMOGENEITY OF TIME

Time. According to Newton, *'Absolute true and mathematical time of itself and from its own nature, flows equally without relation to anything exertnal and is otherwise called duration'.*

Properties. **(*i*) One dimensionality.** Time is one dimensional. It flows only in one direction. It is therefore, specified by a single variable *t*. It is independence of space.

(*ii*) **Homogeneity.** Time is homogenous, *i.e.,* it flows uniformly. The results of an experiment do not change when we change the time of an experiment. In other words, '*The result of an experiment is independent of the change in the origin of time*' This property is known as homogenity of time.

(*iii*) **Isotropy.** Theoretically time is isotropic, *i.e.* the laws of Physics remain unaltered by changing +*t* to −*t*.

Law of conservation of energy from homogeneity of time. Time is homogenous *i.e.* it flows uniformly.

The total energy E of a system is given by $E = U + K$

where U is the potential energy and K the kinetic energy.

The Coulomb force between two charges q_1 and q_2, a distance r apart, is given by

$$\vec{F}_c = \frac{1}{4\pi\varepsilon_0}\left(\frac{q_1 . q_2}{r^2}\right)\hat{r} = k\frac{q_1 . q_2}{r^2}\hat{r}$$

where k is a constant = $1/4\,\pi\varepsilon_0$

Similarly, the graviational force between two masses m_1 and m_2, a distance r apart, is also given by

$$\vec{F}_g = G\left(\frac{m_1 . m_2}{r^2}\right)\hat{r}$$

where G is the gravitational constant.

We find that in the above expressions for force, time does not appear *explicitly.* By *explicit* dependence on time we mean that time should occur as such directly in the expression for force.

Now, for a conservative system $\vec{F} = -\left(\frac{\partial U}{\partial r}\right)\hat{r}$

In other words, U is a function of $\vec{F}$ and $\vec{r}$. If time flows uniformly, the force acting at a point do not depend explicitly on time *i.e.* force $\vec{F}$ is a function of $\vec{r}$ only. Since U is a function of $\vec{F}$ and $\vec{r}$, it is also function of $\vec{r}$ only. This principle of homogeneity of time, therefore, states $\frac{\partial U}{\partial t} = 0$

The expression for kinetic energy $K = \frac{1}{2}mv^2$ also does not depend explicitly on time

$$\frac{\partial K}{\partial t} = 0$$

If total energy E is a function of $\vec{r}$ and t, then $E = E(\vec{r}, t)$

and
$$dE = \frac{\partial E}{\partial r}dr + \frac{\partial E}{\partial t}dt = \frac{\partial}{\partial t}(U+K)\,dr + \frac{\partial}{\partial r}(U+K)dt$$

$$= \left(\frac{\partial U}{\partial r} + \frac{\partial K}{\partial r}\right)dr + \left(\frac{\partial U}{\partial t} + \frac{\partial K}{\partial t}\right)dt = \left[\left(\frac{\partial U}{\partial r}\right) + \left(\frac{\partial K}{\partial r}\right)\right]dr \quad ...(i)$$

as $\frac{\partial U}{\partial t} = 0$ and $\frac{\partial K}{\partial t} = 0$

From (*i*), we have $\frac{dE}{dt} = \left(\frac{\partial U}{\partial r} + \frac{\partial K}{\partial r}\right)\frac{dr}{dt}$

Now $\frac{\partial U}{\partial r} = -F$ and $\frac{\partial K}{\partial r} = \frac{\partial}{\partial r}\left(\frac{1}{2}mv^2\right) = mv\frac{\partial v}{\partial r}$

Since v does not depend explicitly on time, $\frac{\partial v}{dr} = \frac{dv}{dr}$

$$\therefore \quad \frac{\partial K}{\partial r} = mv\frac{dv}{dr} = m\frac{dr}{dt}.\frac{dv}{dr} = m\frac{dv}{dt} = ma$$

Hence $\frac{dE}{dt} = (-F + ma)\frac{dr}{dt}$

The expression within brackets is zero as $F = ma$, accoridng to Newton's second law of motion

$$\therefore \quad \frac{dE}{dt} = 0 \text{ or } E = \text{a constant}$$

Thus we find that the homogeneity of time (or time symmetry) (*ie.* if time flows uniformly) and Newton's second law of motion lead to the principle of conservation of energy.

1.10 GALELIAN TRANSFORMATIONS

When a physical phenomenon is observed in two inertial frames moving with uniform velocity relative to each other and the time interval registered in both the frames is the same, then the datas of results in one frame of reference can be transformed to those in the second frame. This process is known as Galelian transformation.

The equations which connect the position vectors of a particle (or event) its two-inertial frames are known as Galelian transformations.

Galelian transformation equations. We can derive Galelian transformation equation as under. We shall discuss two cases:

Case (*i*). When one inertial frame is moving relative to the other along positive direction of *x*-axis. Let S and S' be the two inertial frames, whose origins O and O' coincide at $t = t' = 0$. Their X-axes OX and $O'X'$ are along the same line and Y and Z-axes are parallel to each other as shown in Fig. 1.4. The frame S' is moving with a uniform velocity v with respect to the frame S along the positive direction of X-axis. Now consider an event P in the frame S specified by space co-ordinate x, y, z and time co-ordinate t. An event is completely known from the space co-ordinates of the point of its occurrence and from the time co-ordinate of its happening. An event can be a collision between two particles.

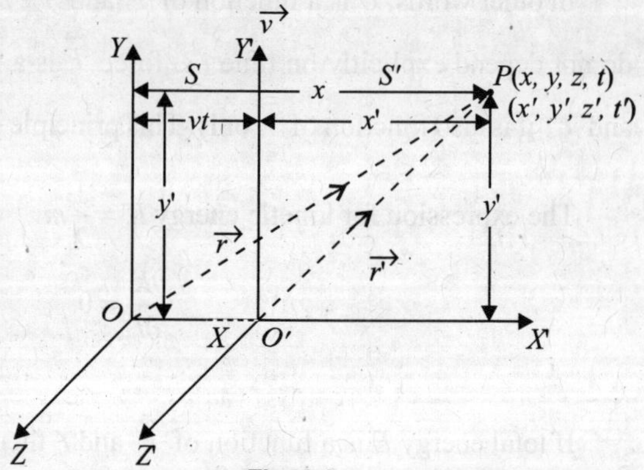

Fig. 1.4

Let the corresponding space and time co-ordinates of the same event in the inertial frame S' be x', y', z' and t'.

Then after a time t, $O'O = vt$ and $x' = x - vt$ $y' = y$ $z' = z$ $t' = t$

The last equation is based upon the universal nature of time as assumed in classical physics. It means that a clock in S measures the same time as a clock in S' provided these were synchronised at time $t = t' = 0$. These equations are known as Galelian transformation equations.

The inverse Galelian transformation equations can be written by changing v to $-v$, x to x', t to t' and so on. These are $x = x' + vt \quad y = y' \quad z = z' \quad t = t'$

If $\vec{r} = \vec{OP}$ is the position vector of the point P in the inertial frames S and $\vec{r} = \vec{OP}$ the position vector of P in the frame S', then $\vec{r} = x\hat{i} + y\hat{j} + z\hat{k}$ and $\vec{r}' = x'\hat{i} + y'\hat{j} + z'\hat{k}$

$$\therefore \quad \vec{r}' = (x - vt)\,\hat{i} + y\hat{j} + z\hat{k} = (x\hat{i} + y\hat{j} + z\hat{k}) - vt\,\hat{i} = \vec{r} - \vec{v}t \text{ where } \vec{v} = v\hat{i}$$

Thus Galelian transformations can also be put as $\vec{r}' = \vec{r} - \vec{v}t \qquad t' = t$

The corresponding inverse Galelian transformations are $\vec{r} = \vec{r}' + \vec{v}t' \qquad t = t'$

Case (*ii*) When one inertial frame is moving relative to the other in any direction. Now let the frame S' move with respect to the frame S with uniform velocity $\vec{v}$ given by $\vec{v} = v_x\,\hat{i} + v_y\,\hat{j} + v_z\,\hat{k}$ in any direction where v_x, v_y and v_z are the components of $\vec{v}$ along x, y and z directions respectively (Fig. 1.5). The origins and the axes of the two frame are so chosen that these coincide at $t = t' = 0$. After a time t the frame S' is separated from the frame S by a distance $v_x t$, $v_y t$ and $v_z t$ along the x, y and z axes respectively. If x, y, z, t are the co-ordinates of an event P in the frame S and x', y', z', t' of the same event in the frame S', then $x' = x - v_x t \quad y' = y - v_y t \quad z' = z - v_z t \quad t' = t$

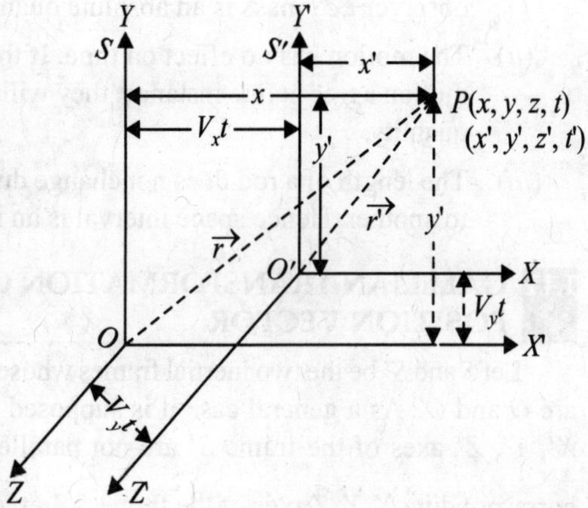

Fig. 1.5

In this case
$$\vec{r}_1' = (x - v_x t)\,\hat{i} + (y - v_y t)\,\hat{j} + (z - v_z t)\,\hat{k}$$
$$= (x\hat{i} + y\hat{j} + z\hat{k}) - (v_x\,\hat{i} + v_y\,\hat{j} + v_z\,\hat{k})\,t = \vec{r} - \vec{v}t$$
$$(\text{where } \vec{v} = v_x\,\hat{i} + v_y\,\hat{j} + v_z\,\hat{k})$$

Thus Galelian transformations can also be put as $\vec{r}' = \vec{r} - \vec{v}t' \qquad t' = t$

We shall, however, restrict our discussion to the simple case discussed in (*i*) where S' is moving along the x-axis relative to the frame S.

Galelian invariance of Length, Mass and Time. It states that '*The basic laws of mechanics do not change under Galelian transformations*'. The basic laws of Newtonian mechanics are the Newton's laws of motion, laws of conservation of momentum and energy etc. We shall prove below that a space interval is invariant under Galelian transformation.

Consider a rod at rest in the inertial frame S having its two ends at the points x_2 and x_1, then for the observer S

The length of the rod $= x_2 - x_1$

The observer S' will consider the rod to be moving with velocity $-v$ from O' to O and will observe the two ends to be at the points x_2' and x_1' at the **same time.**

For the observer S' the length of the rod $= x_2' - x_1'$

$$= (x_2 - vt_2) - (x_1 - vt_1) = (x_2 - x_1) - v(t_2 - t_1)$$

but the ends of the rod have been observed at one and the same time

$$\therefore \quad t_1 = t_2 \text{ or } t_2 - t_1 = 0 \qquad \therefore \; x_2' - x_1' = x_2 - x_1 \qquad ...(i)$$

Relation (*i*) shows that the *space interval or the distance between two points (or length) is in variant under Galelian transformations*.

Thus we find that the space-interval measurement and time-interval measurement are independent of the relative motion of the inertial frames. Classical physics already assumes that mass of a body is constant and independent of its motion with respect to an inertial frame. *Thus all the three fundamental quantities length, mass and time are invariant of the relative motion of the observer*.

This property is known as **Galelian invariance.** Hence,

(*i*) The mass of a body is the same for all observers and it is independent of the motion of the observer *i.e.*, mass is an absolute quantity.

(*ii*) The motion has no effect on time. If the clocks in two inertial frames which are in uniform motion agree at one instance, they will agree at all later times *i.e.*, $t = t'$. Time is an absolute quantity.

(*iii*) The length of a rod does not change due to relative motion of an inertial frame with respect to another. Hence space interval is an invariant quantity.

1.11 GALELIAN TRANSFORMATION OF POSITION VECTOR

Let S and S' be the two inertial frames whose origins are O and O'. As a general case it is supposed that the X', Y', Z' axes of the frame S' are not parallel to the corresponding X, Y, Z axes of the frame S. Let $\vec{R}$ be the position vector of the origin O' with respect to O at $t = t' = 0$. Now consider an event P in the frame S specified by space co-ordinate $\vec{r}$ and time co-ordinate t. An event is completely known from the space co-ordinates of the point of occurrence and from the time co-ordinate of the time of happening.

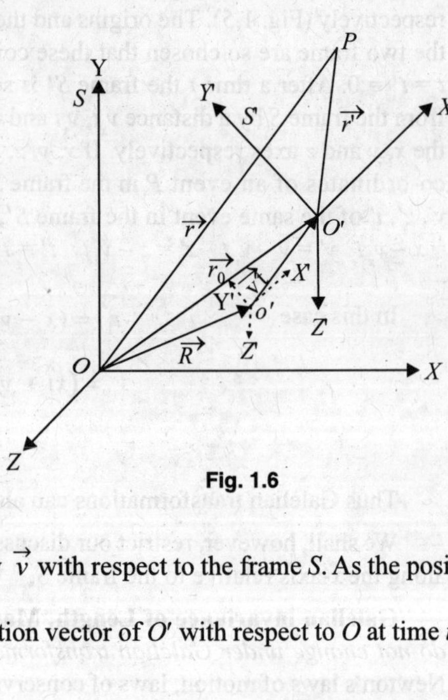

Fig. 1.6

Let the corresponding space and time co-ordinates of the same event in the inertial frame S' be $\vec{r}'$ and t'.

Let the inertial frame S' move with a linear velocity $\vec{v}$ with respect to the frame S. As the position vector of O' with respect to O at $t = 0$ is $\vec{R}$, the position vector of O' with respect to O at time t' in S'

$$OO' = \vec{r_0} = \vec{R} + \vec{v}t'$$

$\therefore$ At time $\qquad t = t'$

$$\vec{r} = \vec{r_0} + \vec{r}' = \vec{R} + \vec{v}t' + \vec{r}' = \vec{r}' + \vec{v}t' + \vec{R} \qquad ...(i)$$

Special cases. (*a*) *When the origins of both the frames coincide initially.* $\vec{R} = 0$

$$\therefore \qquad \vec{r} = \vec{r}' + \vec{v}t' \text{ or } \vec{r}' = \vec{r} - \vec{v}t \qquad ...(ii) \; [\because t = t']$$

(*b*) *When the origins of both the frames coincide initially and their co-ordinate axes are parallel, then* $\vec{r'} = x'\hat{i} + y'\hat{j} + z'\hat{k}$; $\vec{r} = x\hat{i} + y\hat{j} + z\hat{k}$ and $\vec{v} = v_x\,\hat{i} + v_y\,\hat{j} + v_z\,\hat{k}$; $\vec{R} = 0$

Substituting the values in relation (*ii*), we have

$$x'\hat{i} + y'\hat{j} + z'\hat{k} = x\hat{i} + y\hat{j} + z\hat{k} - (v_x\,\hat{i} + v_y\,\hat{j} + v_z\,\hat{k})\,t \quad \text{...(iii)}$$

Equating x, y, z components, we have $x' = x - v_x\,t; y' = y - v_y\,t; z' = z - v_z\,t$

(*c*) *When the origins of both the frames coincide initially, their co-ordinate axes are parallel and the origin of the frames S′ is moving along the* + *x axis with velocity* $\vec{v}$ *relative to O.* Then relation (*iii*) becomes $x'\hat{i} + y'\hat{j} + z'\hat{k} = x\hat{i} + y\hat{j} + z\hat{k} - vt\,\hat{i}$

Equating x, y, z-components on both sides, we have $x' = x - vt \quad y' = y \quad z' = z$

Hence the restricted Galelian transformations are $x' = x - vt$; $y' = y$; $z' = z$ and $t' = t$.

1.12 VELOCITY IS VARIANT TO GALELIAN TRANSFORMATIONS

When an inertial frame S' is moving with a uniform velocity $\vec{v}$ with respect to an inertial frame S and their origins coincide at time $t = t' = 0$, Galelian transformation equation for position vector is

$$\vec{r'} = \vec{r} - \vec{v}t$$

where $\vec{r'}$ is the position vector of a particle in the frame S' and $\vec{r}$ iin the frame S.

Differentiating both sides with respect to time, we get $\dfrac{d\vec{r'}}{dt} = \dfrac{d\vec{r}}{dt} - \vec{v}$

Now $\quad \dfrac{d\vec{r'}}{dt} = \dfrac{d\vec{r'}}{dt'} \cdot \dfrac{dt'}{dt}$

As $\quad t = t' \quad \dfrac{dt'}{dt} = 1 \quad \therefore \quad \dfrac{d\vec{r'}}{dt} = \dfrac{d\vec{r'}}{dt'}$

Hence $\quad \dfrac{d\vec{r'}}{dt'} = \dfrac{d\vec{r}}{dt} - \vec{v}$ or $\vec{u'} = \vec{u} - \vec{v} \quad$ ...(*i*)

where $\vec{u'} = \dfrac{d\vec{r'}}{dt'}$ is the velocity of the particle as observed in frame S' and $\vec{u} = \dfrac{d\vec{r}}{dt}$ is the velocity of the particle as observed in the frame S.

The relation $\vec{u'} = \vec{u} - \vec{v}$ is known as Galelian transformation equation for velocity. This equation clearly shows that velocity in the frame S' is not the same as in the frame S *i.e.*, velocity is a variant under Galelian transformations.

1.13 ACCELERATION IS INVARIANT TO GALELIAN TRANSFORMATIONS

Differentiating equation (*i*), with respect to time, we get

$$\frac{d\vec{u'}}{dt} = \frac{d\vec{u}}{dt} \text{ as } \vec{v} = \text{a constant}$$

But $\quad \dfrac{d\vec{u'}}{dt} = \dfrac{d\vec{u'}}{dt'}\,\dfrac{dt'}{dt} = \dfrac{d\vec{u'}}{dt'} \quad \ldots\left[\because \dfrac{dt'}{dt} = 1\right]$

$$\therefore \quad \frac{d\vec{u'}}{dt'} = \frac{d\vec{u}}{dt} \text{ or } \vec{a'} = \vec{a} \qquad ...(ii)$$

where $\vec{a'} = \frac{d\vec{u'}}{dt'}$ is the acceleration of the particle in the frame S' and $\vec{a} = \frac{d\vec{u}}{dt}$ is the acceleration in the frame S. This shows that the acceleration of a particle is independent of the uniform velocity of one inertial frame of reference relative to another. In other words *acceleration is invariant to Galelian transformations*.

SOLVED EXAMPLES

Example 1.1. *A moving particle has co-ordinates (6t + 3), 8t, 5 m in frame S at any time t. The frame S' is moving relative to S with a velocity* $3\hat{i} + 4\hat{j}$ *m/s. Find the co-ordinates and velocity of the particle in frame S'.*

Solution. Galelian transformation equations for space co-ordinates are

$$x' = x - v_x t \quad y' = y - v_y t \quad z' = z - v_z t$$

Now $\quad x = 6t + 3 \quad y = 8t \quad z = 5; \quad v_x = 3 \quad v_y = 4 \quad v_z = 0$

$$\therefore \quad x' = x - v_x t = 6t + 3 - 3t = 3t + 3$$

$$y' = y - v_y t = 8t - 4t = 4t\,; \; z' = z - v_z t = 5 - 0 = 5$$

$\therefore$ Co-ordinates of the particle in S' are $3t + 3$, $4t$, 5.

Velocity. $x' = 3t + 3 \therefore \frac{dx'}{dt'} = \frac{dx'}{dt} \cdot \frac{dt}{dt'} = \frac{dx'}{dt} \qquad ... \begin{bmatrix} \therefore t' = t \\ dt' = dt \end{bmatrix}$

Hence $\quad \frac{dx'}{dt} = 3 \text{ or } \vec{u'_x} = 3\hat{i}$

Similarly $\quad \frac{dy'}{dt'} = \frac{dy'}{dt} = 4 \text{ or } \vec{u'_y} = 4\hat{j} \quad \frac{dz'}{dt'} = \frac{dz'}{dt} = 0$

$\therefore$ Velocity in frame S', $\vec{u'} = \vec{u'_x} + \vec{u'_y} + \vec{u'_z} = 3\hat{i} + 4\hat{j}$

Example 1.2. *The position vector of a point in the frame S' moving with constant velocity of 10 cm/sec along X-axis is given by (11, 9, 8) cm. Calculate the position with respect to the frame S if the two frames were coincident only* $\frac{1}{2}$ *sec. earlier.*

Solution. As the position co-ordinates are given in the frame S' which is moving with constant velocity of $v = 10$ cm s^{-1} along $+X$-axis, the position co-ordinates in the frame S are given by inverse Galelian transformations

$$x = x' + vt; \quad y = y'; \quad z = z'; \quad t = t'$$

Here $\quad x' = 11 \text{ cm} \quad y' = 9 \text{ cm} \quad z' = 8 \text{ cm} \quad t' = \frac{1}{2} \text{ sec.}$

$$\therefore \quad x = x' + vt = 11 + 10 \times \frac{1}{2} = 16 \text{ cm; } y' = 9 \text{ cm; } z' = 8 \text{ cm}$$

$\therefore$ The position co-ordinates of the point in the frame S are (16, 9, 8) cm.

Example 1.3. *A frame S' is moving with velocity* $5\hat{i} + 7\hat{j}$ *m/s relative to an inertial frame S. A particle is moving with velocity* $(t+5)\hat{i} + 9\hat{j}$ *m/s with respect to S. Find the acceleration of the particle in the frame S'*

Solution. Velocity of the particle in frame S, $\vec{u} = (t+5)\hat{i} + 9\hat{j}$ ms^{-1}

Velocity of frame S' relative to S, $\vec{v} = 5\hat{i} + 7\hat{j}$

According to Galelian transformation equations for velocity,

Velocity of the particle with respect to frame S',

$$\vec{u}' = \vec{u} - \vec{v} = (t+5)\,\hat{i} + 9\hat{j} - (5\hat{i} + 7\hat{j}) = t\,\hat{i} + 2\hat{j} \text{ ms}^{-1}$$

$\therefore$ Acceleration of the particle in the frame S', $\vec{a}' = \dfrac{d\vec{u}'}{dt'} = \dfrac{d\vec{u}'}{dt}$ ... [$\because$ $dt' = dt$]

$$= \frac{d}{dt}[t\,\hat{i} + 2\hat{j}] = \hat{i} \quad \therefore \vec{a}' = \hat{i} \text{ ms}^{-2}$$

1.14 INVARIANCE OF NEWTON'S LAWS OF MOTION IN GALELIAN TRANSFORMATIONS

To prove that Newton's laws of motion are invariant to Galelian transformations we shall first prove that acceleration is invariant to these transformations. We shall consider the case when the co-ordinates axes of the two frames are parallel, their origins coincide at $t' = t = 0$ and the frame S' moves with a uniform velocity $\vec{v}$ with respect to the frame S in the + X direction. In such a case Galelian transformation equations are $x' = x - vt$ $\quad y' = y \quad z' = z \quad t' = t$

Differentiating we get, $dx' = dx - vdt$, $\quad dy' = dy$

$dz' = dz \quad dt' = dt$

Dividing left hand side of each equation by dt' and right hand side by dt we have

$$\frac{dx'}{dt'} = \frac{dx}{dt} - v, \quad \frac{dy'}{dt'} = \frac{dy}{dt}, \quad \frac{dz'}{dt} = \frac{dz}{dt}$$

Now $\dfrac{dx'}{dt'} = u'_x = x$ − component of the velocity $\vec{u}'$ as measured in the frame S'

and $\dfrac{dx}{dt} = u_x = x$ − component of the velocity $\vec{u}$ as measured in the frame S

$\therefore$ $u'_x = u_x - v$

Similarly $u'_y = u_y$, $\quad u'_z = u_z$

Combining the equations for x, y and z components we have

$$u'_x\,\hat{i} + u'_y\,\hat{j} + u'_z\,\hat{k} = (u_x - v)\,\hat{i} + u_y\,\hat{j} + u_z\,\hat{k} = u_x\,\hat{i} + u_y\,\hat{j} + u_z\hat{k} - v\,\hat{i}$$

or $$\vec{u}' = \vec{u} - \vec{v} \qquad ...(i)$$

Differentiating equation (*i*) again we have $du'_x = d(u_x - v) = du_x$

Since $v = a$ constant

Similarly $du'_y = du_y$ and $du'_z = du_z$

Dividing left hand side of each equation by dt' and right hand side by dt, we have

$$\frac{du'_x}{dt'} = \frac{du_x}{dt}; \quad \frac{du'_y}{dt'} = \frac{du_y}{dt}; \quad \frac{du'_z}{dt'} = \frac{du_z}{dt}$$

or $$a'_x = a_x, \; a'_y = a_y, \; a'_z = a_z$$

where a'_x, a'_y, a'_z are the x, y and z components of acceleration in the frame S' and a_x, a_y and a_z in the frame S.

Combining the equations for x, y and z components we have

$$a'_x\hat{i} + a'_y\hat{j} + a'_z\hat{k} = a_x\hat{i} + a_y\hat{j} + a_z\hat{k} \quad \text{or} \quad \vec{a'} = \vec{a} \qquad ...(ii)$$

Hence the components of acceleration of a particle are independent of the uniform velocity of one inertial frame of reference relative to another. In other words **acceleration is an invariant** for a transformation from one inertial frame to another which is having a uniform relative motion of translation.

Newton's laws of motion. Newton's second law of motion is the real law of motion and includes the first and the third law. In mathematical form the second law is stated as $\vec{F} = m\vec{a}$ where $\vec{F}$ is the force acting on a mass m in the frame S. If $\vec{F'}$ is the force acting on the same mass in the frame S' and $\vec{a'}$ is the acceleration produced in the mass as observed in the frame S', then

$$\vec{F'} = m\vec{a'}$$

As proved above in equation (*ii*) $\vec{a'} = \vec{a}$

$$\therefore \quad \vec{F'} = \vec{F}$$

showing thereby the law is invariant to Galelian transformations. Hence **Newton's laws of motion** are invariant to Galelian transformations, and *the relation F = ma holds true under Galelian transformations.*

Thus, the quantities invariant under Galelian transformations are (*i*) mass (*ii*) length (*iii*) time (*iv*) acceleration (*v*) force.

1.15 INERTIAL FRAME IN UNIFORM TRANSLATORY MOTION

A frame of reference having a uniform translatory motion (or moving with constant velocity) relative to an inertial frame is also inertial.

To prove this, consider on inertial frame S' moving relative to another inertial frame S with a uniform velocity $\vec{v}$, along a straight line, then $\vec{a'} = \vec{a}$ i.e. the acceleration of a particle is the same in the two inertial frames. In other words, if a particle is unaccelerated in frame S, it is unaccelerated in S' *i.e.*, a particle at rest or in uniform motion in a straight line (zero acceleration) in S will also be at rest or in uniform motion in a straight line in S'. Hence if S is an inertial frame then S' having a uniform translatory motion relative to it is also an inertial frame.

Force same in two inertial frames. As stated above for two inertial frames connected by Galelian transformations the acceleration of a particle is invariant *i.e.*, $\vec{a'} = \vec{a}$ ...(*i*)

In Newtonian mechanics the mass of a particle is also invariant. Hence multiplying both sides of relation (*i*) by m we have $m\vec{a'} = m\vec{a}$ or $\vec{F'} = \vec{F}$ where $\vec{F}$ is the force on the particle in frame S and $\vec{F}$ the force on the same particle in frame S'.

Thus force acting on a particle is the same in two frames of reference connected by Galelian transformations. In other words, "*If the relation $\vec{F} = m\vec{a}$ is used to define force observers in all frames of reference would agree on the magnitude and direction of force $\vec{F}$ independent of relative velocity of reference frames.*"

1.16 INVARIANCE OF CONSERVATION LAWS IN GALELIAN TRANSFORMATIONS

The two important laws of physics which are invariant to Galelian transformations are (*i*) Law of conservation of linear momentum and (*ii*) Law of conservation of energy.

1.16.1 Law of Conservation of Linear Momentum

The principle of conservation of momentum states,

"When a number of particles interact and there is no external force acting on them, the total momentum of the system is conserved."

The law of conservation of momentum is an invariant *i.e.*, remains unchanged in the two frames of reference connected by Galelian transformations. To prove the same consider two particles of mass m_1 and m_2 moving with velocities $\vec{u_1}$ and $\vec{u_2}$ respectively in the inertial frame *S*. Let their velocities after collision be $\vec{v_1}$ and $\vec{v_2}$ respectively. If we assume that the velocities of the particles are **small** as compared to the velocity of light, then

(*i*) Newton's second law $\vec{F} = m\vec{a} = m\dfrac{d\vec{v}}{dt}$ and

(*ii*) Law of conservation of mass

$m_1 + m_2$ (before collision) $= m_1 + m_2$ (after collision) hold good.

Hence according to the law of conservation of momentum

$$m_1 \vec{v_1} + m_2 \vec{v_2} = m_1 \vec{u_1} + m_2 \vec{u_2} \quad \text{...}(i)$$

Let $\vec{u_1'}$ and $\vec{u_2'}$ be the velocities of m_1 and m_2 before collision and $\vec{v_1'}$ and $\vec{v_2'}$ after collision as seen by an observer in the frame S' moving relative to frame *S* with a constant velocity $\vec{V}$. According to inverse Galelian transformation equations as applied to velocities

$$\vec{u_1} = \vec{u_1'} + \vec{V} \qquad \vec{u_2} = \vec{u_2'} + \vec{V} \qquad \vec{v_1} = \vec{v_1'} + \vec{V} \qquad \vec{v_2} = \vec{v_2'} + \vec{V}$$

Substituting these values in equation (*i*), we have

$$m_1 (\vec{v_1'} + \vec{V}) + m_2 (\vec{v_2'} + \vec{V}) = m_1 (\vec{u_1'} + \vec{V}) + m_2 (\vec{u_2'} + \vec{V})$$

or $$m_1 \vec{v_1'} + m_2 \vec{v_2'} = m_1 \vec{u_1'} + m_2 \vec{u_2'}$$

i.e., the law of conservation of momentum holds good in the moving system also.

1.16.2 Law of Conservation of Energy

According to the principle of conservation of energy, in the above case of collision between two particles, we have

$$\frac{1}{2} m_1 v_1^2 + \frac{1}{2} m_2 v_2^2 = \frac{1}{2} m_1 u_1^2 + \frac{1}{2} m_2 u_2^2 - E$$

where *E* is the part of the energy which appears as heat, light etc. or as potential energy Applying the Galelian transformation equation for velocities we have, for the frame S'

$$\frac{1}{2} m_1 (\vec{v_1'} + \vec{V})^2 + \frac{1}{2} m_2 (\vec{v_2'} + \vec{V})^2 = \frac{1}{2} m_1 (\vec{u_1'} + \vec{V})^2 + \frac{1}{2} m_2 (\vec{u_2'} + \vec{V})^2 - E$$

or $$\frac{1}{2} m_1 (\vec{v_1'} + \vec{V}) \cdot (\vec{v_1'} + \vec{V}) + \frac{1}{2} m_2 (\vec{v_2'} + \vec{V}) \cdot (\vec{v_2'} + \vec{V})$$

$$= \frac{1}{2} m_1 (\vec{u_1'} + \vec{V}) \cdot (\vec{u_1'} + \vec{V}) + \frac{1}{2} m_2 (\vec{u_2'} + \vec{V}) \cdot (\vec{u_2'} + \vec{V}) - E$$

or $\frac{1}{2} m_1 (v_1'^2 + 2\vec{v_1'} . \vec{V} + V^2) + \frac{1}{2} m_2 (v_2'^2 + 2\vec{v_2'} . \vec{V} + V^2)$

$= \frac{1}{2} m_1 (u_1'^2 + 2\vec{u_1'} . \vec{V} + V^2) + \frac{1}{2} m_2 (u_2'^2 + 2\vec{u_2'} . \vec{V} + V^2) - E$

or $\frac{1}{2} m_1 v_1'^2 + \frac{1}{2} m_2 v_2'^2 + (m_1 \vec{v_1'} + m_2 \vec{v_2'}) . \vec{V}$

$= \frac{1}{2} m_1 u_1'^2 + \frac{1}{2} m_2 u_2'^2 + (m_1 \vec{u_1'} + m_2 \vec{u_2'}) . \vec{V} - E$

but $m_1 \vec{v_1} + m_2 \vec{v_2} = m_1 \vec{u_1} + m_2 \vec{u_2}$ according to the principle of conservation of momentum

$\therefore \quad \frac{1}{2} m_1 v_1'^2 + \frac{1}{2} m_2 v_2'^2 = \frac{1}{2} m_1 u_1'^2 + \frac{1}{2} m_2 u_2'^2 - E$

which is the law of conservation of energy as observed in the moving frame S'. Hence the law of conservation of energy is invariant under Galelian transformations.

Example 14. Show that the electromagnetic wave equation

$$\frac{\partial^2 \phi}{dx^2} + \frac{\partial^2 \phi}{\partial y^2} + \frac{\partial^2 \phi}{\partial z^2} - \frac{1}{c} \frac{\partial^2 \phi}{\partial t^2}$$

is non invariant in Galelian transformation equations where ϕ is scalar potential.

Solution. Let an inertial frame S' move along x-direction with respect to the frame S with a speed v. If x, y, z, t are the co-ordinates of a point in frame S and x', y', z', t' the co-ordinates of the same point in frame S', then

$$\frac{\partial \phi}{\partial x} = \frac{\partial \phi}{\partial x'}\frac{\partial x'}{\partial x} + \frac{\partial \phi}{\partial y'}\frac{\partial y'}{\partial x} + \frac{\partial \phi}{\partial z'}\frac{\partial z'}{\partial x} + \frac{\partial \phi}{\partial t'}\frac{\partial t'}{\partial x} \quad ...(i)$$

$$\frac{\partial \phi}{\partial y} = \frac{\partial \phi}{\partial x'}\frac{\partial x'}{\partial y} + \frac{\partial \phi}{\partial y'}\frac{\partial y'}{\partial y} + \frac{\partial \phi}{\partial z'}\frac{\partial z'}{\partial y} + \frac{\partial \phi}{\partial t'}\frac{\partial t'}{\partial y} \quad ...(ii)$$

$$\frac{\partial \phi}{\partial z} = \frac{\partial \phi}{\partial x'}\frac{\partial x'}{\partial z} + \frac{\partial \phi}{\partial y'}\frac{\partial y'}{\partial z} + \frac{\partial \phi}{\partial z'}\frac{\partial z'}{\partial z} + \frac{\partial \phi}{\partial t'}\frac{\partial t'}{\partial z} \quad ...(iii)$$

$$\frac{\partial \phi}{\partial t} = \frac{\partial \phi}{\partial x'}\frac{\partial x'}{\partial t} + \frac{\partial \phi}{\partial y'}\frac{\partial y'}{\partial t} + \frac{\partial \phi}{\partial z'}\frac{\partial z'}{\partial t} + \frac{\partial \phi}{\partial t'}\frac{\partial t'}{\partial t} \quad ...(iv)$$

For the motion of S' along the x-direction of S, the Galelian transformation equations are

$$x' = x - vt, \quad y' = y, \quad z' = z \quad \text{and} \quad t' = t \quad ...(v)$$

From Eq. (v), we get

$$\frac{\partial x'}{\partial x} = 1, \quad \frac{\partial y'}{\partial x} = 0, \quad \frac{\partial z'}{\partial x} = 0, \quad \frac{\partial t'}{\partial x} = 0 \quad ...(vi)$$

Thus Eq. (i) becomes,

$$\frac{\partial \phi}{\partial x} = \frac{\partial \phi}{\partial x'} \quad \text{or} \quad \frac{\partial}{\partial x} = \frac{\partial}{\partial x'}$$

$$\therefore \quad \frac{\partial^2 \phi}{\partial x^2} = \frac{\partial}{\partial x}\left(\frac{\partial \phi}{\partial x}\right) = \frac{\partial}{\partial x'}\left(\frac{\partial \phi}{\partial x'}\right) = \frac{\partial^2 \phi}{\partial x'^2} \quad ...(vii)$$

Similarly, we can prove that

$$\frac{\partial^2 \phi}{\partial y^2} = \frac{\partial^2 \phi}{\partial y'^2} \quad ...(viii)$$

$$\frac{\partial^2 \phi}{\partial z^2} = \frac{\partial^2 \phi}{\partial z'^2} \quad ...(ix)$$

Again from Eq. (*v*), we get

$$\frac{\partial x'}{\partial t} = -v, \frac{\partial y'}{\partial t} = 0, \frac{\partial z'}{\partial t} = 0, \frac{\partial t'}{\partial t} = 1$$

$\therefore$ Eq. (*iv*), becomes $\dfrac{\partial \phi}{\partial t} = -v\dfrac{\partial \phi}{\partial x'} + \dfrac{\partial \phi}{\partial t'}$

or $\dfrac{\partial}{\partial t} = -v\dfrac{\partial}{\partial x'} + \dfrac{\partial}{\partial t'}$

$$\therefore \quad \frac{\partial^2 \phi}{\partial t^2} = \frac{\partial}{\partial t}\left(\frac{\partial \phi}{\partial t}\right) = \left(-v\frac{\partial}{\partial x'} + \frac{\partial}{\partial t'}\right)\left(-v\frac{\partial \phi}{\partial x'} + \frac{\partial \phi}{\partial t'}\right)$$

or $$\frac{\partial^2 \phi}{\partial t^2} = v^2\frac{\partial^2 \phi}{\partial x'^2} - v\frac{\partial^2 \phi}{\partial x' \partial t'} - v\frac{\partial^2 \phi}{\partial t' \partial x'} + \frac{\partial^2 \phi}{\partial t'^2}$$

or $$\frac{\partial^2 \phi}{\partial t^2} = v^2\frac{\partial^2 \phi}{\partial x'^2} - 2v\frac{\partial^2 \phi}{\partial x' \partial t'} + \frac{\partial^2 \phi}{\partial t'^2} \quad ...(x)$$

Using Eqs. (*vii*), (*viii*), (*ix*) and (*x*) and substituting these values in electromagnetic wave equation, we get

$$\frac{\partial^2 \phi}{\partial x^2} + \frac{\partial^2 \phi}{\partial y^2} + \frac{\partial^2 \phi}{\partial z^2} - \frac{1}{c^2}\frac{\partial^2 \phi}{\partial t^2} = \frac{\partial^2 \phi}{\partial x'^2} + \frac{\partial^2 \phi}{\partial y'^2} + \frac{\partial^2 z}{\partial z'^2} - \frac{v^2}{c^2}\frac{\partial^2 \phi}{\partial x'^2} + \frac{2v}{c^2}\frac{\partial^2 \phi}{\partial x' \partial t'} - \frac{1}{c^2}\frac{\partial^2 \phi}{\partial t'^2}$$

$$= \left(1 - \frac{v^2}{c^2}\right)\frac{\partial^2 \phi}{\partial x'^2} + \frac{\partial^2 \phi}{\partial y'^2} + \frac{\partial^2 \phi}{\partial z'^2} - \frac{1}{c^2}\frac{\partial^2 \phi}{\partial t'^2} + \frac{2v}{c^2}\frac{\partial^2 \phi}{\partial x' \partial t'}$$

The right hand side is the equation of *E.M.* wave in the inertial frame *S′*. It does not have the same form as it has in the frame *S*. This proves that electromagnetic wave equation is not invariant under Galelian transformation.

1.17 PROPAGATION OF LIGHT

Sound waves are longitudinal and require a material medium to travel. These waves, therefore, can travel through a gas, a liquid or a solid but not through vacuum. On the other hand light is a form of transverse wave motion. It can pass through material media like glass, water etc. as also through vacuum. But a wave motion always requies a material medium to travel. In order to put the '*wave theory of light*' on a sound footing it was supposed that a hypothetical medium which was given the name '*ether*' exists everywhere, even in vacuum. Light waves thus were a result of the vibtrations of ether particles. The velcoity of wave motion is given by $v = \sqrt{\frac{E}{\rho}}$ where *E* is the *elasticity* of the medium and ρ the *density*. The velocity of light has a large value of 3×10^8 m/sec. Hence 'ether' must have a very large elasticity and very low density. These two properties are opposed to each other, and, therefore, there were grave doubts regarding the actual existence of '*ether*'.

Electromagnetic theory. Electromagnetic theory of light was predicted by Maxwell. According to this theory the velocity of light or that of electromagnetic waves in free space is given by

$$c = \frac{1}{\sqrt{\mu_0 \varepsilon_0}} = 3 \times 10^8 \text{ ms}^{-1}$$

where μ_0 and ε_0 are the permeability and permittivity of free space. *Since μ_0 and ε_0 are absolute quantities releated only to empty space. the velocity of light appears to be an absolute constant.* But Galelian transofmation equations predict different velocities for observers in different internal frames.

Ether hypothesis. In the second half of the 19th century scientists thought that whole space was filled with same luminescent substance - *ether.* They thought that electromagneitc waves needed same medium to propogate in as they imagined them as some kind of mechanical waves. In order to let the waves propogate as quickly as they do (that is with the speed of light) ether had to be very inflexible, but besides it had not to react with celestial bodies moving in it. Creating and using the hypothesis of such a strange substance can be justified by the fact that such a model explained many of the experimental accurances in the field of electromagnetism. Even Maxwell himself in his researches on electricity and magnetism assumed the existence of the incompressible and elastic ether. In *1905 Albert Einstein* finally dicarded the ether hypothesis, categorically stating that *no medium is needed at all for light to travel.*

1.18 GALELIAN TRANSFORMATIONS NOT VALID FOR VELOCITY OF LIGHT

Consider two inertial frames S and S' such that S' is moving with respect to S in the +ve x-direction with a velocity $\vec{v}$ so that $\vec{v} = v\hat{i} = v\hat{i}_x$

If a beam of light travels in the frame S with a velocity $\vec{c}$ along the $+X$ direction, then $\vec{c} = c\hat{i}$.

According to Galelian transformations in Newtonian mechanics the velocity of light as observed by an observer in $\hat{i}$ will be $\vec{c'} = \vec{c} - \vec{v} = (c - v)$

Thus the speed of light in frame S' will be $(c - v)$.

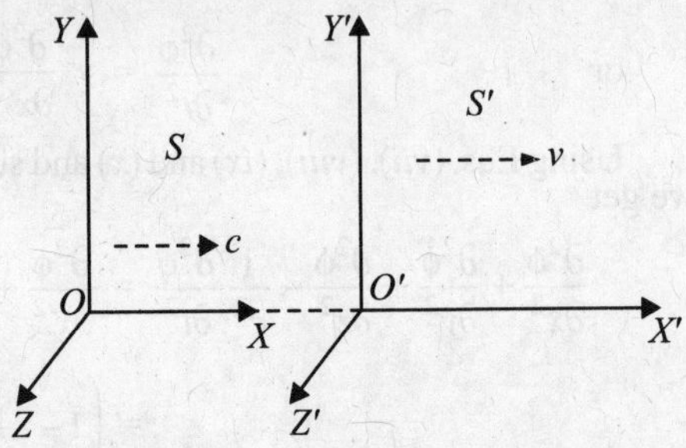

Fig.1.7

This result is absolutely different from that predicted by Maxwell's equations of electromagnetic field. Thus Galelian transformation equations are not valid for electromagnetic radiations or light waves.

1.19 MICHELSON -MORLEY EXPERIMENT

Aim. The aim of Michelson-Morley experiment was to prove the existence of '*ether*' by finding the velocity of drift between the *ether* and the earth.

To prove the existence of '*ether*' attempt was made to find the velocity of drift between the ether and the earth. If ether is assumed to be at rest, the velocity of the earth relative to ether will be its orbital velocity which is about 30 km per second and the velocity of the *ether drift* relative to the earth will be equal in magnitude but opposite in direction.

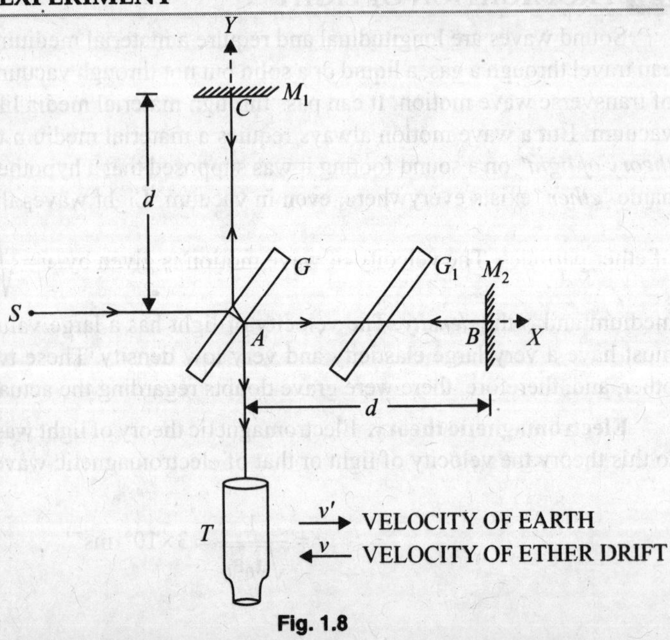

Fig. 1.8

To detect the presumed motion of the earth through ether

Michelson devised an ingenious experiment with the help of Morley using his interferometer. The experiment is classically known as **Michelson-Morley experiment.** In Michelson's interferometer a beam of monochromatic light is split into two parts by a half silvered glass plate G. A part known as *longitudinal part* is transmitted through it along the X-direction and reflected at the front silvered mirror M_2. It retraces its path and after reflection from the back surface of G enters the telescope T.

The other part known as *transverse part* after reflection from the silvered back surface of G, falls on the mirror M_1 along the Y-direction, retraces its path and after refraction through G enters the telescope T. The two beams, therefore, produce interference if the optical path $AB = AC = d$.

Time for light to go from A to B and back to A. If the velocity of the apparatus (or earth) relative to *ether* is v from left to right in the direction AB, then the velocity of *ether* drift is also v but from right to left. If c is the velocity of light, then time taken by light to go from A to $B = \dfrac{d}{c-v}$

and time taken by light to go from B to $A = \dfrac{d}{c+v}$

$\therefore$ Time taken by light to go from A to B and back to A

$$T_x = \frac{d}{c-v} + \frac{d}{c+v} = \frac{2cd}{c^2 - v^2} = \frac{2cd}{c^2\left(1 - \frac{v^2}{c^2}\right)} = \frac{2d}{c\left(1 - \frac{v^2}{c^2}\right)}$$

Time taken by light to go from A to C and back to A. The beam of light reflected from A must leave the half silvered plate G along the path AD, so that the resultant of velocity of light and the velocity of ether drift may act in the direction AC as shown in Fig. 1.9.

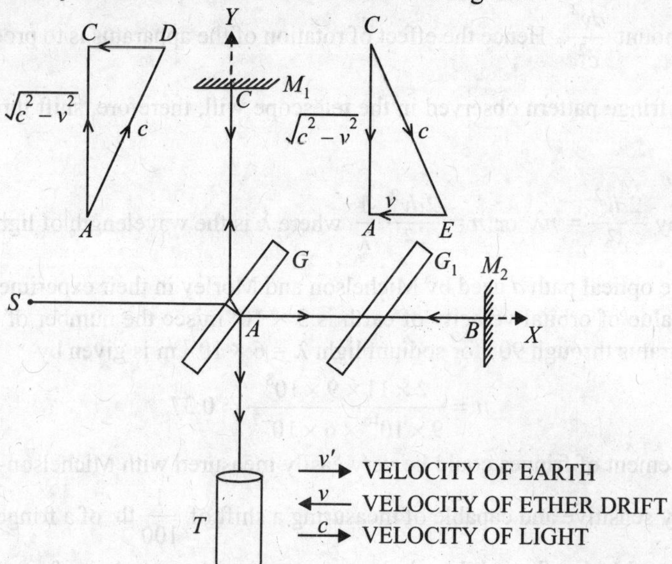

Fig. 1.9

$\therefore$ The resultant velocity of light along $AC = \sqrt{c^2 - v^2}$

After reflection at C from the mirror M_1, the light is reflected along CE, so that the resultant of the velocity of light and the velocity of *ether* drift again acts in the direction CA as shown in Fig. 1.9.

$\therefore$ The resultant velocity of light $= \sqrt{c^2 - v^2}$ along CA

Hence time taken by light to go from A to C and back from C to A

$$T_y = \frac{d}{\sqrt{c^2 - v^2}} + \frac{d}{\sqrt{c^2 - v^2}} = \frac{2d}{\sqrt{c^2 - v^2}} = \frac{2d}{c\sqrt{1 - \frac{v^2}{c^2}}}$$

Expected classical fringe shift. Thus due to *ether drift*, time taken by light to go from A to B and back again to A will be greater than the time taken by light to go from A to C and back again to A by an amount

$$\Delta T = T_x - T_y$$

$$= \frac{2d}{c\left[1 - \frac{v^2}{c^2}\right]} - \frac{2d}{c\sqrt{1 - \frac{v^2}{c^2}}} = \frac{2d}{c}\left\{1 + \frac{v^2}{c^2} - 1 - \frac{1}{2}\frac{v^2}{c^2}\right\}$$

$$= \frac{2d}{c}\left\{\frac{v^2}{2c^2}\right\} = \frac{d}{c}\frac{v^2}{c^2}$$

to a first approximation neglecting higher powers of $\frac{v^2}{c^2}$.

During this time light travels through a distance

$$c\,\Delta T = c \cdot \frac{d}{c} \cdot \frac{v^2}{c^2} = \frac{dv^2}{c^2}$$

If the whole apparatus is turned through 90° the arm AC will become optically longer than the arm AB by an amount $\frac{dv^2}{c^2}$. Hence the effect of rotation of the apparatus is to produce a path difference $\frac{2dv^2}{c^2}$. The fringe pattern observed in the telescope will, therefore, shift through a number of fringes n given by $\frac{2dv^2}{c^2} = n\lambda$ or $n = \frac{2dv^2}{c^2} \cdot \frac{1}{\lambda}$ where λ is the wavelength of light used.

The effective optical path d used by Michelson and Morley in their experiment in 1881 was 11 metres. As the value of orbital velocity of earth is 3×10^4 m/sec the number of fringes shifted by rotating the apparatus through 90° for sodium light $\lambda = 6 \times 10^{-7}$ m is given by

$$n = \frac{2 \times 11 \times 9 \times 10^8}{9 \times 10^{16} \times 6 \times 10^{-7}} = 0.37$$

This displacement of fringes could be very easily measured with Michelson-Morley apparatus which was highly sensitive and capable of measuring a shift of $\frac{1}{100}$th of a fringe.

Michelson and Morley floated the whole apparatus in a large trough of mercury and kept it in slow rotation at the rate of about 10 complete rotations per hour. The results indicated that no significant shift occurred. This indicates that the relative velocity between the earth and *ether* is zero.

Explanation of null results. (*i*) Michelson tried to explain this null result on the hypothesis that the earth dragged the ether along with it, so that there was no relative motion between the earth and the ether. Lodge in 1892 measured the speed of light near rapidly rotating bodies and found that not more than 0.5% of the velocity of the body was communicated to the ether. Michelson's explanation was, therefore, not tenable.

***(ii)* Lorentz-Fitzgerald Hypothesis.** An alternative explanation of the null result was given by Lorentz and Fitzgerald in 1892. According to Lorentz all matter is made up of atoms and atoms are made up of charged particles that produce electric and magnetic fields. These fields must exert forces on the electromagnetic ether thus causing the atoms and molecules in moving matter to be pushed closer together. They pointed out that due to this interaction between *ether* and matter, an object moving relative to the ether becomes shorter in all its dimensions in the direction parallel to the relative velocity, in the ratio

$$1 : \sqrt{1 - \frac{v^2}{c^2}}$$

In other words, an object of length l_0 when at rest, on moving with a velocity v will contract and have a length l given by $l = l_0 \sqrt{1 - \frac{v^2}{c^2}}$ along the direction of motion. This is known as **Lorentz-Fitzgerald contraction hypothesis.**

If this hypothesis is accepted, it would mean that the value of d along AB will become

$$d \sqrt{1 - \frac{v^2}{c^2}}$$

and the time taken by light to go from A to B and back again to A

$$T_x = \frac{2d \sqrt{1 - \frac{v^2}{c^2}}}{c\left(1 - \frac{v^2}{c^2}\right)} = \frac{2d}{c\sqrt{1 - \frac{v^2}{c^2}}}$$

which is the same as T_y. Hence $\Delta T = 0$ so that there is no path difference and hence no fringe-shift. The contraction could not, however, be logically derived from theory.

***(iii)* Einstein's explanation.** In 1905 Einstein proposed an explanation of the null results of Michelson-Morley experiment in a very simple manner. He argued that the reason why it is not possible to detect relative motion through ether is that the *velocity of light is an invariant.* In other words, the velocity of light is constant and does not depend upon the motion of the source, observer or the medium. This is one of the postulates of the special theory of relativity which mathematically explains why the length of an *object* should contract in the ratio $1 : \sqrt{1 - \frac{v^2}{c^2}}$ when it is in motion with a velocity v.

Conclusion of Michelson-Morley experiment.

Michelson-Morley experiment conclusively proved that relative velocity between the earth and ether is zero. Therefore, it is concluded that a medium like ether does not exist.

Light waves are electromagnetic waves and require no material medium for their propagation.

Light travels through vacuum (or free space) with a velocity $c = \frac{1}{\sqrt{\mu_0 \varepsilon_0}}$ where μ_0 is the permeability and ε_0 the permittivity of free space.

As $\mu_0 = 4\pi \times 10^{-7}$ kg m/coul^2 and $\varepsilon_0 = \frac{1}{36\pi \times 10^9}$ coul^2 sec^2/kg m^3, the value of

$$c = \sqrt{\frac{1}{4\pi \times 10^{-7}} \times 36\pi \times 10^9} = 3 \times 10^8 \text{ ms}^{-1}$$

Light is propagated as transverse wave in which electric field vector $\vec{E}$ and magnetic field vector $\vec{H}$ oscillate or vibrate at right angles to the direction of propagation of the wave. The vectors $\vec{E}$ and $\vec{H}$ are also at right angles to each other. If the wave propagates along $\vec{Z}$ axis and $\vec{E}$ is directed along $\vec{X}$ -axis, then $\vec{H}$ is directed along $\vec{Y}$ -axis and vice-versa.

1.20 REASON FOR OBLIQUE TRANSVERSE PATH

Consider Michelson-Mo rely interferometer as moving along with the earth relative to ether frame with a velocity $\vec{v}$ from left to right in the direction AB or AM_2 along the $+X$ direction of its longitudinal arm. A parallel beam of monochromatic light coming from the source S is split into two parts by a half silvered glass plate G. A part is transmitted through it along longitudinal $+X$ direction and reflected at B from the front silvered mirror M_2. It retraces its path and after reflection from the back surface of G at A enters the telescope T. In the mean time the mirror M_2 occupies the position M_2' and the glass plate G, the position G'.

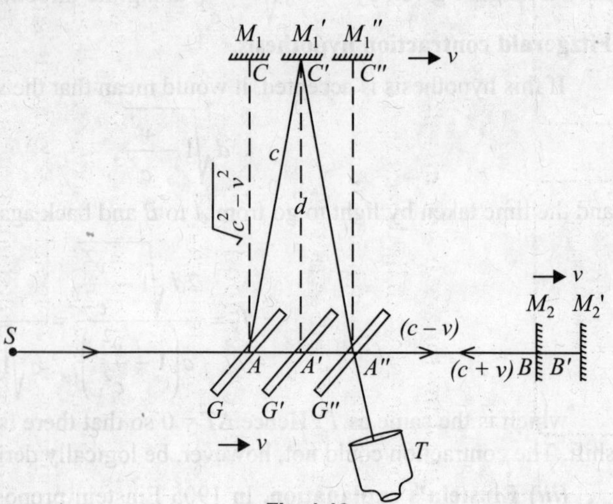

Fig. 1.10

The transverse part of the beam after reflection from the silvered back surface of G falls on the mirror M_1 (Fig. 1.10). In order that the resultant of the velocity of light and velocity of ether drift may act in the transverse direction AC, the beam of light must leave the plate G along the oblique path AC'. After reflection at C' from the mirror M_1 in the position M_1'(when the plate G is in the position G'), the light is reflected along the oblique path C' A'' so that the resultant of the velocity of light and the velocity of ether drift again acts in the transverse direction $C'A'$ and enters the plate G at A'' because in the mean time the mirror M_1 occupies the position M_1'' and the plate G the position G''.

Thus when viewed from ether frame the ray of light travelling along the longitudinal direction (AB) falls on the mirror M_2 at B and is reflected back to A whereas the transverse ray of light travelling along the oblique direction AC' falls on the mirror M_1 at C' in the position M_1' and after reflection travels along the oblique path $C'A''$ to meet the glass plate G at A'' (in the position G'') when the mirror M_1 is in the position M_1''.

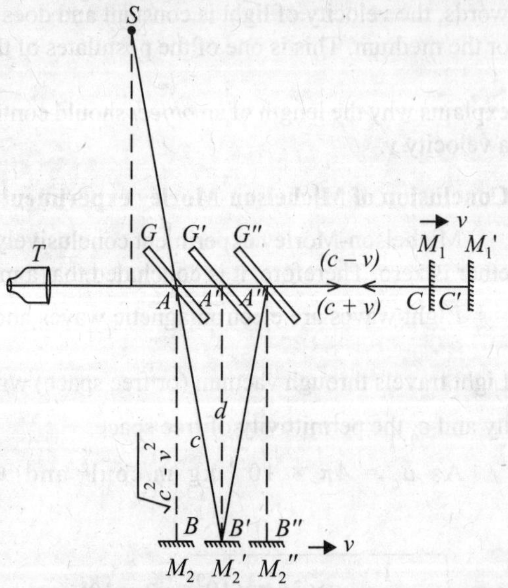

Fig. 1.11

After 90° rotation. After 90° rotation of Michelson-Morley apparatus as shown in Fig. 1.11, he arms interchange their positions and roles. The transverse arm becomes the longitudinal arm and the longitudinal arm becomes the transverse arm. The half silvered glass plate *G* also turns through 90°. In the transverse arm in this orientation light is rendered oblique by the motion of the source *S* (or the collimating lens) which also has been rotated through 90° as also the telescope *T*.

Example 1.5. Find fringe shift when effective length of earth arm is 20 m, orbital velocity of earth about sun is 3 × 10⁴ m/s and wavelength of light used is 6000 Å ***(NagpurUW/2010)***

Solution. Given $V = 3 \times 10^4$ m/s, $d = 20$ m, $\lambda = 6000 \times 10^{-10}$ m $= 6\times 10^{-7}$m

$$\therefore \text{ Number of Fringes shifted, } n = \frac{2dv^2}{c^2}.\frac{1}{\lambda}$$

$$= \frac{2\times20\times9\times10^8}{9\times10^{16}\times6\times10^{-7}}$$

$$= \mathbf{0.67}$$

Example 1.6. Calculate the expected fringe shift in Michelson-Morely experiment if the distance of each path is 2 metres and light is of wavelenght 6000 Å. ***(NagpurU.S/2011)***

Solution. Given $d = 2$m ; $\lambda = 6000$Å $= 6 \times 10^{-7}$ m;

Orbital velocity of earth around the sun, $v = 3\times 10^4$ m/s

$$\therefore \text{ Expected fringe shift, } n = \frac{2dv^2}{c^2}.\frac{1}{\lambda}$$

$$= \frac{2\times2\times9\times10^8}{9\times10^{16}}.\frac{1}{6\times10^{-7}}$$

$$= \mathbf{0.067}$$

Example 1.7. Calculate the expected fringe shift in Michelson-Morley experiment, if the effective length of each path is 6 m, velocity of earth 3×10⁴ m/s and wavelength of monochromatic light used is 5000 A.U ***(NagpurU.S/2010)***

Solution. Given $V = 3 \times 10^4$ m/s; $d = 6$ m, $\lambda = 5000$Å $= 5\ 10^{-7}$

$$\therefore \text{ Expected fringe shift , } n = \frac{2dv^2}{c^2} = \frac{1}{\lambda}$$

$$= \frac{2\times6\times9\times10^8}{9\times10^{16}\times5\times10^{-7}} = \mathbf{0.24}$$

EXERCISE

LONG QUESTIONS

1. What are Newton's notion of space? What are the classical properties of space? What is Eucledian flatness of space? What is the meaning of isotropy of space? Distinguish between homogeneity and isotropy of space. ***(P.U. 2000)***
2. Explain the terms 'frame of reference' 'event' and 'observer' in relativity. ***(MDU 2001; P.U. 2000)***
3. Distinguish between inertial and non-inertial frames of reference. Give one example of each. Is earth an inertial frame? Give reasons. ***(HPU 2003, 2001; Bhopal U. 2004, Bangalore U.2004, OSM U. 2004; M.D.U. 2008, 2003; P.U. 2006, 2004, 2001, Guwahati U. 2000, G.N.D.U. 2007, 2006, Nagpur U. 2011)***

4. What is Newtonian principle of relativity? Discuss with examples why should laws of nature be the same in all inertial frames of reference? *(MDU 2008; H.P.U. 2000)*

5. (a) What does the term translation invariance of space imply? Explain *(Pbi. U. 2002, 2001)*
   (b) Show that the linear uniformity of space leads to Newton's third law of motion and hence to the law of conservation of linear momentum. *(P.U. 2008, M.D.U. 2003)*

6. Prove that law of conservation of linear momentum is a consequence of homogeneity of space. *(Pbi. U. 2007, 2008 , G.N.D.U. 2007, 2004)*

7. (a) What does the term rotational invariance (isotropy) of free space imply?
   (b) Show that the property of rotational invariance of space (isotropy) leads to the law of conservation of angular momentum.
   (M.D.U. 2008, G.N.D.U. 2008, 2007, 2004, 2002, 2001, 2000; PBI. U. 2005, 2001, 2000; P.U. 2009, 2006, 2000, H.P.U. 2003)

8. (a) Give Newton's classical definition of time and its properties.
   (b) What is homogeneity of time? Show that homogeneity of time and Newton's second law of motion result in law of conservation of energy.
   (G.N.D.U. 2007, 2004; P.U. 2005, 2003, 2001, 2000, Pbi. U. 2006)

9. What are Galelian transformations? Derive Galelian transformation equations for two inertial frames. State and prove Galelian invariance.
   (Pbi. U. 2007, 2001, K.U. 2001, M.D.U. 2007; P.U. 2000, G.N. D.U. 2006, 2000; Guwahati U. 2000; Calicut U. 2003)

10. Prove that the Galelian transformation of position vector is expressed as $\vec{r} = \vec{r} + \vec{v}t + \vec{R}$ where $\vec{v}$ is the linear velocity of the frame S′ and $\vec{R}$ is the position vector of the origin O′ as measured by the frame S′ at time t' = 0.

11. (*a*) Show that the distance between two points (or length) is invariant under Galelian transformation. *(P.U. 2001, G.N.D.U. 2008, 2000)*
   (*b*) Show that under Galelian transformations velocity is variant and acceleration is invariant.
   (Pbi. U. 2007, 2006; P.U. 2001, K.U. 2001, M.D.U. 2008, 2003, 2001, G.N.D.U. 2006, 2000)

12. (*a*) What are Galilean transformations? Prove that Newton's laws of motion are invariant under Galelian transformations? *(P.U. 2006, 2000, M.D.U. 2003)*
   (*b*) What are the quantities which are invariant under Galelian transformations?
   (P.U. 2006, 2003; Pbi. U., 2002)

13. (*a*) Show that a frame of reference having a uniform translatory motion (or moving with constant velocity) relative to an inertial frame is also inertial. *(P.U. 2001)*
   (*b*) Show that the force on the particle is the same in two frames connected by Galelian transformations. Hence justify that if the relation $\vec{F} = m\vec{a}$ is used to define force observers in all frames of reference would agree on the magnitude and direction of the force $\vec{F}$ independent of relativity velocity of reference frames. *(Guwahati U. 2000)*

14. Show that the laws of conservation of Momentum and Energy are invariant to Galelian transformations. *(Pbi. U. 2008, 2001, P.U. 2007, 2000; G.N.D.U. 2004, 2003)*

15. (a) Show that the electromagnetic wave equation $\frac{\partial^2\phi}{\partial x^2}+\frac{\partial^2\phi}{\partial y^2}+\frac{\partial^2\phi}{\partial z^2}-\frac{1}{c}\frac{\partial^2\phi}{\partial t^2}$ is non-invariant in Galelian transformation equations where ϕ is scalar potential. *(P.U. 2006)*

16. (a) What was the aim of Michelson-Morley experiment? Describe Michelson-Morley experiment and obtain the expected fringe shift under classical assumptions. Explain how the negative results obtained therefrom are interpreted. Give Lorentz Fitzgerald explanations.

(Pbi. U. 2008, 2007, 2006, 2001, 2000; P.U. 2009, 2008, 2007, 2006, 2005, 2004, 2001, 2000; K.U. 2002, 2001, 2000; M.D.U. 2007, 2006, 2000; H.P.U.2001, G.N.D.U. 2009, 2008, 2007, 2006, Calicut U, 2003, Meerut U. 2007)

17. (a) What efforts were made to explain the null results of Michelson-Morley experiment on the basis of ether hypothesis?

(b) Explain what makes the transverse path of rays oblique when viewed from ether frame.

(c) Draw a ray diagram in ether frame after 90° rotation of the apparatus.

18. (a) Describe Michelson-Morely Experiment.

(b) Discuss theory of the experiment for transmitted rays and reflected rays. Hence show that the fringe shift in fringes due to the rotation of apparatus through 90° is $n = \frac{2dv^2}{xc^2}$, where the symbols have their usual meaning. ***(Nagpur U. 2010 (W)***

19. (a) Explain briefly the necessity of stationary ether hypothesis.

(b) Obtain an expression for fringe-shift in Michelson-Morley experiment due to presence of ether, when the apparatus is rotated through 90°. ***(Nagpur U. S/2010)***

20. Describe Michelson-Morley experiment and show how the negative results obtained from this experiment were interpreted? ***(Nagpur U. S/2011)***

SHORT QUESTIONS

1. Define inertial and non-inertial frames of reference. ***(Nagpur U. 2010)***

2. Why the apparatus for Michelson-Morley experiment was rotated through 90°. ***(G.N.D.U. 2003)***

Ans. The apparatus in Michelson-Morley experiment was rotated through 90° to make the arm *AC* as shown in Fig. 1.8 become optically longer than the arm *AB* by an amount $d\frac{v^2}{c^2}$ where *d* is the distance between half silvered glass plate and the mirror M_1 or M_2. The effect of rotation is to produce a path difference of $2d\ \frac{v^2}{c^2}$ due to which the fringe pattern observed in the telescope shifts through a number of fringes *n* given by

$$2d\frac{v^2}{c^2} = n\lambda$$

where λ is the wavelength of light used. However, no significant shift was observed byMichel-son- Morley.

3. Why did Michelson and Morley repeat the experiment during day and night and during all seasons of the year?

Ans. Michelson-Morley and other scientists repeated the experiment during day and night and during all seasons of the year to find out if there was any effect due to the change in the direction of orbital velocity of the earth as it was different at different times of the year. The experiment was also repeated at different locations of the earth for the same reason.

The earth revolves round the sun and also rotates about its own axis causing change in seasons and producing day and night. The orbital as well as the rotational velocity of earth is non-uniform and varies in magnitude as well as direction. So to avoid the effect of variation in direction and magnitude of the velocity of the earth the experiment was repeated during day and night and all seasons of the year.

MULTIPLE CHOICE QUESTIONS

1. Law of conservation of linear momentum is a consequence of: ***(HPU 2001)***
(*a*) homogeneity of space (*b*) isotropy of space
(*c*) homogeneity of space and time (*d*) homogeneity of time

2. Law of conservation of angular momentum is a consequence of:
(*a*) homogeneity of space (*b*) isotropy of space
(*c*) homogeneity of space and time (*d*) homogeneity of time

3. Homogeneity of time leads to consequence of:
(*a*) linear momentum (*b*) angular momentum
(*c*) total energy (*d*) kinetic energy

4. Which is not explicit function of time?
(*a*) velocity (*b*) acceleration
(*c*) potential energy (*d*) momentum

5. Number of dimensions space has is
(*a*) one (*b*) two
(*c*) three (*d*) four

6. Newton's laws of motion are based on the assumption that space is
(*a*) homogenous (*b*) isotropic
(*c*) homogeneous and isotropic (*d*) invariant under rotation

7. Michelson-Morley experiment was performed to
(*a*) measure speed of light (*b*) prove existence of ether
(*c*) measure of speed of earth relative to ether (*d*) test the isotropy of space

8. Velocity of light in free space is
(*a*) $\sqrt{\mu_o \varepsilon_o}$ (*b*) $(\mu_o \varepsilon_o)^{-\frac{1}{2}}$
(*c*) $\sqrt{\mu_o / \varepsilon_o}$ (*d*) $\sqrt{\varepsilon_o / \mu_o}$

9. Which one is invariant under Galelian transformation?
(*a*) Velocity (*b*) Length
(*c*) Momentum (*d*) Potential energy

ANSWERS

1. (*a*)	**2.** (*b*)	**3.** (*c*)	**4.** (*c*)	**5.** (*c*)
6. (*c*)	**7.** (*c*)	**8.** (*b*)	**9.** (*b*)	

SPECIAL THEORY OF RELATIVITY

2

Chapter

INTRODUCTION

Special theory of relativity is the physical theory of measurement in inertial frames of reference proposed in 1905 by Albert Einstein in the paper '*On the electrodynamics of moving Bodies*'. It generalises Galileo's principle of relativity—that all uniform motion is relative and there is no absolute and well-defined state of rest. Special theory incorporates the principle that the speed of light is the same for all inertial observers regardless of state of motion of the source. This theory has wide range of consequences which have been experimentally verified, including length contraction, time dilation and relativity of simultaneity, contradicting the classical notion that the duration of the time interval between two event is equal for all observers. Combined with other laws of physics, the two postulates (mentioned below) of special relativity predicts the *equivalence of matter and energy* given by formula $E= mc^2$, where c is the speed of light in vacuum. One of the consequences of this theory is that it is impossible for any particle that has rest mass to be accelerated to the speed of light.

The theory was originally termed 'special' because it applied the principle of relativity only to the special case of inertial reference frames, *i.e.* frames of reference in uniform relative motion with respect to each other. Einstein developed general relativity to apply the principle in the more general case *i.e.* to any frame of reference. Special relativity is a valid approximation to general relativity.

2.1. POSTULATES OF SPECIAL THEORY OF RELATIVITY

The special theory of relativity was enunciated in 1905 by Albert Einstein. It has two fundamental postulates These are:

(*i*) *The laws of Physics are invariant in all inertial systems.*

An inertial system is defined as a co-ordinate frame of reference within which the law of inertia *i.e.*, Newton's first law of motion holds. A body on which no net external force acts will move with a uniform velocity if it is in an inertial system. Hence according to this postulate the mathematical form of a physical law remains the same for any two observers moving with constant linear velocity relative to each other. It is, therefore, not possible to distinguish one inertial system from another by an experiment in Physics, as the laws of Physics are the same for an inertial system. In other words, there is no preferred inertial system.

(*ii*) *The speed of light in vacuum is a constant independent of the inertial system, the source and the observer.*

In other words the velocity of light is an **invariant.**

2.2 LORENTZ TRANSFORMATIONS

The equations in relativity Physics which relate the space and time co-ordinates of two co-ordinate systems moving with a uniform velocity relative to one another are called **Lorentz transformations.**

Consider two observers O and O' located in two separate inertial co-ordinate system S and S' respectively. The system S' moves with a uniform velocity v to the *right* along the X-axis relative to S. This is equivalent to the motion of S to the left with a velocity v relative to S'. Suppose each observer carries a *metre rod* (*i.e.*, an arrangement for measuring length) and a *clock* (*i.e.*, an arrangement for measuring time) to measure the position and time of a particle relative to an inertial system. By specifying the *position* and *time* of a physical phenomenon, the observer describes what is called an **event**.

The space and time co-ordinates of an *event* at P described by the observer O are (x, y, z, t) and the co-ordinates of the *same event* as described by the observer O' are $(x'\ y'\ z'\ t')$. The space co-ordinates x, y, z give the distances from the origin O along the X, Y and Z directions as measured by the metre stick of the observer O and t gives the time that he reads on his clock. Suppose both the observers are temporarily at rest with respect to each other when they compare their metre sticks and synchronise their clocks. The system S' is then set in motion with respect to system S. When the origin of S' passes the origin of S both the clocks read zero *i.e.*, $t = 0$ and $t' = 0$ and at this instant $x = x'$.

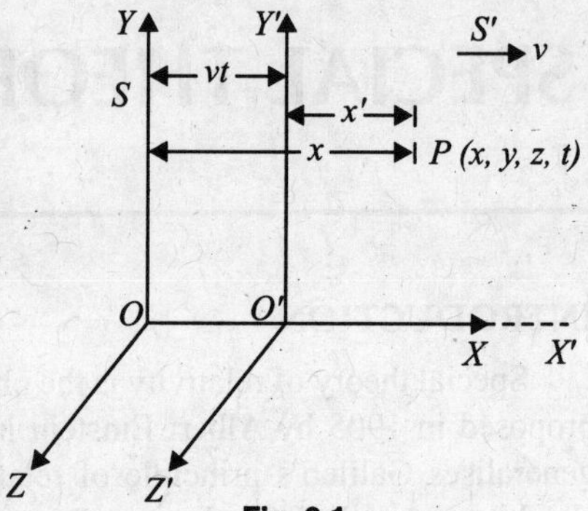

Fig. 2.1

It is evident from Fig. 2.1 that after a time t as measured by O the origin of the system S' is at a distance vt from the origin of the system S,

$$\therefore \quad x' = x - vt$$

As the relative motion between S' and S is at right angles to Y and Z axes, the position co-ordinates

$$y' = y \quad z' = z$$

Further, according to classical Physics the two observers O and O' will compute the same time for any signal originating from P after giving allowance for the velocity of S'

$$\therefore \quad t' = t$$

Hence non-relativistic Galelian transformation equations are

$$x' = x - vt, \quad y' = y, \quad z' = z, \quad t' = t$$

Lorentz transformation equations which satisfy the relativity requirements according to the two postulates must also be *linear* and not quadratic or of a higher order. A quadratic equation has two roots and a higher order equation even more. In order that an event (x, y, z, t) in the inertial system S may correspond to a *single* event (x', y', z', t') in the inertial system S' and *vice-versa* there must be a *one to one correspondence* and the transformation equations must be linear in space co-ordinates as well as in time co-ordinates.

As the inertial system S' is moving with respect to the inertial system S along the X-direction, a ray of light parallel to the X-axis remains always parallel to X'-axis.

$$\therefore \quad y = y' \quad \text{and} \quad z = z'$$

Hence x and t will both depend upon x' and t'.

We shall, therefore, assume the simplest *linear* equations for Lorentz transformations *i.e.*,

$$x = Ax' + Bt', \quad y = y', z = z', \quad t = Gx' + Ht'$$

It has been assumed in the above equations that it is possible that the time interval for the two observers O and O' in the inertial systems S and S' may not be identical.

Imagine that at the time $t = t' = 0$ when the origin O' coincides with the origin O a spherical pulse of light leaves the common origin of S and S'. As the velocity of light is *invariant* each observer sees a spherical wave expanding outwards with the speed c in his own system as measured by his own metre stick and clock.

$\therefore$ For the observer O in the system S, $c = \dfrac{\sqrt{x^2 + y^2 + z^2}}{t}$

or $\quad x^2 + y^2 + z^2 = c^2 t^2 \quad$ or $\quad x^2 + y^2 + z^2 - c^2t^2 = 0 \qquad$ ...(*i*)

Similarly for the observer O' in the system S', $c = \dfrac{\sqrt{x'^2 + y'^2 + z'^2}}{t}$

or $$x'^2 + y'^2 + z'^2 = c^2t'^2 \quad \text{or} \quad x'^2 + y'^2 + z'^2 - c^2t'^2 = 0 \qquad ...(ii)$$

From (*i*) and (*ii*) we have $x^2 + y^2 + z^2 - c^2t^2 = x'^2 + y'^2 + z'^2 - c^2t'^2$

$$\therefore \quad y = y' \text{ and } z = z' \quad \therefore \; x^2 - c^2t^2 = x'^2 - c^2t'^2$$

Substituting $x = Ax' + Bt'$ and $t = Gx' + Ht'$ we have

$$(Ax' + Bt')^2 - c^2 (Gx' + Ht')^2 = x'^2 - c^2t'^2$$

Comparing coefficients on both sides we have

$$x'^2 [A^2 - c^2G^2] = x'^2 \quad \text{or} \quad A^2 - c^2G^2 = 1 \qquad ...(iii)$$

$$t'^2 [B^2 - c^2H^2] = - c^2t'^2 \quad \text{or} \quad B^2 - c^2H^2 = - c^2 \qquad ...(iv)$$

and $$2x't' [AB - c^2 GH] = 0 \quad \text{or} \quad AB - c^2GH = 0 \qquad ...(v)$$

Now when $x' = 0, x = vt$

Substituting these values in the relations

$$x = Ax' + Bt' \quad \text{and} \quad t = Gx' + Ht' \text{ we get } vt = Bt' \text{ and } t = Ht'$$

$$\therefore \quad vHt' = Bt' \quad \text{or} \quad B = vH$$

Substituting $B = vH$ in (*iv*) we have $v^2H^2 - c^2H^2 = -c^2$

or $$H^2 = \frac{c^2}{c^2 \quad v^2} \qquad \boxed{\therefore H = \frac{1}{\sqrt{1 - \frac{v^2}{c^2}}}}$$

Substituting this value of H in $B = vH$, we have $\boxed{B = \dfrac{v}{\sqrt{1 - \frac{v^2}{c^2}}}}$

Substituting $B = vH$ in (*v*), we have

$$AvH - c^2GH = 0 \quad \text{or} \quad Av - c^2G = 0$$

$$\therefore \quad G = \frac{Av}{c^2}$$

Substituting the value of $G = \dfrac{Av}{c^2}$ in (*iii*), we have

$$A^2 - A^2 \frac{v^2}{c^2} = 1$$

or $$\boxed{A = \frac{1}{\sqrt{1 - \frac{v^2}{c^2}}}}$$

Hence $$\boxed{G = \frac{\frac{v}{c^2}}{\sqrt{1 - \frac{v^2}{c^2}}}}$$

Substituting these values of A B G and H, the Lorentz transformation equation, are

$$\boxed{x = \frac{x' + vt'}{\sqrt{1 - \frac{v^2}{c^2}}}}$$

$$\boxed{t = \frac{t' + \frac{v}{c^2}x'}{\sqrt{1 - \frac{v^2}{c^2}}}} \quad \text{... (2.1)}$$

$$\boxed{y = y'} \text{ and } \boxed{z = z'}$$

Inverse Lorentz transformations. To get the inverse transformation equations we note that O is moving with a velocity $-v$ relative to O'. So interchanging x, y, z, t with x', y', z', t' and putting $-v$ in place of v we get

$$x' = \frac{x - vt}{\sqrt{1 - \frac{v^2}{c^2}}}$$

$$t' = \frac{t - \frac{v}{c^2}x}{\sqrt{1 - \frac{v^2}{c^2}}} \quad \text{...(2.2)}$$

$$y' = y;\ z' = z.$$

Galelian transformations special case of Lorentz transformations. When the velocity of moving frame S' *i.e.*, v is much smaller than the velocity of light c, then $v/c << 1$ so that $\frac{v}{c^2}$ and $\frac{v^2}{c^2}$ tend to zero and in the limit

$$x' = x - vt$$

$$y' = y;\ z' = z \text{ and } t' = t \quad \text{...(2.3)}$$

i.e., the *Lorentz transformation equations reduce to Galelian transformations.*

SOLVED EXAMPLES

Example 2.1. *Show by means of Lorentz transformation equations that*

$$x'^2 - c^2 t'^2 = x^2 - c^2 t^2.$$ ***(Kerala U. 2001)***

Solution. Suppose x', y', z', t' and x, y, z, t are the space-time co-ordinates of an event in the inertial frames S' and S respectively and S' is moving with a velocity along the X-direction relative to S, then Lorentz transformation equations are

$$x' = \frac{x - vt}{\sqrt{1 - \frac{v^2}{c^2}}};\quad y' = y;\ z' = z \text{ and } t' = \frac{t - \frac{v}{c^2}x}{\sqrt{1 - \frac{v^2}{c^2}}}$$

$$x'^2 = \frac{x^2 + v^2t^2 - 2xvt}{1 - \frac{v^2}{c^2}}$$

and $$t'^2 = \frac{t^2 + \frac{v^2}{c^4}x^2 - \frac{2xvt}{c^2}}{1-\frac{v^2}{c^2}}$$

or $$c^2t'^2 = \frac{c^2t^2 + \frac{v^2}{c^2}x^2 - 2xvt}{1-\frac{v^2}{c^2}}$$

$\therefore$ $$x'^2 - c^2t'^2 = \frac{x^2 + v^2t^2 - 2xvt - c^2t^2 - \frac{v^2}{c^2}x^2 + 2xvt}{1-\frac{v^2}{c^2}}$$

$$= \frac{x^2\left(1-\frac{v^2}{c^2}\right) - c^2t^2\left(1-\frac{v^2}{c^2}\right)}{1-\frac{v^2}{c^2}} = x^2 - c^2t^2$$

Hence, $x'^2 - c^2t'^2 = x^2 - c^2t^2$ **Proved.**

Example 2.2. *Show that electromagnetic wave equation* $\nabla^2\phi - \frac{1}{c^2}\frac{\partial^2\phi}{\partial t^2} = 0$ *is invariant under Lorentz transformations where φ is a scalar potential.*

Solution. Consider an inertial frame S' moving with a uniform velocity $\vec{v}$ along the X-direction with respect to frame S. The scalar potential ϕ is a function of space time co-ordinates (x', y', z', t') in the frame S' and (x, y, z, t) in the frame S.

The wave equation

$$\nabla^2\phi - \frac{1}{c^2}\frac{\partial^2\phi}{\partial t^2} = \frac{\partial^2\phi}{\partial x^2} + \frac{\partial^2\phi}{\partial y^2} + \frac{\partial^2\phi}{\partial z^2} - \frac{1}{c^2}\frac{\partial^2\phi}{\partial t^2} = 0 \quad ...(i)$$

In the frame S', $\phi = \phi(x', y', z', t')$. Now

$$\frac{\partial\phi}{\partial x} = \frac{\partial\phi}{\partial x'}\frac{\partial x'}{\partial x} + \frac{\partial\phi}{\partial y'}\frac{\partial y'}{\partial x} + \frac{\partial\phi}{\partial z'}\frac{\partial z'}{\partial x} + \frac{\partial\phi}{\partial t'}\frac{\partial t'}{\partial x} \quad ...(ii)$$

$$\frac{\partial\phi}{\partial y} = \frac{\partial\phi}{\partial x'}\frac{\partial x'}{\partial y} + \frac{\partial\phi}{\partial y'}\frac{\partial y'}{\partial y} + \frac{\partial\phi}{\partial z'}\frac{\partial z'}{\partial y} + \frac{\partial\phi}{\partial t'}\frac{\partial t'}{\partial y} \quad ...(iii)$$

$$\frac{\partial\phi}{\partial z} = \frac{\partial\phi}{\partial x'}\frac{\partial x'}{\partial z} + \frac{\partial\phi}{\partial y'}\frac{\partial y'}{\partial z} + \frac{\partial\phi}{\partial z'}\frac{\partial z'}{\partial z} + \frac{\partial\phi}{\partial t'}\frac{\partial t'}{\partial z} \quad ...(iv)$$

and $$\frac{\partial\phi}{\partial t} = \frac{\partial\phi}{\partial x'}\frac{\partial x'}{\partial t} + \frac{\partial\phi}{\partial y'}\frac{\partial y'}{\partial t} + \frac{\partial\phi}{\partial z'}\frac{\partial z'}{\partial t} + \frac{\partial\phi}{\partial t'}\frac{\partial t'}{\partial t} \quad ...(v)$$

According to Lorentz transformation equations

$$x' = \frac{x - vt}{\sqrt{1-\frac{v^2}{c^2}}} = \beta(x - vt),\; y' = y, z' = z, \text{ and } t' = \frac{t - vx/c^2}{\sqrt{1-\frac{v^2}{c^2}}} = \beta\left(t - \frac{vx}{c^2}\right) \quad ...(vi)$$

where $\beta = \dfrac{1}{\sqrt{1-\frac{v^2}{c^2}}}$

Differentiating Eq. (*vi*) w.r.t x, we get

$$\frac{\partial x'}{\partial x} = \beta,\; \frac{\partial y'}{\partial x} = 0,\; \frac{\partial z'}{\partial x} = 0,\; \frac{\partial t'}{\partial x} = -\beta\frac{v}{c^2}$$

Substituting in Eq. (*ii*), we have

$$\frac{\partial \phi}{\partial x} = \beta\frac{\partial \phi}{\partial x'} - \beta\frac{v}{c^2}\frac{\partial \phi}{\partial t'} = \beta\left(\frac{\partial \phi}{\partial x'} - \frac{v}{c^2}\frac{\partial \phi}{\partial t'}\right)$$

or $$\frac{\partial}{\partial x} = \beta\left(\frac{\partial}{\partial x'} - \frac{v}{c^2}\frac{\partial}{\partial t'}\right)$$

so that $$\frac{\partial^2 \phi}{\partial x^2} = \frac{\partial}{\partial x}\left(\frac{\partial \phi}{\partial x}\right) = \beta^2\left(\frac{\partial}{\partial x'} - \frac{v}{c^2}\frac{\partial}{\partial t'}\right)\left(\frac{\partial \phi}{\partial x'} - \frac{v}{c^2}\frac{\partial \phi}{\partial t'}\right)$$

$$= \beta^2\left(\frac{\partial^2 \phi}{\partial x'^2} + \frac{v^2}{c^4}\frac{\partial^2 \phi}{\partial t'^2} - \frac{2v}{c^2}\frac{\partial^2 \phi}{\partial x'\,\partial t'}\right) \quad ...(vii)$$

Again differentiating Eq. (*vi*) w.r.t y, we get

$$\frac{\partial x'}{\partial y} = 0, \frac{\partial y'}{\partial y} = 1, \frac{\partial z'}{\partial y} = 0, \frac{\partial t'}{\partial y} = 0$$

Substituting in Eq. (*iii*), we have

$$\frac{\partial \phi}{\partial y} = \frac{\partial \phi}{\partial y'} \text{ or } \frac{\partial}{\partial y} = \frac{\partial}{\partial y'}$$

so that $$\frac{\partial^2 \phi}{\partial y^2} = \frac{\partial}{\partial y}\left(\frac{\partial \phi}{\partial y}\right) = \frac{\partial}{\partial y'}\left(\frac{\partial \phi}{\partial y'}\right) = \frac{\partial^2 \phi}{\partial y'^2} \quad ...(viii)$$

Similarly from Eq. (*iv*), we get

$$\frac{\partial^2 \phi}{\partial z^2} = \frac{\partial^2 \phi}{\partial' z^2} \quad ...(ix)$$

Now differentiating Eq. (*vi*) w.r. to t we get

$$\frac{\partial x'}{\partial t} = -\beta v,\; \frac{\partial y'}{\partial t} = 0,\; \frac{\partial z'}{\partial t} = 0 \text{ and } \frac{\partial t'}{\partial t} = \beta \quad ...(x)$$

Substituting in Eq. (*v*), we have

$$\frac{\partial \phi}{\partial t} = -\beta v\frac{\partial \phi}{\partial x'} + \beta\frac{\partial \phi}{\partial t'} = \beta\left(\frac{\partial \phi}{\partial t'} - v\frac{\partial \phi}{\partial x'}\right)$$

$$\frac{\partial^2 \phi}{\partial y^2} = \frac{\partial}{\partial y}\left(\frac{\partial \phi}{\partial y}\right) = \frac{\partial}{\partial y'}\left(\frac{\partial \phi}{\partial y'}\right) = \frac{\partial^2 \phi}{\partial y'^2}$$

or

$$\frac{\partial}{\partial t} = \beta\left(\frac{\partial}{\partial t'} - v\frac{\partial}{\partial x'}\right)$$

so that

$$\frac{\partial^2 \phi}{\partial t^2} = \frac{\partial}{\partial t}\left(\frac{\partial \phi}{\partial t}\right) = \beta^2\left(\frac{\partial}{\partial t'} - v\,\frac{\partial}{\partial x'}\right)\left(\frac{\partial \phi}{\partial t'} - v\,\frac{\partial \phi}{\partial x'}\right)$$

$$= \beta^2\left(\frac{\partial^2 \phi}{\partial t'^2} + v^2\frac{\partial^2 \phi}{\partial x'^2} - 2v\frac{\partial^2 \phi}{\partial t' \partial x'}\right) \quad ...(xi)$$

Substituting the values of $\frac{\partial^2 \phi}{\partial x^2}$ from (*vii*), $\frac{\partial^2 \phi}{\partial y^2}$ from (*viii*), $\frac{\partial^2 \phi}{\partial z^2}$ from (*ix*) and $\frac{\partial^2 \phi}{\partial t^2}$ from (*xi*) in wave equation (*i*), we get

$$\frac{\partial^2 \phi}{\partial x^2} + \frac{\partial^2 \phi}{\partial y^2} + \frac{\partial^2 \phi}{\partial z^2} - \frac{1}{c^2}\left(\frac{\partial^2 \phi}{\partial t^2}\right) = \beta^2\left[\frac{\partial^2 \phi}{\partial x'^2} + \frac{v^2}{c^4}\frac{\partial^2 \phi}{\partial t'^2} - \frac{2v}{c^2}\frac{\partial^2 \phi}{\partial x' \partial t'}\right]$$

$$+ \frac{\partial^2 \phi}{\partial y'^2} + \frac{\partial^2 \phi}{\partial z'^2} - \frac{\beta^2}{c^2}\left[\frac{\partial^2 \phi}{\partial t'^2} + \frac{v^2 \partial^2 \phi}{\partial x'^2} - 2v\frac{\partial^2 \phi}{\partial t' \partial x'}\right]$$

$$= \beta^2\left(1 - \frac{v^2}{c^2}\right)\frac{\partial^2 \phi}{\partial x'^2} + \frac{\partial^2 \phi}{\partial y'^2} + \frac{\partial^2 \phi}{\partial z'^2} - \frac{\beta^2}{c^2}\left[1 - \frac{v^2}{c^2}\right]\frac{\partial^2 \phi}{\partial t'^2}$$

$$= \frac{\partial^2 \phi}{\partial x'^2} + \frac{\partial^2 \phi}{\partial y'^2} + \frac{\partial^2 \phi}{\partial z'^2} - \frac{1}{c^2}\frac{\partial^2 \phi}{\partial t'^2} \quad \left[\because 1 - \frac{v^2}{c^2} = \frac{1}{\beta^2}\right]$$

Hence

$$\frac{\partial^2 \phi}{\partial x^2} + \frac{\partial^2 \phi}{\partial y^2} + \frac{\partial^2 \phi}{\partial z^2} - \frac{1}{c^2}\frac{\partial^2 \phi}{\partial t^2} = \frac{\partial^2 \phi}{\partial x'^2} + \frac{\partial^2 \phi}{\partial y'^2} + \frac{\partial^2 \phi}{\partial z'^2} - \frac{1}{c^2}\frac{\partial^2 \phi}{\partial t'^2}$$

i.e., the wave equation $\nabla^2 \phi - \frac{1}{c^2}\frac{\partial^2 \phi}{\partial t^2} = 0$ is invariant to Lorentz transformations.

Note: Maxwell's equations are also invariant under Lorentz transformations.

2.3 VELOCITY OF LIGHT IS ABSOLUTE

The statement '*velocity of light is absolute*' means that the speed of light in vacuum is a constant, independent of the inertial system, the speed of source and the observer. In other words, the velocity of light is invariant. It has the same value in all inertial frames regardless of the motion of the source of light relative to the observer *i.e.*, it is an absolute quantity.

2.4 CONSEQUENCES OF SPECIAL THEORY OF RELATIVITY

The classical or Newtonian mechanics is applicable to the macroscopic bodies moving with the velocities very much less than the velocity of light ($v \ll c$). Whereas the Einstein's relativistic mechanics is applicable to high velocity particles ($v \to c$) and can also be applied to the cases where $v \ll c$ but this becomes Newtonian limit of relativity, hence then classical mechanics is applied. Under this limiting cases Lorentz transformations are reduced to Galelian transformations, as seen in article 2.2. Thus, Einstein's relativistic effects require the high velocities near to *c i.e*, comparable with the velocity of light *c*.

In special theory of relativity, the length of a moving body gets contracted in the direction of motion. For example, a meter scale is moving, it will appear shorter in the direction of motion *w.r.* to the observer at rest. Similarly time interval between two events measured by the observer moving with uniform velocity (or at rest) will be more, *i.e.*, the clock which is in motion with a velocity near c will appear to run slow as compared to the stationary clock. Further, the events will not be simultaneous w.r. to observer at rest or in uniform motion. Thus, the length, time and simultaneity which are absolute in Newtorian relativity will not be absolute in special theory of relativity. Let us examine these cases in detail.

2.5 LENGTH CONTRACTION

The special theory of relativity rejects the absoluteness of space. Hence length of a body measured at rest and in the state of uniform motion will be different in the direction of motion, when its speed approaches to the speed of light c. In other words, in classical physics length of an object is the same for all observers irrespective of their velocities relative to the object. However, according to the theory of relativity the length of an object depends upon the velocity of the observer with respect to the object.

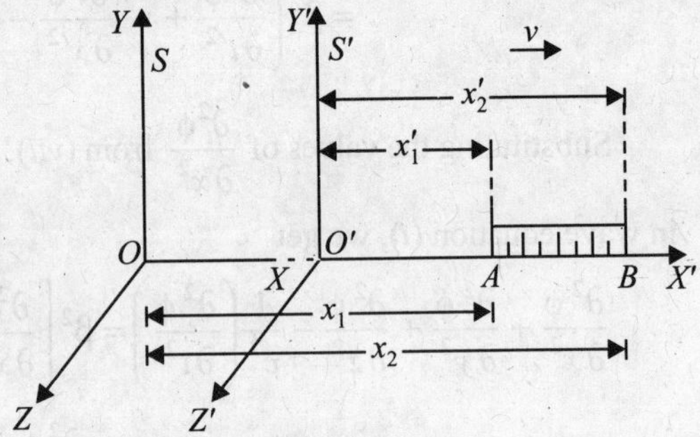

Fig. 2.2

Proof. If two inertial frames S and S' are in relative motion as that S' moves with a uniform velocity v to the right along the X-axis relative to S.

Suppose the observers O and O' in the inertial systems S and S' are at rest with respect to each other with their origins coinciding at some time. They compare their measuring rods and find that both give the same values for the length of an object AB and let it be $= L_0$. Now S' moves to the right along the X-axis with a uniform velocity v with respect to S with the object AB lying along its X-axis. The observer in the system S' places his measuring rod along the X-axis with its left end at $A = x_1'$ and the right end at $B = x_2'$, as shown in Fig. 2.2.

$$\therefore \qquad \text{For } S', \quad AB = x_2' - x_1' = L_0$$

Let us now find the length of the object AB moving with the frame S' as measured by the observer O in the frame S. Suppose the position of the object AB in the reference frame S is x_1 for the end A and x_2 for the end B, then the length of the object lying in the moving frame as measured by S

$$= x_2 - x_1 = L \text{ (say)}$$

As S' is moving to the right along the X-axis with a velocity v with respect to S, therefore according to Lorentz transformations

$$x_1' = \frac{x_1 - vt_1}{\sqrt{1 - \frac{v^2}{c^2}}} \quad \text{and} \quad x_2' = \frac{x_2 - vt_2}{\sqrt{1 - \frac{v^2}{c^2}}}$$

$$\therefore \qquad x_2' - x_1' = \frac{(x_2 - x_1) - v(t_2 - t_1)}{\sqrt{1 - \frac{v^2}{c^2}}}$$

Now in order that S may be able to measure the length of the object AB of S' he should mark the two ends A and B of the object simultaneously *i.e. at one and the same time,*

$$t_2 = t_1$$

$$\therefore \quad x_2' - x_1' = \frac{x_2 - x_1}{\sqrt{1 - \frac{v^2}{c^2}}} \quad ...(i)$$

The proper length *of an object is defined as the length measured by a scale at rest with respect to the object.*

In relation (i) $x_2' - x_1'$ is *actual* or *proper* length of the object as measured by S' himself and is equal to L_0. The *observed* length of the same object as measured by S when S' is in motion is equal to $x_2 - x_1 = L$.

$$\therefore \quad L_0 = \frac{L}{\sqrt{1 - \frac{v^2}{c^2}}}$$

or

$$L = L_0 \sqrt{1 - \frac{v^2}{c^2}}$$

Hence when measuring the length of an object in the system S' moving with a velocity v with respect to an observer in the system S the observer finds the length to be contracted by the factor $\sqrt{1 - \frac{v^2}{c^2}}$. The contraction is *reciprocal i.e.*, the observer in the system S' also finds the length of an object lying in the system S whose *proper* length is L_0 to be equal to

$$L = L_0 \sqrt{1 - \frac{v^2}{c^2}} \quad ...(ii)$$

This phenomenon is known as **length contraction** or space contraction. As it was suggested by Lorentz and Fitzgerald, it is also called Lorentz. Fitzgerald contraction.

Thus, according to the theory of relativity, the length of an object is not absolute, but depends upon the relative motion of the object with respect to the observer. It is maximum when the object is at rest in the observer's inertial system. In other words, *the length of an object measured in a frame moving with respect to an observer is smaller than the length of the same object as observed in stationary frame.*

If the rod is placed parallel to Y-axis or Z-axis perpendicular to the direction of motion along x- axis then,

$L_o = y_2 - y_1$, in S parallel to *Y-axis*

$L = y_2' - y_1'$ in S'

or $L_o = z_2 - z_1$ in S parallel to Z-axis

$L = z_2' - z_1'$ in S'

According to Lorentz inverse transformations,

$L_o = y_2 - y_1 = y_2' - y_1' = L \qquad$ *i.e.*, $L_o = L$

Similarly, $L_o = z_2 - z_1 = z_2' - z_1' = L \qquad$ *i.e.*, $L_o = L$

This shows that length perpendicular to the direction of motion remains unchanged.

Example 2.3. *A circle and square are moving along X-axis. How will they appear to a stationary observer?*

(P.U. 2007, 2006, 2001, 2000; K.U. 2001, Bong. U. 2001, G.N.D.U. 2004, 2002, 2000, M.D.U. 2003, Osm.U. 2004)

Solution. Square and circle in motion. When an inertial frame S' is moving with a velocity $\vec{v}$ along the X-axis with respect to an inertial frame S, then according to relation (i)

$$x_2 - x_1 = (x_2' - x_1')\sqrt{1-\frac{v^2}{c^2}}$$

or
$$L = L_0\sqrt{1-\frac{v^2}{c^2}}$$

i.e. there is a contraction in length of the object in a direction parallel to X-axis. Further according to Lorentz transformations $y_1' = y_1, y_2' = y_2$ and $z_1' = z_1, z_2' = z_2$ *i.e.*, $y_2' - y_1' = y_2 - y_1$ and $z_2' - z_1' = z_2 - z_1$, *i.e.* there is no length contraction in a direction perpendicular to the direction of motion. Thus a moving square appears as a rectangle with the shorter side along its direction of motion (X-axis) and a circle appears as an ellipse with its minor axis along its direction of motion as shown in Fig. 2.3.

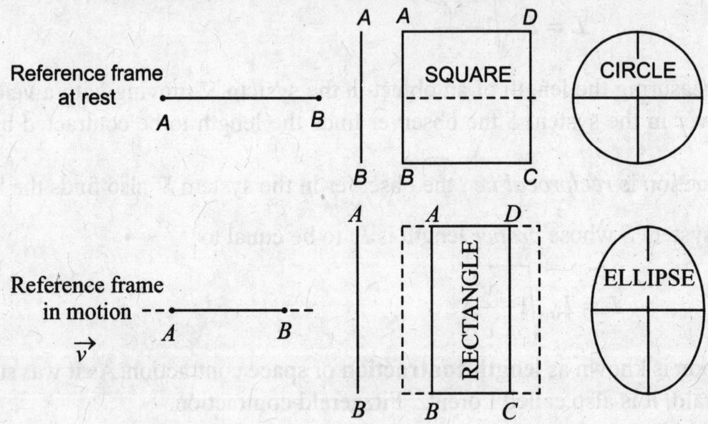

Fig. 2.3

When v is very small as compared to c, $\frac{v^2}{c^2}$ is negligible and therefore Eq. (i) reduces to $L = L_0$ *i.e.*, for ordinary velocities length of an object can be taken to be a constant.

Example 2.4. *The length of a rod as measured in different frames is 10.0 cm., 10.1 cm., 9.8 cm., 9.9 cm. The rod is at rest in one of the frames. Which one? Explain your answer.*

(P.U. 2003)

Solution. The rod is at rest in the frame in which its length is measured as 10.1 cm.

It is because, in a moving frame length contraction takes place and its apparent length decreases. It is not possible for the length to increase in a moving frame. So the frame at rest is that in which the length has the highest value.

Example 2.5. *A rod 1 m long is moving along its length with a velocity 0.6 c. Calculate its length as it appears to (i) an observer on the earth (ii) moving with the rod itself.*

(Nagpur U. 2011)

Solution. The proper length of the rod in the moving (S') frame

$$L_0 = 1 \text{ m}$$

∴ To an observer moving with the rod itself, the length appears to be 1 metre.

Let L be the length of the rod as it appears to an observer on earth, then

$$L = L_0\sqrt{1-\frac{v^2}{c^2}} = 1\sqrt{1-(0.6)^2} = \sqrt{1-0.36}$$

$$= \sqrt{0.64} = 0.8\,\text{m}$$

Example 2.6. *A rod has a length of 2m. Find its length when carried in a rocket with a speed of 2.7×10^8 ms^{-1}.* ***(K.U. 2002, 2000)***

Solution. $$L = L_0\sqrt{1-\frac{v^2}{c^2}} = 2\sqrt{1-\left(\frac{2.7\times10^8}{3\times10^8}\right)^2} = 2\times\sqrt{1-0.81}$$

$$= 2\times\sqrt{0.19} = 2\times0.436 = 0.872\,\text{m}$$

Example 2.7. *The apparent length of a rod in the direction of motion moving with a velocoty of 2×10^8 m/sec is 1 metre. what is the length of the rod at rest?* ***(P.U. 2006)***

Solution. Length of rod at rest , $$L_o = \frac{L}{\sqrt{1-\frac{v^2}{c^2}}}$$

$$= \frac{1}{\sqrt{1-\frac{(2\times10^8)^2}{(3\times10^8)^2}}} = 1.34\ \text{m}$$

Example 2.8. *The length of a spaceship when at rest on the ground measures 100 m. Its length in flight as measured by an observer on the ground is 81 m. Calculate the velocity with which the space-ship is moving?* ***(G.N.D.U. 2008, 2004)***

Solution. $$L = L_o\sqrt{1-\frac{v^2}{c^2}} \text{ or } \frac{81}{100} = \sqrt{1-\frac{v^2}{c^2}} = 0.81$$

$$1-\frac{v^2}{c^2} = (0.81)^2 = 0.6561$$

∴ $$\frac{v^2}{c^2} = 1-0.6561 = 0.3439$$

or $$\frac{v}{c} = 0.5864$$

∴ $$v = 0.5864 \times c = 0.5864 \times 3 \times 10^8 = 1.759 \times 10^8 \text{ m/sec.}$$

Example 2.9. *Calculate the percentage contraction of a rod moving with a velocity 0.8 times the velocity of light in a direction inclined at (a) 60° (b) 45° to its own length.*

Solution. The vector length L of the rod can be resolved into two components:

(*i*) Along the direction of motion $L_x = L\cos 60° = \frac{1}{2}L$

(*ii*) Perpendicular to the direction of motion $L_y = L\sin 60° = \frac{\sqrt{3}}{2}L$

The relativistic contraction occurs only along the direction of motion. There will be a *decrease* in the component L_x. Let its value be L_x'.

$$\therefore \quad L_x' = L_x\sqrt{1-\frac{v^2}{c^2}} = \frac{1}{2}L\sqrt{1-(0.8)^2} = 0.3\,L$$

The component L_y remains unchanged $\quad \therefore L'_y = L_y$

$$\therefore \text{ Length of the rod in the moving frame } L' = \sqrt{\left(L_x'\right)^2 + \left(L_y'\right)^2} = \sqrt{(0.3\,L)^2 + \left(\frac{\sqrt{3}}{2}L^2\right)} = 0.916\,L$$

$$\text{Percentage contraction } \frac{L-0.916\,L}{L}\times 100 = 0.084\times 100 = 8.4\%$$

(*b*) Here $L_x = L\cos 45° = \dfrac{L}{\sqrt{2}};\; L_x' = L_x\sqrt{1-\dfrac{v^2}{c^2}} = \dfrac{L}{\sqrt{2}}\sqrt{1-(0.8)^2} = 0.3\sqrt{2}\,L$

$$L_y = L\sin 45° = \frac{L}{\sqrt{2}};\; L_y' = L_y$$

$$L' = \sqrt{\left(L_x'\right)^2 + \left(L_y'\right)^2} = \sqrt{(0.3\sqrt{2}\,L)^2 + \left(\frac{L}{\sqrt{2}}\right)^2} = \sqrt{0.68\,L} = 0.825\,L$$

$$\text{Percentage contraction } = \frac{L-0.825\,L}{L}\times 100 = 0.175\times 100 = 17.5\%$$

Example 2.10. *Calculate the length and orientation of a rod of length 5 metres in a frame of reference which is moving with a velocity 0.6 c in a direction making an angle of 30° with the rod.* ***(P.U. 2001)***

Solution. The vector length L of the rod can be resolved into two components:

(*i*) Along the direction of motion $L_x = L\cos 30°$

(*ii*) Perpendicular to the direction of motion $L_y = L\sin 30°$

The relativistic contraction occurs only along the direction of motion. There will thus be a *decrease* in the component L_x. Let its value be L_x'.

$$\therefore \quad L_x' = L_x\sqrt{1-\frac{v^2}{c^2}} = L\cos 30°\sqrt{1-(0.6)^2}$$

$$= 5\times\frac{\sqrt{3}}{2}\times 0.8 = 2\sqrt{3} = 3.464 \text{ metres}$$

The component L_y remains unchanged.

$$\therefore \quad L_y' = L_y = L\sin 30° = 5\times\frac{1}{2} = 2.5 \text{ metre}$$

$$\therefore \quad \text{Length of the rod in the moving frame } L' = \sqrt{(L_x')^2 + (L_y')^2}$$

$$= \sqrt{(2\sqrt{3})^2 + (2.5)^2} = 4.27 \text{ metres}$$

Orientation of the rod is given by the angle θ that the rod makes with the direction of velocity v.

$$\therefore \quad \tan\theta = \frac{L_y'}{L_x'} = \frac{2.5}{3.464} = 0.72$$

or $$\theta = \tan^{-1}(0.72) = 35°\,45'$$

Example 2.11. *The length of a rod is found to be half of its length when at rest. What is the speed of the rod relative to the observer?*

Solution. Let L_0 be the length of the rod at rest and L its length when moving with a velocity v, then

$$L = L_0\sqrt{1-\frac{v^2}{c^2}}$$

But $$L = \frac{1}{2}L_0$$

$$\therefore \quad \frac{1}{2}L_0 = L_0\sqrt{1-\frac{v^2}{c^2}} \quad \text{or} \quad \frac{1}{2} = \sqrt{1-\frac{v^2}{c^2}}$$

$$\therefore \quad \frac{1}{4} = 1-\frac{v^2}{c^2} \text{ or } \frac{v^2}{c^2} = 1-\frac{1}{4} = \frac{3}{4}$$

$$\therefore \quad \frac{v}{c} = \frac{\sqrt{3}}{2} \text{ or } v = \frac{\sqrt{3}}{2}c = 0.866\,c$$

Example 2.12. *The length of a rod is 100 metres. If length of this rod is measured by the observer moving parallel to its length is 51 metres, find the speed of the observer.* ***(P. U. 2004)***

Solution. $$L = L_0\sqrt{1-\frac{v^2}{c^2}} \text{ or } 51 = 100\sqrt{1-\frac{v^2}{c^2}} \textbf{ or } (0.51)^2 = 1-\frac{v^2}{c^2}$$

or $$\frac{v^2}{c^2} = 1-0.2601 = 0.7399 \qquad \therefore \quad \frac{v}{c} = 0.86 \textbf{ or } v = 0.86\,c$$

Example 2.13. *Calculate the percentage contraction in the length of a rod in a frame of reference moving with velocity 0.8c in a direction parallel to the length.* ***(Nagpur U. 2010)***

Solution. Given: $v = 0.8c$, $\frac{v}{c} = 0.8$, $\frac{v^2}{c^2} = 0.64$; $\sqrt{1-\frac{v^2}{c^2}} = 0.6$

$$\therefore \quad L = L_o\sqrt{1-\frac{v^2}{c^2}} = 0.6\ L_o$$

$$\therefore \text{ Percentage contraction} = \frac{L_o - 0.6L_o}{L_o}\times 100 = (1-0.6)\times 100 = 0.4\times 100 = 40\%$$

Example 2.14. *The area of a square as measured from the frame attached to it is A_0. What will be its area with respect to an observer who is moving with velocity $\vec{v}$ along one side of the square?*

Solution. Consider a square of side l_0 lying in the X-Y plane of the co-ordinate system S attached to it with one side parallel to X-axis and the other side parallel to Y-axis. Suppose another co-ordinate system S' is moving with a velocity $\vec{v}$ along X-axis. To an observer in S' the side of the square along X-axis will appear to be contracted to l given by

$$l = l_0\sqrt{1-\frac{v^2}{c^2}}$$

The side of the square along Y-axis as measured by the observer in the frame S' will remain unaffected and $= l_0$

$\therefore$ The area of the square with respect to an observer in the frame S' moving with velocity along one side of the square (X-axis)

$$= A = l l_0 = l_0^2 \sqrt{1-\frac{v^2}{c^2}} = A_0 \sqrt{1-\frac{v^2}{c^2}}$$

Example 2.15. *With what speed should a circular disc move parallel to its diameter for its arrea to become half the area at rest?* ***(Kerala U. 2001)***

Solution. Let a be the radius of the circular disc. Suppose the disc moves along X-axis with a velocity v parallel to the X-Y plane of the co-ordinate system at rest, then its diameter $2a$ along X-axis appears to contract to the value $2b$ such that $b = a\sqrt{1-\frac{v^2}{c^2}}$ and the diameter along Y-axis appears to be the same. Thus the circular disc of radius a at rest appears as an ellipse of semi-major axis a along Y-axis (perpendicular to the direction of motion) and semi-minor axis b along X-axis (parallel to the direction of motion).

$\therefore$ The observed area (of the ellipse) = πab

Now, the area of the disc moving with velocity v parallel to its diameter = half its area at rest

$$\therefore \quad \pi a b = \frac{1}{2}\pi a^2$$

But $$b = a\sqrt{1-\frac{v^2}{c^2}} \quad \therefore \quad \pi a^2 \sqrt{1-\frac{v^2}{c^2}} = \frac{1}{2}\pi a^2$$

or $$\sqrt{1-\frac{v^2}{c^2}} = \frac{1}{2} \quad \text{or } 1-\frac{v^2}{c^2} = \frac{1}{4}$$

or $$\frac{v^2}{c^2} = 1-\frac{1}{4} = \frac{3}{4} \quad \therefore \quad \frac{v}{c} = \frac{\sqrt{3}}{2} \text{ or } v = \frac{\sqrt{3}}{2}c = 0.866\,c$$

Example 2.16. *Transform the displacement vector* $(4\hat{i}+3\hat{j})$ *cm in a system S to system S' when S' has a velocity* $0.8\,c\hat{i}$ *relative to S'.* ***(P.U. 2001)***

Solution. In the frame S'

Displacement along X-axis $L_x = 4$ cm

Displacement along Y-axis $L_y = 3$ cm.

S' is moving with a velocity $0.8\,c\hat{i}$ relative to S. This velocity is along X-axis.

$\therefore$ Observed displacement along X-axis, $L_x' = L_x\sqrt{1-\frac{v^2}{c^2}}$

$$= 4\sqrt{1-(0.8)^2} = 4 \times 0.6 = 2.4\,\text{cm}$$

There is no effect on the displacement in a perpendicular direction *i.e.* along Y-axis.

$$\therefore \quad L_y' = L_y = 3 \text{ cm}$$

$\therefore$ Displacement vector with respect to system S' = $(2.4\hat{i} + 3\hat{j})$ cm.

2.6 TIME DILATION

Like space interval, the time interval between two events is also different for two observers moving relative to one another.

Suppose two observers O and O' in the inertial systems S and S' respectively are at rest with respect to each other. They synchronise their respective clocks and they agree that the time interval between any *two events* as measured by their own clocks is the same. Let the clock in the system S give signals at regular intervals and suppose the system S' moves to the right along the X-axis with a uniform velocity v with respect to S. The observer O in the frame S keeps his clock at a fixed point x_1 in his system and measures the time interval T_0 that elapses between two events (say two signals) that occur at times t_1 and t_2 in his frame.

$$\therefore \qquad T_0 = t_2 - t_1 \qquad \text{...}(i)$$

Let the times registered by the observer O' in the inertial system S' between the *same* two events be t_1' and t_2'. As S' is moving to the right with a uniform velocity v with respect to S', therefore, according to Lorentz transformations.

$$\therefore \qquad t_1' = \frac{t_1 - \frac{v}{c^2}x_1}{\sqrt{1-\frac{v^2}{c^2}}}$$

and

$$t_2' = \frac{t_2 - \frac{v}{c^2}x_1}{\sqrt{1-\frac{v^2}{c^2}}}$$

$$\therefore \qquad t_2' - t_1' = \frac{(t_2-t_1) - \frac{v}{c^2}(x_2-x_1)}{\sqrt{1-\frac{v^2}{c^2}}}$$

The clock in the system S remains fixed at the point x_1.

$$x_2 = x_1$$

Hence

$$t_2' - t_1' = \frac{t_2 - t_1}{\sqrt{1-\frac{v^2}{c^2}}}$$

Thus the moving observer O' in the inertial system S' measures the time interval $t_2' - t_1' = T$ between the same two events for which the observer O in the system S measures the time interval

$$t_2 - t_1 = T_0$$

$$\therefore \qquad T = \frac{T_0}{\sqrt{1-\frac{v^2}{c^2}}}$$

Why moving clock appears to run slow? As discussed above the moving observer O' in the inertial system S' measures the time interval T between the same two events for which the observer O in the system S measures time interval T_0 and

$$T = \frac{T_0}{\sqrt{1-\frac{v^2}{c^2}}}$$

As v is less than c, $T > T_0$. Hence the observer S' measures a longer interval of time between the two events with his clock at rest with respect to him. In other words, a clock in the frame S appears to **go slow** to an observer in the frame S' who is in motion with respect to the frame S. **This phenomenon is known as time dilation.** If there are two observers in relative motion with respect to each other, each observer would find the other moving observer's clock to run slow. Thus the consequence of time dilation is reciprocal. As an example of the phenomenon of time dilation, suppose that the velocity of the observer

$$v = 0.98\ c, \text{ then}$$

$$\sqrt{1-\frac{v^2}{c^2}} = \sqrt{1-\frac{0.98^2 c^2}{c^2}} = 0.2$$

$$\therefore \qquad T = \frac{T_0}{0.2} = 5T_0$$

In other words, if the moving observer in the inertial frame S' measures a time interval of 5 seconds between two events by his own clock, the interval between the same two events, as measured by him on the clock of the stationary observer in the inertial frame S, will be 1 second and *vice-versa*.

Proper and non-proper time. We have seen that an observer O in the inertial frame S measures a time interval $T_0 = t_2 - t_1$ between two events by a clock fixed in his frame at a position, say x_1, whereas the time interval between the same two events is measured as $T = t_2' - t_1' = \dfrac{t_2 - t_1}{\sqrt{1-\frac{v^2}{c^2}}}$ by the observer O' in the inertial frame S' moving with a velocity v in the $+X$ direction with respect to the clock in the frame S'.

The time interval between two events which occur at the same position measured by a clock in the inertial frame in which the events occur is called proper time.

The time interval between the same two events measured by an observer from an inertial frame which is moving with respect to the clock is known as non-proper or relativistic time.

When $v << c$. When $v << c$ Lorentz transformations for time reduce to Galelian transformations.

2.7 EXPERIMENTAL VERIFICATION OF TIME DILATION

The high energy particles produced by a synchro-cyclotron give rise to π^+ mesons moving with a velocity 0.99 c when they strike a target. To verify experimentally time dilation, as given by the theory of relativity, the flux of these π^+ mesons was measured at two places 30 metres apart. The π^+ mesons decay with a half-life of $t = 1.8 \times 10^{-8}$ sec. The time taken by π^+ mesons to travel a distance of 30 metres, as measured by an observer at rest in the laboratory –8 is given by

$$T = \frac{30}{0.99 \times 3 \times 10^8} = 10 \times 10^{-8} \text{ sec. (Approx.)}$$

If N_0 is the number of π^+ mesons in the beginning and N after travelling a distance of 30 metres *i.e.*, after a time, T, then

$$\text{fractional flux} \quad \frac{N}{N_0} = \left(\frac{1}{2}\right)^{T/t}$$

$$\text{Now} \qquad \frac{T}{t} = \frac{10 \times 10^{-8}}{1.8 \times 10^{-8}} = 5.6$$

$$\therefore \quad \frac{N}{N_0} = \frac{1}{2^{5.6}} = \frac{1}{48.5} = 2\% \text{ (Approx.)}$$

i.e., the flux should fall to 2% of the original value. On actual observation, the flux was found to be nearly 60% of the actual value. This discrepancy has been explained by calculating the *proper time* T_0 in the moving frame of π^+ meson, according to the theory of relativity, which gives

$$T = \frac{T_0}{\sqrt{1-\frac{v^2}{c^2}}}$$

or

$$T_0 = T\sqrt{1-\frac{v^2}{c^2}} = 10 \times 10^{-8}\,[1-(0.99)^2]^{1/2} = 1.4 \times 10^{-8} \text{ sec.}$$

During this time T_0, $\frac{T_0}{t} = \frac{1.4\times 10^{-8}}{1.8\times 10^{-8}} = 0.78$

and the fractional flux is given by

$$\frac{N}{N_0} = \frac{1}{2^{0.78}} = \frac{1}{1.714} = 60\% \text{ (Approx.)}$$

It is clear from the above experiment that in the laboratory measurement, time taken by π^+ mesons to travel a distance of 30 m is 10×10^{-8} sec, whereas π^+ mesons themselves measure the time as 1.4×10^{-8} secs. This verifies experimentally the time dilation.

2.8. SIMULTANEITY OF EVENTS

This is another important consequnce of Lorentz transformations. It say that any two events accuring simultaneously at different places in one frame of reference will not appear to be simultaneous in another frame of reference. This means simultaneity is not absolute but is relative.

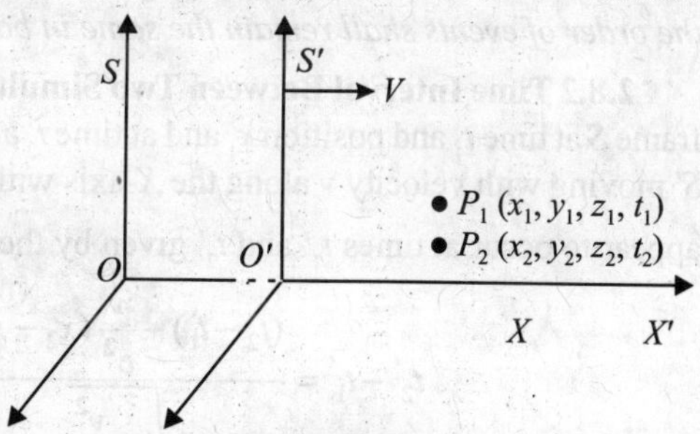

Fig. 2.4

Consider two events characterised by the space and time co-ordinates (x_1, y_1, z_1, t_1) and (x_2, y_2, z_2, t_2) in the frame S and by (x_1', y_1', z_1', t_1') and (x_2', y_2', z_2', t_2') in the frame S'. The frame S' is moving with a velocity relative to S *i.e.*, along $+X$ axis, fig. 2.4.

The time co-ordinates of the two events are connected by Lorentz transformations

$$t_1' = \frac{t_1 - \frac{v}{c^2}x_1}{\sqrt{1-\frac{v^2}{c^2}}} \quad \text{and} \quad t_2' = \frac{t_2 - \frac{v}{c^2}x_2}{\sqrt{1-\frac{v^2}{c^2}}} \qquad \ldots(i)$$

If the two events are *simultaneous* in the frame S, $t_1 = t_2$ and therefore from (*i*) we have $t_1' \neq t_2'$ because $x_1 \neq x_2$. Hence the two events are not simultaneous in S'. Similarly events which are simultaneous in S' will not be so in S.

Thus two events which are simultaneous at *different positions* in reference frame S are not simultaneous in reference frame S' moving with velocity relative to S.

2.8.1 Order of events. Let us now consider the *order of events* as observed in the two frames. Let event t_1 occurs first and t_2 later in the frame S. The time difference between the same two events in S' will be given by

$$t_2' - t_1' = \frac{t_2 - \frac{v}{c^2}x_2}{\sqrt{1-\frac{v^2}{c^2}}} - \frac{t_1 - \frac{v}{c^2}x_1}{\sqrt{1-\frac{v^2}{c^2}}}$$

$$= \frac{1}{\sqrt{1-\frac{v^2}{c^2}}}\left[(t_2 - t_1) - \frac{v}{c^2}(x_2 - x_1)\right]$$

If the term within brackets is **zero,** the two events will be *simultaneous* in S'.

If the term within brackets is **positive** *i.e.,* greater than zero, the *order of events is the same* in both the frames.

However, if the term within brackets is **negative** *i.e.,* less than zero, then the events in S' will be observed to be in the reverse order, but this is an *impossible case.*

The last case will happen if $\quad \frac{v}{c^2}(x_2 - x_1) > (t_2 - t_1)$

As v is always less than c $\quad \frac{1}{c}(x_2 - x_1) > (t_2 - t_1)$

or $\quad (x_2 - x_1) > c\,(t_2 - t_1)$

But this is not possible, as no signal can travel with a velocity greater than that of light. *Hence the order of events shall remain the same in both the inertial frames.*

2.8.2 Time Interval Between Two Simultaneous Events. When two events occur in an inertial frame S at time t_1 and position x_1 and at time t_2 and position x_2, then to an observer in the inertial frame S' moving with velocity v along the X-axis with respect to the inertial frame S, these two events will appear to occur at times t_1' and t_2' given by the relation

$$t_2' - t_1' = \frac{(t_2 - t_1) - \frac{v}{c^2}(x_2 - x_1)}{\sqrt{1-\frac{v^2}{c^2}}}$$

When the events are simultaneous in the frame S, $(t_2 - t_1) = 0$ and if the two points are separated by a distance Δx, then

$$\Delta x = (x_2 - x_1)$$

The two events will, therefore, appear separated by a time interval

$$t_2' - t_1' = \Delta t' = \frac{-\frac{v}{c^2}\Delta x}{\sqrt{1-\frac{v^2}{c^2}}}$$

2.9. FOUR DIMENSIONAL VOLUME ELEMENT INVARIANT UNDER LORENTZ TRANSFORMATIONS

Lorentz transformation equations lead to the formulae of length contraction and time dilation. These are

(*i*) $L = L_0 \sqrt{1-\frac{v^2}{c^2}}$ where L is the observed length in the moving frame S' and L_0 the proper length in the stationary frame S.

(*ii*) $T = \frac{T_0}{\sqrt{1-\frac{v^2}{c^2}}}$ where T is observed time in the moving frame S' and T_0 the proper time in the stationary frame S.

Denoting the observed length in the frame S' as dx' and proper length in the frame S' as dx, we have

$$dx' = dx \sqrt{1-\frac{v^2}{c^2}}$$

The moving frame S' is taken to move with velocity along the X-axis. Therefore $dy' = dy$; $dz' = dz$.

Similarly denoting the observed time in the frame S' as dt', and proper time in the frame S' as dt, we have

$$dt' = \frac{dt}{\sqrt{1-\frac{v^2}{c^2}}}$$

∴ The four dimensional volume element in the moving frame is

$$dx'\, dy'\, dz'\, dt' = dx \sqrt{1-\frac{v^2}{c^2}}\, dy\, dz\, \frac{dt}{\sqrt{1-\frac{v^2}{c^2}}}$$

$$= dx\, dy\, dz\, dt$$

Hence the four dimensional volume element $dx\, dy\, dz\, dt$ is invariant under Lorentz transformations.

Example 2.17. *Half-life of a particle at rest is 17.8 nanosecond. What will be the half-life when its speed is 0.8 c?* ***(Pbi.U. 2001)***

Solution. Half-life of the particle in the rest frame $T_0 = 17.8 \times 10^{-9}$ s

Let T be the half-life of the particle when its speed v is 0.8 c, then

$$T = \frac{T_0}{\sqrt{1-\frac{v^2}{c^2}}} = \frac{17.8 \times 10^{-9}}{\sqrt{1-\frac{(0.8c)^2}{c^2}}} = \frac{17.8 \times 10^{-9}}{\sqrt{1-0.64}}$$

$$= \frac{17.8 \times 10^{-9}}{0.6} = 2.97 \times 10^{-8}\text{ s}$$

Example 2.18. *With what velocity should a rocket move so that every year spent on it corresponds to 4 years on earth?* ***(Nagpur U. 2010; Pbi.U. 2007)***

Solution. The rocket is the moving frame and earth is the rest frame. The time of 1 year on the rocket is the proper time T_0 and the corresponding time on earth is the relativistic time $T = 4$ years.

Now $$T = \frac{T_0}{\sqrt{1-\frac{v^2}{c^2}}}$$

where v is the velocity with which the rocket should move.

Hence $$4 = \frac{1}{\sqrt{1-\frac{v^2}{c^2}}} \text{ or } \sqrt{1-\frac{v^2}{c^2}} = \frac{1}{4}$$

$$\therefore \quad 1-\frac{v^2}{c^2} = \frac{1}{16} \text{ or } \frac{v^2}{c^2} = 1-\frac{1}{16} = \frac{15}{16}$$

$$\therefore \quad v = \sqrt{\frac{15}{16}}\, c = 0.968\, c$$

Example 2.19. *Determine the time (as measured by a clock at rest on the rocket) taken by a rocket to reach a distant star and return to earth with a constant velocity v equal to* $\sqrt{0.9999}$ *c, if the distance to the star is 4 light years.*

Solution. Distance of the star = 4 light years

Time taken by light to reach the star = 4 years

Velocity of the rocket, $v = \sqrt{0.9999}\, c$

Time taken by the rocket to go to the star and back, as measured by an observer on the rocket on a stationary clock on the earth

$$T = \frac{2 \times 4c}{\sqrt{0.9999}\, c} = \frac{8}{\sqrt{0.9999}} \text{ years}$$

Let T_0 be the time measured by the observer on the rocket on the clock in the rocket, then

$$T = \frac{T_0}{\sqrt{1-\frac{v^2}{c^2}}} = \frac{T_0}{\sqrt{1-0.9999}} = \frac{T_0}{0.01}$$

$$\therefore \quad T_0 = T \times 0.01 = \frac{8 \times 0.01}{\sqrt{0.9999}} = 0.08 \text{ year.}$$

Example 2.20. *A clock keeps correct time. With what speed should it travel relative to an observer so that it appears to lose 1 minute in 24 hours?* ***(P.U. 2000)***

Solution. Time as measured by the clock $T = 24 \times 60$ minutes

Time as measured by the observer when the clock is in motion $T_0 = (24 \times 60 - 1)$ minute.

Let v be the velocity of the clock, then

$$T = \frac{T_0}{\sqrt{1-\frac{v^2}{c^2}}}$$

or $$1-\frac{v^2}{c^2} = \left(\frac{T_0}{T}\right)^2 = \left(\frac{24 \times 60 - 1}{24 \times 60}\right)^2 = \left(\frac{1439}{1440}\right)^2 = (0.9993)^2$$
$$= 0.9986$$

or $$\frac{v^2}{c^2} = 1 - 0.9986 = 0.0014$$

or $$\frac{v}{c} = 0.0374$$

$\therefore$ $$v = c \times 0.0374 = 3 \times 10^8 \times 0.0374 = 0.1122 \times 10^8 \text{ ms}^{-1}$$

Example 2.21. *A clock keeps correct time on earth. It is put on a spaceship moving uniformly with a speed of 1 × 10⁸ ms⁻¹. How many hours does it appear to lose per day?* **(P.U. 2003)**

Solution. Time measured by an observer moving with the clock $T = 24$ hrs

Time measured by an observer on earth = T_0

Now $$T = \frac{T_0}{\sqrt{1-\frac{v^2}{c^2}}} = \frac{T_0}{\sqrt{1-\left(\frac{1\times 10^8}{3\times 10^8}\right)^2}} = \frac{3T_0}{\sqrt{8}}$$

or $$T_0 = \frac{24\times\sqrt{8}}{3} = 8\sqrt{8} = 22.627 \text{ hrs}$$

Time lost per day = 24 – 22.627 = 1.373 hrs = 1 hr 22 min 23 sec.

Example 2.22. *If a clock moving along with S' frame with a velocity 0.95c shows an interval of 1 hour, how much time might have elapsed in frame S at rest on earth?*

Solution. $$T = \frac{T_0}{\sqrt{1-\frac{v^2}{c^2}}} = \frac{60}{\sqrt{(1-0.95)^2}} = 192\,\text{m} = 3\,\text{hrs}\ 12\,\text{min}$$

Example 2.23. *Calculate the velocity of a watch when it seems to be slowed down by 1 minute in one hour.* **(Meerut U. 2007)**

Solution. $$T = T_o \Big/ \sqrt{1-\frac{v^2}{c^2}}$$

$\therefore$ $$\left(\frac{T_o}{T}\right)^2 = 1 - \frac{v^2}{c^2} = \left(\frac{60-1}{60}\right)^2 = 0.9669$$

$\therefore$ $$\frac{v^2}{c^2} = 1 - 0.9669 = 0.0331 \text{ or } \frac{v}{c} = 0.1819$$

$\therefore$ $$v = 0.1819\text{ c}$$

Example 2.24. *In the laboratory, the lifetime of particle moving with speed 2.8 × 10⁸ m/sec is found to be 2 × 10⁻⁷ sec. Calculate the proper life time of the particle.* **(P.U. 2001)**

Solution. The proper life time of a particle is the life time in a frame of reference in which the particle is at rest *i.e.*, the frame of reference is moving relative to the laboratory with a velocity equal to the speed of the particle.

Let the proper life time = T_0

Observed life time in Lab. frame $T = 2 \times 10^{-7}$ sec.

Velocity of particle $v = 2.8 \times 10^8$ m/sec

Now $$T = \frac{T_0}{\sqrt{1-\frac{v^2}{c^2}}}$$

$$T_0 = T\sqrt{1-\frac{v^2}{c^2}} = 2\times 10^{-7}\left[1-\left(\frac{2.8}{3}\right)^2\right]^{1/2}$$

$\therefore$ $= 2 \times 10^{-7} \times 0.3588 = 7.176 \times 10^{-3}$ sec.

Example 2.25. *What is the mean life of a burst of π^+ mesons travelling with $\beta = 0.73$ (Proper mean life time $T_0 = 2.5 \times 10^{-8}$ sec)? What is the distance travelled at this velocity during one mean life?*

Solution. The mean life of a moving meson is given by

$$T = \frac{T_0}{\sqrt{1-\frac{v^2}{c^2}}}$$

where T_0 is the proper mean life and v the velocity of the meson.

$$\therefore \quad T = \frac{2.5}{[1-(0.73)^2]^{1/2}} = 3.658 \times 10^{-8} \text{ sec.}$$

The distance travelled by this meson in time T

$$= vT = 0.73 \times 3 \times 10^8 \times 3.658 \times 10^{-8} = 8 \text{ metres}$$

Example 2.26. *A beam of μ-mesons, produced at a height of 20 kms in the earth's atmosphere, travels downwards with a velocity of 0.99c. If 99% of the original mesons decay before reaching the earth's surface, find the mean lifetime of the μ-mesons.*

Solution. Let T be the time noted by an observer in the earth's frame for a μ-meson to travel a distance of 20 km with a velocity 0.99 c with respect to the earth, then

$$T = \frac{20 \times 10^3}{0.99 \times 3 \times 10^8} = 67.34 \times 10^{-6} \text{ sec.}$$

If T_0 is the corresponding time noted by an observer moving with the μ-meson *i.e.*, in an inertial frame in which the μ-meson is at rest, then

$$T = \frac{T_0}{\left[1-\frac{v^2}{c^2}\right]^{1/2}} \quad \text{or} \quad T_0 = T\left(1-\frac{v^2}{c^2}\right)^{1/2}$$

$$= 67.34 \times 10^{-6}[1-(0.99)^2]^{1/2} = 67.34 \times 10^{-6} \times 0.141$$
$$= 9.495 \times 10^{-6} \text{ sec.}$$

During this time T_0 measured in the rest frame of the μ-meson 99% of the mesons decay. If λ is the decay constant, N_0 the number of μ-mesons in the beginning and N at the end of this time, then

$$N = N_0 e^{\lambda T_0} \quad \text{or} \quad \frac{N_0}{N} = e^{+\lambda T_0}$$

or
$$\lambda T_0 = \log_e \frac{N_0}{N} = 2.3026 \times \log_{10} \frac{N_0}{N}$$

But N the number of μ-mesons at the end of time T_0 is only 1% as 99% of the original mesons decay before reaching the earth's surface.

$$\therefore \quad \frac{N_0}{N} = \frac{100}{1} = 100$$

Hence $\lambda T_0 = 2.3026 \log_{10} 100 = 2 \times 2.3026 = 4.6052$

or $$\lambda = \frac{4.6052}{T_0} = \frac{4.6052}{9.495 \times 10^{-6}}$$

But mean lifetime $$= \frac{1}{\lambda} = \frac{9.495 \times 10^{-6}}{4.6052} = 2.06 \times 10^{-6} \text{ sec}$$

Example 2.27. *A burst of 10^4 π^+ mesons travels in a circular path of radius 20 m at a speed $v =$ 0.99 c. The proper mean life of π^+ meson is 2.5×10^{-8} sec.*

(i) How many mesons survive when the burst returns to the point of origin?

(ii) How many mesons would be left in a burst that had remained at rest at the origin for the same period of time?

Solution. Length of the circular path $= 2\pi r = 2\pi \times 20 = 40\ \pi$ metres

Let T be the time noted by an observer in the laboratory frame for a π^+ meson to travel the circular path of 40 π metres with a velocity $v = 0.99\ c$, then

$$T = \frac{40\,\pi}{0.99 \times 3 \times 10^8} = 4.232 \times 10^{-7} \text{ s}$$

If T_0 is the corresponding time noted by an observer moving with the π^+ meson, *i.e.*, in the inertial frame in which π^+ meson is at rest, then

$$T = \frac{T_0}{\sqrt{1 - \frac{v^2}{c^2}}}$$

or $$T_0 = T\left(1 - \frac{v^2}{c^2}\right)^{1/2} = 4.232 \times 10^{-7} \times [1 - (0.99)^2]^{1/2}$$

$$= 4.232 \times 0.141 \times 10^{-7}\text{s} = 0.5967 \times 10^{-7}\text{s}$$

The proper mean life of π^+ meson $= 2.5 \times 10^{-8}$ s

$$\therefore \text{ Decay constant } \lambda = \frac{1}{\text{mean life}} = \frac{1}{2.5 \times 10^{-8}} = 4 \times 10^7.$$

(i) Let N be the number of π^+ mesons out of $N_0 = 10^4$ which survive after the burst returns to the point of origin, then

$$\therefore \quad N = N_0 e^{\lambda T_0}$$

or $$\frac{N_0}{N} = e^{+\lambda T_0} = e^{4 \times 10^7 \times 0.5967 \times 10^{-7}} = 10.88$$

$$\therefore \quad N = \frac{N_0}{10.88} = \frac{10^4}{10.88} = 920$$

(ii) If the burst had remained at rest at the origin, then

$$T_0 = T = 4.232 \times 10^{-7} \text{ s}$$

and $$\frac{N_0}{N} = e^{+\lambda T} = e^{4 \times 10^{-7} \times 4.232 \times 10^{7}} = e^{16.928} = 2.248 \times 10^7$$

$$\therefore \quad N = \frac{N_0}{2.248 \times 10^7} = \frac{10^4}{2.248 \times 10^7} << 1$$

i.e., No π^+ meson would survive.

Example 2.28. *The proper life of π^+ meson is 2.5×10^{-8} sec. If a beam of these mesons of velocity 0.8 c is produced, calculate the distance the beam can travel before the flux of the meson beam is reduced to (i) $\frac{1}{e}$ and (ii) $\frac{1}{e^2}$ times the initial flux.*

Solution. Proper mean lifetime of π^+ meson $T_0 = 2.5 \times 10^{-8}$ *s*

Observed mean lifetime when the π^+ meson beam has a velocity 0.8 *s*

$$T = \frac{T_0}{\sqrt{1-\frac{v^2}{c^2}}} = \frac{2.5 \times 10^{-8}}{\sqrt{1-(0.8)^2}} = \frac{2.5 \times 10^{-8}}{0.6} = 4.166 \times 10^{-8} s$$

If λ is the decay constant, N_0 the flux of the meson beam in the beginning and N at the end of time t, then

$$N = N_0 \, e^{-\lambda t}$$

Now $$\lambda = \frac{1}{\text{mean life time}} = \frac{1}{T} = \frac{1}{4.166 \times 10^{-8}} \text{s}^{-1}$$

Hence $$N = N_0 \, e^{-t/T}$$

(*i*) If t is the time during which the flux reduces to $1/e$, then

$$\frac{N}{N_0} = \frac{1}{e} \text{ or } \frac{1}{e} = e^{-t/T} \text{ or } e^{-1} = e^{-t/T}$$

$$\therefore \quad \frac{t}{T} = 1 \text{ or } t = T = 4.166 \times 10^{-8} \text{ s}$$

Hence the distance travelled by the beam in this time = $vT = 0.8 \times 3 \times 10^8 \times 4.166 \times 10^{-8} = 10$m

(*ii*) If t is the time during which the flux reduces to $1/e^2$, then

$$\frac{N}{N_0} = \frac{1}{e^2} \text{ or } \frac{1}{e^2} = e^{-t/T}$$

or $$e^{-2} = e^{-\frac{t}{T}}$$

$$\therefore \quad \frac{t}{T} = 2 \text{ or } t = 2T = 2 \times 4.166 \times 10^{-8} \text{ s}$$

Hence the distance travelled by the beam in this time = $2vt$

$$= 0.8 \times 3 \times 10^8 \times 2 \times 4.166 \times 10^{-8} = 20 \text{ m}$$

Example 2.29. *The proper life time of an unstable particle is 2.5×10^{-8} sec. At what velocity will it appear to have mean lifetime 2.5×10^{-7} sec?* ***(G.N.D.V. 2007)***

Solution. Proper life time $T_0 = 2.5 \times 10^{-8}$ sec

Observed life time $T = 2.5 \times 10^{-7}$ sec

Velocity of the particle $v = ?$

$$\text{Now } T = \frac{T_o}{\sqrt{1-\frac{v^2}{c^2}}} \quad or \quad \frac{T_o}{T} = \sqrt{1-\frac{v^2}{c^2}}$$

$$\text{or} \qquad \frac{2.5\times10^{-8}}{2.5\times10^{-7}} = \sqrt{1-\frac{v^2}{c^2}} \quad \text{or} \quad \frac{1}{10} = \sqrt{1-\frac{v^2}{c^2}}$$

$$\frac{1}{100} = 1-\frac{v^2}{c^2} \quad \text{or} \quad \frac{v^2}{c^2} = 1-\frac{1}{100} = \frac{99}{100}$$

$$\therefore \qquad \frac{v}{c} = \frac{\sqrt{99}}{10} = \frac{9.9499}{10} = 0.99499$$

$$\therefore \qquad v = 0.99499\text{ c} \cong 0.995\text{ c}$$

2.10 TWIN PARADOX

The time dilation effect is reciprocal in nature. If an inertial frame S' is moving with a velocity $\vec{v}$ with respect to a frame S, then the S-frame observer feels that S' - clock is running slow whereas S'-frame observer feels that S-clock is running slow. Does any of the two clocks actually run slow?

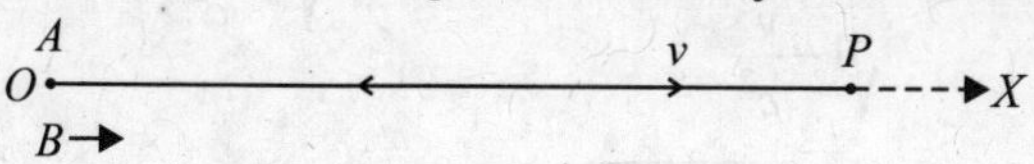

Fig. 2.5

Suppose A and B are twin brothers born at the origin O of the co-ordinate system at a time $t = 0$. The brother B is immediately set in motion along the X-axis with a high speed $v = \frac{\sqrt{3}}{2}c$. Suppose he travels for 15 years as measured by his own clock up to the point P and then returns back to O with the same speed and therefore in the same time. In the opinion of the brother B who is in the moving frame S', he has been travelling for 30 years since his birth and therefore his age is 30 years. This is the *proper time* T_0 as it has been measured by a clock in the inertial frame in which the event takes place, the event being the ageing of B.

The age of B or the corresponding *observed time* T as measured by his twin brother A who is staying at O in the stationary frame S with respect to which brother B is moving is given by

$$T = \frac{T_0}{\sqrt{1-\frac{v^2}{c^2}}} = \frac{30}{\sqrt{1-\frac{3}{4}}} = 60 \text{ years}$$

Therefore the age of B is 60 years from the point of view of the twin brother A who has stayed at O. But when B returns back to O he is only 30 years old.

If we examine the situation from the point of view of the twin brother B who is in the moving frame S', the brother A who has stayed at O is moving with a velocity $\frac{-\sqrt{3}}{2}c$ in the opposite direction. Therefore we might expect A to be 30 years old upon the return of twin B (who is in the moving frame S') who himself will be 60 years old at this time—the precise *opposite* of the result obtained in the preceding paragraph.

This problem is known as *twin paradox*.

To resolve this paradox we must note that the twin B who is in the moving frame S' is in an *accelerated* reference frame, because he undergoes an acceleration at various times in his journey; when he takes off; when he turns round; and when he finally comes to a stop. The results of special theory of relativity which hold only for frames of reference in relative motion at *constant velocity* cannot be applied at all to the above situation. We must apply the formulas of *general* theory of relativity which holds for accelerated frames of reference. According to the principle of equivalence, in this theory a large acceleration produces effects similar to those produced by strong gravitaitonal fields due to which the clocks tick more slowly.

Hence the brother B in the moving frame S' is *indeed younger* on his return back to the point O, than his twin brother A in the stationary frame S. It should, however, be clearly noted that B's life span has not been *extended to him* since however long his 30 years in the moving frame may have seemed to his brother A, it has been 30 years as far as B is concerned.

Example 2.30. *How much younger an astronaut will appear to earth observer if he returns after one year having moved with a velocity 0.5 c ?*

Solution. Let T_0 be the proper time in the spaceship of the astronaut corresponding to observed (or ordinary) time T of 1 year on earth.

The ordinary time T on earth is more than the proper time T_0 in the spaceship of the astronaut and

$$T = \frac{T_0}{\sqrt{1-\frac{v^2}{c^2}}} \quad \text{or} \quad T_0 = T\sqrt{1-\frac{v^2}{c^2}}.$$

or $$T_0 = 1 \times \sqrt{1-(0.5)^2} = \sqrt{0.75} = 0.866 \text{ year} = 10 \text{ months } 12 \text{ days}$$

∴ The astronaut is younger by

$$T - T_0 = 12 - (10 \text{ m} + 12 \text{ d}) = 1 \text{ month } 18 \text{ days}.$$

∴ The astronaut is younger by 1 month and 18 days.

2.11 RELATIVISTIC ADDITION OF VELOCITIES

The Lorentz transformation equations enable us to transform the velocity from one frame of reference to another, in relative motion with respect to it and lead to a relativistic formula for the addition of velocities, known as **Einstein's velocity addition Theorem.**

Let S and S' be the two inertial frames in relative motion, so that S' moves with a uniform velocity v to the right along the X-axis relative to S.

Let $\vec{u}$ and $\vec{u}'$ be the velocities of a particle respectively measured in the inertial frames S and S'. The components of these velocities are (u_x, u_y, u_z) and (u_x', u_y', u_z') given by

$$u_x = \frac{dx}{dt}, u_y = \frac{dy}{dt}, u_z = \frac{dz}{dt}$$

and $$u_x' = \frac{dx'}{dt'}, u_y' = \frac{dy'}{dt'}, u_z' = \frac{dz'}{dt'}$$

According to Lorentz transformation equations

$$x' = \frac{x - vt}{\sqrt{1-\frac{v^2}{c^2}}};\ y' = y;\ z' = z \ \text{ and } t' = \frac{t - \frac{v}{c^2}x}{\sqrt{1-\frac{v^2}{c^2}}}$$

Taking the differentials of above transformation equations, we have

$$dx' = \frac{dx - v.dt}{\sqrt{1-\frac{v^2}{c^2}}};\ dy' = dy;\ dz' = dz;\ dt' = \frac{dt - \frac{v}{c^2}dx}{\sqrt{1-\frac{v^2}{c^2}}}$$

$$\therefore \quad \frac{dx'}{dt'} = \frac{dx - v.dt}{dt - \frac{v}{c^2}dx} = \frac{\frac{dx}{dt} - v}{1 - \frac{v}{c^2}\frac{dx}{dt}}$$

or

$$\boxed{u_x' = \frac{u_x - v}{1 - \frac{v}{c^2}u_x}} \qquad ...(i)$$

and

$$\frac{dy'}{dt'} = \frac{dy\sqrt{1-\frac{v^2}{c^2}}}{dt - \frac{v}{c^2}dx} = \frac{\frac{dy}{dt}\sqrt{1-\frac{v^2}{c^2}}}{1 - \frac{v}{c^2}\cdot\frac{dx}{dt}}$$

or

$$\boxed{u_y' = \frac{u_y\sqrt{1-\frac{v^2}{c^2}}}{1 - \frac{v}{c^2}u_x}} \qquad ...(ii)$$

Similarly

$$\frac{dz'}{dt'} = \frac{dz\sqrt{1-\frac{v^2}{c^2}}}{dt - \frac{v}{c^2}dx} = \frac{\frac{dz}{dt}\sqrt{1-\frac{v^2}{c^2}}}{1 - \frac{v}{c^2}\frac{dx}{dt}}$$

or

$$\boxed{u_z' = \frac{u_z\sqrt{1-\frac{v^2}{c^2}}}{1 - \frac{v}{c^2}u_x}} \qquad ...(iii)$$

Inverse transformation equations. The inverse transformation equations are obtained by putting $v = -v$ and $u_x' = u_x$, $u_y' = u_y$, $u_z' = u_z$. Hence

$$\boxed{u_x = \frac{u_x' + v}{1 + \frac{v}{c^2}u_x'}} \qquad ...(iv)$$

$$\boxed{u_y = \frac{u_y'\sqrt{1-\frac{v^2}{c^2}}}{1 + \frac{v}{c^2}u_x'}} \qquad ...(v)$$

$$u_z = \frac{u_z' \sqrt{1-\frac{v^2}{c^2}}}{1+\frac{v}{c^2}u_x'} \qquad ...(vi)$$

If the velocity of the particle is along the X-axis

$u_x = u \quad u_y = 0 \quad u_z = 0$ then

$$u_x' = \frac{u-v}{1-\frac{v}{c^2}u}; \; u_y' = 0; \; u_z' = 0$$

$$\therefore \qquad u' = \frac{u-v}{1-\frac{v}{c^2}u} \qquad ...(vii)$$

Conversely

$$u = \frac{u'+v}{1+\frac{v}{c^2}u'} \qquad ...(viii)$$

No particle can have velocity greater than light. Now suppose a particle is moving with the velocity of light. (It is possible for a particle of zero rest mass like the photon) as observed by the observer O, then $u = c$. The velocity of the same particle, as observed by the observer O', is given by

$$u' = \frac{c-v}{1-\frac{v}{c}} = c$$

i.e., the observer in the inertial frame S' also finds the particle moving with the velocity c. In other words, it is not possible for a particle to move faster than light. *This also shows that Lorentz transformation equations are in accordance with the constancy of velocity of light.*

Formula for non-relativistic velocities. When v and u are very small as compared to c, $\frac{vu}{c^2}$ and $\frac{v^2}{c^2}$ are nearly equal to zero and in such a case $\sqrt{1-\frac{v^2}{c^2}}$ and $\left(1-\frac{vu}{c^2}\right)$ are very nearly equal to one.

$$\therefore \qquad u_x' = u_x - v, \; u_y' = u_y, \; u_z' = u_z$$

which are similar to Gaielian velocity transformation equations in non-relativistic or classical Newtonian mechanics.

2.12 CONSTANCY OF SPEED OF LIGHT

Einstien's formula. Consider a light photon moving parallel to the X-axis in the inertial frame S with velocity c. If S' is an inertial frame moving with uniform velocity v in the positive X-direction, then relativistic transformation equations for velocity (Einstein's velocity addition theorem) are

$$u_x' = \frac{u_x - v}{1-\frac{v}{c^2}u_x}$$

$$u_y' = \frac{u_y\sqrt{1-\frac{v^2}{c^2}}}{1-\frac{v}{c^2}u_x}$$

$$u_z' = \frac{u_z\sqrt{1-\frac{v^2}{c^2}}}{1-\frac{v}{c^2}u_x}$$

Constancy of speed of light. As the light photon is moving parallel to X-axis with a velocity c

$$u_x = c \qquad u_y = 0 \qquad u_x = 0$$

$$\therefore \qquad u_x' = \frac{c-v}{1-\frac{v}{c}} = c;\ u_y' = 0; \qquad u_z' = 0$$

$\therefore$ The velocity of the photon in the inertial frame $S' = c$.

Similarly if we assume the velocity of light photon in the inertial frame S' to be c along the $+X$ direction and apply inverse velocity transformation equations by substituting $u_x' = c$, we get

$$u_x = \frac{c+v}{1+\frac{v}{c}} = c;\ u_y = 0;\ u_z = 0$$

If the light photon is moving in the $+Y$ direction with a velocity c, in the inertial frame S, then

$$u_x = 0\ ;\ u_y = c\ ;\ u_z = 0$$

The corresponding velocity components, as observed in the inertial frame S', are

$$u_x' = \frac{u_x - v}{1-\frac{v}{c^2}u_x} = -v$$

$$u_y' = \frac{u_y\sqrt{1-\frac{v^2}{c^2}}}{1-\frac{v}{c^2}u_x}$$

$$= u_y\sqrt{1-\frac{v^2}{c^2}} = c\sqrt{1-\frac{v^2}{c^2}}$$

$$u_z' = 0$$

$\therefore$ The speed of light, as measured in S'

$$u' = \sqrt{u_x'^2 + u_y'^2 + u_z'^2} = \sqrt{v^2 + c^2 - v^2} = c$$

Similarly if the light photon is moving in the $+Z$ direction with a velocity c, then $u_x = 0$, $u_y = 0$, $u_z = c$ and we get, as in the above case, $u' = c$.

In other words, the observers in both the frames measure the same value of the speed of light. *Thus the velocity of light is independent of the frame of reference and is, therefore an absolute constant.* We can also say that the velocity addition theorem is in conformity with principle of constancy of speed of light.

Example 2.31. *Two particles are moving along a straight line towards each other with uniform velocities 0.8c and 0.5 c respectively. Calculate the relative velocity of approach between them.*

Solution. As the particles are moving along a straight line towards each other, they are moving in opposite directions. If, therefore, the velocity of one of these (u) is taken as *positive*, the velocity of the other (v) is taken as *negative*.

The relative velocity of approach between the two particles is the velocity of one of the particles which an observer stationed on the other particle will measure. Let this velocity be u', then according to Lorentz velocity transformation equation.

$$u' = \frac{u - v}{1 - \frac{uv}{c^2}} = \frac{0.8c - (-0.5c)}{1 - \frac{.8c(-0.5c)}{c^2}} = \frac{1.3}{1.4}c = 0.93\,c$$

Example 2.32. *In the laboratory two particles are observed to travel in opposite directions each with a velocity of 2.8 × 10⁸ m/s. Find their relative velocity.* ***(G.N.D.U., 2000)***

Solution. As the particles travel in opposite directions, the velocity of one of them (u) is taken as *positive* and the velocity of the other (v) is taken as *negative*. According to Lorentz velocity transformation equations.

$$\text{Relative velocity } u' = \frac{u - v}{1 - \frac{uv}{c^2}} = \frac{2.8 \times 10^8 - (-2.8 \times 10^8)}{1 + \frac{2.8 \times 10^8 \times 2.8 \times 10^8}{3 \times 10^8 \times 3 \times 10^8}}$$

$$= \frac{5.6}{1.87} \times 10^8 = 2.995 \times 10^8 \text{ ms}^{-1}$$

Example 2.33. *Two spacecrafts A and B are moving away from earth in the same direction with speed 0.8 c and 0.6 c respectively. Find the velocity of B with respect to A.* ***(G.N.D.U. 2009)***

Solution. Velocity of B, $u = 0.6c$, Velocity of A, $v = 0.8c$

$$\therefore \text{ Velocity of } B \text{ with respect to } A = \frac{u - v}{1 - \frac{uv}{c^2}} = \frac{(0.6c - 0.8c)}{1 - \frac{(0.6 \times 0.8)c^2}{c^2}} = -\,0.38c$$

Example 2.34. *A spaceship moving away from the earth with velocity 0.6 c fires a rocket (whose velocity relative to the spaceship is 0.7 c)*

(i) away from the earth *(ii) towards the earth.*

What will be the velocity of the rocket, as observed from the earth in the two cases?

(P.U. 2001)

Solution. Velocity of the rocket relative to the earth is given by

$$u = \frac{u' + v}{1 + \frac{v}{c^2}u'}$$

where u' = velocity of the rocket relative to spaceship

v = velocity of the spaceship relative to the earth

The velocity away from the earth is taken as +ve and that towards the earth as –ve.

Here $u' = 0.7c$ and $v = 0.6\,c$

(*i*) *Rocket fired away from the earth*: $u' = +\,0.7\,c$

$$\therefore \quad u = \frac{+0.7c + 0.6c}{1 + 0.6 \times 0.7} = 0.91c$$

(*ii*) *Rocket fired towards the earth*: $u' = -\,0.7\,c$

$$u = \frac{-0.7c + 0.6c}{1 - 0.6 \times 0.7} = 0.17c$$

Example 2.35. *A rocket is sent with a velocity 0.9c. A light pulse is also sent along the same path. What is the velocity of the light pulse relative to the rocket?* ***(G.N.D.U.2008)***

Solution. Velocity of light pulse $u = c$

Velocity of the rocket $v = 0.9c$

Velcoity of light pulse relative to the rocket

$$u' = \frac{u-v}{1-\frac{uv}{c^2}} = \frac{c-v}{1-\frac{cv}{c^2}} = \frac{c-v}{1-\frac{v}{c}} = c$$

∴ Velocity of light pulse relative to the rocket is also c as the *velocity of lilght is an invariant.*

Example 2.36. *A Spacecraft moving away from the earth with velocity 0.5c fires a rocket towards the earth. The velocity of the rocket relative to the spacecraft is 0.8c.*

Find the velocity of the rocket as observed from the earth. ***(G.N.D.U.2009; P.U. 2009)***

Solution. Velocity of the rocket relative to the earth is given by

$$u = \frac{u'+v}{1+\frac{vu'}{c^2}}$$

where, u' = Velocity of rocket relative to the spacecraft = $0.8c$

v = velocity of spacecraft-relative to the earth = $0.5\,c$

As the rocket is fired towards the earth, $u' = -0.8\,c$

$$\therefore \quad u = \frac{-0.8c+0.5c}{1-\frac{(0.5\times 0.8)c^2}{c^2}} = \frac{-0.3c}{1-0.4} = \frac{-0.3c}{+0.6} = -0.5c$$

Example 2.37. *A radioactive atom moves with a velocity $v = 0.1\,c$ along the X-axis of a system S. It emits a β-particle of velocity 0.95c relative to the system S' in which the radioactive atom is at rest. If the β particle is emitted along the X'-axis of S', find its speed relative to S. Compare the values as obtained by Galelian transformations and comment on the results.* ***(Pbi.U. 2000)***

Solution. Let the velocity of the β-particle with respect to the system $S = u_x$ and its velocity with respect to the system $S' = u_x' = 0.95\,c$

As the radioactive atom is at rest in the system S'.

∴ Velocity of the system $S' = v = 0.1\,c$

(*i*) According to Lorentz velocity transformation equations

$$u_x = \frac{u_x' + v}{1+\frac{v}{c^2}u_x'} = \frac{0.95c + 0.1c}{1 + 0.1\times 0.95}$$

$$= \frac{1.05}{1.095}c = 0.903\,c$$

(*ii*) According to Galelian transformation equations

$$u_x = u_x' + v$$

$$= 0.95c + 0.1\,c = 1.05\,c$$

Comments. According to Galelian transformation equations the velocity of the β-particle in the system S comes out to be greater than the velocity of light, which is against the postulate of special theory of relativity that velocity of light is an invariant. This means that Galelian transformation equations no longer hold valid when the velocities concerned are of the order of the velocity of light. For smaller values of v, the difference is very small and negligible. This is why Lorentz transformation equations reduce to Galelian transformation equations for values of $v << c$.

Example 2.38. *Two electron beams travel along the same straight line but in opposite directions with velocities $V = 0.9\ c$ relative to the laboratory frame. Find the relative velocity of the electrons according to Newtonian mechanics. What will be the velocity measured by an observer moving with one of the electron beams?* **(M.D.U. 2001)**

Solution. As the two electron beams travel along the same straight line in opposite directions with a velocity 0.9c, with respect to the laboratory frame, the velocity of one of these (u) is taken as *positive* and that of the other (v) is taken as *negative.*

According to Newtonian mechanics, relative velocity of the electrons

$$u' = u - v = 0.9\ c - (-\ 0.9\ c) = 1.8\ c$$

Velocity measured by an observer moving with one of the electron beams

$$u' = \frac{u-v}{1-\frac{vu}{c^2}} = \frac{0.9c-(-0.9c)}{1-\frac{(-0.9c)\ 0.9c}{c^2}} = \frac{1.8}{1+0.81} = 0.994\ c$$

NOTE: The readers will observe that calculation of relative velocity according to Newtonian Mechanics, gives erroneous results as velocity cannot be greater than the velocity of light.

Example 2.39. *A particle moves in XY plane with a velocity 6×10^9 cm s^{-1} at an angle of 60° with X-axis in system S'. Find the magnitude and direction of its velocity as observed by an observer in system S' when S' has a velocity 3×10^9 cm s^{-1} along positive direction of X-axis.* **(P.U. 2000)**

Solution. Velocity of the frame S' relative to the frame S *along the X-axis* $v = 3 \times 10^9$ cm s^{-1}

Velocity of the particle in frame S', $u = 6 \times 10^9$ cm s^{-1}

The velocity of the particle u can be resolved into two components

Components along the X-axis, $u_x = u \cos 60° = 6 \times 10^9 \times \frac{1}{2} = 3 \times 10^9$ cm s^{-1}

Component along the Y-axis, $u_y = u \sin 60° = 6 \times 10^9 \times \frac{\sqrt{3}}{2} = 3\sqrt{3} \times 10^9$ cm s^{-1}

Let u_s' and u_y' be the value of the components as observed by a person in the frame S', then according to Lorentz velocity transformation equations

$$u_x' = \frac{u_x - v}{1-\frac{u_x v}{c^2}} = \frac{3\times 10^9 - 3\times 10^9}{1-\frac{3\times 10^9\times 3\times 10^9}{3\times 10^{10}\times 3\times 10^{10}}} = 0$$

$$u_y' = \frac{u_y\sqrt{1-\frac{v^2}{c^2}}}{1-\frac{u_x v}{c^2}} = \frac{3\sqrt{3}\times 10^9\sqrt{1-\frac{3\times 10^9\times 3\times 10^9}{3\times 10^{10}\times 3\times 10^{10}}}}{1-\frac{3\times 10^9\times 3\times 10^9}{3\times 10^{10}\times 3\times 10^{10}}} = \frac{3\sqrt{3}\times 10^9}{\sqrt{1-0.01}}$$

$$= \frac{3\sqrt{3}\times 10^9}{\sqrt{0.99}} = \frac{3\times 1.732\times 10^9}{0.995} = 5.22\times 10^9 \text{ cm s}^{-1}$$

The resultant velocity u', as observed by a person in the frame S', is given by

$$u' = \sqrt{u_x'^2 + u_y'^2} = 5.22 \times 10^9 \text{ cm s}^{-1} \text{ along } Y\text{-axis.}$$

Example 2.40. *Two photons approach each other, what is their relative velocity?*

(G.N.D.U., 2000, P.U. 2008)

Solution. Velocity of each photon = c

Suppose the photons move along the direction of X-axis.

As they are approaching each other, applying inverse velocity transformation equation

$$u = \frac{u' + v}{1 + \frac{v}{c^2}u'}$$

we have
$$u = \frac{c + c}{1 + \frac{c^2}{c^2}} = c$$

Hence the relative velocity of two photons approaching each other is equal to the velocity of light.

2.13 RELATIVISTIC DOPPLER EFFECT

The apparent change in the frequency of light due to the motion of the observer relative to the source of the light is known as Doppler effect.

Longitudinal Doppler effect. Consider two inertial frames S and S' such that S' is moving with respect to S with a constant speed v in the $+x$ direction (Fig. 2.6). Suppose a plane wave of light of frequency ν is travelling in the x-direction in the frame S. We represent this wave by the relation

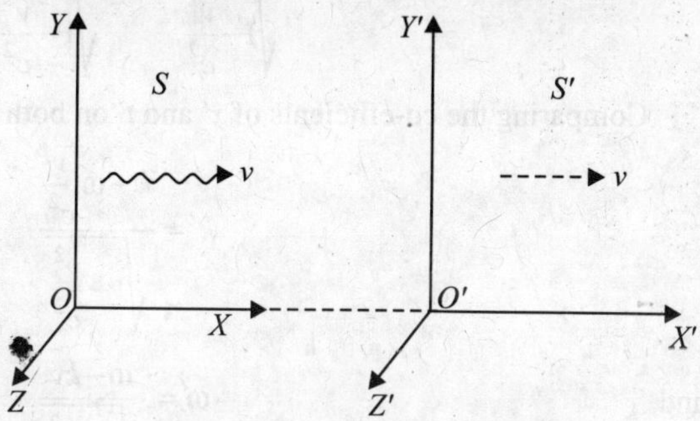

Fig. 2.6

$$a = a_0 \cos \frac{2\pi}{\lambda}(x - ct)$$
$$= a_0 \cos (kx - \omega t)$$

where $k = \frac{2\pi}{\lambda}$, the *propagation constant,* λ being the wavelength of light, $\omega = \frac{2\pi c}{\lambda} = 2\pi\nu$, c being the velocity of light wave.

The wave as seen by an observer in the frame S' is given by

$$a = a_0 \cos (k'x' - \omega' t')$$

where $k' = \frac{2\pi}{\lambda'}$ is the propagation constant, λ' being the wavelength as seen in the frame S' and $\omega' = 2\pi\nu'$, ν' being the frequency as seen in the frame S'.

The quantity $kx - \omega t = \phi_s$ is known as the *phase* of the wave in frame S and $k'x' - \omega't' = \phi_s'$ is the phase of the wave as seen in the frame S'.

In the S' frame the wave will still be a plane wave because Lorentz transformation equations are linear and transform a plane wave into a plane. The phase of the wave is, therefore, invariant under Lorentz transformation *i.e.*, $\phi_s = \phi_{s'}$

or $$kx - \omega t = k' x' - \omega' t' \qquad ...(i)$$

The Lorentz transformation equations are

$$x = \frac{x' + vt'}{\sqrt{1 - \frac{v^2}{c^2}}}$$

$$t = \frac{t' + \frac{v}{c^2} x'}{\sqrt{1 - \frac{v^2}{c^2}}}$$

Substituting the values of x and t in terms of x' and t' in Eq. (i), we have

$$k \frac{(x' + vt')}{\sqrt{1 - \frac{v^2}{c^2}}} - \frac{\omega(t' + \frac{v}{c^2} x')}{\sqrt{1 - \frac{v^2}{c^2}}} = k' x' - \omega' t'$$

or $$\frac{k - \omega \frac{v}{c^2}}{\sqrt{1 - \frac{v^2}{c^2}}} x' - \frac{\omega - kv}{\sqrt{1 - \frac{v^2}{c^2}}} t' = k' x' - \omega' t'$$

Comparing the co-efficients of x' and t' on both sides, we get

$$k' = \frac{k - \omega \frac{v}{c^2}}{\sqrt{1 - \frac{v^2}{c^2}}} \qquad ...(ii)$$

and $$\omega' = \frac{\omega - kv}{\sqrt{1 - \frac{v^2}{c^2}}} \qquad ...(iii)$$

From Eqs. (ii) and (iii), we find that the propagation constant k' and angular frequency ω' of the wave observed in the frame S' are different from the corresponding values in the frame S.

Now $\omega' = 2\pi\nu'$ and $\omega = 2\pi\nu$

Substituting these values in Eq. (iii), we have

$$2\pi\nu' = \frac{2\pi\nu - kv}{\sqrt{1 - \frac{v^2}{c^2}}} = \frac{2\pi\nu - \frac{2\pi}{\lambda} v}{\sqrt{1 - \frac{v^2}{c^2}}}$$

$$\nu' = \frac{\nu - \frac{v}{\lambda}}{\sqrt{1 - \frac{v^2}{c^2}}}$$

For a wave propagating in vacuum $c = \nu\lambda$ or $\frac{1}{\lambda} = \frac{\nu}{c}$

$$\therefore \quad \nu' = \frac{\nu - \frac{v\nu}{c}}{\sqrt{1-\frac{v^2}{c^2}}} = \frac{\nu\left(1-\frac{v}{c}\right)}{\sqrt{1-\frac{v^2}{c^2}}} = \nu\sqrt{\frac{1-\frac{v}{c}}{1+\frac{v}{c}}}$$

or
$$\nu' = \nu\sqrt{\frac{c-v}{c+v}} \qquad \text{...(iv)}$$

In terms of wavelength $\lambda' = \lambda\sqrt{\frac{c+v}{c-v}}$ ...(v)

Eqs. (*iv*) and (*v*) are known as *Longitudinal Doppler effect formulae as the light travels parallel to the direction of motion of S′ frame.*

Eq. (*iv*) gives the apparent frequency of light as observed from a frame moving with respect to the source of light. Eq (*v*) gives the corresponding values of the wavelength of light.

Red shift. From Eqs. (*iv*) and (*v*) we find that when the source of light and the observer are *moving away* from each other

$$\nu' = \nu\sqrt{\frac{c-v}{c+v}} \quad \text{and} \quad \lambda' = \lambda\sqrt{\frac{c+v}{c-v}}$$

where ν and λ are the frequency and wavelength respectively emitted by the source in the frame S and ν' and λ' the corresponding values as observed by the observer in the frame S'. It is clear from relation (*iv*) that ν' is *less* than ν. Hence when the source is receding away from the observer the frequency of the emitted light appears to have decreased *i.e.*, the *wavelength appears to increase* as is also evident from relation (*v*).

This is known as *red shift* since the red colour is on the longer wavelength side of the visible spectrum and explains the red shift of the spectral lines of light received from distant stars. The red shift implies the stars (or nebulae) showing red shift are moving away from us.

Blue shift. Eqs. (*iv*) and (*v*) hold good when the source and the observer are moving away from each other. When the source and observer are moving *towards* each other *v* is taken as *negative*.

In such a case the observed frequency appears to increase *i.e.*, the wavelength appears to decrease. This effect is sometimes called *blue shift.*

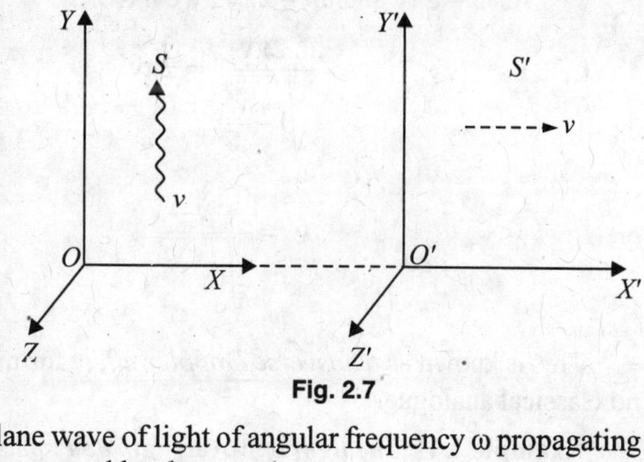

Fig. 2.7

Transverse Doppler effect. Consider a plane wave of light of angular frequency ω propagating in the Y direction (Fig. 2.7). Such a wave is represented by the equation

$$a = a_0 \cos (ky - \omega t)$$

where $k = \frac{2\pi}{\lambda}$ and $\omega = 2\pi\nu$ are the propagation constant and angular frequency in the frame S. The

phase of the wave in the frame $S = \phi_s = ky - \omega t$

Let the phase of the wave in the frame $S' = \phi_s' = k'y' - \omega' t'$ where $k' = \dfrac{2\pi}{\lambda'}$ and $\omega' = 2\pi\nu'$ are the propagation constant and the angular frequency as observed in the frame S'.

As the phase is invariant under Lorentz transformations, we have $ky - \omega t = k'y' - \omega' t'$

Using Lorentz transformation equations $y = y'$ and

$$t = \frac{t' + \frac{v}{c^2}x'}{\sqrt{1-\frac{v^2}{c^2}}}, \text{ we get}$$

$$k\,y' - \frac{\omega(t' + \frac{v}{c^2}x')}{\sqrt{1-\frac{v^2}{c^2}}} = k'\,y' - \omega' t'$$

or

$$ky' - \frac{\omega t'}{\sqrt{1-\frac{v^2}{c^2}}} - \frac{\frac{\omega v}{c^2}x'}{\sqrt{1-\frac{v^2}{c^2}}} = k'\,y' - \omega' t'$$

Comparing co-efficients of y', t' and x' on both sides, we get

$$k = k'; \quad \frac{\omega}{\sqrt{1-\frac{v^2}{c^2}}} = \omega'; \quad \frac{\frac{\omega v}{c^2}x'}{\sqrt{1-\frac{v^2}{c^2}}} = 0$$

As $\omega = 2\pi\nu$ and $\omega' = 2\pi\nu'$, we have

$$\frac{2\pi\nu}{\sqrt{1-\frac{v^2}{c^2}}} = 2\pi\nu'$$

or

$$\nu' = \frac{\nu}{\sqrt{1-\frac{v^2}{c^2}}} \qquad \text{...}(vi)$$

This is known as *transverse Doppler effect* formula. This is a purely relativistic relation and has no classical analogue.

Example 2.41. *The light of wavelength 434 nm is emitted by a star. The light is observed to have a wavelength 532 nm on the earth. Find the speed with which the star is receding from the earth.*

(P.U. 2006: Pbi. U. 2006)

Solution. According to longitudinal Doppler effect given in relation (v), we have

$$\lambda' = \lambda\sqrt{\frac{c+v}{c-v}}$$

where λ is the wavelength of light emitted by the source and λ′ the wavelength measured by the observer, v is the speed with which the star is receding with respect to the earth.

$$\therefore \qquad \frac{\lambda'}{\lambda} = \sqrt{\frac{c+v}{c-v}}$$

Here λ = 434 nm and λ′ = 532 nm. Substituting, we get

$$\frac{532}{436} = 1.226 = \sqrt{\frac{c+v}{c-v}}$$

or $$\frac{c+v}{c-v} = (1.226)^2 = 1.503$$

or $$c + v = 1.503\,(c - v) \qquad \text{or} \qquad v = \frac{0.503}{2.503}c$$

$$\therefore \qquad v = 0.201\text{c} = 6.03 \times 10^7 \text{ m/sec}$$

2.14 RELATIVITY OF MASS

Variation of mass with velocity. Consider two observers O and O' in the inertial frames S and S' at rest with respect to each other with their X-axis lying at a distance apart. The observer O in the system S has a small *hard* sphere A and observer O' in the system S' has an *exactly similar* small hard sphere B (Fig. 2.8). The masses of A and B are the *same* when compared at rest in the same inertial frame. The frame S' now moves to the right along the X-axis with a uniform velocity v. The observer O throws the sphere A in his $+Y$ direction with a speed u as measured by his own measuring scale and his own clock.

At the same time, the observer O' throws the sphere B in his $-Y$ direction with the same speed u, as measured by his own measuring scale and his own clock.

It is supposed that the motion of the particles is confined in the X-Y plane.

The two spheres have an elastic collision midway between the two X-axes. For this to happen the observer O' must project his sphere from a point P. This sphere will appear to the observer O to travel along PB with the resultant velocity of u along $-Y$ direction and v along $+X$-direction. As the velocity along the Y-direction is independent of the velocity along the X-direction the sphere B will have an *elastic* collision with the sphere A at a point exactly midway between the two X-axes, when the observers O and O' are exactly opposite to each other. As the X-component of the velocity of B is not affected by collision, it will appear to the observer O to move back along BQ.

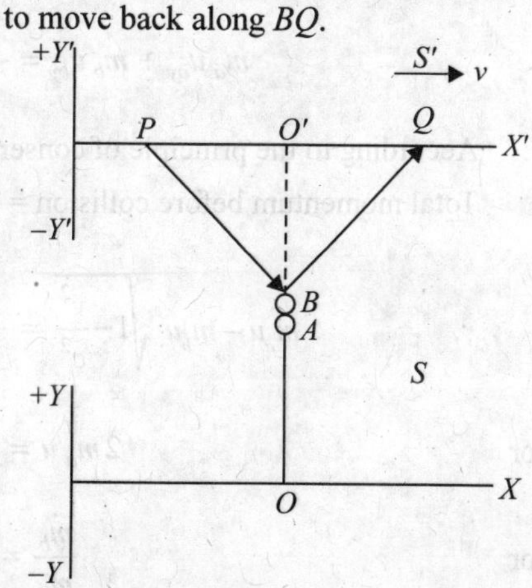

Fig. 2.8

Now let us denote the velocity of the sphere A as measured by the observer O in the frame S as u_a and let its two components in the X-Y plane be u_{ax} and u_{ay}. As the velocity of the sphere A is along $+Y$-direction. $u_{ax} = 0$ and $u_{ay} = u_a = +u$

Similarly let us denote the velocity of the sphere B as measured by the observer O' in the frame S' as $u_b' = -u$ (being along the direction of $-Y$-direction) and let its two components in the X-Y plane be u_{bx}' and u_{by}'.

$$\therefore \qquad u_{bx}' = 0 \text{ and } u_{by}' = u_b' = -u$$

Let us now denote the velocity of the sphere B as measured by the observer O in the frame S as u_b and let its two components in the X-Y plane be u_{bx} and u_{by}. To find the value of u_{bx} and u_{by} in terms of u_{bx}' and u_{by}' we make use of *Lorentz velocity transformation equations*, which are

$$u_x = \frac{u_x' + v}{1 + \frac{v}{c^2} u_x'}$$

$$u_y = \frac{u_y' \sqrt{1 - \frac{v^2}{c^2}}}{1 + \frac{v}{c^2} u_x'}$$

$$u_{bx} = -\frac{u_{bx}' + v}{1 + \frac{v}{c^2} u_{bx}'} = +v \qquad [\because u'_{ox} = 0]$$

and
$$u_{by} = u_{by}' \frac{\sqrt{1 - \frac{v^2}{c^2}}}{1 + \frac{v}{c^2} u_{bx}'} = -u \sqrt{1 - \frac{v^2}{c^2}}$$

The mass of the ball *A*, as measured by *O* and that of the ball *B* as measured by *O'* is the same but let us denote the mass of the ball *A* as measured by $O = m_a$ and the mass of the ball *B* as measured again by *O* in the frame *S* as m_b, then

The total momentum of the two balls along *Y*-axis **before collision** as observed by *O* in the system *S*, is given by

$$m_a u_{ay} + m_b u_{by} = m_a u - m_b u \sqrt{1 - \frac{v^2}{c^2}}$$

After collision, each ball *reverses* its *Y*-component of velocity. Therefore the *Y*-component of the velocity of the two balls along the *Y*-axis **after collision** as observed by *O* in the system *S* is given by

$$u_{ay} = -u \text{ and } u_{by} = +u \sqrt{1 - \frac{v^2}{c^2}}$$

Therefore, total momentum of the two balls along *Y*-axis **after collision,** as observed by *O* in the system *S*, is given by

$$m_a u_{ay} + m_b u_{by} = -m_a u + m_b u \sqrt{1 - \frac{v^2}{c^2}}$$

According to the principle of conservation of momentum:

Total momentum before collision = Total momentum after collision

$$\therefore \qquad m_a u - m_b u \sqrt{1 - \frac{v^2}{c^2}} = -m_a u + m_b u \sqrt{1 - \frac{v^2}{c^2}}$$

or
$$2 m_a u = 2 m_b u \sqrt{1 - \frac{v^2}{c^2}}$$

or
$$\frac{m_b}{m_a} = \frac{1}{\sqrt{1 - \frac{v^2}{c^2}}}$$

where m_a and m_b respectively are the masses of the spheres A and B as measured by the observer O. It is evident that this equation cannot be satisfied unless m_a and m_b are different. But these masses were equal when both of these were measured at rest with respect to stationary observers. Hence the mass also depends upon velocity. As m_a is the mass of the sphere A, in the inertial system S at rest we denote $m_a = m_0$ and as m_b is the mass of the sphere B in the inertial system S' moving with respect to S with a velocity v, we denote

$$m_b = m$$

$$\therefore \quad \frac{m}{m_0} = \frac{1}{\sqrt{1-\frac{v^2}{c^2}}} \quad \text{or} \quad m = \frac{m_0}{\sqrt{1-\frac{v^2}{c^2}}}$$

Hence the mass of a particle, when it is moving with a velocity v, is greater than its mass when it is at rest. In other words, the mass of a body increases with its velocity of motion.

Material particle cannot have velocity equal to or greater than velocity of light.

When the velocity of a moving body is equal to the velocity of light, its mass is *infinite*. On the other hand, if its velocity is greater than the velocity of light its mass is *negative*. Both these things are, however, impossible. It, therefore, means that material particle cannot have a velocity equal to or greater than the velocity of light.

Example 2.42. *An electron moves with a velocity of 0.6 × 10⁸ m/s. Calculate its mass.* $m_0 = 9 \times 10^{-31}$ *kg.*

Solution. $m_0 = 9 \times 10^{-31}$ kg; $v = 0.6 \times 10^8$ m/s

$$\text{Now} \quad m = \frac{m_0}{\sqrt{1-\frac{v^2}{c^2}}}$$

$$= \frac{9\times10^{-31}}{\sqrt{1-\frac{0.6\times10^8\times0.6\times10^8}{3\times10^8\times3\times10^8}}} = 9.183\times10^{-31} \text{ kg.}$$

Example 2.43. *Calculate the velocity at which electron mass is* $\sqrt{3}$ *times the rest mass.* **(*P.U.* 2001)**

Solution. $m = \sqrt{3}\, m_0 \quad \text{or} \quad \frac{m_0}{m} = \frac{1}{\sqrt{3}}$

$$\text{Now} \quad m = \frac{m_0}{\sqrt{1-\frac{v^2}{c^2}}}$$

$$\therefore \quad \frac{m_0}{m} = \sqrt{1-\frac{v^2}{c^2}} = \frac{1}{\sqrt{3}} \quad \text{or} \quad 1-\frac{v^2}{c^2} = \frac{1}{3}$$

$$\therefore \quad \frac{v^2}{c^2} = 1-\frac{1}{3} = \frac{2}{3} \quad \text{or} \quad \frac{v}{c} = \sqrt{\frac{2}{3}} \quad \therefore v = \sqrt{\frac{2}{3}}\, c$$

$$\text{or} \quad v = 0.816\, c = 0.816 \times 3 \times 10^8 = 2.448 \times 10^8 \text{ ms}^{-1}$$

Example 2.44. *What should be the speed of electron so that its relativistic mass is twice its rest mass?* ***(P.U., 2004)***

Solution. Relativistic mass $m = \dfrac{m_0}{\sqrt{1-\dfrac{v^2}{c^2}}}$ where m_0 is the rest mass or $\dfrac{m}{m_0} = 2$

$$\therefore \quad 1 - \frac{v^2}{c^2} = \left(\frac{m_0}{m}\right)^2 = \frac{1}{4} \quad \text{or} \quad \frac{v^2}{c^2} = 1 - \frac{1}{4} = \frac{3}{4} \quad \text{or} \quad \frac{v}{c} = \frac{\sqrt{3}}{2} = 0.866 \text{ or } v = 0.866\,c.$$

Example 2.45. *At what speed must a particle move for its mass to be four times its rest mass?* ***(H.P.U. 2000)***

Solution. Let v be the velocity at which the mass m has a value four *times* the rest mass m_0.

$$\therefore \quad m = 4m_0$$

Now $$m = \frac{m_0}{\sqrt{1-\dfrac{v^2}{c^2}}} \quad \text{or} \quad 4m_0 = \frac{m_0}{\sqrt{1-\dfrac{v^2}{c^2}}}$$

Squaring we get $$16\left(1 - \frac{v^2}{c^2}\right) = 1 \quad \text{or} \quad \frac{v^2}{c^2} = 1 - \frac{1}{16} = \frac{15}{16}$$

$$\therefore \quad v^2 = \frac{15}{16}c^2 \quad \text{or} \quad v = \sqrt{\frac{15}{16}}\,c = 0.968\,c$$

$$= 0.968 \times 3 \times 10^8 = 2.904 \times 10^8 \text{ ms}^{-1}$$

Example 2.46. *With what velocity a particle should move so that its mass appears to increase by 20% of its rest mass?* ***(Pbi. U. 2006)***

Solution. Let m_0 be the rest mass of the particle. If v is the velocity of the particle at which the mass m of the particle appears to increase by 20%, then

$$\frac{m}{m_0} = \frac{100+20}{100} = \frac{6}{5}$$

Now $$m = \frac{m_0}{\sqrt{1-\dfrac{v^2}{c^2}}} \quad \text{or} \quad \frac{m_0}{m} = \sqrt{1-\frac{v^2}{c^2}} \quad \text{or} \quad \frac{5}{6} = \sqrt{1-\frac{v^2}{c^2}}$$

$$\therefore \quad \frac{25}{36} = 1 - \frac{v^2}{c^2} \quad \text{or} \quad \frac{v^2}{c^2} = 1 - \frac{25}{36} = \frac{11}{36} \quad \text{or} \quad \frac{v}{c} = \frac{\sqrt{11}}{6} = 0.553$$

$$\therefore \quad v = 0.553\,c$$

Example 2.47. *An electron having rest mass 9×10^{-31} kg moves with a velocity equal to 80% of that of light. Find its mass. What will be the mass of electron to an observer who is moving with it.* ***(Pbi. U. 2007; P.U. 2007)***

Solution. Velocity of electron $v = \dfrac{80}{100}c = 0.8c$ or $\dfrac{v}{c} = 0.8$

Rest mass of the electron $m_0 = 9 \times 10^{-31}$ kg

$\therefore$ Relativistic mass of electron when it is moving with a velocity v,

$$m=\frac{m_o}{\sqrt{1-\frac{v^2}{c^2}}}=\frac{9\times10^{-31}}{\sqrt{1-0.64}}=\frac{9\times10^{-31}}{0.6}=1.5\times10^{-30}\,\text{kg}$$

To an observer, who is moving with the electron, the electron appears to be at rest. So the mass measured by the observer, who is moving with the electron, is the rest mass of the electron = 9×10^{-31} kg

Example 2.48. *A body of rest mass 100 kg changes its mass by 1 kg in a rocket ship. Calculate the speed of rocket.* **(Nagpur U. 2010)**

Solution. Rest mass of a body m_o = 100 kg

The relativistic mass m = (100 + 1) = 101 kg

Using relation, $$m=\frac{m_o}{\sqrt{1-\frac{v^2}{c^2}}}$$

$$\therefore \quad \frac{m_o}{m}=\frac{100}{101}=0.99=\sqrt{1-\frac{v^2}{c^2}}$$

or $$1-\frac{v^2}{c^2}=(0.99)^2=0.98 \quad \text{or} \quad \frac{v^2}{c^2}=1-0.98=0.0198$$

or $$\frac{v}{c}=\sqrt{0.0198}=0.1407 \quad \text{or} \quad v=0.1407\,c$$

or $$v=0.1407\times3\times10^8$$
$$=4.2\times10^7\ \text{m/s}$$

Example 2.49. *Calculate the mass of a body when its speed is 0.8 times the speed of light. Assume the rest mass of the body to be one gram.* **(Nagpur U. 2011)**

Solution. Given: m_o = 1 gm = 1×10^{-3} kg

v = 0.8 c

$$\therefore \quad \frac{v}{c}=0.8 \quad \text{or} \quad \left(\frac{v}{c}\right)^2=0.64$$

Relativistic mass of the body $$m=\frac{m_o}{\sqrt{1-\frac{v^2}{c^2}}}=\frac{10^{-3}}{\sqrt{1-0.64}}$$

$$=\frac{10^{-3}}{0.6}=1.667\times10^{-3}\,\text{kg}$$

2.15 EINSTEINS' MASS-ENERGY RELATIONSHIP

According to the classical theory, the mass of a moving body is constant and independent of its velocity. According to the theory of relativity, however, it can be proved that the mass of a moving body varies with velocity. According to this theory the *effective mass* of a body when it is moving with a velocity v relative to an observer, is given by

$$m=\frac{m_0}{\sqrt{1-\frac{v^2}{c^2}}}$$

where m_0 is the *rest mass i.e.*, the mass of body when it is at rest relative to the observer and c the velocity of light.

Mass energy relation. According to Newton's second law of motion, the force acting on a body is equal to the rate of change of momentum it produces. If a body of mass m moving with a velocity v has a force F applied to it, then

$$F = \frac{d}{dt}(mv) = m\frac{dv}{dt} + v\frac{dm}{dt}$$

as m and v are both variables.

If the force F acts for a small distance dx, the work done $F\,dx$ is stored in the body as its kinetic energy given by

$$dE_k = Fdx = m\frac{dv}{dt}dx + v\frac{dm}{dt}dx = m\,dv\frac{dx}{dt} + v\,dm\frac{dx}{dt}$$

$$= mv\,dv + v^2dm \qquad ...(i)$$

Since $\dfrac{dx}{dt} = v$

Now $$m = \frac{m_0}{\sqrt{1-\frac{v^2}{c^2}}} \text{ or } m^2 = \frac{m_0^2c^2}{c^2-v^2}$$

or $$m^2c^2 - m^2v^2 = m_0^2c^2$$

Differentiating both sides, we get

$$2m\,dm\,c^2 - 2m\,dm\,v^2 - 2v\,dv\,m^2 = 0$$

or $$dm\,c^2 = (mv\,dv + v^2\,dm)$$

or $$dm\,c^2 = dE_k$$

Integrating both sides, we get

$$c^2\int_{m_0}^{m} dm = \int dE_k$$

or $$c^2(m - m_0) = E_k \qquad ...(ii)$$

or $$E_k = \Delta mc^2$$

where $\Delta m = (m - m_0)$ and is the mass converted into kinetic energy. The equation $E_k = \Delta mc^2$ shows that the kinetic energy of a moving mass is c^2 times the gain in mass. It thus points out that the increase in mass is an indication and measure of the kinetic energy. The mass m_0 is often termed as the rest-mass of the body and the term m_0c^2 as the rest-energy of the body. The rest-energy is regarded as a form of internal energy inherent in the nature of particles constituting matter. Rewriting the equation $c^2(m - m_0) = E_k$ we get

$$mc^2 = E_k + m_0c^2 \qquad ...(iii)$$

or Total energy = Rest mass energy + kinetic energy

If we denote the total energy by E, we get the famous mass energy relation

$$E = mc^2 \qquad ...(iv)$$

This is known as Einstein's mass energy relationship.

Physical significance The relation $E = mc^2$ expressing mass energy equivalence is the most important result of the special theory of relativity. This relation shows that under suitable conditions mass can be converted into energy and *vice-versa.* In classical Newtonian mechanics there are two separate conservation laws namely (*i*) The law of conservation of mass and (*ii*) The law of conservation of energy. Einstein's mass energy relationship in special theory of relativity combines these two laws into one *i.e.*, law of conservation of mass and energy. In nuclear phenomenon in which energy is released there is a corresponding decrease in total mass and in such phenomenon in which energy is absorbed there is an equivalent increase in mass. But as the velocity of light in free space is very large 3×10^8 ms^{-1}, a very small decrease in mass results in the release of a very large amount of energy. For example, the conversion of rest mass of 1 a.m.u. (1.65×10^{-27} kg) releases 931 MeV of energy.

Supporting nuclear phenomenon. (*i*) Nuclear phenomena like *fission* and *fusion* are very important examples of conversion of mass into energy and are most popular present-day methods of producing energy.

(*ii*) The mass energy relationship helps us to determine the binding energy of various nuclei on the basis of mass defect.

(*iii*) Energy released in the annihilation of matter, *e.g.*, annihilation of electron-positron pair also supports the mass-energy relationship.

Example 2.50. *Calculate the ratio of mass of the electron to its rest mass when it is moving with a kinetic energy of 9.5 MeV. Rest mass of electron 0.5 MeV.* ***(K.U. 2001)***

Solution. Let m_0 be the rest mass of the electron and m the mass when moving with a kinetic energy of 9.5 MeV.

$$\text{Kinetic energy} = (m - m_0)\,c^2 = 9.5 \text{ MeV}$$

Also rest-mass (energy) of the electron (Given) $= m_0c^2 = 0.5$ MeV

$$\therefore \quad mc^2 = m_0c^2 + 9.5 \text{ MeV} = 0.5 + 9.5 = 10 \text{ MeV}$$

$$\text{Hence} \quad \frac{m}{m_0} = \frac{mc^2}{m_0\,c^2} = \frac{10}{0.5} = 20$$

Example 2.51. *Calculate the ratio of the mass of the electron to its rest mass when it is moving with a kinetic energy of 1000MeV. Rest mass of the electron = 0.51 MeV.*

Solution. $(m - m_0)c^2 = 1000$ MeV; $m_0c^2 = 0.51$MeV; $mc^2 = m_0c^2 + 1000$ MeV

$$= 1000.51 \text{ MeV} \therefore \frac{m}{m_0} = \frac{1000.51}{0.51} = 1961.78$$

Example 2.52. *Calculate the velocity of 1 MeV electron.* ***(Pbi.U. 2008, Meerut 2006)***

Solution. 1 MeV electron means the kinetic energy of the electron = 1 MeV = 1.6×10^{-13} Joule

Let m_0 be the rest mass of the electron, m its mass when moving with a velocity v and E_k its kinetic energy, then

$$E_k = (m - m_0)\,c^2$$

$$= \left[\frac{m_0}{\sqrt{1-\frac{v^2}{c^2}}} - m_0\right]c^2 = m_0\,c^2\left[\frac{1}{\sqrt{1-\frac{v^2}{c^2}}} - 1\right]$$

$$\text{or} \quad \frac{1}{\sqrt{1-\frac{v^2}{c^2}}} = \frac{E_k}{m_0\,c^2} + 1 = \frac{1.6 \times 10^{-13}}{9 \times 10^{-31} \times 9 \times 10^{16}} + 1 = \frac{160}{81} + 1 = \frac{241}{81}$$

(*iii*) Kinetic energy = Rest mass energy

or $$(m - m_0)\, c^2 = m_0 c^2 \text{ or } m = 2\, m_0$$

But $$m = \frac{m_0}{\sqrt{1 - \frac{v^2}{c^2}}} \quad \therefore \quad 2\, m_0 = \frac{m_0}{\sqrt{1 - \frac{v^2}{c^2}}} \text{ or } \frac{1}{\sqrt{1 - \frac{v^2}{c^2}}} = 2$$

$\therefore$ $$1 - \frac{v^2}{c^2} = \frac{1}{4} \quad \text{or} \quad \frac{v^2}{c^2} = 1 - \frac{1}{4} = \frac{3}{4}$$

or $$\frac{v}{c} = \sqrt{\frac{3}{4}} = 0.866$$

Hence $v = 0.866\, c = 0.866 \times 3 \times 10^8 = 2.598 \times 10^8\ \text{ms}^{-1}$

Example 2.53. *Show that if the variation of mass with velocity is taken into account the kinetic energy of a particle of rest mass m_o moving with velocity v is given by*

$$E = m_o c^2 \left[\left(1 - \frac{v^2}{c^2} \right)^{-\frac{1}{2}} - 1 \right]$$ **(P.U. 2008)**

Solution. Using Einstein's mass-energy relationship, the kinetic energy E_k of a moving particle is given by

$$E_k = c^2 (m - m_o)$$ ($\because$ Eq. (*ii*) of article 2.15)

Putting $E_k = T$, we have $T = c^2 (m - m_0)$

Now $$m = m_o \Big/ \sqrt{1 - \frac{v^2}{c^2}} = m_o \left(1 - \frac{v^2}{c^2} \right)^{-\frac{1}{2}}$$

$\therefore$ $$T = c^2 (m - m_o) = mc^2 - m_o c^2 = m_o \left(1 - \frac{v^2}{c^2} \right)^{-\frac{1}{2}} - m_o c^2$$

$$= m_o c^2 \left[\left(1 - \frac{v^2}{c^2} \right)^{-\frac{1}{2}} - 1 \right]$$

Example 2.54. *Calculate the rest mass of an electron in MeV. Also calculate the speed at which the total relativisitc energy is 1.25 times the rest mass energy $m_o = 9.1 \times 10^{-31}$ kg.* **(G.N. D.U. 2007)**

Solution. Rest mass energy in MeV = $m_o c^2 = 9.1 \times 10^{-31} \times 3 \times 10^8 \times 3 \times 10^8$ Joule

$$= \frac{9.1 \times 10^{-31} \times 9 \times 10^{16}}{1.6 \times 10^{-19}}\ \text{eV}$$

$$= 51.2 \times 10^4\ \text{eV} = 0.512\ \text{MeV}$$

Total relativistic energy = K.E. + Rest mass energy

or $$mc^2 = \frac{m_o c^2}{\sqrt{1 - \frac{v^2}{c^2}}} = 1.25\, m_o c^2$$

$$\therefore \qquad \frac{1}{\sqrt{1-\frac{v^2}{c^2}}}=1.25 \qquad \text{or} \qquad \sqrt{1-\frac{v^2}{c^2}}=\frac{1}{1.25}$$

$$\text{or} \qquad 1-\frac{v^2}{c^2}=\left(\frac{1}{1.25}\right)^2 \text{ or } \frac{v^2}{c^2}=1-\left(\frac{1}{1.25}\right)^2=1-0.64=0.36$$

$$\therefore \qquad \frac{v}{c}=0.6 \text{ or } v=0.6\,c$$

Example 2.55. *Kinetic energy of a particle is (i) 3 times, (ii) twice and (iii) equal to its rest mass energy. What is its velocity?*

(Pbi.U. 2001, 2003; P.U. 2007, 2003, 2001, G.N.D.U. 2009, 2007)

Solution. ***(i)*** Kinetic energy $E_k=(m-m_0)\,c^2$

Rest mass energy $= m_0c^2$

Now Kinetic energy = 3 × Rest mass energy

$$\text{or} \qquad (m-m_0)\,c^2=3m_0\,c^2$$

$$\text{or} \qquad m=4\,m_0$$

$$\text{But} \qquad m=\frac{m_0}{\sqrt{1-\frac{v^2}{c^2}}} \qquad \therefore \qquad 4m_0=\frac{m_0}{\sqrt{1-\frac{v^2}{c^2}}}$$

$$\text{or} \qquad \frac{1}{\sqrt{1-\frac{v^2}{c^2}}}=4$$

$$\therefore \qquad 1-\frac{v^2}{c^2}=\frac{1}{16}$$

$$\text{or} \qquad \frac{v^2}{c^2}=1-\frac{1}{16}=\frac{15}{16}$$

$$\therefore \qquad \frac{v}{c}=\sqrt{\frac{15}{16}}=0.968$$

Hence $v = 0.968\,c = 0.968 \times 3 \times 10^8 = 2.904 \times 10^8$ ms^{-1}

(*ii*) Kinetic energy = 2 × Rest mass energy

$$\text{or} \qquad (m-m_0)\,c^2=2\,m_0\,c^2 \text{ or } m=3\,m_0$$

$$\text{But} \qquad m=\frac{m_0}{\sqrt{1-\frac{v^2}{c^2}}} \qquad \therefore \qquad 3\,m_0=\frac{m_0}{\sqrt{1-\frac{v^2}{c^2}}} \quad \text{or} \quad \frac{1}{\sqrt{1-\frac{v^2}{c^2}}}=3$$

$$\therefore \qquad 1-\frac{v^2}{c^2}=\frac{1}{9} \qquad \text{or} \qquad \frac{v^2}{c^2}=1-\frac{1}{9}=\frac{8}{9}$$

$$\text{or} \qquad \frac{v}{c}=\sqrt{\frac{8}{9}}=0.943$$

Hence $v = 0.943\,c = 0.943 \times 3 \times 10^8 = 2.829 \times 10^8$ ms^{-1}

(*iii*) Kinetic energy = Rest mass energy

or $(m - m_0)\, c^2 = m_0 c^2$ or $m = 2\, m_0$

But $$m = \frac{m_0}{\sqrt{1 - \frac{v^2}{c^2}}} \qquad \therefore\ 2\, m_0 = \frac{m_0}{\sqrt{1 - \frac{v^2}{c^2}}} \text{ or } \frac{1}{\sqrt{1 - \frac{v^2}{c^2}}} = 2$$

$$\therefore \qquad 1 - \frac{v^2}{c^2} = \frac{1}{4} \text{ or } \frac{v^2}{c^2} = 1 - \frac{1}{4} = \frac{3}{4}$$

or $$\frac{v}{c} = \sqrt{\frac{3}{4}} = 0.866$$

Hence $v = 0.866\, c = 0.866 \times 3 \times 10^8 = 2.598 \times 10^8\ \text{ms}^{-1}$

Example 2.56. *What is the speed of the particle whose kinetic energy is equal to half of its rest mass energy?* **(P.U.2006)**

Solution. $$(m - m_o)c^2 = \frac{1}{2} m_o c^2$$

$$\therefore \qquad m = \frac{3}{2} m_o \text{ or } \frac{m_o}{\sqrt{1 - \frac{v^2}{c^2}}} = \frac{3}{2} m_o$$

or, $$\sqrt{1 - \frac{v^2}{c^2}} = \frac{2}{3};\ 1 - \frac{v^2}{c^2} = \frac{4}{9};\ \frac{v^2}{c^2} = \frac{5}{9}$$

$$\therefore \qquad \frac{v}{c} = \frac{\sqrt{5}}{3} = 0.745;\ v = 0.745c$$

Example 2.57. *Find the increase in mass of 2 kg metal piece of specific heat 0.2 when heated through 800°C.* **(H.P.U., 2001)**

Solution. Heat supplied to the metal $E = 2000 \times 0.2 \times 800 = 3.2 \times 10^5$ cal

$$= 3.2 \times 10^5 \times 4.2 = 13.44 \times 10^5 \text{ J}$$

Increase in mass of metal $= \dfrac{E}{c^2} = \dfrac{13.44 \times 10^5}{9 \times 10^{16}} = 1.493 \times 10^{-11}$ kg.

Example 2.58. *Calculate the increase in mass of 500 gm of metal of specific heat 0.15 cal g^{-1} K^{-1} when heated through 900°C.*

Solution. $E = 500 \times 0.15 \times 900 = 6.75 \times 10^4$ cal $= 28.35 \times 10^4$ J

Increase in mass $= \dfrac{28.35 \times 10^4}{9 \times 10^{16}} = 3.15 \times 10^{-12}$ kg.

Example 2.59. *The rest mass of electron is 9.028×10^{-31} kg. Calculate the energy equivalent in electron volts.*

Solution. Rest mass of the electron $m_0 = 9.028 \times 10^{-31}$ kg

Energy equivalent $m_0 c^2 = 9.028 \times 10^{-31} \times (3 \times 10^8)^2$ Joule $= 81.252 \times 10^{-15}$ J

Now one electron volt $= 1.6 \times 10^{-19}$ J

$\therefore$ Energy in electron volts $= \dfrac{81.252 \times 10^{-15}}{1.6 \times 10^{-19}} = 0.5078 \times 10^6 \text{ eV} = 0.5078 \text{ MeV}$

Example 2.60. *The atomic mass unit (a.m.u.) is 1.6558 × 10^{-27} kg. Find the corresponding energy in electron volts.* ***(G.N.D.U. 2006)***

Solution. Atomic mass unit = 1.6558 × 10^{-27} kg

$$\therefore \text{ Energy equivalent } = \frac{1.6558 \times 10^{-27} \times (3\times10^8)^2}{1.6\times10^{-19}} = 931\times10^6 \text{ eV} = 931 \text{ MeV}$$

Example 2.61. *If one gram of a substance is fully converted into energy in one second, how many calories of heat will be produced and how much power will be generated?* ***(G.N.D.U. 2001)***

Solution. Mass of the substance = m_0 = 1 gm = 10^{-3} kg

∴ Energy produced = $m_0c^2 = 10^{-3} \times (3 \times 10^8)^2 = 9 \times 10^{13}$ J

$$= \frac{9\times10^{+13}}{4.2} = 2.14\times10^{13} \text{ calories}$$

As 9×10^{13} J of energy is produced in one second,

Power generated = 9×10^{13} Watts = 9×10^7 Mega Watts

Example 2.62. *A proton has a total relativistic energy 900 MeV. If the rest mass of the proton is 1.6 × 10^{-27} kg, find its speed and kinetic energy.* ***(Pbi. U. 2002, P.U. 2009)***

Solution. Total relativistic energy of the proton = 900 MeV

= $900 \times 10^6 \times 1.6 \times 10^{-19}$ J = 14.4×10^{-11} J

Now relativistic energy $E = mc^2$

∴ Relativistic mass of the proton

$$m = \frac{E}{c^2} = \frac{14.4\times10^{-11}}{(3\times10^8)^2} = \frac{14.4\times10^{-11}}{9\times10^{16}} = 1.6\times10^{-27} \text{ kg}$$

Rest mass of the proton = 1.6×10^{-27} kg

As the rest mass of the proton = Relativistic mass

∴ The proton is at rest and hence its speed as well as kinetic energy is zero.

Example 2.63. *Dynamite liberates 5.4 × 10^6 J/kg when it explodes. What fraction of total energy content is this?*

Solution. Total energy content of 1 kg. mass = $m_0\, c^2 = 1 \times (3 \times 10^8)^2 = 9 \times 10^{16}$ Joule

Energy liberated per kg in explosion = 5.4×10^6 Joule

$$\text{Fraction of total energy } = \frac{5.4\times10^6}{9\times10^{16}} = 0.6\times10^{-10}$$

Example 2.64. *Give relation for relativistic kinetic energy of a particle. Show that for v << c the kinetic energy of the particle is given by classical relation* $\frac{1}{2}m_0v^2$. ***(P.U. 2004, 2001)***

Solution. Relativistic kinetic energy. The kinetic energy of a particle of rest mass m_0 is given by

$$E_k = (m - m_0)\, c^2 \quad ...(i)$$

where m is the mass of the particle when it is moving with a velocity v and is given by

$$m = \frac{m_0}{\sqrt{1 - \frac{v^2}{c^2}}}$$

$$\therefore \quad E_k = \left\{ \frac{m_0}{\sqrt{1 - \frac{v^2}{c^2}}} - m_0 \right\} c^2 = m_0 c^2 \left[\frac{1}{\sqrt{1 - \frac{v^2}{c^2}}} - 1 \right] \quad ...(ii)$$

This expression gives the kinetic energy of a particle of rest mass m_0 moving with a velocity v.

When $v << c$. Relation (*ii*) is the expression for *relativistic kinetic energy* of a particle of rest-mass m_0 moving with a velocity v. When the *speed* v is very small as compared to c,

$$\left(1 - \frac{v^2}{c^2}\right)^{-1/2} = 1 + \frac{1}{2}\frac{v^2}{c^2}$$

Neglecting higher powers of $\frac{v^2}{c^2}$ In such a case

$$E_k = m_0 c^2 \left(\frac{1}{2}\frac{v^2}{c^2}\right) = \frac{1}{2} m_0 v^2$$

This is the usual *non-relativistic* expression for the kinetic energy of a moving particle.

Hence we find that the relativistic expression for kinetic energy of a moving particle reduces to the classical expression for *small* velocitieis.

Example 2.65. *An electron of rest mass 9 × 10^{-31} kg moves with a speed of 80% of that of light. Find its kinetic energy.* ***(Kerala U. 2001)***

Solution. Velocity of the electron $v = \frac{80}{100} c = \frac{4}{5} c$

Kinetic energy of the electron $E_k = (m - m_0)c^2 = m_0 c^2 \left[\frac{1}{\sqrt{1 - \frac{v^2}{c^2}}} - 1 \right]$

$$= m_0 c^2 \left[\frac{1}{\sqrt{1 - \frac{16}{25}}} - 1 \right] = m_0 c^2 \times \frac{2}{3}$$

$$\therefore \quad E_k = 9 \times 10^{-31} \times 9 \times 10^{16} \times \frac{2}{3} = 54 \times 10^{-15}\text{ J} = 5.4 \times 10^{-14}\text{ J}$$

Example 2.66. *Calculate the speed of the electron which has kinetic energy 1.02 MeV. Given rest-mass of the electron = 0.51 MeV.*

Solution. Total energy of the electron is given by

$$E = E_k + m_0 c^2$$

where E_k is the kinetic energy and m_0c^2 the rest mass energy.

Now E_k = 1.02 MeV = 2 × 0.51 = 2 × rest mass energy = $2\, m_0\, c^2$

Total energy $E = mc^2 = \frac{m_0 c^2}{\sqrt{1 - \frac{v^2}{c^2}}}$ where v is the velocity (speed) of the electron.

$$\therefore \qquad \frac{m_0 c^2}{\sqrt{1-\frac{v^2}{c^2}}} = 2\, m_0\, c^2 + m_0\, c^2$$

or
$$\frac{1}{\sqrt{1-\frac{v^2}{c^2}}} = 3$$

$$\therefore \qquad 1-\frac{v^2}{c^2} = \frac{1}{9}$$

or
$$v = \frac{c}{3}\sqrt{8} = 0.943\, c = 2.829 \times 10^8 \text{ m/sec.}$$

Example 2.67. *Given that the total energy of a particle is exactly equal to two times its rest energy, what is the velocity of the particle?*

Solution. Total energy $E = mc^2 = \frac{m_0 c^2}{\sqrt{1-\frac{v^2}{c^2}}} = 2\, m_0\, c^2$

where m_0 is the rest mass and v the velocity of the particle when its energy is twice the rest energy.

$$\therefore \qquad \frac{1}{\sqrt{1-\frac{v^2}{c^2}}} = 2 \qquad \text{or} \qquad 1-\frac{v^2}{c^2} = \frac{1}{4} \qquad \text{or} \qquad \frac{v^2}{c^2} = 1-\frac{1}{4} = \frac{3}{4}$$

$$\therefore \qquad \frac{v}{c} = \frac{\sqrt{3}}{2} \qquad \text{or} \quad v = \frac{\sqrt{3}}{2}c = 0.866\, c = 2.598 \times 10^8 \text{ m/s.}$$

Example 2.68. *The earth receives 1400 W/m² of solar energy. Distance between the earth and the sun is 1.5 × 10¹¹ m. Estimate the rate of decrease of the mass of the sun.* ***(Kerala U. 2001)***

Solution. Solar energy received by the earth = 1400 W/m² = 1400 Joules / m² s

Distance of the earth from the sun $R_0 = 1.5 \times 10^{11}$ m

∴ Total energy radiated by the sun per second

$$E = 4\pi\, R_0^2 \times 1400 = 4\pi \times 1.5 \times 1.5 \times 10^{22} \times 1400 \text{ Joules s}^{-1}$$

If m is the rate of decrease of mass of the sun, then

$$mc^2 = 4\pi \times 1.5 \times 1.5 \times 10^{22} \times 1400 \text{ J/s}$$

or
$$m = \frac{4 \times 22 \times 1.5 \times 1.5 \times 10^{22} \times 1400}{7 \times 3 \times 3 \times 10^{16}} = 4.4 \times 10^9 \text{ kg/s.}$$

Example 2.69. *What is the error involved in calculating the kinetic energy of a particle according to (i) Classical Physics (ii) Relativistic Physics when it is moving at v = 0.5c?*

Solution. Let the rest mass of the particle be = m_0

Velocity of the particle $v = 0.5\, c = 1.5 \times 10^8$ ms⁻¹

(*i*) Kinetic energy of the particle according to Classical Physics

$$= \frac{1}{2} m_0\, v^2 = \frac{1}{2} m_0 \times 1.5 \times 1.5 \times 10^8 \times 10^8 = m_0 \times 1.125 \times 10^{16} \text{ J}$$

(*ii*) Kinetic energy according to Relativistic Physics

$$E_k = (m - m_0)\, c^2$$

$$= \left(\frac{m_0}{\sqrt{1-\frac{v^2}{c^2}}} - m_0\right) c^2 = \left(\frac{1}{\sqrt{1-\frac{v^2}{c^2}}} - 1\right) m_0\, c^2$$

$$= m_0 \left(\frac{1}{\sqrt{1-\frac{1}{4}}} - 1\right) 3 \times 3 \times 10^{16}$$

$$= m_0 \left(\frac{2-\sqrt{3}}{\sqrt{3}}\right) \times 9 \times 10^{16} = m_0 \times 1.395 \times 10^{16}\ \text{J}$$

$$\therefore \text{ Percentage error } = m_0 \times 10^{16} \times \frac{[1.395 - 1.125]}{m_0 \times 1.395 \times 10^{16}} \times 100 = \frac{1.395 - 1.125}{1.395} \times 100 = 19.4\%$$

2.16 ENERGY-MOMENTUM RELATION

The momentum of a particle of mass m (rest mass m_0) moving with a velocity u is given by

$$\vec{p} = m\vec{u} = \frac{m_0\, \vec{u}}{\sqrt{1-\frac{u^2}{c^2}}}$$

$$\therefore \qquad p^2 = \vec{p}.\vec{p} = \left(\frac{m_0\, \vec{u}}{\sqrt{1-\frac{u^2}{c^2}}}\right)\left(\frac{m_0\, \vec{u}}{\sqrt{1-\frac{u^2}{c^2}}}\right)$$

$$= \frac{m_0^2\, u^2}{1-\frac{u^2}{c^2}}$$

or
$$p^2 = \frac{m_0^2\, u^2\, c^2}{c^2 - u^2}$$

Multiplying Equation (*i*) by c^2, we have

$$p^2\, c^2 = \frac{m_0^2\, u^2\, c^4}{c^2 - u^2} \qquad ...(ii)$$

The total energy E of the particle is given by

$$E = mc^2 = \frac{m_0\, c^2}{\sqrt{1-\frac{u^2}{c^2}}}$$

$$\therefore \qquad E^2 = \frac{m_0^2\, c^4}{1-\frac{u^2}{c^2}} = \frac{m_0^2\, c^6}{c^2 - u^2} \qquad ...(iii)$$

Subtracting (*ii*) from (*iii*) we have

$$E^2 - p^2c^2 = \frac{m_0^2\, c^4\, (c^2 - u^2)}{c^2 - u^2} = m_0^2\, c^4$$

$$\therefore \quad E^2 = p^2c^2 + m_0^2\, c^4 \qquad ...(iv)$$

or $$E = \sqrt{p^2\, c^2 + m_0^2\, c^4} = c\sqrt{p^2 + m_0^2\, c^2} \qquad ...(v)$$

Energy and momentum of a photon. For a photon the rest mass $m_0 = 0$.

$\therefore$ Energy of the photon $E = pc$

The energy of a photon of frequency $\nu = h\nu$

$\therefore$ Momentum of the photon $p = \frac{E}{c} = \frac{h\nu}{c} = \frac{h}{\lambda}$

2.17 MOMENTUM-KINETIC ENERGY RELATION

The kinetic energy of a particle of relativistic mass m is given by

$$T = (m - m_0)\, c^2 \;\therefore\; T = mc^2 - m_0c^2$$

$mc^2 = E$ is known as total energy or relativistic energy of the particle.

$$\therefore \quad mc^2 = E = T + m_0\, c^2$$

Also $$E = \sqrt{p^2c^2 + m_0^2\, c^4}$$

$$\therefore \quad \sqrt{p^2c^2 + m_0^2\, c^4} = T + m_0\, c^2$$

Squaring both sides, we get

or $$p^2c^2 + m_0^2\, c^4 = T^2 + m_0^2\, c^4 + 2\, m_0 c^2\, T$$

$$p^2\, c^2 = T^2 + 2\, m_0\, c^2\, T$$

$$\therefore \quad p^2 = \frac{T^2}{c^2} + 2m_0 T$$

or $$p = \sqrt{\frac{T^2}{c^2} + 2\, m_0\, T}$$

Example 2.70. *Calculate the effective mass of a photon of wavelength 1Å.*

Solution. If m is the effective mass of the photon, then

$$h\nu = mc^2$$

where h is Planck's constant and ν the frequency of the photon.

$$\therefore \quad m = \frac{h\nu}{c^2} = \frac{h}{\lambda c} = \frac{6.63 \times 10^{-34}\ \text{J.s}}{10^{-10} \times 3 \times 10^8} = 2.21 \times 10^{-22}\ \text{kg}$$

Example 2.71. *What is the momentum of a proton having kinetic energy 1 BeV? Rest mass of proton = 938.3 MeV.*

Solution. According to the theory of relativity the total energy E of a particle of rest mass m_0 is related to its momentum p by the relation

$$E^2 = p^2c^2 + m_0^2\, c^4$$

The total energy E = Rest mass energy + K.E. = $m_0 c^2$ + K.E.

The rest mass energy of a proton $m_0c^2 = 938.3$ MeV

K.E. of the proton = 1 BeV = 10^9 eV = 10^3 MeV

$$\therefore \quad (10^3 + 938.3)^2 \text{ (MeV)}^2 = p^2c^2 + (938.3)^2 \text{ (MeV)}^2$$

$$\therefore \quad pc = \sqrt{10^6 (1 + 1.877)} \text{ MeV} = 1.696 \times 10^3 \text{ MeV}$$

or $$p = 1.696 \times 10^3 \text{ MeV}/c$$

$$= \frac{1.696 \times 10^3 \times 1.6 \times 10^{-13} \text{ J}}{3 \times 10^8 \text{ ms}^{-1}}$$

$$= 9.045 \times 10^{-19} \text{ Newton sec (kg ms}^{-1})$$

Example 2.72. *Calculate the linear momentum of a 25 GeV proton assuming that rest mass energy of the proton is approximately 1 GeV.*

Solution. This problem is a little different from the problem given in example 2.71. In this problem total energy of the proton including its rest mass energy is given whereas in the problem given in example 2.71 above only kinetic energy of the proton is given.

$E = 25$ GeV $= 25 \times 10^9$ eV; $m_0 c^2 = 1$ GeV $= 10^9$ eV

Now total energy E of a particle of rest mass m_0 is related to its momentum p by the relation

$$E^2 = p^2c^2 + m_0^2 c^4$$

$$\therefore \quad (25 \times 10^9)^2 = p^2c^2 + (1 \times 10^9)^2$$

or $$625 \times 10^{18} = p^2c^2 + 10^{18}$$

or $$p^2 c^2 = 624 \times 10^{18}$$

or $$pc = \sqrt{624} \times 10^9 \text{ eV}$$

$$= \sqrt{624} \times 10^9 \times 1.6 \times 10^{-19} \text{ J}$$

$$\therefore \quad p = \frac{\sqrt{624} \times 1.6 \times 10^{-10}}{3 \times 10^8} \text{ Newton Sec.}$$

$$= 13.3 \times 10^{-18} \text{ Newton Sec.}$$

Example 2.73. *Calculate the wavelength of the radiation emitted by the annihilation of an electron with a positron each of rest mass 9.1×10^{-31} kg.*

Solution. When an electron-positron pair is annihilated, the total rest mass of both the particles is converted into energy and a *pair* of photons of high frequency radiations is given out. The energy will not appear in the form of a single photon, because to conserve the total linear momentum two photons travelling in opposite directions and each sharing half the energy are required. If m is the mass of the electron as well as that of the positron and a pair of photons of frequency ν are emitted, then

$$mc^2 + mc^2 = 2h\nu = \frac{2hc}{\lambda}$$

$$\therefore \quad \lambda = \frac{hc}{mc^2} = \frac{h}{mc}$$

where h is Planck's constant 6.62×10^{-34} Joule sec. and λ the wavelength of the radiation emitted corresponding to the frequency ν of the photon. Hence

$$\lambda = \frac{6.62 \times 10^{-34}}{9.1 \times 10^{-31} \times 3 \times 10^8} = 0.024 \times 10^{-10} \text{ m}$$

2.18 MINKOWSKI SPACE

According to Newtonian classical mechanics *space* and *time* are independent of each other. But from the study of special theory of relativity, we learn that *space* and *time* are interdependent and can be treated on equal basis. In three dimensional space we talk of *position* of a point and denote it by three space co-ordinates x, y, z but in *four dimensional space time continuum* we speak of an *event* and the place and time of its occurrence is denoted by three space co-ordinates x, y, z and fourth time co-ordinate t. To give the time co-ordiante the *same dimension* of length as of x, y, z co-ordinates, the time co-ordinate is multiplied by the velocity of light c and is chosen as an *imaginary* quantity ict where $i = \sqrt{-1}$. A vector in this four dimensional space will be represented by components along four mutually perpendicular axes.

The four dimensional space time continuum is known as *Minkowski's space* and the four co-ordinates in Minkowski's space are x, y, z, ict.

The position vector $\vec{r}$ or the vector distance between the origin and a point in three dimensional space having co-ordinates x, y, z is given by

$$r^2 = x^2 + y^2 + z^2$$

It is not invariant to Lorentz transformations *i.e.*,

$$x^2 + y^2 + z^2 \neq x'^2 + y'^2 + z'^2$$

Similarly the vector interval between an event at the origin and an event in four dimensional space having co-ordinates x, y, z, ict is given by

$$s^2 = x^2 + y^2 + z^2 + (ict)^2$$
$$= x^2 + y^2 + z^2 - c^2 t^2$$

It can be shown that this value of s^2 is *invariant* to Lorentz transformations.

Four dimesnsional element invariant under Lorentz transformations

Lorentz transformation equations are

$$x' = \frac{x - vt}{\sqrt{1 - \frac{v^2}{c^2}}}\,;\; y' = y;\; z' = z \text{ and } t' = \frac{t - \frac{v}{c^2}x}{\sqrt{1 - \frac{v^2}{c^2}}}$$

In the frame S, $s^2 = x^2 + y^2 + z^2 - c^2 t^2$
In the frame S', $s'^2 = x'^2 + y'^2 + z'^2 - c^2 t'^2$

Substituting the values of x', y', z' and t' in terms of x, y, z and t, we get

$$x'^2 + y'^2 + z'^2 - c^2 t'^2 = \frac{(x - vt)^2}{1 - \frac{v^2}{c^2}} + y^2 + z^2 - c^2 \frac{\left(t - \frac{v}{c^2}x\right)^2}{1 - \frac{v^2}{c^2}}$$

$$= \frac{x^2 + v^2 t^2 - 2xvt - c^2 t^2 - \frac{v^2}{c^2}x^2 + 2xvt}{1 - \frac{v^2}{c^2}} + y^2 + z^2$$

$$= \frac{x^2\left(1-\frac{v^2}{c^2}\right) - c^2t^2\left(1-\frac{v^2}{c^2}\right)}{1-\frac{v^2}{c^2}} + y^2 + z^2$$

$$= x^2 + y^2 + z^2 - c^2t^2$$

Thus the quantity $s^2 = x^2 + y^2 + z^2 - c^2t^2 = x'^2 + y'^2 + z'^2 - c^2t'^2$

and is *invariant* under Lorentz transformations. In other words, the four dimensional element is invariant under Lorentz transformations.

Time like and space like interval. The interval s in Minkowski space is given by

$$s^2 = x^2 + y^2 + z^2 - c^2t^2 = x^2 + y^2 + z^2 + (ict)^2$$

If $x^2 + y^2 + z^2 < c^2t^2$ or $r < ct$, r being given by

$$r = \sqrt{x^2 + y^2 + z^2}$$

The interval is called *time like interval*. In this case two events can be connected by a light signal.

If $r > ct$, the interval is known as *space like interval*. In this case two events cannot be connected by a light signal. In order to relate the two events having a space like interval, we require signals moving faster than light which is not possible in accordance with the special theory of relativity.

2.19 TACHYONS

Tachyons are, as yet, hypothetical particles that would move faster than light, if they exist. According to special theory of relativity, the mass of a particle moving with a velocity v is given by $m = \frac{m_0}{\sqrt{1-\frac{v^2}{c^2}}}$. As such, no material particle can move with a velocity greater than the velocity of light because the mass of the particle would become infinite even at $v = c$. For v greater than c, the mass of the particle would become imaginary.

However, physicists like Fienberg have suggested that it is possible that a particle may be produced moving with a velocity greater than the velocity of light from its very birth. These particles have been named by Fienberg as Tachyons, meaning '*swiftly moving*'.

It has been postulated that Tachyons cannot move with a velocity less than the velocity of light. They have an infinite mass when moving with the velocity of light and as the velocity increases the mass decreases and becomes zero at infinity.

The energy equation governing the motion of Tachyons may be written as

$$E = \frac{\mu c^2}{\sqrt{\frac{v^2}{c^2} - 1}}$$

The energy of such a particle will be real if its mass is imaginary and may be related to mass in real universe by the equation

$$\mu = \frac{m}{\sqrt{-1}}$$

So far, it has not been possible to experimentally prove the existence of such a particle.

EXERCISE

LONG QUESTIONS

1. (*a*) State and explain the (basic) fundamental postulates of the special theory of relativity.
(M.D.U. 2008, Pbi. U. 2009, 2007, 2006; G.N.D.U. 2007; P.U. 2006, H.P.U. 2003)

(*b*) A reference frame S′ moves with respect to another frame S with uniform velocity $\vec{v}$. Derive Lorentz space and time transformation equations giving x′, y′, z′, t′ in terms of x, y, z, t, the moving frame coincides with stationary one at t′ = t = 0. Prove that when $\vec{v}$ is much smaller than velocity of light Lorentz transformations reduce to Galelian transformations.
(P.U. 2003, 2001, 2000; G.N.D.U. 2009, 2008, 2001, 2000; K.U. 2002, 2001, 2000, Pbi.U. 2002, 2001, 2000; Bang.U. 2001, 2000; Kerala U. 2001; Meerut U. 2000; M.D.U. 2002)

2. (*a*) Show by means of Lorentz transformation equations that

$$x'^2 - c^2t'^2 = x^2 - c^2t^2$$

(Kerala U. 2001)

(*b*) What do you understand by the statement that velocity of light is absolute? ***(P.U. 2005)***

3. What is the principle of constancy of light? ***(G.N. D.U. 2008)***

4. (*a*) Show that electromagnetic wave equation $\nabla^2\phi - \frac{1}{c^2}\frac{\partial^2\phi}{\partial t^2} = 0$ is invariant under Lorentz transformations where ϕ is a scalar potential.

(*b*) Name the transformations under which Maxwell's equations are invariant.

5. (*a*) On the basis of Lorentz transformations derive an expression for length contraction.

(*b*) Define proper length. ***(G.N.D.U. 2006)***

6. Prove that the length of an object measured in a frame moving with respect to an observer is smaller than the length of the same object as observed in stationary frame.

7. Derive the expression for length contraction using Lorentz transformations. Give its interpretation. ***(Nagpur U. 2010)***

8. States the postulates of special theory of relativity ***(Nagpur U. 2011)***

9. A circle and a square are moving along x-axis. How will they appear to a stationary observer?
(P.U. 2007, 2006, 2001, 2000; K.U. 2001, Bang. U. 2001, G.N. D.U. 2004, 2002, 2000, M.D.U. 2003, OSM. U. 2004)

10. (*a*) What do you understand by time dilation? On the basis of Lorentz transformations discuss the variation of time with velocity according to special theory of relativity. Explain why does a moving clock appear to run slow. Explain the terms proper time and non-proper time. Show that when v << c Lorentz transformations for time reduce to Galelian transformations.
(Keraia U. 2001; Bang. U. 2001; G.N.D.U. 2006, 2003, 2000; K.U. 2002, 2001; P. U. 2007, 2006, 2004; Osm.U. 2004; M.D.U. 2007, 2006)

(*b*) Describe an experiment to verify time dilation. ***(Kerala U. 2001)***

11. What is meant by time dilation? Derive an expression for it. ***(Nagpur U. 2010, 2011)***

12. (*a*) Explain simultaneity of two events in relativity.

(*b*) Show from Lorentz transformation equations that two events simultaneous at different positions in reference frame S are not generally simultaneous in reference frame S′ moving with a velocity $v\hat{i}$ relative to S. ***(Pbi. U. 2005)*** (*c*) Two events occur at different points on X-axis separated by a distance Δx in an inertial frame at the same instant. Show that in another inertial frame moving with uniform velocity v along X-direction with respect to it, these events are not simultaneous but separated by a time interval $\Delta t' = \dfrac{v\Delta x}{\sqrt{1-v^2/c^2}}$.

13. Prove that the four dimensional volume element *dx, dy, dz, dt* is invariant under Lorentz transformations.

14. What do you understand by the term twin paradox? Explain. ***(Pbi. U. 2002)***

15. (a) Starting from Lorentz transformation equations for space and time co-ordinates derive equations for relativistic addition of velocities. Hence prove that no material particle can move with a velocity greater than that of light .

(b) Show that the law agrees with velocity addition formula for non-relativistic velocities.

(P.U. 2008, 2007, 2001; G.N.D.U. 2009, 2007, 2000; Meerut U. 2000; K.U. 2000; M.D.U. 2007; Pbi. U. 2007)

16. Starting with Einstein's velocity addition formula show that it is in conformity with principle of constancy of speed of light. ***(K.U. 2000)***

17. Discuss relativistic Doppler effect and derive relations for longitudinal and transverse Doppler effect. What is meant by the terms red shift and blue shift?

(Meerut U. 2003, 2001, 2000; P.U. 2006, 2003, 2000; G.N.D.U. 2007; H.P.U. 2001, Pbi.U. 2007, 2006)

18. Derive the formula for relativistic variation of mass with velocity, $m = \dfrac{m_0}{\sqrt{1-\dfrac{v^2}{c^2}}}$.

(P.U. 2006, 2005, 2004, 2001, 2000; M.D.U. 2002, 2001, 2000; Indore U. 2001, Pbi.U. 2003; K.U. 2002, 2000; H.P.U. 2003; G.N.D.U. 2008)

(*b*) Hence prove that it is not possible for a material particle to have a velocity equal to or greater than the velocity of light. ***(P.U. 2001, 2000; G.N.D.U. 2004)***

19. Deduce the formula for relativistic variation of mass with velocity. ***(Nagpur U.2010)***

20. Establish mathematically Einstein's mass energy relationship. Explain physical significance of this relation. Mention two nuclear phenomena supporting this relation.

(G.N.D.U. 2004, 2001, 2000; P.U. 2001; Bang. U. 2004, 2001; Osm.U. 2004; Pbi. U. 2003, 2001; M.D.U. 2002, 2000; Meerut U. 2003, 2002;)

21. Obtain Einstein's mass energy equivalence relation. ***(Nagpur U. 2010)***

22. (*a*) Show that the velocity u of a relativisitic particle is given by

$$u = \frac{pc}{\sqrt{p^2 + m_o^2c^2}} \text{ and } E = \sqrt{p^2c^2 + m_o^2c^4}$$

where p is the relativistic momentum.

(*b*) Hence write down formulae for energy and momentum of a photon.

(P.U. 2006, 2001, H.P.U. 2000, Nagpur U. 2011)

(*c*) What should be the speed for a particle at rest mass m_0 in order that its relativistic momentum m_0c? Find its total relativistic energy. ***(P.U. 2009, 2005)***

23. (*a*) Derive relation between relativistic momentum and energy of a free particle. ***(P.U. 2003)***

(*b*) Prove that the momentum *p* of a body of rest mass m_0 and kinetic energy T is given by

$$p = \sqrt{\frac{T^2}{c^2} + 2m_0 T}.$$

24. Comment on the statement that special theory of relativity ascribes energies to all masses and masses to all energies. ***(G.N.D.U. 2004)***

25. (a) What do you mean by Minkowski space? Explain.

(b) Show that $x^2 + y^2 + z^2 - c^2t^2$ is invariant under Lorentz transformations. ***(Meerut U. 2003)***

26. What are Tachyons ? ***(Pbi. U. 2005, 2003)***

27. A rod has a length of 1.2 metre. It is placed in a space ship moving with a velocity 0.6 c relative to earth. Calculate its length as measured by an observer on (i) spaceship and (ii) earth. ***(A.U., 1993) [Ans. (i) 1.2 m (ii) 0.96 m]***

28. Calculate the percentage contraction of a rod moving with a velocity 60% that of light in a direction 60° to its own length. ***(H.P.U., 2000) [Ans. 5%]***

29. Atomic particles in the form of a beam have a velocity 98% of the velocity of light. What is the relativistic mass of an atomic particle as compared with its rest mass? ***[Ans. 5 m_0]***

30. A radioactive atom moving along X-axis with a velocity 0.5 c in the laboratory frame of reference emits an electron with a speed 0.95 c along the X-axis of a system in the radioactive atom. Calculate the speed of the electron in the laboratory system. ***[Ans. 0.983]***

31. Taking velocity of light as 3×10^8 m/s, calculate energy corresponding to a mass of 1.2 gm. ***[Ans. 10.8 × 10^{13} Joule]***

32. Find the velocity at which the mass of a particles is (i) double and (ii) $\sqrt{3}$ times its rest-mass. ***(P.U. 2000) [Ans. (i) 2.59 × 10^8 ms^{-1} (ii) 2.8284 × 10^8 ms^{-1}]***

33. How much mass does an electron gain when it is accelerated to a potential energy of 500 MeV? ***[Ans. 1.77 × 10^{-28} kg]***

34. At what speed should a particle move so that its kinetic energy becomes 1.3 times its rest energy? ***[Ans. 0.9 c]***

35. Calculate the velocity which 1 a.m.u. will have if it has a kinetic energy three times the rest mass energy. ***[Ans. 0.968 c]***

36. If 1000 kg of water is heated from 0 to 100°C, calculate the increase in mass of water. ***(K.U. 2000) [Ans. 4.67 × 10^{-8} kg]***

SHORT QUESTIONS

1. What is the role of Einstein's mass-energy equivalence relation in the development of atomic energy?

Hint: Nuclear phenomena like nuclear fission and fusion in methods of producing nuclear energy.

2. State the postulates of special theory of relativity. ***(Nagpur U. W. 2010)***

3. How does photon differ from material particle? ***(Nagpur U. S/2011)***

4. State the principle of constancy of light?

5. What do you understand by the statement that the velocity of light is absolute?

6. Name the transformations under which Maxwell's equations are invariant.

[**Ans.** Under Lorentz transformation]

7. What do you mean by 'Proper length'

8. Why length contraction is not observerd in daily life?

Ans. Because in our daily life, speed of object is very small as compared to the velocity of light.

9. What is meant by twin paradox?

10. No particle can have velocity greater than light. Explain.

11. State Einstein's velocity addition formula.

12. What is blue shift and red shift?

13. Prove that the rest mass of photon is zero. ***(P.U. 2006)***

Hint. According to mass variation formula, $m = m_o / \sqrt{1 - \frac{v^2}{c^2}}$ or $m_o = m\sqrt{1 - \frac{v^2}{c^2}}$ where m_o is the rest mass and v is velocity with which the mass is moving. A photon moves with a velocity of light. Therefore, for a photon $v=c$. Hence the rest mass of a photon, $m_o = m\sqrt{1 - \frac{c^2}{c^2}} = m \times 0 = 0.$ However, we can not speak about the rest mass of a photon because there exists no frame of reference in which the photons are at rest.

14. Comment on the statement that the rest mass of a particle can be zero. Give examples. ***(P.U. 2006, 2005)***

Hint. The mass of a particle in a frame of reference in which the particle is at rest is called its *rest mass*. The rest mass of a particle can be zero. For examples, the rest mass of particles like a photon and neturino is zero.

15. Is the Newton's second law of motion $\vec{F} = m\vec{a}$ always valid in a special theory of relativity? ***(P.U. 2005)***

Hint. No. The second law of motion $\vec{F} = m\vec{a}$ holds in special theory of relativity only if the velocity $\vec{v}$ is perpendicularto the accleration $\vec{a}$.

16. What are absolute quantitities? Name three absolute quantitites in (*a*) Newtonian mechanics, (*b*) relativity mechanics. ***(P.U. 2005)***

Hint. The quantities which have the same value in all frames of reference are known as absolute quantities.

Examples (a) In Newtonian mechanics, length, mass and time are absolute quantities.

(b) In relativitiy mechanics, velocity of light (c), phase of a wave and interval between two events are invariant.

17. Why electrons can not be acclerated by a cyclotron while protons can be acclerated? ***(G.N.D.U. 2008)***

Hint. When electrons are accelrated by a cyclotron, the mass of electron increases with increase in its velocity. In order that the electron may be acclerated the frequency of the applied electric field across the two Dees has to be constantly adjusted so that the electrons reach the gap after a fixed time interval . Since the mass of the electron is very small, it varies so rapidly with increase in its speed that it is not possible to keep its motion in phase with the frequency applied electric field. However, proton being 1836 times heavier than electron the increase in its mass with velocity is rather slow, so that it becomes possible to keep the motion of the proton in phase with the applied alternating electric field.

MULTIPLE CHOICE QUESTIONS

1. The postulates of special theory of relativity are applicable to
   (a) stationary frame (b) accelerated frames
   (c) inertial frames (d) none of these *(P.U. 1994)*
2. Frame S' moves along the positive X-direction of S frame. A rod placed along Y'-axis if observed from S frame appears
   (a) contracted (b) elongated (c) unchanged *(P.U. 1994)*
3. Rest volume L_0^3 is connected to relativistic volume as
   (a) $L_0^3(1-\beta^2)^{3/2}$ (b) $L_0^3(1-\beta^2)$
   (c) $L_0^3\sqrt{1-\beta^2}$ (d) $\dfrac{L_0^3}{\sqrt{1-\beta^2}}$ *(P.U. 1993)*
4. Relativistic time is given by
   (a) $\dfrac{t+vx'}{\sqrt{1-\beta^2}}$ (b) $\dfrac{t-\dfrac{vx}{c^2}}{\sqrt{1-\beta^2}}$
   (c) $\left(t-\dfrac{v}{c^2}\right)(1-\beta^2)^{-1/2}$ (d) $\left(t+\dfrac{vc^2}{x}\right)(1-\beta^2)^{-1/2}$ (P.U. 1993)
5. Decay of μ-meson supports
   (a) length contraction (b) time dilation
   (c) relativity of mass (d) relativity of energy
6. Of the two twin brothers, one goes on a relativistic tour and comes back. The brother on tour will
   (a) become younger (b) become older
   (c) be of the same age. *(P.U. 1994)*
7. A rocket has a velocity 0.6c. Velocity of light with respect to rocket is *(P.U. 1993)*
   (a) 0.5 c (b) 1.6 c (c) 0.4 c (d) c
8. Two photons approach each other. Their relative velocity will be
   (a) 0 (b) less than c (c) more than c (d) c
   (H.P.U. 1993)
9. Two photons recede from each other. Their relative velocity will be
   (a) zero (b) $\dfrac{c}{2}$ (c) $2c$ (d) c
   (H.P.U. 1994)
10. When one kgm of mass is fully converted into energy, it produces
    (a) 9×10^{16} J (b) 3×10^{10} J (c) 10^{20} J (d) 10^{16} J
    (P.U. 1994)
11. If one gm of a substance is converted into energy, the heat produced is
    (a) 2.15×10^{13} cal (b) 21.5×10^{13} cal (c) 215×10^{13} cal.
12. A stone is raised to the top of Mount Everest. Its rest mass has
    (a) increased (b) decreased (c) remained same *(P.U. 1994)*

13. A body of mass m falls through h metres. The decrease in its mass is equivalent to

(*a*) mgh (*b*) $mghc^2$ (*c*) mgh/c (*d*) mgh/c^2

(P.U. 1994)

14. At what velocity the kinetic energy of a body is equal to its rest mass energy?

(*a*) $\sqrt{2}c$ (*b*) $c/3$ (*c*) $c/2$ (*d*) $\sqrt{3}/2\, c$

(P.U. 1994)

15. A body moves with $0.2c$ velocity, the ratio of the moving mass to rest mass is

(*a*) 1.2 (*b*) 1.02 (*c*) 0.2 (*d*) 1.0

(P.U. 1994)

16. Two bodies of same momentum have different masses. The total mass energy will be greater for ***(P.U. 1994)***

(*a*) lighter body (*b*) heavier body (*c*) none

17. The rest mass of a photon is

(*a*) 0 (*b*) p/c (*c*) $\frac{E}{c^2}$ ***(H.P.U. 1992 1991)***

18. Relativistic transformations were suggested by

(*a*) Newton (*b*) Einstein

(*c*) Huygen (*d*) H.A. Lorentz ***(H.P.U. 2001)***

19. The apparent length of a metre rod moving parallel to its length with velocity $0.8c$ will be

(*a*) 0.5 m (*b*) 0.6 m

(*c*) 1 m (*d*) 1.5 m

ANSWERS

1. (*c*)	**2.** (*c*)	**3.** (*c*)	**4.** (*b*)	**5.** (*b*)
6. (*a*)	**7.** (*d*)	**8.** (*d*)	**9.** (*d*)	**10.** (*a*)
11. (*a*)	**12.** (*a*)	**13.** (*d*)	**14.** (*d*)	**15.** (*b*)
16. (*b*)	**17.** (*b*)	**18.** (*d*)	**19.** (*b*)	

SECTION II

Quantum Mechanics

3. ORIGIN OF QUANTUM THEORY

4. WAVE-PARTICLE DUALITY

5. HEISENBERG'S UNCERTAINTY PRINCIPLE

6. SCHRÖDINGER'S THEORY OF QUANTUM MECHANICS

7. SCHRÖDINGER'S THEORY OF HYDROGEN ATOM

DEVELOPMENT OF QUANTUM MECHANICS

The failure of Classical Mechanics (which includes Newton's laws of motion, inverse law of gravitational attraction between two bodies, Coulomb's inverse law of attraction or repulsion between two charged bodies, law of force on a moving charge in a magnetic field given by Lorentz force etc.) led to the development of *Quantum Mechanics*. To overcome this discrepancy, German scientist Max Planck in 1900, proposed his new theory, now known as old Quantum theory. This theory was applied to many crucial problems of microscopic dimensions and got the satisfactory solution. According to his theory, the exchange of energy between electromagnetic field and matter occurs through a discrete packets or ***quanta*** each of definite value (i.e., $E = h\nu$). The radiation emitted or absorbed by atom or molecule is given by discrete quanta of energy $h\nu$.

During the advent, two quantum concepts began from two distinctly different mathematical procedures: Matrix Mechanics and Wave Mechanics. Matrix mechanics was introduced by Heisenberg while Wave mechanics was developed by Schrödinger. Usually quantum mechanics includes both the concepts and their final results are found to be equivalent. Many distinguished physicists Albert Einstein, Niels Bohr, Louis de Broglie, Max Born, Paul Dirac, Werner Heisenberg, Wolfgang Pauli Erwin Schrödinger, Richard Feynman, Compton contributed to the development of quantum mechanics. The quantum mechanics has thunderous success in solving and explaining the microscopic results of molecular, atomic and nuclear dimensions and also predicting new results. This was the step towards the development of new Quantum theory.

Left to right: *Max Planck, Albert Einstein, Niels Bohr, Louis de Broglie, Max Born, Paul Dirac, Werner Heisenberg, Wolfgang Pauli, Erwin Schrödinger, Richard Feynman.*

ORIGIN OF QUANTUM THEORY

3

Chapter

INTRODUCTION

Quantum mechanics is the theory which attempts to explain the behaviour of *matter* and interaction with *energy* on the scale of atoms and atomic particles i.e., particles of the size of the order of 10^{-10} m. Phenomena such as motion of mechanical objects involving distances larger than about 10^{-6} m can be explained satisfactorily by laws of *classical theoretical physics* which is based on the following basic laws: (i) Newton's laws of motion, (ii) Newton's inverse square law of gravitational attraction between two bodies, (iii) Coulomb's inverse square law of attraction or repulsion between two electrically charged bodies, (iv) The law of force on a moving charge in a magnetic field, i.e., the Lorentz force. However, certain phenomena such as spectral distribution of energy in black body radiation, photo-electric effect etc; and phenomena involving distances of the order of 10^{-10} m could not be explained by classical mechanics. These limitations or failure of classical mechanics led to the development of quantum mechanics, the major revolution in physics. The above limitation i.e., the energy distribution in the spectrum of a black body was successfully explained by *Max Planck* in 1900 who proposed the revolutionary concept of *quantum hypothesis*. This was the origin of the quantum theory.

3.1 FAILURE OF CLASSICAL MECHANICS

There are many important experimental results which could not be explained on the basis of classical mechanics. Some of them are:

1. Stability of atoms and molecules. According to classical mechanics, the atom is stable. When the electron revolves around the nucleus in a circular orbit, it has acceleration. We know that any accelerating charge emits the radiation. The electron during its motion round the nucleus should have emitted energy continuously and as a result the radius of revolution would have reduced continuously and ultimately the electron would have merged into the nucleus. But this is against the stability of the atom. In other words, the classical theory could not explain the stability of the atoms and molecules. The classical theory, has no explaination to the behavior in the region of atomic dimensions (explained in article 3.2 in detail).

2. **The atomic spectra.** Characteristic frequency emission or absorption by the elements.

3. The regularities among the spectral lines.

4. The specific heat of solids at low temperatures.

5. **The black body radiation.** The quantity and quality of the emitted radiation depends only on the temperature of the body and not on the nature of the material of the body. Moreover, the radiation emitted at a particular temperature consists of wide range of frequencies with different intensities.

6. **The photo-electric effect.** The photoelectric emission and scattering of free electrons (photo-electric effect and Compton effect).

7. The emission of X-rays, radioactivity and other nuclear activities.

Some of them are explained below:

3.2 BLACK BODY RADIATION

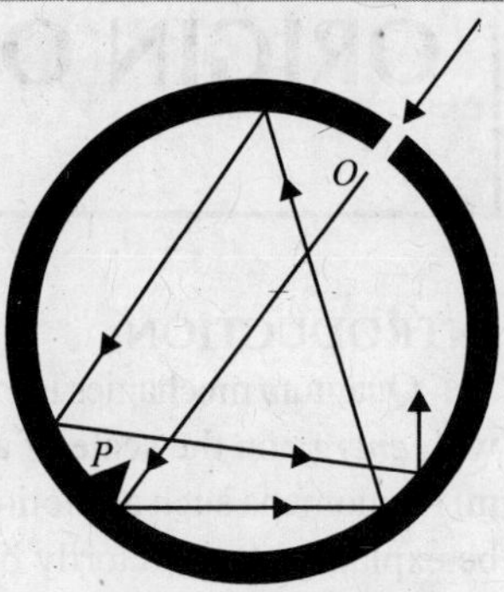

Fig. 3.1 Black body

Black body. *A perfectly black body is one which absorbs all the radiations of all the wavelengths which fall on it*. Since it neither reflects nor transmits any radiation it appears black whatever the colour of incident radiation may be. On the other hand when such a body is heated it emits radiations of all possible wavelengths.

A closed chamber, say a hollow sphere (known as hollow spherical cavity) whose inner surface is coated with platinum black or lamp black having a small opening O in the surface and a projection P opposite to the opening, acts as a *black body*. This is shown in Fig. 3.1.

It should be noted that perfectly blackbody is an ideal concept similar to a perfect gas which does not exist, i.e. it can not be 100% realized in actual practice. The platinum black surface absorbs 98% and lamp black 96% the incident radiation.

Black body radiation. A body emits radiation when it is heated. The emitted radiation from the hot body is called *thermal radiation*. When a black body is heated, it emits more radiations per unit area than any other body at a particular temperature.

Since a perfectly black body absorbs all the radiations falling upon it, has coefficient of absorption i.e., *absorptivity* equal to 1. At any temperature, it emits radiations of all wavelengths from $\lambda = 0$ to $\lambda = \infty$. Hence a perfectly black body has coefficient of emissions i.e. *emissivity* is also equal to 1. However, the amount of radiation is different for different wavelengths.

Distribution of energy in black body spectrum

Lummer and Pringesheim investigated the distribution of energy among the radiation emitted by a black body at different temperatures.

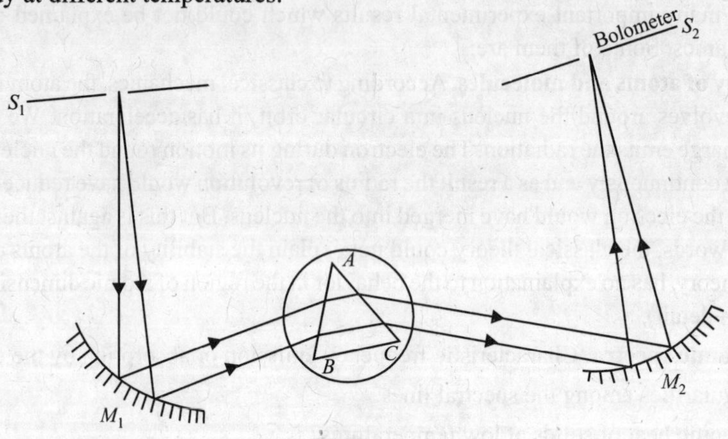

Fig. 3.2

The Experimental arrangement of Lummer and Pringesheim is shown in fig. 3.2. The radiation from a black body pass through the slit S_1 and falls on the reflector M_1, which passes through rock-salt or fluorspar prism ABC kept on a lotating (turn) table of spectrometer. The emergent light is focused by the concave reflector M_2 on a line bolometer placed behind the slit S_2. The bolometer is connected to a sensitive galvanometer. The turn table is slowly rotated so that the different parts of the radiation spectrum successively falls on the bolometer and the corresponding deflection in galvanometer are recorded. The intensity of each line is proportional to the deflection in the galvanometer. Then the curve is drawn between intensity and the wavelength (λ).

Then the body is heated to different temperatures and the curves are drawn for various temperatures. The curves obtained in the above experiment are as shown in Fig. 3.3.

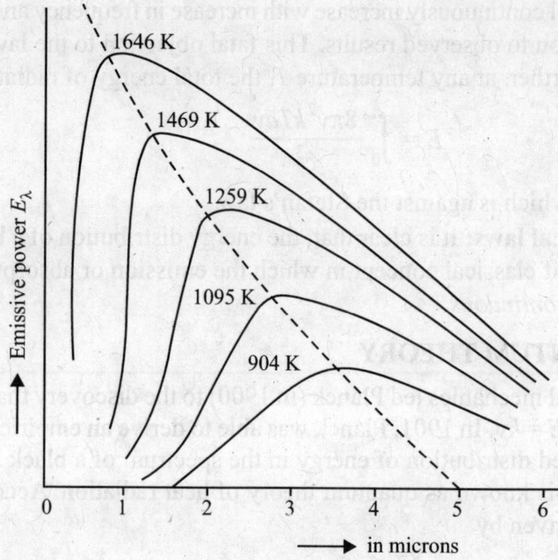

Fig. 3.3

Results. From above curves, we conclude that

(i) At a given temperature the energy is not uniformly distributed in the radiation spectrum of a black body.

(ii) At a given temperature, the intensity of heat radiation increase with increase in wavelength and at a particular wavelength its value is maximum. With further increase in wavelength, the intensity of heat radiation decreases.

(iii) An increase in temperature causes an increase in the emission for all wavelengths.

(iv) An increase in temperature causes a decrease in λ_m where λ_m is the wavelength for which the energy emitted is maximum, such that $\lambda_m T$ = constant. This is called Wien's displacement law.

Explanation: Classical Laws

Wien's Law. The nature of experimental curve of black body spectrum obtained by lummer and Pringsheim was tried to explain by Wien and Rayleigh-Jean applying the existing classical mechanics. According to Wien's, the amount of energy dE associated with a spectral region lying in the wavelength range λ and $(\lambda + d\lambda)$ emitted by a black body at a temperature T is given by

$$dE = K\lambda^{-5} e^{-a/\lambda T} d\lambda \quad \text{...(3.1)}$$

This law, however, as found not to accord with the experimental curves of Lummer and Pringsheim. The success of Wien's law is that it hold good only in the region of shorter wavelengths at lower temperatures. It does not hold good at longer wavelengths at higher temperatures. It is satisfactory for visible region upto 2000 K. But for longer wavelength when $\lambda T > 3 \times 10^{-3}$ m°K, it gives low value of E_λ.

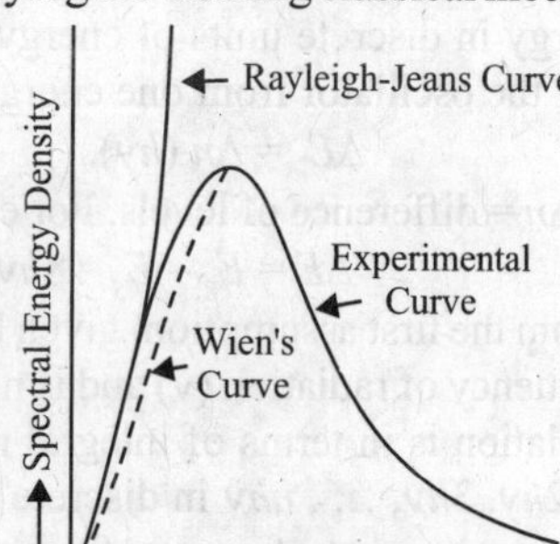

Fig. 3.4

Rayleigh-Jean's law. Rayleigh and Jeans independently assumed that the body consists of large number of atoms each of which is a radiation generator. Each atomic oscillator, oscillates with its own frequency and can emit or absorb the energy in the form of continuous spectrum. They used the method of classical mechanics and showed

$$dE = \frac{8\pi v^2 kTdv}{c^3} \quad \text{...(3.2)}$$

The energy density will continuously increase with increase in frequency and approaches ∞ when $\nu \to \infty$. This is contradiction to observed results. This fatal objection to the law has been known as 'ultraviolet catastrope.' Further, at any temperature T, the total energy of radiation.

$$E = \int_0^\infty \frac{8\pi\nu^2 kT d\nu}{c^3} \quad ...(3.3)$$

is predicted to be infinite which is against the Stefan's law.

Failures of the classical laws: It is clear that, the energy distribution of a black-body could not be explained on the basis of classical concept in which the emission or absorption of energy by the atom was supposed to be *continuous*.

3.3 PLANCK'S QUANTUM THEORY

This failure of classical mechanics led Planck (in 1900) to the discovery that radiation is emitted in quanta whose energy is $E = h\nu$. In 1901, Planck was able to derive an empirical formula to explain the experimentally observed distribution of energy in the spectrum of a black body, on the basis of his revolutionary hypothesis known as quantum theory of heat radiation. According to this theory, the energy distribution is given by

$$E_\lambda d_\lambda = \frac{8\pi hc}{\lambda^5[e^{hc/\lambda kT} - 1]} d\lambda \quad ...(3.4)$$

This relation agrees (and hence completely fit) with the experimental curves obtained in Lummer and Pringesheim experiment.

This formula of distribution of energy with wavelengths, on the basis of his quantum concept, was deduced using following assumptions, which may be called as Planck's quantum postulates. These are:

Planck's quantum postulates

1. A black body contains the atomic oscillators capable of vibrating with all possible frequencies. An oscillator of frequency ν can not have any energy but only discrete values of energy given by

$$E_n = nh\nu \quad ...(3.5)$$

where $n = 0, 1, 2, 3, \ldots$

ν = oscillator frequency

h = Planck's constant.

2. The atomic oscillator can not emit or absorb energy continuously. But it can emit or absorb the energy in discrete units of energy, a tiny packet called 'quanta'. The quantum energy is due to jump of the oscillator from one energy level to the other and is given by

$$\Delta E_n = \Delta n\,(h\nu). \quad ...(3.6)$$

where Δn = difference of levels. For example, if the transition energy levels are $n = 8$ and $n = 5$, then

$$\Delta E = E_8 - E_5 = 8h\nu - 5h\nu = 3h\nu$$

From the first assumption, given by Eq. 3.5 it is clear that each quantum of energy depends upon the frequency of radiation (ν) and is not divisible. Moreover, the exchange of energy between matter and radiation is in terms of integral multiple of $h\nu$. Therefore, the total energy exchanged may be $0, h\nu, 2h\nu, 3h\nu, \ldots, nh\nu$ in discrete manner. However, according to classical theory, the atom can possess any amount of energy from zero and infinity, hence exchange are energies are continuous i.e. from 0 to ∞ which is inadequate to quantum concept.

According to Planck's theory, the atom is associated with the allowed energy states called quantum states and its energies are *quantized* as shown in fig. 3.5. The atom can emit or absorb the radiation i.e. quantum of energy by jumping from one energy state to the other.

The *oscillator* emits energy only when it passes from a higher energy state to a lower energy state and absorbs energy when it goes from a lower energy state to a higher energy state. No emission or absorption of energy takes place when the oscillator is in a given state. The smallest amount of energy which can be emitted or absorbed by the oscillator is $h\nu$. In other words a radiation of frequency ν

is emitted as quantum of energy $E = h\nu$ where h is Planck's universal constant having a value equal to 6.62×10^{-27} erg. sec. (or 6.62×10^{-34} Joule sec.) The quantum is the basic unit of energy and cannot be subdivided. It is the quantum of energy and Einstein named it as a **photon.**

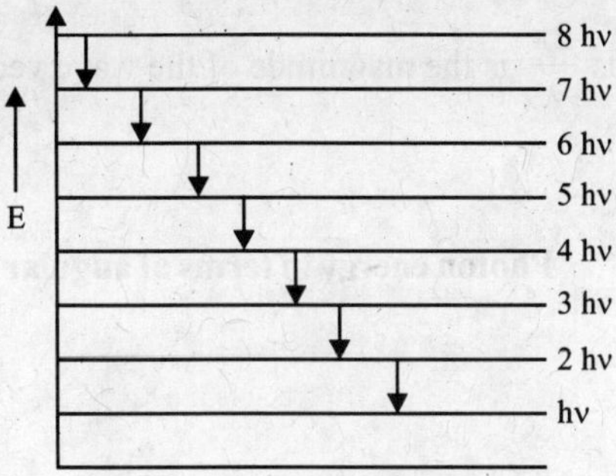

Fig. 3.5. Energy States of atom.

This theory has been able to explain a number of phenomena which could not be explained according to the classical conception of radiation. For example, it gives a very satisfactory explanation of variation of specific heat of solids with temperature, photo-electric effect, Compton effect etc. and has been successfully applied by Bohr in the theory of the hydrogen spectrum.

Photon. If λ is the wavelength of light, its frequency

$$\nu = c/\lambda$$

where c is the velocity of light. Each quantum of energy of this radiation is known as *photon.*

Energy of a photon. The energy E of a photon is given by

$$E = h\nu = \frac{hc}{\lambda} \qquad ...(3.6)$$

Mass. According to the theory of relativity a mass m has an energy equivalent to mc^2 where c is the velocity of light

Hence $$E = mc^2$$

or $$m = \frac{E}{c^2}$$

$\therefore$ Mass of the photon of energy $h\nu = \dfrac{h\nu}{c^2}$

Momentum. As momentum = mass × velocity

$$\textit{Momentum} \text{ of the photon } p = \frac{h\nu}{c^2}c = \frac{h\nu}{c} = \frac{h}{\lambda} \qquad ...(3.7)$$

where λ is the wavelength of the radiation.

A photon is a particle of *zero rest mass*. Its *dynamical variables* are *energy* $(h\nu)$ and *momentum* (h/λ). A photon has *zero charge* and *spin* equal to *one* quantum unit $(h/2\pi = \hbar)$. Two charges or two magnetic poles exert a force on each other due to *exchange of photons*. A photon interacts with all the charged particles and can interact with some neutral particles.

A photon behaves as a small elastic sphere for all purposes. When a photon collides with an electron both momentum and energy are conserved as in the case of collision between two elastic spheres. This fact has been verified by Compton in his experiments on X-rays. When X-rays of frequency ν collide with an electron of mass m, then

$$h\nu = \frac{1}{2}mv^2 + h\nu'$$

where v is the velocity with which the electron moves and ν' is the frequency of the new radiation given out.

Photon linear momentum in terms of wave vector. The magnitude of the linear momentum of a photon of wavelength λ is given by

$$p = \frac{h}{\lambda} = \frac{h}{2\pi}\cdot\frac{2\pi}{\lambda} = \frac{h}{2\pi}|\vec{k}|$$

as $\frac{2\pi}{\lambda}$ is the magnitude of the wave vector $\vec{k}$

$$\therefore \quad p = \hbar |\vec{k}| \quad \left[\because \hbar = \frac{h}{2\pi}\right]$$

Photon energy in terms of angular velocity. The energy of a photon of wavelength λ is given by

$$E = \frac{hc}{\lambda} = h\nu \quad \left[\text{where } \nu = \frac{c}{\lambda}\right]$$

$$= \frac{h}{2\pi} 2\pi\nu = \frac{h}{2\pi}\omega = \hbar\omega \quad ...(3.8)$$

3.4 PLANCK'S RADIATION LAW

Planck's modified Rayleigh-Jeans law with his assumption that a black body is associated with large number of harmonic oscillators. The oscillators have energies.

$$E = nh\nu, n = 0, 1, 2, \ldots \quad ...(i)$$

According to Boltzmann distribution law, the number of oscillators with energies, 0, $h\nu$, $2h\nu$, $3h\nu$, ... would be $N_o, N_e e^{-h\nu/kT}, N_o e^{-2h\nu/kT}$ ... respectively. Thus, the total number of oscillators is given by,

$$N = N_o + N_o e^{-h\nu/kT} + N_o e^{-2h\nu/kT} + \ldots + N_o e^{-nh\nu/kT} \quad ...(ii)$$

where N_o = the number of oscillators having zero energy

k = Boltzmann constant.

Putting $\frac{h\nu}{kT} = \alpha$ in Eqn. (ii), we get,

$$N = N_o + N_o e^{-\alpha} + N_o e^{-2\alpha} + N_o e^{-3\alpha} + \ldots + N_o e^{-n\alpha}$$

$$= N_o (1 + e^{-\alpha} + e^{-2\alpha} + \ldots + e^{-n\alpha})$$

$$= \frac{N_o}{(1 - e^{-\alpha})}$$

or

$$\frac{N_o}{N} = (1 - e^{-\alpha}) \quad ...(iii)$$

The total energy of all the oscillators is,

$$E = 0 + h\nu N_o e^{-\alpha} + 2h\nu N_o e^{-2\alpha} + 3h\nu N_o e^{-3\alpha} + \ldots + nh\nu e^{-n\alpha}$$

$$E = h\nu N_o e^{-\alpha} (1 + 2e^{-\alpha} + 3e^{-2\alpha} + \ldots + ne^{-(n-1)\alpha}) \quad ...(iv)$$

and

$$E e^{-\alpha} = h\nu N_o e^{-\alpha} (e^{-\alpha} + 2e^{-2\alpha} + 3e^{-3\alpha} + \ldots + ne^{-n\alpha}) \quad ...(v)$$

Subtracting Eqn. (v) from (iv), we have,

$$E(1 - e^{-\alpha}) = h\nu N_o e^{-\alpha} (1 + e^{-\alpha} + e^{-2\alpha} + e^{-3\alpha} + \ldots)$$

$$= \frac{h\nu N_o e^{-\alpha}}{1 - e^{-\alpha}}$$

$$\therefore \quad E = \frac{h\nu N_o e^{-\alpha}}{(1 - e^{-\alpha})^2} \quad ...(vi)$$

The average energy per oscillator is thus,

$$\bar{E} = \frac{E}{N} = \frac{N_o}{N} \cdot \frac{h\nu e^{-\alpha}}{(1 - e^{-\alpha})^2}$$

From Eqn. (iii), we have

$$\overline{E} = (1-e^{-\alpha})\cdot\frac{h\nu e^{-\alpha}}{(1-e^{-\alpha})^2} = \frac{h\nu e^{-\alpha}}{(1-e^{-\alpha})}$$

or
$$\overline{E} = \frac{h\nu}{e^{-\alpha}-1} = \frac{h\nu}{e^{h\nu/kT}-1} \qquad \text{...(vii)}$$

The number of oscillators per unit volume in the frequency range ν and ν + dν is

$$N_\nu = \frac{8\pi\nu^3}{c^3}\cdot d\nu$$

Therefore, the energy density i.e. the total energy per unit volume in the frequency range ν and ν+ dν is product of $\overline{E}$ and *Nν*,

$$E_\nu d\nu = N_\nu \cdot \overline{E} = \frac{8\pi\nu^2}{c^3}\cdot d\nu \cdot \frac{h\nu}{e^{\alpha}-1}$$

or
$$E_\nu d\nu = \frac{8\pi h\nu^3}{c^3}\left[\frac{1}{e^{h\nu/kT}-1}\right]d\nu \qquad \text{...(3.9)}$$

Eqn. 3.9 is called **Planck's radiation law.** This law can explain all the experimental results for black body radiation.

The energy density $E_\nu d\nu$ can be written in terms of the wavelength range λ and λ + d λ from

$$\nu = \frac{c}{\lambda}$$

$$|d\nu| = \left|-\frac{c}{\lambda^2}d\lambda\right| \qquad \text{or} \qquad d\nu = \frac{c}{\lambda^2}d\lambda$$

$$\therefore \qquad E_\lambda d\lambda = \frac{8\pi hc}{\lambda^5}\left[\frac{1}{e^{hc/\lambda kT}-1}\right]d\lambda \qquad \text{...(3.10)}$$

Wein's law and Rayleigh-Jeans law can be shown as approximations of Planck's law as follows:

1. Wein's law:

At shorter wavelengths or larger frequencies, $\frac{hc}{\lambda kT} \gg 1$ i.e. $\exp(hc/\lambda kT) \gg 1$, therefore neglecting 1 in the denominator of Eqn. 3.10

$$E_\lambda d\lambda = \frac{8\pi hc}{\lambda^5}\cdot e^{hc/\lambda kT}d\lambda \qquad \text{...(3.11)}$$

This is Wein's law of radiation.

2. Rayleigh-Jeans law:

At longer wavelengths, $\frac{hc}{\lambda kT} \ll 1$, expanding

$$\exp(hc/\lambda kT) \approx 1 + \frac{hc}{\lambda kT} + \frac{1}{2!}\left(\frac{hc}{\lambda kT}\right)^2 + \ldots$$

as $\frac{hc}{\lambda kT} \ll 1$, neglecting higher power terms, we get

$$\exp(hc/\lambda kT) \approx 1 + \frac{hc}{\lambda kT}$$

$\therefore$ Eqn. (3.10) becomes

$$E_\lambda d\lambda = \frac{8\pi hc}{\lambda^5} \cdot \frac{1}{hc/\lambda kT} d\lambda = \frac{8\pi}{\lambda^4} \cdot kTd\lambda \quad ...(3.12)$$

Eqn. is Rayleigh-Jeans law.

Thus Planck's law of radiation is applicable to all the wavelenghs in the black body spectrum.

3. Wein's displacement law:

Differentiating Planck's equation

$E_\lambda = 8\pi hc\,(\lambda^{-5}) \cdot (e^{hc/\lambda kT} - 1)^{-1}$, we have

$$\frac{d(E_\lambda)}{\lambda} = 0 = 8\pi hc\,[-5\lambda^{-6}\,(e^{hc/\lambda kT} - 1)] + (-1)\lambda^{-5}\left[(e^{hc/\lambda kT} - 1)^{-2} \times e^{hc/\lambda kT}\left(-\frac{1}{\lambda^2} \cdot \frac{hc}{kT}\right)\right]$$

or
$$5 + (e^{hc/\lambda kT} - 1)^{-1} \cdot e^{hc/\lambda kT} \cdot \left(-\frac{hc}{\lambda kT}\right) = 0$$

$$5 = \frac{1}{(e^{hc/\lambda kT} - 1)} \cdot e^{hc/\lambda kT} \cdot \frac{hc}{\lambda kT}$$

Let $\frac{hc}{\lambda kT} = x \;\therefore\; 5x + (e^x - 1) = 0$

The only solution of this equation is, $x = 4.965$

$$\therefore \quad \frac{hc}{\lambda kT} = 4.965 \text{ or } \frac{hc}{4.965k} = \lambda_{\max} T = \text{constant} \quad ...(3.13)$$

Eqn. 3.13 is a Wein's displacement law.

3.5 PHOTOELECTRIC EFFECT

The photoelectric effect was discovered by Hertz in 1887. He observed that, when a radiation (light in UV region, X-rays, γ-rays etc.) of high and suitable frequency incident on a metal surface, electrons are emitted from the surface.

Thus, the photoelectric effect is the phenomenon of ejection of electrons from a metal plate when light of a suitable wavelength falls on it.

When ultra-violet light is incident on a negatively charged zinc plate it loses its charge slowly. The loss of charge is due to the ejection of electrons from the surface of the zinc plate under the effect of light. All substances, metals as well as non-metals give out slow moving electrons when illuminated with light of a suitable wavelength.

The electrons emitted are called *photo-electrons* to indicate their mode of production. The value of *e/m* of these electrons is the same as that for electrons produced by other methods.

The alkali metals like sodium, potassium and ceasium are sensitive even to rays from the visible part of the spectrum, whereas zinc, cadmium, etc., are sensitive to only ultra-violet light. *Light of shorter wavelength has more energy and is thus more effective in producing the photo-electrons.*

The effect is also observed in non-metals but as they are highly electro-negative, waves of very short wavelength are required. Photo-electric effect is also observed in liquids and gases. *In gases,* the phenomenon results in *ionisation*.

An electron escaping just from the surface of a plate has the maximum velocity. An electron produced deep down will lose part of its energy in coming out to the surface and hence possesses a lesser velocity.

3.5.1 Experimental arrangement. The experimental arrangement to demonstrate photo-electric effect and obtain the important results is shown in Fig. 3.6.

It consists of an evacuated glass tube *T* fitted with a photo-sensitive plate *A* and a quartz window *W* for allowing the ultra-violet light to fall on the plate *A* from the source *S*. Another plate *B*, is also sealed at the other end of the tube. The middle point of the *H.T.* battery and the plate *A* are earth connected, so that the potential of plate *A* is zero. The plate *B* can be given a positive or a negative potential by the potential dividing arrangement as shown. A voltmeter *V* is connected across the plates *A* and *B* to measure the potential difference whereas a micro-ammeter measures the current flowing in the circuit.

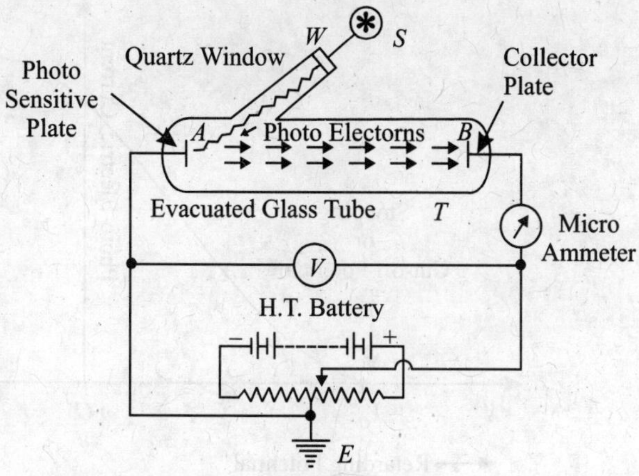

Fig. 3.6

When light of a suitable frequency is allowed to fall on the plate *A* from a source *S* through the quartz window *W* and the plate *B* is kept at a positive potential an electric current flows in the outer circuit as indicated by the micro-ammeter. This is due to photo-electrons emitted from the plate *A* which are attracted towards the plate *B* as shown. Some important results are as under.

1. Effect of intensity of incident light. Ultra-violet light of a *fixed* frequency is allowed to fall on the plate *A*. The collector plate *B* is kept at a positive potential with respect to *A*. The value of the positive potential is adjusted so that the micro-ammeter gives the *maximum* reading *i.e.* all the photo-electrons ejected from the plate *A* are collected by the plate *B*. Keeping the *frequency* of incident light and the accelerating *potential constant* the intensity of incident light is varied and the corresponding values of photo-electric current are measured. If a graph is plotted by taking *intensity of light* along the *X*-axis and the *photo-electric current* along the *Y*-axis, we get a straight line as shown in Fig. 3.7. Thus *the photo-electric current is directly proportional to the intensity of light.* In other words, the number of photo-electrons ejected from the photo-sensitive plate is proportional to the intensity of incident light.

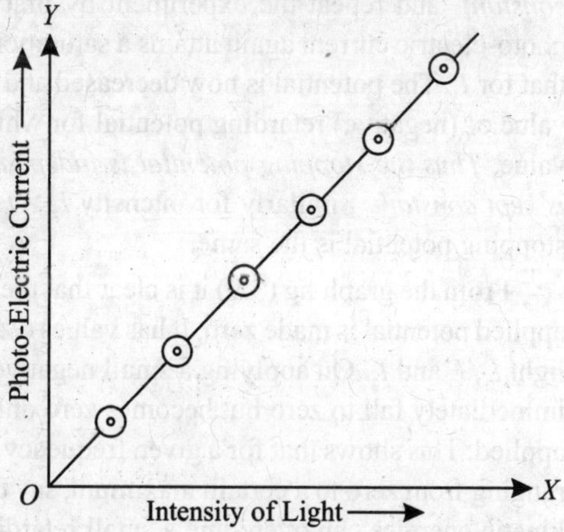

Fig. 3.7

2. Effect of potential. Let us suppose that the collector plate *B* is at zero potential. When light is made to fall on the photo-sensitive plate *A*, photo-electrons are emitted. Some of the photo-electrons which have sufficiently high kinetic energy reach the plate *B* and thus a photo-electric current starts flowing in the external circuit.

Keeping the *intensity* and *frequency* of the incident light constant, slowly increase the positive potential of the electrode *B* and measure the corresponding photo-electric current. It is found that the photo-electric current increases with increase in accelerating potential, till for a certain value of positive potential applied to *B* the current becomes *maximum* beyond which it does not increase further for any increase in positive potential. The maximum value of current is called *saturation current.*

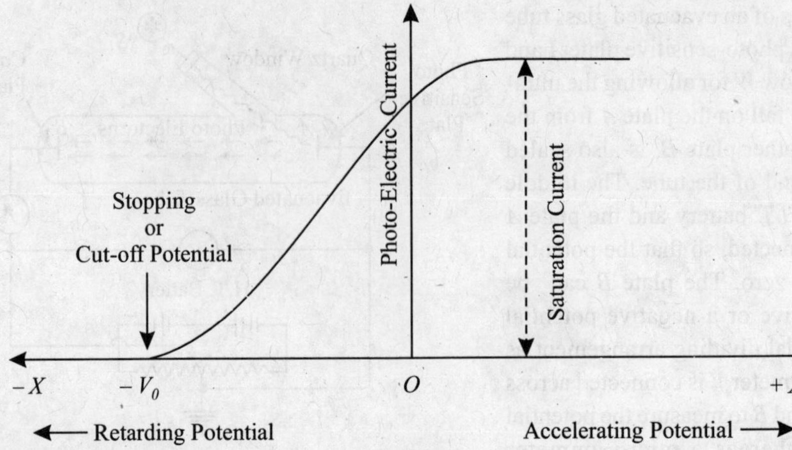

Fig. 3.8

Now decrease the potential applied to the collector plate B to zero and then make it negative with respect to the plate A. As the negative potential applied to B is increased the photo-electric current goes on decreasing. For a sharply defined negative collector potential V_0 the photo-electric current becomes zero.

The retarding negative potential for which the photo-electric current just becomes zero is called the *cut off* or *stopping potential* for the particular frequency of the incident light. At stopping potential no electrons reach the collecting electrode B and hence photo-electric current is zero.

3. Effect of intensity on stopping potential. Let I_1 be the intensity of light used in the last experiment. If we now increase the intensity from I_1 to I_2 such that $I_2 > I_1$, keeping the *frequency constant* and repeat the experiment by first increasing the accelerating voltage we find that the photo-electric current again attains a saturation value. This saturation value of current is higher than that for I_1. The potential is now decreased and then made negative till the current becomes zero. The value of (negative) retarding potential for which current is zero *i.e.* stopping potential has the *same* value. *Thus the stopping potential is independent of the intensity of incident light if the frequency is kept constant.* Similarly for intensity $I_3 > I_2$ the saturation current has a still higher value but the stopping potential is the same.

From the graph fig (3.9) it is clear that the photo-electric current does not become zero when the applied potential is made zero. It has values OA, OB and OC corresponding to the intensity of incident light I_1, I_2 and I_3. On applying a small negative potential to the collector plate B the current does not immediately fall to zero but becomes zero only when a negative potential V_0 = stopping potential is applied. This shows that for a given frequency of light the photo-electrons are emitted with velocities ranging from zero to a certain maximum, say v_{max}. Electrons with higher velocities and hence greater kinetic energies can overcome a small retarding potential and reach the plate B till at the stopping potential even the electron with highest energy and maximum velocity is unable to reach the plate B.

If m is the mass of the electron, e the charge, v_{max} the maximum velocity with which it is released, then V_0 the stopping potential is given by

$$\frac{1}{2} m v_{max}^2 = eV_0$$

or

$$v_{max} = \sqrt{\frac{2eV_0}{m}} \qquad ...(3.14)$$

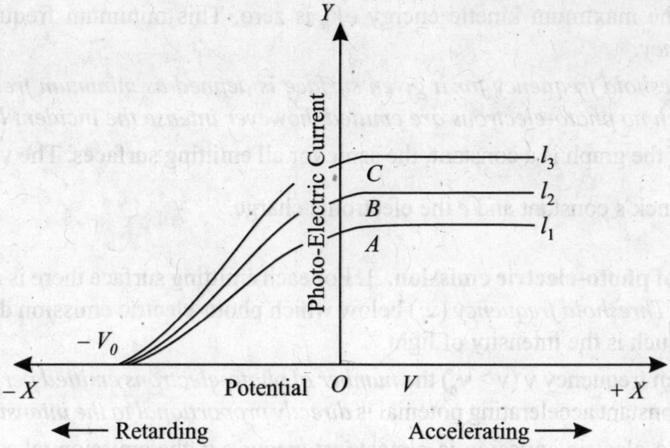

Fig. 3.9

4. Effect of Frequency. Let us have sources of light of the same intensity but different frequencies. Consider a source of frequency ν_1. Adjust the positive potential applied to the collecting plate B till the photo-electric current *just* attains the saturation value. Now first decrease the potential and then make it more negative till the current is zero *i.e.* we get stopping potential say V_1. Now consider source of frequency ν_2. Adjust the positive potential applied to the collecting plate B till the photo-electric cirrent *just* attains the *same* saturation value. Again find the stopping potential as before. Let it be V_2. Similarly find the stopping potential V_3 for a light of frequency ν_3 of the same intensity *i.e.* which gives *just* the same saturation current. For each frequency ν_1, ν_2 and ν_3 plot a graph between

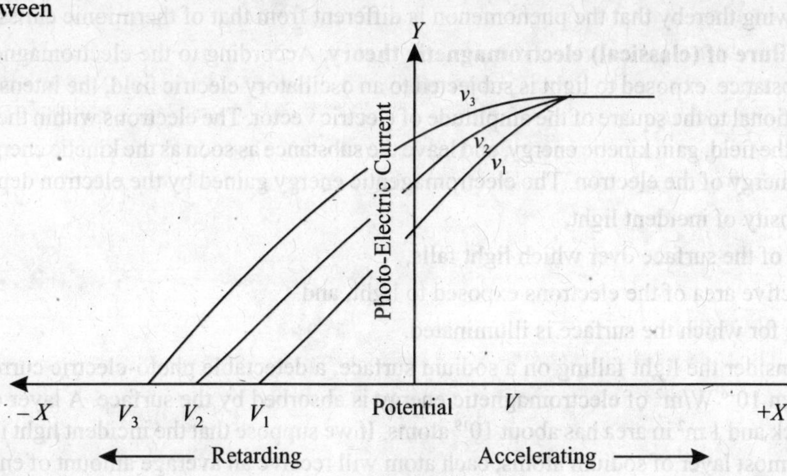

Fig. 3.10

potential and current as shown in Fig. 3.10. If the frequencies are in the increasing order $\nu_3 > \nu_2 > \nu_1$ then the stopping potentials are in the increasing order $V_3 > V_2 > V_1$.

Effect of frequency on stopping potential. If a graph is plotted between the frequency (ν) of the incident light and the corresponding stopping potential, then for any emitting surface we get a straight line not passing through the origin as shown in Fig. 3.11.

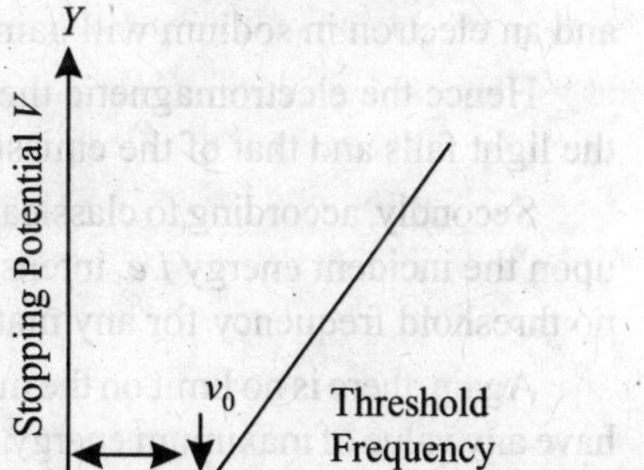

Fig. 3.11

Threshold frequency. The intercept on the frequency axis gives the minimum frequency ν_0 for which the stopping potential V_0 or the maximum kinetic energy eV_0 is zero. This minimum frequency is known as *threshold frequency*.

Thus the threshold frequency for a given surface is defined as minimum frequency of incident light, below which no photo-electrons are emitted, however intense the incident light may be.

The slope of the graph is a constant, the same for all emitting surfaces. The value of the slope = $\frac{h}{e}$ where h is Planck's constant and e the electronic charge.

3.5.2 Laws of photo-electric emission. 1. For each emitting surface there is a certain minimum frequency called *Threshold frequency* (ν_0) below which photo-electric emission does not take place, no matter how much is the intensity of light.

2. For a given frequency ν ($\nu > \nu_0$) the *number of photo-electrons emitted per second* for a given surface under a constant accelerating potenial is *directly proportional to the intensity of incident light.*

3. The photo-electric emission is *almost instaneous e.g.* the emission takes place in less than 3×10^{-9} sec even when the intensity of incident light is as low as 10^{-10} Watt/m^2.

4. The *maximum kinetic energy* and hence the maximum velocity of the ejected photo-electrons *increases with increase in frequency* of the incident light provided it is greater than the threshold frequency. The maximum kinetic energy (or maximum velocity) is independent of the intensity of incident light.

5. The *stopping potential* for a given metal surface is *proportional to the frequency* of incident light but independent of the intensity of light.

6. The rate of emission of photo-electrons is independent of the temperature of the photo-electric material showing thereby that the phenomenon is different from that of thermionic emission.

3.5.3 Failure of (classical) electromagnetic theory. According to the electromagnetic theory of light, a substance exposed to light is subjected to an oscillatory electric field, the intensity of light being proportional to the square of the amplitude of electric vector. The electrons within the substance interact with the field, gain kinetic energy, and leave the substance as soon as the kinetic energy exceeds the binding energy of the electron. The electromagentic energy gained by the electron depends upon

(*i*) intensity of incident light,

(*ii*) area of the surface over which light falls,

(*iii*) effective area of the electrons exposed to light, and

(*iv*) time for which the surface is illuminated.

If we consider the light falling on a sodium surface, a detectable photo-electric current will be obtained when 10^{-6} W/m^2 of electromagnetic energy is absorbed by the surface. A layer of sodium, one atom thick and 1 m^2 in area has about 10^{19} atoms. If we suppose that the incident light is absorbed by the upper most layer of sodium atoms, each atom will receive an average amount of energy equal to 10^{-25} W so that 1.6×10^6 sec (nearly 2 weeks) will be required by an atom to have 1 eV of energy and an electron in sodium will gain sufficient energy to be liberated in about 2 months.

Hence the electromagnetic theory fails to explain why there is no time lag between the instant the light falls and that of the emission of photo-electrons.

Secondly, according to classical electromagentic theory, the ejection of an electron should depend upon the incident energy *i.e.* intensity of light and not on frequency. In other words, there should be no threshold frequency for any material. This is contrary to the observed facts.

Again, there is no limit on the maximum energy to be transferred to the electron *i.e.* the electron can have any value of maximum energy. This is contrary to observed facts. Hence classical electromagnetic theory of light fails to explain the basic facts of photo electric effect.

3.6 EINSTEIN'S EQUATION : QUANTUM THEORY OF PHOTOELECTRIC EFFECT

In 1905, Einstein was successful in explaining the photoelectric effect and the laws of photoelectric effect making use of quantum theory of radiation. He assumed that:

Assumptions. (*i*) A light of frequency ν consists of quanta of energy or photons. Each photon has an energy $h\nu$ and travels with the velocity of light.

(*ii*) In photo electric effect *one* photon is completely absorbed by *one* electron in the photo-cathode.

Derivation of equation. Consider a photon of energy $h\nu$ which ejects a photo-electron from a material, with a velocity υ, then

$$h\nu = \frac{1}{2}mv^2 + \omega \qquad ...(3.15)$$

where ω is the work done or energy spent in just bringing the electron outside the surface. Thus a part of the energy of the incident photon is used by the electron to come out of the metal surface and the remainder is stored in it as kinetic energy. The fastest electron is, therefore, produced nearest to the surface because it requires the least energy to be 'freed' from this surface.

If ω_0 is the **photo-electric work function** *i.e., the energy required to eject an electron just out of the surface* then the frequency of light ν_0 required for the purpose is given by

$$\omega_0 = h\,\nu_0$$

The frequency ν_0 is known as **threshold frequency.**

It is that frequency of light which when falling on a photo sensitive surface is ***just*** *able to liberate electrons without giving them any additional energy.*

It is clear that a frequency less that ν_0 will not be able to produce the photo-electric effect as an electron can absorb only one photon and no more. The **threshold frequency** ν_0 is thus given by

$$\nu_0 = \frac{\omega_0}{h}$$

When a radiation of frequency ν greater than the threshold frequency is incident on the substance the maximum velocity v_{max} with which the electron is ejected is given by the relation

$$h\nu = \frac{1}{2}mv_{max}^2 + \omega_0 \qquad ...(3.16)$$

which is known as **Einstein's photo-electric equation.**

Now $\omega_0 = h\nu_0$

Hence the above equation can be put in the form

$$h\nu = \frac{1}{2}mv_{max}^2 + h\nu_0$$

or

$$\frac{1}{2}mv_{max}^2 = h(\nu - \nu_0)$$

$$= hc\left(\frac{1}{\lambda} - \frac{1}{\lambda_0}\right) \qquad ...(3.17)$$

where λ and λ_0 are wavelengths corresponding to frequencies ν and ν_0.

Thus, an increase in the frequency of the incident light increases the velocity with which a photo-electron is ejected.

Why all the photo-electrons do not have same energy? In the case of surface photo-electric effect in metallic conductors, it is generally assumed that ω_0 is the work done on the electron in bringing it out of the surface of the metal where there exists a *potential barrier.* But according to the theory of electron conduction in metals the conduction electrons have a wide range of energy inside the metal. The possible energy values of a free (conduction) electron near the surface of a metallic conductor is as shown in Fig. 3.12.

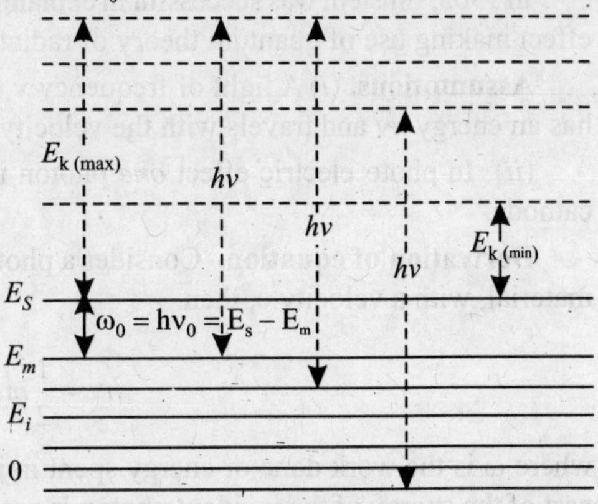

Fig. 3.12

Let the free electron inside the metal have energies ranging from 0 to a maximum value E_m. Then the energy required to remove the electron from the energy level E_s will be the photo-electric work function ω_0 given by

$$\omega_0 = h\nu_0 = E_s - E_m$$

where ν_0 is the threshold frequency.

Now, consider an electron having an energy E_i less than E_m. If a photon of energy $h\nu$ incident on the surface of the metallic conductor ejects this electron from the metal it will have a kinetic energy given by

$$E_k = \frac{1}{2}mv^2 = h\nu - (E_s - E_i)$$

But E_i can have any value between 0 and E_m. Therefore, the kinetic energies and velocities of the electrons ejected by the action of photon of frequency ν will vary over a wide range. The maximum kinetic energy is given by

$$E_{k(\text{max})} = \frac{1}{2}mv_{\text{max}}^2 = h\nu - (E_s - E_m) = (h\nu - E_s) + E_m \qquad ...(3.18)$$

The minimum kinetic energy is given by

$$E_{k(\text{min})} = \frac{1}{2}mv_{\text{min}}^2 = h\nu - (E_s - 0) = (h\nu - E_s) \qquad ...(3.19)$$

Thus, when a monochromatic beam of light is incident on the surface of a metal, the photo-electrons emitted will not all of them, have the same kinetic energy and velocity. The kinetic energy of the emitted electron will vary between $E_{k\,(\text{max})}$ and $E_{k\,(\text{min})}$. It is for this reason that when a potential difference is applied in the opposite direction the photo-electric current reduces to zero gradually and not abruptly as the opposing potential difference is increased and becomes zero only when V is given by

$$Ve = \frac{1}{2}mv^2 = h\nu - \omega_0 = h\nu - h\nu_0$$

Explanation of the laws of photo-electric emission. (*i*) It is clear from Einstein's photo-electric equation

$$\frac{1}{2}mv^2 = h(\nu - \nu_0)$$

that if the frequency of the incident photon $\nu < \nu_0$, the *threshold frequency* the kinetic energy of the emitted electron is *negative*, which is meaningless.

Hence no-photo electrons are emitted if the frequency of incident light is less than the threshold frequency.

(*ii*) For a frequency $\nu > \nu_0$ the kinetic energy of the photo-electron is directly proportional to $(\nu - \nu_0)$. As ν_0 is a constant the kinetic energy of the photo-electron increases with increase in frequency of the incident light. *Thus the maximum velocity of the photo-electron depends upon the frequency of incident light and is independent of intensity of light.*

(*iii*) Since photo-electrons are absorbed in single units, there is a localisation of energy. The photons are, therefore, immediately absorbed and *there is no significant time delay in the emission of photo-electrons.*

(*iv*) If the intensity of light is increased keeping frequency the same, then more photons are incident on the metal surface each photon having the same energy. Hence more electrons are ejected. As an electron can absorb only one photon each electron will have the same maximum energy and will be ejected with the same maximum velocity. Hence an increase in the intensity of incident light *only increases the number of photo-electrons ejected and not their velocity.*

(*v*) The stopping potential V is given by

$$Ve = h\nu - h\nu_0$$

As $h\nu_0$ is a constant, *the stopping potential is proportional to the frequency of the incident photon.* V is independent of intensity.

(*vi*) In Einstein's photo-electric equation $\frac{1}{2}mv^2 = h\nu - h\nu_0$, no factor depends upon temperature. *Hence the rate of emission of photo electrons is independent of temperature.*

SOLVED EXAMPLES

Example 3.1. *The threshold frequency for photo-electric emission in copper is 1.1×10^{15} Hz. Find the maximum energy in eV when light of frequency 1.2×10^{15} Hz is directed on the copper surface.*

Solution. According to Einstein's photo-electric equation

Maximum energy of the ejected electron $\frac{1}{2}mv_{max}^2 = h(\nu - \nu_0)$

Now $\nu_0 = 1.1 \times 10^{15}$ Hz and $\nu = 1.2 \times 10^{15}$ Hz

Taking $h = 6.62 \times 10^{-34}$ Js $= 4.14 \times 10^{-15}$ eVs we have

Maximum energy in eV $= 4.14 \times 10^{-15}\,[1.2 \times 10^{15} - 1.1 \times 10^{15}]$

$= 4.14 \times 0.1 = 0.414$ eV.

Example 3.2. *Calculate the work function in electron volts of a metal, given that photo electric threshold is 6800A°.*

Solution. The work function ω_0 is given by

$$\omega_0 = h\nu_0 = \frac{hc}{\lambda_0}$$

where ν_0 is the threshold frequency and λ_0 the wavelength corresponding to threshold frequency.

$$\lambda_0 = 6800\text{ A°} = 6800 \times 10^{-10}\text{ m}$$

$$\therefore \quad \omega_0 = \frac{6.63 \times 10^{-34} \times 3 \times 10^8}{6800 \times 10^{-10}} = 2.925 \times 10^{-19}\text{ J}$$

$$= \frac{2.925 \times 10^{-19}}{1.6 \times 10^{-19}} = 1.828\text{ eV}$$

Example 3.3. *Is it possible to liberate an electron from a metal surface having work function 4.8 eV with an incident radiation of wavelength (i) 5000 Å and (ii) 2000 Å?*

Solution. For a wavelength of radiation λ, the energy is given by Energy $= \dfrac{hc}{\lambda}$

(*i*) $\lambda = 5000 \text{ Å} = 5000 \times 10^{-10}$ m

$$\therefore \quad \text{Energy} = \frac{hc}{\lambda} = \frac{6.63\times10^{-34}\times3\times10^{8}}{5000\times10^{-10}} \text{ Joules}$$

$$= \frac{6.63\times10^{-34}\times3\times10^{8}}{5000\times10^{-10}\times1.6\times10^{-19}} = 2.486 \text{ eV}$$

As this energy is less than the work function (4.8 eV) for the metal, radiation of wavelength λ = 5000 Å will not be able to liberate an electron.

(*ii*) $\lambda = 2000 \text{ Å} = 2000\times10^{-10}$ m

$$\therefore \quad \text{Energy} = \frac{hc}{\lambda} = \frac{6.63\times10^{-34}\times3\times10^{8}}{2000\times10^{-10}\times1.6\times10^{-19}} = 6.21 \text{ eV}$$

As this energy is greater than the work function (4.8 eV) for the metal, radiation of wavelength λ = 2000 Å will be able to liberate an electron.

Example 3.4. *The threshold wavelength for photo-electric emission in tungsten is 230 nm. What wavelength of incident light must be used in order to eject electrons with a maximum velocity of 5×10^5 ms^{-1}.*

Solution. Threshold wavelength $\lambda_0 = 230$ nm $= 230 \times 10^{-9}$ m.

Maximum velocity of ejected electron $v_{max} = 5 \times 10^5$ ms^{-1}

Mass of the electron $m = 9.1 \times 10^{-31}$ kg

Kinetic energy of the ejected electron

$$\frac{1}{2}mv_{max}^2 = \frac{1}{2}\times9.1\times10^{-31}\times(5\times10^{5})^2$$

$$= 113.75\times10^{-21} \text{ J}$$

Now $$\frac{1}{2}mv_{max}^2 = hc\left(\frac{1}{\lambda}-\frac{1}{\lambda_0}\right)$$

or $$113.75\times10^{-21} = 6.626\times10^{-34}\times3\times10^{8}\left(\frac{1}{\lambda}-\frac{1}{230\times10^{-9}}\right)$$

$$\therefore \quad \left(\frac{1}{\lambda}-\frac{1}{230\times10^{-9}}\right) = \frac{113.75\times10^{-21}}{6.626\times10^{-34}\times3\times10^{8}} = 5.72\times10^{5}$$

or $$\frac{1}{\lambda} = 5.72\times10^{5} + \frac{1}{230\times10^{-9}}$$

$$= 5.72\times10^{5} + 4.348\times10^{6}$$

$$= 49.2\times10^{5}$$

$$\therefore \quad \lambda = \frac{1}{49.2}\times10^{-5} = 20.3\times10^{-8} \text{ m}$$

$$= 203\times10^{-9} \text{ m} = 203 \text{ nm.}$$

Example 3.5. *For a copper surface irradiated by light of wavelength 1849 Å stopping potential is 2.72 volts. Calculate the work function and threshold frequency.*

Solution. According to Einstein's photo-electric equation

$$h\nu = Ve + \omega_0$$

where ν is the frequency of the incident light, V the stopping potential and ω_0 the photo-electric work function.

Now $$\nu = \frac{c}{\lambda};\ e = 1.6 \times 10^{-19} \text{ Coulomb}$$

and $$h = 6.62 \times 10^{-34} \text{ Joule sec.}$$

$$\therefore \quad \frac{hc}{\lambda} = Ve + \omega_0$$

$$\frac{6.62 \times 10^{-34} \times 3 \times 10^8}{1849 \times 10^{-10}} = 1.6 \times 10^{-19} \times 2.72 + \omega_0$$

or $$\omega_0 = (10.740 - 4.352)\ 10^{-19}$$

$$= 6.388 \times 10^{-19} \text{ Joule} = \frac{6.388 \times 10^{-19}}{1.6 \times 10^{-19}} = 3.99 \text{ eV}$$

Threshold frequency $$\nu_0 = \frac{\omega_0}{h} = \frac{6.388 \times 10^{-19}}{6.62 \times 10^{-34}} = 0.965 \times 10^{16} \text{ Hz.}$$

Example 3.6. *In an experiment, the work function of potassium surface is found to be 2.1 eV. Calculate the threshold wavelength. What should be the wavelength of incident radiation if the stopping potential for electrons is 0.43 V?*

Given $h = 6.63 \times 10^{-34}$ Js and $e = 1.6 \times 10^{-19}$ C.

Solution. Work function $\omega_0 = 2.1 \text{ eV} = 2.1 \times 1.6 \times 10^{-19}$ J

If λ_0 is the threshold wavelength, then

$$\omega_0 = h\nu_0 = \frac{hc}{\lambda_0}$$

$$\therefore \quad \lambda = \frac{hc}{Ve + \omega_0}$$

$$= \frac{6.63 \times 10^{-34} \times 3 \times 10^8}{(0.43 \times 1.6 \times 10^{-19}) + (2.1 \times 1.6 \times 10^{-19})}$$

$$= 4970 \times 10^{-10} \text{ m} = 4970 \text{ Å.}$$

Example 3.7. *The work function of sodium metal is 2.3 eV. What is the longest wave length of light that can cause photo-electric emission from sodium?* ***(Gauhati U. 2000; Luck. U. 2001)***

Solution. Longest wavelength = Threshold wavelength $\lambda = \dfrac{hc}{\omega_0}$

$$\omega_0 = 2.3 \text{ eV} = 2.3 \times 1.6 \times 10^{-19} \text{ J}$$

$$\therefore \quad \lambda = \frac{6.63 \times 10^{-34} \times 3 \times 10^8}{2.3 \times 1.6 \times 10^{-19}} = 5405 \times 10^{-10} \text{ m} = 5405 \text{ Å.}$$

Example 3.8. *Light of wavelength 3000Å falls on a metal surface having a work function of 2.3 eV. Calculate the maximum velocity of ejected electron. Given h = 6.6 × 10^{-34} Js, electron mass = 9.1 × 10^{-31} Kg, 1 eV = 1.6 × 10^{-19} J.* **(Nagpur U. 2011/W, Meerut. U. 2005)**

Solution. $E = \frac{1}{2} mv^2 = h\nu - \omega_0 = \frac{hc}{\lambda} - \omega_0 = \frac{6.6 \times 10^{-34} \times 3 \times 10^8}{3000 \times 10^{-10}} - 2.3 \times 1.6 \times 10^{-19}$

$$= 6.6 \times 10^{-19} - 3.68 \times 10^{-19} = 2.92 \times 10^{-19} \text{ J}$$

$$v = \sqrt{\frac{2E}{m}} = \left[\frac{2 \times 2.92 \times 10^{-19}}{9.1 \times 10^{-31}}\right]^{\frac{1}{2}} = 8 \times 10^5 \text{ ms}^{-1}$$

Example 3.9. *A metal surface when illuminated by light of frequency 0.90 × 10^{15} s^{-1} emits electrons which can be stopped by a retarding potential of 0.60 volts. When the same surface is illuminated by light of frequency 1.26 × 10^{15} s^{-1} the required retarding potential is 2.1 volt. Using the data calculate the value of Planck's constant and work function of the metal.*

Solution. When radiation of frequency ν is incident on a metal surface having work function ω_0 and the stopping potential for the emitted photo-electron is V then

$$h\nu = eV + \omega_0$$

(*i*) For $\nu = 0.90 \times 10^{15}$ s^{-1} stopping potential $V = 0.60$ volt.

$\therefore \quad h \times 0.90 \times 10^{15} = 1.6 \times 10^{-19} \times 0.60 + \omega_0$ ...(*i*)

(*ii*) For $\nu = 1.26 \times 10^{15}$ s^{-1} stopping potential $V = 2.1$ volt

$\therefore \quad h \times 1.26 \times 10^{15} = 1.6 \times 10^{-19} \times 2.1 + \omega_0$ ...(*ii*)

Subtracting (*i*) from (*ii*), we have

$$h\,(1.26 - 0.90)\,10^{15} = 1.6 \times 10^{-19}\,(2.1 - 0.60)$$

or $\quad 0.36 \times 10^{15}\, h = 1.5 \times 1.6 \times 10^{-19}$

$\therefore \quad h = \frac{1.5 \times 1.6 \times 10^{-19}}{0.36 \times 10^{15}} = 6.67 \times 10^{-34}$ Joule sec.

Substituting the value of $h = 6.67 \times 10^{-34}$ in (*i*), we have

$$6.67 \times 10^{-34} \times 0.90 \times 10^{15} = 1.6 \times 10^{-19} \times 0.60 + \omega_0$$

or $\quad$ Work function $\quad \omega_0 = (6.67 \times 0.90 - 1.6 \times 0.60)\, 10^{-19} = 5.04 \times 10^{-19}$ **J**

$$= \frac{5.04 \times 10^{-19}}{1.6 \times 10^{-19}} = 3.15 \text{ eV.}$$

3.7 STABILITY OF ATOM

Stability of atom, as discussed in 3.1 was the problem that was troubling the physicist at the end of 19th century. Different models were proposed to resolve the complexity of the atom. Let us have a brief look.

1. J. J. Thomson Model

In 1898, J. J. Thomson proposed a simple model of atom. According to him, atom is spherical in shape with radius about 10^{-10} m. The whole mass of the atom is uniformly distributed and positive charges are also uniformly distributed inside the entire body of the atom (Fig. 3.13).

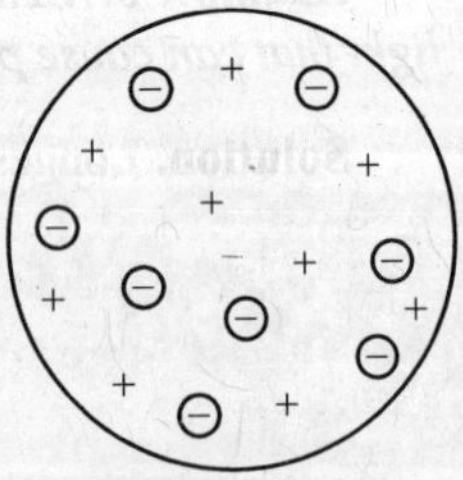

Fig. 3.13. Thomson's model of atom

The electrons are embeded in the atom like plums in the pudding (and hence also named as plum-pudding model of atom). The total positive charge is equal to the total negative charge in the atom, therefore the atom remains stable. The charges may oscillate about their mean position, but keeping the balance of attractive and repulsive forces.

Failure: The short comings of this model are:

(i) Hydrogen atom can emit a single spectral line as the atom consists of only one electron. Experimentally several series and each series consists of many discrete spectral lines were observed.

(ii) Positive charges are distributed over entire body the atom. But according to Rutherford's α-scattering, such distribution is impossible.

2. Rutherford Model

On the basis of α-scattering, Rutherford suggested a model of atom based on the solar system. According to him, atom has a small posively charged (+ze) nucleus of size about 10^{-14} m. The nucleus carry almost entire mass of the atom. The electrons revolve around the nucleus, like planets kee revolving around the sun, in our solar system.

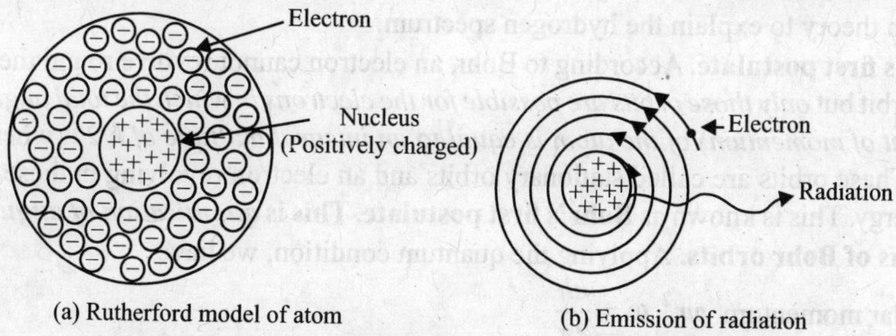

Fig. 3.14.

Failure: Rutherford tried to explain structure and the stability of atom with the help of classical theories (only known as that time). However, the model failed to explain:

(i) Distribution of electrons in the atom,

(ii) Stability of atom and

(iii) Atomic spectrum of hydrogen

According to this model, atom is stable due to balance of attractive and repulsive forces. The radiation comes out of the atom due to accelerating electron in a spiral path [fig. 3.14 (b)]. During this, the electron ultimately falls into the nucleus and destroys the atom giving continuous spectrum. The stability of atom and continuous spectrum are not correct because hydrogen atom gives line spectrum and the atom can never destroy.

3.8 BOHR'S THEORY OF HYDROGEN ATOM

Niel Bohr in 1913 applied the quantum concept to explain the stability of atom and emission and absorption of spectral lines. He improved the Rutherford model of atom. In Bohr's model of atom, stationary electron orbits are quantised and definite spectral line is emitted or absorbed due to jump of electron from one orbit to the other.

Bohr, therefore, assumed that

(i) The atom consists of a central nucleus having a net positive charge and (ii) The electrons revolve round the nucleus in circular orbits. The Coulomb's attractive force on these negatively charged revolving electrons provides the necessary centripetal force for dynamically stable circular orbits.

If r is the radius of the electronic orbit, v the velocity of the electron in its orbit, Ze the charge on the nucleus, Z being the atomic number and e the charge on a proton, then

$$\frac{mv^2}{r} = \frac{kZe \cdot e}{r^2}$$

or $$mv^2 r = k\,Z\,e^2 \qquad ...(i)$$

where k is a constant, the value of which in C.G.S. (Gaussian) units is *one* and in S.I. units is $\frac{1}{4\pi\varepsilon_0} = 9\times 10^9\ Nm^2/C^2$

According to this relation it is possible to have an infinite number of orbits in which the electrons can rotate.

The electron will be stable if it keeps on revolving in its particular orbit. But an electron revolving in a circular orbit, has an acceleration towards the centre and must radiate energy. As a result, the electron will approach the nucleus by a spiral path, giving out radiations of constantly increasing frequency and finally will fall into the nucleus.

Quantum concept

A solution out of this difficulty was found by Niels Bohr in 1913 by making a bold application of quantum theory to explain the hydrogen spectrum.

Bohr's first postulate. According to Bohr, an electron cannot revolve round the nucleus in any arbitrary orbit but *only those orbits are possible for the electrons in which the total angular momentum (or moment of momentum) of the atom is equal to an integral multiple of h/2* π where h is Planck's constant. These orbits are called stationary orbits and an electron revolving in these orbits does not radiate energy. This is known as **Bohr's first postulate. This is *quantization of angular momentum***

Radius of Bohr orbits. Applying the quantum condition, we have

$$\text{Angular momentum } mr^2\omega = \frac{nh}{2\pi}$$

or
$$mvr = \frac{nh}{2\pi} \qquad [\because\ v = r\omega] \quad ...(i\,a)$$

or
$$m^2v^2r^2 = \frac{n^2h^2}{4\pi^2} \quad ...(ii)$$

Dividing (*ii*) by (*i*), we get

$$\frac{m^2v^2r^2}{mv^2r} = \frac{n^2h^2}{4\pi^2}\frac{1}{kZe^2}$$

or
$$r = \frac{n^2h^2}{4\pi^2 mkZe^2} \quad ...(iii)$$

$$\text{In S.I. units} \quad r = \frac{n^2h^2}{4\pi^2\frac{1}{4\pi\varepsilon_0}mZe^2} = \frac{n^2h^2\varepsilon_0}{\pi mZe^2}$$

$$\text{For hydrogen } Z=1, \quad \therefore\ r = \frac{n^2h^2\varepsilon_0}{\pi me^2} = \frac{4\pi\varepsilon_0\hbar^2n^2}{me^2} \quad ...(3.20)$$

i.e. the radius of the orbits varies as the square of n, known as the *principal quantum number.*

Quantisation of Energy

Total energy of the electron. The energy of an electron in any given orbit is the sum of its kinetic and potential energies. The kinetic energy is equal to $\frac{1}{2}mv^2$.

But according to equation (*i*)

$$mv^2 = \frac{kZe^2}{r}$$

$$\therefore \qquad \frac{1}{2}mv^2 = \frac{kZe^2}{2r}$$

[**Note.** Although Z for hydrogen is one, the value of Z is taken in the calculation to include other hydrogen like atoms, *e.g.*, singly ionised helium or doubly ionised lithum.]

The potential energy is given by the work done in bringing the electron of charge $-e$ from infinity to a distance r from the nucleus of charge Ze, which is given by

$$\text{Potential energy} = \int_{\infty}^{r} \frac{kZe\cdot e}{r^2}\,dr = -\frac{kZe^2}{r}$$

$$\text{Hence the total energy} = \frac{kZe^2}{2r} - \frac{kZe^2}{r}$$

or
$$W_n = -\frac{kZe^2}{2r} \qquad \text{...}(iv)$$

Substituting the value of r from (*iii*) in (*iv*), we have

$$W_n = -\frac{2\pi^2k^2Z^2me^4}{n^2h^2}$$

In S.I. units
$$W_n = -2\pi^2\left(\frac{1}{4\pi\varepsilon_0}\right)^2\frac{Z^2me^4}{\pi^2h^2} = \frac{-Z^2me^4}{8\varepsilon_0^2n^2h^2} \qquad \text{...}(v)$$

For hydgrogen Z = 1;
$$W_n = -\frac{me^4}{8\varepsilon_0^2n^2h^2} = -\frac{me^4}{32\pi^2\varepsilon_0^2\hbar^2n^2} \qquad \text{...}(3.21)$$

Thus, the total energy of the electron is quantized and in an orbit varies inversely as the square of n.

Significance of negative energy. The total energy of the electron in a quantised Bohr orbit is negative, which simply means that energy is required to take an electron away from the nucleus and when the electron is at infinite distance from it, its energy is zero. In other words, the electron is bound to the nucleus and motion of the electron is bound one. If E were greater than zero, (or positive) the electron will not follow a closed orbit around the nucleus. In fact the energy E is not a property of the electron alone. It is the property of the system consisting of a negatively charged electron and positively charged nucleus between which there is a force of attraction.

Bohr's second postulate. When an electron is revolving in one of the stationary orbits given by Bohr's quantum condition, it does not radiate energy. According to Bohr, *the radiation of energy only takes place when an electron jumps from one stationary orbit into the other.* The frequency of radiation is given by the relation.

$$W_1 - W_2 = hv$$

where W_1 is the energy of the electron in the first orbit and W_2 that in the second orbit. This is known as **Bohr's second postulate.**

Bohr's explanation of hydrogen spectrum. Bohr's second postulate is known as *radiation condition.*

Applying the radiation condition we can calculate the frequency of the emitted radiation. For instance, when an electron falls from the outer n_2 orbit into the inner n_1 orbit, the difference in energies of the two orbits, is emitted as light of frequency v, given by

$$hv = W_{n1} - W_{n2}$$

$$= \frac{2\pi^2k^2Z^2me^4}{h^2}\left(\frac{1}{n_1^2} - \frac{1}{n_2^2}\right)$$

or $$\nu = \frac{2\pi^2 k^2 Z^2 me^4}{h^3}\left(\frac{1}{n_1^2}-\frac{1}{n_2^2}\right)$$

For hydrogen $Z = 1$

$\therefore$ The frequency $$\nu = \frac{2\pi^2 k^2 me^4}{h^3}\left(\frac{1}{n_1^2}-\frac{1}{n_2^2}\right)$$

and Wave number $$\frac{1}{\lambda} = \frac{2\pi^2 k^2 me^4}{ch^3}\left(\frac{1}{n_1^2}-\frac{1}{n_2^2}\right)$$

or $$\frac{1}{\lambda} = \bar{\nu} = R\left(\frac{1}{n_1^2}-\frac{1}{n_2^2}\right) \quad ...(3.22)$$

where R is known as **Rydberg's constant, and $\bar{\nu}$ as the wave number of the spectral line.**

For a hydrogen like atom, like ionised Helium, where Z is not unity, we have

$$\frac{1}{\lambda} = RZ^2\left(\frac{1}{n_1^2}-\frac{1}{n_2^2}\right) \quad ...(3.23)$$

Rydberg's constant. The value of Rydberg's constant R in S.I. units is given by

$$R = \frac{2\pi^2\left(\frac{1}{4\pi\varepsilon_0}\right)^2 me^4}{ch^3} = \frac{me^4}{8\varepsilon_0^2 ch^3} \quad ...(3.24)$$

Substituting the values of m, e, ε_0, c and h, we have

$$R = \frac{me^4}{8\varepsilon_0^2 ch^3} = \frac{(9.1\times10^{-31}\text{ kg})\ (1.6\times10^{-19}\text{ C})^4}{(8)\ (8.85\times10^{-12}\ F/m)^2 (3\times10^8\text{ ms}^{-1})\ (6.63\times10^{-34}\text{ J.s})^3}$$

$= 1.097 \times 10^7\ \text{m}^{-1}$

The exact value of R is taken to be

$R = 1.09737 \times 10^{-3}\ \text{Å}^{-1} = 1.09737 \times 10^7\ \text{m}^{-1}$

3.9 EMISSION SPECTRUM OF HYDROGEN

The excieted hydrogen atom may return to ground state by the emission of radiations of different frequencies. Long before Bohr formulated his theory of hydrogen spectrum, Balmer, Lyman and Paschen made experimental studies of the hydrogen spectrum in the visible, ultraviolet and infra red regions. They gave an empirical formula representing $\bar{\nu}$, the wave number of the spectral lines as

$$\frac{1}{\lambda} = \bar{\nu} = R\left(\frac{1}{n_1^2}-\frac{1}{n_2^2}\right) \quad \text{(from Eq. 3.22)}$$

When hydrogen atom is excited, the electron absorbs the energy of the exciting agent and jumps into an outer orbit. As an electron in the outer orbit is unstable, it jumps back into the inner orbit thereby radiating the difference in the energy of the two orbits. This energy gives rise to a particular line in the spectrum.

The formation of various series in the hydrogen spectrum is shown by an energy level diagram. The energy corresponding to $n = 1$ is shown along the lowermost line (Fig. 3.15).

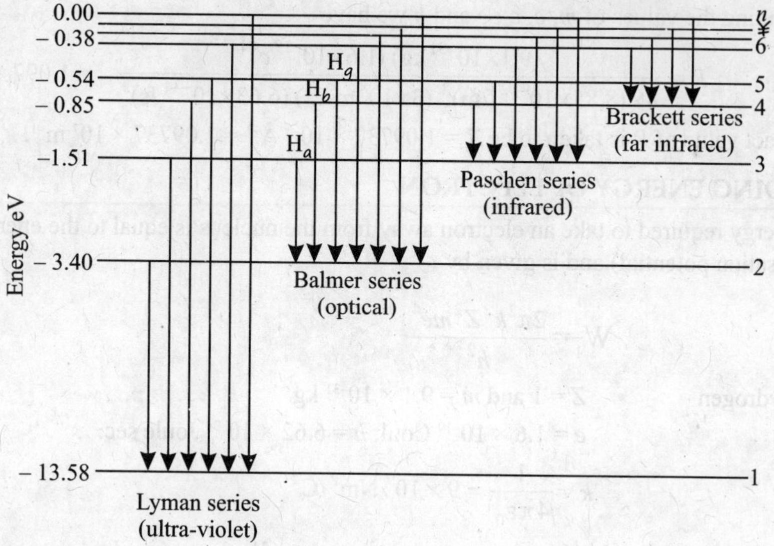

Fig. 3.15

The frequencies of lines in different series remarkably coincide with the values calculated from *Bohr's formula.* These are given in table 3.1.

Serial	*Region*	*Orbit*	*Bohr's Formula*
1. Lyman	Ultra-Violet (UV)	$(n = 1)\ K$	$\bar{v} = R\left(\frac{1}{1^2} - \frac{1}{n_2^2}\right),\ n_2 = 2, 3, 4 \ldots$
2. Balmer	Visible	$(n = 2)\ L$	$\bar{v} = R\left(\frac{1}{2^2} - \frac{1}{n_2^2}\right),\ n_2 = 3, 4, 5, \ldots$
3. Paschen	Infrared	$(n = 3)\ M$	$\bar{v} = R\left(\frac{1}{3^2} - \frac{1}{n_2^2}\right),\ n_2 = 4, 5, 6, \ldots$
4. Brackett	Far-infra red	$(n = 4)\ N$	$\bar{v} = R\left(\frac{1}{4^2} - \frac{1}{n_2^2}\right),\ n_2 = 5, 6, 7, \ldots$
5. Pfund	Extreme. infra red	$(n = 5)\ O$	$\bar{v} = R\left(\frac{1}{5^2} - \frac{1}{n_2^2}\right),\ n_2 = 6, 7, 8, \ldots$

The value of Rydberg's constant, as calculated by the formula $R = \dfrac{2\pi^2 k^2 me^4}{ch^3}$ and the value determined experimentally by Balmer show excellent agreement.

$$\therefore \quad R = \frac{2\pi^2\left(\dfrac{1}{4\pi\varepsilon_0}\right)^2 me^4}{ch^3} = \frac{me^4}{8\varepsilon_0^2 ch^3}.$$

Substituting the values of m, e, ε_0, c and h we have

$$R = \frac{(9.1\times10^{-31}\,\text{kg})\,(1.6\times10^{-19}\,c)^4}{8\times(8.85\times10^{-12}\,\text{f/m})^2\,(3\times10^8\,\text{ms}^{-1})\,(6.63\times10^{-34}\,\text{Js})^3} = 1.097\times10^7\ \text{m}^{-1}$$

The exact value of R is taken to be $R = 1.09737 \times 10^{-3}$ Å^{-1} $= 1.09737 \times 10^7$ m^{-1}.

3.10 BINDING ENERGY OF ELECTRON

The energy required to take an electron away from the nucleus is equal to the energy binding it (or the ionisation potential) and is given by

$$W_1 = \frac{2\pi^2 k^2 Z^2 m e^4}{h^2}$$

For Hydrogen $\quad Z = 1$ and $m = 9.1 \times 10^{-31}$ kg

$e = 1.6 \times 10^{-19}$ Coul; $h = 6.62 \times 10^{-34}$ Joule sec.

$$k = \frac{1}{4\pi\varepsilon_0} = 9\times10^9\ \text{Nm}^2/\text{C}^2$$

$$\therefore \quad W_1 = \frac{2\pi^2\times(9\times10^9)^2\times9.1\times10^{-31}\times(1.6\times10^{-19})^4}{(6.62\times10^{-34})^2}$$

$$= 21.75 \times 10^{-19}\ \text{Joule}$$

$$= \frac{21.75\times10^{-19}}{1.6\times10^{-19}} = 13.6\,\text{eV} \qquad ...(3.25)$$

This energy is also known as *ionisation potential* of hydrogen and represents the *potential energy in the ground state* of hydrogen atom.

3.11 MERITS OF BOHR'S THEORY

The following facts are noted in favour of Bohr's theory:

1. Stability of atom is explained with the Bohr's theory.

2. The atoms in higher excited states (higher values of n) are explained by classical mechanics while lower excited states (low temperature near absolute zero) are explained by quantum mechanics. The classical and quantum theories are bridged together.

3. Size of electron orbit is determined theoretically and it agreed well with the experimental value. For example: Bohr radius $r_1 = 0.53$Å and experimental value = 0.5Å.

4. The radius of electron orbit is calculated with classical mechanics while quantum mechanics is applied to the orbits for quantization.

5. There is excellent agreement between theoretical and experimental value of Rydber constant R = 1.097×10^7m^{-1} and 1.09737×10^7 m^{-1} respectively.

6. Ionization energy $\dfrac{me^4}{8\varepsilon_o^2 h^2} = 13.61$ eV is found to be closer to the experimental value.

7. Emission and absorption spectra of hydrogen are explained well. The spectra of hydrogen like atoms He^+, Li^{++} etc. are also explained.

3.12 LIMITATIONS OF BOHR'S THEORY

Bohr's theory explained successfully the spectrum of hydrogen in terms of only one quantum number, but it could not explain many other aspects of the spectra.

Some facts that could not be explained. (*i*) *Fine structure*. According to Bohr's theory, each spectral line consists of a single wavelength, but a careful study of the spectral lines indicates that each line is not really single, but consists of several lines. These lines are close but separate and distinct. This is known as *fine structure*. For example, H_α line in the Balmer series has been found to consist of 5 component lines. Bohr's theory does not explain fine structure.

(*ii*) *Simultaneous application of classical and quantum theory.* Bohr applied the classical theory to find the radius of the electronic orbit and quantum theory to explain the existence of stationary orbits and calculation of emitted frequency.

(*iii*) *Many electron atoms.* Bohr's theory can be applied to calculate the spectra emitted by hydrogen like atoms like He^+, Li^{++} etc, but could not be applied to atoms with more than one electron.

(*iv*) *Intensity of lines.* Bohr's theory does not give any information regarding intensity of spectral lines.

(*v*) *Selection rules.* Bohr's theory does not predict the selection rules which apply to transition of an electron from one energy level to another.

(*vi*) *Elliptic orbits.* Bohr's theory does not deal with elliptic orbits.

Example 3.10. *All hydrogen like atoms have one electron in the outermost orbit, still their ionisation energy is different. Explain.*

Solution. The ionisation energy of a hydrogen like atom is given by

$$W_1 = -\frac{2\pi^2 k^2 Z^2 me^4}{h^2}$$

where $k = \dfrac{1}{4\pi\varepsilon_0}$. Thus the ionisation energy is proportional to Z^2, Z being the atomic number of the element and hence the ionisation energy of all hydrogen like atoms is different. For example, the ionisation energy of Hydrogen (Z = 1) is – 13.6 eV and the ionisation energy of ionised Helium He^+, (Z = 2) is given by $E = -2^2 \times 13.6 = -54.4$ eV.

Example 3.11. *In a singly ionised Helium atom the electron is in the third orbit. A photon of energy 10.04 eV knocks out the electron. Calculate the stopping potential of the electron. Ionisation potential of H-atom is 13.6 eV.*

Solution. Ionisation potential of hydrogen atom = – 13.6 e V

∴ Ionisation potential of ionised Helium (He^+), Z = 2

$$E_{He} = -Z^2 \times 13.6 = -4 \times 13.6 = -54.4 \text{ eV}$$

For the third orbit $n = 3$

For the ground state orbit $n = 1$

Energy of the electron in the nth orbit $E_n = -\dfrac{2\pi^2 k^2 m Z^2 e^4}{n^2 h^2}$

For $n = 1$, it represents the ionisation energy $= -\dfrac{2\pi^2 k^2 m Z^2 e^4}{h^2}$

$$E_1 = -54.4 \text{ eV for Ionised Helium}$$

∴ Energy of the electron in the third orbit $E_3 = \dfrac{E_1}{3^2} = -\dfrac{54.4}{9} = -6.04$ eV

Energy of the incident photon = 10.04 e V

∴ Energy of ejected electron = 10.04 – 6.04 = 4 eV.

Hence stopping potential required = 4 volt.

Example 3.12. *The ionisation energy of H-like atom is 4 Rydberg. What is the wavelength of radiation emitted when the electron jumps from the excited state to the ground state? Given Rydberg = R = 2.2 × 10^{-18} J.*

Solution. Ionisation energy of H-like atom = 4 Rydberg

or $$E_1 = 4 \times 2.2 \times 10^{-18} = 8.8 \times 10^{-18}\text{ J}$$

This is the energy of the atom in the ground state ($n = 1$)

The energy of the atom in the first excited state (n =2) is given by

$$E_2 = \frac{E_1}{n^2} = \frac{E_1}{2^2} = \frac{8.8 \times 10^{-18}}{4} = 2.2\times 10^{-18}\text{ J}$$

If λ is the wavelength of the radiation emitted when the electron jumps from first excited state to the ground state, then

$$E_1 - E_2 = h\nu = \frac{hc}{\lambda}$$

$$\therefore \quad \lambda = \frac{hc}{E_1 - E_2} = \frac{6.63\times 10^{-34} \times 3\times 10^8}{(8.8 - 2.2)\times 10^{-18}} = 3.01\times 10^{-8}$$

$$= 301 \times 10^{-10}\text{ m} = 301\text{ Å}.$$

Example 3.13. *Calculate the radius of the first Bohr orbit of Hydrogen atom. Given e = 1.6 × 10^{-19} C, h = 6.63 × 10^{-34} Joule sec, k = 9 × 10^9 Nm²/ C² and m = 9.1 × 10^{-31} kg.*

(Gauhati, U., 2000)

Solution. For hydrogen Z =1 ; For the first orbit $n = 1$, $e = 1.6 \times 10^{-19}$ C; $h = 6.63 \times 10^{-34}$ Js; $k = 9 \times 10^9$ Nm²/ C; $m = 9.1 \times 10^{-31}$ kg.

$$\therefore \quad r = \frac{(6.63\times 10^{-34})^2}{4\pi^2 \times 9.1\times 10^{-31} \times 9\times 10^9 \times (1.6\times 10^{-19})^2}$$

$$= 0.53 \times 10^{-10}\text{ m} = 0.53\text{ Å}$$

Example 3.14. *Suppose the ionised lithium ion in its ground state is like H-atom. Calculate the radius of ionised lithium in its ground state. Radius of hydrogen atom is 0.54 Å.*

(Pbi, U. 2001)

Solution. According to the value of r, given by relation (i), for the same value of n, $r \propto \frac{1}{Z}$, all other factors being constants.

For Hydrogen $Z_H = 1$; For Lithium $Z_{Li} = 3$

$$\therefore \quad \frac{r_{Li}}{r_H} = \frac{Z_H}{Z_{Li}} = \frac{1}{3}$$

$$\therefore \quad r_{Li} = \frac{r_H}{3} = \frac{0.54}{3}\text{ Å} = 0.18\text{ Å}$$

$$= 0.18 \times 10^{-10}\text{ m}$$

Example 3.15. *The first exitation energy of hydrogen like atom is 40. 8 eV. Find the energy required to remove the electron from the ion.* ***(P.U. 2001)***

Solution. The energy of the electron in the nth state of a hydrogen like atom of atomic number Z

$$W_n = -\frac{2\pi^2 k^2 m e^4 Z}{n^2 h^2}$$

Energy required to remove the electron from the ion = energy in the ground state $n = 1$

$$\frac{2\pi^2 k^2 m e^4 Z}{h^2} = W$$

First excitation energy of the atom = energy in the first excited state $n = 2$

$$= \frac{2\pi^2 k^2 m e^4 Z}{2^2 h^2} = \frac{W}{4} = 40.8\,\text{eV}$$

$$\therefore \qquad W = 4 \times 40.8 = 163.2 \text{ eV.}$$

Example 3.16. *The wave length of first line of Balmer series of hydrogen is 6562. 8 Å. Calculate the ionisation potential and first excitation potential for hydrogen. (Given h = 6.63 × 10⁻³⁴ J)*

Solution. The wave length λ of the spectral line of hydrogen spectrum is given by

$$\frac{1}{\lambda} = \frac{2\pi^2 k^2 m e^4}{ch^3}\left[\frac{1}{n_1^2} - \frac{1}{n_2^2}\right]$$

or

$$\frac{ch}{\lambda} = \frac{2\pi^2 k^2 m e^4}{h^2}\left[\frac{1}{n_1^2} - \frac{1}{n_2^2}\right]$$

For the first line of Balmer series $n_1 = 2$ and $n_2 = 3$

$$\therefore \qquad \frac{ch}{\lambda} = \frac{2\pi^2 k^2 m e^4}{h^2}\left[\frac{1}{2^2} - \frac{1}{3^2}\right]$$

$$= \frac{2\pi^2 k^2 m e^4}{h^2} \times \frac{5}{36}$$

or

$$\frac{2\pi^2 k^2 m e^4}{h^2} = \frac{ch}{\lambda} \times \frac{36}{5} = \frac{3\times10^8 \times 6.63\times10^{-34} \times 36}{6562.8\times10^{-10}\times5}$$

$$= 21.82 \times 10^{-19}\text{ J} = \frac{21.82\times10^{-19}}{1.6\times10^{-19}} = 13.63 \text{ eV.}$$

Energy of the electron in the nth orbit of hydrogen atom is given by $W_n = \frac{2\pi^2 k^2 m e^4}{n^2 h^2}$

∴ Ionisation potential = Energy required to remove the electron from $n = 1$ orbit

$$= \frac{2\pi^2 k^2 m e^4}{h^2} = 13.63 \text{ eV.}$$

First excitation potential = Energy required to remove the electron from $n = 2$ orbit

$$= \frac{2\pi^2 k^2 m e^4}{2^2 h^2} = \frac{13.63}{4} = 3.41\, eV.$$

3.13 RITZ COMBINATION PRINCIPLE

The Rydberg-Ritz Combination Principle is the theory proposed by Walter Ritz in 1908 to explain relationship of the spectral lines for all atoms. The principle states that *the spectral lines of any element include frequencies that are either the sum or the difference of the frequencies of two other lines.*

Ritz found that Balmer's formula was just one prominent example of a more widespread phenomenon, in "line spectra" emitted by hot gas. If one formed differences between the values of (1/λ) in pairs spectral lines of some atom, sometimes *two different pairs gave the same difference*. That meant of course that the differences in frequency (c/λ) also matched (where c is the speed of *light*).

Suppose that one spectral line, of frequency v_1 resulted from a jump from energy A to energy B, and another line, or frequency v_2, originated in a jump from energy C to energy D. Then

$$hv_1 = A - B$$

$$h\nu_2 = C - D$$

In such atoms, it might perhaps be possible for transitions to also occur from A to C, and from B to D, causing the emission of photons with frequencies ν_3 and ν_4. In those cases

$$h\nu_3 = A - C$$
$$h\nu_4 = B - D$$

If such transitions are possible, then (as may be checked)

$\nu_3 - \nu_4 = (1/h)\,[(A - C) - (B - D)] = \nu_1 - \nu_2$

As noted earlier, the wavelengths of spectral lines (and hence their frequencies) could be established with very high precision. If such equalities as the one written above occurred by blind chance, they should be quite rare. Ritz and others found matching differences (of various sorts) much more prevalent, and that supported the idea that atoms would exist (at least briefly) at different energy levels.

Importance of Ritz combination principle:

The Ritz combination principle allowed physicists to translate the jumble of observed spectral lines into a more orderly (and smaller) scheme of energy levels. Based on understanding based on the quantum theory of the atom, such levels can be sorted into meaningful families, and various questions can be addressed, e.g. why certain transitions exist while others seem "forbidden," even through they too are expected to release energy.

One question was, why did levels sometimes split into two or more narrowly separated levels. For instance, the yellow light of sodium actually contains two closely spaced wavelengths.

A related effect was the **splitting of a single spectral wavelength** into several closely spaced ones, when light was emitted from the region of a strong **magnetic field**. This is the **Zeeman** effect, discovered 1896 by the Dutchman Pieter Zeeman, and the separation of the wavelengths often gives an indication of the strength of the magnetic field in the source region of the light. It was the Zeeman splitting of spectral lines emitted from **sunspots** that led George Ellery Hale in 1908 to realize that sunspots were in fact strongly magnetized, to about 1500 gauss (0.15 Tesla).

Radioactive nuclei emit gamma rays (γ rays), photons with energies about a million times higher than those of atoms. It took a while to measure their wavelength with any precision, but by about 1949 it was realized that they too displayed the Ritz principle. That confirmed the long-standing suspicion that nuclei too had energy levels, and that the associated transitions were the source of gamma rays.

3.14 COMPTON EFFECT

When *X*-rays fall on matter, a part is scattered without any change in wavelength. This is known as **unmodified or coherent or classical scattering.** In 1922, Compton using Bragg's crystal spectrometer discovered that in addition to classical scattering in which the wavelength of *X*-rays remains unchanged the secondary radiation contains *X*-rays of lower frequency or longer wavelength than those of the incident beam. This is known as **modified or incoherent scattering.** The phenomenon is known as **Compton effect.** In such a case an electron is also ejected with an energy depending upon its direction. The electron is known as **Compton recoil electron.**

Assumptions. Compton calculated mathematically the increase in wavelength making use of quantum theory and making the following assumptions:

1. The X-rays of frequency ν (wavelength λ) consist of photons of energy $h\nu$ and momentum $\dfrac{h}{\lambda} = \dfrac{h\nu}{c}$.

2. The electrons are assumed to be *free* and *stationary*. It is, because the binding energy of the electron to the atom is of the order of 10 eV which is very small as compared to the energy of *X*-ray

photon which is of the order of 10 keV. The kinetic energy of the electron being of the same order as the binding energy it is comparatively at rest.

3. The collision between the high energy photon and the electron is *elastic*. The energy and momentum carried by the scattered photon and recoil electron is governed by the laws of conservation of energy and momentum respectively.

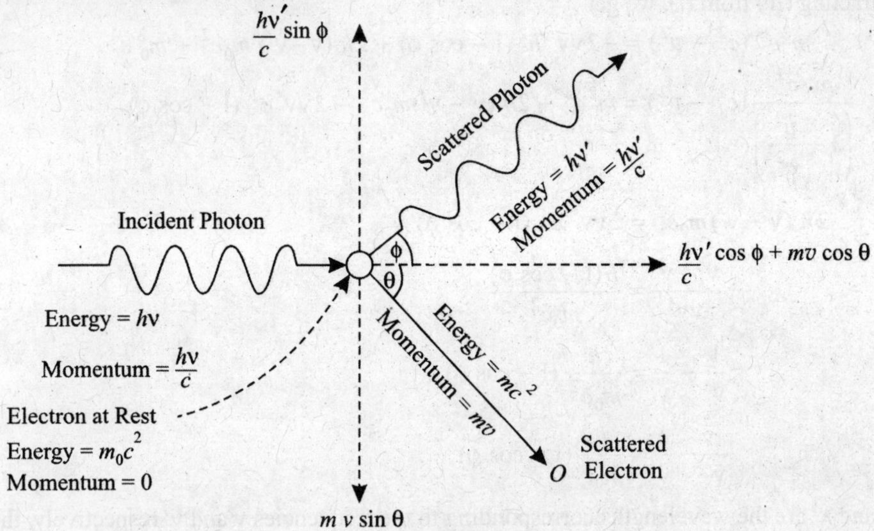

Fig. 3.16

3.14.1 Compton shift. Suppose an *X*-ray photon of frequency ν and energy $h\nu$ collides with an electron supposed to be at rest in the target. A part of its energy is imparted to the electron which is ejected with a velocity v in a direction making an angle θ with that of the incident *X*-ray photon. The remaining energy is given out as a scattered *X*-ray photon of lower frequency ν', moving in the direction making an angle ϕ with that of the incident photon. If m_0 is the rest mass of the electron, then its mass while moving with a velocity v according to the theory of relativity is given by

$$m = \frac{m_0}{\sqrt{1-\frac{v^2}{c^2}}}$$

where c is the velocity of light.

(*a*) Applying the *principle of conservation of energy*, we have

$$h\nu + m_0c^2 = h\nu' + mc^2$$

or
$$mc^2 = h(\nu - \nu') + m_0c^2$$

Squaring, we get

$$m^2c^4 = h^2(\nu^2 - 2\nu\nu' + \nu'^2) + 2h(\nu - \nu')\,m_0c^2 + m_0^2c^4 \quad ...(i)$$

(*b*) Applying the *principle of conservation of momentum*, we have in the direction of the incident photon *i.e.* along *X*-axis.

$$\frac{h\nu}{c} = \frac{h\nu'}{c}\cos\phi + mv\cos\theta$$

or
$$mvc\cos\theta = h(\nu - \nu'\cos\phi) \quad ...(ii)$$

and in a direction perpendicular to the direction of the incident photon *i.e.*, along *Y*-axis

$$0 = \frac{h\nu'}{c}\sin\phi - mv\sin\theta$$

or $$mvc \sin\theta = h\nu' \sin\phi \qquad ...(iii)$$

Squaring (*ii*) and (*iii*) and adding, we get

$$m^2v^2c^2 = h^2(\nu^2 - 2\nu\nu'\cos\phi + \nu'^2\cos^2\phi + \nu'^2\sin^2\phi)$$

or $$m^2v^2c^2 = h^2(\nu^2 - 2\nu\nu'\cos\phi + \nu'^2) \qquad ...(iv)$$

Subtracting (*iv*) from (*i*), we get

$$m^2c^2(c^2 - v^2) = -2\nu\nu' h^2(1 - \cos\phi) + 2h(\nu - \nu')m_0c^2 + m_0^2c^4$$

or $$\frac{m_0^2c^2}{\left(1 - \frac{v^2}{c^2}\right)}(c^2 - v^2) = m_0^2c^4 + 2h(\nu - \nu')m_0c^2 - 2\nu\nu'h^2(1 - \cos\phi)$$

or $$2h(\nu - \nu')m_0c^2 = 2\nu\nu'h^2(1 - \cos\phi)$$

or $$\frac{\nu - \nu'}{\nu\nu'} = \frac{h(1 - \cos\phi)}{m_0c^2}$$

or $$\frac{1}{\nu'} - \frac{1}{\nu} = \frac{h}{m_0c^2}(1 - \cos\phi)$$

or $$\frac{c}{\nu'} - \frac{c}{\nu} = \frac{h}{m_0c}(1 - \cos\phi)$$

If λ and λ' are the wavelengths corresponding to the frequencies ν and ν' respectively, then

$$\lambda' - \lambda = \Delta\lambda = \frac{h}{m_0c}(1 - \cos\phi) \qquad ...(3.26)$$

Factors on which Compton shift depends. The relation given by Eq. (3.26) known as *Compton equation* shows that the increase in wavelength or Compton shift $\Delta\lambda$ *is independent of the wavelength of the incident radiations as well as the nature of the scattering substance, but depends upon the angle of scattering.*

3.14.2 Wavelength of scattered photon. The wavelength of scattered photon

$$\lambda' = \lambda + \Delta\lambda = \lambda + \frac{h}{m_0c}(1 - \cos\phi)$$

Compton wavelength: Substituting the values

$$h = 6.624 \times 10^{-34} \text{ Joule sec.}$$

$$m_0 = 9.107 \times 10^{-31}\ kg,\ c = 2.998 \times 10^8 \text{ m/s}$$

in relation (3.26) we get

$$\frac{h}{m_0c} = \frac{6.624 \times 10^{-34}}{9.107 \times 10^{-31} \times 2.998 \times 10^8}$$

$$= 0.0242 \times 10^{-10}\ m = 0.0242\ \text{Å}$$

$$\therefore \qquad \Delta\lambda = 0.0242\,(1 - \cos\phi)\ \text{Å}.$$

This shows that when the incident radiation of wavelength λ is scattered through an angle ϕ by a free electron, the scattered wavelength λ' of *X*-rays is greater than that of the incident radiation by an amount 0.0242 $(1 - \cos\phi)$ Å.

For a given value of the angle ϕ, the shift in wavelength is independent of the wavelength of incident radiation.

For $\phi = 90°$.

$$\Delta\lambda = \frac{h}{m_0 c} = 0.0242 \text{ Å}$$

This is known as **Compton wavelength** and is denoted by λ_c.

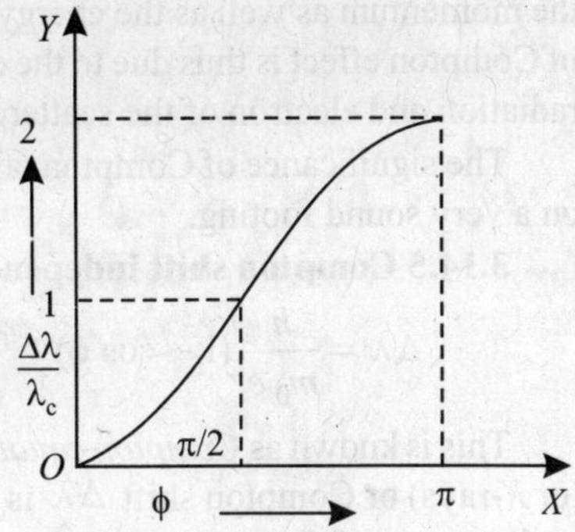

Fig. 3.17

Distinction between Compton wavelength and Compton shift. Compton wavelength is a constant quantity equal to 0.0242 Å. It is the Compton shift for an angle of scattering $\phi = 90°$. Compton shift is a variable quantity depending upon the angle of scattering. Compton shift $\Delta\lambda = \frac{h}{m_0 c}(1 - \cos\phi)$.

As $\frac{h}{m_0 c} = \lambda_c$ the Compton wavelength

Compton shift = Compton wavelength $(1 - \cos\phi)$

or $\Delta\lambda = \lambda_c\ (1 - \cos\phi)$

As the Compton shift is proportional to $(1 - \cos\phi)$, it increases from zero for $\phi = 0$ to $\Delta\lambda = 2\lambda_c$ for $\phi = \pi$. The variation of $\Delta\lambda$ (in terms of Compton wavelength λ_c *i.e.* $\left.\frac{\Delta\lambda}{\lambda_c}\right)$ with ϕ is shown in Fig. 3.17.

3.14.3 Verification of results. The schematic arrangement of the apparatus used by Compton to verify the results is shown in Fig. 3.18. A monochromatic stream of incident *X*-rays is collimated into a fine beam by lead blocks *BB* and falls on a scatterer *S*. The *X*-rays are scattered at an angle ϕ and detected with the help of Bragg's spectrometer *BS* which can move along the arc of a circle.

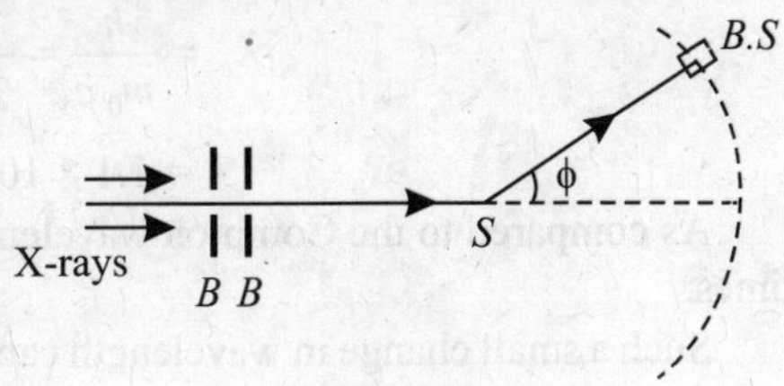

Fig. 3.18

The results obtained by Compton have been verified expreimentally. The wavelength displacement for the K_α line from molybdenum target scattered by carbon for values of $\phi = 0$, 45°, 90° and 135° is shown in Fig. 3.19. At each angle the scattered beam consists of two distinct lines, the unmodified line which has the same wavelength as the incident radiation and the modified line which has a longer wavelength depending upon the angle of scattering. The displacement $\Delta\lambda$ has been found to be in fair aggrement with the values calculated from Equation (*i*), For $\phi = \pi$ (not shown in the figure).

$$\Delta\lambda = 0.048 \text{ Å}.$$

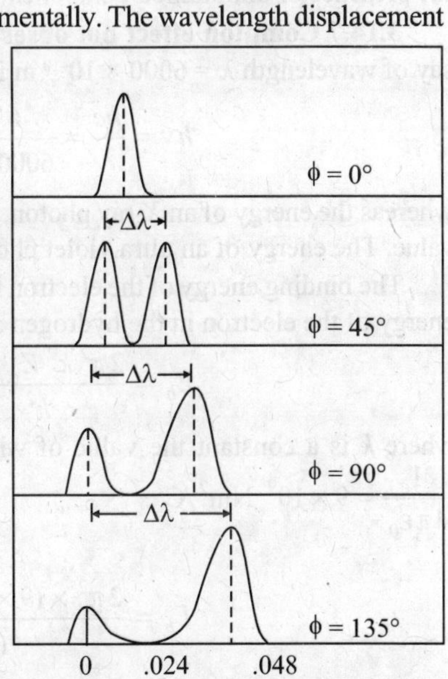

Fig. 3.19

3.14.4 Significance of Compton effect. The Compton effect clearly shows the particle like nature of electro-magnetic radiation. Not only a precise quantum of energy $h\nu$ can be assigned to a photon but also a precise quantum of momentum $\frac{h\nu}{c} = \frac{h}{\lambda}$. The total momentum of a monochromatic radiation cannot have *any* value but is only an exact multiple of the linear momentum of a single photon. In other words the momentum as well as the energy of the electromagentic radiation is quantised. The phenomenon of Compton effect is thus due to the elastic collision between two particles, the photon of the incident radiation and electron of the scatterer.

The significance of Compton effect lies in the fact that it has put the quantum theory of radiation on a very sound footing.

3.14.5 Compton shift independent of incident wavelength. Compton shift is given by

$$\Delta\lambda = \frac{h}{m_0 c}(1 - \cos\phi) \qquad \text{(from Eq. 3.26)}$$

This is known as *Compton equation*. It shows that increase in wavelength of the incident radiation (or *X*-rays) or Compton shift $\Delta\lambda$ is independent of the wavelength of the incident radiation as well as the nature of the scattering substance. It only depends upon the angle of scattering.

3.14.6 Presence of unmodified component in Compton effect. In deriving the relation for Compton shift it is assumed that the electron is *free* and *stationary*. But all the electrons do not satisfy this condition. When Compton scattering takes place from an electron tightly bound to the atom, the collision is in fact with the atom to which the electron is tightly bound. As the Compton wavelength $\lambda_c = \frac{h}{m_0 c}$ is inversely proportional to the mass of the scattering particle its magnitude for an atom is nearly 10^{-4} times the value for an electron as calculated below for an atom of carbon of mass $m_0 = 2 \times 10^{-26}$ kg for which

$$\lambda_c = \frac{h}{m_0 c} = \frac{6.6 \times 10^{-34}}{2 \times 10^{-26} \times 3 \times 10^8} = 1.1 \times 10^{-16} \text{ m}$$

$$= 1.1 \times 10^{-6} \text{ Å}$$

As compared to the Compton wavelength for electron *i.e.* 2.4×10^{-2} Å this value is nearly 10^{-4} times.

Such a small change in wavelength cannot be resolved by an *X*-ray spectrometer and hence such photons are recorded without change of wavelength for all values of the scattering angle. This explains the presence of unmodified component of *X*-ray in the study of Compton effect.

3.14.7 Compton effect not observed with visible light. The energy of a visible light photon say of wavelength $\lambda = 6000 \times 10^{-10}$ m is given by

$$h\nu = \frac{hc}{\lambda} = \frac{6.6 \times 10^{-34} \times 3 \times 10^8}{6000 \times 10^{-10} \times 1.6 \times 10^{-19}} = 2\,\text{eV (Approx.)}$$

whereas the energy of an *X*-ray photon, say of wavelength 1 Å will be more than 10^3 times the above value. The energy of an ultra violet photon is of the order of 10 eV.

The binding energy of the electron in the atoms is of the order of 10 eV. For example, the binding energy of the electron in the hydrogen atom

$$E_b = \frac{2\pi^2 k^2 Z^2 me^4}{h^2}$$

where k is a constant the value of which in C.G.S. (Gaussian) system is 1 and in S.I. units is $\frac{1}{4\pi\varepsilon_0} = 9 \times 10^9 \text{ Nm}^2/\text{C}^2$.

$$\therefore \quad E_b = \frac{2\pi^2 \times (9 \times 10^9)^2 \times (9.1 \times 10^{-31}) \times (1.6 \times 10^{-19})^4}{(6.62 \times 10^{-34})^2 \times 1.6 \times 10^{-19}}$$

$$= 13.6 \text{ eV.}$$

Hence these electrons can be treated as *free* when *X*-rays or gamma rays are incident but these cannot be treated as *free* for visible or ultra-violet light. Thus for visible or ultra-violet light the whole of the atom takes part in Compton scattering and not the electron alone.

The Compton wavelength for the electron $\lambda_c = \dfrac{h}{m_0 c} = 2.4 \times 10^{-2}$ Å and for an atom of carbon it is 1.1×10^{-6} Å. To detect a change in wavelength of this order, the optical instrument should have resolving power $\dfrac{\lambda}{\Delta\lambda} = \dfrac{6000 \text{ Å}}{1.1 \times 10^{-6} \text{ Å}} = 5 \times 10^9$ which is much beyond the limit of resolution of these instruments. Hence Compton effect cannot be observed with visible or ultraviolet light.

Thus, Out of visible light and γ-rays, γ-rays can most easily show Compton effect.

3.14.8 Relation between φ and θ. Referring to Fig. 3.16 in accordance with the law of conservation of momentum, we have

For horizontal direction $mvc \cos\theta = h(\nu - \nu' \cos\phi)$ ...(*i*)

and for vertical direction $mvc \sin\theta = h\nu' \sin\phi$ ...(*ii*)

where ν and ν′ are the frequencies of the incident and scattered photon respectively, φ and θ the angle of scattering for the photon and the electron respectively, *m* the mass and *v* the velocity of the electron.

Dividing (*ii*) by (*i*), we have

$$\tan\theta = \frac{\nu' \sin\phi}{\nu - \nu' \cos\phi} = \frac{\sin\phi}{\dfrac{\nu}{\nu'} - \cos\phi} = \frac{\sin\phi}{\dfrac{\lambda'}{\lambda} - \cos\phi} \qquad ...(iii)$$

Now $$\Delta\lambda = \lambda' - \lambda = \frac{h}{m_0 c}(1 - \cos\phi)$$

$$\therefore \quad \frac{\lambda'}{\lambda} = 1 + \frac{h}{m_0 c\lambda}(1 - \cos\phi)$$

Substituting in (*iii*), we have

$$\tan\theta = \frac{\sin\phi}{\left[1 + \dfrac{h(1 - \cos\phi)}{m_0 c\lambda}\right] - \cos\phi} = \frac{\sin\phi}{(1+\alpha)(1 - \cos\phi)}$$

where $$\alpha = \frac{h}{m_0 c\lambda} = \frac{h\nu}{m_0 c^2}$$

or $$\tan\theta = \frac{2\sin\phi/2 \cos\phi/2}{(1+\alpha)\, 2\sin^2\phi/2} = \frac{\cot\phi/2}{1+\alpha} \qquad ...(iv)$$

$$\therefore \quad \tan\theta = \frac{\cot\phi/2}{1 + \dfrac{h\nu}{m_0 c^2}}. \qquad ...(3.27)$$

3.14.9 Behaviour of photon and electron when φ varies from 0 to π. We shall discuss the behaviour of photon and electron for φ = 0, φ = π/2 and φ = π.

1. When $\phi = 0$; $\cot\phi/2 = \infty$ ∴ $\tan\theta = \infty$ or $\theta = \pi/2$

Also $\Delta\lambda = 0$.

This means that the photon goes unscattered, while the electron moves at right angles to the direction of the photon.

2. When $\phi = \dfrac{\pi}{2}$; $\cot\dfrac{\phi}{2} = 1$ ∴ $\tan\theta = \dfrac{1}{1+\alpha}$

This being positive θ lies between 0 and 90°

Also $$\Delta\lambda = \frac{h}{m_0 c} = 0.0242 \text{ Å}.$$

This means that the photon gets scattered and moves at right angles to the direction of incidence. The change in wavelength is 0.0242 Å while the electron moves in any direction making an angle θ less than $\pi/2$ with the direction of incidence.

3. When $\phi = \pi$; $\cot \phi/2 = 0$ $\therefore$ $\tan\theta = 0$ or $\theta = 0$

Also $$\Delta\lambda = \frac{2h}{m_0 c} = 0.0484 \text{ Å}. \quad ...(3.28)$$

This means that the photon reverses its direction with change in wavelength of 0.0484 Å while the electron moves in the direction of the incident photon. Thus, maximum change in wavelength takes place when the photon gets scattered in the direction of incidence.

3.14.10 Electron cannot be scattered at greater than 90°. The angle of scattering for the electron θ is related to angle of scattering for the photon ϕ as under

$$\tan\theta = \frac{\cot\phi/2}{1 + \dfrac{h\nu}{m_0 c^2}}$$

Now $1 + \dfrac{h\nu}{m_0 c^2}$ = a constant = k (suppose)

$$\therefore \quad k \tan\theta = \cot\phi/2.$$

The maximum value of $\cot\phi/2 = \infty$

Hence the maximum value of $\tan\theta = \infty$

or $$\theta = \pi/2$$

In other words, the electron cannot be scattered at an angle greater than 90°.

Example 3.17. *An X-ray photon is found to have doubled its wavelength on being scattered by 90°. Find the energy and wavelength of incident photon.*

Solution. When an *X*-ray photon is Compton scattered through an angle ϕ, the increase in wavelength

$$\Delta\lambda = \frac{h}{m_0 c}(1 - \cos\phi)$$

For $\phi = 90°$, $\cos\phi = 0$

$$\therefore \quad \Delta\lambda = \frac{h}{m_0 c} = \frac{6.63 \times 10^{-34}}{9.1 \times 10^{-31} \times 3 \times 10^8} = 0.02425 \times 10^{-10} \text{ m}$$

If λ is the wavelength of incident photon, then wavelength of scattered photon = $\lambda + \Delta\lambda$

But $$\lambda + \Delta\lambda = 2\lambda$$

$$\therefore \quad \lambda = \Delta\lambda = 0.02425 \times 10^{-10} \text{ m} = 0.02425 \text{ Å}$$

Energy of the incident photon = $h\nu$

$$= \frac{hc}{\lambda} = \frac{6.63 \times 10^{-34} \times 3 \times 10^8}{0.02425 \times 10^{-10}}$$

$$= \frac{19.89 \times 10^{-16}}{0.2425} \text{ J} = \frac{19.89 \times 10^{-16}}{0.02425 \times 1.6 \times 10^{-19}}$$

$$= 0.51 \times 10^6 \text{ eV} = 0.51 \text{ MeV}.$$

Example 3.18. *Calculate the wavelength of incident X-ray photon which produces recoil electron of energy 4 keV in Compton effect. The electron recoils in the direction of incident photon and photon is scattered at an angle of 180°.*

Solution. Let $m = 9.1 \times 10^{-31}$ kg be the mass and v the velocity of the electron, then

Kinetic energy of the recoil electron $= \frac{1}{2} mv^2 = 4$ keV

$$= 4000 \times 1.6 \times 10^{-19} = 6.4 \times 10^{-16} \text{ J}$$

Multiplying by $2m$, we have, $m^2v^2 = 2m \times 6.4 \times 10^{-16} = 2 \times 9.1 \times 10^{-31} \times 6.4 \times 10^{-16}$

or $$mv = \sqrt{2 \times 9.1 \times 10^{-31} \times 6.4 \times 10^{-16}} = 34.13 \times 10^{-24} \text{ kg m s}^{-1}$$

Let λ be the wavelength of incident X-ray photon and λ' of the scattered photon, then according to the law of conservation of energy

$$\frac{hc}{\lambda} - \frac{hc}{\lambda'} = 6.4 \times 10^{-16} \text{ J} \quad ...(i)$$

As the electron recoils in the direction of incident photon, the angle of scattering for the electron $\theta = 0$

As the incident photon is scattered through an angle of 180°, $\phi = 180°$.

According to the principle of conservation of linear momentum, in the direction of incident photon

$$\frac{h}{\lambda} = \frac{h}{\lambda'} \cos \phi + mv \cos \theta$$

$$= \frac{h}{\lambda'} \cos 180° + mv \cos 0 = -\frac{h}{\lambda'} + mv$$

$$\therefore \quad \frac{h}{\lambda} + \frac{h}{\lambda'} = mv$$

The initial and final momentum in a direction perpendicular to the direction of incident photon is zero.

Hence $$\frac{h}{\lambda} + \frac{h}{\lambda'} = 34.13 \times 10^{-24}$$

Multiplying both sides by c the velocity of light

$$\frac{hc}{\lambda} + \frac{hc}{\lambda'} = 34.13 \times 10^{-24} \times 3 \times 10^8 = 102.39 \times 10^{-16} \text{ J} \quad ...(ii)$$

Adding (*i*) and (*ii*), we have

$$\frac{2hc}{\lambda} = 6.4 \times 10^{-16} + 102.39 \times 10^{-16} = 108.79 \times 10^{-16} \text{ J}$$

$$\therefore \quad \lambda = \frac{2hc}{108.79 \times 10^{-16}} = \frac{2 \times 6.3 \times 10^{-34} \times 3 \times 10^8}{108.79 \times 10^{-16}} = 0.347 \times 10^{-10} \text{ m}$$

$$= 0.347 \text{ Å}.$$

Example 3.19. *In an experiment on Compton scattering X-rays of wavelength 10^{-10} m are incident. Calculate the wavelength of X-rays scattered in the backward direction.* ***(Gharwal U. 2000)***

Solution. When a photon is scattered in the backward direction, the angle of scattering for the photon $\phi = 180° = \pi$ and change in wavelength $\Delta\lambda = \frac{2h}{m_0 c} = 0.0484$ Å (see Eq. 3.28)

Wavelength of incident X-ray photon $\lambda = 1 \times 10^{-10}$ m $= 1$ Å

$\therefore$ Wavelength of X-ray photon scattered $\lambda' = \lambda + \Delta\lambda$

$$= 1 + 0.0484 = 1.0484 \text{ Å}.$$

3.15 OLD QUANTUM THEORY AND ITS LIMITATIONS

The old quantum theory based on Planck's hypothesis could very sucessfully explain phenomenon like black body radiation, photo electric effect, Compton effect, variation of specific heat of solids with temperature and hydrogen spectrum, it has a number of draw backs and limitations. Some of these are :

(*i*) Bohr's quantisation rules are arbitrary. The theory does not provide any physical explanation for making this assumption.

(*ii*) Bohr's postulate of discrete non radiating energy states on which old quantum theory is based were empirical without any theoretical background.

(*iii*) The theory could not explain the spectra of Helium and other more complex atoms.

(*iv*) This theory could not explain why certain spectral lines are more intense than others. In other words, why certain transitions between energy levels have greater probability of occurrence than others.

(*v*) It also does not explain fine structure of spectral lines *i.e.* why certain spectral lines consist of more than one line having slightly different wavelengths.

(*vi*) The theory does not explain the process connected with spin of the electron and Pauli's exclusion principle.

(*vii*) It is unable to explain dispersion of light.

(*viii*) Above all it does not explain how individual atoms interact with one another to give the physical and chemical properties of matter in aggregate of macroscopic size.

All these and other difficulties have been explained on the basis of quantum mechanics, called wave mechanics or matrix mechanics.

EXERCISE

LONG QUESTIONS

1. Explain briefly the origin of quantum theory in view of failure of classical mechanics.
2. Draw the experimental curves of the black body radiation for different temperatures. Discuss how classical approaches failed to account for the spectral distribution of energy density in a black body radiation. How did Planck's theory overcome this difficulty?
3. Obtain an expression of Planck's distribution law. Deduce Wein's displacement law from Planck's law. Did Planck's theory justify the Rayleigh-Jeans law? How?
4. Discuss the black body radiation spectrum. How are different classical theories failed to explain the spectrum? How is the problem solved? Explain. Derive Planck's radiation law.
5. Explain why classical mechanics can not be directly applied to atomic phenomenon? Give a brief account of Planck's quantum theory of radiation.
6. What is a black body? Discuss the distribution of energy in the spectrum of a black body. Describe and explain Planck's quantum theory of radiation.
7. Discuss the inadequacy of classical theory to account for the photoelectric effect. How did Einstein explain the phenomenon? Derive Einstein's photoelectric equation.
8. Explain the phenomenon of photoelectric emission. Derive Einstein's photoelectric equation. Show that the photoelectric effect depends upon frequency of incident radiation and not on the intensity. ***(H.P.U. 2002; Meerut U. 2001, Luck U. 2001, Nagpur U. 2002)***

9. What is photoelectric effect? Define photoelectric work function and threshold frequency. What are the laws of photoelectric emission? State Einstein's photoelectric equation.

10. Give the historical development of quantum mechanics with spacial reference to the black body problem. What modifications were made by Planck in the established theories to solve the black body problem?

11. What are the failures of classical theory to explain the photoelectric effect? Describe the experimental arrangement to study this effect. Give the important observations in this experiment. Draw the necessary graphs.

12. What is quantum theory of electromagnetic energy? What major deviation from wave theory was postulated in quantum theory? How did Einstein use quantum theory to explain the photoelectric effect?

13. State and explain the fundamental postulates of Bohr's theory of hydrogen atom. Obtain an expression for the energy of electron in the nth orbit. Explain the emission of spectral lines. Determine the number of possible transitions.

14. On the basis of Bohr's theory, obtain an expression for the radius, frequency and energy of the first orbit of hydrogen atom. Draw the energy level diagram of hydrogen atom. Discuss the spectral series emitted by hydrogen atom.

15. Explain how Bohr's model successfully accounts for the hydrogen spectra. The energy of electron in the nth Bohr's orbit is negative. Explain.

16. How is the classical mechanics failed to explain the stability of atom? Explain. To what extent Bohr was successful to explain this stability? How the problem of emission of radiation solved by Bohr?

17. (*a*) Explain briefly quantum theory of radiation. What is photon? State its properties. Express linear momentum in terms of wave vector $|\vec{k}|$ and energy of photon in terms of angular velocity ω.

(*b*) What are limitations of old quantum theory? ***(P.U. 2003, MDU 2000)***

18. What is photo-electric effect ? Draw a labelled diagram of the apparatus you will use to demonstrate photoelectric effect. Write down its important results and show them graphically. ***(Pbi U. 2001, Gharwal U. 2000, Nagpur U. 2002 P.U. 2004)***

19. (*a*) State the laws of photoelectric emission. ***(H.P.U. 2003, 2002; Meerut U. 2001; Gharwal U. 2000)***

(*b*) In what way classical electromagnetic theory of light fails to explain the basic facts of photo-electricity. ***(P.U. 2003, G.N.D.U. 2001, Bang. U. 2001)***

20. (*a*) What is meant by work function of a material? Show how you will measure it experimentally?

(*b*) Draw a curve showing stopping potential against frequency of a photo sensitive material. How do you determine the following with the help of the curve.

(i) Threshold frequency, (ii) Work function and (iii) planck's constant.

21. (*a*) What is Compton effect? Derive an expression for Compton shift and wavelength of scattered photon.

(*b*) What is Compton wavelength? Determine its value. Distinguish between Compton shift and Compton wavelength. What are the factors on which Compton shift depends? ***(H.P.U. 2003, 2000, M.D.U. 2008, 2006, 2002, 2000; Pbi. U. 2004, 2002, 2001, 2000; Meerut U. 2001, Madurai U. 2002; P.U. 2000, Luck U. 2002)***

22. (*a*) A beam of monochromatic X-rays of wavelength λ is scattered by a light element. Obtain an expression for the wavelength of scattered radiation. Which apparatus will you set up to verify your results? ***(Meerut U. 2000, Pbi. U. 2001)***

(*b*) Discuss the physical significance of compton effect.

23. How do you account for the presence of unmodified component of X-rays in the study of Compton effect?

24. (*a*) Why Compton effect can not be observed with visible light?

(Pbi. U. 2003, 2000; H.P.U. 2001, 2000; P.U. 2004, 2000; G.N.D.U. 2004)

(*b*) Out of visible light and γ-rays which can most easily show Compton effect and why?

(Pbi. U. 2000)

25. (*a*) Derive a relation between the angle of scattering of photon and that of electron in Compton effect.

(*b*) Discuss the behaviour of the photon and electron when angle of scattering for the photon is 0, $\frac{\pi}{2}$ and π. Show that the change in wavelength of the scattered radiation in Compton scattering varies from 0 to 0.484 Å if the angle of scattering varies from 0 to 180°.

(Luck U., 2002)

(*c*) Explain why the electron cannot be scattered at an angle greater than 90°.

26. (*a*) State Bohr's postulates and use them to determine the expression for (*i*) radius of Bohr orbit and (*ii*) total energy of the electron in a hydrogen atom in the nth state. What is the significance of negative energy of the electron in the quantized Bohr Orbit?

(*b*) Explain the hydrogen spectrum on the basis of this theory. What is the value of Rydberg's constant?

(*c*) What are the limitations of Bohr's theory? State the fact which could not be explained by it. ***(Luck. U. 2002. 2001; Bang. U. 2001, Gauhati U. 2000, Cal. U. 2002)***

27. (*a*) What are the various series of hydrogen spectrum? Calculate the wavelength of the first member and limiting member of each series.

SHORT QUESTIONS

1. What happens when radiations are incident on material medium? Explain your answer.
2. Draw the black body radiation curves at different temperatures. State the relation between maximum wavelength and temperature.
3. Give the characteristic observations of experimental black body radiation curves.
4. As temperature of a black body is increased, how does λ_m change? Also comment upon total radiations emitted at different temperatures.
5. What is a black body? Draw the graphs of black body spectrum at three temperatures?
6. State Planck's law of radiation. How does it explain black body spectrum?
7. Derive Planck's radiation law. Show that Planck's radiation law and Rayleigh-Jeans law become identical at very high temperature.
8. Deduce Wein's displacement law from Planck's law. Did Planck's theory justify the Rayleigh-Jeans law? How?
9. What is photoelectric effect? Discuss the experimental arrangement to study the photoelectric effect.
10. How does the photocurrent change with intensity of incident light? Draw the necessary graphs.
11. How does the photocurrent change with frequency of incident light? Draw the necessary graphs.
12. What is cut off potential ? Why should such potential exist?
13. Derive Einstein's photoelectric equation. Are all photoelectrons emitted with same kinetic energy? ***(Nagpur U. 2011)***

14. Establish a relation between stopping potential and maximum K.E. of photoelectrons. Why should the photoelectric effect not be observed with infrared light?
15. What are the failures of classical wave theory in explaining photoelectric effect?
16. How does Einstein's equation explain the observations in photoelectric effect?
17. Why should the current in photoelectric effect reach a saturation value? How does this value change with intensity of light?
18. Show by taking suitable example that there must be measurable time lag between incident light and observation of photocurrent.
19. How is classical mechanics failed to explain stability of atom?
20. State the laws of photoelectric emission and explain them.
21. Why classical mechanics can not be directly applied to the phenomenon of atomic dimensions?
22. What is basic difference between the postulates of classical and quantum theory? What are limitations of old quantum theory?
23. Discuss the difference between Planck's postulates and Einstein's postulates about quantization.
24. Describe in detail the experimental arrangement to study photoelectric effect. State the conclusions.
25. State and explain the postulates of Bohr's theory of hydrogen atom.
26. Obtain an expression for the frequencies of spectral lines in hydrogen atom using Bohr's theory.
27. Derive an expression for energy of nth orbit of hydrogen atom.
28. Draw the energy level diagram of hydrogen atom. Explain the emission of spectral lines.
29. Discuss the spectral series emitted by hydrogen atom.
30. Obtain an equation for the radius of nth Bohr orbit of hydrogen atom.
31. Explain how Bohr's model successfully accounts for the hydrogen atom.
32. What are drawbacks of Bohr's theory of hydrogen atom?
33. What are limitations of old quantum theory?
34. Show that compton shift is independent of wavelength of incident X-rays.
35. Calculate the ionisation potential or potential energy in the ground state for a hydrogen atom. ***(Meerut U. 2002, Bang. U. 2000)***
36. Calculate the wavelength of the first line of the Paschen series if $R = 1.097 \times 10^7$ m. ***(Nagpur U. 2011)***
37. How does photon differ from material particle? ***(Nagpur U. 2011)***
38. What is the relation between work function and threshold frequency?
39. Name and explain different series of hydrogen spectra. ***(Nagpur U. 2011)***
40. State the limitations of Bohr's theory. ***(Nagpur U. 2011)***
41. State the range of wavelength of infra red radiations.
42. What type of spectrum is obtained from a black body at high temperature?
43. How will you explain the stability of atom?
44. Calculate the velocity of electron in hydrogen atom in its ground state. **(Ans. 2.18×10^6 m/s)**
45. State Planck's hypothesis of quantum of energy. If frequency is doubled what will be energy of quantum?
46. What is photoelectric effect? How does the current change when intensity of light is increased keeping its frequency constant?
47. What is threshold frequency? Is it same for all metals?
48. Draw a graph between stopping potential and frequency of incident radiation.

49. State Einstein's photoelectric equation. Explain work function.

50. Explain why the glass tube in the photoelectric experiment should be evacuated?

51. If a distance of a monochromatic source from the photo cathode is increased, how will the cut off potential change? Explain .

52. Why can not insulators exhibit photoelectric effect under visible or ultraviolet light?

53. Show the straight line graphs of two metals A and B whose work functions are 2eV and 3eV in only one figure. Mark them as A and B.

54. Show the photoelectric effect is frequency dependent phenomenon.

55. State laws of photoelectric emission.

56. State Planck's radiation law. How is Wein law obtained from it?

57. Calculate the radius of the first Bohr orbit of hydrogen atom.

58. State the characteristics of photoelectrons.

59. State the properties of photon.

60. Define photoelectric work function and threshold frequency.

61. What is a black body? Draw the energy distribution curves.

62. Show the classical theory is inadequate to explain the photoelectric effect.

63. For a monochromatic beam of radiations of intensity $10^5 W/m^2$, find average number of photons per cubic meter in (i) 10 KHz radio waves and (ii) 10 MeV, gamma rays.

[Hint: Intensity = Energy / sq. m/s = Energy / m^3 × volume of cylinder of length c meter and unit area of cross-section and $I = Nh\nu \cdot c$ where N = photon density / m^3]

[Ans. (i) 5.03 × 10^{25} photon/m^3 and (ii) 2.083 × 10^8 photon/m^3]

64. A metal of work function 1.5 eV is irradiated by radiation of wavelength 2000 A⁰. Calculate the maximum K.E. of photoelectrons. What will be the threshold wavelength of the metal?

(Ans. K.E.$_{(max)}$ = 4.7 eV, λ_0 = 8280 Å)

65. The energy required to remove an electron from sodium is 2.3 eV. Will the metal show photo-electric effect for light of wavelength 6800 A⁰? **(Ans. No. effect, λ_0 = 5404 Å < 6800 Å)**

66. Light of wavelength 2000A⁰ falls on an aluminium surface whose work function is 4.2 eV. What is the K.E. of (i) the fastest and (ii) the slowest emitted electrons? (iii) what is the stopping potential? (iv) what is the threshold wavelength for aluminium?

(Ans. (i) 2eV (ii) Zero (iii) 2V (iv) 3000 Å)

67. A light of wavelength 6000A⁰ incident on a metal surface emits photoelectrons of maximum velocity 4 × 10^5 m/s. Find the threshold wavelength. **(Ans. 7631 Å)**

68. Find the difference of K.E. of the fastest photoelectrons emitted from a given metal surface for light of wavelengths 2540 A⁰ and 4360A⁰. Is this difference depends upon the surface of metal and intensity of incident light?

69. A human eye with normal light can detect 10^{-18} joule of electromagnetic energy of wavelength 6000 A°. How many photons must there be in the incident beam?

[Hint. $E = Nh\nu = nhc/\lambda$ where N = no. of photons
$N = E\lambda/hc = 10^{-18} \times 6000 \times 10^{-19} / 6.63 \times 10^{-34} \times 3 \times 10^8 = 3$]

70. The adjacent figure shows a graph between a stopping potential and frequency of light for a metal. Calculate the photoelectric work function of the metal. Threshold frequency 4.5 × 10^{14} Hz workfunction.

[Hint: Workfunction $\phi = h\nu_0 = \dfrac{6.63\times10^{-34}\times4.5\times10^{14}}{1.6\times10^{-19}}$ = 1.86 eV.]

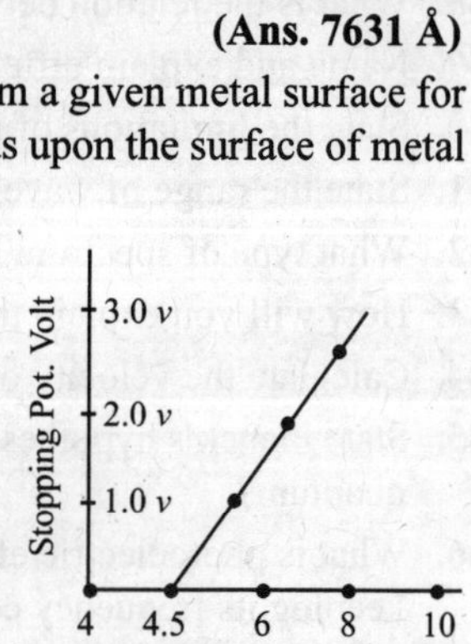

Fig. 3.20

71. Photoelectric and Compton effect both arise due to the action of photon on electron but the two effects are not the same. Explain. ***(Luck.U. 2001; P.U. 2001; Bang. U. 2001; Pbi.U. 2000)***

[HINT. In photo electric effect the energy of the incident photon is completely absorbed by the bound electron. This electron is then emitted with an energy equal to the energy absorbed by it minus the work function which is the minimum energy required to just eject the electron by photo electric emission.

In Compton effect, the photon strikes a free or loosely bound electron and gives a part of its energy to the electron thereby causing it to recoil. The photon still exists with a reduced energy (or greater wave length) and is also scattered. The directions in which the incident photon is scattered and that in which the electron recoils are such that both energy and momentum are conserved.]

72. All hydrogen like atoms have one electron in outer most orbit, still their ionisation energy is different. Explain.

[Hint: Hydrogen ($z = 1$), He^+ ($z = 2$), Li^{++} ($z = 3$) and $w_1 \propto z^2$].

73. What is Ritz combination principle ? ***(Nagpur U. 2011)***

MULTIPLE CHOICE QUESTIONS

Mark the correct answer.

1. The energy corresponding to a wavelength λ is

(*a*) $\frac{h}{\lambda c}$ (*b*) $\frac{hc}{\lambda}$ (*c*) $\frac{\lambda}{hc}$ (*d*) $\frac{\lambda c}{h}$

2. The momentum of a photon is given by

(*a*) $\frac{h}{\lambda}$ (*b*) $\hbar k$ (*c*) $\frac{h\nu}{c}$ (*d*) $\frac{\hbar\omega}{c}$

3. The photoelectric effect takes place only when the wavelength of incident light is

(*a*) above threshold wavelength (*b*) below threshold wavelength

(*c*) equal to threshold wavelength (*d*) Any wavelength

4. The photo-electric effect involves only

(*a*) bound electron. (*b*) free electron (*c*) Both bound and free electron

5. If an electron can be stopped by a potential of 5 volts, its kinetic energy is

(*a*) 5 volts (*b*) 5 Joules

(*c*) 5 cals (*d*) 5 eV

6. Relative to the original photon, the photon that emerges from a collision with an electron has

(*a*) more frequency (*b*) more wavelength

(*c*) more energy (*d*) more velocity

7. The shift in wavelength $\Delta\lambda$ in Compton effect is equal to Compton wavelength when

(*a*) $\phi = 0$ (*b*) $\phi = \pi/2$ (*c*) $\phi = \pi$ (*d*) $\phi = \frac{\pi}{4}$

8. The phenomenon which points towards corpuscular nature of electro magnetic waves is

(*a*) Interference (*b*) diffraction

(*c*) polarisation (*d*) Compton effect.

ANSWERS

1. (*b*)	**2.** All are correct	**3.** (*b*)	**4.** (*a*)	**5.** (*d*)
6. (*b*)	**7.** (*b*)	**8.** (*d*)		

QUANTUM THEORY : WAVE-PARTICLE DUALITY

4
Chapter

INTRODUCTION

The old quantum theory has explained some of the phenomena like photoelectric effect, Compton effect, black body spectrum, emission of spectral lines etc. as discussed in previous chapter, by considering particle nature of a wave. However, the difficulty with quantum theory was how to explain the optical effects i.e. phenomena like interference, diffraction etc; which are exhibited by the very same light. For this, Einstein suggested the light waves merely serves as 'guide' for photons, each carrying energy $h\nu$. This idea was difficult to accept at that time because particle nature and wave nature are apparently different for the reason the particle has a definite mass, velocity, momentum etc. in the space while a wave is associated with wavelength, frequency, phase, amplitude, intensity etc.

To resolve this, Victor de Broglie, in 1924 suggested the duality of matter for which he was awarded Nobel Prize in 1929. He proposed that the matter possess 'wave' as well as 'particle' characteristics. The motion of electrons in atom is guided by peculiar kind of waves similar to the electromagnetic waves, which serves to guide photons. According to him, electrons as well as protons which ordinarily behave like particle, under certain conditions behave like a train a waves. The wavelength of these waves depends upon the momentum which in turn depends upon the mass and velocity of the particles. This idea was immediately accepted and latter it was extended by Heisenberg, Schrodinger, Dirac, Max Born and others. This led to the foundation of new branch called 'Wave Mechanics' or 'New Quantum Mechanics'. In this chapter we shall study the experimental evidence of duality of matter and Heisenberg's uncertainty principle.

The waves associated with a moving particle are called *matter waves*.

4.1 DE-BROGLIE HYPOTHESIS FOR MATTER WAVES

According to quantum theory, radiation of frequency ν consists of quanta or photons, each of energy $E = h\nu$ where h is Planck's constant the value of which is $= 6.62 \times 10^{-34}$ Joule sec.

The equivalent mass of the photon $m = \dfrac{E}{c^2} = \dfrac{h\nu}{c^2}$

Since the speed of the photon in free space is c, the equivalent momentum of the photon

$p = mc = \dfrac{E}{c^2}\,c = \dfrac{h\nu}{c} = \dfrac{h}{\lambda}$ where λ is the wavelength of the radiation of frequency ν.

By analogy with this, de-Broglie suggested that a moving particle is associated with a wave. The frequency of the wave is taken to be $\nu = \dfrac{E}{h} = \dfrac{mc^2}{h}$ where m is the mass of the particle. The wavelength of the wave $\lambda = \dfrac{h}{p} = \dfrac{h}{mv}$ where v is the velocity of the particle.

Thus, a particle of mass m moving with a velocity v has an associated wavelength

$$\lambda = \frac{h}{p} = \frac{h}{mv} \qquad ...(4.1)$$

This is known as de-Broglie *wave equation* and λ is called de-Broglie *wavelength*. The associated wave is termed *matter*, *guide*, *pilot* or de-Broglie wave.

de-Broglie wave velocity or phase velocity of the wave is given by

$$v_p = \nu\lambda = \frac{mc^2}{h}.\frac{h}{mv} = \frac{c^2}{v}$$

Since the particle velocity v, must be less than the velocity of light c, de-Broglie waves travel faster than velocity of light.

Like other waves the de-Broglie wave can also be represented by a function $\Psi(\vec{r},t)$ *i.e.* a function of position $\vec{r}$ and time t. It is known as *wave function* representing *matter wave* guiding the particle. The wave function can be positive, negative or a complex quantity. It has an amplitude or modulus and a phase like other complex functions. The wave vector $\vec{k}$ for de-Broglie wave is given by

$$|\vec{k}| = k = \frac{2\pi}{\lambda} = \frac{2\pi}{h/p} = \frac{p}{h/2\pi} = \frac{p}{\hbar}$$

where $$\hbar = \frac{h}{2\pi}.$$

4.1.1 de-Broglie wavelength in terms of energy and temperature. de-Broglie wavelength according to relation (4.1) is given by

$$\lambda = \frac{h}{p} = \frac{h}{mv}$$

If E is the kinetic energy of the moving particle, then

$$E = \frac{1}{2}mv^2 = \frac{1}{2}\frac{m^2v^2}{m} = \frac{p^2}{2m}$$

or $$p = \sqrt{2mE} \qquad ...(4.2)$$

Hence de-Broglie wavelength $$\lambda = \frac{h}{p} = \frac{h}{\sqrt{2mE}} \qquad ...(4.3)$$

According to kinetic theory of gases, the average kinetic energey of the material particle is given by $$E = \frac{1}{2}mv^2 = \frac{3}{2}kT$$

where k is Boltzmann's constant = 1.38×10^{-23} JK^{-1} and T is the absolute temperature.

$$\therefore \qquad \lambda = \frac{h}{\sqrt{2mE}} = \frac{h}{\sqrt{3mkT}}. \qquad ...(4.4)$$

Thus, wavelength associated with a moving particle is proportional to $1/\sqrt{T}$.

SOLVED EXAMPLES

Example 4.1. *Calculate the de-Broglie wavelength of an average Helium atom in a funace of 400 K. Planck's constant h = 6.6 × 10^{-34} Js and Boltzmann's constant K = 1.38 × 10–23 JK^{-1} and mass of a neutron = 1.67 × 10^{-27} kg.* **(Gauhati U. 2000)**

Solution. de-Broglie wavelength

$$\lambda = \frac{\lambda}{\sqrt{3mkT}} = \frac{6.6\times10^{-34}}{[3\times4\times1.67\times10^{-27}\times1.38\times10^{-23}\times300]}$$

$$= 0.62 \times 10^{-10} \text{ m} = 0.62 \text{ A°}$$

Example 4.2. *Calculate the wavelength of thermal neutrons at 27°C, given mass of neutron = 1.67 × 10⁻²⁷ kg, Planck's constant h = 6.6 × 10⁻³⁴ Js and Boltzmann's constant k = 1.376 × 10⁻²³ JK⁻¹.* **(K.U. 2001)**

Solution. de-Broglie wavelength

$$\lambda = \frac{h}{\sqrt{3mkT}}$$

$$= \frac{6.6 \times 10^{-34}}{[3 \times 1.67 \times 10^{-27} \times 1.376 \times 10^{-23} \times (273 + 27)]^{1/2}}$$

$$= 1.452 \times 10^{-10} \text{ m} = 1.452 \text{ Å}$$

Example 4.3. *Calculate the energy in eV corresponding to a wavelength of 1Å for electron and neutron. Given h = 6.6 × 10⁻³⁴ Js⁻¹*

Electron mass = 9.1 × 10⁻³¹ kg; Neutron mass = 1.7 × 10⁻²⁷ kg.

1 eV = 1.6 × 10⁻¹⁹ J. **(M.D.U. 2000)**

Solution. According to de-Broglie, wavelength $\lambda = \dfrac{h}{mv}$

For an electron $v = \dfrac{h}{m\lambda} = \dfrac{6.6 \times 10^{-34}}{9.1 \times 10^{-31} \times 10^{-10}} = 7.2 \times 10^{6}\ ms^{-1}$

For a neutron $v = \dfrac{h}{m\lambda} = \dfrac{6.6 \times 10^{-34}}{1.7 \times 10^{-27} \times 10^{-10}} = 3.88 \times 10^{3}\ ms^{-1}$

As the velocity in both the cases is much less than the velocity of light, it is a non-relativistic case.

Hence $\lambda = \dfrac{h}{\sqrt{2mE}}$ or $\lambda^2 = \dfrac{h^2}{2mE}$ or $E = \dfrac{h^2}{2m\lambda^2}$

For an electron $E = \dfrac{6.6 \times 6.6 \times 10^{-34} \times 10^{-34}}{2 \times 9.1 \times 10^{-31} \times 10^{-10} \times 10^{-10}} = 2.39 \times 10^{-17}$ **J**

$$= \frac{2.39 \times 10^{-17}}{1.6 \times 10^{-19}} = 1.49 \times 10^2 \text{ eV} = 149 \text{ eV}$$

For a neutron $E = \dfrac{6.6 \times 6.6 \times 10^{-34} \times 10^{-34}}{2 \times 1.7 \times 10^{-27} \times 10^{-10} \times 10^{-10}} = 12.8 \times 10^{-21}$ **J**

$$= \frac{12.8 \times 10^{-21}}{1.6 \times 10^{-19}} = 8 \times 10^{-2} \text{ eV} = 0.08 \text{ eV}.$$

Example 4.4. *Calculate the de-Broglie wave length of 1 MeV proton. Do we require relativistic calculation ?*

Solution. If we calculate the velocity of proton using the relation $E = \frac{1}{2} mv^2$, we get

$$v = \sqrt{\frac{2E}{m}} = \sqrt{\frac{2 \times 10^6 \times 1.6 \times 10^{-19}}{1.67 \times 10^{-17}}} = 1.38 \times 10^7\ m/s$$

which is nearly less than one twentieth of the velocity of light. So we do not require relativistic calculation.

Hence de-Broglie wavelength $\lambda = \dfrac{h}{\sqrt{2mE}}$

Mass of the proton $m = 1.67 \times 10^{-27}$ kg

Energy of the proton $= 1$ MeV $= 10^6 \times 1.6 \times 10^{-19}$ **J**

$$\therefore \quad \lambda = \frac{6.62 \times 10^{-34}}{\sqrt{2 \times 1.67 \times 10^{-27} \times 1.6 \times 10^{-19}}} = 2.86 \times 10^{-14} \ m.$$

Example 4.5. (*a*) *Find the de-Broglie wave length in angstrom for an electron of energy V electron volts.* ***(P.U. 2005)***

(*b*) *Calculate de-Broglie wave length of an electron whose kinetic energy is*

(*i*) *500 eV* ***(M.D.U. 2000)***

(*ii*) *300 eV* ***(M.D.U. 2008)***

(*iii*) *250 eV* ***(G.N.D.U. 2007)***

(*iv*) *50 eV* ***(Meerut U. 2003; Poorvanchal U. 2001)***

(*v*) *1 eV.* ***(Gharwal U. 2000)***

Solution. (*a*) de-Broglie wavelength $\lambda = \dfrac{h}{\sqrt{2mE}}$

Energy $E = V$ electron volts $= V \times 1.6 \times 10^{-19}$ J

Electron mass $m = 9.1 \times 10^{-31}$ kg; $h = 6.62 \times 10^{-34}$ Js

$$\therefore \quad \lambda = \frac{h}{\sqrt{2mE}} = \frac{6.62 \times 10^{-34}}{\sqrt{2 \times 9.1 \times 10^{-31} \times 1.6 \times 10^{-19}\ V}} = \frac{6.62 \times 10^{-34}}{[2 \times 9.1 \times 10^{-31} \times 1.6 \times 10^{-19}]^{\frac{1}{2}}\ V^{\frac{1}{2}}} m$$

$$= \frac{6.62 \times 10^{-34} \times 10^{10}}{[2 \times 9.1 \times 10^{-31} \times 1.6 \times 10^{-19}]^{\frac{1}{2}}\ V^{\frac{1}{2}}} \text{Å} = \frac{12.25}{V^{\frac{1}{2}}} \text{Å}$$

(*b*) (*i*) $E = 500$ eV

$$\text{de-Broglie wavelength } \lambda = \frac{h}{\sqrt{2mE}} = \frac{6.62 \times 10^{-34}}{\sqrt{2 \times 9.1 \times 10^{-31} \times 500 \times 1.6 \times 10^{-19}}}$$

$$= 0.55 \times 10^{-10}\ m = 0.55 \text{ Å}.$$

(*ii*) $E = 300$ eV

$$\text{de-Broglie wavelength } \lambda = \frac{h}{\sqrt{2mE}} = \frac{6.62 \times 10^{-34}}{\sqrt{2 \times 9.1 \times 10^{-31} \times 300 \times 1.6 \times 10^{-19}}}$$

$$= 0.71 \times 10^{-10}\ m = 0.71 \text{ Å}$$

(*iii*) $E = 250$ eV

$$\text{de-Broglie wavelength } \lambda = \frac{h}{\sqrt{2mE}} = \frac{6.62 \times 10^{-34}}{\sqrt{2 \times 9.1 \times 10^{-31} \times 250 \times 1.6 \times 10^{-19}}}$$

$$= 0.77 \times 10^{-10}\ m = 0.77 \text{ Å}$$

(*iv*) $E = 50$ eV

$$\text{de-Broglie wavelength } \lambda = \frac{h}{\sqrt{2mE}} = \frac{6.62 \times 10^{-34}}{\sqrt{2 \times 9.1 \times 10^{-31} \times 50 \times 1.6 \times 10^{-19}}}$$

$$= 1.73 \times 10^{-10}\ m = 1.73 \text{ Å}.$$

(v) $E = 1$ eV

de-Broglie wavelength $\lambda = \dfrac{h}{\sqrt{2mE}} = \dfrac{6.62 \times 10^{-34}}{\sqrt{2 \times 9.1 \times 10^{-31} \times 1 \times 1.6 \times 10^{-19}}}$

$= 12.2 \times 10^{-10}\ m = 12.2$ Å.

Example 4.6. *Find the de-Broglie wavelength of an electron in the first Bohr orbit of hydrogen atom.* ***(P.U. 2000)***

Solution. Energy of electron in the *n*th Bohr orbit of hydrogen atom $E_n = -\dfrac{13.6}{n^2}$ eV

For the first orbit $n = 1$

$\therefore \quad E_1 = \dfrac{13.6}{1^2} = -13.6 \text{ eV} = -13.6 \times 1.6 \times 10^{-19}$ **J**

$\therefore$ de-Broglie wavelength $\lambda = \dfrac{h}{\sqrt{2mE}}$

$$= \frac{6.62 \times 10^{-34}}{\sqrt{2 \times 9.1 \times 10^{-31} \times 13.6 \times 1.6 \times 10^{-19}}} = 3.3 \times 10^{-10}\ m = 3.3 \text{ Å}.$$

Example 4.7. *Ultra violet light of wave length 1500Å is incident on Molybdenum whose work function is 4.15 eV. Find the de-Broglie wave length associated with the photo-electron.* ***(Bang. U. 2000)***

Solution. Energy of ultra-violet light of wave length $\lambda = 1500$ Å will be

$$\frac{hc}{\lambda} = \frac{6.62 \times 10^{-34} \times 3 \times 10^{8}}{1500 \times 10^{-10}}$$

$$= 1.324 \times 10^{-18} \text{ J}$$

$$= \frac{1.324 \times 10^{-18}}{1.6 \times 10^{-19}} = 8.275 \text{ eV}$$

Work function of molybdenum $\omega = 4.15$ eV

$\therefore$ Energy of photo-electron emitted $E = 8.275 - 4.150 = 4.125$ eV

de-Broglie wavelength associated with photo-electron

$$\lambda = \frac{h}{\sqrt{2mE}} = \frac{6.62 \times 10^{-34}}{\sqrt{2 \times 9.1 \times 10^{-31} \times 4.125 \times 1.6 \times 10^{-19}}}$$

$$= 6 \times 10^{-10}\ m = 6 \text{ Å}.$$

Example 4.8. *The wavelength of yellow spectral line of Sodium in 5890 Å. At what kinetic energy an electron will have the same de-Broglie wavelength?* ***(Pbi. U. 2001)***

Solution. Let *E eV* be the energy of the electron to have a wavelength of 5890 Å $= 5890 \times 10^{-10}\ m$

Now $\quad \lambda = \dfrac{h}{\sqrt{2mE}}$ or $\lambda^2 = \dfrac{h^2}{2mE}$ or $E = \dfrac{h^2}{2m\lambda^2}$

$$= \frac{6.62 \times 10^{-34} \times 6.62 \times 10^{-34}}{2 \times 9.1 \times 10^{-31} \times 5890 \times 10^{-10} \times 5890 \times 10^{-10}} = 6.94 \times 10^{-25} \text{ J}$$

$$= \frac{6.94 \times 10^{-25}}{1.6 \times 10^{-19}} = 4.34 \times 10^{-6} \text{ eV}.$$

Example 4.9. *An electron and a proton have a wavelength of 2 Å. What are their momenta and total energies ?* ***(Pbi. U. 2003; Meerut U. 2000)***

Solution. (*i*) **Electron** $\lambda = 2$ Å $= 2 \times 10^{-10}$ *m*

de-Broglie wavelength $\lambda = \dfrac{h}{\sqrt{2mE}} \therefore \lambda^2 = \dfrac{h^2}{2mE}$

or $$E = \frac{h^2}{2m\lambda^2} = \frac{6.62 \times 6.62 \times 10^{-68}}{2 \times 9.1 \times 10^{-31} \times 2 \times 2 \times 10^{-20}}$$

$$= 0.6 \times 10^{-17} \text{ J} = \frac{0.6 \times 10^{-17}}{1.6 \times 10^{-19}} = 37.5 \text{ eV}$$

Also de-Broglie wavelength $\lambda = \dfrac{h}{mv}$

$\therefore$ Momentum $mv = \dfrac{h}{\lambda} = \dfrac{6.62 \times 10^{-34}}{2 \times 10^{-10}} = 3.31 \times 10^{-24}$ *kg ms*$^{-1}$.

(*ii*) **Proton** $\lambda = 2$ Å $= 2 \times 10^{-10}$ *m*

$$E = \frac{h^2}{2m\lambda^2} = \frac{6.62 \times 6.62 \times 10^{-68}}{2 \times 1.67 \times 10^{-27} \times 2 \times 2 \times 10^{-20}}$$

$$= 3.28 \times 10^{-21} \text{ J} = \frac{3.28 \times 10^{-21}}{1.6 \times 10^{-21}} = 0.02\, eV$$

Momentum $= \dfrac{h}{\lambda} = \dfrac{6.62 \times 10^{-34}}{2 \times 10^{-10}} = 3.31 \times 10^{-24}$ *kg ms*$^{-1}$.

Example 4.10. *An electron and a proton have the same de-Broglie wavelength. Prove that the kinetic energy of the electron is greater than that of proton.* ***(M.D.U. 2006)***

Solution. de-Broglie wave length of electron $\lambda_e = \dfrac{h}{\sqrt{2m_e E_e}}$

de-Broglie wave length of proton $\lambda_p = \dfrac{h}{\sqrt{2m_p E_h}}$

But $\lambda_e = \lambda_p$ or $\dfrac{h}{\sqrt{2m_e E_e}} = \dfrac{h}{\sqrt{2m_p E_h}}$ or $m_e E_e = m_p E_p$

or $$\frac{E_e}{E_p} = \frac{m_p}{m_e}$$

As $m_p > m_e$; $E_e > E_p$ *i.e.* the K.E. of the electron is greater than that of proton.

Example 4.11. *A proton and a deuteron have the same kinetic energy. Which has longer wavelength?* ***(Meerut U. 2001)***

Solution. Let *m* be the mass of proton and v_p its velocity.

$\therefore$ Mass of deuteron = 2 *m*. Let v_d be its velocity.

Kinetic energy of proton $= \frac{1}{2} m v_p^2$

Kinetic energy of deuteron $= \frac{1}{2} 2m v_d^2 = m v_d^2$

As the proton and deuteron have the same kinetic energy

$$= \frac{1}{2} m v_p^2 = m v_d^2 \quad \text{or} \quad v_d^2 = \frac{1}{2} v_p^2$$

$$\therefore \quad v_d = \frac{1}{\sqrt{2}} v_p$$

de-Broglie wavelength of moving proton $\lambda_p = \dfrac{h}{m v_p}$

de-Broglie wavelength of moving deuteron $\lambda_d = \dfrac{h}{2 m v_d} = \dfrac{h}{2m \frac{1}{\sqrt{2}} v_p} = \dfrac{h}{\sqrt{2} m v_p}$

$$\therefore \quad \frac{\lambda_d}{\lambda_p} = \frac{h \times m v_p}{\sqrt{2} m v_p \times h} = \frac{1}{\sqrt{2}}$$

$\therefore$ $\lambda_p = \sqrt{2}\, \lambda_d$ *i.e.* proton has a higher de-Broglie wavelength.

Example 4.12 *A proton and an electron have equal kinetic energies. Compare their de-Broglie wave lengths.* ***(Meerut U. 2005)***

Solution. Let *M* be the mass of the proton, *V* its velocity and *m* the mass of electron and *v* its velocity, then

Kinetic energy of the proton $= \frac{1}{2} M V^2$

and kinetic energy of electron $= \frac{1}{2} m v^2$

$$\therefore \quad \frac{1}{2} M V^2 = \frac{1}{2} m v^2 \quad \text{or} \quad \frac{v^2}{V^2} = \frac{M}{m} \quad \text{or} \quad \frac{v}{V} = \left(\frac{M}{m}\right)^{\frac{1}{2}}$$

de-Broglie wavelength of proton $\lambda_p = \dfrac{h}{MV}$

de-Broglie wavelength of electron $\lambda_e = \dfrac{h}{mv}$

$$\therefore \quad \frac{\lambda_p}{\lambda_e} = \frac{h}{MV} \cdot \frac{mv}{h} = \frac{m}{M} \cdot \frac{v}{V} = \frac{m}{M}\left(\frac{M}{m}\right)^{\frac{1}{2}} = \left(\frac{m}{M}\right)^{\frac{1}{2}}$$

$$= \left[\frac{9.1 \times 10^{-31}}{1.67 \times 10^{-27}}\right]^{\frac{1}{2}} = 0.0233$$

4.1.2 de-Broglie Wavelength of Electron. de-Broglie wavelength of an electron in terms of potential difference (V) through which it is accelerated is calculated as under.

Let *v* be the velocity acquired by the electron when it is accelerated through potential differnce *V*, then

Work done on the electron = eV

Kinetic energy gained by the electron $= \frac{1}{2} mv^2$

$$\therefore \quad \frac{1}{2} mv^2 = Ve$$

or

$$v = \sqrt{\frac{2Ve}{m}} \quad ...(4.5)$$

The wavelength associated with the electron (for a non-relativistic case)

$$\lambda = \frac{h}{mv}$$

$$\therefore \quad \lambda = \frac{h}{m\sqrt{\frac{2Ve}{m}}} = \frac{h}{\sqrt{2Vem}}. \quad ...(4.6)$$

In S.I. units. In S.I. units V is in volts, e in coulomb, m in kg and h in Joule sec. Various values are

$$e = 1.6 \times 10^{-19}\ C;\ m = 9.11 \times 10^{-31}\ \text{kg}$$

$$h = 6.62 \times 10^{-34}\ \text{Joule sec}.$$

$$\therefore \quad \lambda = \frac{6.62 \times 10^{-34}}{\sqrt{V}\ \sqrt{2 \times 1.6 \times 10^{-19} \times 9.11 \times 10^{-31}}}$$

$$= \frac{12.26 \times 10^{-10}}{\sqrt{V}}\ \text{metre}$$

$$= \frac{12.26}{\sqrt{V}}\ \text{Å} = \sqrt{\frac{150}{V}}\ \text{Å}. \quad ...(4.7)$$

Example 4.13. *In a diode the anode has a potential of 100 volts relative to the cathode. Find the wavelength associated with the electron as it reaches the anode. Also calculate the wavelength when the anode has a potential of 200 volts.* ***(Bang. U. 2001)***

Solution. Here V = 100 volts. Substituting, we have

$$\lambda = \frac{12.26}{\sqrt{V}} = \frac{12.26}{\sqrt{100}} = 1.226\ \text{Å}$$

For V = 200 volts

$$\lambda = \frac{12.26}{\sqrt{V}} = \frac{12.26}{\sqrt{200}} = 0.867\ \text{Å}.$$

Example 4.14 *What is the wavelength associated with the electron accelerated through 1000 volts.*

Solution. Here V = 1000 volts. Substituting, we have

$$\lambda = \frac{12.26}{\sqrt{V}} = \frac{12.26}{\sqrt{1000}} = 0.338\ \text{Å}.$$

Example 4.15 *The electron beam in a television receiver tube is accelerated by 12000 volts. Calculate de-Broglie wavelength of the electron.*

Solution. Here $V = 12000$ volts. Substituting, we have

$$\lambda = \frac{12.26}{\sqrt{V}} = \frac{12.26}{\sqrt{1200}} = 0.1119 \text{ Å}.$$

Example 4.16. *What voltage must be applied to an electron microscope to produce electron of wavelength $\lambda = 1$ Å?* **(Meerut U. 2006)**

Solution. $\lambda = \sqrt{\frac{150}{V}}$ Å $\quad \therefore \quad$ For $\lambda = 1$Å, $V = 150$ Volt

Example 4.17. *Find the ratio of the wavelength of deuteron and proton accelerated through the same potential difference.* **(Meerut U. 2007)**

Solution. $\lambda = \frac{h}{\sqrt{2Vem}}$. The mass of deuteron = 2 × mass of proton

Charge on deuteron = Charge on proton

$$\therefore \frac{\lambda_D}{\lambda_P} = \frac{\dfrac{h}{\sqrt{2\,Ve\,2m}}}{\dfrac{h}{\sqrt{2Ve\,m}}} = \frac{1}{\sqrt{2}} \qquad \therefore \quad \lambda_p = \sqrt{2}\,\lambda_D$$

Example 4.18. *Calculate de-Broglie wavelength of an α-particle accelerated through a potential difference of 4 kV.*

Solution. Mass of α-particle $m = 4$ a.m.u. $= 4 \times 1.66 \times 10^{-27}$ kg

$= 6.64 \times 10^{-27}$ kg

Charge on the α-particle $q = 2e = 2 \times 1.6 \times 10^{-19}$ C

$= 3.2 \times 10^{-19}$ C

Let v be the velocity of the α-particle when it is accelerated through a potential difference V, then work done on the α-particle $= qV$

Kinetic energy gained by the α-particle $= \frac{1}{2} mv^2$

$$\therefore \quad = \frac{1}{2} mv^2 = qV$$

$$\therefore \quad v = \sqrt{\frac{2qV}{m}}$$

The wavelength associated with α-particle (non-relativistic case)

$$\lambda = \frac{h}{mv} = \frac{h}{m\sqrt{\dfrac{2qV}{m}}} = \frac{h}{\sqrt{2\,q\,mV}}$$

Here $\qquad V = 4$ kV $= 4000$ Volts

$$\therefore \quad \lambda = \frac{6.62 \times 10^{-34}}{\sqrt{2 \times 3.2 \times 10^{-19} \times 6.64 \times 10^{-27} \times 4000}}$$

$$= \frac{6.62}{41.23} \times 10^{-12} = 0.16 \times 10^{-12} \text{ m}.$$

4.2 WAVE NATURE OF MICROSCOPIC PARTICLES

The de-Broglie wavelength associated with a moving particle is given by the relation $\lambda = h/p$. When the momentum of the particle is very large, the associated wavelength is very small. For macroscopic particles such as a marble, a cricket ball etc momentum $|\vec{p}| \sim 10^{-5}$ kg ms^{-1}.(Suppose m = 1 gm = 10^{-3} kg and v = 1 cm s^{-1} = 10^{-2} m s^{-1} so that momentum $|\vec{p}| = 10^{-5}$ kg ms^{-1}) and therefore the corresponding de-Broglie wavelength $\lambda = \frac{6.6 \times 10^{-34}}{10^{-5}} = 6.6 \times 10^{-29}$ m. This wavelength is so small that its measurement is not at all possible. The de-Broglie wavelength can be measured if it is of the order of the size of the atom (~ 10^{-10} m). For this the momentum of the particle should be about 10^{-24} kg ms^{-1}. Such a small magnitude of momentum can be had only for micro-particles like electrons, protons, neutrons, atoms etc because their masses are very small.

For an electron moving with a speed $v = 10^7$ ms^{-1} the de-Broglie wavelength.

$$\lambda = \frac{h}{mv} = \frac{6.6 \times 10^{-34}}{9 \times 10^{-31} \times 10^7}$$

$$= 0.73 \times 10^{-10}\ m$$

Thus it is clear that wave properties associated with particles can be observed only for particles of microscopic (very small) and not with those of macroscopic size.

This also explains why we cannot observe de-Broglie wavelength associated with a fast moving cricket ball.

4.3 WAVELENGTH OF ELECTRON MOVING WITH RELATIVISTIC VELOCITY

According to the theory of relativity, the mass m of a particle moving with velocity v is given by

$$m = \frac{m_0}{\left(1 - \frac{v^2}{c^2}\right)^{1/2}}$$

where m_0 is the rest mass of the particle and c the velocity of light.

The momentum of the particle $p = mv = \frac{m_0 v}{\left(1 - \frac{v^2}{c^2}\right)^{1/2}}$

According to de-Broglie wave equation the wavelength λ associated with the moving particle is given by

$$\lambda = \frac{h}{p} = \frac{h}{\frac{m_0 v}{\left(1 - \frac{v^2}{c^2}\right)^{1/2}}} = \frac{h}{m_0} \frac{\left(1 - \frac{v^2}{c^2}\right)^{1/2}}{v}$$

$$= \frac{h}{m_0 c} \frac{\left(1 - \frac{v^2}{c^2}\right)^{1/2}}{\frac{v}{c}} \qquad \text{...(4.8)}$$

Example 4.19. *Show that de-Broglie wavelength of a particle is equal to its Compton wavelength when its velocity is 0.707c.*

Solution. As proved in (4.8) de-Broglie wavelength associated with a moving particle

$$\lambda = \frac{h}{m_0 c} \frac{\left(1 - \frac{v^2}{c^2}\right)^{1/2}}{\frac{v}{c}}$$

For $v = 0.707\ c = \frac{c}{\sqrt{2}}$, we have

$$\lambda = \frac{h}{m_0 c} \frac{\left(1 - \frac{1}{2}\right)^{1/2}}{\sqrt{\frac{1}{2}}} = \frac{h}{m_0 c}$$

which is the Compton wavelength for the particle.

Example 4.20. *Calculate de-Broglie wavelength of an electron moving with velocity* $\frac{3}{5}c$.

(P.U. 2000)

Solution. $$\lambda = \frac{h}{m_0 c} \frac{\left(1 - \frac{v^2}{c^2}\right)^{1/2}}{\frac{v}{c}} = \frac{h}{m_0 c} \frac{\left(1 - \frac{9}{25}\right)^{1/2}}{\frac{3}{5}} = \frac{h}{m_0 c} \times \frac{4}{5} \times \frac{5}{3}$$

$$= \frac{4}{3} \frac{h}{m_0 c} = \frac{4}{3} \times \frac{6.62 \times 10^{-34}}{9.1 \times 10^{-31} \times 3 \times 10^8} = 0.032 \times 10^{-10}\ m$$

$$= 0.032\ \text{Å}$$

Example 4.21. *Calculate the energy of a proton having de-Broglie wavelength 0.5×10^{-15} m.*

Solution. According to de-Broglie

$$\lambda = \frac{h}{p} = \frac{h}{mv}$$

or $$v = \frac{h}{m\lambda} = \frac{6.6 \times 10^{-34}}{1.6 \times 10^{-27} \times 0.5 \times 10^{-15}} = 8.25 \times 10^8\ ms^{-1}$$

which is greater than the velocity of light. Therefore, it is a relativistic case.

Hence $$\lambda = \frac{h}{p} \text{ or } p = \frac{h}{\lambda} = \frac{6.6 \times 10^{-34}}{0.5 \times 10^{-15}} = 13.2 \times 10^{-19}\ m$$

According to the theory of relativity

$$E^2 = (pc)^2 + E_0^2 = (pc)^2 + (m_0 c^2)^2$$

where E is the total energy and $E_0 = m_0 c^2$ is the rest mass energy of the particle

$$\therefore \quad E^2 = (13.2 \times 10^{-19} \times 3 \times 10^8)^2 + (1.6 \times 10^{-27} \times 9 \times 10^{16})^2$$

$$= (39.6 \times 10^{-11})^2 + (14.4 \times 10^{-11})^2$$

Hence $$E = 42.2 \times 10^{-11}\ \text{J}$$

$\therefore$ Kinetic energy of the proton

$$= E - E_0 = 42.2 \times 10^{-11} - 14.4 \times 10^{-11} \text{ J}$$

$$= 27.8 \times 10^{-11} \text{ J} = 1730 \text{ MeV}.$$

Example 4.22. *Find the de-Broglie wavelength of an electron moving with velocity 10^7 ms^{-1}.* ***(Pbi. U. 2001; Luck.U.,2003)***

Solution. Velocity of the electron $v = 10^7$ ms^{-1}

As the velocity of the electron is small as compared to the velocity of light, it is a non-relativistic case

$\therefore$ de-Broglie wavelength of the electron

$$\lambda = \frac{h}{mv} = \frac{6.62 \times 10^{-34}}{9.11 \times 10^{-31} \times 10^7} = 0.7266 \times 10^{-10} \; m$$

$$= 0.7266 \text{ Å}$$

4.4 DE-BROGLIE PHASE (OR WAVE) VELOCITY AND PARTICLE VELOCITY

Wave motion is a form of disturbance which travels through a medium due to the repeated periodic motion of the particles of the medium about their mean positions, the motion being handed over from one particle to the next. The individual oscillators which make up the medium only execute simple harmonic motion about their mean positions and *do not* themselves travel through the medium with the wave. Every particle begins its vibrations a little later than its predecessor and there is a progressive change of phase as we travel from one particle to the next. *It is the phase relationship of these particles that we observe as a wave and the velocity with which the plane of equal phase travels through the medium is known as phase velocity (or wave velocity).*

If λ is wavelength and ν the frequency of the wave, then

Phase velocity $\quad v_p = \nu\lambda$ ...(*i*)

Now $\quad \omega = 2\pi\nu$ or $\nu = \dfrac{\omega}{2\pi}$ and

$$k = \frac{2\pi}{\lambda} \text{ or } \lambda = \frac{2\pi}{k}$$

$\therefore$ Phase velocity $v_p = \nu\lambda = \dfrac{\omega}{2\pi} \dfrac{2\pi}{k} = \dfrac{\omega}{k}$ ...(*ii*)

According to quantum theory, the energy of the wave

$$E = h\nu \qquad \therefore \nu = \frac{E}{h}$$

and momentum $p = \dfrac{h}{\lambda} \quad \therefore \lambda = \dfrac{h}{p}$

Substituting in (*i*), we have, the phase velocity

$$v_p = \frac{E}{h} \cdot \frac{h}{p} = \frac{E}{p} \qquad ...(4.9)$$

This relation also holds true in the case of de-Broglie waves associated with a moving particle.

de-Broglie phase velocity exceeds the velocity of light. In a *non-relativistic* case for a particle of mass m, moving with a velocity v, the total energy $E = mc^2$ and momentum $p = mv$.

$\therefore$ Phase velocity of the associated de-Broglie wave

$$v_p = \frac{E}{p} = \frac{mc^2}{mv} = \frac{c^2}{v} \qquad ...(4.10)$$

As $v < c$ *the phase velocity of the associated wave is always greater than c, the velocity of light in vacuum.*

In a *relativistic case* for de-Broglie waves associated with a moving particle of rest mass m_0, having a momentum p, the total energy is given by

$$E = [m_0^2 c^4 + p^2 c^2]^{1/2}$$

$$\therefore \quad v_p = \frac{E}{p} = \left[\frac{m_0^2 c^4 + p^2 c^2}{p^2}\right]^{1/2}$$

$$= c\left[\frac{m_0^2 c^2}{p^2} + 1\right]^{1/2} = c\left[\frac{m_0^2 c^2 \lambda^2}{h^2} + 1\right]^{1/2} \qquad ....(4.11)$$

As the term $\frac{m_0^2 c^2 \lambda^2}{h^2}$ is always a *positive* quantity for non-zero values of m_0 or λ, the *phase velocity of the associated wave is always greater than the velocity of light in vacuum*. But according to the theory of relativity it is not possible that the velocity of matter wave be equal to or greater than the velocity of light. If the velocity of the matter wave is equal to or greater than the velocity of light, then the wave associated with the particle will move faster than the particle itself thus leaving the particle far behind. *Hence a moving material particle cannot be equivalent to a single wave train*. It is always equal to a group of waves and the *particle velocity is equal to the group velocity* of a group of waves.

Phase velocity exceeds the particle velocity. According to relation 4.10, $v_p = \frac{c^2}{v}$, *the phase velocity* v_p *of the associated wave is always greater than c the velocity of light.*

As a material particle can never have even the velocity of light, *the phase velocity also exceeds the particle velocity.*

De-Broglie waves show dispersion in vacuum. Further, even in free space or vacuum the velocity of de-Broglie waves depends upon wavelength. As a result, de-Broglie waves of different wavelengths have different phase velocities. **Thus, vacuum is a dispersive medium** so far as de-Broglie waves are concerned but is not a dispesive medium for electro magnetic waves as in their case the waves of all wavelength travel with the same velocity in vacuum.

Relation between Particle Velocity and Wave Velocity

Phase velocity (or wave velocity) $v_p = \frac{E}{p}$, according to relation (*iii*). For a particle of mass m moving with a velocity v, $E = \frac{1}{2} mv^2$ and $p = mv$

$$\therefore \quad v_p = \frac{E}{p} = \frac{1}{2}\frac{mv^2}{mv} = \frac{1}{2} v \qquad ...(4.12)$$

Thus *phase velocity* (or *wave velocity*) *is equal to half the particle velocity.*

4.5 GROUP VELOCITY

A group consists of a number of waves of different frequencies superimposed upon each other. For example, white light consists of a continuous visible wavelength spectrum ranging from about 3000 Å in the violet to about 7000 Å in the red region. *The velocity with which the slowly varying envelop of the modulated pattern due to a group of waves travels in a medium is known as the group velocity.* The importance of the group velocity lies in the fact that this is the velocity with which the energy in the wave group is transmitted.

To derive an expression for group velocity consider a group of waves consisting of only two components of equal amplitude and having angular frequencies ω_1 and ω_2 differing by a small amount and represented by the equations

$$y_1 = a \cos (\omega_1 t - k_1 x)$$

and
$$y_2 = a \cos (\omega_2 t - k_2 x)$$

where $\frac{\omega_1}{k_1}$ and $\frac{\omega_2}{k_2}$ represent their respective phase (wave) velocities. ω being equal to $2\pi\nu$ and $k = \frac{2\pi}{\lambda}$. It is further supposed that $\frac{\omega_1}{k_1} \neq \frac{\omega_2}{k_2}$ *i.e.* it is a *dispersive medium*. The resultant amplitude is given by

$$y = y_1 + y_2 = a [\cos (\omega_1 t - k_1 x) + \cos (\omega_2 t - k_2 x)]$$

$$= 2a \cos \left[\frac{(\omega_1 + \omega_2)}{2} t - \frac{(k_1 + k_2)}{2} x\right] \cos \left[\frac{(\omega_1 - \omega_2)}{2} t - \frac{(k_1 - k_2)}{2} x\right]$$

$$= 2a \cos (\omega t - kx) \cos \left(\frac{\Delta \omega t}{2} - \frac{\Delta kx}{2}\right)$$

where
$$\omega = \frac{\omega_1 + \omega_2}{2},\ k = \frac{k_1 + k_2}{2}$$

$$\Delta \omega = \omega_1 - \omega_2 \text{ and } \Delta k = k_1 - k_2$$

The resultant wave has two parts.

(*i*) **Phase velocity.** A wave of frequency ω, propagation constant k and velocity

$$v_p = \frac{\omega}{k} = \frac{2\pi\nu}{2\pi/\lambda} = \nu\lambda \quad ...(i)$$

This is the *phase velocity* or wave velocity.

(*ii*) **Group velocity.** A second wave of frequency $\frac{\Delta\omega}{2}$, propagation constant $\frac{\Delta k}{2}$ and velocity $v_g = \frac{\Delta\omega}{\Delta k}$

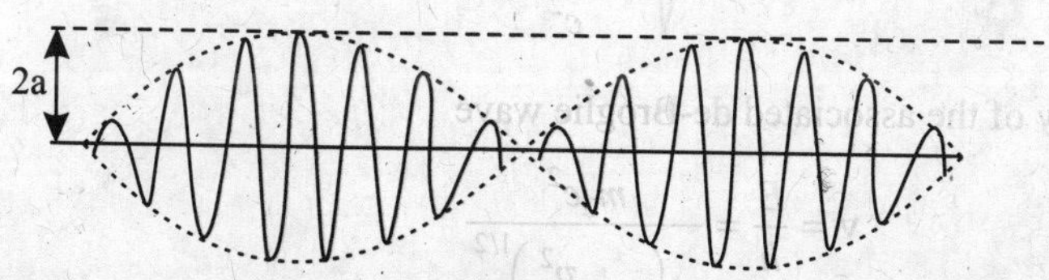

Fig. 4.1

It consists of a group of waves of the first type and is a very slowly moving envelop of frequency $\frac{\Delta\omega}{2}$ and propagation constant $\frac{\Delta k}{2}$. The modulated pattern moves with a velocity

$$v_g = \frac{\Delta\omega}{\Delta k} \text{ known as the } \textit{group velocity}.$$

If $\Delta\omega$ and Δk are very small, we can put

Group velocity
$$v_g = \frac{\partial\omega}{\partial k} \qquad ...(ii)$$

Also
$$v_g = \frac{2\pi\,\partial\nu}{2\pi\,\partial\left(\frac{1}{\lambda}\right)} = -\lambda^2\frac{\partial\nu}{\partial\lambda}. \qquad ...(4.13)$$

Relation between phase velocity and group velocity. We have seen that

$$v_p = \frac{\omega}{k} \quad \therefore \omega = kv_p$$

and
$$v_g = \frac{d\omega}{dk} = \frac{dkv_p}{dk} = v_p + \frac{k\,dv_p}{dk}$$

But
$$k = \frac{2\pi}{\lambda} \quad \therefore dk = -\frac{2\pi}{\lambda^2}d\lambda$$

and
$$\frac{k}{dk} = -\frac{\lambda}{d\lambda}$$

Hence
$$v_g = v_p - \lambda\frac{dv_p}{d\lambda} \qquad ...(4.14)$$

Particle velocity. *It is the velocity with which the particle travels.*

Relation between group velocity and particle velocity. A particle moving with a velocity v is supposed to consist of a group of waves according to de-Broglie hypothesis. For a material particle of rest mass m_0 moving with a velocity v and having an effective mass m the total energy E is given by

$$E = mc^2 = \frac{m_0c^2}{\sqrt{1-\frac{v^2}{c^2}}}$$

and momentum
$$p = mv = \frac{m_0v}{\sqrt{1-\frac{v^2}{c^2}}}$$

The frequency of the associated de-Broglie wave

$$\nu = \frac{E}{h} = \frac{m_0c^2}{h\left(1-\frac{v^2}{c^2}\right)^{1/2}}$$

and
$$\omega = 2\pi\nu = \frac{2\pi m_0 c^2}{h\left(1-\frac{v^2}{c^2}\right)^{1/2}}$$

$$\therefore \quad d\omega = \frac{2\pi m_0}{h\left(1-\frac{v^2}{c^2}\right)^{3/2}} v\, dv \qquad ..(4.15)$$

The wavelength of the associated de-Broglie wave

$$\lambda = \frac{h}{p} = \frac{h\left(1-\frac{v^2}{c^2}\right)^{1/2}}{m_0 v}$$

But
$$k = \frac{2\pi}{\lambda} = \frac{2\pi m_0 v}{h\left(1-\frac{v^2}{c^2}\right)^{1/2}}$$

$$\therefore \quad dk = \frac{2\pi m_0}{h}\left[\left(1-\frac{v^2}{c^2}\right)^{-1/2} dv + v\frac{v}{c^2}\left(1-\frac{v^2}{c^2}\right)^{-3/2} dv\right]$$

$$= \frac{2\pi m_0}{h\left(1-\frac{v^2}{c^2}\right)^{3/2}}\left[dv - \frac{v^2}{c^2}dv + \frac{v^2}{c^2}dv\right]$$

$$= \frac{2\pi m_0\, dv}{h\left(1-\frac{v^2}{c^2}\right)^{3/2}}. \qquad ...(4.16)$$

Dividing (4.15) by (4.16), we have

Group velocity $v_g = \frac{d\omega}{dk} = v,$ the particle velocity.

Thus, the de-Broglie wave group associated with a particle travels with the same velocity as the particle itself.

Example 4.23. *Find the phase and group velocity of an electron whose de-Broglie wavelength is 1.2 Å. (neglect relativity effect).* ***(G.N.D.U., 2002)***

Solution. Group velocity $v_g = v$ the particle velocity

Now
$$\lambda = \frac{h}{mv}$$

$\therefore$ Velocity of the moving particle

$$v = \frac{h}{m\lambda} = \frac{6.62 \times 10^{-34}}{9.1 \times 10^{-31} \times 1.2 \times 10^{-10}}$$

$$= 6.06 \times 10^6 \text{ ms}^{-1}$$

$\therefore$ Group velocity $\quad v_g = 6.06 \times 10^6 \text{ ms}^{-1}$

Phase velocity $$v_p = \frac{\omega}{k}$$

Using the relation $$E = h\nu = \frac{h}{2\pi} 2\pi\nu = \hbar\omega$$

and $$p = \frac{h}{\lambda} = \frac{h}{2\pi}.\frac{2\pi}{\lambda} = \hbar k$$

We have $$v_p = \frac{\omega}{k} = \frac{\hbar\omega}{\hbar k} = \frac{E}{p} \quad ...(i)$$

For a particle of mass m moving with a velocity v

$$E = \frac{1}{2} mv^2 \text{ and } p = mv$$

$$\therefore \quad E = \frac{p^2}{2m}$$

Substituting $E = \frac{p^2}{2m}$ in (*i*), we get

$$v_p = \frac{p^2}{2m.p} = \frac{p}{2m} = \frac{h}{2m\lambda} = \frac{6.62 \times 10^{-34}}{2 \times 9.1 \times 10^{-31} \times 1.2 \times 10^{-10}}$$

$$= 3.03 \times 10^6 \text{ ms}^{-1}.$$

Note. From the above calculation we find that phase velocity v_p is one half of the group velocity or the particle velocity.

In order that a moving localised particle may be represented as a wave it is necessary that the phase (or wave) velocity should be equal to the particle velocity, otherwise the matter wave would not be able to keep pace with the particle whose motion it guides. Thus a localised particle cannot be represented by a single wave. It can only be represented by a group of waves and in such a case group velocity of the wave group is equal to the particle velocity.

4.6 REST MASS OF PHOTON

In a *relativistic case* the de-Broglie waves associated with the moving particle of rest mass m_0, having a momentum p, the total energy is given by

$$E = \left[m_0^2 c^4 + p^2 c^2\right]^{1/2}$$

$$\therefore \text{ Phase (or wave velocity) } v_p = \frac{E}{p} = \left[\frac{m_0^2 c^4 + p^2 c^2}{p^2}\right]^{1/2}$$

$$= c\left[\frac{m_0^2 c^2}{p^2} + 1\right]^{1/2} = c\left[\frac{m_0^2 c^2 \lambda^2}{h^2} + 1\right]^{1/2} \quad ...(i)$$

Now consider de-Broglie wave travelling with velocity c. This corresponds to the propagation of electro-magnetic waves. The velocity of the associated particle *i.e.* the *photon* is also c. Substituting $v_p = c$ in Eq. (*i*), we have

$$c = c\left[\frac{m_0^2 c^2 \lambda^2}{h^2} + 1\right]^{1/2}$$

$$\text{or} \qquad \frac{m_0^2 c^2 \lambda^2}{h^2} = 0 \quad \text{or} \quad m_0 = 0 \qquad [\because c, \lambda \text{ and } h \neq 0]$$

Thus the rest mass of the photon is zero.

4.7 EVIDENCES OF MATTER WAVES: DIFFRACTION AND INTERFERENCE OF PARTICLES

An essential property of a *wave* is that it can undergo *diffraction*. According to classical physics, this property of diffraction is not shown by 'particles'. However, direct evidence of the existence of de-Broglie waves *associated* with moving *micro-particles* has been furnished by experiments on *electron diffraction*.

Theory. A free electron starting from rest, on being accelerated through a potential difference V acquires a kinetic energy given by

$$\frac{1}{2} m_0 v^2 = eV$$

where m_0, e and v are the rest mass, charge and velocity of the electron respectively.

$$\therefore \qquad v = \sqrt{\frac{2Ve}{m_0}}$$

The de-Broglie wavelength associated with the electron (moving non-relativistically) is given by

$$\lambda = \frac{h}{m_0 v} = \frac{h}{\sqrt{2Vem_0}} \qquad \ldots(4.17)$$

In S.I. units

$$\lambda = \frac{6.62 \times 10^{-34}}{\sqrt{V}\ (2 \times 1.6 \times 10^{-19} \times 9.1 \times 10^{-31})^{1/2}}$$

$$= \frac{12.27 \times 10^{-10}}{\sqrt{V}}\ m = \frac{12.27}{\sqrt{V}}\ \text{Å} \qquad \ldots(4.18)$$

Thus, the wavelength of de-Broglie waves associated with an electron accelerated through a potential difference of 100 volts in vacuum is $\lambda = 1.227$Å. Since the wavelength is comparable with the wavelength of *X*-rays, the de-Broglie waves should also be diffracted by crystals just like *X-rays*.

4.7.1 Davisson and Germer's Experiment. Davisson and Germer experimentally demonstrated the diffraction of electrons.

The electrons are produced by heating a tungsten filament F and are accelerated through a known potential difference V by maintaining a steady potential difference between the filament F and a plate P.

The electrons are collimated by passing through narrow holes and finally emerge through a fine hole H in the form of a very narrow beam. The whole arrangement is known as an *Electron gun*. The fine beam of electrons falls normally on the surface of a *single* Nickel crystal N. This is prepared by baking the block of Nickel in a high temperature oven, when *very small* individual crystals of which the block is normally composed of form a *large single* crystal, all the atoms of which are arranged in a regular lattice. The electrons are scattered by the crystal in all directions and the intensity of the scattered beam in any direction is measured by allowing the beam to enter the ionisation chamber I set at appropriate angle and noting the deflection in a sensitive galvanometer connected to I.

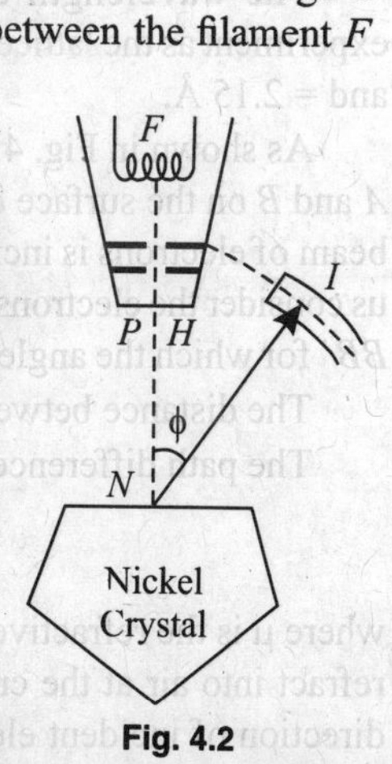

Fig. 4.2

The chamber can be rotated with the crystal as the centre and the intensity of the scattered beam can be determined as a function of scattering angle ϕ, the crystal being held in a fixed position throughout the experiment. The results of the experiment are represented in the form of a *polar graph*. In the polar graph, a line is drawn from the origin O inclined at an angle ϕ with Y-axis which represents the direction of incident beam. The length of the line drawn from O is taken proportional to the intensity of the diffracted electron beam. The end points of these lines drawn for different values of ϕ are joined to get a smooth, continuous curve. This curve is known as the *polar graph*.

Effect of increasing electron energy on the scattering angle. To study the effect of increasing electron energy on the scattering angle ϕ the observations were taken for increasing values of accelerating voltages. The graph is at first fairly *smooth* but at a potential difference of 44 V and corresponding electron energy a *spur* or *hump* appears on the curve at a scattering angle $\phi = 60°$. As the accelerating potential and hence the corresponding electron energy is increased the length of the *spur* increases and a *sharp maxima* is obtained for electron energy corresponding to 54V at a scattering angle $\phi = 50°$. With further increase in accelerating potential and electron energy the *spur* decreases in the length and finally disappears at 68V at a scattering angle $\phi = 40°$. For further increase in electron energy or accelerating voltage the graph is again fairly smooth.

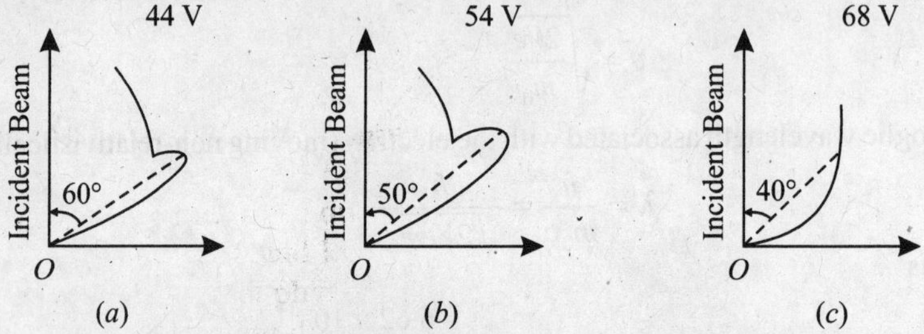

Fig. 4.3

Intensity of electron beam against scattering angle. If we plot the intensity of electron beam against the scattering angle ϕ for a potential difference of 54 V, we get a curve as shown in Fig. 4.4. This shows that electrons with kinetic energy 54 eV suffer maximum scattering at $\phi = 50°$.

Discussion. The de-Broglie wave-length associated with the electron accelerated through a potential difference of 54 volts according to relation 4.18 is given by

$$\lambda = \frac{12.27}{\sqrt{54}} \text{ Å} = 1.66 \text{ Å}$$

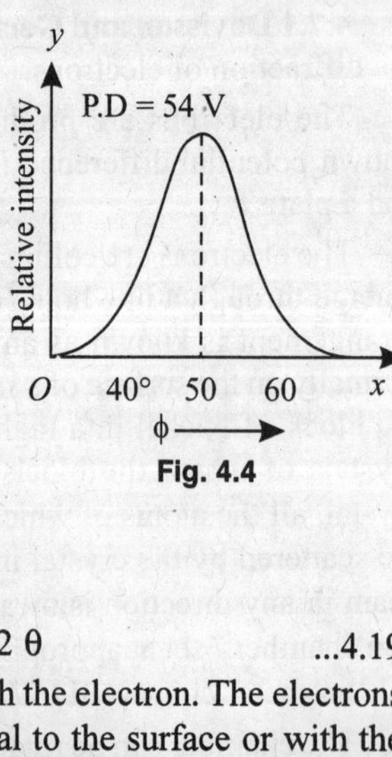

Fig. 4.4

The wavelength can also be calculated from the diffraction experiment as the lattice constant for a single crystal of Nickel is known and = 2.15 Å.

As shown in Fig. 4.5 the distance between the two adjacent atoms A and B on the surface of the crystal D = lattice constant. The incident beam of electrons is incident normally on the surface of the crystal. Let us consider the electrons diffracted from the two adjacent planes AA' and BB' for which the angle of incidence as also the angle of reflection is θ.

The distance between these two planes $= d = D \sin \theta$.

The path difference between the two beams reflected at P and Q

$$= MQ + QN$$

$$= 2\mu d \cos \theta = 2\mu D \sin \theta \cos \theta = \mu D \sin 2\theta \qquad \text{...4.19}$$

where μ is the refractive index of the crystal for the wave associated with the electron. The electrons refract into air at the crystal surface making an angle ϕ with the normal to the surface or with the direction of incident electron beam. The angle of incidence of the reflected electron on the crystal surface $= 2\theta$.

$$\therefore \qquad \mu = \frac{\sin \phi}{\sin 2\theta}$$

Substituting in Eq. 4.19 we have

Path difference = $D \sin \phi$

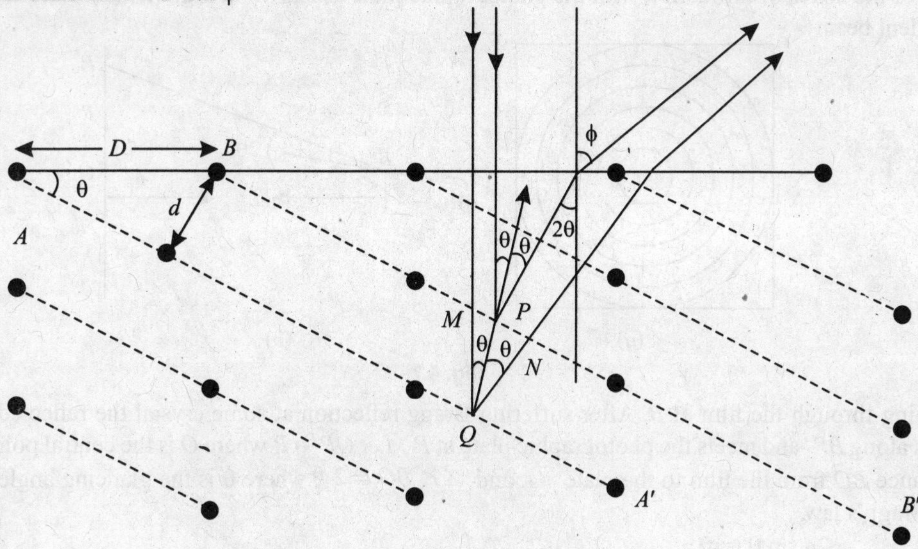

Fig. 4.5

For maximum intensity $D \sin \phi = n \lambda$ where $n = 1, 2, 3, ...$

For the first order maximum $n = 1$

$$\therefore \qquad \lambda = D \sin \phi = 2.15 \sin 50° = 1.65 \text{ Å}$$

This result shows excellent agreement with calculated value of 1.66 Å from de-Broglie hypothesis.

4.7.2 G.P. Thomson's Experiment. The wave like nature of electrons was confirmed by G.P. Thomson by his famous *Electron diffraction* experiment.The diffraction pattern obtained by passing a beam of electrons of velocity v through a thin gold foil was exactly similar to that obtained from X-rays of wavelength $\lambda = h/mv$.

The electrons are produced by passing an electric current through a filament F and accelerated by giving a high positive potential of the order of 50,000 volts to the anode A (fig. 4.6). The whole apparatus is kept highly evacuated. The electron beam passes through a fine hole in a metal block B and falls on a gold foil G of thickness 10^{-8} m. The elctrons suffer diffraction and are recorded on a photographic plate P giving rise to diffraction rings exactly similar to those obtained by diffraction through a small circular hole.

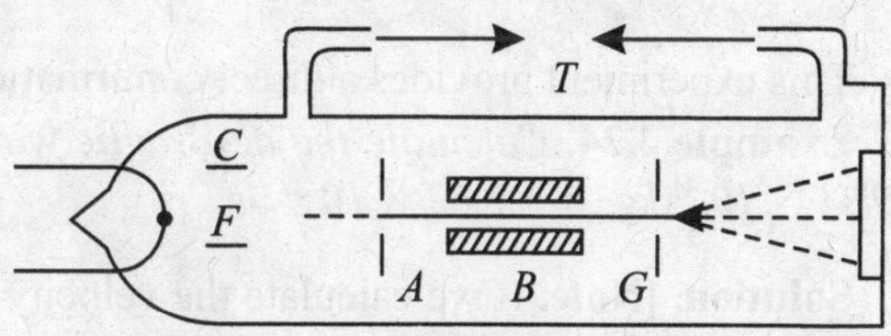

Fig. 4.6

To rule out the possibility that the cathode ray beam produced X-rays on striking the film and these X-rays produced the diffraction pattern, the experiment was repeated by introducing a magnetic induction field between the film and the photographic plate. It was found that diffraction pattern was exactly similar but moved as a whole to one side due to the effect of the mgnetic field, showing that the diffraction pattern was due to charged particles (electrons), all of them having the same veocity.

On removing the film only central spot is obtained on the photographic plate without any diffraction pattern.

Further experiments showed that not only electrons but also protons, neutrons and molecules behave like waves under suitable conditions.

To calculate λ from radii of rings. Gold foil is a poly-crystalline film. In such a film there are some crystals set at the correct angle to give a Bragg reflection. If there are sufficient number of crystals distributed at random, Bragg reflection will form a series of rings arising from the intersection of the cones of diffraction with the photographic plate as shown in Fig. 4.7. (*a*). Let *AB* be the incident beam

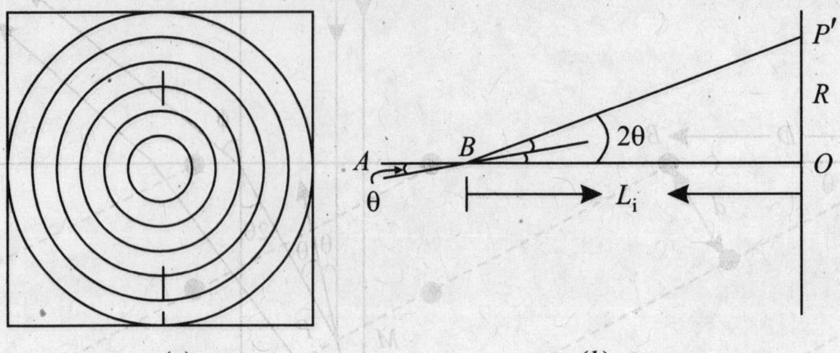

Fig. 4.7

passing through the film at *B*. After suffering Bragg reflection at some crystal the reflected beam goes along *BP'* and meets the photographic plate at *P'*. *Let OP' = R* where *O* is the central point. The distance *BO* from the film to the plate = *L* and $\angle P'BO = 2\theta$ where θ is the glancing angle given by Bragg's law,

$$2d \sin\theta = n\lambda$$

where *d* is the perpendicular distance between two nearest lattice planes of gold crystal.

Now $$\frac{R}{L} = \tan 2\theta = 2\theta$$

Also $$2d \sin\theta = 2d\theta \qquad [\because \theta \text{ is small}]$$

$$\therefore \quad d\frac{R}{L} = n\lambda \quad \text{or} \quad \lambda = \frac{dR}{nL}$$

From this relation wavelength associated with the accelerated electron can be calculated. This agrees with the calculated value $\lambda = \dfrac{12.27}{\sqrt{V}}$

This experiment provides a direct confirmation of the wave nature of the electron.

Example 4.24. *Calculate the de-Broglie wavelength of 1 MeV electron. Mass of the electron* $= 9.11 \times 10^{-31}$ *kg;* $h = 6.62 \times 10^{-34}$ *Js*

Solution. [Note. If we calculate the velocity of the electron using the relation $E = \frac{1}{2}mv^2$, we get

$$v = \sqrt{\frac{2E}{m}} = \sqrt{\frac{2 \times 10^6 \times 1.6 \times 10^{-19}}{9.11 \times 10^{-31}}} = 5.927 \times 10^8 \text{ ms}^{-1}$$

which is not possible as it is greater than the velocity of light. We should, therefore, use relativistic form of the equation.]

Kinetic energy of the electron $= 10^6 \; eV = 10^6 \times 1.6 \times 10^{-19}$ **J**

$= 1.6 \times 10^{-13}$ **J**

Let m_0 be the rest mass of the electron and m the mass of the moving electron.

∴ Total energy of the moving electron

$$mc^2 = m_0c^2 + K.E.$$

$$= 9.11\times10^{-31}\times9\times10^{16} + 1.6\times10^{-13}$$

∴ $$m = \frac{9.11\times10^{-31}\times9\times10^{16} + 1.6\times10^{-13}}{9\times10^{16}}$$

$$= 26.88\times10^{-31}\text{ kg}$$

Now $$m = \frac{m_0}{\sqrt{1-\dfrac{v^2}{c^2}}} \quad \text{or} \quad 1-\frac{v^2}{c^2} = \left(\frac{m_0}{m}\right)^2 = \left(\frac{9.11\times10^{-31}}{26.88\times10^{-31}}\right)^2$$

$$= (0.3389)^2 = 0.1149$$

∴ $$\frac{v^2}{c^2} = 1 - 0.1149 = 0.8851$$

or $$\frac{v}{c} = 0.9407$$

∴ $$v = 3\times10^8\times0.9407 = 2.822\times10^8\text{ ms}^{-1}$$

Hence de-Broglie wavelength

$$\lambda = \frac{h}{mv} = \frac{6.62\times10^{-34}}{26.88\times10^{-31}\times2.822\times10^8}$$

$$= 0.0087\times10^{-10}\text{ m} = 0.0087\text{ Å}.$$

Example 4.25. *Which has a shorter wavelength; a 10 eV photon or a 10 eV electron? Explain your answer.* ***(P.U. 2007)***

Solution. Energy of a 10 eV photon $E = 10\times1.6\times10^{-19}$ Joule

For a photon $$E = h\nu = \frac{hc}{\lambda}$$

or $$\lambda = \frac{hc}{E} = \frac{6.62\times10^{-34}\times3\times10^8}{10\times1.6\times10^{-19}}$$

$$= 1240\times10^{-10}\text{ m} = 1240\text{ Å}$$

For an electron $$E = \frac{1}{2}mv^2$$

∴ $$v = \sqrt{\frac{2E}{m}}$$

and momentum $$p = mv = \sqrt{2mE}$$

∴ $$\lambda = \frac{h}{p} = \frac{h}{\sqrt{2mE}}$$

$$= \frac{6.62\times10^{-24}}{(2\times9.1\times10^{-31}\times10\times1.6\times10^{-19})^{1/2}}$$

$$= 3.88\times10^{-10}\text{ m} = 3.88\text{ Å}.$$

Explanation. Form the above calculations we find that for the electron (any particle) the de-Broglie wavelength is inversely proportional to the square root of the energy (in electron volts) or $\lambda \propto \frac{1}{\sqrt{E}}$ and for a photon the wavelength is inversely proportional to the energy or $\lambda \propto \frac{1}{E}$.

Hence for the same energy the de-Broglie wavelength of the electron is much smaller as compared to the wavelength of the photon.

Example 4.26. *Show that the expression for de-Broglie wavelength of a particle with rest mass energy E_0, and kinetic energy K both in eV can also be written as*

$$\lambda = \frac{hc}{[K(2m_0c^2 + K)]^{1/2}} = \frac{1.241 \times 10^{-6}}{[K\,(2E_0 + K)]^{1/2}} \text{ metre.} \qquad \textit{(K.U. 2002)}$$

Solution. Rest mass energy $E_0 = m_0c^2$ where m_0 is the rest mass of the particle and c the velocity of light. If m is the mass of the particle in motion then

$$E_0 + K = mc^2$$

If v is the velocity of the particle, then

$$m = \frac{m_0}{\sqrt{1 - \frac{v^2}{c^2}}}$$

$$\therefore \quad \frac{m}{m_0} = \frac{1}{\sqrt{1 - v^2/c^2}} = \frac{E_0 + K}{E_0}$$

or

$$\frac{1}{1 - \frac{v^2}{c^2}} = \frac{(E_0 + K)^2}{E_0^2}$$

$$E_0^2 = (E_0 + K)^2 - \frac{v^2}{c^2}(E_0 + K)^2$$

$$= (E_0 + K)^2 - m^2c^2v^2$$

$$\therefore \quad m^2c^2v^2 = 2KE_0 + K^2 = K(2E_0 + K)$$

or

$$mv = \frac{\sqrt{K(2E_0 + K)}}{c}$$

Hence

$$\lambda = \frac{h}{mv} = \frac{hc}{[K\,(2E_0 + K)]^{1/2}} = \frac{hc}{[K\,(2m_0c^2 + K)]^{1/2}} \qquad [\because E_0 = m_0c^2]$$

As E_0 and K are in eV, we have

$$h = 6.626 \times 10^{-34} \text{ Js} = 4.136 \times 10^{-15} \text{ eVs}$$

and

$$c = 3 \times 10^8 \text{ ms}^{-1}$$

$$\therefore \quad hc = 4.136 \times 10^{-15} \times 3 \times 10^8 = 1.241 \times 10^{-6} \text{ eV}m$$

Hence

$$\lambda = \frac{[1.241 \times 10^{-6}]}{[K\,(2E_0 + K)^{1/2}]} \text{ metre.}$$

Example 4.27. *An electron and a proton have the same de-Broglie wavelength. Prove that the energy of the electron is greater.*

Solution. Let m be the mass and v the velocity of the electron and M and V the corresponding values for the proton.

∴ de-Broglie wavelength for the electron

$$\lambda_e = \frac{h}{mv}$$

and for the proton $\lambda_p = \frac{h}{MV}$

But $\lambda_e = \lambda_p$

$$\therefore \quad \frac{h}{mv} = \frac{h}{MV} \text{ or } mv = MV$$

$$\text{Energy of the electron} = \frac{1}{2}mv^2 = \frac{1}{2}MV \times v$$

$$= \frac{1}{2}MV\frac{MV}{m} = \frac{1}{2}MV^2\frac{M}{m}$$

$$= \frac{M}{m} \times \text{Energy of the proton.}$$

As $M > m$, the energy of the electron is greater than that of the proton of same de-Broglie wavelength.

4.8 FREE AND LOCALIZED PARTICLE

4.8.1 Free particle. *A particle is known as a free particle if there is no force acting on it.*

As force F is the time rate of change of momentum p, $F = \frac{dp}{dt}$. For a free particle $F = 0$ *i.e.* $\frac{dp}{dt} = 0$ or p = a constant

Also kinetice energy $E = \frac{1}{2}mv^2 = \frac{m^2v^2}{2m} = \frac{p^2}{2m}$

As p is a constant E is also a constant.

According to de-Broglie hypothesis a material particle has a wave associated with it. The displacement of matter wave is represented by Ψ, which is a function of space and time co-ordinates. This function which describes the matter wave associated with a material particle is known as its wave function.

As a free particle has a constant momentum $p = \frac{h}{\lambda} = \frac{h}{2\pi}.\frac{2\pi}{\lambda} = \hbar k$ and constant energy $E = h\nu = \frac{h}{2\pi}.2\pi\nu = \hbar\omega$ the matter wave travelling in the positive X-direction is given by a complex wave function of the type

$$\Psi(x,t) = Ae^{i(kx - \omega t)}$$

This function repeats itself after one differentiation. As such the matter wave associated with the particle describes the particle completely because a knowledge of Ψ at $t = 0$ is sufficient to know its values any time later.

Characteristic parameters 1. A free particle has a constant definite momentum.

2. It has a constant definite energy.

3. It is characterised by a complex wave function of the type $\Psi(x,t) = Ae^{i(kx - \omega t)}$

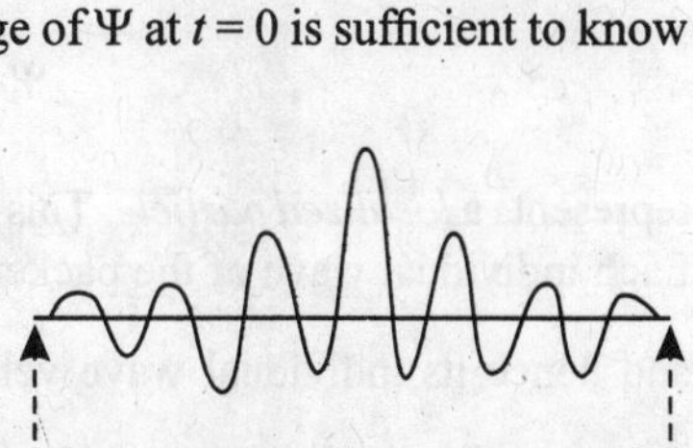

Fig. 4.8

4.8.2 Localized Particle and its Wave function

According to de-Broglie hypothesis a material particle has a wave associated with it. The wave function for a free particle is given by

$$\Psi\ (x,t) = A\ e^{i(kx - \omega t)}$$

This wave function exists at all points in space and is no where zero. In other words, the matter wave represented by this wave function and hence the particle must be present every where from $x = -\infty$ to $x = +\infty$. But this is not possible because at a given time the particle must be *localised in space at just one point.*

Thus a monochromatic wave *i.e.,* a wave of only one wavelength which spreads out over the whole of one dimensional space cannot represent a localised particle. Ideally, a localised particle should occupy just a point space but as the nearest approach let the particle be localised in a region of length Δx in one dimensional space *i.e.,* the wave function is zero every where except in the region Δx where the particle is localised. In the region Δx the wave function should have oscillatory character and a wavelength. *Such a wave of finite extent is called a wave packet* and is shown in Fig. 4.8. *Thus a localized particle is represented by a wave packet and not by a single wave of infinite extent.*

A wave packet consists of a number of waves, say N of slightly different frequencies ω_1, ω_2 ..... ω_i.......ω_n and appropriate amplitudes A_1, A_2.......A_i.....A_n superimposed on each other in such a way that the resultant amplitude is non-zero in the region Δx and zero outside it. The wave function can now be written as

$$\Psi\ (x, t) = \sum_{i=1}^{N} A_i\ e^{i(k_i x - \omega_i t)}$$

The value of Δx depends upon the range of frequencies. Longer the range smaller is Δx. If we have an infinite number of frequencies, the summation changes to integral and

$$\Psi\ (x,t) = \int_{-\infty}^{+\infty} A(k)\ e^{i(kx - \omega t)}\ dk$$

Thus we find that—

(i) A localised particle is always represented by a wave packet consisting of a linear combination of an infinite number of waves each having an amplitude $A\ (k)$

(ii) An important point to be noted is in connection with uncertainty principle. The wave packet representing a localised particle consists of an infinite number of waves of different values of k. As $p = \hbar k$, different values of k mean that the wave packet does not have a precise momentum (or velocity). In other words localisation of particle in as small a space as possible has been achieved at the cost of bringing about a large uncertainty in momentum.

4.9 WAVE PACKET: AMPLITUDE AND ITS VELOCITY

4.9.1 Amplitude of wave packet. The wave packet

$$\Psi\ (x,t) = \int_{-\infty}^{\infty} A(k)\ e^{i(kx - \omega t)}\ dk \qquad ...(i)$$

represents a *localized particle.* This wave packet is formed by combining an infinite number of waves. Each individual wave of the packet is known as *phase wave.* Each phase wave has its own k and ω and hence its individual wave velocity $v_p = \dfrac{\omega}{k}$. Here $A\ (k)$ is the amplitude of the plane wave corresponding to the wave vector k. The individual waves have wide range of velocities. The wave packet, however, moves with a constant velocity known as *group velocity*.

Suppose the wave packet given in Eq. (*i*), instead of having an infinite number of waves only consists of waves for which k lies between $k_0 - \Delta k$ and $k_0 + \Delta k$ and $A(k) = 1$, Then

$$A(k) = 1 \text{ for } k_0 - \Delta k < k < k_0 + \Delta k$$

and $$A(k) = 0 \text{ outside.}$$

Under these conditions Eq. (*i*) becomes

$$\Psi(x,t) = \int_{k_0 - \Delta k}^{k_0 + \Delta k} e^{i(kx - \omega t)}\, dk \qquad \text{...(ii)}$$

As the nearest approximation, if $\frac{d\omega}{dk}$ is the rate of change of ω with k, ω_0 the value of ω at $k = k_0$ and ω at $k = k$, then

$$\omega = \omega_0 + (k - k_0)\left(\frac{d\omega}{dk}\right)$$

$$\therefore \quad kx - \omega t = kx - \omega_0 t - (k - k_0)\left(\frac{d\omega}{dk}\right)t$$

$$= kx + k_0 x - \omega_0 t - k_0 x - (k - k_0)\left(\frac{d\omega}{dk}\right)t$$

$$= (k_0 x - \omega_0 t) + (k - k_0)x - (k - k_0)\left(\frac{d\omega}{dk}\right)t$$

$$= (k_0 x - \omega_0 t) + (k - k_0)\left[x - \left(\frac{d\omega}{dk}\right)t\right]$$

Substituting this value of $kx - \omega t$ in Eq. (*ii*), we have

$$\Psi(x,t) = \int_{k_0 - \Delta k}^{k_0 + \Delta k} e^{i(k_0 x - \omega_0 t) + i(k - k_0)\left[x - \left(\frac{d\omega}{dk}\right)t\right]}\, dk$$

$$= e^{i(k_0 x - \omega_0 t)} \int_{k_0 - \Delta k}^{k_0 + \Delta k} e^{i(k - k_0)\left[x - \left(\frac{d\omega}{dk}\right)t\right]}\, dk \qquad \text{...(iii)}$$

Put $x - \left(\frac{d\omega}{dk}\right)t = f$, then the integral portion of Eq. (*iii*) becomes $I = \int_{k_0 - \Delta k}^{k_0 + \Delta k} e^{if(k - k_0)}\, dk$

Now let $k - k_0 = k'$ Then $dk = dk'$

When $k = k_0 - \Delta k$ (lower limit)

$$k' = k - k_0 = k_0 - \Delta k - k_0 = -\Delta k$$

When $k = k_0 + \Delta k$ (upper limit)

$$k' = k - k_0 = k_0 + \Delta k - k_0 = +\Delta k$$

$$\therefore \quad I = \int_{-\Delta k}^{+\Delta k} e^{ifk'}\, dk' \qquad \text{...(iv)}$$

$$= \left[\frac{e^{ifk'}}{if}\right]_{-\Delta k}^{+\Delta k}$$

$$= \frac{e^{if\Delta k} - e^{-if\Delta k}}{if}$$

$$= \frac{2\sin f\,\Delta k}{f}$$

or
$$I = \frac{2\sin\left[\Delta k\left\{x - \left(\frac{d\omega}{dk}\right)t\right\}\right]}{x - \left(\frac{d\omega}{dk}\right)t}$$

Substituting the value of the integral (I) in Eq (*iii*), we get

$$\Psi(x,t) = \frac{2\sin\left[\Delta k\left\{x - \left(\frac{d\omega}{dk}\right)t\right\}\right]}{x - \left(\frac{d\omega}{dk}\right)t}\, e^{i(k_0 x - \omega_0 t)} \quad \text{...}(v)$$

Comparing with the given equation (*i*), we find that the above equation represents a plane wave travelling in the $+X$ direction having an amplitude

$$A(x,t) = \frac{2\sin\left[\Delta k\left\{x - \left(\frac{d\omega}{dk}\right)t\right\}\right]}{x - \left(\frac{d\omega}{dk}\right)t} \quad \text{...}(vi)$$

The amplitude varies with time as well as with x and therefore, itself corresponds to a wave motion.

4.9.2 Velocity of the wave packet. The amplitude of the wave packet is given by Eq. (*vi*),

A function of the type $\frac{\sin ax}{x}$ has a maximum at $x = 0$. Therefore, A (x,t) will have a maximum at

$$x - \left(\frac{d\omega}{dk}\right)t = 0 \quad \text{or } x = \left(\frac{d\omega}{dk}\right)t \quad \text{...}(vii)$$

Eq. (*vii*) gives the position x of amplitude peak (maximum) A at time t.

At $t = 0$; $x = x_1 = 0$

At $t = \Delta t$; $x = x_2 = \left(\frac{d\omega}{dk}\right)\Delta t$

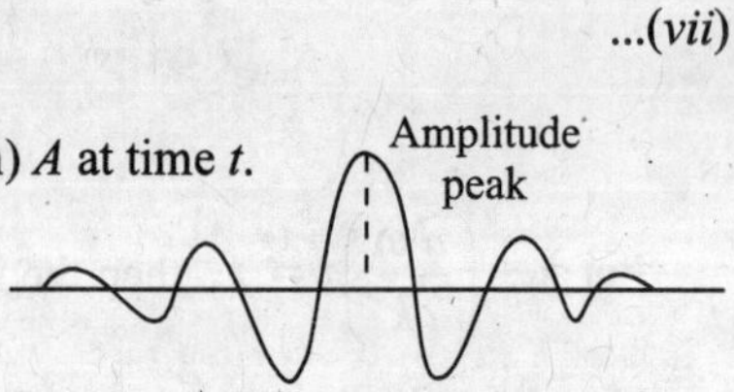

Fig. 4.9

Thus the peak of the wave packet amplitude moves through a distance $x_2 - x_1 = \left(\frac{d\omega}{dk}\right)\Delta t$ in a time Δt

$$\therefore \text{ Velocity of the amplitude peak } = \frac{x_2 - x_1}{\Delta t}$$

$$= \frac{\left(\frac{d\omega}{dk}\right)\Delta t}{\Delta t} = \frac{d\omega}{dk}$$

The velocity of the amplitude peak is known as the *group velocity* and gives the velocity of the group of waves or wave packet

$$\therefore \quad v_g = \frac{d\omega}{dk}.$$

Example 4.28. *A one dimensional wave packet is given by*

$$\Psi(x,t) = \frac{1}{\sqrt{2\pi\hbar}} \int_{-\infty}^{\infty} a(p)\, e^{(i/\hbar)(px - Et)} dp$$

Write the relation for amplitude of the component of p. ***(G.N.D.U. 2002)***

Solution. The Fourier transform can be written as

$$\Psi(x,t) = \frac{1}{\sqrt{2\pi}} \int_{-\infty}^{\infty} a(k)\, e^{i(kx - \omega t)}\, dk$$

$$a(k) = \frac{1}{\sqrt{2\pi}} \int_{-\infty}^{\infty} \Psi(x,t)\, e^{-i(kx - \omega t)}\, dx$$

The propagation constant k and angular frequency ω in terms of momentum and energy are

$$k = \frac{p}{\hbar} \quad \text{and} \quad \omega = \frac{E}{\hbar} = \frac{p^2}{2m\hbar}$$

∴ The exponential term would become $\exp\left[\frac{i}{\hbar}\left(px - \frac{p^2}{2m}t\right)\right]$

or $\exp\left[\frac{i}{\hbar}(px - Et)\right]$ and hence the wave packet $\Psi(x,t)$ would become

$$\Psi(x,t) = \frac{1}{\sqrt{2\pi\hbar}} \int_{-\infty}^{\infty} a(p)\, e^{i/\hbar\,(px - Et)}\, dp$$

Now, applying Fourier transform the amplitude of the component of momentum p is given by

$$a(p) = \frac{1}{\sqrt{2\pi\hbar}} \int_{-\infty}^{\infty} \Psi(x,t)\, e^{-i/\hbar\,(px - Et)}\, dx.$$

4.10 QUANTIZATION IN HYDROGEN ATOM (Bohr quantum condition)

In Bohr atom model, the concept of stationary electron orbit of any radius is not totally correct (Fig. 4.10). According to de-Broglie, the electron moving in a stationary orbit is associated with the matter waves which propogate along the circumference repeatedly (Fig. 4.11).

In fact, by considering wave like behaviour of electron, the quantization of angular momentum of an electron in atom can be derived in a more natural way. According to Bohr, the angular momentum L of an electron should be

$$L = mvr = n\hbar \qquad \ldots(i)$$

where $n = 1, 2, 3, \ldots$

When the electron moves in a circular stationary orbit, it is assumed to generate a series of waves. The electron wave meets itself for one complete circumference (Fig. 4.11) is said as allowed orbit. But when it is incompleted (Fig. 4.10), then there is destructive interference, such *an orbit is not allowed*. Thus, de-Broglie waves of electrons in the orbit behave as a standing or stationary wave. The only necessary condition is that the circumference or length of the orbit should be whole number multiple of the de Broglie wavelength i.e.

$$2\pi r = n\lambda \qquad \ldots(ii)$$

From Eq. (i) and (ii), we have

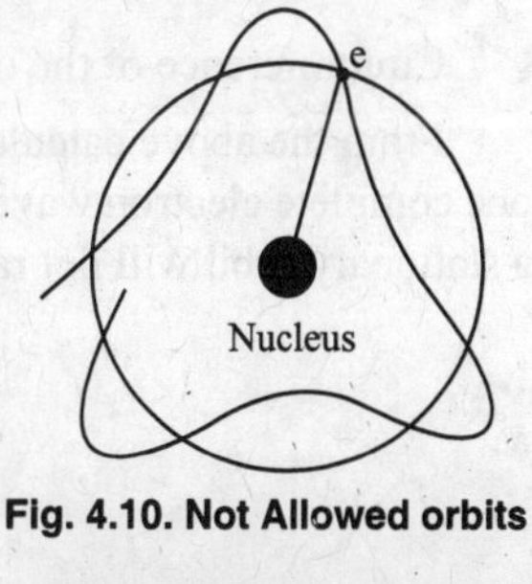

Fig. 4.10. Not Allowed orbits

$$2\pi r = n \cdot \frac{h}{mv} \qquad \left(\because \lambda = \frac{h}{mv}\right)$$

or $$mvr = n \cdot \frac{h}{2\pi} = n\hbar$$

$$\therefore \qquad L = n\hbar \qquad ...(iii)$$

where n is principle quantum number. Therefore,

if $n = 1$, circumference = λ

if $n = 2$, circumference = 2λ

if $n = 3$, circumference = 6λ and so on.

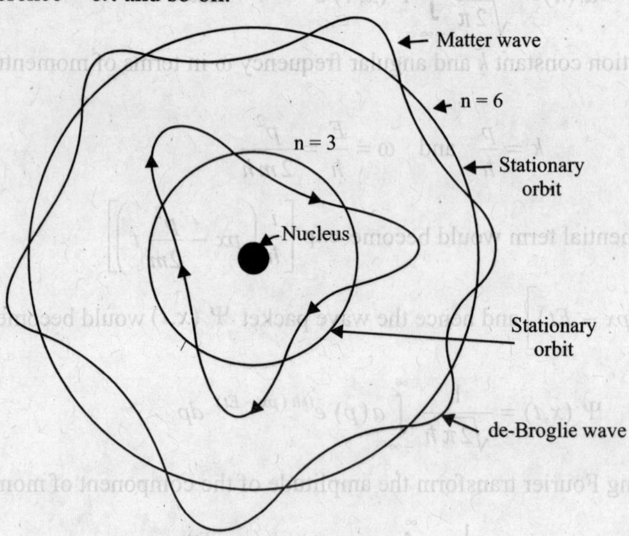

Fit. 4.11. Allowed orbits

Let us examine the wave like behaviour of electron in the orbit of hydrogen atom in ground state.

The de Broglie wavelength, $\lambda = \dfrac{h}{mv}$

where, $v = \dfrac{e}{\sqrt{4\pi\varepsilon_0 mr}}$ $\qquad \therefore \lambda = \dfrac{h}{me} \cdot \sqrt{4\pi\varepsilon_0 mr} = \dfrac{h}{e}\sqrt{\dfrac{4\pi\varepsilon_0 r}{m}}$

$$= \frac{6.62\times10^{-34}}{1.6\times10^{-19}} \times \sqrt{\frac{4\times3.14\times8.85\times10^{-12}\times5.3\times10^{-11}}{9.1\times10^{-31}}}$$

$$\lambda = 3.3 \times 10^{-10} \text{ m} = 3.3 \text{A}°$$

Here, $r = 0.53$ A°

Circumference of the orbit, $2\pi r = 2 \times 3.14 \times 0.53 \times 10^{-10}$ m $= 3.3 \times 10^{-10}$ m $= 3.3$ A°

From the above calculations, we conculde that the electron in a hydrogen atom corresponds to one complete electron wave joined on itself. This wave mechanical concept of electron revolving in a stationary orbit will not radiate energy. Hence it works as allowed orbit.

EXERCISE

LONG QUESTIONS

1. (*a*) Explain the concept of wave-particle dualism. What led de-Broglie to suggest that matter has wave characteristics? ***(G.N.D.U. 2001, M.D.U. 2000, Madurai 2003)***

(*b*) State the de-Broglie hypothesis of matter waves. Derive an expression for de-Broglie wavelength of matter particle in terms of kinetic energy and temperature.

(P.U. 2002, 2001, 2000; Luck. U. 2001, G.N.D.U. 2001, Bang. U. 2004, 2000; Pbi 2000, Nag. U. 2002, Madurai U. 2003)

2. What is de-Broglie hypothesis? Show that the wavelength λ associated with an electron of mass *m* and kinetic energy E is given by $\lambda = \dfrac{h}{\sqrt{2mE}}$ ***(Nag. U. 2011)***

3. Derive a formula expressing de-Broglie wavelength (in Angstrom units) of an electron in tems of potential difference V (in volts) through which it is accelerated.

(Pbi. U. 2004; Luck.U. Meerut. U. 2002; H.P.U.)

4. Why are wave properties of particles normally observed only when we study very small particles? ***(Pbi. U. 2002; H.P.U. 2002; 2001; P.U., 2005, 1990)***

OR

Why cannot we observe de-Broglie wavelength with a fast moving cricket ball? ***(P.U., 2001)***

5. Show that the wavelength of an electron of rest mass m_0 moving with a velocity v is given by

$$\lambda = \frac{h}{m_0 c}\frac{\left(1-\dfrac{v^2}{c^2}\right)}{\dfrac{v}{c}}.$$

(Meerut U. 2003)

6. (a) Describe with necessary theory Davisson and Germer experiment for establishing wave nature of the electron.

(Madurai U. 2003; Pbi. U. 2003; M.D.U. 2002; K.U. 2001; Gauhati U. 2000; Bang.U. 2000; Meerut U. 2007, 2006; Nagpur U. 2011, 2002)

(*b*) What is the effect of increasing the electron energy on the scattering angle in Davisson and Germer experiment?

7. Discuss the two experiments of electron diffraction which established the existence of matter waves. ***(Meerut U. 2001; Gauhati U. 2000)***

OR

Describe G.P. Thomson's experiment on electron diffraction and explain the important conclusion. ***(Meerut U. 2003)***

8. (*a*) Define de-Broglie phase velocity and particle velocity. Show that phase velocity exceeds the velocity of light as well as the particle velocity and de-Broglie waves show dispersion even in vacuum. ***(P.U. 2001, 2000)***

(*b*) Derive the relation between wave velocity and particle velocity. ***(K.U. 2001)***

9. (*a*) Define phase velocity, particle velocity and group velocity of a wave packet. Derive a relation between group velocity and phase velocity.

(*b*) Show that the particle velocity is equal to the group velocity of the wave packet.

(Pbi.U. 2004, 2003, 2002, 2001; G.N.D.U. 2008, 2006, 2004, 2003; Meerut U. 2003, 2000, K.U. 2002, 2001, P.U. 2005, 2001, 2000; M.D.U. 2007, 2001; H.P.U. 2001; Gharwal U. 2000)

10. What is a free particle ? Which parameters characterize it?

(Pbi. U. 2003, 2002; Nagpur U. 2011, 2002)

11. What do you mean by localized particle? *(Pbi. U. 2001)*

12. Consider the wave function representing the wave packet

$$\Psi(x,t) = \int_{-\infty}^{\infty} A(k)\, e^{i(kx-\omega t)}\, dk$$

and find the expression for amplitude and group velocity of the wave packet. *(P.U. 2003)*

13. Show that the rest mass of a photon is zero *(P.U. 2000)*

14. Find the phase and group velocity of an electron whose de-Broglie wavelength is 1.2A° (neglect relativity effect)

(Ans. Phase velocity = 3.03×10^6 ms^{-1}, group velocity = 6.06×10^6 ms^{-1})

SHORT QUESTIONS

1. State and prove the de-Broglie hypothesis.

2. Show that wavelength of matter wave is $\lambda = h/p$, where p is momentum of a particle.

3. If E is the K.E. of material particle, what will be its de Broglie wavelength?

4. Describe an experiment in short in support of de-Broglie hypothesis.

5. Explain Bohr's quantum rule using de-Broglie hypothesis.

6. Write a note on dual nature of matter.

7. Establish a relation between phase velocity and group velocity.

8. How is de-Broglie's hypothesis applied for the quantization of hydrogen atom? What is Bohr quantum condition.

9. Describe G.P. Thomson experiment on diffraction of electron beam. Explain the result.

10. Show that velocity of a particle is equal to the group velocity.

11. What is wave-packet? Explain.

12. What are matter waves? Explain with experimental evidence.

MULTIPLE CHOICE QUESTIONS

1. The wavelength λ associated with a particle of mass *m* moving with a velocity *v* is given by

(*a*) $\lambda = \dfrac{h}{mv}$ (*b*) $\lambda = \dfrac{hv}{m}$ (*c*) $\lambda = \dfrac{m}{hv}$ (*d*) $\lambda = \dfrac{mv}{h}$

2. Davisson and Germer experiment relates to

(*a*) Interference (*b*) Electron diffraction

(*c*) Polarisation (*d*) Fluorescence

3. The relation between de-Broglie phase velocity v_p and velocity of light *c* is

(*a*) $v_p > c$ (*b*) $v_p = c$ (*c*) $v_p < c$

4. The relation between de-Broglie wave group velocity v_g and particle velocity *v* is

(*a*) $v_g > v$ (*b*) $v_g = v$ (*c*) $v_g < v$

5. Which of the following is correct statement of Heisenberg's uncertainty principle?

(*a*) $\Delta p\, \Delta x \geq \frac{\hbar}{2}$ (*b*) $\Delta L\, \Delta \theta \geq \frac{\hbar}{2}$ (*c*) $\Delta E\, \Delta t \geq \frac{\hbar}{2}$

6. The wave length of de-Broglie waves associated by an electron accelerated with 100 V is

(*a*) 1Å (*b*) 1.5 Å

(*c*) 1.27 Å (*d*) 1.28 Å ***(Meerut U. 2003)***

ANSWERS

1. (*a*) **2.** (*b*) **3.** (*a*) **4.** (*b*)

5. all statements are correct. **6.** (*c*)

HEISENBERG'S UNCERTAINTY PRINCIPLE

5

Chapter

5.1 HEISENBERG'S UNCERTAINTY PRINCIPLE

The de-Broglie wavelength associated with a moving particle travelling with a uniform velocity v is given by

$$\lambda = \frac{h}{mv}$$

This is a monochromatic wave of *infinite* extent. The phase velocity v_p of such a monochromatic de-Broglie wave is given by

$$v_p = \nu\lambda = \frac{h\nu}{h/\lambda} = \frac{E}{p} = \frac{mc^2}{mv} = \frac{c^2}{v}$$

and is thus always greater than the velocity of light in vacuum. It is, therefore, theoretically impossible for a monochromatic de-Broglie wave train to transport a particle or carry energy. The phase velocity, merely represents the rate at which a given phase of a mono-chromatic wave train advances and is purely a mathematical concept.

Further, the stability of a material particle like an electron demands that it should be concentrated over a small region of space at any given instant of time. Thus, the mass and charge of a particle like the electron are localised entities whereas the de-Broglie wave with which we represent the moving particle is of infinite extent by virtue of its being monochromatic.

By taking a group of waves of nearly equal wavelengths centered round the de-Broglie wavelength $\lambda = \frac{h}{p}$ we obtain a *wave packet* the amplitude of which is different from zero only in a certain small region of space having dimensions of the particle. This small region of space is associated with the position of the particle and this wave packet travels with the velocity known as group velocity v_g which is equal to the particle velocity v.

Hence, instead of associating a single monochromatic de-Broglie wave with a particle we associate with it a wave packet consisting of a group of waves of nearly equal amplitude centred round de-Broglie wavelength $\lambda = \frac{h}{p}$ of the particle. The wave packet has the dimensions of the localised particle and travels with the same velocity as the particle.

The association of a group of waves with a moving particle means that the position of the particle at any instant of time cannot be specified with any desired degree of accuracy. All that we can say is that the particle is somewhere within the group of waves i.e. within a small region of space. Thus, a certain inaccuracy will always creep in ascertaining the position of a moving particle like the electron, proton etc. With a similar reasoning it can be argued that a certain amount of inaccuracy will creep in determining the velocity and hence the momentum of the particle.

Heisenberg's uncertainty principle states that in any simultaneous determination of the position and momentum of a particle, the product of the uncertainties is equal to, or greater than Planck's constant h, i.e. $\Delta x . \Delta p_x \geq h$

for motion along X-axis, where Δx is the fundamental error or uncertainty in the measurement of the position and Δp_x the fundamental error or uncertainty in the measurement of momentum (velocity).

A more accurate calculation gives

$$\Delta p_x \; \Delta x \geq \frac{1}{2} \frac{h}{2\pi} = \frac{1}{2} \hbar$$

According to classical ideas, it is possible for a particle to occupy a fixed position and have a definite momentum and we can predict exactly its position and momentum at any time later. According to uncertainty principle, it is not possible to determine accurately the simultaneous values of position and momentum of a particle at any time. If it is desired to reduce the error, in the determination of position *i.e.*, to reduce Δx, then this can only be done at the expense of accuracy in determining the momentum and the error involved in Δp will increase.

It should be born in mind that, uncertainties are not due to inperfections of the measuring instruments or way of measurements. These are inherent due to wave nature of the particle.

5.2 TIME-ENERGY UNCERTAINTY RELATION

Heisenberg's uncertainty principle is aplicable to all conjugate or complementary pairs of physical variables the product of which has the dimensions of Planck's constant i.e. action (energy × time).

The most common of these pairs are; position-linear momentum, time-energy, and angular momentum-angular displacement.

The time energy uncertainty principle states:

In any simultaneous determination of the time and energy of a particle, the product of the uncertainties is equal to or greater than Planck's constant h.

Thus, $\Delta E . \Delta t \geq h$

where ΔE is the uncertainty in the measurement of energy and Δt the corresponding uncertainty in the measurement of time.

Derivation from position momentum uncertainty principle. To obtain the time energy uncertainty principle from position momentum uncertainty relation consider the case of an electron of mass m moving with a velocity v so that its kinetic energy

$$E_k = \frac{1}{2} mv^2$$

Taking the mass of the electron to be constant, the uncertainty in the measurement of E_k is given by

$$\Delta E_k = \Delta \left(\frac{1}{2} mv^2 \right) = m\text{v} \; \Delta v = v \Delta p$$

Now the velocity $\qquad v = \dfrac{\Delta x}{\Delta t}$

$\therefore \qquad \Delta E_k = \dfrac{\Delta x}{\Delta t} \Delta p$

or $\qquad \Delta x . \Delta p = \Delta E_k . \Delta t \geq h$

More accurately $\qquad \Delta x . \Delta p = \Delta E_k . \Delta t \geq \dfrac{1}{2} . \dfrac{h}{2\pi} = \dfrac{1}{2} \hbar.$

5.3 ANGULAR MOMENTUM – ANGULAR POSITION UNCERTAINTY

To verify the uncertainty principle $\Delta L . \Delta\theta \geq \frac{\hbar}{2}$ consider a particle of mass m moving in a circle of radius R with a velocity v. As the particle moves in a circular path uncertainty principle will apply in a direction tangent to the circle.

$$\therefore \qquad \Delta p_s \; \Delta s = \hbar / 2$$

where s is the linear displacement measured along the circle and p_s the linear momentum.

Now angular displacement θ is related to the arc length s by

$$\theta = \frac{s}{R}$$

or

$$\Delta s = R \Delta\theta$$

The angular momentum L is related to the linear momentum by

$$L = m v R = p_s R$$

$$\therefore \qquad \Delta p_s = \frac{\Delta L}{R}$$

Hence

$$\Delta p_s \; \Delta s = \frac{\Delta L}{R} R \Delta\theta = \Delta L \Delta\theta = \frac{\hbar}{2}.$$

5.4 HEISENBERG'S UNCERTAINTY PRINCIPLE FROM DE-BROGLIE WAVE CONCEPT

Heisenberg's uncertainty principle can be proved with the help of de-Broglie's wave concept that a particle in motion can be taken as a group of waves, the group velocity being equal to the particle velocity.

As a simple case, consider a group of waves consisting of only two components of equal amplitude and having frequencies ω_1 and ω_2 differing by a small amount and represented by the equations

$$\psi_1 = A \cos(\omega_1 t - k_1 x)$$

and

$$\psi_2 = A \cos(\omega_2 t - k_2 x)$$

where $\frac{\omega_1}{k_1}$ and $\frac{\omega_2}{k_2}$ represent their respective phase velocities

$$\text{as} \left(\frac{\omega}{k} = \frac{2\pi\nu}{\frac{2\pi}{\lambda}} = \nu\lambda = v_p \right)$$

The resultant amplitude is given by

$$\psi = \psi_1 + \psi_2 = 2A \cos(\omega t - kx) \cos\left(\frac{\Delta\omega}{2} t - \frac{\Delta k}{2} x \right)$$

where

$$\omega = \frac{\omega_1 + \omega_2}{2} \quad ; \quad k = \frac{k_1 + k_2}{2}$$

$$\Delta\omega = \omega_1 - \omega_2 \quad ; \quad \Delta k = k_1 - k_2 \qquad \text{[For proof, See 4.5]}$$

The resultant wave is shown in Fig. 4.1 where the loop formed travels with the group velocity

$$v_g = \frac{\Delta\omega}{\Delta k}$$

As the group velocity of the de-Broglie wave group associated with the moving particle is equal to the particle velocity, the loop formed is equivalent to the position of the particle. Thus, the position of the particle cannot be given with certainty. It will lie somewhere between the two consecutive nodes. In other words, the fundamental error in the measurement of the position of the particle is equal to the distance between two consecutive nodes.

The condition for the formation of a node is given by

$$\cos\left(\frac{\Delta\omega}{2}t - \frac{\Delta k}{2}x\right) = 0$$

$$\therefore \quad \frac{\Delta\omega}{2}t - \frac{\Delta k}{2}x = \frac{\pi}{2}, \frac{3\pi}{2} \ldots (2n+1)\pi/2$$

Thus, if x_1 and x_2 are the positions of two consecutive nodes

$$\frac{\Delta\omega}{2}t - \frac{\Delta k}{2}x_1 = (2n+1)\pi/2$$

and
$$\frac{\Delta\omega}{2}t - \frac{\Delta k}{2}x_2 = (2n+3)\pi/2$$

$$\therefore \quad \frac{\Delta k}{2}(x_1 - x_2) = \pi$$

or error in the measurement of position of the particle $(x_1 - x_2)$

$$= \Delta x = \frac{2\pi}{\Delta k}$$

Also
$$k = \frac{2\pi}{\lambda} = \frac{2\pi p}{h}$$

$$\therefore \quad \Delta k = \frac{2\pi}{h}\Delta p$$

Hence
$$\Delta x = \frac{2\pi h}{2\pi\,\Delta p} = \frac{h}{\Delta p}$$

or
$$\Delta x . \Delta p = h$$

where Δp is the fundamental error in the measurement of momentum p.

The above treatment is very elementary. If we consider a group consisting of a very large number of waves of continuously varying frequencies we get the product of the fundamental errors in the measurement as

$$\Delta p\,\Delta x \geq \frac{1}{2}\frac{h}{2\pi} = \frac{1}{2}\hbar$$

This value is used for more accurate calculations. The constant $\hbar = h/2\pi = 1.054 \times 10^{-34}$ Js

$= 1.054 \times 10^{-27}$ erg. sec

$= 6.58 \times 10^{-16}$ eV sec.

5.5 ILLUSTRATION OF UNCERTAINTY PRINCIPLE

The uncertainty principle can be illustrated by a few imaginary or thought experiments which are logically performed on the basis of known principles. One of such experiments is Heisenberg's γ-ray microscope.

Heisenberg's hypothetical gamma (γ) ray microscope:

The principle can be illustrated by considering the inevitable error in finding the position of an electron in a hypothetical experiment using a γ-ray microscope. This method was suggested by Bohr and fully developed by Heisenberg.

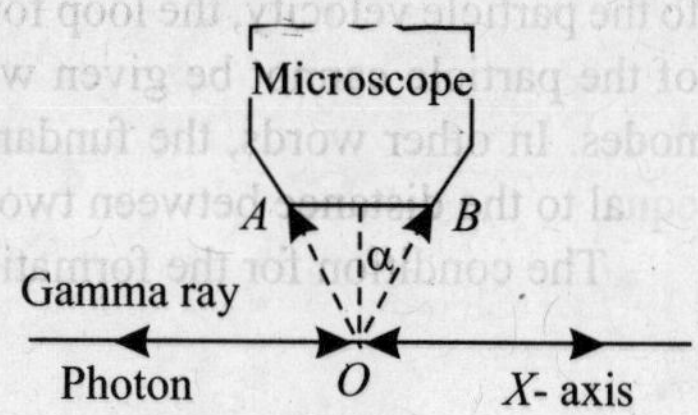

Fig. 5.1

Let O be the electron, λ the wavelength of γ-rays and α the semi-vertical angle of the cone of rays that enter the microscope objective from this object. The resolving power of the microscope is given by

$$\Delta x = \frac{\lambda}{2 \sin \alpha}$$

where Δx is the shortest distance between two points in the field of view which can be distinguished as separate. In other words, if the position of the electron O changes by an amount equal to or less than Δx, the microscope will not be able to detect it or the uncertainty in the measurement of the position of electron is Δx.

The electron is seen by the γ-radiation scattered by it into the microscope. During this scattering, the electron will recoil and there will be a change in its momentum. In order that this change may be as small as possible consider a single γ-ray photon incident on it along the X-axis, then a scattered photon of wavelength λ will enter the objective anywhere between OA and OB. The momentum of the scattered photon is $\frac{h}{\lambda}$. If it enters the objective along OA, its component along the X-axis is $\frac{h}{\lambda}$ sin α. Hence momentum imparted to the electron in the X-direction

$$= \frac{h}{\lambda'} - \frac{h}{\lambda} \sin \alpha \qquad ...(i)$$

where λ' is the wavelength of the γ-radiation before scattering.

If the scattered photon enters the microscope along OB, then its component along the X-axis will be $-\frac{h}{\lambda} \sin \alpha$.

Hence momentum imparted to the electron in the X-direction

$$= \frac{h}{\lambda'} - \left(-\frac{h}{\lambda} \sin \alpha \right)$$

$$= \frac{h}{\lambda'} + \frac{h}{\lambda} \sin \alpha \qquad ...(ii)$$

The momentum given to the electron can, therefore, have any value between those given by equations (*i*) and (*ii*). The uncertainty or the error involved in the measurement of the momentum of the electron, therefore, is given by

$$\Delta p_x = \left(\frac{h}{\lambda'} + \frac{h}{\lambda} \sin \alpha \right) - \left(\frac{h}{\lambda'} - \frac{h}{\lambda} \sin \alpha \right) = \frac{2h}{\lambda} \sin \alpha$$

$$\therefore \qquad \Delta x.\Delta p_x = \frac{\lambda}{2 \sin \alpha} \cdot \frac{2h}{\lambda} \sin \alpha = h$$

However, a more rigorous treatment gives $\Delta x.\Delta p_x = \hbar / 2$.

5.6 SIGNIFICANT FOR ONLY MICROSCOPIC PARTICLES

The uncertainty principle gives the minimum absolute uncertainty in the measurement of conjugate quantities like (x, p_x), (y, p_y), (z, p_z), (t, E) and (θ, L). The actual uncertainties are much larger and include contributions on account of errors due to measuring instruments, conditions of performance of experiment and certain other causes. The uncertainty principle is applicable to all objects but its consequences are significant only for such minute particles of microscopic size as electrons, protons, neutrons, etc., because in the case of *macroscopic* (large size) objects the uncertainties in the measurement of these quantities are so large that their products exceed the value of Planck's constant h by many orders of magnitude. Thus due to the smallness of Planck's constant the uncertainty principle becomes meaningless for measurements in connection with macroscopic bodies.

5.7 HEISENBERG'S UNCERTAINTY PRINCIPLE BY DIFFRACTION OF ELECTRONS

Consider a narrow slit AB of width Δx as shown in Fig. 5.2. Let a beam of electrons fall on this slit. After passing through the slit the electron beam spreads out i.e. it suffers diffraction. The first minimum of the diffraction pattern is given by the relation $d \sin \theta = n \lambda$ where $d = \Delta x$, $n = 1$, and θ is the angle through which the electron beam is diffracted

$$\therefore \qquad \Delta x \sin \theta = \lambda \qquad \text{...}(i)$$

It is very difficult to locate the exact position of the slit at which the beam of electrons enters in the slit. Therefore, the uncertainty in the measurement of the position of the electron beam in the slit in a direction perpendicular to the direction of incident beam is equal to the width of the slit

$$= \Delta x = \frac{\lambda}{\sin \theta}. \qquad \text{...}(ii)$$

Let $\vec{p}$ be the momentum of the electron. If the electron moves in the direction EC after diffraction then momentum of the electron in a direction perpendicular to the initial direction of the electron *i.e.* along $OC = \vec{p} \sin \theta$. If the electron after diffraction goes in the direction ED, then momentum of the electron in the direction $OD = -\vec{p} \sin \theta$. As the electron may be any where within the diffraction pattern from angle $-\theta$ to $+\theta$, the perpendicular component of the momentum of the electron may be any where between $-\vec{p} \sin \theta$ to $+\vec{p} \sin \theta$.

$\therefore$ Uncertainty in the measurement of momentum of the electron

$$\Delta p = \vec{p} \sin \theta - (-\vec{p} \sin \theta) = 2 \vec{p} \sin \theta$$

$$= 2 p \sin \theta$$

But $$\lambda = \frac{h}{p} \text{ or } p = \frac{h}{\lambda} \therefore \Delta p = 2 \frac{h}{\lambda} \sin \theta \qquad \text{...}(iii)$$

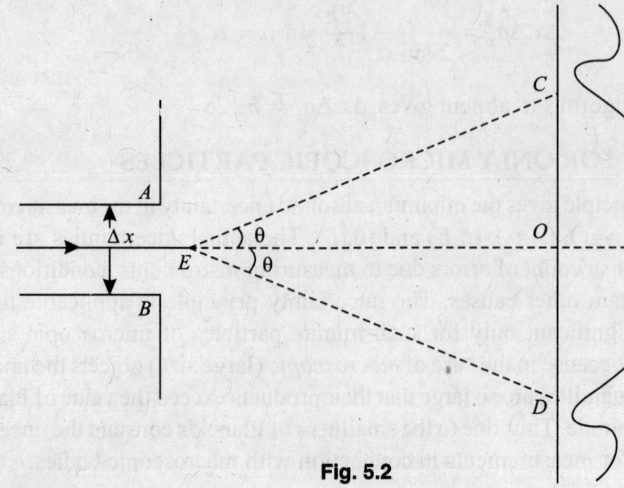

Fig. 5.2

From (*ii*) and (*iii*), we have

$$\Delta x . \Delta p = \frac{\lambda}{\sin \theta} . \frac{2h}{\lambda} \sin \theta = 2h$$

which is in agreement with Heisenberg's uncertainty principle.

SOLVED EXAMPLES

Example 5.1. *The position and momentum of 0.5 keV electron are simultaneously determined. If its position is located within 0.2 nm, what is the percentage uncertainty in its momentum?*

Solution. Energy of the electron

$$E = 0.5 \times 10^{3} \text{ eV} = 0.5 \times 10^{+3} \times 1.6 \times 10^{-19} \text{ J}$$

$$= 8 \times 10^{-17} \text{ J}$$

As it is a non-relativistic case, momentum of the electron

$$p = \sqrt{2mE} = (2 \times 9 \times 10^{-31} \times 8 \times 10^{-17})^{1/2}$$

$$= 1.2 \times 10^{-23} \text{ kg ms}^{-1}$$

Uncertainty in position $\Delta x = 0.2 \times 10^{-9}$ m $= 2 \times 10^{-10}$ m

Uncertainty in momentum Δ_p is given by

$$\Delta p . \Delta x = \frac{\hbar}{2}$$

$$\therefore \qquad \Delta p = \frac{\hbar}{2\Delta x} = \frac{h}{2 \times 2\pi \times \Delta x} = \frac{7 \times 6.6 \times 10^{-34}}{2 \times 2 \times 22 \times 2 \times 10^{-10}}$$

$$= 2.6 \times 10^{-25} \text{ kg ms}^{-1}$$

$\therefore$ Percentage uncertainty in momentum $= \dfrac{\Delta p}{p} \times 100$

$$= \frac{2.6 \times 10^{-25}}{1.2 \times 10^{-23}} \times 100 = 2.2\%.$$

Example 5.2. *The position and momentum of a 1 keV electron are simultaneously measured. If the position is located within 1 Å, what is the percentage of uncertainty in momentum?*

(Gharwal U. 2000)

Solution. Momentum $p = \sqrt{2mE} = \sqrt{2 \times 9 \times 10^{-31} \times 1000 \times 1.6 \times 10^{-19}}$

$$= 16.97 \times 10^{-24} \text{ kg ms}^{-1}$$

$$\Delta x = 1 \times 10^{-10} \text{ } m; \therefore \Delta p = \frac{\hbar}{2\,\Delta x} = \frac{6.6 \times 10^{-34}}{2 \times 2 \times 3.142 \times 10^{-10}}$$

$$= 5.2 \times 10^{-25} \text{ kg ms}^{-1}$$

$$\text{Percentage uncertainty in momentum} = \frac{5.2 \times 10^{-25}}{16.97 \times 10^{-24}} \times 100 = 3\%$$

Example 5.3. *Calculate the uncertainty in measurement of momentum of an electron if the uncertainty in locating it is 1Å.*

Solution. Uncertainty in locating the electron $\Delta x = 1\text{Å} = 10^{-10}$ m

$\therefore$ Uncertainty in measurement of momentum $\Delta p = \dfrac{\hbar}{2\,\Delta x}$

$$= \frac{6.62 \times 10^{-34}}{2 \times 2 \times 3.142 \times 10^{-10}} = 5.2 \times 10^{-25} \text{ kg ms}^{-1}.$$

Example 5.4. *An electron has a momentum 5.4×10^{-26} kg ms^{-1} with an accuracy of 0.05%. Find the minimum uncertainty in the location of the electron.* ***(P.U. 2002)***

Solution. Momentum of the electron $= 5.4 \times 10^{-26}$ kg ms^{-1}

$\therefore$ Uncertainty in the measurement of momentum

$$\Delta p = \frac{5.4 \times 10^{-26} \times .05}{100} = 27 \times 10^{-30} \text{ kg ms}^{-1}$$

Hence uncertainty in the location of electron $\Delta x = \dfrac{\hbar}{2\,\Delta p}$

$$= \frac{6.62 \times 10^{-34}}{2 \times 2 \times 3.142 \times 27 \times 10^{-30}} = 1.95 \times 10^{-6} \text{ } m.$$

Example 5.5. *An experiment designed to know the position of a proton, can result in changing the proton's kinetic energy by at the most 1 KeV. Find the maximum accuracy in the position of proton.*

(P.U. 2008)

Solution. Maximum change in kinetic energy $\Delta E = 1$ KeV $= 1000$ eV $= 1000 \times 1.6 \times 10^{-19}$ J

Mass of proton $= 1.67 \times 10^{-27}$ kg

$\therefore$ Maximum change in momentum $\Delta p = \sqrt{2\, m\, \Delta E}$

$$= [2 \times 1.67 \times 10^{-27} \times 1000 \times 1.6 \times 10^{-19}]^{1/2}$$

$$= [2 \times 1.67 \times 1.6 \times 10]^{1/2}\, 10^{-12} = 7.3 \times 10^{-22} \text{ kg ms}^{-1}$$

But $\Delta p\, \Delta x = \dfrac{\hbar}{2} = \dfrac{1}{2}\dfrac{h}{2\pi} = \dfrac{h}{4\pi}$ where Δx is the maximum accuracy in the position of the proton

$$\therefore \quad \Delta x = \frac{h}{4\pi\Delta p} = \frac{6.62 \times 10^{-34}}{4 \times 3.142 \times 7.3 \times 10^{-22}} = 7.2 \times 10^{-14} \text{ m.}$$

Example 5.6. *An electron has speed of 6.6×10^4 m/s with an accuracy of 0.01%. Calculate uncertainty in position of electron. Given mass of electron = 9.1×10^{-31} kg; Planck's constant h = 6.6×10^{-34} Js.* **(Meerut U. 2001)**

Solution. Uncertainty in the measurement of momentum

$$\Delta p = \Delta\,(mv) = m\,\Delta v = 9.1\times10^{-31}\times6.6\times10^{4}\times\frac{0.01}{100}$$

$$= 60\times10^{-31} = 6\times10^{-30}\ \text{kg ms}^{-1}$$

∴ Uncertainty in the measurement of position $\Delta x = \dfrac{\hbar}{2\Delta p}$

$$= \frac{6.6\times10^{-34}}{2\times2\times\frac{22}{7}\times6\times10^{-30}} = \frac{6.6\times7\times10^{-4}}{2\times2\times22\times6}$$

$$= 8.75 \times 10^{-6}\ \text{m}.$$

Example 5.7. *An electron has a speed of 1.05×10^4 ms^{-1} within an accuracy of 0.01%. Calculate uncertainty in position of electron. Given $\hbar$ = 1.05×10^{-34} Js and m = 9×10^{-31} kg.* **(M.D.U. 2007)**

Solution. Uncertainty in velocity $\Delta v = 1.05 \times 10^{+4} \times \dfrac{.01}{100} = 1.05\ \text{ms}^{-1}$

$$\therefore \text{ Uncertainty in position} = \frac{\hbar}{2\Delta p} = \frac{\hbar}{2m\,\Delta v} = \frac{1.05\times10^{-34}}{2\times9\times10^{-31}\times1.05} = 5.5\times10^{-5}\ \text{m}$$

Example 5.8. *An electron of mass 9.1×10^{-31} kg has a speed of 1 km s^{-1} with an accuracy of 0.05%. Calculate the uncertainty with which the position of the electron can be located.* **(P.U. 2001)**

Solution. $\Delta p = m\Delta v = 9.1\times10^{-31}\times1000\times\dfrac{0.05}{100} = 4.55\times10^{-31}\ \text{kg ms}^{-1}$

$$\therefore \quad \Delta x = \frac{\hbar}{2\Delta p} = \frac{6.62\times10^{-34}}{2\times2\times3.142\times4.55\times10^{-31}} = 1.15\times10^{-4}\ m.$$

Example 5.9. *A particle of mass 10^{-6} gm has a speed of 1 m/s. If the speed is uncertain by 0.01% what is the minimum uncertainty in the position of the particle.* **(Pbi. U. 1993)**

Solution. $\Delta p = m\,\Delta v = 10^{-9}\times1\times\dfrac{01}{100} = 10^{-13}\ \text{kg ms}^{-1}$

$$\Delta x = \frac{\hbar}{2\Delta p} = \frac{6.62\times10^{-34}}{2\times2\times3.142\times10^{-13}} = 0.52\times10^{-21}\ m = 5.2\times10^{-22}\ m.$$

Example 5.10. *Calculate the uncertainty in the momentum and velocity of an electron confined in a box of length 1Å.* **(Meerut U. 2000; Cal. U. 1992)**

Solution. Uncertainty in the position of electron = length of the box Δx = 1 Å = 1×10^{-10} m

∴ Uncertainty in the momentum of the electron

$$\Delta p = \frac{\hbar}{2\Delta x} = \frac{h}{4\pi\Delta x} = \frac{6.62\times10^{-34}}{4\times3.142\times10^{-10}} = 5.27\times10^{-25}\ \text{kg ms}^{-1}$$

Now $\Delta p = \Delta (mv) = m\Delta v$

$\therefore$ Uncertainty in velocity $\Delta v = \dfrac{\Delta p}{m} = \dfrac{5.27 \times 10^{-25}}{9.1 \times 10^{-31}} = 5.8 \times 10^5$

Example 5.11. *Compare the uncertainties in velocity of a proton and an electron contained in a 20 Å box.* **(Pbi. U. 2001)**

Solution. Uncertainty in the position of the particle (proton or electron) = length of the box, $\Delta x = 20$ Å $= 20 \times 10^{-10}$ m

Uncertainty in the momentum of the particle

$$\Delta p = \frac{\hbar}{2\Delta x} = \frac{h}{4\pi\Delta x}$$

As Δp does not depend upon mass or charge of the particle, Δp (proton) = Δp (electron)

Now $$\Delta p = \Delta (mv) = m\Delta v$$

$\therefore$ Uncertainty in the velocity of the particle $\Delta v = \dfrac{\Delta p}{m}$

Hence uncertainty in velocity of proton Δv (proton) $= \dfrac{\Delta p}{m_p}$

and uncertainty in velocity of electron Δv (electron) $= \dfrac{\Delta p}{m_e}$

$$\therefore \quad \frac{\Delta v \text{ (proton)}}{\Delta v \text{ (electron)}} = \frac{\Delta p}{m_p} \times \frac{m_e}{\Delta p} = \frac{m_e}{m_p}$$

$$= \frac{9.1 \times 10^{-31}}{1.67 \times 10^{-27}} = 5.45 \times 10^{-4}.$$

Example 5.12. *The electron in hydrogen atom may be thought of as confined to a nucleus of radius = 5 × 10⁻¹¹ metre. Calculate the minimum-uncertainty in the momentum of the electron. Also calculate the minimum kinetic energy of the electron.*

Given $m_e = 9 \times 10^{-31}$ kg; 1 eV $= 1.6 \times 10^{-19}$ J, $h = 6.62 \times 10^{-34}$ Js.

Solution. According to Heisenberg's uncertainty principle

$$\Delta x . \Delta p \geq \frac{\hbar}{2} \geq \frac{h}{4\pi}$$

where Δx is the uncertainty in position and Δp_x the minimum uncertainty in the momentum of a particle. As the electron is confined to a radius of 5×10^{-11} metre, the uncertainty in its position $\Delta x = 5 \times 10^{-11}$ metre

$$\therefore \quad \Delta p_x \geq \frac{h}{4\pi\,\Delta x} \geq \frac{6.62 \times 10^{-34}}{4\pi \times 5 \times 10^{-11}} = 1.054 \times 10^{-24} \text{ kg.ms}^{-1}$$

Now kinetic energy $E = \dfrac{p^2}{2m}$

$$\Delta E = \frac{\Delta p^2}{2m} \geq \frac{(1.054 \times 10^{-24})^2}{2 \times 9 \times 10^{-31}} \geq 6.142 \times 10^{-19} \text{ J}$$

$$\geq \frac{6.142 \times 10^{-19}}{1.6 \times 10^{-19}} = 3.84 \text{ eV}.$$

Example 5.13. *Life time of a nucleus in the excited state is 10^{-12} s. Calculate the probable uncertainty in energy and frequency of a γ-ray photon emitted by it.* ***(K.U. 2003)***

Solution. According to Heisenberg's time energy uncertainty principle

$$\Delta E \,.\, \Delta t = \frac{h}{2\pi} = \hbar$$

As the life time of a nucleus in the excited state is 10^{-12} s, we have $\Delta t = 10^{-12}$ s

$$\therefore \quad \Delta E = \frac{h}{2\pi\, \Delta t} \geq \frac{6.62 \times 10^{-34}}{2\pi \times 10^{-12}} = 1.054 \times 10^{-22} \text{ J}$$

The frequency of γ-ray photon is uncertain by an amount

$$\Delta\nu = \frac{\Delta E}{h} = \frac{1.054 \times 10^{-22}}{6.62 \times 10^{-34}} = 1.59 \times 10^{11} \text{ Hz.}$$

Example 5.14. *The speed of a bullet of mass 50 gm is measured to be 300 m/s with an uncertainty of 0.01%. With what accuracy can we locate the position of the bullet if it is measured simultaneously with its speed.*

Solution. Uncertainty in the measurement of momentum

$$\Delta p \geq \Delta\,(mv) \geq m\,\Delta\, v \geq \frac{50}{1000} \times 300 \times \frac{0.01}{100}$$

$$\geq 15 \times 10^{-4} \text{ kg ms}^{-1}$$

$\therefore$ Uncertainty in the measurement of position $\Delta x \geq \dfrac{h}{\Delta p}$

$$\geq \frac{6.62 \times 10^{-34}}{15 \times 10^{-4}} \geq 4.41 \times 10^{-31}\ m$$

Note. If we take $\Delta x \cdot \Delta p \geq \hbar = \dfrac{h}{2\pi}$, then

$$\Delta x \geq 7 \times 10^{-32} \text{ m}$$

5.8 APPLICATIONS OF UNCERTAINTY PRINCIPLE

5.8.1 Non-existence of Electrons in Nucleus

If a particular type of particle is to exist within the nucleus then the uncertainty in its *position* in the nucleus should not exceed the radius of the nucleus. The radius of a typical atomic nucleus is of the order of 10^{-14} m. Hence if the electron is to exist within the nucleus the uncertainty in its position is given by

$$\Delta x \geq 10^{-14} \text{ m}$$

According to Heisenberg's uncertainty principle,

$\Delta x . \Delta p_x = \dfrac{\hbar}{2}$ where Δp_x is the uncertainty in the x-component of momentum or the momentum itself and $\hbar = \dfrac{h}{2\pi}$

∴ Uncertainty in the momentum of the electron if it is to exist in the nucleus

$$\Delta p_x \geq \frac{\hbar}{2\Delta x} = \frac{6.6 \times 10^{-34}}{2 \times 2\pi \times 10^{-14}} = 5.3 \times 10^{-21} \text{ kg ms}^{-1}$$

This means that the total momentum of the *electron* in the nucleus must be at least of the order of 5.3×10^{-21} kg ms^{-1}.

The rest mass of the electron = 9.1×10^{-31} kg.

Hence the above value of the momentum is *relativistic* as a non-relativistic relation for momentum gives a value of $v > c$.

According to the theory of relativity, the total relativistic energy of the electron

$$E = [m_0^2 c^4 + p^2 c^2]^{1/2}$$

$$= [(9.1 \times 10^{-31})^2 \times (3 \times 10^8)^4 + (5.3 \times 10^{-21})^2 \times (3 \times 10^8)^2]^{1/2}$$

$$= 3 \times 10^8 \, [9.1 \times 9.1 \times 9 \times 10^{-46} + 5.3 \times 5.3 \times 10^{-42}]^{1/2}$$

$$= 16 \times 10^{-13} \text{ Joule} = \frac{16 \times 10^{-13}}{1.6 \times 10^{-19}} = 10 \text{ MeV}.$$

Thus, the kinetic energy of a free electron confined within a nucleus should have a minimum value of 10 MeV. The rest mass energy of an electron is only about 0.5 MeV and experimentally, the highest value of the kinetic energy of a β-particle emitted spontaneously by a nucleus is found to be about 4 MeV. Since this is much less than the calculated value of the minimum kinetic energy which the electron, if it exists inside the nucleus, should possess, we come to the conclusion that electrons cannot be constituents of the atomic nucleus.

Other reasons for electrons not constituents of the nucleus. It has been observed that electrons are ejected from the nucleus as β-particles. This lead to the conclusion that electrons were constituents of the nucleus. But there are a number of reasons to believe that *electrons are not present within the nucleus* and the ejection of a β-particle from the nucleus is due to the conversion of a neutron into a proton within the nucleus.

The arguments against the presence of an electron in the nucleus are:

(*i*) Taking the radius of the nucleus as 10^{-14} m it can be proved on the basis of Heisenberg's uncertainty principle that the minimum energy the electron must possess to be within the nucleus is about 10 MeV.

But the electrons emitted during β-decay do not have energy more than 2 or 3 MeV. An electron having energy of this order cannot be a constituent of the nucleus.

(*ii*) The spin angular momentum of the proton as well as the electron is $\frac{1}{2}\hbar$ as both these are Fermi particles. Thus nucleus with *even* number of protons + electrons will have an *integral spin* and the nuclei with *odd* number of protons + electrons will have *half integral spin.* This rule is not obeyed by nuclei. For example, deuteron which is a heavier isotope of hydrogen with atomic number (charge) 1, and mass number 2, if composed of protons and electrons must have 2 protons and 1 electron giving a nuclear spin of $\frac{1}{2}$ or $\frac{3}{2}$ depending upon the orientation of the particles.

The observed spin of the deuteron is, however, 1 from which it is again seen that the electrons are not the constituents of the nucleus.

(*iii*) The spin magnetic moment of the proton is about 0.15% of that of the electron and thus almost negligible. If the electrons exist in the nucleus the spin magnetic moment of the nucleus should be nearly the same as that of the electron. But the observed magnetic moment of the nuclei are comparable to that of the proton and are thus much smaller than the spin magnetic moment of the electron. This again shows that electrons are not the constituents of the nucleus.

(*iv*) On the basis of the forces between nuclear particles, the binding energy per nuclear particle comes out to be of the order of about 8 MeV. If the elctrons exist within the nucleus, these nuclear electrons should interact with protons very strongly.

The orbital electrons, on the other hand, only interact electrically with the proton which is comparatively a weaker interaction. Even when fast electrons are scattered by nuclei, these behave in such a way as if these were acted upon only by electric forces. On the contrary when protons are scattered by nuclei, these show departure from electric forces, which is supposed to be due to strong nuclear forces. It is not possible for the electron to behave in two different ways, i.e. for one-half to have a strong interaction and for the other half to have only weak interaction. This again indicates that the electrons are not the constituents of the nucleus.

5.8.2 Co-Existence of Protons, Neutrons and α-particles within the nucleus

If a particular type of particle is to exist within the nucleus, then the uncertainty in its position in the nucleus should not exceed the radius of the nucleus. The radius of the typical atomic nucleus is of the order of 10^{-14} m. Hence if a proton, or a neutron or an α-particle is to exist within the nucleus, the uncertainty in their position is given by

$$\Delta x \geq 10^{-14} \text{ m}$$

According to Heisenberg's uncertainty principle

$$\Delta x \, . \, \Delta p_x \approx \frac{\hbar}{2}$$

where Δp_x is the uncertainty in the x-component of momentum or the momentum itself and $\hbar = h/2\pi$

∴ Uncertainty in the momentum of the proton, or neutron or α-particle, if it is to exist in the nucleus

$$\Delta p_x \geq \frac{\hbar}{2\Delta x} = \frac{6.6 \times 10^{-34}}{2 \times 2\pi \times 10^{-34}} = 5.3 \times 10^{-21} \text{ kg ms}^{-1}$$

This means that the total momentum of the proton or neutron or α-particle in the nucleus must be at least of the order of 5.3×10^{-21} kg ms^{-1}

The rest mass of the proton or the neutron is of the order of $m_0 = 1.67 \times 10^{-27}$ kg. Hence the value of momentum 5.3×10^{-21} kg ms^{-1} as calculated above is *non-relativistic* as the value of v comes out to be

$$\frac{5.3 \times 10^{-21}}{1.67 \times 10^{-27}} = 3.1 \times 10^{6} \text{ ms}^{-1}$$

The corresponding value of kinetic energy is given by

$$E_k = \frac{p^2}{2m_0} = \frac{(5.3 \times 10^{-21})^2}{2 \times 1.67 \times 10^{-27}} = 8.4 \times 10^{-15} \text{ Joule}$$

$$= \frac{8.4 \times 10^{-15}}{1.6 \times 10^{-19}} = 52 \text{ keV}$$

The rest mass of α-particle is nearly four times the proton mass. Thus, the α-particle should have a minimum kinetic energy of 13 keV. Since the energy carried by the protons or neutrons emitted by the nuclei are greater than 52 keV and for α-particles more than 13 keV, these particles can exist in the nuclei.

5.8.3. The Zero Point Energy

Consider a particle confined to move under the influence of a potential of linear dimensions *l*. Since the particle can exist *any where* in this region, the uncertainty in its position $\Delta x = l/2$

According to uncertainty principle Δp_x if is the uncertainty in the *x*-component of its momentum, then

$$\Delta p_x . \Delta x = \frac{\hbar}{2}$$

or

$$\Delta p_x = \frac{\hbar}{2\,\Delta x} = \frac{\hbar}{l}$$

Suppose the minimum value of the momentum of the particle is equal to its uncertainty Δp_x, then

$$p_x = \Delta p_x = \frac{\hbar}{l}$$

If m_0 is the rest mass of the particle, then the lowest value of kinetic energy in a potential of finite dimension

$$E = \frac{p^2}{2\,m_0} = \frac{p_x^2}{2\,m_0} = \frac{\hbar^2}{2m_0\,l^2}$$

This is the minimum energy possessed by a particle confined to a box of finite length *l*. *Thus we conclude that a particle confined to a region of space cannot have zero kinetic energy.*

Further, a physical system is in its lowest energy state when its temperature is close to absolute zero. Therefore, it is clear that even at 0K, the kinetic energy of a system or a particle will have some *finite value* and not zero. This energy is called *zero point energy*. In other words a particle confined to a finite dimensional potential is never at rest, not even at 0 K because it means $p_x = 0$ and $\Delta p_x = 0$, which is not possible. Such a motion is called *zero point motion*.

Some typical examples of zero point energy are (*i*) particle in a box and (*ii*) a harmonic oscillator. The typical behaviour of liquid helium at temperatures close to 0 K is also due to zero point energy.

5.8.4. Binding Energy of an Electron in an Atom

Consider an electron revolving round the nucleus of an atom in an orbit of radius *r*. As the electron is inside the atom, it is located within a distance *r*. The uncertainty in the position of the electron Δx is equal to the radius of the orbit.

$$\therefore \qquad \Delta x = r$$

According to uncertainty principle $\Delta p\,\Delta x = \hbar/2$

$$\therefore \qquad \Delta p = \frac{\hbar}{2r}$$

Hence minimum value of the momentum of the electron in its orbit

$$p = \Delta p = \hbar/2r$$

Taking *r* to be of the order of 10^{-10} m, the value of

$$p = \frac{\hbar}{2r} = \frac{h}{4\pi r} = \frac{6.6 \times 10^{-34}}{4\pi \times 10^{-10}} = 5.2 \times 10^{-25} \text{ kg ms}^{-1}$$

As this momentum is non-relativistic, the kinetic energy of the electron

$$K = \frac{p^2}{2m} = \frac{(5.2 \times 10^{-25})^2}{2 \times 9.1 \times 10^{-31}} = 1.5 \times 10^{-19} \text{ J} = 1 \text{ eV (App.)}$$

The potential energy of the electron in the field of the nucleus with atomic number Z,

$$V = \frac{-Ze^2}{4\pi\varepsilon_0 r} = -Z\frac{(1.6 \times 10^{-19})^2 \times 9 \times 10^9}{10^{-10}}$$

$$= -Z \times 14.4\text{eV}$$

$\therefore$ Total energy $E = K + V = 1 - 14.4$ ZeV

For hydrogen $Z = 1$

$\therefore$ $E = 1 - 14.4 = -13.4$ eV

For helium $Z = 2$

$\therefore$ $E = 1 - 28.8 = -27.8$ eV.

The two values agree closely with the binding energy of the outer most electron in Hydrogen (–13.6 eV) and helium (– 24.6 eV) respectively.

Our precise calculations give the binding energy of an electron in a hydrogen atom to be 13.4 eV. Thus roughly speaking it is of the order of 15 eV.

5.8.5. Size of the Elementary Cell in Phase Space

In quantum statistical Physics, the state of a microsystem is defined by *six* variables-three components of position vector and three components of momentum vector–in a six dimensional position momentum space, known as phase space. If Δx and Δp_x be the uncertainty in the *x*-component of position and momentum respectively, then

$$\Delta x . \Delta p_x = \hbar / 2$$

Similarly for *y* and *z* components

$$\Delta y . \Delta p_y = \hbar / 2$$

$$\Delta z . \Delta p_z = \hbar / 2$$

Multiplying we have

$$\Delta x \ \Delta y \ \Delta z \ \Delta p_x \ \Delta p_y \ \Delta p_z = \left(\frac{\hbar}{2}\right)^3$$

The product $\Delta_x \Delta_y \Delta_z \Delta p_x \Delta p_y \Delta p_z$ is called the volume of the elmentary cell in the phase space. Thus the volume of the elementary cell in the phase space $= (\hbar / 2)^3$. A macroscopic system consisting of *N* micro-systems is described by 6*N* variables. The graphical representation for the macro-system has 6*N* dimensions and is called Γ-phase space. In Γ-phase space the size of the elementary cell $= \left(\frac{\hbar}{2}\right)^{3N}$

5.8.6. Uncertainty in Frequency of Radiation Emitted by an Atom

On transition from a higher energy state to a lower energy state the nuclei, atoms and molecules emit characteristic electromagnetic radiation. In the case of nucleus, these radiations are gamma rays and in the case of atoms, these are in the ultra violet and visible regions. For molecules these fall in the infra-red region.

An excited system has a small finite life time ΔT and the uncertainty for this interval is of the same value. The energy of the excited state is therefore also uncertain. If ΔE is the uncertainty in energy of the state, then

$$\Delta E.\Delta T \geq \hbar/2$$

or

$$\Delta E \geq \frac{\hbar}{2\Delta T} = \frac{h}{4\pi\Delta T}$$

As $E = h\nu$ the uncertainty in the frequency of the emitted photon $\Delta\nu$ is given by $\Delta E = h\,\Delta\nu$

5.9 MINIMUM ENERGY OF HARMONIC OSCILLATOR USING UNCERTAINTY PRINCIPLE

A simple example of a harmonic oscillator is the motion of a particle of mass m attached to a spring of force constant k. According to classical mechanics the energy of the oscillator is minimum when it is at rest. Quantum mechanically an oscillator at rest will have *zero* uncertainty in its position and hence the uncertainty in its momentum will be infinite. An infinite momentum uncertainty leads to an infinite amount of kinetic energy which is impossible. Thus quantum mechanically a harmonic oscillator cannot be at rest and cannot have zero energy.

Suppose Δx is the uncertainty in the position of the oscillator and Δp the uncertainty in its momentum, then

$$\Delta x.\Delta p = \frac{1}{2}\hbar$$

or

$$\Delta p = \frac{\hbar}{2\Delta x}$$

The total energy of the oscillator = kinetic energy + potential energy

$$E = \frac{(\Delta p)^2}{2m} + \frac{1}{2}k(\Delta x)^2$$

The angular frequency of an oscillator $\omega = \sqrt{\frac{k}{m}}$

$$\therefore \quad k = m\,\omega^2$$

Hence

$$E = \frac{\hbar^2}{2m(2\Delta x)^2} + \frac{1}{2}m\omega^2(\Delta x)^2$$

Putting $\Delta x = r$, we have

$$E = \frac{\hbar^2}{8mr^2} + \frac{1}{2}m\omega^2 r^2$$

The value of E will be minimum when $\frac{dE}{dr} = 0$

Now

$$\frac{dE}{dr} = -\frac{\hbar^2}{4mr^3} + m\omega^2 r$$

When

$$\frac{dE}{dr} = 0$$

$$\frac{\hbar^2}{4mr^3} = m\omega^2 r \quad \text{or} \quad r^4 = \frac{\hbar^2}{4m^2\omega^2}$$

or
$$r = \sqrt{\frac{\hbar}{2m\omega}}$$

Thus E is minimum for
$$r = \sqrt{\frac{\hbar}{2m\omega}}$$

$$\therefore \quad E_{min} = \frac{\hbar^2}{8m.\dfrac{\hbar}{2m\omega}} + \frac{1}{2}m\omega^2 \frac{\hbar}{2m\omega} = \frac{1}{4}\hbar\omega + \frac{1}{4}\hbar\omega$$

$$= \frac{1}{2}\hbar\omega$$

This gives the *minimum energy of the oscillator*. We get the same result by solving Schrodinger's wave equation for a harmonic oscillator.

5.10 SUPERIORITY OF AN ELECTRON MICROSCOPE

The resolving power of a microscope is given by

$$\text{Resolving power} = \frac{A}{\lambda}$$

where A is the aperture (diameter) of the lens used as objective and λ the wavelength of light. Thus smaller the wavelength higher is the resolving power. If we use ultra-violet light, it has a wavelength 4000 Å = 4000×10^{-10} m = 4×10^{-7} m.

But, if we use an electron microscope and accelerate electrons through a potential difference of 64 kV = 64000 V, the velocity v attained by the electron is given by

$$eV = \frac{1}{2}mv^2 \text{ or } v = \sqrt{\frac{2eV}{m}}$$

$$= \sqrt{\frac{2 \times 1.6 \times 10^{-19} \times 64 \times 10^3}{9.1 \times 10^{-31}}} = 1.5 \times 10^8 \text{ ms}$$

$$\therefore \quad \frac{v}{c} = \frac{1.5 \times 10^8}{3 \times 10^8} = \frac{1}{2}$$

As this velocity is only half the velocity of light, we use relativity formula for de-Brogle wavelength of the electron

$$\therefore \quad \lambda = \frac{h}{m_0 c} \frac{\left(1 - \dfrac{v^2}{c^2}\right)^{1/2}}{\dfrac{v}{c}}$$

Substituting $\frac{v}{c} = \frac{1}{2}$ and values of other constants, we get

$$\lambda = \frac{\dfrac{6.62\times10^{-34}}{9.1\times10^{-31}\times3\times10^{8}}\left(1-\dfrac{1}{4}\right)^{1/2}}{\dfrac{1}{2}}$$

$$= \frac{6.62\times10^{-34}\times\sqrt{3}}{9.1\times10^{-31}\times3\times10^{8}} = \frac{6.62\times10^{-34}}{9.1\times10^{-31}\times1.732\times10^{8}} = 4.2\times10^{-12}\text{ m.}$$

This wavelength is 10^5 times smaller than the wavelength of ultra violet / visible light. Thus an electron-microscope has a resolving power 10^5 times the resolving power of an optical microscope. *An electron-microscope can, therefore, produce sharp images with higher magnification and is superior to the optical microscope.*

5.11 COMPLEMENTARITY PRINCIPLE

As a consequence of uncertainty principle both particle and wave aspect of matter cannot be measured in the same experiment. If an experiment is so designed as to measure particle properties of matter, then in this experiment the error in the measurement of position co-ordinate Δx and time co-ordinate Δt should be zero. In turn it means that momentum and energy and therefore, the wave aspects like wavelength λ and frequency ν given by $\lambda = \dfrac{h}{p}$ and $\nu = \dfrac{E}{h}$ will be completely unknown.

Similarly, if an experiment is designed to accurately measue the wave aspects $\Delta\lambda$ and $\Delta\nu$ or Δp and ΔE, the position and time co-ordinates will be completely unknown. We, therefore, see that when the particle aspect of matter is observed, the wave aspect is suppressed and vice-versa.

The inability to observe the wave and particle aspect of matter at the same time is known as principle of complementarity.

5.12 DIFFERENCE BETWEEN QUANTUM MECHANICS AND CLASSICAL MECHANICS

1. The classical mechanics deals with particles of *macroscopic size* whreas quantum mechanics deals with particles of *microscopic size*.

2. The laws of classical mechanics were developed on the basis of Newton's laws of motion having the basic equation $F = ma$. This is why classical mechanics is also known as Newtonian Mechanics. On the other hand one method of approach to obtain the laws of quantum mechanics is on the basis of Schrodinger's wave equation which in one dimensional time dependent form is stated as

$$-\frac{\hbar^2}{2m}\frac{\partial^2\psi}{\partial x^2} + V\psi = i\hbar\frac{\partial\psi}{\partial t}$$

where ψ is the wave function associated with and V the potential energy of the particle.

3. In classical mechanics (or Newtonian Mechanics), the future position and momentum of a paticle can be completely determined if its initial position and momentum as well as the forces acting on it are known. In quantum mechanics there is an inherent uncertainty (in accordance with Heisenberg's uncertainty principle) in the determination of initial position and momentum of a particle. As a result it is not possible to completely know the future position and momentum at a later time.

4. Quantum mechanics deals with *probabilities* whereas classical mechanics deals with '*certainties*'. For example, in classical mechanics the radius of the orbit of an electron in the ground state of a hydrogen atom is *always exactly* 5.3×10^{-11} m whereas according to quantum mechanics this is the *most probable* value of the radius. In other words, if we measure the radius of the orbit of the electron by suitable experiments, the results obtained will be different, may be more or less than the above value but when a very large number of observations are taken the value most likely to be found will be 5.3×10^{-11} m.

EXERCISE

1. State Heisenberg's uncertainty principle and explain it in view of de-Broglie wave concept.
(G.N.D.U. 2007; Meerut U. 2002; Luck. U. 2001, 1996; H.P.U. 2001, 1996; P.U. 2007, 2006, 2004, 2001; Pbi. U. 2008; Poorvanchal U. 2002; M.D.U. 2001; Gauhati U. 2000)

2. (*a*) State time energy uncertainty relation and obtain it from position momentum uncertainty principle. *(K.U. 2001; H.P.U. 2001; Pbi.U. 1999; P.U., 1997, Luck.U., 1996)*

   (*b*) Verify that uncertainty principle can be expressed in the form $\Delta L \cdot \Delta\theta \geq \frac{h}{2}$ where L is the angular momentum and θ the angular position.
   (P.U. 2001, 1999; Luck.U. 2001, G.N.D.U., 1991)

3. Starting from de-Broglie's wave concept obtain Heisenberg's uncertainty principle.
(P.U 2001; 1995; Pbi. U. 1999; Luck.U., 1996)

4. (*a*) State Heisenberg's uncertainty principle and derive it from a hypothetical gamma ray microscope.
   (Meerut U. 2007, 2005, 2003, 2001, 2000; K.U. 2003, 2000; P.U. 2002, 1999; Bang. U. 2001, 2000; Gharwal U. 2002, 1999; G.N.D.U. 2000, Pbi.U. 2000)

   (*b*) Why is uncertainty principle important for microscopic particles but insignificant in practical life? *(P.U. 2001)*

5. (*a*) Illustrate Heisenberg's uncertainty principle by diffraction of a beam of electrons by a narrow slit. *(Meerut. U. 2006; Madurai U. 2003; G.N.D.U.2001, 1999; Pbi.U.2001)*

6. Prove position momentum uncertainty principle using particle approach. (P.U., 1995)

7. (*a*) By applying uncertainty principle explain non-existence of electrons in the atomic nucleus.
   (Pbi. U. 2008, 2002, 2000; M.D.U. 2006, 2003; Meerut U. 2003; Gharwal U. 2000; Poorvanchal U. 2002; Bang.U. 2000; P.U. 1997, 1995; G.N.D.U. 2004, 1996)

   (*b*) What other reasons show why electrons cannot exist inside the nucleus? *(P.U., 1991)*

8. Using uncertainty principle show that, protons, neutrons and α-particles can exist within the nucleus. *(G.N.D.U. 1997, 1996; P.U. 1997)*

9. Apply Heisenberg's uncertainty principle to explain the following:—
   (*i*) Non-existence of electrons within the nucleus *(G.N.D.U. 2007; M.D.U. 2003)*
   (*ii*) Existence of protons, neutrons and α-particles *(G.N.D.U. 2007; Pbi. U. 2006)*
   (*iii*) Existence of finite zero-point energy *(Pbi. U. 2006; G.N.D.U. 2005; P.U. 1999)*
   (*iv*) Binding energy of an electron in a hydrogen atom is of the order of 15 eV *(H.P.U. 2002)*
   (*v*) Size of the elementary cell in phase space *(P.U. 2000)*
   (*vi*) The radiation emitted from an excited atom does not appear with a precise frequency.

10. Show that an electron can not exist in the nucleus of radius in the range 10^{-14} m, on the basis of uncertainty principle. *(Nagpur U. 2011)*

11. By applying uncertainty principle explain minimum energy of harmonic oscillator.
(M.D.U.2003; G.N.D.U. 2000: P.U., 1997)

12. Resolving power of a microscope depends upon wavelength of light used, In what way electron microscope in superior? *(P.U., 1997)*

13. What is complementarity principle? *(Gharwal U. 2000)*

14. Explain the difference between quantum mechanics and classical mechanics. ***(Luck U. 2001)***

15. A beam of short wavelength gives accurately the position of a particle-justify.

[**Hint.** According to Heisenberg's uncertainty principle

$$\Delta p . \Delta x = \frac{\hbar}{2}$$

To determine the position of *a* particle accurately, the value of Δx must be small. As $\Delta p . \Delta x = a$ constant, Δp should be large. The wavelength associated with a moving particle is given by

$$\lambda = \frac{h}{mv} = \frac{h}{p}.$$

$$p = \frac{h}{\lambda}$$

Therefore, for large p (or Δp) λ should be small. So short wavelength beam will give more accurately the position of the particle.

16. The concept of trajectory is meaningless in quantum mechanics. Explain.

(Banglaore U. 2001)

[**Hint.** To determine the path of a particle *i.e.* its trajectory it is necessary that its *position* and *momentum* must be known accurately at every point of its motion. But according to Heisenberg's uncertainty principle both position and momentum cannot be known exactly at one and the same time. Thus, according to quantum mechanics, it is impossible to predict the exact trajectory and hence it is a meaningless concept.]

17. An electron is confined to a nucleus of radius 10^{-14}cm. Calculate the uncertainty in the momentum of an electron if it is in the nucleus (**Ans.** 5.3×10^{-21} kg ms^{-1})

MULTIPLE CHOICE QUESTIONS

1. Heisenberg's uncertainty principle is

(*a*) $\Delta E . \Delta x \geq \frac{h}{2\pi}$ (*b*) $\Delta E . \Delta t \geq \frac{h}{4\pi}$

(*c*) $\Delta p . \Delta t \geq \frac{h}{4\pi}$ (*d*) $\Delta x . \Delta t \geq \frac{h}{4\pi}$

2. Heisenberg's uncertainty Principle is

(*a*) $\Delta p_x . \Delta x \geq \frac{\hbar}{2}$ (*b*) $\Delta p_x . \Delta x \geq \hbar$

(*c*) $\Delta p_x . \Delta x \geq 2\hbar$ (*d*) $\Delta p_x . \Delta x \geq \hbar^2$

3. Heisenberg's uncertainty principle in term of angular position and angular momentum is

(*a*) $\Delta L . \Delta\theta \geq \hbar$ (*b*) $\Delta L . \Delta\theta \geq \hbar^2/2$

(*c*) $\Delta L . \Delta\theta \geq \hbar^2$ (*d*) $\Delta L . \Delta\theta \geq \frac{\hbar}{2}$

ANSWERS

1. (*b*)	**2.** (*a*)	**3.** (*d*)

SCHRÖDINGER'S THEORY OF QUANTUM MECHANICS

6
Chapter

6.1 WAVE FUNCTION (Ψ) OF A MOVING PARTICLE

The relation $\lambda = h/p$ between momentum p and wavelength λ which has been experimentally verified for both photons and particles (like electrons) suggests that it is possible to use a *concentrated group of waves* or a *wave packet* to describe localised particles and photons (or quantas) of radiation. The associated de-Broglie wave function $\Psi\ (\vec{r},t)$ is a function of space variables (x, y, z) or $\vec{r}$ and time t and it can give nearly complete information about the state of a physical system at a particular time in accordance with the rules of quantum mechanics. It is also known as *state function.* The value of the wave function associated with a moving particle at the particular point x, y, z in space at the time t is related to the likelihood of finding the particle there at that time.

Physical significance of wave function Ψ and $|\Psi|^2$. The wave function Ψ is assigned three basic properties:

(*i*) *It can interfere with itself.* This property explains the phenomenon of electron diffraction.

(*ii*) *It is large in magnitude where the particle* (*electron*) or *photon is likely to be located and small elsewhere.*

In other words, the probability of experimentally finding the particle described by the wave function Ψ at the point x, y, z at the time t is proportional to $|\Psi|^2$ there at t.

Thus, $|\Psi|^2$ *i.e.*, **the square of the magnitude** (amplitude) **of the wave function** $\Psi\ (\vec{r},t)$ gives the *probability density* of finding the physical system (particle or photon) at a particular place at a given time.

The wave function is, in general, a complex quantity and thus consists of two parts, the amplitude and phase.

If we write the complex quantity $\Psi = A + iB$ where A and B are real functions, then its complex conjugate $\Psi^* = A - iB$ and $\Psi\ \Psi^* = A^2 - i^2 B^2 = A^2 + B^2$ which is always a **positive real** quantity (or zero).

$$\therefore \qquad \text{Probability density} = |\Psi|^2$$

In case Ψ is complex, the probability density is given by the product $\Psi\Psi^*$ of Ψ and its complex conjugate Ψ^*.

(*iii*) *The wave function describes the behaviour of a single particle or photon* and not the statistical distribution of a number of such quanta.

6.2 CONDITIONS FOR THE WAVE FUNCTION TO BE WELL BEHAVED

Due to the interpretation of $|\Psi|^2$ as a probability density, the wave function Ψ must obey the following conditions.

(*i*) Ψ must be finite for all values of x, y and z.

(*ii*) Ψ must be single valued *i.e.* for each set of values of x, y and z, Ψ must have one value only.

(*iii*) Ψ must be continuous in all regions except in those regions where the potential energy $V(x, y, z) = \infty$.

(*iv*) Ψ must vanish at infinity *i.e.* $\Psi = 0$ as $x \to \pm\infty$, $y \to \pm\infty$ or $z \to \pm\infty$. In other words $\int_{-\infty}^{+\infty} \psi^* \psi \, dx < \infty$.

(*v*) The partial derivatives of Ψ *i.e.* $\frac{\partial \Psi}{\partial x}, \frac{\partial \Psi}{\partial y}, \frac{\partial \Psi}{\partial z}$ must also be finite, single valued and continuous at all points, except at points where the potential $V(x)$ is infinite.

These requirements are necessary in order that measurable quantities like momentum, energy etc. obtained from the wave function Ψ are well defined.

6.3 NEED FOR DIFFERENTIAL WAVE EQUATION

The motion of a particle of atomic dimensions cannot be described by Newton's laws of motion which are applicable to bodies of macroscopic dimensions. The behaviour of particles of microscopic size is governed by the associated de-Broglie waves or the wave function $\Psi(\vec{r}, t)$, but Ψ is a complex quantity and square of its modulus $|\Psi|^2$ or $\Psi\Psi^*$ gives the probability of finding the particle at the position vector $\vec{r}$ corresponding to Cartesian co-ordinates *x, y, z* and time *t*.

To describe the motion of the particle in space and time we must know the value of $\Psi(\vec{r}, t)$ at all times and for this purpose it is necessary to find a differential equation which controls the space time behaviour of the wave function $\Psi(\vec{r}, t)$. The solution of such a differential equation will gives the *possible* motion of the particle. Hence the necessity of such a differential equation.

6.4 TIME-DEPENDENT SCHRÖDINGER'S WAVE EQUATION

A plane wave moving in the *X*-direction is represented in the exponential form as

$$y = Ae^{i(\omega t - kx)} \qquad \text{...}(i)$$

where $\omega = 2\pi\nu$ (ν = frequency of the wave) and k the propagation constant $= \frac{2\pi}{\lambda}$ (λ = wavelength of the wave).

According to quantum theory, the **energy** of a wave

$$E = h\nu$$

and therefore

$$\nu = \frac{E}{h}$$

and

$$\omega t = 2\pi\nu t = 2\pi \frac{E}{h} t = \frac{Et}{h/2\pi} = \frac{Et}{\hbar}$$

and **momentum**

$$p_x = \frac{h}{\lambda} \text{ or } \frac{1}{\lambda} = \frac{p_x}{h}$$

$$\therefore \quad kx = \frac{2\pi}{\lambda} x = 2\pi \frac{p_x x}{h} = \frac{p_x}{h/2\pi} x = \frac{p_x}{\hbar} x$$

where

$$\hbar = \frac{h}{2\pi} = 1.054 \times 10^{-34} \text{ Js}.$$

Substituting in (*i*) we can express a plane wave moving in the *X*-direction as

$$y = Ae^{i\left(\frac{E}{\hbar}t - \frac{p_x x}{\hbar}\right)}$$

$$= Ae^{\frac{i}{\hbar}(Et - xp_x)} \qquad \text{...}(ii)$$

By analogy between photons and electron beams, we assume that the electrons in a beam travelling in the X-direction with constant momentum p_x and energy E are controlled by *pilot* waves with a wave function Ψ which plays the same role for electrons and other particles as the displacement y for sound waves or electric field vector E for electro-magnetic waves.

Thus $\Psi(x, t)$ as a function of x and t denotes the amplitude of matter wave at a point x at a time t and is given by an equation similar to equation (*ii*) *i.e.*,

$$\Psi(x,t) = Ae^{\frac{i}{\hbar}(Et - xp_x)}$$

However, in wave mechanics, we use the exponential form with a *negative sign i.e.*,

$$\psi(x,t) = Ae^{\frac{-i}{\hbar}(Et - xp_x)}$$

$$= Ae^{\frac{i}{\hbar}(xp_x - Et)} \quad ...(iii)$$

for a wave moving along the positive direction of X-axis. We represent a wave moving along the negative direction of X-axis by the equation

$$\Psi(x,t) = Ae^{\frac{-i}{\hbar}(xp_x - Et)}$$

This merely introduces a phase difference of π from the notation used for sound waves, electromagnetic and other waves.

Differentiating equation *(iii)* with respect to time and writing only Ψ for $\Psi(x, t)$, we have

$$\frac{\partial \Psi}{\partial t} = -\frac{i}{\hbar} EAe^{\frac{i}{\hbar}(xp_x - Et)} = -\frac{i}{\hbar} E\Psi \quad ...(iv)$$

or $$E\Psi = -\frac{\hbar}{i}\frac{\partial \Psi}{\partial t} \quad ...(v)$$

Differentiating equation *(iii)* with respect to x, we have

$$\frac{\partial \Psi}{\partial x} = \frac{i}{\hbar} p_x Ae^{\frac{i}{\hbar}(xp_x - Et)}$$

and $$\frac{\partial^2 \Psi}{\partial x^2} = -\frac{p_x^2}{\hbar^2} Ae^{\frac{i}{\hbar}(xp_x - Et)} = -\frac{p_x^2}{\hbar^2}\Psi$$

or $$p_x^2 \Psi = -\hbar^2 \frac{\partial^2 \Psi}{\partial x^2} \quad ...(vi)$$

Particle subject to force. When a particle is not free but is subject to a force, its total energy is the sum of its kinetic energy and potential energy.

For a particle of mass m, moving with a velocity v piloted by the wave function Ψ, the total energy in a *non-relativistic* case is given by

$$E = \frac{1}{2}mv^2 + V$$

where $\frac{1}{2}mv^2$ is the kinetic energy and V the potential energy of the particle—a function of both x and t. If p_x is the momentum of the particle, then

$$p_x = mv \text{ and } \frac{1}{2}mv^2 = \frac{1}{2m}m^2v^2 = \frac{p_x^2}{2m}$$

$$\therefore \qquad E = \frac{p_x^2}{2m} + V \qquad \text{...(vii)}$$

Multiplying both sides by Ψ, we have

$$\frac{p_x^2}{2m}\Psi + V\Psi = E\Psi$$

Substituting the value of $E\Psi$ from *(v)* and $p_x^2\Psi$ from *(vi)*, we have

$$-\frac{\hbar^2}{2m}\frac{\partial^2\Psi}{\partial x^2} + V\Psi = -\frac{\hbar}{i}\frac{\partial\Psi}{\partial t} = i\hbar\frac{\partial\Psi}{\partial t} \qquad \text{...(viii)}$$

As Ψ is a function of x and t, we write it explicitly as

$$-\frac{\hbar^2}{2m}\frac{\partial^2}{\partial x^2}\Psi(x,t) + V\Psi(x,t) = i\hbar\frac{\partial}{\partial t}\Psi(x,t) \qquad \text{...(6.1)}$$

This is Schrödinger's *one dimensional time dependent* wave equation for waves associated with a moving particle in a potential field V.

Three dimensional wave equation. For motion of the particle in three dimensions Ψ is a function of x, y, z and t (or $\vec{r}$ and t). Extending to the three dimensional case, Schrödinger's time dependent wave equation is given by

$$-\frac{\hbar^2}{2m}\left[\frac{\partial^2\Psi}{\partial x^2} + \frac{\partial^2\Psi}{\partial y^2} + \frac{\partial^2\Psi}{\partial z^2}\right] + V\Psi = i\hbar\frac{\partial\Psi}{\partial t}$$

or

$$-\frac{\hbar^2}{2m}\nabla^2\Psi(\vec{r},t) + V\Psi(\vec{r},t) = i\hbar\frac{\partial\Psi(\vec{r},t)}{\partial t} \qquad \text{...(6.2)}$$

This is Schrödinger's *time dependent, three dimensional, quantum mechanical* wave equation for waves associated with a moving particle in a potential field V. This equation can be used only for non-relativistic problems. A more elaborate formulation is required when particle speeds comparable with that of light are involved.

Free particle. A particle is said to be *free* when no force acts on it, in such a case the potential energy V of the particle is zero and the total energy E is only kinetic.

Putting $V = 0$ in equation (6.1), we have for a one dimensional case

$$-\frac{\hbar^2}{2m}\frac{\partial^2}{\partial x^2}\Psi(x,t) = i\hbar\frac{\partial\Psi}{\partial t}(x,t)$$

or simply

$$-\frac{\hbar^2}{2m}\frac{\partial^2\Psi}{\partial x^2} = i\hbar\frac{\partial\Psi}{\partial t} \qquad \text{...(6.3)}$$

This is Schrödinger's *one dimensional time dependent* wave equation associated with a *free* particle.

Three dimensional wave equation for a free particle. For a three dimensional case, we have from equation (6.2)

$$-\frac{\hbar^2}{2m}\left[\frac{\partial^2\Psi}{\partial x^2} + \frac{\partial^2\Psi}{\partial y^2} + \frac{\partial^2\Psi}{\partial z^2}\right] = i\hbar\frac{\partial\Psi}{\partial t}$$

or $$-\frac{\hbar^2}{2m}\nabla^2\Psi(\vec{r},t) = i\hbar\frac{\partial\Psi(\vec{r},t)}{\partial t}$$

or simply $$-\frac{\hbar^2}{2m}\nabla^2\Psi = i\hbar\frac{\partial\Psi}{\partial t} \quad ...(6.4)$$

This is Schrödinger's *time dependent, three dimensional* wave equation for waves associated with a moving *free* particle.

6.5 SCHRÖDINGER'S WAVE EQUATION IN TERMS OF HAMILTONIAN OPERATOR

For derivation of Schrödinger's time dependent three dimensional, quantum mechanical wave equation

$$-\frac{\hbar^2}{2m}\nabla^2\Psi(\vec{r},t) + V\Psi(\vec{r},t) = i\hbar\frac{\partial\Psi(\vec{r},t)}{\partial t} \quad [\because \text{Eq. 6.2}]$$

or $$-\frac{\hbar^2}{2m}\nabla^2\Psi + V\Psi = i\hbar\frac{\partial\Psi}{\partial t}$$

or $$\left(-\frac{\hbar^2}{2m} + V\right)\Psi = i\hbar\frac{\partial}{\partial t}\Psi$$

or $$\widehat{H}\Psi = \widehat{E}\Psi \quad ...(6.5)$$

where $\widehat{H}$ is the Hamiltonian operator $\left(-\frac{\hbar^2}{2m} + V\right)$ and $\widehat{E}$ is the energy operator $i\hbar\frac{\delta}{\delta t}$.

6.6 EIGEN FUNCTION AND EIGEN VALUE

A function f is called an eigen function of the operator $\hat{A}$ if when the operator $\hat{A}$ operates on the function f, we get the same function multiplied by a constant C *i.e.*

$$\hat{A}f = Cf \quad ...(i)$$

The constant C is called the **eigen value** of the operator $\hat{A}$

Equation (*i*) is known as **eigen value equation.**

For example, if the operator $\hat{A} = \frac{d}{dx}$ and f is a function of x given by $f(x) = e^{mx}$, then

$$\hat{A}f(x) = \frac{d}{dx}e^{mx} = me^{mx} = mf(x)$$

or $$\hat{A}f = mf$$

Therefore $f = e^{mx}$ is an eigen function of the operator $\hat{A} = \frac{d}{dx}$ with eigen value $= m$

SOLVED EXAMPLES

Example 6.1. *Show that $u(x) = A\,e^{iax}$ is an eigen function of the momentum operator $\widehat{p}_x$ and determine the corresponding eigen value.* ***(Pbi. U. 2008)***

Solution. The momentum operator $\widehat{p}_x = -i\hbar\frac{d}{dx}$

$$\widehat{p}_x\, u(x) = -i\hbar\frac{d}{dx}[A\,e^{iax}] = -i\hbar\, i\, a\, A\, e^{iax}$$
$$= \hbar\, a\, A\, e^{iax} = \hbar a\, u(x)$$

As the above equation satisfies the relation $\hat{A}f = Cf$, $u(x)$ is an eigen function of momentum operator $\hat{p}_x$ with eigen value $\hbar a$.

Example 6.2 *The operator for Z-component of angular momentum is* $\hat{L}_z = -i\hbar\frac{\partial}{\partial\phi}$, *determine whether or not sin mϕ is its eigen function.* ***(Pbi.U. 2000)***

Solution. Now $f = \sin m\phi$ and operator $\hat{L}_z = -i\hbar\frac{\partial}{\partial\phi}$

$$\therefore \quad \hat{A}f = \hat{L}_z \sin m\phi = -i\hbar\frac{\partial}{\partial\phi}\sin m\phi = -i\hbar\, m \cos\phi$$

As $L_z \sin m\phi \neq$ a constant × $\sin m\phi$

$\sin m\phi$ is not an eigen function of angular momentum operator $\hat{L}_z = -i\hbar\frac{\partial}{\partial\phi}$.

6.7 TIME INDEPENDENT SCHRÖDINGER'S WAVE EQUATION

The time dependent, three dimensional Schrödinger's wave equation is given by

$$-\frac{\hbar^2}{2m}\nabla^2\Psi + V\Psi = i\hbar\frac{\partial\Psi}{\partial t} \quad ...(i)$$

and explicitly in terms of $\vec{r}$ and t, we have

$$-\frac{\hbar^2}{2m}\nabla^2\Psi(\vec{r},t) + V\Psi(\vec{r},t) = i\hbar\frac{\partial\Psi(\vec{r},t)}{\partial t} \quad ...(ii)$$

The one dimensional Schrödinger's wave equation is

$$-\frac{\hbar^2}{2m}\frac{\partial^2\Psi}{\partial x^2} + V\Psi = i\hbar\frac{\partial\Psi}{\partial t} \quad ...(iii)$$

and explicitly in terms of x and t, we have

$$-\frac{\hbar^2}{2m}\frac{\partial^2\Psi(x,t)}{\partial x^2} + V\Psi(x,t) = i\hbar\frac{\partial\Psi(x,t)}{\partial t} \quad ...(iv)$$

When the potential energy V does not depend explicitly on time and is a function of x only and the total energy E is constant, the wave function $\Psi(x, t)$ can be written as the product of two separate functions $\psi(x)$ a function only of x and $f(t)$ a function only of t.

$$\therefore \quad \Psi(x, t) = \psi(x) f(t)$$

The above relation is briefly written as $\Psi = \psi f$

Hence $\quad \frac{\partial\Psi}{\partial x} = f\frac{\partial\psi}{\partial x}$ and $\frac{\partial^2\Psi}{\partial x^2} = f\frac{\partial^2\psi}{\partial x^2}$ and $\frac{\partial\Psi}{\partial t} = \psi\frac{\partial f}{\partial t}$

Now $\quad \Psi = Ae^{\frac{i}{\hbar}(xp_x - Et)}$

$$\frac{\partial\Psi}{\partial t} = -\frac{i}{\hbar}EAe^{\frac{i}{\hbar}(xp_x - Et)} = -\frac{i}{\hbar}E\Psi = \frac{-i}{\hbar}E\psi f \quad ...(v)$$

Substituting in equation (*iii*), we have

$$-\frac{\hbar^2}{2m} f \frac{\partial^2 \psi}{\partial x^2} + V f \psi = i\hbar \left(\frac{-i}{\hbar} \right) E \psi f \qquad ...(vi)$$

Dividing both sides by *f*, we have

$$\frac{-\hbar^2}{2m} \frac{\partial^2 \psi}{\partial x^2} + V \psi = E \psi$$

or $$\frac{\partial^2 \psi}{\partial x^2} + \frac{2m}{\hbar^2}(E - V)\psi = 0 \qquad ...(6.6)$$

As ψ is a function of x only it may be stated explicitly

as $$\frac{\partial^2 \psi(x)}{\partial x^2} + \frac{2m}{\hbar^2}(E - V)\psi(x) = 0 \qquad ...(viii)$$

Further ψ is a function of *only one* variable x, we can replace *partial derivative* by total derivative $\frac{d\psi}{dx}$ and $\frac{d^2\psi}{dx^2}$.

We can, therefore, write equation (*viii*) as

$$\frac{d^2 \psi}{dx^2} + \frac{2m}{\hbar^2}(E - V)\psi = 0 \qquad ...(6.7)$$

This is *one dimensional, time independent* Schrödinger's equation.

Three dimensional time independent equation. For motion of the particle in three dimensions ψ is a function of x, y, z or $\vec{r}$. In such a case the *time independent* or *stationary state* form of Schrödinger's quantum mechanical wave equation is given by

$$\left(\frac{\partial^2 \psi}{\partial x^2} + \frac{\partial^2 \psi}{\partial y^2} + \frac{\partial^2 \psi}{\partial z^2} \right) + \frac{2m}{\hbar^2}(E - V)\psi = 0$$

or $$\nabla^2 \psi + \frac{2m}{\hbar^2}(E - V)\psi = 0 \qquad ...(6.8)$$

More explicitly $$\nabla^2 \psi(\vec{r}) + \frac{2m}{\hbar^2}(E - V)\psi(\vec{r}) = 0 \qquad ...(6.9)$$

Equation for time dependent part. From the relation (*v*)

$$\frac{\partial \Psi}{\partial t} = \frac{-i}{\hbar} E \psi f \text{, we have}$$

$$\frac{\psi \partial f}{\partial t} = \frac{-i}{\hbar} E \psi f \qquad [\because \Psi = \psi(x) f(t)]$$

or $$i\hbar \frac{\partial f}{\partial t} - Ef = 0$$

Again replacing partial derivatives by total derivative and more explicitly

$$i\hbar \frac{df(t)}{dt} - Ef(t) = 0 \qquad ...(6.10)$$

or $$\frac{d\,f(t)}{f\,(t)} = \frac{Edt}{i\,\hbar} = -\frac{i}{\hbar}Edt$$

Integrating both sides, we get

$$\ln f(t) = -\frac{i}{\hbar}Et$$

or $$f\,(t) = e^{-\frac{i}{\hbar}Et}$$

This is the equation for $f\,(t)$ the *time dependant part* of the wave equation Ψ

$\therefore$ $$\Psi = \Psi(x)\;e^{-\frac{i}{\hbar}Et} \quad ...(6.11)$$

Eigen functions and Eigen values. For a particle of mass m moving with a velocity v, the total energy in a *non-relativistic* case $E = \frac{1}{2}mv^2 + V$ where $\frac{1}{2}mv^2$ is the kinetic energy and V the potential energy of the particle. If p_x is the momentum of the particle, then

$$p_x = mv \text{ and } \frac{1}{2}mv^2 = \frac{p_x^2}{2m}$$

$\therefore$ $$E = \frac{p_x^2}{2m} + V$$

or $$E - V = \frac{p_x^2}{2m} = \frac{p^2}{2m} \quad \text{(writing only } p \text{ for } p_x\text{)}$$

Eq.(6.6) can now be written as

$$\frac{\partial^2\psi}{\partial x^2} + \frac{2m}{\hbar^2}\frac{p^2}{2m}\psi = 0$$

or $$\frac{\partial^2\psi}{\partial x^2} + \frac{p^2}{\hbar^2}\psi = 0 \quad ...(6.12)$$

We will get several solutions of equation (6.6) or (6.11) for sets of values of total energy E_1, E_2,............E_n and momentum p_1, p_2,............p_n. The wave function ψ_1 is obtained when momentum is p_1 and total energy E_1, a wave function ψ_2 is obtained when momentum is p_2 and total energy is E_2..... and in general a wave function ψ_n is obtained when momentum is p_n and total energy E_n. The wave functions ψ_1, ψ_2.....ψ_n are called *Eigen functions* for *Eigen values* of momentum p_1, p_2,............ p_n *and for* Eigen values *of total energy* E_1, E_2,............E_n respectively.

Note. Schrödinger's wave equation is not valid for relativistic particles because in deriving this equation we use classical (non-relativistic) expression for total energy $E = \frac{p^2}{2m} + V$. We also take the momentum of the particle as non-relativistic and equal to mv. The kinetic energy is taken as $\frac{1}{2}mv^2$, which is also a non-relativistic expression.

Example 6.3. *A particle moving along the positive direction of X-axis in a region of positive potential V (x) is represented by a wave packet given by*

$$\Psi\,(x,t) = \frac{1}{\sqrt{2\pi\hbar}}\int A(p)\;e^{i/\hbar(px-Et)}\;dp$$

Obtain the time dependent Schrödinger's wave equation for a non-relativistic total energy.

(Pbi. U. 2005; P.U. 2000)

Solution. Schrödinger's wave equation. The wave function representing the wave packet is

$$\Psi(x, t) = \frac{1}{\sqrt{2\pi\hbar}} \int A(p)\, e^{i/\hbar (p_x - Et)}\, dp \qquad ...(i)$$

Differentiating Eq. (*i*) with respect to time and writing only Ψ for $\Psi(x, t)$, we have

$$\frac{\partial \Psi}{\partial t} = \frac{1}{\sqrt{2\pi\hbar}} \left(\frac{-i}{\hbar} E\right) \int A(p)\, e^{i/\hbar (p_x - Et)}\, dp$$

$$= \frac{-i}{\hbar} E\Psi$$

or

$$E\Psi = -\frac{\hbar}{i} \frac{\partial \Psi}{\partial t} \qquad ...(ii)$$

Differentiating Eq (*i*) with respect to *x*, we have

$$\frac{\partial \Psi}{\partial x} = \frac{1}{\sqrt{2\pi\hbar}} \left(\frac{ip}{\hbar}\right) \int A(p) e^{i/\hbar (p_x - Et)}\, dp$$

and

$$\frac{\partial^2 \Psi}{\partial x^2} = \frac{1}{\sqrt{2\pi\hbar}} \left(\frac{-p^2}{\hbar^2}\right) \int A(p)\, e^{i/\hbar (p_x - Et)}\, dp$$

$$= \frac{-p^2}{\hbar^2} \Psi$$

or

$$p^2 \Psi = -\hbar^2 \frac{\partial^2 \Psi}{\partial x^2} \qquad ...(iii)$$

For a particle of mass *m*, moving with a velocity *v* piloted by the wave packet Ψ, the total energy in a non-relativistic case is given by

$$E = \frac{1}{2} mv^2 + V$$

where $\frac{1}{2} mv^2$ is the kinetic energy and *V* the potential energy of the particle - a function of both *x* and *t*. The momentum of the particle $p = mv$

$$\therefore \quad E = \frac{1}{2} mv^2 + V = \frac{p^2}{2m} + V$$

Multiplying both sides by Ψ, we have

$$\frac{p^2}{2m} \Psi + V\Psi = E\Psi \qquad ...(iv)$$

substituting the value of $E\Psi$ from *(ii)* and $p^2\Psi$ from *(iii)* in *(iv)*, we get

$$-\frac{\hbar^2}{2m} \frac{\partial^2 \Psi}{\partial x^2} + V\Psi = -\frac{\hbar}{i} \frac{\partial \Psi}{\partial t}$$

or

$$-\frac{\hbar^2}{2m} \frac{\partial^2 \Psi}{\partial x^2} + V\Psi = i\hbar \frac{\partial \Psi}{\partial t}$$

As Ψ is a function of *x* and *t*, we write it explicitly as

$$\frac{-\hbar^2}{2m} \frac{\partial^2}{\partial x^2} \Psi(x,t) + V\Psi(x,t) = i\hbar \frac{\partial}{\partial t} \Psi(x, t).$$

Example 6.4. *Show that the wave equation Ψ (x, t) = A cos (kx – ω t) does not satisfy time dependent Schrödinger's wave equation for a free particle.* **(Pbi. U. 2001)**

Or

Can we represent matter waves associated with a free particle by wave function Ψ(x, t) = A cos (ωt – kx). **(P.U. 2006)**

Solution. Schrödinger's time dependent one dimensional wave equation for a free particle is given by

$$-\frac{\hbar^2}{2m}\frac{\partial^2\Psi}{\partial x^2} = i\hbar\frac{\partial\Psi}{\partial t} \qquad ...(i)$$

The given wave equation is $\Psi(x, t) = A\cos(kx - \omega t)$

$$\therefore \qquad \frac{\partial\Psi}{\partial t} = \omega A\sin(kx - \omega t)$$

and $$\frac{\partial\Psi}{\partial x} = -Ak\sin(kx - \omega t) \text{ or } \frac{\partial^2\Psi}{dx^2} = -Ak^2\cos(kx - \omega t)$$

Now $$\frac{-\hbar^2}{2m}\frac{\partial^2\Psi}{\partial x^2} = \frac{-\hbar^2}{2m}\times\left[-Ak^2\cos(kx - \omega t)\right]$$

$$= \frac{\hbar^2 k^2}{2m}A\cos(kx - \omega t) \qquad ...(ii)$$

and $$i\hbar\frac{\partial\Psi}{\partial t} = i\hbar\omega A\sin(kx - \omega t) \qquad ...(iii)$$

As the right hand sides of Eq. (*ii*) and (*iii*) are not equal Eq. (*i*) i.e. Schrodinger's wave equation for a free particle is not satisfied. Therefore, we cannot represent matter waves associated with a free particle by the wave equation given above.

6.8 SCHRÖDINGER'S WAVE EQUATION IN MOMENTUM REPRESENTATION

The three dimensional time dependent Schrödinger's wave equation Ψ is given by

$$\frac{-\hbar^2}{2m}\nabla^2\Psi + V\Psi = i\hbar\frac{\partial\Psi}{\partial t} \qquad ...(i)$$

The wave function for any plane wave is given by

$$\Psi = e^{i/\hbar(\vec{p}.\vec{r} - Et)}$$

Any superposition of such waves forms a wave packet whose wave equation using Fourier's theorem is written as

$$\Psi = \frac{1}{(\sqrt{2\pi\hbar})^3}\int a(p)\, exp.\left[\frac{i}{\hbar}(\vec{p}.\vec{r} - Et)\right] dp \qquad ...(ii)$$

Differentiating Eq. (*ii*) with respect to *t*, we get

$$\frac{\partial\Psi}{\partial t} = \frac{1}{(\sqrt{2\pi\hbar})^3}\int\left(\frac{-iE}{\hbar}\right)a(p)\, exp.\left[\frac{i}{\hbar}(\vec{p}.\vec{r} - Et)\right] dp \qquad ...(iii)$$

Further $\vec{p}.\vec{r} = p_x x + p_y y + p_z z$

$$\frac{\partial\Psi}{\partial x} = \frac{1}{(\sqrt{2\pi\hbar})^3}\int\frac{ip_x}{\hbar}a(p)\, exp.\left[\frac{i}{\hbar}(\vec{p}.\vec{r} - Et)\right] dp$$

and
$$\frac{\partial^2\Psi}{\partial x^2}=\frac{1}{\left(\sqrt{2\pi\hbar}\right)^3}\int\left(\frac{-p_x^2}{\hbar^2}\right)a(p)exp.\left[\frac{i}{\hbar}(\vec{p}.\vec{r}-Et)\right]dp$$

Similarly
$$\frac{\partial^2\Psi}{\partial y^2}=\frac{1}{\left(\sqrt{2\pi\hbar}\right)^3}\int\left(\frac{-p_y^2}{\hbar^2}\right)a(p)exp.\left[\frac{i}{\hbar}(\vec{p}.\vec{r}-Et)\right]dp$$

and
$$\frac{\partial^2\Psi}{\partial z^2}=\frac{1}{\left(\sqrt{2\pi\hbar}\right)^3}\int\left(\frac{-p_z^2}{\hbar^2}\right)a(p)exp.\left[\frac{i}{\hbar}(\vec{p}.\vec{r}-Et)\right]dp$$

$\therefore$
$$\nabla^2\Psi=\frac{\partial^2\Psi}{\partial x^2}+\frac{\partial^2\Psi}{\partial y^2}+\frac{\partial^2\Psi}{\partial z^2}$$
$$=\frac{1}{\left(\sqrt{2\pi\hbar}\right)^3}\int\left(-\frac{p_x^2+p_y^2+p_z^2}{\hbar^2}\right)a(p)exp.\left[\frac{i}{\hbar}(\vec{p}.\vec{r}-Et)\right]dp$$
$$=\frac{1}{\left(\sqrt{2\pi\hbar}\right)^3}\int\left(-\frac{p^2}{\hbar^2}\right)a(p)exp.\left[\frac{i}{\hbar}(\vec{p}.\vec{r}-Et)\right]dp \qquad ...(iv)$$

Substituting the value of $\nabla^2\Psi$ from Eq. *(iv)* and $\frac{\partial\Psi}{\partial t}$ from Eq. *(iii)* in Eq. *(i),* we get.
$$-\frac{\hbar^2}{2m}\left(-\frac{p^2}{\hbar^2}\right)\frac{1}{\left(\sqrt{2\pi\hbar}\right)^3}\int a(p)\,exp.\left[\frac{i}{\hbar}(\vec{p}.\vec{r}-Et)\right]dp+V\Psi$$
$$=\frac{1}{\left(\sqrt{2\pi\hbar}\right)^3}\,i\hbar\left(\frac{-iE}{\hbar}\right)\int a(p)\,exp.\left[\frac{i}{\hbar}(\vec{p}.\vec{r}-Et)\right]dp$$

or
$$\frac{1}{\left(\sqrt{2\pi\hbar}\right)^3}\left(\frac{p^2}{2m}-E\right)\int a(p)exp.\left[\frac{i}{\hbar}(\vec{p}.\vec{r}-Et)\right]dp+V\Psi=0 \qquad ...(6.13)$$

This is Schrödinger's three dimensional wave equation in momentum representation.

One dimensional wave equation. For a one dimensional case Schrödinger's wave equation is given by
$$\frac{-\hbar^2}{2m}\frac{\partial^2\Psi}{\partial x^2}+V\Psi=i\hbar\frac{\partial\Psi}{\partial t} \qquad ...(v)$$

The equation for a wave packet in one dimension is
$$\Psi=\frac{1}{\sqrt{2\pi\hbar}}\int a(p)exp.\left[\frac{i}{\hbar}(p_x x-Et)\right]dp$$

$\therefore$
$$\frac{\partial\Psi}{\partial t}=\frac{1}{\sqrt{2\pi\hbar}}\int\left(\frac{-iE}{\hbar}\right)a(p)exp.\left[\frac{i}{\hbar}(p_x x-Et)\right]dp \qquad ...(vi)$$

and
$$\frac{\partial\Psi}{\partial x}=\frac{1}{\sqrt{2\pi\hbar}}\int\left(\frac{i}{\hbar}p_x\right)a(p)exp.\left[\frac{i}{\hbar}(p_x x-Et)\right]dp$$
$$\frac{\partial^2\Psi}{\partial x^2}=\frac{1}{\sqrt{2\pi\hbar}}\int-\left(\frac{p_x}{\hbar}\right)^2a(p)exp.\left[\frac{i}{\hbar}(p_x x-Et)\right]dp \qquad ...(vii)$$

Substituting the value of $\frac{\partial^2\Psi}{\partial x^2}$ from Eq *(vii)*, and $\frac{\partial\Psi}{\partial t}$ from Eq *(vi)* in Eq. (6.13), we get.

$$\frac{1}{\sqrt{2\pi\hbar}}\left(\frac{p_x^2}{2m}-E\right)\int a(p)\,exp.\left[\frac{i}{\hbar}(p_x x-Et)\right]dp+V\Psi=0 \quad ...(6.14)$$

This is Schrödinger's one dimensional wave equation in momentum representation.

Significance of momentum wave function. According to Born's interpretation of the wave function $\Psi(x, t)$

$$P(\vec{r})=\Psi\,\Psi^*=|\Psi|^2$$

represents the probability density of the particle in the state Ψ. Similarly the interpretation of momentum wave function $\phi(k)$ is that

$$P(\vec{p})=P(k)=\phi^*(\vec{k})\phi(\vec{k})$$

represents the probability density in $\vec{k}$ space or momentum space. Thus, the probability that the momentum of a particle lies between $\hbar k$ and $\hbar(k+dk)$ is given by

$$P(k)\,dk=\phi^*(k)\,\phi(k)\,dk$$
$$=|\phi(k)|^2\,dk \quad ...(6.15)$$

6.9 BORN'S INTERPRETATION

Schrödinger's time dependent wave equation for a particle states

$$\frac{-\hbar^2}{2m}\nabla^2\Psi(\vec{r},t)+V\Psi(\vec{r},t)=i\hbar\frac{\partial\Psi(\vec{r},t)}{\partial t}$$

The one dimensional Schrödinger's wave equation is

$$\frac{-\hbar^2}{2m}\frac{\partial^2\Psi(x,t)}{\partial x^2}+V\Psi(x,t)=i\hbar\frac{\partial\Psi(x,t)}{\partial t}.$$

Significance of wave function. In this equation, the wave function $\Psi(\vec{r},t)$ gives the behaviour of the particle at a given position $\vec{r}$ and at a given time t or the wave function $\Psi(x, t)$ gives the behaviour of the particle at a given point x *in one dimension* at a given time t. Its magnitude is *large* in regions where the probability of finding the particle is *high* and its magnitude is *small* in regions where the probability of finding the particle is *low. Thus, we can regard* Ψ *as a measure of the probability of the particle around a particular position.*

Born's interpretation. Born interpreted the relation between the wave function $\Psi(x, t)$ and the location of the particle x by drawing an analogy between intensity of light or photon beam and intensity of electron beam.

Meaning of E^2 for electromagnetic waves. Consider a beam of light (electromagnetic wave) incident normally on a screen. The electric field vector $\vec{E}$ of the beam is given by

$$E=A\sin(kx-\omega t)$$

where A is the value of the instantaneous electric field on the screen.

For an electromagnetic wave, the intensity I at a point due to a monochromatic beam of frequency v is given by $I=c\varepsilon_0\overline{E^2}$ where $\overline{E^2}$ is the average of the square of the instantaneous magnitudes of the electric field vector of the wave over a complete cycle, c the velocity of light and ε_0 the permittivity of free space. The intensity may also be interpreted as the number *N of photons each of energy hv crossing unit area per second* at the point under consideration normal to the direction of motion of the photons.

Thus $$I=N\,hv$$

And, we have $$I=c\varepsilon_0\overline{E^2}=N\,hv$$

or $$N=\frac{c\varepsilon_0}{hv}\overline{E^2}$$

or $$N \propto \overline{E^2} \qquad ...(6.16)$$

This relation is valid only when a large number of photons are involved *i.e.*, the beam has a large intensity. It is not possible to observe the usual pattern of intensity variation or diffraction if we consider the scattering of only a single photon by a crystal or passage of only a single photon through a narrow slit. Statistically speaking the probability of striking of a photon on a screen is very large at points where the intensity in the diffraction pattern is a *maximum* and very small at points where the intensity in the diffraction pattern is a *minimum.* In the case of a single photon, we can only say that the probability of the photon striking the screen is highest at places where the wave theory predicts a maximum and lowest at places where the wave theory predicts a minimum.

As $N \propto \overline{E^2}$, $\overline{E^2}$ is a measure of the probability of photon crossing unit area in unit time at the point under consideration. Therefore, in one dimension, $\overline{E^2}$ is a measure of the probability per unit length of finding the photon at the position x at time t.

Meaning of $|\Psi|^2$ for matter waves. In accordance with the above line of argument as suggested by Born, we interpret the square of the amplitude of the wave function Ψ at a point as a measure of *probability density i.e.*, the probability per unit volume, of the particles being at that point. Now, the probability of finding a particle at a particular region must be *real* and *positive* but the wave function Ψ is in general a complex quantity. Thus, the probability density is given by

$$P(\vec{r},t) = |\Psi(\vec{r},t)|^2 = \Psi^*\Psi \qquad ...(6.17)$$

where Ψ^* is the complex conjugate of Ψ.

Thus there is one fundamental difference between Ψ which represents a matter wave associated with a material particle and E which represents a wave associated with a photon. The difference is that $\Psi(\vec{r},t)$ is a complex quantity whereas the electric field E is a real and observable quantity. So Ψ itself has no physical significance but $|\Psi|^2$ has a meaning and as stated above represents the probability density. In a one dimensional case the probability of finding a particle in the length dx is

$$P\,dx = \Psi(x,t)\,\Psi^*(x,t)\,dx$$

To understand how the probability density is related to wave function Ψ, we shall explain normalisation, conservation of probability and steady state.

6.10 NORMALISATION

If we consider a small element of volume dv defined by the co-ordinates $(x, x + dx)$; $(y, y + dy)$; and $(z, z + dz)$ or $(\vec{r}, \vec{r} + d\vec{r})$, then the probability of finding the particle existing within this element of volume dv is given by

$$P(dv) = \Psi^*\Psi\,dv = \Psi\Psi^*\,dx\,dy\,dz$$

The probability of finding the particle in a finite volume v is given by

$$\int_v P dv = P(\text{v}) = \iiint_v \Psi^*\Psi\,dx\,dy\,dz$$

In one dimension only, we have

$$\int_{x_1}^{x_2} P dx = \int_{x_1}^{x_2} \Psi^*\Psi\,dx$$

The particle must always be somewhere in space so that extending the integral over all space, the probability becomes a certainty *i.e.*, it equals *unity*

$$\therefore \qquad \iiint_{All\ space} \Psi^*\Psi\,dx\,dy\,dz = 1$$

or $$\iiint_{All\ space} |\Psi|^2\, dx\, dy\, dz = 1 \qquad ...(6.18)$$

The process of integration over all possible locations to give unity is called **normalisation.**

6.11 CONSERVATION OF PROBABILITY

If it is further assumed that the particles are neither created nor destroyed, the total probability of locating a system in the region continues to be *one* for all time *i.e.* the *normalisation is preserved in time* or is independent of time. Further, just as matter is conserved in hydrodynamics, charge in electrodynamics, the probability density is also conserved.

In other words, if the probability of finding a particle in some *bounded* region decreases, the possibility of finding it outside that region increases by the same amount. *This principle is known as conservation of probability.*

6.12 STEADY STATE

When a system does not change with time, it is said to be in a *steady state* or *a stationary state.* Such a system can be represented by a wave function ψ of space co-ordinates only and it is represented as $\psi(\vec{r})$ in the three dimensional case and $\psi(x)$ in a one dimensional case. The function $\psi(x)$ as well as $\frac{d\psi(x)}{dx}$ must be *finite, single valued* and *continuous* for all values of x. If $\psi(x)$ is infinite at any point in space, the probability $|\psi(x)|^2$ of locating the system at that point is infinite *i.e. the particle is completely localised.* This is against the wave property of the system. If $\psi(x)$ is multi-valued, then the probability of finding the system between x and $x + dx$ given by $|\psi(x)|^2$ will also have more than one values. This is physically unacceptable. The derivative of a discontinuous function is always infinite. Therefore if $\psi(x)$ and $\frac{d\psi(x)}{dx}$ are to be finite, they must be continuous.

6.13 PROBABILITY DENSITY

For a particle in a state Ψ where $\Psi(x, t)$ is the normalised wave function associated with the particle, probability density is defined as the product of Ψ and its complex conjugate Ψ^*

$\therefore$ Probability desity $P = \Psi^*(x, t)\, \Psi(x, t)$

In a one dimesioal case the probability of locatign the particle i a distace dx is given by

$$P\, dx = \Psi^*\, \Psi\, dx$$

In three dimensional space

$$Pdv = \Psi^*\, \Psi\, dv = \Psi^*\, \Psi\, dxdydz.$$

Probability density current. In a one dimensional case probability density current or flux is defined by the expression

$$S(x, t) = \frac{i\hbar}{2m}\left[\Psi \frac{\partial \Psi^*}{\partial x} - \Psi^* \frac{\partial \Psi^*}{\partial x}\right] \qquad ...(6.19)$$

where Ψ and Ψ^* have the same meaning as for probability density, $\frac{\partial \Psi^*}{\partial x}$ and $\frac{\partial \Psi}{\partial x}$ are the partial derivatives of Ψ^* and Ψ with respect to x.

In a three dimensional case probability current

$$S = \frac{i\hbar}{2m}[\Psi\, \Delta\, \Psi^* - \Psi^* \nabla \Psi] \qquad ...(6.19)$$

The value of probability density P varies with time from element to element. This flow probability density is known as probability density current or simply probability current.

6.14 NORMALISATION FACTOR

For a one dimensional case, we have

$$P = \Psi(x, t)\, \Psi^*(x, t)$$

where P is the probability per unit length of finding the particle at the position x at time t. Therefore probability of finding the particle in the length dx is

$$P\, dx = \Psi(x, t)\, \Psi^*(x, t)\, dx$$

and the total probability of finding the particle *somewhere* along X-axis is

$$P_T = \int_{-\infty}^{+\infty} P\, dx = \int_{-\infty}^{+\infty} \Psi(x,t)\, \Psi^*(x,t)\, dx$$

But total probability of finding the particle must be **unity**

$$\therefore \qquad \int_{-\infty}^{+\infty} \Psi(x,t)\Psi^*(x,t)\, dx = 1 \qquad \text{...(6.20)}$$

A wave function which satisfies the above condition is said to be normalised.

Equation (6.20) gives the mathematical expression for the normalisation condition.

Normalisation factor. Some times Ψ is not a normalised wave function. In such a case we can multiply Ψ by a constant A such that the new wave function $A\Psi$ is a normalised wave function because if Ψ is a solution of Schrödinger's wave equation $A\Psi$ is also a solution.

If $A\Psi$ is a normalised wave function then

$$\int_{\text{All space}} (A\Psi)\, (A\Psi)^*\, dx\, dy\, dz = 1$$

or

$$\int_{\text{All space}} (A\Psi)\, (A\Psi)^*\, dv = 1 \quad \text{where } dv = dx\, dy\, dz$$

or in a one dimensional case

$$\int_{-\infty}^{+\infty} A\Psi(x, t)\, A\Psi^*(x,t)\, dx = 1$$

$$|A^2| = \frac{1}{\int_{-\infty}^{+\infty} \Psi\Psi^* dx}$$

In a three dimensional case

$$|A^2| = \frac{1}{\int_{\text{All space}} \Psi\Psi^* dv} \qquad \text{...(6.21)}$$

A is known as *normalising constant* or *normalising factor.*

Normalisation is independent of time. The wave function $\Psi(x, t)$ can be written as a product of two separate functions $\Psi(x)$ a function of only x and $f(t)$ a function of only t, and is given by

$$\Psi(x,t) = \psi(x)\, f(t) = \psi(x)\, e^{-\frac{iEt}{\hbar}}$$

Substituting this value of $\Psi(x, t)$ and corresponding value of $\Psi^*(x, t)$ in equation (i), we have

$$\int_{-\infty}^{+\infty} \psi(x) e^{-iEt/\hbar}\, \psi^*(x) e^{iEt/\hbar}\, dx = 1$$

or

$$\int_{-\infty}^{+\infty} \psi(x)\, \psi^*(x)\, dx = 1 \qquad \text{...(6.22)}$$

As $\psi(x)$ and $\psi^*(x)$ are functions of x only and are independent of time, the probability of locating a system in the region $-\infty$ to $+\infty$ continues to be *one* for all times *i.e.* normalisation is preserved in time or is *independent* of time.

6.15 IMPORTANCE OF NORMALISATION

The amplitude of a wave function is arbitrary. It is because Schrödinger's wave equation is a linear equation and it can be multiplied by a constant of any arbitrary magnitude. The wave function so obtained still remains a solution of the Schrödinger's wave equation. After normalisation the amplitude of the wave function has a fixed constant value. Thus normalisation is important to give Schrödinger's wave equation a constant fixed amplitude.

Example 6.5. *Find the normalisation constant of the particle described by the Gaussian wave packet wave function.*

$$\psi(x) = A\, e^{-\frac{\alpha^2 x^2}{2}}\, e^{ikx}$$ **(P.U. 2007)**

Solution. The normalisation condition is

$$\int_{-\infty}^{+\infty} \psi(x)\, \psi^*(x)\, dx = 1$$

where $\psi^*(x)$ is the complex conjugate of $\psi(x)$

Now $\quad \psi(x) = A\, e^{-\frac{\alpha^2 x^2}{2}}\, e^{ikx}$

$\therefore \quad \psi^*(x) = A\, e^{\frac{-\alpha^2 x^2}{2}}\, e^{-ikx}$

and $\quad \psi(x)\psi^*(x) = A^2\, e^{-\alpha^2 x^2}$

Hence $\displaystyle\int_{-\infty}^{+\infty} \psi(x)\, \psi^*(x)\, dx = A^2 \int_{-\infty}^{+\infty} e^{-\alpha^2 x^2}\, dx = 1$

Now $\displaystyle\int_{-\infty}^{+\infty} e^{-\alpha^2 x^2}\, dx$ is a standard integral the value of which is equal to $\dfrac{\sqrt{\pi}}{\alpha}$

$$\therefore \quad A^2 \frac{\sqrt{\pi}}{\alpha} = 1 \quad \text{or} \quad A^2 = \frac{\alpha}{\sqrt{\pi}} \quad \text{or} \quad A = \frac{\alpha^{1/2}}{\pi^{1/4}}$$

Hence the normalisation constant $A = \dfrac{\alpha^{1/2}}{\pi^{1/4}}$

Example 6.6. *Normalise the one dimensional wave function $\psi(x) = e^{-|x|} \sin \alpha x$.* **(P.U. 2006)**

Solution. Let the normalised wave function be

$$\psi'(x) = A\, \psi(x)$$

where A is the normalisation constant.

Then, according to normalisation condition

$$\int_{-\infty}^{+\infty} \psi'^*\, \psi'\, dx = 1$$

or

$$|A|^2 \int_{-\infty}^{+\infty} \psi^*\, \psi\, dx = 1$$

Now $$\int_{-\infty}^{+\infty} \psi^* \psi\, dx = \int_{-\infty}^{0} \psi^* \psi\, dx + \int_{0}^{+\infty} \psi^* \psi\, dx$$

Put $$\int_{-\infty}^{0} \psi^* \psi\, dx = I_1 \quad \text{and} \quad \int_{0}^{+\infty} \psi^* \psi\, dx = I_2\text{, then}$$

$$I_1 = \int_{-\infty}^{0} \psi^* \psi\, dx = \int_{-\infty}^{0} (e^{+x} \sin \alpha x)(e^{+x} \sin \alpha x)\, dx = \int_{-\infty}^{0} e^{2x} \sin^2 \alpha x\, dx$$

and $$I_2 = \int_{0}^{+\infty} \psi^* \psi\, dx = \int_{0}^{+\infty} (e^{-x} \sin \alpha x)(e^{-x} \sin \alpha x)\, dx = \int_{0}^{+\infty} e^{-2x} \sin^2 \alpha x\, dx$$

Let us first solve I_1,

$$I_1 = \int_{-\infty}^{0} e^{2x} \sin^2 \alpha x\, dx = \int_{-\infty}^{0} e^{2x} \left(\frac{1 - \cos 2\alpha x}{2}\right) dx$$

$$= \frac{1}{2}\int_{-\infty}^{0} e^{2x}\, dx - \frac{1}{2}\int_{-\infty}^{0} e^{2x} \cos 2\alpha x\, dx$$

$$= \frac{1}{2} \times \frac{1}{2}[e^{2x}]_{-\infty}^{0} - \frac{1}{2} I_1' = \frac{1}{4} - \frac{1}{2} I_1'$$

where $$I_1' = \int_{-\infty}^{0} e^{2x} \cos 2\alpha x\, dx$$

Integrating, we have

$$I_1' = \left[\left(\frac{e^{2x}}{2} \cos 2\alpha x\right)_{-\infty}^{0} + \int_{-\infty}^{0} 2\alpha \sin 2\alpha x \frac{e^{2x}}{2}\, dx\right]$$

$$= \frac{1}{2} + \alpha \left[\left(\sin 2\alpha x \frac{e^{2x}}{2}\right)_{-\infty}^{0} - 2\alpha^2 \int_{-\infty}^{0} \cos 2\alpha x \frac{e^{2x}}{2}\, dx\right]$$

$$= \frac{1}{2} + 0 - \alpha^2 \int_{-\infty}^{0} \cos 2\alpha x\, e^{2x}\, dx = \frac{1}{2} - \alpha^2 I_1'$$

$$\therefore \quad I_1' (1 + \alpha^2) = \frac{1}{2} \quad \text{or} \quad I_1' = \frac{1}{2(1 + \alpha^2)}$$

Hence $$I_1 = \frac{1}{4} - \frac{1}{2}\frac{1}{2(1 + \alpha^2)} = \frac{1}{4}\left[1 - \frac{1}{1 + \alpha^2}\right] = \frac{1}{4}\left[\frac{\alpha^2}{1 + \alpha^2}\right]$$

Similarly $$I_2 = \int_{0}^{+\infty} e^{-2x} \sin^2 \alpha x\, dx = \frac{1}{4}\left[\frac{\alpha^2}{1 + \alpha^2}\right]$$

Hence $$|A|^2 \int_{-\infty}^{+\infty} \psi^* \psi\, dx = |A|^2 \left[\frac{1}{4}\frac{\alpha^2}{1 + \alpha^2} + \frac{1}{4}\frac{\alpha^2}{1 + \alpha^2}\right] = |A|^2 \frac{1}{2}\left[\frac{\alpha^2}{1 + \alpha^2}\right] = 1$$

or $$|A|^2 = \frac{2(1 + \alpha^2)}{\alpha^2} \quad \text{or} \quad A = \sqrt{\frac{2(1 + \alpha^2)}{\alpha^2}}$$ and the normalised wave function is

$$\psi'(x) = A\,\psi(x) = \sqrt{\frac{2\,(1+\alpha^2)}{\alpha^2}}\; e^{-|x|} \quad \sin \alpha\, x$$

Example 6.7. *Normalise the one dimensional wave function $\psi(x) = e^{-|x|} \sin x$.* ***(P.U. 2005)***

Solution. This problem is the same as solved above except that here we have $\sin x$ instead of $\sin \alpha x$ *i.e.* $\alpha = 1$.

$\therefore$ Proceeding as above, we get normalisation constant

$$A = \sqrt{\frac{2\,(1+\alpha^2)}{\alpha^2}} = \sqrt{\frac{2\,(1+1)}{1}} = \sqrt{4} = 2$$

Hence normalised wave function is $\psi'(x) = 2e^{-|x|} \sin x$.

Example 6.8. *If $\Psi_1(x, t)$ and $\Psi_2(x, t)$ are both solutions of Schrödinger's wave equation for a given potential $V(x, t)$ show that $\Psi = a_1 \Psi_1 + a_2 \Psi_2$ in which a_1 and a_2 are arbitrary constants is also a solution.* ***(G.N.D.U. 2006; Pbi.U. 2003)***

Solution. The time dependent Schrödinger's wave equation in one dimension is

$$-\frac{\hbar^2}{2m}\frac{\partial^2}{\partial x^2}\Psi(x,t) + V(x,t)\Psi(x,t) = i\hbar\frac{\partial}{\partial t}\Psi(x,t) \quad \text{...(i)}$$

As $\Psi_1(x, t)$ and $\Psi_2(x, t)$ are both solutions of this equation

$$\therefore \quad -\frac{\hbar^2}{2m}\frac{\partial^2}{\partial x^2}\Psi_1(x,t) + V(x,t)\Psi_1(x,t) = i\hbar\frac{\partial}{\partial t}\Psi_1(x,t) \quad \text{...(ii)}$$

and

$$-\frac{\hbar^2}{2m}\frac{\partial^2}{\partial x^2}\Psi_2(x,t) + V(x,t)\Psi_2(x,t) = i\hbar\frac{\partial}{\partial t}\Psi_2(x,t) \quad \text{...(iii)}$$

Multiplying equation *(ii)* by a_1 and equation *(iii)* by a_2, adding, rearranging and putting Ψ_1 for $\Psi_1(x, t)$, Ψ_2 for $\Psi_2(x, t)$ and V for $V(x, t)$, we have

$$-\frac{\hbar^2}{2m}\frac{\partial^2}{\partial x^2}(a_1\Psi_1 + a_2\Psi_2) + V(a_1\Psi_1 + a_2\Psi_2)$$

$$= i\hbar\frac{\partial}{\partial t}(a_1\Psi_1 + a_2\Psi_2) \quad \text{...(iv)}$$

Comparing equation *(iv)* with equation *(i)*, we get

$$\Psi(x, t) = a_1\Psi_1 + a_2\Psi_2$$

and therefore $\Psi = a_1 \Psi_1 + a_2 \Psi_2$ is also a solution of Schrödinger's time dependent wave function.

6.16 SUPERPOSITION OF STATES

Schrödinger's equation is a linear equation. The solutions of this equation, therefore, obey the principle of superposition. The principle states *'If two possible states of a system (or particle) are represented by wave functions ψ_1 and ψ_2, then corresponding to their linear combination there exists another state ψ of the system'*.

Thus $\quad \psi = c_1\psi_1 + c_2\psi_2$

where c_1 and c_2 are ordinary complex numbers independent of space co-ordinates but possibly functions of time.

The function ψ will satisfy the same conditions of continuity and integrability which are satisfied by ψ_1 and ψ_2.

In a general case, for more than two states

$$\psi = c_1 \psi_1 + c_2 \psi_2 + c_3 \psi_3 + \ldots\ldots\ldots c_n \psi_n$$

$$= \sum_{i=1}^{n} c_i \psi_i \qquad \ldots(6.23)$$

According to the principle of superposition of states, if we multiply a wave function by an arbitrary non-zero complex number, the new function will correspond to the same state of the particle. This principle further shows that quantum mechanics is a linear theory having no correspondence in classical mechanics. The explanation of the phenomenon of interference of light waves is based upon the principle of superposition.

6.17 OPERATOR

An operator is a mathematical rule or prescription. Mathematical operations in algebra and calculus like adding, subtracting, multiplying, dividing, finding the square root, differentiating with respect to a variable, or integration are represented by characteristic symbols like, +, –, ×, ÷, $\sqrt{\ }$, $\frac{\partial}{\partial t}$, and $\int f dx$ can be considered as operators.

Thus if A is an operator generally represented as $\hat{A}$ and stands for the operation $\frac{\partial}{\partial x}$, then

$$\hat{A} x^3 = \frac{\partial}{\partial x}(x^3) = 3x^2$$

and
$$4x^2 \hat{A} x^3 = 4x^2 \frac{\partial}{\partial x}(x^3) = 12x^4.$$

Commuting and non-commuting operators. Consider two operators $\hat{A}$ and $\hat{B}$ where $\hat{A}$ stands for $\frac{\partial}{\partial x}$ and $\hat{B}$ stands for the operation $x^2\, \times$, then if the two operators $\hat{A}$ and $\hat{B}$ operate together on a function say y which is a function of variable x, then

$$\hat{A}\hat{B}\, y = \frac{\partial}{\partial x}(x^2 \times)\, y = \frac{\partial}{\partial x} x^2 y = 2xy + x^2 \frac{\partial y}{\partial x}$$

On the other hand

$$\hat{B}\hat{A}\, y = (x^2 \times) \frac{\partial}{\partial x}\, y = x^2 \frac{\partial y}{\partial x}$$

Thus if there are two operators applied together $\hat{A}\,\hat{B}$ means, first apply $\hat{B}$, then $\hat{A}$ and $\hat{B}\,\hat{A}$ means first apply $\hat{A}$ then $\hat{B}$. From the above we find that $\hat{A}\hat{B} \neq \hat{B}\hat{A}$ because $\hat{A}\hat{B} - \hat{B}\hat{A} = 2xy$.

Now consider two operators $\hat{A}$ and $\hat{B}$ where $\hat{A}$ represents $\frac{\partial}{\partial x}$ and $\hat{B}$ represents $a\, \times$; a being a constant, then

$$\hat{A}\hat{B}\, y = \frac{\partial}{\partial x}(a \times) y = \frac{\partial}{\partial x}\, ay = a \frac{\partial y}{\partial x}$$

and
$$\hat{B}\hat{A}\, y = (a \times) \frac{\partial}{\partial x}\, y = a \frac{\partial y}{\partial x}$$

Thus
$$\hat{A}\hat{B} = \hat{B}\hat{A} \text{ and } \hat{A}\hat{B} - \hat{B}\hat{A} = 0.$$

Two operators $\hat{A}$ and $\hat{B}$ for which $\hat{A}\hat{B} = \hat{B}\hat{A}$ or for which $\hat{A}\hat{B} - \hat{B}\hat{A} = 0$ are said to be *commuting.*

Commutator. The commutator of two operators $\hat{A}$ and $\hat{B}$ denoted as $[\hat{A}, \hat{B}]$ is defined as $[\hat{A}, \hat{B}] = \hat{A}\hat{B} - \hat{B}\hat{A}$.

Linear operator. An operator $\hat{A}$ is said to be *linear* if it satisfies the following two equations

$$\hat{A}(\alpha f) = \alpha \hat{A} f \quad \text{...(i)}$$

and

$$\hat{A}(f+g) = \hat{A}f + \hat{A}g \quad \text{...(ii)}$$

where α is a constant

Linear momentum operator see article 6.19 (*iii*) **Energy operator** see article 6.19 (iv) and (v).

Hermitian operator. An operator $\hat{A}$ is called *Hermitian* if it satisfies the equation

$$\int_{-\infty}^{+\infty} f^*(\hat{A}g)\, dx = \int_{-\infty}^{+\infty} (\hat{A}f)^* g\, dx$$

where $f(x)$ and $g(x)$ are any two well behaved functions of x.
Hermitian operator may also be written as

$$\int_{-\infty}^{\infty} f^* \hat{A} g\, dx = \int_{-\infty}^{\infty} g \hat{A}^* f^*\, dx.$$

Operator applied to a wave function. *In quantum mechanics an operator when applied to a wave function gives the corresponding observable quantity of the system multiplied by the wave function.* For example, consider the wave function

$$\Psi = Ae^{\frac{i}{\hbar}(xp_x - Et)}$$

then

$$\frac{\partial \Psi}{\partial x} = \frac{i}{\hbar} p_x Ae^{\frac{i}{\hbar}(xp_x - Et)} = \frac{i}{\hbar} p_x \Psi$$

or

$$\frac{\hbar}{i}\frac{\partial}{\partial x}\Psi = p_x \Psi$$

If we represent the operator p_x denoted as $\hat{p}_x$ corresponding to the observable property x-component of momentum p_x by $\frac{\hbar}{i}\frac{\partial}{\partial x}$ and if it operates on wave function Ψ, we get

$$\hat{p}_x \Psi = \frac{\hbar}{i}\frac{\partial}{\partial x}\Psi = p_x \Psi$$

Thus, the linear momentum operator $\hat{p}_x$ when applied to the wave function Ψ gives the corresponding observable quantity *i.e.* linear momentum p_x multiplied by the wave function.

Similarly,

$$\frac{\partial \Psi}{\partial t} = \frac{-i}{\hbar} EAe^{i/\hbar(xp_x - Et)} = \frac{-i}{\hbar} E\Psi$$

or

$$i\hbar\frac{\partial \Psi}{\partial t}\Psi = E\Psi$$

If we represent the operator E (denoted by $\hat{E}$) corresponding to the observable property 'total energy' E by $i\hbar\frac{\partial}{\partial t}$ and if it operates on the wave function Ψ, we get

$$\hat{E}\Psi = i\hbar\frac{\partial}{\partial t}\Psi = E\Psi \qquad ...(6.24)$$

Thus, the total energy operator $\hat{E}$ when applied to the wave function Ψ gives the corresponding observable quantity *i.e.* the total energy multiplied by the wave function.

6.18 EXPECTATION VALUE

Dynamical variable. Any quantity which depends upon the co-ordinates of a particle is called its dynamical variable e.g. position, momentum or energy in terms of which all properties of motion can be expressed.

Expectation value. Consider a variable say *x*- the positon co-ordinate of a particle along X-axis. Suppose it takes *n* discrete *values* $x_1, x_2, x_3 ......x_n$, then

$$\text{Average value of } x = <x> = \frac{x_1 + x_2 + ... + x_i ....x_n}{n}$$

$$= \frac{\sum_{i=1}^{n} x_i}{n}$$

Thus to find the *average value* we add all the values of the variable and divide by the total number of values.

If the variable *x* takes the value x_1, m_1 times, the value x_2, m_2 times and so on, then

$$\text{Average value of } x = <x> = \frac{m_1 x_1 + m_2 x_2 .... + m_i x_i ...... + m_n x_n}{m_1 + m_2 + ... + m_i + ...m_n}$$

$$= \frac{\sum_{i=1}^{n} m_i x_i}{\sum_{i}^{n} m_i} = \frac{\sum_{i=1}^{n} m_i x_i}{n}$$

where $n = m_1 + m_2 + ... + m_i + ...m_n$

or
$$<x> = \sum_{i=1}^{n} x_i \frac{m_i}{n}$$

Now $\frac{m_i}{n} = P_i$ the *probability* of occurrence of the value x_i

$$\therefore \qquad <x> = \sum_{i=1}^{n} x_i P_i$$

If however, the variable *x* is continuous, instead of discrete, then in place of summation we substitute the integral sign and get

$$<x> = \int x P\, dx . \qquad ...(6.25)$$

Relation with probability: Now, consider a particle described by the wave function Ψ *(x, t)*. The probability of finding the particle in a distance *dx* around *x* is given by

$$P\, dx = \Psi\ (x,t)\ \Psi^*\ (x,t)\ dx$$

As *P dx* is only the *probability,* it means it is not possible to predict *exactly* the position *x* of the particle. In other words; if we make a large number of observations, then each measurement will give a different result. Hence we use the average value of a large number of measurements of *x* to predict the position of the *expectation value.*

Expectation value in operator notation. To define the expectation value in terms of corresponding operators, we find that when a given system is in a state Ψ which need not be an eigen function of the operator $\hat{A}$, the average of a large number of independent measurements of an observable corresponding to the operator $\hat{A}$ is given by

$$< A > = \frac{\int\limits_{\text{Available region}} \Psi^* \hat{A} \Psi \, d\vec{r}}{\int\limits_{\text{Available region}} \Psi^* \Psi \, d\vec{r}}$$

$< A >$ is called the *expectation value* of the observable.

For a normalised function $\int\limits_{\text{Available region}} \Psi^* \Psi \, d\vec{r} = 1$

$$\therefore \quad < A > = \int\limits_{\text{Available region}} \Psi^* \hat{A} \Psi \, d\vec{r}$$

The above result may also be put in the form

$$<A> = \int\limits_{\text{Available region}} \Psi^* \hat{A} \Psi \, dv$$

$$= \int \Psi^* \hat{A} \Psi \, dv.$$

(*i*) **Expectation value of position** x. In a one dimensional case the expectation value of position co-ordinate x given by

$$< x > = \int_{-\infty}^{+\infty} \Psi^* (x, t) \, \hat{x} \, \Psi (x, t) \, dx = \int_{-\infty}^{+\infty} x |\Psi|^2 \, dx. \quad \text{...(6.26)}$$

(*ii*) **Expectation value of momentum.** The expectation value for the momentum vector $\vec{p}$ is given by

$$<\vec{p}> = \int \Psi^* \, \hat{p} \, \Psi \, d\vec{r} = \int\limits_{\text{Available region}} \Psi^* \, (-i\hbar \vec{\nabla}) \, \Psi \, d\vec{r}$$

$$= -i\hbar \int\limits_{\text{Available region}} \Psi^* \, \vec{\nabla} \Psi \, d\vec{r}$$

In a one dimensional case

$$p_x = -i\hbar \frac{\partial}{\partial x}$$

$$\therefore \quad <p_x> = -i\hbar \int_{-\infty}^{+\infty} \Psi^* (x, t) \frac{\partial}{\partial x} \Psi (x, t) \, dx = -i\hbar \int_{-\infty}^{+\infty} \Psi^* \frac{\partial \Psi}{\partial x} dx. \quad \text{...(6.27)}$$

(*iii*) **Expectation value of energy.** The expectation value of energy is given by

$$<E> = \int_{-\infty}^{+\infty} \Psi^* \hat{E} \Psi \, dx = \int_{-\infty}^{+\infty} \Psi^* \left(i\hbar \frac{\partial}{\partial t} \right) \Psi \, dx$$

$$= i\hbar \int_{-\infty}^{+\infty} \Psi^* \frac{\partial \Psi}{\partial t} dx. \quad \text{...(6.28)}$$

Example 6.9. *A particle is described by a wave function* $\psi = \sqrt{\frac{2}{L}} \sin \frac{n\pi}{L} x$ *for* $0 < x < L$ *and* $= 0$ *elsewhere. Find the expectation value of the momentum of the particle.* **(G.N.D.U. 2005)**

Solution. Given $\psi = \sqrt{\frac{2}{L}} \sin \frac{n\pi}{L} x$

$$= \left(\frac{2}{L}\right)^{\frac{1}{2}} \sin \frac{n\pi}{L} x$$

The expectation value of momentum is gven by

$$<p> = -i\hbar \int_{-\infty}^{+\infty} \psi^* \frac{\partial}{\partial x}[\psi]\, dx \qquad ...(i)$$

The wave function exists in the region $0 < x < L$ and is equal to zero elsewhere

$\therefore$ Eq. (*i*) can be put in the form

$$<p> = -i\hbar \int_0^L \psi^* \frac{\partial \psi}{\partial x} dx$$

or

$$<p> = -i\hbar \int_0^L \left(\frac{2}{L}\right)^{\frac{1}{2}} \sin \frac{n\pi}{L} x \frac{\partial}{\partial x}\left[\left(\frac{2}{L}\right)^{\frac{1}{2}} \sin \frac{n\pi}{L} x\right] dx$$

$$= -\frac{2i\hbar}{L} \int_0^L \sin \frac{n\pi}{L} x \frac{\partial}{\partial x}\left[\sin \frac{n\pi}{L} x\right] dx$$

$$= -\frac{2i\hbar}{L}\frac{n\pi}{L} \int_0^L \sin \frac{n\pi}{L} x \cos \frac{n\pi}{L} x\, dx$$

$$= -\frac{2i\hbar}{L}\frac{n\pi}{L}\frac{1}{2} \int_0^L 2\sin \frac{n\pi}{L} x \cos \frac{n\pi}{L} x\, dx$$

$$= -\frac{i\hbar}{L}\frac{n\pi}{L} \int_0^L \sin \frac{2n\pi}{L} x\, dx$$

$$= \frac{i\hbar}{L}\frac{n\pi}{L}\left[\frac{\cos \frac{2n\pi}{L} x}{\frac{2\pi n}{L}}\right]_0^L = \frac{i\hbar}{2L}[\cos 2n\pi - \cos 0]$$

$$= \frac{i\hbar}{2L}[1 - 1] = 0 \quad \text{since } \cos 2n\pi = 1, \cos 0 = 1.$$

$\therefore$ Expectation value of momentum $<p> = 0$

6.19 POSTULATES OF QUANTUM MECHANICS

On the basis of the wave nature of a moving particle, we have developed Schrodinger's wave equation. In order to use the wave equation and to obtain a quantitative description of matter on the atomic or nuclear scale like the values of linear momentum, angular momentum and energy etc. of a micro-particle, we have to make some assumptions which are known as postulates of wave mechanics. The wave function $\Psi(\vec{r}, t)$ gives an almost complete knowledge of the behaviour of the particle in space and time consistent with Heisenberg's uncertainty principle. Similarly, the wave function Ψ

$(\vec{r})$ gives the behaviour of the particle in the stationary state which is independent of time. The fundamental postulates are;

Postulate 1. *There is a linear quantum mechanical operator corresponding to every observable which describes the motion of a particle.*

(*i*) **Cartesian co-ordinate operator.** These operators are represented as $\hat{x}, \hat{y}, \hat{z}$ and correspond to *x, y, z* co-ordinates respectively. When any of these operators operates on a wave function, the result is the function multiplied by the corresponding co-ordinate.

Thus $\quad \hat{x}\Psi = x\Psi.$

These operators are also called *position operators*.

(*ii*) **Time operator:** The time operator is represented as $\hat{t}$. When this operator operates on a wave function ψ, the result is function multiplied by time.

Thus $\quad \hat{t}\,\psi = t\psi$

(*iii*) **Linear momentum operator.** These operators are represented as $\hat{p}_x, \hat{p}_y, \hat{p}_z$ and $\hat{p}$ corresponding to *x, y or* z component or total value of linear momentum respectively. Their values are

$$\hat{p}_x = -i\hbar\frac{\partial}{\partial x};\quad \hat{p}_y = -i\hbar\frac{\partial}{\partial y};\quad \hat{p}_z = -i\hbar\frac{\partial}{\partial z}$$

and $\quad \hat{p} = -i\hbar\vec{\nabla}$

When any of the first three linear momentum operators operates on a wave function, the result is the function multiplied by corresponding components of momentum. Thus

$$\hat{p}_x\Psi = -i\hbar\frac{\partial\Psi}{\partial x} = p_x\Psi$$

When the operator $\hat{p}$ operates on the function, we get the function multiplied by total momentum. Thus

$$\hat{p}\Psi = -i\hbar\vec{\nabla}\,\Psi = p\Psi.$$

(*iv*) **Potential energy operator.** This operator is represented as $\hat{V}(\vec{r})$ and corresponds to the potential energy $V(\vec{r})$. Thus

$$\hat{V}(\vec{r})\Psi = V(\vec{r})\,\Psi.$$

(*v*) **Total energy opertor.** This operator is represented as $\hat{E}$ and corresponds to the total energy *E* of the system. Its values is $\hat{E} = i\hbar\frac{\partial}{\partial t}$. When it operates on a function the result is the function multiplied by the observable *E*. Thus

$$\hat{E}\Psi = i\hbar\frac{\partial\Psi}{\partial t} = E\Psi. \qquad ...(6.29)$$

(*vi*) **Hamiltonian operator.** In classical mechanics, Hamiltonian is defined as $H = \frac{p^2}{2m} + V$ and it gives the total energy *E* of the system. Hence the total energy operator for a non-relativistic particle is also represented by Hamiltonian operator $\hat{H}$. As $\hat{p} = i\hbar\vec{\nabla}$, the Hamiltonian operator

$$\hat{H} = -\frac{\hbar^2}{2m}\nabla^2 + V$$

When it operates on the function, the result is the function multiplied by total energy E. Thus

$$\hat{H}\Psi = \left(-\frac{\hbar^2}{2m}\nabla^2 + V\right)\Psi = E\Psi. \qquad ...(6.30)$$

Postulate 2. Eigen value

The only possible values that can be obtained from the measurement of an observable of a system, whose operator is $\hat{A}$ *are the eigen values* A_n *of the equation.* Thus

$$\hat{A}\Psi = A_n \Psi_n$$

The index n used to identify the eigen values and eigen functions is known as the *quantum number.* A system described by eigen function Ψ_n is said to be in the quantum state *n.*

For example, *(i)* If a particle is somewhere on the *X*-axis we have the momentum operator

$$\hat{p}_x = -i\hbar\frac{\partial}{\partial x}, \text{ then the equation is}$$

$$-i\hbar\frac{\partial}{\partial x}\Psi = (p_x)_n \Psi_n$$

Hence $(p_x)_n$ gives the n values that can be obtained from the measurement of the observable *x*-component of linear momentum whose operator is $\hat{p}_x$. Similarly for the total momentum operator $\hat{p} = -i\hbar\vec{\nabla}$ the equation is

$$-i\hbar\vec{\nabla}\Psi = p_n \Psi_n$$

Ψ_n is called the eigen function and p_n an eigen value of the linear momentum.

(ii) For the total energy $\hat{E} = i\hbar\frac{\partial}{\partial t}$, the equation is

$$i\hbar\frac{\partial\Psi}{dt} = E_n \Psi_n$$

Similarly for the Hamiltonian operator $\hat{H} = -\frac{\hbar^2}{2m}\nabla^2 + V$, the equation is

$$\left(-\frac{\hbar^2}{2m}\nabla^2 + V\right)\Psi = E_n \Psi_n \qquad ...(6.31)$$

Again Ψ_n is an eigen function and E_n an eigen value of total energy.

Postulate 3.

The probability of finding the particle in the space, between $\vec{r}$ *and* $\vec{r} + d\vec{r}$, *is* $\Psi^*\Psi\, d\vec{r}$ *where* Ψ *is the function and* Ψ^* *its complex conjugate.*

Thus the probability of finding the particle in a finite interval between $\vec{r}_1$ and $\vec{r}_2$ is given by $\int_{r_1}^{r_2} \Psi^*\Psi\, d\vec{r}$. When there is only one particle which is present somewhere, the total probability of finding the particle somewhere is **unity.**

$$\therefore \qquad \int_{-\infty}^{+\infty} \Psi^*\Psi\, d\vec{r} = 1. \qquad ...(6.32)$$

Postulate 4. Expectation Value

When a given system is in a state Ψ which need not be an eigen function of the operator $\hat{A}$, the average of a large number of independent measurements of an observable corresponding to the operator $\hat{A}$ is given by

$$<A> = \frac{\int_{Available\ region} \Psi^* \hat{A} \Psi \, d\vec{r}}{\int_{Available\ region} \Psi^* \Psi \, d\vec{r}}$$

<A> is called the *expectation value* of the observable.

For a normalised function $\int_{Available\ region} \Psi^* \Psi \, d\vec{r} = 1$

$$\therefore \quad <A> = \int_{Available\ region} \Psi^* \hat{A} \Psi \, d\vec{r}.$$

(*i*) **Expectation value of position *x*.** In a one dimensional case the expectation value of position co-ordinate x is given by

$$<x> = \int_{-\infty}^{+\infty} \Psi^*(x,t)\, \hat{x}\, \Psi(x,t)\, dx = \int_{-\infty}^{+\infty} x|\Psi|^2\, dx \qquad ...(6.33)$$

and (*ii*) the **expectation value for the momentum** vector $\vec{p}$ is given by

$$<\vec{p}> = \int_{Available\ region} \Psi^*(-i\hbar \vec{\nabla}) \Psi d\vec{r}$$

$$= -i\hbar \int_{Available\ region} \Psi^* \vec{\nabla} \Psi d\vec{r}$$

In a one dimensional case

$$p_x = -i\hbar \frac{\partial}{\partial x}$$

$$\therefore \quad <p_x> = -i\hbar \int_{-\infty}^{+\infty} \Psi^*(x,t) \frac{\partial}{\partial x} \Psi(x,t)\, dx. \qquad ...(6.34)$$

6.20 SCHRÖDINGER'S EQUATION FROM POSTULATES OF QUANTUM MECHANICS

Schrödinger's wave equation can be obtained from the postulates of quantum mechanics.

Suppose a particle of rest mass m moving with a non-relativistic velocity $\vec{v}$ has a linear momentum $\vec{p}(\vec{r}) = m\vec{v}$ and a kinetic energy

$$E_k = \frac{1}{2} m v^2 = \frac{\left|\vec{p}\,(\vec{r}, t)\right|^2}{2m}$$

If $V(\vec{r}, t)$ is the potential energy of the particle, then the total energy $E(\vec{r},t)$ is given by

$$E(\vec{r},t) = \frac{\left|\vec{p}(\vec{r},t)\right|^2}{2m} + V(\vec{r},t)$$

Solution. All the quantities in the above equation are *observable.* Therefore, expressing the equation in terms of corresponding operators in accordance with postulate 1, we have

$$\hat{E}(\vec{r},t) = \frac{|\hat{p}(\vec{r},t)|^2}{2m} + \hat{V}(\vec{r},t)$$

$$= \frac{\hat{p}(\vec{r},t).\hat{p}(\vec{r},t)}{2m} + \hat{V}(\vec{r},t)$$

or $$i\hbar\frac{\partial}{\partial t} = \frac{-i\hbar\vec{\nabla}\,.\,-i\hbar\vec{\nabla}}{2m} + \hat{V}(\vec{r},t) \qquad \left[\because \hat{p} = -i\hbar\vec{\nabla} \text{ and } \hat{E} = i\hbar\frac{\partial}{\partial t}\right]$$

$$= -\frac{\hbar^2}{2m}\nabla^2 + \hat{V}(\vec{r},t)$$

Applying the operator equation to the wave function $\Psi(\vec{r}, t)$ of the particle, we have

$$-\frac{\hbar^2}{2m}\nabla^2\Psi(\vec{r},t) + V(\vec{r},t)\,\Psi(\vec{r},t) = i\hbar\frac{\partial\Psi(\vec{r},t)}{\partial t} \qquad ...(6.35)$$

This is Schrödinger's time dependent quantum mechanical wave equation.

APPLICATIONS OF QUANTUM MECHANICS

6.21 FREE PARTICLE IN ONE DIMENSIONAL BOX

Consider a particle of rest mass *m* confined to move in a rectangular box under the following boundary conditions:

Boundary conditions

(*i*) *The walls of the box are rigid, hard and elastic*. Thus when the particle strikes the wall it rebounds with the same kinetic energy. Hence its total energy *E* remains constant.

(*ii*) *The walls of the box are non-penetrable.* This means that outside the box the potential energy is so large that the total energy of the particle can never exceed it. The particle, therefore, always remains within the box. There is no chance of finding the particle outside the box.

(*iii*) *The box is one dimensional*. Thus, the particle moves only along a straight line, say along the *X*-axis. The length of the box = *l*.

(*iv*) *The particle is free*. Thus, the potential energy *V* of the particle inside the box is zero. As the potential energy is independent of time, we shall use Schrödinger's time independent wave equation to study the state of the system. The potential energy at various points is shown in Fig. 6.1 and because of its appearance it is known as infinite rectangular or square potential well.

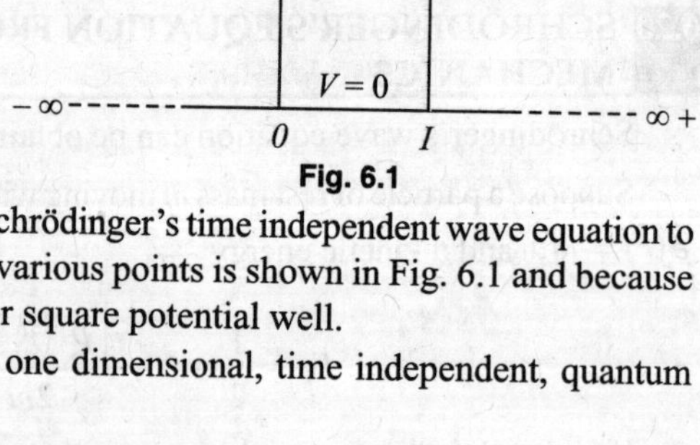

Fig. 6.1

Schrödinger's Wave Equation. Schrödinger's one dimensional, time independent, quantum mechanical wave equation gives

$$\frac{\partial^2\psi(x)}{\partial x^2} + \frac{2m}{\hbar^2}(E - V)\,\psi(x) = 0$$

For a free particle $V = 0$. Therefore, the Eigen value equation for such a particle in a one dimensional box under non-relativistic conditions is given by

$$\frac{\partial^2\psi(x)}{\partial x^2} + \frac{2m}{\hbar^2}E\,\psi(x) = 0 \qquad ...(i)$$

Solution. Since the particle remains within the box of length l, $\psi(x)$ has a finite value *i.e.* $\psi(x) \neq 0$ for values of x lying between 0 and l. Mathematically

$$\psi(x) \neq 0 \text{ for } 0 < x < l$$

In other words,

$$\psi(x) = 0 \text{ for } x < 0 \text{ and for } x > l.$$

As the potential energy of the particle is zero, the total energy E is the kinetic energy and is always a positive quantity.

Now $$E = \frac{1}{2} mv^2 = \frac{m^2 v^2}{2m} = \frac{p_x^2}{2m}$$

$\therefore$ $$\frac{2m}{\hbar^2} E = \frac{p_x^2}{\hbar^2} = k^2$$

Since $$\frac{p_x}{\hbar} = \frac{p_x \cdot 2\pi}{h} = \frac{2\pi}{\lambda} = k$$

where k is the *propagation constant*.

$\therefore$ Equation (*i*), may be written as

$$\frac{\partial^2 \psi(x)}{\partial x^2} + k^2 \psi(x) = 0 \qquad ...(ii)$$

The general solution of equation (*ii*) is

$$\psi(x) = A \sin kx + B \cos kx \qquad ...(iii)$$

where A and B are arbitrary constants.

To prove that $\psi(x) = A \sin kx + B \cos kx$ is the solution of the equation $\frac{\partial^2 \psi(x)}{\partial x^2} + k^2 \psi(x) = 0$, differentiate equation (*iii*) for $\psi(x)$ twice and we get

$$\frac{\partial \psi(x)}{\partial x} = kA \cos kx - kB \sin kx$$

and $$\frac{\partial^2 \psi(x)}{\partial x^2} = -k^2 A \sin kx - k^2 B \cos kx$$

$$= -k^2 (A \sin kx + B \cos kx)$$

$$= -k^2 \psi(x)$$

or $$\frac{\partial^2 \psi(x)}{\partial x^2} + k^2 \psi(x) = 0.$$

Therefore equation (*iii*) is the solution of the differential equation (*ii*).

To find the values of A and B we apply the boundary conditions. For continuity of the wave function at the boundary, we have

$$\psi(x) = 0 \text{ when } x = 0 \text{ and when } x = l$$

Substituting $\psi(x) = 0$ when $x = 0$ in (*iii*), we have

$$0 = B$$

$\therefore$ Equation (*iii*) now becomes

$$\psi(x) = A \sin kx \qquad ...(iv)$$

Substituting $\psi(x) = 0$ when $x = l$ in (*iv*), we have

$$0 = A \sin kl$$

$\therefore$ $kl = n\pi$ where $n = 1, 2, 3$ ....

or $$k = \frac{n\pi}{l}. \qquad ...(v)$$

We do not take $n = 0$, which means the momentum and hence the kinetic energy of the particle is zero *i.e.*, the particle is at rest for all times which is not physically acceptable. For $k = 0$, $\psi(x) = 0$ everywhere in the box, which means that the probability density $|\psi|^2 = 0$ *i.e., the particle is not in the box*. The negative values of n simply repeat the corresponding values for positive n and do not give independent solutions.

$$\therefore \qquad \psi(x) = A \sin \frac{n\pi}{l} x. \qquad \text{...(vi)}$$

Evaluation of A and normalisation. To find the value of A, we *normalise* the Eigen function $\psi(x)$. The probability of finding the particle anywhere inside the box is given by $\psi^*(x)\,\psi(x)\,dx = |\psi(x)|^2$ and is **one** as there is only one particle and there is **no** chance of its being located outside the box.

$$\therefore \qquad \int_0^l |\psi(x)|^2\, dx = 1$$

or
$$A^2 \int_0^l \sin^2 \frac{n\pi x}{l}\, dx = 1$$

or
$$\frac{A^2}{2} \int_0^l 2\sin^2 \frac{n\pi x}{l}\, dx = 1$$

But
$$\int_0^l 2\sin^2 \frac{n\pi x}{l}\, dx = \int_0^l \left(1 - \frac{\cos 2\pi n x}{l}\right) dx = l$$

$$\therefore \qquad \frac{A^2}{2}\, l = 1 \qquad \text{or} \qquad A = \left(\frac{2}{l}\right)^{1/2}$$

Hence
$$\psi(x) = \left(\frac{2}{l}\right)^{1/2} \sin \frac{n\pi}{l} x \qquad \text{...(6.36)}$$

Eq. (6.36) is the *normalised wave function* for a particle inside a box of finite width l.

Eigen function. From the wave equation (6.36), the various eigen functions are

$$\psi_1 = \left(\frac{2}{l}\right)^{1/2} \sin \frac{\pi}{l} x,$$

$$\psi_2 = \left(\frac{2}{l}\right)^{1/2} \sin \frac{2\pi}{l} x, \ldots \psi_n = \left(\frac{2}{l}\right)^{1/2} \sin \frac{n\pi}{l} x$$

with $p_1, p_2, \ldots p_n$ as eigen values of momentum, $k_1, k_2, \ldots k_n$ as eigen values of propagation constant and $E_1, E_2, \ldots E_n$ as eigen values for the total energy of the particle.

Eigen values of propagation constant k. According to relation (*v*)

$$k = \frac{n\pi}{l}$$

$$\therefore \qquad \text{Eigen values of } k \text{ are} \quad k_1 = \frac{\pi}{l},\ k_2 = \frac{2\pi}{l} \ldots.\ k_n = \frac{n\pi}{l}. \qquad \text{...(vii)}$$

Eigen values of momentum. Now

$$k = \frac{2\pi}{\lambda} = \frac{2\pi p}{h} = \frac{p}{\hbar}$$

or $$p = \hbar\, k = \frac{n\pi\hbar}{l}$$

$\therefore$ Eigen values of the momentum p of the particle are

$$p_1 = \frac{\pi\hbar}{l},\ p_2 = \frac{2\pi\hbar}{l}\ \ldots\ p_n = \frac{n\pi\hbar}{l}. \qquad ...(viii)$$

Eigen values of energy. If we consider the energy of the particle non-relativistically

$$E = \frac{p^2}{2m} = \frac{n^2\pi^2\hbar^2}{2ml^2}$$

$\therefore$ Eigen values of the energy E of the particle are

$$E_1 = \frac{\pi^2\hbar^2}{2ml^2},\ E_2 = \frac{2^2\pi^2\hbar^2}{2ml^2}\ \ldots\ E_n = \frac{n^2\pi^2\hbar^2}{2ml^2}. \qquad ...(ix)$$

Difference from classical results. The results based on quantum mechanical considerations differ from classical results in following respects.

(1) According to classical considerations the particle in a box can have any value of energy from zero to infinity, but according to quantum mechanics, only *discrete values of energy* (and hence momentum) are permissible *i.e.* momentum and energy are quantised.

Quantisation of momentum and energy. The momentum of the particle can only have values which are integral multiples of

$$\frac{\pi\hbar}{l} = \frac{\pi h}{2\pi l} = \frac{h}{2l}$$

or $$p_n = \frac{nh}{2l} \qquad ...(x)$$

and no other values. Thus, the linear momentum is quantised, as the only values it can have are multiples of $\frac{h}{2l}$. We also find that the quantum of momentum is independent of the mass of the particle *i.e.* in a box of length l, the momentum of any particle, proton, neutron etc. will be intergal multiple of the value $\frac{h}{2l}$. Thus the momentum of a particle inside the box is

(*i*) directly proportional to the natural number *i.e.* n

(*ii*) inversely proportional to the length l of the box

and (*iii*) independent of the mass of the particle.

Under non-relativistic conditions, the energy of the particle can *only* have values given by

$$E_n = \frac{n^2\pi^2\hbar^2}{2ml^2} \text{ where } n = 1, 2, 3 \qquad ...(6.37)$$

The energy can, therefore, have only discrete values and is thus quantised, as the only values it can have is n^2 times $\frac{\pi^2\hbar^2}{2ml^2}$.

Thus, the energy of a particle inside the box is

(*i*) directly proportional to the square of natural number *i.e.* n^2

(*ii*) inversely proportional to the square of the length of the box *i.e.* $\frac{1}{l^2}$ and

(*iii*) inversely proportional to the mass of the particle.

(2) According to quantum mechanics, the minimum energy that the particle can have is $E_1 = \dfrac{\pi^2 \hbar^2}{2ml^2}$ and not zero, as required classically. [See article **6.24**]

(3) Classically the particle is equally likely to be found at any point within the box from 0 to l but according to quantum mechanics, the probability of finding the particle is maximum at the *antinodes* and minimum at the *nodes*. [See article **6.23**]

Zero and Negative Values of *n* not Permissible

The normalised wave function for a free particle inside a box of finite width l is given by

$$\psi(x) = \left(\frac{2}{l}\right)^{1/2} \sin \frac{n\pi}{l} x$$

where n is the quantum number having values $n = 1, 2, 3....$

Now $\dfrac{n\pi}{l} = k$ where k is *the propagation constant*

and
$$k = \frac{2\pi}{\lambda} = \frac{p_x\, 2\pi}{h} = \frac{p_x}{\hbar}$$

$\therefore$
$$p_x = \hbar k$$

Also
$$E = \frac{1}{2} mv^2 = \frac{m^2 v^2}{2m} = \frac{p_x^2}{2m} = \frac{\hbar}{2m} k^2$$

We only take positive values of quantum number n. We do not take $n = 0$ because it would mean $k = 0$ and hence momentum p_x and kinetic energy E of the particle will be zero *i.e.* the particle is at rest for all times which is not physically acceptable. For $k = 0$, $\psi(x) = 0$ everywhere in the box, which means that the probability density $|\psi|^2 = 0$ *i.e. the particle is not in the box*, which again is not possible.

The negative values of n simply repeat the corresponding values for positive n and do not give independent solutions.

6.22 ENERGY LEVELS

The normalised wave function for a particle in a box of finite width l is given by

$$\psi(x) = \left(\frac{2}{l}\right)^{1/2} \sin \frac{n\pi}{l} x$$

Now $\dfrac{n\pi}{l} = k$ the *propagation constant*

and momentum
$$p = \frac{h}{\lambda} = \frac{h}{2\pi}\frac{2\pi}{\lambda} = \hbar k$$

$\therefore$
$$p = \frac{n\hbar\pi}{l}$$

Also kinetic energy
$$E = \frac{1}{2} mv^2 = \frac{m^2 v^2}{2m} = \frac{p^2}{2m}$$
$$= \frac{n^2 \pi^2 \hbar^2}{2ml^2}$$

Thus we find that for a free particle in a one dimensional box *i.e.* an infinite rectangular or square potential well, the energy of the particle can have only discrete values given by

$$E_n = \frac{n^2 \pi^2 \hbar^2}{2ml^2} \quad \text{where } n = 1, 2, 3 \text{ ...etc.} \qquad ...(6.38)$$

These discrete values of energy are known as energy states or *energy levels.*

Ground state. The minimum value of energy is for $n = 1$ and is given by

$$E_1 = \frac{\pi^2 \hbar^2}{2ml^2}$$

It is called the *ground state energy* of the particle. The state of the particle for $n = 1$ is called the *ground state.*

Excited states. The states of the particle for $n = 2, 3, 4$ etc are called *excited states*. For $n = 2$ we have the *first* excited state with energy $E_2 = 4\dfrac{\pi^2 \hbar^2}{2ml^2} = 4E_1$ and for $n = 3$ we have the *second* excited state with energy $E_3 = 9\dfrac{\pi^2 \hbar^2}{2ml^2} = 9\,E_1$ and so on.

Energy level diagram. From the above discussion we find that the particle can have only *discrete* values of energy given by E_1, E_2, E_3 ..... *i.e.* the energy of the particle is quantised. These discrete allowed values of energy E_1, E_2, E_3 .... etc are called *energy levels* and are shown in Fig. 6.2.

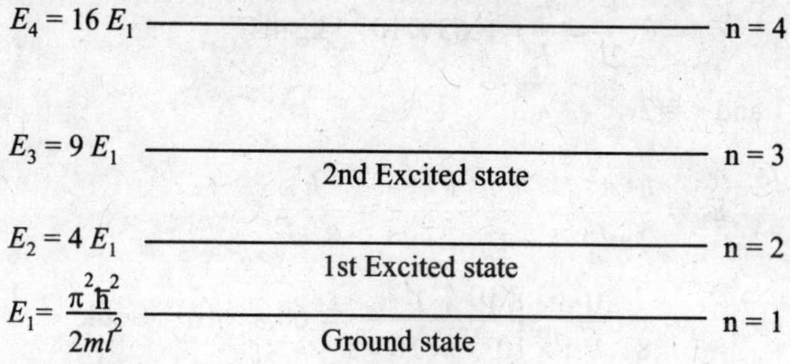

Fig. 6.2

Example 6.10. *Show that the wavelength of de-Broglie wave associated with a particle in a box and in the first excited state is equal to the length of the box.*

(Pbi.U. 2003, P.U. 2003; H.P.U. 2002)

Solution. For a particle in a box, the momentum of the particle can only have discrete values which are integral multiples of $\dfrac{h}{2l}$ *i.e.*

$$p_n = \frac{nh}{2l} \quad \text{where } n = 1, 2, 3 \text{ .....} \qquad \textbf{[See Eq. } x \textbf{ of article 6.21]}$$

For $n = 1$, we have the *ground* state. For the *first excited* state $n = 2$

$$\therefore \quad p = \frac{2h}{2l} = \frac{h}{l} \quad \text{or} \quad l = \frac{h}{p} \qquad ...(i)$$

The wave length of de-Broglie wave associated with a particle is given by

$$\lambda = \frac{h}{mv} = \frac{h}{p} \qquad ...(ii)$$

Comparing (*i*) and (*ii*), we have

$$\lambda = l$$

This shows that '*the wave length of de-Broglie wave associated with a particle in a box in the first excited state is equal to the length of the box*'.

Example 6.11. *State the values of momentum and energy of a particle in one dimensional box with impenetrable walls. Find their values for an electron in a box of length 1Å for n = 1 and n = 2. Mass of the electron = 9.1 × 10⁻³¹ kg and h = 6.63 × 10⁻³⁴ Js.* ***(P.U. 2001; K.U. 2000)***

Solution. For a particle in one dimensional box of length *l* with impenetrable walls

Momentum $$p_n = n\frac{\hbar\pi}{l} \text{ where } n = 1, 2, 3, ...$$

and Energy $$E_n = \frac{p^2}{2m} = \frac{n^2\hbar^2\pi^2}{2ml^2}$$

Momentum for $n = 1$ and $n = 2$

For $n = 1$ $$p_1 = \frac{\hbar\pi}{l} = \frac{h}{2\pi}\frac{\pi}{l} = \frac{h}{2l}$$

$$= \frac{6.63\times10^{-34}}{2\times10^{-10}} = 3.315\times10^{-24}\text{ kg ms}^{-1}$$

For $n = 2$ $$p_2 = 2\frac{h}{2l} = \frac{h}{l} = 6.63\times10^{-24}\text{ kg ms}^{-1}$$

Energy for $n = 1$ and $n = 2$

For $n = 1$ $$E_1 = \frac{\hbar^2\pi^2}{2ml^2} = \frac{h^2\pi^2}{4\pi^2 . 2ml^2} = \frac{h^2}{8ml^2}$$

$$= \frac{(6.63\times10^{-34})^2}{8\times9.1\times10^{-31}\times10^{-20}} = 60.4\times10^{-19}\text{ Joule}$$

$$= \frac{60.4\times10^{-19}}{1.6\times10^{-19}} = 37.7\text{ eV}$$

For $n=2$ $$E_2 = 2^2\frac{h^2}{8ml^2} = 4\times37.7 = 150.8\text{ eV}.$$

Example 6.12. *An electron is constrained to move in a one dimensional box of length 0.1 nm Find the first three energy eigen values and the corresponding de-Broglie wave length.* ***(Bang.U. 2001; Nagpur U. 2002)***

Solution. 0.1 n m = 10^{-10} m = 1Å. The first two energy levels for $n = 1$ and $n = 2$ have been calculated above.

For $n = 3$, $$E_3 = 3^2\frac{h^2}{8ml^2} = 9\times37.7 = 339.3\text{ eV}$$

and $$p_3 = 3\frac{h}{2l} = 3\times3.315 = 9.945\text{ kg ms}^{-1}$$

de-Broglie wave length $$\lambda_n = \frac{h}{mv} = \frac{h}{p} = \frac{h}{n}\cdot\frac{2l}{h} = \frac{2l}{n}$$

For $n = 1$, $\lambda_1 = 2\,l = 2$Å

$$n = 2,\ \lambda_2 = \frac{2l}{2} = l = 1\text{Å}$$

$$n = 3,\ \lambda_3 = \frac{2l}{3} = 0.667l = 0.667\text{Å}.$$

Example 6.13. *Find minimum energy of an electron constrained to move linearly in a infinitely high potential box of length 10^{-11} m. (Given mass of electron = 9.11×10^{-31} kg, h = 6.63×10^{-34} Js)* ***(Nagpur U. 2011, G.N.D.U. 2008)***

Solution. For minimum energy $n = 1$; $l = 10^{-11}$ m

$$\therefore \quad E_1 = \frac{h^2}{8\,ml^2} = 3780 \text{ eV}$$

Example 6.14. *Calculate the energy difference between the ground state and the first excited state for an electron in a box of length 1Å.* ***(Meerut U. 2006, 2005, 2003)***

Solution. $E_n = \dfrac{n^2 h^2}{8ml^2}$ For ground state $n = 1$; For first excited state $n = 2$,

$$\therefore \quad E_2 - E_1 = (2^2 - 1^2)\frac{h^2}{8ml^2} = 3 \times 37.7 = 113.1\text{eV}.$$

Example 6.15. *Calculate the value of lowest energy of an electron in one dimensional force free region of length 4 Å.* ***(H.P.U., 2002)***

Solution. For a particle in one dimensional force free region of length l, the energy is given by

$$E_n = \frac{p^2}{2m} = \frac{n^2 \hbar^2 \pi^2}{2ml^2} = \frac{n^2 h^2 \pi^2}{4\pi^2\, 2ml^2} = \frac{n^2 h^2}{8ml^2}$$

The lowest energy state is for $n = 1$

$$\therefore \quad E_1 = \frac{h^2}{8ml^2} = \frac{(6.63 \times 10^{-34})^2}{8 \times 9.1 \times 10^{-31} \times (4 \times 10^{-10})^2}$$

$$= 3.78 \times 10^{-19} \text{ J}$$

$$= \frac{3.78 \times 10^{-19}}{1.6 \times 10^{-19}} = 2.36\text{eV}.$$

Example 6.16. *Compute the energy of the lowest three levels for an electron in a square well of width 3 Å.* ***(Pbi.U. 2002)***

Solution. $E_n = \dfrac{n^2 h^2}{8ml^2}$. For $n = 1$, $E_1 = \dfrac{h^2}{8ml^2} = \dfrac{6.63 \times 6.63 \times 10^{-68}}{8 \times 9.1 \times 10^{-31} \times 3 \times 3 \times 10^{-20}}$

$$= 6.71 \times 10^{-19} \text{ J} = \frac{6.71 \times 10^{-19}}{1.6 \times 10^{-19}} = 4.19\text{eV}$$

$$E_2 = 4\frac{h^2}{8ml^2} = 16.76\text{eV}$$

$$E_3 = 9\frac{h^2}{8ml^2} = 37.71\text{eV}.$$

Example 6.17. *The eigen value of energy of an electron confined to a one dimensional box of length 0.4 nm is 604 eV. What is the order of the excited state? Calculate the momentum of the electron in that state.*

Solution. $E_n = n^2 \dfrac{h^2}{8ml^2}$. As calculated in the question $E_1 = \dfrac{h^2}{8ml^2} = 2.36\,\text{eV}$

$$\therefore \quad \frac{E_n}{E_1} = \frac{604}{2.36} = 256 = n^2 \quad \text{or} \quad n = 16$$

$$\text{Momentum} \quad p_n = n\frac{h}{2l} = \frac{16 \times 6.63 \times 10^{-34}}{2 \times 4 \times 10^{-10}} = 13.26 \times 10^{-24}\ \text{kg ms}^{-1}.$$

Example 6.18. *The lowest energy possible for a certain particle entrapped in a box is 40 eV. What are the next three higher energies the particle can have?*

Solution. For a particle in a one dimensional box

$$E_n = \frac{n^2 \hbar^2 \pi^2}{2ml^2}$$

The lowest possible energy corresponds to $n = 1$ and is given by

$$E_1 = \frac{\hbar^2 \pi^2}{2ml^2} = 40\,\text{eV}$$

The next higher energy state corresponds to $n = 2$ and is given by

$$E_2 = 2^2 \frac{\hbar^2 \pi^2}{2ml^2} = 4 \times 40 = 160$$

Similarly for $n = 3$ $\quad E_3 = 3^2 E_1 = 9 \times 40 = 360$ eV

and $\quad$ for $n = 4$ $\quad E_4 = 4^2 E_1 = 16 \times 40 = 640$ eV.

Example 6.19. *The minimum energy possible for a particle entrapped in a one dimensional box is 3.2 × 10*$^{-18}$ *J. What are the next three energies in eV, the particle can have?*

Solution. Energy of a particle trapped in a one dimensional box

$$E_n = \frac{n^2 \hbar^2 \pi^2}{2ml^2}$$

i.e. $E_n \propto n^2$ where for minimum energy $n = 1$

$$\text{Now minimum energy of the particle in eV} = \frac{3.2 \times 10^{-18}\ \text{J}}{1.6 \times 10^{-19}} = 20$$

∴ Next three values of energy are $20 \times 2^2 = 80$ eV

$20 \times 3^2 = 180$ eV

$20 \times 4^2 = 320$ eV.

Example 6.20. *Find the probability that a particle in a one dimensional box of length L can be found between 0.4 L to 0.6 L.*

(i) for the ground state

(ii) for the first excited state and

(iii) for the second excited state. **(P.U., 2001; G.N.D.U., 2000)**

Solution. The wave function for a free particle in a one dimensional box of length L is given by

$$\psi_n = \left(\frac{2}{L}\right)^{1/2} \sin \frac{n\pi}{L} x$$

The probability of finding a particle in a distance Δx about the mean position x is

$$P_n \Delta x = \psi_n \psi_n^* \Delta x$$

$$= \left(\frac{2}{L}\right)^{1/2} \sin \frac{n\pi x}{L} \left(\frac{2}{L}\right)^{1/2} \sin \frac{n\pi x}{L} \Delta x = \frac{2}{L} \sin^2 \frac{n\pi}{L} x \, \Delta x$$

Here mean position $x = \dfrac{0.4L + 0.6L}{2} = 0.5L = \dfrac{L}{2}$

and $\Delta x = 0.6\,L - 0.4\,L = 0.2\,L$

For the ground state $n = 1$

$$\therefore \quad P_1 \,\Delta x = \frac{2}{L} \sin^2 \left(\frac{\pi}{L}\frac{L}{2}\right) \times 0.2\,L = 2 \sin^2 \frac{\pi}{2} \times 0.2 = 0.4$$

For the first excited state $n = 2$

$$\therefore \quad P_2 \,\Delta x = \frac{2}{L} \sin^2 \left(\frac{2\pi}{L}\frac{L}{2}\right) \times 0.2\,L = 2 \sin^2 \pi \times 0.2 = \text{Zero}$$

For the second excited state $n = 3$

$$\therefore \quad P_3 \,\Delta x = \frac{2}{L} \sin^2 \left(\frac{3\pi}{L}\frac{L}{2}\right) \times 0.2\,L = 2 \sin^2 \left(\frac{3\pi}{2}\right) \times 0.2 = 0.4$$

Thus for $n = 1, 3, 5$ (odd values) etc., the probability of finding the particle within a distance $\Delta x = 0.2\,L$ in the mean position $x = 0.5\,L$ is 0.4.

For $n = 2, 4, 6$ (even values) etc., the probability is zero.

Example 6.21. *A particle is moving in one dimensional potential box of infinite height and width 10 Å. Calculate the probability of finding the particle within an interval of 4 Å at the centre of the box when it is in the state of least energy.* ***(Nagpur U. 2002)***

Solution. The state of least energy is the ground state for $n = 1$

Width of one dimensional potential box $L = 10$ Å $= 10 \times 10^{-10}$ m

Position of the centre of the box $x = \dfrac{L}{2} = \dfrac{10 \times 10^{-10}}{2} = 5 \times 10^{-10}$ m

Interval $\Delta x = 4$ Å $= 4 \times 10^{-10}$ m

$\therefore$ Probability of finding the particle within an interval Δx about the mean position x

$$P_n \,\Delta x = \frac{2}{L} \sin^2 \left(\frac{n\pi x}{L}\right) \Delta x$$

$$= \frac{2}{10 \times 10^{-10}} \sin^2 \left(\frac{\pi \times 5 \times 10^{-10}}{10 \times 10^{-10}}\right) 4 \times 10^{-10}$$

$$= 0.8 \sin^2 \frac{\pi}{2} = 0.8.$$

6.23 PROBABILITY DENSITY IN ONE DIMENSIONAL BOX

In the case of a free particle in a one dimensional box the value of normalised wave function $\psi(x)$ is given by the relation

$$\psi_n(x) = \left(\frac{2}{l}\right)^{1/2} \sin \frac{n\pi}{l} x \qquad \text{(From Eq. 6.36)}$$

where l is the length of the box.

The probability of finding the particle in the one dimensional space between the points x and $x + dx$ is given by $\psi\psi^* \, dx$ where ψ is the wave function and ψ^* its complex conjugate. Thus the probability of finding the particle in a finite interval between x_1, and x_2 is given by

$$P = \int_{x_1}^{x_2} \psi\psi^* dx$$

The probability of finding a particle in a distance dx about the mean position x is

$$P_n\, dx = \psi_n\, \psi_n{}^* dx = \left(\frac{2}{l}\right)^{\frac{1}{2}} \sin\frac{n\pi x}{l}\left(\frac{2}{l}\right)^{1/2} \sin\frac{n\pi x}{l}\, dx$$

$$= \frac{2}{l}\sin^2\frac{n\pi}{l}\, x\, dx$$

$\therefore$ Probability density $P_n = \psi_n \psi_n{}^* = |\psi_n|^2 = \dfrac{2}{l}\sin^2\dfrac{n\pi}{l}x$. ...(6.39)

Graph. For $n = 1, 2, 3$ and 4 the normalised wave function ψ_1, ψ_2, ψ_3 and ψ_4 and corresponding probability densities $|\psi_1^2|$, $|\psi_2^2|$, $|\psi_3^2|$ and $|\psi_4^2|$ are plotted as shown in Fig. 6.3 (*a*) and (*b*). It is clear from the figure that ψ_n has *positive* as well as *negative* values whereas $|\psi_n^2|$ is always *positive*.

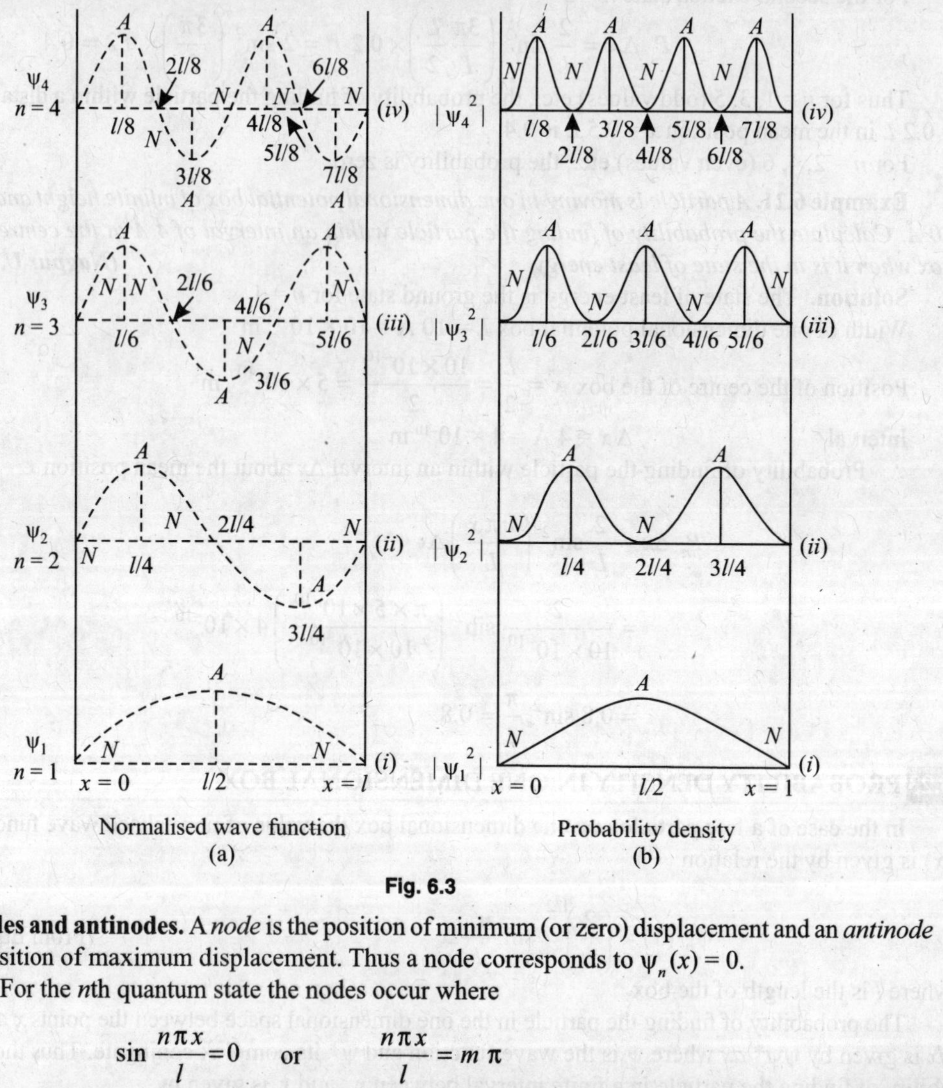

Fig. 6.3

Nodes and antinodes. A *node* is the position of minimum (or zero) displacement and an *antinode* is the position of maximum displacement. Thus a node corresponds to $\psi_n(x) = 0$.

$\therefore$ For the *n*th quantum state the nodes occur where

$$\sin\frac{n\pi x}{l} = 0 \quad \text{or} \quad \frac{n\pi x}{l} = m\pi$$

where $m = 0, 1, 2, \ldots n$

$$\therefore \qquad x = m\frac{l}{n}$$

Thus the nodes are located at, $0, \frac{l}{n}, \frac{2l}{n} \ldots \frac{nl}{n}$

An antinode corresponds to the maximum value of $\psi_n(x)$. Since the maximum value of $\sin \frac{n\pi}{l} x = 1$, for an antinode

$$\psi_n(x) = \left(\frac{2}{l}\right)^{1/2}$$

$\therefore$ For the nth quantum state the antinodes occur where

$$\sin \frac{n\pi x}{l} = 1 \qquad \text{or} \qquad \frac{n\pi x}{l} = (2m+1)\frac{\pi}{2}$$

$$\therefore \qquad x = \frac{2m+1}{2n} l$$

Thus, the antinodes are located at $x = \frac{l}{2n}, \frac{3l}{2n} \ldots \frac{(2n-1)}{2n} l$

When $n = 1$, nodes occur at $x = 0$ and $x = l$

antinodes occur at $x = \frac{l}{2}$ [Fig. 6.3 (*a*)(*i*)]

When $n = 2$, nodes occur at $x = 0$; $x = \frac{l}{2}$; and $x = l$

antinodes occur at $x = \frac{l}{4}$; $x = \frac{3l}{4}$ [Fig. 6.3 (*a*)(*ii*)]

When $n = 3$, nodes occur at $x = 0$; $x = \frac{l}{3}$; $x = \frac{2l}{3}$; and $x = l$

antinodes occur at $x = \frac{l}{6}$; $x = \frac{3l}{6}$; $x = \frac{5l}{6}$ [Fig. 6.3 (*a*)(*iii*)]

When $n = 4$, nodes occur at $x = 0$; $x = \frac{l}{4}$; $x = \frac{2l}{4}$; $x = \frac{3l}{4}$; and $x = l$

antinodes occur at $x = \frac{l}{8}$; $x = \frac{3l}{8}$; $x = \frac{5l}{8}$; $x = \frac{7l}{8}$. [Fig. 6.3 (*a*)(*iv*)]

In terms of probability density $|\psi_n^2|$. In terms of probability density, an *antinode* is a point of *maximum* probability density and a *node* is a point of *zero* probability density. We, therefore, find that for the *n*th quantum state there are *n antinodes i.e.* there are *n* equally probable positions in the box at which the particle is most likely to be found. Similarly, there are $n - 1$ nodes, each node lying in between two antinodes *i.e.* there are $n - 1$ equally spaced positions where the probability of finding the particle is zero. Of course, the two positions $x = 0$ and $x = l$ are permanent positions of nodes.

In the ground state $n = 1$, the position of maximum probability is given by $x = \frac{l}{2}$. For the first excited state $n = 2$, the probability is maximum for $x = \frac{l}{4}$ and $x = \frac{3l}{4}$ but the probability for $x = \frac{l}{2}$

is zero. According to the classical theory, the probability of finding the particle is the same at every point in the box but according to the quantum theory, there is a greater probability of finding the particle at some points than at others. The probability of locating the particle at $x = 0$ or $x = l$, *i.e.* near the wall is always zero.

Example 6.22. *Evaluate the first three energy levels of an electron enclosed in a box of width 10 Å. Compare it with those of glass marble of mass 1 gm, contained in a box of width 20 cm. Can these levels of the marble be measured experimentally?* ***(Meerut U. 2000; P.U., 1994)***

Solution. Energy of a particle in a one dimensional box

$$E_n = \frac{p^2}{2m} = \frac{n^2 \hbar^2 \pi^2}{2ml^2} = \frac{n^2 h^2}{8ml^2}$$

For **electron** $n = 1$; $l = 10\text{Å} = 10 \times 10^{-10}\ m$; $m = 9.1 \times 10^{-31}$ kg

$$E_1 = \frac{1 \times (6.63 \times 10^{-34})^2}{8 \times 9.1 \times 10^{-31} \times 100 \times 10^{-20}} = 0.604 \times 10^{-19}\ \text{J}$$

$$= \frac{0.604 \times 10^{-19}}{1.6 \times 10^{-19}} = 0.377\ \text{eV}$$

For $n = 2$; $\quad E_2 = 2^2 \dfrac{h^2}{8ml^2} = 4 \times 0.377 = 1.508\ \text{eV}$

For $n = 3$; $\quad E_3 = 3^2 \dfrac{h^2}{8ml^2} = 9 \times 0.377 = 3.393\ \text{eV}$

For glass **marble** $n = 1$; $l = 20$ cm $= 0.2$m ; $m = 1$ gm $= 10^{-3}$ kg

$$E_1 = \frac{1 \times (6.63 \times 10^{-34})^2}{8 \times 10^{-3} \times (0.2)^2} = 1.374 \times 10^{-63}\ \text{J}$$

$$= \frac{1.374 \times 10^{-63}}{1.6 \times 10^{-19}} = 0.858 \times 10^{-44}\ \text{eV}$$

For $n = 2$; $\quad E_2 = 2^2 \dfrac{h^2}{8ml^2} = 4 \times 0.858 \times 10^{-44} = 3.432 \times 10^{-44}\ \text{eV}$

For $n = 3$; $\quad E_3 = 3^2 \dfrac{h^2}{8ml^2} = 9 \times 0.858 \times 10^{-44} = 7.722 \times 10^{-44}\ \text{eV}$

These levels of the marble cannot be measured experimentally as the energy involved is very small and is almost zero.

6.24 PARTICLE WITH ZERO ENERGY

Quantum mechanically, a particle will not exist in a box if its energy is zero. From this, we will discuss the zero point energy.

For a free particle in a one dimensional box of length l, the wave equation is

$$\psi_n(x) = \left(\frac{2}{l}\right) \sin \frac{n\pi}{l} x \qquad \text{...}(i)$$

and the given values of energy are

$$E_n = \frac{n^2 \pi^2 \hbar^2}{2ml^2} \qquad \text{...}(ii)$$

For $n = 0$, $E_0 = 0$ and $\psi_0(x) = 0$

It means that for $E_0 = 0$ the probability of the particle being within the box is nil *i.e.*, there can be no particle in the box. Hence if the particle is in the box, n cannot be zero. In other words, a particle will not exist in a box if its energy is zero.

Zero point energy. The smallest possible value of n in relation (*i*) and (*ii*) is **one**, for which

$$\psi_1(x) = \left(\frac{2}{l}\right)^{1/2} \sin\frac{\pi x}{l} \text{ and } E_1 = \frac{\pi^2 \hbar^2}{2ml^2}$$

Thus, the minimum energy that the particle can have is $E_1 = \dfrac{\pi^2 \hbar^2}{2ml^2}$. In other words, even when the temperature of the box is reduced to 0 K, the total energy will still be E_1 and will not be zero. Therefore, $E_1 = \dfrac{\pi^2 \hbar^2}{2ml^2}$ is called the *zero point energy* of the particle.

The zero point energy is (*i*) inversely proportional to m and (*ii*) inversely proportional to the square of the length of the box.

Hence if m is very large, *i.e.*, for a heavy particle or if l is very large *i.e.*, for a box of the laboratory size, zero point energy is *almost* zero. If $l \to \infty$ the particle behaves as a *free particle* without any box and does not possess any zero point energy.

6.25 ORTHOGONAL WAVE FUNCTIONS

The two wave functions $\psi_m(x)$ and $\psi_n(x)$ are said to be orthonormal or orthogonal when

$$\int_{\text{Available region}} \psi_m^*(x)\, \psi_n(x)\, dx = 0$$

Wave functions of two different states are orthogonal. The wave functions for two different states m and n of a free particle in a box are given by

$$\psi_m(x) = \left(\frac{2}{l}\right)^{1/2} \sin\frac{m\pi}{l} x \text{ and } \psi_n(x) = \left(\frac{2}{l}\right)^{1/2} \sin\frac{n\pi}{l} x$$

$$\therefore \quad \int_0^l \psi_m^*(x)\, \psi_n(x)\, dx = \frac{2}{l}\int_0^l \sin\frac{m\pi}{l} x \sin\frac{n\pi}{l} x$$

$$= \frac{2}{l}\,\frac{1}{2}\int_0^l \left[\frac{\cos(m-n)\pi x}{l} - \frac{\cos(m+n)\pi x}{l}\right] dx$$

$$= \frac{1}{l}\left[\frac{\sin(m-n)\pi x/l}{(m-n)\pi/l}\right]_0^l - \left[\frac{\sin(m+n)\pi x/l}{(m+n)\pi/l}\right]_0^l = 0 \quad \text{...(6.40)}$$

since both the functions within the brackets are each equal to zero.

Hence the wave functions for two different states are orthogonal or orthonormal.

Two lowest states. For the two lowest states $n = 1$, $m = 2$ and the result holds good.

6.26 PARITY

The behaviour of a wave function under reflection in a line or point of symmetry is called parity.

Consider the case of a wave function under inversion of co-ordinates through the origin *i.e.* x is replaced by $-x$, y by $-y$ and z by $-z$. If the potential energy function is unchanged in the process *i.e.*, $V(x, y, z) = V(-x, -y, -z)$, then the wave function either remains the same or changes sign. If the sign remains the same it is said to have *even parity and if the sign changes it is said to have odd parity.*

For example, Schrödinger's wave function in one dimension i.e. x only is

$$-\frac{\hbar^2}{2m}\frac{d^2\psi(x)}{dx^2}+V(x)\psi=E\psi(x) \quad ...(i)$$

If we change the variable in Eq. (i) from x to x' where $x' = -x$ then $dx' = -dx$ and $dx'^2 = dx^2$.

Hence $\quad \frac{d\psi}{dx}=\frac{d\psi}{dx'}\cdot\frac{dx'}{dx}=-\frac{d\psi}{dx'}$

and $\quad \frac{d^2\psi}{dx^2}=\frac{d}{dx}\left(\frac{d\psi}{dx}\right)=\frac{d}{dx'}\left(\frac{d\psi}{dx}\right)\frac{dx'}{dx}=\frac{d}{dx'}\left(\frac{d\psi}{dx'}\right)=\frac{d^2\psi}{dx'^2}$

∴ Eq (i) becomes

$$-\frac{\hbar^2}{2m}\left(\frac{d^2\psi}{dx'^2}\right)+V(x')\psi(x')=E\psi(x') \quad ...(ii)$$

If the potential energy function $V(x) = V(x') = V(-x)$ *i.e.* the potentials are symmetric about the origin $x = 0$ and we change the sign of x in (i), we get

$$-\frac{\hbar^2}{2m}\frac{d^2\psi(-x)}{dx^2}+V(-x)\psi(-x)=E\psi(-x). \quad ...(iii)$$

Even and odd parity. $\psi(x)$ and $\psi(-x)$ are the solutions of the same wave equation with the same eigen value E. Suppose there is only one linearly independent solution for a given eigen value E, then $\psi(x)$ and $\psi(-x)$ can only differ by a multiplicative constant

∴ $\quad \psi(-x) = \varepsilon\,\psi(x)$

Changing x to $-x$ in this equation, we have

$$\psi(x) = \varepsilon\,\psi(-x)$$

Hence $\quad \psi(-x)\,\psi(x) = \varepsilon^2\,\psi(x)\,\psi(-x)$

or $\quad \varepsilon^2 = 1 \quad ∴ \quad \varepsilon = \pm 1$

The constant ε is known as the parity of the state. It is also known as the parity operator for the function $\psi(x)$ and then denoted as Π. If $\varepsilon = +1$, $\psi(x)$ is symmetric about the origin and the wave function has *even* parity. If $\varepsilon = -1$, $\psi(x)$ is antisymmetric about the origin and the wave function has *odd* parity.

The $\psi(x)$, x curves of Fig. 6.3 (a) show that for odd values of n the eigen functions are symmetric with respect to $x = \frac{l}{2}$ line, whereas for even values of n the eigen functions are antisymmetric. Thus, for quantum states with $n = 1, 3, 5$ etc. the parity is even and for values of $n = 2, 4, 6$ etc. the parity is odd in the case of a free particle in a one dimensional box.

Parity operator. *The parity operator Π for the function ψ of the variable x is defind by the relation*

$$\Pi\,\psi(x) = \psi(-x)$$

We have proved above $\varepsilon^2 = 1$. Similarly it can be proved that

$$\Pi^2 = 1 \quad \text{or} \quad \Pi = \pm 1$$

The eigen function belonging to the eigen values for $+1$ of the operator are the *even function* ψ (even) which satisfy the relation $\psi_e(x) = \psi_e(-x)$. The eigen function belonging to the eigen values for -1 of the operator are *odd functions* ψ (odd) which satisfy the relation $\psi_0(x) = -\psi_0(-x)$.

6.27 PARITY OF ONE DIMENSIONAL HARMONIC OSCILLATOR WAVE FUNCTION

Schrödinger's wave equation is given by

$$-\frac{\hbar^2}{2m}\frac{d^2\psi}{dx^2}+[V(x)-E]\psi=0$$

For a one dimensional harmonic oscillator $V(x) = \frac{1}{2} kx^2$

$\therefore$ Schrödinger's wave equation becomes

$$\frac{\hbar^2}{2m} \frac{d^2 \psi}{dx^2} + \left[\frac{1}{2} kx^2 - E\right] \psi = 0. \qquad ...(6.41)$$

Now $\frac{d^2 \psi}{dx^2}$ and $V(x) = \frac{1}{2} kx^2$ are both invariant to symmetry transformation $x \to (-x)$. Harmonic oscillator Schrödinger's wave equation, therefore, remains unchanged for a transformation x to $(-x)$.

Hence the basic reason for the parity of a one dimensional harmonic oscillator wave function is that potential energy function $V(x) = \frac{1}{2} kx^2$ is symmetric about the origin and is invariant under transformation $x \to (-x)$.

6.28 PARTICLE IN THREE DIMENSIONAL RECTANGULAR BOX

Consider a *free* particle of rest mass m confined to move non-relativistically in a rectangular box of sides l_x, l_y and l_z, the sides being rigid and parallel to the X, Y and Z co-ordinate axes respectively. As the particle is *free*, the potential energy $V(x, y, z)$ of the particle inside the box is zero. Therefore, the total energy E of the particle is equal to the kinetic energy and is always a *positive* quantity. For non-relativistic motion

$$E = \frac{1}{2} mv^2 = \frac{m^2 v^2}{2m} = \frac{p^2}{2m}$$

where p is the magnitude of total momentum of the particle. Further, the potential energy being independent of time, we can use Schrödinger's time independent, three dimensional wave equation, given by

$$\nabla^2 \psi + \frac{2m}{\hbar^2} (E - V) \psi = 0$$

or

$$\nabla^2 \psi + \frac{2m}{\hbar^2} E \psi = 0$$

or

$$\left(\frac{\partial^2 \psi}{\partial x^2} + \frac{\partial^2 \psi}{\partial y^2} + \frac{\partial^2 \psi}{\partial z^2}\right) + \frac{p^2}{\hbar^2} \psi = 0 \qquad ...(i)$$

Now ψ is a function of x, y, z co-ordinates. We can, therefore, put

$$\psi(x, y, z) = \psi_x \psi_y \psi_z$$

where ψ_x, ψ_y, ψ_z are the functions respectively of x, y and z co-ordinates only.

$$\therefore \quad \frac{\partial^2 \psi}{\partial x^2} = \psi_y \psi_z \frac{\partial^2 \psi_x}{\partial x^2}; \quad \frac{\partial^2 \psi}{\partial y^2} = \psi_x \psi_z \frac{\partial^2 \psi_y}{\partial y^2}$$

and

$$\frac{\partial^2 \psi}{\partial z^2} = \psi_x \psi_y \frac{\partial^2 \psi_z}{\partial z^2}$$

Also

$$p^2 = p_x^2 + p_y^2 + p_z^2$$

where p_x, p_y and p_z are the x, y and z components of the total linear momentum $\vec{p}$ respectively.

Substituting in (*i*), we have

$$\psi_y \psi_z \frac{\partial^2 \psi_x}{\partial x^2} + \psi_x \psi_z \frac{\partial^2 \psi_y}{\partial y^2} + \psi_x \psi_y \frac{\partial^2 \psi_z}{\partial z^2} + \frac{p_x^2 + p_y^2 + p_z^2}{\hbar^2} \psi_x \psi_y \psi_z = 0$$

or $$\frac{1}{\psi_x}\frac{\partial^2 \psi_x}{\partial x^2}+\frac{1}{\psi_y}\frac{\partial^2 \psi_y}{\partial y^2}+\frac{1}{\psi_z}\frac{\partial^2 \psi_z}{\partial z^2}+\frac{p_x^2+p_y^2+p_z^2}{\hbar^2}=0 \quad ...(ii)$$

As the three co-ordinates are independent of each other, we can separate equation (*ii*) into three independent equations, each equation being a function of *one* co-ordinate only.

$$\therefore \quad \frac{\partial^2 \psi_x}{\partial x^2}+\frac{p_x^2}{\hbar^2}\psi_x=0; \quad \frac{\partial^2 \psi_y}{\partial y^2}+\frac{p_y^2}{\hbar^2}\psi_y=0 \text{ and } \frac{\partial^2 \psi_z}{\partial z^2}+\frac{p_z^2}{\hbar^2}\psi_z=0$$

The solutions of these equations, respectively are

$$\psi_x = A_x \sin\frac{p_x}{\hbar}x + B_x \cos\frac{p_x}{\hbar}x \quad ...(iii)$$

$$\psi_y = A_y \sin\frac{p_y}{\hbar}y + B_y \cos\frac{p_y}{\hbar}y \quad ...(iv)$$

$$\psi_z = A_z \sin\frac{p_z}{\hbar}z + B_z \cos\frac{p_z}{\hbar}z \quad ...(v)$$

Applying the boundary conditions, we have

$\psi_x = 0$ when $x = 0$ and $x = l_x$

$\psi_y = 0$ when $y = 0$ and $y = l_y$

$\psi_z = 0$ when $z = 0$ and $z = l_z$

Substituting $\psi_x = 0$ when $x = 0$ in (*iii*), we get $B_x = 0$.

Again substituting $B_x = 0$ and $\psi_x = 0$ when $x = l_x$, we have

$$\sin\frac{p_x}{\hbar}l_x = 0$$

or $$\frac{p_x}{\hbar}l_x = n_x\pi$$

where n_x can have only integral values 1, 2, 3 ... etc.

Similarly, substituting $\psi_y = 0$ when $y = 0$ and $y = l_y$ in (*iv*), we get

$$B_y = 0 \text{ and } \frac{p_y}{\hbar}l_y = n_y\pi$$

where n_y has only integral values 1, 2, 3 ... etc.

and substituting $\psi_z = 0$ when $z = 0$ and $z = l_z$ in (*v*), we get

$$B_z = 0 \text{ and } \frac{p_z}{\hbar}l_z = n_z\pi$$

where n_z has only intergal values 1, 2, 3 ... etc.

Combining all these, we get

$$\psi = \psi_x\psi_y\psi_z = A_x A_y A_z \sin\frac{n_x\pi}{l_x}x \sin\frac{n_y\pi}{l_y}y \sin\frac{n_z\pi}{l_z}z$$

$$= A \sin\frac{n_x\pi}{l_x}x \sin\frac{n_y\pi}{l_y}y \sin\frac{n_z\pi}{l_z}z \quad ...(vi)$$

where $A = A_x A_y A_z$ and n_x, n_y and n_z are all independent integers.

The probability of finding the particle anywhere inside the box is equal to one as there is only one particle and there is no chance of it being located outside the box.

$$\therefore \quad \int_0^{l_x}\int_0^{l_y}\int_0^{l_z} |\psi|^2\, dx\, dy\, dz$$

$$= A^2 \int_0^{l_x}\int_0^{l_y}\int_0^{l_z} \sin^2 \frac{n_x \pi}{l_x} x \sin^2 \frac{n_y \pi}{l_y} y \sin^2 \frac{n_z \pi}{l_z} z\, dx\, dy\, dz$$

$$= A^2 \frac{l_x}{2}\frac{l_y}{2}\frac{l_z}{2} = \frac{A^2}{8} l_x l_y l_z = 1 \qquad \text{...(vii)}$$

as $\int_0^{l_x} \sin^2 \frac{n_x \pi}{l_x} x = \frac{l_x}{2}$; $\int_0^{l_y} \sin^2 \frac{n_y \pi}{l_y} y = \frac{l_y}{2}$ and $\int_0^{l_z} \sin^2 \frac{n_z \pi}{l_z} z = \frac{l_z}{2}$

$\therefore$ Amplitude A from equation (*vii*) is given by

$$A = \sqrt{\frac{8}{l_x l_y l_z}} = \sqrt{\frac{8}{V}}$$

where $V = l_x l_y l_z$ = the volume of the box.

Thus the **eigen functions are**

$$\psi(n_x, n_y, n_z) = \sqrt{\frac{8}{l_x l_y l_z}} \sin \frac{n_x \pi}{l_x} x \sin \frac{n_y \pi}{l_y} y \sin \frac{n_z \pi}{l_z} z$$

The momentum $p = [p_x^2 + p_y^2 + p_z^2]^{1/2}$. Therefore, the **eigen values of momentum** are

$$p\,(n_x, n_y, n_z) = \left[\left(\frac{n_x \pi \hbar}{l_x}\right)^2 + \left(\frac{n_y \pi \hbar}{l_y}\right)^2 + \left(\frac{n_z \pi \hbar}{l_z}\right)^2\right]^{\frac{1}{2}}$$

$$= \pi \hbar \left[\frac{n_x^2}{l_x^2} + \frac{n_y^2}{l_y^2} + \frac{n_z^2}{l_z^2}\right]^{\frac{1}{2}} \qquad \text{...(6.42)}$$

The energy $E = \frac{p^2}{2m}$. Therefore, the **eigen values of energy** are

$$E\,(n_x, n_y, n_z) = \frac{\pi^2 \hbar^2}{2m}\left[\frac{n_x^2}{l_x^2} + \frac{n_y^2}{l_y^2} + \frac{n_z^2}{l_z^2}\right] \qquad \text{...(6.43)}$$

Thus, we find that instead of only one quantum number required in the case of a particle in one dimensional box, we require three quantum numbers, one corresponding to each co-ordinate x, y, z.

If any of the three quantum number n_x, n_y, n_z, is zero, $\psi = 0$, which implies the absence of the particle. This is not physically acceptable. Therefore, none of the quantum numbers can have zero value. Hence the *ground state* corresponds to $n_x = n_y = n_z = 1$

$$\therefore \qquad \psi_{1,1,1} = \sqrt{\frac{8}{l_x l_y l_z}} \sin \frac{\pi x}{l_x} \sin \frac{\pi y}{l_y} \sin \frac{\pi z}{l_z}$$

and the value of *zero point energy* is given by

$$E_{1,1,1} = \frac{\pi^2 \hbar^2}{2m}\left[\frac{1}{l_x^2} + \frac{1}{l_y^2} + \frac{1}{l_z^2}\right]$$

The first excited state is obtained when one of the quantum numbers is 2 and the other two are one each.

$$\therefore \qquad E_{2,1,1} = \frac{\pi^2 \hbar^2}{2m}\left[\frac{4}{l_x^2}+\frac{1}{l_y^2}+\frac{1}{l_z^2}\right]$$

$$E_{1,2,1} = \frac{\pi^2 \hbar^2}{2m}\left[\frac{1}{l_x^2}+\frac{4}{l_y^2}+\frac{1}{l_z^2}\right]$$

$$E_{1,1,2} = \frac{\pi^2 \hbar^2}{2m}\left[\frac{1}{l_x^2}+\frac{1}{l_y^2}+\frac{4}{l_z^2}\right]$$

Degeneracy. We now suppose that the box is *cubical i.e.*, $l_x = l_y = l_z = l$.

The *eigen functions* are now given by

$$\psi\,(n_x, n_y, n_z) = \sqrt{\frac{8}{l^3}}\,\sin\frac{n_x\,\pi}{l}\,x\,\sin\frac{n_y\,\pi}{l}\,y\,\sin\frac{n_z\,\pi}{l}\,z$$

where as already stated $n_x = 1, 2, 3$ ...etc., $n_y = 1, 2, 3$ ... etc., $n_z = 1, 2, 3$ ... etc., are independent integral quantum numbers.

The *eigen values of momentum* now are

$$p\,(n_x, n_y, n_z) = \frac{\pi\hbar}{l}\left[n_x^2 + n_y^2 + n_z^2\right]^{1/2}$$

The *eigen values of energy* now are

$$E\,(n_x, n_y, n_z) = \frac{\pi^2\,\hbar^2}{2ml^2}\left[n_x^2 + n_y^2 + n_z^2\right]$$

The lowest state or ground state is given by $n_x = n_y = n_z = 1$ and the corresponding values of momentum and energy are

$$\text{Momentum} \qquad p_{1,1,1} = \frac{\pi\hbar}{l}\sqrt{3}$$

$$\text{Energy} \qquad E_{1,1,1} = \frac{\pi^2\hbar^2}{2\,ml^2}\,3$$

The eigen state will be

$$\psi_{1,1,1} = \sqrt{\frac{8}{l^3}}\,\sin\frac{\pi x}{l}\,\sin\frac{\pi y}{l}\,\sin\frac{\pi z}{l}$$

The next higher state or first excited state will have 3 sets of quantum numbers 2, 1, 1 or 1, 2, 1 or 1, 1, 2. For all these three quantum states the values of momentum and energy are

$$\text{Momentum} \qquad p_{2,1,1} = \frac{\pi\hbar}{l}\left[2^2 + 1^2 + 1^2\right]^{1/2} = \frac{\pi\hbar}{l}\sqrt{6}$$

$$p_{1,2,1} = \frac{\pi\hbar}{l}\left[1^2 + 2^2 + 1^2\right]^{1/2} = \frac{\pi\hbar}{l}\sqrt{6}$$

$$p_{1,1,2} = \frac{\pi\hbar}{l}\left[1^2 + 1^2 + 2^2\right]^{1/2} = \frac{\pi\hbar}{l}\sqrt{6}$$

$$\therefore \qquad p_{2,1,1} = p_{1,2,1} = p_{1,1,2} = \frac{\pi \hbar}{l}\sqrt{6}$$

Similarly $\quad E_{2,1,1} = E_{1,2,1} = E_{1,1,2} = \dfrac{\pi^2 \hbar^2}{2ml^2}\, 6$

The *eigen function* corresponding to the three sets of quantum number are, however, different. Thus

$$\psi_{2,1,1} = \sqrt{\frac{8}{l^3}} \sin \frac{2\pi x}{l} \sin \frac{\pi y}{l} \sin \frac{\pi z}{l}$$

$$\psi_{1,2,1} = \sqrt{\frac{8}{l^3}} \sin \frac{\pi x}{l} \sin \frac{2\pi y}{l} \sin \frac{\pi z}{l}$$

and $\quad \psi_{1,1,2} = \sqrt{\dfrac{8}{l^3}} \sin \dfrac{\pi x}{l} \sin \dfrac{\pi y}{l} \sin \dfrac{2\pi z}{l}$

we, therefore, find that for a single value of energy or momentum; there are three different quantum states. *The property of two or more quantum states of a particle, having different sets of quantum numbers and different eigen functions, to have the same value of momentum and energy, is called* **degeneracy.**

A free particle in a *cubical* box is thus a case of degeneracy. The lowest state is not degenerate as there is only *one* eigen function for $n_x = n_y = n_z = 1$. The next higher state is said to be *three fold* degenerate as three sets of quantum numbers (2, 1, 1), (1, 2, 1) and (1, 1, 2) and three eigen functions $\psi_{2,1,1}$; $\psi_{1,2,1}$; $\psi_{1,1,2}$ have the same value of energy and momentum.

The next higher state (2, 2, 1) has also *three fold* degeneracy, because the three different eigen function $\psi_{2,2,1}$; $\psi_{1,2,2}$; $\psi_{2,1,2}$ will have the same value of momentum given by

$$p_{2,2,1} = p_{1,2,2} = p_{2,1,2} = \frac{\pi \hbar}{l}\left[2^2 + 2^2 + 1\right]^{1/2} = \frac{\pi \hbar}{l}\, 3$$

and the same value of energy, given by

$$E_{2,2,1} = E_{1,2,2} = E_{2,1,2} = \frac{\pi^2 \hbar^2}{2ml^2}\left[2^2 + 2^2 + 1^2\right] = \frac{\pi^2 \hbar^2}{2ml^2}\, 9$$

The next higher state (2, 2, 2) is not degenerate, but if we proceed further we get three fold degeneracy, except when all the three quantum numbers are the same.

Energy Levels of a Particle

The energy levels of a particle in a cubical box of each side l is given by

$$E(n_x, n_y, n_z) = \frac{\pi^2 \hbar^2}{2ml^2}\left[n_x^2 + n_y^2 + n_z^2\right]$$

(*i*) For the *lowest energy state* $n_x = n_y = n_z = 1$

$$\therefore \qquad E_{1,1,1} = \frac{\pi^2 \hbar^2}{2ml^2}(1^2 + 1^2 + 1^2) = \frac{\pi^2 \hbar^2}{2ml^2} \times 3$$

The lowest state is non-degenerate

(*ii*) For the *first excited state.* There are three sets quantum number — (2,1,1) or (1,2,1) or (1,1,2) and

$$E_{2,1,1} = E_{1,2,1} = E_{1,1,2} = \frac{\pi^2 \hbar^2}{2ml^2}(2^2 + 1^2 + 1^2) = \frac{\pi^2 \hbar^2}{2ml^2} \times 6$$

The first excited state has three fold degeneracy

(*iii*) For the *second excited state* there are again three sets of quantum number — (2,2,1) or (2,1,2) or (1,2,2) and

$$E_{2,2,1} = E_{2,1,2} = E_{1,2,2} = \frac{\pi^2 \hbar^2}{2ml^2}(2^2 + 2^2 + 1^2) = \frac{\pi^2 \hbar^2}{2ml^2} \times 9$$

The second excited state also has three fold degeneracy.

(*iv*) For the *third excited state* $n_x = n_y = n_z = 2$ and

$$E_{2,2,2} = \frac{\pi^2 \hbar^2}{2ml^2}(2^2 + 2^2 + 2^2) = \frac{n^2 \hbar^2}{2ml^2} \times 12$$

This state is also non-degenerate.

Example 6.23. *An electron is put in a cubical box of each side 1 Å. Find the values of its momentum and energy for the ground state and the first excited state.*

Solution. For the ground state $n_x = n_y = n_z = 1$

$$\therefore \text{ Momentum } p_{1,1,1} = \frac{\pi\hbar}{l}\sqrt{3} = \frac{\pi h}{2\pi l}\sqrt{3} = \frac{h}{2l}\sqrt{3} = \frac{6.63 \times 10^{-34}}{2 \times 10^{-10}}\sqrt{3}$$

$$= (3.315 \times 10^{-24})\sqrt{3} = 5.74 \times 10^{-24} \text{ kgms}^{-1}$$

$$\text{Energy } E_{1,1,1} = \frac{\pi^2 \hbar^2}{2ml^2} 3 = \frac{\pi^2 h^2}{4\pi^2 2 ml^2} 3 = \frac{h^2}{8ml^2} 3$$

$$= \frac{(6.63 \times 10^{-34})^2 \times 3}{8 \times 9.1 \times 10^{-31} \times 10^{-20}}$$

$$= 60.4 \times 3 \times 10^{-19} = 181 \times 10^{-19} \text{ Joule}$$

$$= 113 \text{ eV}$$

For the first excited state, one of the quantum numbers is = 2

$$\therefore \text{ Momentum } p_{2,1,1} = p_{1,2,1} = p_{1,1,2} = \frac{\pi\hbar}{l}\sqrt{6}$$

$$= \frac{h}{2l}\sqrt{6} = 3.315 \times 10^{-24} \times \sqrt{6}$$

$$= 8.13 \times 10^{-24} \text{ kg. ms}^{-1}$$

$$\text{Energy } E_{2,1,1} = E_{1,2,1} = E_{1,1,2} = \frac{\pi^2 \hbar^2}{2ml^2} 6 = \frac{h^2}{8ml^2} 6$$

$$= 60.4 \times 6 \times 10^{-19} = 362 \times 10^{-19} \text{ Joule.}$$

$$= 226 \text{ eV.}$$

Example 6.24. *Think of the nucleus as a box with a size of 10^{-14} m across. Compute the energy of a neutron confined to the nucleus. ($h = 6.26 \times 10^{-34}$ Js, $m = 1.6 \times 10^{-27}$ kg.)*

Solution. Take the nucleus to be a cubical box of size 10^{-14} m across.

$$\therefore \quad l_x = l_y = l_z = l = 10^{-14} \text{ m}$$

The neutron will have the lowest energy in the quantum state given by

$$n_x = n_y = n_z = 1$$

Now energy $= E_{1,1,1} = \dfrac{\pi^2 \hbar^2}{2m}\left[\dfrac{n_x^2}{l_x^2} + \dfrac{n_y^2}{l_y^2} + \dfrac{n_z^2}{l_z^2}\right] = \dfrac{\pi^2 \hbar^2}{2m}\left[\dfrac{3}{l^2}\right]$

$$= \frac{3\pi^2 h^2}{4\pi^2 \times 2ml^2} = \frac{3h^2}{8ml^2}$$

$$= \frac{3 \times [6.62 \times 10^{-34}]^2}{8 \times 1.6 \times 10^{-27} \times 10^{-28}} = 10.29 \times 10^{-13} \text{ J} = 6.43 \text{ MeV}$$

For the first excited state, one of the quantum numbers is = 2

$\therefore$ Energy $= E_{2,1,1} = E_{1,2,1} = E_{1,1,2} = \dfrac{\pi^2\hbar^2}{2m}\left[\dfrac{1+4+1}{l^2}\right] = \dfrac{\pi^2 \hbar^2}{2ml^2} \times 6$

$$= \frac{h^2}{8ml^2} 6 = 12.86 \text{ MeV}$$

Thus the first excited state is three fold degenerate.

For the second excited state two of the quantum number are = 2

$\therefore$ Energy $= E_{2,2,1} = E_{2,1,2} = E_{1,2,2} = \dfrac{\pi^2\hbar^2}{2m}\left[\dfrac{4+4+1}{l^2}\right] = \dfrac{\pi^2 \hbar^2}{2ml^2} \times 9$

$$= \frac{h^2}{8ml^2} 9 = 19.29 \text{ MeV}$$

The second excited state is also three fold degenerate.

For the third excited state all the quantum number are = 2

$\therefore$ Energy $= E_{2,2,2} = \dfrac{\pi^2\hbar^2}{2m}\left[\dfrac{4+4+4}{l^2}\right] = \dfrac{\pi^2 \hbar^2}{2ml^2} \times 12$

$$= \frac{h^2}{8ml^2} 12 = 25.72 \text{ MeV}$$

The third excited state is not degenerate.

Example 6.25. *Find the potential barrier for the α-particle emitted by ${}_{86}Rn^{222}$ assuming effective nuclear radius $r_o = 1.5\ A^{1/3}$ Fermi.* **(Bang. U. 2004 similar; P.U. 2000)**

Solution. **Radius of the nucleus $r_o = 1.5\ A^{1/3}$ Fermi**

For ${}_{86}Rn^{222}$, $A = 222$; 1 Fermi $= 10^{-15}$ m

$\therefore$ **$r_o = 1.5 \times (222)^{1/3} \times 10^{-15}$ m $= 1.5 \times 6.055 \times 10^{-15} = 9.08 \times 10^{-15}$ m**

Mass of the α-particle $m = 4 \times 1.67 \times 10^{-27}$ kg

Take the nucleus to be a cubical box of each side $l = 9.08 \times 10^{-15}$ m

The α-particle will have the lowest energy in the quantum state given by $n_x = n_y = n_z = 1$ and its value is

$$E_{1,1,1} = \frac{\pi^2 \hbar^2}{2ml^2}(n_x^2 + n_y^2 + n_z^2) = \frac{\pi^2 \hbar^2}{2ml^2} \times 3$$

$$= \frac{\pi^2 h^2}{4\pi^2 \times 2ml^2} \times 3 = \frac{3h^2}{8ml^2} = \frac{3 \times (6.62 \times 10^{-34})^2}{8 \times 4 \times 1.67 \times 10^{-27} \times (9.08 \times 10^{-15})^2}$$

$$= 2.98 \times 10^{-13} \text{ J} = 1.86 \text{ MeV}$$

$\therefore$ **Minimum potential barrier for the α-particle emitted by ${}_{86}Rn^{222} = 1.86$ MeV.**

6.29 POTENTIAL STEP AND POTENTIAL BARRIER

Potential step. *A potential step corresponds to the force field acting on a particle being zero every where except in a limited region.* A single step potential is shown in Fig. 6.4 (*a*) and is represented by

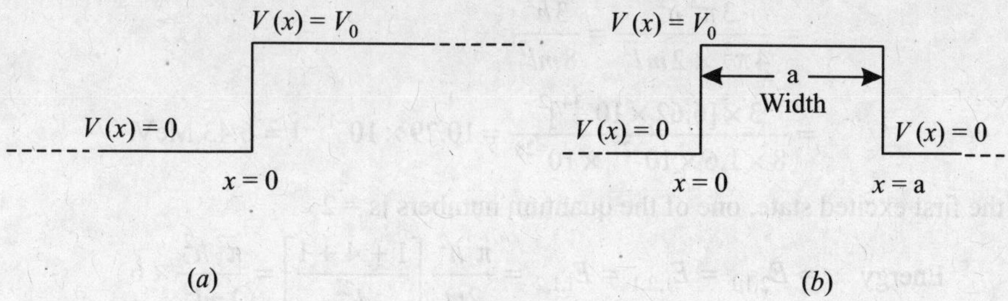

Fig. 6.4

Potential $V(x) = 0$ for $x < 0$

Potential $V(x) = V_0$ for $x \geq 0$

The height of the potential step is said to be V_0.

Potential barrier. A potential barrier has a *double step*, first from $V(x) = 0$ to $V(x) = V_0$ and the second from $V(x) = V_0$ to $V(x) = 0$, as shown in Fig. 6.4 (*b*)

The distance between the two steps is known as the *width of the barrier*. A potential barrier is, therefore, represented as

$$V(x) = 0 \text{ for } x < 0$$
$$V(x) = V_0 \text{ for } 0 \leq x \leq a$$
$$V(x) = 0 \text{ for } x > a$$

6.30 ONE DIMENSIONAL POTENTIAL STEP (OR BARRIER)

Assuming that the total energy E of the incident particle is greater than V_0, we shall establish Schrödinger's wave equation for the particle and we will solve it. From this we shall calculate the co-efficient of reflection and transmission. The result will be discussed from both, the classical and quantum point of view.

The one dimensional function of a potential step (or barrier)

$$V_x = 0 \text{ for } x < 0$$
$$= V_0 \text{ for } x \geq 0$$

The particle moves like a free particle in the region $x < 0$ but as it approaches the step potential towards the right *i.e.* for $x > 0$ it has to face a potential hill or step or barrier of height (potential energy) $V_x = V_0$. A one dimensional potential step (or barrier) is shown in Fig. 6.5. There is a discontinuity at $x = 0$.

6.30.1 Total Energy $E > V_0$; the Step Potential

Let a particle of mass *m* and total energy *E* be incident from the left (Region I) on the potential step $V(x) = V_0$ at $x = 0$, the origin of the *X*-axis. This is equivalent to a quantum particle striking a stationary wall of finite height *h*.

The potential energy of the particle is *zero* when it is to the left of the step *i.e.* left of the origin *O* and the potential energy $V(x) = V_0$ to the right of the step or the origin.

$\therefore \quad V(x) = 0$ for $x < 0$ Region I

and $\quad V(x) = V_0$ for $x > 0$ Region II

Suppose that the total energy of the incident particle is *greater* than V_0, the value of step potential *i.e.* $E > V_0$

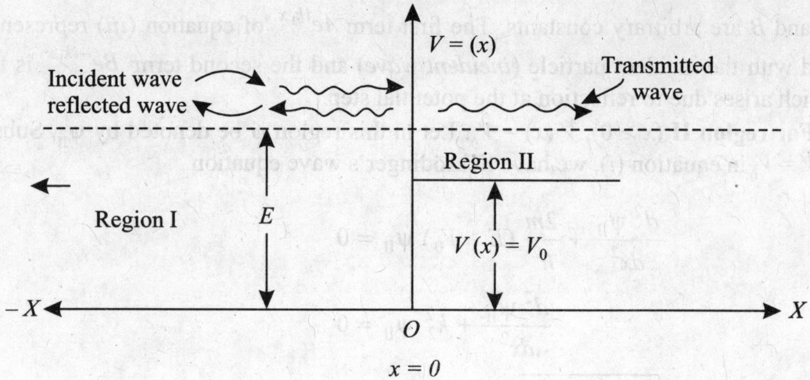

Fig. 6.5

Classical view point. According to classical mechanics, the particle having energy $E > V_0$ travelling in region I in the $+X$ direction will experience a retarding force $F = \frac{-dV(x)}{dx}$ at the point $x = 0$. This force will slow down the motion of the particle in the direction of increasing x but the particle will keep moving and enter the region II *i.e.*, $x > 0$.

The momentum of the particle in region I ($x < 0$) is given by

$$E = \frac{p_1^2}{2m} \quad \text{or} \quad p_1 = \sqrt{2mE}$$

In region II ($x > 0$) the momentum of the particle is given by

$$E - V_0 = \frac{p_2^2}{2m} \quad \text{or} \quad p_2 = \sqrt{2m(E - V_0)}$$

As $p_2 < p_1$ the velocity of the particle in region II will be less than that in region I but the particle with $E > V_0$ *will always be transmitted*. The change in potential from 0 to V_0 at $x = 0$ only provides a retarding force which reduces the speed of the particle.

Quantum mechanical view point. According to quantum mechanics, the motion of the particle is governed by Schrödinger's wave equation for regions I and II. As the potential V_0 is independent of time, we have to solve Schrödinger's time independent wave equation to analyse the behaviour of the particle. It is given by

$$\frac{d^2\psi}{dx^2} + \frac{2m}{\hbar^2}(E - V)\psi = 0 \qquad ...(i)$$

(*a*) **For region I** ($x < 0$) $V = 0$. Let in this region ψ be denoted by ψ_1. Substituting $V(x) = 0$ in equation (*i*), we have Schrödinger's wave equation

$$\frac{d^2\psi_I}{dx^2} + \frac{2m}{\hbar^2}E\psi_I = 0$$

or

$$\frac{d^2\psi_I}{dx^2} + k_1^2\psi_I = 0 \qquad ...(ii)$$

where

$$k_1 = \frac{\sqrt{2mE}}{\hbar} = \frac{p_1}{\hbar} = \frac{2\pi}{\lambda_1}$$

The general solution of equation (*ii*) is given by

$$\psi_I = Ae^{ik_1x} + Be^{-ik_1x} \qquad ...(iii)$$

where A and B are arbitrary constants. The first term $Ae^{ik_1 x}$ of equation (*iii*) represents the wave associated with the incident particle (*incident wave*) and the second term $Be^{-ik_1 x}$ is the *reflected wave*, which arises due to reflection at the potential step.

(*b*) **For region II** ($x > 0$), $V(x) = V_0$. Let in this region ψ be denoted by ψ_{II}. Substituting the value of $V = V_0$ in equation (*i*), we have Schrödinger's wave equation

$$\frac{d^2 \psi_{II}}{dx^2} + \frac{2m}{\hbar^2}(E - V_0)\,\psi_{II} = 0$$

or
$$\frac{d^2 \psi_{II}}{dx^2} + k_2^2\,\psi_{II} = 0 \qquad \text{...}(iv)$$

where
$$k_2 = \frac{\sqrt{2m(E - V_0)}}{\hbar} = \frac{p_2}{\hbar} = \frac{2\pi}{\lambda_2}$$

The general solution of equation (*iv*) is given by

$$\psi_{II} = Ce^{ik_2 x} + De^{-ik_2 x} \qquad \text{...}(v)$$

where C and D are arbitrary constants. The function $Ce^{-ik_2 x}$ gives the *transmitted wave*. The function $De^{-ik_2 x}$ represents a wave moving to left in –X direction. But there is neither a sources of particles in region II nor a further potential step to produce a reflected wave, which means there is no reflected wave and therefore $D = 0$. Substituting $D = 0$ in equation (v), we get

$$\psi_{II} = Ce^{ik_2 x} \qquad \text{...}(vi)$$

Determination of constants. To evaluate the constants B and C in terms of A, we apply the boundary conditions which are:–

1. The wave function ψ must be continuous at the boundary $x = 0$

∴
$$\psi_I(0) = \psi_{II}(0)$$

or
$$Ae^{ik_1 \times 0} + Be^{-ik_1 \times 0} = Ce^{ik_2 \times 0}$$

or
$$A + B = C \qquad \text{...}(vii)$$

2. The first order derivatives of the wave function ψ *i.e.* $\frac{d\psi}{dx}$ must also be continuous at $x = 0$. According to this condition

$$\left(\frac{d\psi_I}{dx}\right)_{x=0} = \left(\frac{d\psi_{II}}{dx}\right)_{x=0}$$

or
$$\left[ik_1 Ae^{ik_1 x} - ik_1 Be^{-ik_1 x}\right]_{x=0} = \left[ik_2 Ce^{ik_2 x}\right]_{x=0}$$

or
$$k_1(A - B) = k_2 C$$

or
$$A - B = \frac{k_2}{k_1} C \qquad \text{...}(viii)$$

Adding equations (*vii*) and (*viii*), we get

$$2A = C + \frac{k_2}{k_1} C = C\left(\frac{k_1 + k_2}{k_1}\right)$$

or
$$C = \frac{2k_1}{k_1 + k_2} A \qquad \text{...}(ix)$$

Subtracting (*viii*) from, (*vii*), we get

$$2B = C - \frac{k_2}{k_1} C = C \left(\frac{k_1 - k_2}{k_1} \right)$$

or
$$B = \frac{k_1 - k_2}{2k_1} C = \frac{k_1 - k_2}{2k_1} \cdot \frac{2k_1}{k_1 + k_2} A$$

$$= \frac{k_1 - k_2}{k_1 + k_2} A \qquad \text{...}(x)$$

Substituting the values of C and B from equations (*ix*) and (*x*) in equations (*iii*) and (*vi*), we get

$$\psi_I = Ae^{ik_1 x} + A \frac{k_1 - k_2}{k_1 + k_2} e^{-ik_1 x} \qquad \text{...}(xi)$$

and
$$\psi_{II} = A \frac{2k_1}{k_1 + k_2} e^{ik_2 x} \qquad \text{...}(xii)$$

The first term in equation (*xi*) gives the *incident wave function* and the second term the *reflected wave function.*

∴ Incident wave function

$$\psi_{in} = Ae^{ik_1 x} \qquad \text{...}(xiii)$$

and Reflected wave function

$$\psi_{re} = A \frac{k_1 - k_2}{k_1 + k_2} e^{-ik_1 x} \qquad \text{...}(xiv)$$

From (*xiii*) and (*xiv*), we get

$$\psi_I = \psi_{in} + \psi_{re}$$

Similarly equation (*xii*) gives the *transmitted wave function.*

∴ Transmitted wave function

$$\psi_{tr} = A \frac{2k_1}{k_1 + k_2} e^{ik_2 x} \qquad \text{...}(xv)$$

Thus
$$\psi_{II} = \psi_{tr}$$

Reflection Co-efficient. The reflection co-efficient R is defined as

$$R = \frac{\text{Reflected probability density current}}{\text{Incident probability density current}} = \frac{S_{re}}{S_{in}}$$

Now, *probability density current for a free particle in a given region is defined as the product of its velocity and probability density.*

∴
$$S_{re} = v_1 \psi_{re} \psi_{re}{}^*$$
$$S_{in} = v_1 \psi_{in} \psi_{in}{}^*$$

where v_1 is the velocity of the particle in region I.

Hence
$$R = \frac{S_{re}}{S_{in}} = \frac{v_1 \psi_{re} \psi_{re}{}^*}{v_1 \psi_{in} \psi_{in}{}^*} = \frac{\psi_{re} \psi_{re}{}^*}{\psi_{in} \psi_{in}{}^*} \qquad \text{...}(xv)\ (a)$$

$$= \frac{A \frac{k_1 - k_2}{k_1 + k_2} e^{-ik_1 x} \cdot A \frac{k_1 - k_2}{k_1 + k_2} e^{+ik_1 x}}{Ae^{+ik_1 x} \cdot Ae^{-ik_1 x}}$$

∴
$$R = \left(\frac{k_1 - k_2}{k_1 + k_2} \right)^2 \qquad \text{...}(6.44)$$

Interpretation of the result. *The reflection co-efficient R is a measure of the probability that the incident particle may be reflected.* From relation (6.44) we find that the value of R is not zero unless $k_1 = k_2$ which means $V_0 = 0$ *i.e.* there is no potential barrier. In other words *there will always be some reflection at the step potential*. This is contrary to our classical view point where we have explained that for $E > V_0$, the particle is *always transmitted* and it would not at all be reflected.

Transmission co-efficient. The transmission co-efficient T is defined as

$$T = \frac{\text{Transmitted probability density current}}{\text{Incident probability density current}} = \frac{S_{tr}}{S_{in}}$$

Now $S_{tr} = v_{II}\,\psi_{tr}\,\psi_{tr}^*$ where v_{II} is the velocity of the particle in region II.

$$\therefore \quad T = \frac{v_{II}\,\psi_{tr}\,\psi_{tr}^*}{v_I\,\psi_{in}\,\psi_{in}^*} = \frac{v_{II}}{v_I}\,\frac{A\dfrac{2k_1}{k_1+k_2}\,e^{ik_2x}\,.\,A\dfrac{2k_1}{k_1+k_2}\,e^{-ik_2x}}{Ae^{ik_1x}\,.\,Ae^{-ik_1x}} \qquad \text{...(xvi)}$$

$$= \frac{v_{II}}{v_I}\left(\frac{2k_1}{k_1+k_2}\right)^2 \qquad \text{...(xvii)}$$

Again $$v_I = \frac{p_1}{m} = \frac{\hbar\,k_1}{m}$$

and $$v_{II} = \frac{p_2}{m} = \frac{\hbar\,k_2}{m}$$

$$\therefore \quad \frac{v_{II}}{v_I} = \frac{k_2}{k_1} \qquad \text{...(xviii)}$$

Substituting the value of $\dfrac{v_{II}}{v_I} = \dfrac{k_2}{k_1}$ in relation (*xvii*) we have

$$T = \frac{k_2}{k_1}\cdot\frac{4k_1^2}{(k_1+k_2)^2} = \frac{4k_1\,k_2}{(k_1+k_2)^2} \qquad \text{...(6.45)}$$

Interpretation of the result. *The transmission co-efficient is a measure of the probability that the incident particle will be transmitted across the potential step.* We find that the value of T is not **one**, *i.e.* the particle has some probability of not being transmitted, which is again contrary to classical view point, where the particle with $E > V_0$ *will always be transmitted.* Also R and T remain unchanged if k_1 and k_2 are exchanged in equations (6.44) and (*xvi*). In other words we shall get the same value of R and T if the particle is incident on the step potential from the right.

From equations (6.44) and (*xviii*) we find that

$$R + T = 1$$

Thus we find that the behaviour of a particle under quantum conditions is very much different from that of a particle under classical conditions.

Sum of Reflection and Transmission Co-Efficients is One

When a particle of mass m and energy E meets a one-dimensional potential barrier of height V_0 ($E > V_0$). The co-efficient of reflection is given by

$$R = \left[\frac{k_1 - k_2}{k_1 + k_2}\right]^2 \qquad \text{...(iii)}$$

where $$k_1 = \frac{\sqrt{2mE}}{\hbar} \text{ and } k_2 = \frac{\sqrt{2m(E - V_0)}}{\hbar}$$

and co-efficient of transmission is given by

$$T = \frac{4k_1 k_2}{(k_1 + k_2)^2} \qquad \text{...(iv)}$$

Adding (iii) and (iv), we have

$$R + T = \frac{(k_1 - k_2)^2}{(k_1 + k_2)^2} + \frac{4k_1 k_2}{(k_1 + k_2)^2} = \frac{(k_1 + k_2)^2}{(k_1 + k_2)^2} = 1 \qquad \text{...(6.46)}$$

Thus the sum of reflection and transmission co-efficients is equal to **one.**

Example 6.26. *What ratio of* $\frac{E}{V_0}$ *is necessary for scattering from a one dimensional step potential, so that the transmission probability is 50%?*

Solution. Transmission co-efficient for step potential $T = \frac{4k_1 k_2}{(k_1 + k_2)^2}$

Now $\qquad T = \frac{1}{2}$

$$\therefore \quad \frac{1}{2} = \frac{4k_1 k_2}{(k_1 + k_2)^2} \quad \text{or } (k_1 + k_2)^2 = 8k_1 k_2 \qquad \text{...(i)}$$

$$\therefore \quad k_1^2 + k_2^2 = 6k_1 k_2$$

Substituting the value of $k_1 = \frac{\sqrt{2mE}}{\hbar}$ and $k_2 \frac{\sqrt{2m(E - V_0)}}{\hbar}$ in (*i*), we have

$$\frac{2mE}{\hbar^2} + \frac{2m(E - V_0)}{\hbar^2} = \frac{6\sqrt{2mE}\sqrt{2m(E - V_0)}}{\hbar^2}$$

or

$$E + E - V_0 = 6\sqrt{E(E - V_0)}$$

Squaring both sides, we get

$$4E^2 - 4EV_0 + V_0^2 = 36E^2 - 36EV_0$$

or

$$32E^2 - 32EV_0 - V_0^2 = 0$$

or

$$32\frac{E^2}{V_0^2} - 32\frac{E}{V_0} - 1 = 0$$

This is quadratic equation in $\frac{E}{V_0}$ and its roots are

$$\frac{32 \pm \sqrt{32^2 + 128}}{64} = \frac{1 \pm \sqrt{\frac{9}{8}}}{2} = \frac{1 \pm 1.06}{2}$$

$$\therefore \qquad \frac{E}{V_0} = 1.03 \text{ or } -0.03$$

But negative value of $\frac{E}{V_0}$ is not allowed

Hence $\frac{E}{V_0} = 1.03$.

6.30.2 Total Energy $E < V_0$; the Step Potential

Let a particle of mass m and total energy E be incident from left (region I) on the potential step $V(x) = V_0$ at $x = 0$, the origin of the X-axis.

The potential energy of the particle is *zero* when it is to the left of the step *i.e.* left of the origin O and the potential energy $V(x) = V_0$ to the right of the step or the origin.

$\therefore$ $V(x) = 0$ for $x < 0$ Region I

$V(x) = V_0$ for $x \geq 0$ Region II

As $0 < E < V_0$ the total energy of the incident particle is positive and *less* than V_0, the value of step potential *i.e.*, $E < V_0$

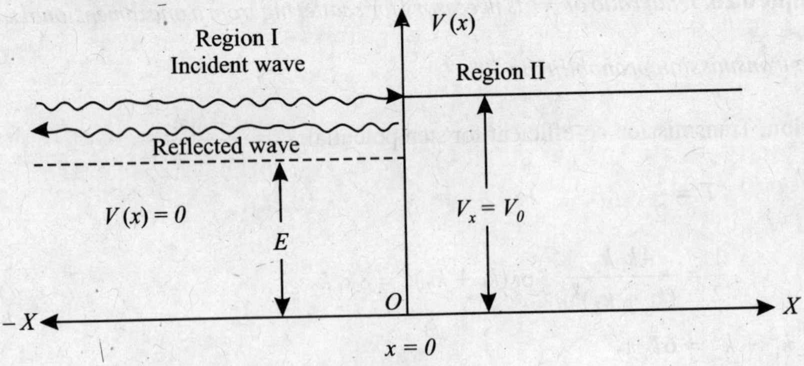

Fig. 6.6

Classical view point. According to classical mechanics, when the particle having energy E incident from region I reaches the point $x = 0$, it experiences a force $F = \dfrac{-dV}{dx}$ in the $-X$ direction.

Since the total energy E is constant and less than V_0, so according to classical mechanics, the particle cannot penetrate into the region II ($x > 0$). This is because in region II

$$E = \frac{p^2}{2m} + V(x) < V(x)$$

or
$$\frac{p^2}{2m} < 0 \qquad ...(i)$$

Thus the kinetic energy in region II will be *negative*, which is meaningless in classical mechanics. Relation (*i*) also leads to the conclusion that p^2 is negative *i.e.* the momentum of the particle is an imaginary quantity, which is again meaningless. So the particle cannot enter into region II and is completely reflected at the step potential, with the *same magnitude* of momentum, but in the opposite direction.

Quantum mechanical view point. According to quantum mechanics, the motion of the particle is governed by Schrödinger's wave equation for regions I and II. As the potential V_0 is independent of time, we have to solve Schrödinger's time independent wave equation to anaylse the behaviour of the particle. It is given by

$$\frac{d^2\psi}{dx^2} + \frac{2m}{\hbar^2}(E - V)\psi = 0$$

(*a*) **For region I** ($x < 0$) $V = 0$. Let in this region ψ be denoted by ψ_I. Substituting $V(x) = 0$ in the above equation we have

$$\frac{d^2\psi_I}{dx^2} + \frac{2m}{\hbar^2}E\psi_I = 0$$

or $$\frac{d^2 \psi_I}{dx^2} + k_1^2 \psi_I = 0 \qquad ...(ii)$$

where $$k_1 = \frac{\sqrt{2mE}}{\hbar} = \frac{p_1}{\hbar} = \frac{2\pi}{\lambda_1}$$

The general solution of equation (*ii*) is given by

$$\psi_I = Ae^{ik_1 x} + Be^{-ik_1 x} \qquad ...(iii)$$

where A and B are arbitrary constants. The first term $Ae^{ik_1 x}$ of equation (*iii*) represents the wave associated with the incident particle (*incident wave*) and the second term $Be^{-ik_1 x}$ is the *reflected wave*, which arises due to reflection at the potential step.

(*b*) **For region II** $(x > 0)$, $V(x) = V_0$. Let in this region ψ be denoted by ψ_{II}. Substituting the value of $V = V_0$ in the equation

$$\frac{d^2 \psi}{dx^2} + \frac{2m}{\hbar^2}(E - V)\psi = 0, \text{ we get}$$

$$\frac{d^2 \psi_{II}}{dx^2} + \frac{2m}{\hbar^2}(E - V_0)\psi_{II} = 0$$

but $E < V_0$ and hence $(E - V_0)$ is a negative quantity. Putting $(E - V_0) = -(V_0 - E)$, we have

$$\frac{d^2 \psi_{II}}{dx^2} - \frac{2m}{\hbar^2}(V_0 - E)\psi_{II} = 0$$

or $$\frac{d^2 \psi_{II}}{dx^2} - k_2^2 \psi_{II} = 0 \qquad ...(iv)$$

where $$k_2^2 = \frac{2m}{\hbar^2}(V_0 - E)$$

The solution of equation (*iv*) is given by

$$\psi_{II} = Ce^{-k_2 x} + De^{k_2 x} \qquad ...(v)$$

where C and D are arbitrary constants. The wave function ψ_{II} becomes infinite as $x \to \infty$ due to the presence of the term $De^{k_2 x}$. In order that ψ_{II} may remain *finite*, we must have $D = 0$. Substituting $D = 0$ in equation (*v*), we get

$$\psi_{II} = Ce^{-k_2 x} \qquad ...(vi)$$

Determination of constants. To evalute the constant B and C in terms of A, we apply the boundary conditions which are:–

1. The wave function ψ must be continuous at the boundary $x = 0$

$\therefore$ $$\psi_I(0) = \psi_{II}(0)$$

or $$Ae^{ik_1 \times 0} + Be^{-ik_1 \times 0} = Ce^{-k_2 \times 0}$$

or $$A + B = C \qquad ...(vii)$$

2. The first order derivative of the function ψ *i.e.* $\frac{d\psi}{dx}$ must also be continuous at $x = 0$. According to this condition

$$\left(\frac{d\psi_I}{dx}\right)_{x=0} = \left(\frac{d\psi_{II}}{dx}\right)_{x=0}$$

or $$\left[ik_1 Ae^{ik_1 x} - ik_1 Be^{-k_1 x}\right]_{x=0} = \left[-k_2 Ce^{-k_2 x}\right]_{x=0}$$

or $$ik_1(A - B) = -k_2 C$$

$\therefore$ $$A - B = \frac{-k_2}{ik_1} C = i\frac{k_2}{k_1} C \qquad ...(viii)$$

Adding (*vii*) and (*viii*), we get

$$2A = C + i\frac{k_2}{k_1} C$$

or $$C = \frac{2k_1}{k_1 + ik_2} A = \frac{2ik_1}{ik_1 - k_2} A \qquad ...(ix)$$

Subtracting (*viii*) from (*vii*), we get

$$2B = C - i\frac{k_2}{k_1} C = \left(\frac{k_1 - ik_2}{k_1}\right) C$$

$$= \left(\frac{ik_1 + k_2}{ik_1}\right) C = \frac{ik_1 + k_2}{ik_1} \cdot \frac{2ik_1}{ik_1 - k_2} A$$

or $$B = \frac{ik_1 + k_2}{ik_1 - k_2} A \qquad ...(x)$$

Substituting the values of B and C from equations (*ix*) and (*x*) in equations (*iii*) and (*vi*), we get

$$\psi_I = Ae^{ik_1 x} + A\frac{ik_1 + k_2}{ik_1 - k_2} e^{-ik_1 x} \qquad ...(xi)$$

and $$\psi_{II} = A\frac{2ik_1}{ik_1 - k_2} e^{-k_2 x} \qquad ...(xii)$$

The first term in equation (*xi*) gives the *incident wave function* and the second term the *reflected wave function*.

$\therefore$ Incident wave function

$$\psi_{in} = Ae^{ik_1 x} \qquad ...(xiii)$$

and Reflected wave function

$$\psi_{re} = A\frac{ik_1 + k_2}{ik_1 - k_2} e^{-ik_1 x} \qquad ...(xiv)$$

Reflection Co-efficient. The reflection co-efficient R is defined as

$$R = \frac{\text{reflected probability density current}}{\text{Incident probability density current}} = \frac{S_{re}}{S_{in}}$$

Now *probability density current for a free particle in a given region is defined as the product of its velocity and probability density.*

If v_1 is the velocity of the particle in region I, then

$$R = \frac{S_{re}}{S_{in}} = \frac{v_1 \psi_{re} \psi_{re}^*}{v_1 \psi_{in} \psi_{in}^*} = \frac{\psi_{re} \psi_{re}^*}{\psi_{in} \psi_{in}^*}$$

$$= \frac{A\dfrac{ik_1 + k_2}{ik_1 - k_2} e^{-ik_1 x} \cdot A\dfrac{-ik_1 + k_2}{-ik_1 - k_2} e^{ik_1 x}}{Ae^{ik_1 x} \cdot Ae^{-ik_1 x}}$$

$$= \frac{(ik_1 + k_2)(-ik_1 + k_2)}{(ik_1 - k_2)(-ik_1 - k_2)} = 1 \qquad ...(6.47)$$

The reflection co-efficient is **one**. This means that the particle incident on the step potential with $E < V_0$ is *always reflected*. This result is in agreement with the predictions of classical mechanics.

Discussion. Quantum mechanically and according to relation (*xii*) there is a finite probability of locating the particle in the region which is forbidden classically.

Penetration (or leakage) through the potential barrier. Although the reflection co-efficient is **unity**, *i.e.* incident particle with $E < V_0$ is always reflected, the wave function ψ_{II} in region II $x > 0$ has a finite value given by $\psi_{II} = Ce^{-k_2 x}$ where $C = \frac{2ik_1}{ik_1 - k_2} A$. This wave function does not become zero at $x = 0$ but decays exponentially. The particle thus penetrates into region II, but is always *finally* reflected.

Leakage probability. The probability per unit length of finding the particle in region II is given by

$$P = \psi_{II}\,\psi_{II}^* = Ce^{-k_2 x}\; C^* e^{-k_2 x} = CC^* e^{-2k_2 x}$$

Now $$CC^* = A\,\frac{2ik_1}{ik_1 - k_2}\,.\,A\,\frac{-2ik_1}{-ik_1 - k_2} = A^2\,\frac{4k_1^2}{k_1^2 + k_2^2}$$

$$\therefore \quad P = A^2\,\frac{4k_1^2}{k_1^2 + k_2^2}\,e^{-2k_2 x} \qquad ...(6.48)$$

Thus there is a finite probability of finding the particle in the region $x > 0$ or *leakage*. This is in sharp contrast to the classical mechanics. This phenomenon is known as *penetration of the classically excluded or forbidden region*. Equation (6.48) shows that probability of finding the particle in region II decreases exponentially with increasing x as shown in Fig. 6.7

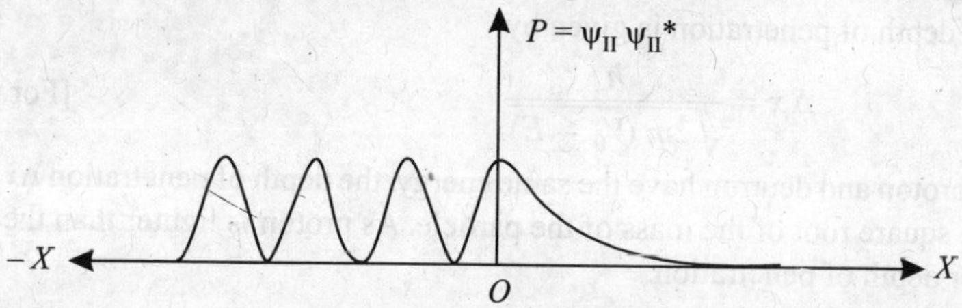

Fig. 6.7

Penetration distance. *The penetration distance or depth of penetration is defined as the distance in which the wave function* ψ_{II} *reduces to 1/e of its value at* $x = 0$.

The wave function ψ_{II} at $x = 0$, is $Ce^{-k_2 x} = Ce^{-k_2 \times 0} = C$. Let Δx be the distance from $x = 0$ where the wave function is $1/e$ of its value at $x = 0$.

$$\therefore \quad \psi_{II}\,\Delta x = \frac{1}{e}\,\psi_{II}\,(x = 0)$$

or $$Ce^{-k_2 \Delta x} = \frac{1}{e}\,C$$

or $$e^{-k_2 \Delta x} = \frac{1}{e}$$

Taking natural logs on both sides

$$-k_2 \Delta x = -1$$

or $$\Delta x = \frac{1}{k_2} = \frac{\hbar}{\sqrt{2m(V_0 - E)}} \qquad ...(6.49)$$

Thus penetration distance is inversely proportional to the square root of the mass of the incident particle.

Example 6.27. *An electron and a proton of same energy (E) approach a potential barrier of height V, greater than E. Do they have the same probability of getting through? Explain.*

(G.N.D.U., 2000)

Solution. As shown in the probability of a particle getting through the barrier and finding it on the other side is given by

$$P = A^2 \frac{4k_1^2}{k_1^2 + k_2^2} e^{-2k_2 x} = 4A^2 \frac{1}{1 + \frac{k_2^2}{k_1^2}} e^{-2k_2 x}$$

Now, $\frac{k_2^2}{k_1^2} = \frac{2m(V_0 - E)}{\hbar^2} \cdot \frac{\hbar^2}{2mE} = \frac{V_0 - E}{E}$ and is independent of the mass of the particle. The only factor which depends upon mass is $e^{-2k_2 x}$ because

$$e^{-2k_2 x} = e^{-2\sqrt{2m(V_0 - E)}\, x}$$

Larger the value of m smaller is the value of $e^{-2\sqrt{2m(V_0 - E)}\, x}$ and smaller is the probability. As the proton has a much greater (1836 times) mass as compared to the mass of the electron, *the probability of the proton getting through the barrier is much smaller than that of the electron.*

Example 6.28. *A proton and a deutron having the same energy penetrate a given rectangular barrier. Which particle has greater depth of penetration?* ***(P.U., 2000)***

Solution. When a particle of mass m having an energy E meets a potential barrier of height V_0 $(E < V_0)$ the depth of penetration is given by

$$\Delta x = \frac{\hbar}{\sqrt{2m(V_0 - E)}}$$ [For proof See Eq. 6.49]

As the proton and deutron have the same energy, the depth of penetration Δx is inversely proportional to the square root of the mass of the particle. As proton is lighter than the deutron, the proton has a greater depth of penetration.

6.31 TRANSMISSION ACROOSS POTENTIAL BARRIER

Potential barrier of finite height and width. Consider a particle of mass m incident on a potential barrier of height V_0 and width a as shown in Fig. 6.7. The energy of the particle is E and $E < V_0$. On both sides of the barrier $V = 0$, which means that no forces act upon the particle in regions I and III. The barrier is thus represented by a potential of the form

$V(x) = 0$ for $x < 0$ Region I

$V(x) = V_0$ for $0 \le x \le a$ Region II

$V(x) = 0$ for $x > a$ Region III

Classical view point. According to classical mechanics, the particle of total energy $E < V_0$ in region I $(x < 0)$ cannot overcome a barrier of height V_0 and penetrate into region II. It cannot, therefore, appear in region III. This means that the *particle will always be reflected back.*

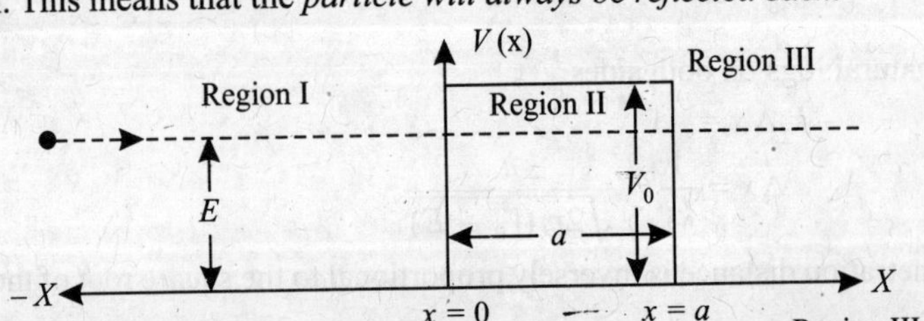

Fig. 6.7

Quantum mechanical view point. According to quantum mechanics, the motion of the particle is governed by Schrödinger's wave equation for regions I, II, III. As the potential V_0 is independent of time, we have to solve Schrödinger's time independent wave equation to analyse the behaviour of the particle. It is given by

$$\frac{d^2\psi}{dx^2} + \frac{2m}{\hbar^2}(E - V)\psi = 0 \qquad ...(i)$$

For region I ($x < 0$), $V(x) = 0$. Let in this region ψ be denoted by ψ_I. Substituting $V(x) = 0$ in equation (*i*), we have

$$\frac{d^2\psi_I}{dx^2} + \frac{2m}{\hbar^2}E\psi_I = 0$$

or
$$\frac{d^2\psi_I}{dx^2} + k_1^2\psi_I = 0 \qquad ...(ii)$$

where
$$k_1^2 = \frac{2mE}{\hbar^2}$$

The general solution of equation (*ii*) is given by

$$\psi_I = Ae^{ik_1x} + Be^{-ik_1x} \qquad ...(iii)$$

where A and B are arbitrary constants.

This first term in equation (*iii*) represents the incident wave ψ_{in} and the second term the reflected wave ψ_{re}

$\therefore$
$$\psi_{in} = Ae^{ik_1x} \qquad ...(iv)$$

and
$$\psi_{re} = Be^{-ik_1x} \qquad ...(v)$$

For region II ($0 \le x \le a$), $V_{(x)} = V_0$. Let in this region ψ be denoted by ψ_{II}. Substituting the value of $V = V_0$ in equation (*i*), we have

$$\frac{d^2\psi_{II}}{dx^2} + \frac{2m}{\hbar^2}(E - V_0)\psi_{II} = 0$$

but $E < V_0$ and hence $(E - V_0)$ is a negative quantity. Putting $(E - V_0) = -(V_0 - E)$, we have

$$\frac{d^2\psi_{II}}{dx^2} - \frac{2m}{\hbar^2}(V_0 - E)\psi_{II} = 0$$

or
$$\frac{d^2\psi_{II}}{dx^2} - k_2^2\psi_{II} = 0 \qquad ...(vi)$$

where
$$k_2^2 = \frac{2m}{\hbar^2}(V_0 - E)$$

The general solution of equation (*vi*) is given by

$$\psi_{II} = Ce^{-k_2x} + De^{k_2x} \qquad ...(vii)$$

where C and D are arbitrary constants.

For region III ($x > a$), $V(x) = 0$. Let in this region ψ be denoted by ψ_{III}. Substituting $V = 0$ in equation (*i*), we have

$$\frac{d^2\psi_{III}}{dx^2} + \frac{2mE}{\hbar^2}\psi_{III} = 0$$

or
$$\frac{d^2\psi_{III}}{dx^2} + k_1^2\psi_{III} = 0 \qquad ...(viii)$$

where $$k_1^2 = \frac{2mE}{\hbar^2} \quad \text{(as for region I)}$$

The general solution of equation (*viii*) is given by

$$\psi_{III} = Fe^{ik_1 x} + Ge^{-ik_1 x}$$

where F and G are arbitrary constants.

The function $Ge^{-ik_1 x}$ represents a wave moving to the left in $-x$ direction. But there is neither a source of particles in region III nor a further potential step to produce a reflected wave. This means, there is no reflected wave and, therefore, $G = 0$

Substituting $G = 0$ in equation (*ix*), we get

$$\psi_{III} = Fe^{ik_1 x} \qquad ...(x)$$

This is the *transmitted wave* in region III and may be represented as

$$\psi_{tr} = Fe^{ik_1 x} \qquad ...(xi)$$

Determination of constants. To evaluate the constants A, B, C and D in terms of F, we apply the boundary conditions which are:–

1. The wave function ψ must be continuous at the boundary $x = 0$

$$\therefore \quad \psi_I(0) = \psi_{II}(0)$$

or $$Ae^{ik_1 \times 0} + Be^{-ik_1 \times 0} = Ce^{-k_2 \times 0} + De^{k_2 \times 0}$$

or $$A + B = C + D \qquad ...(xii)$$

2. The wave function ψ must be continuous at the boundary $x = a$

$$\therefore \quad \psi_{II}(a) = \psi_{III}(a)$$

or $$Ce^{-k_2 a} + De^{k_2 a} = Fe^{ik_1 a} \qquad ...(xiii)$$

3. The first order derivatives of the wave function ψ *i.e.* $\frac{d\psi}{dx}$ must be continuous at $x = 0$

$$\therefore \quad \left|\frac{d\psi_I}{dx}\right|_{x=0} = \left|\frac{d\psi_{II}}{dx}\right|_{x=0}$$

or $$\left[ik_1 Ae^{ik_1 x} - ik_1 Be^{-ik_1 x}\right]_{x=0} = \left[-k_2 Ce^{-k_2 x} + k_2 De^{k_2 x}\right]_{x=0}$$

or $$ik_1 (A - B) = -k_2 (C - D)$$

or $$A - B = \frac{-k_2}{ik_1}(C - D) = \frac{ik_2}{k_1}(C - D) \qquad ..(xiv)$$

4. The first order derivative of the wave function ψ *i.e.* $\frac{d\psi}{dx}$ must be continuous at $x = a$

$$\therefore \quad \left|\frac{d\psi_{II}}{dx}\right|_{x=a} = \left|\frac{d\psi_{III}}{dx}\right|_{x=a}$$

or $$\left[-k_2 Ce^{-k_2 x} + k_2 De^{k_2 x}\right]_{x=a} = \left|ik_1 Fe^{ik_1 x}\right|_{x=a}$$

or $$-k_2 Ce^{-k_2 a} + k_2 De^{k_2 a} = ik_1 Fe^{ik_1 a}$$

or $$-Ce^{-k_2 a} + De^{k_2 a} = i\frac{k_1}{k_2} Fe^{ik_1 a} \qquad ...(xv)$$

Adding equations (*xiii*) and (*xv*), we have

$$2De^{k_2 a} = \left(1 + i\frac{k_1}{k_2}\right) Fe^{ik_1 a}$$

or
$$D = \frac{F}{2}\, e^{ik_1 a}\left(1 + \frac{i k_1}{k_2}\right) e^{-k_2 a} \quad ...(xvi)$$

Subtracting equation (*xv*) from (*xiii*), we get

$$2Ce^{-k_2 a} = \left(1 - \frac{i k_1}{k_2}\right) Fe^{ik_1 a}$$

or
$$C = \frac{F}{2}\, e^{ik_1 a}\left(1 - \frac{i k_1}{k_2}\right) e^{k_2 a} \quad ...(xvii)$$

Again adding equations (*xvi*) and (*xvii*), we have

$$C + D = \frac{F}{2}\, e^{ik_1 a}\left[\left(e^{k_2 a} + e^{-k_2 a}\right) - i\frac{k_1}{k_2}\left(e^{k_2 a} - e^{-k_2 a}\right)\right]$$

$$= Fe^{ik_1 a}\left[\left(\frac{e^{k_2 a} + e^{-k_2 a}}{2}\right) - i\frac{k_1}{k_2}\left(\frac{e^{k_2 a} - e^{-k_2 a}}{2}\right)\right]$$

$$= Fe^{ik_1 a}\left[\cosh(k_2 a) - i\frac{k_1}{k_2}\sinh(k_2 a)\right] \quad ...(xviii)$$

Also subtracting equation (*xvi*) from (*xvii*), we have

$$C - D = \frac{F}{2}\, e^{ik_1 a}\left[(e^{k_2 a} - e^{-k_2 a}) - i\frac{k_1}{k_2}(e^{k_2 a} + e^{-k_2 a})\right]$$

$$= Fe^{ik_1 a}\left[\sinh(k_2 a) - i\frac{k_1}{k_2}\cosh(k_2 a)\right] \quad ...(xix)$$

Substituting the value of $(C + D)$ from equation (*xviii*) in (*xii*), we have

$$A + B = Fe^{ik_1 a}\left[\cosh(k_2 a) - i\frac{k_1}{k_2}\sinh(k_2 a)\right] \quad ...(xx)$$

Also substituting the value of $(C - D)$ from equation (*xix*) in (*xiv*), we have

$$A - B = i\frac{k_2}{k_1}\left[Fe^{ik_1 a}\left\{\sinh(k_2 a) - i\frac{k_1}{k_2}\cosh(k_2 a)\right\}\right]$$

$$= Fe^{ik_1 a}\left[\cosh(k_2 a) + i\frac{k_2}{k_1}\sinh(k_2 a)\right] \quad ...(xxi)$$

Adding (*xx*) and (*xxi*), we get

$$A = Fe^{ik_1 a}\left[\cosh(k_2 a) + \frac{i}{2}\left(\frac{k_2}{k_1} - \frac{k_1}{k_2}\right)\sinh(k_2 a)\right] \quad ...(xxii)$$

and subtracting (*xxi*) from (*xx*), we get

$$B = \frac{-i}{2}\, Fe^{ik_1 a}\left[\left(\frac{k_1}{k_2} + \frac{k_2}{k_1}\right)\sinh(k_2 a)\right] \quad ...(xxiii)$$

Reflection co-efficient. The reflection co-efficient R is defined as

$$R = \frac{\text{Reflected probability density current}}{\text{Incident probability density current}} = \frac{S_{re}}{S_{in}}$$

Now, *the probability density current for a free particle in a given region is defined as the product of its velocity and probability density.*

$$S_{re} = v_1 \psi_{re} \psi_{re}^*$$

and $$S_{in} = v_1 \psi_{in} \psi_{in}^*$$

where v_1 is the velocity of the particle in region I.

$$\therefore \quad R = \frac{v_1 Be^{-ik_1 x} \; B^* e^{+ik_1 x}}{v_1 Ae^{ik_1 x} \; A^* e^{-ik_1 x}} = \frac{BB^*}{AA^*}$$

$$= \frac{-\frac{i}{2} Fe^{ik_1 a}\left[\left(\frac{k_1}{k_2}+\frac{k_2}{k_1}\right)\sinh(k_2 a)\right] \times \frac{i}{2} F^* e^{-ik_1 a}\left[\left(\frac{k_1}{k_2}+\frac{k_2}{k_1}\right)\sinh(k_2 a)\right]}{Fe^{ik_1 a}\left[\cosh(k_2 a)+\frac{i}{2}\left(\frac{k_2}{k_1}-\frac{k_1}{k_2}\right)\sinh(k_2 a)\right] \times F^* e^{-ik_1 a}\left[\cosh(k_2 a)-\frac{i}{2}\left(\frac{k_2}{k_1}-\frac{k_1}{k_2}\right)\sinh(k_2 a)\right]}$$

$$= \frac{\frac{1}{4}\left(\frac{k_1}{k_2}+\frac{k_2}{k_1}\right)^2 \sinh^2(k_2 a)}{\cosh^2(k_2 a)+\frac{1}{4}\left(\frac{k_2}{k_1}-\frac{k_1}{k_2}\right)^2 \sinh^2(k_2 a)}$$

$$= \frac{\left[\frac{k_2^2+k_1^2}{k_1 k_2}\right]^2 \sinh^2(k_2 a)}{4\cosh^2(k_2 a)+\left(\frac{k_2^2-k_1^2}{k_1 k_2}\right)^2 \sinh^2(k_2 a)} \qquad \text{...(xxiv)}$$

The reflection co-efficient R is a measure of the probability that the particle is reflected from the barrier.

Transmission co-efficient. The transmission co-efficient T is defined as

$$T = \frac{\text{Transmitted probability density current}}{\text{Incident probability density current}} = \frac{S_{tr}}{S_{in}}$$

Now $$S_{tr} = v_{III} \psi_{tr} \psi_{tr}^* = v_1 \psi_{tr} \psi_{tr}^* \qquad [\because v_{III} = v_I]$$

$$\therefore \quad S_{tr} = v_1 Fe^{ik_1 x} \; F^* e^{-ik_1 x} = v_1 FF^*$$

and $$S_{in} = v_1 Ae^{ik_1 x} \; A^* e^{-ik_1 x} = v_1 AA^*$$

$$\therefore \quad T = \frac{FF^*}{AA^*} = \frac{1}{\cosh^2(k_2 a)+\frac{1}{4}\left(\frac{k_2}{k_1}-\frac{k_1}{k_2}\right)^2 \sinh^2(k_2 a)}$$

$$= \frac{1}{1+\sinh^2(k_2 a)+\left(\frac{k_2^2-k_1^2}{2k_1 k_2}\right)^2 \sinh^2 k_2 a}$$

$$= \frac{1}{1+\sinh^2(k_2 a)\left[1+\left(\frac{k_2^2-k_1^2}{2k_1 k_2}\right)^2\right]}$$

$$= \frac{1}{1 + \left(\frac{k_2^2 + k_1^2}{2k_1 k_2}\right)^2 \sinh^2 (k_2 a)} \qquad ...(xxv)$$

Substituting the values of k_1^2 and k_2^2, we get

$$T = \frac{1}{1 + \frac{V_0^2}{4E(V_0 - E)} \sinh^2 \frac{\sqrt{2m(V_0 - E)}}{\hbar} a} \qquad ...(6.50)$$

6.32 TUNNEL EFFECT

The transmission co-efficient T is a measure of the probability that the particle will be transmitted through the barrier. From relation (*xxv*), we find that T is not zero. This shows that there is *some* probability that the particle may penetrate through the barrier and appear in region III even when its energy E is less than V_0 the barrier height.

This prediction of the quantum theory is contrary to the classical behaviour of the particle. A schematic representation of the wave function in regions I, II and III is shown in Fig. 6.8.

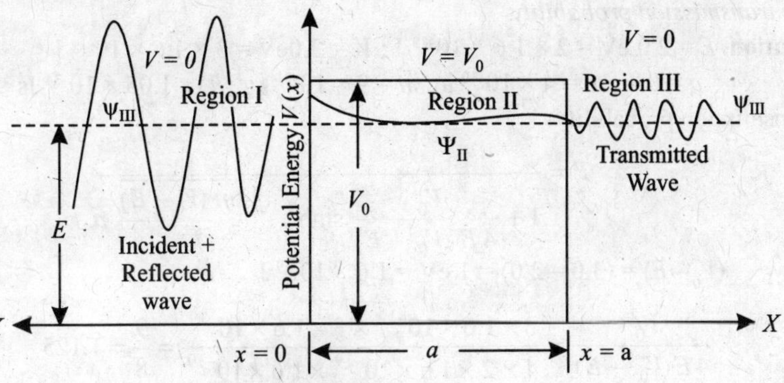

Fig. 6.8

The phenomenon of transmission of a particle through a potential barrier of finite width and height, even when its energy is less than the barrier height, is called quantum mechanical tunneling.

The probability of transmission depends upon V_0 and a. If the barrier is very high or quite thick, the probability of penetration becomes small. The theory of tunnel effect is used to explain α-decay. There are many examples of tunneling of electrons in Solid State Physics.

6.33 CONDITION FOR PERFECT TRANSMISSION

For perfect transmission, that is, for the particle to tunnel through the barrier without any reflection the transmission co-efficient T should be equal to one or $T = 1$. From Eq. 6.50 we find that T can take the value unity only if

$$\frac{V_0^2}{4E(V_0 - E)} \sinh^2 \frac{\sqrt{2m(V_0 - E)}}{\hbar} a = 0$$

or

$$\sinh^2 \frac{\sqrt{2m(V_0 - E)}}{\hbar} a = 0 \quad \text{or} \quad \sinh^2 k_2 a = 0$$

where $k_2 = \frac{\sqrt{2m(V_0 - E)}}{\hbar}$ as $\frac{V_0^2}{4E(V_0 - E)} \neq 0$

This is possible if $\sinh k_2 a = 0$ or $k_2 a = n\pi$ or $a = \frac{n\pi}{k_2}$ ...(i)

Now $k_2 = \frac{2\pi}{\lambda_2}$ because $\sqrt{2m(V_0 - E)} = p$ the momentum of the particle in region II [Fig. 6.8] of the barrier and $\frac{p}{\hbar} = \frac{2\pi p}{h} = \frac{2\pi}{\lambda_2}$ where λ_2 is the de-Broglie wave length associated with the particle in region II. Substituting the value of $k_2 = \frac{2\pi}{\lambda_2}$ in Eq. (i), we have

$$a = \frac{n\pi}{k_2} = \frac{n\pi}{2\pi}\lambda_2 = n\frac{\lambda_2}{2} \qquad ...(6.51)$$

Hence the transmission through the rectangular barrier will be perfect if the length of the barrier is equal to an integral or half integral mutliple of de-Broglie wavelength associated with the particle in region II.

Example 6.29. *Electrons of energy 2.0 eV are incident on a barrier 3.0 eV high and 0.4 nm wide. Find the transmission probability.* ***(P.U., 2001)***

Solution. $E = 2.0$ eV $= 2 \times 1.6 \times 10^{-19}$ J ; $V_0 = 3.0$eV $= 3 \times 1.6 \times 10^{-19}$ J

$a = 0.4$ nm $= 4 \times 10^{-10}$ m; $m = 9 \times 10^{-31}$ kg ; $\hbar = 1.05 \times 10^{-34}$ Js

Transmission probability

$$T = \frac{1}{1 + \frac{V_0^2}{4E(V_0 - E)} \sinh^2 \frac{\sqrt{2m(V_0 - E)}}{\hbar} a}$$

Now $(V_0 - E) = (3.0 - 2.0) = 1$ eV $= 1.6 \times 10^{-19}$ J

$$\therefore \quad \frac{V_0^2}{4E(V_0 - E)} = \frac{3 \times 1.6 \times 10^{-19} \times 3 \times 1.6 \times 10^{-19}}{4 \times 2 \times 1.6 \times 10^{-19} \times 1.6 \times 10^{-19}} = \frac{9}{8} = 1.125$$

and $$\frac{\sqrt{2m(V_0 - E)}}{\hbar} a = \frac{(2 \times 9 \times 10^{-31} \times 1.6 \times 10^{-19})^{1/2}}{1.05 \times 10^{-34}} \times 4 \times 10^{-10} = 2.045$$

$\sinh(2.045) = 3.805$

$$\therefore \quad T = \frac{1}{1 + 1.125 \times (3.805)^2} = 0.058$$

Example 6.30. *Electrons of energy 2.0 eV are incident on a barrier 4.0 eV high and 0.4 nm wide. Find the transmission probability.* ***(G.N.D.U., 2000)***

Solution. The question is same as above, except that $V_0 = 4.0$ eV instead 3.0 eV

$$\therefore \quad \frac{V_0^2}{4E(V_0 - E)} = 1 \text{ and } \frac{\sqrt{2m(V_0 - E)}}{\hbar} a = 2.892$$

$\sinh(2.892) = 8.990$ and $\sinh^2(2.892) = 80.83$

$$\therefore \quad T = \frac{1}{1 + 80.83} = \frac{1}{81.83} = 0.0122$$

6.34 HARMONIC OSCILLATOR

When a system vibrates about an equilibrium position such that (i) the magnitude of the force acting on it is proportional to its displacement from the equilibrium position and (ii) the direction of the force is opposite to that of displacement, it is called a harmonic oscillator.

Examples. Familiar examples are; a simple pendulum, an object floating in a liquid, a diatomic molecule and an atom in a crystal lattice. There are many more examples in the macroscopic and microscopic field.

Physical importance. The importance of a harmonic oscillator in both classical and quantum physics, lies in the fact that if any force, which is a function of displacement x, acts upon a particle or system, it can be shown to be proportional to x for *small* displacements and hence the particle acts as a simple harmonic oscillator.

Any force which is a function of x can be expressed in a Maclaurin's series about the equilibrium position $x = 0$ as

$$F(x) = F_{x=0} + \left(\frac{dF}{dx}\right)_{x=0} x + \frac{1}{2}\left(\frac{d^2 F}{dx^2}\right)_{x=0} x^2 + \frac{1}{6}\left(\frac{d^3 F}{dx^3}\right)_{x=0} x^3$$

Since $x = 0$ is the equilibrium position $F_{x=0} = 0$

Also when x is small x^2, x^3 etc are *very very* small and the terms containing x^2, x^3 etc can be neglected, as compared to the term containing x. The only term of significance, when x is small is, therefore,

$$F(x) = \left(\frac{dF}{dx}\right)_{x=0} x$$

This is the statement of Hooke's law when $\left(\frac{dF}{dx}\right)_{x=0}$ is negative *i.e.* it is a restoring force.

Thus the importance of a harmonic oscillator lies in the fact that all oscillations are simple harmonic in character, when their amplitude is sufficiently small.

Energy of a Harmonic oscillator according to classical mechanics. Consider a particle of mass m constrained to move along X-axis. If we take $x = 0$ as the equilibrium position, the restoring force F acting on the oscillating particle, at a displacement x is given by

$$F \propto -x \text{ or } F = -kx \qquad ...(i)$$

The potential energy of the particle is related to the force by the equation

$$F = -\frac{dV}{dx} \qquad ...(ii)$$

From (*i*) and (*ii*), we have

$$F = \frac{-dV}{dx} = -kx \text{ or } dV = k\,x\,dx$$

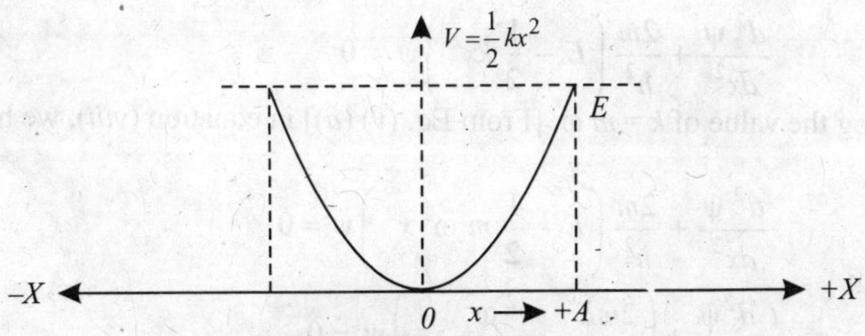

Fig. 6.9

∴ Potential energy V for a displacement x from the equilibrium position $x = 0$ is given by

$$V = \int_0^x k\,x\,dx = \frac{1}{2}\,k\,x^2 \qquad ...(i)$$

As $V \propto x^2$ a graph between potential energy V of the harmonic oscillator and displacement x is a parabola, as shown in Fig. 6.9. According to classical mechanics

$$F = ma = m\frac{d^2 x}{dt^2} = -\,k\,.\,x$$

or
$$\frac{d^2 x}{dt^2} + \frac{k}{m}x = 0 \qquad ...(iv)$$

This is the equation of motion of the particle and its solution is

$$x = A \sin(\omega t + \phi) \qquad ...(v)$$

where A is the amplitude of the oscillations, ϕ the initial phase and ω the *classical angular frequency* of oscillation, given by

$$\omega = \sqrt{\frac{k}{m}} \qquad ...(v)\,(a)$$

When the particle vibrates and is at the mean position O ($x = 0$) its total energy E is all kinetic and potential energy is zero. As the particle moves outwards, its kinetic energy decreases and potential energy increases (Total energy remaining constant = E). At the extreme positions $x = \pm A$, the kinetic energy is zero and the potential energy $V = E$, the total energy. The particle, therefore, cannot go beyond $x = \pm A$, because then the potential energy will be greater then V and hence kinetic energy will have to be negative, which is a meaningless concept. Thus, at the points $x = \pm A$ the velocity becomes zero, changes sign and the particle reverses its direction. These points are called classical turning points. At these points,

Total energy E = potential energy V

∴ Potential energy (or potential) $V = \frac{1}{2}\,k\,x^2 = \frac{1}{2}\,kA^2$; at $x = \pm A$

The total energy E of the oscillator is proportional the square of the amplitude A. As A can have any value, the total energy can also have any value from zero to ∞.

Schrödinger's equation. Schrödinger's one dimensional time independent equation is

$$\frac{d^2\psi}{dx^2} + \frac{2m}{\hbar^2}(E - V)\,\psi = 0 \qquad ...(vii)$$

For a harmonic oscillator $V = \frac{1}{2}\,k\,x^2$

∴
$$\frac{d^2\psi}{dx^2} + \frac{2m}{\hbar^2}\left(E - \frac{1}{2}kx^2\right)\psi = 0 \qquad ...(viii)$$

Substituting the value of $k = m\,\omega^2$ [From Eq. (*v*) (*a*)] in equation (*viii*), we have

$$\frac{d^2\psi}{dx^2} + \frac{2m}{\hbar^2}\left(E - \frac{1}{2}\,m\,\omega^2\,x^2\right)\psi = 0 \qquad ...(ix)$$

or
$$\frac{d^2\psi}{dx^2} + \left(\frac{2mE}{\hbar^2} - \frac{m^2\,\omega^2}{\hbar^2}\,x^2\right)\psi = 0 \qquad ...(x)$$

Energy levels or Energy eigen values. To find the eigen values of energy

put $\frac{m\omega}{\hbar} = \alpha$ and $\frac{2mE}{\hbar^2} = \beta$ in equation (*x*), so that we get

$$\frac{d^2\psi}{dx^2} + \left(\beta - \alpha^2 x^2\right)\psi = 0 \qquad ...(xi)$$

To solve equation (*xi*) it is convenient to substitute $\sqrt{\alpha}\, x = y$, a *dimensionless* constant.

When $y = \sqrt{\alpha}\, x$, $\frac{dy}{dx} = \sqrt{\alpha}$

and
$$\frac{d\psi}{dx} = \frac{d\psi}{dy} \cdot \frac{dy}{dx} = \sqrt{\alpha}\,\frac{d\psi}{dy}$$

$$\frac{d^2\psi}{dx^2} = \frac{d}{dx}\left(\frac{d\psi}{dx}\right) = \frac{d}{dy}\left(\sqrt{\alpha}\,\frac{d\psi}{dy}\right)\frac{dy}{dx} = \alpha\frac{d^2\psi}{dy^2}$$

Substituting the above value of $\frac{d^2\psi}{dx^2}$ in equation (*xi*), we have

$$\alpha\frac{d^2\psi}{dy^2} + \left(\beta - \alpha^2\frac{y^2}{\alpha}\right)\psi = 0$$

or
$$\frac{d^2\psi}{dy^2} + \left(\frac{\beta}{\alpha} - y^2\right)\psi = 0 \qquad ...(xii)$$

To get an acceptable solution of Eq. (*xii*) we first try to get an asymptotic solution i.e. when $y >> \beta/\alpha$. In such a case, the term β/α can be neglected, as compared to the term y^2 and Eq. (*xii*) reduces to

$$\frac{d^2\psi}{dy^2} - y^2\psi = 0 \qquad ...(xiii)$$

The solution of eq. (*xiii*) is given by $\psi = e^{\pm y^2/2}$ *i.e.* there are two possible solutions $\psi = e^{+y^2/2}$ and $\psi = e^{-y^2/2}$. The quantity $|\psi|^2$ gives the probability of finding the particle some where along *X*-axis. Therefore, it must decrease continuously to zero as $x \to -\infty$ and $x \to +\infty$. But the solution $\psi = e^{+y^2/2}$ increases rapidly with increasing *y* and hence with increasing *x*. Therefore, it is not an acceptable solution of the wave function.

Thus
$$\psi = e^{-y^2/2};\ |y| \to \infty$$
is the *only acceptable solution*.

The standard procedure for solving differential equation like Eq. (*xii*) valid for all values of *y* is to try a solution of the form

$$\psi = Ae^{-y^2/2}\,H(y) \qquad ...(xiii)(a)$$

where *A* is a constant and *H*(*y*) a polynomial in *y*, known as *Hermite polynomial*

To determine the function *H*(*y*) we calculate

$$\frac{d\psi}{dy} = -\,Aye^{-y^2/2}\,H + Ae^{-y^2/2}\,\frac{dH}{dy}$$

and $$\frac{d^2\psi}{dy^2} = -Ae^{-y^2/2}H + Ay^2 e^{-y^2/2}H - Aye^{-y^2/2}\frac{dH}{dy}$$

$$-Aye^{-y^2/2}\frac{dH}{dy} + Ae^{-y^2/2}\frac{d^2H}{dy^2}$$

Substituting this value of $\frac{d^2\psi}{dy^2}$ in Eq. (*xii*), we have

$$Ae^{-y^2/2}\left(-H + y^2 H - 2y\frac{dH}{dy} + \frac{d^2H}{dy^2}\right) + \left(\frac{\beta}{\alpha} - y^2\right)Ae^{-y^2/2}H = 0$$

or $$\frac{d^2H}{dy^2} - 2y\frac{dH}{dy} + \left(\frac{\beta}{\alpha} - 1\right)H = 0 \qquad ...(xiv)$$

Writing $(\beta/\alpha - 1) = 2n$ or $\beta/\alpha = 2n + 1$, equation (*xiv*) becomes

$$\frac{d^2H}{dy^2} - 2y\frac{dH}{dy} + 2nH = 0 \qquad ...(xv)$$

Eigen values of energy or eigen states. This is a standard mathematical equation, known as *Hermite equation*. The solutions of Eq. (*xv*) are called *Hermite polynomials* given by

$$H_n(y) = (-1)^n e^{y^2}\frac{d^n}{dy^n}\left[e^{-y^2}\right]$$

Only those solution of Eq. (*xv*) are acceptable for all values of y for which

$$\beta/\alpha = 2n + 1$$

where n is an integral number 0, 1, 2, 3, etc. This number n is called *quantum number*.

Substituting the value of

$$\beta = \frac{2mE}{\hbar^2} \text{ and } \alpha = \frac{\omega m}{\hbar}$$

we have $$\beta/\alpha = \frac{2mE}{\hbar^2}\cdot\frac{\hbar}{\omega m} = \frac{2E}{\hbar\omega}$$

Hence $$2E = \hbar\omega(\beta/\alpha) = \hbar\omega(2n+1)$$

or $$E = \hbar\omega\left(n + \frac{1}{2}\right)$$

More generally $$E_n = \hbar\omega\left(n + \frac{1}{2}\right)$$

If v is the classical frequency

$$\hbar\omega = \frac{h}{2\pi}\cdot 2\pi v = hv$$

and $$E_n = hv\left(n + \frac{1}{2}\right) \qquad ...(6.52)$$

Eq. (6.52) gives the *eigen values* (or permitted values) of total energy for a harmonic oscillator for $n = 0, 1, 2, 3...$ These energies are discrete and their corresponding energy levels are shown in Fig. 6.10.

Energy level diagram. The energy of a harmonic oscillator is quantised in steps of $h\nu$. The energy levels are equally spaced with the separation between adjacent levels being $h\nu$. When $n = 0$ $E_0 = \frac{1}{2} h\nu$ which is the lowest value of energy known as zero point energy. Thus we find that according to quantum theory, the energy of a harmonic oscillator is quantised in steps of $h\nu$, with lowest energy $\frac{1}{2} h\nu$. This is in contrast to classical theory, where energy of the oscillator is continuous, from zero to infinity.

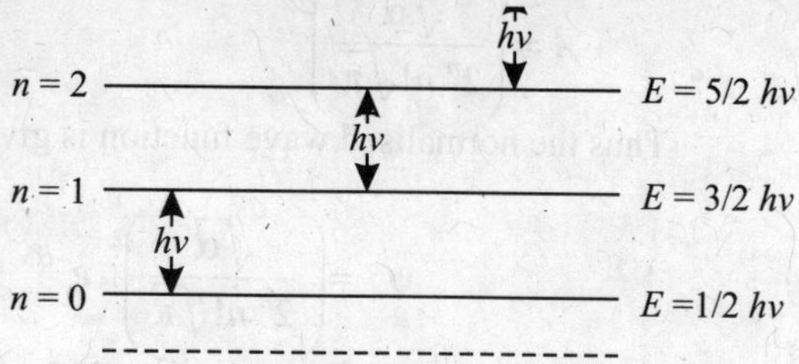

Fig. 6.10

For comparison between energy levels of a harmonic oscillator and energy levels of a hydrogen atom. (Refer Ch. 7).

QUANTISATION OF ENERGY OF A HARMONIC OSCILLATOR

The energy of a harmonic oscillator is given by

$$E_n = h\nu\left(n + \frac{1}{2}\right) \text{ where } n = 0, 1, 2, 3, ... \text{ etc.} \quad (\because \text{Eq. 6.52}) ...(i)$$

Thus a harmonic oscillator cannot have any value of energy, as predicted by the classical theory, but only discrete values of energy, given by Eq. (*i*) i.e. the energy is quantised in steps of $h\nu$.

Zero point energy. The energy of a harmonic oscillator has the smallest value for $n = 0$. Thus the minimum energy of the oscillator is

$$E_0 = \frac{1}{2} h\nu$$

This is known as *zero point energy,* because a harmonic oscillator in equilibrium with its surroundings would approach an energy $E = E_0 = \frac{1}{2} h\nu$ and not $E = 0$, as the temperature approaches 0K.

Zero point energy in conformity with uncertainty principle. The existence of zero point energy is required by the uncertainty principle, according to which

$$\Delta E \, \Delta t \geq \hbar/2$$

or $\Delta E \geq \frac{\hbar}{2\Delta t}$ which is not zero.

6.35 WAVE FUNCTIONS OF A HORMONIC OSCILLATOR

The general wave function of a harmonic oscillator is given by

$$\psi_n = Ae^{-y^2/2} H_n(y)$$

where A is a constant $H_n(y)$ is the Hermite polynomial, given by

$$H_n(y) = (-1)^n e^{y^2} \frac{d_n}{dy^n}\left(e^{-y^2}\right) \quad ...(i)$$

and $y = \sqrt{\alpha}\, x$ with $\alpha = m\omega/\hbar$

Normalised wave function. On normalising the wave function, we get

$$A = \left(\frac{\sqrt{\alpha}}{2^n\, n!\sqrt{\pi}}\right)^{\frac{1}{2}}$$

Thus the normalised wave function is given by

$$\psi_n = \left(\frac{\sqrt{\alpha}}{2^n\, n!\sqrt{\pi}}\right)^{\frac{1}{2}} e^{-\alpha x^2/2}\, H_n\left(\sqrt{\alpha}\, x\right) \qquad \text{...(ii)}$$

or
$$\psi_n = \left(\frac{2mv}{\hbar}\right)^{1/2} \left(2^n\, n!\right)^{-1/2} H_n(y)\, e^{-y^2/2} \qquad \text{...(iii)}$$

for $n = 0, 1, 2, 3$ etc. because $\sqrt{\dfrac{\alpha}{\pi}} = \sqrt{\dfrac{m\omega}{\hbar\pi}} = \sqrt{\dfrac{2mv\pi}{\hbar\pi}} = \left(\dfrac{2mv}{\hbar}\right)^{1/2}$

First three normalised wave functions. The first three normalized wave functions or *harmonic oscillator eigen functions* are (*i*) ground state for $n = 0$, (*ii*) first excited state for $n = 1$ and (*iii*) second excited state for $n = 2$.

1. Ground state wave function. For the ground state wave function $n = 0$ and $H_n(y)$ for $n = 0$ from equation (*i*) is given by

$$H_1(y) = (-1)^0\, e^{y^2} e^{-y^2} = 1 \text{ ; Also } n! = 0! = 1$$

∴ From equation (*ii*) we get

$$\psi_0 = \left(\frac{\alpha}{\pi}\right)^{1/4} e^{-\alpha x^2/2} \qquad \text{...(iv)}$$

The plot of the ground state wave function is shown in Fig. 6.11 (*a*). It has the shape of a Gaussian curve and even parity.

2. First excited state wave function. For the first excited state $n = 1$ and $H_n(y)$ for $n = 1$ from equation (*i*) is given by

$$H_1(y) = (-1)^1\, e^{y^2} \frac{d}{dy}\left(e^{-y^2}\right) = -1e^{y^2}\left(-2\,ye^{-y^2}\right)$$
$$= 2\,y = 2\sqrt{\alpha}\, x$$

∴ From equation (*ii*), we get

$$\psi_1 = \left(\frac{\sqrt{\alpha}}{2\sqrt{\pi}}\right)^{1/2} e^{-\alpha x^2/2} \times 2\sqrt{\alpha}\, x$$
$$= \left(\frac{\alpha}{\pi}\right)^{1/4} \sqrt{2}\sqrt{\alpha}\; xe^{-\alpha x^2/2} \qquad \text{...(v)}$$

The plot of the first excited state wave function is shown in Fig. 6.11 (*b*). It has odd parity.

3. Second excited state wave function. For the second excited state $n = 2$ and $H_n(y)$ for $n = 2$ from equation (*i*) is given by

$$H_2(y) = (-1)^2\, e^{y^2} \frac{d^2}{dy^2}\left(e^{-y^2}\right) = e^{y^2} \frac{d}{dy}\left(-2ye^{-y^2}\right)$$
$$= e^{y^2}\left(4y^2 - 2\right) e^{-y^2} = 2\left(2y^2 - 1\right) = 2\left(2\alpha x^2 - 1\right)$$

∴ From equation (*ii*), we get

$$\psi_2 = \left(\frac{\sqrt{\alpha}}{2^2 \times 2!\sqrt{\pi}}\right)^{1/2} \times 2(2\alpha x^2 - 1)e^{-\alpha x^2/2}$$

$$= \left[\frac{\sqrt{\alpha}}{2\sqrt{\pi}}\right]^{1/2} \left(2\alpha x^2 - 1\right)e^{-\alpha x^2/2} \qquad ...(vi)$$

The plot of the second excited state wave function is shown in Fig. 6.11 (*c*). It has even parity.

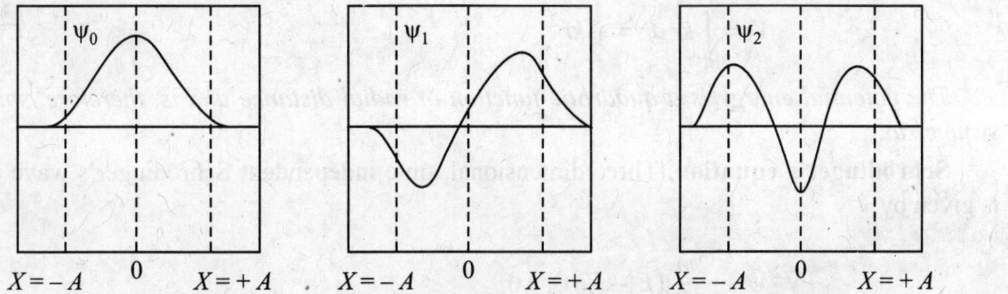

(*a*) (*b*) (*c*)

Fig. 6.11

The vertical lines show the limits at $X = +A$ and $X = -A$ between which a classical oscillator, with the same energy, would vibrate. Quantum mechanically, we find that the harmonic oscillator slightly exceeds the amplitude A with an exponentially decreasing probability and penetrates into classically forbidden regions.

Example 6.31. *Find the expectation value of energy when the state of a harmonic oscillator is described by the following wave functions.*

$$\psi(x, t) = \frac{1}{\sqrt{2}}[\psi_0(x, t) + \psi_1(x, t)]$$

where $\psi_0(x, t)$ and $\psi_1(x, t)$ are the wave functions for the ground state and the first excited state respectively. **(P.U. 2008)**

Solution. The normalised wave function for the ground state of a harmonic oscillator is given by

$$\psi_0(x, t) = \left(\frac{\alpha}{\pi}\right)^{1/4} e^{-\alpha x^2/2}$$

and for the first excited state

$$\psi_1(x, t) = \left(\frac{\alpha}{\pi}\right)^{1/4} \sqrt{2}\sqrt{\alpha}\, x\, e^{-\alpha x^2/2}$$

The expectation value of energy is given by

$$< E > = \int_{-\infty}^{+\infty} \psi^*(x, t)\, E\, \psi(x, t)\, dx$$

Now $$\psi_0^*(x, t) = \left(\frac{\alpha}{\pi}\right)^{1/4} e^{-\alpha x^2/2} = \psi_0(x, t)$$

and $$\psi_1^*(x, t) = \left(\frac{\alpha}{\pi}\right)^{1/4} \sqrt{2}\sqrt{\alpha}\, x\, e^{-\alpha x^2/2} = \psi_1(x, t)$$

6.36 THREE DIMENSIONAL HARMONIC OSCILLATOR

When a particle vibrates in space about an equilibrium position such that the magnitude of the force F acting on it is proportional to its displacement r from the equilibrium position and the direction of the force is opposite to that of displacement, it is called a three dimensional harmonic oscillator.

$$\therefore \qquad F = -kr$$

or $$F = -\frac{dV}{dr} = -kr \text{ or } dV = kr\,dr$$

∴ Potential energy for a displacement r from the equilibrium position $r = 0$ is given by

$$V = \int_0^r kr\,dr = \frac{1}{2}kr^2$$

The potential energy is a quadratic function of radial distance and is, therefore, spherically symmetric.

Schrödinger's equation. Three dimensional, time independent Schrödinger's wave equation is given by

$$\nabla^2 \psi + \frac{2m}{\hbar^2}(E - V)\psi = 0$$

or $$\nabla^2 \psi + \frac{2m}{\hbar^2}\left(E - \frac{1}{2}kr^2\right) = 0 \qquad ...(i)$$

This equation is similar to Eq. (*i*), of article 6.21 where we have

$$\frac{d^2\psi}{dx^2} \text{ for } \nabla^2 \psi \text{ and } \frac{1}{2}kx^2 \text{ for } \frac{1}{2}kr^2$$

Proceeding in the same manner we get

$$\nabla^2 \psi + (\lambda - r^2)\psi = 0 \qquad ...(ii)$$

To solve Eq. (*ii*) $\psi = \psi\,(r)$ may be taken as the product of three functions $X(x)$ – *a* function of x only, $Y(y)$ a function of y only and $Z(z)$ a function of z only.

$$\therefore \qquad \psi(r) = \psi(x, y, z) = X(x)\,Y(y)\,Z(z) = XYZ$$

Putting Eq. (*ii*) in the from

$$\frac{\partial^2 \psi}{\partial x^2} + \frac{\partial^2 \psi}{\partial y^2} + \frac{\partial^2 \psi}{\partial z^2} + (\lambda - x^2 - y^2 - z^2)\psi = 0$$

and substituting $\psi = XYZ$, we have

$$\frac{\partial^2 (XYZ)}{\partial x^2} + \frac{\partial^2 (XYZ)}{\partial y^2} + \frac{\partial^2 (XYZ)}{\partial z^2} + (\lambda - x^2 - y^2 - z^2)XYZ = 0$$

or $$YZ\frac{\partial^2 X}{\partial x^2} + XZ\frac{\partial^2 Y}{\partial y^2} + XY\frac{\partial^2 Z}{\partial z^2} + (\lambda - x^2 - y^2 - z^2)XYZ = 0$$

Dividing both sides by XYZ, we get

$$\frac{1}{X}\frac{\partial^2 X}{\partial x^2} + \frac{1}{Y}\frac{\partial^2 Y}{\partial y^2} + \frac{1}{Z}\frac{\partial^2 Z}{\partial z^2} + (\lambda - x^2 - y^2 - z^2) = 0$$

or $$\left(\frac{1}{X}\frac{\partial^2 X}{\partial x^2} - x^2\right) + \left(\frac{1}{Y}\frac{\partial^2 Y}{\partial y^2} - y^2\right) + \left(\frac{1}{Z}\frac{\partial^2 Z}{\partial z^2} - z^2\right) = -\lambda \qquad ...(iii)$$

When x changes only the *first* term in Eq. (*iii*) changes, when y changes only the *second* term and when z changes only the *third* term changes.

$$\therefore \qquad -\lambda = -(\lambda_x + \lambda_y + \lambda_z)$$

where λ_x, λ_y and λ_z respectively represent the value of λ corresponding to only a change in x, y and z.

$$\therefore \qquad \frac{1}{X}\frac{\partial^2 X}{\partial x^2} - x^2 = -\lambda_x \quad \text{or} \quad \frac{d^2 X}{dx^2} + (\lambda_x - x^2) = 0 \qquad ...(iv)$$

$$\frac{1}{Y}\frac{\partial^2 Y}{dy^2} - y^2 = -\lambda_y \quad \text{or} \quad \frac{d^2 Y}{dy^2} + (\lambda_y - y^2) = 0 \qquad ...(v)$$

and

$$\frac{1}{Z}\frac{\partial^2 Z}{dz^2} - z^2 = -\lambda_z \quad \text{or} \quad \frac{d^2 Z}{dz^2} + (\lambda_z - z^2) = 0 \qquad ...(vi)$$

Each equation (*iv*), (*v*) or (*vi*) represents the equation for a one dimensional harmonic oscillator along the respective co-ordinate axis. These equations have an acceptable solution only when

$$\lambda_x = 2n_x + 1;\ \lambda_y = 2n_y + 1;\ \lambda_z = 2n_z + 1$$

where n_x, n_y, n_z have positive integral values

0, 1, 2, 3 ....

$$\therefore \qquad \lambda = \lambda_x + \lambda_y + \lambda_z = 2n_x + 1 + 2n_y + 1 + 2n_z + 1$$
$$= 2(n_x + n_y + n_z) + 3$$

or $\qquad \lambda = 2n + 3 \qquad$ where $n = n_x + n_y + n_z$

Allowed energies Total energy $E = \frac{1}{2}\hbar\omega\lambda$

$$= \frac{1}{2}(2n+3)\hbar\omega$$

$$= \left(n + \frac{3}{2}\right)\hbar\omega$$

Zero point energy. The lowest value of energy for a three dimensional harmonic oscillator is given by

$$E_0 = \frac{3}{2}\hbar\omega \qquad (\text{For } n = 0)$$

Thus the zero point energy of a three dimensional harmonic oscillator is *three* times the zero point energy of a one dimensional (or linear) harmonic oscillator.

Wave function. The acceptable wave functions are given by

$$X = A_x\, e^{-x^2/2}\, H_{n_x}(x)$$

$$Y = A_y\, e^{-y^2/2}\, H_{n_y}(y)$$

and

$$Z = A_z\, e^{-z^2/2}\, H_{n_z}(z)$$

where H is Hermite polynomial.

$$\therefore \qquad \psi(x, y, z) = XYZ = A_x A_y A_z\, e^{-x^2/2}\, e^{-y^2/2}\, e^{-z^2/2}\, H_{n_x}(x)\, H_{n_y}(y)\, H_{n_z}(z)$$

$$= Ae^{-(x^2+y^2+z^2)/2}\, H_{n_x}(x)\, H_{n_y}(y)\, H_{n_z}(z)$$

or

$$\psi_{n_x, n_y, n_z} = Ae^{-r^2/2}\, H_{n_x}(x)\, H_{n_y}(y)\, H_{n_z}(z) \qquad ...(vii)$$

where $A = A_x A_y A_z$ and $r^2 = x^2 + y^2 + z^2$

Normalising the wave function, we get

$$A = \left[\frac{1}{2^n\, n_x\,!\, n_y\,!\, n_z\,!\,\pi^{3/2}}\right]^{\frac{1}{2}}$$

Normalized wave function, for a three dimensional harmonic oscillator is

$$\psi_{n_x,\,n_y,\,n_z} = \left[\frac{1}{2^n\, n_x\,!\, n_y\,!\, n_z\,!\,\pi^{3/2}}\right]^{\frac{1}{2}} e^{-r^2/2} H_{n_x}(x)\, H_{n_y}(y)\, H_{n_z}(z) \qquad ...(6.53)$$

Degeneracy. The state of a three dimensional harmonic oscillator is determined by three quantum numbers, n_x, n_y and n_z which satisfy the condition

$$n = n_x + n_y + n_z$$

For $n = 0$; $n_x = 0, n_y = 0, n_z = 0$ and $\psi = \psi_{000}$

For $n = 1$; $\begin{cases} n_x = 1, n_y = 0, n_z = 0 \text{ and } \psi = \psi_{100} \\ n_x = 0, n_y = 1, n_z = 0 \text{ and } \psi = \psi_{010} \\ n_x = 0, n_y = 0, n_z = 1, \text{ and } \psi = \psi_{001} \end{cases}$

Similarly for $n = 2$, there are 6 states ψ_{110}, ψ_{011}, ψ_{101} and ψ_{200}, ψ_{020} and ψ_{002}

In genergal, the degree of degeneracy or total number of states for a given value of $n = \frac{1}{2}(n+1)(n+2)$

Thus for $n = 0$ we have only one state

For $n = 1$, we have 3 states

For $n = 2$, we have 6 states

It is only the state for $n = 0$, which is non–degenerate

Example 6.32. *A pendulum with a length of one metre has a bob with a mass of 0.1 kg. What is the zero point energy?* ***(P.U. 2006; Pbi.U., 2003)***

Solution. Length of the pendulum $l = 1$m

$$\therefore \quad \text{Time period } T = 2\pi\sqrt{\frac{l}{g}} = 2\pi\sqrt{\frac{1}{9.81}}$$

or $$\text{frequency } \nu = \frac{1}{T} = \frac{\sqrt{9.81}}{2\pi} = \frac{3.13}{2\pi}$$

$$\therefore \quad \text{zero point energy} = \frac{1}{2}h\nu = \frac{1}{2} \times \frac{6.63\times10^{-34}\times 3.13}{2\pi}$$

$$= 1.65 \times 10^{-34} \text{ J} = 1.03 \times 10^{-15} \text{ eV.}$$

Example 6.33. *Calculate zero point energy of a harmonic oscillator of frequency 2×10^{15} Hz.* ***(Nag. U., 2002)***

Solution. $E = \frac{1}{2}h\nu = \frac{6.63\times10^{-34}\times2\times10^{15}}{2}$

$$= 6.63 \times 10^{-19} \text{ J} = 4.14 \text{ eV}$$

Example 6.34. *Find the zero point energy in electron volts of a pendulum whose period is 1 second.* ***(Pbi. U. 2008)***

Solution. $T = 1$ sec. $\therefore$ Frequency $\nu = \frac{1}{T} = 1$ Hertz

$$\text{Zero point energy} = \frac{1}{2} h\nu = \frac{6.63 \times 10^{-34} \times 1}{2 \times 1.6 \times 10^{-19}} = 2.07 \times 10^{-15} \text{ eV}$$

Example 6.35. *A particle of mass 1 mg is attached to a spring of spring constant 10^{-3} Nm^{-1}. Calculate its zero point energy and classical value of amplitude of zero point vibration.*

Solution. Mass of the particle $m = 1$ mg $= 10^{-6}$ kg

Spring constant $s = 10^{-3}$ Nm^{-1}

$$\text{Frequency of vibration } \nu = \frac{1}{2\pi}\sqrt{\frac{s}{m}} = \frac{1}{2\pi}\sqrt{\frac{10^{-3}}{10^{-6}}} = 5.03 \text{ s}^{-1}$$

$$\therefore \text{ Zero point energy } E_0 = \frac{1}{2} h\nu = \frac{1}{2} \times 6.63 \times 10^{-34} \times 5.03$$

$$= 16.67 \times 10^{-34} \text{ J} = \frac{16.67 \times 10^{-34}}{1.6 \times 10^{-19}} = 10.4 \times 10^{-15} \text{ eV}$$

In classical physics total energy of the oscillator is given by

$$E = \frac{1}{2} m a^2 \omega^2$$

$$\text{But } \omega^2 = \frac{s}{m} \quad \therefore \quad E = \frac{1}{2} ma^2 \frac{s}{m} = \frac{1}{2} sa^2$$

or

$$a = \sqrt{\frac{2E}{s}} = \sqrt{\frac{2 \times 16.67 \times 10^{-34}}{10^{-3}}}$$

$$= 1.826 \times 10^{-15} \text{ m.}$$

Example 6.36. *Calculate the zero point energy of a system consisting of a mass of 0.5 gm connected to a fixed point by a spring which is stretched by 1 cm by a force of 1 k. dyne. The particle is constrained to move along X-axis.* ***(G.N.D.U., 2006)***

Solution. Spring constant $s = \frac{1000}{1} = 1000$ dyne cm^{-1}

$$\therefore \quad \nu = \frac{1}{2\pi}\sqrt{\frac{s}{m}} = \frac{1}{2\pi}\sqrt{\frac{1000}{0.5}} = 7.1 \text{ s}^{-1}$$

$$\text{Zero point energy} = \frac{1}{2} h\nu = \frac{1}{2} \times 6.63 \times 10^{-34} \times 7.1 = 23.5 \times 10^{-34} \text{ J} = 14.7 \times 10^{-15} \text{ eV}$$

Example 6.37. *Show that if the linear harmonic oscillator is in its normal state, then probability of finding it outside the classical limits is approximately 16%.*

(P. U., 2001; Poorvanchal U. 2002)

Solution. The ground state wave function of a harmonic oscillator is given by

$$\psi_o = \left[\frac{\alpha}{\pi}\right]^{\frac{1}{4}} e^{-\alpha x^2/2}$$

$$\text{where} \qquad \alpha = \frac{m\omega}{\hbar}$$

According to quantum theory the energy of the oscillator in the ground state $= \frac{1}{2}\hbar\omega$

According to classical theory the total energy of the particle undergoing harmonic oscillations $= \frac{1}{2} m\, \omega^2 A^2$

where A is the classical amplitude.

$$\therefore \qquad \frac{1}{2} m\, \omega^2 A^2 = \frac{1}{2} \hbar\omega$$

or $$A^2 = \frac{\hbar}{m\omega} = \frac{1}{\alpha}$$

or $$A = \pm \frac{1}{\sqrt{\alpha}}$$

In the normal state of the oscillator the probability of finding the particle in the classical region $-A$ to $+A$ is given by $P = \int_{-A}^{+A} \psi^* \psi dx = \int_{-\frac{1}{\sqrt{\alpha}}}^{+\frac{1}{\sqrt{\alpha}}} \left[\frac{\alpha}{\pi}\right]^{1/2} e^{-\alpha x^2} dx$

Substituting $\sqrt{\alpha}\, x = q$, we have $dq\ \sqrt{\alpha}\, dx$ and

$$P = \int_{-1}^{+1} \frac{1}{\sqrt{\alpha}} \left[\frac{\alpha}{\pi}\right]^{1/2} e^{-q^2} dq = \int_{-1}^{+1} \frac{1}{\sqrt{\pi}} e^{-q^2} dq = \frac{2}{\sqrt{\pi}} \int_0^1 e^{-q^2} dq$$

$$= \frac{2}{\sqrt{\pi}} \int_0^1 \left[1 - q^2 + \frac{q^4}{2!} - \frac{q^6}{3!} + \frac{q^8}{4!} + \ldots\right] dq$$

$$= \frac{2}{\sqrt{\pi}} \left[q - \frac{q^3}{3} + \frac{q^5}{10} - \frac{q^7}{42} + \frac{q^9}{216} + \ldots\right]_0^1$$

$$= \frac{2}{\sqrt{\pi}} \left[1 - \frac{1}{3} + \frac{1}{10} - \frac{1}{42} + \frac{1}{216} + \ldots\right] = 0.84 = 84\%$$

As the probability of finding the particle from $-\infty$ to $+\infty$ is unity and as proved above the probability of finding the particle in the normal state of the oscillator with in the classical limits is 0.84 or 84%, the probability of finding the particle outside the classical limits in the normal state of the harmonic oscillator is 1–0.84 = 0.16 = 16%.

EXERCISE

1. (*a*) What do you understand by the wave function Ψ of a moving particle?

   (P.U. 2006; Meerut. U. 2002)

   (*b*) Give the physical significance of wave function. What does the square of wave function signify?

   (G.N.D.U. 2007, 2006; M.D.U. 2006, 2003, 2000, K.U. 2002, 2001; P.U. 2000; Nagpur U. 2002; Gharwal. U. 2000; Meerut. U. 2003)

   (*c*) What are the conditions and limitations that the wave function must obey?

2. What are the conditions for a wave function to be called well behaved wave function?

   (P.U. 2008, 2007; G.N.D.U. 2007, 2005; Meerut. U. 2007; H.P.U. 2003, 2001)

3. Explain the need for differential wave equation. ***(H.P.U. 2002)***

4. (*a*) Starting from the wave equation and introducing energy and monentum of the particle obtain expression for three dimensional Schrödinger's equation in time dependent form.

(M.D.U. 2007, 2002; P.U. 2006, 2003, 2001; Pbi. U. 2006; H.P.U. 2003, 2002, 2000, K.U. 2003, 2002; G.N.D.U. 2003, 2002, Luck U. 2002, 2001; Meerut. U. 2001, Bang. U. 2001; Nagpur. U. 2011)

(*b*) Derive Schrödinger's wave equation and write it in terms of Hamiltonian operator. ***(P.U. 2008)***

5. Obtain three dimensional time independent Schrödinger's wave equation from time dependent Schrödinger's equation.

(Meerut. U. 2003, 2001; H.P.U. 2003, 2002, 2000; G.N.D.U. 2006, 2002; Bang. U. 2001; P.U. 2007, 2001, 2000; K.U. 2000; M.D.U. Nagpur U. 2002)

6. Why the Schrödinger's wave equation is not valid for relativistic particles? ***(P.U.2002; Pbi. U. 2001)***

7. Obtain Schrödinger's wave equation in momentum representation. What is significance of momentum wave function? ***(P.U. 2001, 2000)***

8. (*a*) Write Schrödinger's wave equation for a particle. What is the physical interpretation of the wave function Ψ? Explain Born's interpretation.

[H.P.U. 2003; G.N.D.U. 2002; Bang.U. 2001, 2000; Luck.U. 2002]

(*b*) What do you mean by normalisation and probability value of a wave function? ***(G.N.D.U. 2006)***

9. (*a*) Explain probability current and probability density. ***(Poorvanchal U. 2002; G.N.D.U., 2004, 2001)***

(*b*) Show that change in probability density in a region of space is equal to the net change in probability current into that region. ***(P.U. 2003)***

10. What is normalisation of a wave function? How is it mathematically expressed? Find out the normalisation factor of Schrödinger's wave function. Prove that normalisation is independent of time. ***(G.N.D.U. 2008; K.U. 2000,1991, P.U. 2005; Nagpur U. 2002)***

11. State the conditions that the wave function must obey or limitations of wave function.

12. State the conditions for the wave function to be well behaved.

13. What do you understand by superposition of states? ***(P.U. 2007, 2006, 2005; Pbi. U. 2000)***

14. What is the importance of normalising a wave function? ***(H.P.U. 2002)***

15. (*a*) Explain the terms 'operator'. Derive expression for (i) linear momentum operator (ii) energy operator (iii) What is a Hermitian operator?

(G.N.D.U. 2008, 2007; Pbi. U. 2005; Nagpur U. 2002; Meerut. U. 2006)

(*b*) Explain the terms 'expectation value' and 'eigen value'. ***(P.U. 2006; Meerut. U. 2005, 2006)***

(*c*) State the postulates of quantum mechanics and hence derive Schrödinger's wave equation. ***(Nagpur U. 2002; G.N.D.U. 2001)***

16. What is an eigen function and eigen value?

(G.N.D.U. 2006, 2005; Nagpur, U. 2002; P.U. 2006, 2001; Luck.U. 2001)

17. (*a*) Define expectation value of a dynamical variable. How is it related to probability? Give expression for expectation value of position, momentum and energy in terms of corresponding operators.

(Nagpur U. 2002; Pbi.U. 2002; G.N.D.U. 2007, 2001; Bang.U. 2000; Gauhati.U. 2000)

(*b*) Why $\Psi = ax^2$ is not acceptable wave finction in quantum mechanics ?

18. What are operators ? Obtain an expression for quantum mechanical operator for energy. ***(Nagpur U. 2011)***

19. (*a*) State the postulates of quantum mechanics.

(*b*) Obtain Schrodinger's equation from postulates of quantum mechanics.

20. State Schrödinger's wave equation for a free particle in a one dimensional closed box with infinitely hard walls. State the boundary conditions and solve it to obtain normalised wave function for the particle. Calculate the eigen functions and corresponding eigen values of propagation constant, momentum and energy for different states of the particle. How do the results differ from classical?

(Nagpur U. 2001; G.N.D.U. 2003, 2002; H.P.U. 2002, 2001;M.D.U. 2002; Pbi.U. 2002;Luck. U. 2002, K.U. 2002, 1991; P.U. 2001, Gharwal U. 2000; Meerut U. 2000; Bang. U. 2004)

21. Show that free particle in one dimensional box can have positive energy values which are discrete. ***(Nagpur U. 2002)***

22. Show that free particle in one dimensional box n can not have zero and negative values.

23. (*a*) Show that the linear momentum and energy of a particle in a one dimensional box is quantised. Discuss the comparison between classical and quantum ideas. ***(Bang. U. 2001)***

(*b*) A particle is moving in one dimensional box and its wave function is given by

$\psi(x) = A \sin \dfrac{n\pi}{a} x$. Normalise the wave function. ***(H.P.U. 2001; P.U. 2001)***

24. (*a*) Obtain energy levels for a particle inside a box of finite width l. ***(P.U. 2000)***

(*b*) What is ground state, excited state and energy level? Draw the energy level diagram of a particle in a box. ***(Pbi.U. 2002, K.U. 2000)***

25. (*a*) A particle of mass m is confined to a one dimensional box of length l. Derive expression for (i) wave function (ii) probability density of the particle. Show these on separate graphs. ***(P.U. 2000)***

(*b*) (i). What are nodes and antinodes ? Where do they occur? Explain their presence in terms of probability density.

26. Prove quantum mechanically that a particle will not exist in a box if its energy is zero. Discuss zero point energy. ***(G.N.D.U. 2000)***

27. Why a particle in a box cannot have zero energy quantum mechanically? How does it change when the length of the box l tends to infinity? ***(G.N.D.U. 2007, 2005)***

28. (*a*) What are orthogonal wave functions? ***(G.N.D.U., 2002)***

(*b*) For a particle in one dimensional box, show that the wave functions for two different states are orthogonal (ortho-normal). Hence show that the wave functions of the two lowest states are orthogonal. ***(Pbi.U., 2003)***

29. (*a*) Explain briefly the concept of parity as applied to wave function. When will the wave function have odd or even parity?

(*b*) What is parity operator? ***(Pbi.U., 2006, 2001; P.U., 2000)***

30. What is the basic reason for the parity of one dimensional harmonic oscillator wave function? ***(Pbi.U., 2003)***

31. (*a*) Write Schrödinger's equation for a particle in a rigid box and solve it. Find the eigen functions when a particle is kept in a rectangular box of dimensions $l_x \times l_y \times l_z$. Find the eigen values of momentum and energy.

(*b*) Show that for a single value of energy or momentum, different quantum states are possible. Explain the term degeneracy. ***(G.N.D.U., 2001; Pbi.U., 2000; 1999, P.U., 2000)***

32. Find the energies of eight lowest energy levels of a particle in a cubical box. Which of these levels are degenerate? ***(P.U., 1999)***

33. Differentiate between a potential step and potential barrier. Give physical example in each case. ***(G.N.D.U., 2001)***

34. (*a*) What is one dimensional potential barrier or potential step? ***(K.U., 2001)***

(*b*) A particle is moving in one dimensional potential given by

$$V = 0 \text{ For } x < 0$$
$$V = V_0 \text{ for } x \geq 0$$

Assuming that total energy E of the incident particle is greater than V_0, write down Schrödinger's wave equation for the particle and solve it. Calculate the co-efficient of reflection and transmission. Also discuss the case classically.

(P.U., 2003; M.D.U., 2008, 2006, 2003; G.N.D.U., 2005, 2002; K.U., 2002, 2001; Meerut U., 2002; H.P.U., 2001; Luck.U. 2001; Pbi.U., 2005; Poorvanchal U. 2003, 2001)

35. Show that the sum of the reflection and transmission coefficient is one. ***(M.D.U. 2003; Pbi. U. 2002)***

36. What is potential step? Find the reflection and transmission co-efficient for potential step of the form $0 < E < V_0$. Show that there is a finite probability of locating the particle in the region which is forbidden classically. What is the penetration distance? Give its relation to the mass of the incident particle. ***(Pbi. U. 2008; K.U., 2002, 2001; Meerut.U., 2002; Luck.U., 2001; M.D.U., 2001, 2000; P.U., 2008, 2001; G.N.D.U., 2007)***

37. (*a*) A particle of mass m and energy $E < V_0$ travelling along x-axis has a potential barrier defined by

$$V(x) = \begin{cases} 0 & \text{for } x < 0 \\ V_0 & \text{for } 0 < x < a \\ 0 & \text{for } x > a \end{cases}$$

Derive the expression for reflection and transmission co-efficient of the particle.

(*b*) What do you mean by tunneling through the barrier?

(Meerut U., 2003, 2001, 2000; K.U., 2003; H.P.U., 2003, 2000; M.D.U., 2000; P.U., 2005, 2000; Poorvanchal U. 2003; Gharwal U., 2001; Pbi.U., 2006)

Or

What is tunnel effect? ***(G.N.D.U. 2006, P.U. 2006, 2005)***

38. (*a*) What do you mean by tunneling effect? Show how it is encountered in quantum system of a particle moving along X-axis with an energy E when it faces a rectangular barrier of height V_0, such that $E < V_0$? ***(G.N.D.U., 2008)***

(*b*) What is the condition for perfect transmission through a rectangular potential barrier? ***(G.N.D.U. 2005)***

39. What is a harmonic oscillator? Give examples. What is its physical importance? ***(P.U. 2007; Pbi.U., 2002)***

40. (*a*) What do you understand by an harmonic oscillator? What does classical mechanics say about its energy? What are the modifications introduced into classical model by Quantum mechanics? Obtain an expression for the energy levels of the harmonic oscillator applying Schrödinger's equation.

(*b*) Plot the energy levels of the oscillator and compare with energy levels of hydrogen atom.

(M.D.U., 2007, 2003, 2002, 2001; Meerut U., 2006, 2003, 2002, 2001; Pbi.U., 2006, 2002; P.U., 2001; K.U., 2001, Gauhati U., 2000; Gharwal U., 2000; G.N.D.U., 2006, 2005, 1996; Nagpur U. 2002; Poorvanchal U. 2002)

41. Write potential for one dimensional harmonic oscillator and use it to build up a time independent Schrödinger's wave equation. Solve the equation for its eigen energies and eigen wave functions. ***(G.N.D.U. 2008; P.U. 2007; Meerut U. 2006)***

42. Solve the Schrödinger's equation for a linear harmonic oscillator for its energy eigen values. ***(P.U. 2007)***

43. (*a*) Show quantitatively that energy of an harmonic oscillator is quantised in steps of hν and approaches $\frac{1}{2}h\nu$ as temperature approaches 0 K. How do you explain the existence of zero point energy?

(*b*) Draw energy level diagram of a harmonic oscillator and compare with those of particle in a box. ***(M.D.U. 2007; H.P.U. 2003; Gharwal U. 2003, 2000; P.U. 2005, 2001; Nagpur U. 2011, 2002; G.N.D.U. 2005)***

44. Write the general and normalised wave function of a harmonic oscillator. State the first three normalised wave functions of the oscillator and show them graphically. ***(P.U. 2006; H.P.U. 2003; K.U. 2001)***

45. Show that the Schrödinger's equation for a linear harmonic oscillator can be written as $\psi'' + (\lambda - x^2)\,\psi = 0$ where distance and energy are expressed in units of $\sqrt{\frac{\hbar}{m\omega}}$ and $\frac{1}{2}\hbar\omega$ respectively. Solve the equation to obtain eigen values. ***(G.N.D.U. 2003)***

46. Solve Schrödinger's equation for three dimensional harmonic oscillator in Cartesian co-ordinates and obtain its energy levels. Which state of three dimensional harmonic oscillator is non-degenerate? ***(Pbi.U., 2003)***

47. Show that the quantity $y = \sqrt{\alpha}\,x$ where $\alpha = \frac{\omega m}{\hbar}$ as used in harmonic oscillator is dimensionless. ***(Pbi. U., 2002)***

48. Is the function $\psi = a \cos m\phi$ an eigen function of Z-component of angular momentum operator? If so find the eigen value. ***(H.P.U. 2001)***

Hint. $\hat{A} f = \hat{L}_z\, a \cos m\phi = -i\hbar \frac{\partial}{\partial \phi} a \cos m\phi = i\hbar\, am \sin m\phi$.

As $L_z\, a \cos m\phi \neq a$ constant $\times\, a \cos m\phi$, $a \cos m\phi$ is not an eigen function of angular momentum operator.

49. A particle of kinetic energy 9 eV is incident on a potential step of height 5 eV. Find the value of reflection co-efficent *R*.

Hint. $R = \left(\frac{k_1 - k_2}{k_1 + k_2}\right)^2$; $k_1 = \frac{\sqrt{2mE}}{\hbar}$ and $k_2 = \frac{\sqrt{2m(E - V_0)}}{\hbar}$

$$\therefore \quad k_1 - k_2 = \frac{\sqrt{2m}}{\hbar}(\sqrt{E} - \sqrt{E - V_0})$$

$$k_1 + k_2 = \frac{\sqrt{2m}}{\hbar}(\sqrt{E} + \sqrt{E - V_0})$$

or $$\frac{k_1 - k_2}{k_1 + k_2} = \frac{\sqrt{E} - \sqrt{E - V_0}}{\sqrt{E} + \sqrt{E - V_0}} = \frac{\sqrt{9} - \sqrt{9 - 5}}{\sqrt{9} + \sqrt{9 - 5}} = \frac{1}{5}$$

$$\therefore \quad R = \frac{1}{25} = 0.04$$

[Ans. 0.04]

MULTIPLE CHOICE QUESTIONS

1. Which of the following wave functions represent a free particle moving along + X axis?

   (a) $A \sin (kx - \omega t)$ (b) $A \cos (kx - \omega t)$ (c) $Ae^{i(kx - \omega t)}$ (d) $Ae^{-i(kx - \omega t)}$

2. A free particle has

   (a) definite energy but indefinite momentum
   (b) definite momentum but indefinite energy
   (c) definite energy and definite momentum
   (d) indefinite energy and indefinite momentum

3. Schrödinger's wave equation for a moving particle contains

   (a) first order time derivative (b) second order time derivative
   (c) third order time derivative (d) no time derivative.

4. For a stationary state the probability density is

   (a) function of time
   (b) independent of time
   (c) independent of space co-ordinates

5. The probability of finding a particle in a distance dx around a point x is

   (a) ψ^* (b) $\psi\psi^* dx$ (c) $\psi\,\psi^*$ (d) ψ

6. For a three dimensional case, which is the normalisation condition

   (a) $\int \psi_n{}^* \psi_m \, dV = 1$ for $n = m$ (b) $\int \psi_n{}^* \psi_m \, dV = 1$ for $n \neq m$
   (c) $\int \psi_n{}^* \psi_m \, dV = 0$ for $n = m$ (d) $\int \psi_n{}^* \psi_m \, dV = 0$ for $n \neq m$

7. The energy of a particle in a one dimensional closed box is

   (a) directly proportional to quantum number n (b) directly proportional to n^2
   (c) inversely proportional to n (d) inversely proportional to n^2

8. The energy of a particle in a one dimensional closed box of length l is

   (a) directly proportional to l (b) directly proportional to l^2
   (c) inversely proportional to l (d) inversely proportional to l^2

9. The number of nodes in the ground state wave function of a particle in an infinite square well is

   (a) 1 (b) 2 (c) 3 (d) 4

10. Two wave functions $\psi_m(x)$ and $\psi_n(x)$ are orthogonal if

    (a) $\int \psi_m(x)\psi_n(x)\,dx = 0$ (b) $\int \psi_{m^*}(x)\psi_n(x)\,dx = 0$
    (c) $\int \psi_m(x)\psi_n(x)\,dx = 1$ (d) $\int \psi_{m^*}(x)\psi_n(x)\,dx = 1$

11. For a particle trapped in a box of length l the average value of momentum $\langle p \rangle$ is

    (a) $\frac{h}{l}$ (b) $\frac{h}{2l}$ (c) 0 (d) 1

12. The reflection co-efficient for a particle incident on a potential step with energy E less than height of the step is

    (a) 1 (b) 0 (c) $\frac{1}{2}$ (d) $\frac{1}{3}$

13. The relation between co-efficient of reflection R and transmission T for a particle incident on a potential step with energy E greater than height of the step is

(a) $R = T$ (b) $R + T = 0$ (c) $R + T = 1$ (d) $R = \frac{1}{T}$

14. The energy of a one dimensional harmonic oscillator in the ground state is

(a) 0 (b) $\frac{1}{2}\hbar\omega$ (c) $\frac{3}{2}\hbar\omega$ (d) $\frac{5}{2}\hbar\omega$

15. The energy of a one dimensional harmonic oscillator in the first exicited state is

(a) 0 (b) $\frac{1}{2}\hbar\omega$ (c) $\frac{3}{2}\hbar\omega$ (d) $\frac{5}{2}\hbar\omega$

ANSWERS

1. (c)	**2.** (c)	**3.** (a)	**4.** (b)	**5.** (b)
6. (a)	**7.** (b)	**8.** (d)	**9.** (b)	**10.** (b)
11. (c)	**12.** (a)	**13.** (c)	**14.** (b)	**15.** (c)

SCHRÖDINGER'S THEORY OF HYDROGEN ATOM

7
Chapter

INTRODUCTION

For hydrogen atom the potential energy of the electron

$$V(r) = -\frac{e^2}{4\pi \epsilon_0 r}$$

i.e., the potential energy of the electron only depends upon r the distance between the electron and the nucleus and is independent of any other factor. In other words, the potential is *spherically symmetric* and hence conservative in nature. The expression for V also does not involve time explicitly. Therefore, Schrödinger's time independent equation is applicable.

Because of spherical symmetry it is more convenient to deal with the problem in spherical polar co-ordinates defined by r, θ and ϕ. As these three co-ordinates are orthogonal to each other it is possible to express Schrödinger's time independent wave function ψ as the product of three separate functions each of only one independent variable r, θ or ϕ.

The spherical co-ordinates r, θ and ϕ have the following ranges

For r the range is $0 \rightarrow \infty$

For θ the range is $0 \rightarrow \pi$

For ϕ the range is $0 \rightarrow 2\pi$

Hence, in order to understand the meaning and the occurance of n, l and m quantum numbers, let us derive Schrödingers three dimensional time independent wave equation in spherical polar co-ordinates r, θ and ϕ.

7.1 SCHRÖDINGER'S THREE DIMENSIONAL TIME-INDEPENDENT WAVE EQUATION IN SPHERICAL POLAR CO-ORDINATES

The relation between Cartesian co-ordinates and spherical polar co-ordinates is

$$x = r\sin\theta\cos\phi$$

$$y = r\sin\theta\sin\phi$$

$$z = r\cos\theta$$

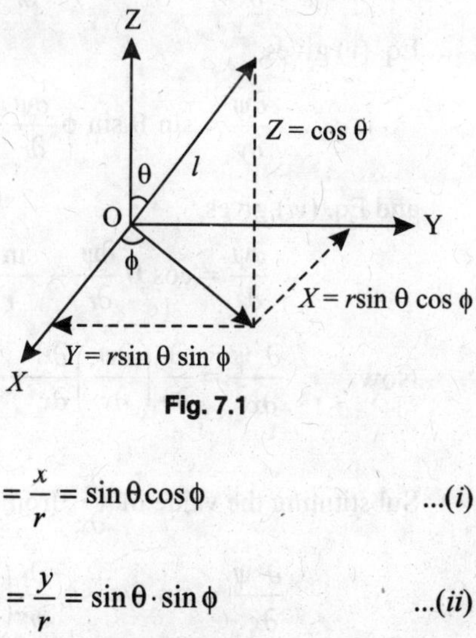

Fig. 7.1

$$\therefore \qquad r = \sqrt{x^2 + y^2 + z^2}$$

$$\theta = \tan^{-1}\sqrt{x^2 + y^2 / z}$$

$$\phi = \tan^{-1}\frac{y}{x}$$

$$\frac{\partial r}{\partial x} = \frac{1}{2}\left(x^2 + y^2 + z^2\right)^{-1/2} \cdot 2x = \frac{x}{r} = \sin\theta\cos\phi \qquad ...(i)$$

$$\frac{\partial r}{\partial y} = \frac{1}{2}\left(x^2 + y^2 + z^2\right)^{-1/2} \cdot 2y = \frac{y}{r} = \sin\theta\cdot\sin\phi \qquad ...(ii)$$

$$\frac{\partial r}{\partial z} = \frac{1}{2}\left(x^2+y^2+z^2\right)^{-1/2}\cdot 2z = \frac{z}{r} = \cos\theta \qquad ...(iii)$$

Now
$$\frac{\partial}{\partial x} = \frac{\partial r}{\partial x}\frac{\partial}{\partial r} + \frac{\partial \theta}{\partial x}\frac{\partial}{\partial \theta} + \frac{\partial \phi}{\partial x}\frac{\partial}{\partial \phi}$$

$$= \sin\theta\cos\phi\,\frac{\partial}{\partial r} + \frac{1}{r}\cos\theta\cos\phi\frac{\partial}{\partial \theta} - \frac{\sin\phi}{r\sin\theta}\frac{\partial}{\partial \phi} \qquad ...(vi)$$

$$\frac{\partial}{\partial y} = \frac{\partial r}{\partial y}\frac{\partial}{\partial r} + \frac{\partial \theta}{\partial y}\frac{\partial}{\partial \theta} + \frac{\partial \phi}{\partial y}\frac{\partial}{\partial \phi}$$

$$= \sin\theta\sin\phi\frac{\partial}{\partial r} + \frac{1}{r}\cos\theta\sin\phi\frac{\partial}{\partial \theta} + \frac{\cos\phi}{r\sin\theta}\frac{\partial}{\partial \phi} \qquad ...(v)$$

$$\frac{\partial}{\partial z} = \frac{\partial r}{\partial z}\frac{\partial}{\partial r} + \frac{\partial \theta}{\partial z}\frac{\partial}{\partial \theta} + \frac{\partial \phi}{\partial z}\frac{\partial}{\partial \phi}$$

$$= \cos\theta\frac{\partial}{\partial r} - \frac{\sin\theta}{r}\frac{\partial}{\partial \theta} \qquad ...(vi)$$

Schrödinger's three dimensional time independent wave equation is

$$\nabla^2\psi + \frac{2m}{\hbar^2}(E-V)\psi = 0$$

In Cartesian co-ordinates $\nabla^2\psi = \frac{\partial^2\psi}{\partial x^2} + \frac{\partial^2\psi}{\partial y^2} + \frac{\partial^2\psi}{\partial z^2}$ and hence the equation becomes

$$\frac{\partial^2\psi}{\partial x^2} + \frac{\partial^2\psi}{\partial y^2} + \frac{\partial^2\psi}{\partial z^2} + \frac{2m}{\hbar^2}(E-V)\psi = 0$$

From Eq. (*iv*), we get

$$\frac{\partial\psi}{\partial x} = \sin\theta\cdot\cos\phi\,\frac{\partial\psi}{\partial r} + \frac{1}{r}\cos\theta\cdot\cos\phi\,\frac{\partial\psi}{\partial\theta} - \frac{\sin\phi}{r\sin\theta}\frac{\partial\psi}{\partial\phi} \qquad ...(vii)$$

Eq. (*v*) gives,

$$\frac{\partial\psi}{\partial y} = \sin\theta\sin\phi\,\frac{\partial\psi}{\partial r} + \frac{1}{r}\cos\theta\sin\phi\,\frac{\partial\psi}{\partial\theta} + \frac{\cos\phi}{r\sin\theta}\frac{\partial\psi}{\partial\phi} \qquad ...(viii)$$

and Eq. (*vi*) gives,

$$\frac{\partial\psi}{\partial z} = \cos\theta\,\frac{\partial\psi}{\partial r} - \frac{\sin\theta}{r}\frac{\partial\psi}{\partial\theta} \qquad ...(ix)$$

Now
$$\frac{\partial^2\psi}{\partial x^2} = \frac{\partial}{\partial r}\left(\frac{\partial\psi}{\partial x}\right)\frac{\partial r}{\partial x} + \frac{\partial}{\partial\theta}\left(\frac{\partial\psi}{\partial x}\right)\frac{\partial\theta}{\partial x} + \frac{\partial}{\partial\phi}\left(\frac{\partial\psi}{\partial x}\right)\frac{\partial\phi}{\partial x}$$

Substituting the value of $\frac{\partial r}{\partial x}$ from (*i*), $\frac{\partial\theta}{\partial x}$ from (*ii*) and $\frac{\partial\phi}{\partial x}$ form (*iii*), we have

$$\frac{\partial^2\psi}{\partial x^2} = \sin\theta\cos\phi\,\frac{\partial}{\partial r}\left(\frac{\partial\psi}{\partial x}\right) + \frac{\cos\theta\cos\phi}{r}\frac{\partial}{\partial\theta}\left(\frac{\partial\psi}{\partial x}\right) - \frac{\sin\phi}{r\sin\theta}\frac{\partial}{\partial\phi}\left(\frac{\partial\psi}{\partial x}\right)$$

Substituting the value of $\frac{\partial\psi}{dx}$ from (*vii*) in the above equation we have

$$\frac{\partial^2\psi}{\partial x^2} = \sin\theta\cos\phi\frac{\partial}{\partial r}\left[\sin\theta\cos\phi\frac{\partial\psi}{\partial r} + \frac{1}{r}\cos\theta\cos\phi\frac{\partial\psi}{\partial\theta} - \frac{\sin\phi}{r\sin\theta}\frac{\partial\psi}{\partial\phi}\right]$$

$$+\frac{\cos\theta\cos\phi}{r}\frac{\partial}{\partial\theta}\left[\sin\theta\cos\phi\frac{\partial\psi}{\partial r} + \frac{1}{r}\cos\theta\cos\phi\frac{\partial\psi}{\partial\theta} - \frac{\sin\phi}{r\sin\theta}\frac{\partial\psi}{\partial\phi}\right]$$

$$-\frac{\sin\phi}{r\sin\theta}\frac{\partial}{\partial\phi}\left[\sin\theta\cos\phi\frac{\partial\psi}{\partial r} + \frac{1}{r}\cos\theta\cos\phi\frac{\partial\psi}{\partial\theta} - \frac{\sin\phi}{r\sin\theta}\frac{\partial\psi}{\partial\phi}\right]$$

$$= \sin^2\theta\cos^2\phi\frac{\partial^2\psi}{\partial r^2} + \sin\theta\cos\theta\cos^2\phi\frac{\partial}{\partial r}\left(\frac{1}{r}\frac{\partial\psi}{\partial\theta}\right)$$

$$-\sin\phi\cos\phi\frac{\partial}{\partial r}\left(\frac{1}{r}\frac{\partial\psi}{\partial\phi}\right) + \frac{\cos\theta\cos^2\phi}{r}\frac{\partial}{\partial\theta}\left(\sin\theta\frac{\partial\psi}{\partial r}\right)$$

$$+\frac{\cos\theta\cos^2\phi}{r^2}\frac{\partial}{\partial\theta}\left(\cos\theta\frac{\partial\psi}{\partial\theta}\right) - \frac{\cos\theta\cos\phi\sin\phi}{r^2}\frac{\partial}{\partial\theta}\left(\frac{1}{\sin\theta}\frac{\partial\psi}{\partial\phi}\right)$$

$$-\frac{\sin\phi}{r}\frac{\partial}{\partial\phi}\left(\cos\phi\frac{\partial\psi}{\partial r}\right) - \frac{\sin\phi\cot\theta}{r^2}\frac{\partial}{\partial\phi}\left(\cos\phi\frac{\partial\psi}{\partial\theta}\right)$$

$$+\frac{\sin\phi}{r^2\sin^2\theta}\frac{\partial}{\partial\phi}\left(\sin\phi\frac{\partial\psi}{\partial\phi}\right) \quad \text{...}(x)$$

Similarly $$\frac{\partial^2\psi}{\partial y^2} = \frac{\partial}{\partial r}\left(\frac{\partial\psi}{\partial y}\right)\frac{\partial r}{\partial y} + \frac{\partial}{\partial\theta}\left(\frac{\partial\psi}{\partial y}\right)\frac{\partial\theta}{\partial y} + \frac{\partial}{\partial\phi}\left(\frac{\partial\psi}{\partial y}\right)\frac{\partial\phi}{\partial y}$$

Substituting the value of $\frac{\partial r}{\partial y}$, $\frac{\partial\theta}{\partial y}$ and $\frac{\partial\phi}{\partial y}$, we have

$$\frac{\partial^2\psi}{dy^2} = \sin\theta\sin\phi\frac{\partial}{\partial r}\left(\frac{\partial\psi}{\partial y}\right) + \frac{1}{r}\cos\theta\sin\phi\frac{\partial}{\partial\theta}\left(\frac{\partial\psi}{\partial y}\right) + \frac{\cos\phi}{r\sin\theta}\frac{\partial}{\partial\phi}\left(\frac{\partial\psi}{\partial y}\right)$$

Substituting the value of $\frac{\partial\psi}{\partial y}$ from (*viii*) in the above equation, we have

$$\frac{\partial^2\psi}{\partial y^2} = \sin\theta\sin\phi\frac{\partial}{\partial r}\left[\sin\theta\sin\phi\frac{\partial\psi}{\partial r} + \frac{1}{r}\cos\theta\sin\phi\frac{\partial\psi}{\partial\theta} + \frac{\cos\phi}{r\sin\theta}\frac{\partial\psi}{\partial\phi}\right]$$

$$+\frac{1}{r}\cos\theta\sin\phi\frac{\partial}{\partial\theta}\left[\sin\theta\sin\phi\frac{\partial\psi}{\partial r} + \frac{1}{r}\cos\theta\sin\phi\frac{\partial\psi}{\partial\theta} + \frac{\cos\phi}{r\sin\theta}\frac{\partial\psi}{\partial\phi}\right]$$

$$+\frac{\cos\phi}{r\sin\theta}\frac{\partial}{\partial\phi}\left[\sin\theta\sin\phi\frac{\partial\psi}{\partial r} + \frac{1}{r}\cos\theta\sin\phi\frac{\partial\psi}{\partial\theta} + \frac{\cos\phi}{r\sin\theta}\frac{\partial\psi}{\partial\phi}\right]$$

$$= \sin^2\theta\sin^2\phi\frac{\partial^2\psi}{\partial r^2} + \sin\theta\cos\theta\sin^2\phi\frac{\partial}{\partial r}\left(\frac{1}{r}\frac{\partial\psi}{\partial\theta}\right)$$

$$+\cos\phi\sin\phi\frac{\partial}{\partial r}\left(\frac{1}{r}\frac{\partial\psi}{\partial\theta}\right) + \frac{\cos\theta\sin^2\phi}{r}\frac{\partial}{\partial\theta}\left(\sin\theta\frac{\partial\psi}{\partial r}\right)$$

$$+\frac{\cos\theta\sin^2\phi}{r^2}\frac{\partial}{\partial\theta}\left(\cos\theta\frac{\partial\psi}{\partial\theta}\right)+\frac{\cos\theta\sin\phi\cos\phi}{r^2}\frac{\partial}{\partial\theta}\left(\frac{1}{\sin\theta}\frac{\partial\psi}{\partial\phi}\right)$$

$$+\frac{\cos\phi}{r}\frac{\partial}{\partial\phi}\left(\sin\phi\frac{\partial\psi}{\partial r}\right)+\frac{\cos\phi\cot\theta}{r^2}\frac{\partial}{\partial\phi}\left(\sin\phi\frac{\partial\psi}{\partial\theta}\right)$$

$$+\frac{\cos\phi}{r^2\sin^2\theta}\frac{\partial}{\partial\phi}\left(\cos\phi\frac{\partial\psi}{\partial\phi}\right) \quad ...(xi)$$

Adding equations (*i*) and (*xi*), we get

$$\frac{\partial^2\psi}{\partial x^2}+\frac{\partial^2\psi}{\partial y^2}=\sin^2\theta\frac{\partial^2\psi}{\partial r^2}+\sin\theta\cos\theta\frac{\partial}{\partial r}\left(\frac{1}{r}\frac{\partial\psi}{\partial\theta}\right)+\frac{\cos\theta}{r}\frac{\partial}{\partial\theta}\left(\sin\theta\frac{\partial\psi}{dr}\right)$$

$$+\frac{\cos\theta}{r^2}\frac{\partial}{\partial\theta}\left(\cos\theta\frac{\partial\psi}{\partial\theta}\right)+\frac{1}{r}\frac{\partial\psi}{\partial r}+\frac{\cot\theta}{r^2}\frac{\partial\psi}{\partial\theta}+\frac{1}{r^2\sin^2\theta}\frac{\partial^2\psi}{\partial\phi^2} \quad ...(xii)$$

Similarly $\dfrac{\partial^2\psi}{dz^2}=\dfrac{\partial}{\partial r}\left(\dfrac{\partial\psi}{\partial z}\right)\dfrac{\partial r}{\partial z}+\dfrac{\partial}{\partial\theta}\left(\dfrac{\partial\psi}{\partial z}\right)\dfrac{\partial\theta}{\partial z}+\dfrac{\partial}{\partial\phi}\left(\dfrac{\partial\psi}{\partial z}\right)\dfrac{\partial\phi}{\partial z}$

Substituting the value of $\dfrac{\partial r}{\partial z}$, $\dfrac{\partial\theta}{\partial z}$ and $\dfrac{\partial\phi}{\partial z}$, we have

$$\frac{\partial^2\psi}{dz^2}=\cos\theta\frac{\partial}{\partial r}\left(\frac{\partial\psi}{\partial z}\right)-\frac{1}{r}\sin\theta\frac{\partial}{\partial\theta}\left(\frac{\partial\psi}{\partial z}\right)$$

Substituting the value of $\dfrac{\partial\psi}{\partial z}$ from (*ix*) in the above equation, we have

$$\frac{\partial^2\psi}{\partial z^2}=\cos\theta\frac{\partial}{\partial r}\left[\cos\theta\frac{\partial\psi}{\partial r}-\frac{\sin\theta}{r}\frac{\partial\psi}{\partial\theta}\right]$$

$$-\frac{\sin\theta}{r}\frac{\partial}{\partial\theta}\left[\cos\theta\frac{\partial\psi}{\partial r}-\frac{\sin\theta}{r}\frac{\partial\psi}{\partial\theta}\right]$$

$$=\cos^2\theta\frac{\partial^2\psi}{\partial r^2}-\sin\theta\cos\theta\frac{\partial}{\partial r}\left(\frac{1}{r}\frac{\partial\psi}{\partial\theta}\right)-\frac{\sin\theta}{r}\frac{\partial}{\partial\theta}\left(\cos\theta\frac{\partial\psi}{\partial r}\right)$$

$$+\frac{\sin\theta}{r^2}\frac{\partial}{\partial\theta}\left(\sin\theta\frac{\partial\psi}{\partial\theta}\right) \quad ...(xiii)$$

Adding eqs. (*xii*) and (*xiii*), we have

$$\frac{\partial^2\psi}{\partial x^2}+\frac{\partial^2\psi}{\partial y^2}+\frac{\partial^2\psi}{\partial z^2}=\frac{\partial^2\psi}{\partial r^2}+\frac{2}{r}\frac{\partial\psi}{\partial r}+\frac{1}{r^2}\frac{\partial^2\psi}{\partial\theta^2}+\frac{\cot\theta}{r^2}\frac{\partial\psi}{\partial\theta}+\frac{1}{r^2\sin^2\theta}\frac{\partial^2\psi}{\partial\phi^2}$$

Thus $$\nabla^2\psi=\frac{\partial^2\psi}{\partial r^2}+\frac{2}{r}\frac{\partial\psi}{\partial r}+\frac{1}{r^2}\frac{\partial^2\psi}{\partial\theta^2}+\frac{\cot\theta}{r^2}\frac{\partial\psi}{\partial\theta}+\frac{1}{r^2\sin^2\theta}\frac{\partial^2\psi}{\partial\phi^2} \quad ...(7.1)$$

From which we have

$$\nabla^2=\frac{\partial^2}{\partial r^2}+\frac{2}{r}\frac{\partial}{\partial r}+\frac{1}{r^2}\frac{\partial^2}{\partial\theta^2}+\frac{\cot\theta}{r^2}\frac{\partial}{\partial\theta}+\frac{1}{r^2\sin^2\theta}\frac{\partial^2}{\partial\phi^2}$$

$\therefore$ Schrödinger's wave equation in spherical polar co-ordinates is

$$\frac{\partial^2 \psi}{\partial r^2} + \frac{2}{r}\frac{\partial \psi}{\partial r} + \frac{1}{r^2}\frac{\partial^2 \psi}{\partial \theta^2} + \frac{\cot\theta}{r^2}\frac{\partial \psi}{\partial \theta} + \frac{1}{r^2 \sin^2\theta}\frac{\partial^2 \psi}{\partial \phi^2} + \frac{2m}{\hbar^2}(E - V)\,\psi = 0. \quad ...(7.2)$$

7.2 SCHRÖDINGER'S WAVE EQUATION FOR HYDROGEN ATOM

A hydrogen atom consist of a *single* electron going round the nucleus which contains only *one* nucleon *i.e.* a proton. It is the simplest atom and thus provides the simplest application of Schrödinger's wave equation to an atom. The results obtained can be extended to similar systems, such as

(*i*) singly ionised Helium (He^+), doubly ionised Lithium (Li^{++}), triply ionised Beryllium (Be^{+++}) i.e. ions having only one electron going round the nucleus, which may consist of more than one nucleon and hence more than one proton (*ii*) atoms or ions having closed shells and only one electron in the outermost shell, such as alkali atoms and alkaline earth ions.

These systems are known as hydrogen-like systems or *hydrogenic atoms.*

Thus a hydrogenic atom consists of Z protons and N neutrons in the nucleus and only one *electron* in the surrounding space. The electron, which carries a negative charge, is bound to the nucleus, which has a positive charge due to the Coulomb force of electrostatic attraction. Hence the electrostatic potential energy of an electron with charge $-e$ in the field of the nucleus with charge $+Ze$ is given by

$$V(r) = -\frac{kZe^2}{r} \quad ...(i)$$

where r is the distance between the electron and the nucleus both assumed to be point charges and k is a constant the value of which in S.I. units is $\frac{1}{4\pi\varepsilon_0} = 9 \times 10^9\ \text{Nm}^2/\text{C}^2$.

For hydrogen $Z = 1$

$$\therefore \quad V(r) = -\frac{e^2}{4\pi\varepsilon_0 r} \quad ...(i)\,(a)$$

This is, therefore, a two body problem, but it can be reduced to a one-body problem by

(*i*) replacing the electron mass m_e, by its reduced mass m, given by

$$m = \frac{Mm_e}{M + m_e}$$

where M is the mass of the nucleus and (*ii*) by using the *centre of mass* as the origin of the co-ordinate system or by assuming the nucleus to have an infinite mass.

As given by equation (*i*), the potential energy of the electron only depends upon the distance between the electron and the nucleus and is independent of any other factor, like the angular position of the electron with respect to the nucleus. The value of V is the same at every point on the surface of a sphere of radius r. In other words, the potential is spherically symmetric and hence conservative in nature. The expression for V also does not involve time *explicitly*, the system is stationary and Schrödinger's time independent form of wave equation applies to it.

Because of spherical symmetry, it is more convenient to deal with the problem in spherical polar co-ordinates defined by the variables r, θ and ϕ.

Schrödinger's time independent, three dimensional wave equation in Cartesian Co-ordinates is given by

$$\nabla^2 \psi + \frac{2m}{\hbar^2}(E - V)\,\psi = 0 \quad ...(ii)$$

where m is the reduced mass of the electron, E its total energy, V the potential energy and $\hbar = \frac{h}{2\pi}$.

The value of ∇^2 in cartesian co-ordinates is

$$\nabla^2 = \frac{\partial^2}{\partial x^2} + \frac{\partial^2}{\partial y^2} + \frac{\partial^2}{\partial z^2}$$

and in spherical polar co-ordinates is

$$\nabla^2 = \frac{\partial^2}{\partial r^2} + \frac{2}{r}\frac{\partial}{\partial r} + \frac{1}{r^2}\frac{\partial^2}{\partial \theta^2} + \frac{\cot\theta}{r^2}\frac{\partial}{\partial \theta} + \frac{1}{r^2 \sin^2\theta}\frac{\partial^2}{\partial \phi^2}$$

$$= \frac{1}{r^2}\frac{\partial}{\partial r}\left(r^2 \frac{\partial}{\partial r}\right) + \frac{1}{r^2 \sin\theta}\frac{\partial}{\partial \theta}\left(\sin\theta \frac{\partial}{\partial \theta}\right) + \frac{1}{r^2 \sin^2\theta}\frac{\partial^2}{\partial \phi^2} \quad ...(ii)a$$

Substituting the value of V from (*i*), relation (*ii*) can be put in spherical polar co-ordinates as

$$\frac{1}{r^2}\frac{\partial}{\partial r}\left(r^2 \frac{\partial \psi}{\partial r}\right) + \frac{1}{r^2 \sin\theta}\frac{\partial}{\partial \theta}\left(\sin\theta \frac{\partial \psi}{\partial \theta}\right) + \frac{1}{r^2 \sin^2\theta}\frac{\partial^2 \psi}{\partial \phi^2} + \frac{2m}{\hbar^2}\left(E + kZ\frac{e^2}{r}\right)\psi = 0 \quad ...(iii)$$

The three co-ordinates r, θ and ϕ are orthogonal to each other. It is, therefore, possible to express the wave function ψ, which is a function of all the three co-ordiantes r, θ and ϕ, as the product of three separate functions, each of only one independent variable r, θ or ϕ. Thus

$$\psi(r, \theta, \phi) = R(r)\ \Theta(\theta)\ \Phi(\phi)$$

and

$$\frac{\partial \psi}{\partial r} = \Theta\,\Phi\,\frac{dR}{dr}$$

$$\frac{\partial \psi}{\partial \theta} = R\,\Phi\,\frac{d\Theta}{d\theta}$$

$$\frac{\partial \psi}{\partial \phi} = R\,\Theta\,\frac{d\Phi}{d\phi}$$

$$\frac{\partial^2 \psi}{\partial \phi^2} = R\,\Theta\,\frac{d^2\Phi}{d\phi^2}$$

Substituting in (*iii*), dividing by $R\ \Theta\ \Phi$, multiplying by $r^2 \sin^2\theta$ and re-arranging, we get

$$\frac{\sin^2\theta}{R}\frac{d}{dr}\left(r^2 \frac{dR}{dr}\right) + \frac{\sin\theta}{\Theta}\frac{d}{d\theta}\left(\sin\theta \frac{d\Theta}{d\theta}\right)$$

$$+ \frac{2mr^2 \sin^2\theta}{\hbar^2}\left[E + \frac{kZe^2}{r}\right] = -\frac{1}{\Phi}\frac{d^2\Phi}{d\phi^2} \quad ...(iv)$$

The left hand side of equation (*iv*) is a function of θ and r, whereas the right hand side is a function of ϕ only *i.e.*, the two sides of the equation are indepedent of each other. Therefore, each side must be equal to a constant. Let this constant be m_l^2.

$$\therefore \qquad -\frac{1}{\Phi}\frac{d^2\Phi}{d\phi^2} = m_l^2$$

or

$$\frac{d^2\Phi}{d\phi^2} + m_l^2\,\Phi = 0 \quad ...(v)$$

This is the independent equation for the variable ϕ. Equation (*iv*) can now be put as

$$\frac{\sin^2\theta}{R}\frac{d}{dr}\left(r^2\frac{dR}{dr}\right)+\frac{\sin\theta}{\Theta}\frac{d}{d\theta}\left(\sin\theta\frac{d\Theta}{d\theta}\right)+\frac{2mr^2\sin^2\theta}{\hbar^2}\left[E+\frac{kZe^2}{r}\right]=m_l^2$$

Dividing both sides by $\sin^2\theta$ and separating the variables r and θ, we have

$$\frac{1}{R}\frac{d}{dr}\left(r^2\frac{dR}{dr}\right)+\frac{2mr^2}{\hbar^2}\left[E+\frac{kZe^2}{r}\right]$$

$$=\frac{m_l^2}{\sin^2\theta}-\frac{1}{\Theta\sin\theta}\frac{d}{d\theta}\left(\sin\theta\frac{d\Theta}{d\theta}\right)$$

Here again, the left hand side of the equation is a function of r only, whereas the right hand side is a function of θ only. Therefore, each side must be equal to a constant. This constant is taken as $l\,(l+1)$, where l is greater than $|m_l|$ *i.e.*, the magnitude of m_l. This value of the constant is taken so that we get only finite values of the function Θ for all values of θ.

Therefore, we have two more equations, one for variable θ and the other for variable r given by

$$\frac{m_l^2}{\sin^2\theta}-\frac{1}{\Theta\sin\theta}\frac{d}{d\theta}\left(\sin\theta\frac{d\Theta}{d\theta}\right)=l\,(l+1)$$

or
$$\frac{m_l^2}{\sin^2\theta}-\frac{1}{\Theta}\left[\frac{d^2\Theta}{d\theta^2}+\cot\theta\frac{d\Theta}{d\theta}\right]=l\,(l+1) \qquad ...(vi)$$

and
$$\frac{1}{R}\frac{d}{dr}\left(r^2\frac{dR}{dr}\right)+\frac{2mr^2}{\hbar^2}\left[E+\frac{kZe^2}{r}\right]=l\,(l+1)$$

or
$$\frac{d^2R}{dr^2}+\frac{2}{r}\frac{dR}{dr}+\frac{2m}{\hbar^2}\left[E+\frac{kZe^2}{r}\right]R=\frac{l\,(l+1)}{r^2}R \qquad ...(7.3)$$

where $\frac{kZe^2}{r}=\frac{e^2}{4\pi\varepsilon_0 r}$ for hydrogen. $\left[\because k=\frac{1}{4\pi\varepsilon_0} \text{ and } Z=1\right]$

Thus we get three independent equations, one a function of r, the second a function of θ and the third a function of ϕ.

These can be solved to obtain R, Θ and Φ. Once these are known we can get

$$\psi=R\,(r)\;\Theta\;(\theta)\;\Phi\;(\phi) \qquad ...(7.4)$$

It means to solve Schrödinger's equation for hydrogen, we must find a solution of above three equations (*v*), (*vi*) and (7.3).

7.3 SOLUTION OF Θ (θ) FOR HYDROGEN ATOM

The equation is

$$\frac{m_l^2}{\sin^2\theta}-\frac{1}{\Theta\sin\theta}\frac{d}{d\theta}\left(\sin\theta\frac{d\Theta}{d\theta}\right)=l\,(l+1)$$

or
$$\frac{1}{\sin\theta}\frac{d}{d\theta}\left(\sin\theta\frac{d\Theta}{d\theta}\right)+\left[l\,(l+1)-\frac{m_l^2}{\sin^2\theta}\right]\Theta=0 \qquad ...(i)$$

To solve equation (*i*) put $x=\cos\theta$ ...(*ii*)

$$\therefore \qquad \frac{dx}{d\theta}=-\sin\theta$$

Now $$\frac{d\Theta}{d\theta} = \frac{d\Theta}{dx} \cdot \frac{dx}{d\theta} = -\sin\theta \frac{d\Theta}{dx}$$

Multiplying both sides by sin θ , we have –

$$\sin\theta \frac{d\Theta}{d\theta} = -\sin^2\theta \frac{d\Theta}{dx} \qquad ...(iv)$$

$$\therefore \quad \frac{d}{d\theta}\left(\sin\theta \frac{d\Theta}{d\theta}\right) = \frac{d}{d\theta}\left(-\sin^2\theta \frac{d\Theta}{dx}\right)$$

or $$\frac{d}{d\theta}\left(\sin\theta \frac{d\Theta}{d\theta}\right) = \frac{d}{dx}\left[(\cos^2\theta - 1)\frac{d\Theta}{dx}\right]\frac{dx}{d\theta}$$

Substituting $\frac{dx}{d\theta} = -\sin\theta$ and $\cos\theta = x$, we have

$$\frac{d}{d\theta}\left(\sin\theta \frac{d\Theta}{d\theta}\right) = -\sin\theta \frac{d}{dx}\left[(x^2 - 1)\frac{d\Theta}{dx}\right] \qquad ...(v)$$

Substituting this value in equation (*i*), we get

$$\frac{d}{dx}\left[(1 - x^2)\frac{d\Theta}{dx}\right] + \left[l(l+1) - \frac{m_l^2}{\sin^2\theta}\right]\Theta = 0$$

or $$\frac{d}{dx}\left[(1 - x^2)\frac{d\Theta}{dx}\right] + \left[l(l+1) - \frac{m_l^2}{(1 - x^2)}\right]\Theta = 0 \qquad (7.5)$$

This equation is known as *Legendre equation*. The solutions of this equation are known as *associated Legendre polynomials* of degree l which can be written as

$$\Theta_{lml} = C_{lml} P_{lml} (\cos\theta)$$

where $l = |m_l|, |m_l| + 1, |m_l| + 2, ...$

It is evident from equation (*vi*) that the solution of this equation will depend upon two quantum numbers l and m_l. These solutions are finite and well behaved for all values of Θ (*i.e.*, for all values of x between – 1 and + 1) only if the constant l is an integer and $l \geq |m_l|$. This constant l is called orbital quantum number.

The constant $C_{l m_l}$ can be calculated by normalising the function Θ_{l, m_l}

A few normalised Θ_{l, m_l} functions are as under.

l	m_l	Θ_{l, m_l}
0	0	$\Theta_{00} = \frac{1}{\sqrt{2}}$
1	0	$\Theta_{10} = \sqrt{\frac{3}{2}} \cos\theta$
1	+1	$\Theta_{1+1} = \frac{\sqrt{3}}{2} \sin\theta$
1	–1	$\Theta_{1-1} = \frac{\sqrt{3}}{2} \sin\theta$

7.4 AZIMUTHAL COMPONENT

The (azimuthal) component wave function Φ for the variable (ϕ) is given by the equation

$$\frac{d^2\Phi}{d\phi^2} + m_l^2\ \Phi = 0 \qquad ...(i)$$

The solution of equation (*i*) is $\Phi = Ae^{i\,m_l\,\phi}$
where A is an arbitrary constant. For the wave function ψ (r, θ, ϕ) to be single valued over the available region, the function $\Phi(\phi)$ should also be single valued over the ϕ space. Hence when the value for angle ϕ is increased by 2 π (radian) the function Φ should have the same value.

$$\therefore \quad Ae^{im_l\phi} = Ae^{im_l(\phi+2\pi)} = Ae^{i(m_l\phi+m_l 2\pi)}$$

or $$Ae^{i\,m_l\phi} = Ae^{i\,m_l\phi} \times e^{i\,m_l 2\pi}$$

$$\therefore \quad e^{im_l 2\pi} = 1$$

Now $$e^{im_l 2\pi} = \cos 2\,\pi\, m_l + i \sin 2\pi\, m_l = 1$$

Equating real and imaginary parts, we have

$$\cos 2\,\pi\, m_l = 1$$

This is possible only if $m_l = 0, \pm 1, \pm 2, ...$ etc.

Thus ϕ space is quantised. The constant m_l is called magnetic quantum number of electron.

Normalised eigen functions. To find the value of A, normalising the function Φ over ϕ space from 0 to 2 π , we have

$$\int_0^{2\pi} \Phi^*_{m_l}\ \Phi_{m_l} = 1 \text{ or} \int_0^{2\pi} Ae^{\ im_l\phi}\ Ae^{\ -im_l\phi}\ d\phi = 1$$

or $$A^2\,[\phi]_0^{2\pi} = 1 \therefore A = \frac{1}{\sqrt{2\pi}}$$

Hence normalized eigen functions for Φ are

$$\Phi_{m_l} = \frac{1}{\sqrt{2\pi}} e^{im_l\phi} \text{ where } m_l = 0, \pm 1, \pm 2... \qquad ...(7.6)$$

The zero value of m_l corresponds to no rotation, positive integral values to rotation in anticlockwise direction and negative integral values to rotation in clockwise direction.

Significance of the magnetic quantum number m_l. The significance of the magnetic quantum number m_l predicted by the component wave function Φ lies in the fact that $m_l\,\hbar$ gives the Z-component of the angular momentum vector.

The operator for the Z-component of the angular momentum, in spherical polar co-ordinates is given by

$$\hat{L}_z = -\ i\hbar \frac{\partial}{\partial\phi}$$

If the operator $\hat{L}_z$ operates upon the function ψ , we have

$$L_z\ \psi = -\ i\hbar \frac{\partial\psi}{\partial\phi} = -\ i\hbar \frac{\partial}{\partial\phi}(R\,\Theta\,\Phi)$$

$$= -\ i\hbar\ R\,\Theta \frac{\partial\Phi}{\partial\phi} = -\ i\hbar\ R\,\Theta(i\,m_l\ \Phi)$$

as $$\Phi = Ae^{im_l\phi} \text{ and } \frac{\partial\Phi}{\partial\phi} = i\,m_l\ Ae^{\ im_l\phi} = i\,m_l\ \Phi$$

$\therefore \quad L_z \psi = -i\hbar\,(i\,m_l)\,R\Theta\Phi = \hbar\,m_l\,\psi$

$\therefore$ Z-component of the angular momentum

$$L_z = m_l \hbar$$

The value of $|m_l| \le l$ and $m_l = 0, \pm 1 \pm 2$ etc. ...(7.7)

For given value of orbital quantum number l, the total angular momentum $L = \sqrt{l\,(l+1)}\,\hbar$, but its component along the Z-axis

$$L_z = m_l \hbar$$

$$\therefore \quad L_z = 0, \pm\hbar, \pm 2\hbar \text{ ...etc.} \qquad \text{...(7.8)}$$

Thus, we find that though the total angular momentum is not an integral multiple of $\hbar$, the Z-component is an integral multiple of $\hbar$. For a given value of l, there are $(2l + 1)$ different values of m_l. Hence there are only $(2l+1)$ values of L_z. The Z-component is, therefore, quantised and is said to have spatial quantisation.

Values of quantum number *m*. The quantum number m or m_l is known as *'magnetic quantum number*. Its possible values are 0, ±1, ±2...... *The zero value of m_l (or m) corresponds to no rotation, positive integral values to rotation in the anti-clockwise direction and negative integral values to rotation in clock wise direction.*

7.5 ORBITAL QUANTUM NUMBER

The angular momentum operator in spherical polar co-ordinates is given by

$$\hat{L}^2 = -\hbar^2 \left[\frac{\partial^2}{\partial\theta^2} + \cot\theta\frac{\partial}{\partial\theta} + \frac{1}{\sin^2\theta}\frac{\partial^2}{\partial\phi^2}\right]$$

In a hydrogen atom, the wave function $\psi\,(r, \theta, \phi)$ can be put as a product of three independent functions, one of only one variable.

$$\therefore \quad \psi\,(r, \theta, \phi) = R\,(r)\,\Theta\,(\theta)\,\Phi\,(\phi)$$

If, therefore, the angular momentum operator $\hat{L}^2$ operates upon the function ψ, we have

$$\hat{L}^2\,\psi = -\hbar^2\left[\frac{\partial^2}{\partial\theta^2} + \cot\theta\frac{\partial}{\partial\theta} + \frac{1}{\sin^2\theta}\frac{\partial^2}{\partial\phi^2}\right] R\,\Theta\,\Phi$$

$$= -\hbar^2\left[\frac{\partial^2\Theta}{\partial\theta^2} + \cot\theta\frac{\partial\Theta}{\partial\theta}\right] R\,\Phi + \left[\frac{1}{\sin^2\theta}\frac{\partial^2\Phi}{\partial\phi^2}\right] R\,\Theta$$

Again in a hydrogen atom

$$\frac{\partial^2\Phi}{\partial\phi^2} = -m_l^2\,\Phi$$

$$\therefore \quad \hat{L}^2\,R\,\Theta\,\Phi = -\hbar^2\left[\frac{\partial^2\Theta}{\partial\theta^2} + \cot\theta\frac{\partial\Theta}{\partial\theta} - \frac{m_l^2}{\sin^2\theta}\Theta\right] R\,\Phi \qquad \text{...}(i)$$

But

$$\frac{\partial^2\Theta}{\partial\theta^2} + \cot\theta\frac{\partial\Theta}{\partial\theta} - \frac{m_l^2}{\sin^2\theta}\Theta = -[l\,(l+1)\Theta]$$

Substituting in equation (*i*), we get

$$\hat{L}^2 R\ \Theta\Phi = \hbar^2 l(l+1) R\,\Theta\Phi$$

or $$\hat{L}^2 \psi = \hbar^2 l(l+1)\psi \qquad ...(7.9)$$

Orbital quantum number. Equation (7.9) shows that the wave function ψ is an eigen function of $\hat{L}^2$ with eigen values $l(l+1)\hbar^2$. The eigen values of the operator $\hat{L}^2$ measure the experimental values of square of the orbital angular momentum.

Thus the magnitude of the angular momentum vector is given by

$$L^2 = \hbar^2 l(l+1)$$

or $$|\vec{L}| = \hbar\sqrt{l(l+1)} \qquad ...(7.10)$$

For $l = 0$, angular momentum $L = 0$ for $l = 1$, $L = \sqrt{2}\ \hbar$ for $l = 2$, $L = \sqrt{6}\ \hbar$ and so on. The integral l is, therefore, called orbital quantum number. It can have values 0, 1, 2, 3 etc., and the corresponding values of the angular momentum of the electron are 0, $\sqrt{2}\ \hbar$, $\sqrt{6}\,\hbar$ and so on.

This means that the magnitude of orbital angular momentum $\vec{L}$ of the electron is quantised. The unit of angular momentum $\hbar = h/2\pi$.

7.6 SPHERICAL HARMONICS

Schrödinger's wave equation for Hydrogen atom in spherical polar co-ordinates can be resolved into three independent equations, one $R(r)$ a function of r, the second $\Theta(\theta)$ a function of θ and the third $\Phi(\phi)$ a function of ϕ. The solution of $\Phi(\phi)$ equation is

$$\Phi_{m_l} = A\, e^{i m_l \phi} \quad \text{or} \quad \Phi_m = A_m\, e^{im\phi}$$

where m_l or $m = 0, \pm 1, \pm 2$, etc. The solution of $\Theta(\theta)$ equation is

$$\Theta_{l\, m_l} = C_{l\, m_l}\, P_{l\, m_l}(\cos\theta) \quad \text{or} \quad \Theta_{lm} = C_{lm}\, P_l^m(\cos\theta)$$

The product of function Φ_m *and function* Θ_{lm} *is known as spherical harmonic function.* It is denoted by $Y_l^m(\theta, \phi)$ and depends upon two quantum numbers m and l

$$\therefore\quad Y_l^m(\theta,\phi) = \Theta_{lm}\,\Phi_m = C_{lm}\,P_l^m(\cos\theta)\,A_m\,e^{im\phi}$$

$$= B_{lm}\,P_l^m(\cos\theta)\,e^{im\phi} \quad \text{where } B_{lm} = A_m\,C_{lm}.$$

Normalisation of the above function gives the value of B_{lm} as

$$B_{lm} = \sqrt{\frac{(2l+1)}{4\pi}\,\frac{(l-m)!}{(l+m)!}} \qquad ...(7.11)$$

and normalised spherical harmonic is given by

$$Y_l^m(\theta,\phi) = \sqrt{\frac{(2l+1)}{4\pi}\,\frac{(l-m)!}{(l+m)!}}\,P_l^m(\cos\theta)\,e^{im\phi} \qquad ...(7.12)$$

The first few spherical harmonics are

l	m	$Y_l^m(\theta, \phi)$
0	0	$Y_0^0 = \frac{1}{\sqrt{4\pi}}$
1	0	$Y_1^0 = \sqrt{\frac{3}{4\pi}} \cos\theta$
1	1	$Y_1^1 = -\sqrt{\frac{3}{8\pi}} \sin\theta\, e^{i\phi}$
1	–1	$Y_1^{-1} = \sqrt{\frac{3}{8\pi}} \sin\theta\, e^{-i\phi}$
2	0	$Y_2^0 = \sqrt{\frac{5}{16\pi}} (3\cos^2\theta - 1)$

For solution of Schrödinger's wave equation (azimuthal component) for H-atom (See article 7.4) For solution of Schrödinger's wave equation $\Theta(\theta)$ for H-atom (See article 7.3)

It has been proved in article 7.5 that the wave function $\psi(r, \theta, \phi)$ for hydrogen atom is an eigen function of operator $\hat{L}_z$ with eigen values $l(l+1)h^2$ where l is the orbital quantum number.

We can prove that wave function ψ_{nlm} or $\psi(r, \theta, \phi)$ for hydrogen atom is an eigen function of z-component of angular momentum operator $\hat{L}_z$ with eigen values $m_l h$ where m_l (or m) is the magnetic quantum number as under:

To prove $\hat{L}_z\, \psi_{nlm} = (\hbar\, m_l)\, \psi_{nlm}$

In spherical polar co-ordinates $\hat{L}_z = -i\hbar \frac{\partial}{\partial\phi}$ and $\psi_{nlm} = R_{nl}\, \Theta_{lm_l}\, \Phi_{m_l}$. Also $\Phi_{m_l} = \frac{1}{\sqrt{2\pi}} e^{im_l\phi}$

$$\therefore \quad \hat{L}_z\, \psi_{nlm} = -i\hbar \frac{\partial}{\partial\phi} \psi_{nlm}$$

$$= -i\hbar \frac{\partial}{\partial\phi} \left[R_{nl}\ {}_{lm_l}\ \Phi_{m_l} \right]$$

$$= -i\hbar\, R_{nl} \Theta_{lm_l} \frac{\partial}{\partial\phi} \Phi_{m_l}$$

using $\qquad \Phi_{m_l} = \frac{1}{\sqrt{2\pi}} e^{im_l\phi}$ we have

$$\frac{\partial}{\partial\phi} \Phi_{m_l} = i\, m_l \frac{1}{\sqrt{2\pi}} e^{im_l\phi} = i\, m_l\, \Phi_{m_l}$$

Hence $\qquad \hat{L}_z\, \psi_{nlm} = (-i\,\hbar)\,(im_l)\, R_{nl}\, \Theta_{lm_l}\, \Phi_{m_l}$

$$= \hbar m_l\, \psi_{nlm} \qquad ..(7.13)$$

It means ψ_{nlm} is an *eigen function of* $\hat{L}_z$ *with eigen value* $m_l \hbar$

Since operator $\hat{L}_z$ gives the Z-component of orbital angular momentum $\vec{L}$ of the electron

$$L_z = m_l \hbar$$

Thus magnetic quantum number m_l is a measure of the Z-component of angular momentum.

7.7 PRINCIPAL QUANTUM NUMBER (n) AND RADIAL WAVE FUNCTION $R(r)$

The $R(r)$ equation or the radial equation for hydrogen (Z = 1) is given by Eq. (7.3) as

$$\frac{d^2R}{dr^2} + \frac{2}{r}\frac{dR}{dr} + \frac{2m}{\hbar^2}\left[E + \frac{e^2}{4\pi\varepsilon_0 r}\right]R = \frac{l(l+1)}{r^2}R$$

or
$$\frac{d^2R}{dr^2} + \frac{2}{r}\frac{dR}{dr} + \left[\frac{2mE}{\hbar^2} + \frac{2me^2}{4\pi\varepsilon_0 \hbar^2 r} - \frac{l(l+1)}{r^2}\right]R = 0 \qquad ...(i)$$

Putting $\frac{-2mE}{\hbar^2} = \beta^2$ in (i) we get

[**Note.** Negative sign with energy E indicates that the electron is bound to the nucleus]

$$\frac{d^2R}{dr^2} + \frac{2}{r}\frac{dR}{dr} + \left[-\beta^2 + \frac{2me^2}{4\pi\varepsilon_0 \hbar^2 r} - \frac{l(l+1)}{r^2}\right]R = 0$$

or
$$\frac{1}{r^2}\frac{d}{dr}\left(r^2\frac{dR}{dr}\right) + \left[-\beta^2 + \frac{2me^2}{4\pi\varepsilon_0 \hbar^2 r} - \frac{l(l+1)}{r^2}\right]R = 0 \qquad ...(ii)$$

Now put $\rho = 2\beta r$ $\qquad ...(iii)$

where ρ is a dimensionless variable.

$$\therefore \qquad \frac{d\rho}{dr} = 2\beta$$

Also
$$\frac{dR}{dr} = \frac{dR}{d\rho}\cdot\frac{d\rho}{dr} = 2\beta\frac{dR}{d\rho}$$

or
$$r^2\frac{dR}{dr} = 2\beta r^2\frac{dR}{d\rho} = 2\beta\frac{\rho^2}{4\beta^2}\frac{dR}{d\rho} = \frac{\rho^2}{2\beta}\frac{dR}{d\rho}$$

Differentiating both sides of this equation w. r. t. r, we get

$$\frac{d}{dr}\left(r^2\frac{dR}{dr}\right) = \frac{d}{dr}\left(\frac{\rho^2}{2\beta}\frac{dR}{d\rho}\right)$$

$$= \frac{d}{d\rho}\left(\frac{\rho^2}{2\beta}\frac{dR}{d\rho}\right)\frac{d\rho}{dr}$$

$$= 2\beta\frac{d}{d\rho}\left(\frac{\rho^2}{2\beta}\frac{dR}{d\rho}\right)$$

$$= \frac{d}{d\rho}\left(\rho^2\frac{dR}{d\rho}\right) \qquad ...(iv)$$

Substituting the value $r = \rho/2\beta$ from (iii) and $\frac{d}{dr}\left(r^2\frac{dR}{dr}\right)$ form (iv) in equation (ii), we get

$$\frac{4\beta^2}{\rho^2}\frac{d}{d\rho}\left(\rho^2\frac{dR}{d\rho}\right)+\left[-\beta^2+\frac{me^2\beta}{\pi\varepsilon_0\hbar^2\rho}-\frac{4\,l(l+1)\,\beta^2}{\rho^2}\right]R=0$$

or
$$\frac{1}{\rho^2}\frac{d}{d\rho}\left(\rho^2\frac{dR}{d\rho}\right)+\left[-\frac{1}{4}+\frac{me^2}{4\pi\varepsilon_0\hbar^2\rho\beta}-\frac{l\,(l+1)}{\rho^2}\right]R=0$$

or
$$\frac{d^2R}{d\rho^2}+\frac{2}{\rho}\frac{dR}{d\rho}+\left[-\frac{1}{4}+\frac{me^2}{4\pi\varepsilon_0\hbar^2\rho\beta}-\frac{l\,(l+1)}{\rho^2}\right]R=0$$

or
$$\frac{d^2R}{d\rho^2}+\frac{2}{\rho}\frac{dR}{d\rho}+\left[\frac{n}{\rho}-\frac{l\,(l+1)}{\rho^2}-\frac{1}{4}\right]R=0 \qquad ...(v)$$

where
$$n=\frac{me^2}{4\pi\varepsilon_0\hbar^2\beta} \qquad ...(v)\,(a)$$

Asymptotic solution. To find the asymptotic solution of Eq. (*v*) when $\rho\to\infty$, we can neglect terms containing $\frac{1}{\rho}$ and $\frac{1}{\rho^2}$ for large values of ρ and get

$$\frac{d^2R}{d\rho^2}-\frac{R}{4}=0 \qquad ...(vi)$$

Two solutions of Eq. (*vi*) are

$$R=Ae^{+\rho/2} \text{ and } R=Ae^{-\rho/2}$$

Since ρ varies from zero to infinity, the solution $R=Ae^{+\rho/2}$ goes on increasing as ρ increases. Therefore $Ae^{+\rho/2}$ is not an acceptable solution. Thus $R=Ae^{-\rho/2}$ is the acceptable solution because $Ae^{-\rho/2}$ decreases as ρ (and hence *r* the distance of the electron from the nucleus) increases to infinity. With this asymptotic solution, the possible solution of Eq.(*v*) is

$$R=Ae^{-\rho/2}F(\rho) \qquad ...(vii)$$

where $F(\rho)$ is a function of ρ yet to be determined. Differentiating Eq. (*vii*) with respect of ρ, we get

$$\frac{dR}{d\rho}=-\frac{1}{2}Ae^{-\rho/2}F+Ae^{-\rho/2}\frac{dF}{d\rho}$$

and
$$\frac{d^2R}{d\rho^2}=\frac{A}{4}e^{-\rho/2}F-\frac{A}{2}e^{-\rho/2}\frac{dF}{d\rho}-\frac{A}{2}e^{-\rho/2}\frac{dF}{d\rho}+Ae^{-\rho/2}\frac{d^2F}{d\rho^2}$$

or
$$\frac{d^2R}{d\rho^2}=Ae^{-\rho/2}\frac{d^2F}{d\rho^2}-Ae^{-\rho/2}\frac{dF}{d\rho}+\frac{A}{4}e^{-\rho/2}F$$

Substituting the values of $\frac{dR}{d\rho}$ and $\frac{d^2R}{d\rho^2}$ in Eq. (*v*), we get

$$Ae^{-\rho/2}\frac{d^2F}{d\rho^2}-Ae^{-\rho/2}\frac{dF}{d\rho}+\frac{A}{4}e^{-\rho/2}F-\frac{A}{\rho}e^{-\rho/2}F+\frac{2A}{\rho}e^{-\rho/2}\frac{dF}{d\rho}$$
$$+\left(\frac{n}{\rho}-\frac{l\,(l+1)}{\rho^2}-\frac{1}{4}\right)AFe^{-\rho/2}=0$$

Dividing throughout by $Ae^{-\rho/2}$ and simplifying we get

$$\frac{d^2F}{d\rho^2}+\left(\frac{2}{\rho}-1\right)\frac{dF}{d\rho}+\left[\frac{n-1}{\rho}-\frac{l(l+1)}{\rho^2}\right]F=0 \qquad ...(viii)$$

In order that the solution of Eq. (*viii*) *remains finite* for all values of ρ, the condition is

$$n-1=l,\ l+1,\ l+2,\ l+3\ ....\text{etc.}$$

or

$$n=l+1,\ l+2,\ l+3,\ ....\text{etc.} \qquad ...(viii\ a)$$

n is called the principal quantum number. The condition in n is expressed as a condition in l as follows

$$l=0,1,2...(n-1)$$

Energy levels. The value of $n=\dfrac{me^2}{4\pi\varepsilon_0\hbar^2\beta}$ as given in equation (*v*) (*a*)

$$\therefore \qquad \beta=\frac{me^2}{4\pi\varepsilon_0^2\hbar^2}\cdot\frac{1}{n}$$

or

$$\beta^2=\frac{m^2e^4}{16\pi^2\varepsilon_0\hbar^4}\cdot\frac{1}{n^2} \qquad ...(ix)$$

But

$$\beta^2=\frac{-2mE}{\hbar^2}$$

$$\therefore \qquad \frac{-2mE}{\hbar^2}=\frac{-m^2e^4}{16\pi^2\varepsilon_0^2\hbar^4}\cdot\frac{1}{n^2}$$

or

$$E=\frac{-me^4}{32\pi^2\varepsilon_0^2\hbar^2}\cdot\frac{1}{n^2} \qquad ...(x)$$

where n has the values given in Eq. (*viii a*).

This shows that the energy of the electron in a hydrogen atom depends upon principal quantum number n. We, therefore write

$$E_n=-\frac{me^4}{32\pi^2\varepsilon_0^2\hbar^2}\cdot\frac{1}{n^2} \qquad ...(xi)$$

This equation gives the energy levels of the hydrogen atom which are indentical with those predicted by Bohr.

Radial wave function. The radial wave function of the hydrogen atom is given by Eq. (*vii*)

$$R=Ae^{-\rho/2}F(\rho) \qquad ...(xii)$$

It is also clear from equation (*viii*) that the solution of this equation will depend upon two quantum numbers n and l. Eq. (*xii*) can, therefore, be written as

$$R_{nl}(\rho)=A_{nl}e^{-\rho/2}F_{nl}(\rho)$$

Substituting the value of $\rho=2\beta r$ from Eq. (*iii*), we get

$$R_{nl}(r)=A_{nl}e^{-\beta r}F_{nl}(2\beta r) \qquad (xiii)$$

Now

$$\beta=-\frac{me^2}{4\pi\varepsilon_0\hbar^2}\cdot\frac{1}{n} \qquad \text{[according to Eq. (ix)]}$$

$$=\frac{1}{a_0 n} \text{ where } a_0=\frac{4\pi\varepsilon_0\hbar^2}{me^2}$$

is the radius of the first Bohr orbit. Substituting the value of $\beta=\dfrac{1}{a_0 n}$ in Eq. (*xiii*), we have

$$R_{nl} = A_{nl}\, e^{-r/a_0 n}\, F_{nl}\left(\frac{2r}{a_0\, n}\right) \qquad ...(xiv)$$

It can be shown that the polynomial $F_{nl}\left(\frac{2r}{a_0\, n}\right)$ is the product of $\left(\frac{2r}{a_0\, n}\right)^l$ and the *associated Laguerre polynominal* $\overset{2l+1}{\underset{n+1}{L}}$. Thus $F_{nl}\left(\frac{2r}{a_0\, n}\right) = \left(\frac{2r}{a_0\, n}\right)^l \overset{2l+1}{\underset{n+1}{L}}\left(\frac{2r}{n\, a_0}\right)$

Thus the acceptable radial wave functions are

$$R_{nl} = A_{nl}\, e^{-r/na_0}\left(\frac{2r}{n\, a_0}\right)^l \overset{2l+1}{\underset{n+1}{L}}\left(\frac{2r}{n\, a_0}\right) \qquad ...(xv)$$

It is seen that two quantum numbers n and l are required to identifiy different radial wave functions. A few normalised radial wave functions are given below.

n	l	R_{nl}
1	0	$R_{10} = \frac{2}{a_0^{3/2}}\, e^{-r/a_0}$
2	0	$R_{20} = \frac{1}{(2\, a_0)^{3/2}}\left(2 - \frac{r}{a_0}\right) e^{-r/2a_0}$
2	1	$R_{21} = \frac{1}{(2\, a_0)^{3/2}} \frac{1}{\sqrt{3}} \frac{r}{a_0} e^{-r/2a_0}$

7.8 COMPLETE WAVE FUNCTION FOR THE HYDROGEN ATOM

The values of component wave functions are

$$\Phi_{m_l}(\phi) = \frac{1}{\sqrt{2\pi}}\, e^{im_l\phi}$$

$$\Phi_{lml}(\theta) = C_{lml}\, P_{lml}(\cos\theta)$$

and

$$R_{nl}(r) = A_{nl}\, e^{-r/na_0}\left(\frac{2r}{na_0}\right)^l \overset{2l+1}{\underset{n+1}{L}}\left(\frac{2r}{na_0}\right)$$

Substituting the values R_{nl}, Θ_{lml} and Φ_{ml}, the complete wave function for the hydrogen atom becomes

$$\psi_{nlm_l}(r,\theta,\phi) = A_{nl}\, e^{-r/na_0}\left(\frac{2r}{na_0}\right)^l \overset{2l+1}{\underset{n+1}{L}}\left(\frac{2r}{na_0}\right) C_{lm_l}\, P_{lm_l}(\cos\theta)\, \frac{1}{\sqrt{2\pi}}\, e^{im_l\phi}$$

or

$$\psi_{nlm_l} = N_{nlm_l}\, e^{-r/na_0}\left(\frac{2r}{na_0}\right)^l \overset{2l+1}{\underset{n+1}{L}}\left(\frac{2r}{na_0}\right) P_{lm_l}(\cos\theta)\, e^{im_l\phi} \qquad ...(7.14)$$

where $N_{nlm_l} = A_{nl}\, C_{lm_l}\, \frac{1}{\sqrt{2\pi}}$ is called the normalisation constant. To find the value of N_{nlml} apply the normalisation condition

$$\int_{\text{all space}} \psi_{nlm_l}\, \psi^*_{nlm_l}\, dV = 1$$

Thus the complete wave function for the hydrogen atom is specified by three quantum numbers n, l and m_l, n, the principal quantum number denotes the *shell* or *orbit* in which the electron is moving, l the orbital quantum number denotes the *subshell*, the wave function of the subshell being called the *orbital. Although the complete wave function* ψ *depends upon all the three quantum numbers, the energy of the hydrogen atom only depends upon the quantum number n.*

(*i*) **Ground state wave function.** For the ground state $n = 1$. Therefore $l = 0$ and $m_l = 0$. Substituting these values in Eq. (*xvi*), we have

$$\psi_{100} = R_{10}\, \Theta_{00}\, \Phi_0 = \frac{1}{\sqrt{2\pi}} \cdot \frac{1}{\sqrt{2}} \frac{2}{a_0^{3/2}} e^{-r/a_0}$$

$$= \frac{1}{\sqrt{\pi}\, a_0^{3/2}} e^{-r/a_0} \qquad \text{...(7.15)}$$

(*ii*) **First excited state wave function.** For the first excited state $n = 2$. Therefore, the two possible values of l are, $l = 1$ and $l = 0$. For $l = 1$, $m_l = +1, -1$ and for $l = 0$; $m_l = 0$. Substituting these values in Eq. (*xvi*) we have

$$\psi_{200} = R_{20}\, \Theta_{00}\, \Phi_0 = \frac{1}{\sqrt{2\pi}} \cdot \frac{1}{\sqrt{2}} \frac{1}{(2a_0)^{3/2}} \left(2 - \frac{r}{a_0}\right) e^{-r/2a_0}$$

$$= \frac{(2 - r/a_0)}{4\sqrt{2\pi}\, a_0^{3/2}} e^{-r/2a_0}$$

$$\psi_{210} = R_{21}\, \Theta_{10}\, \Phi_0 = \frac{1}{\sqrt{2\pi}} \sqrt{\frac{3}{2}} \cos\theta \frac{1}{(2a_0)^{3/2}} \frac{1}{\sqrt{3}} \left[\frac{r}{a_0}\right] e^{-r/2a_0}$$

$$= \frac{1}{4\sqrt{2\pi}\, a_0^{3/2}} \frac{r}{a_0} e^{-r/2a_0} \cos\theta$$

$$\psi_{211} = R_{21}\, \Theta_{11}\, \Phi_1 = \frac{1}{\sqrt{2\pi}} e^{i\phi} \frac{\sqrt{3}}{2} \sin\theta \frac{1}{(2a_0)^{3/2}} \frac{1}{\sqrt{3}} \left[\frac{r}{a_0}\right] e^{-r/2a_0}$$

$$= \frac{1}{8\sqrt{\pi}} \frac{1}{a_0^{3/2}} \left(\frac{r}{a_0}\right) e^{-r/2a_0} \sin\theta\, e^{+i\phi}$$

$$\psi_{21-1} = R_{21}\, \Theta_{1-1}\, \Phi_{-1} = \frac{1}{\sqrt{2\pi}} e^{-i\phi} \frac{\sqrt{3}}{2} \sin\theta \frac{1}{(2a_0)^{3/2}} \frac{1}{\sqrt{3}} \left[\frac{r}{a_0}\right] e^{-r/2a_0}$$

$$= \frac{1}{8\sqrt{\pi}} \frac{1}{a_0^{3/2}} \left(\frac{r}{a_0}\right) e^{-r/2a_0} \sin\theta\, e^{-i\phi} \qquad \text{...(7.16)}$$

SOLVED EXAMPLES

Example 7.1. *The ground state wave function for hydrogen atom* $\psi = \frac{1}{\sqrt{\pi}} \frac{1}{a_0^{3/2}} e^{-r/a_0}$. *Show that the average distance of the electron from the nucleus is 1.5* a_0. ***(P.U. 2008)***

Solution. The average value of r is given by

$$\langle r \rangle = \int_0^\infty \psi\, r\, \psi^*\, dV$$

$$= \int_0^\infty \frac{1}{\sqrt{\pi}} \frac{1}{a_0^{3/2}} e^{-r/a_0}\; r \frac{1}{\sqrt{\pi}} \frac{1}{a_0^{3/2}} e^{-r/a_0}\; dV$$

$$= \frac{1}{\pi a_0^3} \int_0^\infty r\, e^{-2r/a_0}\; dV$$

Now, in spherical polar co-ordinates the volume element

$$dV = r^2\, dr \sin\theta\, d\theta\, d\phi$$

$$\therefore \qquad < r > = \frac{1}{\pi a_0^3} \int_0^\infty r^3\, e^{-2r/a_0}\, dr \int_0^\pi \sin\theta\, d\theta \int_0^{2\pi} d\phi \qquad ...(i)$$

To find the value of $\int_0^\infty r^3\, e^{-2r/a_0}\, dr$ we use the standard integral

$\int_0^\infty e^{-bx}\, x^n\, dx = \frac{n!}{b^{n+1}}$ and get

$$\int_0^\infty r^3\, e^{-2r/a_0}\, dr = \frac{3!}{\left(\frac{+2}{a_0}\right)^4} \qquad \left[\because n = 3 \text{ and } b = \frac{2}{a_0}\right]$$

Also $\int_0^\pi \sin\theta\, d\theta = [-\cos\theta]_0^{2\pi} = 2$

and $\int_0^{2\pi} d\phi = [\phi]_0^{2\pi} = 2\pi$

Substituting in (*i*), we get

$$\langle r \rangle = \frac{1}{\pi a_0^3} \frac{3!}{\left(\frac{2}{a_0}\right)^4} \times 2 \times 2\pi = \frac{3}{2}\, a_0 = 1.5\, a_0$$

Example 7.2. *Find the expectation value of the electron from the nucleus in H-atom for the 2p state.* ***(P. U. 2001)***

Solution. One of the 2*p* states for *H*-atom is given by

$$\psi_{210} = \frac{1}{4\sqrt{2\pi}} \frac{1}{a_0^{3/2}} \frac{r}{a_0} e^{-r/2a_0} \cos\theta \qquad \text{(For this state } l = 1;\ m_l = 0\text{)}$$

$\therefore$ Expectation value of the electron from the nucleus

$$< r > = \int \psi_{210}^*\; r\; \psi_{210}\; dV$$

$$= \frac{1}{16\,(2\pi)} \frac{1}{a_0^3} \frac{1}{a_0^2} \int r^2\, e^{-r/a_0}\; r \cos^2\theta\; r^2\, dr \sin\theta\, d\theta\, d\phi$$

$$= \frac{1}{32\pi a_0^5} \int r^5\, e^{-r/a_0}\, dr \int \cos^2\theta \sin\theta\, d\theta \int d\phi$$

Now the first integral $I_1 = \int \left(\frac{r}{a_0}\right)^5 e^{-r/a_0}\, dr$

To solve it put $r = a_0 t \quad \therefore dr = a_0\, dt$

and $I_1 = \int t^5 e^{-t} a_0\, dt = a_0 \int_0^\infty t^5 e^{-t}\, dt$

$= a_0\, 5!$

The second integral $I_2 = \int_0^\pi \cos^2\theta \sin\theta\, d\theta$

To solve it put $\cos\theta = q$ $\therefore dq = -\sin\theta\, d\theta$

$\therefore$ $I_2 = -\int_1^{-1} q^2\, dq = \left[\frac{q^3}{3}\right]_{-1}^{1} = \frac{1}{3} + \frac{1}{3} = \frac{2}{3}$

The third integral $I_3 = \int_0^{2\pi} d\phi = 2\pi$

Thus $\langle r \rangle = \frac{1}{32\pi} a_0 \times 5! \times \frac{2}{3} \times 2\pi$

$= \left[\frac{1}{32\pi}\frac{4\pi}{3} 5 \times 4 \times 3 \times 2 \times 1\right] a_0$

or $< r > = 5\, a_0$

Example 7.3. *Starting from the radial equation for R (r) of the electron in the hydrogen atom show that the orbital angular momentum $L = \hbar\sqrt{l\,(l+1)}$ where l is the orbital quantum number.*

(G.N.D.U., 2000)

Solution. Orbital angular momentum. The *radial equation* is given by

$$\frac{d^2R}{dr^2} + \frac{2}{r}\frac{dR}{dr} + \frac{2m}{\hbar^2}\left[E + \frac{kZe^2}{r}\right]R - \frac{l(l+1)}{r^2}R = 0 \qquad ...(i)$$

The above equation is solely dependent on the radial motion of the electron *i.e.*, the motion of the electron along the co-ordinate $\vec{r}$ towards or away from the nucleus. It does not at all depend upon the orbital motion due to which the co-ordinates θ and ϕ undergo a change. The total energy E includes the kinetic energy due to radial motion as well as kinetic energy due to orbital motion and potential energy.

$\therefore$ Total energy $E = E_k\,(\text{orbital}) + E_k\,(\text{radial}) - \frac{kZe^2}{r}$

where $-\frac{kZe^2}{r}$ is the potential energy, k being a constant $= \frac{1}{4\pi\varepsilon_0} = 9 \times 10^9\ \text{Nm}^2/\text{C}^2$ in S.I. units.

$\therefore$ $E + \frac{kZe^2}{r} = E_k\,(\text{orbital}) + E_k\,(\text{radial})$

Substituting in (*i*), we have

$$\frac{d^2R}{dr^2} + \frac{2}{r}\frac{dR}{dr} + \frac{2m}{\hbar^2}[E_k \text{ (orb.)} + E_k \text{ (rad.)}]\, R - \frac{l(l+1)}{r^2} R = 0$$

This equation will depend only on radial motion if

$$E_k \text{ (orb.)} - \frac{l(l+1)\hbar^2}{2mr^2} = 0$$

If v is the orbital velocity of the electron, then

$$E_k \text{ (orb.)} = \frac{1}{2} mv^2$$

$$\therefore \quad \frac{l(l+1)\hbar^2}{2mr^2} = \frac{1}{2} mv^2$$

or $$(mvr)^2 = \hbar^2 l(l+1)$$

But mvr gives the magnitude of the orbital angular momentum $|\vec{L}|$

$$\therefore \quad |\vec{L}| = \hbar\sqrt{l(l+1)}$$

where l is the orbital quantum number.

7.9 SIGNIFICANCE OF QUANTUM NUMBERS *n*, *l* AND *m*

In the discussion of Schrödinger's theory of hydrogen atom we came across three quantum number n, l and m_l. We shall now discuss their physical significance.

(*i*) Principal quantum number. The quantum number n is called the principal (or radial) quantum number. It can have only positive integral values 1, 2, 3, ... n.

A well behaved finite and physically acceptable solution of the component function $R(r)$ of the variable r in Schrödinger's wave function for a hydrogenic atom is possible only if n is an integer (greater than l). In this discussion, n is a dimensionless constant given by

$$E_n = -\frac{me^4}{32\pi^2\varepsilon_0^2\hbar^2} \cdot \frac{1}{n^2} \qquad ...(i)$$

or $$n = \frac{e^2}{4\pi\varepsilon_0\hbar^2}\left(-\frac{m}{2E_n}\right)^{1/2} \qquad ...(ii)$$

Thus, the principal quantum number defines the total energy of the electron in the orbit and shows that the energy is quantised in accordance with Eq. (i) *i.e.*

$$E_n \propto \frac{1}{n^2}$$

as predicted by Bohr's theory.

The orbit of the electron may be a circle or an ellipse, the energy will depend only on the principal quantum number. For a circular orbit, the radius of the nth orbit r_n is given by

$$E_n = \frac{me^4}{32\pi^2\varepsilon_0^2\hbar^2} \cdot \frac{1}{n^2} = \frac{e^2\, me^2}{4\pi\varepsilon_0 \times 2 \times 4\pi\varepsilon_0\hbar^2 n^2} = \frac{e^2}{4\pi\varepsilon_0 \times 2r_n}$$

where $$r_n = \frac{4\pi\varepsilon_0\hbar^2 n^2}{me^2} \qquad ...(iii)$$

Thus the radius of the electronic orbit is proportional to the square of the principal quantum number.

For an elliptic orbit $r_n = a_n$ gives the semi-major axis of the ellipse.

However, the wave function $\psi(r, \theta, \phi)$ in Schrödinger's equation does not give the exact orbit of the electron. It only gives the probability of finding the electron at a certain position (r,θ,ϕ). Therefore, in terms of wave mechanics the maximum probability of finding the electron would be at any point along a circular or elliptical orbit round the nucleus.

Principal quantum number cannot be zero. Principal quantum number n cannot be zero. If $n = 0$, then according to equation (*i*) $E_n = \infty$, which is an impossibility.

Again, if $n = 0$, $r_n = 0$, which means the electron will be located at the centre of the nucleus, which is again impossible.

(*ii*) **Orbital quantum number.** The quantum number l is called the orbital quantum number. It defines the orbital angular momentum of the electron given by

$$L = \sqrt{l(l+1)}\,\hbar$$

It can have only positive integral values, 0, 1, 2, ... $(n - 1)$

For example, for $n = 3$, the possible values of l are 0, 1, 2 only.

In Bohr's theory, quantisation of angular momentum was introduced as a postulate without giving any justification, but in quantum mechanics, the quantisation of angular momentum appears as a result of the condition that the solution of the component function $\Theta(\theta)$ of the variable θ should be finite.

The orbital quantum number also defines the shape of the electron orbit. The semi-minor axis of the ellipse is given by $b_n = a_n \dfrac{l+1}{n}$.

When $l = n - 1$ [*e.g.* $n = 3, l = 2$] $b_n = a_n$ the orbit will be a circle. As the value of l decreases the ellipticity increases. The value of l cannot be $\geq n$ because then $b_n > a_n$ which is not possible. Hence the accepted values of l are 0, 1, 2,... $(n - 1)$ only. For given value of n *i.e.* for a given value of total energy, the electron can be in n different states having n different values of angular momentum. For example, for $n = 3$, $l = 0, 1, 2$ which corresponds to three different values of angular momentum given by $L = 0, L = \sqrt{2}\,\hbar, L = \sqrt{6}\,\hbar$. Historically $l = 0, 1, 2, 3$ etc. states of electron are represented by letters *s, p, d, f* respectively after the classification *sharp, principal, diffuse, fundamental* etc. which was adopted to designate the hydrogen spectra in the early stages of the development of the atomic theory. The state of the electron is thus represented as

1 *s*	for $n = 1$	$l = 0$
2 *s*	for $n = 2$	$l = 0$
2 *p*	for $n = 2$	$l = 1$
3 *s*	for $n = 3$	$l = 0$
3 *p*	for $n = 3$	$l = 1$
3 *d*	for $n = 3$	$l = 2$ and so on.

(*iii*) **Magnetic quantum number m_l.** The magnetic quantum number decides the Z-component of the orbital angular momentum given by $L_z = m_l \hbar$. Since m_l can have integral values $-l, -(l - 1)$, 0, ... $(l - 1)$, l, the Z-component of the angular momentum is also quantised.

Why magnetic quantum number is so called. An atom behaves like a small magnetic dipole of magnetic moment $\vec{\mu} = -\left(\dfrac{e}{2m}\right)\vec{L}$ where m is the reduced mass of the electron and $\vec{L}$ the angular momentum. Its Z-component is given by

$$\mu_z = -\frac{e}{2m} L_z = -\frac{e}{2m} m_l \hbar$$

Hence μ_z is also quantised and can have only integral values of $\frac{e\hbar}{2m}$. The quantity $\frac{e\hbar}{2m}$ is known as a Bohr magneton. States corresponding to different values of m_l can be observed in an external magnetic field, the direction of the applied field is taken as the direction of Z-axis. m_l is, therefore, called *magnetic (orbital) quantum number*.

How are they inter-related. *The principal quantum number* has the integral values $n = 1, 2, 3,..$etc. For one value of n of principal quantum number, the orbital quantum number l has n values given by $l = (n-1), (n-2) \ldots 2, 1, 0$.

For each value of orbital quantum number l, the magnetic quantum number m_l has $(2l + 1)$ values given by

$$m_l = \pm l, \pm (l-1) \ldots \pm 2, \pm 1, 0$$

Maximum number of energy states in a shell of principal quantum number *n*. From the above we find that for any value of principal quantum number n there are n states having different values of orbital quantum from 0 to $(n-1)$ and for each value of orbital quantum number l, there are $(2l+1)$ states with different values of magnetic quantum number m_l.

$$\therefore \text{ Total number of states } N = \sum_{l=0}^{l=n-1} (2l+1) = 2\sum_{l=0}^{l=n-1} l + n \qquad \left[\because \sum_{l=0}^{l=n-1} = n\right]$$

$$= 2\frac{(n-1)(n)}{2} + n = n^2$$

In other words, for a given n the total number of eigen functions or states $= n^2$ or the wave function ψ for the hydrogen atom has a degeneracy $= n^2$. The above relationship is shown in a tabular form for $n = 1$, $n = 2$ and $n = 3$.

n	l	m_l	$2l+1$	$N = n^2$
1	0	0	1	1
2	1 0	−1, 0, + 1 0	3 1	4
3	2 1 0	−2, −1, 0 + 1, +2 −1, 0, + 0	5 3 1	9

7.10 DEGENERACY

It is known that for a given value of n, the principal quantum number, the total number of eigen functions or states is n^2 i.e. there are n^2 different wave functions. The energy of a quantum state only depends upon n. *Thus for a given n there are n^2 different states having the same energy.* As shown in the table given above for $n = 2$ there are four states with ($l = 1$, $m_l = +1$ or ψ_{21+1}) ($l = 1$, $m_l = 0$ or ψ_{210}), ($l = 1$, $m_l = -1$ or ψ_{21-1}), and ($l = 0$, $m_l = 0$ or ψ_{200}).

A state of energy E_n is said to be degenerate if there are two or more than two different eigen functions corresponding to the same energy. The number of eigen functions corresponding to a state gives the *degree of degeneracy*. Thus in the above example the state $n = 2$ is *four fold degenerate i.e.* there are four coincident levels with the same energy.

Example 7.4. *List the set of quantum numbers for n = 3, for hydrogen atom.*

(P. U. 2001, GNDU 2001)

Solution. Refer to the table shown above and note that for $n = 3$; $l = 2, 1, 0$ and corresponding to $l = 2$; $m_l = +2, +1, 0, -1, -2$, corresponding to $l = 1$; $m_l = +1, 0, -1$ and corresponding to $l = 0$; $m_l = 0$.

Example 7.5. *Energy eigen values of hydrogen atom are given by* $E_n = \frac{-k m e^4}{2n^2 \hbar^2}$. *Discuss the significance of 'n' and the degeneracy of each state.* **(P.U. 2007; G.N.D.U. 2003)**

Solution. The energy eigen values of hydrogen atom $E_n = \frac{-k m e^4}{2n^2 \hbar^2}$

Susbstituting the value of $k = \frac{1}{4\pi\varepsilon_0}$, we get

$$E_n = -\frac{me^4}{16\pi^2 \varepsilon_0^2 2n^2 \hbar^2} = -\frac{me^4}{32\pi^2 \varepsilon_0^2 \hbar^2} \frac{1}{n^2} \quad ...(i)$$

The quantum number n is called the principal (or radial) quantum number. It can have only positive integral values 1, 2, 3...n. Thus the principal quantum number n defines the total energy of the electron in the nth orbit and shows that the energy is quantised in accordance with Eq. (*i*) $E_n \propto \frac{1}{n^2}$

7.11 SPACE QUANTIZATION OF $\vec{L}$ AND PRECESSION

The importance of the role of orbital magnetic quantum number lies in the fact that the Z-component of angular momentum $\vec{L}$ is given by

$$L_z = m_l \hbar \quad ...(i)$$

where m_l is the orbital magnetic quantum number. The value of $|m_l| \le l$ the orbital quantum number and $m_l = 0, \pm 1, \pm 2$ etc. For a given value of l, the total orbital angular momentum

$$L = \sqrt{l(l+1)}\, \hbar$$

but its component along Z-axis is

$$L_z = m_l \hbar \quad ...(ii)$$

$$\therefore \quad L_z = 0, \pm \hbar, \pm 2\hbar \text{ ...etc.}$$

Space quantisation of Z-component. From relations (*i*) and (*ii*) we find that though the total angular momentum is not an integral multiple of $\hbar$, the *Z-component is an integral multiple of $\hbar$.*

The Z-component is, therefore, quantised and is said to have spatial quantisation. For a given value of L, there are $(2l + 1)$ different values of L_z. *Thus the magnetic quantum number plays the major role in spatial quantisation of angular momentum and tells us the number and magnitude of different quantised values of L_z for a given value of l.*

Why L does not have a definite direction? The direction of orbital angular momentum is not fixed in space. It is continuously changing but it always makes the same angle with the Z-axis *i.e.* the direction of the applied magnetic field. The angular momentum vector, therefore, describes a cone, with constant vertex angle θ around the direction of magnetic field. In other words, the vector $\vec{L}$ is continuously precessing around the Z-axis. If the angular momentum vector $\vec{L}$ were fixed in space, then its components L_x, L_y as well as L_z will have definite values. In other words, the electron will be confined to a definite plane. For example, if $\vec{L}$ is always in the Z-direction, the electron will have to be in the $x - y$ plane at all times. But according to uncertainty principle, the electron cannot be confined to a single plane. There is, therefore, a built in uncertainty in the electron's Z-co-ordinate. Hence, the direction of $\vec{L}$ is continuously changing, such that the average values of L_x and L_y are zero, whereas L_z has specific values $m_l \hbar$.

This explains why only one component i.e. Z-component of L is quantised and why the angular momentum vector L cannot have a definite direction.

Quantisation of magnitude of $\vec{L}$. The magnitude of angular momentum vector $\vec{L}$ cannot have any value. It can have only values given by $|\vec{L}| = \sqrt{l(l+1)}\ \hbar$ where $l = 0, 1, 2...$ *i.e.* the magnitude of $\vec{L}$ is quantised.

The Z-component of $\vec{L}$ is given by

$$L_z = m_l \hbar \text{ where } m_l = 0, \pm 1, \pm 2, ...$$

Thus the magnitude of L_z is also quantised and is an integral multiple of $\hbar$ i.e. $m_l \hbar$ (m_l being positive as well as negative)

Quantisation of direction of L and precession. The Z-component of the angular momentum $L_z = L \cos\theta$. Hence the angular momentum vector L can have only $(2l+1)$ discrete orientations in space, given by

$$\cos\theta = \frac{L_z}{L} = \frac{m_l \hbar}{[l(l+1)]^{1/2}\hbar} = \frac{m_l}{[l(l+1)]^{1/2}}$$

where $m_l = -l, -(l-1), ... 0, ... (l-1), l$

From the above it is clear that θ cannot have any value between 0 and π, as given by classical physics, but can have only discrete values given by the above relation.

The angular momentum vector $\vec{L}$ can neither be oriented parallel nor anti-parallel to the direction of the magnetic field. As $\cos\theta = \dfrac{m_l}{\sqrt{l(l+1)}}$ and maximum value of $m_l = l$, $\cos\theta < 1$, which means $\theta > 0$. Similarly, the minimum value being $-l$, $\cos\theta < -1$ and $\theta < \pi$.

Therefore, the value of θ is > 0 but $< \pi$. In classical physics, however, it is assumed that a magnetic dipole completely aligns itself along the direction of the magnetic field. The experimental results confirm the quantum mechanical values.

For $l = 3$, $m_l = +3, +2, +1, 0, -1, -2, -3$ and

$$L = \sqrt{3(3+1)}\hbar = \sqrt{12}\hbar = 2\sqrt{3}\hbar \text{ and}$$

$$L_z = +3\hbar, +2\hbar, +1\hbar, -1\hbar, -2\hbar, -3\hbar.$$

All these values are shown in Fig. 7.2. The circles show that the vector $\vec{L}$ is precessing round the Z-axis, and has only a definite direction in space *i.e.* the direction of angular momentum vector L is also quantised.

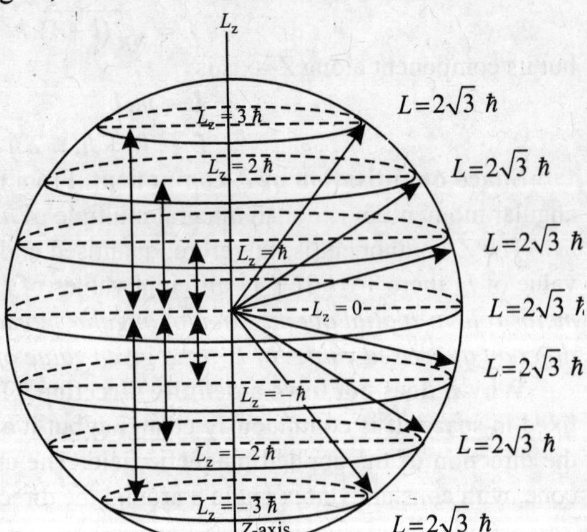

Fig. 7.2

Example 7.6. *Calculate the allowed angles between the angular momentum vector $\vec{L}$ of an electron in the state $l = 2$ and $l = 1$ and its Z-component. Show these graphically.*

Solution. (*i*) $l = 1$. For $l = 1$, the magnitude of the angular momentum vector

$$L = \sqrt{l(l+1)}\hbar = \sqrt{2}\hbar$$

For $l = 1$, the allowed values of m_l are $m_l = +1, 0, -1$

$\therefore$ The orientations of angular momentum vector in space are given by for $m_l = +1$; $\cos\theta_1 = \dfrac{m_l \hbar}{\sqrt{l(l+1)}\hbar} = \dfrac{1}{\sqrt{2}}$ $\therefore \theta_1 = 45°$

for $m_l = +0$; $\cos\theta_2 = 0$ $\therefore \theta_2 = 90°$

for $m_l = -1$; $\cos\theta_3 = \dfrac{-1}{\sqrt{2}}$ $\therefore \theta_3 = 135°$

This set of 3 allowed orientations is shown in Fig. 7.3 (*a*).

(*ii*) **$l = 2$.** For $l = 2$ the magnitude of the angular momentum vector

$$L = \sqrt{l(l+1)}\,\hbar = \sqrt{2(2+1)}\,\hbar = \sqrt{6}\,\hbar$$

For $l = 2$ the allowed values of m_l are $m_l = +2, +1, 0, -1, -2$.

$\therefore$ The required angles are

for $m_l = 2$; $\cos\theta_1 = \dfrac{m_l \hbar}{\sqrt{l(l+1)}\hbar} = \dfrac{2}{\sqrt{6}}$ $\therefore \theta_1 = 35°\,9'$

for $m_l = 1$; $\cos\theta_2 = \dfrac{1}{\sqrt{6}}$ $\therefore \theta_2 = 65°\,54'$

for $m_l = 0$; $\cos\theta_3 = 0$ $\therefore \theta_3 = 90°$

for $m_l = -1$; $\cos\theta_4 = -\dfrac{1}{\sqrt{6}}$ $\therefore \theta_4 = 114°\,6'$

for $m_l = -2$; $\cos\theta_5 = -\dfrac{2}{\sqrt{6}}$ $\therefore \theta_5 = 144°\,51'$

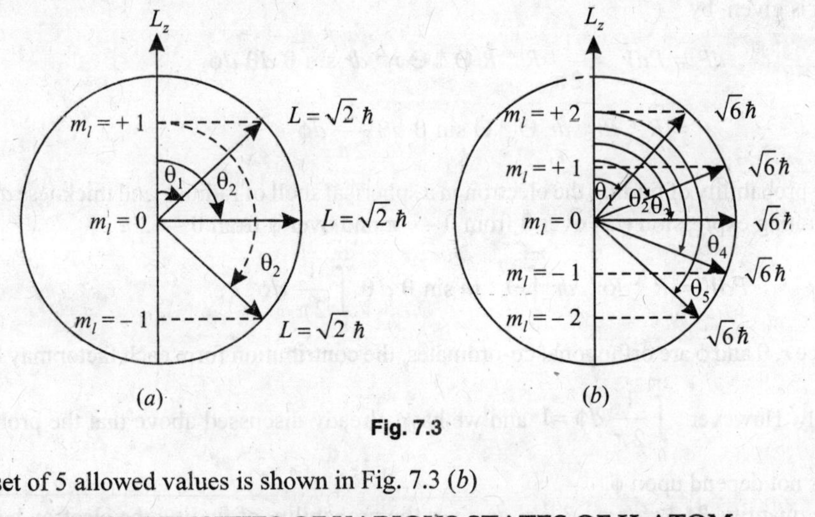

Fig. 7.3

This set of 5 allowed values is shown in Fig. 7.3 (*b*)

7.12 PROBABILITY DENSITIES OF VARIOUS STATES OF H-ATOM

Probability densities. According to quantum theory, the electron wave function ψ in a hydrogen atom is given by $\psi = R\,\Theta\,\Phi$ where $R = R_{nl}(r)$ describes how ψ varies with r when the principal (or total) and orbital quantum numbers have the values n and l respectively; $\Theta = \Theta_{lm_l}(\theta)$ describes how ψ varies with θ when the orbital and magnetic quantum numbers have the values l and m_l respectively and $\phi_{m_l}(\phi)$ describes how ψ varies with ϕ when the magnetic quantum number is m_l.

The probability density is defined as the probability of finding the electron in a unit volume around a given point. It is given by

$$P = \psi\psi^* = |\psi|^2$$

$$= R_{nl}^* (r) R_{nl} (r) \Theta_{lml}^* (\theta) \Theta_{lml}(\theta) \Phi_{ml}^* (\phi) \Phi_{ml} (\phi)$$

$$= | R_{nl} (r) |^2 | \Theta_{lml} (\theta) |^2 | \Phi_{ml} (\phi) |^2 \quad ...(i)$$

Now $\Phi_{m_l} = \frac{1}{\sqrt{2\pi}} e^{i m_l \phi}$

$\therefore \quad \Phi^*_{m_l} \Phi_{m_l} = \frac{1}{\sqrt{2\pi}} e^{i m_l \phi} \frac{1}{\sqrt{2\pi}} e^{-i m_l \phi} = \frac{1}{2\pi}$

Substituting value of $\Phi^*_{m_l} \Phi_{m_l} = \frac{1}{2\pi}$ in equation (*i*), we have

$$P = \frac{1}{2\pi} R^*_{nl} R_{nl} \Theta^*_{lm_l} \Theta_{lm_l} = \frac{1}{2\pi} |R_{nl}|^2 |\Theta_{lm_l}|^2 \quad ...(ii)$$

This equation shows that the probability density does not depend upon the azimuthal co-ordinate ϕ i.e. the distribution of electron is completely symmetric about Z-axis and hence there is equal chance of the electron being found at one angle ϕ as at any other.

Thus the probability density depends upon the product of two quantities $(R^*_{nl} R_{nl})$ and $(\Theta^*_{lm_l} \Theta_{lm_l})$. The factor $(R^*_{nl} R_{nl})$, (henceforth written as $R^* R$) determines the probability density of the electron as a function of *r* along a given direction and is known as radial probability density. Similarly $(\Theta^*_{lm_l} \Theta_{lm_l})$ (henceforth written as $\Theta^* \Theta$) determines the probability density as a function of θ along a given radius and is known as *directional* (*or polar*) *probability density.*

Radial probability density. The probability of finding an electron in a volume $dV = r^2\, dr \sin\theta\, d\theta\, d\phi$ is given by

$$dP = PdV = \frac{1}{2\pi} R^* R\, \Theta^* \Theta\, r^2\, dr \sin\theta\, d\theta\, d\phi$$

$$= R^* R r^2\, dr\, \Theta^* \Theta \sin\theta\, d\theta \frac{1}{2\pi} d\phi \quad ...(iii)$$

The probability of finding the electron in a spherical shell of radius *r* and thickness *dr* is obtained by integrating expression (*iii*) over θ from $0 \rightarrow \pi$ and over ϕ from $0 \rightarrow 2\pi$

$$\therefore \quad PdV = R^* R r^2\, dr \int_0^{\pi} \Theta^* \Theta \sin\theta\, d\theta \int_0^{2\pi} \frac{1}{2\pi} d\phi$$

Since *r*, θ and ϕ are orthogonal co-ordinates, the contribution form each factor may be discussed separately. However $\int_0^{2\pi} \frac{1}{2\pi} d\phi = 1$ and we have already discussed above that the probability density does not depend upon ϕ.

The quantity $R^* R r^2 dr = | R |^2 r^2\, dr$ gives the probability of finding the electron between *r* and $(r + dr)$ regardless of the direction θ.

Ground state. For the ground (1*s*) state $n = 1, l = 0$

$\therefore$ Radial probability density $P_{10} = | R_{10} |^2 r^2 \quad ...(iv)$

Substituting the value of $R_{10} = \frac{2}{a_0^{3/2}} e^{-r/a_0}$ in (*iv*), we have

For the 1*s* state, $P_{10} = r^2 \left| \frac{2}{a_0^{3/2}} e^{-r/a_0} \right|^2 = \frac{4r^2}{a_0^3} e^{-2r/a_0}$

First excited state. For the first excited state $n = 2$. So there are two values of l i.e. $l = 0$ and $l = 1$. The radial probability for $n = 2$, $l = 0$ i.e. $2s$ state is given by

$$P_{20} = r^2 \, |R_{20}|^2 = r^2 \left[\frac{1}{(2a_0)^{3/2}} \left(2 - \frac{r}{a_0} \right) e^{-r/2a_0} \right]^2$$

$$= \frac{r^2}{8a_0^3} \left(2 - \frac{r}{a_0} \right)^2 e^{-r/a_0}$$

The radial probability for $n = 2$ and $l = 1$ *i.e.* $2p$ state is given by

$$P_{21} = r^2 \, |R_{21}|^2 = r^2 \left[\frac{1}{(2a_0)^{3/2}} \frac{1}{\sqrt{3}} \frac{r}{a_0} e^{-r/2a_0} \right]^2$$

$$= \frac{r^4}{24a_0^5} e^{-r/a_0}$$

2nd excited state. For the second excited state $n = 3$. So there are three values of l *i. e.* $l = 0$, $l = 1$ and $l = 2$. We can calculate the radial probability P_{30} for $n = 3$, $l = 0$ *i.e.* $3s$ state, P_{31} for $n = 3$, $l = 1$ *i.e.* $3p$ state and P_{32} for $n = 3$, $l = 2$ *i. e* $3d$ state, by substituting the corresponding values of radial wave function R_{30}, R_{31} and R_{32} respectivley in the expression $P_n = |R_{nl}|^2 r^2$.

The variation of radial probability density $P(r)\,dr$ with the radial distance measured in units of Bohr radius a_0 is shown in Fig. 7.4 (*a*), (*b*) and (*c*). Fig. 7.4(*a*) gives the plot of 1s, $2s$ and $3s$ states, Fig. 7.4(*b*) gives the plot of $2p$ and $3p$ states and Fig. 7.4(*c*) gives the plot of $3d$ state.

From Fig. 7.4. (*a*) it is clear that for $1s$ state $P(r)\,dr = 0$ at $r = 0$ *i.e.* at the nucleus itself. The probability density is maximum at $r = a_0$ which is in agreement with the radius of the orbit for the given state, as calculated by Bohr's theory. For $2s$ state $P(r)\,dr = 0$ at $r = 0$. There are two maxima at $r = a_0$ and at $r = 5\,a_0$. The second maximum is higher than the first. For $3s$ state there are three maximum at $r = 0$, $r = 4\,a_0$ and $r = 12\,a_0$. The maxima become higher as r increases.

As shown in Fig. 7.4 (*b*) the $2p$ electron has a maximum radial probability at $r = 4\,a_0$ and $3p$ electron has two maxima at $r = 2\,a_0$ and $r = 12\,a_0$.

As shown in Fig. 7.4 (c) the $3d$ state has a maximum at about $10\,a_0$.

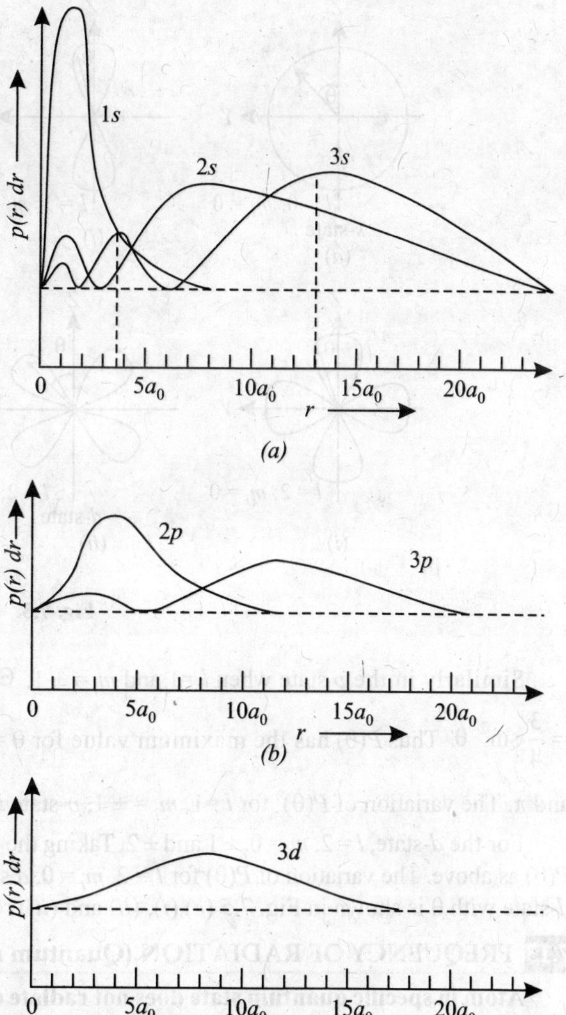

Fig. 7.4

Thus there is a higher probability of finding the electron at one of those distances where the maxima take place. In each case, the probability of finding the electron at $r = 0$ is zero.

Directional (or polar) probability density. The value of Θ_{lml} varies with the polar angle θ except for the s-electron for which $l = 0$ and $m_l = 0$. The value of $\Theta_{00} = \frac{1}{\sqrt{2}}$ = a constant. Therefore, the probability $P(\theta) = \Theta^*_{00}\ \Theta_{00} = \frac{1}{2}$ and is thus independent of θ. This fact, combined with symmetry about Z-axis, shows that the distribution is spherical, as shown in Fig. 7.5 (a) i.e. the probability of locating an s-electron has the same value at a given r in all directions.

In the p-state when $l = 1$ and $m_l = 0$, $\Theta_{10} = \sqrt{\frac{3}{2}}\cos\theta$. Therefore, $P(\theta) = \frac{3}{2}\cos^2\theta$. Thus $P(\theta)$ has the maximum value for $\theta = 0$ [$\cos\theta = 1$] and $\theta = \pi$ [$\cos\theta = -1$] and the minimum value (zero) for $\theta = \pm\pi/2$. The variation of $P(\theta)$ with θ for $l = 1$, $m_l = 0$, p state is shown in Fig. 7.5 (b) (i). The probability distribution has two lobes symmetrical about the Z-axis.

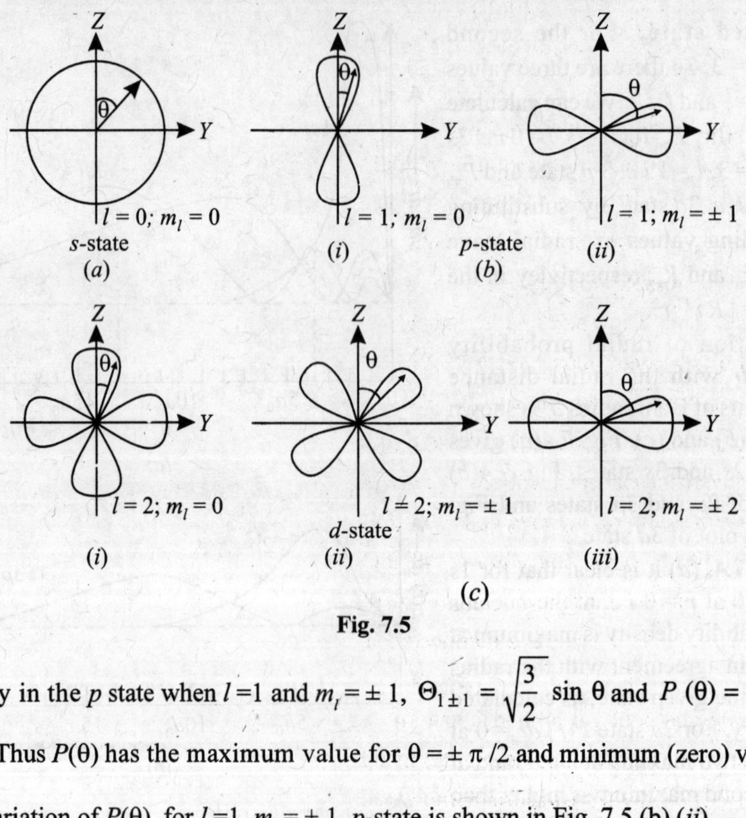

Fig. 7.5

Similarly in the p state when $l = 1$ and $m_l = \pm 1$, $\Theta_{1\pm1} = \sqrt{\frac{3}{2}}\sin\theta$ and $P(\theta) = \Theta^*_{1\pm1}\ \Theta_{1\pm1} = \frac{3}{4}\sin^2\theta$. Thus $P(\theta)$ has the maximum value for $\theta = \pm\pi/2$ and minimum (zero) value for $\theta = 0$ and π. The variation of $P(\theta)$ for $l = 1$, $m_l = \pm 1$, p-state is shown in Fig. 7.5 (b) (ii).

For the d-state, $l = 2$, $m_l = 0, \pm 1$ and ± 2. Taking the values of Θ_{20}, $\Theta_{2\pm1}$ and $\Theta_{2\pm2}$ we can calculate $P(\theta)$ as above. The variation of $P(\theta)$ for $l = 2$, $m_l = 0$, d state, $l = 2$, $m_l = \pm 1$ d-state and $l = 2$, $m_l = \pm 2$ d-state with θ is shown in Fig. 7.5 (c) (i), (ii) and (iii) respectively.

7.13 FREQUENCY OF RADIATION (Quantum mechanical treatment)

Atom in specific quantum state does not radiate energy. To explain the spectrum of hydrogen atom, Bohr postulated that an electron revolving round the nucleus in a stationary orbit does not radiate energy. A radiation of frequency ν is emitted only when the electron jumps from a higher energy state E_{n2} to a lower energy state E_{n1} given by

$$\nu = \frac{E_{n_2} - E_{n_1}}{h} \qquad ...(7.17)$$

Although Bohr made this assumption without assigning any reason, the above relation can be derived on the basis of the quantum mechanical theory of the atom.

The underlying principle is that an atomic electron, whose average position relative to the nucleus is constant in time, does not radiate energy; but if the average position relative to the nucleus oscillates, electromagnetic radiations are emitted, the frequency of which is the same as natural frequency of oscillation of the electron.

For the sake of simplicity, we shall consider the motion of the electron only along the X-direction. The time dependent wave function Ψ_n of an electron in a state of quantum number n and energy E_n is given by

$$\Psi_n = \psi_n e^{-iE_n t/\hbar} \qquad ...(i)$$

where ψ_n is the time independent part of wave function and $e^{-iE_n t/\hbar}$ is the time dependent part.

The average posititon (expectation value) $< x >$ of such an electron is given by

$$< x > = \int_{-\infty}^{+\infty} \Psi_n^* \hat{x} \Psi_n \, dx = \int_{-\infty}^{+\infty} \Psi_n^* x \Psi_n \, dx$$

where $\hat{x}$ is the x-co-ordinate operator and $\Psi_n^* = \psi_n^* e^{+iE_n t/\hbar}$

$$\therefore \qquad < x > = \int_{-\infty}^{+\infty} \psi_n^* e^{+iE_n t/\hbar} x \psi_n e^{-iE_n t/\hbar} \, dx$$

$$= \int_{-\infty}^{+\infty} \psi_n^* x \psi_n \, dx \qquad ...(ii)$$

The above expression is constant in time, as ψ_n and ψ_n^* are functions of position only. The electron does not oscillate and no radiation occurs.

Thus quantum mechanics predicts that an electron in a specific quantum state does not radiate energy.

Frequency of radiation. Now consider an electron changing from one energy state n_2 to another energy state n_1. For example, let the atom be in the ground state n_1 at $t = 0$. Suppose the electron in the ground state is excited by a process of some kind, say due to the energy absorbed from a beam of incident radiation and moves to a higher (excited) energy state n_2. Subsequently the atom emits radiation corresponding to a transition of the electron from the excited energy state E_{n2} to the ground state E_{n1}. At some time during the intervening period, the atom existed in the state n_2, whereas finally it is in the state n_1. The wave function Ψ of an electron capable of existing in the states n_1 and n_2 is given by

$$\Psi = a\Psi_{n1} + b\Psi_{n2}$$

where $\quad a^* a$ = probability of electron in state n_1

and $\quad b^* b$ = probability of electron in state n_2

Also $a^* a + b^* b = 1$

$a^* a = a^2$ and $b^* b = b^2$ [as the electron must exist either in the state n_1 or in the state n_2]

At $t = 0$ the electron is in the state n_1

$\therefore \qquad a = 1$ and $b = 0$

when the electron is in the state n_2

$\qquad a = 0$ and $b = 1$

Finally, after de-excitation, again the electron is in the state n_1

and $$a = 1 \text{ and } b = 0$$

When the electron is either in the state n_1 or in the state n_2 it does not radiate energy, but when it is in the midst of the transition from state n_2 to state n_1 where $a \neq 0, b \neq 0$, electromagnetic radiation are given out. At this stage, the expectation or average value of the position co-ordinate x is given by

$$< x > = \int_{-\infty}^{+\infty} \Psi^* \; x \; \Psi \; dx$$

$$= \int_{-\infty}^{+\infty} \left(a^* \Psi^*_{n_1} + b^* \Psi^*_{n_2}\right) x \left(a \Psi_{n_1} + b \Psi_{n_2}\right) dx$$

$$= \int_{-\infty}^{+\infty} x \left[a^2 \Psi^*_{n_1} \Psi_{n_1} + a^* b \Psi^*_{n_1} \Psi_{n_2} + b^* a \Psi_{n_1} \Psi^*_{n_2} + b^2 \Psi^*_{n_2} \Psi_{n_2}\right] dx \quad ...(iii)$$

As x is not an operator, its position in the integral is of no significance.

Substituting the value of

$$\Psi_{n_1} = \psi_{n_1} e^{-E_{n_1} t/\hbar}, \; \Psi^*_{n_1} = \psi^*_{n_1} e^{+iE_{n1} t/\hbar}$$

$$\Psi_{n_2} = \psi_{n_2} e^{-E_{n_2} t/\hbar}, \; \Psi^*_{n_2} = \psi^*_{n_2} e^{+iE_{n_2} t/\hbar}$$

in equation (*iii*), we have

$$< x > = a^2 \int_{-\infty}^{+\infty} x \psi^*_{n_1} \psi_{n_1} \; dx + a^* b \int_{-\infty}^{+\infty} x \psi^*_{n_1} e^{+E_{n_1} t/\hbar} \psi_{n_2} e^{-iE_{n_2} t/\hbar} dx$$

$$+ b^* a \int_{-\infty}^{+\infty} x \psi_{n_1} e^{-E_{n_1} t/\hbar} \psi^*_{n_2} e^{+iE_{n_2} t/\hbar} dx + b^2 \int_{-\infty}^{+\infty} x \psi^*_{n_2} \psi_{n_2} \; dx \quad ...(iv)$$

As proved above, the first and the last integrals are constant in time. It is only the second and the third integrals which contribute to the time variation in x. For a finite bound system of two states, as in the present case

$$b^* a = a^* b \quad \text{and} \quad \psi^*_{n_2} \psi_{n_1} = \psi^*_{n_1} \psi_{n_2}$$

Substituting in equation (*iv*), the time varying term becomes

$$a^* b \int_{-\infty}^{+\infty} x \psi^*_{n_1} \psi_{n_2} \left[e^{-i \frac{(E_{n_2} - E_{n_1})}{\hbar} t} + e^{+i \frac{(E_{n_2} - E_{n_1})}{\hbar} t} \right] dx$$

But $$\frac{e^{-i \frac{(E_{n_2} - E_{n_1})}{\hbar} t} + e^{+i \frac{(E_{n_2} - E_{n_1})}{\hbar} t}}{2} = \cos \frac{(E_{n_2} - E_{n_1})}{\hbar} t$$

$\therefore$ The time varying term is

$$= 2 a^* b \cos \frac{(E_{n_2} - E_{n_1})}{\hbar} t \int_{-\infty}^{+\infty} x \psi^*_{n_1} \psi_{n_2} \; dr \quad ...(v)$$

The above term (*v*) contains the time varying factor

$$\cos \frac{(E_{n_2} - E_{n_1})}{\hbar} t = \cos 2\pi \frac{(E_{n_2} - E_{n_1})}{h} t$$

Thus we find that position of the electron oscillates sinusoidally with a frequency ν *given* by

$$\nu = \frac{E_{n_2} - E_{n_1}}{h} \quad \text{...(vi)}$$

and the amplitude of oscillations is given by

$$2a^* b \int_{-\infty}^{+\infty} x \, \psi_{n_1}^* \, \psi_{n_2} \, dx \quad \text{...(vii)}$$

Finally $$<x> = a^2 \int_{-\infty}^{+\infty} x \, \psi_{n_1}^* \, \psi_{n_1} \, dx + b^2 \int_{-\infty}^{+\infty} x \, \psi_{n_2}^* \, \psi_{n_2} \, dx$$

$$+ 2\, a^* b \cos 2\pi \nu t \int_{-\infty}^{+\infty} x \, \psi_{n_1}^* \, \psi_{n_2} \, dx \quad \text{...(7.18)}$$

When the electron is in state n_1 or state n_2 the probabilities b^2 or a^2 are respectively zero and the electron is, on the average stationary *i.e.* $<x>$ = a constant. When the electron undergoes a transition between two states, its average position oscillates with a frequency ν. This frequency is the same as that given by Bohr.

Allowed and forbidden transitions. We have seen that the amplitude of the oscillations produced when an electron in an atom jumps from a higher energy state n_2 to a lower energy state n_1 is given by

$$2\, a^* b \int_{-\infty}^{+\infty} x \, \psi_{n_1}^* \, \psi_{n_2} \, dx$$

If the motion of the electron, instead of being restricted along the *X*-axis, can take place in any direction, then x is replaced by $\vec{r}$ and dx by dV in relation (*vii*) and the integral becomes

$$M_{n_2 n_1} = \int_{\text{all space}} \vec{r} \, \psi_{n_1}^* \, \psi_{n_2} \, dV$$

This quantity is known as matrix element of the position vector $\vec{r}$ taken between the initial state n_1 and final state n_2.

In order that a transition between the two states n_2 and n_1 may take place, the integral $M_{n_2 n_1} = \int_{\text{all space}} \vec{r} \, \psi_{n_1}^* \, \psi_{n_2} \, dV$ should not be zero, as the intensity of radiation is proportional to $M_{n_2 n_1}^2$ and if $M_{n_2 n_1}^2 = 0$, the intensity of the spectral line is zero *i.e.* there is no radiation.

Thus the transitions for which $M_{n_2 n_1} = \int_{\text{all space}} \vec{r} \, \psi_{n_1}^* \, \psi_{n_2} \, dV = 0$ are called **forbidden transitions.**

The only **allowed transitions** are those for which the matrix $M_{n_2 n_1}^2$ is non-zero *i.e.* has a finite value, or

$$M_{n_2 n_1} = \int_{\text{all space}} \vec{r} \, \psi_{n_1}^* \, \psi_{n_2} \, dV \neq 0 \quad \text{...(viii)}$$

Selection rules. The condition that the matrix element

$$M_{n_2 n_1} = \int_{\text{all space}} \vec{r}\ \psi^*_{n_1}\ \psi_{n_2}\ dV \neq 0$$

for an allowed transition leads to the selection rules for radiative transitions between two quantum states, depending upon their various quantum numbers.

In the case of hydrogen atom, there are three quantum numbers, n, l and m_l which specify the initial and final states involved in a radiative transition. Only those transitions are allowed which obey the following rules.

1. The magnetic quantum number m_l changes by unity (+ 1 or – 1) or remains unchanged *i.e.*

$$\Delta m_l = \pm 1 \text{ or } 0$$

2. The orbital quantum number l changes by unity (+ 1 or – 1) *i.e.*

$$\Delta l = \pm 1$$

3. The change in total or (principal) quantum number n is not restricted.

Derivation of selection rule for m_l. To explain how the condition

$M_{n_2 n_1} = \int_{\text{all space}} \vec{r}\ \psi^*_{n_1}\ \psi_{n_2}\ dV \neq 0$ leads to the above selection rules, we shall derive the selection rule for magnetic quantum number m_l.

For hydrogen atom $\psi_n = R_{nl}(r)\ \Theta_{lml}(\theta)\ \Phi_{ml}(\phi)$

Substituting in (*viii*), with proper subscripts, we have

$$M_{n_2 n_1} = \int_{\text{all space}} \vec{r}\ R^*_{n_1 l_1}\ R_{n_2 l_2}\ \Theta^*_{l_1}\ \Theta_{l_2}\ \Phi^*_{m_{l1}}\ \Phi_{m_{l2}}\ r^2\ dr \sin\theta\ d\theta\ d\phi$$

because $dV = r^2\ dr \sin\theta\ d\theta\ d\phi$. Writing the above as product of three separate integrals, we have

$$M_{n_2 n_1} = \int_{r=0}^{\infty} R^*_{n_1 l_1}\ R_{n_2 l_2}\ r^2\ dr \int_{\theta=0}^{\pi} \Theta_{l_1}{}^*\ \Theta_{l_2}\ \sin\theta\ d\theta \int_{\phi=0}^{2\pi} \vec{r}\ \Phi^*_{ml_1}\ \Phi_{ml2}\ d\phi$$

To derive the selection rules for magnetic quantum number m_l we consider the integral

$$\int_{\phi=0}^{2\pi} \vec{r}\ \Phi^*_{ml_1}\ \Phi_{ml2}\ d\phi$$

It is a vector integral and has three components, say I_x, I_y and I_z along X, Y and Z axes respectively, corresponding to three components x, y, and z of $\vec{r}$.

$$\therefore \qquad I_x = \int_0^{2\pi} x\ \Phi^*_{ml_1}\ \Phi_{ml_2}\ d\phi\ ,\ I_y = \int_0^{2\pi} y\ \Phi^*_{ml_1}\ \Phi_{ml2}\ d\phi$$

and

$$I_z = \int_0^{2\pi} z\ \Phi^*_{ml_1}\ \Phi_{ml_2}\ d\phi \qquad \text{...}(ix)$$

The relation between cartesian and polar co-ordinates gives

$$x = r \sin\theta \cos\phi,\ y = r \sin\theta \sin\phi,\ z = r\cos\theta$$

Substituting these values in Eq.(*ix*), we have

$$I_x = \int_0^{2\pi} x\ \Phi^*_{ml_1}\ \Phi_{ml_2}\ d\phi = r \sin\theta \int_0^{2\pi} \Phi^*_{ml_1}\ \Phi_{ml2}\ \cos\phi\ d\phi$$

Now $\Phi^*_{ml_1} = \dfrac{1}{\sqrt{2\pi}} e^{-im_{l1}\phi}, \Phi_{m_{l2}} = \dfrac{1}{\sqrt{2\pi}} e^{+im_{l2}\phi}$

and $\cos\phi = \dfrac{e^{i\phi} + e^{-i\phi}}{2}$

$$\therefore \quad I_x = \frac{r\sin\theta}{4\pi}\left[\int_0^{2\pi} e^{-im_{l1}} e^{+im_{l2}\phi} e^{i\phi}\, d\phi + \int_0^{2\pi} e^{-im_{l1}\phi} e^{+im_{l2}\phi} e^{-i\phi}\, d\phi\right]$$

$$= \frac{r\sin\theta}{4\pi}\left[\int_0^{2\pi} e^{i(m_{l2} - m_{l1} + 1)\phi}\, d\phi + \int_0^{2\pi} e^{i(m_{l2} - m_{l1} - 1)\phi}\, d\phi\right]$$

where $(m_{l2} - m_{l1} + 1)$ and $(m_{l2} - m_{l1} - 1)$ are equal to either zero or to an integral value $k = \pm 1, \pm 2, \pm 3$ ... because m_{l_2} and m_{l_1} are respectively limited to $0, \pm 1, \pm 2 \ldots \pm l_2$ and $0, \pm 1, \pm 2 \ldots \pm l_1$.

Further $e^{ik\phi} = \cos k\phi + i \sin k\phi$

and hence $\displaystyle\int_0^{2\pi} e^{ik\phi}\, d\phi = \int_0^{2\pi} \cos k\phi\, d\phi + i\int_0^{2\pi} \sin k\phi\, d\phi = 0$

for $k = \pm 1, \pm 2, \pm 3$ ... etc.

Hence $m_{l2} - m_{l1} + 1 = 0$ or $m_{l2} - m_{l1} = -1$ or $\Delta m_l = -1$

or $m_{l2} - m_{l1} - 1 = 0$ or $m_{l2} - m_{l1} = +1$ or $\Delta m_l = +1$

because only under the condition $\Delta m_l \pm 1$ the above integral is non-zero and has a definite value- the condition for allowed transition. If $(m_{l2} - m_{l1} \pm 1)$ has a finite integral value k, then whole of the integral is zero and the condition for allowed transition is not fulfilled.

The same rule will be obtained by evaluating the integral

$$I_y = \int_0^{2\pi} y\, \Phi^*_{m_{l1}}\, \Phi_{m_{l_2}}\, d\phi$$

But if we consider the integral $I_z = \displaystyle\int_0^{2\pi} z\, \Phi^*_{m_{l1}}\, \Phi_{m_{l2}}\, d\phi$, then

$$I_z = r\cos\theta \int_0^{2\pi} \Phi^*_{m_{l1}}\, \Phi_{m_{l2}}\, d\phi = \frac{r\cos\theta}{4\pi}\int_0^{2\pi} e^{i(m_{l2} - m_{l_1})\phi}\, d\phi$$

This integral $I_z = 0$ for all values of $(m_{l2} - m_{l1})$ except for $m_{l2} - m_{l1} = 0$ or $\Delta m_l = 0$ for which the integral is non-zero and finite.

$\therefore$ The rules for allowed transitions in a hydrogen atom for magnetic quantum number m_l are

$$\Delta m_l = 0, \pm 1$$

Although the above derivation has been made for hydrogen atom, the selection rule $\Delta m_l = 0, \pm 1$ as well as $\Delta l = \pm 1$ holds true for transitions that involve single outer shell electron in any atom.

Energy level diagram. Energy level diagrams of some transitions allowed by the selection rule $\Delta l = \pm 1$ for hydrogen are shown in Fig. 7.6.

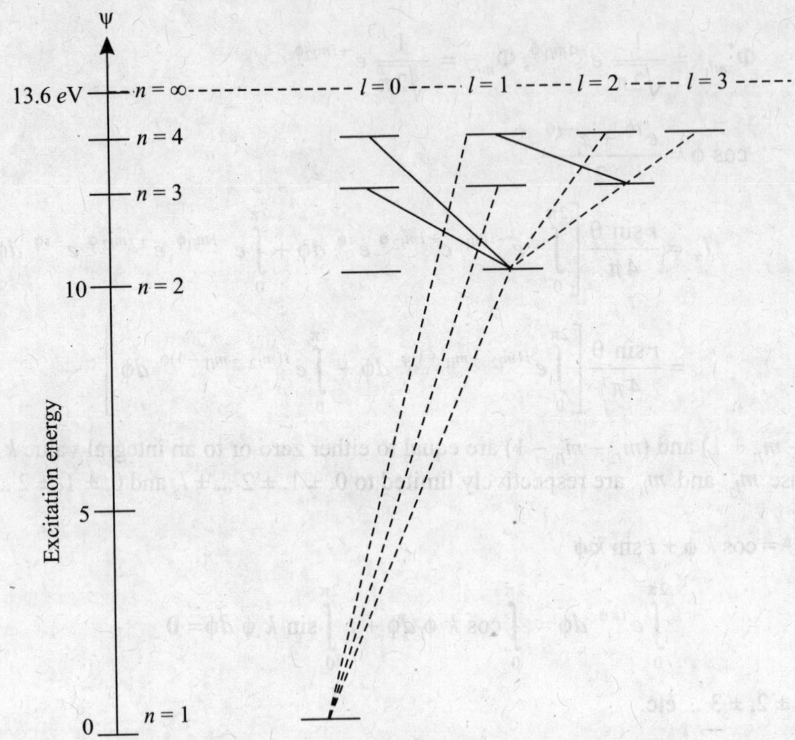

Fig. 7.6

The excitation energy for $n = 1$, $n = 2$, $n = 3$, $n = 4$... and $n = \infty$ is shown along the Y-axis.

For $n = 1$, l has only one value *i.e.* $l = n - 1 = 0$. This energy level is shown aganist $n = 1$ under $l = 0$.

For $n = 2$, l has two values *i.e.* $l = n - 1 = 1$ and $l = n - 2 = 0$. These energy levels are shown aganist $n = 2$ under $l = 0$ and $l = 1$. For $n = 3$, l has three values *i.e.* $l = n - 1 = 2$, $l = n - 2 = 1$ and $l = n - 3 = 0$. These energy levels are shown aganist $n = 3$ under $l = 0$, $l = 1$ and $l = 2$. For $n = 4$, l has four values *i.e.* $l = n - 1 = 3$, $l = n - 2 = 2$, $l = n - 3 = 1$ and $l = n - 4 = 0$. These energy levels are shown aganist $n = 4$ under $l = 0$, $l = 1$, $l = 2$ and $l = 3$.

Transitions allowed by the selection rule $\Delta l = +1$ are shown by dotted lines from $l = 1 \rightarrow l = 0$; $l = 2 \rightarrow l = 1$ and $l = 3 \rightarrow l = 2$. Transitions allowed by the selection rule $\Delta l = -1$ are shown by full lines from $l = 0 \rightarrow l = 1$ and $l = 1 \rightarrow l = 2$.

Example 7.7. *Is the transition $\psi_{210} \rightarrow \psi_{100}$ forbidden? Explain.*

Solution. This is transition from the state $n = 2$, $l = 1$, $m_l = 0$ to the state $n = 1$, $l = 0$ $m_l = 0$.

This satisfies the selection rule $\Delta l = \pm 1$ and $\Delta m_l = 0$. So this transition is *allowed* [There is no selection rule for n]

Example 7.8. *Is the transition $\psi_{200} \rightarrow \psi_{100}$ allowed or forbidden? Justify your answer.*

Solution. This is a transition from the state $n = 2$; $l = 0$, $m_l = 0$ to the state $n = 1$, $l = 0$, $m_l = 0$. This transition satisfies the rules $\Delta m_l = 0$ but does not satisfy the selection rule for orbital angular momentum quantum number $\Delta l = \pm 1$ because in this case $\Delta l = 0$. The transition $\psi_{200} \rightarrow \psi_{100}$ is therefore forbidden (not allowed).

7.14 COMPARISON OF QUANTUM MODEL AND BOHR MODEL OF HYDROGEN ATOM

The two models compare with regard to following points, which also give in what respects the predictions of quantum mechanics for angular momentum differ form those based on Bohr's theory.

Orbital angular momentum. 1. In Bohr's theory, quantisation of angular momentum was introduced as a postulate without giving any justification, but in quantum mechanics quantisation of angular momentum appears as a result of the condition that the solution of the component function $\Theta(\theta)$ of the variable θ should be finite. The orbital quantum number also defines the shape of the electron orbit. The semi-minor axis of the ellipse is given by $b_n = a_n \dfrac{l+1}{n}$.

2. In Bohr's theory, the same quantum number n is involved in the expression for angular momentum and energy, but in quantum mechanics angular momentum is represented by a quantum number different from that used for the expression of total energy.

3. In Bohr's theory, n is not equal to *zero i.e.* the angular momentum can never be zero, because zero angular momentum in the theory means no motion of the electron and its consequent collapse into the nucleus. But in quantum mechanics, angular momentum can have zero value when $l = 0$. Physically it means that the wave function ψ does not change with θ for these states for which $l = 0$.

4. In Bohr' theory, the electron in the nth quantum state has only one value of angular momentum $= n\hbar$, but in quantum mechanics, there are n such values for $l = (n-1)(n-2) \ldots 2, 1, 0$. The maximum value of $L = \hbar\sqrt{l(l+1)} = \hbar\sqrt{(n-1)\,n}$ which is less than $n\hbar$ the value given by Bohr's theory. The spectroscopic data are in complete agreement with quantum mechanical values.

5. According to quantum mechanics, the radius of the first Bohr orbit gives the distance at which the probability of finding the electron is maximum. Similarly the radii of other Bohr orbits are also the distances at which the probability of finding the electron is maximum. Thus the quantum mechanics deals with the radii of the electronic orbits in terms of probability density.

6. Bohr's theory introduces only one quantum number n, whereas quantum mechanics introduces three quantum numbers n, l and m_l. Bohr's theory does not give the selection rules, according to which the electron can jump from one energy level to another. These are correctly predicted by quantum mechanics, which also sets the condition for allowed and forbidden transitions.

Energy. In quantum model as well as in Bohr's theory the energy of the electron in hydrogen atom is the same and equal to $\dfrac{-13.6 \text{ eV}}{n^2}$.

According to Bohr's theory

$$E_n = \frac{-2\pi^2 k^2 Z^2 m e^4}{h^2} \cdot \frac{1}{n^2}$$

where $k = \dfrac{1}{4\pi\varepsilon_0}$ and for hydrogen Z = 1. Substituting we get

$$E_n = -2\pi^2 \frac{1}{(4\pi\varepsilon_0)^2} \frac{me^4}{h^2} \frac{1}{n^2}$$

$$= -\frac{me^4}{32\pi^2 \varepsilon_0^2 \hbar^2} \cdot \frac{1}{n^2} = -\frac{13.6 \text{ eV}}{n^2}$$

According to Quantum theory, the radial part of Schrödinger's wave equation for hydrogen atom also gives the energy value as

$$E_n = -\frac{me^4}{32\pi^2 \varepsilon_0^2 \hbar^2} \frac{1}{n^2} = -\frac{13.6 \text{ eV}}{n^2} \qquad \ldots(7.18)$$

EXERCISE

1. State Schrödinger's three dimensional time independent wave equation in Cartesian co-ordinates. Derive from it the wave equation in spherical polar co-ordinates.
(Pbi. U., 2002; P.U., 2007, 2006, 2001, 1995; Bang. U., 2001; G.N.D.U., 1997)

2. Coulomb's potential energy of an electron at a distance r from the nucleus of a hydrogen atom $V = \frac{-e^2}{4\pi\varepsilon_0 r}$. Substituting this value of V in Schrödinger's steady state equation obtain three independent equations in three different spherical polar co-ordinates r, θ and ϕ.
(Pbi, U., 2003; H.P.U. 2003, 2002, 2000, 1997; G.N.D.U., 1995, 1992; P.U., 2006, 1995, 1991, 1990; Poorvanchal U. 2002)

3. Derive Schrödinger's time-independent wave equation for hydrogen atom in spherical polar co-ordinates r, θ and ϕ.

4. Solve Θ (θ) equation for Hydrogen atom.

5. (*a*) State azimuthal component of Schrödinger's wave equation of H-atom. Obtain its solution and normalised eigen functions. Explain the significance of the quantum number predicted by it. *(Cal. U. 2003; Pbi. U., 2006, 2002; 1993; H.P.U., 1991; G.N.D.U., 1992)*

   (*b*) The ϕ part of Schrödinger's equation expressed in spherical polar co-ordinates has the solution $\Phi = e^{im\phi}$. Write with justification the possible values of quantum number m.
(G.N.D.U. 2003)

6. Define orbital angular momentum and orbital quantum number. State the relation connecting them. *(P.U. 1997; G.N.D.U., 1996)*

7. (*a*) What are spherical harmonics? *(G.N.D.U. 2008)*

   (*b*) Solve the angular parts of Schrödinger's equation for a spherically symmetric potential and show that the solutions are eigen functions of $\hat{L}_z$ and $\hat{L}^2$ operator. *(Pbi.U. 2005)*

8. Prove that $\hat{L}_z \psi_{nlm} = (\hbar m_l) \psi_{nlm}$ where the symbols have their usual meanings. *(P. U. 2002)*

9. (*a*) Solve radial part [R (r)] of Schrödinger's equation of hydrogen atom obtaining its energy values and radial wave function *(G.N.D.U. 2007; Poorvanchal U. 2002; P.U., 2008, 2007, 2002, 2001, 1999; Pbi. U. 2002, 1995; Gharwal U. 1999; H.P.U. 1992)*

   (*b*) Substituting the values of R_{nl}, Θ_{lm_l} and Φ_{m_l} state the complete wave function of the hydrogen atom. Find the value of the complete ground state and first excited state wave functions. *(H.P. U., 2003; P.U. 2000, 1999)*

10. (*a*) Explain the physical significance of quantum number n, l and m_l. How are they inter-related?

    (*b*) Can principal quantum number be zero? Why is the magnetic quantum number so called?

    (*c*) Obtain an expression for the maximum number of energy states in the shell of total quantum number n.
(Madurai U. 2002, G.N.D.U. 2007; H.P.U. 2003, 1995; Meerut U. 2002; M.D.U. 2001; Pbi. U. 2008, 2001; P.U. 2005, 2001, 2000)

11. (*a*) Write a short note on degeneracy. Show that hydrogen has n^2 fold degeneracy for the nth state. *(Nagpur U. 2002; P.U. 2008, 2007, 2006, 2005, 2003, 2002, 2001, 2000, Pbi. U. 2008, 2005, 2001; G.N.D.U. 2005, 2001; H.P.U. 2000)*

    (*b*) List the set of quantum numbers for $n = 3$; for hydrogen atom.
(P.U. 2001; G.N.D.U., 2001)

12. What role magnetic quantum number plays in space quantisation of angular momentum? Show that the Z-component of $\vec{L}$ is quantised. What prohibits the angular momentum vector $\vec{L}$ in having a definite direction? Discuss the space quantisation of angular momentum clearly explaining the meaning of quantisation of its magnitude and direction. Show the precession of L about Z-axis. ***(Pbi. U. 2006, 2005; H.P.U. 2002)***

13. Explain the term probability density. Calculate the probability densities of various states in a hydrogen atom (n = 1, 2 and 3) and represent them graphically. ***(H.P.U. 2001; G.N.D.U., 2002)***

14. (*a*) On the basis of quantum mechanics prove that a radiation of frequency ν is emitted when an electron in an atom jumps from a higher energy state E_{n2} to a lower energy state E_{n1} given by

$$\nu = \frac{E_{n2} - E_{n1}}{h}$$

(*b*) What is meant by allowed transitions and forbidden transitions for an atom in the excited state? How do they lead to selection rules? Explain. ***(Pbi.U., 2001; G.N.D.U.,2007, 2006, 2005; P.U.,2008, 2007)***

(*c*) Give energy level diagrams of some transitions allowed by selection rule $\Delta l = \pm 1$. ***(Pbi. U., 2001)***

15. Compare the quantum model of hydrogen atom and Bohr's model. In what respects do the predictions of quantum mechanics for energy and orbital angular momentum differ from those based on Bohr's theory? ***(P.U., 2001; G.N.D.U., 2004; H.P.U., 2005)***

16. The wavelength of sodium D_1 line is 590 nm. Calculate the difference in energy levels involved in the emission or absorption of this line. **(Ans. 3.37×10^{-19} J)**

17. Consider an electron in the excited state (n = 3) in hydrogen atom. List all the possible values of l and m_l for the electron. **[Ans. l = 2, 1, 0; $m_l = \pm 2, \pm 1, 0$]**

MULTIPLE CHOICE QUESTIONS

1. The velocity of electron in the first Bohr orbit is

(*a*) $\frac{1}{10}c$ (*b*) $\frac{1}{13.7}c$ (*c*) $\frac{1}{137}c$ (*d*) $\frac{1}{1370}c$

2. The Rydberg constant for hydrogen is R. Which of the following values give the short wavelength limit of Lyman series?

(*a*) R (*b*) $\frac{1}{R}$ (*c*) $2R$ (*d*) $\frac{2}{R}$

3. Rydberg constant for hydrogen is $1.097 \times 10^7\ m^{-1}$. What is the longest wavelength in Lyman series?

(*a*) 1612 Å (*b*) 1216 Å (*c*) 911 Å (*d*) 119 Å

4. The ratio of proton mass to electron mass is

(*a*) 1.837 (*b*) 18.37 (*c*) 183.7 (*d*) 1837

5. The wave function of hydrogen atom depends only on

(*a*) n (*b*) l (*c*) n and l (*d*) n, l and m_l

6. The value of principal quantum number n corresponding to ground state of hydrogen is

(*a*) Zero (*b*) one (*c*) two (*d*) ∞

7. The energy of hydrogen atom depends only on.

(*a*) l (*b*) m (*c*) n (*d*) l and m

8. The orbital angular momentum of first excited state of hydrogen is

(*a*) $\sqrt{2}\,\hbar$ (*b*) $\sqrt{2}\,h$ (*c*) $2\,h$ (*d*) $2\hbar$

9. The maximum value allowed for orbital quantum number for a given principal quantum number n is

(*a*) n (*b*) $n+1$ (*c*) $n-1$ (*d*) $n-2$

10. The Z-component of orbital angular momentum in the ground state of hydrogen is

(*a*) 0 (*b*) $\hbar$ (*c*) $2\hbar$ (*d*) $3\,\hbar$

11. The number of allowed orientations for orbital angular momentum for the state $l = 1$ are

(*a*) 0 (*b*) 1 (*c*) 2 (*d*) 3

12. The selection rules for the allowed transitions between energy levels of hydrogen atom are

(*a*) $\Delta l = 0\ \Delta m = \pm 1$ (*b*) $\Delta l = 0\ \Delta m = 0$

(*c*) $\Delta l = \pm 1\ \Delta m = \pm 1$ (*d*) $\Delta l = \pm 1\ \Delta m = 0$

ANSWERS

1. (*c*)	**2.** (*b*)	**3.** (*b*)	**4.** (*d*)	**5.** (*d*)
6. (*b*)	**7.** (*c*)	**8.** (*a*)	**9.** (*c*)	**10.** (*a*)
11. (*d*)	**12.** (*c*)			

SECTION III

ATOMIC PHYSICS

8. ATOMS WITH ONE ELECTRON
9. ATOMS WITH MANY ELECTRONS
10. X-RAYS

ATOMS WITH ONE ELECTRON

8

Chapter

INTRODUCTION

Here we make an entry into the mysterious world of atomic physics through the behaviour of atoms having only one electron in their outermost orbit. The simplest example is that of hydrogen atom. In this, we will study about the discovery of spin, space quantization, spin-orbit coupling, vector atom model of atom required to explain the complex spectra of atoms, fine and hyperfine structure of spectral lines and Stern-Gerlach experiment leading to some of the above facts. Lastly we shall discuss about the normal and anomalous Zeeman effect.

8.1 CRITICAL, IONISATION AND EXCITATION POTENTIALS

Critical potential. *The least energy, in electron volts, which is necessary to excite a free neutral atom from its ground state to a higher energy state, is called a critical potential of the atom.*

Ionisation potential. A critical potential is called an *ionisation potential*, when the absorption of energy by the excited atom just removes an electron from it *i.e.*, the electron is raised from the level $n = 1$ to $n = \infty$.

Excitation potential. A critical potential is called an *excitation potential*, when the absorption of energy by the unexcited atom just raises it to a higher energy state (but dose not ionise it). On the basis of Bohr's theory, the various energy states in a hydrogenic atom are given by

$$E_n = -\frac{2\pi^2 k^2 Z^2 me^4}{n^2 h^2} = -\frac{13.6 \text{ eV}}{n^2} \qquad \text{(for Hydrogen Z = 1)}$$

Thus the energy of the electron in a hydrogen atom for the First, Second, Third etc., orbits are – 13.6 eV ($n = 1$), – 3.4 eV ($n = 2$), – 1.51 eV ($n = 3$), and 0 eV ($n = \infty$).

Hence the energy required to raise the hydrogen atom from the ground state to the first excited state is equal to (13.6 – 3.4) = 10.2 eV, that required to raise it to the second excited state is (13.6 – 1.51) = 12.09 eV and the energy required to ionise it is (13.6 – 0) = 13.6 eV.

In other words, discrete amounts of energy are required to excite the hydrogen atom to different energy states, which are thus quantised. The first and second excitation potentials are 10.2 and 12.09 volts respectively and the ionisation potential is 13.6 volts.

8.2 FRANK-HERTZ EXPERIMENT

Frank-Hertz experiment indicates the presence of quantised energy states in an atom and gives a method of measuring their values.

The apparatus consists of a glass tube filled with the gas vapours of the atoms under study, fitted with an indirectly heated cathode *C*, a metal grid *G* and a plate *P* as shown in fig. 8.1. The cathode gives a copious supply of electrons when heated. These electrons are accelerated towards the grid *G* maintained at a positive potential V_a with respect to the cathode. The value of V_a can be varied by a potentiometer arrangement. Therefore, the electrons reaching *G* have a kinetic energy = $e\,V_a$. The plate *P* is maintained at a small negative (or retarding) potential of about 0.5 V with respect to *G*, due to which only those electrons reach *P* after crossing *G* which have sufficient kinetic energy to overcome V_r. The electrons reaching *P* give rise to an electric current i_p in the plate grid circuit. This current is measured by a sensitive milli-ammeter *mA*.

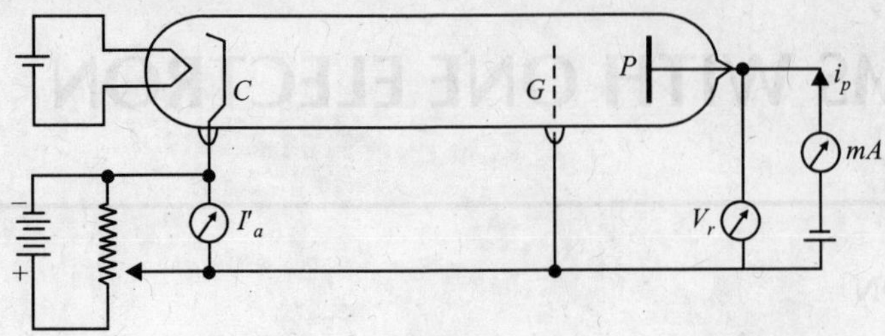

Fig 8.1

As the electrons proceed form C to G, these collide with the atoms of the vapour or gas in the tube. The electrons lose *little* energy so long as their collisions with *heavy* atoms of the gas are *elastic*. The electrons are thus accelerated through a potential difference V_a and will reach the plate P if $V_a > V_r$. As V_a approaches the value of a critical potential, the electrons lose most of their energy in *inelastic* collisions and fail to reach the plate P, as on reaching G their own energy is too small to overcome the opposing potential difference V_r. Thus, if V_a is increased from zero upward, the energy of the electrons reaching G goes on increasing and more and more reach P, thereby giving rise to an increasing current i_p. But for a certain potential V_{a1}, the current suddenly drops. This corresponds to the critical potential, when most of the energy of the electron is lost to the atoms of the gas, which get excited and the electrons, on reaching G are left with insufficient energy to overcome the retarding potential difference V_r. As V_a is further increased, the plate current again begins to increase, as the electrons now have sufficient energy left, even after experiencing an inelastic collision, to reach the plate P. The current goes on increasing till a second value $V_{a_2} = 2V_{a_1}$ is reached. At this stage the current again drops suddenly. This is due to the fact that the electron loses its energy again by exciting *another* gas atom to the same energy level through an inelastic collision. The electrons which have lost their energy by multiple inelastic scattering are themselves left with insufficient energy to overcome the potential V_r on reaching G.

If the applied voltage V_a is further increased, a series of critical potentials for a particular atomic sample can be obtained. Thus the higher potentials are due to more than one inelastic collision and are multiples of the lowest value.

A graph between V_a — the accelerating potential and i_p the plate current for mercury vapour is shown in Fig. 8.2.

Discussion of graph between plate current and accelerating voltage. Frank and Hertz observed that with mercury vapour there is a sudden fall in i_p for values of $V_a = 4.9\ V, 9.8\ V$ and $14.7\ V$ as shown in Fig. 8.2. Thus the first excitation state for mercury is 4.9 V higher than the ground state, and 9.8 V and 14.7 V correspond to twice and three time this value.

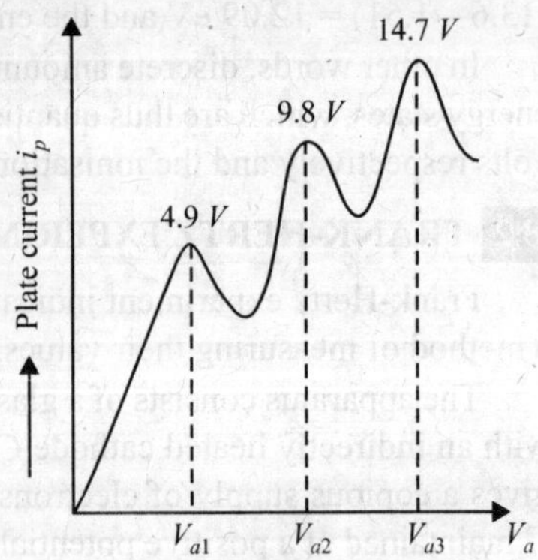

Fig. 8.2

If the pressure of mercury vapour is low, the mean free path of the electrons in the vapour will be large. In such a case, it is possible that the electrons accelerated by a higher value of V_a may undergo the first inelastic collision when it has acquired energy corresponding to first excitation potential of mercury *i.e.*, 4.9 V and then may travel towards G without undergoing any further collision till it acquires 6.7 eV of energy, which corresponds to the next critical potential of mercury.

Determination of critical potentials. The regular spacing of peaks in the $i_p - V_a$ graph shows that mercury atoms accept energy from the bombarding electrons only when these electrons have an energy at least equal to 4.9 eV. This implies that electrons lose energy in discrete amounts in inelastic collisions with the atoms. Hence the first excited state of a mercury atom is 4.9 eV which gives the value of *first critical potential*.

Verification. The Frank-Hertz experiment not only gives a clear indication of the existence of energy levels in an atom, but also a method of measuring the same. The values are in close agreement with spectroscopic data, as the wave length corresponding to first excitation potential of 4.9 V comes out to be 2540 Å as calculated below, whereas the actual value is 2536 Å.

$$\lambda = \frac{c}{v} = \frac{hc}{hv} = \frac{hc}{\Delta E} = \frac{hc}{e\Delta V}$$

$$= \frac{12.4 \times 10^3 \ eV \ \text{Å}}{4.9 \ eV} = 2540 \ \text{Å}.$$

SOLVED EXAMPLES

Example 8.1. *In the spectrum of sodium a line of wave length 5896 Å is obtained during transition of an electron from the first excited state to the ground state. At what different voltages there would be drop in current in Frank Hertz experiment ?* ***(Pbi. U. 2002)***

Solution. Suppose in Frank Hertz experiment first drop in current takes place at a voltage ΔV. Then this voltage corresponds to the energy difference between the first excited state and the ground state. The wave length of light obtained in this transition is given by

$$\lambda = \frac{hc}{e\Delta V} \quad \text{or} \quad \Delta V = \frac{hc}{e\lambda}$$

$$= \frac{6.63 \times 10^{-34} \times 3 \times 10^8}{1.6 \times 10^{-19} \times 5896 \times 10^{-10}} = 2.1 \text{ Volt}$$

Further drop in current will occur at 4.2 V, 6.3 V etc.

Example 8.2. *Show that for a charged particle rotating about a point the magnetic dipole moment is proportional to its angular momentum.* ***(Pbi. U. 2000)***

Solution. Consider a charged particle of mass *m* having a charge *e* rotating about a point with a linear velocity *v* in a circular path of radius *r*, then it is equivalent to an electric current given by

$$I = \frac{\text{Charge}}{\text{Time takan to once complete the orbit}}$$

$$= \frac{e}{2\pi r/v} = \frac{ev}{2\pi r}$$

The magnitude of the dipole magnetic moment of the rotating charged particle, which is equivalent to a current loop

$$p_m = IA = \frac{ev}{2\pi r} \cdot \pi r^2 = \frac{1}{2} evr$$

The magnitude of the angular momentum of the charged particle of mass *m* rotating with a velocity *v* in a circular path of radius *r* is given by

$$L = mvr$$

$$\therefore \quad p_m = \frac{1}{2}\frac{e}{m} L \quad \text{or} \quad p_m \propto L \qquad [\because e/m = \text{a constant}]$$

Hence the magnitude of the dipole moment of a charged particle rotating about a point is proportional to its angular momentum.

8.3 QUANTISATION OF MAGNETIC MOMENT

Orbital magnetic moment. Consider an electron of mass m, charge $-e$ revolving round the nucleus of an atom in a circular orbit of radius r in $x-y$ plane with a velocity (of magnitude) v.

The charge circulating in a loop is equivalent to current

$$I = -\frac{e}{T} = \frac{-ev}{2\pi r}$$

A current loop has a magnetic moment.

∴ Orbital magnetic moment of the electron

$$\mu_l = \text{current} \times \text{area} = I\,A$$

$$= \frac{-ev}{2\pi r}\,\pi r^2 = -\frac{evr}{2}$$

Multiplying and dividing by m, we have

$$\mu_l = \frac{-emvr}{2m} = \frac{-e}{2m} L$$

where $L = mvr$, the orbital angular momentum of the electron.

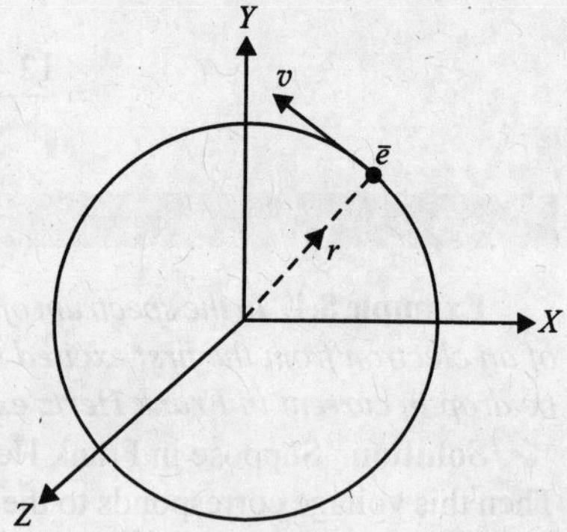

Fig. 8.3

In vector notation

$$\vec{\mu}_l = \frac{-e}{2m}\vec{L} \qquad \text{...(i)}$$

The negative sign shows that the direction of the orbital magnetic moment is opposite to that of the orbital angular momentum.

Ratio of orbital magnetic moment to angular momentum. From equation (*i*), we have

$$\frac{\text{Orbital magnetic moment}}{\text{Orbital angular momentum}} = \frac{\vec{\mu}_l}{\vec{L}} = \frac{-e}{2m}.$$

Magnetic moment of the atom. Again equation (*i*) is generally written as

$$\vec{\mu}_l = -g_l\frac{e}{2m}\vec{L}$$

where $g_l = 1$ and is called *orbital g-factor of the electron.*

According to quantum mechanics

$$\vec{L} = \sqrt{l(l+1)}\,\hbar$$

where l is the orbital quantum number, having values 0, 1, 2 ..... etc.

$$\therefore \qquad \vec{\mu}_l = -g_l\frac{e}{2m}\sqrt{l(l+1)}\,\hbar = \frac{-e\hbar}{2m}\sqrt{l(l+1)} \qquad \text{...(ii)}$$

For $l = 0$, $\mu_l = 0$; For $l = 1$, $\mu_l = \sqrt{2}\,\frac{e\hbar}{2m}$; For $l = 2$, $\mu_l = \sqrt{6}\,\frac{e\hbar}{2m}$.

Z-component of orbital magnetic moment. The Z-component of orbital angular momentum, $\vec{L} = L_z = m_l\hbar$ where m_l is the orbital magnetic quantum number. For a given value of l

$$m_l = l, (l-1) \ldots 0, 1 \ldots. (-l+1), -l$$

The Z-component of orbital magnetic moment $(\mu_l)_z = \frac{-e}{2m}L_z = \frac{-e\hbar}{2m}m_l$

For $m_l = 1,\ (\mu_l)_z = \dfrac{-e}{2m}\hbar$

Bohr magneton. *A Bohr magneton is the unit of magnetic moment* and has the value

$$\mu_B = \frac{e\hbar}{2m}$$

Substituting the value of $e = 1.6 \times 10^{-19}$ coulomb, $m = 9.1 \times 10^{-31}$ kg and

$$\hbar = \frac{h}{2\pi} = 1.055 \times 10^{-34} \text{ Js we have}$$

$$\mu_B = \frac{1.6 \times 10^{-19} \times 1.055 \times 10^{-34}}{2 \times 9.1 \times 10^{-31}} = 9.27 \times 10^{-24} \text{ Am}^2$$

Physical significance. The presence of smallest unit of magnetic moment equal to one Bohr magneton $= \dfrac{e\hbar}{2m}$ shows that the magnetic moment of the atom, like charge, mass etc. is quantised.

Quantisation of magnetic moment. The Z-component of orbital magnetic moment

$$(\mu_l)_z = -\frac{e\hbar}{2m} m_l$$

where $-\dfrac{e\hbar}{2m}$ is the smallest unit of magnetic moment known as a *Bohr magneton* and m_l is an integral number given by

$$m_l = l, (l-1) \ldots (-l+1), -l$$

Thus the magnetic moment is always an integral multiple of the smallest unit Bohr magneton. *This shows the quantisation of magnetic moment.*

8.4 LARMOUR PRECESSION (Atomic Dipole in a Strong Magnetic Field $\vec{B}$)

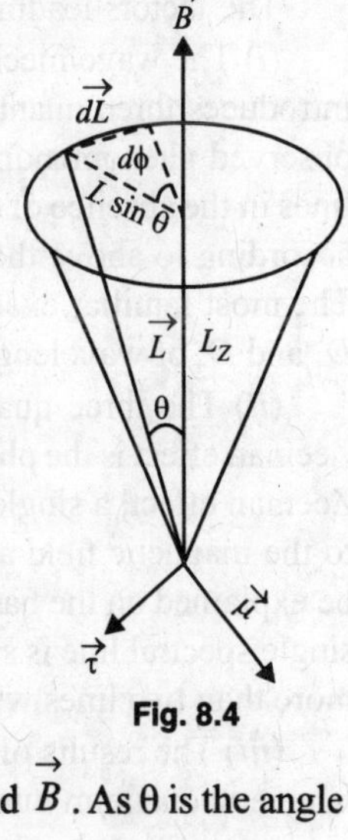

Fig. 8.4

An electron with an orbital magnetic moment $\vec{\mu}$ will experience a torque $\vec{\tau}$ given by

$$\vec{\tau} = \vec{\mu} \times \vec{B} = -\frac{e}{2m}\vec{L} \times \vec{B}$$

This torque will cause a change in angular momentum, given by

$$\vec{\tau} = \frac{d\vec{L}}{dt} = -\frac{e}{2m}\vec{L} \times \vec{B}$$

The change in $\vec{L}$, $d\vec{L}$ is perpendicular to both $\vec{L}$ and $\vec{B}$ resulting in a precession of $\vec{L}$ about the direction of $\vec{B}$.

or $$\frac{|d\vec{L}|}{dt} = \frac{e}{2m} LB \sin\theta$$

where θ is the angle between the direction of vector $\vec{L}$ and the vector $\vec{B}$.

The vector $d\vec{L}$ lies in the plane of precession perpendicular to both $\vec{L}$ and $\vec{B}$. As θ is the angle between the direction of $\vec{L}$ and that of $\vec{B}$, the component of $\vec{L}$ in the plane of precession $= L \sin\theta$.

Now let $d\phi$ be the angle that $d\vec{L}$ subtends at the point where the vector $\vec{B}$, passes through the plane of precession, then

$$d\phi = \frac{|d\vec{L}|}{L \sin\theta}$$

The angular velocity of precession

$$\omega_p = \frac{d\phi}{dt} = \frac{\left|\frac{d\vec{L}}{dt}\right|}{L \sin\theta} = \frac{\frac{e}{2m} LB \sin\theta}{L \sin\theta}$$

$$= \frac{e}{2m} B.$$

This is known as *Larmour precession* and ω_p is equal to angular frequency difference observed in the normal Zeeman effect between the displaced and the undisplaced spectral lines.

Frequency of Larmour precession. $\nu_p = \frac{\omega_p}{2\pi}$

$$= \frac{1}{2\pi} \frac{e}{2m} B$$

$$= \frac{e}{4\pi m} B$$

Example 8.3 *An atom is placed is a magnetic field of strength 0.1 Tesla. Calculate the rate of precession. Given : m = 9 × 100*$^{-31}$ *kg.* ***(M.D.U. 2001)***

Solution. Rate of precession $\nu = \frac{eB}{4\pi m} = \frac{1.6 \times 10^{-19} \times 0.1}{4\ \pi \times 9 \times 10^{-31}}$

$$= 14.14 \times 10^8\ s^{-1}$$

8.5 FACTORS LEADING TO DISCOVERY OF SPIN

The factors leading to the discovery of spin are:

(*i*) The wave mechanical treatment of hydrogenic atoms, on the basis of Schrödinger's theory, introduces three quantum numbers *n, l* and m_l. These three quantum numbers explain most of the observed phenomenon for hydrogen atom, but fail to account for the *fine structure* of the spectral lines in the absence of external magnetic field. Fine structure means that certain spectral lines, which according to above theory should be *singlets* consist of two or more closely spaced spectral lines. The most familiar example is *D* line of sodium of wavelength 5893 Å which has two components, D_1 and D_2 of wavelengths 5890 Å and 5896 Å.

(*ii*) The three quantum numbers were also insufficient to explain *anomalous Zeeman effect*. Zeeman effect is the phenomenon of splitting up of spectral lines in a strong magnetic field. In normal. Zeeman effect, a single spectral line is split up into *three* spectral lines, when viewed at right angles to the magnetic field and into two lines when viewed longitudinally. Normal Zeeman effect could be explained on the basis of (orbital) magnetic quantum number, m_l. In anomalous Zeeman effect, a single spectral line is split up into *more than three* spectral lines, when viewed transversely and into more than two lines when viewed longitudinally.

(*iii*) The results of Stern – Gerlach experiment could not be explained on the basis of (orbital), magnetic quantum number m_l. For a given value of *l*, m_l has $(2l + 1)$ values, so that for hydrogenic atoms, the magnetic moment vector $\vec{\mu}$ should have an odd number of orientations with respect to the applied magnetic field. But in Stern-Gerlach experiment for hydrogen, silver etc., there are only two traces, showing thereby that there were only *two* orientations of the observed magnetic moment. These could not be due to orbital angular momentum, but there is some other cause behind it. This necessitated the introduction of the concept of spin in quantum theory.

8.6 SPIN ANGULAR MOMENTUM

The idea of intrinsic spin angular momentum was proposed by Goudsmit Uhelnbeck. According to this hypothesis, an electron '*spins*' about an axis passing through its centre of mass. It has, therefore, an *inherent intrinsic* or *internal* angular momentum and a magnetic moment associated with spin rotation. The spin motion of the electron is in addition to its orbital motion around the nucleus, just as the earth spins about its own axis and also revolves round the sun.

The correct interpretation of spin momentum was given by Dirac in his theory of relativistic quantum mechanics, where a new quantum number corresponding to the fourth time co-ordinate was introduced as a natural requirement and was related to spin. Spin is essentially a quantum mechanical phenomenon and has no analog in classical physics.

Magnitude. The spin angular momentum $\vec{S}$ of the electron has a constant *magnitude* given by

$$S = \sqrt{s(s+1)}\,\hbar = \sqrt{\frac{1}{2}\left(\frac{1}{2}+1\right)}\,\hbar = \frac{\sqrt{3}}{2}\,\hbar$$

where s is a quantum number for '*spin*' motion, similar to quantum number l for '*orbital*' motion and can have only one value $\frac{1}{2}$. A free electron has only intrinsic spin angular momentum. Hence the total angular momentum of a free electron $S = \frac{\sqrt{3}}{2}\,\hbar$

8.7 SPACE QUANTISATION

The component of the spin angular momentum along the direction of the external magnetic field (parallel to *Z*-axis) is also *space quantised* like the *Z*-component of orbital angular momentum and is given by $S_z = m_s\,\hbar = \pm\frac{1}{2}\,\hbar$. m_s is called the *spin magnetic quantum number* for which the allowed values are $+\frac{1}{2}$ and $-\frac{1}{2}$. These are generally known as *up* and *down* spins respectively.

Hence we find that the spin angular momentum cannot be aligned completely with the applied magnetic field. It can align itself only in two directions for which the Z-component is $\pm 1/2\,\hbar$. Thus by virtue of its spin, the electron possesses two *internal* (or intrinsic) quantum states for two values of $m_s = +\frac{1}{2}$ and $m_s = -\frac{1}{2}$. This shows the space quantisation of electron spin.

Two orientations of spin angular momentum. If θ is the angle between the spin vector $\vec{S}$ and Z-axis, then

$$\cos\theta = \frac{S_z}{S} = \frac{m_s\,\hbar}{\sqrt{s(s+1)}\,\hbar}$$

$$= \frac{m_s}{\sqrt{s(s+1)}}$$

For *spin up*, $m_s = +\frac{1}{2}$

$$\cos\theta_1 = \frac{\frac{1}{2}}{\sqrt{\frac{1}{2}\left(\frac{1}{2}+1\right)}} = \frac{1}{\sqrt{3}} = 0.577$$

∴ $\theta_1 = 54°45'$

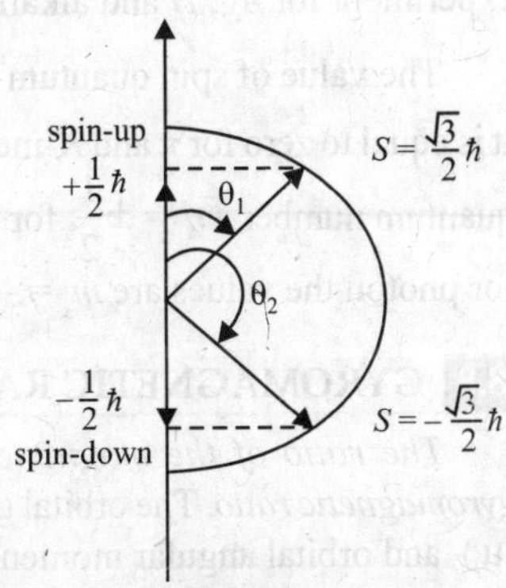

Fig. 8.5

For spin down, $m_s = -\frac{1}{2}$

$$\cos\theta_2 = \frac{-\frac{1}{2}}{\sqrt{\frac{1}{2}\left(\frac{1}{2}+1\right)}} = -\frac{1}{\sqrt{3}} = -0.577$$

$$\therefore \qquad \theta_2 = 125°15'$$

θ_1 and θ_2 are shown in Fig. 8.5.

Spin-magnetic moment. The magnetic moment associated with spin angular momentum is given by

$$\vec{\mu}_s = -2\frac{e}{2m}\vec{S} = -2\frac{e}{2m}\sqrt{s(s+1)}\,\hbar$$

$$= -g_s\frac{e\hbar\sqrt{s(s+1)}}{2m} = -g_s\,\mu_B\sqrt{s(s+1)}$$

$$= \frac{-g_s\,\mu_B\,\vec{S}}{\hbar}$$

where $g_s = 2$ is called the *spin g-factor* and $\frac{e\hbar}{2m} = \mu_B$ the Bohr magneton.

As $s = \frac{1}{2}$ $\qquad \mu_s = -g_s\,\mu_B\sqrt{\frac{1}{2}\left(1+\frac{1}{2}\right)} = -g_s\,\mu_B\frac{\sqrt{3}}{2}$

or $\qquad |\mu_s| = 2\mu_B\frac{\sqrt{3}}{2} = \sqrt{3}\,\mu_B = 1.73\mu_B$

The Z-component of magnetic moment

$$(\mu_s)_z = 2\frac{e}{2m}S_z = 2\frac{e}{2m}m_s\,\hbar \qquad [\because S_z = m_s\,\hbar]$$

$$= \pm\frac{e}{2m}\hbar \qquad \left[\because m_s = \pm\frac{1}{2}\right]$$

The fact that $(\mu_s)_z$ has only two values $+\frac{e}{2m}\hbar$ and $-\frac{e}{2m}\hbar$ has been verified by Stern Gerlach experiment for *Ag*, *H* and alkalie metals i.e. for those atoms for which in ground state $l = 0$.

The value of spin quantum number $s = \frac{1}{2}$ for electron, proton, neutron, neutrino, μ meson etc. It is equal to zero for π and *K*-mesons and has a value 1 for photon. Hence the values of spin magnetic quantum number $m_s = \pm\frac{1}{2}$ for electron, photon, neutrino and μ-meson, zero for π and *K*-meson and for photon the values are $m_s = -1$, and $+1$.

8.8 GYROMAGNETIC RATIO

The ratio of the magnetic moment to angular momentum is a constant and is known as gyromagnetic ratio. The orbital gyromagnetic ratio between Z-component of orbital magnetic moment $(\mu_l)_z$ and orbital angular momentum L_z is given by

$$g\,(\text{orbital}) = \frac{(\mu_l)_z}{L_z} = \frac{\frac{e}{2m}\, m_l\, \hbar}{m_l\, \hbar} = \frac{e}{2m}$$

The spin gyromagnetic ratio between Z-component of *spin* magnetic moment $(\mu_s)_z$ and spin angular momentum S_z in given by

$$g\,(\text{spin}) = \frac{(\mu_s)_z}{S_z} = \frac{\frac{e}{2m}\hbar}{\frac{1}{2}\hbar} = 2\frac{e}{2m} = 2g\ (\text{orb.})$$

Thus, the spin gyromagnetic ratio is *twice* the orbital gyromagnetic ratio.

The correct interpretation of Z-component of spin magnetic moment $= \frac{e}{2m}\hbar$ was also given by Dirac on the basis of relativistic quantum mechanics.

Example 8.4. *What is the total angular momentum of a free electron?* ***(P.U. 2006)***

Solution. A free electron has no orbital angular momentum. It has only an intrinsic spin angular momentum with spin quantum number $s = \frac{1}{2}$. So its total angular momentum

$$J = s = \sqrt{s\,(s+1)}\ \hbar = \sqrt{\frac{1}{2}\times\frac{3}{2}}\ \hbar = \frac{\sqrt{3}}{2}\ \hbar$$

Example 8.5. *A beam of electrons enters a uniform magnetic field of 1.2 Tesla along the direction of the field. Find the energy difference between electrons whose spin is parallel and anti parallel to the field.* ***(P.U. 2001)***

Solution. The spin magnetic moment of the electron is given by

$$\vec{\mu}_s = -\frac{g_s\, \mu_B}{\hbar}\vec{S}$$

where $g_s = 2$, $\mu_B = 1$ Bohr magneton $= 9.27 \times 10^{-24}$ Am^2

When placed in a magnetic field $\vec{B}$, the energy of the electron is given by

$$E = -\vec{\mu}_s\,.\vec{B}$$

$$= \frac{g_s\, \mu_B}{\hbar}\vec{S}\,.\,\vec{B}$$

Suppose the magnetic field acts in the Z-direction, then $\vec{B} = B\,\hat{k}$ and hence

$$E = \frac{g_s\, \mu_B}{\hbar} B\, S_z$$

Now $S_z = m_s\ \hbar$ where m_s is the magnetic spin quantum number.

$$\therefore \qquad E = g_s\, \mu_B\, B\, m_s$$

(*i*) Electron spin parallel to magnetic field. In this case $m_s = +\frac{1}{2}$. The energy of the electron

$$E(\uparrow) = +\frac{1}{2}\, g_s\, \mu_B\, B$$

(*ii*) Electron spin anti-parallel to magnetic field. In this case $m_s = -\frac{1}{2}$. The energy of the electron

$$E\,(\downarrow) = -\frac{1}{2}\, g_s\, \mu_B\, B$$

$\therefore$ Energy difference $E\,(\uparrow) - E\,(\downarrow) = \frac{1}{2}\, g_s\, \mu_B\, B + \frac{1}{2}\, g_s\, \mu_B\, B$

$$= g_s \mu_B B$$

Substituting the values, we get

$$E(\uparrow) - E(\downarrow) = 2\times9.27\times10^{-24}\times1.2 = 22.248\times10^{-24}\text{ J}$$

$$= \frac{22.248\times10^{-24}}{1.6\times10^{-19}} = 13.9\times10^{-5}\text{ eV.}$$

8.9 SPIN-ORBIT COUPLING

Generally an electron in an atom possesses both the orbital and spin angular momentum. Since these are vector quantities, the resultant angular momentum is given by

$$\vec{J} = \vec{L} + \vec{S} \qquad ...(i)$$

where $\vec{J}$ is known as *total angular momentum vector.*

The combination of orbital and spin angular momentum vectors $\vec{L}$ and $\vec{S}$ to form a single total angular momentum vector $\vec{L}$ is known as spin orbit coupling.

Total angular momentum vector $\vec{J}$. *The total angular momentum vector $\vec{J}$ for an atom is the vector sum of the resultant orbital angular momentum $\vec{L}$ and the resultant spin angular momentum $\vec{S}$.*

$$\therefore \qquad \vec{J} = \vec{L} + \vec{S}$$

The vector $\vec{J}$ is also quantised and its magnitude is given by

$$J = \sqrt{j(j+1)}\,\hbar$$

where j is the *total angular momentum quantum number.* Further $\vec{J}$ is also apace quantised, similar to $\vec{L}$ and $\vec{S}$. In other words $\vec{J}$ can have only some particular orientations.

Z-component of $\vec{J}$. The component of $\vec{J}$ in the direction of the applied magnetic field (Z-axis) is given by

$$J_z = m_j\,\hbar \qquad ...(ii)$$

where m_j is the *magnetic total quantum number* and has $(2j + 1)$ values given by $-j$, $(-j + 1)$, $(-j + 2)$, ... $+ (j - 1)$, $+j$

Quantisation of Z-component. From relation (*ii*), we find that the Z-component of total angular momentum J_z is quantised in half integral multiples of $\hbar$ since m_j has half integral values as proved below.

From equation (*i*) for Z-components of $\vec{J}$, $\vec{L}$ and $\vec{S}$, we have

$$J_z = L_z \pm S_z$$

or $$m_j\,\hbar = m_l\,\hbar \pm m_s\,\hbar$$

or $$m_j = m_l \pm m_s \qquad ...(iii)$$

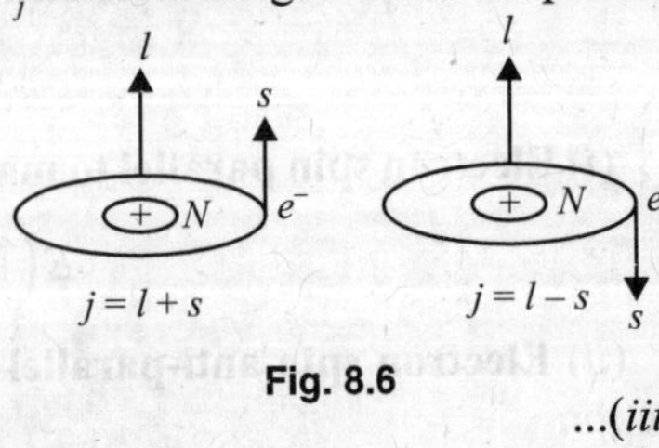

Fig. 8.6

The possible values of m_l range from $+ l$ through 0 to $-l$ and those of m_s are $\pm$ s where $s = \frac{1}{2}$. The quantum number l is always an integer or 0. Therefore, m_j is always half integral and its possible values range from $+ j$ to $-j$ in integral steps.

$$\therefore \qquad j = l \pm s \qquad ...(iv)$$

Total angular momentum quantum number j for a one electron atom. From the relation $j = l \pm s$ we find that for a given value of orbital angular momentum quantum number l for an electron in a *one electron atom*, two possible values of total angular momentum quantum number j are possible.

These are $j = l + \frac{1}{2}$ (spin up) when s is parallel to l, as shown in Fig. 8.6 (a) and $j = l - \frac{1}{2}$ (spin down) when s is anti-parallel to l as shown in Fig. 8.6 (b).

For $l = 0$, the total angular momentum quantum number $j = \frac{1}{2}$.

For $l = 1$, $j = l + \frac{1}{2} = 1 + \frac{1}{2} = \frac{3}{2}$ and $j = l - \frac{1}{2} = 1 - \frac{1}{2} = \frac{1}{2}$

$\therefore$ Magnitude of $\vec{L} = \sqrt{l(l+1)}\,\hbar = \sqrt{1(1+1)}\,\hbar = \sqrt{2}\,\hbar$

$$\vec{S} = \sqrt{s(s+1)}\,\hbar = \sqrt{\frac{1}{2}\left(1+\frac{1}{2}\right)}\,\hbar = \frac{\sqrt{3}}{2}\,\hbar$$

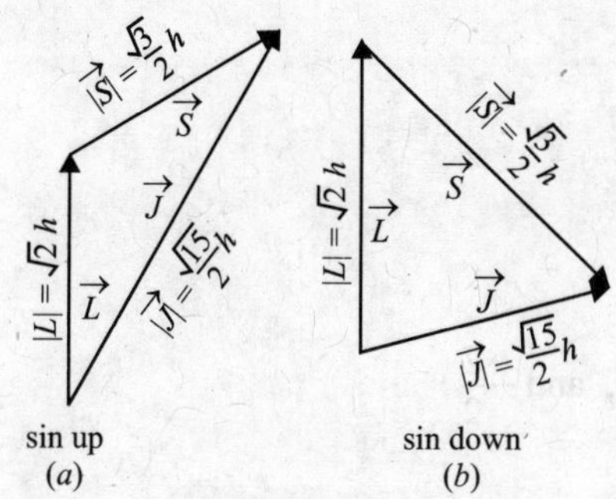

Fig. 8.7

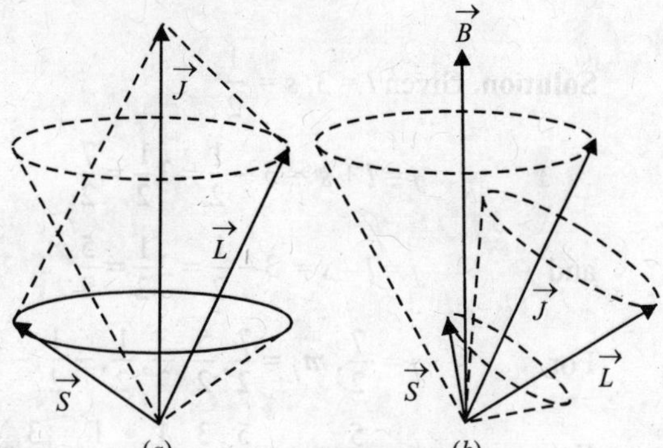

Fig. 8.8

For $j = \frac{3}{2}$ the magnitude of vector

$$\vec{J} = \sqrt{j(j+1)}\,\hbar = \sqrt{\frac{3}{2}\left(\frac{3}{2}+1\right)}\,\hbar$$

$$= \frac{\sqrt{15}}{2}\,\hbar$$

For $j = \frac{1}{2}$ the magnitude of vector

$$\vec{j} = \sqrt{\frac{1}{2}\left(\frac{1}{2}+1\right)}\,\hbar = \frac{\sqrt{3}}{2}\,\hbar$$

The magnitude and direction of $\vec{J}$ for $j = 3/2$ (spin up) *i.e.* $l = 1$ and $s = \frac{1}{2}$ and for $j = \frac{1}{2}$ (spin down) *i.e.* $l = 1$ and $s = -\frac{1}{2}$ with corresponding magnitude and direction of $\vec{L} = \sqrt{2}\,\hbar$ and $\vec{S} = \frac{\sqrt{3}}{2}\,\hbar$ is shown in Fig. 8.7(a) and (b).

Precession of $\vec{L}$ and $\vec{S}$ about $\vec{J}$. The orbital and spin angular momenta $\vec{L}$ and $\vec{S}$ of the atomic electron *interact magnetically* which is known as *spin-orbit interaction*. Due to this spin-orbit interaction, $\vec{L}$ and $\vec{S}$ precess independently around the resultant vector $\vec{J}$ as shown in Fig. 8.8(a).

When an atom is placed in an external magnetic field $\vec{B}$, then $\vec{J}$ precesses about direction of $\vec{B}$ where as $\vec{L}$ and $\vec{S}$ continue to precess about $\vec{J}$ as shown in Fig. 8.8(b). The vector $\vec{J}$ is *space quantised* with respect to the external magnetic field $\vec{B}$, i.e. it can have only $(2j + 1)$ discrete orientations with respect to the external magnetic field $\vec{B}$ such that m_j (the component of j along the magnetic field vector $\vec{B}$) can have values given by

$$m_j = j, (j-1), ... (-j+1), -j$$

These discrete orientations of $\vec{J}$ with respect to $\vec{B}$ give rise to slightly different energies and explain the anomalous Zeeman effect.

Example 8.6. *Find the possible values of j and m_j for states in which $l = 3$ and $s = \frac{1}{2}$.*

(P.U. 2007)

Solution. Given $l = 3$, $s = \frac{1}{2}$

$$\therefore \qquad j = l + s = 3 + \frac{1}{2} = 3\frac{1}{2} = \frac{7}{2}$$

$$\text{and} \qquad j = l - s = 3 - \frac{1}{2} = 2\frac{1}{2} = \frac{5}{2}$$

$$\text{For} \qquad j = \frac{7}{2}, m_j = \frac{7}{2}, \frac{5}{2}, \frac{3}{2}, \frac{1}{2}, -\frac{1}{2}, -\frac{3}{2}, -\frac{5}{2}, \text{ and } -\frac{7}{2}$$

$$\text{For} \qquad j = \frac{5}{2}, m_j = \frac{5}{2}, \frac{3}{2}, \frac{1}{2}, -\frac{1}{2}, -\frac{3}{2} \text{ and } -\frac{5}{2}$$

Example 8.7. *Find the possible orientations of total angular momentum vector $\vec{J}$ corresponding o $j = 3/2$ with respect to a magnetic field along Z-axis.* ***(P.U. 2005)***

Solution. The vector $\vec{J}$ is space quantised with respect to the external magnetic field $\vec{B}$.

$$\text{Magnitude of } \vec{J} = J = \sqrt{j(j+1)}\,\hbar = \sqrt{\frac{3}{2} \times \frac{5}{2}}\,\hbar = \sqrt{\frac{15}{2}}\,\hbar$$

where j is total angular momentum quantum number. $\vec{J}$ can have only $(2j + 1)$ discrete orientations with respect to the external magnetic field such that J_z (the Z-component of $\vec{J}$ along the direction of magnetic field) $= m_j\,\hbar$, where m_j is the total angular momentum magnetic quantum number having values

$$m_j = j, (j-1), ... (-j+1), (-j).$$

Also $\frac{J_z}{J} = \cos\theta$, where θ is the angle that vector $\vec{J}$ makes with magnetic field vector $\vec{B}$.

$$\therefore \qquad \cos\theta = \frac{J_z}{J} = \frac{m_j\,\hbar}{\sqrt{j(j+1)}\,\hbar} = \frac{m_j}{\sqrt{j(j+1)}}$$

For $j = 3/2$, the values of m_j are 3/2, 1/2, –1/2 and –3/2

$$\therefore \qquad \cos\theta_1 = \frac{\frac{3}{2}}{\sqrt{\frac{3}{2}\times\frac{5}{2}}} = \frac{3}{\sqrt{15}} = \frac{3}{3.873} = 0.7746$$

$$\text{or} \qquad \theta_1 = 39.2°$$

$$\cos\theta_2 = \frac{\frac{1}{2}}{\sqrt{\frac{3}{2}\times\frac{5}{2}}} = \frac{1}{\sqrt{15}} = \frac{1}{3.873} = 0.2582$$

$$\text{or } \theta_2 = 75°$$

Similarly $\theta_3 = 180° - 75° = 105°$ and $\theta_4 = 180° - 39.2 = 140.8°$

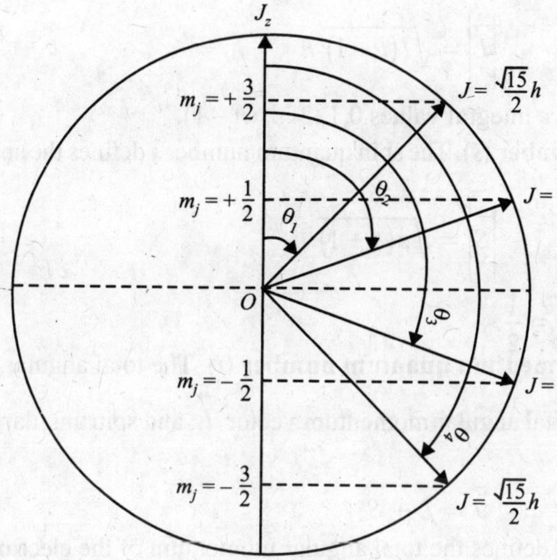

Fig. 8.9

So the possible orientations of $\vec{J}$ with respect to $\vec{B}$ are 39.2°, 75°, 105°, 140.8° as shown in Fig. 8.9.

8.10 VECTOR MODEL OF ATOM

Difficulties with Bohr-Sommerfeld model Bohr's theory was able to explain only the series spectra of the simplest hydrogen atom. It could not, however, explain the multiple structure of spectral lines even in the case of hydrogen atom. Sommerfeld's modification could explain the fine structure of spectral lines of hydrogen but failed to predict the correct number of fine structure lines. Both the theories could not give any information about the relative intensities of spectral lines and also could not explain the complex spectra of alkali metals like *sodium*. These theories also failed to explain phenomenon like Zeeman effect, Stark effect etc.

Vector model of the atom. The vector model of the atom was introduced to explain the complex spectra of atoms and their relation to atomic structure. The two distinct features of the vector model are

(*i*) Spatial quantisation and

(*ii*) Electron spin

Quantum numbers associated with vector atom model. The quantum numbers associated with vector atom model are

(*i*) *Principal quantum number* (n)
(*ii*) *Orbital quantum number* (l)
(*iii*) *Spin quantum number* (s)
(*iv*) *Total angular momentum quantum number* (j)
(*v*) *Magnetic orbital quantum number* (m_l)
(*vi*) *Magnetic spin quantum number* (m_s)
(*vii*) *Magnetic total angular momentum quantum number* (m_j)

(*i*) **Principal quantum number (*n*).** This quantum number is the same as used in Bohr-Sommerfeld theory of Hydrogen spectrum, n denotes the serial number of the shell starting from the inner most taken as 1. It can take only *positive integral values (excluding zero)* i.e. 1, 2, 3, 4 .... n etc.

(*ii*) **Orbital quantum number (*l*).** The quantum number l defines the orbital angular momentum of the electron in the atom, given by

$$\left|\vec{L}\right| = \sqrt{l(l+1)}\,\hbar$$

It can have only *positive* integral values 0,1, 2 .... $(n-1)$.

(*ii*) **Spin quantum number (*s*).** The spin quantum number s defines the spin angular momentum of the electron given by

$$\left|\vec{S}\right| = \sqrt{s(s+1)}\,\hbar$$

It has only *one* value $s = \frac{1}{2}$.

(*iv*) **Total angular momentum quantum number (*j*).** The total angular momentum vector $\vec{J}$ is the vector sum of the orbital angular momentum vector $\vec{L}$ and spin angular momentum vector $\vec{S}$ given by

$$\vec{J} = \vec{L} + \vec{S}$$

The quantum number j defines the total angular momentum of the electron given by

$$\left|\vec{J}\right| = \sqrt{j(j+1)}\,\hbar$$

Total angular momentum quantum number j for a one electron atom is given by

$$j = l \pm s$$

Thus for a given value of orbital quantum number l for an electron in a one electron atom two values of total angular momentum quantum number j are possible. These are $j = l + \frac{1}{2}$ (spin – up) where s is parallel to l and $j = l - \frac{1}{2}$ (spin down) where s is anti-parallel to l.

For $l = 0$, $j = \frac{1}{2}$ (only one value)

To explain the splitting of spectral lines in a magnetic field there are three more quantum numbers known as magnetic orbital quantum number m_l, magnetic spin quantum number m_s and magnetic total angular momentum quantum number m_j.

(*v*) **Magnetic orbital quantum number m_l.** The component of orbital angular momentum $\vec{L}$ in the direction of the magnetic field, known as the Z-component is given by

$$L_z = m_l \hbar$$

where m_l is the orbital magnetic quantum number. For a given value of l, $m_l = l, (l-1) \ldots 0, 1, -1 \ldots (-l+1), -l$. The Z-component of orbital angular momentum is *space-quantised i.e.* it can be inclined to the direction of external magnetic field or Z-axis only along directions given by

$$\cos\theta = \frac{m_l}{l}$$

(*vi*) **Magnetic spin quantum number.** The component of the spin angular momentum along the direction of the external magnetic field or the Z-direction is also *space quantised* and is given by

$$S_z = m_s \hbar = \pm\frac{1}{2}\hbar$$

m_s is called the spin magnetic quantum number for which the allowed values are $+\frac{1}{2}$ and $-\frac{1}{2}$.

(*vii*) **Magnetic total angular momentum quantum number** m_j. The component of the total angular momentum $\vec{J}$ in the direction of the applied magnetic field or the Z-direction is given by

$$J_z = m_j \hbar$$

where m_j is the magnetic total angular momentum quantum number and has $(2j+1)$ values given by $m_j = -j, (-j+1), (-j+2) \ldots\ldots (j-1), j$ ....(*i*)

Relation : To obtain the relationship between various angular momentum quantum numbers, we have for the Z-component of $\vec{J}, \vec{L}$ and $\vec{S}$ that

$$J_z = L_z \pm S_z$$

or
$$m_j \hbar = m_l \hbar \pm m_s \hbar$$

or
$$m_j = m_l \pm m_s$$

Precession of $\vec{L}$ and $\vec{S}$ about $\vec{J}$. The orbital and spin angular momenta $\vec{L}$ and $\vec{S}$ of the atomic electron *interact magnetically* which is known as *spin-orbit interaction*. Due to this spin,orbit interaction $\vec{L}$ and $\vec{S}$ precess independently around the resultant vector $\vec{J}$ as shown in Fig 8.8 (*a*). When an atom is placed in an external magnetic field $\vec{B}$, then $\vec{J}$ precesses about the direction of $\vec{B}$ whereas $\vec{L}$ and $\vec{S}$ continue to precess about $\vec{J}$ as shown in Fig. 8.8 (*b*). The vector $\vec{J}$ is *space quantised* with respect to the external magnetic field $\vec{B}$ i.e. it can have only $(2j+1)$ discrete orientations with respect to the external magnetic field $\vec{B}$ such that m_j, the component of j along the magnetic field vector $\vec{B}$ can have values given by

$$m_j = j, (j-1) \ldots\ldots (-j+1), -j$$

If no external torque acts on the atom, then the total angular momentum $\vec{J}$ is conserved. *This is the vector model of the one electron atom.*

8.11 SPIN-ORBIT INTERACTION

The electron, in an atom, has two angular momenta (*i*) Orbital angular momentum due to orbital motion of the electron around the nucleus and (*ii*) Spin angular momentum due to the intrinsic property of the electron.

The orbital motion of the electron around the nucleus gives rise to a large *internal magnetic field*. The intrinsic magnetic moment of the electron due to its spin interacts with this internal magnetic field, which leads to the modification of the energy levels of the atom.

The interaction between the orbital magnetic moment and spin magnetic moment is called spin-orbit interaction.

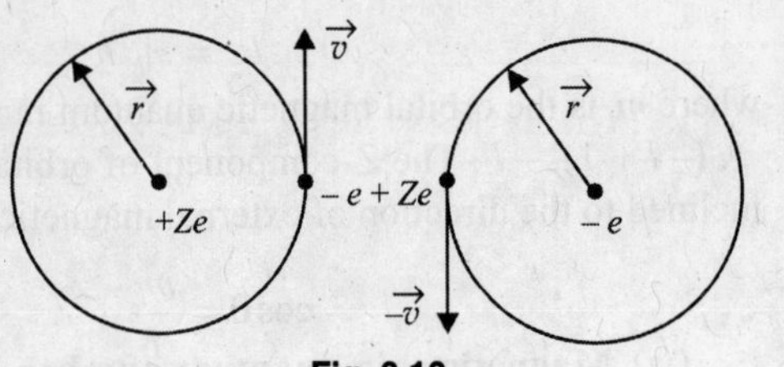

Fig. 8.10

Internal magnetic field. We shall first calculate the value of *internal magnetic field.*

Suppose the electron of charge $-e$ and mass m is moving around the nucleus of charge $+Ze$ in a circular orbit of radius r with a velocity $\vec{v}$ as shown in Fig. 8.10 (*a*). Since the motion is relative, we can consider the electron to be at rest and the nucleus to be moving around the electron in an orbit of radius r and velocity $-\vec{v}$, as shown in Fig. 8.10 (*b*).

The motion of the nucleus with positive charge in a circular orbit is equivalent to the flow of current in a circular loop of radius r. The value of current

$$I = \frac{Ze}{T} = \frac{Ze}{2\pi r/v} = \frac{Zev}{2\pi r}$$

The magnitude of the magnetic field due to this current I at the centre of the circular loop

$$B = \frac{\mu_0 I}{2r} = \frac{\mu_0 Zev}{4\pi r^2} \quad \text{...(i)}$$

As $\vec{r}$ and $\vec{v}$ are vectors and the direction of $\vec{B}$ is perpendicular to both $\vec{r}$ and $\vec{v}$, relation (*i*) can be put in the vector form as

$$\vec{B} = \frac{\mu_0 Ze}{4\pi r^2} \frac{(\vec{r} \times \vec{v})}{r}$$

Multiplying and dividing by m, we have

$$\vec{B} = \frac{\mu_0 Ze}{4\pi m r^3}(\vec{r} \times m\vec{v})$$

Now $\vec{r} \times m\vec{v}$ is the angular momentum of the electron $= \vec{L}$

$$\therefore \quad \vec{B} = \frac{\mu_0 Ze}{4\pi m r^3} \vec{L} \quad \text{...(ii)}$$

Also $$c = \frac{1}{\sqrt{\mu_0 \varepsilon_0}} \text{ or } \mu_0 \varepsilon_0 = \frac{1}{c^2}$$

$$\therefore \quad \mu_0 = \frac{1}{\varepsilon_0 c^2}$$

Substituting in eq. (*ii*), we have

$$\vec{B} = \frac{Ze}{4\pi\varepsilon_0 mc^2 r^3} \vec{L} \quad \text{...(iii)}$$

This result shows that *the magnetic field seen by the electron depends upon its own orbital angular momentum.*

Spin-Orbit interaction energy. We have supposed that the electron is at rest with respect to the nucleus. Therefore, it possesses only spin magnetic moment $\vec{\mu}_s$. When a dipole of magnetic moment $\vec{\mu}_s$ is placed in a field of magnetic flux density $\vec{B}$, the potential energy of the magnetic dipole is given by

$$E'_M = -\vec{\mu}_s \cdot \vec{B} \quad \text{...(iv)}$$

The spin magnetic moment is given by $\vec{\mu}_s = -g_s \frac{\mu_B}{\hbar} \vec{S}$

where g_s is the spin g factor, μ_B the Bohr magneton and $\vec{S}$ spin angular momentum vector.

Substituting in relation (*iv*), we get

$$E'_M = g_s \frac{\mu_B}{\hbar} \vec{S} \cdot \vec{B}$$

But from Eq. (*iii*) $\vec{B} = \dfrac{Ze}{4\pi\varepsilon_0 mc^2 r^3} \vec{L}$

$$\therefore \quad E'_M = \frac{g_s \mu_B Ze}{4\pi\varepsilon_0 mc^2 r^3 \hbar} \vec{S} \cdot \vec{L}$$

Substituting the value of $\mu_B = \dfrac{e\hbar}{2m}$ and spin g–factor = 2, we get

$$E'_M = \frac{Ze^2}{4\pi\varepsilon_0 m^2 c^2 r^3} \vec{S} \cdot \vec{L}$$

This energy has been calculated in the frame of reference in which the electron is at rest. If we transform it into a frame of reference in which the nucleus is at rest, then the above value of potential energy is reduced by a factor 2 and the net potential energy due to spin orbit interaction is given by

$$E_M = \frac{1}{2} E'_M = \frac{Ze^2}{8\pi\varepsilon_0 m^2 c^2 r^3} \vec{S} \cdot \vec{L} \quad \text{...(v)}$$

Value of $\vec{S} \cdot \vec{L}$. The total angular momentum $\vec{J}$ of the electron is given by

$$\vec{J} = \vec{L} + \vec{S}$$

$$\therefore \quad \vec{J} \cdot \vec{J} = J^2 = \left(\vec{L} + \vec{S}\right) \cdot \left(\vec{L} + \vec{S}\right)$$

$$= L^2 + S^2 + \vec{L} \cdot \vec{S} + \vec{S} \cdot \vec{L}$$

But $\quad \vec{L} \cdot \vec{S} = \vec{S} \cdot \vec{L}$

$$\therefore \quad J^2 = L^2 + S^2 + 2\vec{S} \cdot \vec{L}$$

or $$\vec{S} \cdot \vec{L} = \frac{J^2 - L^2 - S^2}{2} \quad \text{...(v)(a)}$$

As $\quad j = \sqrt{j(j+1)}\,\hbar; \; L = \sqrt{l(l+1)}\,\hbar; \; S = \sqrt{s(s+1)}\,\hbar$

$$\vec{S}.\vec{L} = \frac{[j(j+1) - l(l+1) - s(s+1)]\hbar^2}{2} \quad ...(vi)$$

Substituting in Eq. (*v*), we have

$$E_M = \frac{Ze^2\hbar^2}{16\pi\varepsilon_0 m^2 c^2 r^3}[j(j+1) - l(l+1) - s(s+1)]$$

If E_n is the energy of the atom in the *n*th state, then total energy of the atom $E = E_n + E_M$ *Thus* E_M *gives the change in total energy of the atom due to spin orbit coupling.*

Example 8.8. *Find the magnitude of spin orbit energy for the state (i)* $2p_{1/2}$ *and (ii)* $2p_{3/2}$ *of the hydrogen atom. The radius of the orbit is* $3a_0 = 1.5$ *Å.* ***(G.N.D.U. 2004)***

Solution. The spin orbit interaction energy is given by

$$E_M = \frac{Ze^2\hbar^2}{16\pi\varepsilon_0 m^2 c^2 r^3}[j(j+1) - l(l+1) - s(s+1)]$$

$$= A[j(j+1) - l(l+1) - s(s+1)]$$

where $$A = \frac{Ze^2\hbar^2}{16\pi\varepsilon_0 m^2 c^2 r^3}$$

For hydrogen $Z = 1$; $r = 3a_0 = 1.5$ Å $= 1.5 \times 10^{-10}$ m

$$e = 1.6 \times 10^{-19} \text{ C}, \hbar = \frac{h}{2\pi} = \frac{6.63 \times 10^{-34}}{2 \times 3.142} = 1.05 \times 10^{-34} \text{ Js}$$

$$\varepsilon_0 = 8.854 \times 10^{-12} \text{ C}^2/\text{Nm}^2; m = 9 \times 10^{-31} \text{ kg}; c = 3 \times 10^8 \text{ ms}^{-1}$$

$$\therefore \quad A = \frac{1 \times 1.6 \times 1.6 \times (10^{-19})^2 \times 1.05 \times 1.05 \times (10^{-34})^2}{16 \times 3.142 \times 8.854 \times 10^{-12} \times 9 \times 9 \times 10^{-62} \times 9 \times 10^{16} \times (1.5 \times 10^{-10})^3}$$

$$= 2.577 \times 10^{-34} \text{ J} = \frac{2.577 \times 10^{-24}}{1.6 \times 10^{-19}} = 1.61 \times 10^{-5} \text{ eV}$$

(*i*) For the state $2p_{1/2}$; $j = \frac{1}{2}, l = 1, s = \frac{1}{2}$

$$\therefore \quad [j(j+1) - l(l+1) - s(s+1)] = \frac{1}{2} \times \frac{3}{2} - 1 \times 2 - \frac{1}{2} \times \frac{3}{2} = -2$$

$\therefore$ Spin orbit interaction energy for the state $2p_{1/2} = -2A$

$$= -2 \times 1.61 \times 10^{-5} = -3.22 \times 10^{-5} \text{ eV}$$

(*ii*) For the state $2p_{3/2}$; $j = \frac{3}{2}, \; l = 1, \; s = \frac{1}{2}$

$$\therefore \quad [j(j+1) - l(l+1) - s(s+1)] = \frac{3}{2} \times \frac{5}{2} - 1 \times 2 - \frac{1}{2} \times \frac{3}{2} = +1$$

$\therefore$ Spin orbit interaction energy for the state $2p_{3/2} = +A$

$$= 1.61 \times 10^{-5} \text{ eV}$$

8.12 FINE STRUCTURE OF HYDROGEN ATOMIC SPECTRUM

The spectral lines in the spectrum of atomic hydrogen are not single lines; but each line consists of two closely spaced lines. *This doubling of spectral lines in hydrogen spectrum is known as fine structure*. For example, the 6563 Å Balmer line in hydrogen spectrum is found to consist of two lines separated by 1.4 Å.

This can be explained on the basis of spin-orbit interaction.

The energy-levels in atomic spectroscopy are labelled as $(n)^{2s+1} l_j$

where n = principal quantum number

s = spin angular momentum quantum number

j = total angular momentum quantum number

l = orbital angular momentum quantum number

denoted as s for $l = 0$; p for $l = 1$; d for $l = 2$; f for $l = 3$ and so on.

Ground state. The ground state of hydrogen atom has $n = 1$.

Therefore, for this state $l = 0$ (s-state) $j = l + s = \frac{1}{2}$

$\therefore$ Ground state of hydrogen is represented as $1^2s_{1/2}$

First excited state. For first excited state $n = 2$

$\therefore$ For this state $l = 0$ or 1

For $l = 0$ (s-state), $j = l + s = \frac{1}{2}$

$\therefore$ This state is written as $2^2s_{1/2}$.

For $l = 1$ (p-state) $j = l \pm s = \left(1 + \frac{1}{2}\right) = \frac{3}{2}$ and $\left(1 - \frac{1}{2}\right) = \frac{1}{2}$

Thus we have two states $2\,^2p_{1/2}$ and $2\,^2p_{3/2}$.

In all, the first excited state has three sub-states $2\,^2s_{1/2}$ and $2\,^2p_{1/2}$ and $2\,^2p_{3/2}$.

All these states have the same principal quantum number $n = 2$.

If we do not consider the effect of spin-orbit interaction, these states *are degenerate* and have the same energy – 3.4 eV and coincide exactly.

Second excited state. For second excited state $n = 3$

$\therefore$ For this state $l = 0$, 1 or 2

For $l = 0$ (s-state) $j = \frac{1}{2}$ and this state is written as $3\,^2s_{1/2}$.

For $l = 1$ (p-state) $j = \frac{3}{2}$ and $\frac{1}{2}$

and these states are written as $3\,^2p_{1/2}$ and $3\,^2p_{3/2}$.

For $l = 2$ (d-state) $j = \frac{5}{2}$ and $\frac{3}{2}$

and these states are written as $3\,^2d_{3/2}$ and $3\,^2d_{5/2}$.

All these *five* states are *degenerate*, have the same energy –1.51 eV and coincide exactly. All these states are shown in Fig. 8.11 (*a*).

Effect of Spin-orbit interaction. Now let us consider the effect of spin-orbit interaction energy due to spin orbit coupling on these energy states. The spin-orbit interaction energy is given by

$$E_M = \frac{Ze^2 \hbar^2}{16\pi\varepsilon_0 m^2 c^2 r^3}\left[\, j(j+1) - l(l+1) - s(s+1)\right]$$

$$= A\left[\, j(j+1) - l(l+1) - s(s+1)\right]$$

where $$A = \frac{Ze^2 \hbar^2}{16\pi\varepsilon_0 m^2 c^2 r^3}$$

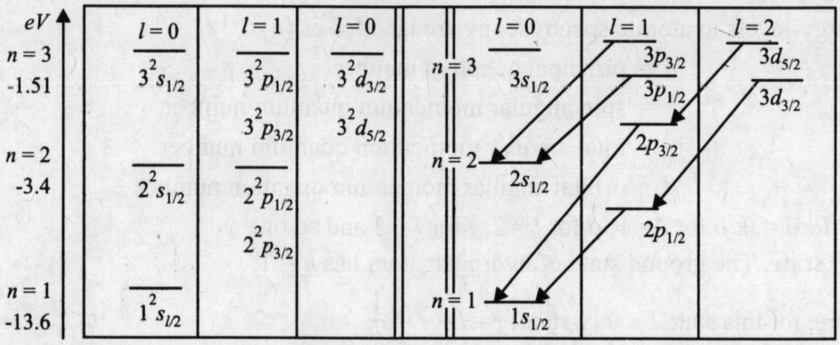

Without spin-orbit coupling (a) Without spin-orbit interaction (b)

Fig. 8.11

For *s*-states, $l = 0,\ s = \frac{1}{2}$ and $j = \frac{1}{2}$

$\therefore$ Spin-orbit interaction energy for *s*-state

$$E_M = A\,[j\,(j+1) - l\,(l+1) - s\,(s+1)]$$

$$= A\left[\frac{1}{2}\left(\frac{1}{2}+1\right) - 0 - \frac{1}{2}\left(\frac{1}{2}+1\right)\right] = 0$$

Thus the total energy of the atom does not change even in the presence of spin-orbit interaction. Hence *all s-states remain single*.

For *p*-state $l = 1,\ s = \frac{1}{2},\ j = \frac{1}{2}$ or $\frac{3}{2}$

For $$j = \frac{1}{2},\ E_M = A\left[\frac{1}{2}\left(\frac{1}{2}+1\right) - 1(1+1) - \frac{1}{2}\left(\frac{1}{2}+1\right)\right] = -\,2A$$

For $$j = \frac{3}{2},\ E_M = A\left[\frac{3}{2}\left(\frac{3}{2}+1\right) - 1(1+1) - \frac{1}{2}\left(\frac{1}{2}+1\right)\right] = +\,A$$

Thus the spin-orbit interaction splits each *p*-level into *two* energy states, raising the energy for $j = \frac{3}{2}$ by $+A$ and lowering the energy for $j = \frac{1}{2}$ by $-2A$.

For *d*-state. $l = 2,\ s = \frac{1}{2},\ j = \frac{3}{2}$ or $\frac{5}{2}$

For $$j = \frac{3}{2},\ E_M = A\left[\frac{3}{2}\left(\frac{3}{2}+1\right) - 2(2+1) - \frac{1}{2}\left(\frac{1}{2}+1\right)\right] = -\,3A$$

For $$j = \frac{5}{2},\ E_M = A\left[\frac{5}{2}\left(\frac{5}{2}+1\right) - 2(2+1) - \frac{1}{2}\left(\frac{1}{2}+1\right)\right] = +\,2A$$

Thus the spin-orbit interaction splits the *d*-level into *two* energy states raising the energy for $j = \frac{5}{2}$ by $+2A$ and lowering the energy for $j = \frac{3}{2}$ by $-3A$.

The effect of spin-orbit interaction on various levels is shown in Fig. 8.11(*b*). All levels except *s*-level are doublet levels. The doubling of levels gives rise to a doubling of spectral lines. The spectral lines appear in pairs, having very close frequency (or wavelength). The separation between the lines is too small and is known as *fine-structure.*

The doublet lines formed by transition from $3p_{3/2}$ and $3p_{1/2}$ states to $2s_{1/2}$ state, from $3d_{5/2}$ to $2p_{3/2}$, from $3d_{3/2}$ to $2p_{1/2}$ and from $2p_{3/2}$ and $2p_{1/2}$ to $1s_{1/2}$ states is shown in Fig. 8.11(*b*).

Selection rules for *j* in doublet spectra

In doublet spectra arising from atomic system containing but one valence electron *i.e.* one electron in the outer most incomplete shell, the transition from one energy state to another the selection rule for total angular momentum quantum number *j* are

j changes by 0, + 1 or – 1 only

i.e. $\Delta j = 0, \pm 1$

However, the transition from $j = 0$ to $j = 0$ (0–0 transition) is forbidden.

Example 8.9 *The ground state of Sodium is* $3\,^2s_{\frac{1}{2}}$. *Find n, l,s, j.* ***(Luck, U. 2001)***

Solution. The energy level in atomic spectroscopy is labelled as $(n)^{(2s+1)}l_j$.

The level with $l = 0$ is denoted as *s*-level

$\therefore$ For $3\,^2s_{\frac{1}{2}}$

$$n = 3$$
$$l = 0$$
$$(2s + 1) = 2 \quad \text{or} \quad s = \frac{1}{2}$$
$$j = \frac{1}{2}$$

Example 8.10 *Calculate the values of l, s and j and L, S and J for a d electron in one electron atomic system.* ***(P.U. 2008)***

Solution. For a *d* electron $l = 2$ and $s = \frac{1}{2}$ as it is one electron atomic system.

$\therefore$ Possible values of *j* are

$$j = l + s \text{ and } j = l - s$$

or
$$j = 2 + \frac{1}{2} = \frac{5}{2} \quad \text{and} \quad j = 2 - \frac{1}{2} = \frac{3}{2}$$

$\therefore$ *j* values are $\frac{5}{2}, \frac{3}{2}$

Now spin angular momentum

$$S = \sqrt{s\,(s+1)}\,\hbar$$
$$= \sqrt{\frac{1}{2}\left(\frac{1}{2}+1\right)}\,\hbar = \sqrt{\frac{3}{2}}\,\hbar$$

Orbital angular momentum

$$L = \sqrt{l\,(l+1)}\,\hbar$$
$$= \sqrt{2 \times (2+1)}\,\hbar = \sqrt{6}\,\hbar$$

Total angular momentum

$$J = \sqrt{j\,(j+1)}\,\hbar$$
$$= \sqrt{\frac{5}{2} \times \left(\frac{5}{2}+1\right)}\,\hbar$$

$$= \sqrt{\frac{35}{2}}\,\hbar$$

and $$J = \sqrt{\frac{3}{2} \times \left(\frac{3}{2} + 1\right)}\,\hbar$$

$$= \frac{\sqrt{15}}{2}\,\hbar$$

Example 8.11. *Spin orbit coupling splits all states except s-state into two sub states. Why are s-states exception to this rule?*

Solution. The spin orbit interaction energy is given by

$$E_M = A\,[j\,(j+1) - l\,(l+1) - s\,(s+1)]$$

where $$A = \frac{Ze^2\,\hbar^2}{16\pi\varepsilon_0\,m^2\,c^2\,r^3} = \text{a constant}$$

For *s*-state; $l = 0,\quad s = \frac{1}{2}$ and $j = \frac{1}{2}$

$$\therefore \left[j(j+1) - l(l+1) - s(s+1)\right] = \left[\frac{1}{2}\left(\frac{1}{2} + 1\right) - 0 - \frac{1}{2}\left(\frac{1}{2} + 1\right)\right] = 0$$

Thus the total energy of the atom in the *s*-state does not change even in the presence of spin-orbit interaction. *Hence all s-states remain single.*

Example 8.12. *Consider p-electron in one electron atom and calculate the values of L,S and J.* ***(P.U. 2001; H.P.U. 2000; G.N.D.U.; 2003)***

Solution. For a *p*-electron $l = 1$ and $s = \frac{1}{2}$

$\therefore$ Possible values of *j* are $j = l + s$ and $j = l - s$

or $$j = 1 + \frac{1}{2} = \frac{3}{2} \text{ and } j = 1 - \frac{1}{2} = \frac{1}{2}$$

$\therefore$ *j* values are $\frac{3}{2}, \frac{1}{2}$.

Now Orbital angular momentum $L = \sqrt{l(l+1)}\,\hbar$

$$= \sqrt{1(1+1)}\,\hbar = \sqrt{2}\,\hbar \qquad [\text{For } l = 1]$$

Spin angular momentum $S = \sqrt{s(s+1)}\,\hbar$

$$= \sqrt{\frac{1}{2}\left(\frac{1}{2} + 1\right)}\,\hbar = \frac{\sqrt{3}}{2}\,\hbar \qquad \left[\text{For } s = \frac{1}{2}\right]$$

Total angular momentum $J = \sqrt{j(j+1)}\,\hbar$

$$= \sqrt{\frac{3}{2}\left(\frac{3}{2} + 1\right)}\,\hbar = \frac{\sqrt{15}}{2}\,\hbar \qquad \left[\text{For } j = \frac{3}{2}\right]$$

and $$J = \sqrt{\frac{1}{2}\left(\frac{1}{2} + 1\right)}\,\hbar = \frac{\sqrt{3}}{2}\,\hbar \qquad \left[\text{For } j = \frac{1}{2}\right]$$

8.13 TOTAL MAGNETIC MOMENT AND LANDE'S FACTOR

Total magnetic moment. An atom possesses a magnetic moment, because its electrons have orbital and spin angular momenta and associated magnetic moments.

The *vector sum* of the orbital and spin angular momenta $\vec{L}$ and $\vec{S}$ of an electron is called its total angular momentum $\vec{J}$ given by

$$\vec{J} = \vec{L} + \vec{S} \qquad ...(i)$$

The magnitude of the total angular momentum

$$J = \sqrt{j(j+1)}\,\hbar$$

where j is the total angular momentum quantum number. The Z-component of $\vec{J}$ is space quantised similar to the Z-component of $\vec{L}$ and $\vec{S}$ and is given by

$$J_z = m_j \hbar$$

where m_j is the *total magnetic quantum number.* It has $(2j + 1)$ values given by $-j, -j+1 \ldots +j-1, +j$

The orbital magnetic moment vector is given by

$$\vec{\mu}_l = -g_l \frac{e}{2m}\vec{L} = -g_l \frac{e\hbar}{2m}\frac{\vec{L}}{\hbar} = -\frac{\mu_B}{\hbar}\vec{L} \qquad ...(i)(a)$$

and its magnitude $\mu_l = -\mu_B\sqrt{l(l+1)}$ where g_l = orbital g factor = 1, $\mu_B = \frac{e\hbar}{2m} = 1$. Bohr magneton and the magnitude of the orbital angular momentum $L = \sqrt{l(l+1)}\,\hbar$, l being the *orbital quantum number*.

The spin magnetic moment vector is given by

$$\vec{\mu}_s = -g_s \frac{e}{2m}\vec{S} = -\frac{e\hbar}{2m}\frac{\vec{S}}{\hbar} = -2\frac{\mu_B}{\hbar}\vec{S} \qquad ...(i)(b)$$

and its magnitude $\mu_s = -2\mu_B\sqrt{s(s+1)}$ where g_s = spin g factor = 2 and the magnitude of the spin angular momentum $S = \sqrt{s(s+1)}\,\hbar$; s being the *spin quantum number*.

∴ Total magnetic moment vector

$$\vec{\mu} = \vec{\mu}_l + \vec{\mu}_s = -\frac{\mu_B}{\hbar}\left(\vec{L} + 2\vec{S}\right)$$

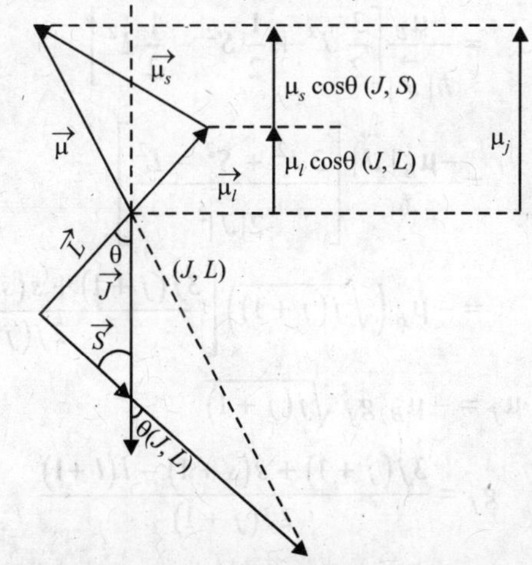

Fig. 8.12

The *effective total magnetic moment vector* $\vec{\mu}_j$ is obtained by adding components of $\vec{\mu}_l$ and $\vec{\mu}_s$ along the direction of vector $\vec{J}$ as shown in Fig 8.12.

$$\therefore \quad \mu_j = \mu_l \cos\theta(\vec{J}, \vec{L}) + \mu_s \cos\theta(\vec{J}, \vec{S})$$

or
$$\mu_j = \vec{\mu}_l \cdot \frac{\vec{J}}{|\vec{J}|} + \vec{\mu}_s \cdot \frac{\vec{J}}{|\vec{J}|} = (\vec{\mu}_l + \vec{\mu}_s) \cdot \frac{\vec{J}}{|\vec{J}|}$$

$$= -\frac{\mu_B}{\hbar} \frac{(\vec{L} + 2\vec{S}) \cdot (\vec{J})}{|\vec{J}|}$$

$$= \frac{-\mu_B}{\hbar} \frac{(\vec{L} \cdot \vec{J} + 2\vec{S} \cdot \vec{J})}{|\vec{J}|} \qquad \text{...(ii)}$$

But according to Eq. (*i*); $\vec{J} = \vec{L} + \vec{S}$. Substituting in Eq. (*ii*) we have

$$\mu_j = \frac{-\mu_B}{\hbar} \left[\frac{\vec{L} \cdot \left(\vec{L} + \vec{S}\right) + 2\vec{S} \cdot \left(\vec{L} + \vec{S}\right)}{|\vec{J}|} \right]$$

$$= \frac{-\mu_B}{\hbar} \left[\frac{L^2 + 2S^2 + 3\left(\vec{L} \cdot \vec{S}\right)}{|\vec{J}|} \right]$$

But $\vec{L} \cdot \vec{S} = \frac{1}{2}\left(J^2 - L^2 - S^2\right)$ ...[See **article. 8.11** Eq. (*v*) (*a*)]

$$\therefore \quad \mu_j = \frac{-\mu_B}{\hbar} \left[\frac{L^2 + 2S^2 + \frac{3}{2}\left(J^2 - L^2 - S^2\right)}{|\vec{J}|} \right]$$

$$= \frac{-\mu_B}{\hbar |\vec{J}|} \left[\frac{3}{2} J^2 + \frac{1}{2} S^2 - \frac{1}{2} L^2 \right]$$

$$= \frac{-\mu_B |\vec{J}|}{\hbar} \left[\frac{3J^2 + S^2 - L^2}{2|\vec{J}|^2} \right]$$

$$= -\mu_B \left(\sqrt{j(j+1)}\right) \left[\frac{3j(j+1) + s(s+1) - l(l+1)}{2j(j+1)} \right]$$

or Total magnetic moment $\mu_j = -\mu_B \, g_j \sqrt{j(j+1)}$ ...(*iii*)

where
$$g_j = \frac{3j(j+1) + s(s+1) - l(l+1)}{2j(j+1)} \qquad \text{...(iv)}$$

$$= 1 + \frac{j(j+1) + s(s+1) - l(l+1)}{2j(j+1)}$$

is called the g factor or *Lande's splitting factor*.

Lande's g factor for S electron. For an S-electron, the orbital quantum number $l = 0$ and spin quantum number $s = \frac{1}{2}$.

$\therefore$ Total quantum number $j = l + s = \frac{1}{2}$

Lande's g factor $= 1 + \frac{j(j+1) + s(s+1) - l(l+1)}{2j(j+1)}$

$$= 1 + \frac{\frac{1}{2} \times \frac{3}{2} + \frac{1}{2} \times \frac{3}{2}}{2 \times \frac{1}{2} \times \frac{3}{2}} = 1 + 1 = 2$$

Example 8.13. *Calculate Lande's g factor for a p-electron.* ***(P.U. 2007, 2002)***

Solution. For a p-electron $l = 1$ and $s = \frac{1}{2}$, therefore j has two values $j = 1 + \frac{1}{2} = 3/2$ and $j = l - \frac{1}{2} = \frac{1}{2}$.

(*i*) For $j = \frac{3}{2}$, $l = 1$ and $s = \frac{1}{2}$

$$g = 1 + \frac{j(j+1) + s(s+1) - l(l+1)}{2j(j+1)}$$

$$= 1 + \frac{\frac{3}{2} \times \frac{5}{2} + \frac{1}{2} \times \frac{3}{2} - 1 \times 2}{2 \times \frac{3}{2} \times \frac{5}{2}} = 1 + \frac{1}{3} = \frac{4}{3}$$

(*ii*) For $j = \frac{1}{2}$, $l = 1$ and $s = \frac{1}{2}$

$$g = 1 + \frac{\frac{1}{2} \times \frac{3}{2} + \frac{1}{2} \times \frac{3}{2} - 1 \times 2}{2 \times \frac{1}{2} \times \frac{3}{2}} = 1 - \frac{1}{3} = \frac{2}{3}$$

Example 8.14. *Calculate Lande's g factor and total magnetic moment (i) for $^2D_{3/2}$ state. (ii) for $^2P_{3/2}$ state and (iii) $^2F_{5/2}$ state.* ***(M.D.U. 2008, 2003)***

Solution. (*i*) For $^2D_{3/2}$ state $l = 2$

$$(2s + 1) = 2 \quad \therefore s = \frac{1}{2}$$

$$j = \frac{3}{2}$$

$$\therefore \quad g_j = 1 + \frac{j(j+1) + s(s+1) - l(l+1)}{2j(j+1)}$$

$$=1+\frac{\frac{3}{2}\times\frac{5}{2}+\frac{1}{2}\times\frac{3}{2}-2\times3}{2\times\frac{3}{2}\times\frac{5}{2}}=1-\frac{1}{5}$$

$$=\frac{4}{5}$$

$$\mu_j=-\mu_B\ g_j\ \sqrt{j(j+1)}$$

$$=-\mu_B\times\frac{4}{5}\times\sqrt{\frac{3}{2}\times\frac{5}{2}}=-\mu_B\ \frac{2\sqrt{15}}{5}$$

$$=-\frac{2\sqrt{15}}{5}\ \mu_B$$

where μ_B = 1 Bohr magneton = 9.27×10^{-24} Am^2

(*ii*) For $^2P_{3/2}$ $l=1$, $(2s+1)=2$ $\therefore s=\frac{1}{2}$; $j=\frac{3}{2}$

$$g_j=\frac{4}{3}$$ [For calculation See **Example 8.13(*i*)**]

$$\therefore\qquad \mu_j=-\mu_B\ g_j\ \sqrt{j(j+1)}$$

$$=-\mu_B\times\frac{4}{3}\times\sqrt{\frac{3}{2}\times\frac{5}{2}}$$

$$=-\frac{2}{3}\sqrt{15}\ \mu_B$$

where μ_B = 1 Bohr magneton = 9.27×10^{-24} Am^2

(*iii*) For $^2F_{5/2}$ state, $l=3$

$$2(s+1)=2\quad\therefore\quad s=\frac{1}{2}$$

$$j=5/2$$

$$\therefore\qquad g_j=1+\frac{j(j+1)+s(s+1)-l(l+1)}{2j\,(j+1)}$$

$$=1+\frac{\frac{5}{2}\times\frac{7}{2}+\frac{1}{2}\times\frac{3}{2}-3\times4}{2\times\frac{5}{2}\times\frac{7}{2}}$$

$$=1-\frac{1}{7}=\frac{6}{7}$$

$$\mu_j=\mu_B\ g_j\ \sqrt{j\,(j+1)}$$

$$=-\mu_B\times\frac{6}{7}\times\sqrt{\frac{5}{2}\times\frac{7}{2}}$$

$$=-\mu_B\times\frac{3}{7}\times\sqrt{35}$$

$$=-\frac{3\sqrt{35}}{7}\ \mu_B$$

8.14 HYPERFINE STRUCTURE OF SPECTRAL LINES

Atomic nuclei have radii of the order of 10^{-4}Å which are small compared with typical distance of an electron from the nuclei which is equal to 1 Å. The nuclei are also much smaller than electron. It is, therefore, a very good approximation to consider the nuclei to be positive point charges of infinite mass. The assumption is insufficient to explain many experimental results e.g. the observation of *hyperfine structure* of atomic energy levels using high resolution spectrographs. The *hyperfine* shift in electronic energy levels is usually much smaller than that corresponding to fine structure. The following factors contribute towards hyperfine structure of spectral lines:

(*i*) The nucleus has an intrinsic angular momentum given by

$$|\vec{I}| = \sqrt{I(I+1)}\,\hbar$$

where $\vec{I}$ is the *spin of the nucleus* and I the *nuclear angular momentum quantum number*.

(*ii*) Associated with $\vec{I}$ is a magnetic moment $\vec{\mu}_N$ given by

$$\vec{\mu}_N = g\frac{e}{m_p}\vec{I}$$

where m_p is the mass of the proton and g the Lande's g-factor for the nucleus. For proton the value of $g = 2.79$.

(*iii*) The Z-component $\vec{I}$ (in the direction of the external magnetic field) is space quantised and is given by

$$I_z = m_i\,\hbar$$

(*iv*) The atomic states are designated by total angular momentum $\vec{F}$ given by

$$\vec{F} = \vec{J} + \vec{I}$$

where

$$\vec{F} = \sqrt{F(F+1)}\,\hbar$$

F is known as total *angular momentum quantum number* and has integral values from $(J - I)$ to $(J + I)$.

This means that each level with a given value of J has a multiplicity of $(2I + 1)$ if $J > I$ and a multiplicity of $(2J + 1)$ if $J \leq I$.

(*v*) The Z-component of $\vec{F}$ (in the direction of the external magnetic field) is also space quantised and has values given by

$$F_z = m_F\,\hbar = (m_i + m_j)\,\hbar$$

(*vi*) The allowed electronic dipole transitions satisfy the following selection rules

$$\Delta I = 0 \quad \text{(Zero)}$$

$$\Delta F = \pm 1, 0 \qquad (F = 0 \rightarrow F = 0 \text{ not allowed})$$

$$\Delta m_F = \pm 1, 0$$

(*vii*) The nuclear magnetic moment interacts with the magnetic field created at the nucleus by the electron. This magnetic field is due to (*i*) the orbital motion of the electron about the nucleus and (*ii*) the intrinsic (or spin) magnetic moment of the electron. The interaction energy due to the magnetic field of the orbital motion of the electron is smaller than the corresponding fine structure term by a factor of about m_e/m_p which is of the order of 1/1000. Taking into consideration the interaction energy due to both the factors, the hyperfine splitting between $F = 1$ and $F = 0$ levels of the ground state of hydrogen atom comes out to be

$$E_1(F=1) - E_1(F=0) = \frac{16\,m_e}{2\,m_p}\left(g\alpha^2\right)|E_I|$$

As transition between $F = 1 \rightarrow \vec{F} = 0$ i.e. $\Delta F = +1$ is permitted, this leads to a spectral line with frequency

$$\nu = \frac{E_1 (F = 1) - E_1 (F = 0)}{h} = 1.420 \times 10^9 s^{-1}$$

which corresponds to a wave length of about 21.1 cm. This is the famous 21cm line observed by the radio astronomers in the spectrum of interstellar or intergalaxic hydrogen.

Conclusion. Nuclear spin introduces an additional multiplicity of atomic levels. For the hydrogen atom each energy level acquires an additional multiplicity of 2. The magnetic moment associated with nuclear spin introduces a hyperfine splitting between the levels. These splittings are about 1000 times smaller than the fine structure splitting.

The $\nu = 1.420 \times 10^9 s^{-1}$ frequency radiation (known as the 21 cm hydrogen line) corresponding to the hyperfine transition between the ground state levels with $F = 1$ and $F = 0$ is used to study the distribution and motion of hydrogen in interstellar and inter-galaxic space using Doppler's effect.

8.15 RELATIVISTIC SOMMERFIELD THEORY OF HYDROGEN SPECTRUM

Bohr's theory predicted with great accuracy the spectral lines of hydrogen and singly ionised helium. According to Bohr, the lines in the hydrogen spectrum should each have a well defined wave length. Studies with high resolving power spectrographs, however, showed that H_α, H_β and H_γ lines in the hydrogen spectrum were not single but each line consisted of a number of lines very close to each other. For example, H_α line was found by Michelson to consist of two lines with a wave length difference of only 0.13 Å.

This is known as *fine structure of spectral lines*. In terms of energy levels, the existence of fine structure means that instead of a single energy level corresponding to a given value of quantum number n there exist more than one energy levels close to each other. Bohr's theory could not explain this *fine structure* and existence of more than one energy levels corresponding to a single quantum number n.

To explain the observed fine structure Sommerfeld made following modifications in Bohr's theory.

(*i*) The path of the electron around the nucleus, in general is an ellipse with the nucleus as one of the foci. The circular orbit of Bohr being only a special case.

(*ii*) The velocity of an electron moving in an elliptic orbit varies considerably at different points of the orbit. The velocity is greater when the electron is near the nucleus but smaller when it is far away from the nucleus. Due to the variation in velocity of the electron, its mass varies in accordance with the special theory of relativity. Sommerfeld took into account this variation of mass of the electron with velocity to explain the fine structure.

Elliptic orbits. When the electron in a hydrogen like atom having nuclear charge + Ze moves in an elliptic orbit, with nucleus at one of the foci, there are two co-ordinates which may change

(*i*) The distance r of the electron from the nucleus known as *radius vector* and

(*ii*) The angle ϕ that the radius vector makes with the major axis of the ellipse taken along X-axis.

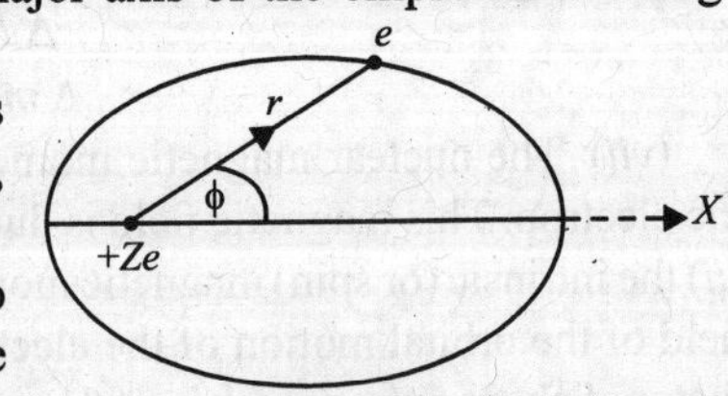

Fig. 8.13

In the case of a circular orbit r remains constant and it is only the co-ordinate ϕ that varies periodically. Sommerfeld, therefore, introduced a new postulate.

'The momenta associated with the co-ordinates r and ϕ extending over one period of the motion are an integral multiple of h'.

These conditions are expressed mathematically as

$$\oint p_\phi \, d\phi = n_\phi \, h$$

and

$$\oint p_r \, dr = n_r \, h$$

where p_ϕ and p_r are the momenta associated with the co-ordinates ϕ and r respectively. n_ϕ is called the *angular* or *azimuthal quantum number* and n_r is called the *radial quantum number*. n_ϕ and n_r can take only positive integral values and $n_\phi + n_r = n$, the principal quantum number which can take integral values 1, 2, 3 ..... etc.

Evaluation of $\oint p_\phi \, d\phi$. The angular momentum p_ϕ correrponding to the co-ordinate ϕ is the angular momentum of the electron in the elliptic orbit and is a constant according to Kepler's law.

$$\therefore \quad \int_0^{2\pi} p_\phi \, d\phi = n_\phi h$$

or $$p_\phi \, [\phi]_0^{2\pi} = n_\phi h$$

or $$p_\phi \, 2\pi = n_\phi h \;\; or \;\; p_\phi = n_\phi \frac{h}{2\pi} = n_\phi \hbar \qquad ...(i)$$

Evaluation of $\oint p_r \, dr$. The equation of an ellipse in polar co-ordinates is

$$\frac{1}{r} = \frac{1 + \varepsilon \cos\phi}{a\left(1-\varepsilon^2\right)} \qquad ...(ii)$$

where a is the semi-major axis and ε the eccentricity of ellipse. Differentiating both sides of Eq. (*ii*) with respect to ϕ, we get

$$-\frac{1}{r^2}\frac{dr}{d\phi} = -\frac{\varepsilon \sin\phi}{a\left(1-\varepsilon^2\right)}$$

or $$\frac{1}{r}\frac{dr}{d\phi} = \frac{r\,\varepsilon \sin\phi}{a\left(1-\varepsilon^2\right)}$$

Substituting the value of r from Eq, (*ii*), we have

$$\frac{1}{r}\frac{dr}{d\phi} = \frac{a\left(1-\varepsilon^2\right)}{1+\varepsilon\cos\phi} \cdot \frac{\varepsilon\sin\phi}{a\left(1-\varepsilon^2\right)} = \frac{\varepsilon\sin\phi}{1+\varepsilon\cos\phi} \qquad ...(iii)$$

Now
$$\begin{aligned} p_r\, dr &= m\frac{dr}{dt}\, dr \\ &= m\left(\frac{dr}{d\phi}.\frac{d\phi}{dt}\right)\left(\frac{dr}{d\phi}\right) d\phi \\ &= m\left(\frac{dr}{d\phi}\right)^2 \frac{d\phi}{dt}\, d\phi \\ &= m\left(\frac{d\phi}{dt}\right)\left(\frac{dr}{d\phi}\right)^2 d\phi \end{aligned}$$

Also $p_\phi = mvr = mr^2\omega = mr^2 \dfrac{d\phi}{dt}$

or $$m\frac{d\phi}{dt} = \frac{p_\phi}{r^2}$$

Hence $$n_r\, h = \oint p_r\, dr = \oint \left(m\frac{d\phi}{dt}\right)\left(\frac{dr}{d\phi}\right)^2 d\phi$$

$$= \oint \frac{p_\phi}{r^2}\left(\frac{dr}{d\phi}\right)^2 d\phi$$

$$= \oint p_\phi \left(\frac{1}{r}\frac{dr}{d\phi}\right)^2 d\phi \qquad ...(iv)$$

Now $\oint p_\phi \, d\phi = n_\phi \frac{h}{2\pi}$ [According to Eq. (*i*)]

and $\frac{1}{r}\frac{dr}{d\phi} = \frac{\varepsilon \sin\phi}{1+\varepsilon\cos\phi}$ [According to Eq. (*iii*)]

Substituting in Eq. (*iv*), we get

$$n_r\, h = n_\phi \frac{h}{2\pi} \oint \left(\frac{\varepsilon \sin\phi}{1+\varepsilon\cos\phi}\right)^2 d\phi$$

$$= n_\phi \frac{h}{2\pi} \int_0^{2\pi} \frac{\varepsilon^2 \sin^2\phi}{(1+\varepsilon\cos\phi)^2} d\phi \qquad ...(v)$$

But $\int_0^{2\pi} \frac{\varepsilon^2 \sin^2\phi}{(1+\varepsilon\cos\phi)^2} d\phi$ is a standard integral the value of which

$$= 2\pi\left[\frac{1}{\left(1-\varepsilon^2\right)^{1/2}} - 1\right]$$

Substituting in Eq. (*v*), we get

$$n_r\, h = n_\phi \frac{h}{2\pi} 2\pi \left[\frac{1}{\left(1-\varepsilon^2\right)^{1/2}} - 1\right]$$

or
$$n_r = n_\phi \left[\frac{1}{\left(1-\varepsilon^2\right)^{1/2}} - 1\right]$$

$$= \frac{n_\phi}{\left(1-\varepsilon^2\right)^{1/2}} - n_\phi$$

or
$$n_r + n_\phi = \frac{n_\phi}{\left(1-\varepsilon^2\right)^{1/2}}$$

or
$$n = \frac{n_\phi}{\left(1-\varepsilon^2\right)^{1/2}}$$

or
$$1-\varepsilon^2 = \frac{n_\phi^2}{n^2} \qquad ...(vi)$$

For an ellipse $1-\varepsilon^2 = \frac{b^2}{a^2}$ where *a* and *b* are the semi-major and semi-minor axes respectively.

$$\therefore \quad \frac{b^2}{a^2} = \frac{n_\phi^2}{n^2} \quad or \quad \frac{b}{a} = \frac{n_\phi}{n} \qquad ...(vii)$$

Total energy. The total energy of the electron in an elliptic orbit of principal quantum number n.

E_n = Potential energy P.E. + Kinetic energy K.E.

and Potential energy $= \dfrac{-Ze^2}{4\pi\varepsilon_0 r}$

To find the kinetic energy the tangential velocity of the electron at any instant is resolved into two components

(*i*) radial velocity $\dfrac{dr}{dt}$ along the radius vector r and (*ii*) transverse velocity $r\dfrac{d\phi}{dt}$ at right angles to the radius vector.

$$\therefore \quad \text{K.E.} = \frac{1}{2}m\left(\frac{dr}{dt}\right)^2 + \frac{1}{2}m\left(r\frac{d\phi}{dt}\right)^2$$

$$= \frac{p_r^2}{2m} + \frac{p_\phi^2}{2mr^2}$$

$$\text{Total energy } E_n = \frac{p_r^2}{2m} + \frac{p_\phi^2}{2mr^2} - \frac{Ze^2}{4\pi\varepsilon_0 r}$$

It can be shown that the above expression gives the value of

$$E_n = \frac{-me^4Z^2}{8\varepsilon_0^2 h^2}\left[\frac{1}{n_r + n_\phi}\right]^2$$

$$= -\frac{me^4Z^2}{8\varepsilon_0^2 h^2}\frac{1}{n^2} \qquad ...(viii)$$

We, therefore, find that the expression for total energy is the same as that obtained by Bohr for a given value of n irrespective of the value of n_ϕ. This means that the theory of elliptic orbits introduces no new energy levels and is unable to explain the fine structure of spectral lines.

Summerfeld solved the fine structure problem on the basis of variation of mass of the electron with velocity according to the theory of relativity.

Relativistic theory. The velocity of the electron in a circular orbit is constant but the velocity of the electron in an elliptic orbit varies, being maximum at the perihilion and minimum at the apehilim. This velocity is quite large, being about 1/137 of the velocity of light. Sommerfeld, therefore, modified his theory by taking into account the variation of mass of the electron with velocity. When the relativity effect is taken into account the energy of the electron is found to depend on both, the principal quantum number n and azimuthal quantum number n_ϕ. The value of energy is now given by

$$E_n = -\frac{mZ^2e^4}{8\varepsilon_0^2 h^2 n^2}\left[1 + \frac{\alpha^2 Z^2}{n}\left(\frac{1}{n_\phi} - \frac{3}{4n}\right)\right] \qquad ...(ix)$$

where $\alpha = \dfrac{e^2}{2\varepsilon_0 hc} = \dfrac{1}{137}$ and is known as fine structure constant.

The first term in Eq. (*ix*) is the energy of the electron in the orbit with principal quantum number n. The second term is the *Sommerfeld relativity correction.* As this term depends upon azimuthal quantum number n_ϕ, it results in splitting of energy levels of the atom into the number of permitted values of n_ϕ.

Fine structure of H_α – line. The $H\alpha$ line of Balmer series of Hydrogen spectrum is due to the transition for $n = 3$ to $n = 2$ state. For $n = 3$, there are three possible values of n_ϕ i.e. 1, 2 and 3. Similarly there are two possible energy levels for $n = 2$ as shown in Fig. 8.14.

Theoretically following six transitions are possible.

$3_3 \to 2_2$; $3_3 \to 2_1$; $3_2 \to 2_2$; $3_1 \to 2_1$; $3_1 \to 2_2$; $3_1 \to 2_1$

Sommerfeld also applied a *selection rule* that

'*Only those transitions are allowed for which* $\Delta n_\phi = \pm 1$ *ie.* n_ϕ *can change only by + 1 or – 1.* When this rule is applied $3_3 - 2_1$, $3_2 - 2_2$ and $3_1 \to 2_1$ for which $\Delta n_\phi = 2$ or zero become forbidden and only the remaining three are permitted. The forbidden lines are indicated by *dotted lines* and permitted lines are indicated by *full lines*.

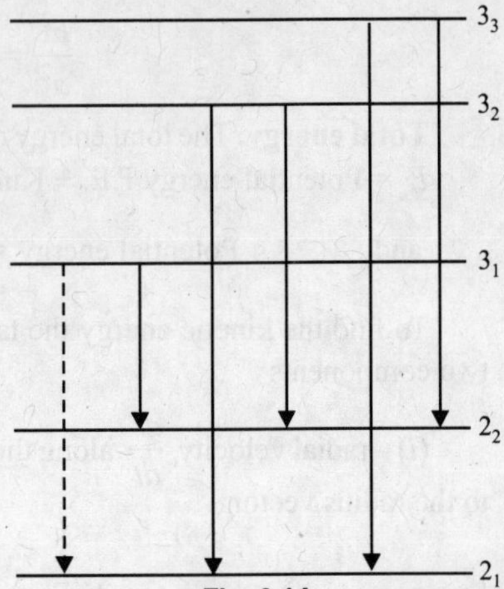

Fig. 8.14

This agrees with the actually observed three components of the H_α - line and this explains the fine structure.

Zero Angular Momentum

According to classical dynamics the length of the semi-major axis a of the elliptic orbit is determined solely by the principal quantum number n, while the length of the semi-minor axis b depends upon the azimuthal quantum number n_ϕ. Now,

$$\frac{b}{a} = \frac{n_\phi}{n}$$

and $n = n_\phi + n_r$ where n_r is the radial quantum number. Therefore, $n_\phi \le n$.

For the normal state of the hydrogen atom, the principal quantum number $n = 1$. Thus corresponding to $n = 1$, there are two possible values of n_r and n_ϕ i.e. $n_r = 0$, $n_\phi = 1$ and $n_r = 1$, $n_\phi = 0$. When $n_\phi = 0$, the value of semi - minor axis $b = 0$ and the ellipse degenerates into a straight line passing through the nucleus. This implies that the electron will have to pass through the nucleus twice during every period. This type of motion is not possible. Hence n_ϕ cannot be zero i.e. the lowest possible value of $n_\phi = 1$. When $n_\phi = 0$, the value of angular momentum $p_\phi = n_\phi h = 0$ since n_ϕ cannot be zero, p_ϕ also cannot be zero.

This is why according to Sommerfeld's elliptic treatment of hydrogen atom zero angular momentum was considered impossible.

8.16 CORRECTION TERM IN FINE STRUCTURE OF BALMER LINES OF HYDROGEN ATOM

On the basis of elliptic orbits and taking into consideration the relativistic variation of mass of the electron with velocity Sommerfeld obtained the value of energy of the electron in the orbit with principal quantum number n and azimuthal quantum number n_ϕ as

$$E_n = -\frac{mZ^2e^4}{8\varepsilon_0^2 h^2 n^2}\left[1 + \frac{\alpha^2 Z^2}{n}\left(\frac{1}{n_\phi} - \frac{3}{4n}\right)\right]$$

where $-\dfrac{m Z^2 e^4}{8\varepsilon_0^2 h^2 n^2}$ is the value of energy according to Bohr and $\dfrac{-m Z^4 e^4 \alpha^2}{8\varepsilon_0^2 h^2 n^3}\left(\dfrac{1}{n_\phi} - \dfrac{3}{4n}\right)$ is the correction which one makes to obtain the fine structure of the Balmer lines of hydrogen atom. α is known as fine structure constant and has a value $\dfrac{1}{137}$.

Example 8.15. *How many elliptical and how many circular orbits are there for n = 1, n = 2 and n = 3 according to Sommerfeld's model of the atom ?* ***(Bang. U. 2004, 2000)***

Solution. $\boldsymbol{n = 1}$. For the first orbit $n = 1$. As $n_r + n_\phi = 1$ and $n_\phi \neq 0$, the only value of n_ϕ can have is $n_\phi = 1$. When $n = n_\phi = 1$; $b = a$ and the first orbit is a circle.

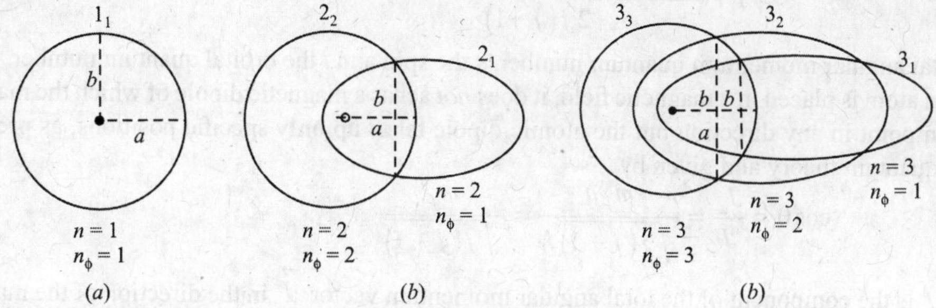

Fig. 8.15

$\boldsymbol{n = 2}$. For $n = 2$, n_ϕ can have the value 2 or 1 so that there are two possible orbits, $n = 2$, $n_\phi = 2$ and $n = 2$, $n_\phi = 1$. $n = 2$, $n_\phi = 2$ orbit is a circle and $n = 2$, $n_\phi = 1$ orbit is an ellipse as shown in Fig. 8.15 (*b*).

$\boldsymbol{n = 3}$. For $n = 3$, n_ϕ can have the value 3, 2 and 1 so that there are three possible orbits; $n = 3$, $n_\phi = 3$; $n = 3$, $n_\phi = 2$; $n = 3$, $n_\phi = 1$. $n = 3$, $n_\phi = 3$ orbit is a circle, $n = 3$, $n_\phi = 2$ is an ellipse and $n = 3$, $n_\phi = 1$ is also an ellipse as shown in Fig. 8.15 (*c*).

The allowed orbits are indicated by assigning the values of n and n_ϕ. Thus the three orbits of $n = 3$ are represented by symbols 3_3, 3_2 and 3_1, the subscript being the azimuthal quantum number n_ϕ.

8.17 STERN-GERLACH EXPERIMENT

Aim. Stern and Gerlach studied the deflection of atomic beams in non-uniform magnetic fields and thus proved the existence of electron spin and associated magnetic moment.

Apparatus. The apparatus consisted of an air-tight metal box B in which very high vacuum was produced. Metallic silver was evaporated by heating it electrically in the furnace F Fig. 8.16(*a*) The stream of silver vapours coming out of the furnace was rendered into a narrow beam by the slits S_1 and S_2. When no magnetic field was applied, a fine line was obtained on the glass plate G due to the deposit of metallic silver. When a non-uniform magnetic field having a very steep gradient in the vertical direction was applied with the help of a strong electromagnet NS having wedge and groove type pole-pieces, the beam was found to split up into two components, one on either side of the original line. The trace of silver particles now consisted of two lines except at the ends, as shown in Fig 8.16 (*b*) because here the silver atoms pass through the region where the field is very weak.

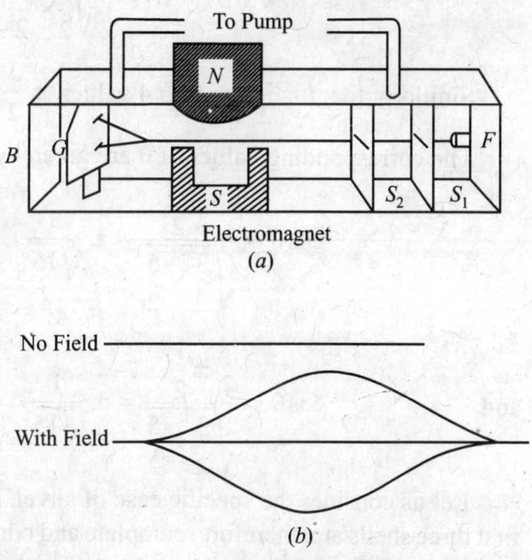

Fig. 8.16

Explanation on the basis of Quantum theory. An atom possesses a magnetic moment because its electrons have orbital and spin angular momenta and associated magnetic moments. Due to the combined effect of both, the magnitude of the total magnetic moment of the electron is given by

$$\mu_j = -\mu_B\, g_j \sqrt{j(j+1)}$$ [For proof see **article. 8.13** Eq. (*iii*)]

where $\mu_B = \dfrac{e\hbar}{2m} = 1$ Bohr magneton, g_j is Lande's splitting factor

$$= 1 + \frac{j(j+1) + s(s+1) - l(l+1)}{2j(j+1)}$$

j the total angular momentum quantum number, s the spin and l the orbital quantum number. Hence when an atom is placed in a magnetic field, it does *not* act as a magnetic dipole of which the magnetic axis can point in any direction but the atomic dipole takes up **only** specific positions, as predicted by the quantum theory and given by

$$\cos\theta = \frac{J_z}{J} = \frac{m_j \hbar}{\sqrt{j(j+1)}\,\hbar} = \frac{m_j}{\sqrt{j(j+1)}} \qquad ...(i)$$

where J_z is the component of the total angular momentum vector $\vec{J}$ in the direction of the magnetic induction vector $\vec{B}$, m_j being the *total magnetic quantum number* and θ the angle between the vectors $\vec{J}$ and $\vec{B}$.

For any value of j, m_j has $(2j + 1)$ values from $-j$ to $+j$ given by $(-j)$, $(-j+1)$, ... $(+j-1)$, $(+j)$.

For example, if $j = \frac{1}{2}$, m_j has 2 values $+\frac{1}{2}$ and $-\frac{1}{2}$. In such a case, there are only two values of θ given by

$$\cos\theta = \frac{m_j}{\sqrt{j(j+1)}} = \frac{\pm\frac{1}{2}}{\sqrt{\frac{1}{2}\times\frac{3}{2}}} = \pm\frac{1}{\sqrt{3}}$$

Similarly for $j = \frac{3}{2}$, m_j has 4 values $\pm\frac{3}{2}$ and $\pm\frac{1}{2}$.

The corresponding values of θ are given by

$$\cos\theta = \frac{\pm\frac{3}{2}}{\sqrt{\frac{3}{2}\times\frac{5}{2}}} = \pm\frac{3}{\sqrt{15}}$$

and
$$\cos\theta = \frac{\pm\frac{1}{2}}{\sqrt{\frac{3}{2}\times\frac{5}{2}}} = \pm\frac{1}{\sqrt{15}}$$

Let us consider the specific case of silver. Its atomic number is 47 and it has 47 electrons. The first three shells are, therefore, complete and contain 2, 8 and 18 electrons respectively. The first three sub-shells of the fourth shell are also complete and respectively contain 2, 6 and 10 electrons. This makes in all 46 electrons. The last electron is 5 s electron *i.e.*, it occupies the first s sub-shell of the 5th shell $n = 5$ and the quantum state corresponding to it is represented as $5\,{}^2s_{\frac{1}{2}}$. For this electron $n = 5$, $l = 0$, $j = \frac{1}{2}$ and $s = \frac{1}{2}$ and therefore Lande's splitting factor $g_j = 2$ and $m_j = \pm\frac{1}{2}$. The component of μ_j along the direction of magnetic induction.

$$\mu_j \cos\theta = \mu_B\, 2\sqrt{j(j+1)}\,\frac{m_j}{\sqrt{j(j+1)}} = 2m_j\,\mu_B \qquad ...(ii)$$

Let the silver atoms enter the magnetic field with a speed v moving along X-axis. Let the magnetic induction B vary at the rate $-\frac{\partial B}{\partial z}$ along the Z-axis, then the magnitude of the force acting on atomic dipole along the Z-axis is given by

$$F_z = 2m_j \mu_B \frac{\partial B}{\partial z} \quad ...(iii)$$

$\therefore$ Acceleration $a = \frac{F_z}{M} = 2m_j \mu_B \frac{\partial B}{\partial z}\frac{1}{M}$

where M is the mass of silver atom.

If L is the length of the path of the beam through the magnetic field, then

Time taken to travel through the field $= \frac{L}{v}$

and displacement of the atom along Z-axis on coming out of the field

$$z = \frac{1}{2}at^2 = \frac{1}{2}2m_j \mu_B \frac{\partial B}{\partial z}\frac{1}{M}\frac{L^2}{v^2} \quad ...(iv)$$

According to Maxwell's law of distribution of velocities, the most probable velocity of silver atoms at a temperature T K is given by

$$v = \sqrt{\frac{3kT}{M}} \text{ where } k \text{ is Boltzmann's constant}$$

Substituting in Eq. (*iv*), we get

$$z = \frac{1}{2}2m_j \mu_B \frac{\partial B}{\partial Z}\frac{L^2}{M}\frac{M}{3kT}$$

$$= \frac{1}{6}2m_j \mu_B \frac{L^2}{kT}\frac{\partial B}{\partial z} \quad ...(v)$$

Now $\quad m_j = \pm\frac{1}{2}$

$\therefore \quad 2m_j \mu_B = \pm \mu_B$

Substituting in Eq. (*v*), we have

$$z = \pm\frac{1}{6}\mu_B \frac{L^2}{kT}\frac{\partial B}{\partial z} \quad ...(vi)$$

The maximum displacement at the centre of the trace is given by the above relation.

Conclusion and significance. The value of z, the displacement of the line deposit from its original position has been calculated on the basis of quantum theory and is in close agreement with experimental values. The results of this experiment, therefore, give an experimental confirmation of space orientation of total magnetic moment vector m_j and spin magnetic moment vector m_s.

In the normal (ground) state $l = 0$. Therefore $j = \frac{1}{2}$ and $s = \frac{1}{2}$

For $j = \frac{1}{2}$, $m_j = \pm\frac{1}{2}$ and for $s = \frac{1}{2}$, $m_s = \pm\frac{1}{2}$

Thus the whole magnetic moment of silver atom is due to spin of one of its electrons.

∴ Eq. (*v*) may be put as

$$z = \frac{1}{6} 2 m_s \mu_B \frac{L^2}{kT} \frac{\partial B}{\partial z} \quad \text{...(vii)}$$

$$= \pm \frac{1}{6} \mu_B \frac{L^2}{kT} \frac{\partial B}{\partial z} \quad \text{...(viii)}$$

This result is the same as given in Eq. (*vi*). Thus there are two values of *z*-corresponding to two values of $m_s = \pm \frac{1}{2}$ which give rise to two opposite spin orientations and hence instead of a single central trace, we have two distinct traces-one displaced above and the other below the normal trace. It also gives an experimental verification of the interaction between orbital angular momentum and spin angular momentum (spin-orbit interaction) and their associated magnetic moments, as well as *space quantisation* of total magnetic moment vector μ_j.

Stern Gerlach experiment was designed to detect the magnetic dipole moment of atoms. This is done by passing a beam of *neutral* atoms possessing intrinsic magnetic moment through an inhomogeneous magnetic field. If, instead of neutral atoms, ions (which have a charge) are used, then moving charged ions will directly interact with whole of the magnetic field whereas the inhomogeneous part of the field will only interact with the intrinsic dipole moment of the atom. The first effect will be much greater than the second effect and will completely mask it. It will, therefore, not be possible to draw any direct logical conclusion.

Example 8.16. *Determine the maximum separation of a beam of hydrogen atoms that moves a distance of 20 cm with a speed of 2×10^5 m/s perpendicular to a magnetic field, whose gradient is 2×10^{-24} T/m and $M_H = 1.67 \times 10^{-27}$ kg.*

Solution. In the ground state, hydrogen atoms have zero orbital angular momentum. Therefore orbital angular momentum quantum number $l = 0$ and hence $m_l = 0$.

Now the *z*-component in the direction of the magnetic induction field of the total, orbital and spin angular momenta are connected by the relation

$$j = l + s$$

$$\therefore \quad j = s = \frac{1}{2} \qquad [\because l = 0]$$

and $\quad m_j = m_l + m_s$

$$\therefore \quad m_j = m_s = \pm \frac{1}{2}$$

$$\mu_B = 9.27 \times 10^{-24} \text{ J/T}$$

Magnetic field gradient $\frac{\partial B}{\partial z} = 2 \times 10^2$ T/m

$M = M_H$ = Mass of hydrogen atom = 1.67×10^{-27} kg

Length of the path L = 20 cm = 0.2 m

Velocity $\quad v = 2 \times 10^5$ m/s

Now $$z = \pm \frac{1}{2} 2 m_j \mu_B \frac{\partial B}{\partial z} \frac{1}{M} \frac{L^2}{v^2}$$

$$= \pm \frac{1}{2} \times 9.27 \times 10^{-24} \times 2 \times 10^2 \times \frac{1}{1.67 \times 10^{-27}} \left(\frac{0.2}{2 \times 10^5} \right)^2$$

$$= \pm 5.55 \times 10^{-7} \text{ m}$$

$\therefore$ Maximum separation $= 2z = 1.11 \times 10^{-6}$ m.

Example 8.17. *In Stern Gerlach experiment silver atoms traverse a distance of 0.1 m through a non-homogeneous field of gradient 60 T m⁻¹. If the separation between the two traces on the collector plate is 0.15 mm, find the velocity of the silver atoms. Mass of silver atom = 1.79 × 10⁻²⁵ kg, Bohr magneton = 9.27 × 10⁻²⁴ J T⁻¹.* **(Bang. U. 2001)**

Solution. $$2z = 2 \times \frac{1}{2} \times 2\, m_j\, \mu_B \frac{\partial B}{\partial Z} \frac{1}{M} \frac{L^2}{v^2} \quad \left[\text{For silver } m_j = \frac{1}{2}\right]$$

$$\therefore \quad v^2 = \mu_B \frac{\partial B}{\partial Z} \frac{1}{M} \frac{L^2}{2z} = \frac{9.27 \times 10^{-24} \times 60 \times 0.1 \times 0.1}{1.79 \times 10^{-25} \times 0.15 \times 10^{-3}}$$

or $$v^2 = 207150$$

or $$v = \sqrt{207150} = 455\ m s^{-1}$$

8.18 ZEEMAN EFFECT : NORMAL AND ANOMALOUS

Zeeman effect. In 1896, Zeeman discovered that if a source of light producing line spectrum is placed in very strong magnetic field the spectral lines are split up into components-dou-blets or triplets or even more complex arrangements. This is called *Zeeman effect*.

Normal Zeeman effect. *The splitting of a spectral line into three components in a magnetic field when viewed in a direction perpendicular to the magnetic field is called normal Zeeman effect.*

Anomalous Zeeman effect. *When a single spectral line is split up into four or more lines, when observed in a direction perpendicular to that of the magnetic field, it is known as* **anomalous Zeeman effect.** It is called anomalous Zeeman effect, because it could not be explained on the basis of classical physics.

Normal Zeeman effect is obtained from sources of elements like *Ca, Cu, Zn, Cd* etc., and anomalous Zeeman pattern is obtained from elements like *Na, Cr* etc.

Experimental arrangement. To study Zeeman effect, experimentally, a gas discharge tube is placed between the pole-pieces of a strong electro-magnet. Light from this source is studied with the help of a spectrograph of high resolving power. The Zeeman effect may be observed in two ways:

(*i*) *Perpendicular to the direction of the magnetic field.* In this case, a single spectral line observed when no field is applied, splits up into three components when the magnetic field is applied *i.e.*, a triplet consisting of three spectral lines is observed. The central line occupies the undisturbed position and is plane polarised with vibrations parallel to the field. The outer two lines are also plane polarised, having vibrations in a direction perpendicular to the field. This is known as **normal transverse Zeeman effect.**

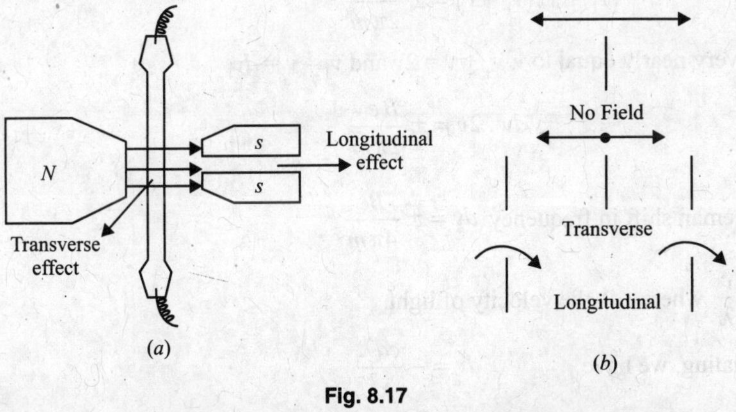

Fig. 8.17

(*ii*) *Parallel to the direction of the magnetic field.* To observe the effect, a hole is bored through one of the pole-pieces and the light coming from the source through the hole is examined with the spectrograph. In this case, a single spectral line is split up into two components when the magnetic field is applied *i.e.*, a doublet consisting of two spectral lines is observed. Both the lines are displaced from the original position and are circularly polarised in opposite directions. This is known as **normal longitudinal Zeeman effect.** The displacement in the two cases are proportional to the strength of the magnetic field and are equal for the same value of magnetic field, as shown in Fig. 8.17(*b*).

Classical theory of normal Zeeman effect. The phenomenon has been explained by Lorentz on the consideration of the electronic theory of matter.

The centripetal force acting on the revolving electron is given by

$$F = \frac{mv^2}{r} \qquad ...(i)$$

where v is the velocity of the electron in its orbit of radius r.

But $v = \frac{2\pi r}{T} = 2\pi r \nu$ where ν is the numer of revolutions made by the electron per second.

Thus $$F = \frac{m}{r} 4\pi^2 r^2 \nu^2 = 4\pi^2 mr\nu^2 \qquad ...(ii)$$

Now if a magnetic field is applied in a direction perpendicular to the plane of the electronic orbit, an additional force = $Be\,v = Be\,2\,\pi\,r\,\nu$ acts on the electron.

According to Fleming's left-hand rule, the direction of this force will be towards or away from the centre, according as the electronic motion is clockwise or counter-clockwise. If the additional force is towards the centre of the atom, the orbit contracts and thus frequency of rotation increases.

If the additional force is away from the centre, the orbit expands and hence frequency decreases. This force is added to or subtracted from the attractive force, so that in the new state of dynamic equilibrium—

$$F \pm Be 2\pi r \nu = 4\pi^2 mr\nu_1^2 \qquad ...(iii)$$

where ν_1 is the frequency in the new orbit.

Subtracting (*ii*) from (*iii*), we have

$$4\pi^2 mr\left(\nu_1^2 - \nu^2\right) = \pm Be 2\pi r \nu$$

$$\therefore \qquad \nu_1^2 - \nu^2 = \frac{Be\nu}{2\pi m}$$

or $$(\nu_1 + \nu)(\nu_1 - \nu) = \pm \frac{Be\nu}{2\pi m}$$

But ν_1 is very nearly equal to ν, $\nu_1 + \nu \doteq 2\nu$ and $\nu_1 - \nu = d\nu$

$$\therefore \qquad d\nu \,.\, 2\nu = \pm \frac{Be\nu}{2\pi m} \qquad ...(v)$$

or Zeeman shift in frequency $$d\nu = \pm \frac{Be}{4\pi m} \qquad ...(vi)$$

Now $\nu = \frac{c}{\lambda}$ where c is the velocity of light.

Differentiating, we have $$d\nu = -\frac{c\,d\lambda}{\lambda^2}$$

Hence expressing in terms of wavelength, equation (*vi*) becomes

$$d\lambda = \pm \frac{\lambda^2}{c} \frac{Be}{4\pi m}$$

where λ is the wavelength of the original line and $d\lambda$ the change in wavelength or *Zeeman shift in wavelength* brought about by the application of the field.

The value of e/m determined by this method is 1.77×10^{11} C/kg. It is the same as determined by J.J. Thomson for cathode rays.

The diversity in the nature and positions of the lines, when observed in a direction perpendicular or parallel to the magnetic field can be explained as follows:

The emission of light by a glowing gas is due to the vibratory motion of the electrons, which is of the simple harmonic type. Any linear motion of an electron can be resolved into a linear motion along the field and another linear component perpendicular to the field. The latter can be considered as equivalent to two opposite circular motions perpendicular to the field. Now, if this is observed in a direction perpendicular to the field, the linear motion parallel to the field will send out un-disturbed plane polarised vibrations of the original frequency v. The two circular motions perpendicular to the field will send in the above-mentioned directions two plane polarised vibrations of frequencies $v + dv$ and $v - dv$. Hence when examined with a spectrometer, three plane polarised lines are observed, the central in the undisturbed position and other two also plane polarised, but in a perpendicular plane. The frequency difference is given by equation (*vi*).

When light is viewed parallel to the direction of the magnetic field, two circularly (but in the opposite sense) polarised lines are observed with a frequency difference given by (*vi*). The vibrations parallel to the field do not send any energy in this direction.

8.19 NORMAL ZEEMAN EFFECT

The splitting of a spectral line into three components under the effect of a magnetic field is known as normal Zeeman effect.

All the three components can be seen when viewed in a direction perpendicular to the direction of the magnetic field (*transverse Zeeman effect*) but only the outer two components are seen when viewed in a direction parallel to the magnetic field (*longitudinal Zeeman effect*). The central line is missing.

The central line has the same wavelength as that of the original spectral line and the other two lines are equally spaced on either side of the central line. The shift in wavelength of other lines depends upon the original wavelength and the strength of the magnetic field.

Normal Zeeman effect is observed for light emitted from sources of elements like *Ca, Cu* and *Zn* etc. which have two electrons in the outermost valence shell. These two electrons have anti-parallel *spin* (one has *spin up* ↑ and the other *spin down* ↓) so that the net spin $S = 0$. For quantum mechanical treatment of normal Zeeman effect we shall, therefore, consider the orbital angular momentum only.

Effect of Magnetic field on energy levels on quantum theory. The magnetic moment of a current loop of area $\vec{A}$ through which a current i is flowing, is given by

$$\vec{\mu} = i\vec{A}$$

Considering magnitudes only $\mu = i A$.

An electron revolving in a circular orbit of radius r and making n revolutions is equivalent to a current ($-ne$) in a circular loop of area πr^2. The magnetic moment of the electron

$$\mu = i A = -n e \pi r^2 \qquad ...(i)$$

The angular momentum of the electron

$$\vec{L} = \vec{r} \times m\vec{v}$$

where m is the mass of the electron and $\vec{v}$ its linear velocity, the magnitude of which is given by

$$v = 2\pi\, n\, r$$

As $\vec{v}$ is perpendicular to $\vec{r}$, the magnitude of angular momentum

$$L = 2\pi\, m\, n\, r^2 \qquad \text{...(ii)}$$

From (i) and (ii), we have

$$\mu = -\frac{e}{2m} L$$

where μ is the magnetic moment of the electron due to its orbital motion. Let it be denoted by μ_l.

$$\therefore \qquad \vec{\mu}_l = -\left(\frac{e}{2m}\right)\vec{L} \qquad \text{...(iii)}$$

When a magnetic field of flux density $\vec{B}$ is applied, the torque tending to align the magnetic moment vector $\vec{\mu}_l$ of the magnetic dipole in the direction of the field vector $\vec{B}$ is given by

$$\vec{\tau} = \vec{\mu}_l \times \vec{B}$$

The magnitude of the torque is $\tau = \mu_l B \sin\theta$

where θ is the angle the vector $\vec{\mu}_l$ makes with the field vector $\vec{B}$.

The potential energy gained by the *dipole* is given by

$$\delta E = \int_{\pi/2}^{\theta} \mu_l B \sin\theta\, d\theta = -\mu_l B \cos\theta$$

The lower limit is taken as $\pi/2$ because the potential energy of the dipole is minimum when it lies in a direction at right angles to the field.

Substituting the value of μ_l from (*iii*) we have

$$\delta E = \frac{e}{2m} LB\cos\theta \qquad \text{...(iv)}$$

But $$L = \sqrt{l\,(l+1)}\hbar = \sqrt{l\,(l+1)}\,\frac{h}{2\pi}$$

where l is the *orbital quantum number* of the electron in the atom, and h the Planck's constant.

Also $$\cos\theta = \frac{L_z}{L} = \frac{m_l}{\sqrt{l(l+1)}}$$

where L_z is the component of the angular momentum vector $\vec{L}$ in the direction of the magnetic field vector $\vec{B}$ and m_l the magnetic quantum number which has $(2l + 1)$ values from $-l$ to $+l$ through zero.

Substituting the values of L and $\cos\theta$ in (*iv*), we have

$$\delta E = m_l\left(\frac{eh}{4\pi m}\right)B = m_l\left(\frac{e\hbar}{2m}\right)B$$

The quantity $\left(\frac{eh}{4\pi m}\right) = \left(\frac{e\hbar}{2m}\right)$ is called the *Bohr magneton* and has a value 9.27×10^{-24} J/T (Joule per Tesla).

Suppose an electron in the atom goes from an initial higher energy state E_{oi} to final lower energy state E_{of} when no magnetic field is applied, then the frequency of the emitted photon (spectral line).

$$\nu_0 = \frac{E_{oi} - E_{of}}{h}$$

When a magnetic induction field $\vec{B}$ is applied, the energy of the initial and final states is given by

$$E_i = E_{oi} + m_{li}\left(\frac{eh}{4\pi m}\right)B \qquad ...(v)$$

and

$$E_f = E_{of} + m_{lf}\left(\frac{eh}{4\pi m}\right)B \qquad ...(vi)$$

Splitting of spectral lines–normal Zeeman effect. As the value of energy of the higher as well as lower energy state (E_{oi} and E_{of}) has been modified, the frequency of the emitted photon (spectral line) is now given by

$$\nu = \frac{E_i - E_f}{h} = \frac{E_{oi} - E_{of}}{h} + (m_{li} - m_{lf})\left(\frac{e}{4\pi m}\right)B$$

$$= \nu_0 + \Delta m_l \left(\frac{e}{4\pi m}\right)B \qquad ...(vii)$$

As only those transitions are possible according to selection rules for which $\Delta m_l = +1$, 0 and -1, we respectively get *three* values of ν, given by

$$\nu_1 = \nu_0 + \left(\frac{e}{4\pi m}\right)B = \nu_0 + \Delta\nu \qquad ...(viii)$$

$$\nu_2 = \nu_0 \qquad ...(ix)$$

and

$$\nu_3 = \nu_0 - \left(\frac{e}{4\pi m}\right)B = \nu_0 - \Delta\nu \qquad ...(x)$$

where

$$\Delta\nu = \left(\frac{e}{4\pi m}\right)B \qquad ...(xi)$$

As

$$\nu = \frac{c}{\lambda};\ d\nu = -c\frac{d\lambda}{\lambda^2} \text{ or } d\lambda = -\frac{\lambda^2}{c}d\nu$$

∴ Zeeman shift in terms of wavelength, is given by

$$d\lambda = \Delta\lambda = \pm\frac{\lambda^2}{c}\left(\frac{e}{4\pi m}\right)B$$

Detailed analysis of transition probabilities and polarisation shows that radiation emitted during the transition $\Delta m_l = 0$ is polarised in a plane parallel to B and that emitted during the transition $\Delta m_l = \pm 1$ is polarised in a plane perpendicular to B. Thus, ν_1 and ν_2 are ⊥ or σ components and ν_0 is ∥ or π component.

For longitudinal Zeeman effect $\Delta m_l = 0$ transition is forbidden and hence ν_0 component is not observed. The ν_1 and ν_2 lines corresponding to $\Delta m_l = \pm 1$are circularly polarised in opposite senses.

Energy level diagram showing transitions. As an example, consider the case of a spectral line emitted due to the transition of the electron from the *d*-state ($l = 2$) to the *p*-state ($l = 1$) for an atom for which $S = 0$.

Before the application of the mangnetic field $\nu_0 = \dfrac{E_{oi} - E_{of}}{h}$

On the application of the magnetic field, the energy level corresponding to $l = 2$ splits up into $(2l + 1) = 5$ sub-levels for $m_l = +2, +1, 0, -1$ and -2 and the energy level corresponding to $l = 1$ splits up into $(2l + 1) = 3$ sub-levels for values of $m_l = +1, 0$ and -1. According to transition rules, only those transitions are possible for which $\Delta m_l = 0$ or ± 1. For $\Delta m_l = 0$, three transitions are possible *i.e.* from $m_l = +1, 0$ and -1 for $l = 2$ to $m_l = +1, 0$ and -1 for $l = 1$. All these transitions give the original frequency $\nu_0 = \dfrac{E_{of} - E_{oi}}{h}$ and, therefore, combine to give only *one* line. For $\Delta m_l = +1$, again three transitions are possible *i.e.* from $m_l = +2, +1$ and 0 for $l = 2$ to $m_l = +1, 0$ and -1 for $l = 1$ respectively. All these three transitions give a frequency $\nu_1 = \nu_0 + \Delta\nu$ where $\Delta\nu = \dfrac{\Delta E}{h} = \dfrac{e}{4\pi m} B$ and combine to give only one line.

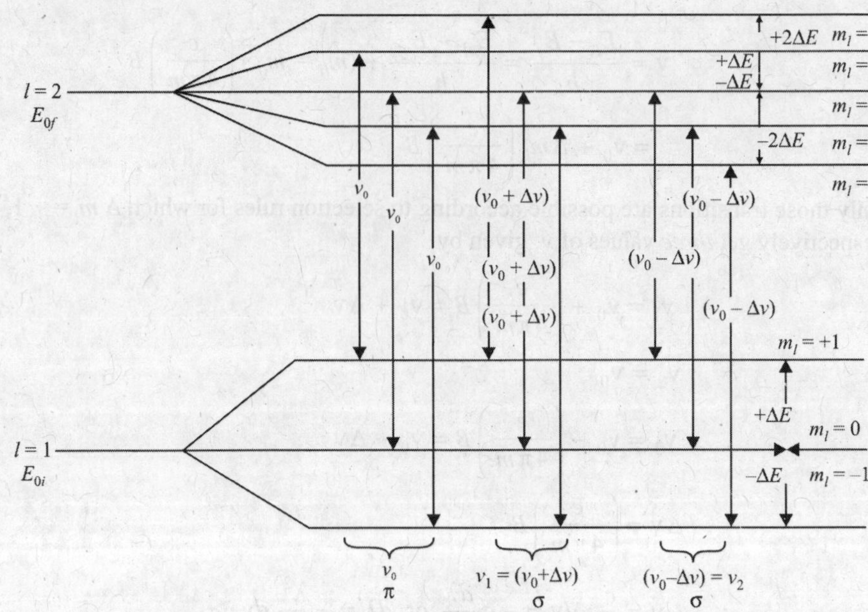

Fig. 8.18

For $\Delta m_l = -1$ again three transitions are possible *i.e.* from $m_l = 0, -1$ and -2 for $l = 2$ to $m_l = +1$, 0 and -1 for $l = 1$ respectively. All these three transitions give a frequency $\nu_0 = \nu_0 - \Delta\nu$ and combine to give only a *single* line.

Splitting into only three spectral lines. From the above discussion we find that corresponding to each of the three selection rules $\Delta m_l = +1, 0$, and -1, there are three possible transitions, but all three give rise to only one frequency as all the frequencies are equal. Thus there are only three spectral lines, one each for $\Delta m_l = +1$, $\Delta m_l = 0$ and $\Delta m_l = -1$.

In other words, we get a *triplet* consisting of the original frequency ν_0 and two frequencies on either side displaced by $\Delta\nu = \dfrac{e}{4\pi m} B$. The state of polarisation of the lines is also indicated in Fig. 8.18.

Example 8.18. *A certain transition in an atom, of wave length 4226 Å when observed in a magnetic field of 4 Tesla exhibits a normal Zeeman pattern whose components are separated by 5.6 × 10^{10} cycles per second. Calculate the value of e/m.* **(P.U. 2002)**

Solution. When placed in a magnetic field *B*, the frequency difference for normal Zeeman pattern is

$$\Delta v = \left(\frac{e}{4\pi m}\right) B = \frac{e}{m}\frac{B}{4\pi}$$

or

$$\frac{e}{m} = \frac{4\pi\,\Delta v}{B} = \frac{4 \times 3.142 \times 5.6 \times 10^{10}}{4}$$

$$= 1.759 \times 10^{11}\ C/kg$$

8.20 NORMAL ZEEMAN EFFECT IN *H, He, Li, Mg, Na* ATOMS

Normal Zeeman effect occurs only in atoms which have a total spin $S = 0$ so that the spin magnetic moment of the atom is zero. For *S* to be zero, the number of electrons in the outermost valence orbit should be even. The number of electrons in the *n*th orbit can have a maximum value equal to $2n^2$, i.e. the first orbit can have $2 \times 1^2 = 2$ electrons, second $2 \times 2^2 = 8$ electrons and so on with the provison that the outermost orbit can have a maximum of 8 electrons. Thus all orbits except the outermost orbit, which are full contain an even number of electrons. When the number of electrons in the outermost orbit is even, the total number of electrons is even.

As normal Zeeman effect occurs between those states which have zero spin angular momentum, for these states total angular momentum *J* is equal to orbital angular momentum *L*. This condition is satisfied only by singlet states. Hence normal Zeeman effect occurs for spectral lines produced by transition between singlet states.

The electronic configuration of *He* is $1s^2$ and of magnesium is $(1s^2\ 2s^2\ 2p^6)\ 3s^2$, so that helium has two electron in 1*s* orbit and magnesium 2 electrons in 3s orbit with anti-parallel spin and $S = 0$. All other elements given above have only one electron in outermost valence orbit, since their electronic configuration is $H = 1s^1$, $Li = 1s^2, 2s^1$ and $Na = (1s^2\ 2s^2\ 2p^6)\,3s^1$ and a spin $S = \frac{1}{2}$.

Therefore *He* and *Mg* exhibit normal Zeeman effect and *H*, *Li* and *Na* do not exhibit normal Zeeman effect.

8.21 ANOMALOUS ZEEMAN EFFECT : QUANTUM THEORY

The quantum mechanical explanation of anomalous Zeeman effect is based upon the concept of '*spin*'. An atom possesses a magnetic moment, because its electrons have orbital and spin angular momenta and associated magnetic moments. The magnitude of the total magnetic moment vector is given by

$$\mu_j = -\,\mu_B\, g_j \sqrt{j\,(j+1)} \qquad \text{...(}i\text{) [see Equation (}iii\text{) article } \mathbf{8.13}]$$

where

$$\mu_B = \frac{e\hbar}{2m},\quad g_j = 1 + \frac{j\,(j+1) + s\,(s+1) - l\,(l+1)}{2\,j\,(j+1)}$$

known as Lande's splitting factor, *j* the total angular momentum quantum number, *s* spin quantum number and *l* orbital quantum number.

When a magnetic induction field $\vec{B}$ is applied, the torque tending to align the total magnetic moment vector $\vec{\mu}_j$ in the direction of the field vector $\vec{B}$ is given by

$$\vec{\tau} = \vec{\mu}_j \times \vec{B}$$

The magnitude of the torque $\tau = \mu_j B \sin\theta$, where θ is the angle that the vector $\vec{\mu}_j$ makes with the field vector $\vec{B}$. The magnetic energy gained by the atomic dipole is given by

$$\delta E = \int_{\pi/2}^{0} \mu_j B \sin\theta \, d\theta = -\mu_j B \cos\theta = -\vec{\mu}_j . \vec{B} \qquad ...(ii)$$

The lower limit is taken as $\pi/2$ because the potential energy of the dipole is minimum when it lies in a direction at right angles to the field.

Substituting the value of μ_j from (*i*) in (*ii*), we have

$$\delta E = \mu_B g_j B \sqrt{j(j+1)} \cos\theta = \mu_B g_j m_j B \qquad ..(ii)(a)$$

Since $$\cos\theta = \frac{J_z}{J} = \frac{m_j \hbar}{\sqrt{j(j+1)}\hbar} = \frac{m_j}{\sqrt{j(j+1)}}$$

Suppose an electron in the atom goes from an initial higher energy state E_{oi} to a final lower energy state E_{of}, when no magnetic field is aplied, then the frequency of the emitted photon

$$\nu_0 = \frac{E_{oi} - E_{of}}{h}$$

When a magnetic induction field $\vec{B}$ is applied, the energy of the initial and final energy states respectively is given by

$$E_i = E_{oi} + \mu_B g_{ji} m_{ji} B$$

and $$E_f = E_{of} + \mu_B g_{jf} m_{jf} B$$

where g_{ji}, m_{ji} and g_{jf}, m_{jf} are the values of Lande's splitting factor g_j and total magnetic quantum number m_j for the initial and final energy states respectively.

The frequency of the emitted photon is now given by

$$\nu = \frac{E_i - E_f}{h} = \frac{E_{oi} - E_{of}}{h} + \frac{\mu_B B}{h}[g_{ji} m_{ji} - g_{jf} m_{jf}]$$

$$= \frac{E_{oi} - E_{of}}{h} + \frac{\Delta E}{h}[g_{ji} m_{ji} - g_{jf} m_{jf}]$$

$$= \nu_0 + \Delta\nu [g_{ji} m_{ji} - g_{jf} m_{jf}] \qquad ...(iii)$$

where $$\Delta E = \mu_B B \text{ and } \Delta\nu = \frac{\mu_B B}{h} = \frac{eh}{4\pi m} . \frac{B}{h} = \frac{e}{4\pi m} B$$

Zeeman pattern for D_1 and D_2 lines of sodium. Anomalous Zeeman effect is observed if the atomic states involved in producing the spectral line have $\vec{S} \neq 0$. In this case, a single spectral line is split up into four or more lines. As $\vec{S} \neq 0$ the values of g_{ji} and g_{jf} are different for initial and final energy states of the atom.

As an example, consider the splitting up of D_1 and D_2 lines of *Na*. The D_1 line of $\lambda_1 = 5896$ Å is emitted by $3^2S_{1/2} - 3^2P_{1/2}$ transition of the electron and D_2 line of $\lambda_2 = 5890$ Å by $3\,^2S_{1/2} - 3\,^2P_{3/2}$ transition of the electron. Since the atomic number of $Na = 11$, it has two electrons in the first orbit, $n = 1$, 8 in the second $n = 2$ and one in the third $n = 3$.

$\therefore$ $3\ ^2S_{1/2}$ level means $n = 3$, $l = 0$, $j = \frac{1}{2}$ and $(2s + 1) = 2$ i.e., $s = \frac{1}{2}$

$3\ ^2P_{1/2}$ level means $n = 3$, $l = 1$, $j = \frac{1}{2}$ and $s = \frac{1}{2}$

$3\ ^2P_{3/2}$ level means $n = 3$, $l = 1$, $j = \frac{3}{2}$ and $s = \frac{1}{2}$

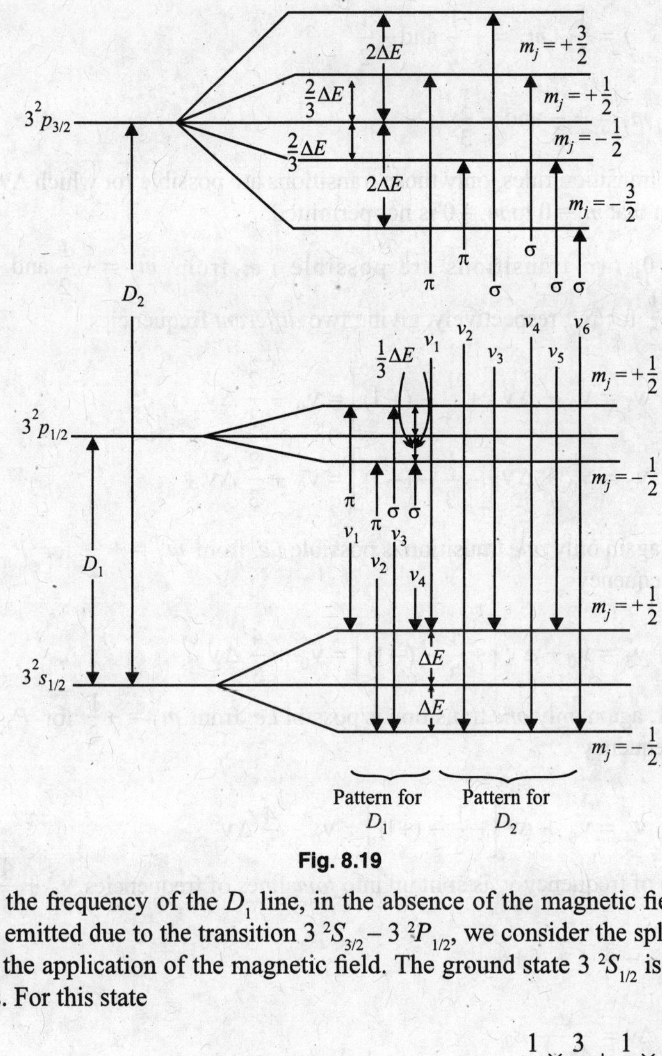

Fig. 8.19

D_1-line. Let the frequency of the D_1 line, in the absence of the magnetic field, be denoted by ν_0. As the line is emitted due to the transition $3\ ^2S_{3/2} - 3\ ^2P_{1/2}$, we consider the splitting of these two energy states on the application of the magnetic field. The ground state $3\ ^2S_{1/2}$ is split up into $(2j + 1) = 2$ sub-levels. For this state

$$g_j = 1 + \frac{j\,(j+1) + s\,(s+1) - l\,(l+1)}{2\,j\,(j+1)} = 1 + \frac{\frac{1}{2}\times\frac{3}{2} + \frac{1}{2}\times\frac{3}{2}}{2\times\frac{1}{2}\times\frac{3}{2}} = 2$$

$\because$ $\quad j = \frac{1}{2};\ m_j = +\frac{1}{2}$ and $-\frac{1}{2}$

$\therefore$ $\quad g_j m_j = +1$ and -1

The state $3^2P_{1/2}$ is also split up into $(2j + 1) = 2$ sub-levels.

For this state

$$g_j = 1 + \frac{\frac{1}{2}\times\frac{3}{2}+\frac{1}{2}\times\frac{3}{2}-1\times 2}{2\times\frac{1}{2}\times\frac{3}{2}} = \frac{2}{3}$$

$$\because \quad j = \frac{1}{2};\ m_j = +\frac{1}{2} \text{ and} -\frac{1}{2}$$

$$\therefore \quad g_j m_j = +\frac{1}{3} \text{ and } -\frac{1}{3}$$

According to transition rules, only those transitions are possible for which $\Delta m_j = 0, +1$, or -1, with the exception that $m_j = 0$ to $m_j = 0$ is not permitted.

For $\Delta m_j = 0$, *two* transitions are possible *i.e.* from $m_j = +\frac{1}{2}$ and $-\frac{1}{2}$ for $^2P_{1/2}$ to $m_j = +\frac{1}{2}$ and $-\frac{1}{2}$ for $^2S_{1/2}$ respectively, giving two *different* frequencies.

$$\nu_1 = \nu_0 + \Delta\nu\left[+\frac{1}{3}-(+1)\right] = \nu_0 - \frac{2}{3}\Delta\nu$$

and

$$\nu_2 = \nu_0 + \Delta\nu\left[-\frac{1}{3}-(-1)\right] = \nu_0 + \frac{2}{3}\Delta\nu$$

For $\Delta m_j = 1$, again only *one* transition is possible *i.e.* from $m_j = +\frac{1}{2}$ for $^2P_{1/2}$ to $m_j = -\frac{1}{2}$ for $^2S_{1/2}$, giving the frequency

$$\nu_3 = \nu_0 + \Delta\nu\left[+\frac{1}{3}-(-1)\right] = \nu_0 + \frac{4}{3}\Delta\nu$$

For $\Delta m_j = -1$, again only *one* transition is possibl *i.e.* from $m_j = -\frac{1}{2}$ for $^2P_{1/2}$ to $m_j = +\frac{1}{2}$ for $^2S_{1/2}$, giving the frequency

$$\nu_4 = \nu_0 + \Delta\nu\left[-\frac{1}{3}-(+1)\right] = \nu_0 - \frac{4}{3}\Delta\nu$$

So the D_1 line of frequency ν_0 is split up into *four* lines of frequencies $\nu_0 + \frac{4}{3}\Delta\nu, \nu_0 + \frac{2}{3}\Delta\nu$, $\nu_0 - \frac{2}{3}\Delta\nu$ and $\nu_0 - \frac{4}{3}\Delta\nu$, where

$$\Delta\nu = \frac{e}{4\pi m}B$$

D_2-line. Let the frequency of the D_2 line, in the absence of the magnetic field, be denoted by ν_0. As this line is emitted due to the transition $3\,^2S_{1/2} - 3\,^2P_{3/2}$, we consider the splitting of these two energy states on the application of the magnetic field.

For the ground state $3\,^2S_{1/2}$, number of sub-levels = 2, $g_j = 2$ and $m_j = +\frac{1}{2}$ and $-\frac{1}{2}$, so that $g_j m_j = +1$ and -1. The $3\,^2P_{3/2}$ state is split up into $(2j+1) = \left(2\times\frac{3}{2}+1\right) = 4$ sub-levels.

For this state

$$g_j = 1 + \frac{\frac{3}{2}\times\frac{5}{2}+\frac{1}{2}\times\frac{3}{2}-1\times 2}{2\times\frac{3}{2}\times\frac{5}{2}} = \frac{4}{3}$$

$\because \quad j = \frac{3}{2} \qquad m_j = +\frac{3}{2}, +\frac{1}{2}, -\frac{1}{2}, -\frac{3}{2}$

$\therefore \quad g_j m_j = +2, +\frac{2}{3}, -\frac{2}{3}, -2$

For $\Delta m_j = 0$, *two* transitions are possible from $m_j = +\frac{1}{2}$ and $-\frac{1}{2}$ for $^2P_{3/2}$ to $m_j = +\frac{1}{2}$ and $-\frac{1}{2}$ for $^2S_{1/2}$ respectively, giving two *different* frequencies–

$$\nu_1 = \nu_0 + \Delta\nu\left[+\frac{2}{3}-(+1)\right] = \nu_0 - \frac{1}{3}\Delta\nu$$

and $$\nu_2 = \nu_0 + \Delta\nu\left[-\frac{2}{3}-(-1)\right] = \nu_0 + \frac{1}{3}\Delta\nu$$

For $\Delta m_j = +1$, *two* transitions are possible from $m_j = +\frac{3}{2}$ and $+\frac{1}{2}$ for $^2P_{3/2}$ to $m_j = +\frac{1}{2}$ and $-\frac{1}{2}$ for $^2S_{1/2}$ respectively, giving two *different* frequencies–

$$\nu_3 = \nu_0 + \Delta\nu\,[+2-(+1)] = \nu_0 + \Delta\nu$$

and $$\nu_4 = \nu_0 + \Delta\nu\left[+\frac{2}{3}-(-1)\right] = \nu_0 + \frac{5}{3}\Delta\nu$$

For $\Delta m_j = -1$, *two* transitions are possible for $m_j = -\frac{1}{2}$ and $-\frac{3}{2}$ for $^2P_{3/2}$ to $m_j = +\frac{1}{2}$ and $-\frac{1}{2}$ for $^2S_{1/2}$ respectively, giving two *different* frequencies–

$$\nu_5 = \nu_0 + \Delta\nu\left[-\frac{2}{3}-(+1)\right] = \nu_0 - \frac{5}{3}\Delta\nu$$

and $$\nu_6 = \nu_0 + \Delta\nu\,[-2-(-1)] = \nu_0 - \Delta\nu$$

Thus D_2 line of frequency ν_0 is split up into *six* lines of frequencies $\nu_0 + \frac{5}{3}\Delta\nu, \nu_0 + \Delta\nu,$ $\nu_0 + \frac{1}{3}\Delta\nu, \nu_0 - \frac{1}{3}\Delta\nu, \nu_0 - \Delta\nu$ and $\nu_0 - \frac{5}{3}\Delta\nu$, where $\Delta\nu = \frac{e}{4\pi m}B$.

The spectral lines corresponding to $\Delta m_j = 0$ give electric polarisation or π component parallel to B while $\Delta m_j = \pm 1$ gives $\perp$ or σ component. The same are also shown in Fig. 8.19.

Example 8.19. *Evaluate the Lande's g factor for $^3P_1$ level and use the result to predict the splitting (in eV) of the level when the atom is in an external magnetic field of 0.1 Tesla. Given Bohr magneton = 9.3 × 10^{-24} Am^2.* ***(G.N.D.U., 2002)***

Solution. For $^3P_1$ level $\quad l = 1$

$(2s+1) = 3 \quad \therefore s = 1$

$j = 1$

$$\therefore \qquad g_j = 1 + \frac{j(j+1) + s(s+1) - l(l+1)}{2j(j+1)}$$

$$= 1 + \frac{1\times 2 + 1\times 2 - 1\times 2}{2\times 1\times 2} = 1 + \frac{1}{2} = \frac{3}{2}$$

Now $\delta E = \mu_B \, g_j \, m_j \, B$ ...**[Eq. (*ii*) (*a*) article 8.21]**

For $j = 1$, the possible values of m_j are $+1, 0, -1$.

For $m_j = +1$

$$\delta E = 9.3\times 10^{-24} \times \frac{3}{2} \times 1 \times 0.1 = 1.395\times 10^{-24} \text{ J}$$

$$= \frac{1.395\times 10^{-24}}{1.6\times 10^{-19}} = 0.872\times 10^{-5} \text{ eV} = +8.72\times 10^{-6} \text{ eV}$$

For $m_j = 0$ $\qquad \delta E = 0$

For $m_j = -1$ $\qquad \delta E = -8.72\times 10^{-6}$ eV.

The $^3P_1$ level will split into three levels.

EXERCISE

LONG QUESTIONS

1. (*a*) What are critical potentials? Distinguish between critical potential, ionisation potential and excitation potential. ***(Meerut U. 2003, 2002, 2000; Gauhati U. 2000)***

   (*b*) Describe briefly Frank-Hertz experiment which clearly demonstrated the existence of discrete energy levels in an atom. Discuss the graph obtained between the current and accelerating voltage. How does this method help in determining the critical potentials of the atom? ***(Utkal U. 2003; Pbi; U. 2005, 2003, 2000; P.U. 2006, 2005, 2002; G.N.D.U. 2005, 2004, 2001; H.P.U. 2000; Meerut U. 2003, 2002, 2000; Gharwal U. 2000).***

2. (*a*) Explain what is magnetic moment of an atom and derive an expression for it. Hence show that the ratio of the orbital magnetic moment to angular momentum of the atom is $e/2m$. ***(G.N.D.U. 2008, 2007; Gujrat U. 2004)***

   (*b*) What is Bohr magneton? Calculate its value. What is its physical significance? Is magnetic moment quantised? ***(G.N.D.U. 2008, 2007)***

3. Calculate the angular frequency at which an electron's orbital magnetic moment $\vec{\mu}$ precesses in a magnetic field $\vec{B}$. OR An atomic dipole is subjected to very strong magnetic field B so that it begins to process about the field. Calculate the frequency of Larmour precession. ***(Pbi. U. 2001)***

4. (*a*) Name the factors which lead to the discovery of spin. When and by whom the idea of intrinsic spin angular momentum of the electron was proposed? What is its value? What is meant by space quantisation of electron spin? Calculate two orientations of electron spin angular momentum $\vec{S}$ with respect to the magnetic field direction. ***(P.U. 2007; M.D.U. 2007; Meerut U. 2003; Pbi. U. 2002; G.N.D.U., 2007)***

   (*b*) What is gyro-magnetic ratio? Calculate its orbital and spin value.

5. (*a*) What is spin orbit coupling? Define total angular momentum vector J and show that its Z-component is quantised. Find the value of total angular momentum $\vec{J}$ for one electron atom and the value of J for $l = 1$.

(*Gujrat U. 2004; G.N.D.U., 2008, 2005; Pbi. U. 2000*)

(*b*) Show in a diagram how orbital and spin angular momentum vectors $\vec{L}$ and $\vec{S}$ precess about $\vec{J}$. What happens when a magnetic field is applied?

6. What are the difficulties with Bohr-Sommerfeld model of the atom? Describe the main features of the vector atom model. Explain the various quantum numbers associated with it.

(*Gujrat U. 2004; M.D.U. 2003; Meerut U. 2001, 2000; K.U. 2001*)

7. What do you mean by spin-orbit interaction? Calculate the change in total energy of the atom due to spin-orbit coupling. **(*H.P.U. 2002; G.N.D.U. 2006, 2002, 2001, 1995; M.D.U. 2002; K.U. 2002; Pbi. U., 2008; P.U. 2006*)**

8. Show that $\cos\theta = \dfrac{j(j+1) - l(l+1) - s(s+1)}{2\sqrt{l(l+1)\,s(s+1)}}$, where θ is the angle between $\vec{L}$ and $\vec{S}$.

(*G.N.D.U. 2001*)

9. What is meant by fine structure of hydrogen spectra? How is it explained on the basis of spin orbit interaction?

(*H.P.U. 2003, 2002, 2001; Pbi.U. 2006, 2005, 2003; P.U. 2008, 2006, 2003, 2002, 2001, 2000; Luck. U. 2002, 2001; G.N.D.U. 2007, 2006, 2005*)

10. What are the selection rules for j for observing doublet spectra in single electron system?

(*G. N. D.U 2002*)

11. (*a*) Derive an expression for the total magnetic moment of an electron in an atom due to the interaction of its orbital and spin angular momenta. Hence find the value of Lande's g factor.

(*b*) Calculate Lande's g factor for s electron.

(*G.N.D.U. 2007, Pbi. U. 2006; P.U. 2006, 2005, 2001*)

12. (*a*) Explain hyperfine sturcture of spectral lines.

(*b*) Explain how different factors are responsible for hyperfine structure of spectral lines.

(*M.D.U. 2003, 2002, 2000; K.U. 2001*)

13. Discuss relativistic Sommerfeld theory of hydrogen atomic spectrum. How does it explain fine structine of Hα-line? **(*Meerut U. 2003, 2001; Luck .U. 2002, 2001; Madurai U. 2002*)**

14. Why was zero angular momentum considered impossible in the Sommerfeld elliptic treatment of hydrogen atom? **(*G.N.D.U. 2003*)**

15. Describe Stern Gerlach experiment with necessary theory. What was the aim of the experiment? Discuss its significance. **(*P.U. 2007, 2006, 2003, 2002, 2001, 2000, K.U. 2002; Madurai U. 2002; Meerut U. 2003, 2002; Pbi. U. 2004, 2002, 2000; G.N.D.U. 2007, 2006, 2005, 2002; H.P.U. 2000; Bang. U. 2000; Luck. U. 2002*)**

16. (*a*) What is Zeeman effect? Distinguish between normal and anomalous Zeeman effect. Discuss experimental arrangement for observing normal Zeeman effect.

(*b*) Give the explanation of normal Zeeman effect on classical theory and obtain an expression for Zeeman shift.

(*P.U. 2006, 2002, 2000; K.U. 2001; Luck. U. 2001, Bang. U. 2001, 2000; Gauhati U. 2000; Pbi. U. 2000*)

17. Discuss the general quantum mechanical theory of anomalaus Zeeman effect, with special reference to Zeeman pattern for D_1 and D_2 lines of sodium. Draw a neat diagram to illustrate the Zeeman splitting of D_1 and D_2 lines of sodium.

(Osm. U. 2004, G.N.D.U. 2003, 2001, H.P.U. 2001, K.U. 2002; M.D.U. 2001, 2000; P.U. 2001, Luck U. 2003, 2001, Purvanchal U. 2003, 2001)

18. Determine the normal Zeeman splitting of Cadmium red line 6438 Å when the atoms are placed in a magnetic field of 0.009 T. **[Ans. d λ = 1.74 × 10^{-3} Å]**

19. In a normal Zeeman experiment the Calcium 4226 Å line splits into 3 lines separated by 0.25 Å in a magnetic field of 3 T. Detemine e/m for the electron. **[Ans. e/m = 1.76 × 10^{11} C/kg]**

SHORT QUESTIONS

1. What is orbital magnetic moment of an electron ?
2. Show that ratio of orbital magnetic moment to angular momentum is $-\frac{e}{2m}$.
3. What is Bohr magneton ? Explain its physical significance.
4. Explain the quantisation of magnetic moment.
5. What are the selection rules for j for observing doublet spectra in single electron system ? ***(G.N.D.U. 2002)***
6. What do you mean by hyperfine structure of spectral lines ?
7. What is hyperfine shift in electronic energy levels in spectral lines. ?
8. Write down the correction which one makes to obtain the fine structure of the Balmer lines of hydrogen. ***(G.N.D.U. 2002)***
9. Can Stern Gerlach experiment be performed with ions rather than neutral atoms? ***(Pbi. U. 2005; G.N.D.U. 2004; P.U. 2003; H.P.U. 2002; Luck .U. 2001)***
10. A beam of silver atoms in a Stern Gerlach experiment obtained from an oven heated to a temperature 1500 K passes through an inhomogeneous magnetic field having a field gradient of 2 Wb/m² per cm perpendicular to the beam. The pole pieces are 10 cm long. What is the separation between the components of the beam on a photographic plate placed at a distance of 50 cm ? **(Ans. 0.3278 cm)**
11. What is normal Zeeman effect? On the basis of quantum theory, explain the effect of magnetic field on energy levels of an atom. Hence explain the splitting of spectral lines in normal Zeeman effect and show that a spectral line splits into three spectral lines. ***(Gujrat U. 2004; P.U., 2007, 2006, 2003; G.N.D.U. 2008, 2004, 2003; Pbi. U. 2008, 2003, 2001; H.P.U. 2003, 2000; Madurai U. 2002; Bang. U., 2004, 2001, 2000; M.D.U. 2007; Poorvanchal U. 2001)***
12. (*a*) Explain why normal Zeeman effect occurs only in atoms with even number of electrons. ***(H.P.U. 2003, 2000; Pbi. U. 2002; G.N.D.U., 2005)***

    (*b*) Which of the elements H, He, Li, Mg and Na might exhibit normal Zeeman effect and why? ***(G.N.D.U. 2000; Pbi. U. 2002)***

MULTIPLE CHOICE QUESTIONS

1. The value of Bohr magneton is

    (*a*) 9.27 × 10^{-27} Am² (*b*) 9.27 × 10^{-24} Am²
    (*c*) 9.27 × 10^{-19} Am² (*d*) 9.27 × 10^{-7} Am²

2. The direction of orbital magnetic moment in hydrogen atom is
   (*a*) same as the direction of orbital angular momentum
   (*b*) opposite to that of orbital angular momentum
   (*c*) pependicular to the direction of orbital angular momentum vector.
   (*d*) in a direction making some angle with the orbital angular momentum vector.
3. The energy of a hydrogen atom in the first excited state is
   (*a*) – 13.6 eV (*b*) – 1.51 eV (*c*) – 3.4 eV (*d*) – 0.85 eV
4. The value of spin quantum number of an electron in hydrogen atom is
   (*a*) $-\frac{1}{2}$ (*b*) $+\frac{1}{2}$ (*c*) 1 (*d*) 0
5. The number of allowed orientations of spin angular momentum of the elctron in hydrogen atom are
   (*a*) 0 (*b*) 1 (*c*) 2 (*d*) 3
6. The total number of allowed values for total magnetic quantum number m_j for a given j are
   (*a*) j (*b*) $2j$ (*c*) $2j+1$ (*d*) $2j-1$
7. The number of values allowed to total quantum number j of an electron in hydrogen atom for a given non-zero l are
   (*a*) 0 (*b*) 1 (*c*) 2 (*d*) 3
8. On which of the following levels of hydrogen the spin-orbit interaction has no effect?
   (*a*) *s*-level (*b*) *p*-level (*c*) *d*-level (*d*) *f*-level.
9. The principal series of hydrogen atom is observed during the transition of the electron from
   (*a*) *s* to *p* states (*b*) *p* to *s* states (*c*) *d* to *p* states (*d*) *p* to *d* states
10. The value of Lande's splitting factor g_j for *s*-state is
   (*a*) 0 (*b*) 1 (*c*) 2 (*d*) $\frac{1}{2}$
11. In normal Zeeman effect a level of given l splits into
   (*a*) l levels (*b*) $2l$ levels (*c*) $(2l+1)$ levels (*d*) $(2l-1)$ levels

ANSWERS

1. (*b*)	2. (*b*)	3. (*c*)	4. (*b*)	5. (*c*)
6. (*c*)	7. (*c*)	8. (*a*)	9. (*b*)	10. (*c*)
11. (*c*)				

ATOMS WITH MANY ELECTRONS

9

Chapter

INTRODUCTION

Having studied atoms with single electron in outermost orbit, let us study now atoms with many electrons leading to singlet and triplet spectra in fine structure of alkaline elements due to L–S and J–J couplings. To understand this, let us start with wave functions which are symmetric and anti-symmetric depending upon the type of particles under consideration, say bosons, fermions etc.

9.1 TOTAL WAVE FUNCTION OF TWO IDENTICAL PARTICLES

In the study of quantum theory of hydrogen atom, we found that the complete wave function ψ of the electron in the atom can be expressed as the product of three separate, independent wave functions R, Θ, Φ, each describing that part of ψ which is a function of one of the three co-ordinates r, θ and ϕ respectively.

Similarly, the total wave function of n non-interacting particles can be written as product of n functions *i.e.*

$$\psi(1, 2, 3...n) = \psi(1)\,\psi(2)\,\psi(3)\,.....\,\psi(n)$$

Let us consider a system consisting of only two identical, non-interacting particles—*i.e. particles having the same mass, charge and spin.* Let particle 1 be in the quantum state (a), denoted as $\psi_a(1)$ and particles 2 be in the quantum state (b), denoted as $\psi_b(2)$. The total wave function of the two particles is then given by

$$\psi(1, 2) = \psi_a(1)\,\psi_b(2) \qquad ...(i)$$

Now if the particles are exchanged, so that particle 1 is in the state (b) and particle 2 in the state (a) then total wave function $\psi(2, 1)$ of the two particles is given by

$$\psi(2, 1) = \psi_a(2)\,\psi_b(1) \qquad ...(ii)$$

Condition for acceptable wave function. The time independent Schrödinger's wave equation is given by

$$H\psi = E\psi \qquad ...(iii)$$

where H is the Hamiltonian and E the total energy.

When the total wave function of the two identical particles is $\psi(1, 2)$, equation (iii) becomes

$$H(1, 2)\,\psi(1, 2) = E\,\psi(1, 2) \qquad ...(iv)$$

When the particles are exchanged and the total wave function is $\psi(2, 1)$, equation (iii) becomes

$$H(2, 1)\,\psi(2, 1) = E\,\psi(2, 1) \qquad ...(v)$$

Since the particles are identical, the Hamiltonian remains the same under exchange of particles

$\therefore \qquad H(1, 2) = H(2, 1)$

Substituting $\qquad H(2, 1) = H(1, 2)$ in Eq. (v) we have

$$H(1, 2)\,\psi(2, 1) = E\,\psi(2, 1) \qquad ...(vi)$$

Comparing Eqs. (iv) *and* (vi), *we find that* $\psi(1, 2)$ *and* $\psi(2, 1)$ *are both solutions of Schrödinger's wave equation.*

So these functions must be related by a constant

or $$\psi(2, 1) = \lambda\,\psi(1, 2) \qquad ...(vii)$$

In other words, when the two particles are interchanged, the wave function gets multiplied by a constant λ.

Similarly, interchanging the labels of the particles in Eq. (*vii*), we have

$$\psi(1, 2) = \lambda\, \psi(2, 1) \quad ...(viii)$$

Substituting the value of $\psi(2, 1) = \lambda\, \psi(1, 2)$ in Eq. (*viii*), we get

$$\psi(1, 2) = \lambda^2\, \psi(1, 2)$$

or $\lambda^2 = 1$

$$\lambda = \pm 1$$

Substituting this value of λ in Eq. (*vii*), we get

$$\psi(2, 1) = \pm\, \psi(1, 2) \quad ...(ix)$$

The condition given by equation (ix) must be satisfied by the acceptable wave function of a two identical particle system.

Symmetric and anti-symmetric wave functions. From Eq. (*ix*) we find that only two types of wave function are acceptable.

(*i*) For which $\psi(2, 1) = +\psi(1, 2)$

and (*ii*) For which $\psi(2, 1) = -\psi(1, 2)$

Symmetric wave function. *If a wave function does not change sign on interchanging the particles, it is said to be a symmetric wave function,*

For a symmetric wave function

$$\psi(2, 1) = +\psi(1, 2)$$

or $\psi_a(1)\,\psi_b(2) = +\psi_b(1)\,\psi_a(2)$

Since the particles are identical, there is equal probability for both the wave functions $\psi(1, 2)$ and $\psi(2, 1)$. Therefore, the actual probability is obtained by a linear combination of the two wave functions.

∴ For symmetric wave function

$$\psi_s = \frac{1}{\sqrt{2}}\left[\psi_a(1)\,\psi_b(2) + \psi_b(1)\,\psi_a(2)\right] \quad ...(x)$$

where $\frac{1}{\sqrt{2}}$ is the normalisation factor. It is seen that ψ_s remains unchanged by an exchange of particles.

Anti-symmetric wave function. *If a wave function changes sign on interchanging the particles, it is said to be an anti-symmetric wave function.*

For an anti-symmetric wave function

$$\psi(2, 1) = -\psi(1, 2)$$

or $\psi_a(1)\,\psi_b(2) = -\psi_b(1)\,\psi_a(2)$

The linear combination of these two wave functions gives the anti-symmetric wave function.

$$\psi_A = \frac{1}{\sqrt{2}}\left[\psi_a(1)\,\psi_b(2) - \psi_b(1)\,\psi_a(2)\right] \quad ...(xi)$$

ψ_A changes sign when the particles 1 and 2 are exchanged.

This distinction between symmetric and anti-symmetric wave functions leads to two kinds of particles *i.e. Fermions* and *Bosons*.

Fermions. From Eq. (*xi*) we find that, when the wave functions of two particles are anti-symmetric, the two cannot exist in the same quantum state, because if $a = b$, $\psi_A = 0$. Comparing this quantum mechanical statement with Pauli's exclusion principle, which states that no two electrons in an atom can be in the same quantum state, we conclude that the system of electrons is described by an anti-symmetric wave function, which changes sign upon exchange of any pair of electrons.

Further the experimental results show that *all particles which have odd half integral* spin $\left(\frac{1}{2}, \frac{3}{2} ...\right)$ have anti-symmetric wave functions. Such particles obey Pauli's exclusion principle, when they are in the same system *i.e.* move in a common force-field, so that each member of the system is in a different quantum state. These particles include electron, proton, neutron, μ-meson and neutrino. All particles with spin $\frac{1}{2}\hbar$ are referred to as Fermi particles or *Fermions* as they obey Fermi-Dirac Statistics.

Bosons. From Eq. (*x*), we find that when $a = b$, $\psi_s \neq 0$. Thus the particles, which have symmetric wave functions, do not obey Pauli's exclusion principle *i.e.* more than one such particle can exist in the same quantum state. *These particles have zero or integral spin* and obey Bose-Einstein statistics. These are known as Bose particles or *Bosons.* These include photons, π and *K* mesons, α-particle and Helium atom He_2^4.

Exchange degeneracy. We have seen that both the wave functions ψ (1, 2) and ψ (2, 1) correspond to the same value of total energy *E*. As ψ_s and ψ_A are linear combinations of ψ (1, 2) and ψ (2, 1), these also correspond to the same value of total energy *E*. But ψ_s and ψ_A are two different wave functions. Thus the two-particle system is *degenerate.* This degeneracy arises due to exchange of particle labels and is, therefore, called *exchange degeneracy*.

Degenerate states. Whenever in a dynamical system there are several different states which correspond to the same total energy, these states are said to be degenerate states.

9.2 SYMMETRIC AND ANTI-SYMMETRIC WAVEFUNCTIONS LEAD TO PAULI'S EXCLUSION PRINCIPLE

Pauli's exclusion principle. An electron in an atom can exist in a quantum state defined by 4 quantum numbers.

(*i*) *Principal quantum number n*. This decides the total energy of the electron. It has integral values 1, 2, 3, 4 etc.

(*ii*) *Orbital quantum number l*. This decides the orbital angular momentum and magnetic moment of the electron due to its orbital motion. For one particular value of the principal quantum number *n*, it has *n* different integral values 0, 1, 2 ...... (*n* – 1).

(*iii*) *Magnetic quantum number* m_l. This decides the spatial quantisation of the angular momentum and magnetic moment of the electron, due to its orbital motion, in the direction of the applied magnetic field (taken as *Z*-axis). For one particular value of orbital quantum number *l*, it has (2*l* + 1) different integral values – *l*, – (*l* – 1), – 1, 0, + 1 ... (*l* – 1), + 1.

(*iv*) *Spin magnetic quantum number* m_s. The spin quantum number *s* decides the spin angular momentum and magnetic moment of the electron due to spin. Its value for the electron $= \frac{1}{2}$. The spin magnetic quantum number decides the spatial quantisation of the magnetic moment of the electron due to spin in the direction of the applied magnetic field. It has two values $m_s = \pm\frac{1}{2}$.

In the case of an atom having more than one electron, the electrons may arrange themselves in different quantum states. The principle which determines the arrangement of electrons in different quantum states in an atom was enunciated by W. Pauli and is known as *Pauli's exclusion principle.*

It states that no two electrons in an atom can exist in the same quantum state. In other words, each electron in the atom must have a different set of quantum numbers *n*, *l*, m_l and m_s.

A simple explanation of Pauli's exclusion principle can be given, using quantum mechanics. To describe a system of two identical particles, there are two different wave functions.

(*i*) **Symmetric wave function ψ_s.** It is given by

$$\psi_s = \frac{1}{\sqrt{2}}[\psi_a(1)\psi_b(2) + \psi_b(1)\psi_a(2)] \quad ...(i)$$

where ψ_a (1) denotes that particle 1 is in the quantum state (*a*) and ψ_b (2) denotes that particle 2 is in the quantum state (*b*). Similarly ψ_b (1) denotes that particle 1 is in the quantum state (*b*) and ψ_a (2) denotes that particle 2 is in the quantum state (*a*). $\frac{1}{\sqrt{2}}$ is the normalising factor.

(*ii*) **Anti-symmetric wave function ψ_A.** It is given by

$$\psi_A = \frac{1}{\sqrt{2}}[\psi_a(1)\psi_b(2) - \psi_b(1)\psi_a(2)] \quad ...(ii)$$

An electron is a particle with spin $\frac{1}{2}\hbar$. All identical particles with half integral spin (*Fermions*) are governed by anti-symmetric wave functions.

If we try to place two electrons in the same quantum state *i.e.* $a = b$, equation (*ii*) becomes

$$\psi_A = \frac{1}{\sqrt{2}}[\psi_a(1)\psi_a(2) - \psi_a(1)\psi_a(2)]$$

$$\psi_A = 0$$

The wave function ψ_A is identically equal to zero. As a result, two electrons cannot exist simultaneously in the same quantum state. In the language of quantum mechanics, Pauli's exclusion principle can be stated as

'*The wave function of a many electron system is anti-symmetric to exchange of co-ordinates of any two electrons*'.

On the other hand, if the wave function is symmetric—as in the case of Bosons—which include photons, π and K mesons and α-particles and have 0 or integral spin, ψ_s is not equal to zero, even when the two such particles are in the same quantum state. Therefore, more than one such particle can stay in the same quantum state *i.e.* Pauli's exclusion principle is not obeyed.

Thus we find that a knowledge of symmetric and anti-symmetric wave functions leads to Pauli's exclusion principle.

No two electrons have same quantum state

Electrons are Fermi particles having spin $\frac{1}{2}$. The total wave function of two electrons is antisymmetric and is given by

$$\psi_A = \frac{1}{\sqrt{2}}[\psi_a(1)\psi_b(2) - \psi_b(1)\psi_a(2)] \quad ...(i)$$

where ψ_a (1) denotes that the particle (1) is in the quantum state (*a*) and ψ_b (2) denotes that the particle (2) is in the quantum state (*b*). Similarly ψ_b (1) denotes that the particle (1) is in the quantum state (*b*) and ψ_a (2) denotes that particle (2) is in the quantum state (*a*). $\frac{1}{\sqrt{2}}$ is the normalising factor.

If we try to place two electrons in the same quantum state *i.e.* $a = b$, then equation (*i*) becomes

$$\psi_A = \frac{1}{\sqrt{2}}[\psi_a(1)\psi_a(2) - \psi_a(1)\psi_a(2)]$$

or $\quad \psi_A = 0$

The wave function ψ_A is identically equal to zero. This implies that *no two electrons can have the same quantum state.*

In other words, each electron in the atom must have a different set of quantum numbers, *n, l* m_l and m_s.

9.3 WAVE FUNCTIONS FOR BOSONS AND FERMIONS

To describe a system of two identical particles there are two different wave functions

symmetric wave function $$\psi_s = \frac{1}{\sqrt{2}}[\psi_a(1)\psi_b(2) + \psi_b(1)\psi_a(2)] \quad ...(i)$$

and *anti-symmetric wave function* $$\psi_A = \frac{1}{\sqrt{2}}[\psi_a(1)\psi_b(2) - \psi_b(1)\psi_a(2)] \quad ...(ii)$$

where ψ_a (1) denotes that the particle (1) is in the quantum state (*a*) and ψ_b (2) denotes that the particle (2) is in the quantum state (*b*). Similarly ψ_b (1) denotes that the particle (1) is in the quantum state (*b*) and ψ_a (2) denotes that the particle (2) is in the quantum state (*a*). $\frac{1}{\sqrt{2}}$ is the normalising factor.

Fermions-antisymmetric wave function. Fermions are particles having half integral spin $\left[\frac{1}{2}, \frac{3}{2} ....\right]$ and obey Pauli's exclusion principle according to which no two *identical fermions* e.g. electrons in an atom can be in the same quantum state.

Now, if we try to place two identical Fermions (say electrons) in the *same quantum state* i.e. $a = b$, then equation (*ii*) becomes

$$\psi_A = \frac{1}{\sqrt{2}}[\psi_a(1)\psi_a(2) - \psi_a(1)\psi_a(2)]$$

or $$\psi_s = 0$$

The wave function ψ_A is identically equal to zero. This implies that two identical Fermi particles cannot exists simultaneously in the same quantum state. Thus Pauli's exclusion principle is obeyed only when two identical Fermions have antisymmetric wave function i.e. the summetry character of the wave function of multiparticle system of Fermions is *antisymmetric.*

Bosons-symmetric wave function. Bosons are particles which have zero or integral spin in units of $\hbar$ e.g. photons, π or *K*-mensons and α-particles.

In the case of Bosons Pauli's exclusion principle is not obeyed and more than one such particle can stay in the same quantum state. A multi particle system of Bosons can, therefore, be represented by the symmetric wave function

$$\psi_s = \frac{1}{\sqrt{2}}[\psi_a(1)\,\psi_b(2) + \psi_b(1)\,\psi_a(2)]$$

9.4 SINGLET AND TRIPLET STATES

To explain the difference between singlet and triplet states, we consider a system of only two electrons. Let the explicit interaction, like Coulomb interaction between the two electrons, be zero. The wave function for this system will be anti-symmetric, given by

$$\psi_A = \frac{1}{\sqrt{2}}[\psi_a(1)\psi_b(2) - \psi_a(2)\psi_b(1)]$$

The symbol (*a*) specifies the space co-ordinates and spin of one electron and (*b*) the space co-ordinates and spin of the other electron.

If ψ^o is the orbital or *space eigen function* and σ the *spin eigen function* of the electron, the total wave function

$$\psi_A = (\text{space eigen function}) \times (\text{spin eigen function})$$

or $$\psi_A = \psi^o \sigma$$

If we exchange the labels of the particles, the total wave function ψ_A will be anti-symmetric *only* if

(*i*) *Either the space function ψ^o is symmetric and spin function σ is anti-symmetric* or

(*ii*) *The space function ψ^o is anti-symmetric and spin function σ is symmetric.*

If the space function ψ^o and spin function σ are *both symmetric* or are *both anti-symmetric,* the total wave function will be symmetric. Therefore, for total wave function of two electrons to be anti-symmetric, ψ_A is given by

either $$\psi_A = \psi_s^o \sigma_A$$

or $$\psi_A = \psi_A^o \sigma_s$$

where ψ_s^o and ψ_A^o are the space and σ_s and σ_A the spin symmetric and anti-symmetric wave function respectively.

Spin wave function of a single electron. Each electron has a spin quantum number $s = \frac{1}{2}$. The magnitude of the spin angular momentum vector $\vec{s}$ of each electron is given by

$$|\vec{s}| = \sqrt{s(s+1)}\,\hbar$$

or $$|\vec{s}| = \sqrt{\frac{1}{2}\left(\frac{1}{2}+1\right)}\,\hbar = \frac{\sqrt{3}}{2}\,\hbar$$

The magnetic spin quantum number m_s for an electron can have only two values $+\frac{1}{2}$ and $-\frac{1}{2}$. Thus the spin angular momentum vector $\vec{s}$ of an electron *can have two orientations with respect to Z-axis,*

When $$m_s = +\frac{1}{2} \qquad s_z = +\frac{1}{2}\hbar$$

This orientation is known as **spin-up** state and is represented by spin wave function σ_+.

When $$m_s = -\frac{1}{2} \qquad s_z = -\frac{1}{2}\hbar$$

This orientation is known as **spin-down** state and is represented as σ_-. σ_+ and σ_- are spin wave functions of a single electron.

Spin wave function for two electrons. If s_1 and s_2 are the spin quantum numbers of electron 1 and electron 2 in a two-electron system, then the total spin quantum number S can take values

$$(s_1 + s_2),\ (s_1 + s_2 - 1) \ldots\ldots\ldots |s_1 - s_2|$$

The magnitude of the total spin angular momentum of the two electrons

$$|\vec{S}| = \sqrt{S(S+1)}\,\hbar$$

As $s_1 = s_2 = \frac{1}{2}$, the resultant spin S of the two electrons can have values

$$S = \left(\frac{1}{2} + \frac{1}{2}\right) = 1 \ or \ S = \left(\frac{1}{2} - \frac{1}{2}\right) = 0$$

When $S = 1$, the spins are said to be **parallel.**

When $S = 0$, the spins are said to be **anti-parallel.**

The Z-component of total spin angular momentum of the two electrons is given by

$$S_z = M_s \hbar$$

where M_s is the total spin magnetic quantum number and can take values

$$M_s = (S), (S-1), (S-2) .... (-S)$$

Triplet state. When the spins are parallel, $S = 1$. Corresponding to $S = 1$, the permissible values of M_s are

$$M_s = +1, 0, -1$$

i.e. the total angular momentum vector $\vec{S}$ can take *three* orientations in space.

In Fig. 9.1 (*a*), both the electrons are in the *spin up* state, so that the total spin magnetic number

$$M_s = m_{s_1} + m_{s_2} = \frac{1}{2} + \frac{1}{2} = 1$$

The spin function σ_1 for two electrons, corresponding to this state ($M_s = +1$) is given by the product of individual electron spin up states

$$\therefore \qquad \sigma_1 = \sigma_+(1)\,\sigma_+(2) \quad (M_s = +1)$$

In Fig. 9.1 (*b*), electron 1 is in the *spin up* state and electron 2 in the *spin-down* state, so total magnetic spin quantum number $M_s = m_{s_1} - m_{s_2} = 0$. The spin function for two electrons corresponding to this state is given by

$$\sigma' = \sigma_+(1)\,\sigma_-(2)$$

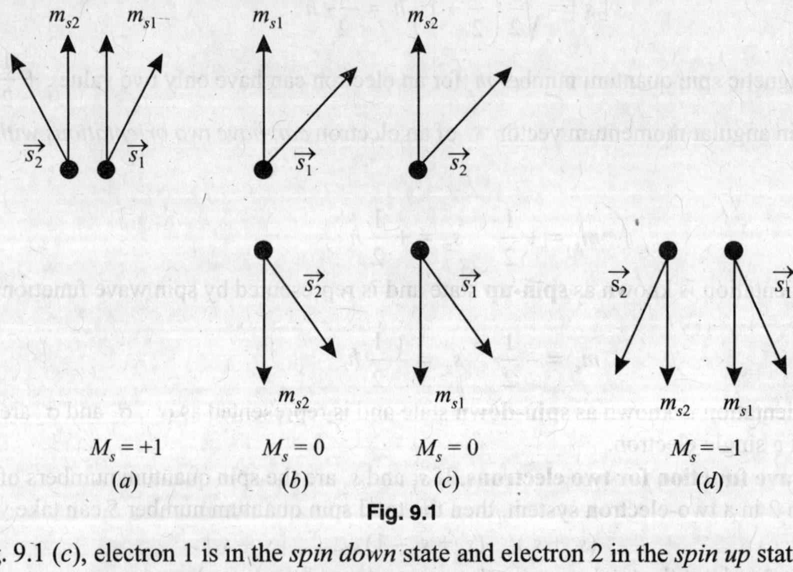

Fig. 9.1

In Fig. 9.1 (*c*), electron 1 is in the *spin down* state and electron 2 in the *spin up* state.

The spin function σ'', for two electrons corresponding to this state is given by

$$\sigma'' = \sigma_-(1)\,\sigma_+(2)$$

Since the electrons are identical, the probability that σ' describes the system at any time is equal to the probability that σ'' describes the system at the same time. So the complete spin function to describe this state is a linear combination of σ' and σ'', given by

$$\sigma_2 = \frac{1}{\sqrt{2}}(\sigma' + \sigma'')$$

or
$$\sigma_2 = \frac{1}{\sqrt{2}}[\sigma_+(1)\sigma_-(2) + \sigma_-(1)\sigma_+(2)] \quad (M_s = 0)$$

In Fig. 9.1 (*d*), both the electrons are in the *spin down* state, so the total spin magnetic quantum number $M_s = m_{s_1} + m_{s_2} = -\frac{1}{2} - \frac{1}{2} = -1$. This spin function σ_3 for two electrons corresponding to this state ($M_s = -1$) is given by

$$\sigma_3 = \sigma_-(1)\,\sigma_-(2)$$

Thus, corresponding to spin quantum number $S = 1$, there are three spin functions σ_1, σ_2 and σ_3. *All these three states are symmetric with respect to exchange of particles*. Therefore, for the triplet state, the space part of the wave function ψ^o will be anti-symmetric.

There are, therefore, *three symmetric* spin functions for $S = 1$ corresponding to three values of $M_s = +1$, 0 and -1. These are–

$$\sigma_s = \begin{bmatrix} \sigma_1 = \sigma_+(1)\sigma_+(2) & (M_s = +1) \\ \sigma_2 = \frac{1}{\sqrt{2}}[\sigma_+(1)\sigma_-(2) + \sigma_-(1)\sigma_+(2)] & (M_s = 0) \\ \sigma_3 = \sigma_-(1)\sigma_-(2) & (M_s = -1) \end{bmatrix}$$

The state with $S = 1$ is called a *triplet* state.

Singlet state. When the spin of two electrons are anti-parallel, $S = 0$. Corresponding to $S = 0$, the permissible value of $M_s = 0$ *i.e.* the total spin angular momentum vector $\vec{S}$ can have only one orientation in space. This orientation is produced when the spin angular momenta of the two electrons are in opposite directions. As shown in Fig. 9.2(*a*), electron 1 is in the *spin up* ($m_s = +1/2$) state and electron 2 is in the *spin down* state ($m_s = -1/2$).

This state can be described as

$$\sigma' = \sigma_+(1)\,\sigma_-(2)$$

In Fig. 9.2 (*b*), electron 2 is in *spin up* state and electron 1 in *spin down* state. This state can be described as

$$\sigma'' = \sigma_+(2)\,\sigma_-(1)$$

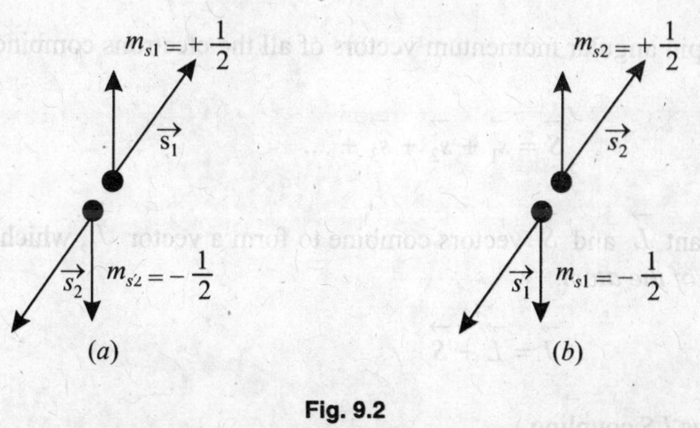

Fig. 9.2

The probability that the system of electrons is described by σ' or by σ'' is the same. Therefore, complete spin wave function to describe this state is again a linear combination of σ' and σ'' given by

$$\sigma_A = \frac{1}{\sqrt{2}}(\sigma' - \sigma'')$$

$$= \frac{1}{\sqrt{2}}[\sigma_+(1)\sigma_-(2) - \sigma + (2)\,\sigma_-(1)]$$

This spin function, corresponding to S = 0, is anti-symmetric.

The space part of the singlet wave function ψ_s^o is, therefore, symmetric.

Thus corresponding to $S = 0$, there is only one anti-symmetric spin function σ_A. This state with $S = 0$ *is called a singlet state.*

Elements giving singlets and triplets

For energy levels to be divided into singlets and triplets, the total spin quantum number S should be zero or one.

For $S = 0$, we have a singlet state

For $S = 1$, we have a triplet state

Now $S = 0$ or 1, only when there are two electrons in the valence (outermost) orbit of the atom, according as their spins are anti-parallel or parallel. Out of the given list, the following will have singlet and triplet state:– *Mg, Ca, Ba*. The other elements *Ne, Cl, Cu* and *Ag* will not have singlet and triplet states.

9.5 COUPLING SCHEMES : LS AND JJ COUPLINGS

Coupling schemes. The electron in an atom has an orbital angular momentum and a spin angular momentum. In atoms having two or more electrons, the orbital and spin angular momenta of all the electrons can be added together in two ways, depending upon the *interaction* or *coupling* between the two kinds of angular momenta. These schemes are:–

(*i*) *LS* coupling or Russel-Saunder's coupling and (*ii*) *JJ* coupling.

***(i) LS* coupling.** *LS* coupling is known as *normal coupling* as it takes place more frequently. In this type of coupling, the orbital angular momentum vectors of all the electrons combine to form a resultant vector given by

$$\vec{L} = \vec{l_1} + \vec{l_2} + \vec{l_3} + \ldots\ldots$$

Similarly, the spin angular momentum vectors of all the electrons combine to form a resultant vector, given by

$$\vec{S} = \vec{s_1} + \vec{s_2} + \vec{s_3} + \ldots.$$

Then the resultant $\vec{L}$ and $\vec{S}$ vectors combine to form a vector $\vec{J}$, which represents the *total angular momentum of the atom.*

$$\therefore \quad \vec{J} = \vec{L} + \vec{S}$$

This is known as *LS* coupling.

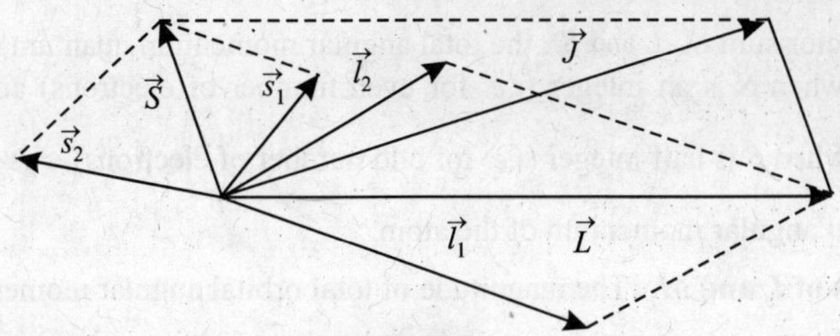

Fig. 9.3

Vector diagram of *LS* coupling. Consider an atom containing two electrons. Suppose $\vec{l_1}$ and $\vec{l_2}$ are their orbital angular momentum vectors and $\vec{s_1}$ and $\vec{s_2}$ their spin angular momentum vectors.

Then, the total angular momentum vector $\vec{L}$ of the two electrons is the vector sum of $\vec{l_1}$ and $\vec{l_2}$. Similarly, the total spin vector $\vec{S}$ is the vector sum of $\vec{s_1}$ and $\vec{s_2}$. Thus the resultant total angular momentum vector $\vec{J}$ is the vector sum of $\vec{L}$ and $\vec{S}$, as shown in Fig. 9.3.

or $$\vec{J} = \vec{L} + \vec{S}$$

Explanation. *LS* coupling is governed by the following rules:–

(*i*) All the three vectors $\vec{L}, \vec{S}$ and $\vec{J}$ are quantised. The quantum number corresponding to $\vec{L}, \vec{S}$ and $\vec{J}$ are denoted by L, S, and J respectively.

(*ii*) The total orbital angular momentum quantum number L is an integer, including zero *i.e.*

$$L = 0, 1, 2, ... \text{etc.}$$

For an atom with two electrons with orbital quantum numbers l_1 and l_2, the total orbital quantum number L has the values

$|l_1 + l_2|, |l_1 + l_2 - 1| ..... |l_1 - l_2|$

(*iii*) The total spin quantum number S is an integer or half integer, depending upon the number of electrons and the direction of their spin vectors. In the case of *one electron atom*, the spin quantum number of the electron $s = \frac{1}{2}$. Therefore the total spin quantum number $S = \frac{1}{2}$. In the case of *two electron atoms*, S is either 1 or zero, depending upon whether the spin vectors are parallel or anti-parallel. Similarly in case of three-electron system, S is either 3/2 or 1/2, depending upon whether spins are parallel or anti-parallel and for 4 electrons $S = 2, 1, 0$ for the same reason. This is shown in Fig. 9.4.

One Electron	Two Electrons		Three Electrons		Four Electrons		
$\uparrow s_1$	$s_1\uparrow$ $s_2\uparrow$	$s_1\uparrow\downarrow s_2$	$\uparrow s_1$ $\uparrow s_2$ $\uparrow s_3$	$\uparrow s_1 \downarrow s_3$ $\uparrow s_2$	$\uparrow s_1$ $\uparrow s_2$ $\uparrow s_3$ $\uparrow s_4$	$\uparrow s_1$ $\uparrow s_2 \downarrow s_4$ $\uparrow s_3$	$\uparrow s_1 \downarrow s_3$ $\uparrow s_2 \downarrow s_4$
$S = \frac{1}{2}$	$S = 1$	$S = 0$	$S = \frac{3}{2}$	$S = \frac{1}{2}$	$S = 2$	$S = 1$	$S = 0$

Fig. 9.4

(*iv*) As $\vec{J}$ is vector sum of $\vec{L}$ and $\vec{S}$, the total angular momentum quantum number J = integer (0, 1, 2, 3 .... etc) when S is an integer (*i.e.* for even number of electrons) and J = half integer $\left(\frac{1}{2}, \frac{3}{2}, \frac{5}{2} \text{.... etc.}\right)$ when S is half integer (*i.e.* for odd number of electrons). J is always *positive*, as it represents the total angular momentum of the atom.

Allowed values of L and M_L. The magnitude of total orbital angular momentum is given by

$$|\vec{L}| = \sqrt{L(L+1)}\,\hbar$$

For an atom with two electrons, the total angular momentum quantum number L can have the values

$$|l_1 + l_2|\,|l_1 + l_2 - 1| \,....\, |l_1 - l_2|$$

where l_1 and l_2 are the orbital quantum numbers of electron 1 and 2 respectively.

The Z-component of the total orbital angular momentum is given by

$$L_z = M_L$$

where M_L is the total orbital magnetic quantum number. M_L can have the values given by

$$M_L = L, L-1, L-2, ... -L$$

Allowed values of S and M_s. The magnitude of the total spin angular momentum is given by

$$|\vec{S}| = \sqrt{S(S+1)}\,\hbar$$

where S can take values

$$S = |s_1 + s_2|, |s_1 + s_2 - 1| \,....\, |s_1 - s_2|$$

for an atom with two electrons, having spin quantum number s_1 *for* electron 1 and s_2 for electron 2.

The Z-component of total spin angular momentum is given by

$$S = M_s\,\hbar$$

where M_s can have the values given by

$$M_s = S, S-1, S-2, ... -S$$

Allowed values of J and M_j. The magnitude of the resultant total angular momentum vector of the atom is given by

$$|\vec{J}| = \sqrt{J(J+1)}\,\hbar$$

where the possible values of J are

$$J = |L+S|, |L+S-1| \,.....\, |L-S|$$

The Z-component of resultant total angular momentum is given by

$$J_z = M_j\,\hbar$$

where M_j can have the values given by

$$M_J = J, J-1, J-2 .... -J$$

Both L and M_L have always an integral value or zero, while the other quantum numbers are half integral, if an odd number of electrons is involved and integral or zero, if an even number of electrons is involved. When $L > S$, J can have $2S + 1$ values, but when $L < S$, J can have $2L + 1$ values.

Why coupling of l's and spins occur? *LS* coupling generally takes place in the case of light atoms, because the interaction between orbital angular momentum vectors $\vec{l}_1, \vec{l}_2, \vec{l}_3$ etc. and that between the spin angular momentum vectors $\vec{s}_1, \vec{s}_2, \vec{s}_3$ etc. among different electrons in the atom is much greater than the spin orbit interactions of the same electrons *i.e.* between $\vec{l}_1$, and s_1 of electron 1 or between l_2 and $\vec{s}_2$ of electron 2 and so on.

Due to the asymmetrical distribution of charge densities of electrons in an atom the electric force between the electrons varies with relative orientations of the angular momentum vector. Only those orientations are stable for which total orbital angular momentum is quantised according to the relation

$$|\vec{L}| = \sqrt{L(L+1)}\,\hbar$$

The coupling between various $\vec{l}$ vectors of different electrons takes place in this way, because it leads to the *lowest energy state* of the atom. This is obtained when the total orbital angular momentum $\vec{L}$ is a maximum.

To understand this point, consider two electrons in the same Bohr orbit. As the electrons repel each other, they tend to revolve round the nucleus in the same direction, which makes $\vec{L}$ maximum. If, however, the electrons revolve in opposite directions, they would cross each other very frequently and would need energy to overcome the Coulomb repulsion while approaching each other, thus leading to a higher energy of the system. In quantum mechanical terms, the wave functions of various electrons in an atom overlap the least when $\vec{L}$ is a maximum. As a system tends to be in the lowest energy state, we have

$$\vec{L} = \vec{l}_1 + \vec{l}_2 + \vec{l}_3 \ldots$$

It is a little difficult to understand coupling between spins of different electrons, as spin is a purely quantum mechanical phenomenon with no classical analog. The spin angular momenta of various electrons $\vec{s}_1, \vec{s}_2$ etc combine to form a resultant $\vec{S}$ *i.e.*

$$\vec{S} = \vec{s}_1 + \vec{s}_2 + \vec{s}_3 + \ldots$$

The *lowest energy state* is again obtained when total spin $\vec{S}$ is a maximum. We know that this happens when the *spins are parallel* and with two electrons in the atom, the total spin quantum number

$$S = s_1 + s_2 = \frac{1}{2} + \frac{1}{2} = 1$$

In such a case, there are three different possible values of M_s the total spin magnetic quantum number *i.e.* there are three different spin wave functions which means a greater average separation in space of the electrons and accordingly a lower total energy.

When *LS* coupling breaks down? The electric forces which couple various $\vec{l}$ vectors into a single resultant $\vec{L}$ vector and $\vec{s}$ vectors into a single resultant $\vec{S}$ vector, are stronger than the forces of magnetic spin-orbit interaction that couple $\vec{L}$ and $\vec{S}$ to form $\vec{J}$ in *light atoms*. The same situation prevails even when a moderate external magnetic field is applied.

But in *heavy atoms*, the nuclear charge is very large and produces a spin-orbit interaction comparable in magnitude to the Coulomb interaction between various $\vec{l}$ vectors and between various $\vec{s}$ vectors so that *LS* coupling begins to break down. In a strong magnetic field (> 1 *T*) also a similar breakdown of the scheme takes place, giving rise to Paschen-Back effect in atomic spectra. As *LS* coupling starts failing, the resultant angular momentum vectors j_1, j_2 etc. of individual electrons add together directly to form the total angular momentum vector $\vec{J}$ of the whole atom. This is known as $j-j$ coupling.

(*ii*) $j-j$ coupling. In $j-j$ coupling, each electron is considered separately and its contribution to the total angular momentum of the atom is obtained by coupling first the individual spin and orbital angular momentum vectors, using the relation $\vec{j} = \vec{l} + \vec{s}$. Then the total angular momentum vector $\vec{J}$ of the atom will be the vector sum of all the individual $\vec{j}$ vectors of the electrons.

This is given by

$$\vec{J} = [\vec{j_1} + \vec{j_2} + \vec{j_3} + ...]$$

$$= (\vec{l_1} + \vec{s_1}) + (\vec{l_2} + \vec{s_2}) + (\vec{l_3} + \vec{s_3}) + ...$$

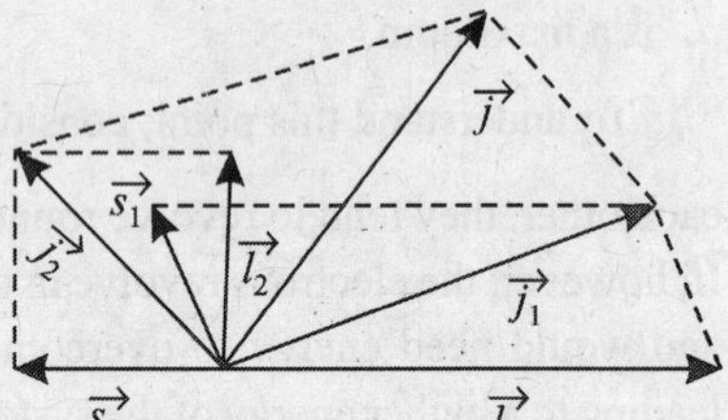

Fig. 9.5

A vector diagram showing the $j-j$ coupling of a two-electrons atom is shown in Fig. 9.5. $\vec{j_1}$ is the vector sum of $\vec{l_1}$ and $\vec{s_1}$, $\vec{j_2}$ is the vector sum of $\vec{l_2}$ and $\vec{s_2}$, $\vec{J}$ is the vector sum of $\vec{j_1}$ and $\vec{j_2}$.

Selection rules. All the possible transitions between different energy states of an atom do not appear as spectral lines. These are restricted according to the following selection rules.

For LS coupling. *Rule for L.* Only those spectral lines are observed for which

$$\Delta L = \pm 1$$

i.e. L changes by ± 1.

Rule for S. Only those spectral lines are observed for which

$$\Delta S = 0$$

Rule for J. Only those spectral lines are observed for which

$$\Delta J = \pm 1 \text{ or zero}$$

but 0 – 0 transition is not allowed.

Presence of magnetic field. In the presence of magnetic field following selection rules apply:–

Rule for M_L. For the orbital magnetic quantum number

$$\Delta M_L = 0 \text{ or} \pm 1$$

i.e. either M_L does not change or changes by 1.

Rule for M_s. The spin magnetic quantum number never changes *i.e.*

$$\Delta M_s = 0$$

Rule for M_J. For the total magnetic quantum number

$$\Delta M_J = 0 \text{ or } \pm 1 \qquad \text{(Same as for } M_L\text{)}$$

For *j–j* coupling. For $j - j$ coupling $\vec{L}$ and $\vec{S}$ remain unidentified and the selection rules for L and S no longer apply. The selection rules are

$$\left.\begin{matrix}\Delta j_1 = 0, \pm 1 \\ \Delta j_2 = 0\end{matrix}\right] \text{ or vice versa } i.e. \left[\begin{matrix}\Delta j_2 = 0, \pm 1 \\ \Delta j_1 = 0\end{matrix}\right.$$

$$\Delta J = 0, \pm 1 \qquad [0\text{–}0 \text{ transition not allowed}]$$

Thus, for lighter atoms LS coupling scheme holds and for heavier atoms JJ coupling scheme holds.

9.6 COMPARISON OF *LS* AND JJ COUPLING

***LS* coupling.** In *LS* coupling the orbital angular momentum vectors of all the electrons in the atom combine to form a resultant vector given by

$$\vec{L} = \vec{l_1} + \vec{l_2} + \vec{l_3} + \ldots.$$

Similarly the spin angular momentum vectors of all the electrons combine to form a resultant vector given by

$$\vec{S} = \vec{s_1} + \vec{s_2} + \vec{s_3} + \ldots.$$

Then the resultant $\vec{L}$ and $\vec{S}$ vectors combine to form a vector $\vec{J}$ which represents the *total angular momentum of the atom*

$$\vec{J} = \vec{L} + \vec{S}$$

***J–J* Coupling.** In $j - j$ coupling each electron is considered separately and its contribution to the total angular momentum of the atom is obtained first by considering the individual spin and orbital angular momentum vector of each electron, using the relation

$$\vec{j} = \vec{l} + \vec{s}$$

Then the total angular momentum vector $\vec{J}$ of the atom is given by vectorially combining the total angular momentum of all the electrons *i.e.*

$$\vec{J} = \vec{j_1} + \vec{j_2} + \vec{j_3} + \ldots. = (\vec{l_1} + \vec{s_1}) + (\vec{l_2} + \vec{s_2}) + (\vec{l_3} + \vec{s_3}) + \ldots$$

For light atoms. *LS* coupling generally takes place in the case of *light atoms* because the interaction between orbital angular momentum vectors $\vec{l_1}, \vec{l_2}, \vec{l_3} \ldots$ etc among different electrons in the atom is much greater than spin orbit interactions of the same electrons *i.e.* between $\vec{l_1}$ and $\vec{s_1}$ of electron 1 or between $\vec{l_2}$ and $\vec{s_2}$ of electron 2 and so on.

However, *in heavy atoms,* the nuclear charge is very large and produces a spin orbit interaction comparable in magnitude to the coulomb interaction between various $\vec{l}$ vectors and between various $\vec{s}$ vectors so that *LS* coupling breaks down. As *LS* coupling breaks down, the resultant angular momentum vectors $\vec{j_1}$, $\vec{j_2}$ etc of individual electrons add directly to form the total angular momentum vector $\vec{J}$ of the whole atom.

Thus, LS coupling scheme holds for light atoms.

SOLVED EXAMPLES

Example 9.1. *What are the possible values of L, S and J (total angular momentum quantum number) under LS coupling for two atomic electrons having orbital quantum number?*

(*i*) $l_1 = 1$ *and* $l_2 = 1$ ***(P.U. 2006)***

(*ii*) $l_1 = 1$ *and* $l_2 = 2$ ***(Pbi.U. 2000; G.N.D.U., 2000)***

(*ii*) $l_1 = 1$ *and* $l_2 = 3$

Solution. Possible values of *L*. The resultant orbital quantum number *L* has the values $L = (l_1 + l_2)$, $(l_1 + l_2 - 1)$, ... $|l_1 - l_2|$

$l_1 = 1;\ l_2 = 1$	$l_1 = 1;\ l_2 = 2$	$l_1 = 1;\ l_2 = 3$
$\therefore\ L = 2, 1, 0$	$\therefore\ L = 3, 2, 1$	$\therefore\ L = 4, 3, 2$

Possible values of *S*. The spin quantum numbers of individual electrons are $s_1 = \frac{1}{2}$ and $s_2 = \frac{1}{2}$. So the values of resultant spin quantum number *S* are

$$S = (s_1 + s_2), (s_1 + s_2 - 1), ... |s_1 - s_2|$$

$\therefore$ $S = 1, 0$ $\qquad$ $S = 1, 0$ $\qquad$ $S = 1, 0$

Possible values of *J*. Allowed values of *J* are $(L + S)$, $(L + S - 1)$ .... $|L - S|$

For $L = 2, S = 1$	For $L = 3, S = 1$	For $L = 4, S = 1$
$J = 3, 2, 1$	$J = 4, 3, 2$	$J = 5, 4, 3$
For $L = 2, S = 0$	For $L = 3, S = 0$	For $L = 4, S = 0$
$J = 2$	$J = 3$	$J = 4$
For $L = 1, S = 1$	For $L = 2, S = 1$	For $L = 3, S = 1$
$J = 2, 1, 0$	$J = 3, 2, 1$	$J = 4, 3, 2$
For $L = 1, S = 0$	For $L = 2, S = 0$	For $L = 3, S = 0$
$J = 1$	$J = 2$	$J = 3$
For $L = 0, S = 1$	For $L = 1, S = 1$	For $L = 2, S = 1$
$J = 1$	$J = 2, 1. 0$	$J = 3, 2, 1$
For $L = 0, S = 0$	For $L = 1, S = 0$	For $L = 2, S = 0$
$J = 0$	$J = 1$	$J = 2$

9.7 DISTRIBUTION OF ELECTRONS IN SHELLS AND SUB-SHELLS

The distribution of electrons in various shells and sub-shells is determined in accordance with the following rules:

(*i*) Only one electron can exist in one particular quantum state, given by one set of values for n, l, m_l and m_s

(*ii*) A system of particles is stable when its total energy is minimum.

The energy of an electron in any quantum state is mainly decided by the principal quantum number n. For the same value of n, energy depends upon the value of l. Higher the values of l, higher is the energy.

For the same values of l, the energy for $m_s = +\frac{1}{2}$ is slightly greater than that for $m_s = -\frac{1}{2}$.

(*iii*) The electrons having the same value of principal quantum number n form a shell. The number of electrons in a shell = $2n^2$. The shells are named K, L, M, N, O .... for values of n = 1, 2, 3, 4, 5 ... respectively.

n	*Shell*	l	*Sub-shell*	m_l	m_s	*No. of electrons in sub-shell with spectroscopic notation*		*Total no. of electrons in shell*
1	K	0	s	0	$+\frac{1}{2}, -\frac{1}{2}$	2	$1s^2$	2
2	L	0	s	0	$+\frac{1}{2}, -\frac{1}{2}$	2	$2s^2$	
2	L	1	p	$-1, 0, +1$	$+\frac{1}{2}, -\frac{1}{2}$	6	$2p^6$	8
3	M	0	s	0	$+\frac{1}{2}, -\frac{1}{2}$	2	$3s^2$	
3	M	1	p	$-1, 0, +1$	$+\frac{1}{2}, -\frac{1}{2}$	6	$3p^6$	
3	M	2	d	$-2, -1, 0, +1, +2$	$+\frac{1}{2}, -\frac{1}{2}$	10	$3d^{10}$	18
4	N	0	s	0	$+\frac{1}{2}, -\frac{1}{2}$	2	$4s^2$	
4	N	1	p	$-1, 0, +1$	$+\frac{1}{2}, -\frac{1}{2}$	6	$4p^6$	32
4	N	2	d	$-2, -1, 0, +1, +2$	$+\frac{1}{2}, -\frac{1}{2}$	10	$4d^{10}$	
4	N	3		$\begin{bmatrix} -3, -2, -1, 0 \\ +1, +2, +3 \end{bmatrix}$	$+\frac{1}{2}, -\frac{1}{2}$	14	$4f^{14}$	

and so on.

(*iv*) The electrons belonging to a certain shell and having the same value of the orbital quantum number l form a **sub-shell**. The number of electrons in a sub-shell = 2 (2l + 1). The sub-shells are named *s*, *p*, *d*, *f*, *g* for values of l = 0, 1, 2, 3, 4. The energy of all the electrons in a sub-shell is almost the same, as the change in energy due to difference in the values of m_l and m_s is extremely small.

The number of electrons in various shells and sub-shells with their quantum numbers are given in the table on the last page.

Generally, a shell having 2 n^2 electrons is said to be *completely closed* or *filled*. Similarly, a sub-shell having 2 (2l + 1) electrons is *filled* or *completed*.

Every electron in a closed sub-shell is matched by another having the value of m_l of opposite sign. Similarly, the values of m_s for the electrons in a closed sub-shell are also paired. The resultant value of $\Sigma (m_l + m_s)$ for a closed sub-shell is, therefore, equal to zero. Hence the total angular momentum of the atom due to the orbital and spin angular momenta of the electrons in a closed sub-shell is zero. As a result, a closed shell, which is made up of closed sub-shells, does not contribute anything towards the total angular momentum of the atom.

Electronic configuration. *The electronic configuration of an atom is the distribution of electrons in various shells and sub-shells around the nucleus of the atom.*

To represent the electronic configuration of an atom, the value of the *principal quantum number* n is written as a *prefix* to the letter representing its l-value. For example, a state in which n = 2. l = 0 is a 2 *s* state. A state, in which n = 4, l = 2 is a 4*d* state and so on. The number of electrons having the same n and l values is indicated by an index written at the upper right of the letter representing the l-value. For example, a state with n = 2, l = 1 is a 2*p* state. If all the six electrons are there, it is a 2 p^6 state, but if out of six only 4 electrons are present, it is a 2 p^4 state.

We shall illustrate the electronic configuration of first three atoms and sodium.

Hydrogen (Z = 1). In Hydrogen atom there is only one electron. In the normal state, it is in the lowest possible energy level *i.e.* ground state with n = 1, l = 0. Its electronic configuration is $1s^1$.

Helium (Z = 2). Helium has two electrons. In the normal state, both the electrons are in the state $n = 1, l = 0, m_s = +\frac{1}{2}$ for one and $m_s = -\frac{1}{2}$ for the other electron. Its electronic configuration is $1s^2$.

For Example: (*i*) Lithium *Li* (Z = 3). Lithium has three electrons. In the normal state one of the electrons is in the 2*s* sub-shell (n = 2, l = 0) as the capacity of the 1*s* sub-shell (n = 1, l = 0) is only 2, being given by 2(2l + 1). The electronic configuration is $1s^2\ 2s^1$.

(*ii*) Sodium *Na* (Z = 11). Sodium has atomic number 11. The eleven electrons of sodium in the normal state are designated as

$1\,s^2\ 2\,s^2\ 2\,p^6\ 3\,s^1$

meaning thereby that the *first* *K*-shell (n = 1) contains two electrons (l = 0), the *second* *L*-shell (n = 2) contains two electrons in *s*-sub-shell (l = 0) 6 in the *p*-subshell (l = 1) and the third *M* shell (n = 3) contains only one electron in the *s*-sub-shell (l = 0).

The filling of different energy levels or states occurs in such a way that the resulting configuration has minimum energy. For atoms with a large number of electrons, the state or shells are filled according to the order

$1s^2, 2s^2, 2p^6, 3s^2\ 3p^6, 4s^2, 3d^{10}, 4p^6, 5s^2, 4d^{10}, 5p^6, 6s^2, 4f^{14}, 5d^{10}, 6p^6, 7s^2, 6d^{10}, 5f^{14}$. The figures as superscripts give the maximum occupancy of the sub-shell.

9.8 ELECTRONIC CONFIGURATION OF SOME ELEMENTS WITH SPINS

The spin configurations of *p*-electrons of the elements having Z = 7 to Z = 10 are as under:

For Z = 7, electronic configuration is $1s^2$, $2s^2$, $2p^3$

Spin orientations of *p*-electrons are ↑ ↓ ↑

For Z = 8, electronic configuration is $1s^2$, $2s^2$, $2p^4$

Spin orientations of *p*-electrons are ↑ ↓ ↑ ↓

For Z = 9. electronic configuration is $1s^2$, $2s^2$, $2p^5$

Spin orientations of *p*-electrons are ↑ ↓ ↑ ↓ ↑

For Z = 10, electronic configuration is $1s^2$, $2s^2$, $2p^6$

Spin orientations of *p*-electrons are ↑ ↓ ↑ ↓ ↑ ↓

Electronic configuration of some elements

Copper (*Cu*) (Z = 29) has 29 electrons. 2 electrons occupy the first (*K*-shell), 8-the two sub shells of second (*L*-shell) and 18-the three subshells of third (*M*-shell). Remaining one electron-the valency electron occupies the 4th (*N*-shell). The electronic configuration is

$$\underset{K}{\underline{1s^2}},\ \underset{L}{\underline{2s^2\,2p^6}},\ \underset{M}{\underline{3s^2\,3p^6\,3d^{10}}},\ \underset{N}{\underline{4s^1}}$$

Cobalt (*Co*) (Z = 27) has 27 electrons. 2 electrons occupy the first (*K*-shell), 8-the two sub-shells of second (*L*-shell), the third (*M*-shell) has two electrons is *s*-sub-shell, 6 in *p*-sub-shell and only 7 in *d*-sub-shell. The last two electrons occupy the 4th (*M*-shell) in *s*-sub-shell before the third (*M*-shell) is completed. The electronic configuration is

$$\underset{K}{\underline{1s^2}},\ \underset{L}{\underline{2s^2\,2p^6}},\ \underset{M}{\underline{3s^2\,3p^6\,3d^7}},\ \underset{N}{\underline{4s^2}}$$

Thallium (*Tl*) (Z = 81) has 81 electrons. It has the first four shells complete with 2 + 8 + 18 + 32 = 60 electrons. The fifth (*O*-shell) takes up 18 electrons in the first three sub-shells but before this shell is completed the sixth (*P*-shell) starts being filled up. Of the last 3 electrons two are *s*-electrons and one is *p*-electron.

The complete electronic configuration is

$$\underset{K}{\underline{1s^2}},\ \underset{L}{\underline{2s^2\,2p^6}},\ \underset{M}{\underline{3s^2\,3p^6\,3d^{10}}},\ \underset{N}{\underline{4s^2\,4p^6\,4d^{10}\,4f^{14}}},\ \underset{O}{\underline{5s^2\,5p^6\,5d^{10}}},\ \underset{P}{\underline{6s^2\,6p^1}}$$

Aluminium. The ground state electronic configuration for Aluminium (*Al*) (*Z* = 13) is

$$1s^2\ 2s^2\ 2p^6\ 3s^2\ 3p^1$$

C, Ne, Fe. Applying Pauli's exclusion principle to synthesise the normal state, we have for

Carbon	**(*C*) (*Z* = 6)**	$1s^2\ 2s^2\ 2p^2$
Neon	**(*Ne*) (*Z* = 10)**	$1s^2\ 2s^2\ 2p^6$
Iron	**(*Fe*) (*Z* = 26)**	$1s^2\ 2s^2\ 2p^6\ 3s^2\ 3p^6\ 3d^6\ 4s^2$

Example 9.2. *Consider an atom with electronic configuration $1s^2\ 2s^2\ 2p^1$ and find possible values of l, s and j. Also find the magnitude of angular momenta L, S, J.*

Solution. Electronic configuration of the atom = $1s^2\ 2s^2\ 2p^1$. $1s$ and $2s$ sub-shells both contain two electrons each and are, therefore, closed sub shells. These have zero angular momentum. The angular momentum of the atom is only due to 2 p electron. For this p electron

$$n = 2;\ l = 1;\ s = \frac{1}{2}$$

∴ The allowed values of j are

$$j = l + s = 1 + \frac{1}{2} = \frac{3}{2}$$

and
$$j = l - s = 1 - \frac{1}{2} = \frac{1}{2}$$

Magnitude of orbital angular momentum $L = \sqrt{l\,(l+1)}\ \hbar$

$$= \sqrt{1\,(1+1)}\ \hbar = \sqrt{2}\ \hbar$$

Magnitude of spin angular momentum $S = \sqrt{s\,(s+1)}\ \hbar$

$$= \sqrt{\frac{1}{2}\left(\frac{1}{2}+1\right)}\ \hbar = \frac{\sqrt{3}}{2}\ \hbar$$

Magnitude of total angular momentum $J = \sqrt{j\,(j+1)}\ \hbar$

For
$$j = \frac{3}{2},\quad J = \sqrt{\frac{3}{2}\left(\frac{3}{2}+1\right)}\ \hbar = \frac{\sqrt{15}}{2}\ \hbar$$

For
$$j = \frac{1}{2},\quad J = \sqrt{\frac{1}{2}\left(\frac{1}{2}+1\right)}\ \hbar = \frac{\sqrt{3}}{2}\ \hbar$$

9.9 SPECTRAL NOTATION REPRESENTATION

Spectral notation. The specification of n and l for each electron in an atom is called the electronic configuration. In this configuration, the orbital angular momentum states of individual electrons are described by a lower case (small) letter as follows:–

$$l = 0\quad 1\quad 2\quad 3\quad 4\quad 5$$

$$\text{letter} = s\quad p\quad d\quad f\quad g\quad h$$

A similar scheme, using capital letters, is used to designate the entire electronic state of the atom, according to its total orbital angular momentum quantum number as follows—

$$L = 0\quad 1\quad 2\quad 3\quad 4\quad 5$$

$$\text{Letter} = S\quad P\quad D\quad F\quad G\quad H$$

The state of an atom is determined by three quantum numbers L, S and J, where

$$L = l_1 + l_2 + ...$$

and
$$S = s_1 + s_2 + ...$$

for all the individual electrons of the atom and

$$J = L + S$$

The quantum number J has the possible values

$$J = (L + S), (L + S - 1), ... (L - S)$$

The state of an atom obeying LS coupling is given by the term

$$^{2s+1}L_j$$

This term is called *spectroscopic term*. In this term, the factor $(2S + 1)$ represents the *multiplicity of states* when $L > S$, because J ranges from $(L + S)$ through 0 to $(L - S)$. Thus when $S = 0$, the multiplicity is 1 (*singlet state*) and $J = L$; when $S = \frac{1}{2}$, the multiplicity is 2 (doublet state) and $J = L \pm \frac{1}{2}$; when $S = 1$, the multiplicty is 3 (triple state) and $J = L + 1, L, L - 1$ and so on.

(When $S > L$, the multiplicty is given by $(2L + 1)$. The total angular momentum quantum number J is used as a subscript after L, representing the total orbital quantum number.

Examples. The state of an atom with $L = 1$, $S = \frac{1}{2}$ and $J = 3/2$ is written as $^2P_{3/2}$. It is read as '*doublet P three half state*'.

The state of an atom with $L = 2$, $S = 1$ and $J = 2$ is written as $^3D_2$. It is read as '*triplet D two state*'.

If we want to describe all the states of an atom with $L = 2$ and $S = 1$, then all the possible values of J are taken. These are 3, 2, 1 and the corresponding terms are $^3D_3$, $^3D_2$, $^3D_1$.

Example 9.3. *Explain the meaning of each symbol in spectroscopic notation $^{2s+1}L_j$ used for describing the state of an atom.* ***(Luck. U., 2003)***

Solution. The spectroscopic notation $^{2s+1}L_j$ contains the three quantum numbers S, L and J which are necessary to determine the state of an atom.

In the pre-superscript $(2S + 1)$, S represents the total spin quantum number given by

$$S = s_1 + s_2 + s_3 \ldots$$

for all individual electrons in the atom.

L represents the total orbital angular momentum given by

$$L = l_1 + l_2 + l_3 \ldots$$

for all individual electron in the atom.

The post-subscript J represents the total angular momentum of the atom and is given by

$$J = L + S$$

The quantum number J has the possible values

$$J = |L + S|, \quad |L + S| - 1, \ldots\ldots\ldots\ldots\ldots\ldots |L - S|$$

The factor $(2S + 1)$ represents the multiplicity of states.

Example 9.4. *Find the S, L and J values that correspond to each of the following states* $^1P_1$, $^1S_0$, $^2P_2$, $^3P_2$, $^2D_{3/2}$, $^2D_{5/2}$, $^2P_{3/2}$, $^2S_{1/2}$, $^3S_1$, $^3D_0$, $^3D_2$, $^5F_5$, $^6H_{5/2}$.

(Pbi. U., 2008, 2005; P.U. 2007, 2006; G.N.D.U., 2001, 2000)

Solution. (*i*) For $^1P_1$; $L = 1$; $(2S + 1) = 1$; $\therefore$ $S = 0$. Hence $J = 1$

[As $J = 1$ is an admissible value for $L = 1$ and $S = 0$, $^1P_1$ represents an allowed state with $J = 1$]

(*ii*) For $^1S_0$; $L = 0$; $(2S + 1) = 1$; $\therefore S = 0$. Hence $J = 0$

[As $J = 0$ is an admissible value for $L = 0$ and $S = 0$, $^1S_0$ represents an allowed state with $J = 0$]

(*iii*) For $^2P_2$; $L = 1$; $(2S + 1) = 1$; $\therefore S = \frac{1}{2}$. Hence $J = \frac{3}{2}$ and $\frac{1}{2}$

[As $J = 2$ is not an admissible value with $L = 1$ and $S = \frac{1}{2}$, $^2P_2$ term does not represent an allowed state]

(*iv*) For $^3P_2$; $L = 1$; $(2S + 1) = 3$; $\therefore S = 1$. Hence $J = 2$ and zero

[As $J = 2$ is an admissible value for $L = 1$ and $S = 1$, $^3P_2$ represents an allowed state with $J = 2$]

(*v*) For $^2D_{3/2}$; $L = 2$; $(2S + 1) = 2$; $\therefore S = \frac{1}{2}$. Hence $J = \frac{5}{2}$ and $\frac{3}{2}$

[As $J = 3/2$ is an admissible value with $L = 2$ and $S = 1/2$, $^2D_{3/2}$ term represents an allowed state with $J = 3/2$]

(*vi*) For $^2D_{5/2}$; $L = 2$; $2(S + 1) = 2$ $\therefore S = \frac{1}{2}$. Hence $J = \frac{5}{2}$ and $\frac{3}{2}$

[As $J = 5/2$ is an admissible vlaue with $L = 2$ and $S = 1/2$, $^2D_{5/2}$ term represents an allowed state with $J = 5/2$]

(*vii*) For $^2P_{3/2}$; $L = 1$; $(2S + 1) = 2$; $\therefore S = \frac{1}{2}$. Hence $J = \frac{3}{2}$ and $\frac{1}{2}$

[As $J = 3/2$ is an admissible value for $L = 1$ and $S = \frac{1}{2}$, $^2P_{3/2}$ term represents an allowed state with $J = 3/2$]

(*viii*) For $^2S_{1/2}$; $L = 0$; $(2S + 1) = 2$; $\therefore S = \frac{1}{2}$. Hence $J = \frac{1}{2}$

[As $J = 1/2$ is an admissible value with $L = 0$ and $S = 1/2$, $^2S_{1/2}$ is an allowed state with $J = 1/2$]

(*ix*) For $^3S_1$; $L = 0$, $(2S + 1) = 3$ $\therefore S = 1$. Hence $J = 1$.

[As $J = 1$ is an admissible value with $L = 0$ and $S = 1$; $^3S_1$ is an allowed state with $J = 1$]

(*x*) For $^3D_0$; $L = 2$; $(2S + 1) = 3$; $\therefore S = 1$. Hence $J = 3, 2, 1$.

[As $J = 0$ is not an admissible value with $L = 2$ and $S = 1$, $^3D_0$ is not an allowed state.]

(*xi*) For $^3D_2$; $L = 2$; $(2S + 1) = 3$; $\therefore S = 1$. Hence $J = 3, 2, 1$.

[As $J = 2$ is an admissible value with $L = 2$ and $S = 1$, $^3D_2$ is an allowed state]

(*xii*) For $^5F_5$; $L = 3$; $(2S + 1) = 5$; $\therefore S = 2$. Hence $J = 5, 4, 3, 2, 1$

[As $J = 5$ is an admissible value for $L = 3$ and $S = 2$, $^5F_5$ represents an allowed state with $J = 5$]

(*xiii*) For $^6H_{5/2}$; $L = 5$; $(2S + 1) = 6$; $\therefore S = \frac{5}{2}$. Hence $J = \frac{15}{2}, \frac{13}{2}, \frac{11}{2}, \frac{9}{2}, \frac{7}{2}, \frac{5}{2},$

[As $J = \frac{5}{2}$ is an admissible value for $L = 5$ and $S = \frac{5}{2}$; ${}^6H_{5/2}$ represents an allowed state with $J = \frac{5}{2}$]

9.10 SPECTRA OF ALKALIE ATOMS

Alkalie atoms. *The atoms which have only one electron outside the completely filled (or closed) shells are known as one valence electron atoms*. Hydrogen (Z = 1) and alkalie atoms *Li* ($Z = 3$), *Na* ($Z = 11$), *K* ($Z = 19$), *Rb* ($Z = 37$) and *Cs* ($Z = 55$) belong to this class.

Since the complete shells do not contribute either to the orbital or spin angular momentum of the atom, the total angular momentum of alkalie atoms is entirely due to the valence electron. So their optical spectra is characterised by the states of the valence electron. We shall take the example of sodium as a typical case.

Energy levels of sodium atom. The electronic configuration of sodium in its ground state is $1s^2\ 2s^2\ 2p^6\ 3s^1$. The electrons in the $1s^2\ 2s^2\ 2p^6$ states do not contribute to angular momentum of sodium atom and constitute the core. The *valence electron* is in the 3*s*-state.

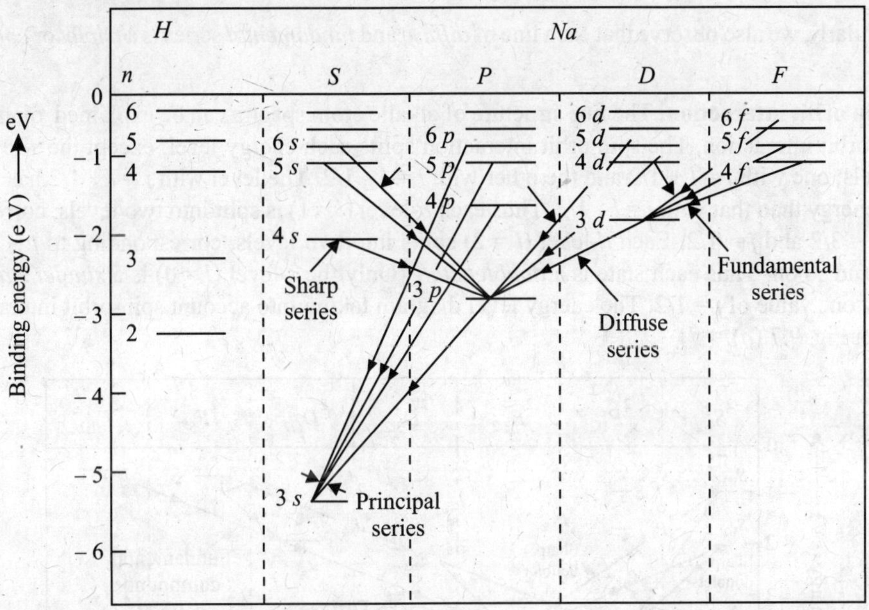

Fig. 9.6

The valence (or the optical) electron moves in the net field of the nucleus [charge + *Ze*] and the core of electrons [charge − ($Z - 1$) *e*] where $Z = 11$. This core of electrons acts as a shield between the nucleus and the optical electron. The *shield effect* makes the energy levels of sodium dependent upon *l*- the orbital quantum number, in addition to *n*- the principal quantum number. In the case of hydrogen, there is no such core of electrons. Therefore 3*s*, 3*p* and 3*d* levels of hydrogen have the same energy. Similarly its 4*s*, 4*p*, 4*d* and 4*f* levels also have the same energy. But in the case of sodium, all levels such as 3*s*, 3*p*, 3*d* and 4*s*, 4*p*, 4*d*, 4*f* have different energies.

Allowed transitions. When the 3*s* electron in sodium is excited, it goes to 3*p*, 3*d*, 4*s*, 4*p*, 4*d*, 4*f* energy levels. The energy level diagram of sodium is shown in Fig. 9.6. When this excited electron returns to a state of lower energy, radiations are emitted, which appear as spectral lines, when seen through a spectrograph. Only those transitions are allowed for which orbital quantum number l changes by 1 or $\Delta l = \pm 1$.

Spectrum and series obtained. The spectrum of sodium, like that of hydrogen consists of a number of different series. Each series consists of transitions to a particular level. (Fig. 9.6)

(*i*) ***Principal series.*** The series of spectral lines produced by transitions from *p*-states to 3s-state is called principal series.

(*ii*) ***Sharp series.*** The series of spectral lines produced by transitions from *s*-states to 3*p*-state is called sharp series.

(*iii*) ***Diffuse series.*** The series of spectral lines produced by transitions from *d*-states to 3*p*-state is called diffuse series.

(*iv*) ***Fundamental series.*** The series of spectral lines produced by transitions from *f*-states to 3*d*-state is called *fundamental series*.

Fine structure in sodium spectrum. When we observe the spectrum of sodium with the help of a high resolving power spectrograph, we find that each line of *sharp* and *principal* series is a *doublet* *i.e.* it consists of two lines very close to each other.

Similarly, we also observe that each line of *diffuse* and *fundamental* series is a *triplet* or *compound doublet*.

Spin orbit interaction. The fine structure of alkalie atom spectra can be explained on the basis of spin orbit interaction. The spin orbit interaction splits each energy level, except the *s*-level-into two levels; one with $j = l + 1/2$ and the other with $j = l - 1/2$. The level with $j = l + 1/2$ has slightly higher energy than that with $j = l - 1/2$. Thus each *p*-level ($l = 1$) is split into two levels, corresponding to $j = 3/2$ and $j = 1/2$. Each *d*-level ($l = 2$) splits into two levels, corresponding to $j = 5/2$ and $j = 3/2$ and so on. Thus each state is a *doublet state*. Only the *s*-level ($l = 0$) is a *singlet state*, as it has only one value of $j = 1/2$. The energy level diagram taking into account spin-orbit interaction is shown in Fig. 9.7 (*a*).

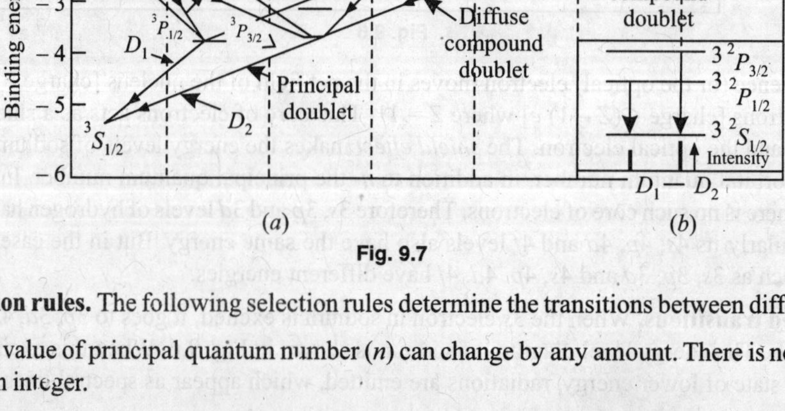

Fig. 9.7

Selection rules. The following selection rules determine the transitions between different states.

(*i*) The value of principal quantum number (n) can change by any amount. There is no restriction *i.e.* Δn = an integer.

(*ii*) The total quantum number j can change by zero or unity *i.e.*

$$\Delta j = \pm 1 \text{ or zero}$$

However zero j to zero j i.e. $j_1 = 0$ to $j_2 = 0$ transition is not allowed.

(*iii*) The emitted photon in a given transition carries away angular momentum = 1 $\hbar$. Therefore, orbital angular momentum quantum number l must change by unity *i.e.*

$$\Delta l = \pm 1$$

***D*-lines of sodium.** The principal series of sodium atom is produced by transition from *p*-states to 3*s*-state. These lines are all *doublets*, because they start from *doublet p-level* and end on *singlet s-level*. The first two members of principal series arise from the transitions

$$D_1 \text{ line} - 3\,^2P_{1/2} \rightarrow 3\,^2S_{1/2} — \lambda_1 = 5896 \text{ Å}$$

$$D_2 \text{ line} - 3\,^2P_{3/2} \rightarrow 3\,^2S_{1/2} — \lambda_2 = 5890 \text{ Å}$$

For the first transition $\Delta l = -1$ and $\Delta j = 0$. For the second $\Delta l = -1$ and $\Delta j = -1$. These transitions are shown separately in Fig. 9.7(*b*). The intensity of D_2 line is twice that of D_1 line, as shown. All other lines of the principal series are in the ultraviolet region.

Compound doublets of sodium. The transition from *d*-levels to 3*p*-state gives rise to *diffuse series*. The transition form *f*-level to 3*d*-state gives rise to *fundamental series*. Both these series start from doublet levels and end on doublet levels. So, there are four possible transitions, but one of these transitions is not allowed, because it does not satisfy the selection rule $\Delta j = \pm 1$ or 0 ($j = 0$ to $j = 0$ not allowed).

In diffuse series, the transitions $3d_{5/2} \rightarrow 3p_{3/2}$, $3d_{3/2} \rightarrow 3p_{3/2}$, $3d_{3/2} \rightarrow 3p_{1/2}$, are allowed, but transition $3d_{5/2} \rightarrow 3p_{1/2}$, is not allowed because $\Delta j = 2$ is forbidden. The lines of diffuse and fundamental series are, therefore, *triplets*. These triplets are known as *compound doublets*.

The principal and sharp doublets and diffuse and fundamental compound doublets are shown in Fig. 9.7(*a*) with corresponding transitions.

Thus the spectrum of sodium consists of doublets and compound doublets or triplets.

9.11 SPECTRA OF ALKALINE EARTH ELEMENTS

Alkaline earth elements. The elements of group II of the periodic table are known as alkaline earth elements. These elements have a closed shell of electrons, plus two electrons in the outermost *s*-shell. Since the angular momentum of a closed shell is zero, the total angular momentum of an alkaline earth atom is due to the angular momenta of two valence electrons.

The elements of alkaline earth group are *Be* ($Z = 4$), *Mg* ($Z = 12$), *Ca* ($Z = 20$), *Sr* ($Z = 38$), *Ba* ($Z = 56$) and *Ra* ($Z = 88$). To discuss the spectra of alkaline earth elements, we shall take the example of magnesium.

Energy levels. (*i*) ***Ground state.*** The ground state electronic configuration of magnesium is ($1s^2$ $2s^2$, $2p^6$) $3s^2$. In the ground state, both the valence electrons have the same principal quantum number (n) and orbital quantum number (l) i.e. these are *equivalent electrons*. Their spins are, therefore, *anti-parallel* and total spin quantum number $S = 0$. Hence the ground state is a singlet and denoted as $^1S_0$.

(*ii*) ***Excited states.*** Suppose only one of the two electrons of magnesium is excited to a higher energy state, say (np). The configuration of magnesium atom is now ($3s$, np). The two electrons are *non-equilavent*. The orbital quantum number of the two electrons—one in the *s*-state and the other in the *p*-state--are respectively

$$l_1 = 0 \quad \text{and} \quad l_2 = 1$$

$\therefore$ Total orbital quantum number $L = 1$.

The total spin quantum number S can have two values

$S = \frac{1}{2} + \frac{1}{2} = 1$ (parallel spin) and $S = \frac{1}{2} - \frac{1}{2} = 0$ (Anti-parallel spin).

For $L = 1$; $S = 1$; $J = (L + S), (L + S - 1) ... (L - S) = 2, 1, 0$.

The corresponding spectroscopic terms are ${}^3P_2$, ${}^3P_1$, and ${}^3P_0$. Thus ${}^3P_{2,1,0}$ is a *triplet state.*

Similarly if one of the valence electrons is excited to *d*-state, we again get two distinct sets of energy levels. The singlet state ${}^1D_2$, corresponding to $S = 0$ (anti-parallel spin of valence electrons), $L = 2$ and $J = 2$ and triplet state ${}^3D_{3,2,1}$, corresponding to $S = 1$ (parallel spin) $L = 2$ and $J = 3, 2, 1$.

The energy levels of magnesium, therefore, fall into two sets; *singlets* ($S = 0$) and *triplet* ($S = 1$) as shown in Fig. 9.8.

If there is no mutual force of repulsion between the two electrons the singlet state ${}^1P_1$ and the triplet state ${}^3P_{2,1,0}$ should have the same energy. But due to the electrostatic repulsion, the energy of each singlet level is greater than that of the corresponding triplet level, the difference being of the order of 1 eV. Further, due to spin orbit interaction, the triplet states with different J values have slightly different energy, such that the energy level with smallest values of J lies lowest.

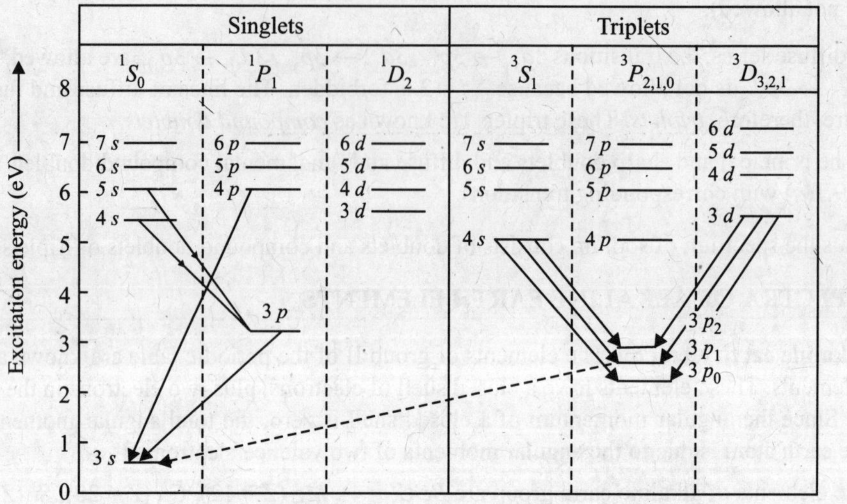

Fig. 9.8

Selection rules. For the spectra of two electron system only those transitions are allowed which obey the following selection rules:–

$\Delta L = 0, \pm 1$; $\Delta J = 0, \pm 1$, (0 → 0 transition not allowed) and $\Delta S = 0$.

As $\Delta S = 0$ transition is only permitted, singlet states only combine with other singlet states and triplet states only combine with other triplet states *i.e.* singlet and triplet states do not mix.

However, some transitions for which $\Delta S \neq 0$ take place and these give rise to inter-combination lines, the transition from ${}^3p_1$ (triplet state) to ${}^1S_0$ (singlet state) is a typical example. Inter-combination lines are rare and have very low intensity.

Metastable states. The transition from $^3P_0$, $\rightarrow$ $^1S_0$ is from $J = 0 \rightarrow J = 0$ and is not allowed. The transition from $^3P_2$, $\rightarrow$ $^1S_0$ is from $J = 2 \rightarrow J = 0$ *i.e.* $\Delta J = 2$ and is also not allowed. These states $^3P_0$ and $^3P_2$ are, therefore, known as *metastable states* and in the absence of collisions, an atom can persist in either of these states for a relatively long time (a second or more).

Example 9.5. *Given the energy difference between two p-energy levels in Sodium atom is 0.00214 eV, calculate the separation of the D-lines taking $\lambda_0 = 589.3$ n.m.* ***(Bang. U., 2003)***

Solution. Energy difference $\Delta E = 0.00214$ eV $= 2.14 \times 10^{-3} = 2.14 \times 10^{-3} \times 1.6 \times 10^{-19}$, J

$$\therefore \text{ Frequency difference } \Delta\nu = \frac{\Delta E}{h} = \frac{2.14 \times 1.6 \times 10^{-22}}{6.62 \times 10^{-34}}$$

Now $$\nu = \frac{c}{\lambda} \qquad \therefore d\nu = \frac{-c}{\lambda^2} d\lambda$$

or $$d\lambda = -\frac{d\nu\lambda^2}{c} \quad \therefore \Delta\lambda = \frac{\Delta\nu\lambda^2}{c} \text{ (ignoring the negative sign)}$$

or $$\Delta\lambda = \frac{2.14 \times 1.6 \times 10^{-22}}{6.62 \times 10^{-34}} \times \frac{589.3 \times 10^{-9} \times 589.3 \times 10^{-9}}{3 \times 10^8} = 5.987 \times 10^{-10}\ m = 5.987\ \text{Å}$$

Example 9.6. *Calculate the angle between the total and orbital angular momentum vectors for the $^4D_{3/2}$ state.*

Solution. For $^4D_{3/2}$ state $2S + 1 = 4 \qquad \therefore S = \frac{3}{2}$

$$L = 2 \qquad J = \frac{3}{2}$$

Now $\therefore$ $$\vec{J} = \vec{L} + \vec{S} \text{ or } \vec{J} - \vec{L} = \vec{S}$$

$$\vec{S}\cdot\vec{S} = \left(\vec{J} - \vec{L}\right)\cdot\left(\vec{J} - \vec{L}\right)$$

or $$S^2 = J^2 + L^2 - 2\vec{J}\cdot\vec{L} \text{ or } \vec{J}\cdot\vec{L} = \frac{J^2 + L^2 - S^2}{2}$$

But $$\vec{J}\cdot\vec{L} = \left|\vec{J}\right|\cdot\left|\vec{L}\right| \cos\theta = \sqrt{J(J+1)}\ \sqrt{L(L+1)}\ \hbar^2 \cos\theta$$

and $$J^2 + L^2 - S^2 = J(J+1)\hbar^2 + L(L+1)\hbar^2 - S(S+1)\hbar^2$$

$\therefore$ $$\cos\theta = \frac{J(J+1)\hbar^2 + L(L+1)\hbar^2 - S(S+1)\hbar^2}{2\sqrt{J(J+1)}\ \sqrt{L(L+1)}\ \hbar^2}$$

Substituting the values of J, L and S

$$\cos\theta = \sqrt{\frac{2}{5}} \text{ or } \theta = 50°52'$$

EXERCISE

1. (*a*) What do you mean by symmetric and anti-symmetric wave functions? What is the difference between them? How do they lead to the concept of Fermions and Bosons?
(P.U. 2006, 2005, 2002, 2001, 2000; H.P.U. 2002; Pbi. U. 1995; G.N.D.U. 2008, 2005)
(*b*) What is exchange degeneracy? How does symmetry of wave function correspond to degeneracy of states? ***(G.N.D.U. 2006; H.P.U., 2002)***
2. State and explain Pauli's exclusion principle. How does a knowledge of symmetric and anti-symmetric wave functions lead to this principle?
(Meerut. U. 2003, 2000; Pbi. U. 2005, 2004, 2002; Luck. U. 2002, 2001; P. U. 2007, 2006; G. N. D. U. 2007, 2006, 2004, 2003; Gujrat U. 2004)
3. Show that no two electrons have the same quantum state. ***(P. U. 2008, 2001; Luck. U. 2001)***
4. Describe the symmetry character of the wave function of multiparticle system of bosons and fermions separately. ***(G.N.D.U. 2007)***
5. (*a*) Explain the difference between singlet and triplet states. Also discuss the physical significance of these states.
(*b*) Which of the following elements would you expect to have energy levels divided into singlet and triplet states? Ne, Mg, Cl, Ca, Cu, Ag, Ba.
6. (*a*) Describe and explain LS coupling. Under what conditions does it hold? Why does a coupling of *L's* and spins occur? Under what condition *LS* coupling breaks down and what kind of new coupling takes over? Describe and explain *JJ* coupling. Illustrate *LS* and *JJ* coupling with the help of vector diagram. Give the selection rules for *L*, *S* and *J*. ***(H.P.U. 2003; M.D.U. 2002; P.U. 2007, 2005, 2001, 2000; Pbi. U. 2001, G.N.D.U. 2008, 2004, 2000; Gauhati. U. 2000)***
(*b*) (*i*) Which scheme holds for lighter atoms? ***(P.U. 2005)***
(*ii*) Which scheme holds for heavier atoms? ***(P.U. 2007)***
7. What is *JJ* coupling? Explain *JJ* coupling in case of two electron spectra. Draw vector model diagram depicting *JJ* coupling. ***(G.N.D.U. 2002, 2000)***
8. (*a*) Compare *LS* and *JJ* coupling schemes. Which scheme holds for light atoms? ***(H.P.U. 2002)***
(*b*) State the selection rules for *LS* coupling for light atoms. What happens for heavier atoms? ***(H.P.U. 2000)***
9. (*a*) Show how the electrons are distributed in various shells and sub-shells? State their quantum number. ***(Meerut. U., 2002, 2001)***
(*b*) How is the electronic configuration of an atom represented? Give the electronic configuration of Na. ***(Luck. U. 2002; G.N.D.U., 2002)***
(*c*) Write down the electronic configuration of Li atom. Why is the third electron not accommodated in the shell corresponding to $n = 1$? ***(Vid. S. U., 2004)***
(*d*) How are electrons distributed in various sub-shells for $n = 3$? Give quantum numbers of electrons in first sub-shell. **(Ans: See table in Q. 9.7, under n = 3) *(Pbi. U. 2008)***
10. (*a*) Give the electronic configuration of elements with Atomic numbers $Z = 7$ to $Z = 10$, indicating parallel and anti-parallel spins of *p*-electrons.
(*b*) Give electronic configuration for Cu (Z = 29), Co (Z = 27) and Thallium Tl (Z = 81) in their ground state.
11. (*a*) Aluminium has two 3*s* electrons and one 3*p* electron outside the filled inner shells. What is the ground state configuration? ***(Pbi. U. 2000)***

(*b*) Apply Pauli's exclusion principle to synthesise the normal state of C, Ne and Fe. **(*G.N.D.U. 2001, 2000*)**

12. Explain qualitatively the general features of the spectra of alkalie atoms. Draw energy level diagrams of sodium and explain principal features of its spectrum. Discuss the various series observed. **(*M.D.U. 2007*)**

13. What is meant by fine structure of spectral lines? Describe how electron spin coupled with orbital motion explains the fine structure of alkalie spectra. Explain why each line of principal series of Na is a doublet whereas each line of diffuse series is a triplet.

(*H.P.U. 2003, 2000; K.U. 2001, 1996; G.N.D.U. 2001, 2000; Luck U. 2002; Pbi.U. 2001; M.D.U. 2006, 2000*)

14. (*a*) Show that s-levels are singlet and all other levels are doublet. **(*M.D.U. 2000*)**

(*b*) What are the selection rules for observing doublet spectra in single electron atoms?

15. (*a*) Discuss the spectra of alkaline earth elements qualitatively. **(*Pbi. U. 2003; H.P.U. 2003*)**

(*b*) Give selection rules for spectra of two electron system. **(*G.N.D.U., 2002*)**

16. Two angular momenta with quantum numbers, $J_1 = 3/2$ and $J_2 = 2$ combine. How many sets of J and M_j are there? **(Ans. 20)**

MULTIPLE CHOICE QUESTIONS

1. Wave functions which change sign under exchange of particles are called

(*a*) symmetric (*b*) anti-symmetric

2. A system of electrons is described by a wave function which should be

(*a*) anti-symmetric (*b*) symmetric

3. The energy levels of a multi electron atom depend upon

(*a*) n (*b*) l (*c*) both n and l

4. Electrons which have same n and l quantum numbers are known as

(*a*) equivalent (*b*) non-equivalent

5. Fermions are particles with

(*a*) integral spin (*b*) half integral spin (*c*) zero spin

6. Bosons are particles with

(*a*) zero and integral spin (*b*) half integral spin

7. Helium atom in a singlet state is known as

(*a*) ortho helium (*b*) para helium

8. Spin wave functions are symmetric with respect to exchange of particle for

(*a*) singlet state (*b*) doublet state (*c*) triplet state

9. The spectral lines in the principal series of sodium are

(*a*) singlet (*b*) doublet (*c*) triplet

10. The yellow *D*-lines of Sodium spectrum correspond to

(*a*) Sharp series (*b*) Fundamental series

(*c*) Principal series (*d*) Diffuse Series

ANSWERS

1. (*b*)	**2.** (*a*)	**3.** (*c*)	**4.** (*a*)	**5.** (*b*)
6. (*a*)	**7.** (*b*)	**8.** (*a*)	**9.** (*b*)	**10.** (*c*)

X-RAYS

10

Chapter

INTRODUCTION

X-rays discovered by William Roentgen in 1895 for which he was honoured by Nobel Prize in 1901. Basically X-rays are high energy photons (1 – 100 KeV) or electro-magnetic radiation, having a very short wavelength of the order of 1 Å. Typical X-ray wavelength lies between the range of 1 Å and 100 Å. Accordingly they are classified into *hard X-rays* and *soft X-rays*. Uses of X-rays are multifold. Usually they are used in surgery, radiotherapy, engineering, industry, detective departments, scientific research and so many. Hence, the name X-rays is popularly known even to a Common man.

10.1 PRODUCTION OF X-RAYS

The modern X-ray tube was designed by Dr. Coolidge and is known after his name as *Coolidge tube*. It is widely used for commercial and medical purposes.

CONSTRUCTION AND WORKING OF COOLIDGE TUBE

The construction of Coolidge tube to produce X-rays is schematically shown in Fig. 10.1.

The cathode C consists of a tungsten filament, heated by passing a current through it. It then emits a large number of electrons called *thermions* which travel towards the *anti-cathode A,* also called the *target* with a velocity depending upon the potential difference applied between A and C. By surrounding the cathode with a molybdenum shield M, maintained at a negative potential with respect to the filament circuit, the rays can be focussed to a fine spot on A, as without the shield the electrons would fly away in all directions. The tube is completely exhausted to the no-discharge stage, so that no ionisation can take place.

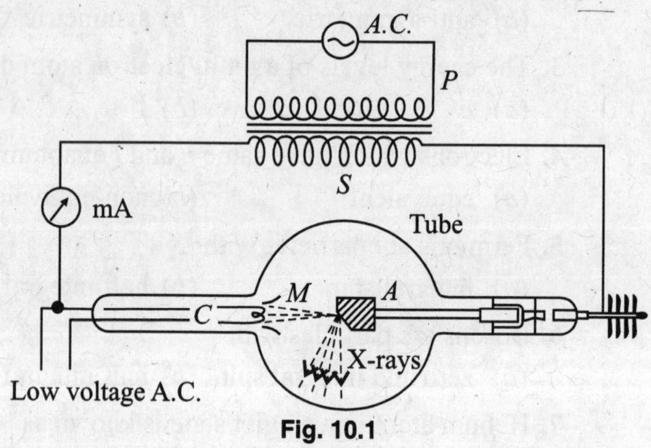

Fig. 10.1

A current of 30 mA at a voltage of 100,000 volts is used; the high voltage is obtained from a step-up transformer. The tube acts as its own rectifier, *unless the anode attains such a high temperature that it begins to emit electrons.* In large plants, the current from the transformer is rectified by thermionic rectifiers and high voltage capacitors.

The anti-cathode in a Coolidge tube must have the following characteristics:

(*i*) High atomic weight, to produce hard X-rays.

(*ii*) High melting point to withstand the high temperature developed, as most of the energy of the impinging electron is converted into heat.

(*iii*) High thermal conductivity, to get rid of the heat produced.

(*iv*) Low vapour pressure at high temperatures.

Tantalum, platinum and tungsten fulfil these requirements, but tungsten is the best and is widely used. In actual practice, a heavy block of this metal is mounted on a thick copper rod and the face is sloped at about 45° to the electron beam. The heat generated by the impact of electrons is removed by special devices. In some cases, the anode is water-cooled.

Control of intensity and quality. (*i*) The higher the potential difference between the cathode and the anti-cathode, the greater is the velocity of electrons emitted. The electrons possessing greater energy produce more penetrating X-rays. The *quality of X-rays,* therefore, varies with the *difference of potential between the cathode and the anti-cathode.*

(*ii*) The number of electrons given out by the filament is proportional to its temperature, which can be adjusted by varying the current in the filament circuit. The *intensity* of X-rays, therefore, varies with *filament current.* Thus separate control of quality and intensity can be achieved.

Properties of X-rays. (*i*) They are not deflected by electric and magnetic fields. This property distinguishes them from the cathode rays and indicates that the X-rays do not consist of charged particles.

(*ii*) They are highly penetrating and can pass through many solids, which are opaque to ordinary light, *e.g.,* wood, flesh etc. The transparency depends upon the density of the material. The denser the substance the less transparent it is to X-rays. A sheet of lead one millimetre thick, absorbs these rays, whereas the rays can pass through the same thickness of aluminium.

The penetrating power of X-rays depends upon

(*a*) The applied potential difference.

(*b*) The atomic weight of the material of the anti-cathode.

The greater the potential difference and higher the atomic weight, the more penetrating are the X-rays produced. The X-rays having a high penetrating power are known as **hard** X-rays and those with low penetrating power are known as **soft X-rays.**

(*iii*) They cause fluorescence in many substances, *e.g.,* barium platino-cyanide, cadmium tungstate, zinc sulphide, etc.

(*iv*) They produce a photo-chemical action and effect a photographic plate and they are even more effective than light.

(*v*) They ionise a gas and also eject electrons from certain metals on which they fall (photoelectric effect)

(*vi*) They are propagated in straight lines with the velocity of light *i.e.* 3×10^8 metres/sec.

(*vii*) Like light, X-rays consist of electromagnetic waves of very short wavelength and show reflection, refraction, interference, diffraction and polarisation in a similar way. These effects can be demonstrated by suitable arrangements.

(*viii*) When X-rays are incident on matter, they give rise to a complex phenomenon known as *secondary radiations,* which consist of three different types of rays:

(*a*) **Scattered X-rays.** They are practically of the same nature and wavelength as the original or the primary rays. Their properties do not depend upon the nature of the scattering substance.

(*b*) **Corpuscular rays.** They consist of fast moving electrons produced by the photoelectric process. Their properties are independent of nature of the scattering substance, but depend upon the quality of primary X-rays.

(*c*) **Characteristic X**-rays. The wavelength of the characteristic X-rays is equal to or less than the primary X-rays, but it is *characteristic* of the substance upon which the primary X-rays are incident and is independent of the wavelength of the primary X-rays.

(*ix*) They have a destructive effect on living tissues. Exposure of human bodv to X-rays causes reddening of the skin and surface sores.

10.2 NEED OF HIGH ATOMIC NUMBER MATERIAL AS TARGET

For most of the purposes we need *hard* X-rays which have a high penetrating power. The penetrating power depends upon the energy of X-rays and hence on the frequency. According to Moseley's law the frequency of X-rays is given by

$$\nu = Rc\,[\,Z - b\,]^2 \left[\frac{1}{n_1^2} - \frac{1}{n_2^2}\right]$$

where R is Rydberg's constant, c the velocity of light, Z the atomic number of material of the target and b the screening constant. Thus frequency is proportional to the square of the atomic number. Materials of high atomic number have high atomic weight and hence are more useful for producing highly penetrating X-rays.

High melting point. At the anticathode most of the energy of the impinging electrons is converted into heat due to which a very high temperature is developed at the anticathode. In order that the anticathode may withstand this very high temperature, it should have a high melting point.

10.3 USES OF X-RAYS

(*i*) **Surgery.** X-rays can pass through flesh and not through bones. They are of extensive use in surgery to detect fractures, foreign bodies, diseased organs etc. The patient is made to stand between an X-ray source and a fluorescent screen. The luminosity of the screen depends upon the stopping power of different parts of the body. Thus a deep shadow of the bones and a light shadow of the flesh will be obtained. A photograph may also be obtained. It is called a *radiograph.* If the various parts do not show sufficient contrast, artificial means are adopted, *e.g.,* in radiographing the portions of the alimentary tract, the patient is fed with a meal of barium sulphate.

(*ii*) **Radiotherapy.** X-rays kill the diseased tissues of the body. Hence they are used to cure intractable skin diseases, malignant tumours etc. If the affected part is superficial soft rays are applied and for deep seated organs hard rays are used.

(*iii*) **Engineering.** Because of their high penetrating power, they are used to investigate the structure of metals, testing of castings for flaws and gas pockets, causes of weakness in structures and to detect cracks and blow holes in metal plates. The method has the advantage that unlike some mechanical tests, it does not destroy the piece. Moreover, it often suggests alterations in mould design and heat treatment to avoid a recurrence of the detected faults.

(*iv*) **Industry.** X-rays are employed in industry to detect defects in motor tyres, golf and tennis balls, wood and wireless valves. They are used in testing the uniformity of insulating material and for detecting the presence of pearls in oysters.

(*v*) **Detective departments.** X-rays are used at customs posts for detection of contraband goods like explosives, opium, etc. concealed in leather or wooden cases, in examining of parcels without opening them and in the detection of precious metals like gold and silver in the body of smugglers. They are also used in distinguishing real diamonds, gems and pearls from artificial ones, pure ghee from vegetable ghee and real documents from forged ones.

(*vi*) **Scientific research.** X-rays have even been employed to investigate the *structure of crystals,* structure and properties of atoms and arrangement of atoms and molecules in matter.

10.4 ORIGIN AND MECHANISM OF X-RAYS (DUANE-HUNT LAW)

X-ray spectrum. The dependence of intensity of X-rays emitted by a source on wavelength (or frequency) is known as its X-ray spectrum.

It has been observed that the X-ray spectrum emitted by all the target material consists of a *continuous spectrum,* having radiations of all possible wavelength, within a certain range and superimposed upon it a sharp *line spectrum* of definite wavelength known as *characteristic spectrum,* the lines being characteristic of the material used as target.

Continuous spectrum. The X-rays emitted by all the target materials have a continuous distribution of intensity above a certain minimum wavelength λ_m given by

$$\lambda_m = \frac{1.242 \times 10^{-6}}{V} m = \frac{12420 \times 10^{-10}}{V} m = \frac{12420}{V} \text{Å}$$

where V is the applied voltage. This relation is known as *Duane-Hunt law.* The minimum wavelength, known as *continuous spectrum boundary or Duane-Hunt limit,* is independent of the nature of the target.

Origin and mechanism of production. Continuous X-rays are produced by the phenomenon of *Bremsstrahlung,* which is a German word meaning *'braking'* or *'slowing down'* radiation. Electrons emitted from the cathode in the X-ray tube are accelerated towards the target, strike it, penetrate deep into the interior of its atoms and are attracted by the nuclei due to strong electrostatic interaction. The electron, therefore, deviates from its original path, as shown in Fig. 10.2.

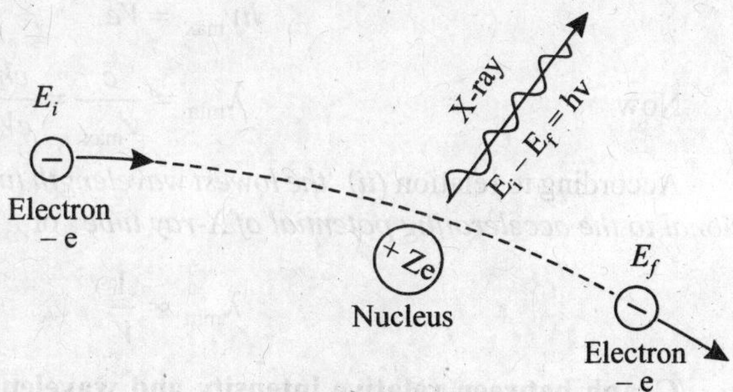

Fig. 10.2

Source of energy. The deviation of the electron from its straight line path is equivalent to its collision with the nucleus. So the electron loses some of its energy due to the collision. This energy appears as an X-ray photon. If E_i is the initial kinetic energy of the electron and E_f is the final kinetic energy, then the energy of the X-ray photon = $E_i - E_f$. If ν is the frequency of the X-ray photon emitted, then

$$h\nu = E_i - E_f$$

The electron may suffer several collisions with various nuclei before coming to rest. Each collision is accompanied by the emission of an X-ray photon. Hence, a number of photons of different frequencies are emitted. As there are a large number of electrons in the beam and each electron suffers collision in a different way, we get photons of almost all fequencies or wavelength, *thus producing a continuous X-ray spectrum.*

The production of continuous spectrum is the result of inverse photo-electric effect with electron kinetic energy $E_i - E_f$ being transformed into a photon of energy hv.

Lowest wavelength limit. According to classical theory, the continuous X-ray spectrum should consist of all frequencies between zero and infinity. But according to quantum theory, there is a minimum wavelength only, up to which the continuous X-ray spectrum extends. To find the value of λ_{min}, let the velocity of the incident electron decrease from v_i to v_f due to interaction with the nucleus. Then frequency of X-ray photon is given by

$$h\nu = \frac{1}{2} m v_i^2 - \frac{1}{2} m v_f^2$$

When the electron is completely brought to rest by electrostatic interaction *i.e.* $v_f = 0$, the entire kinetic energy of the electron is radiated as an X-ray photon and the photon has maximum frequency or minimum wavelength given by

$$h\nu_{max} = \frac{1}{2} m v_i^2$$

or $$\nu_{max} = \frac{m v_i^2}{2h}$$

$$\therefore \qquad \lambda_{min} = \frac{c}{\nu_{max}} = \frac{2ch}{mv_i^2} \qquad ...(i)$$

λ_{min} inversely proportional to accelerating velocity. If V is the potential difference between the cathode and the target to accelerate the electron, then

$$\frac{1}{2}mv_i^2 = Ve$$

$$\therefore \qquad h\nu_{max} = Ve$$

$$\text{Now} \qquad \lambda_{min} = \frac{c}{\nu_{max}} = \frac{ch}{eV} = \frac{1.242 \times 10^{-6}}{V} m \qquad ...(ii)$$

According to relation *(ii)* *'the lowest wavelength limit of continuous spectra is inversely proportional to the accelerating potential of X-ray tube'* or

$$\lambda_{min} \propto \frac{1}{V}$$

Graph between relative intensity and wavelength. The plot of relative intensity of X-ray spectra versus wavelength at four different potential differences applied to an X-ray tube between the cathode and the target (Tungsten) is shown in Fig. 10.3.

It is clear from the graph that as the potential difference V applied to the X-ray tube increases from 30 *kV to 40* kV to 50 kV and then to 70 kV the minimum wavelength limit decreases from 0.4 Å to 0.3 Å to 0.24 Å and then to 0.17Å showing thereby that $\lambda_m \propto \frac{1}{V}$ or $\lambda_m \times V$ = a constant.

Other special features of the continuous X-ray spectrum are:

(1) *Change of target material.* As $\lambda_{min} = \frac{ch}{eV}$ and *c, h* and *e* are constants, the cut-off wavelength is, therefore, the same for all target materials. In other words, the range of the continuous spectrum does not depend upon the nature cf the target material. Hence, the plots will remain unchanged with change in target material.

(2) *Change in accelerating voltage.* (*i*) The continuous spectrum ends abruptly at a minimum wavelength λ_{min}. The value of λ_{min} decreases with increasing accelerating potential.

(*ii*) The intensity rises to a maximum at a certain wavelength, after which it gradually falls off. The intensity never reaches zero, showing that the radiation contains all possible wavelength, above a certain minimum.

(*iii*) The wavelength corresponding to maximum intensity λ_m shifts gradually to lower wavelength side as the accelerating potential is increased in such a way that $(\lambda_m V^{1/2})$ is almost a constant.

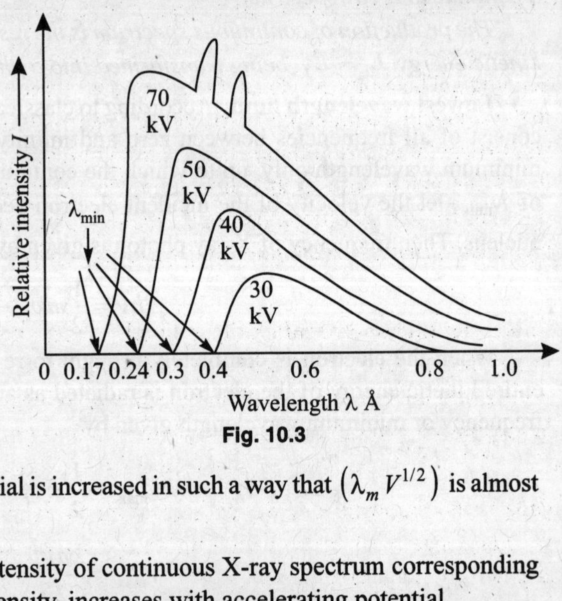

Fig. 10.3

(*iv*) For a particular target material, the intensity of continuous X-ray spectrum corresponding to all wavelengths, as well as the minimum intensity, increases with accelerating potential.

(v) For applied potential difference 70 *kV*, two sharp peaks are seen. The sharp peaks are due to characteristic radiation. The line spectra is absent till the potential difference is greater than a particular value. The smoothly varying curves represent the *continuous spectrum* and the superemposed lines on the continuous background constitute the *characteristic spectrum.*

10.5 SOFT AND HARD X-RAYS

The penetrating power of X-rays depends upon (*i*) the applied potential difference between the cathode and the anticathode and (*ii*) the atomic weight of the material of the anticathode. The greater the potential difference, smaller is the wavelength of X-rays, higher is their energy and more penetrating are the rays. The X-rays having high penetrating power are known as *hard X-rays*. These carry high energy. Those with low penetrating power are known as *soft X-rays*. These are of longer wavelength, carry lower energy and are less penetrating.

SOLVED EXAMPLES

Example 10.1. *The potential difference between the cathode and anode in X-ray tube is doubled and the separation between the cathode and target is also doubled. How will it affect the cut off wavelength?* **(Pbi. U. 2002)**

Solution. The values of cut off wavelength or

$$\lambda_{min} = \frac{ch}{eV} = \frac{1.242 \times 10^{-6}}{V} m$$

The cut off wavelength λ_{min} is thus inversely proportional to the potential difference between the cathode and the anode (or anticathode). When this potential difference is doubled the cut off wavelength λ_{min} becomes one half.

Cut off wavelength does not depend upon the distance between the cathode and the target. Doubling the distance between the cathode and the target will, therefore, have no effect on the cut off or minimum wavelength of X-rays produced.

10.6 CHARACTERISTIC X-RAY SPECTRUM

When the p. d. between the cathode and target exceeds a certain minimum value, approximately 35 KV for molybdenum target, the contimuous spectrum becomes the characteristic X-ray spectrum. It can be seen from Fig. 10.4 that the background of characteristic X-ray spectrum is continuous spectrum where there are high intensity peaks at certain definite wavelengths. The wevelengths at which these peaks appear depend on the target material.

The minimum p. d. at which the continuous spectrum becomes characteristic X-ray spectrum is different for different target meterial. For example, for copper, molybdenum and tungsten, the values are 9 KV, 29 KV and 72 KV respectively.

In the spectrum, first longer wavelength is K_α line, second shorter wavelength is K_β line and so on. These lines are the characteritic of the target atom and different for different elements. When heavy elements are used as target, the other lines L_α, L_β, L_y of L-Series, M_α, M_β etc. of M-series are observed.

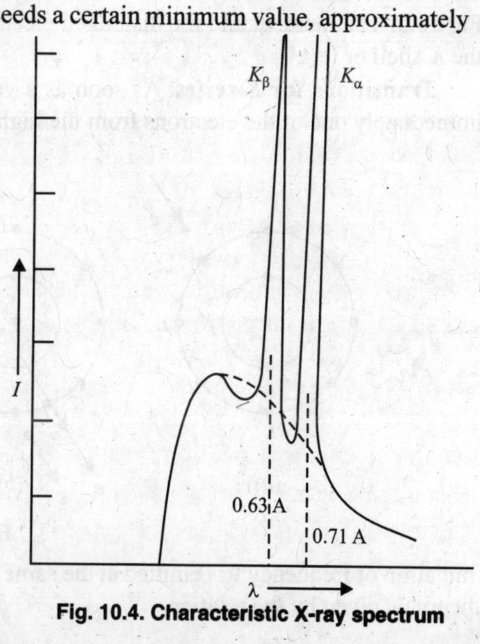

Fig. 10.4. Characteristic X-ray spectrum

The characteristic X-ray spectrum consists of sharp peaks super imposed on the continuous spectrum. Their wavelength is characteristic of the target element. It is highly monochromatic and independent of the applied voltage. The intensity of the characteristic spectrum lines is very large.... sometimes 10^4 times that of the continuous spectrum in that region.

The characteristic X-rays from substances of higher atomic number are of shorter wavelength than those from substances of lower atomic number and hence are more penetrating. The characteristic wavelength is given by Moseley's law

$$\lambda (Z - b)^2 = a \text{ (constant)}$$

where the value of b is different for different groups.

Every substance gives rise to a particular set of characteristic X-rays, consisting of one or more series, known as *K, L, M* etc. series, The *X*- rays in *K* series of a particular element have smaller wavelength than in the *L or M* series, whereas the wavelength of the X-rays in the same series, say *K*, produced from an element of higher atomic weight, is less than that produced from an element of lower atomic weight.

Why it is so called. The characteristic (or discrete) X-ray spectrum consists of sharp lines the wavelength of which only depends upon the target element. The characteristic wavelength is given by Moseley's law; $\lambda(Z - b)^2$ = a constant where Z is the atomic number of the target element.

As the wavelength is characteristic of the target element it is known as characteristic spectrum. Characteristic X-ray spectrum is independent of applied voltage.

Production of characteristic X-rays. An atom consists of a central nucleus having a positive charge and revolving round it, in more or less circular orbits, a suitable number of electrons. The electron in the inner most orbit is attracted by the nucleus with the greatest force and to detach it from the atom maximum energy is required. The innermost orbit is given the name *K* level or shell and the outer orbits *L, M*, etc. levels respectively. Thus more energy is required to detach an electron from an inner level than from an outer level.

When an atom is bombarded by fast moving electrons, an electron coming with a high velocity penetrates through the outer shell containing electrons and removes an electron from one of the inner shells. An incoming electron knocking out an electron from *K* shell of an atom is shown in Fig 10.5. The incident and the dislodged electrons both rush out of the atom, causing a vacancy in the *K* shell or level.

Transitions for *K*-series. As soon as a vacancy in the *K*-shell is caused as explained above, immediately one of the electrons from the higher levels (*e.g. L* level) jumps into the *K* level. A

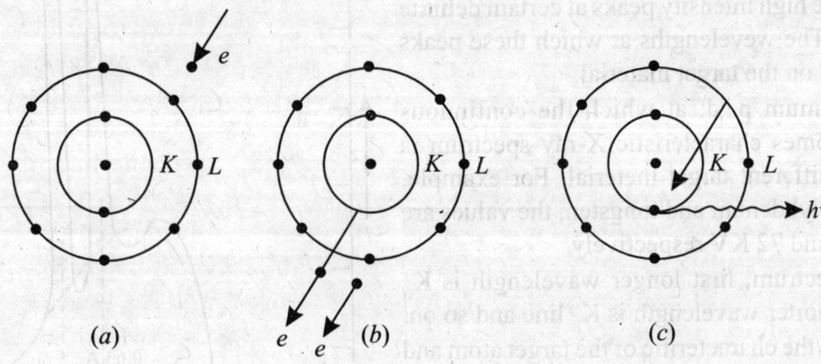

Fig. 10.5

radiation of frequency v is emitted at the same time. The frequency of the radiation, on the quantum theory, is given by the relation

$$E_K - E_L = h\nu$$

or
$$\nu = \frac{E_K - E_L}{h}$$

where E_K is the energy required to extract an electron from the K level and E_L is the energy required to extract an electron from the L level and h is Planck's constant having a value equal to 6.62×10^{-34} Joule-sec. The frequency of the radiation is very high and is of the order of 10^{18} per second.

The wavelength of the corresponding X-ray is given by

$$\lambda = \frac{c}{\nu} = \frac{hc}{E_K - E_L}$$

If an electron from the M level falls into the K level, X-rays of still higher frequency (shorter wavelength) will be produced. The frequencies of X-rays produced by the fall of an electron from an outer level to the K level can be arranged into a series, called K series, consisting of K_a, K_β, K_γ etc. lines, as shown in Fig. 10.6.

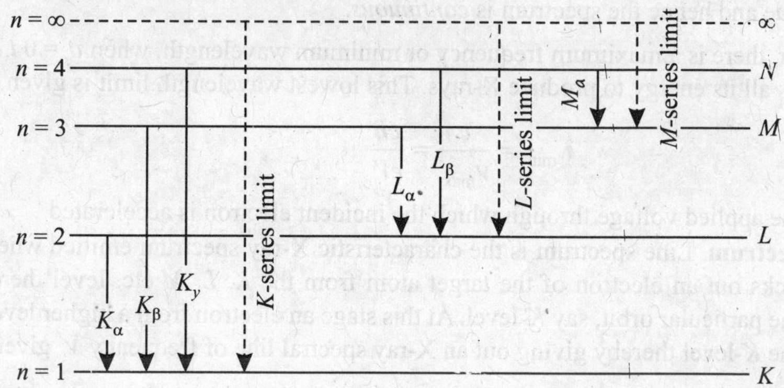

Fig. 10.6

A jump from L level to K level gives rise to K_α line and from M level to K level K_β line and so on, till we get the limiting line, known as *K-series limit.*

Transition for *L* series. Similarly, when the incident electrons carry a lesser amount of energy, an electron is displaced from the L level and an electron from M or other outer level takes its place, so that a radiation of a lower frequency than that of K series is emitted. This corresponds to the L series of the X-ray spectrum. It is evident that from the same element, the L lines of the X-ray spectrum have smaller frequency (larger wavelength) than the K lines. A jump from M shell to L shell gives rise to L_α line and from N shell to L shell gives rise to L_β line and so on, till we get the limiting line, known as *L-series limit.*

The wavelength of these lines in the *K, L,* or *M* series depends upon the atomic weight of the anticathode and these wavelengths are known as **characteristic X-rays** for the material of the given anticathode or target.

However, it should be noted that even a high speed electron bombarding on a hydrogen target, can not emit characteristic X-rays. The shortest wavelength emitted by hydrogen atom corresponds to the limiting line in Lyman series for which the wavelength is 911Å. The wavelength of X-rays lies in the range 0.1Å to 100Å. Thus the radiation emitted by the hydrogen atom will not lie in the X-ray region.

Example 10.2. *Calculate the momentum of photons corresponding to the minimum wavelength emitted from X-ray tube whose operating voltage is 120 kV.* ***(P.U. 2007)***

Solution. Minimum wavelength $\lambda_m = \frac{ch}{eV} = \frac{1.242\times10^{-6}}{V} m = \frac{12420\times10^{-10}}{V} m$

For $V = 120$ KV $= 120 \times 10^3$ V; $\lambda_m = \frac{12420}{120\times10^3} = 0.1035$Å $= 0.1035 \times 10^{-10}$ m

$\therefore$ Momentum $p = \frac{h}{\lambda} = \frac{6.62\times10^{-34}}{0.1035\times10^{-10}} = 63.96 \times 10^{-24}$ Newton metre.

10.7 PHOTONS OF CONTINUOUS AND CHARACTERISTIC X-RAYS

Continuous spectrum. The continuous spectrum is emitted due to the loss of energy of the incident electron due to a *collision* with the nucleus. The frequency of X-rays produced is given by

$$h\nu = \frac{1}{2} mv_i^2 - \frac{1}{2} mv_f^2$$

where v_i is the initial velocity of the incident electron and v_f its final velocity. The loss of energy may have any value and hence the spectrum is *continuous.*

However, there is a maximum frequency or minimum wavelength, when $v_f = 0$ *i.e.* the incident electron loses all its energy to produce X-rays. This lowest wavelength limit is given by

$$\lambda_{\text{min}} = \frac{c}{\nu_{\text{max}}} = \frac{ch}{eV}$$

where V is the applied voltage through which the incident electron is accelerated.

Line spectrum. Line spectrum is the characteristic X-ray spectrum emitted when the incident electron knocks out an electron of the target atom from the *K, L, M* etc. level thereby causing a vacancy in the particular orbit, say *K*-level. At this stage an electron from a higher level, say *L* level, jumps into the *K*-level thereby giving out an X-ray spectral line of frequency ν given by

$$\nu = \frac{E_K - E_L}{h}$$

where E_K is the energy required to extract the electron from the *K*-level and E_L the energy required to extract the electron form *L*-level. This frequency is known as K_α line.

Similarly if a vacancy is caused in the *L*-level and an electron jumps into it from the *M* -level we get the L_α line. Thus the line spectrum is characteristic of the target atom.

Difference. The two photons, one from continuous X-rays and the other from characteristic X-rays, differ only in their mode of creation.

The photon of continuous X-rays is emitted due to loss of energy of the incident electron when it has a collision with the nucleus of the target atom.

The photon of characteristic X-rays is emitted when an electron from a higher inner shell jumps to a lower inner shell after the incident electron knocks out an electron of the target element from the lower inner shell (or level).

Example 10.3. *The X-ray emission spectrum of a substance gives the following values of K_a and K_a lines.*

$$K_a \textit{ line} = 0.2 \text{ Å}; \; L_a \textit{ line} = 1.6 \text{ Å}.$$

Using the data find the wavelength of K_a line. **(P.U. 2005)**

Solution. Wavelength of K_α line $\lambda_1 = \frac{hc}{E_K - E_L} = 0.2$Å $= 0.2 \times 10^{-10}$ m

$$\therefore \quad E_K - E_L = \frac{hc}{\lambda_1} = \frac{6.62 \times 10^{-34} \times 3 \times 10^8}{0.2 \times 10^{-10}} = 99.3 \times 10^{-16}\ J$$

Wavelength of L_α line, $\lambda_2 = \dfrac{hc}{E_L - E_M} = 1.6\ \text{Å} = 1.6 \times 10^{-10}\ m$

$$\therefore \quad E_L - E_M = \frac{hc}{\lambda_2} = \frac{6.62 \times 10^{-34} \times 3 \times 10^8}{1.6 \times 10^{-10}} = 12.4 \times 10^{-16}\ J$$

Wavelength of K_β line, $\lambda = \dfrac{hc}{E_K - E_M}$

Now $\quad E_K - E_M = E_K - E_L + E_L - E_M = 99.3 \times 10^{-16} + 12.4 \times 10^{-16}\ J$

$$= 111.7 \times 10^{-16}\ J$$

$$\therefore \quad \lambda = \frac{hc}{E_K - E_M} = \frac{6.62 \times 10^{-34} \times 3 \times 10^8}{111.7 \times 10^{-16}} = 0.18 \times 10^{-10}\ m = 1.18\ \text{Å}$$

10.8 BREMSSTRAHLUNG

In an X-ray tube, electrons emitted from the cathode and accelerated towards the anticathode or target are *'slowed down'* or undergo *'braking'* due to the strong electrostatic interaction of the nucleus. In the process, the electron loses some of its energy. This energy appears as an X-ray photon. As we get photons of almost all energies up to a certain maximum, it gives rise to a continuous X-ray spectrum having a maximum frequency or minimum wavelength.

Since these continuous X-rays are produced due to *'slowing down' or 'braking'* of the fast moving electron, these are known as *'Bremsstrahlung'* which is the German word for 'slowing down' or 'braking'.

10.9 DIFFERENCE IN ORIGIN OF X-RAYS AND VISIBLE LIGHT

When an atom is excited, the outermost valence electron moves to one of the higher energy excited states (or virtual orbits). The life of the electron in these orbits or excited states is very small. The electron, therefore, jumps back to its ground state or a lower energy state (orbit); the difference in the energy of the two states is radiated as electromagnetic radiation (light).

The transitions of the outer electron of an atom usually involve only a few electron volts of energy. Even removing an electron requires 13.6 eV for hydrogen and 54.6 eV for helium. These transitions accordingly are associated with photons whose wavelengths lie in the near visible part of the electromagnetic spectrum.

X-rays, on the other hand, are produced by transitions among inner electron orbits. For example, if a high energy fast moving electron ejects a *K* electron from an atom, the vacancy in the *K*-shell is filled by an electron jumping into *K*-orbit from *L, M, N* etc. orbit. In the process, the difference in energy is radiated as an X-ray photon.

The inner electrons of an atom, particularly the heavier ones are not well shielded from the full nuclear charge by intervening electron shells and are, therefore, very tightly bound. In Sodium, for example, only 5.13 eV energy is required to remove the outermost 3 *s* (M_I of *M* shell) electron, while the corresponding figures for the inner electrons are 31 eV for each 2 *p* ($L_{II} L_{III}$ of *L* shell), 63 eV for each 2 *s* (L_I of *L* shell) and 104.1 eV for each 1*s* (*K* shell) electron. Transitions involving inner electrons of the atom give rise to high energy photons, lying in the X-ray region of the electro magnetic spectrum, having wavelength much smaller than those of visible light.

Hence production of visible light is due to transitions between the virtual or excited states of the valence electron whereas X-rays are due to the transitions of the inner electrons in *K, L, M* etc. shells.

X-ray production inverse of photo-electric effect. In photo-electric effect the energy of the incident photon is absorbed by an electron in the metal and it is knocked out carrying the energy of the incident photon. If v is the velocity of the electron ejected, E is kinetic energy and ν the frequency of the incident photon, then

$$h\nu = \frac{1}{2}mv^2 = E$$

The production of continuous X-rays spectra is the result of inverse photo-electric effect, in which the kinetic energy of the incident fast moving electron is carried by the emitted X-ray photon. If E_i *is* the initial kinetic energy of the electron, E_f the final kinetic energy, then the energy of the X-ray photon emitted $= E_i - E_f = E$. If ν is the frequency of the photon, then

$$h\nu = E_i - E_f = E$$

Therefore, it is quite appropriate to regard X-ray production as inverse of photo-electric effect.

Example10.4. *An X-ray tube operates at the voltage given below. Find the maximum speed of electrons striking the anticathode and shortest wavelength of X-rays produced.*

(*i*) *40 kV*
(*ii*) *20 kV*
(*iii*) *100 kV* ***(Gauhati U. 2000)***
(*iv*) *50 kV* ***(H.P.U. 2003, 1996; Pbi. U. 2002)***
(*v*) *18 kV* ***(P.U. 2001)***

Solution. Let V be the difference of potential applied to the X-ray tube, then the work done on the electron in moving from the cathode to the anticathode is Ve, where e is the charge on the electron. If ν is the maximum velocity acquired by the electron, then

$$\text{Maximum kinetic energy} = \frac{1}{2}mv^2$$

$$\text{or} \quad \frac{1}{2}mv^2 = Ve$$

$$\text{or} \quad v = \sqrt{\frac{2Ve}{m}} = \sqrt{V}\sqrt{\frac{2e}{m}}$$

$$= \sqrt{V}\sqrt{\frac{2\times1.6\times10^{-19}}{9\times10^{-31}}} = \sqrt{V}\times0.596\times10^6\ ms^{-1}.$$

(*i*) V = 40 kV = 40, 000 V

$$\therefore \quad v = \sqrt{40{,}000}\times0.596\times10^6 = 200\times0.596\times10^6 = 119.2\times10^6\ ms^{-1}$$

(*ii*) V = 20 kV = 20,000 V

$$\therefore \quad v = \sqrt{20{,}000}\times0.596\times10^6 = 84.27\times10^6\ ms^{-1}$$

(*iii*) V = 100 kV = 100,000 V

$$\therefore \quad v = \sqrt{100{,}000}\times0.596\times10^6 = 188.45\times10^6\ ms^{-1}$$

(*iv*) V = 50 kV = 50,000 V

$$\therefore \quad v = \sqrt{50{,}000}\times0.596\times10^6 = 133.27\times10^6\ ms^{-1}$$

(*v*) V = 18 kV = 18000 V

$$\therefore \quad v = \sqrt{18000}\times0.596\times10^6 = 79.96\times10^6\ ms^{-1}$$

Shortest wavelength $\lambda_m = \frac{ch}{eV} = \frac{12420 \times 10^{-10}}{V} m = \frac{12420}{V}$ Å

(*i*) V = 40 kV = 40,000 V

$\therefore \quad \lambda_m = \frac{12420}{40,000} = 0.31$ Å

(*ii*) V = 20 kV = 20, 000 V

$\therefore \quad \lambda_m = \frac{12420}{20,000} = 0.62$ Å

(*iii*) V = 100 kV = 100,000 V

$\therefore \quad \lambda_m = \frac{12420}{100,000} = 0.124$ Å

(*iv*) V = 50 kV = 50,000 V

$\therefore \quad \lambda_m = \frac{12420}{50,000} = 0.248$ Å

(*v*) V = 18 kV = 18,000 V

$\therefore \quad \lambda_m = \frac{12420}{18,000} = 0.69$ Å

Example 10.5. *Electrons are accelerated in a television tube through a potential difference of 9.8 kV. Find the highest frequency and minimum wavelength of the elelctromagnetic waves emitted, when these strike the screen of the tube. In which region of the spectrum will these waves lie?*

(P.U. 2002)

Solution. Minimum wavelength is given by

$$\lambda_{\min} = \frac{hc}{eV} = \frac{12420 \times 10^{-10}}{V} m = \frac{12420}{V} \tilde{A}$$

Here $V = 9.8$ kV

$\therefore \quad \lambda_{mim} = \frac{12420}{9.8 \times 10^3} = 1.26$ Å

$\therefore \quad \nu_{\max} = \frac{c}{\lambda_{\min}} = \frac{3 \times 10^8}{1.26 \times 10^{-10}} = 2.38 \times 10^{18}\ Hz$

The waves lie in the X-ray region of the electromagnetic spectrum.

Example 10.6. *If the K, L, and M energy levels of Pt are 78000, 12000 and 3000 eV respectively, calculate the wavelength of K_α and K_β lines from Pt. (1 eV = 1.6 × 10^{-19} J and h = 6.55 × 10^{-34} Joule-sec.)*

Solution. Energy of *K* level, $E_K = 78000$ eV

$= 78000 \times 1.6 \times 10^{-19}$ Joule

Energy of *L* level, $E_L = 12000 \times 1.6 \times 10^{-19}$ Joule

Energy of *M* level, $E_M = 3000 \times 1.6 \times 10^{-19}$ Joule

Wavelength of K_α line, $\lambda_1 = \frac{hc}{E_K - E_L}$

$$= \frac{6.55 \times 10^{-34} \times 3 \times 10^{8}}{(78000 - 12000) \times 1.6 \times 10^{-19}}$$

$$= 0.186 \times 10^{-10} \text{m} = 0.186 \text{ Å}$$

Wavelength of K_β line, $\lambda_2 = \frac{hc}{E_K - E_M}$

$$= \frac{6.55 \times 10^{-34} \times 3 \times 10^{8}}{(78000 - 3000) \times 1.6 \times 10^{-19}}$$

$$= 0.163 \times 10^{-10} \text{ m} = 0.163 \text{ Å}$$

Example 10.7. *An X-ray tube operating at (i) 30 kV, (ii) 44 kV (iii) 50 kV and (iv) 200 kV emits a continuous spectrum with shortest wavelength (i) 0.414 Å, (ii) 0.284 Å (iii) 0.248 × 10^{-10}m and (iv) 6.2 × 10^{-12}m; calculate Planck's constant. Given e = 1.6 × 10^{-19} C and c = 3 × $10^8 ms^{-1}$.*

Solution. Minimum wavelength is given by

$$\lambda_{min} = \frac{ch}{eV}$$

$$\therefore \quad h = \frac{eV}{c} \lambda_{min}$$

Now $e = 1.6 \times 10^{-19}$ C and $c = 3 \times 10^{8}$ ms^{-1}

$$\therefore \quad \frac{e}{c} = \frac{1.6 \times 10^{-19}}{3 \times 10^{8}} = 0.5333 \times 10^{-27}$$

$$\therefore \quad h = 0.5333 \times 10^{-27} \ V \lambda_{min}$$

(i) $V = 30$ kV $= 30{,}000$ V; $\lambda_{min} = 0.414$ Å $= 0.414 \times 10^{-10}$ m

$\therefore \ h = 0.5333 \times 10^{-27} \times 3 \times 10^{4} \times 0.414 \times 10^{-10} = 0.662 \times 10^{-33} = 6.62 \times 10^{-34}$ Js

(ii) $V = 44$ kV $= 44{,}000$ V; $\lambda_{min} = 0.284$ Å $= 0.284 \times 10^{-10}$ m

$\therefore \ h = 0.5333 \times 10^{-27} \times 44 \times 10^{3} \times 0.284 \times 10^{-10} = 6.66 \times 10^{-34}$ Js

(iii) $V = 50$ kV $= 50 \times 10^{3}$ V ; $\lambda_{min} = 0.248 \times 10^{-10}$ m

$\therefore \ h = 0.5333 \times 10^{-27} \times 50 \times 10^{3} \times 0.248 \times 10^{-10} = 6.61 \times 10^{-34}$ Js

(iv) $V = 200$ kV $= 2 \times 10^{5}$ V , $\lambda_{min} = 6.2 \times 10^{-10}$ m

$\therefore \ h = 0.5333 \times 10^{-27} \times 2 \times 10^{5} \times 6.2 \times 10^{-12} = 6.61 \times 10^{-34}$ Js

Example 10.8. *The K- absorption limit for Uranium is 0.1Å. What is the excitation potential of the tube to give this radiation?* **(P.U. 2004)**

Solution. K-absorption limit = minimum wavelength $\lambda_{min} = \frac{ch}{eV}$

$$\therefore \quad V = \frac{ch}{e\lambda_{min}} = \frac{3 \times 10^{8} \times 6.62 \times 10^{-34}}{1.6 \times 10^{-19} \times 0.1 \times 10^{-10}} = 124 \times 10^{3} \ V = 124 \text{ kV}$$

Example 10.9. *Given that K-absorption edge for lead is 0.14 Å and the minimum voltage required for producing K-lines in lead is 88.6 KV, determine the ratio of h/e.* **(G.N.D.U. 2002)**

Solution. K-absorption edge = minimum wavelength $\lambda_{min} = 0.14$ Å $= 0.14 \times 10^{-10}$ m
$V = 88.6 \times 10^3$ V

$$\lambda_{min} = \frac{ch}{eV} \therefore \frac{h}{e} = \lambda_{min}\frac{V}{c} = \frac{0.14 \times 10^{-10} \times 88.6 \times 10^3}{3 \times 10^8}$$

$$= 4.135 \times 10^{-15} \text{ Joule sec/C}$$

Example 10.10. *What voltage must be applied to an X-ray tube for it to emit X-rays with minimum wavelength of (i) 40 pm (ii) 0.5Å (iii) 1.0Å and (iv) 0.25 Å.*

(G.N.D.U. 2004; P.U., 2001; Luck., 2007; Meerut. U. 2000)

Solution. Minimum wavelength is given by $\lambda_{min} = \frac{ch}{eV}$

$\therefore$ Voltage that must be applied $V = \frac{ch}{e\lambda_{min}} = \frac{ch}{e}\frac{1}{\lambda_{min}}$

Here $c = 3 \times 10^8$ ms^{-1} ; $h = 6.626 \times 10^{-34}$ Js ; $e = 1.6 \times 10^{-19}$ C

$$\therefore \quad \frac{ch}{e} = \frac{3 \times 10^8 \times 6.626 \times 10^{-34}}{1.6 \times 10^{-19}} = 12.42 \times 10^{-7}$$

(i) $\lambda_{min} = 40$ pm $= 40 \times 10^{-12}$ m

$$\therefore \quad V = \frac{ch}{e} \cdot \frac{1}{\lambda_{min}} = \frac{12.42 \times 10^{-7}}{40 \times 10^{-12}} = 31 \times 10^3 V = 31 kV$$

(ii) $\lambda_{min} = 0.5$ Å $= 0.5 \times 10^{-10}$ m

$$\therefore \quad V = \frac{12.42 \times 10^{-7}}{0.5 \times 10^{-10}} = 24.84 \times 10^3 \; V = 24.84 \; kV$$

(iii) $\lambda_{min} = 0.1$ Å $= 0.1 \times 10^{-10}$ m

$$\therefore \quad V = \frac{12.42 \times 10^{-7}}{0.1 \times 10^{-10}} = 124.1 \times 10^3 \; V = 124.1 \; kV$$

(iv) $\lambda_{min} = 0.25$ Å $= 0.25 \times 10^{-10}$ m

$$V = \frac{12.42 \times 10^{-7}}{0.25 \times 10^{-10}} = 49.68 \times 10^3 \; V = 49.68 \; kV.$$

Example 10.11. *An X-ray tube is operated at an anode potential of 10 kV and anode current of 15 mA. Calculate (i) number of electrons hitting the anode per second and (ii) the minimum wavelength produced by the X-ray tube.*

(Luck. U. 2003)

Solution. Current = 15 mA = 15×10^{-3} A

$\therefore$ Charge conveyed to the anode per second = $15 \times 10^{-3} C$

Charge on the electron = 1.6×10^{-19} C

∴ Number of electrons hitting the anode per second

$$= \frac{15 \times 10^{-3}}{1.6 \times 10^{-19}} = 9.4 \times 10^{16} \text{ electrons per second.}$$

Now $$\lambda_{min} = \frac{ch}{eV} = \frac{3 \times 10^8 \times 6.62 \times 10^{-34}}{1.6 \times 10^{-19} \times 10 \times 10^3} = 1.24 \times 10^{-10} = 1.24 \text{ Å}$$

Example 10.12. *An X-ray tube is operated at an anode potential 12.4 KV and current of 15 mA. Calculate*

(i) minimum wavelength produced by X-ray tube

(ii) number of electrons hitting the anode per second ***(Pbi. U. 2001)***

Solution. $\lambda_{min} = 1$Å; Number of electrons hitting the cathode per second $= 9.4 \times 10^{16}$ electrons per second.

Example 10.13. *An X-ray tube is operated at 50 kV. and current through the tube is 1 mA. What is the number of electrons striking the target per second?* ***(P.U. 2000)***

Solution. Number of electrons striking the target per second $= \frac{10^{-3}}{1.6 \times 10^{-19}} = 6.25 \times 10^{15}$

10.10 MOSELEY'S WORK

Moseley made a systematic study of characteristic X-ray spectra of different metallic elements in 1913. He used an X-ray spectrometer with a crystal of potassium ferrocyanide and received the reflected X-rays on a photographic plate, instead of using an ionisation chamber. At first he photographed the spectra of elements from calcium to zinc, in order of increasing atomic weight and later he extended his work upto tin. He calculated the wavelength of X-rays, using Bragg's law and found that the spectral lines emitted by these elements belonged to two distinct series, a short wavelength group belonging to K series and a relatively large wavelength group belonging to L series.

The K series of every element consisted of two lines, one of which was stronger in intensity than the other. The stronger line, denoted as K_α line, is longer in wavelength than the weaker line denoted as K_β for every element. As the atomic number is increased both K_α and K_β lines shift towards shorter wavelength side.

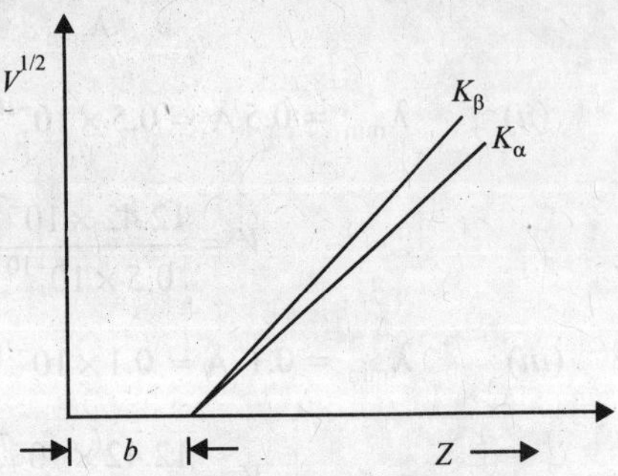

Fig. 10.7: Moseley's diagram

Moseley found a striking regularity in the spectral lines of each series. The spectrum emitted by each element was found to be exactly similar to that of the other, the only difference being that the lines had different wavelength. The frequency of any particular line in a series varied regularly from one element to the next in the periodic table. On plotting a graph between the square root of the frequency of any particular line in a series, say K_α line, which is the most intense in the series and the atomic number of the element, Moseley found that it was a straight line, as shown in Fig. 10.7.

10.11 MOSELEY'S LAW

The frequency of the K lines produced from an element of higher atomic number is greater than that produced from element of lower atomic number.

The higher the atomic number, higher is the frequency of X-rays produced and more penetrating are the rays. It can be shown by the application of Bohr's theory of the atom that if ν is the frequency of the characteristic X-radiation, then

$$\sqrt{\nu} \propto (Z - b)$$

or $$\nu^{1/2} = k(Z - b)$$

where Z is atomic number of the target materials and k and b are constants of which the value for K series, L series etc. are different.. The constant b is known as *screening constant.* This is known as **Moseley's law** and is stated as follows:

The frequency of spectral line in X-ray spectrum varies as the square of the atomic number of the element emitting it.

It is a very good rule to take the screening constant b as the number of extra nuclear electrons inside the orbit to which the electron falls in radiating together with $\frac{1}{2}(n-1)$, where n is the number of electrons in the orbit. For hydrogen $b = 0$, for helium $b = \frac{1}{2}$.

Taking the case of copper for K-lines, $b = \frac{1}{2}$, as there are two electrons in the innermost orbit. In the second orbit there are 8 electrons, so that in falling into this orbit, a L line would be produced and

$$b = 2 + \frac{1}{2}(8 - 1) = 5.5$$

10.12 DERIVATION OF MOSELEY'S LAW ACCORDING TO BOHR'S THEORY

According to Bohr's theory the energy of an electron in the nth orbit of a hydrogen like atom of atomic number Z is given by

$$E_n = -\frac{mZ^2e^4}{8\varepsilon_0^2h^2}\cdot\frac{1}{n^2}$$

For states with principal quantum numbers n_1 and n_2, the corresponding energy of the electron is given by

$$E_{n1} = -\frac{mZ^2e^4}{8\varepsilon_0^2h^2}\frac{1}{n_1^2}$$

and $$E_{n2} = -\frac{mZ^2e^4}{8\varepsilon_0^2h^2}\frac{1}{n_2^2}$$

If we take into consideration the effect of electrons which screen the positive charge on the nucleus, then Z is reduced to $(Z - b)$ where b is a constant. Taking the value of $b = b_1$ for the the orbit with principal quantum number n_1 and $b = b_2$ for the orbit with principal quantum number n_2, we have

$$E_{n1} = -\frac{me^4(Z-b_1)^2}{8\varepsilon_0^2h^2}\cdot\frac{1}{n_1^2}$$

and $$E_{n2} = -\frac{me^4(Z-b_2)^2}{8\varepsilon_0^2h^2}\cdot\frac{1}{n_2^2}$$

$\therefore$ Energy difference $$E_{n1} - E_{n2} = -\frac{me^4}{8\varepsilon_0^2h^2}\left[\frac{(Z-b_2)^2}{n_2^2} - \frac{(Z-b_1)^2}{n_1^2}\right]$$

When Z has a very high value $b_1 = b_2' = b$, then

$$E_{n1} - E_{n2} = \Delta E = \frac{me^4 (Z-b)^2}{8\varepsilon_0^2 h^2}\left[\frac{1}{n_2^2} - \frac{1}{n_1^2}\right]$$

$$\therefore \quad \nu = \frac{\Delta E}{h} = \frac{me^4 (Z-b)^2}{8\varepsilon_0^2 h^3}\left[\frac{1}{n_2^2} - \frac{1}{n_1^2}\right] \qquad ...(i)$$

According to Bohr's theory of spectrum of hydrogen like atoms

$$\frac{1}{\lambda} = RZ^2\left[\frac{1}{n_2^2} - \frac{1}{n_1^2}\right] \qquad ...(ii)$$

where $R = \dfrac{me^4}{8\varepsilon_0^2 ch^3}$ is Rydberg's constant, the value of which is $1.097 \times 10^7\ m^{-1}$. From relation (*ii*) we have

$$\nu = \frac{c}{\lambda} = cRZ^2\left[\frac{1}{n_2^2} - \frac{1}{n_1^2}\right]$$

where $$cR = \frac{me^4}{8\varepsilon_0^2 h^3}$$

Substituting $\dfrac{me^4}{8\varepsilon_0^2 h^3} = cR$ in equation (*i*), we get

$$\nu = \frac{c}{\lambda} = cR(Z-b)^2\left[\frac{1}{n_2^2} - \frac{1}{n_1^2}\right]$$

Frequency of K_α -line. The frequency ν of the K_α line in Moseley's work is obtained by putting $n_1 = 2$ and $n_2 = 1$ so that the above equation becomes

$$\nu = \frac{c}{\lambda} = cR\,(Z-b)^2\left[\frac{1}{1^2} - \frac{1}{2^2}\right]$$

For K_α line, $b = 1$ as explained on next page.

$$\therefore \quad \nu = Rc(Z-1)^2\left[\frac{1}{1^2} - \frac{1}{2^2}\right] = \frac{3}{4}cR(Z-1)^2$$

In other words, the K_α line in the X-ray spectrum is emitted when the electron jumps from the orbit $n = 2$ to the orbit $n = 1$. In the case of hydrogen, this gives a line in the Lyman series in the ultraviolet region. The factor (Z - 1) in place of Z is due to the fact that when one of the two electrons in the *K* shell is knocked out, the remaining one electron lowers the positive charge on the nucleus from Z to (Z - 1) and the electron jumping from the *L* level to fill the vacancy in the *K* level moves in an electric field of magnitude (Z – 1) *e*.

Thus with $b = 1$, we have

$$\nu = \frac{3}{4}cR\,(Z-1)^2$$

or $$\nu^{1/2} = \left(\frac{3}{4}cR\right)^{1/2}(Z-1)$$

$$= k\,(Z-1) \qquad ...(iii)$$

Thus the value of the constant k for K_α line is given by

$$k\,(\text{for } K_\alpha \text{ line}) = \left(\frac{3}{4}cR\right)^{1/2}$$

$$= \left[\frac{3}{4}(3\times10^{8}\ ms^{-1})\times(1.097\times10^{7}\ m^{-1})\right]^{\frac{1}{2}}$$

$$= 4.97\times10^{7}$$

Relation (iii) may also be put as

$$\nu = k^2\ (Z-1)^2 = a(Z-1)^2$$

$$= a(Z-b)^2 \qquad \text{...(iv)}$$

where $a = 0.247\times10^{16}\ s^{-1}$ and $b = 1$

Thus Eq. *(iv)* gives the approximate relation for K_α -line of an element.

Importance and conclusions. According to this law, it is the atomic number and not the atomic weight of the element which determines its characteristic properties, both physical and chemical. According to Moseley's law, the atomic number is the number of positive units of electric charge on the nucleus of an atom. The atoms must be arranged in the periodic table according to their atomic numbers and not according to their atomic weights. Moseley thus corrected the position of argon, cobalt and tellurium in the periodic table and he found that in the new positions the elements were in places corresponding to their chemical properties. The atomic weight of tellurium (Z = 52) is 127.61 and that of iodine (Z = 53) is 126. 92. Hence in the periodic table, tellurium should precede iodine, though its atomic weight is greater.

Similarly Z = 27 for cobalt and Z = 28 for nickel, but their respective atomic masses are 58.93 and 58.71. The order, on the basis of atomic mass, does not explain the chemical properties of cobalt and nickel, but when these are placed in the periodic table according to the atomic numbers they are in the right place. Moseley also pointed out the existence of gaps in the group of rare earth elements at values Z = 43, 61, 72 and 75. These elements have been subsequently discovered. The first two, technetium and promethium, have no stable isotope and were first produced in the laboratory many years later. The last two, hafnium and rhenium, were isolated in 1920's.

Example 10.14. *Determine wavelength of K_α X-rays emitted by an element having $Z = 79$, $a = 2.468\times10^{15}\ s^{-1}$ and $b = 1$.*

Solution. The frequency of an X-ray spectral line, according to Moseley's law, is given by

$$\nu = a(Z-b)^2$$

Here Z = 79, $\quad a = 2.468\times10^{15}\ s^{-1},\ b = 1$

$$\therefore \quad \nu = 2.468\times10^{15}\ (79-1)^2 = 1.502\times10^{19}\ s^{-1}$$

Hence $\quad \lambda = \dfrac{c}{\nu} = \dfrac{3\times10^{8}}{1.502\times10^{19}} = 0.1997\times10^{-10}\ m$

$$= 0.1997\text{Å}$$

Example 10.15. *Calculate the frequency of K_α line when atomic number of the anticathode is 79. Given $R = 1.097\times10^{7}\ m^{-1}$.*

Solution. The frequency of the K_α line is given by

$$\nu = Rc\ (Z-1)^2\left[\frac{1}{1^2}-\frac{1}{2^2}\right] = \frac{3}{4}cR\ (Z-1)^2$$

Substituting the values, we have

$$\nu = \frac{3\times 3\times 10^8 \times 1.097\times 10^7\ (79-1)^2}{4} = 1.502\times 10^{19}\ s^{-1}$$

Example 10.16. *If the K_α line of Mo (Z = 42) has a wavelength 0.7Å, find the wavelength of corresponding line of Cu (Z = 29).* **(P.U. 2006)**

Solution. The frequency of K_α line is given by

$$\nu = Rc\,(Z-1)^2\left[\frac{1}{1^1}-\frac{1}{2^2}\right] = \frac{3}{4}cR\,(Z-1)^2$$

For Molybdenum Mo, $Z = 42$

$$\therefore \quad \nu_{Mo} = \frac{3}{4}cR\,(42-1)^2 = \frac{3}{4}cR\,41^2$$

For Copper Cu, $\quad Z = 29$

$$\nu_{Cu} = \frac{3}{4}cR\,(29-1)^2 = \frac{3}{4}cR\,28^2$$

Now $\nu = \dfrac{c}{\lambda}$ $\quad\therefore\quad \dfrac{\lambda_{Cu}}{\lambda_{Mo}} = \dfrac{c}{\nu_{Cu}}\times\dfrac{\nu_{Mo}}{c} = \dfrac{\nu_{Mo}}{\nu_{Cu}} = \dfrac{41^2}{28^2}$

Hence $\quad \lambda_{Cu} = \dfrac{41^2}{28^2}\lambda_{Mo} = \dfrac{41^2}{28^2}\times 0.7\text{Å} = 1.5\text{Å}$

Example 10.17. *Calculate the frequency of K_α X-ray line of Pb atom. Given Z for Pb = 82 and $R' = 3.289\times 10^{15}\ s^{-1}$.* **(H.P.U. 2002)**

Solution. The frequency of the K_α line is given by

$$\nu = Rc\,(Z-1)^2\left(\frac{1}{1^2}-\frac{1}{2^2}\right) = \frac{3}{4}cR(Z-1)^2 \quad ...(i)$$

Here $R' = 3.289\times 10^{15}\ s^{-1}$. This value of R' has dimensions of s^{-1}. It is in fact $= cR$ in the above relation. The dimension of R in Eq. (*i*) is m^{-1}. Therefore dimension of $R'\ ms^{-1}\ m^{-1} = s^{-1}$

Hence $\quad \nu = \dfrac{3}{4}R'(Z-1)^2 = \dfrac{3}{4}\times 3.289\times 10^{15}\times(82-1)^2$

$$= 1.618\times 10^{19}\ s^{-1}$$

Example 10.18. *Find the energy in eV and wavelength of K_α X-ray line of Co-27.* **(G.N.D. U., 2001)**

Solution. The frequency of a K_α X-ray spectral line, according to Moseley's law, is given by

$$\nu = \frac{3}{4}Rc(Z-1)^2$$

Here $Z = 27$; $\quad R = 1.097\times 10^7\ m^{-1}$; $c = 3\times 10^8\ ms^{-1}$

$$\therefore \quad \nu = \frac{3}{4}\times 1.097\times 3\times 10^8\ (27-1)^2$$

$$= 1.67\times 10^{18}\ Hz$$

Energy $E = h\nu$, where $h = 6.626\times 10^{-34}\ Js = 4.136\times 10^{-15}\ eVs$

$$\therefore \quad E = 4.136 \times 10^{-15} \times 1.67 \times 10^{18} \text{ eV}$$
$$= 6.907 \times 10^{3} \text{ eV} = 6.907 \text{ KeV}$$

Wavelength $\lambda = \dfrac{c}{\nu} = \dfrac{3 \times 10^{8}}{1.67 \times 10^{18}} = 1.79 \times 10^{-10}$
$$= 1.79 \text{ Å}$$

Example 10.19. *Calculate ionisation potential of K-shell electron of copper. Given Z for copper is 29 and Rydberg's constant for hydrogen* $R = 1.097 \times 10^{7}\ m^{-1}$. **(H.P.U.2000)**

Solution. The frequency of K_α X-ray spectral line according to Moseley's law is given by

$$\nu = \frac{3}{4} Rc(Z-1)^2$$

Here $R = 1.097 \times 10^{7}\ m^{-1}$; $Z = 29$; $c = 3 \times 10^{8}\ ms^{-1}$

$$\therefore \quad \nu = \frac{3}{4} \times 1.097 \times 3 \times 10^{8}\ (29-1)^2$$
$$= 1.936 \times 10^{18}\ Hz$$

Energy of K_α line $E = h\nu$ and $h = 6.626 \times 10^{-34}$ Js

$$= 4.136 \times 10^{-15}\ eVs$$

$$\therefore \quad E = 4.136 \times 10^{-15} \times 1.936 \times 10^{18} = 8.011 \times 10^{3} \text{ eV}$$
$$= 8.011 \text{ KeV}$$

Now energy of K_α line, $E = E_k - E_L$ where E_k is the ionisation energy of K-shell

$$\therefore \quad E_K = E + E_L$$

Taking $E_L = 0.931$ KeV, we have

$$E_K = 8.011 + 0.931 = 8.942 \text{ KeV}$$

Example 10.20. *The wavelength of* $\boldsymbol{K_\alpha}$ *line in copper is 1.54 Å. Calculate the ionisation potential of K-shell electron in copper. Given the energy of L-shell is 0.923 KeV.* **(P.U., 2007)**

Solution. Wavelength of K_α line = 1.54Å $= 1.54 \times 10^{-10}$ m

$\therefore$ Energy of X-rays corresponding to wavelength of K_α-line

$$E_\alpha = h\nu = \frac{hc}{\lambda} = \frac{6.626 \times 10^{-34} \times 3 \times 10^{8}}{1.54 \times 10^{-10}} \text{ J}$$

$$= \frac{6.626 \times 10^{-34} \times 3 \times 10^{8}}{1.54 \times 10^{-10} \times 1.6 \times 10^{-19}} = 8.067 \times 10^{3}\ eV$$
$$= 8.067 \text{ KeV}$$

The K_α line is obtained when the electron jumps from L-shell to K-shell

$$\therefore \quad E_\alpha = E_L - E_K \text{ or } E_K = E_\alpha + E_L$$
$$= (8.067 + 0.923)\ 10^{3} \text{ eV} = 8.99 \text{ KeV}$$

$\therefore$ Ionisation energy of K-shell copper electron $= 8.99 \times 10^{3}$ eV

The ionisation energy in eV = Ionisation potential in volts (numerically)

$\therefore$ Ionisation potential of K-shell copper electron

$$= 8.99 \times 10^{3} \text{ volt.}$$

Example 10.21. *The wavelength of L_α X-ray line of Platinum (Z = 78) is 1.321 Å. An unknown substance emits L_α X-rays of wavelength 4.174 Å. Calculate the atomic number of the unknown substance b = 7.4 for L_α line.*

Solution. The frequency of an X-ray spectral line, according to Moseley's law, is given by

$$\nu = a(Z - b)^2$$

Let ν_1 and ν_2 be the frequencies of the L_α line of Platinum and the unknown substance respectively, then

$$\nu_1 = a\,(Z_1 - b)^2 \text{ and } \nu_2 = (Z_2 - b)^2$$

or
$$\frac{\nu_1}{\nu_2} = \frac{(Z_1 - b)^2}{(Z_2 - b)^2} \text{ or } \sqrt{\frac{\nu_1}{\nu_2}} = \frac{Z_1 - b}{Z_2 - b}$$

$$\therefore \quad (Z_2 - b) = (Z_1 - b)\sqrt{\frac{\nu_2}{\nu_1}}$$

But
$$\nu_2 = \frac{c}{\lambda_2} \text{ and } \nu_1 = \frac{c}{\lambda_1} \quad \therefore \quad \frac{\nu_2}{\nu_1} = \frac{\lambda_1}{\lambda_2}$$

Hence $(Z_2 - b) = (Z_1 - b)\sqrt{\dfrac{\lambda_1}{\lambda_2}}$

Give $b = 7.4$, $Z_1 = 78$; $\lambda_1 = 1.321$ Å and $\lambda_2 = 4.174$ Å

$$\therefore \quad (Z_2 - 7.4) = (78 - 7.4)\sqrt{\frac{1.321}{4.174}} = 70.6 \times 0.5625 = 39.72$$

or
$$Z_2 = 7.4 + 39.72 = 47.12$$

Thus the atomic number of the unkonwn substance is 47.12 or 47.

Example 10.22. *When 0.5 Å X-rays strike a material, the photo-electrons from the K-shell are observed to move in a circle of radius 23 mm in a normal magnetic field of 2×10^{-2} T. What is the binding energy of the K-shell electron ?* ***(Nagpur U. 2002)***

Solution. Let the velocity of the photo-electron from the *K*-shell be *v*. As it enters the magnetic field $B = 2 \times 10^{-2}$ T normal to it, the force acting on the electron = *Bev.*

This force acts at right angles to the direction of motion of the electron. Therefore, the electron moves in a circular path of radius *R*, given by

$$Bev = \frac{mv^2}{R}$$

$$\therefore \quad v = \frac{BeR}{m} = \frac{2 \times 10^{-2} \times 1.6 \times 10^{-19} \times 23 \times 10^{-3}}{9 \times 10^{-31}} = 8.17 \times 10^7 \; ms^{-1}$$

The kinetic energy of the photo electron $= \dfrac{1}{2}mv^2 = \dfrac{1}{2} \times 9 \times 10^{-31} \times (8.17 \times 10^7)^2$

$$= 300 \times 10^{-17} J = \frac{300 \times 10^{-17}}{1.6 \times 10^{-16}} = 18.75 \text{ KeV}$$

Energy of the incident photon $= h\nu = \dfrac{hc}{\lambda} = \dfrac{6.62 \times 3 \times 10^8 \times 10^{-34}}{0.5 \times 10^{-10}} = 39.72 \times 10^{-16}$ J

$$= \frac{39.72 \times 10^{-16}}{1.6 \times 10^{-19}} = 24.8 \text{ KeV}$$

$\therefore$ Binding energy of K-shell election = Energy of incident photon – Kinetic energy of photoelectron

$$= 24.8 - 18.75 = 6.05 \text{ KeV}$$

10.13 DOUBLET STRUCTURE OF X-RAY SPECTRA (FINE STRUCTURE)

Moseley's observations about characteristic X-ray spectra were made in 1913. As spectrometers with higher resolving powers were made, most of the lines in the X-ray spectrum were found to consist of two or more lines. For example, the K_α line was found to be *doublet i.e.* two very close lines, for all elements beyond Sodium (At. No. 11). Similarly K_β line also showed a complex structure.

Thus each X-ray line is found to consist of two or more lines very close together. This is known as fine structure of X-ray lines.

To explain the emission of characteristic X-ray spectra, we considered the total energy difference only due to change of principal quantum number *n.* But to have a complete picture the effect of energy of an electron is determined by its configuration, which depends upon the principal as well as the orbital quantum number. Further, to take into account the sufficiently large influence of spin quantum number, we take the vector sum of the orbital and spin quantum number *j*- the total angular momentum quantum number.

In the K-shell ($n = 1$, $l = 0$) the orbital is spherically symmetrical and only one value of j is possible *i.e.* $j = l + s = +\frac{1}{2}$. The number of levels in K-shell $(2l + 1) = 1$ *i.e.* there is only one *level*. In spectroscopic notation, it is denoted as $1s_{1/2}$. For the L-shell ($n = 2$, $l = 1, 0$) two values of j are possible $j = \frac{3}{2}$ and $j = \frac{1}{2}$ corresponding to $l = 1$ *i.e.* $(l + s) = \left(1 + \frac{1}{2}\right) = \frac{3}{2}$ and $(l - s) = \left(1 - \frac{1}{2}\right) = \frac{1}{2}$ and one value $j = \frac{1}{2}$ corresponding to $l = 0$ *i.e.* $(l + s) = +\frac{1}{2}$. Thus there are *three sub-shells* $L_1\left(n = 2, l = 0, j = +\frac{1}{2}\right)$; $L_{II}\left(n = 2, l = 1, j = \frac{1}{2}\right)$ and $L_{III}\left(n = 2, l = 1, j = \frac{3}{2}\right)$. In spectroscopic notation, these are denoted as $L_1 = 2s_{1/2}$; $L_{II} = 2p_{1/2}$ $L_{III} = 2p_{3/2}$. For the M-shell ($n = 3$, $l = 2, 1, 0$) there are *five sub-shells,* $M_I\left(n = 3, l = 0, j = \frac{1}{2}\right) = 3s_{1/2}$, $M_{II}\left(n = 3, l = 1, j = \frac{1}{2}\right) = 3p_{1/2}$, $M_{III}\left(n = 3, l = 1, j = \frac{3}{2}\right) = 3p_{3/2}$, $M_{IV}\left(n = 3, l = 2, j = \frac{3}{2}\right) = 3d_{3/2}$ and $M_V\left(n = 3, l = 2, j = \frac{5}{2}\right) = 3d_{5/2}$

Similarly for *N-shell* there are *seven sub-shells:*

$N_I = 4s_{1/2}, N_{II} = 4p_{1/2}, N_{III} = 4p_{3/2}, N_{IV} = 4d_{3/2}, N_V = 4d_{5/2}, N_{VI} = 4f_{5/2}, N_{VII} = 4f_{7/2}$.

It should be noted that each sub-shell can have a maximum of $(2j + 1)$ electrons.

This is known as the *multiplicity* of the shell and arises from the fact that the total angular momentum vector can be oriented in $(2j + 1)$ directions, according to the rule of spatial quantisation.

Thus, the multiplicity of the K-shell $\left(j = \frac{1}{2}\right) = 2$ for, $L_1\left(j = \frac{1}{2}\right) = 2$, $L_{II}\left(j = \frac{1}{2}\right) = 2$,

$L_{III}\left(j=\frac{3}{2}\right)=4,\ M_I\left(j=\frac{1}{2}\right)=2,\ M_{II}\left(J=\frac{1}{2}\right)=2,\ M_{III}\left(j=\frac{3}{2}\right)=4,\ M_{IV}\left(j=\frac{3}{2}\right)=4,$

$M_V\left(j=\frac{5}{2}\right)=6$ and so on for N-sub-shells.

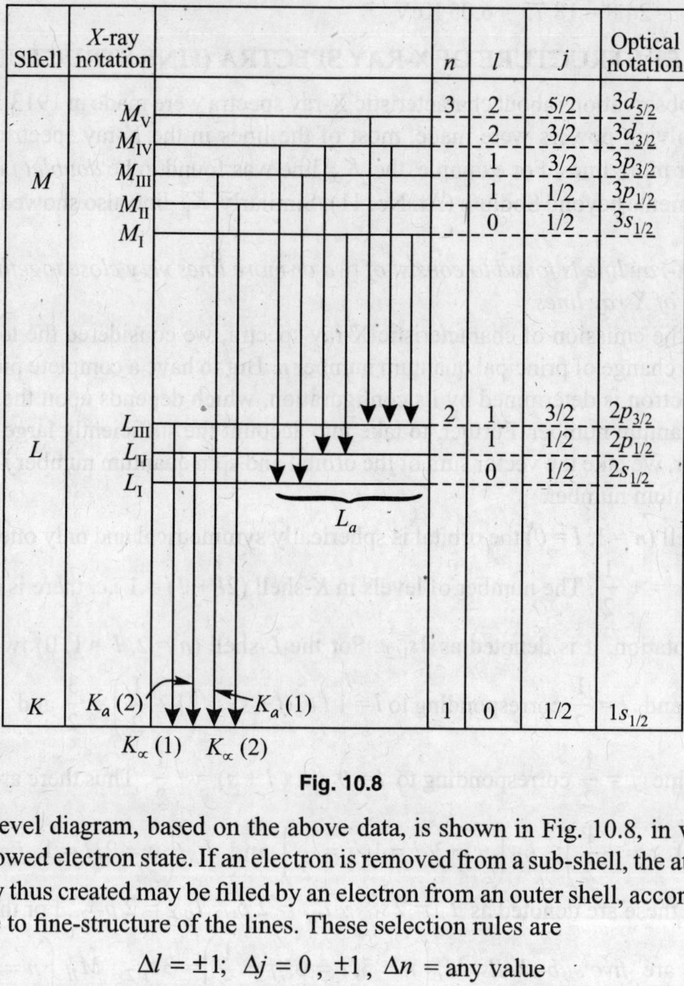

Fig. 10.8

An energy level diagram, based on the above data, is shown in Fig. 10.8, in which each level represents an allowed electron state. If an electron is removed from a sub-shell, the atom gets ionised.

The vacancy thus created may be filled by an electron from an outer shell, according to selection rules, giving rise to fine-structure of the lines. These selection rules are

$$\Delta l = \pm 1;\ \Delta j = 0,\ \pm 1,\ \Delta n = \text{any value}$$

K_{α_1} K_{α_2} K_{β_1} K_{β_2} lines. K_α line which is produced by the transition of an electron from L-level to K- level, will now consist of two lines K_α (1) and K_α (2) for transitions from L_{III} or $2p_{3/2}$ to K or $1s_{1/2}$ and from L_{II} or $2p_{1/2}$ to K or $1s_{1/2}$, level according to $l = 1 \rightarrow l = 0$ or $\Delta l = +1$ and $j=\frac{3}{2} \rightarrow j=\frac{1}{2}$ or $\Delta j = +1$ and $l = 1 \rightarrow l = 0$ or $\Delta l = +1$ and $j=\frac{1}{2} \rightarrow j=\frac{1}{2}$ or $\Delta j = 0$. The transition from L_1 to K (or $2s_{1/2}$ to $1s_{1/2}$) involving $l = 0 \rightarrow l = 0$ or $\Delta l = 0$ is not allowed. Thus K_α line is a doublet. K_β and K_γ lines also consist of two very close lines each. Some allowed transitions giving rise to fine structure in K_α, K_β and L_α lines is shown in Fig. 10.8. It is clear from the figure that K_{β_1} line is produced by transition from M_{III} or $3p_{3/2}$ to K or $1s_{1/2}$ level and K_{β_2} line is produced by transition from M_{II} or $3p_{1/2}$ to K or $1s_{1/2}$ level.

There are other weak lines observed in X-ray spectra which cannot be expressed as difference of any two levels given in the figure. These lines are called *non-diagram lines* and generally lie close to one of diagram lines. These are known as *satellite lines* and are produced due to transitions in atoms from which two or more electrons have been removed.

10.14 AUGER EFFECT

Photo-electric effect. *The ejection of an electron from a metal by the action of ultraviolet or visible light is known as photo-electric effect.*

These photo-electrons are the conduction electrons ejected from the outermost shell of the atom. The minimum energy required to eject a photo-electron, known as photo-electric work function or threshold energy, has only one value for a given metal.

When X-rays are used to produce photo-electric effect, it is the inner electrons that are ejected. This electron may be from the *K*-shell, *L*-shell or *M*-shell, depending upon the energy of the incident X-rays. The *K*-shell has the highest binding energy. If, therefore, the incident X-ray photon energy is less than the binding energy of the *K*-shell, then only possibility is that an *L* or *M*- electron is ejected. The photo-electric effect with X-rays, therefore, may have more than one value of work function W_K, W_L, W_M etc. corresponding to *K, L, M* etc. shell.

Auger effect. Suppose when a metal is subjected to an X-ray beam, the X-ray photons knock out an electron from the *K*-shell, thus creating a vacancy in the *K*-shell. Now, an electron from one of the higher shells, say *L, M*,... etc., will jump into the *K*-shell and an X-ray photon, corresponding to one of the characteristic *K*-lines, will be emitted.

If all the *K*-shell vacancies are filled by emissions of X-ray photons, then the number of *K*-electrons knocked out will be equal to the number of *K*-photons emitted. But it has been observed experimentally that the number of *K*-photons emitted is less than the *K*-shell electrons knocked out. In other words, the *fluorescence yield,* defined as

$$E = \left(\frac{n_p}{n_e}\right)$$

where n_p is the number of *K*-photons emitted and n_e the number of *K*-electrons knocked out (*i.e.* the number of *K*- shell vacancies) is *less than one*, ranging from 0.1 in case of light nuclei to about 0.95 for uranium.

Auger electron and Auger transition. The explanation of the observations given above was provided by Auger in 1925. When the incident X-ray photon knocks out an electron from one of the inner orbits of an atom, say the *K*-electron, there is a vacancy in the *K*-shell. According to Auger, the ejection of the *K*-shell electron is often accompanied by an alternate process, in which the absorbing atom can get rid of its excitation energy, resulting from the photo-electric effect in an inner (say *K*) shell by dropping an electron from an outer (say *L*) shell into the vacancy in the inner shell and simultaneously ejecting another electron usually from the same outer shell, producing an atom in a state of double ionisation. This emitted electron is called *Auger electron* and this transition is known as *Auger transition.*

In this process, the excitation energy of the atom is utilised to free and eject an electron in the outer shell, while the inner shell gets filled up. The ejection of the electron from the outer shell is not through the photo-electric absorption of any photon emitted by the atom, as the vacancy in the inner shell gets filled up but is a part of the process of re-adjustment of the atom. For example, the vacancy in the *K*-shell may be filled up by an electron from L_I sub-shell and the electron from the L_{II} sub-shell may be knocked out, with the result that there are two vacancies in the *L*-shell. *This non-radiative process is known as Auger effect.*

Thus the de-excitation of the atom may be accompanied either by the emission of a characteristic X-ray photon or an Auger electron. The two processes taken together fully account for the number of vacancies in the K-shell. Auger effect is similar to the process known as *auto-ionisation* in optical spectroscopy and is also called *internal conversion.* The Auger effect competes strongly with X-ray fluorescence in de-exciting the atom. In some cases, more than one Auger electron may be emitted.

Auger electron energy. Under certain conditions, an atom in a state of single ionisation spontaneously undergoes Auger transition, leaving the atom in a state of double ionisation.

When an electron from the L-shell jumps into the K-shell and a photon of energy $h\nu$ is emitted, then

$$h\nu = E_K - E_L$$

where E_K and E_L are the ionisation (binding) energies of the atom for the K and L shell respectively. If E_{L2} is the ionisation (binding) energy of the atom with two L- electrons missing after Auger electron is given by

$$E = \frac{1}{2} mv^2 = h\nu - (E_{L2} - E_L)$$

$$= (E_K - E_L) - (E_{L2} - E_L)$$

$$= E_K - E_{L2} = E_K - 2E_L \qquad [\because E_{L2} = 2E_L]$$

For example in Zn (Z = 30) the ionisation (binding) energy of K-shell = 9.659 KeV and L- shell = 1.021 KeV, the kinetic energy of the Auger electron emitted from the L-shell

$$= E_K - 2E_L = 9.659 - 2(1.021)$$

$$= 7.617 \text{ KeV}$$

Observations of Auger. Auger made an experimental study of photo-electron and Auger electron tracks in a mixture of 95 parts Hydrogen and 5 parts Argon and noted the following points.

(*i*) The photo-electron and accompanying Auger electron tracks have a common origin.

(*ii*) The length of the photo-electron track varies inversely as the wavelength of the X-ray photon, but the length of the Auger electron track is independent of wavelength. The length of the Auger track increases with the atomic number of the element.

(*iii*) The direction of the Auger electron track was random and independent of the direction of photo-electron track.

(*iv*) All the photo-electron tracks are not accompanied by Auger electron tracks.

(*b*) An Auger electron cannot be emitted with visible light. Visible light cannot knock out an electron from K, L or M shell of the atom, a prerequisite for emission of Auger electron.

10.15 ABSORPTION OF X-RAYS

X-rays can pass through matter. The intensity is reduced in the process, the reduction depending upon the nature of the material and frequency of X-rays. High frequency X-rays are generally absorbed less and are known as *hard* X-rays. The low frequency X-rays are absorbed more and are known as *soft* X-rays.

When a beam of X-rays is allowed to pass through a material like aluminium, the part of intensity of X-rays is absorbed and remaining part is transmitted. The transmitted intensity is measured with the help of ionization chamber.

Intensity of X-rays. *The intensity of X-ray beam, at any point, is defined as the amount of energy carried per second per unit area, perpendicular to the direction of flow of energy.*

Let I_0 be the intensity of X-ray beam, incident normally and dI the further fall in its intensity on traversing an additional distance dx then

$$\text{Rate of decrease of intensity} = -\frac{dI}{dx}$$

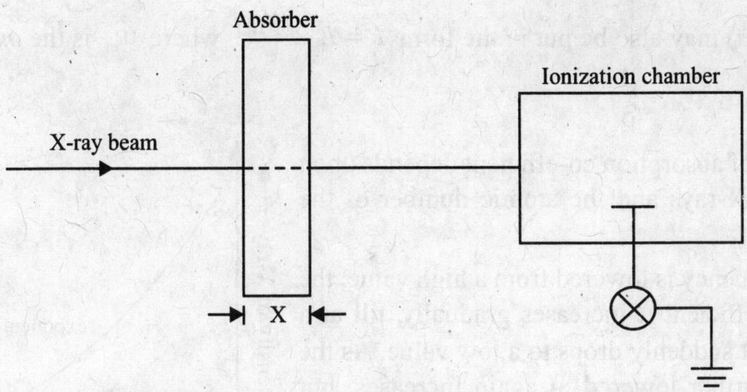

Fig. 10.9: Absorption of X-rays

The rate of decrease of intensity at any point is directly proportional to the intensity at that point

$$\therefore \qquad \frac{dI}{dx} \propto I$$

or
$$\frac{dI}{dx} = -\mu I \qquad \text{...}(i)$$

where μ is a constant of proportionality and is known as *linear absorption co-efficient.*

Eq. *(i)* can be put in the form

$$\frac{dI}{I} = -\mu\, dx$$

Integrating both sides, we get

$$\int \frac{dI}{I} = -\int \mu\, dx$$

or
$$\ln I = -\mu x + C \qquad \text{...}(ii)$$

where C is the constant of integration, the value of which can be determined by the application of boundary conditions

At $x = 0; I = I_0$

$$\therefore \qquad C = \ln I_0$$

Substituting in Eq. *(ii)*, we have

$$\ln I = -\mu x + \ln I_0$$

or
$$\ln I - \ln I_0 = -\mu x$$

or
$$\ln \frac{I}{I_0} = -\mu x$$

$$\therefore \qquad I = I_0\, e^{-\mu x} \qquad \text{...}(iii)$$

This equation shows that the absorption of X-rays through a given material follows the exponential law as shown in Fig. 10.10.

If
$$x = \frac{1}{\mu},\; I = \frac{I_0}{e}$$

The linear absorption co-efficient of the material is, therefore, defined as the reciprocal of the thickness after which the intensity of X-rays falls to 1/e of its original intensity.

Relation (*iii*) may also be put in the form $I = I_0\ e^{-\mu_m \rho x}$ where μ_m is the *mass absorption co-efficient,* given by $\mu_m = \dfrac{\mu}{\rho}$

The value of absorption co-efficient depends upon wavelength of X-rays and the atomic number of the material.

As the frequency is lowered from a high value, the absorption co-efficient μ increases gradually, till at a frequency ν_K it suddenly drops to a low value. As the frequency is further lowered, μ again increases, but drops again at a frequency μ_{L_1}. The edges in the plot of μ against frequency are called *K*-absorption edge L_1, absorption edge etc.

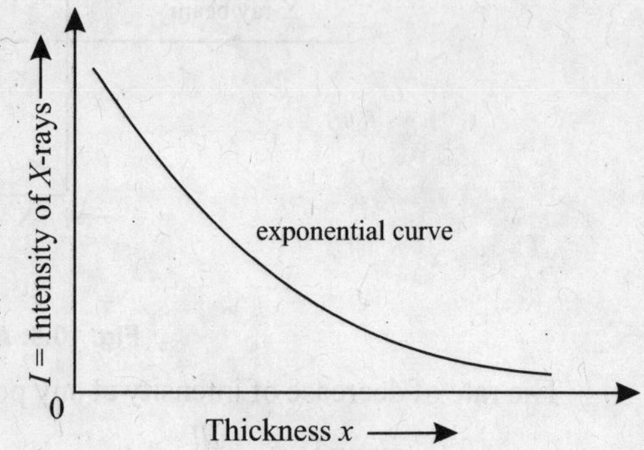

Fig. 10.10

10.16 CHARACTERISTIC ABSORPTION SPECTRUM

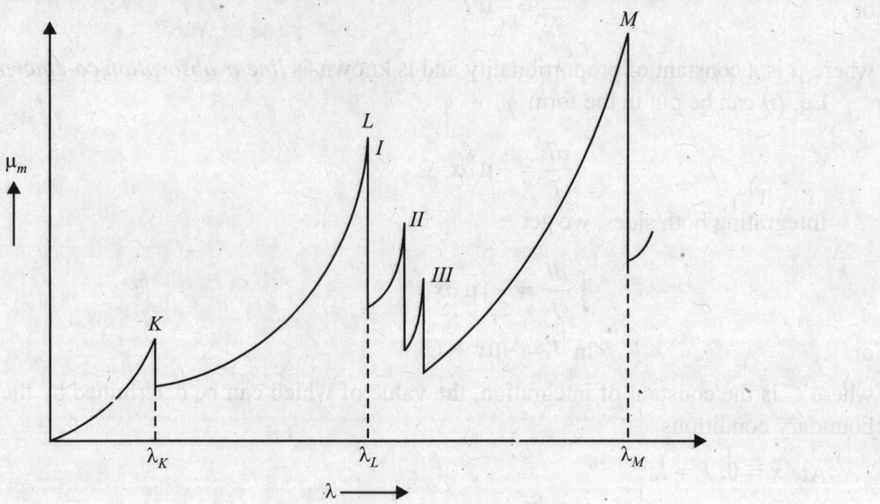

Fig. 10.11: X-ray absorption spectrum

The mass aborption coefficient μ_m is the characteristic of absorbing material. In Fig. 10.11, μ*m* is plotted against λ of X-rays. To study absorption spectrum of an element, the continuous X-ray spectrum is used. Fig. 10.11 shows discontinuities at λ_K, λ_L, λ_M etc. They are called ***absorption edges***. When the position of wavelength λ_K is reached, μ_m drops studdenly, it again increases and gives further absorption edges at λ_{LI}, λ_{LII}, λ_{LIII} etc.

The first break occurs at the wavelength λ_K which is called K absorption limit. The others are at λ_L, (λ_{LI}, λ_{LII}, λ_{LIII}) which are called as L absorption limit and so on. It has been observed that the absorption discontinuities λ_K, λ_L, λ_M etc, are the chareacteristic of the absorbing element. The incident X-rays with the wavelengths shorter than that of a given discontinuity are absorbed by the element to a greater extent than the rays with the wavelengths larger than the critical values λ_K, λ_L, λ_M etc.

The characteristic absorption of X-rays is due to the interaction with the electrons in inner shells of the absorbing atom. Thus, absorption lines are not at the same frequencies corresponding to charcteristic emission lines. Consider an example of K_α radiation : for K_α absorption, there must be transfer of electron from K to L shell but there is no vacancy in the L-shell. Hence there is no transition of electron. Similarly, for absorption of L_α radiation, transition of electron must be from L to M shell, but there is no vacancy in M-shell and so on.

The absorption discontinuities may be explained as follows : $E_K = h\nu_K$ is the energy required to knock out the electron from K-shell. If the frequency of incident X-rays ν is equal to or greater than ν_K then it can eject the electron from K-shell. The energy $h\nu_K$ is utilized in ejection and remaining energy becomes K.E. of ejected electron. If wavelength of incident X-rays is increased, ν becomes lesser than ν_K. Hence there will be no ejection of electron and abosorption will drop suddenly. This is first absorption discontinuity at the wavelength longer than λ_K. On further increase in wavelength, other absorption discontinuities occur. At other discontinuities, λ becomes equal or greater than λ_L, λ_M etc., at which $h\nu_L = E_L$, $h\nu_M = E_M$. As there are three sublevels of L, the absorption edges are three, L_I, L_{II} and L_{III}. Similarly five sublevels of M at which five discontinuities may occur.

As the frequency of absorption edges gives the energy required to knock out the electron from the shell, the difference of ν values of absorption edges give ν values of emission lines in X-ray spectra.

Example 10.23. *The intensity of X-ray beam falls to 25% of its value after penetrating a target 1 cm thick. Calculate the absorption co-efficient of the target.* **(H.P.U. 2001)**

Solution. Here $\frac{I}{I_0} = 0.25; x = 1cm = 0.01\text{m}$. If μ is the absorption co-efficient, then

$$\frac{I}{I_0} = e^{-\mu x} \quad \text{or} \quad 0.25 = e^{-\mu \times .01}$$

$\therefore$ $\quad \ln 0.25 = -0.01\,\mu$

or $\quad -1.3863 = -0.01\,\mu$

or $\quad \mu = \frac{1.3863}{.01} = 138.63\ m^{-1}$

Example 10.24. *The absorption co-efficient of a material is 16.2 m^{-1} for K_a radiation of tungsten. Calculate the percentage of intensity of this line that passes through a plate of thickness 0.184 m.* **(Nagpur 2011)**

Solution. Given $\mu = 16.2\ \text{m}^{-1}, x = 0.184$ m

We know that $\quad I = I_0 e^{-\mu x}$

or $\quad \frac{I}{I_0} = e^{-\mu x} = e^{-(16.2 \times 0.184)}$

$$= 0.05$$
$$= 5\ \%$$

Example 10.25. *An X-ray tube produces 0.1 Å X-rays. Find the accelerating voltage.*

Solution. Given $\lambda = 0.1$ Å

Using Duane-Hunt law, $A_{min} = \frac{12410\text{Å}}{V}$

$$\therefore \quad V = \frac{12410\text{Å}}{\lambda_{min}} = \frac{12410\ \text{Å}}{0.1\ \text{Å}} = 124100\ \text{volt}$$

$$= 124.1\ \text{kV}.$$

EXERCISE

LONG QUESTIONS

1. (a) What are X-rays? What is the wavelength range of X-rays? **(P.U. 2000)**

(b) Describe the construction and working of a Coolidge tube. How can you control (i) the intensity and (ii) the quality of X-rays? Give the properties and uses of X-rays. What are hard and soft X-rays ?

(Pbi. U. 2004; H.P.U. 2002, 2000 G.N.D.U., 2001; P.U. 2001; Burd. U., 2000)

(c) Why should anticathode have high atomic number and high melting point ? **(H.P.U. 2001)**

2. (a) Discuss the origin and mechanism of production of the continuous X-ray spectra. What is the source of energy of photon of continuous X-rays? Show that the lowest wave-length limit of continuous X-ray spectra is inversely proportional to the accelerating potential of the X-ray tube.

(b) Draw four graphs of relative intensity of continuous spectra versus wavelength of X-rays and show that λ_{min} is proportional to $\frac{1}{V}$. **(H. P. U. 2000; P.U. 2007; Pbi. U. 2006, 2004, 2003, 2000; G. N. D. U. 2001; Meerut. U. 2000)**

3. (a) State the difference between soft and hard X-rays. **(Nagpur U. 2011, G..N.D.U. 2008)**

(b) Distinguish between continuous and characteristic (or discrete) X-ray spectra. Why is the characteristic spectra so called? How is the production of characteristic X-ray spectra accounted for? Discuss the transition for K and L series.

(H.P.U. 2003, 2001; G.N.D.U. 2001; Bang. U. 2001; Pbi.U. 2004, 2003, 2002, 2000; Meerut. U. 2001; Gauhati U. 2000, P.U. 2008, 2006, 2000; Cal.U. 2002,)

4. (a) Both line and continuous spectrum emitted by the same target are of different origin. Explain. **(Bang. U. 2003)**

(b) What is the difference between a photon of continuous X-rays and a photon of characteristic X-rays? **(P.U. 2006)**

5. (a) What is Bremsstrahlung? **(Pbi. U. 2001; P.U., 2000)**

(b) Explain the difference in origin of X-rays and visible light. **(Pbi. U., 2001)**

(c) Is it appropriate to regard X-ray production as the inverse of photo-electric effect? **(H.P.U., 1994; P.U. 1999, 1990)**

6. What is the difference between optical spectra and X-ray spectra? Hence discuss their grouping into series. **(G.N.D.U. 2003, 2002; Meerut U. 2001)**

7. (a) Describe Moseley's work on X-rays. State and explain Moseley's law. Show it graphically.

(b) Derive it on the basis of Bohr's theory and state and discuss the approximate relation between frequency of K_α line of X-rays of an element.

(c) Discuss the importance of Moseley's observations of X-ray spectra of different elements. What conclusions were drawn by him?

(Pbi. U. 2008, 2003, 2001, 2000; Luck. U. 2001; G.N.D.U. 2007, 2005, 2004, 2000; H.P.U. 2000; Bang. U. 2001; Gauhati. U. 2000; P.U. 2007, 2005, 2001; Cal. U. 2002)

8. (a) Explain daublet structure of X-ray spectra.

(b) Discuss the fine structure of X-ray spectra. How can it be explained? Indicate the origin of $K_{\alpha 1}$, $K_{\alpha 2}$ and $K_{\beta 1}$ and $K_{\beta 2}$ lines in the energy level diagram.

9. (*a*) Compare photo-electric effect with Auger effect. Explain Auger effect. What is Auger transition and Auger electron? Give the observations made by Auger in the study of photo-electrons emitted by X-rays. ***(H.P.U. 2000 ; Pbi U. 2008, 2006, 2005, 2000 P.U. 2008, 2007, 2005, 2001; G.N.D.U. 2007, 2004)***

(*b*) Can an Auger electron be emitted when a metal is bombarded with visible light? ***(P.U. 2006)***

10. An X-ray beam of intensity I_0 is incident on a substance. Derive an expression for the intensity of the beam when it has passed through a thickness x of the substance. Define absorption co-efficient. On what factors does it depend?

11. (*a*) Discuss the main features of characteristic X-ray spectra.

(*b*) Show that the intensity of X-ray beam decreases exponentially when it passes through a material of thickness x. ***(Nagpur U. 2011)***

SHORT QUESTIONS

1. State the difference between soft and hard X-rays. ***(Nagpur U. 2011, G. N. D. U. 2008)***

2. State the types of X-ray spectra and explain their origin. ***(G. N. D. U. 2007)***

3. Why discrete X-rays from an atom are called characteristic X-rays ?

4. A target of hydrogen is bombarded with high speed electrons. Will the target emit characteristic X-rays ? ***(Pbi. U. 2002)*** **[Ans. No.]**

5. When a Coolidge tube is operated for some time, it becomes hot. Where does the heat come from? ***(Pbi. U. 2003)***

[**Hints.** In a Coolidge tube, the thermionic electrons emitted by the cathode are accelerated through a large potential difference between the cathode and the anode (or anticathode). When these high energy electrons strike the anticathode a part of the energy is converted into X-rays but a large fraction of the energy of the impinging electrons is converted into heat. As a result the Coolidge tube becomes hot when it is operated for some time.]

MULTIPLE CHOICE QUESTIONS

1. Which of the following has the highest frequency ?

(*a*) visible light (*b*) ultraviolet light (*c*) X-rays (*d*) infra-red rays

2. Wavelength of visible light is more than that of X-rays. What is the velocity of X-ray in vacuum?

(*a*) $< c$ (*b*) $= c$ (*c*) $> c$ (*d*) $< or > c$

3. In a Coolidge tube the electrons are focussed on the target by using

(*a*) a concave cathode (*b*) magnetic lens (*c*) an electron gun

4. The quality of X-rays is determined by

(*a*) filament current (*b*) filament voltage

(*c*) potential difference between the cathode and the anode

(*d*) size of the cathode

5. The intensity of X-rays is determined by

(*a*) filament current (*b*) filament voltage

(*c*) potential difference between the cathode and the anode

(*d*) size of the cathode

6. On increasing the number of electrons striking the anode of an X-ray tube, which of the following characteristics of X-rays increase?
   (*a*) frequency (*b*) intensity (*c*) wavelength (*d*) quality
7. An X-ray tube operates at 2000 kV. What is the energy of the most energetic X-ray photon produced?
   (*a*) 2000 eV (*b*) 2000 MeV (*c*) 2000 keV (*d*) 200 keV
8. Continuous X-rays are produced as a result of
   (*a*) transition of electrons of target from lower to higher orbits
   (*b*) transition of electrons of target from higher to lower orbit
   (*c*) loss of kinetic energy of incident electrons
   (*d*) loss of mass of incident electrons.
9. The minimum frequency limit of X-ray spectrum depends upon
   (*a*) nature of target material (*b*) kinetic energy of incident electron
   (*c*) degree of vacuum in Coolidge tube.
10. The frequency of any line in characteristic X-ray spectrum is directly proportional to
   (*a*) atomic number of the target element
   (*b*) square of atomic number of the target element
   (*c*) square root of the atomic number of the target element
   (*d*) is independent of the atomic number of the target element

ANSWERS

1. (*c*)	**2.** (*b*)	**3.** (*c*)	**4.** (*c*)	**5.** (*a*)
6. (*b*)	**7.** (*c*)	**8.** (*c*)	**9.** (*b*)	**10.** (*b*)

SECTION IV

Molecular Physics

11. VIBRATIONAL AND ROTATIONAL SPECTRA
12. RAMAN EFFECT
13. SPECTROSCOPIC TECHNIQUES

VIBRATIONAL AND ROTATIONAL SPECTRA

11
Chapter

11.1 DIATOMIC MOLECULE

A molecule having two atoms separated by a distance which is large as compared to atomic dimensions is called a diatomic molecule. Examples: H_2, O_2, N_2, CO, NO etc. Diatomic molecules are classified as follows.

Homonuclear molecule. *It consists of two identical atoms i.e. the molecule has two identical nuclei*. Such a molecule is called *homonuclear*.

Examples: H_2, O_2, N_2 etc.

Hetro-nuclear molecule. *It consists of two dissimilar atoms i.e. the molecule has two non-identical nuclei*. Such a molecule is called *hetronuclear*.

Examples: NO, HCl, CO etc.

11.2 MOLECULAR SPECTRA

Molecules can remain in the ground state as well as in excited states. *The transition between the allowed energy states of a molecule with the emission or absorption of radiation i.e. photon, gives rise to molecular spectra*. The frequency of the emitted or absorbed photon is given by

$$\nu = \frac{E_2 - E_1}{h} \text{ if } E_2 > E_1 \qquad \text{(absorption)}$$

or

$$\nu = \frac{E_1 - E_2}{h} \text{ if } E_1 > E_2 \qquad \text{(emission)}$$

The energy levels of molecules are more complex than those of the atom. In the case of an atom, we neglect the nuclear motion, because we suppose that the centre of mass is stationary. In a molecule, even when the centre of mass of the molecule remains stationary, various types of motions of individual nuclei can take place.

Cause of excitation. The excited energy levels of a diatomic molecule for a fixed electronic configuration arise from

(*i*) Excitation of the electrons in the molecule from a lower energy state to a higher energy state.

(*ii*) The rotation of the molecule about an axis passing through the centre of mass of the molecule and perpendicular to the line joining the atoms.

(*iii*) The vibration of the atoms about their equilibrium position in the molecule.

Therefore, the total energy of a molecule consists of three parts

(*i*) Energy due to the excitation of electrons in the molecule—Electronic energy = E_e.

(*ii*) Energy due to the rotation of the molecule—rotational energy = E_r

(*iii*) Energy due to the vibration of the molecule—vibrational energy = E_v

∴ Total energy of molecule is

$$E = E_e + E_r + E_v$$

The magnitude of these energies are sufficiently different and

$$E_e > E_v > E_r$$

Energy levels. The energy levels corresponding to electronic energy E_e are known as *electronic energy levels.*

The energy levels corresponding to rotational energy E_r are known as *rotational energy levels.*

The energy levels corresponding to vibration energy E_v are known as *vibrational energy levels.*

11.3 KINDS OF MOLECULAR SPECTRA

Molecular spectra generally fall into three distinct regions of wavelength.

(*i*) **Pure rotational spectrum.** *The transition between the rotational energy states of a molecule with the emission or absorption of a photon gives rise to pure rotational spectrum.*

In this case the electronic configuration of the atoms and the vibrational state of the molecule remain the same. Rotational levels are separated by quite small energy intervals of the order of 10^{-3} eV and pure rotation spectra lie in the *microwave* or far infra red region (10^7 Å).

(*ii*) **Rotational-vibrational spectrum.** *The transition between the vibrational energy states of a molecule with the emission or absorption of a photon, when the electronic configuration of the atoms remains the same, gives rise to rotational vibrational spectrum.*

If the rotational states of the molecule also do not change, we get a *pure vibrational spectrum* Vibrational spectra lies in the *near infrared* region (10^5 Å). The vibrational states are separated by energy intervals of about 0.1 eV.

(*iii*) **Electronic spectrum.** *The transition between two electronic energy states of an atom in a molecule with emission or absorption of photon gives rise to electronic spectrum*. The energy separation between different electronic states is 1eV – 10 eV and the electronic spectra lies in the *visible and ultraviolet* part of the spectrum.

Molecules Show Band Spectrum

The total energy of a molecule in the excited state is given by

$$E = E_e + E_v + E_r$$

where E_e is the *excitation energy of the electrons* in the molecule, E_v the energy due to *vibration* of the molecule and E_r the energy due to *rotation* of the molecule.

The electronic energy levels in a molecule are separated by energy intervals of the order of 1 eV – 10 eV. Each electronic energy level may have a number of vibrational energy levels which are separated by energy intervals of about 0.1 eV. Each vibrational energy level may have a number of rotational energy levels separated by energy intervals of the order of 10^{-3} eV.

Thus each electronic energy level in reality consists of a large number of levels closely spaced together. The transition between different electronic energy levels each of which is a band of a number of levels gives rise to band spectrum rather than line spectrum.

11.4 ROTATIONAL SPECTRA

A diatomic molecule consists of two atoms separated by a distance, which is large as compared to atomic dimensions.

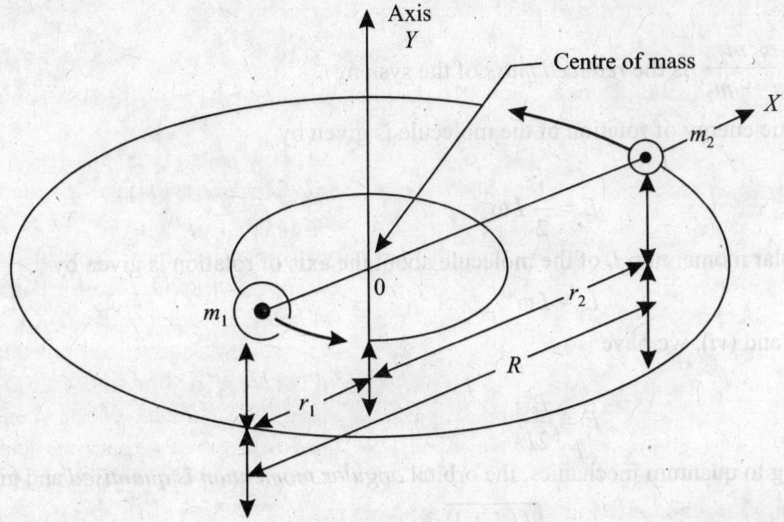

Fig. 11.1

Rotational energy of a diatomic molecule. Consider a diatomic molecule consisting of two atoms of masses m_1 and m_2 separated by a constant distance R, as shown in Fig. 11.1. Suppose the whole molecule formed by these two atoms rotates with an angular velocity ω about an axis passing through the centre of mass of the system and perpendicular to the line joining m_1 and m_2. Let r_1 and r_2 be the distances of the atoms of mass m_1 and m_2 respectively from the centre of mass, then

$$r_1 + r_2 = R \quad \ldots(i)$$

We take the centre of mass as the origin and the line joining the two masses as the X-axis. For a balanced system about the centre of mass

$$m_1 r_1 = m_2 r_2 \quad \ldots(ii)$$

The moment of inertia of the diatomic molecule about an axis passing through the centre of mass is given by

$$I = m_1 r_1^2 + m_2 r_2^2 \quad \ldots(iii)$$

From Eq. (*ii*) $$r_1 = \frac{m_2}{m_1} r_2$$

$$\therefore \quad R = r_1 + r_2 = \frac{m_2}{m_1} r_2 + r_2 = r_2 \left(\frac{m_1 + m_2}{m_1} \right)$$

or $$r_2 = \left(\frac{m_1}{m_1 + m_2} \right) R$$

Similarly $$r_1 = \left(\frac{m_2}{m_1 + m_2} \right) R$$

Substituting the values of r_1 and r_2 in Eq. (*iii*), we get

$$I = \frac{m_1 m_2^2}{(m_1 + m_2)^2} R^2 + \frac{m_2 m_1^2}{(m_1 + m_2)^2} R^2$$

$$= m_1 m_2 \frac{(m_1 + m_2)}{(m_1 + m_2)^2} R^2 = \frac{m_1 m_2}{m_1 + m_2} R^2$$

$$= m' R^2 \qquad ...(iv)$$

where $m' = \frac{m_1 m_2}{m_1 + m_2}$ is the *reduced mass* of the system.

The kinetic energy of rotation of the molecule is given by

$$E = \frac{1}{2} I \omega^2 \qquad ...(v)$$

The angular momentum L of the molecule about the axis of rotation is given by

$$L = I \omega \qquad ...(vi)$$

From (*v*) and (*vi*), we have

$$E = \frac{L^2}{2I} \qquad ...(vii)$$

According to quantum mechanics, the orbital *angular momentum is quantised* and given by

$$L = \sqrt{J(J+1)}\, \hbar$$

where J is known as the *rotational quantum number*, which can take integral values 0, 1, 2, 3...etc

As the energy depends upon quantum number J, we denote is as E_J

$$\therefore \qquad E_J = \frac{L^2}{2I} = J(J+1) \frac{\hbar^2}{2I} \qquad ...(viii)$$

$$= J(J+1) \frac{\hbar^2}{2m' R^2} \qquad ...(ix)$$

Eq. (*viii*) and (*ix*) show that:–

(*i*) A diatomic molecule can not have any arbitrary value of rotational energy. The energy is quantised and is limited by rotational quantum number J.

(*ii*) The rotational energy depends upon moment of inertia I *i.e.* on the mass, shape and size of the molecule.

In deriving relation (*viii*) we have assumed the rotation about an axis perpendicular to the axis of symmetry of diatomic molecule. We have neglected the rotation of the molecule about the axis of symmetry. The reason for the same is that the mass of the atom is concentrated almost entirely at its nucleus, the radius of which is about 10^{-4} of the radius of the atom as a whole. The main contribution to the moment of intertia of a diatomic molecule about the axis of symmetry comes from its electrons. The total mass of the electrons is about 1/4000 of the total molecular mass and these are concentrated in a region whose radius is about half the bond length R. As the allowed rotational energy levels are inversely proportional to I, rotation about the symmetry axis must involve energies of the order of 10^4 times the value of E_J. This means energies of several eV are involved for any rotation about the symmetry axis. The bond energies being of the same order of magnitude, the molecule will very likely dissociate under these conditions.

Frequency of rotational spectra. The rotational spectra arises due to transitions between the rotational energy levels. Just as for atomic spectra, there are *selection rules* for the transitions between rotational energy levels also.

Selection rule. The selection rule is that '*rotational quantum number J must change by one unit, i.e.*

$$\Delta J = \pm 1$$

Suppose the transition takes place from initial energy state $E_i = E_j$ with rotation quantum number $J_i = J$ to final energy state $E_f = E_{j-1}$ with $J_f = J - 1$, then the frequency of the photon emitted is given by

$$h\nu = E_i - E_f = E_J - E_{J-1}$$

$$= J(J+1)\frac{\hbar^2}{2I} - (J-1)J\frac{\hbar^2}{2I}$$

$$= \frac{J\hbar^2}{2I}(J+1-J+1) = \frac{J}{I}\hbar^2 = \frac{Jh^2}{4\pi^2 I}$$

$$\therefore \qquad \nu = J\frac{h}{4\pi^2 I} = J\nu_0 \quad \text{with} \quad J = 0, 1, 2...$$

where $$\nu_0 = \frac{h}{4\pi^2 I}$$

In actual practice rotational spectra is obtained in **absorption** *i.e.* transitions take place form a lower state with initial quantum number J to next higher state with final quantum number $J + 1$

In such a case

$$\nu = \frac{E_{J+1} - E_J}{h} = \frac{h}{4\pi^2 I}(J+1)$$

Representation of rotational energy levels and rotational spectrum. The rotational energy levels for a diatomic molecule and rotational absorption spectra are shown in Fig. 11.2 (*a*) and (*b*).

It should be noted that the energy lebels are not equally spaced however the frequency lines are equally spaced; as discussed below:

Energy level not equally spaced. It is seen from the energy level diagram that the energy separation between successive levels is not the same. The separation goes on increasing with increasing J values. The energy of a rotational level is given by

$$E_J = J(J+1)\frac{\hbar^2}{2I}$$

where J is the rotational quantum number having values 0, 1, 2....

For $\quad J = 0; \quad E_J = E_0 = 0$

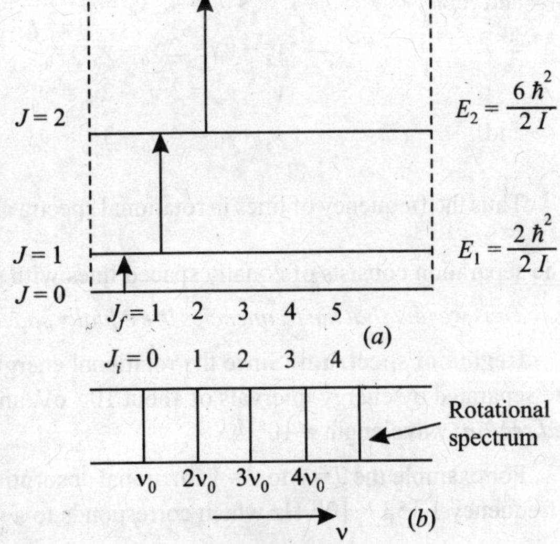

Fig. 11.2

For $$J = 1;\quad E_J = E_1 = \frac{2\hbar^2}{2I}$$

For $$J = 2;\quad E_J = E_2 = \frac{6\hbar^2}{2I}$$

For $$J = 3;\quad E_J = E_3 = \frac{12\hbar^2}{2I}$$

For $$J = 4;\quad E_J = E_4 = \frac{20\hbar^2}{2I}$$

$\therefore$ $$E_1 - E_0 = \frac{2\hbar^2}{2I};\quad E_2 - E_1 = \frac{4\hbar^2}{2I}$$

$$E_3 - E_2 = \frac{6\hbar^2}{2I};\quad E_4 - E_3 = \frac{8\hbar^2}{2I}$$

Thus the difference between two consecutive energy levels is not equal but goes on increasing as the value of J increases.

Frequency lines equally spaced. It is seen from the rotational spectrum diagram that frequency of the lines in the spectrum consist of equally spaced lines with spacing = $\nu_0 = \frac{h}{4\pi^2 I}$. The frequency of a line in rotational spectrum is given by

$$\nu = J\frac{h}{4\pi^2 I} \quad \text{where } J = 0, 1, 2, 3...$$

For $$J = 0,\ \nu = 0 \quad \text{For } J = 1;\ \nu_1 = \frac{h}{4\pi^2 I} = \nu_0$$

For $$J = 2;\ \nu_2 = \frac{2h}{4\pi^2 I} = 2\nu_0;\quad \text{For } J = 3,\ \nu_3 = \frac{3h}{4\pi^2 I} = 3\nu_0$$

and For $$J = 4,\ \nu_4 = \frac{4h}{4\pi^2 I} = 4\nu_0$$

$\therefore$ $$\nu_2 - \nu_1 = 2\nu_0 - \nu_0 = \nu_0$$

$$\nu_3 - \nu_2 = 3\nu_0 - 2\nu_0 = \nu_0$$

and $$\nu_4 - \nu_3 = 4\nu_0 - 3\nu_0 = \nu_0.$$

Thus the frequency of lines in rotational spectra are integral multiples of frequency $\nu_0 = \frac{h}{4\pi^2 I}$.

The separation consists of equally spaced lines with spacing = ν_0

Thus we find that the frequencies are equally spaced, but the energy levels are not equally spaced.

Region of spectrum. Since the rotational energies are very small, the rotational energy levels are separated by energy intervals of about 10^{-3} eV, and the spectra lie in the *micro-wave* or *far infra red region;* wavelength = 10^7 Å.

For example the $J = 0$ to $J = 1$ rotational absorption line in CO (carbon monoxide) molecule has a frequency 1.153×10^{11} Hz which corresponds to a wavelength 2.6×10^{-3} m = 2.6×10^7 Å.

Rotational energy in ground state. The rotational energy of a diatomic molecule is given by

$$E_J = \frac{J(J+1)\hbar^2}{2I}$$

where J is the rotational quantum number. For the ground state $J = 0$. Hence $E_J = 0$.

As the rotational energy in the ground state is zero, it means that *a molecule cannot rotate in the ground state.*

SOLVED EXAMPLES

Example 11.1. *The $J = 0$ to $J = 1$ absorption line in carbon monoxide (CO) occurs at a frequency 1.153×10^{11} Hz; calculate the moment of inertia, bond length, and the lowest energy level of the molecule corresponding to $J = 1$. (Given 1 a.m.u. = 1.66×10^{-27} kg)*

Solution. For absorption spectrum for transition from initial quantum number J to final quantum number $J + 1$, the frequency is given by

$$\nu = \frac{h}{4\pi^2 I}(J+1)$$

Here $J = 0$ and $J + 1 = 1$

$$\therefore \quad \nu = \frac{h}{4\pi^2 I}$$

$$\text{or} \quad I = \frac{h}{4\pi^2 \nu} = \frac{6.63 \times 10^{-34}}{4\pi^2 \times 1.153 \times 10^{11}} = 1.46 \times 10^{-46}\ \text{kg m}^2$$

Bond length. To find the bond length, we first find the reduced mass of the molecule given by

$$m' = \frac{m_1 m_2}{m_1 + m_2} = \frac{16 \times 12}{16 + 12} \times 1.66 \times 10^{-27}\ \text{kg} = 1.14 \times 10^{-26}\ \text{kg}$$

Now $\quad I = m' R^2 \quad$ where R = bond length

$$\text{or} \quad R = \sqrt{\frac{I}{m'}} = \left(\frac{1.46 \times 10^{-46}}{1.14 \times 10^{-26}}\right)^{\frac{1}{2}} = 1.14 \times 10^{-10}\ \text{m.}$$

Energy. For $J = 1$, $E_{J=1} = \dfrac{\hbar^2}{2I} J(J+1) = \dfrac{\hbar^2}{I} = \dfrac{h^2}{4\pi^2 I}$

$$\therefore \quad E_{J=1} = \frac{6.63 \times 6.63 \times 10^{-68}}{4\pi^2 \times 1.46 \times 10^{-46}} = 7.61 \times 10^{-23}\ J$$

$$= 4.76 \times 10^{-4}\ \text{eV}$$

Example 11.2. *Find the energy and angular velocity of a CO molecule in the ground state. Given CO bond length = 0.113 nm and mass of atoms $m_c = 1.99 \times 10^{-26}$ kg and $m_O = 2.66 \times 10^{-26}$ kg.*

Solution. *CO* molecule is a diatomic molecule. As proved in part (*b*) the energy of a diatomic molecule in the ground state ($J = 0$)

$$\therefore \quad E_J = 0$$

Since energy $E_J = \dfrac{1}{2} I \omega^2 = 0$, the angular velocity $\omega = 0$.

Example 11.3. *The moment of inertia of the CO molecule is 1.46 × 10⁻⁴⁶ kg m². Calculate the energy in electron volts in the lowest rotational energy level of the CO molecule.*

(P.U. 2008)

Solution. The lowest rotational energy level is for $J = 1$

$E_{J=1} = 4.76 \times 10^{-4}$ eV. (For calculation refer Example 11.1)

Example 11.4. *Calculate the rotational energy corresponding to J = 1 for hydrogen molecule, given that inter-molecular distance (bond length) is 0.074 nm and mass of H-atom 1.674×10^{-27} kg. (Given $h = 6.63 \times 10^{-34}$ JS.)*

(Kerala. U. 2001)

Solution. Mass of each hydrogen atom $m = 1.674 \times 10^{-27}$ kg

$\therefore$ Reduced mass of hydrogen molecule $m' = \dfrac{m_1 m_2}{m_1 + m_2}$

or
$$m' = \frac{m^2}{2m} = \frac{m}{2} = 0.837 \times 10^{-27} \text{ kg}$$

Inter molecular distance = bond length $R = 0.074$ nm

$= 0.074 \times 10^{-9}\ m$

$\therefore$ Moment of Inertia $I = m'R^2 = 0.837 \times 10^{-27} \times 0.074 \times 0.074 \times 10^{-18}$

$= 4.58 \times 10^{-48}$ kg m²

Rotational quantum number $J = 1$; $h = 6.63 \times 10^{-27}\ Js$

Rotational energy
$$E_J = \frac{J(J+1)\hbar^2}{2I} = \frac{J(J+1)h^2}{4\pi^2 \times 2I}$$

$$= \frac{1 \times 2 \times 6.63 \times 6.63 \times 10^{-68}}{4 \times 3.142 \times 3.142 \times 2 \times 4.58 \times 10^{-48}}$$

$$= 0.243 \times 10^{20}\text{J} = \frac{0.243 \times 10^{-20}}{1.6 \times 10^{-19}} = 0.152 \text{ eV}$$

Example 11.5. *Show that at room temperature the molecules may be in excited rotational states.*

Or

Using example of HCl molecule, show that generally molecules may be in the rotational excited states.

(P.U. 2008)

Solution. The energy E_J corresponding to rotational quantum number J is given by

$$E_J = J(J+1)\frac{\hbar^2}{2I}$$

For molecule like *HCl*, $E_J = 1.3 \times 10^{-3}\, J(J+1)$ eV as proved below.

For *HCl* molecule, mass of hydrogen atom $m_1 = 1.67 \times 10^{-27}$ kg

Mass of chlorine atom $m_2 = 35\, m_1 = 35 \times 1.67 \times 10^{-27}$ kg

The reduced mass of *HCl* molecule $m' = \dfrac{m_1 m_2}{m_1 + m_2}$

$$= \frac{m_1 \times 35m_1}{m_1 + 35m_1} = \frac{35}{36} m_1$$

$$= \frac{35}{36} \times 1.67 \times 10^{-27} \text{ kg}$$

$$R = 1.27 \times 10^{-10} \text{ m}$$

∴ Moment of inertia of HCl molecule

$$I = m' R^2$$

$$= \frac{35}{36} \times 1.67 \times 10^{-27} \times 1.27 \times 1.27 \times 10^{-20} \text{ kg m}^2$$

$$= 2.62 \times 10^{-47} \text{ kg m}^2$$

Now $$E_J = J(J+1)\frac{\hbar^2}{2I} = \frac{J(J+1)\,h^2}{2 \times I \times 4\pi^2} = J(J+1)\frac{6.62 \times 6.62 \times 10^{-68}}{2 \times 2.62 \times 10^{-47} \times 4\pi^2} \text{ Joule}$$

$$= J(J+1)\frac{6.62 \times 6.62 \times 10^{-68}}{2 \times 2.62 \times 10^{-47} \times 4\pi^2 \times 1.6 \times 10^{-19}} \text{ eV}$$

$$= 1.3 \times 10^{-3}\, J(J+1)$$

This is a very small energy. At room temperature the thermal energy of the molecules $\left(\frac{3}{2}kT\right)$ is of the order of 2.5×10^{-2} eV. Thus ordinary collisions between molecules provide necessary energy for excitation. In other words, at room temperature, the molecules may be in a highly excited rotational state.

11.5 VIBRATIONAL SPECTRA

Vibrations of a molecule. While deriving the expression for the energy of rotational levels and the corresponding frequencies of spectral lines for transitions between rotational energy states, it was assumed that the inter-nuclear distance remains constant *i.e.* the diatomic molecule was regarded as a rigid body.

But on being sufficiently excited, the nuclei do not maintain a constant fixed separation. They oscillate back and forth along the line joining them in such a way that the centre of mass remains fixed. The vibratory motion is in addition to the rotation of the molecule.

Expression for vibrational energy levels. Let us consider the vibrational motion of a diatomic molecule. Suppose the two atoms forming the molecule have masses m_1 and m_2 and are separated by an *equilibrium distance* R_0 between them. As the molecule vibrates along the line joining the atoms the distance between them changes. Let at any instant during the vibration the separation between m_1 and m_2 be R *i.e.* there is a change in the distance between the two atoms equal to $(R - R_0)$. As a result, a restoring force comes into play. We suppose that the *restoring force is proportional to the deviation* from the equilibrium position R_0 for oscillations having a *small amplitude*

∴ $$F \propto -(R - R_0)$$

The negative sign indicates that the force is directed towards the centre of mass.

Putting $R - R_0 = x$, we have

$$F \propto -x \quad \text{or} \quad F = -kx \qquad ...(i)$$

where k is called the *force constant* and is a measure of the stiffness of the bond

But $$F = ma = m\frac{d^2 x}{dt^2}$$

Substituting in (*i*), we get

$$m\frac{d^2 x}{dt^2} = -kx$$

or $$\frac{d^2 x}{dt^2} = -\frac{k}{m}x \qquad ...(ii)$$

In the case of a diatomic molecule we have two atoms of mass m_1 and m_2 joined by a bond which may be compared to an elastic spring, as shown in Fig. 11.3 (*a*). As there is no external force acting on the system, the linear momentum remains constant. The oscillations of the two atoms, therefore, do not change the position of centre of mass *i.e.* m_1 and m_2 vibrate back and forth, relative to centre of mass in opposite directions and both reach the extremes of their respective motion at the same time. Such a two body oscillator is equivalent in all respects to the motion of a single harmonic oscillator of mass equal to the reduced mass of the two-body system *i.e.* m in Eq. (*ii*) should be replaced by m'—the reduced mass, given by

$$m' = \frac{m_1 m_2}{m_1 + m_2}$$

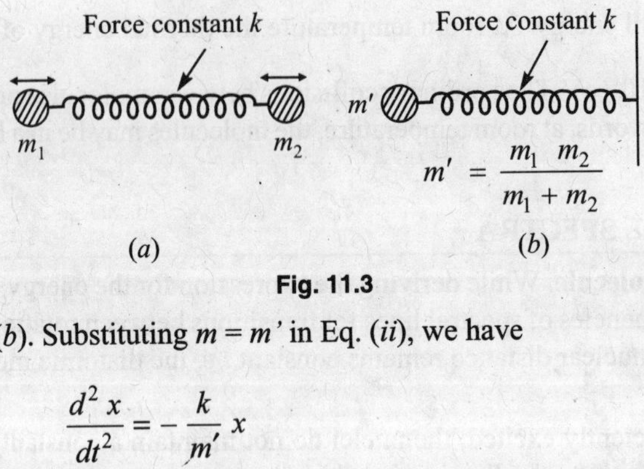

(*a*) (*b*)

Fig. 11.3

as shown in Fig. 11.3 (*b*). Substituting $m = m'$ in Eq. (*ii*), we have

$$\frac{d^2 x}{dt^2} = -\frac{k}{m'}x$$

This is the equation of simple harmonic motion and according to classical physics, the frequency of oscillations is given by

$$\nu_0 = \frac{1}{2\pi}\sqrt{\frac{k}{m'}} \qquad ...(iii)$$

Quantum mechanical treatment. Let us now treat the problem *quantum mechanically*. For this purpose, we first calculate the potential energy of oscillating molecule. According to Eq. (*i*)

$$F = -kx$$

The potential energy of the system is related to force by the relation

$$F = -\frac{dV}{dx}$$

$$\therefore \qquad \frac{dV}{dx} = kx$$

or $$dV = kx\,dx$$

$\therefore$ Potential energy for a displacement x from the equilibrium position $x = 0$ is given by

$$V = \int_0^x k\,x\,dx = \frac{1}{2}kx^2 \qquad \text{...(iv)}$$

Schrodinger's one dimensional, time independent equation is given by

$$\frac{d^2\psi}{dx^2} + \frac{2m}{\hbar^2}(E - V)\,\psi = 0$$

Substituting $m = m'$ and $V = \frac{1}{2}k.x^2$ for the molecular oscillator, Schrodinger's equation becomes

$$\frac{d^2\psi}{dx^2} + \frac{2m'}{\hbar^2}\left(E - \frac{1}{2}kx^2\right)\psi = 0$$

or

$$-\frac{\hbar^2}{2m'}\frac{d^2\psi}{dx^2} + \frac{1}{2}kx^2\,\psi = E\psi \qquad \text{...(v)}$$

The energy levels of the simple harmonic oscillator are now given by the solution of Eq. (*v*), which are

$$E_v = \left(v + \frac{1}{2}\right)h\nu_0 \qquad \text{...(vi)}$$

where $\nu_0 = \frac{1}{2\pi}\sqrt{\frac{k}{m'}}$ and $v = 0, 1, 2...$ etc is the *vibrational quantum number*. For the ground state $v = 0$, and hence ground state vibrational energy $E_0 = \frac{1}{2}h\nu_0$, where ν_0 is the classical frequency.

Vibrational energy levels are equally spaced. The energy levels given by Eq. (*vi*) are called *vibrational levels* of the molecule. Vibrational energy levels corresponding to vibrational quantum number $v = 0, 1, 2, 3...$ are shown in Fig. 11.4.

The energy levels are *equally spaced*. The energy of the ground state is not zero (as in the case of rotational energy levels) but has a value $\frac{1}{2}h\nu_0$.

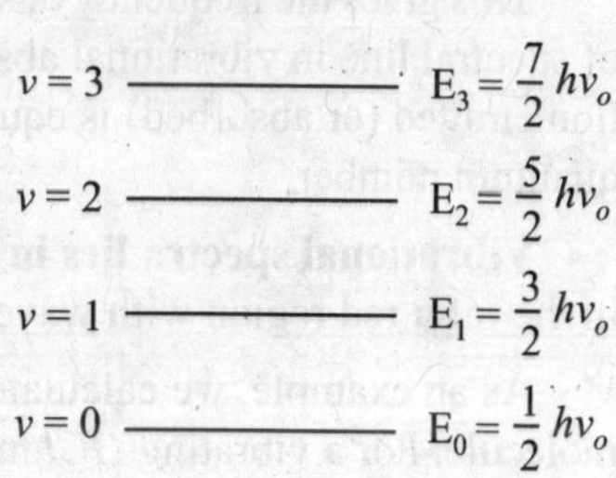

Fig. 11.4

This is in accordance with uncertainty principle, because if the molecule has zero vibrational energy *i.e.* it is at rest, the uncertainty in position $\Delta x = 0$ which means uncertainty in its momentum $\Delta p = \infty$. But a particle with $E = 0$ cannot have infinitely uncertain momentum.

The vibrational energy levels are given by $E_v = \left(v + \frac{1}{2}\right)h\nu_0$ where v is the vibrational quantum number.

$$\text{For } v = 0,\ E_v = E_0 = \frac{1}{2}h\nu_0;\ \text{For } v = 1,\ E_v = E_1 = \frac{3}{2}h\nu_0$$

$$\text{For } v = 2,\ E_v = E_2 = \frac{5}{2}h\nu_0;\ \text{and For } v = 3,\ E_v = E_3 = \frac{7}{2}h\nu_0$$

$$\therefore \qquad E_1 - E_0 = E_2 - E_1 = E_3 - E_2 = h\nu_0$$

Thus the vibrational energy levels are equally spaced with energy interval $= h\nu_0$

Selection rules. The selection rule for transitions between vibrational energy levels is as under:-

Only those transitions are allowed between vibrational energy levels for which the vibrational quantum number v changes by one unit i.e.

$$\Delta v = \pm 1$$

The reason for this rule is given below:-

An oscillating dipole, whose frequency is ν_0 can only absorb or emit electromagnetic radiation of the same frequency. All quanta of frequency ν_0, have the energy $h\,\nu_0$. Thus the oscillating dipole can only absorb an energy $\Delta E = h\nu_0$ at one time, so that its energy increases from $\left(v + \frac{1}{2}\right)h\nu_0$ to $\left(v + \frac{1}{2} + 1\right)h\nu_0$. Similarly it can only emit an energy $\Delta E = h\nu_0$. Hence the selection rule $\Delta v = \pm 1$.

Frequency of lines in vibrational spectra. If the transition occurs from an energy level with quantum number $(v + 1)$ to an energy level with quantum number v, then

$$h\nu = \Delta E = E_{v+1} - E_v$$

$$= \left(v + 1 + \frac{1}{2}\right)h\nu_0 - \left(v + \frac{1}{2}\right)h\nu_0$$

or

$$h\nu = h\nu_0$$

$\therefore$

$$\nu = \nu_0 = \frac{1}{2\pi}\sqrt{\frac{k}{m'}}$$

This gives the frequency of spectral line in vibrational emission spectra. Similarly, the frequency of spectral line in vibrational absorption spectra is also ν_0. Thus we find that the frequency of radiation emitted (or absorbed) is equal to classical frequency ν_0 irrespective of the values of vibrational quantum number.

Vibrational spectra lies in infra red region. The frequency of lines in vibrational spectra lies in the infra red region with wavelength ranging from 8000Å – 50,000Å for a number of molecules.

As an example, we calculate the wavelength corresponding to the vibrational spectrum of HCl molecule. For a vibrating HCl molecule the force constant $k = 470\ \text{Nm}^{-1}$.

The reduced mass
$$m' = \frac{1 \times 35}{1 + 35} \times 1.67 \times 10^{-27} = 1.62 \times 10^{-27}\ \text{kg.}$$

Now, Frequency
$$\nu = \frac{1}{2\pi}\sqrt{\frac{k}{m'}} = \frac{1}{2\pi}\sqrt{\frac{470}{1.62 \times 10^{-27}}} = 8.6 \times 10^{13}\ \text{s}^{-1}$$

Hence wavelength
$$= \frac{3 \times 10^8}{8.6 \times 10^{13}} = 0.35 \times 10^{-5}\ \text{m}$$

$$= 35000\text{Å}$$

This wave length lies in the infra-red region.

A molecule exhibits its vibrational spectra when it is sufficiently excited so that its nuclei do not maintain a constant fixed separation but oscillate back and forth along the line joining them in such a way that the centre of mass remains fixed. This vibrational motion is in addition to the rotation of the molecule.

Example 11.6. *The force constant of CO bond is 187 Nm^{-1}. Find the frequency of vibration of CO molecule and spacing between vibrational levels. Mass of ^{12}C atom = 1.99 × 10^{-26} kg and of ^{16}O atom = 2.66 × 10^{-26}kg.*

(G.N.D.U. 2001)

Solution. Reduced mass of *CO* molecule

$$m' = \frac{m_1 m_2}{m_1 + m_2}$$

$$= \frac{1.99 \times 10^{-26} \times 2.66 \times 10^{-26}}{(1.99 + 2.66) \times 10^{-26}} = 1.14 \times 10^{-26} \text{ kg.}$$

Frequency of vibration

$$\nu = \nu_0 = \frac{1}{2\pi}\sqrt{\frac{k}{m'}}$$

$$= \frac{1}{2\pi}\sqrt{\frac{187}{1.14 \times 10^{-26}}} = 2.04 \times 10^{13} \text{ Hz}$$

Separation between vibrational energy levels in *CO*

$$\Delta E = E_{v+1} - E_v = h\nu_0$$

$$= 6.63 \times 10^{-34} \times 2.04 \times 10^{13}$$

$$= 1.35 \times 10^{-20} \text{ J} = 8.44 \times 10^{-2} \text{ eV.}$$

Example 11.7. *For a diatomic molecule the reduced mass is 1.14 × 10^{-26} kg and force constant of the bond is 167.8 Nm^{-1}. Determine the spacing between the vibrational energy levels in eV.*

(Pbi. U. 2000)

Solution.

$$\Delta E = \frac{h}{2\pi}\sqrt{\frac{k}{m'}} = \frac{6.63 \times 10^{-34}}{2 \times 3.142}\sqrt{\frac{167.8}{1.14 \times 10^{-26}}} = 1.28 \times 10^{-19} \text{ J}$$

$$= 0.08 \text{eV.}$$

11.6 VIBRATIONAL ROTATIONAL SPECTRA

Pure vibrational spectra are observed only in liquids in which no rotation can take place due to interaction between adjacent molecules. Since the excitation energies involved in molecular rotation are much smaller than those involved in vibration, the freely moving molecules in a gas or vapour are almost always rotating, regardless of their vibrational states. The spectra of such molecules do not show isolated lines corresponding to each vibrational transition, but instead a large number of closely spaced lines due to transitions between various rotational states of one vibrational level and the rotational states of the other vibrational level. When these spectra are observed with a spectrometer, the lines appear as broad streaks called *vibration-rotation band.*

Vibration-rotation energy levels. The vibration and rotation of a molecule can take place independently of each other. The vibrational and rotational energies of a diatomic molecule are vastly different. So the rotation of a diatomic molecule does not affect its vibrational motion.

Therefore, the total energy *E* of a molecule is the sum of the vibrational and rotational energy

Hence

$$E_{vJ} = E_v + E_J$$

$$= \left(v + \frac{1}{2}\right)h\nu_0 + J(J+1)\frac{\hbar^2}{2I} \quad ...(i)$$

The total energy of the molecule depends upon two quantum numbers v and J. Therefore we denote this energy as E_{vJ}

The total energy of a molecule is *minimum* when $v = 0$, $J = 0$ and has the value $E_{0,0} = \frac{1}{2} h\nu_0$. It is known as *ground state* energy of the molecule. The higher energy values of the molecule, corresponding to $v = 0, J = 1$; $v = 0, J = 2$; ... are called excited states. The energy of the *first excited state* is

$$E_{0,1} = \frac{1}{2} h\nu_0 + \frac{\hbar^2}{I}$$

The energy of the second excited state is

$$E_{0,2} = \frac{1}{2} h\nu_0 + \frac{3\hbar^2}{I}$$

The energies of the 3rd, 4th ... excited state of the molecule are

$$E_{0,3} = \frac{1}{2} h\nu_0 + \frac{6\hbar^2}{I}$$

$$E_{0,4} = \frac{1}{2} h\nu_0 + \frac{10\hbar^2}{I} \text{ and so on.}$$

The allowed values of energy $E_{0,0}; E_{0,1}; E_{0,2}$... are called *vibration-rotation energy levels* of the molecule.

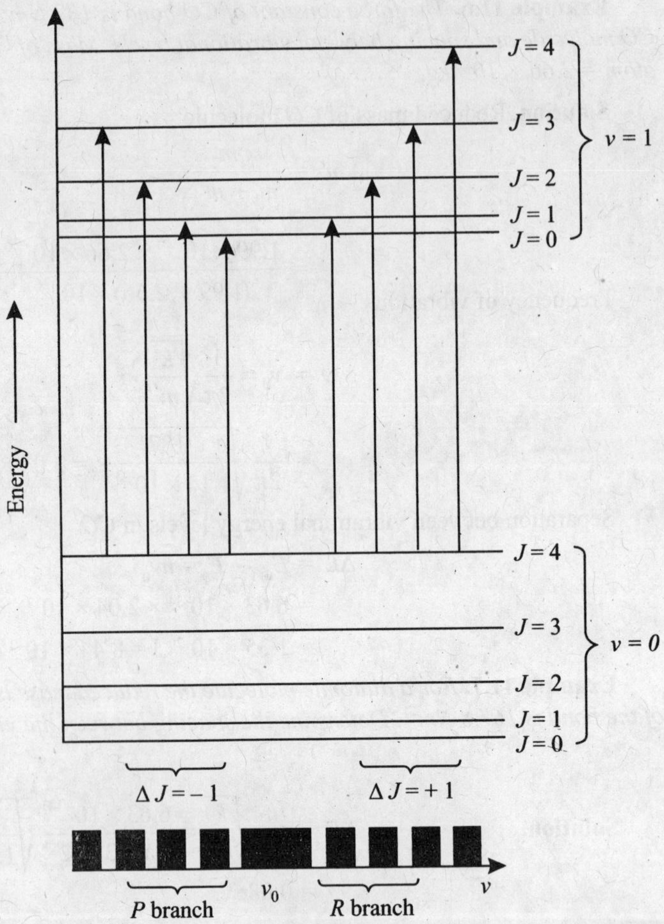

Fig. 11.5

The ground state vibrational energy $\frac{1}{2} h\nu_0$ is greater than kT. Therefore, only a few molecules at room temperature are in an excited state. Further, E_v is very much greater than E_J. So, even at high temperatures only the vibrational states corresponding to $v = 0$ and $v = 1$ are excited.

Energy level diagram. Fig. 11.5 shows $J = 0, 1, 2, 3$ and 4 rotational levels of a diatomic molecule, corresponding to two vibrational levels $v = 0$ and $v = 1$. The spectral lines in absorption are also shown.

Frequency of spectral lines. The selection rules for vibrational rotational spectra are

$\Delta v = \pm 1$ and $\Delta J = \pm 1$

The transition with $\Delta J = 0$ is not allowed.

The selection rule $\Delta v = +1$ applies to absorption spectra and $\Delta v = -1$ applies to emission spectra.

In absorption spectra, the transition form $v = 0 \rightarrow v = 1$ ($\Delta v = +1$) falls into two categories:–

(*i*) ***P*-Branch.** Here $\Delta J = -1$, i.e., transitions are ($J \rightarrow J - 1$). The frequencies of the spectral lines in *P*-branch are given by– $h\nu_p = E_{1, J-1} - E_{0, J}$

$$= \left[\left(1+\frac{1}{2}\right)h\nu_0 + (J-1)\,J\frac{\hbar^2}{2I}\right] - \left[\left(0+\frac{1}{2}\right)h\nu_0 + J\,(J+1)\frac{\hbar^2}{2I}\right]$$

$$= h\nu_0 - J\,\frac{\hbar^2}{I}$$

$$\therefore \qquad \nu_P = \nu_0 - J\,\frac{h}{4\pi^2 I};\ J = 1, 2, 3...$$

(*ii*) ***R*-Branch.** Here $\Delta J = +1$ *i.e.* transitions are $(J \rightarrow J+1)$. The frequency of the spectral lines in *R*-branch are given by– $h\nu_R = E_{1, J+1} - E_{0, J}$

$$= \left[\left(1+\frac{1}{2}\right)h\nu_0 + (J+1)\,(J+2)\,\frac{\hbar^2}{2I}\right] - \left[\left(0+\frac{1}{2}\right)h\nu_0 + J\,(J+1)\,\frac{\hbar^2}{2I}\right]$$

$$= h\nu_0 + (J+1)\frac{\hbar^2}{I}$$

$$\therefore \qquad \nu_R = \nu_0 + (J+1)\,\frac{h}{4\pi^2 I};\ J = 0, 1, 2, 3...$$

(*iii*) ***Q*-Branch.** Here $\Delta J = 0$ *i.e.* transitions are $(J \rightarrow J)$. The frequencies of the spectral lines in *Q*-branch are given by

$$h\nu_Q = E_{1,J} - E_{0,J}$$

$$= h\nu_0$$

$$\therefore \qquad \nu_Q = \nu_0$$

As transition for which $\Delta J = 0$ is *not allowed* in diatomic molecules, there is *no line at* $\nu_Q = \nu_0$ *i.e. Q-branch is missing*. Each line in the rotation-vibration spectrum consists of a series of lines corresponding to different values of *J*. These lines are closely spaced and appear continuous. These are, therefore, said to form bands, as shown at the bottom [Fig. 11.5(*b*)].

The spacing between the lines in both *P* and *R* branch is $\Delta\nu = \dfrac{h}{4\pi^2 I} = \dfrac{\hbar}{2\pi I}$. We can, therefore, find the moment of inertia of the molecule by measuring $\Delta\nu$. Moment of inertia can thus be calculated from the micro-wave pure rotational spectra, as well as from the infra-red vibrational-rotational spectra.

11.7 POTENTIAL ENERGY CURVE

The vibrational-rotational energy levels of a diatomic molecule, on a potential energy versus inter-atomic distance curve, are shown in Fig. 11.6. As the vibrational quantum number *v* increases, the higher vibrational states of the molecule do not obey the relation

$$E = \left(v + \frac{1}{2}\right)h\nu_0 \text{ where } \nu_0 = \frac{1}{2\pi}\sqrt{\frac{k}{m'}}$$

k being the force-constant of the molecule and *m'* its reduced mass.

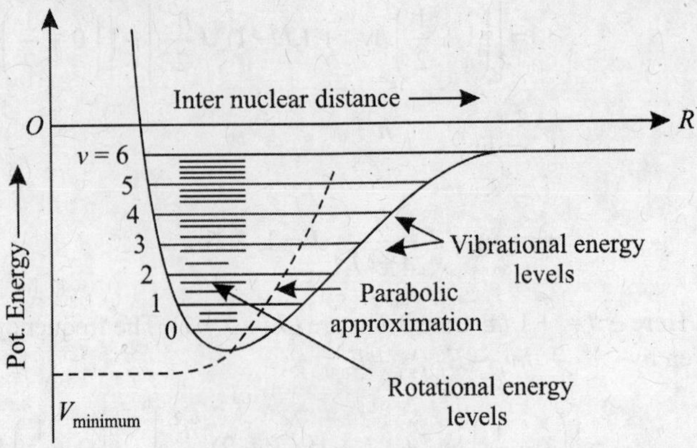

Fig. 11.6

If we look at the potential energy V versus inter-nuclear distance R curve, we find that in the neighbourhood of the minimum of the curve, which corresponds to the normal configuration of the molecule, the shape of the curve is very nearly a parabola. Therefore, for lower vibrational states of the molecule, the potential energy curve is parabolic but for higher vibrational states *i.e.* with increasing energy, the potential energy curve deviates more and more from parabolic form, so that the energy-well is wider than the parabola. As a result, the spacing between adjacent energy levels of higher vibrational quantum number is less than the spacing between adjacent energy levels for lower values of vibrational quantum number, so that for higher quantum numbers, the energy levels are closer.

11.8 SPECTRA OF HOMONUCLEAR DIATOMIC MOLECULE

The vibrational spectra are observed only in absorption. The electromagnetic radiation can induce transition among vibrational energy levels. For transfer of radiation energy to vibrating molecule, an electrical coupling should be present. If the vibraing molecule produces an oscillating dipole moment, then this dipole moment interacts with electric field of radiation and thus gives rise to coupling. Molecules like *HCl, HBr* having dissimilar atoms, have a resultant dipole moment which is a function of internuclear distance and therefore, show vibrational spectra. Molecules like H_2, N_2, O_2, which have similar atoms, do not possess net dipole moment. Such molecules cannot interact with radiation and do not show vibrational spectra. This is why homonuclear diatomic molecules do not show vibrational rotational spectra.

11.9 ELECTRONIC SPECTRA OF DI-AND POLY ATOMIC MOLECULES

The energies of rotation and vibration in a molecule are due to the motion of the atomic nuclei, because the whole mass of the molecule is concentrated in the nuclei.

In a molecule, the electrons can also be excited to higher energy states from their ground state. The spacing of these levels is much greater than the spacing of vibrational or rotational levels. The electronic transitions thus involve much greater energies and give rise to spectra in the visible or ultraviolet region.

Electronic energy levels in a molecule are separated by energies of the order of eV. Each electronic energy level may have a number of vibrational energy levels, with each vibrational energy level further having a number of rotational energy levels, with energy separation of 10^{-2} to 10^{-3} eV. Thus, each electronic level consists of a large number of levels closely spaced together. Each electronic transition, therefore, produces a *series of closely spaced lines, called a band* due to the presence of vibrational and rotational states in each electronic state.

All the molecules exhibit electronic spectra, because a change in dipole moment is always accompanied by a change in electronic configuration of the molecule. So, even *homo-nuclear molecules* like H_2, N_2 and O_2, which do not exhibit either rotational or vibrational spectra because they do not posses a permanent dipole moment do show electronic spectra which possesses rotational and vibrational *fine structure*. The moment of inertia and bond force constant of such molecules can be calculated from the fine structure in electronic spectral bands.

Frequency of electronic spectra. When the electronic states of a molecule are excited, the total energy of the molecule in a given quantum state, is given by

$$E = E_e + E_v + E_J$$

The frequency of radiation, when transition between any two states takes place, is given by

$$\nu = \frac{E_1 - E_2}{h} = \frac{E_{e1} - E_{e2}}{h} + \frac{E_{v1} - E_{v2}}{h} + \frac{E_{J1} - E_{J2}}{h}$$

$$= \frac{\Delta E_e}{h} + \frac{\Delta E_v}{h} + \frac{\Delta E_J}{h}$$

$$= \frac{1}{h}[\Delta E_e + \Delta E_v + \Delta E_J]$$

This is most general type of transition and is known as *rotation-vibration-electronic transition.* Since $E_e > E_v > E_J$ the frequency or position of the electronic spectra is determined by ΔE_e (*i.e.* change in E_e). This lies in the visible or ultraviolet region.

Poly atomic molecule. When electronic excitation takes place in a poly-atomic molecule, it brings about a *change in shape* of the molecule. This can be determined from rotational fine structure in its band spectrum. The change in shape is due to the reason that wave function of electrons in different states have different characters, which give rise to different *bond geometries*. For example, the molecule of beryllium hydride (BeH_2) is *linear* ($H-Be-H$) in one state and *bent* $\left(\begin{array}{c} H - Be \\ \quad\ | \\ \quad\ H \end{array}\right)$ in another state.

11.10 LUMINESCENCE

Luminescence. When a molecule absorbs energy, usually in the form of a high frequency radiation and re-emits it in the form of visible light, the phenomenon is known as *Luminescence*. Luminescence is, therefore, reverse of absorption. If the molecule emits light simultaneously the process is called *fluorescence*. If it takes 10^{-7} sec. or longer to emit light, the process is called *phosphorescence* and the substance is known as *phosphor*.

Fluorescence. A molecule, in an excited state, can lose energy in a number of ways and return to the ground state.

The molecule may emit a photon of the same energy (or frequency) as that of the photon it absorbed and return to the ground state in a *single step*.

On the other hand, the excited molecule may share some of its vibrational energy with other molecules by collision and move down to a lower vibrational quantum state. It may then emit a photon of lower energy (lower frequency) and return to the ground state. The downward radiative transition originating from a lower vibrational level in the excited upper electronic state, as shown in Fig. 11.7.

This phenomenon, in which a molecule absorbs a photon of higher energy and emits a photon of lower energy (lower frequency or longer wavelength) is known as fluorescence.

Fluorescence excited by ultraviolet light is used in the identification of minerals and certain compounds, which are of interest in the study of bio-chemistry. Another application of fluorescence is in a *fluorescent tube*.

Phosphorescence. In atomic spectra, the selection rule for spin quantum number is $\Delta S = 0$. The same selection rule also applies to molecular spectra. This means radiative transitions between electronic states with different total spin are forbidden *i.e.* no transition is allowed between *singlet states* ($S = 0$) and *triplet states* ($S = 1$).

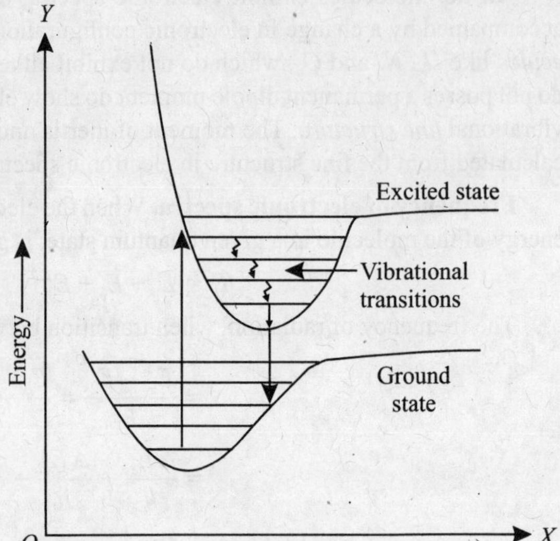

Fig. 11.7

As shown in Fig. 11.8, a molecule in the *singlet ground state* absorbs a photon and is raised to another *singlet excited state*. The molecule in the singlet excited state may have collisions with other molecules and undergo radiation-less transitions to a lower vibrational level. It can happen that this lower vibrational level in the singlet state may have the same energy as one of the vibrational levels in the excited triplet state. There is then a certain probability that a shift to the triplet state with the same vibrational energy may take place. Further, collisions with other molecules may again take place in the triplet state and the molecule may shift to $v = 0$ level (in the triplet state).

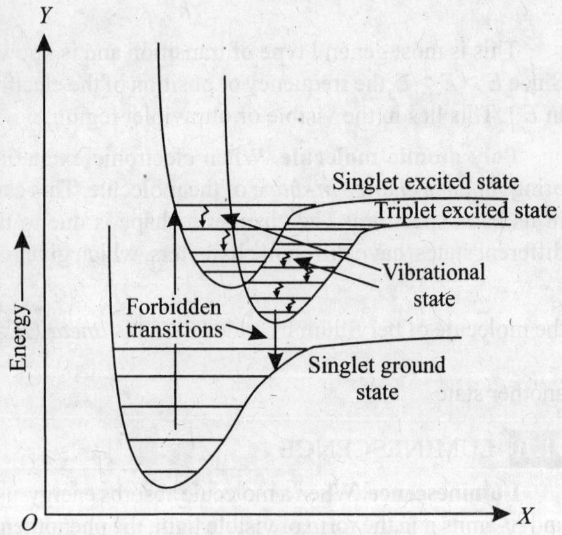

Fig. 11.8

A radiative transition from the triplet state ($S = 1$) to a singlet state ($S = 0$) is *forbidden* by selection rules. It does not, however, mean that such a transition is *impossible* to occur, but it means that there is only a *very small probability* of this to take place. Such transitions have a very long half life and the molecule may stay in the excited state for long time. The *resulting* radiation may be emitted minutes or even hours after the initial absorption of photon. This type of delayed radiation is called *phosphorescence* and the transitions are known as *phosphorescent transitions*.

EXERCISE

LONG QUESTIONS

1. (*a*) Distinguish between homo-nuclear and hetro-nuclear molecules. Give examples of each. ***(P.U., 2002)***

   (*b*) Explain general characteristics of a diatomic molecular spectra. What is the cause of excited energy levels in a molecule? What do you mean by electronic, rotational and vibrational energy level? What are the different kinds of molecular spectra? ***(G.N.D.U. 2007)*** (*c*) Why do molecules show band spectra rather than line spectra?

(H.P.U., 2003, 2002, 2001; Pbi. U. 2005, 2001)

2. (*a*) Obtain an expression for rotational energy levels of a diatomic molecule and the frequency of rotational spectra. State clearly the selection rules. Show energy levels and rotational spectrum of a diatomic molecule. In which region of the electro-magnetic spectrum do the rotational spectra of molecules lie?

(*b*) Show that in rotational spectra the energy levels are not equally spaced whereas the frequencies are equally spaced.

(Pbi. U. 2005, 2004, 2001; Kerala U. 2001; Bang. U. 2000; G.N.D.U. 2005, 2000; M.D.U. 2006)

3. Find the rotational energy of a diatomic molecule in the ground state. Hence show that the molecules do not rotate in the ground state.

(P.U. 2006, 2003; G.N.D.U., 1997; Pbi. U., 1993)

4. What is the rotational energy of a diatomic molecule in the ground state? What inference do you draw from this result? ***(Pbi.U. 2006)***

5. Why all molecules do not show rotational spectra?

6. (*a*) Give theory of vibrational spectra and obtain an expression for energies of various vibrational levels. Show that the energy levels are equally spaced. How is the frequency of various lines obtained in vibrational spectra of molecules? What are the selection rules for vibrational states? In which region of the electro-magnetic spectra do the vibrational spectra of molecules lie?

(H.P.U. 2002, 2000; G.N.D.U. 2004; Pbi. U., 2006, 2002)

(*b*) Under what conditions a molecule exhibits its vibrational spectra? ***(G.N.D.U. 2007)***

7. One cannot obtain vibration rotation spectra of homonuclear diatomic molecules. Why?

(G.N.D.U. 2003)

8. H_2, N_2 and O_2 do not show vibrational spectra. Why?

9. (*a*) Discuss vibrational-rotational spectra of diatomic molecules. How is it explained? Draw energy level diagrams to explain the spectra. What are *P*, *Q* and *R* branches in vibrational rotational spectra?

(M.D.U. 2008; Pbi. U. 2003, P.U., 2008, 2007, 2005; G.N.D.U, 2002)

(*b*) Show vibrational and rotational energy levels of a diatomic molecule on a potential energy versus inter-atomic distance curve. Explain the formation of these levels.

(Kerala. U. 2001; G.N.D.U. 2002)

10. Describe in brief the electronic spectra of diatomic molecules. What is the effect of electronic excitation on a poly-atomic molecule?

(G.N.D.U. 2000; Pbi. U., 2000)

11. What is luminescence? Explain the difference between fluorescence and phosphorescence.

(P.U., 2007, 1993; G.N.D.U., 1992)

12. The spacing between vibrational levels of carbon monoxide (CO) molecule is 0.08 eV. Calculate the value of force constant. Take the mass of Carbon atom = 12 and that of oxygen atom = 16 times the mass of proton = 1.67×10^{-27} kg. ***(G.N.D.U., 1993; P.U., 1992)***

(Ans. 167.8 Nm^{-1})

13. The carbon monoxide (CO) molecule has a bond length *R* of 0.113 nm and the masses of the ^{12}C and ^{16}O atoms are respectively 1.99×10^{-26} kg and 2.66×10^{-26} kg. Find the energy and angular velocity of CO molecule when it is in its lowest rotational state.

Hint. Lowest rotational state has J = 1 and $E = \frac{1}{2} I \omega^2$

[Ans. $E = 7.61 \times 10^{-23}$ J; $\omega = 3.23 \times 10^{11}$ rad s^{-1}]

SHORT QUESTION

1. Do all molecules show rotational spectra?

[Hint. Rotational spectra arise from transitions between rotational energy states. Only molecules which have electric dipole moments can absorb or emit electromagnetic radiation (or photons) in such transitions. The emission of radiation is due to the rotation of electrical dipole moment.

According to classical electrodynamics, the rotation of a molecule leads to radiation of energy, only if a changing dipole moment is associated with its motion. In the case of molecules consisting of *unlike* atoms like *HCl* or *CO*, such a permanent dipole moment does exist, since the centres of gravity of positive and negative charges do not coincide. The component of dipole moment along a fixed direction changes periodically as the molecules rotate, giving rise to a rotational spectrum. For molecules consisting of *like* atoms–say N_2, H_2 etc. no such dipole moment is present and no radiation takes place due to pure rotation. Conversely, energy in the far infrared region is absorbed by molecules having permanent dipole moment. In general, non polar diatomic molecules with identical atoms like H_2, O_2, N_2 etc and symmetric poly-atomic molecules such as CO_2 $(C=O=C)$ and CH_4 $\left\{\begin{matrix} & H & \\ & | & \\ H- & C & -H \\ & | & \\ & H & \end{matrix}\right\}$ do not exhibit rotational spectra.

Note. (Transitions between rotational states in molecules like H_2, CO_2, CH_4 etc can take place during collisions, which may disturb the symmetry of the electrons or the nuclei).]

MULTIPLE CHOICE QUESTIONS

Mark the correct answer

1. The stable configuration in hydrogen molecular ion occurs at a distance of
   (*a*) 0.106 Å (*b*) 1.06 Å (*c*) 10.6 Å (*d*) 106 Å
2. The binding energy of hydrogen molecular ion is
   (*a*) 2.65 MeV (*b*) 2.65 KeV (*c*) 2.65 eV (*d*) 2.65 J
3. Which of the following is homonuclear molecule?
   (*a*) CO (*b*) O_2 (*c*) NO (*d*) CO_2
4. Which of the following states for a diatomic molecule have $m = 2$?
   (*a*) π (*b*) σ (*c*) δ (*d*) ϕ
5. Which type of spin wave function describes a system of electrons?
   (*a*) symmetric (*b*) anti-symmetric (*c*) none of these
6. The rotational energy levels of a molecule are
   (*a*) equally spaced (*b*) unequally spaced (*c*) no energy levels at all.

7. The rotational spectra the frequency lines are

(*a*) equally spaced (*b*) unequally spaced (*c*) no energy levels at all.

8. The vibrational energy level of a molecule are

(*a*) equally spaced (*b*) unequally spaced (*c*) no energy levels at all.

9. The selection rule for transitions in rotational spectra is

(*a*) $\Delta J = 0$ (*b*) $\Delta J = \pm 1$ (*c*) $\Delta J = \pm 2$

ANSWERS

1. (*b*)	**2.** (*c*)	**3.** (*b*)	**4.** (*c*)	**5.** (*b*)
6. (*b*)	**7.** (*a*)	**8.** (*a*)	**9.** (*b*)	

RAMAN EFFECT

12
Chapter

INTRODUCTION

Raman effect, change in wavelength of light that occurs when a light beam is deflected by molecules. This phenomenon is named for *Sir Chandrasekhra Venkata Raman*, who discovered in 1928. When a beam of light traverses a dust-free, transparent sample of a *Chemical compound*, a small fraction of the light emerges in directions other than that of the incident (incoming) beam. Most of this scattered light is considered as of unchanged wavelength *i.e.* of same frequency. A small part, however, has wavelength (or frequency) different that of the incident light, its presence is a result of the *Raman effect*.

Raman scattering is easily understood if the incident light is considered as consisting of particles (*i.e.* particle nature of light) or photons (with energy proportional to frequency, $E = h\nu$), that strikes the molecules of the sample. Most of the encounters are *elastic,* and the photons are scattered with unchanged energy and frequency. On some occasions, however the molecules takes up energy from or gives up energy to the photons, which are thereby scattered with diminised or increased energy, hence with lower or higher frequency. The frequency shifts are thus measures of the amounts of energy involved in the transition between initial and final states of the scattering molecule. Raman showed that the nature of this scattered light was dependent on the type of sample present. Other scientists quickly understood the significance of this phenomenon as an analytical and research tool and called it *Raman effect*. Its current uses range from the non-destructive identification of the minerals to the early detection of life-threatening diseases. For his discovery Raman was awarded the *Nobel Prize in physics* in 1930.

12.1 RAMAN EFFECT

In 1928, Sir C.V. Raman, while studying the phenomenon of scattering of light, found that when a beam of monochromatic light was passed through organic liquids like benzene, toluene etc., the scattered light contained other frequencies, in addition to the frequency of the incident light. Raman observed the scattered light by means of a spectrometer placed at right angles to the incident light and found that in addition to the un-modified original spectral line (main line) a number of new lines were present on both sides of the main line. These lines are known as *Raman lines* and the spectrum formed is called *Roman spectrum*. To sum up

Light scattered from the molecules of the scattering material at right angles to the incident direction contains radiations not only of incident frequency ν *but also radiations of frequencies* $\nu' = \nu - \Delta\nu$*. This scattering of light with change in frequency is known as Raman effect.*

The change in frequency $\Delta\nu$ is known as *Raman shift*. This shift is *different for different substances* producing scattering and is independent of the frequency of incident light.

Stoke's and Anti Stoke's lines. (*i*) If $\Delta\nu$ is **positive** $\nu' < \nu$ *i.e.* the *frequency* of the scattered radiation is less than that of the incident radiation. In other words the wavelength of the scattered radiation is longer than that of the incident radiation. (Fig. 12.1)

These spectral lines appear in the Raman spectrum on the lower frequency or longer wavelength side of the unmodified spectral line and are designated as *Stoke's lines*.

(*ii*) If $\Delta \nu = 0$ *i.e.* $\nu' = \nu$, the frequency of the scattered radiation (hence the wavelength) is the *same* as that of the incident radiation. *This un-modified spectral line in Raman spectrum appears in its original place. This scattering of light by molecules, without change in frequency (or wavelength), is known as Rayleigh scattering or coherent scattering and the spectral line is called Rayleigh line.*

(*iii*) If $\Delta \nu$ is **negative,** $\nu' > \nu$ *i.e.* the *frequency* of the scattered radiation is *greater* than that of the incident radiation. In other words, the wavelength of the scattered radiation is *shorter* than that of the incident radiation. These spectral lines appear in the Raman spectrum on the *higher frequency* (or *shorter wavelength*) side of the un-modified line and are designated as *anti-Stoke's lines*.

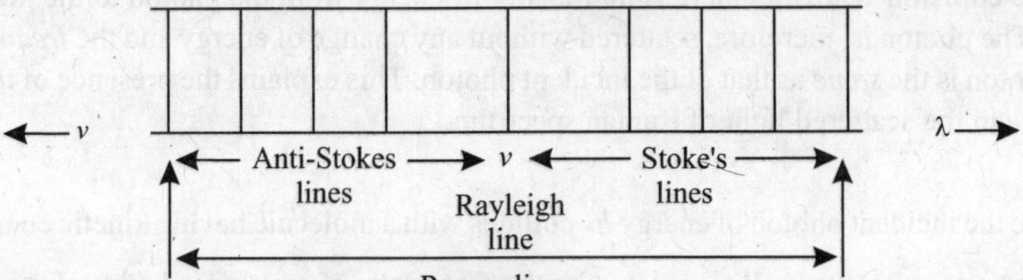

Fig. 12.1

12.2 CHARACTERISTICS OF RAMAN LINES

It is seen in the Raman spectrum that

(*i*) The Stoke's lines are always more intense than anti-Stoke's lines.

(*ii*) The Raman lines are symmetrically displaced with respect to the parent or original line.

(*iii*) The frequency difference between the modified lines and parent line represents the frequency of the corresponding infra-red absorption line.

12.3 EXPERIMENTAL ARRANGEMENT

The apparatus used to study Raman effect in liquids is shown in Fig. 12.2. It was originally developed by Wood.

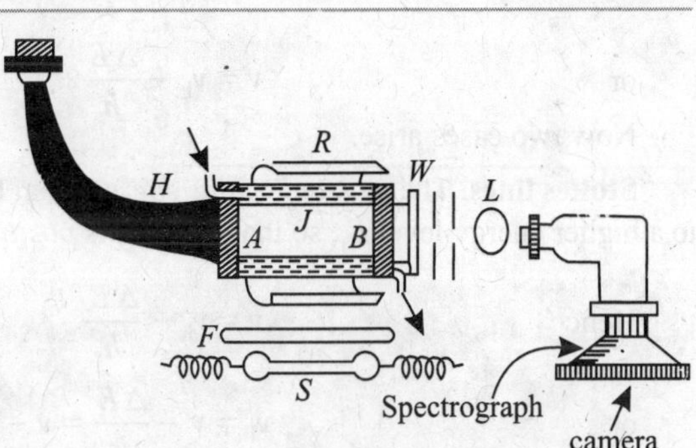

Fig. 12.2

It consists of a glass tube *AB* about 1 to 2 cm in diameter and 10 to 15 cm long, containing the pure experimental liquid free from dust and air bubbles. The tube is closed at one end by an optically plane glass plate which forms the window *W* and the other end is drawn into the shape of a horn *H* and blackened on the outside. It is known as *Raman tube*. Light from a *mercury arc source S* is first passed through a suitable *filter F*, which allows only monochromatic radiation of wavelength λ = 4358 Å to pass through. The tube is surrounded by a water-jacket *J* through which cold water is circulated to prevent overheating of the liquid due to the proximity of the mercury arc. A semi-cylindrical polished aluminium reflector *R* is used to further increase the intensity of illumination.

The scattered light coming out of the window *W* is focused on the slit of a spectrograph with the help of a lens-system *L*. The spectrograph is carefully aligned along the axis of the Raman tube and screened from the direct rays of the arc. The spectrograph must have a *large light gathering power* and a prism of *high resolving power*. A *short focus camera* is used to photograph the spectrum.

If a sufficiently long time of exposure is used, then on developing the photographic plate the Raman spectrum is found to consist of a number of Stoke's lines a few anti-Stoke's lines and a strong un-modified Rayleigh line.

12.4 EXPLANATION OF RAMAN EFFECT ON QUANTUM THOERY

According to quantum theory of light, a wave of frequency ν consists of photons, each of energy $h\nu$. When this light falls on the molecules of a solid, liquid or gas, the photon undergoes a collision with the molecule.

(*i*) If the collision is *elastic*, there is no transfer of energy from the photon to the molecule and vice-versa. The photon is, therefore, scattered without any change of energy and the *frequency* of the scattered photon is the *same* as that of the incident photon. This explains the presence of *un-modified Rayleigh line* in the scattered light of Raman spectrum.

Suppose the incident photon of energy $h\nu$ collides with a molecule having kinetic energy $\frac{1}{2}mv^2$ and intrinsic energy E. After collision, let ν_1 be the frequency of scattered photon, v the velocity of molecule and E_1 its intrinsic energy, then according to the law of conservation of energy—

$$h\nu + \frac{1}{2}mv^2 + E = h\nu_1 + \frac{1}{2}mv_1^2 + E_1$$

Since there is no change in temperature during the collision, there is no change in kinetic energy,

$$\therefore \qquad \frac{1}{2}mv^2 = \frac{1}{2}mv_1^2$$

Hence $$h\nu + E = h\nu_1 + E_1$$

or $$h\nu = h\nu_1 + (E_1 - E) = h\nu_1 + \Delta E$$

or $$\nu = \nu_1 + \frac{\Delta E}{h}$$

Now two cases arise.

Stokes lines. The *molecule gains energy* from the photon and goes from a lower energy level E to a higher energy level E_1, so that $(E_1 - E)$ is *positive* and $= +\Delta E$

Hence $$\nu = \nu_1 + \frac{\Delta E}{h}$$

or $$\nu_1 = \nu - \frac{\Delta E}{h} = \nu - \Delta\nu$$

where $$\Delta\nu = \frac{\Delta E}{h}$$

The *frequency* of the scattered photon is less than that of the incident photon.

This explains the presence of *modified Stoke's lines* on the *lower frequency* or *longer wavelength* side of the Raman spectrum.

Anti-Stokes lines. The molecule may happen to be initially in an excited state. It may *lose energy to the photon*, which may now have a frequency ν_2. The molecule goes from the initial higher energy level E to a lower energy level E_1, so that $(E_1 - E)$ is *negative* and $= -\Delta E$

Hence $$\nu = \nu_2 - \frac{\Delta E}{h}$$

or $$\nu_2 = \nu + \frac{\Delta E}{h} = \nu + \Delta \nu$$

The *frequency* of the scattered photon is *higher* than that of the incident photon. This explains the presence of *Anti-Stoke's lines* on the *higher frequency* or *shorter wavelength* side of the Raman spectrum.

12.5 INTENSITY OF RAMAN LINES

At ordinary temperatures, there are more molecules in the lower energy state, so that there are more transitions from lower energy state to higher energy state, where energy is absorbed by the molecule from the photon than those from higher energy state to lower energy state, where energy is absorbed by the photon from the molecule. Therefore, *anti-Stoke's lines* are generally *fainter* than Stoke's lines. In some cases, anti-Stokes lines are so faint that they are not even observable. However, the *intensity* of anti-Stokes lines *increases with temperature*, as a rise in temperature increases the relative population of molecules in the higher energy state.

Detailed study of Raman spectrum. When a *spectrograph of small dispersion* is used with the scattering substance consisting of diatomic molecules at ordinary (low) temperatures the most intense Raman line is the one displaced to the lower frequency or longer wavelength side of the un-displaced exciting radiation. The *displacement is found to agree with the frequency of the main infra-red vibrational absorption band of the scattering molecules at low temperature*. The production of this Stoke's line is due to the absorption of energy of the incident radiation by the molecule to get itself excited from the vibrational level with vibrational quantum number $v = 0$ of the lowest electronic state, which is most highly populated at ordinary temperatures, to the upper vibrational level with $v = 1$.

If the scattered radiation is examined through a *spectrograph with large dispersion*, it is found that Raman lines with large displacements from the exciting radiation line show a structure which corresponds to that shown by the *rotational structure of the vibrational-rotational spectra* of the molecules of the scattering substance. In addition, we also get a series of lines *close* to and on *either side* of the exciting spectral line, which are almost *equally spaced* with *separation* between successive members *twice* the separation observed in pure rotational spectra of diatomic molecules of the scattering substance.

12.6 RAMAN EFFECT – A TWO STEP PROCESS

Raman effect is in fact a two-step process. To explain this, consider a scattering molecule in the rotational state *A* having rotational quantum number *J*. In the ordinary rotation spectrum, the selection rule is $\Delta J = \pm 1$.

But according to quantum analysis of the process of Raman spectrum, a Raman jump is possible between two levels *A* and *B*, only if there exists a third level *C*, such that ordinary transitions are allowed between *A* and *C* and *C* and *B*. This amounts to saying that the molecule first goes to the state *C* from *A* and then jumps from *C* to *B i.e. it is a two step process*.

During the transition from state *A* with rotational quantum number J_A to the state *C* with rotational quantum number J_C, the *rotational quantum number must change by* ± 1 in accordance with selection rule.

$$\therefore \quad J_C = J_A \pm 1$$

The transition from state *C* to state *B* with rotational quantum number J_B is also governed by the same selection rule.

$\therefore \qquad J_B = J_C \pm 1$

The total change in rotational quantum number—

$$J_B - J_A = \Delta J = J_C \pm 1 - J_A$$
$$= J_A \pm 1 \pm 1 - J_A = \pm 1 \pm 1$$

The possible values of ΔJ are (+ 1, + 1), (– 1, – 1) and (–1, +1) *i.e.*, ± 2, 0.

The case $\Delta J = 0$, involves no change in molecular energy. The scattered Raman radiation, therefore, has the same frequency as that of the incident light and there is no Raman shift.

This corresponds to un-modified line or Rayleigh scattering.

Selection rule for Raman scattering. The case $\Delta J = \pm 2$ gives rise to Raman scattering. So *the selection rule* for Raman scattering is

$$\Delta J = \pm 2$$

Thus we see that transitions which are not allowed in ordinary rotational spectra, are allowed in Raman spectra.

For $\Delta J = +2$ we get *Stoke's lines* and for $\Delta J = -2$ we get *anti-Stoke's lines*.

Since rotational quanta have small value, a large number of rotational energy levels will be occupied by many molecules at ordinary temperatures. Hence several Raman transitions with initial value of $J = 0, 1, 2, 3$ ... upto 10 will be observed.

12.7 PURE ROTATIONAL RAMAN SPECTRA

The frequency of Stoke's lines arising from the transitions from the rotational energy level E_J to E_{J+2} is given by

$$\nu_J = \nu - \frac{E_{J+2} - E_J}{h}$$

$$= \nu - \frac{h^2}{8\pi^2 I}\left[\frac{(J+2)(J+3) - (J(J+1))}{h}\right]$$

because rotational energy $E_J = \dfrac{h^2}{8\pi^2 I} J(J+1)$

or
$$\nu_J = \nu - \frac{h}{8\pi^2 I}(4J+6)$$

$$= \nu - \frac{h}{4\pi^2 I}(2J+3)$$

$\dfrac{h}{4\pi^2 I}(2J+3) = \dfrac{\hbar}{2\pi I}(2J+3)$ is called *Raman shift.*.

12.8 RAMAN SHIFT

Raman shift $\nu_J - \nu = \dfrac{\hbar}{2\pi I}(2J+3)$ where $J = 0, 1, 2, 3...$

For $J = 0$, Raman shift $= \dfrac{3\hbar}{2\pi I} = 3\nu_1$, where $\nu_1 = \dfrac{\hbar}{2\pi I}$

For $J = 1$, Raman shift $= \dfrac{5\hbar}{2\pi I} = 5\nu_1$

For $J = 2$, Raman shift $= \dfrac{7\hbar}{2\pi I} = 7\nu_1$

Thus spacing between the original undisplaced Rayleigh line and the first Raman line is $3\nu_1$

where $\nu_1 = \dfrac{\hbar}{2\pi I}$

Spacing between two successive Strokes line. The frequency of Stokes lines is given by

$$\nu - 3\nu_1, \nu - 5\nu_1, \nu - 7\nu_1 \ldots$$

and spacing between two successive Stoke's lines in Raman spectra

$$[\Delta\nu]_R = 2\nu_1 = 2\frac{\hbar}{2\pi I} = \frac{\hbar}{\pi I} = \frac{h}{2\pi^2 I}$$

Thus spacing between any two successive Raman (or Stoke's) lines (except the first) is $2\nu_1$. The expression for spacing is independent of J. Therefore the lines are equally spaced.

Spacing between two successive Anti-Stokes lines. For Anti-Stokes lines $\Delta J = -2$

$$\therefore \quad \nu_J - \nu = -\frac{h}{4\pi^2 I}(2J+3) = -\frac{\hbar}{2\pi I}(2J+3)$$

$\therefore$ Raman shift for $J = 0, 1, 2\ldots$ is $-3\nu_1, -5\nu_1, -7\nu_1 \ldots$ respectively. The frequency of anti-Stoke's lines is given by

$$\nu + 3\nu_1,\ \nu + 5\nu_1,\ \nu + 7\nu_1 \ldots.$$

The spacing between two successive anti-Stokes line is also

$$[\Delta\nu]_R = 2\nu_1 = \frac{\hbar}{\pi I} = \frac{h}{2\pi^2 I}$$

Spacing between two successive rotational lines. The frequency difference between two successive lines in pure rotational spectra of a diatomic molecule is given by

$$\Delta\nu = \frac{h}{4\pi^2 I}$$

Thus the frequency difference (spacing) in Raman lines $[\Delta\nu]_R$ is twice the frequency difference (spacing) between successive lines in pure rotational spectrum of the molecule.

The homonuclear molecules like H_2, N_2, O_2 etc do not exhibit vibrational and rotational spectra. These molecules exhibit Raman spectra. Therefore, their vibrational and rotational energy levels, as well as molecular constants like bond-length, bond force constant etc. and molecular structure can be determined from Raman effect.

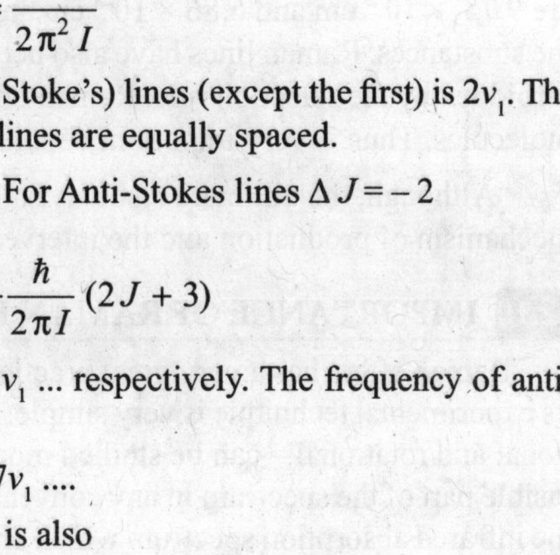

Fig. 12.3

Weaker antistoke's lines will occur at the same distance from, but to higher frequency side of exciting frequency. The rotational-vibrational lines occur for optically anisotropic motecules. The intensity of the spectral lines is greater for greater anisotropy. Moecules like H_2, O_2, N_2 etc. can produce rotational haman spectra due to anisotropy though they have no absorption for the corresponding lines in for-infrared region.

12.9. COMPLIMENTERY CHARACTER OF RAMAN AND INFRARED SPECTRA.

Experimentally it has been observed that Raman spectra occur due to changes in the molecular states and Raman lines mostly occur in far-infrared and near-infrared regions. Thus, it closely resembles rotational-vibrational absorption spectrum of a molecule. Again in many experimental observations Raman frequencies are found to be equal to infrared absorption frequencies of the specimen or to difference of absorption frequencies. But intensity-Variation for Raman lines and infrared absorption lines are found to be different. Moreover, all Raman lines do not have their corresponding infrared bands or vice-versa. Thus, the most intense Raman lines in benzene and toulene whose frequency shifts are 10.3×10^{-4} cm and 10.2×10^{-4} cm respectively, have no corresponding intense infrared absorption bands. On the other hand, intense absorption bands for the above two specimens respectively are 9.75×10^{-4} cm and 6.86×10^{-4} cm and there are no corresponding lines in the Raman spectra of the substances. Raman lines have also been observed corresponding to infrared transitions which are forbidden by the selection rule. Raman effect is the best compliment of infrared absorption effect by molecules. Thus, a very important conclusion can be drawn for these effects as:

"Although, the ulitimate result as far as the molecule is concerned, is same in both the cases, the mechanism of production and the intervening laws are much different in the two effects."

12.10 IMPORTANCE OF RAMAN EFFECT

Raman effect has a very great practical importance. The phenomenon is universal in nature and its experimental technique is very simple. The infra red absorption spectra of molecules—both vibrational and rotational—can be studied more easily with the help of Raman effect by shifting it to the visible part of the spectrum in any convenient position. Raman spectrum contains even those lines of the infrared absorption spectrum which are forbidden in a direct study. Homo-nuclear molecules like H_2 which do not exhibit vibrational and rotational spectra directly, exhibit the same in Raman effect.

Thus it enables us to study the structure of matter, mainly as regards the constitution of molecules, their number, arrangement and motion in gaseous, liquid and solid state.

12.11 APPLICATION OF RAMAN EFFECT

The following are the four important applications of Raman effect in Physics.

1. Molecular structure. Raman effect has been put to very great use in the study of structure of molecules. For this purpose, we make an evaluation of wavelength of Raman lines, their intensities and state of polarisation. Raman spectra include the *vibrational spectrum* of the molecules. The natural frequency of *diatomic molecule* is given by $\nu = \left(v + \frac{1}{2}\right)\nu_0$ where $\nu_0 = \frac{1}{2\pi}\sqrt{\frac{k}{m'}}$, k being the restoring force per unit displacement or force-constant and m' the reduced mass of the molecule. Thus we can find the masses of the atoms, the bond-length and strength of the bond between the atoms.

From this relation, it is clear that a molecule for which k has a large value *i.e.* the binding force between the atoms is large, has a *higher* characteristic frequency than the one in which the force is weak. This force depends upon the nature of inter-atomic bonds. In molecules having a *covalent bond*, the polarisability changes considerably due to nuclear oscillations, which in turn gives rise to intense Raman lines. In molecules having *electro-valent bond*, the binding electrons change over from one nucleus to other in the formation of the molecule. The polarisability of the molecule, therefore, is not effected by nuclear oscillations and hence no Raman lines appear.

Triatomic molecules. In the case of *triatomic* molecules, from a study of Raman effect we can find whether the molecule is *symmetric* or *asymmetric; linear* or *non-linear*. From the number and intensity of the observed lines in Raman effect alongwith the data from the infra-red spectrum important conclusions can be drawn regarding molecular structure. The *rule of mutual exclusion*, derived on theoretical grounds, states that '*for molecules with a centre of symmetry, transitions that are allowed in the infra-red spectrum are forbidden in the Raman spectrum and vice-versa*'. This rule does not imply that all transitions *forbidden in one must occur* in the other *i.e.* some transitions may be *forbidden in both*. On the other hand, certain transitions can occur both in the infra-red and Raman spectra in case of molecules without a centre of symmetry. For example, carbon dioxide CO_2 has two very strong bands in the infra-red absorption spectrum at wave number $= \frac{1}{\lambda} = 66800\,\text{m}^{-1}$ and 234900 m^{-1} while there is only one strong band at wave number $= \frac{1}{\lambda} = 13890\ \text{m}^{-1}$, in Raman spectra. As none of the bands appear both in Raman as well as infra-red spectrum, the CO_2 molecule has a *centre of symmetry*, according to the rule of mutual exclusion. So the CO_2 molecule is *linear* and *symmetric* and has a structure *O–C–O*.

Nitrous oxide (N_2O) has three absorption bands at $\frac{1}{\lambda} = 222400\,\text{m}^{-1}$, 128500 m^{-1}, and 58900 m^{-1} of which the first two appear in Raman spectra. Thus the molecule cannot have a centre of symmetry *i.e.* it is *asymmetric* and its structure is *N–N–O*.

Raman spectrum also tells us that water (H_2O) molecule, though symmetric is *non-linear* or *bent*. There is further evidence to show that the bend in the system is 120° (Fig. 12.4).

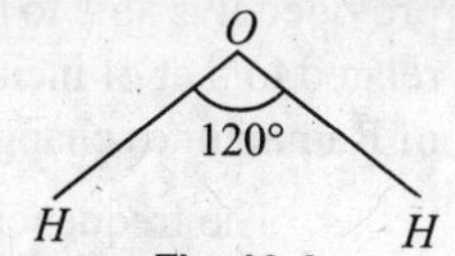

Fig. 12.4

2. Nature of liquids. In the gaseous state, the translatory motion of the molecules obeys Maxwell's law of distribution of velocities. Hence Rayleigh scattering is spread accordingly.

In the solid state, it has been shown that the translatory motions of the molecules may be resolved into a system of elastic waves traversing the media in all directions. On the basis of this theory, the scattered light should only consist of two displaced components, depending on the velocity of elastic waves and the angle of scattering.

The observations of Raman spectra have led to the conclusion that liquid state is an *intermediate* state in which the thermal motion of the liquid molecule is neither wholly disorganised or random nor wholly organised in the form of elastic waves in solids.

3. Crystal Physics. Raman effect in crystals is complimentary to X-ray crystal study. The frequency deduced from Raman spectra and the known positions and masses of the atoms in the crystal enable us to determine the *binding force* in crystals. With the help of Raman effect, the existence of '*mosaic structure*' and '*thermal agitation*' of the crystals has also been studied. These studies show the existence of discrete monochromatic infra red frequencies, which explain the fine structure in the infrared spectrum of the crystals.

Specific heat capacity of solids, brilliance of metals and their molecular structure has been explained by Raman effect.

4. Nuclear Physics. Raman effect has also been applied in the study of some aspects of nuclear Physics, such as *spin, statistics* and *isotopic constituents* of the molecules.

The rotational Raman lines of *homo-nuclear* molecules like H_2, D_2, N_2, O_2 etc have a *characteristic property of alternating intensities*. For example, for hydrogen, Raman lines for transitions between *odd rotational levels are three times* more intense than those arising from transitions between *even rotational levels*. This can be shown to be due to the fact that hydrogen nucleus has a *spin half*, obeys *Fermi-Dirac statistics* and that the *statistical weight* of the odd rotational levels is *three* times that of even rotational levels.

Similarly for Raman spectrum of *deuterium* D_2 on the same lines leads to the conclusion that D_2 nucleus has a *spin one*, obeys Bose-*Einstien statistics* and the *statistical weight of even lines* is *twice* that of *odd lines*.

On the same lines, nuclear spin of oxygen has been found to be zero.

5. Chemical effects. Various chemical effects, like strength of chemical bonds, electrolytic dissociation, hydrolysis etc have been understood through Raman effect. It has also been specially useful in the analysis of complicated mixture of organic molecules.

12.12 RAMAN EFFECT AND FLUORESCENCE

Raman effect and fluorescence resemble each other in the fact that the incident light in both cases suffers a change in wavelength and lines of new wavelengths appear in the spectrum of either. But Raman effect is quite different from fluorescence. The main points of difference are:—

1. The frequencies in fluorescent spectrum are always less than that of incident frequency, while in Raman effect, the scattered light has spectral lines both with higher frequency (anti-Stoke's lines) and lower frequency (Stoke's lines) than that for the incident frequency.

2. The frequencies in fluorescent spectra are independent of the frequency of exciting radiation, provided it is able to produce fluorescence, but in case of Raman lines, their frequencies are directly related to that of incident radiation. If the frequency of incident radiation is changed, the frequency of Raman lines change at the same rate.

3. The frequencies in fluorescent spectra are fixed by the nature of the scattering material, but in case of Raman spectra, it is the frequency shift of the Raman lines that depends upon the scatterer and not the actual frequencies.

4. The lines in fluorescent spectra are *unpolarised*, whereas Raman lines are *strongly polarised*.

5. Raman frequencies are either actual infra red frequencies in the absorption spectrum of the scatterer or difference in such frequencies, but this is not the case in fluorescent scattering.

12.13 RAMAN SCATTERING AND RAYLEIGH SCATTERING

It is true that '*Raman effect deals with interchange of energy between the incident radiation and matter i.e. molecules of scattering substance, during the process of scattering of light*.

The incident photon of energy *hv* shares energy with the molecule having intrinstic energy *E* and is scattered with energy hv_1, leaving the molecule with intrinstic energy E_1, so that

$$hv + E = hv_1 + E_1$$

We have not taken into consideration the kinetic energy of the molecule, because such interchange of energy does not involve a change in temperature and hence there is no change in the kinetic energy of the molecule.

If the molecule gains energy, the photon loses an equal amount and the frequency of the scattered photon is lowered, giving rise to Stoke's lines of Raman spectra. If the molecule is initially in an excited state, it loses energy and the photon gains an equal amount. The frequency of the scattered photon is increased, giving rise to anti-Stokes lines.

Raman effect and Rayleigh scattering. Rayleigh scattering is *coherent scattering i.e.* scattered light has the same frequency as that of incident light. In other words, there is no exchange of energy between the incident photon and the molecule of the scattering substance or the *collision is elastic*.

In Raman effect, as stated above, there is interchange of energy and the collision is *in-elastic*. The scattered photon may have an energy less than or greater than that of the incident photon, giving rise to Stokes lines on the lower frequency (longer wavelength) side or anti-Stokes lines on the higher frequency (shorter wavelength) side of the un-modified spectral line in the Raman spectrum.

12.14 RECENT TECHNIQUES FOR RAMAN SPECTROSCOPY

The invention of LASER and its use as a strong monochromatic source has revolutionzied the technique of Raman spectroscopy, making it more conveniant and more accurate. Raman scattering is inherently a weak process and unless the source is intense the lines obtained are diffused and needed much longer time of exposure. LASER is an ideal source for Raman experiments, because it is narrow, highly monochromatic and highly intense. Raman spectroscopy is particularly well suited to study the biological systems. Today it is used as an analytical and research tool. Computer averaging of multiple scans is commonly used to improve the experimental technique. Multichannel detectors, each sensing a different small part of the frequency range increases the sensitivity of the experiment. Recently, nearinfrared LASER excitation has been successfull using NdYAG LASER operating at 9398 cm^{-1}. Its current uses range from the non-destructive identification of the minerals to the early detection of life threatening diseases.

SOLVED EXAMPLES

Example 12.1. *In an experiment in the study of Raman effect using mercury green radiation of λ = 546.1 nm, a Stokes line of wavelength 554.3 nm was observed. Find Raman shift and wavelength corresponding to Anti-Stokes line.* ***(M.D.U. 2007 similar)***

Solution. Wavelength of exciting radiation

$$\lambda = 546.1 \text{ nm} = 546.1 \times 10^{-9} \text{ m} = 5461 \times 10^{-10} \text{ m}$$

$\therefore$ Frequency of exciting radiation $\nu = \dfrac{c}{\lambda} = c \times \dfrac{1}{5461 \times 10^{-10}}$

$$= c \times 1.8312 \times 10^{-4} \times 10^{10} = c \times 18312 \times 10^{2} \text{ s}^{-1}$$

Wavelength of Stokes line $\lambda_1 = 554.3 \text{ nm} = 554.3 \times 10^{-9} \text{ m} = 5543 \times 10^{-10}$ m.

$\therefore$ Frequency of Stokes line $\nu_1 = \dfrac{c}{\lambda_1} = c \times \dfrac{1}{5543 \times 10^{-10}}$

$$= c \times 1.8041 \times 10^{-4} \times 10^{10} = c \times 18041 \times 10^{2} \text{ s}^{-1}$$

$\therefore$ Raman shift $= \nu - \nu_1 = c \times 18312 \times 10^2 - c \times 18041 \times 10^2$

$$= c \times 10^2 [271] = 3 \times 10^8 \times 10^2 \times 271$$

$$= 813 \times 10^{10} \text{ s}^{-1}$$

As the Stokes and Anti-Stokes line have the same frequency displacement with respect to the exciting line and for Anti-Stokes line $\nu'_1 > \nu$

$\therefore$ Frequency of Anti stokes line $\nu'_1 = \nu + c \times 10^2 \times 271$

$$= c \times 10^2 [18312 + 271]$$

and wavelength of Anti-Stokes line $\lambda'_1 = \dfrac{c}{c \times 10^2 \times 18583}$

$$= 5.381 \times 10^{-5} \times 10^{-2} = 5381 \times 10^{-10} \text{ m}$$

$$= 5381 \text{ Å}.$$

Example 12.2. *In an experiment with Raman scattering the wave length of incident radiation is 436.8 nm. The wave length of the first scattered radiation is 462.4 nm. Find the wave length of the next scattered line.* ***(Bang. U. 2001)***

Solution. In Raman spectrum the frequency of the first scattered radiation on the longer wave length (lower frequency) side i.e. first Stokes line is given by $v - 3v_1$ and the frequency of the next Stokes line is given by $v - 5v_1$ where v is frequency of the incident radiation and $v_1 = \frac{\hbar}{2\pi I} = \frac{h}{4\pi^2 I}$

Here, wavelength of incident radiation $\lambda = 436.8 \times 10^{-9}$ m

$$= 4368 \times 10^{-10} \text{ m} = 4368 \text{ Å}$$

$$\therefore \quad v = \frac{c}{\lambda} = \frac{3 \times 10^8}{4368 \times 10^{-10}} = 6868 \times 10^{11} \text{ Hz}$$

Wavelength of first Stokes line $\lambda' = 462.4 \times 10^{-9}$ m

$$= 4624 \times 10^{-10} \text{ m} = 4624 \text{ Å}$$

$$\therefore \quad v' = \frac{c}{\lambda'} = \frac{3 \times 10^8}{4624 \times 10^{-10}} = 6488 \times 10^{11} \text{ Hz}$$

But $\quad v' = v - 3v_1 \quad \therefore 3v_1 = v - v'$

$$= (6868 - 6488) \times 10^{11} = 380 \times 10^{11} \text{ Hz}$$

or $\quad v_1 = 126.7 \times 10^{11}$ Hz

Frequency of second stokes line $= v - 5v_1$

$$= 6868 \times 10^{11} - 5 \times 126.7 \times 10^{11} = (6868 - 633.5)\, 10^{11}$$

$$= 6234.5 \times 10^{11} \text{ Hz}$$

$$\therefore \text{ Wavelength of second Stokes line} = \frac{3 \times 10^8}{6234.5 \times 10^{11}} = 4812 \times 10^{-10} \text{ m}$$

$$= 4812 \text{ Å}$$

EXERCISE

1. What is Raman effect? What are Stoke's and anti-Stoke's lines in Raman spectrum? Give the experimental set up to study Raman effect with the help of a neat diagram. ***(P. U. 2003; H.P.U. 2003, 2001)***
2. (*a*) How is Raman effect explained on the basis of quantum theory? Explain the origin of Stokes and Anti-Stokes lines in Raman spectrum. Why are the anti-Stokes lines fainter than Stokes lines? Give a detailed study of Raman effect.
3. How is Raman effect a two step process? Give the selection rules for Raman scattering. Explain, showing clearly the transitions within rotational levels of the molecule. Show that the spacing in Raman lines is twice the spacing between successive lines in rotational spectrum of the molecules. ***(G.N.D.U. 2008, 2005, 2004, 2003, 2002, 2001, 2000; H.P.U. 2001, 2000; Pbi. U. 2006, 2001, 2000; Kerala. Ŭ. 2001; P.U. 2006, 2005; Cal. U. 2003)***
4. (*a*) What is Raman effect? Explain its importance. ***(H.P.U. 2000)***
   (*b*) Write applications of Raman effect explaining how it is used in the study of molecular structure and triatomic molecules. ***(M.D.U. 2008)***

5. Bring out clearly the difference between Raman effect and flourescene.
6. Raman effect and fluorescence resemble each other in that the incident light in both the cases suffers a change in wavelength and new lines appear in spectrum, yet they are different. Explain.
7. Give your reasons for or against the statement 'Raman effect deals with interchange of energy between radiation and matter, during scattering of light'. How Raman effect is different from Rayleigh scattering? ***(Gauhati. U., 2000; Kerala. U., 2001)***
8. Distinguish between Raman scattering and Rayleigh scattering. ***(G.N.D.U. 2008)***
9. The carbon monoxide (CO) molecule has a bond length R of 0.113 nm and the masses of the ^{12}C and ^{16}O atoms are respectively 1.99×10^{-26} kg and 2.66×10^{-26} kg. Find the energy and angular velocity of CO molecule when it is in its lowest rotational state.

   Hint. Lowest rotational state has $J = 1$ and $E = \frac{1}{2} I \omega^2$

   [Ans. $E = 7.61 \times 10^{-23}$ J; $\omega = 3.23 \times 10^{11}$ rad s^{-1}]
10. The exciting line in an experiment with Raman effect is 5460 Å. If the Stokes line has $\lambda = 5520$ Å, calculate the wavelength of Anti-Stokes line. **[Ans. $\lambda = 5400$ Å]**

MULTIPLE CHOICE QUESTIONS

1. In Raman effect, Stoke's lines are observed when
   (*a*) $\Delta\nu$ is positive (*b*) $\Delta\nu$ is negative (*c*) $\Delta\nu$ is zero.
2. In Raman effect, anti-Stoke's lines are observed when
   (*a*) $\Delta\nu$ is positive (*b*) $\Delta\nu$ is negative (*c*) $\Delta\nu$ is zero.
3. In Raman effect, when $\Delta\nu$ is 0, spectral lines obtained are called
   (*a*) Stoke's lines (*b*) anti-Stoke's lines (*c*) Rayleigh lines.
4. The relative intensity of spectral lines in Raman effect observed are :
   (*a*) Stoke's lines are always more intense than anti-Stoke's lines.
   (*b*) Stoke's lines are equally intense with anti-Stoke's lines
   (*c*) anti-Stoke's lines are always more intense than Stoke's lines
5. The selection rule for Raman scattering is
   (*a*) $\Delta J = \pm 1$ (*b*) $\Delta J = \pm 2$ (*c*) $\Delta J = \pm 3$ (*d*) $\Delta J = \pm 4$
6. Raman's lines are
   (*a*) strongly polarised (*b*) weakly polarised (*c*) never polarised at all.
7. Stoke's or anti-Stokes line in Raman spectrum are
   (*a*) equally spaced (*b*) unequally spaced (*c*) mixed of both.

ANSWERS

1. (*a*)	2. (*b*)	3. (*c*)	4. (*a*)	5. (*b*)
6. (*a*)	7. (*a*)			

SPECTROSCOPIC TECHNIQUES

13

Chapter

INTRODUCTION

Spectroscopy is a technique that uses the interaction of energy with a sample to perform an analysis. The data that is obtained from spectroscopy is called a **spectrum**. *A spectrum is a plot of the intensity of energy detected versus the wavelength (or mass or momentum or frequency, etc.) of the energy.* A spectrum can be used to obtain information about atomic and molecular energy levels, molecular geometries, chemical bonds, interactions of molecules, and related processes. Often, spectra are used to identify the components of a sample (qualitative analysis). Spectra may also be used to measure the amount of material in a sample (quantitative analysis). There are several instruments and techniques that are used to perform a spectroscopic analysis. In simplest terms, spectroscopy requires an energy source (commonly a laser, but this could be an ion source or radiation source) and a device for measuring the change in the energy source after it has interacted with the sample (often a spectrophotometer or interferometer). Different spectroscopes generally used in practice are vibrational spectroscopy (IR and Raman), UV-vis spectroscopy, Nuclear Magnetic Resonance (NMR), Electron Spin Resonance (ESR), X-ray spectroscopy, electron spectroscopy, Mössbauer spectroscopy, mass spectroscopy etc.

13.1 TYPES OF SPECTROSCOPY

There are as many different types of spectroscopy depending upon the energy sources needed. Here are some examples:

1. Atomic Absorptoion spectroscopy: Energy absorbed by the sample is used to assess its characteristics. Sometimes absorbed energy causes light to be released from the sample, which may be measured by a technique such as fluorescence spectroscopy.

2. Electron Spectroscopy: There are serveral types of electron spectroscopy, all associated with measuring changes in electronic energy levels.

3. UV, Visible and Infra-red Spectroscopy: The infra-red absorption spectrum of a substance is sometimes called its molecular fingerprint. Although frequently used to identify materials, infra-red spectroscopy also may be used to quantify the number of absorbing molecules. UV-Vis-NIR spectroscopy is a technique used to examine features of this part (Ultra-Violet, Visible and Near Infra Red)[1] of the light spectrum. In this region, molecules undergo electronic transitions. These transitions can be seen as either the presence of fluorescence, phosphorescence or even the absorption of light in or at a specific wavelength or wavelength region. UV-Vis and fluorescence spectroscopy is used as part of a general identification tool used for either general property characterization or for the analysis of synthetic organic compounds.

4. Gamma-ray Spectrosecopy: Gamma radiation is the energy source in this type of spectroscopy, which includes activation analysis and Mossbauer spectroscopy.

5. Laser Spectroscopy: Absorption spectroscopy, fluorescence spectroscopy, Raman spectroscopy, and surface-enhanced Raman spectroscopy commonly use laser light as un energy source. Laser spectroscopy provides information about the interaction of coherent light with matter. Laser spectroscopy generally has high resolution and sensitivity.

1"Near UV" and "near IR" means the parts of the UV and IR spectra with are close to the visible spectrum.

6. Mass Spectrometry: A mass spectrometer source produces ions. Information about a sample may be obtained by analyzing the dispersion of ions when they interact with the sample, generally using the mass-to-charge ratio.

7. Electron Paramagnetic Spectroscopy: This is a microwave technique based on splitting electronic energy fields in a magnetic field. It is used to determine structures of samples containing unpaired electrons.

8. Raman Spectroscopy: Raman scattering of light by molecules may be used to provide information on a sample's chemical composition and molecular structure.

9. X-ray Spectroscopy: This technique involves excitation of inner electrons of atoms, which may be seen as x-ray absorption. An *x*-ray fluorescence emission spectrum may be produced when an electron falls from a higher energy state into the vacancy created by the absorbed energy.

10. Astronomical Spectroscopy: Energy from celestial objects is used to analyze their chemical composition, density, pressure, temprature, magnetic fields, velocity, and other characteristics. There are many energy types (spectroscopies) that may be used in astronomical spectroscopy.

11. Attenuated Total Reflectance Spectroscopy: This is the study of substances in thin firms or on surfaces. The sample is penetrated by an energy beam one or more times and the reflected energy is analyzed. Attenuated total reflectance spectroscopy and the related technique called frustrated multiple internal reflection spectroscopy are used to analyze coating and opaque liquids.

12. Fourier Transform Spectroscopy: This is a family of spectroscopic techniques in which the sample is irradiated by all relevant wavelengths simultaneously for a short period of time. The absorption spectrum is obtained by applying a mathematical analysis to the resulting energy pattern.

13.2 UV – VISIBLE SPECTROPHOTOMETRY

The most common spectrophotometers are used in the UV (200 to 400 mm) and visible (400 to 700 nm) regions of the spectrum, and some of these instruments also operate into the near-infra-red regions as well.

Visible region spectrophotometry is used extensively in colorimetry science, Ink manufacturers, printing companies, textiles vendors, and many more, need the data provided through colorimetry. They take readings in the region of every 5–20 nanometers along the visible region, and produce a spectral reflectance curve or a data stream for alternative presentations. These curves can be used to test a new batch of colorant to check if it makes a match to specifications, e.g., ISO printing standards.

Traditional visible region spectrophotometers cannot detect if a colorant or the base material has fluorescence. This can make it difficult to manage color issues if for example one or more of the printing inks is fluorescent. Where a colorant contains fluorescence, a bi-spectral fluorescent spectrophotometer is used. There are two major setups for visual spectrum spectrophotometers, d/8 (spherical) and 0/45. The names are due to the geometry of the light source, observer and interior of the measurement chamber. Scientists use this instrument to measure the amount of compounds in a sample. If the compound is more concentrated more light will be absorbed by the sample. If the compound ismore concentrated more light will be absorbed by the sample; within small ranges, the Beer-Lambert law holds and the absorbance between samples vary with concentration linearly. In the case of printing measurements two alternative settings are commonly used without/with UV filter to control better the effect of UV brighteners within the paper stock.

13.3 IR SPECTROPHOTOMETRY

Spectrophotometers designed for the main infrared region are quite different because of the technical requirements of measurement in that region. One major factor is the type of photosensors that are available for different spectral regions, but infrared measurement is also challenging because virtually everything emits IRT light as thermal radiation, especially at wavelengths beyond about 5 μm.

Another complication is that quite a few materials such as glass and plastic absorb infra-red light, making it incompatible as an optical medium. Ideal optical materials are salts, which do not absorb strongly. Samples for IR spectrophotometry may be smeared between two discs of potassium bromide or ground with potassium bromide and pressed into a pellet. Where aqueous solutions are to be measured, insoluble silver chloride is used to construct the cell.

Spectroradiometers, which operate almost like the visible region spectrophotometers, are designed to measure the spectral density of illuminants in order to evaluate and categorize lighting for sales by the manufacturer, or for the customers to confirm the lamp they decided to purchase is within their specifications.

13.4 SPECTROPHOTOMETERS

Principle

If a white light is passed through a coloured substance, some of the light gets absorbed. A solution containing hydrated copper ions, for example, look pale blue because the solution absorbs light from the red end of the spectrum. The remaining wavelengths in the light combine in the eye and brain to give the appearance of cyan (pale blue). Some colourless substances also absorb light - but in the ultra-violet region. Since we can't see UV light, we don't notice this absorption. Different substances absorb different wavelengths of light, and this can be used to help to identify the substance — the presence of particular metal ions, for example, or of particular functional groups in organic compounds. This amount of absorption is also dependent on the concentration of the substance if it is in solution. Measurement of the amount of absorption can be used to find concentration of very dilute solution. An absorption spectrometer measures the way that the light absorbed by a compound varies across the UV and visible spectrum.

The use of spectrophotometers spans various scientific field, such as physics, materials science, chemistry, biochemistry, and molecular biology. They are widely used in many industries including semiconductors, laser and optical manufacturing, printing and forensic examination, as well in laboratories for the study of chemical substances. Ultimately, a spectrophotometer is able to determine, depending on the control or calibration, what substances are present in a target and exactly how much through calculations of observed wavelengths.

Historically, spectrophotometers use a monochromator containing a diffraction grating to produce the analytical spectrum. The grating can either be movable or fixed. If a single detector, such as a photomultiplier tube or photodiode is used, the grating can be scanned stepwise so that the detector can measure the light intensity at each wavelength (which will correspond to each "step"). Arrays of detectors, such as charge coupled devices (CCD) or photodiode arrays (PDA) can also be used. In such systems, the grating is fixed and the intensity of each wavelength of light is measured by a different detector in the array. Additionally, most modern mid-infrared spectrophotometers use a Fourier transform technique to acquire the spectral information. The technique is called Fourier transform infra-red spectroscopy.

Designing of Spectrophotometer

Spectrophotometry involves the use of a spectrophotometer. A spectrophotometer is a photometer that can measure intennsity as a function of the light source wavelength. Important features of spectrophotometers are spectral bandwidth and linear range of absorption of reflectance measurement.

A spectrophotometer is commonly used for the measurement of transmittance or reflectance of solutions, transparent or opaque solids, such as polished glass, or gases, However they can also be designed to measure the diffusivity on any of the listed light ranges that usually cover around 200 nm – 2500 nm using different controls and calibrations. Within these ranges of light, calibrations are needed on the machine using standards that vary in type depending on the wavelength of the photometric determination.

When making transmission measurements, the spectrophotometer quantitatively compares the fraction of light that passes through a reference solution and a test solution. For reflectance measurements, the spectrophotometer quantitatively compares the fraction of light that reflects from the reference and test samples. Light from the source lamp is passed through a monochromator, which diffracts the light into a "rainbow" of wavelengths and outputs narrow bandwidths of this diffracted spectrum. Discrete frequencies are transmitted through the test sample. Then the photon flux density (watts per metre squared usually) of the transmitted or reflected light is measured with a photodiode, charge coupled device or other light sensor. The transmittance or reflectance value for each wavelength of the test sample is then compared with the transmission (or reflectance) values from the reference sample.

In short, the sequence of events in a modern spectrophotometer is as follows:

1. The light source is imaged upon the sample
2. A fraction of the light is transmitted or reflected from the sample
3. The light from the sample is imaged upon the entrance slit of the monochromator
4. The monochromator separates the wavelengths of light and focuses each of them onto the photodutector sequentially.

13.5 SINGLE AND DOUBLE BEAM SPECTROPHOTOMETERS

There are two major classes of devices: **single beam** and **double beam**. A single beam spectrophotometer (Fig. 13.1) measures the relative light intensity of the beam before and after a test sample is inserted.

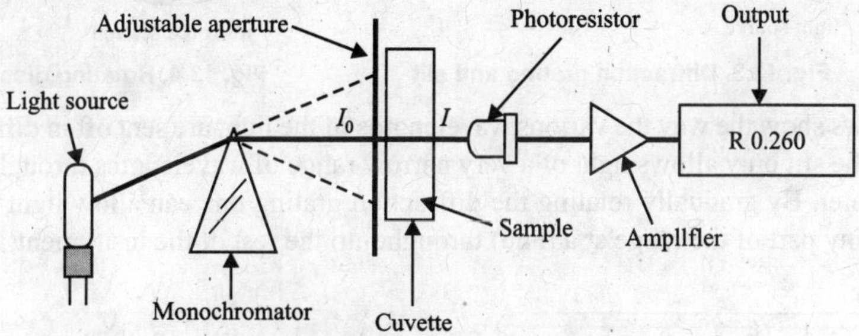

Fig. 13.1 Single beam spectrophotometer

A double beam spectrophotometer (Fig. 13.2) compares the light intensity between two light paths, one path containing a reference sample and the other the test sample. Although comparison

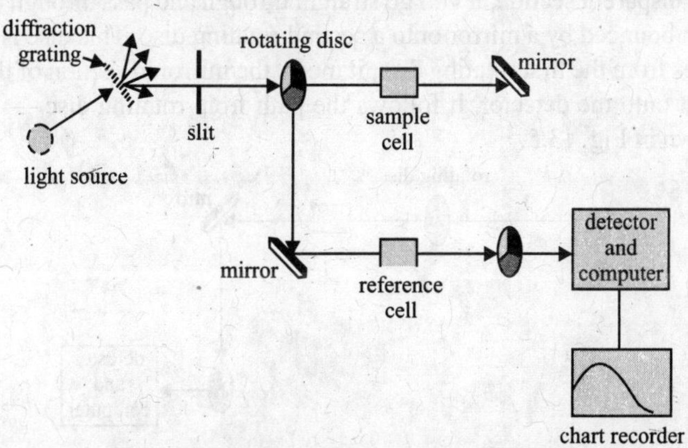

Fig. 13.2. Double beam spectrophotometer

measurements from double beam instruments are easier and more stable, single beam instruments can have a larger dynamic range and are optically simpler and more compact. Additonally, some specialized instruments, such as spectrophotometer built onto microscopes or telescopes, are single beam instruments due to practicality.

Working

Light Source:

We need a light source which gives the entire visible spectrum plus the near ultra-violet so that you are covering the range from about 200 nm to about 800 nm. (This exxtends slightly into the near infrared as well.) You can't get this range of wavelengths from a single lamp, and so a combination of two is used - a deuterium lamp for the UV part of the spectrum, and a tungsten / halogen lamp for the visible part. The combined output of these two bulbs is focussed on to a diffraction grating.

Diffraction grating and the slit:

We are familiar with the way that a prism splits light into its component (constituent) colours. A diffraction grating does the same job, but more efficiently.

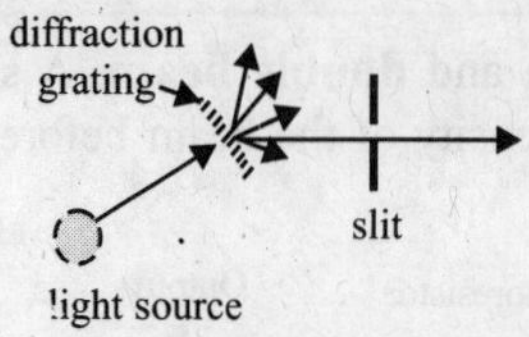

Fig. 13.3. Diffraction grating and slit

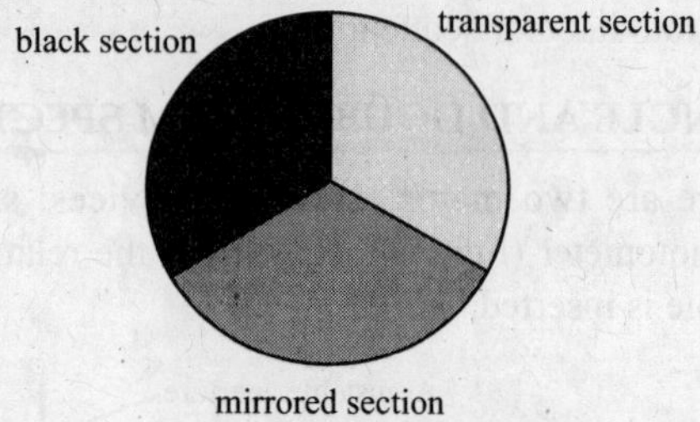

Fig. 13.4. Rotating disc

The arrows show the way the various wavelengths of the light are sent off in different directions (Fig. 13.3). The slit only allows light of a very narrow range of wavelengths through into the rest of the spectrometer. By gradually rotating the diffraction grating one can allow light from the whole spectrum (a tiny part of the range at a time) through into the rest of the instrument.

Rotating disc

Each disc is made up of a number of different segments (Fig. 13.4.). In this figure, we have shown three different sections - other designs may have a different number. The light coming from the diffraction grating and slit will hit the rotating disc and one of three things can happen:

1. If it hits the transparent section, it will go straight through and pass through the cell containing the sample. It is then bounced by a mirror onto a second rotating disc. This disc is rotating such that wnen the light arrives from the first rotating disc, it meets the mirrored section of the second rotating disc. That bounces it onto tne detector. It follows the path from rotating disc — mirror — another rotating disc as shown in Fig. 13.5.

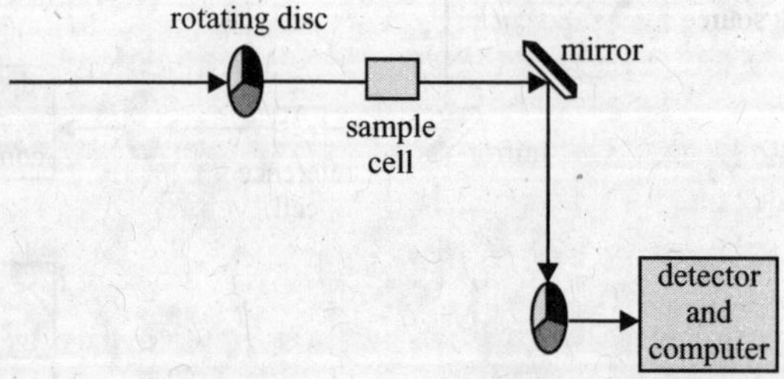

Fig. 13.5. Path of ray passing through sample cell

2. If the original beam of light from the slit hits the mirrored section of the first rotating disc, it is bounced down along the path as shown in Fig. 13.6. After the mirror, it passes through a reference cell. Finally the light gets to the second disc which is rotating in such a way that it meets the transparent section. It goes straight through to the detector.

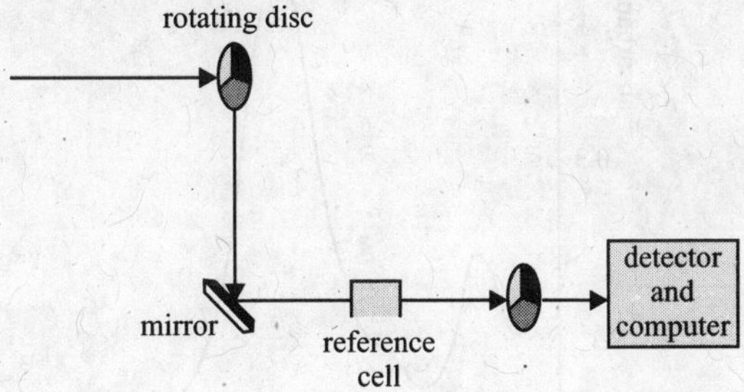

Fig. 13.6. Path of ray passing through reference cell

3. If the light meets the first disc at the black section, it is blocked - and for a very short while no light passes through the spectrometer. This just allows the computer to make allowance for any current generated by the detector in the absence of any light.

Sample and Reference cells

These are small rectangular glass or quartz containers. They are often designed so that the light beam travels a distance of 1 cm through the contents. The sample cell contains a solution of the substance you are testing — usually very dilute. The solvent is chosen so that it doesn't absorb any significant amount of light in the wavelength range we are interested in (200 - 800 nm). The reference cell just contains the pure solvent.

Detector and Computer

The detector converts the incoming light into a current. The higher the current, the greater the intensity of the light. For each wavelength of light passing through the spectrometer, the intensity of the light passing through the reference cell is measured. This is usually refered to as I_0 (here I stands for Intensity). The intensity of the light passing through the sample cell is also measured for that wavelength - denoted by the symbol, I.

If is less than I_0, then obviously the sample has absorbed some of the light. From this the computer convert this into something called the *absorbance* of the sample - given the symbol, A.

The relationship between A and the two intensities is given by :

$$A = \log_{10} \frac{I_0}{I}$$

On most of the diagrams you will come across, the absorbance ranges from 0 to 1, but it can go higher than that.

An absorbance of 0 at some wavelength means that no light of that particular wavelength has been absorbed. The intensities of the sample and reference beam are both the same, so the ratio I_0/I is 1. Log_{10} of 1 is zero. An absorbance of 1 happens when 90% of the light at that wavelength has been absorbed - which means that the intensity is 10% of what it would otherwise be. In that case, I_0/I is 100/10 (= 10) and $\log_{10}$ of 10 is 1.

Chart recorder

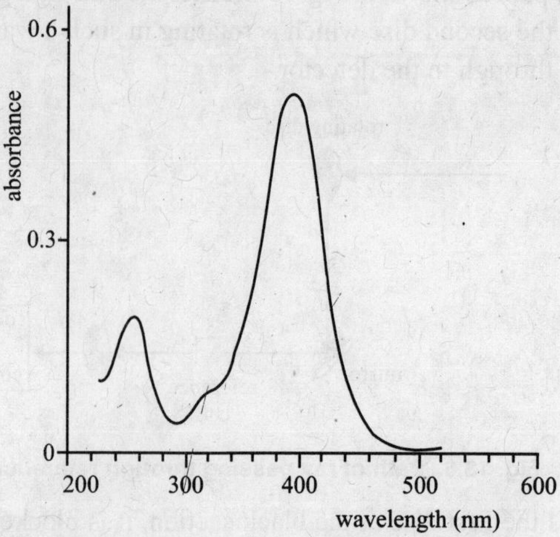

Fig. 13.7 absorbance against wavelength curve

Chart recorders usually plot absorbance against wavelength. The output might look like curve as shown in Fig. 13.7. This particular substance has two absorbance peaks at 255 and 395 nm.

EXERCISE

1. What is Spectroscopy ?
2. What is a spectrum ? What information is obtained from it ?
3. What instruments and techniques are needed ?
4. Mention different types of spectroscopies used ?
5. Explain the principle and working of a double beam spectrophotometer.
6. Explain the IR spectrophotometry.
7. Explain the UV – visible spectrophotometry.
8. How measurements are done in different spectrophotometers ?

SECTION V

Nuclear Physics

14. NUCLEAR DETECTORS

15. STRUCTURE OF NUCLEI

16. NUCLEAR MODELS

17. NUCLEAR REACTIONS

NUCLEAR DETECTORS

14

Chapter

INTRODUCTION

In experimental and applied particle physics, nuclear physics and nuclear engineering, **a particle detector**, also known as a **radiation detector** or **nuclear detector**, is a device used to detect, track and indentify high energy particles, such as those produced by nuclear decay, cosmic radiation, or reactions in a particle accelerator. Detectors designed for modern accelerators are huge, both in size and cost. The term **counter** is often used instead of detector, when the detector counts the particles but does not resolve its energy of ionization. Let us start with the interaction of charged particles and neutrons with matter.

14.1 INTERACTION OF CHARGED PARTICLES WITH MATTER

The charged particles like alpha (α) and beta (β) particles emitted by radioactive substances, protons, electrons, helium nuclei etc. ejected in nuclear reactions collide with the particles of medium elastically and impart their energy and momentum to the particles of the medium. In the process of collision, the particles of the medium get ionized producing pairs of negative and positive ions. Obvioasly, the number of ion-pairs produced depends upon the initial energy of the charged particles colliding. This process in which the number of ion-pairs produced in the medium is called *primary ionization.* In this process the energy of the charged particle decreases but not sufficiently large enough to slow down the particle. So, the charge particles can produce number of successive collisions, loosing energy and momentum in every collision and finally becomes ineffective due to low energy. The rate of loss of energy per unit path travelled in a medium has been calculated by Neil Bohr and then corrected by Bethe-Bloch.

14.2 NEIL-BOHR'S FORMULA

The total energy dE transferred to the medium when the particle travels a distance dx in it *i.e.* the kinetic energy lost by the particle in a length dx of its path is given by the integral of equation between the minimum and maximum values ($b_{\min}$ and $b_{\max}$) of the impact parameter b.

$\therefore$ Energy lost by the particle in traversing a unit distance in the medium

$$\frac{-dE}{dx} = \frac{z^2 e^4 n}{4\pi\varepsilon_0^2 m_e v^2} \int_{b_{\min}}^{b_{\max}} \frac{db}{b}$$

$$= \frac{z^2 e^4 n}{4\pi\varepsilon_0^2 m_e v^2} \log_e \frac{b_{\max}}{b_{\min}} \qquad ...(i)$$

At first it seems natural to put $b_{\min} = 0$ and $b_{\max} = \infty$ but this leads to an absurd result *i.e.*, the rate of loss of energy is infinite and the charged particle will stop instantly.

To find $b_{\max}$. The electrons in the stopping material are not free but revolve round the nucleus in stationary orbits with frequency ν. The force on the electron is an impulse of short duration $= \frac{b}{v}$. If this time is equal to the period of revolution of the electron, then

$$\frac{1}{\nu} = \frac{b}{v} \quad \text{or} \quad b = \frac{v}{\nu}$$

At values of b, greater than $\frac{v}{\nu}$, $\frac{b}{v}$ will be greater than $\frac{1}{\nu}$, the period of revolution of the electron in the atom. In such a case, the electron will absorb no energy and therefore the moving particle will not lose any energy. Hence value of b should not exceed $\frac{v}{\nu}$ *i.e.,*

$$b_{max} = \frac{v}{\nu}$$

To find b_{min}. The lower limit of b can be determined by Heisenberg's uncertainty principle. In an elastic *head-on-collision* (assuming the heavy particle at rest and electron impinging upon it) the maximum change in velocity of the electron = $2v$ and hence the maximum change in its momentum = $2\,m_e\,v$.

$\therefore$ According to uncertainty principle $b_{min} = \dfrac{h}{2 m_e v}$

This is approximately the size of the electron as seen by the moving particle

$$\therefore \quad -\frac{dE}{dx} = \frac{z^2 e^4 n}{4\pi\varepsilon_0^2 m_e v^2} \log_e \frac{2 m_e v^2}{h\nu} \quad ...(ii)$$

where $h\nu = I$, the average *excitation energy* (or *binding* energy) of the atomic electron of the medium.

Equation (*ii*) is known as *Bohr's formula.*

If N is the number of atoms per unit volume of the medium and Z the atomic number of the medium, then

Number of electrons per unit volume $n = NZ$

Hence relation (*ii*) can be put in the form

$$-\frac{dE}{dx} = \frac{z^2 e^4 NZ}{4\pi\varepsilon_0^2 m_e v^2} \log_e \frac{2 m_e v^2}{h\nu} \quad ...(iii)$$

Stopping power. The quantity $\dfrac{-dE}{dx}$ is known as the *stopping power* of the medium.

It is defined as the loss of energy of the incident charged particle per unit length of the medium.

Factors on which stopping power depends. (*i*) It does not depend upon the mass of the incident particle.

(*ii*) It depends upon the charge of the incident particle. It is *directly* proportional to z^2 where ze is the charge on the incident particle. Thus the stopping power of a material for α-particles ($z = 2$) is four times the stopping power for the protons ($z = 1$) for the same velocity.

(*iii*) It is *inversely* proportional to v^2 where v is the velocity of the incident charged particle.

Assumptions. In arriving at relation given by Eq. (*iii*) following assumptions have been made

(*i*) The heavy charged particle transfers energy only to the atomic electrons and energy transferred to the heavy positive nucleus is negligible.

(*ii*) The velocity of the heavy charged particle is so high that the revolving orbital electron does not appreciably change its position during the time the charged particle passes by it.

(*iii*) The velocity of the heavy charged particle remains constant during the collision as the change in velocity due to collision with a single electron is negligible.

Bethe-Bloch Modification (formula)

When the velocity of the incoming charged particle is very large, so that it is comparable to the velocity of light, relativistic correction has to be applied to the value of b_{min} and b_{max}.

To find b_{max}. Due to relativistic effect the electric field of the incident particle is contracted in the direction of motion and increased in a perpendicular direction by a factor $\dfrac{1}{\sqrt{1-\dfrac{v^2}{c^2}}}$. This *sharpens* the impulse given to the electron and its duration is given by $\dfrac{b}{v}\sqrt{1-\dfrac{v^2}{c^2}}$. Hence b_{max} is now given by

$$\frac{1}{\nu} = \frac{b_{max}\sqrt{1-\dfrac{v^2}{c^2}}}{v} \qquad \text{...}(iv)$$

or

$$b_{max} = \frac{v}{\nu\sqrt{1-\dfrac{v^2}{c^2}}}$$

To find b_{min}. To find b_{min} we have assumed the heavy particle at rest and electron impinging upon it. Relativistically, in a co-ordinate system in which the incident particle is at rest and the electron has a velocity v, the maximum change in momentum of the electron is given by

$$\frac{2m_e v}{\sqrt{1-\dfrac{v^2}{c^2}}} \text{ and hence } b_{min} = \frac{h}{2m_e v}\sqrt{1-\frac{v^2}{c^2}} \qquad \text{...}(v)$$

In deriving this relation, it has been assumed that the orbital velocity of the electron is much less than v.

From (*iv*) and (*v*), for a relativistic case

$$\frac{b_{max}}{b_{min}} = \frac{v}{\nu\sqrt{1-\dfrac{v^2}{c^2}}} \times \frac{2m_e v}{h\sqrt{1-\dfrac{v^2}{c^2}}}$$

$$= \frac{2m_e v^2}{h\nu\left(1-\dfrac{v^2}{c^2}\right)}$$

$\therefore$ Applying relativity correction

$$\frac{-dE}{dx} = \frac{z^2e^4NZ}{4\pi\varepsilon_0^2 m_e v^2}\left[\log_e \frac{2m_e v^2}{h\nu\left(1-\dfrac{v^2}{c^2}\right)}\right] \qquad \text{...}(vi)$$

$$= \frac{z^2e^4NZ}{4\pi\varepsilon_0^2 m_e v^2}\left[\log_e \frac{2m_e v^2}{h\nu} - \log_e\left(1-\frac{v^2}{c^2}\right)\right]$$

In 1930, Bethe developed an approximate quantum mechanical relationship and a more exact formula was derived by F. Bloch in 1933. Bethe-Bloch formula is

$$\frac{-dE}{dx} = \frac{z^2 e^4 NZ}{4\pi\varepsilon_0^2 m_e v^2}\left[\log_e \frac{2m_e v^2}{h\nu} - \log_e\left(1 - \frac{v^2}{c^2}\right) - \frac{v^2}{c^2}\right] \quad ...(vii)$$

Relative stopping power. *The ratio of the stopping power of a material to the stopping power of air at N.T.P. for equal energy loss is called relative stopping power of the material.*

Relative atomic stopping power. The relative atomic stopping power is defined as

$$S_a = \frac{\left(1/N\right)_m \left(dE/dx\right)_m}{\left(1/N\right)_{air} \left(dE/dx\right)_{air}}$$

where $(N)_m$ and $(N)_{air}$ are the number of atoms per unit volume of the material and air respectively and $\left(dE/dx\right)_m$ and $\left(dE/dx\right)_{air}$, the stopping power of the material and air respectively.

The relative atomic stopping power for air at N.T.P (*Standard air*) is taken to be *unity.* Comparatively the relative atomic stopping power for H_2 is 0.2, for copper it is 2.4 and for lead 4.3.

14.3 LIMITATIONS TO BETHE-BLOCH FORMULA

Bethe-Bloch formula of interaction of charged particles with matter is not applicable to electrons. This is the limitation of this formula.

The following are the main reasons why Bethe-Bloch formula of interaction of charged particles with matter does not apply to electrons. Bethe-Bloch formula states

$$\frac{-dE}{dx} = \frac{z^2 e^4 NZ}{4\pi\varepsilon_0^2 m_e v^2}\left[\log_e \frac{2m_e v^2}{h\nu} - \log_e\left(1 - \frac{v^2}{c^2}\right) - \frac{v^2}{c^2}\right]$$

where $\frac{-dE}{dx}$ is the loss of energy per unit length, ze the charge on the particle, v its velocity, Z the atomic number of the absorber which contains N atoms per unit volume whose average ionisation energy is $h\nu$ and m_e the rest mass of the electron.

Reasons. (*i*) The derivation of the above formula assumes that the incident particle is practically undeflected. But the incident particle acquires a *transverse* component of momentum per collision approximately equal to that given to the absorbing electron. If the incident particle is itself an electron the transverse velocity corresponding to the momentum will not be negligible as assumed in the case of a *heavy* incident particle. Thus if the incident particle is an electron it will be deflected through much larger angles and its path will be very irregular and not at all straight as is the path of the heavy charged particle.

(*ii*) Electrons have a continuous distribution of initial energy whereas heavy charged particles are mono-energetic.

(*iii*) Collision between identical particles involves an energy exchange phenomenon which has to be taken into account for electron-electron collisions.

(*iv*) Electrons of energy very much less than 1 MeV lose energy mostly by excitation or ionisation of the atoms of the material through which they pass. The charge on the electron never changes and the electron loses energy by ionisation even at very low energies unlike the heavy positively charged particles which lose their charge by recombination at lower energies.

(*v*) **Bremsstrahlung.** Electrons with energies very much greater than 1 MeV radiate considerable fraction of their energy in the form of *X*-rays due to their inelastic collisions with coulomb field of atomic nuclei and also with electrons.

This phenomenon is, therefore, known as *Bremsstrahlung*. Bremsstrahlung is a German term meaning breaking radiation. It represents energy radiated per unit time in the form of electro-magnetic radiation when a charged particle moving with a high speed is accelerated or decelerated. When a charged electron moving with relativistic speed passes near a nucleus of charge *Ze*, its path is bent a little. Such a bending is due to the influence of the nuclear field and gives rise to emission of radiation which is found to be proportional to the square of acceleration. The radiation losses are negligble for heavy charged particles like proton, deutron and α-particle because it is inversely proportional to the square of the mass of incident particle.

(*vi*) High speed of electrons makes it necessary to employ a relativistic treatement of the collision mechanism.

14.4 RANGE AND STRAGGLING OF A CHARGED PARTICLE

Range. When a heavy charged particle moves through a medium it excites and ionises the atoms with which it comes sufficiently close and in the process loses some kinetic energy by transferring it to the orbital electrons of the atoms of the medium. The particle, therefore, goes on losing speed. As the particle slows down to a speed of the same order of magnitude as that of the outer orbital electrons in the atoms of the medium, it picks up an electron which reduces its effective charge and this reduction in the value of *z* results in a decrease in the rate of loss of energy of the particle. This continues until the particle becomes neutral so that the ionisation density in the medium drops to zero by the time the particle stops.

Very seldom does a heavy chargd particle come close enough to the nucleus of an atom of the medium to get deviated from its path through the large angle Rutherford scattering or to be lost through the initiation of a nuclear reaction. The charged particle, therefore, travels in a straight line path losing energy discontinuously in ionising and exciting the atoms of the medium. We may, however, consider the process as almost continuous due to the large number of ions generated by the particle. This leads to the concept of well defined range of the particle in the medium which is related to the total energy of the particle.

The distance travelled by a charged particle in a given medium before it comes to rest is called its range. This is the *average* distance travelled by the charged particle in the medium before it loses all its kinetic energy.

The more accurate formula for range gives a better approximation as

$$R \propto E^{3/2} \propto v^3$$

where E_0 is initial kinetic energy of the particle and v its velocity.

Empirical range energy relations. The range of an **α-particle** in air at 15°C and standard atmospheric pressure is given by empirical relation $R = bE_0^{3/2} = 0.00318\,E_0^{3/2}$ metre where E_0 is the initial energy of the α-particle in MeV. A semi-empirical formula for the **proton** range energy relation is $R = \left(\dfrac{E_0}{9.3}\right)^{1.8}$ where E_0 is in MeV and R in metres of air. The range energy relation for an **electron** valid for aluminium in the energy range 0.15 MeV to 0.8 MeV is given by

$$R = 0.00407\,E_0^{1.38} \text{ metre.}$$

Straggling. Identical charged particles, even if having the same initial velocity do not all have the same range. While passing through matter particles lose energy in small but definite amounts. After undergoing a large number of collisions their energy is reduced by a certain specific amount The number of collisions by which each particle loses energy will, however, be different. Therefore, the value of the range is distributed over a small interval about a mean value. This phenomenon is known as *straggling*.

Consider the case of α-particles emitted by natural radio-active elements. These α-particles can be easily absorbed by a sheet of paper or a few centimeters of air. If the α-particles are counted, say, by a G.M. counter it is found that their number remains constant upto a certain distance R from the source and then drops to zero in a manner shown by the graph between '*Relative Counting rate*' and '*range*' (Fig. 14.1).

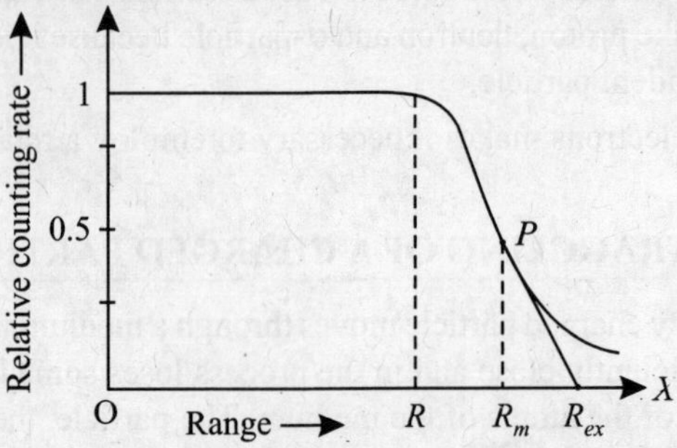

Fig. 14.1

The distance R is called the range of the particle and its value depends upon the initial energy of the particle. The mean (or average) value of the range R_m is defined as the *value of the range at which the number of α-particles falls to one half of the original value.* Let this point on the graph be P corresponding to R_m the mean value of the range.

The *extrapolated range* R_{ex} is defined as *the distance of the point on the X (range) axis where the tangent to the curve at the point P passed by half the particles intersects the X-axis.*

The distance between R_{ex} and R_m is known as '*range straggling*' or '*straggling parameter*'.

When the incident beam of heavy charged particles passes through the stopping material and suffers range straggling, the emergent beam after suffering multiple collisions broadens out giving rise to '*angular straggling*'.

14.5 ENERGY LOSS OF ELECTRONS AND HEAVEY IONS DURING INTERACTION

The energy loss of electron on their passage through matter is a complicated phenomenon as compared to the energy loss by ionisation of heavy ions due to the following reasons.

1. Electrons being small particles suffer large deflections on collision with atomic electrons of the absorbing material. The path of the electron is, therefore, highly irregular.

2. Electrons emitted by radioactive elements have a continuous distribution of energy whereas heavy ions like α-particles ($_2He^4$) are mono-energetic.

3. In the case of electrons, the two colliding particles are identical in character. This requires that exchange phenomenon should be taken into account in the theoretical investigation of electron-electron collision.

4. β-particles have very high velocities–comparable to the velocity of light. This makes it necessary to employ relativistic treatment of the collision mechanism.

14.6 PASSAGE OF A FAST ELECTRON THROUGH MATTER

There are several processes which contribute to dissipation of energy of electrons while passing through matter.

(*i*) **Ionisation.** The loss of energy of electrons due to ionisation may be treated in a manner similar to that for heavy charged particles. However, it is a case of collision between identical particles. When electron energy is low, the loss due to ionisation predominates. Ignoring relativity effect, the loss of energy per unit path length of an electron in the low energy limit is given by the same relation as for a heavy charged particle. *i.e.*,

$$\left(\frac{-dE}{dx}\right)_{ion} = \frac{e^4 NZ}{4\pi\varepsilon_0^2\, m_e v^2}\left[\log_e \frac{2m_e v^2}{h\nu} - 1.2329\right] \qquad ...(i)$$

We have taken $z = 1$ for the electron and there is an additional numerical term -1.2329. $\left(\frac{-dE}{dx}\right)$ due to ionisation is directly proportional to N the number of atoms per unit volume, Z the atomic number of the medium and inversely proportional to $m_e v^2$ or $\frac{1}{2}m_e v^2$ or E the initial kinetic energy of the electron.

$$\therefore \qquad \left(\frac{-dE}{dx}\right)_{ion} \propto \frac{NZ}{E}$$

Hence energy loss due to ionisation is quite high at low energies and decreases as the energy increases. As the energy exceeds the rest mass energy $m_e c^2$ of the electron the particle behaves relativistically and the terms $\log_e \frac{2m_e v^2}{h\nu}$ in Eq. (*i*) becomes dominant and ionisation again starts increasing.

The ionisation loss is proportional to NZ which is nearly a constant for nuclei having small value of Z. The energy loss through ionisation is most important factor for electron energies below 1 MeV.

(*ii*) **Radiation (Bremsstrahlung).** Electrons with energies very much greater than 1 MeV radiate considerable fraction of their energy in the form of X-rays due to their inelastic collisions with coulomb field of atomic nuclei and also with electrons. This phenomenon is known as *Bremsstrahlung.* It is a German term meaning breaking radiation. It represents energy radiated per unit time in the form of electromagnetic radiation when a charged particle moving with a high speed is accelerated or decelerated. When a charged electron moving with relativistic speed passes near a nucleus of charge Ze, its path is bent a little. Such a bending is due to the influence of the nuclear field and gives rise to emission of radiation. The energy loss per unit length due to radiation is direactly proportional to Z^2 and also directly proportional to E *i.e.*,

$$\left(\frac{-dE}{dx}\right)_{rad} \propto Z^2 E$$

Thus the radiation loss is low for small values of energy and is almost zero for very low energy values. It increases with kinetic energy of the electron. Energy loss through radiation depends upon Z^2 while energy loss through ionisation depends upon Z only. Radiation losses are, therefore, more prominent in heavy materials.

(*iii*) **Scattering by atomic electrons and back scattering.** The electrons moving through a medium lose a small amount of energy in scattering by atomic electrons and back scattering by nuclei similar to that of Rutherford scattering of α-particles.

14.7 INTERACTIONS OF NEUTRONS WITH MATTER

Neutrons are uncharged (neutral) particles and do not interact with atomic electrons in the matter through which they are passing, but they do interact with the nuclei of these atoms. The nuclear force, with leads to these interactions, is very short ranged which means the neutrons have to pass close to a nucleus for an interaction to take place. Bacause of the small size of the nucleus in relation to the atoms as a whole, the neutrons will have a low probability of interaction, and could thus travel considerable distances in matter.

The way in which neutrons interact with matter depends to a large extent on their energies, which can range from hundreds of MeV down to fractions of an eV. The most common neutron reactions are:

1. Elastic scattering: This is analogous to a billiard ball type of collision. The neutron collides with a nucleus and rebounds in a different direction. The energy the neutron loses is gained by the target nucleus which moves away at a increased speed. If the neutron collides with a massive nucleus it rebounds with almost the same speed and loses very little energy. Light nuclei, on the other hand, will gain a lot of energy from such a collision and will therefore be more effective for slowing down neutrons.

2. Inelastic scattering: A neutron may strike a nucleus and form a compound nucleus instead of bouncing off as in elastic scattering. This nucleus is unstable and emits a neutron of lower energy together with a gamma photon which takes up the remaining energy. This process, called inelastic scattering, is most effective at high neutron energies in heavy materials, but at lower energies (a few MeV) elestic scattering becomes a more important reaction for energy loss proovided there are light nuclei present.

Elastic scattering is not effective in slowing down neutrons with very high energy (above 150 MeV).

3. Radiative capture: This is one of the most common neutron reactions. The neutron is again captured by a nucleus which emits only a gamma photon. This reaction, which occurs in most materials, is the most important one for neutrons with very low energy. The product nuclei of (n, γ) reactions are usually redioactive and are beta and gamma emitters.

Two of the neutron capture reactions which are important from the radiological standpoint at most accelerator and nuclear reactor facilities are the (n, γ) reaction in Co-59, which is normal stable cobalt metal and quite commonly occurs in steel, to produce Co-60, which is radioactive. The cobalt readily captures neutrons, and Co-60 has a half-life of about 5 years. The other is the neutron capture in Na-23, which is normal, stable sodium. In this case the product is the radioisotope Na-24. Traces of sodium are present in the concrete shielding.

Neutron Detection

Collisions between neutrons and atomic nuclei are rate events, because both are tiny compared to atoms. Collisions can be elastic, or inelastic. Other interactions are (n, γ), (n, α), (n, p), and $(n, 2n)$ nuclear reactions. In collisions with heavy nuclides, neutrons lose little energy. In collisions with light nuclides such as H, D, He, and C, neutrons lose a mojor portion of their energies. A neutron can lose all its kinetic energy in a single collision with proton. Thus, light nuclides are effective moderators, but not heavy nuclides.

Captures of neutron by proton and deuterium lead to the formation of deuterium and tritium respectively. The cross sections are small and many neutrons decay if not captured.

Neutrons are detected indirectly by observing the protons knocked loose by them or by nuclear reactions induced by them. For example, the alpha particles are easily detected in the following reactions.

$$^{14}N + n \longrightarrow {}^{11}B + \alpha$$

$$^{10}B + n \longrightarrow {}^{7}Li + \alpha$$

Slow neutrons are measured by a proportional counter, which is filled with BF_3 gas. The products, ^7Li and α ionize the gas in the proportional counter and the signals are detected. Fission reactions induced by neutrons can also serve for neutron detection.

Neutrons whose kinetic energy is above about 1 keV are generally classified as **fast neutrons.** They give large portions of their kinetic energy to the protons. Energetic protons ionize, and they are detected as such.

Since a neutron has no charge it can easily enter into a nucleus and cause reaction. Neutrons interact primarily with the nucleus of an atom, except in the special case of magnetic scattering where the interaction involves the neutron spin and the magnetic moment of the atom. Neutron reactions can take place at any energy, so one has to pay particular attention to the energy dependence of the interaction cross section. In a nuclear reactor neutrons with energies from 10^{-3} eV (1 meV) to 10^{7} eV (10 MeV) are of interest, this means we will be covering an energy of 10^{10} eV.

Interaction of Neutrons with Human Tissue and its Protections

The interaction of neutrons with human body (particularly human tissues) is important in personal dosimetry frequenctly used in nuclear reactors for human protection from neutrons. The human body is composed largely of water, about 60% by weight, which contain many hydrogen nuclei. Elastic scattering of the neutrons with the hydrogen nuclei will cause the protons to recoil violently. Similarly elastic collisions of neutrons with carbon, oxygen or other heavier nuclei will cause these to recoil. Because the mass of protons and the other recoiling nuclei is much greater than that of electrons, they generate a much denser ion path resulting in more damage to the tissue. Once neutrons have been slowed down by elastic collisions to thermal energy, 0.025 eV, thev are readily captured by some of the reactions described above.

A very common reaction is the (n, γ) reaction, particularly with hydrogen. The gamms photon produced in the reaction always has an energy of 2.2 MeV and will cause indirect ionization as described previously. When neutrons are absorbed by an (n, γ) reaction in the body the tissues will be further damaged by gamma radiation in addition to damage which they receive in slowing down the neutrons. Other radio-nuclides may be formed in the body by the interaction of slow neutrons with stable nuclei. However, the dose contributed by these radionuclides is usually insignificant compared to the dose the neutrons themselves contribute.

Our normal method of protecton from this external radiation hazard is shielding. Shielding will consist of concrete, water, and/or poly-type materials because of their high water content and possibly a borated material to absorb the thermalized neutrons. (Please refer chapter 17.11 and 17.12 for details about neutron)

14.8 NUCLEAR RADIATION AND DETECTORS

Detection instruments are used to detect sub-atomic particles or high energy photons. Sub-atomic particles are too small to be detected directly. Thus the necessity of suitable detecting instruments. The working of various detection instruments is based upon one of the following three main characteristic properties of the particles and rays emitted in nuclear interactions.

(*i*) **Ionisation.** They ionise the gas through which they pass.

(*ii*) **Fluorescence.** They cause fluorescence in certain materials.

(*iii*) **Photographic effect.** They effect a photographic plate.

The following detecting devices depend upon ionisation property

(*a*) Ionisation chamber (*b*) Proportional counter

(*c*) Geiger Mueller counter (*d*) Semi-conductor detector

(*e*) Wilson cloud chamber (*f*) Bubble chamber.

Except for semi-conductor detector all others are *gas filled ionisation based nuclear detectors*. Semi-conductor detector is a solid state device.

The *scintillation counter depends upon fluorescence*. The *nuclear emulsion* detection depends upon *photographic effect*.

14.9 IONISATION CHAMBER

Principle. The ionisation chamber works on the principle that charged sub-atomic particles can ionise a gas. The number of ion pairs formed gives information with regard to the nature of the incident particle as well as its energy.

Ionisation chambers are of two types :-

(*i*) *Pulse ionisation chamber*. In this type of ionisation chamber, the ionisation produced by charged particles traversing the gas is detected as a *single voltage pulse*. This is known as *non-integrating type*.

(*ii*) *Current ionisation chamber*. In this type of ionisation chamber, pulses are not separated but continuous accumulation of charge leads to flow of electric current. This is known as *integrating type*.

The working of the two types depends upon the response time of the system relative to the frequency of arrival of the ionising particles. If the response time is small each pulse produced by the ionising particle can be detected separately and counted. Thus the non-integrating type must have a *small* response time whereas the integrating type must have a *long* response time.

Construction. An ionisation chamber consists of a metallic cylinder C with a thin axial wire W enclosed in a glass envelop in which a suitable inert gas (like Argon) is filled. A high potential difference is established between the cylinder and the wire electrode; the wire electrode being at a positive potential with respect to the cylinder. A resistance R, a capacitor and an amplifier cum pulse counter are connected as shown in Fig. 14.2. The electric field is such that the loss of ions through recombination is kept as low as possible, at the same time ensuring that ionisation by collision does not occur. The ionisation chamber thus operates under *saturation condition*.

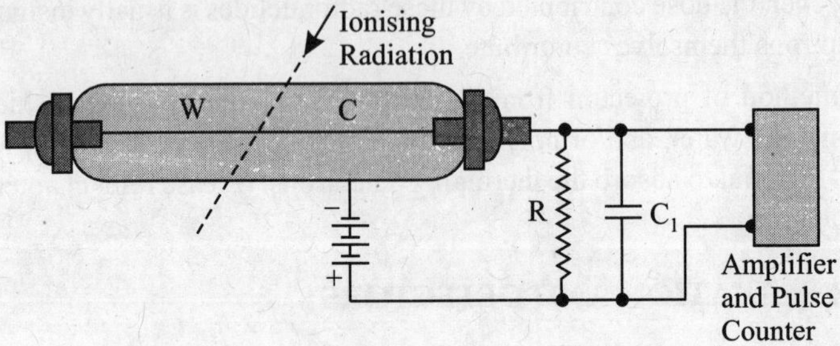

Fig. 14.2

Working. When a charged particle enters the active volume of the chamber, it ionises the gas molecules and produces a large number of ion pairs along its path. The positive ions move towards the cylindrical electrode and the negative ions (electrons) towards the positive wire electrode. The positive ions being heavier will move slowly.

Suppose the number of ion pairs produced = n. Then the height of the resulting voltage pulse ΔV across the external resistor is given by $\Delta V = \dfrac{ne}{C}$ where C is the capacitance of the capacitor formed by cylinder-wire system.

It has been calculated that 35 eV of energy is required to form an ion pair in air. If the incoming particle has an energy of 1 MeV and it loses all its energy in the chamber, then the number of ion pairs produced $n = \dfrac{1 \times 10^6}{35} = 2.86 \times 10^4$. If the capacitance of the ionisation chamber is $10\,pF = 10 \times 10^{-12}\,F$, then

Pulse height $$\Delta V = \frac{ne}{C} = \frac{2.86 \times 10^4 \times 1.6 \times 10^{-19}}{10 \times 10^{-12}} = 4.57 \times 10^{-4}\,V$$

The voltage pulse or the *source e.m.f.* produces a flow of current i in the external resistance R given by

$$i = \frac{ne}{CR}$$

If the time constant of the instrument (CR) is large as compared to the time between ionising events a steady state is reached and a *direct current* may be measured. On the other hand if (CR) is small each pulse produced by an ionising event may be recorded separately. If the individual particles are to be counted, then the pulses are fed to a pulse amplifier which is connected to the ionisation chamber by a coupling capacitor. In the integrating type ionisation chamber, the total ionisation current produced by the incident radiation is measured.

It has been found that if a grid is introduced in between the electrodes of an ionisation chamber and maintained at half the potential of the electrodes, the response becomes very quick and performance greatly improved.

Characteristic Regions

The variation of ionisation current with applied voltage due to the passage of a charged particle through the usual cylindrical ionisation chamber is known is its characteristic curve and is as shown in Fig. 14.3.

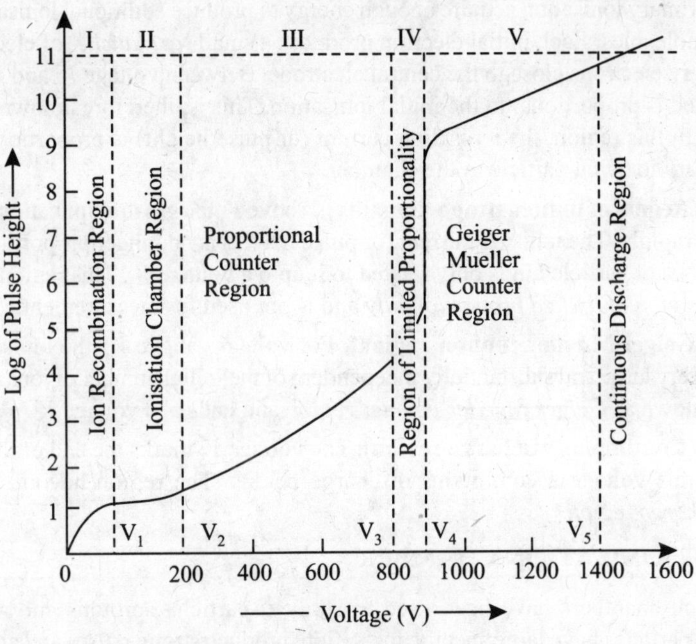

Fig. 14.3

Variation of ionisation current (Pulse height) with voltage. The ionisation chamber, the proportional counter, Geiger Mueller counter and other gas filled particle detectors are based upon the production of ionisation in a gas and the separation and collection of the ions by means of an electric field. The experimental arrangement in all these cases consists of a cylindrical electrode fixed along the axis of the chamber and well insulated from it. The chamber is filled with a gas at a pressure of about one atmosphere. A voltage V is applied between the cylinder and the central electrode through a resistance R and a capacitor C_1, the wire electrode being at a positive potential with respect to the conducting cylinder as shown in Fig. 14.2.

The motion of ion pairs to their respective electrodes constitutes an ionisation current which gives rise to a pulse the height of which is proportional to the ionisation current.

Region I. (Ion-recombination region). Suppose a charged particle enters the cylinder and some ionisation occurs in the gas with the formation of positive and negative ions. If there is no voltage across the electrodes the ions will recombine and no charge will appear across the capacitor. As the voltage is increased some positive ions reach the cylinder which is at a negative potential and an equal number of negative ions reach the central wire electrode which is at a positive potential, while others still recombine. This is, therefore, the *ion-recombination region*. In this region, the number of particles reaching the respective electrodes and hence the ionisation current (or pulse height) is proportional to applied voltage.

Region II (Ionisation chamber region). At a voltage V_1 the loss of ions by recombination becomes negligible and all the negative ions reach the central electrode. The ionisation current (or pulse height) attains a constant value. As V is increased above V_1, the ion pairs move to their respective electrodes so rapidly that almost all of them reach the electrodes. There is virtually no recombination of ions. The ionisation current (or pulse height) remains independent of potential as long as the voltage is increased upto V_2, the point at which secondary electrons are produced by collisions in the gas. This region is known as *Ionisation chamber region or saturation region.*

Region III (Proportional counter region). When the voltage is increased beyond V_2, the electrons released in the primary ionisation acquire enough energy to produce additional ionisation by collision with other gas molecules. Each initial electron produces a small '*avalanche*' of electrons with most of these secondary electrons close to the central electrode. Between voltage V_2 and V_3 the number of ion pairs collected is proportional to the initial ionisation. This is, therefore, known as *proportional counter region*. In this region, the ionisation current (or pulse height) is proportional to the energy lost by the primary ionising particle in the chamber.

Region IV (Region of limited proportionality). Above V_3 the gas multiplication effect continues to increase very rapidly. The ionising current (or pulse height) is no longer proportional to the initial energy of the incident particle but is only related to it upto a voltage V_4. This region between V_3 and V_4 is known as *region of limited proportionality* and is not used for measurement.

Region V (Geiger Mueller counter region). For voltages above V_4, the ionisation current (or pulse height) is very large and substantially independent of the initial number of ions and their energy. This region is known as *Geiger Mueller counter region* and ends at a voltage V_5.

Region VI (Continuous discharge region). The voltage V_5 marks the end of the useful voltage because above this voltage a continuous discharge begins. The region beyond V_5 is known as *continuous discharge region.*

14.10 USES OF IONISATION CHAMBER

1. Ionisation chambers have been used to study α-particles, protons and nuclei of lighter elements. These particles have large specific ionisation, produce strong pulses and are easily counted by the ionisation chamber.

2. For neutron detection, the chamber is filled with Boron triflouride (where boron is enriched with ${}_5B^{10}$).

3. The ionisation chambers are not used for electrons as it is less sensitive to β-particles because of their low primary ionisation.

As β-rays and γ-rays do not produce enough ionisation to give pulses detectable above the background noise, these cannot be detected by ionisation chamber.

14.11 SPECIFIC IONISATION

It is defined as the number of ion pairs per unit distance between the electrodes.

Gas multiplication factor. *It is defined as the total number of ion pairs produced by a single primary ion.*

If the strength of the electric field applied to a proportional counter or G.M. counter is such that an electron knocked out by a primary ion during its passage between two collisions with the gas molecules can gain sufficient energy to ionise the molecule with which it collides, then the number of charge carriers increases with every collision by a factor of 2. In such a case, the multiplication factor $m = 2^n$ where n is the average number of collisions of the primary generated electron on its path to the anode.

SOLVED EXAMPLES

Example 14.1. *A sample of Uranium emitting α-particles of energy 4.18 MeV is placed near an ionisation chamber. Assuming that 12 particles per second enter the chamber, calculate the current produced. An ion pair requires energy of 40 eV. Charge on the electron e = 1.6 × 10⁻¹⁹ Coulomb.* **(K.U. 2002)**

Solution. Energy of the α-particle = 4.18 MeV

Number of α-particles entering ionisation chamber per second = 12

∴ Total energy imparted per sec = 12 × 4.18 MeV

Energy required to produce 1 ion pair = 40 eV

∴ Number of ion pairs produced per second $n = \dfrac{12 \times 4.18 \times 10^6}{40}$

Charge collected per second = current $i = \dfrac{12 \times 4.18 \times 10^6 \times 1.6 \times 10^{-19}}{40}$

$$= 2 \times 10^{-13} \text{ A}$$

Example 14.2. *An α-particle of energy 4.18 MeV is completely stopped in an ionisation chamber. Calculate the pulse voltage and current in an external resistance of 10^6 Ω. Energy required to produce one ion pair = 35 eV. Capacitance of the chamber 25 μμF.*

Solution. Pulse voltage height $\Delta V = \dfrac{ne}{C}$

Number of ion pairs produced $n = \dfrac{4.18 \times 10^6}{35}$

$$\therefore \qquad \Delta V = \frac{4.18 \times 10^6 \times 1.6 \times 10^{-19}}{35 \times 25 \times 10^{-12}} = 7.6 \times 10^{-4} \text{ V}$$

Current
$$i = \frac{ne}{CR} = \frac{7.6 \times 10^{-4}}{10^6} = 7.6 \times 10^{-10} \text{ A}$$

14.12 PROPORTIONAL COUNTER

A proportional counter is basically a *gas filled, ionisation based, nuclear detector.*

Low energy ionising particles cannot be detected by an ionisation chamber as the height of the voltage pulse produced by them is very low. In a proportional counter, the voltage pulse is sufficiently high.

Construction. The construction of a proportional counter is almost the same as that of an ionisation chamber. Its main parts are :

(*i*) *A hollow copper cylindrical tube* about 20 cm long and 2 cm in diameter containing a mixture of 10% methane and 90% argon at a pressure of about one atmosphere.

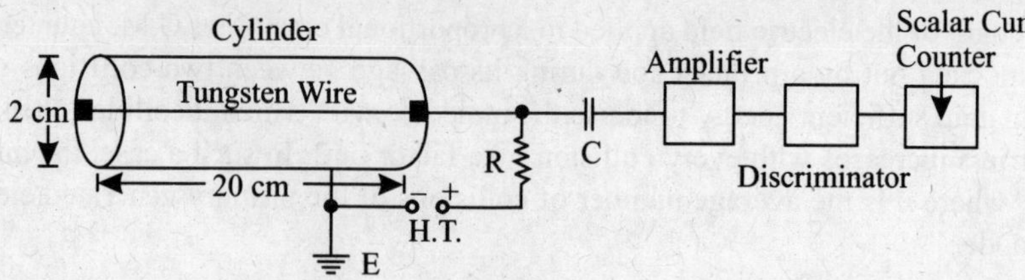

Fig. 14.4

(*ii*) *A tungsten electrode* consisting of a fine tungsten wire of radius about 0.001 cm fixed along the axis of the tube and well insulated from it. A positive potential of about 700 volts is applied to the central electrode with respect to the cylindrical tube so that the counter works in the *proportional counter region.*

(*iii*) The electric pulses generated in the counter are amplified with the help of an *amplifier* and then fed to a counting device known as *scalar cum counter*. To allow pulses of required voltage height and to cut off low voltage pulses which might disturb the working, an arrangement known as *discriminator* is used.

Principle. When the potential difference between the central wire and the metallic cylinder increases past a certain value, the electrons produced due to ionisation are accelerated towards the central wire which is at a positive potential. In this process, the electrons gain large energy and create further ion pairs along their path. The resulting '*avalanche*' of secondary electrons that reaches the anode may represent a gas multiplication factor as high as 1000 or more with correspondingly larger output pulse.

The size of the output pulse is proportional to the number of primary electrons produced by the ionising particle and the number of primaries is proportional to the energy of the ionising particle. Thus the height of the output pulse is proportional to the energy of the ionising particle.

If W is the average energy for generation of one ion pair and ΔE the energy lost by the incident charged particle in the gas between the chamber electrodes, then number of ion pairs produced

$$n = \frac{\Delta E}{W}$$

If m is the multiplication (or amplification) factor, then each primary ion will produce m ion pairs. Hence charge carried to the central electrode $Q = m\dfrac{\Delta E}{W}e$. As Q is proportional to energy ΔE, the pulse height is proportional to the energy of the ionising particle. This is why it is called a *proportional counter*.

Working. If b is the radius of the cylinder and a the radius of the wire, then the radial field E_r at a distance r from the central axis is given by

$$E_r = \frac{V}{r \log_e \left(\dfrac{b}{a}\right)}$$

Since the central wire is very thin and the potential difference is fairly large, the electric field at a distance r from the central wire is very high. Hence electrons travelling towards the wire are rapidly accelerated when near it and produce additional electrons in the region due to ionisation by collision.

The proportional counter, thus gives single pulses of height proportional to the energy of the ionising particles. These voltage pulses filtered by the capacitor are passed through the discriminator circuit. The discriminator level is adjusted in such a way that it does not allow the passage of low voltage pulses which include undesirable spurious and noise pulses. The discriminator sends to the counter only genuine pulses which are duly registered and counted.

Applications. (*i*) *Detection of low energy X-rays, γ-rays,* and *β-rays.*

Proportional counters have been found very useful for detection and measurement of energy of soft radiations like *X*-rays, γ-rays, β-rays, mesons and protons.

For example, electrons from the β-decay of tritium which have a maximum energy of only 18 KeV can be detected by a proportional counter. The procedure is to incorporate tritium in the gas filling the counter and then to record the height of the pulses that are produced.

In experiments on Mossbauer effect, it is required to detect γ-rays of energy 14.4 KeV from Fe^{27} against a strong back ground of 123 KeV γ-rays and 6.3 KeV *X*-rays. A typical proportional counter operated at 2100 volts with single channel pulse height analyser in place of discriminator is used for this purpose.

(*ii*) **Neutron detection.** Proportional counters can be used to detect neutrons although neutrons do not ionise the gas they pass through. However, the absorption of neutrons by a nucleus is usually followed by the emission of a charged particle or a γ-ray and this secondary radiation is detected by the counter. A proportional counter filled with boron trifluoride (BF_3) is generally used for the purpose. The neutrons react with boron according to the reaction

$$_0n^1 + {}_5B^{10} \longrightarrow {}_3Li^7 + {}_2He^4$$

and emit an α-particle. Natural Boron contains 19% of $_5B^{10}$ isotope and a specially sensitive counter is designed by using BF_3 which has been enriched with a higher proportion of B^{10}. BF_3 counters are specially responsive to slow neutrons as the reaction cross-section is inversely proportional to neutron speed. The 1.47 MeV α-particles from the reaction give rise to much higher pulses which are readily distinguished and counted.

Example 14.3. *It is required to operate a proportional counter with a maximum radial field of 108 volt/m. What is the applied voltage required if the radius of the wire and tube are 0.002 cm and 1 cm. respectively ?*

Solution. Radial field

$$E_r = \frac{V}{r \log_e \left(\frac{b}{a}\right)}$$

∴ Radial field at the surface of wire = maximum radial field

$$E = \frac{V}{a \log_e \left(\frac{b}{a}\right)}$$

Here $a = 0.002$ cm $= 2 \times 10^{-5}$ m; $b = 1$ cm $= 10^{-2}$ m; $E = 10^8$ V/m

$$10^8 = \frac{V}{2 \times 10^{-5} \times 2.303 \log_{10}\left(\frac{10^{-2}}{2 \times 10^{-5}}\right)}$$

or $$V = 2 \times 10^{-5} \times 2.303 \times 2.6990 \times 10^8 = 12430 \text{ volt.}$$

Example 14.4. *Estimate the number of ion pairs produced in a proportional counter by a 10 MeV proton if the counter size and pressure are large enough for all proton energy to be absorbed. If the gas multiplication factor is 10^2, how much charge flows in the counter when the proton is* *absorbed. If the pulse current flows for 2×10^{-3} sec, through output resistor of 10^4 ohm, find the height of voltage pulse.*

Solution. Energy required for producing one ion pair $W = 35$ eV

∴ Number of pairs produced by 10 MeV proton

$$= \frac{\Delta E}{W} = \frac{10 \times 10^6}{35}$$

$$= 2.86 \times 10^5$$

Gas multiplication factor $m = 10^{+2} = 100$

∴ Charge carried by the ions to the central wire $Q = m\dfrac{\Delta E}{W}e$

$$= 100 \times 2.86 \times 10^5 \times 1.6 \times 10^{-19} = 4.576 \times 10^{-12} \text{ C}$$

Duration of the pulse $t = 2 \times 10^{-3}$ sec

∴ Rate of flow of charge = current through the resistor

$$i = \frac{Q}{t} = \frac{4.576 \times 10^{-12}}{2 \times 10^{-3}} = 2.288 \times 10^{-9} \text{ A}$$

Height of the voltage pulse

$$\Delta V = iR = 2.288 \times 10^{-9} \times 10^4$$

$$= 2.288 \times 10^{-5} \text{ V}$$

14.13 GEIGER-MUELLER COUNTER

When the voltage applied to a gas filled detector is increased beyond the proportional counter region, the pulse height does not remain proportional to the energy of the incident ionising radiation or charged particle. The counter which operates in this region of higher voltage is known as a G.M. counter.

Specific characteristic of G.M. region. The specific characteristics of G.M. region are :

(*i*) The region starts at the voltage where the region of limited proportionality ends.

(*ii*) In this region, the ionisation current or pulse height is independent of the nature and energy of the incident ionizing particle or radiation.

(*iii*) A magnified view of the curve in the G.M. region shows a flat portion known as *plateau.* It the voltage is increased beyond the plateau region, the *continuous discharge region* begins.

Construction. The construction of a G.M. counter is similar to that of a proportional counter. The main components are :-

(*i*) *Hollow copper cylindrical tube.* It is a cylindrical copper tube *M* of diameter 1.5 cm and length 2–100 cm depending upon the purpose for which it is to be used. It is enclosed in another cylindrical glass tube *G* and partially evacuated. The tube either contains air at a low pressure or a mixture of Argon with 10% vapours of ethyl alcohol at a pressure of 10 cm of mercury column.

(*ii*) *Tungsten wire electrode.* A fine tungsten wire *W* of diameter 0.5 mm is stretched along the axis of the copper tube, is well insulated from it and sealed in the glass tube. The positive terminal of the H.T. is connected to the central wire through a high resistance *R* and negative terminal to the copper cylinder.

(*iii*) *Amplifier and counting device.* The electric pulses produced in the G.M. counter are fed to a pre-amplifier, scalar and finally detected and counted by a counter.

Working. Consider an ionising particle passing through the gas in the G.M. counter. As in the case of a proportional counter an '*avalanche*' of ions will be produced close to the central wire near its midpoint. The condition for avalanche to start is

$$E_a \lambda = \frac{V\lambda}{a \log_e b/a} = I \text{ (Ionisation potential)}$$

where E_a is the electric field at the central wire of radius a, b is the radius of the cathode-cylinder, λ the mean free path and V the applied voltage.

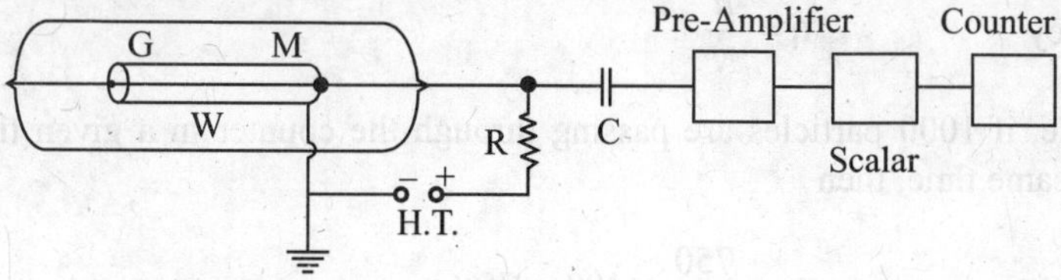

Fig. 14.5

If the resulting positive ions get a chance to capture electrons, photons (in the ultra-violet region), may be emitted. These photons again knock out electrons by collisions with gas atoms.

[**Note.** It should be noted that in the proportional counter region, the population of the positive ions and the electrons reach high values so close to the wire where the field strength and, therefore, the particle velocities are very high, that the chance of their encountering each other before the electrons reach the wire and thus leading to the generation of ultra-violet radiations are almost negligible].

The potential V in the G.M. counter region being sufficiently high, the secondary ionisation process is initiated at a comparatively larger distance from the wire so that the chances of ultra-violet radiation, photon emission are very much increased and thus the ultra-violet radiation is emitted in all directions.

The ultra violet photon travelling axially may also release electrons from the gas at either end of the original avalanche producing further avalanches at sites close to the original one. The effect spreads along the length of the wire in this manner in small steps at the rate of about 10 cm per μs *i.e.* with a velocity of propagation of the order of 10^7 cm/sec. This leads to a multiplication factor m as high as 10^8. With the avalanches spread over the whole length of the wire the output pulse no longer depends upon the number of primary ions initially produced by the ionising particle and proportionality which is the characteristic of proportional counter no longer exists.

Dead time. The wire potential changes with the drift of the positive ions. The rate of change of wire potential is most rapid as the positive ions move away from the immediate neighbour hood of the wire because in this region the velocity is very high and hence the drift time very small due to the high field strength. The counter remains dead till the positive ions have moved sufficiently away from the avalanche site to put the wire back at the *Geiger threshold* potential.

The time for which the counter is dead is called the dead time. During this time interval the counter does not record any event.

Recovery time. After the dead time interval, the counter may operate with a reduced pulse size till the positive ions reach the cylinder and the counter potential regains its original value for operation in the Geiger region.

The recovery time is the time after which the original pulse levels are restored. During the recovery time, the pulses are recorded but are of smaller size.

The dead time and the recovery time are of the order of 100 μs which limits the counting rate of a G.M counter to about 50,000 pulses per second.

True counting rate. The *true counting rate n* is given by

$$n = \frac{\text{Observed counting rate}}{1 - \text{observed counting rate} \times \text{dead time}}$$

∴ The **dead time loss** = True counting rate – observed counting rate

Efficiency of counting. *The efficiency of counting is defined as the ratio of the number of counts observed per unit time (n) to the number of ionising particles which pass through the counter during that time (N).*

∴ Efficiency $\eta = \frac{n}{N}$

For example, if 1000 particles are passing through the counter in a given time and 750 are recorded in the same time, then

$$\eta = \frac{750}{1000} \times 100 = 75\%$$

Quenching and its necessity. The ultra-violet radiation given out in the process described above may release electrons from the gas atoms in the regions far away from the wire or from the inner surface of the metal cylinder. Further as the positive ions of the gas in the counter approach the cylinder, these may extract electrons from its metal surface to become neutral gas atoms, releasing the excess energy left with them after the process as photons which in turn, may also knock out electrons from the metal surface.

In order that the counter should record a chain of ionising particles it is necessary that the discharge caused by the first particle is completely quenched before the arrival of the next particle. In other words, it means that the *sheath* of positive ions should be eliminated. *Elimination of positive sheath around the wire is known as quenching.*

Self-quenching. To bring about automatic internal quenching or self-quenching a small percentage (about 10%) of a polyatomic vapour like ethyl alcohol or methane is added to Argon gas in the counter. Alcohol has low ionisation potential of 11.30 eV as compared to Argon which has an ionisation potential of 15.7 eV. With this the chances of the ultraviolet photons from the gas travelling far from their places of origin well within the avalanche site due to primary ionising event will be considerably diminished as most of them will be lost in dissociating the molecules of alcohol or methane. The ultra-violet photons will, therefore, initiate fresh ionisation only along the axial wire lying *very close* to the site of their production but will not be able to travel radially without encountering the polyatomic molecules, to regions outside the avalanches producing initial ionising events.

From energy considerations also we find that positive Argon ions which encounter thousands of collisions with polyatomic molecules during their passage from the avalanche site close to the wire to the surface of the cylinder will borrow electrons from these molecules to become neutral atoms, the excess energy, if any, being shared as kinetic energy of the colliding particles. Thus only the positive ions of the poly-atomic vapour reach the cylinder surface. These ions again extract electrons from the metal surface and become neutral molecules, the excess energy left after this process goes to dissociate the molecules and emitted as photon. The recycling of the counter by spurious electrons is thus prevented and the counter becomes *self-quenched.*

This means that the polyatomic molecules of ethyl aclcohol or methane perform two functions.

(*i*) It prevents the photo-electric effect of the quanta at the cathode and

(*ii*) It prevents the creation of secondary electrons when the positive ions impinge the cathode.

Self-quenching with halogen vapours. The fact that in a self-quenched counter dissociation of poly-atomic molecules takes place means that the number of such molecules in the counter will go on diminishing and will become ultimately zero. This happens after the counter has recorded about 10^9 pulses. *The life of the counter is thus limited.* G.M. counters using halogen vapours (*Cl* or *Br*) as the quenching agents do not suffer from this draw back. The halogen atoms produced by dissociation have a tendency to recombine to form molecules again. The cylinder of a G.M. counter using halogen vapour should be made up of some material like stainless steel, which is not chemically acted upon by Halogens.

Operation and testing. To test the operation of a G.M. counter a radioactive source whose half-life is very long as compared to the duration of the experiment is taken and placed in a fixed position at a suitable distance from the counter so that the ionising particles enter the counter at a constant rate. The applied voltage is gradually increased in small steps keeping the source-counter assembly undisturbed. When the applied voltage is low, the multiplication factor (m) also has a small value so that the output pulse is not strong enough to operate the mechanical counter even after amplification. The counting rate will, therefore be zero. As the applied potential is raised the multiplication factor increases and more and more of the ionising particles starting with those which produce maximum ionisation in the counter, start getting counted. At a certain value of applied voltage, the number of pulses increases almost linearly with voltage. This is the *proportional counter region.* With further increase in voltage the counting rate rises further and soon the *Geiger region* is reached when the pulses due to all the ionising particles are recorded and these also have the same height. This operating potential is called the *Geiger threshold* of the counter under the given conditions.

Plateau. When the voltage is increased beyond the threshold the counting rate remains the same for a large increase in voltage but the pulse height rises. The pulse counting rate remains the same as all the particles are being counted. This portion of the graph between counting rate and applied potential is, therefore, a straight line, parallel to *X*-axis. This straight portion of the curve is called the *plateau*. The plateau continues until the excess potential over the threshold called *over voltage* becomes so high that the counter breaks down to a continuous discharge.

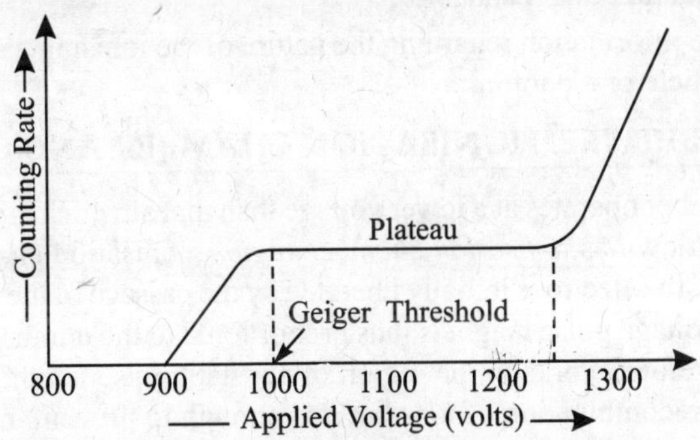

Fig. 14.6

The variation of counting rate with applied voltage is shown in Fig. 14.6. For a good G.M. counter the plateau region extends over about 200 V. The operating potential is chosen so that the value of over voltage is not critical and small fluctuations in it do not effect the counting rate.

The plateau has a slight slope, which corresponds to about 2% increase in counting rate for an operating voltage increase of 100 V. This slope is due to end effects.

While using the counter care should be taken to see that the activity of the source is not so high as to exceed the maximum counting rate of the G.M. counter. If this precaution is not taken the counter gets *choked*.

Applications. G.M. counters are very useful for detecting nuclear radiations and charged particles. It is being largely used for measuring cosmic ray intensities and recording cosmic ray events.

The counter set up with transistorised electronics is light and portable. It serves best for mineral prospecting and other similar work.

14.14 ADVANTAGES AND DISADVANTAGES OF G. M. COUNTER

Advantages of a G.M. counter:

1. G.M. counters are very useful for detecting nuclear particles like α and β-particles and protons, for measuring cosmic ray intensities and recording cosmic ray events.
2. They are sensitive to production of even a single ion pair.
3. The pulse height is sufficiently high which requires little amplification.
4. The pulse size is independent of the nature and energy of the incident ionising particle or radiation.
5. The pulse height is constant over a wide range of applied voltage (known as *plateau width*) so that the power supply need not be precisely regulated as required in a proportional counter.
6. The counter set up with transistorised electronics is light and portable and can be used for mineral prospecting and cosmic ray work.

Disadvantages or Limitations.

1. It has a large dead time and recovery time of about 200–400 μs following each pulse which prevents its use at very high counting rates of above 10^4 particles per second.
2. G.M. counters have very low intrinsic efficiency for detecting γ-radiation.
3. **It cannot detect uncharged particles like neutrons.**

Reason: The working of a G.M. counter depends upon the ionising property of the incident particle; which on passing through the gas in the G.M. counter produces an avalanche of ions. A neutron being uncharged has no ionising property and hence cannot activate the G.M. counter.

4. G.M. tubes have a limited life as the quenching gases dissociate and change the pressure of the inert gas filled inside it. This reduces the flatness of the plateau so that it has a steep slope and makes the counter undependable.
5. It cannot provide information regarding the nature of the ionising particle or even whether it is a charged particle or a photon.

14.15 DIFFERENCE BETWEEN IONISATION CHAMBER AND GM COUNTER

The ionisation chamber operates at a lower voltage than that at which the G.M. counter operates. This voltage region is known as *ionisation chamber region*. Ionisation chamber is able to make a complete collection of all the electrons initially liberated by the passage of the ionising particle without gas amplification. The voltage pulse height is thus proportional to the number of these electrons. The ionisation chamber, therefore, works in the region of constant pulse height. The applied voltage is high enough to prevent recombination of ions but low enough to prevent multiplication. Ionisation chamber is useful for detecting α-particles and fission fragments. β-particles and γ-rays do not produce enough ionisation to produce pulses detectable above the back ground noise.

The G.M. counter works at a much higher voltage than the ionisation chamber. This voltage region is known as *G.M. counter region*. At this stage the discharge spreads throughout the length of the counter tube with the result that the pulse height is independent of ionisation. A G.M. counter, therefore, gives a sufficiently high voltage pulse and does not need further amplification. It is most useful for detecting α and β particles, γ-rays and cosmic rays.

14.16 COMPARISON BETWEEN PROPORTIONAL COUNTER AND G.M. COUNTER

	Proportional Counter		*GM Counter*
1.	It works in the proportional counter voltage region which is lower than the GM counter voltage region.	1.	It works in the G.M. counter voltage region which is higher than the proportional counter voltage region.

2.	It works on the principle of gas amplification (or multiplication).	2.	It works on the principle of individual ionisation.
3.	The pulse height is proportional to the energy of the incident ionising particle.	3.	The pulse height is independent of the energy of the incident ionising particle.
4.	The pulse voltage is not very high so that it requires further amplification.	4.	The pulse voltage is sufficiently high so that it does not ordinarily require further amplification.
5.	It can detect low energy radiation like X-rays and γ-rays.	5.	It cannot detect low energy radiation.
6.	It can detect uncharged particles like neutrons.	6.	It cannot detect uncharged particles as it works on the principle of individual ionisation.
7.	It has a high resolving power.	7.	The resolving power is low. The resolving time of a GM counter is limited due to a large dead time and recovery time.
8.	It can be used to study Mossbauer effect as it can detect γ-rays.	8.	It cannot be used to study Mossbauer effect.
9.	Power supply must be precisely regulated.	9.	The pulse height is constant over a wide range of applied voltage, so the power supply need not be precisely regulated.

Example 14.5. *The dead time of a Geiger Mueller counter is 400 μs. What is the true count rate for measured count rates of 110, 1100, 9890 counts per minute ?*

Solution. Dead time $t = 400\ \mu s = \frac{400}{60} \times 10^{-6} = 6.66 \times 10^{-6}$ min

True count rate
$$n = \frac{\text{observed count rate}}{1 - \text{observed count rate} \times \text{dead time}}$$

$$= \frac{n_{obs}}{1 - n_{obs} \times t}$$

(*i*) $n_{obs} = 110$

$$\therefore \quad n = \frac{110}{1 - 110 \times 6.66 \times 10^{-6}} = \frac{110}{0.9992674} = 110.08$$

(*ii*) $n_{obs} = 1100$

$$\therefore \quad n = \frac{1100}{1 - 1100 \times 6.66 \times 10^{-6}} = \frac{1100}{0.992674} = 1108.1$$

(*iii*) $n_{obs} = 9890$

$$\therefore \quad n = \frac{9890}{1 - 9890 \times 6.66 \times 10^{-6}} = \frac{9890}{0.9341326} = 10587$$

Example 14.6. *A G.M. counter with dead time of 300 μs records 16000 counts per minute. What is dead time loss in counting rate.* ***(Pbi. U. 2002)***

Solution. Dead time $t = 300\ \mu s = \frac{300}{60} \times 10^{-6} = 5 \times 10^{-6}$ min.

$$n = \frac{n_{obs}}{1 - n_{obs} \times t} = \frac{16000}{1 - 16000 \times 5 \times 10^{-6}} = \frac{16000}{0.92}$$

$$= 17391$$

Dead time loss $= n - n_{obs} = 17391 - 16000 = 1391$ counts/min.

Example 14.7. *A G.M. counter wire collects 10^8 electrons during each discharge when the counting rate is 1500 per minute. What is the current in the circuit ? What is the voltage developed across 2×10^6 ohm leak resistor ?* ***(Calicut. U. 2001)***

Solution. Number of electrons per discharge = 10^8

Number of discharges per minute = counting rate = 1500

$\therefore$ Number of discharges per second $= \frac{1500}{60}$

$\therefore$ Current in the circuit = charge conveyed per second

$$i = \frac{1500}{60} \times 10^8 \times 1.6 \times 10^{-19} = 4 \times 10^{-10} \text{ A}$$

Voltage developed across $R = 2 \times 10^6\ \Omega$, $V = iR$

$$= 4 \times 10^{-10} \times 2 \times 10^6 = 8 \times 10^{-4} \text{ volt.}$$

Example 14.8. *A G.M. counter collects 10^8 electrons per discharge when the counting rate is 500 counts per minute. What will be the average current in the curcuit ?* ***(M.D.U. Similar 2001)***

Solution. $$i = \frac{500}{60} \times 10^8 \times 1.6 \times 10^{-19} = 1.33 \times 10^{-10} \text{ A.}$$

Example 14.9. *A G.M. counter collects 10^8 electrons per discharge. The average current in the circuit is 6.65×10^{-6} A. Find the counting rate. $e = 1.6 \times 10^{-19}$ C.* ***(Calicut. U. 2000)***

Solution. Let the counting rate be n per minute.

$$\therefore \quad 6.65 \times 10^{-6} = \frac{n}{60} \times 10^8 \times 1.6 \times 10^{-19}$$

or $$n = \frac{6.65 \times 10^{-6} \times 60}{10^8 \times 1.6 \times 10^{-19}} = 25 \times 10^6 \text{ per minute}$$

14.17 SCINTILLATION COUNTER

A scintillation counter is used for the detection and measurement of energy of atomic radiation.

Principle. It is based upon the principle that when high energy radiation falls on a fluorescent material it produces flashes of light called '*scintillations*'. The earliest form of scintillation counter consists of a small wire the tip T of which is dipped in Radium bromide or any other radioactive material Fig. 14.7. It is placed in front of a screen S coated with Zinc sulphate, Barium platino-cyanide or calcium tungstate and viewed through a low power microscope M in a dark room. The α-particles emitted by the radioactive material excite luminescence in the material coated on the screen. Every time an α-particle is incident on the screen, a bright spot of light becomes visible. In this way the number of particles emitted by the radioactive material in a given time can be counted. The arrangement is known as a *spinthariscope*.

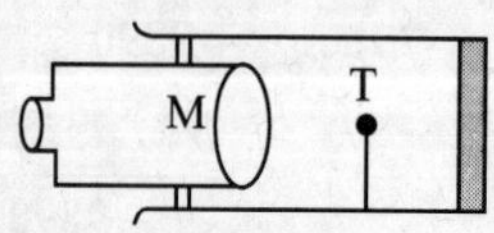

Fig. 14.7

The process is, however, tedious and tiresome. More over the persistence of vision restricts the counting rate at about 100 per minute. The same principle is used in a modern scintillation counter.

Block diagram. The block diagram of a scintillation counter is shown in Fig. 14.8.

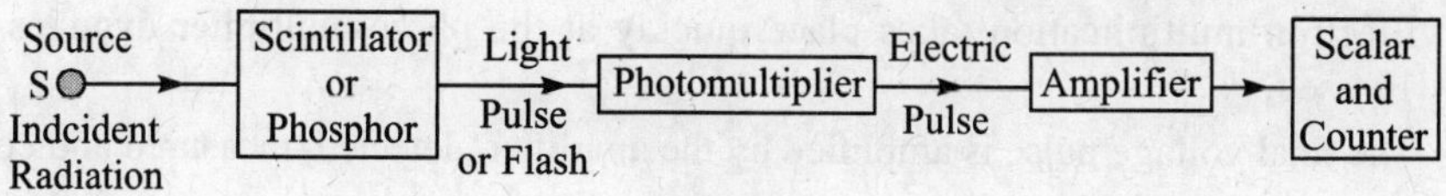

Fig. 14.8

Construction and operation. As shown in the block diagram *S* is the source which emits ionising radiation. These radiations fall on the scintillator and lose energy to excite or ionise the atoms of the '*phosphor*'. These excited atoms emit photons in the blue-violet or ultra-violet region of the optical spectrum when they return to the normal ground state. Thus the *scintillator* converts the energy of ionising radiations into flashes of light or *scintillations*. The phosphor is optically coupled to the envelop of the photo multiplier tube.

The output pulse from the photo-multiplier is fed to a pulse-amplifier followed by scalar and counter unit.

In order to remove the undesired low energy pulses, a *discriminator* circuit is introduced between the amplifier and the counter. Power to the various stages is supplied from a stabilised power supply.

The photomultiplier tube is the most important component of the whole arrangement. We shall describe it in detail.

Photo-multiplier. The light emitted by the flashes falls on the transparent photo cathode of a photo-multiplier shown in Fig. 14.9. The photo-cathode consists of an antimony ceasium layer deposited on glass. It emits a secondary electron when a photon of light-falls on it. This photo electron is accelerated by a potential difference of about 100 volts to the first **dynode** of the photo-multiplier tube. This single electron having a kinetic energy of about 100 eV collides with the surface of the dynode which may emit 2 or more electrons. These secondary electrons are accelerated to the second dynode which is maintained at a potential of + 100 volts with respect to the first. At this surface, electron multiplication by secondary emission again takes place. In a typical photo-multiplier there are nine dynodes and 10 stages of electron amplifications. The electrons emitted by the final dynode (No. 9) are collected at a solid metal disc electrode called the *anode* maintained at + 100 volts with respect to the last dynode. All this takes place in about 10^{-6} sec. The necessary voltage to all the dynodes is supplied from a stabilized power supply provided with a suitable potential dividing arrangement for the necessary subdivision of the voltage. A typical photo-multiplier tube has an overall amplification of the order of 10^6 *i.e.* a single photo-electron emitted by the photo-cathode produces approximately 10^6 electrons at the final collector. If m is the electron multiplication factor of a single dynode stage and there are n stages, then total amplification is m^n.

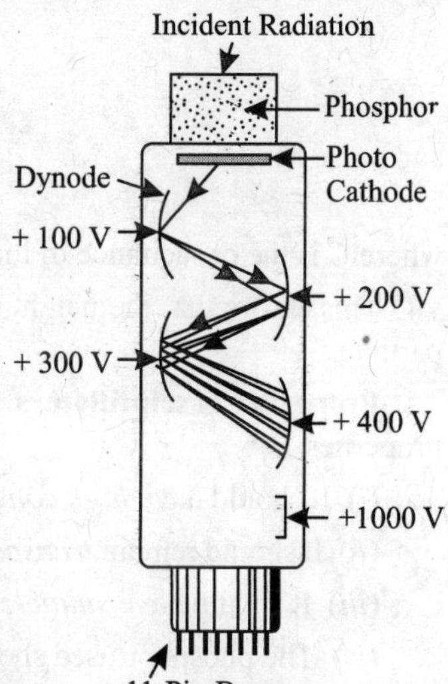

Fig. 14.9

The whole operation is summarised as under and takes place in following distinct steps.

(*i*) The incident radiation is absorbed by the 'phosphor' (or scintillator) and the atoms (or molecules) are excited.

(*ii*) The excited atoms (or molecules) of the fluorescent material of the phosphor de-excite and produce a light flash in a very short time.

(*iii*) The emitted light photons are transmitted to the photo-cathode of the photo-multiplier tube.

(*iv*) Photo-electrons are produced due to absorption of light photons at the photo cathode.

(*v*) Electron multiplication takes place quickly at the photo-multiplier dynodes in about 10^{-6} sec.

(*vi*) The final voltage pulse is amplified by the amplifier, detected, measured and counted by the scalar-counter assembly.

Height of output voltage pulse. Suppose the energy of the incident ionising particle is E_i. If E_v is the mean energy of the photons emitted and η the efficiency of conversion, then number of photons produced in the phosphor by an ionising particle

$$N_1 = \eta \frac{E_i}{E_v} \quad \left[\text{or } \eta = \frac{N_1 E_v}{E_i} \right]$$

The number of photons falling on the photo-cathode of the photo-multiplier tube $N_2 = \alpha\, TN_1$ where α is the fraction of the total number of photons falling on the photo-cathode and T the optical transparency. These photons are converted into photo-electrons with an efficiency η_e.

∴ Number of photo-electrons produced by the cathode $N = \eta_e N_2$

If D be the fraction of the photo-electrons reaching the first dynode, then

Number of photo-electrons reaching the first dynode $= DN$

If m is the multiplication factor of each dynode and there are n dynodes, the number of electrons collected at the anode $= DN\, m^n$

∴ Charge carried to the anode

$$= \frac{E_i}{E_v} \eta\, \alpha\, T\, \eta_e\, D\, m^n\, e$$

and pulse height

$$V = \frac{E_i}{E_v C} \eta\, \alpha\, T\, \eta_e\, D\, m^n\, e$$

where C is the capacitance of the system.

This shows that the height of the output pulse is proportional to the energy E_i of the incident particle.

Properties of scintillators. The scintillator material used in the counter must have the following properties.

(*i*) It should have *high conversion efficiency* to convert the incident radiation into scintillation.

(*ii*) It should remain *transparent to its own light*. It is this light which produces photo-emission.

(*iii*) It should have *suitable value of refractive index* so that there is no total internal reflection.

(*iv*) The phosphor used should have *high resolving power* for incident radiation.

(*v*) It should be *stable* and *long lasting* to withstand prolonged exposure to radiation.

Types of scintillators used. (*i*) **Sodium iodide.** This type of scintillator consists of a single crystal of *NaI*. The size of the crystal depends upon the requirements of the experiment. For use with low energy γ-rays a small quantity of thallium iodide activator is used with *NaI* crystal as impurity. As *NaI* is hygroscopic and moisture can spoil the crystal, it is sealed in air tight aluminium can with a window of pyrex glass to allow the ionising radiations to penetrate without any appreciable absorption. The can is coated on its inside with *MgO*. This coating serves as a reflector. A fluid of silicon or Canada balsam is used for optical coupling between scintillator and photo-multiplier tube. The whole assembly *is called scintillator head*.

Some times *NaI* is replaced by *CsI* which is not hygroscopic and is, therefore, preferred.

(*ii*) **Zinc sulphide.** It is extensively used for detection of those particles which have a short range like α-particles. As it readily becomes opaque to its own radiation it is used only in thin layers.

(*iii*) **Organic phosphors.** Organic materials like *anthracene and stilbeme* are used for detection of β-particles. Anthracene gives the highest yield of 15 photons for a 1000 eV β-particle.

(*iv*) **Plastics and liquid scintillators.** Specially prepared transparent plastic materials cut to the desired shape and size are used in high energy physics.

The liquid scintillator has a solvent like toluene and Xylene in which is added a small percentage of diphenyloxazole or terphenyl as solute. The radiation energy is absorbed by the solvent and transferred to the solute which emits the light.

(*v*) **Gases.** For counting of heavy charged particles in the presence of γ-rays, Xenon is used as a phosphor which emits radiation in ultraviolet region.

Superiority of scintillation counter. Scintillation counters have a high efficiency of detection, short resolving time, linear response over a wide range of energy of incident radiation. Due to these advantages, scintillation counters are considered superior to **G.M. counters.**

Because of its extremely short duration pulse and high resolution, the scintillation counter is far superior to a **proportional counter.**

Example 14.10. *A scintillation detector has 5% proton conversion efficiency and the photo-cathode multiplier has 1% photo-electron conversion efficiency with light collection efficiency of 10%. It has 10 dynodes and secondary emission factor of these dynodes = 4. If the anode capacitance to ground is 30 pF, find the pulse height in volts.* **(P.U. 2002)**

Solution. Proton conversion efficiency = 5%

If E_i is the initial energy of the proton, out of which a fraction η is used to produce photons and E_v is the energy converted into photons, then proton conversion efficiency

$$= \frac{E_v}{\eta E_i} = \frac{5}{100} = \frac{1}{20}$$

or $$N_1 = \frac{\eta E_i}{E_v} = 20$$

Light collection efficiency = 10%

Light collection efficiency = fraction of total number of photons from the phosphor falling on the photo-cathode (α) × optical transparency of the cathode (T)

$$= \alpha T = \frac{10}{100} = \frac{1}{10}$$

photo-electric conversion efficiency $\eta_e = 1\% = \dfrac{1}{100}$

It is supposed that all the electrons from the cathode reach the first dynode *i.e.* $D = 1$.

Secondary emission factor of each dynode $m = 4$

Total number of dynodes $= 10$

Total multiplication factor $= m^n = 4^{10} = 1.05 \times 10^6$

Capacitance $C = 30\ pf = 30 \times 10^{-12}\ F,\ e = 1.6 \times 10^{-19}$ C

Pulse voltage height $$V = \frac{1}{C}\frac{\eta E_i}{E_v}\alpha T \eta_e D m^n e$$

$$= \frac{1 \times 20 \times 1 \times 1 \times 1.05 \times 10^6 \times 1.6 \times 10^{-19}}{30 \times 10^{-12} \times 10 \times 100}$$

$$= 1.12 \times 10^{-4} \text{ V}.$$

14.18 RADIATION DETECTORS BASED ON IMAGE FORMATION

There are three radiation detectors based on image formation. These are:

(*i*) Emulsion detector, (*ii*) Wilson Cloud chamber (*iii*) Bubble Chamber. Let us describe them one by one.

Nuclear Emulsion Detector. For studying certain nuclear reactions photographic emulsions can be used directly as particle track detectors. The photographic emulsion consists of *AgBr* (or any silver halide) in a matrix of gelatine. A heavy charged particle traversing a photographic emulsion interacts with tiny silver halide crystals and produces a latent image of its track, which can be developed and fixed when the deposited black silver in the negative gives a permanent record.

Ordinary optical photographic emulsions are not suitable for quantitative studies. The density of such emulsions is low and the developed crystal grains are too large and widely spaced to give any well defined range to the tracks. These defects are overcome by having emulsions in which the proportion of silver bromide to gelatine is higher than in optical emulsions by about 10 times. Moreover, the silver bromide grain size is made very small with diameters of 0.1 to 0.2 μm. Such fine grain nuclear emulsions with thickness varying from 25–1000 μm (micron) much greater than that of the optical emulsion have been developed by Powell and his group.

The energy required by various particles to travel 1 mm in an average nuclear emulsion are

$$e = 0.7 \text{ MeV}; p = 14.0 \text{ MeV}$$
$$\mu = 5.5 \text{ MeV}; d = 20.0 \text{ MeV}$$
$$\pi = 6.1 \text{ MeV}; \alpha = 55.0 \text{ MeV}$$

The smaller the grain size, the thicker the emulsion, the more sensitive it is to ionising particles. Nuclear emulsions for experimental work are supported on glass plates. Photographic companies have developed a series of nuclear emulsions for detecting particles with different specific ionisation. For example, Ilford D_1 emulsion is suitable for detecting fission fragments and α-particles, Ilford C_2 emulsion for mesons, protons and heavy nuclei, and Ilford G_5 for high energy electrons.

Nuclear emulsion technique. One of the methods to develop thick emulsions as are required for nuclear detection work is to soak them in the developer maintained at a low temperature of 5°C. The rate of development will be very low at such a temperature. Once the whole emulsion has been suffused by the developer the temperature is raised to 28° C so that it is uniformly developed quickly. The nuclear emulsions with their high silver halide content shrink, which poses a serious problem on the accuracy of nuclear emulsion measurements. This shrinkage can be controlled by soaking in glycerine, filling the voids that are left by the removal of silver halide due to shrinkage. A comparison of nuclear and optical emulsions is given below.

Property	*Optical emulsion*	*Nuclear emulsion*
AgBr : Gelatine (mass)	47 : 53	80 : 20
AgBr : Gelatine (volume)	15 : 85	45 : 55
Grain size (micron)	1 – 3.5	0.1 – 0.6
Thickness (micron)	2 – 3	25 – 2000
Sensitivity to light	Very high	Poor
Response to α-particles	Dense blackening	Individual tracks
β-particles	Moderate	Faint fog
γ-rays	Faint	Almost none

The tracks produced in an emulsion are very short. They can be magnified and photographed. High quality microscopes with small depth of view, accurate depth gauges and attached devices for measuring track lengths and angles are used to scan the developed plates. The general procedure used for displaying emulsion tracks is to make enlargements of microscope photographs of neighbouring sections and then to piece the photographic prints together.

Under favourable conditions it is possible to determine the charge, mass and initial kinetic energy of the charged particle brought to rest in the nuclear emulsion. For this purpose, the range of the particle, the grain density and its variation along the track, the small scattering angle due to collision of the particle in the emulsion are all accurately measured.

The number of grains deposited per micron $\frac{dN}{dx}$ of track is closely related to the energy loss per unit length of track $\frac{dT}{dx}$ KeV/micron by the empirical relation

$$\frac{dN}{dx} = a\left\{1 - e^{bz(dT/dx)^{1/2} c^{1/2}}\right\}$$

where a, b and c are empirical constants and Z is the integral number characterising the charge of the ionising particle.

The total range R in microns is connected to the initial kinetic energy of the ionising particle by the empirical relation

$$K.E. = KZ^{2n} m^{1-n} R^n$$

where K and n are empirical constants and m the mass of the particle in units of proton mass.

The total number of silver grains deposited N is also given by an empirical relation

$$N = AZ^{2p} m^{1-p} R^p$$

where A and p are other empirical constants.

Emulsions impregnated with target materials are used for studying different nuclear reactions. Similarly special emulsions are available for the study of α-decay and for the detection of neutrons and γ-rays.

Gelatine emulsion contains a considerable amount of hydrogen so that a neutron on passing through the nuclear emulsion may collide with a proton and give up most of its energy. Nuclear emulsions can, therefore, be used for neutron spectroscopy.

In another method to detect neutrons, the emulsion is impregnated with B^{10}.

Gamma rays of energy above 3 MeV can be detected through the track of protons released in photo-disintegration reaction in emulsion loaded with deutron compounds.

Advantages. 1. Nuclear emulsions give continuous and permanent records of the trajectories of charged particles through them. Particles of different ionising powers produce tracks which are quite different.

2. The emulsion is relatively light and cheap. Because of their lightness emulsion plates are very suitable for high altitude cosmic ray work and can be sent in balloons and space ships. The cosmic ray events once recorded can be studied by developing the exposed plates conveniently in the laboratory. Study of cosmic ray events in emulsion lead to the discovery of π and K-mesons.

3. The emulsion is continuously sensitive and is consequently always available to record an event unlike a cloud chamber which is sensitive only for a fraction of a second after an expansion and remains insensitive for several seconds between successive expansions.

4. The high density of emulsion gives it a high stopping power about 1000 times that of standard air. Many short lived, unstable high energy particles can be brought to rest in the emulsion, before their decay and their decay scheme can be studied. For example, the measurement of the ranges of mesons emitted from stopped pions before they decay into electrons enables us to determine the mean life time of muons.

5. The energy of product particles in a nuclear reaction can be determined from their range in the emulsion and the differential scattering cross-section from counts of the number of tracks formed in each emulsion.

Limitations and drawbacks. 1. The main drawback of the nuclear emulsion is that their sensitivity and thickness are effected by temperature, humidity, age of the emulsion before development and the conditions under which the emulsion plates are developed. The scanning of the plates and the analysis of the tracks obtained are also laborious and tiring when done manually.

2. It is very difficult to determine the sign of the charge and the momentum of a particle from observations of the curvature of the path in a strong magnetic field. The actual path length in the emulsion is small as compared to that in a cloud chamber and the multiple coulumb scattering in the emulsion stimulates the curvature.

3. Another serious disadvantage of emulsion is that they remain always sensitive to ionising particles. There is, therefore, no method of triggering them by the type of particles we wish to study, as in the case of cloud chambers.

14.19 WILSON CLOUD CHAMBER

Wilson cloud chamber is used to photograph the tracks of high energy ions and electromagnetic radiations capable of producing ionisation.

Principle. If there is a sudden expansion of saturated vapour in a chamber, super-cooling of the vapour occurs. If some dust particles are present in the chamber, tiny droplets will be formed by condensation on them. But if we have *completely dust free* saturated air and it is suddenly allowed to expand and thereby cool, no condensation will take place. If instead of dust particles some ions are available in the chamber during expansion, they serve as nuclei for condensation. In other words, if an ionising particle passes through the chamber during an expansion process, ions are produced along its path, droplets condense on these ions and the particle track becomes visible. This is how the cloud chamber makes visible the path of an incoming particle.

Construction and working. It consists of a small glass chamber A resting on a rubber disc in a trough containing water and fitted with a movable piston P as shown in Fig. 14.10. The chamber in this way is completely saturated with water vapour. A large evacuated flask F, fitted with a movable piston p is connected to the chamber by a side tube. The space within the movable piston P is reduced by the blocks BB. It is evident from the figure that as soon as the piston p is moved from the position a to b the piston P moves downwards and causes an adiabatic expansion in the chamber A. If at the same time a high-velocity ion or an electromagnetic radiation passes through the chamber ionisation takes place and the charged particles serve as nuclei for the condensation of water vapour. As the chamber is illuminated the water-vapours will reflect light and the path of the ionising agents will become visible. If a camera with its shutter open is placed on the chamber the path of the ionising agent can be photographed.

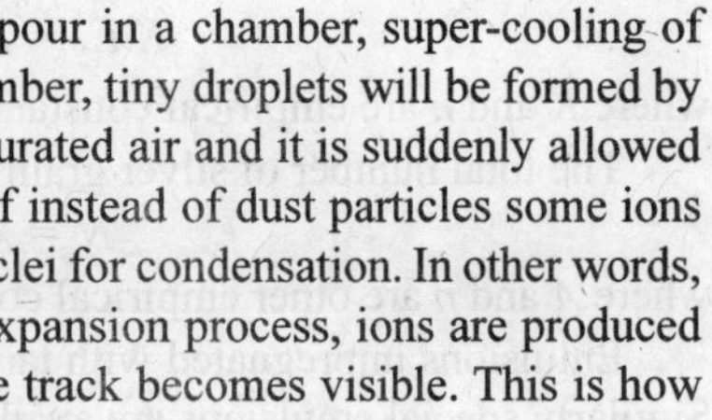

Fig. 14.10

In air saturated with water-vapour but absolutely free from dust particles, expansion in the ratio of 1 : 1.25 causes condensation to occur on negative ions and expansion ratio of 1 : 1.31 gives condensation on positive ions also and expansion ratio of 1 : 1.38 brings about condensation even without any charged particles.

The ionising agent can be easily identified from its path in the cloud chamber.

Alpha particles being comparatively massive go straight and their paths *are thick, straight and sharply defined*.

Beta particles being lighter are easily deflected by collision and their paths *are thin and crooked*.

Gamma rays are seldom observed in a cloud chamber as a single ray produces only *one or two ions in passing through several metres of air*.

The modern cloud chambers are automatic, *i.e.*, as an ionising particle enters the cloud chamber, expansion takes place and at the same time the camera shutter also opens thus enabling the path to be photographed.

Uses. (*i*) Cloud chambers are used to study the specific ionisation along the track of a charged particle and the range of such particles.

(*ii*) The sign of the electric charge and momentum p of the particle can be determined by placing the chamber in a strong magnetic field and noting the radius of curvature of the track.

Suppose a particle of mass m and charge q moves with a velocity v in a direction perpendicular to that of the magnetic field of flux density B, then under the action of the force due to the magnetic field the particle will move in a circular path of radius R such that the force due to the magnetic field is balanced by the centrifugal force

$$\therefore \quad \frac{mv^2}{R} = Bqv$$

$$\text{or} \quad mv = p = Bq\,R$$

Knowing p the kinetic energy of the particle can be calculated by the relation

$$K.E. = E_k = \sqrt{p^2 c^2 + m_0^2 c^4} - m_0 c^2$$

Limitations. (*i*) The range of the particle may exceed the dimensions of the chamber so that only a part of the track is photographed.

(*ii*) When there is a forked track, there is an uncertainty about the nature of the nuclei constituting the branches of the track.

Diffusion cloud chamber. The disadvantage of the cloud chamber lies in the fact that it needs a definite time to recover after an expansion and be ready to record successive events. Thus it cannot maintain a continuous record of events taking place in the chamber. This drawback is over come in the *continuously sensitive diffusion cloud chamber*. It consists of a chamber containing air or other gas in which a vertical temperature gradient is maintained by keeping the top at a higher temperature and bottom at a lower temperature. The liquid used is methyl alcohol. It vaporises in the hot region where the vapour pressure is high. The vapour diffuses downward *continuously* where the vapour pressure is low and condensation takes place. Since the saturation vapour pressure is an exponential function of temperature, a steady condition of super saturation can be achieved and there is a region near the cold base where the super saturation factor is high enough for condensation to take place around the available ions.

The chamber remains continuously sensitive to ionising particles until the supply of volatile liquid is exhausted. The system is illuminated by a strong source of light and the track of the particle is photographed by a camera. The *sensitive region* of the diffusion chamber is about 7.5 cm deep.

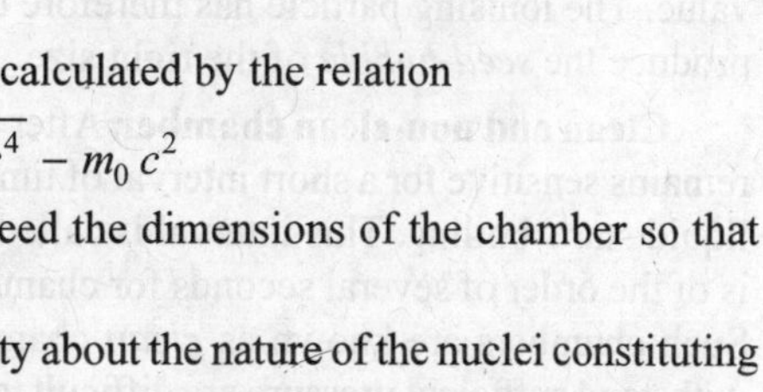

Fig. 14.11

The discovery of the positron, the muons in the secondary cosmic radiation, the K° meson and the hyperon $\mathring{\Lambda}$ as well as the first accurate determination of the muon rest mass, were made with the cloud chamber.

14.20 BUBBLE CHAMBER

A cloud chamber is not suitable to detect energetic particles because the interaction of energetic particles cannot be completely observed in a cloud chamber. The *bubble chamber* invented by Glaser in 1952 combines the precise tracks and high stopping power of emulsions with easy scanning of cloud chamber photographs.

Process of visible track. Under normal atmospheric pressure, a liquid boils at its normal boiling point with the evolution of bubbles. But if the liquid is heated under high pressure to a temperature well above its normal boiling point, a sudden release of pressure will leave the liquid in a *super-saturated* state *i.e.* with its temperature above its normal boiling point for a finite interval of time provided no discontinuities are present in it. If an ionising radiation passes through the liquid within a few milli-seconds after the pressure is released, the string of ions of microscopic size left in the track of the ionising particle act as nuclei or condensation centers for the formation of bubbles. The vapour bubbles, to begin with, are small but rapidly grow in size and attain a *visible size* in a time of about 10 – 100 μs and can be photographed using a short duration flash before they become too large. Thus we get a record of the path of the particle in the liquid. The chamber is reset by applying the original pressure to it.

Thus in a bubble chamber a vapour bubble forms in a super heated liquid whereas in a cloud chamber a liquid drop forms in a super-saturated vapour.

In a bubble chamber, a bubble will grow only if its initial size is larger than a certain critical value. The ionising particle has therefore to deposit sufficient energy along its track in the liquid to produce the *seed-bubble* of the right size.

Clean and non-clean chamber. After the pressure over the liquid is released, the bubble chamber remains sensitive for a short interval of time ready to record an ionising event before the superheated liquid starts boiling. This interval is called sensitive time of the bubble chamber. The *sensitive time* is of the order of several seconds for chambers with no sharp discontinuities of curvature of surface. Such chambers are known as *clean* chambers. Workable clean chambers of suitable size, able to withstand sufficient pressure are difficult to make. The chambers in common use are of the non-clean or *dirty* type.

Construction and working. A schematic diagram of a liquid hydrogen bubble chamber designed by L.W. Alvarez and operating at a temperature of 27 K is shown in Fig. 14.12. The vessel was made

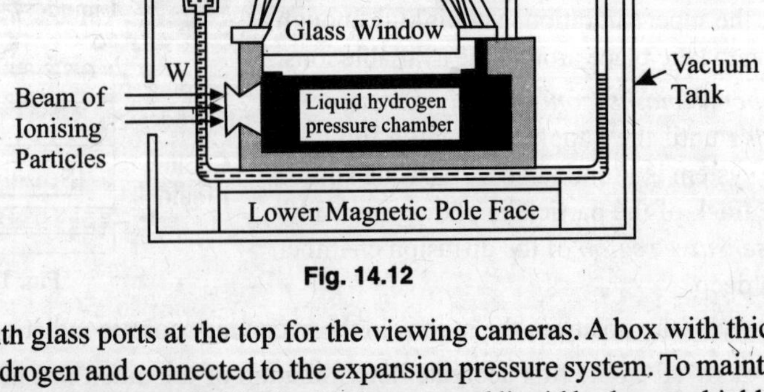

Fig. 14.12

of stainless steel with glass ports at the top for the viewing cameras. A box with thick glass walls is filled with liquid hydrogen and connected to the expansion pressure system. To maintain the chamber at constant temperature it is surrounded by liquid nitrogen and liquid hydrogen shield. A beam of high energy ionising particles is allowed to enter the chamber from a side window *W*. Sudden release of pressure from the expansion valve is followed by a light flash and the camera takes a stereoscopic view of the chamber. The light flash is triggered by the delayed pulse from a counter which is activated by the same ionising particle that traverses the bubble chamber. From the curvature and length of the track when the bubble chamber is operated in a magnetic field, the momentum, energy and sign of charge of the ionising particle is determined.

The *cycling time* of the chamber which is the time that lapses between the release of pressure on the chamber liquid and the next recompression of it is of the order of 30 milli-second.

Other transparent liquids of low surface tension and high vapour pressure are also used in the bubble chamber. In earlier bubble chambers, diethyl ether was used. Now a days in addition to liquid hydrogen, deuterium, helium, propane, Xenon and other liquids have been successfully used for different purposes. For example, liquid hydrogen is replaced by propane (C_2H_3) to avoid the need of low temperature of – 246° C required for liquid hydrogen. Liquid deuterium is used, because it has a fairly lose bound neutron to a proton, for observations on *neutron interaction*. For studying high energy interaction of nucleons, liquid helium is used at 4 K and at 1 atmosphere pressure. For study of interaction of gamma rays and other neutral particles, a liquid containing heavy atoms is preferred because it has a high stopping power. Liquid Xenon is used for studying the interaction with high energy protons and neutral pions.

Advantages. 1. The density of the liquid is very large as compared to that of a gas even at a high pressure. Therefore, the chances of collision of high energy particles with liquid molecules are very much greater. As a result there is much higher chance of their track being recorded.

2. The sensitive time of bubble chambers is very short, the background due to unwanted tracks is, therefore, considerably reduced.

3. There is no need to apply a clearing field to remove the back ground due to old tracks before the chamber is set sensitive because the life time of the bubble nuclei produced by the ionising particle is less than a mili-second which is very short as compared with the time it takes to release the pressure on the liquid.

4. The bubbles grow rapidly and as a result the tracks are not likely to get distorted due to convection currents in the liquid.

5. The bubble chamber is sensitive even to particles of low ionising power.

Drawback. The sensitive time between superheating by expansion and boiling is a few milli-seconds and entry of particles and photography must take place during this time. The bubble chamber therefore, loses the advantage of continuous sensitivity of the diffusion chamber.

14.21 SPARK CHAMBER

The visual techniques of cloud chamber and bubble chamber provide excellent spatial resolution for studies of high energy particles but their time determination is not so good. The electronic devices coupled with ionisation counters provide highly accurate time determination and resolution but give no spatial information about the particle. The two essential features of high resolution in space (cycling time 50 per second) and time (less than 1 μs) were achieved simultaneously in a spark chamber.

Principle. The basic principle of operation is that '*a high voltage is maintained between two plates spaced in a vessel containing a gas but the electric field is not quite enough to permit the passage of a spark. Now if an ionising particle enters the gas space, a spark will pass and it will tend to follow the path of ion pairs produced by the ionising particle*'.

Construction and working. It consists of a series of large, thin metal (Aluminium) plates mounted in an air tight chamber containing a gas like helium or a mixture of 90% neon and 10% helium at a pressure of 1 atmosphere. Alternate plates are connected to a triggered high voltage *d.c.* supply. A typical chamber may consist of 25 to 100 plates, each about 1 *m* square and 1 *mm* thick accurately spaced about 3 – 5 *mm* apart. Alternate plates are connected to a 10,000 – 15000 volt *d.c.* source.

When a high energy charged particle enters such a chamber, the passage of the ionising particles is detected by counters on either side of the chamber. The counter pulses are fed into a coincidence circuit whose output triggers the supply of high voltage pulse between the alternate plates of the spark chamber and also opens the camera shutter. Traversing the chamber, a charged particle produces many ion pairs along its path and sudden application of very high voltages to alternate plates while the others are left at ground potential, results in a very high electric field across the gaps. Electrical breakdown then occurs along the trails of the ion and sparks jump between the pairs of oppositely charged plates as shown in Fig. 14.13. So the trajectory of a given particle through the counter is marked by a series of sparks.

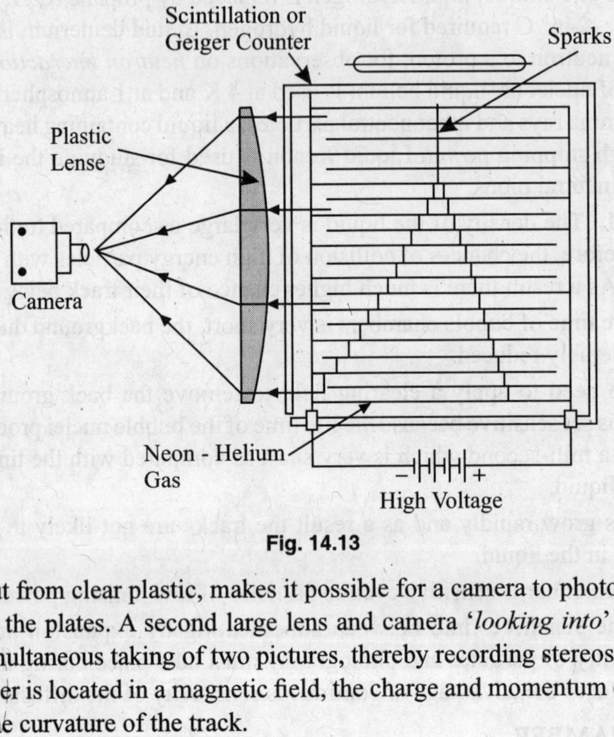

Fig. 14.13

A large lens cut from clear plastic, makes it possible for a camera to photograph the light from sparks between all the plates. A second large lens and camera '*looking into*' an adjascent side of the box permits simultaneous taking of two pictures, thereby recording stereoscopic photos of each event. If the chamber is located in a magnetic field, the charge and momentum of the particle can be determined from the curvature of the track.

Before the spark chamber can record an event it is necessary to remove all the ions from the spaces between the plates, otherwise these may initiate spark breakdown of the gas when the chamber is pulsed with high voltage. This is accomplished by applying a continuous *d.c.* voltage between the plates so that the electrons in the gas between the plates are swept off within a short time after their production. Spark breakdown of the gas between the plates can occur due to an ionising particle only if the time that lapses between the passage of the particle through the chamber and pulsing of the high voltage is less than the time *d.c.* voltage takes to clear off the electrons generated in the gas by the ionising particle.

A spark chamber is capable of rapid operation. It provides excellent definition of direction and can be operated in coincidence with other counters. As it is rendered sensitive for only a very short time, it is used to detect and record rare events.

EXERCISE

1. Explain the interaction of charged particles with matter.
2. (*a*) A charged particle (charge *ze*) having a mass '*M*' and kinetic energy '*E*' is passing through a certain medium having '*n*' electrons per unit volume. If the motion of the particle is along X-direction, derive an expression (using classical logic) for the rate of loss of energy $\left(\frac{-dE}{dx}\right)$. Define stopping power of the medium. On what factors does it depend ? Outline the assumptions made.

   (*b*) State Bethe and Bloch modification and write down Bethe-Bloch formula.

   (P.U. 2004, 2003, 2002, 2001, 2000; G.N.D.U. 2004, 2003, 2002; H.P.U. 2001, 2000; Pbi.U. 2001; K.U. 2002; M.D.U. 2000)
3. (*a*) Define relative atomic stopping power.

   (*b*) The relative atomic stopping power is one in case of (*i*) Standard air (*ii*) H_2 (*iii*) Lead (*iv*) Iron. ***(H.P.U. 2000)***
4. Why does Bethe-Bloch formula of interaction of charged particles with matter not become applicable to electrons ? ***(P.U. 2000, 1992)***
5. Explain the limitation of Bethe-Bloch formula in applying it to the interaction of electron with matter.
6. (*a*) What do you mean by range of a moving charged particle ? Write also the empirical range energy relations for different particles. ***(G.N.D.U. 1991)***

   (*b*) Explain the terms 'Range' and 'Straggling' of a charged particle. ***(K.U. 2000)***
7. (*a*) The energy loss of electrons is a much more complicated phenomenon than energy loss by ionisation of heavy ions. Explain it. ***(P.U. 2000)***

   (*b*) Discuss the motion of high energy electrons through a medium. How does a fast electron lose energy on its passage through matter ?
8. Explain interaction of neutrons with matter.
9. (*a*) Give the three properties of nuclear radiation used in detection instruments. What are gas filled ionisation based nuclear detectors ?

   (P.U. 2001; G.N.D.U. 2003; H.P.U. 2001)

   (*b*) Describe the construction, principle and working of an ionisation chamber. Explain the difference between ionisation chamber and GM counter. Why is an ionisation chamber less sensitive to β-particle ?

   (Calicut. U. 2001; 2000; G.N.D.U. 2004)
10. (*a*) Describe and discuss the variation of ionisation current with applied voltage due to the passage of a charged particle through the usual cylindrical ionisation detector. What is specific characteristic of G.M. region ?
11. (*a*) Discuss the curve between pulse height and applied voltage for a gas filled counter serving as (*i*) an ionisation counter (*ii*) a proportional counter (*iii*) a Geiger Mueller counter.

    (P.U. 2001)

    (*b*) Describe specific ionisation, and gas multiplication factor.
12. Explain the principle, construction and working of a proportional counter (gas filled, ionisation based nuclear detector). Give its application and explain how it can be used for neutron detection.
13. (*a*) Discuss in detail the construction and working of a G.M. counter. What is dead time and recovery time ? What do you mean by quenching of a G.M. counter ? What is its necessity ? How is it achieved internally ?

(*b*) Describe briefly its operation and testing. What is plateau as applied to a G.M. counter? Give its application. What are its limitations ?

(P.U. 2004, 2003, 2002, 2001, 2000; G.N.D.U. 2004, 2003; Pbi. U. 2004, 2002, 2001, 2000; K.U. 2001, 2000; M.D.U. 2001; Calicut. U. 2002 H.P.U. 2004)

14. Give the difference between ionisation chamber and G. M. counter.

15. Give the difference between ionisation chamber, proportional counter and G.M. counter.

(M.D.U. 2002, 2001; H.P.U. 2001; K.U. 2002; P.U. 2003; G.N.D.U. 2002)

16. State the advantages and disadvantages of a G.M. counter. A G.M. counter cannot detect neutrons. Why ? ***(Bang. U. 2002)***

17. Explain the principle, construction and operation of a scintillation radiation counter. Draw a block diagram to show the main components. Briefly describe a few types of scintillators used. In what way a scintillation counter is superior to a G.M. counter ?

(P.U. 2004, 2002, 2001, 2000; Pbi. U. 2001, 2000; G.N.D.U. 2004; Luck. U. 2001; Calicut. U. 2003, 2000)

18. (*a*) Name three radiation detectors based on image formation. ***(G.N.D.U. 2001)***

(*b*) Explain the principle and operation of a nuclear emulsion charged particle detector giving nuclear emulsion technique. What are its advantages and limitations ?

(P.U. 2003, 2001, 2000)

19. Describe Wilson's cloud chamber. How does a cloud chamber make visible the path of an incoming particle ? What is its main disadvantage and how is it removed in a diffusion cloud chamber ? ***(Calicut. U. 2003)***

20. What is a bubble chamber ? Describe the process by which ionising particles make a visible track in a bubble chamber. Give its advantages and drawbacks.

(P.U. 2004, 2000; G.N.D.U. 2004, 2003, 2000; Calicut. U. 2003; Gauhati. U. 2000)

21. Write a short note on Spark Chamber.

22. You have at your disposal two G.M. counters, one has a long and flat plateau and the other has a short and steep plateau. Which would you prefer ? Explain with reasons.

[**Hint.** A G.M. counter with a long and flat plateau is always preferred to one with a short and steep plateau.

A long and flat plateau means that the counting rate remains constant over a fairly large voltage region i.e. the counting rate is independent of the potential difference across the tube which is the main characteristic of a G.M. counter.

With a short and steep plateau the counting rate rises rapidly and steeply in a short voltage interval and the results are no longer dependable. When the operating potential is changed on a short plateau, it is possible that the over voltage may become critical and even small fluctuations may effect the counting rate or the counter may go beyond its plateau region. It fact, when the plateau becomes notably short and steep it is a sign that the counter tube is nearing the end of its useful life.]

MULTIPLE CHOICE QUESTIONS

1. The time during which pulses are recorded but are of smaller size in G.M. counter is called

(*a*) Recovery time (*b*) Dead time

(*c*) Resolving time (*d*) no specific name.

Ans. (*a*)

STRUCTURE OF NUCLEI

15
Chapter

15.1 STRUCTURE OF THE NUCLEUS

An atom is electrically neutral under normal conditions. But some atoms are radioactive which emit negatively or positively charged particles (β-rays and α-rays). Hence it was assumed that the atoms are made up of equal number of positive and negative charges.

In the year 1897, Sir J.J. Thomson discovered the existence of a very small particle which carried a negative charge. This particle was called the **electron.**

Thomson model. Thomson, therefore, assumed that an atom consisted of a sphere of positive charge of uniform density in which an equal and opposite negative charge was symmetrically distributed in the form of electrons.

Rutherford model. The experiments on the scattering of α-particles showed that some of the α-particles were deflected through more than 90° on their passage through a thin gold foil. Rutherford interpreted the observation as an indication of the presence of a very small entity having a strong positive charge on it, which exerts a strong force of repulsion on the α-particle. *This small entity, carrying a positive charge was called the* **nucleus.** Calculations from experimental results showed that the approach of an α-particle to the nucleus of an atom was of the order of 10^{-14} m, whereas from a study of kinetic theory of gases and other phenomena, the diameter of the atom was shown to be of the order 10^{-10} m.

Rutherford, therefore, came to the conclusion that an atom only consists of a small massive nucleus, which carries a positive charge with a comparatively vast region around containing the electrons. The electrostatic force of attraction of the massive positively charged nucleus was supposed to be balanced by the centrifugal force due to the rotation of the electrons around the nucleus. Thus Rutherford transformed the static model due to Thomson into a dynamic model in which the nucleus plays the role of the sun and the electrons that of the orbiting planets in the solar system.

15.2 PROTON-ELECTRON MODEL

Until the discovery of the neutron it was assumed that an atom only contains electrons and protons. The concept of proton-electron constitution of the nucleus arose from the following experimentally observed facts:

(*i*) Some radioactive elements emitted β-rays which were found to be fast moving **electrons.**

(*ii*) The discovery of isotopes and their whole number atomic weights suggested that all elements were built out of hydrogen atoms. The fractional atomic weights are due to the presence of two or more isotopes each of which has an integral mass. The experimental results show that nuclei of different elements can be regarded as being a collection of an integral number of hydrogen nuclei. The nucleus of hydrogen has a positive charge equal in magnitude to that of the electron and is known as a **proton.**

According to *Proton-electron model*, a nucleus of atomic mass number A and atomic charge number Z consisted of A protons and $(A - Z)$ electrons, thus giving it a net charge $+ Ze$. The nucleus was supposed to be surrounded by Z extra-nuclear electrons, thereby making the atom on the whole neutral. For example, an atom of nitrogen of atomic weight 14 and atomic number 7 was supposed to contain 14 protons and 14 electrons. 14 protons and 7 electrons are packed together in the small space of the nucleus and the remaining seven electrons revolve round the nucleus in suitable orbits.

Failure of the electron-proton hypothesis. Although proton-electron model had some satisfactory aspects, yet there were a number of objections against the model. The most serious objection is that *the electron can not exist within the nucleus.*

15.3 PROTON-NEUTRON HYPOTHESIS

Until the discovery of neutron it was assumed that a nucleus only contains electrons and protons. This set up of the nucleus has changed with the discovery of the neutron.

Neutron. In 1930, Bothe and Becker bombarded Beryllium with α-particles and found that a very penetrating radiation was produced. J. Chadwick, in 1932 showed that this radiation consisted of particles of mass slightly greater than that of the proton but carried *no charge* because its path was not deflected by electric and magnetic fields. The particle also did not ionize a gas. This new fundamental particle was given the name **neutron.**

With the discovery of the neutron the set up of the nucleus has changed. To overcome the difficulties experienced by proton-electron model, Heisenberg proposed a new hypothesis that the *nucleus consists of protons and neutrons.* According to proton-neutron hypothesis, a nucleus of atomic mass (number) A and atomic charge (number) Z consisted of Z protons giving it a net charge $+ Ze$ and $N = (A - Z)$ neutrons giving it a total mass $(N + Z) = A$. For example, the nitrogen nucleus contains 7 protons and 7 neutrons, thus making atomic weight 14 and atomic number 7. The number of extra nuclear electrons is 7 *i.e.*, $= Z$, thus making the atom, on the whole, neutral. As the nucleus of an atom contains protons and neutrons a common name has been given to these particles, *i.e.*, **nucleons.** In $_{92}U^{238}$ nucleus there are in all 238 nucleons out of which 92 are protons and $238 - 92 = 146$ are neutrons. The rest mass of the proton is equal to 1.0072766 a.m.u. $= 1.6725 \times 10^{-27}$ kg and that of the neutron is equal to 1.0086654 a.m.u. $= 1.6748 \times 10^{-27}$ kg. The proton and neutron are, therefore, two charge states of the same particle – the nucleon. The proton is the *protonic state* with charge $+ e$ and the neutron is the *neutronic state* with charge zero.

The proton-neutron theory has overcome the objections of the proton-electron theory thereby avoiding the failures of proton-electron hypothesis and paving the way for acceptance of proton-neutron hypothesis.

Avoiding failures of Proton-Electron Hypothesis

(*i*) **Isotopic mass.** The isotopic masses can be easily explained on the basis of proton-neutron hypothesis. Isotopes[1] of the same atom have the same Z *i.e.*, the same number of protons but different atomic masses *i.e.*, A. Thus they have only a different number of neutrons *i.e.*, $N = (A - Z)$.

1 **Isotopes.** *Isotopes are nuclei having the same atomic* (proton) *number Z but different atomic mass A e.g.*, $_8O^{16}$ and $_8O^{17}$. The isotopes of an element contain the same number of protons but different number of neutrons.

Hydrogen has three isotopes, (*i*) Hydrogen $^1H_1$; (*ii*) Deuterium $^2H_1$ and (*iii*) Tritium $^3H_1$.

The nucleus of each isotope of hydrogen contains one proton, but a deuterium nucleus ($^2H_1$) contains a proton as well as a neutron and tritium nucleus ($^3H_1$) contains one proton and two neutrons.

At present about 297 different isotopes, from hydrogen (mass no. 1) to uranium (mass no. 238) including radioactive ones are known. About 300 unstable isotopes have been produced by disintegration. Some elements like As, F, I and Au have only a single isotope whereas others like Xenon and Hg have nine each. The atomic weights of the isotopes of Xenon are 124, 126, 128, 129, 130, 131, 132, 134 and 136. The relative abundance of various isotopes is found from the intensity of lines obtained from Aston's mass spectrograph. In general, elements of odd atomic numbers do not have more than 2 stable isotopes whereas those of even atomic numbers usually have a much larger number.

(*ii*) **Nuclear spin.** It has resolved the difficulty regarding nuclear spin. Proton and neutron are both Fermions and have a spin $\frac{1}{2}\hbar$. If the nucleus contains an even number of nucleons (protons + neutrons) *i.e.*, *A* is even like N_7^{14}, $_8O^{16}$ it has an *integral* or *zero spin*. On the other hand if the nucleus contains an *odd* number of nucleons like $_4Be^9$ having nine (odd) particles it has a half integral spin. This agrees with experimental values and explains the observed spin of deutron equal to *one*.

(*iii*) **Nuclear magnetic dipole moment.** The magnetic dipole moment associated with nuclear spin $= \frac{e\hbar}{2m_p}$ where m_p is the rest mass of the proton. The quantity $\frac{e\hbar}{2m_p}$ is called *nuclear magneton* and is denoted as β_n. It has a value 3.15×10^{-8} eV/T (or 5.05×10^{-27} J/T). The magnetic dipole moment associated with a spinning electron is $\frac{e\hbar}{2m_e}$ where m_e is the rest mass of the electron. The quantity $\frac{e\hbar}{2m_e}$ is called Bohr magneton μ_B and has a value 5.79×10^{-5} eV/T (or 9.27×10^{-24} J/T). The ratio of a nuclear magneton to Bohr magneton is $\frac{m_e}{m_p} = \frac{1}{1836}$ *i.e.*, a nuclear magneton is almost 2000 times smaller than Bohr magneton.

A neutron has a magnetic moment equal to $-1.913\ \beta_n$. As it carries no charge it cannot posses any magnetic moment due to orbital motion. A proton has an intrinsic magnetic moment $2.792\ \beta_n$. Its magnetic moment due to orbital motion is of the order of β_n. Thus the resultant magnetic moment of the nucleus should be of the order of nuclear magneton.

The observed magnetic moment of the nucleus is also of the order of nuclear magneton lying between $-2\beta_n$ to $+4\beta_n$. Thus the proton-neutron hypothesis avoids the difficulty of explaining the nuclear magnetic moment on the basis of proton-electron theory.

Acceptance of proton-neutron theory. (*i*) This theory removes the difficulty of having an electron in the nucleus and is based upon two actually existing fundamental particles, the proton and the neutron.

(*ii*) The presence of protons and neutrons in the nucelus is in accordance with Heisenberg's uncertainty principle, according to which

$$\Delta x \,.\, \Delta p = \frac{\hbar}{2}$$

where Δp is the uncertainty in momentum. As Δx is the nuclear radius $\geq 10^{-14}$ m, the value of

$$\Delta p = \frac{\hbar}{2\Delta x} = \frac{1.055 \times 10^{-34}}{2 \times 10^{-14}} = 5.3 \times 10^{-21} \text{ kg ms}^{-1}.$$

The rest mass of the proton (or neutron) is of the order of $m_p = 1.67 \times 10^{-27}$ kg. Hence the corresponding value of velocity $v = \frac{5.3 \times 10^{-21}}{1.67 \times 10^{-27}} = 3 \times 10^5$ ms^{-1}. It is, therefore, a non relativistic case. The value of kinetic energy is given by

---

Most of the physical and chemical properties of an element are determined by the number and arrangement of the electrons in its atoms. Since the isotopes of an element have almost identical electron structures, the two isotopes of chlorine, for intance, have the same yellow colour, the same suffocating odour, the same bleaching effect, and the same ability to combine chemically with metals. Because boiling and freezing points depend somewhat on atomic mass, these properties differ slightly between the isotopes, as does density.

$$E_k = \frac{p^2}{2m_p} = \frac{(5.3\times10^{-21})^2}{2\times1.67\times10^{-27}} = 8.4\times10^{-15} \text{ Joule} = 52 \text{ KeV}$$

Since the energy carried by the protons or neutrons emitted by the nuclei is greater than 52 KeV the protons and neutrons can exist within the nucleus.

(*iii*) The process of β-emission *i.e.*, emission of electrons from the nuclei is explained on the basis of conversion of a neutron into a proton as these are two charge states of the same particle–the nucleon. An electron does not pre-exist in a nucleus but is formed just at the instant of emission, by the transformation of a neutron into a proton according to the equation

$$n \longrightarrow p + e^-$$

The positron emission is also explained on the same basis by the conversion of proton into a neutron according to the relation

$$n = n + e^+$$

(*iv*) **Classification of elements.** This theory also leads to more refined classification of elements as follows:

(*a*) **Isotopes.** These are nuclei having the same atomic (*proton*) number Z but different atomic masses (A) *e.g.*, ${}_8O^{16}$ and ${}_8O^{17}$.

The isotopes of an element contain the same number of protons but different number neutrons. All the isotopes of an element have identical chemical properties and differ physically only in mass.

(*b*) **Isotones.** These are nuclei having the same neutron number N *e.g.*, ${}_6C^{13}$ and ${}_7N^{14}$. Both have $(13 - 6) = 7$ or $(14 - 7) = 7$ neutrons.

(*c*) **Isobars.** These are nuclei having the same mass number A (total number of nucleons *i.e.*, (protons plus neutrons) but different atomic number Z, *e.g.*, ${}_6C^{14}$ and ${}_7N^{14}$. Both have the same mass number $A = 14$ but $Z = 6$ for ${}_6C^{14}$ and $Z = 7$ for ${}_7N^{14}$.

(*d*) **Isomers.** These are atoms which have the same atomic mass A and same atomic number Z but differ from one another in their nuclear energy states. They also exhibit differences in their internal structure and have different life times. They are also known as *isomeric nuclei*.

(*e*) **Mirror nuclei.** Nuclei with same mass number A but with proton and neutron number interchanged *i.e.*, the number of protons in one is equal to the number of neutrons in the other, are called *mirror nuclei e.g.*, ${}_4B^7$ ($Z = 4, N = 3$) and ${}_3Li^7$ ($Z = 3, N = 4$).

15.4 PROPERTIES OF NUCLEI

(*i*) **Charge.** If Z is the charge number of nucleus *i.e.*, the number of protons in it, then

Charge on the nucleus = + Ze

where e is the positive charge equal to the charge on the electron = 1.6×10^{-19} Coulomb.

(*ii*) **Mass.** If A is the mass number *i.e.*, the total number of nucleons in the nucleus *i.e.*, Z protons and $(A - Z)$ neutrons, then the mass of the nucleus is very nearly equal to A atomic mass units. In terms of atomic units, the mass of carbon atom C^{12} is taken to be = 12 (a.m.u.) and

1 a.m.u. (also written as 1 u) = 1.6604×10^{-27} kg = 931.48 MeV.

∴ Nuclear mass $M_N = A\, m_N$ where m_N is the mass of a nucleon

(*iii*) **Radius.** As the nucleus is approximately spherical, its volume is proportional to the total number of nucleons in it or its mass number A.

$$\therefore \quad \frac{4}{3}\pi r^3 \propto A$$

where r is the radius of the nucleus.

Hence $r \propto A^{1/3}$

or $r = r_0 A^{1/3}$

where $r_0 = 1.3 \times 10^{-15}$ m = 1.3 fermi [1 fermi (fm) = 10^{-15} m]

(*iv*) **Density.** Nuclear density $\rho_N = \dfrac{\text{Nuclear mass}}{\text{Nuclear volume}}$

Nuclear mass = Am_N where A = mass number and m_N = mass of the nucleon = 1.67×10^{-27} kg.

$$\therefore \quad \text{Nuclear volume} = \frac{4}{3}\pi r^3 = \frac{4}{3}\pi\,(r_0 A^{1/3})^3 = \frac{4}{3}\pi r_0^3 A$$

$$\therefore \quad \rho_N = \frac{Am_N}{\frac{4}{3}\pi r_0^3 A} = \frac{m_N}{\frac{4}{3}\pi r_0^3} = \frac{1.67 \times 10^{-27}}{\frac{4}{3}\pi\,(1.3 \times 10^{-15})^3}$$

$$= 1.816 \times 10^{17} \text{ kg m}^{-3}$$

This shows that the nuclear matter is in a highly compressed state.

As the density of the nucleus is independent of A, its value is almost the same for all nuclei.

In other words, nuclear mass is proportional to nuclear volume or size.

(*v*) **Binding energy.** The nucleons exert short range nuclear forces of attraction on each other. When any two particles attract each other, the sum of their masses, when separated, exceeds that of the bound system since energy (or mass) must be added to the system to separate it into component particles.

This energy is called *binding energy* and holds the nucleons together to form a stable nucleus. Thus the stability of the nucleus is due to the decrease in mass of the constituent particles which combine to form the nucleus.

For a nucleus having Z protons and $N = (A - Z)$ neutrons;

$$\text{Assumed nuclear mass} = Zm_p + Nm_n$$

where m_p is the mass of the proton and m_n that of the neutron. The real nuclear mass M_N is, however, less than the assumed nuclear mass. The difference in assumed and real mass $(Zm_p + Nm_n) - M_N = \Delta m$ gives the *decrease* in mass and the *binding energy* is given by

Binding energy B.E = $\Delta mc^2 = c^2\,[(Zm_p + Nm_n) - M_N]$

(*vi*) **Angular momentum of the nucleons. 1. Spin angular momentum.** Each nucleon *i.e.*, each proton and each neutron in the nucleus has an intrinsic angular momentum which may be pictured as being caused by particle's spinning motion about an axis through its centre of mass. This is known as *spin angular momentum* and its magnitude is given by

$$|\vec{s}| = \sqrt{s\,(s+1)}\,\hbar$$

where $s = 1/2$ is the *spin angular momentum quantum number.*

The *magnetic spin quantum number* m_s is the component of s in a specified quantisation direction – that of the applied magnetic field usually taken as Z-axis. With $s = \frac{1}{2}$, $m_s = +\frac{1}{2}$ and $-\frac{1}{2}$ corresponding to *parallel and anti-parallel* alignment with respect to Z-axis.

2. Orbital angular momentum. In addition, each individual nucleon may be pictured as having an angular momentum associated with its orbital motion within the nucleus. This is known as *orbital angular momentum* and its magnitude is given by

$$|\vec{l}| = \sqrt{l\,(l+1)}\,\hbar$$

where l is the *orbital angular momentum quantum number* having only *integral* values 0, 1, 2, 3, ...

The *orbital magnetic quantum number* m_l is the component of l in the specified quantisation direction – that of applied magnetic field unusaually taken as *Z*-axis. m_l can take $(2l + 1)$ possible values such as, $l, (l - 1) \dots + 1, 0, - 1, \dots - (l - 1), - l$.

3. Total angular momentum. The total angular momentum $\vec{j}$ of a nucleon is the vector sum of its orbital and spin angular momenta

$$\therefore \qquad \vec{j} = \vec{l} = \vec{s}$$

The magnitude of the total angular momentum of a nucleon is given by

$$|\vec{j}| = \sqrt{j(j+1)}\,\hbar$$

where j is the *total angular momentum quantum number*. For both proton and neutron which are Fermions $s = 1/2$ and l is an integer. Hence j can take on the values 1/2, 3/2, 5/2, ... *i.e.*, j is always *half integral*.

The *magnetic total angular momentum quantum number* m_j is the component of j in the specified quantisation direction – that of the applied magnetic field usually taken as the *Z*-axis. m_j can take $(2j + 1)$ possible values such as $j, (j - 1) \dots - (j - 1), - j$.

4. Total angular momentum of the nucleus (nuclear spin). The total angular momentum of the nucleus $\vec{I}$ for a particular nuclear state is the resultant of the individual total angular momenta of all the constituent nucleons in the nucleus. The magnitude of the total angular momentum of the nucleus is given by

$$|\vec{I}| = \sqrt{I(I+1)}\,\hbar$$

where I is the *total angular momentum quantum number for the nucleus*. The value of I depends upon the type of interaction or coupling between the nucleons.

I generally called **nuclear spin** is an integral multiple of $\hbar$ $(h/2\pi)$ if A is *even* and an *odd half integral multiple* of $\hbar$ when A is *odd*. If both Z and A are *even* the value of I in the ground state is always *zero*. All *odd-odd* nuclei *i.e.*, nuclei with *odd* number of *protons* and *odd* number of *neutrons* and hence *even A* have an *integral spin I*. On the other hand all *odd-even* nuclei *i.e.*, nuclei with *odd* number of *protons* and *even* number of *neutrons* or *odd* number of *neutrons* and even number of protons and hence *odd A* have *half integral spin* lying between $\hbar/2$ and $9\hbar/2$.

(*vii*) **Magnetic moment.** The magnetic dipole moment associated with nuclear spin $= \dfrac{e\hbar}{2m_p}$

where m_p is the rest mass of the proton. The quantity $\dfrac{e\hbar}{2m_p}$ is called **nuclear magneton** and is denoted as β_n.

$$\therefore \qquad \beta_n = \frac{e\hbar}{2m_p} = \frac{1.6 \times 10^{-19}\ \text{C} \times 6.58 \times 10^{-16}\ \text{eV}}{2 \times 1.673 \times 10^{-27}\ \text{kg}}$$

$$= 3.15 \times 10^{-8}\ \text{eV/T}$$

$$= 5.5 \times 10^{-27}\ \text{J/T}$$

The magnetic dipole moment associated with a spinning electron is $\dfrac{e\hbar}{2m_e}$ where m_e is the rest mass of the electron. The quantity $\dfrac{e\hbar}{2m_e}$ is called *Bohr magneton* μ_B and has a value

$$= 5.79 \times 10^{-5}\ \text{eV/T (or } 9.27 \times 10^{-24}\ \text{J/T).}$$

The ratio of a nuclear magneton to a Bhor magneton

$$\frac{\beta_n}{\mu_B} = \frac{m_e}{m_p} = \frac{1}{1836}$$

For all nuclei magnetic moment μ_N is given by

$$\mu_N = g\,\frac{e\hbar}{2m_p}\,\frac{\vec{I}}{\hbar}$$

where $\vec{I}$ is the total angular momentum of the nucleus.

The g-factor varies from nucleus to nucleus. The magnetic dipole moment of a proton is given by

$$\therefore \qquad \mu_{\text{proton}} = 2.792\,\frac{e\hbar}{2m_p} = 2.792\,\beta_n$$

The magnetic dipole moment of a neutron is given by

$$\mu_{\text{neutron}} = -1.913\,\frac{e\hbar}{2m_p} = -1.913\,\beta_n$$ [For proof See. **Eq. (v) article 15.9**]

Having a positive charge the magnetic field around a proton is parallel to the direction of its total angular momentum and hence its magnetic moment is positive. A negative magnetic moment of a neutron is a clear indication that the neutron is a complex particle containing negative and positive charges in equal amounts and that the negative charge is, on the average farther from the axis of rotation.

As an example, consider the case of a deutron. It is a nuclear particle composed of one proton and one neutron. Its known spin $I = 1$ *i.e.*, the spins of the two nucleons are *parallel*. Since the two magnetic moments are oppositely directed the resultant magnetic moment should be $2.792 - 1.913 = 0.879\ \beta_n$. Precision measurements show that $\mu_d = 0.857\ \beta_n$.

The difference is well outside the experimental error and has been explained by assuming that the ground state of deutron is not a simple $l = 0$ state.

The magnetic moment of *odd-even* and *even-odd* nuclei is due to the unpaired single nucleon. In general the magnetic moment of the nucleus is about one thousandth of the magnetic moment of the electron.

(*viii*) **Electric quadruple moment.** It is found that nuclei do not have electric dipole moment but electric quadruple moments have been observed in a number of nuclei. The electric quadruple moment of a nuclear charge distribution which is symmetric about the Z-axis is given by

$$Q = \frac{1}{e}\int_V (3z^2 - r^2)\,\rho\,(x, y, z)\,dv$$

where $\rho\,(x, y, z)$ is the charge density and $r^2 = x^2 + y^2 + z^2$. For uniformly charged ellipsoid of revolution defined by the equation

$$\frac{x^2 + y^2}{a^2} + \frac{z^2}{b^2} = 1$$

the electric quadruple moment reduces to

$$Q = \frac{2}{5}\,Z\,(b^2 - a^2)$$

where Ze is the total nuclear charge.

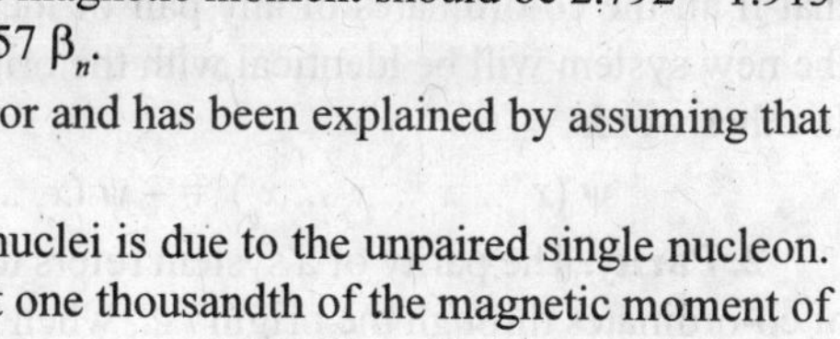

Fig. 15.1

Evidently $Q = 0$ for a spherically symmetric charge distribution ($a = b$). For a charge distribution stretched in the Z-direction (spin axis) ($b > a$) *i.e.*, prolate spheroid Q is **positive** and Q is **negative** for an oblate spheroid ($b < a$) *i.e.*, nucleus flattened along the spin axis Dimension of the quadruple moment is that of area. In nuclear physics area is measured in *barn*. (1 barn = 10^{-28} m^2). Quadruple moment is also measured in fm^2 and its value mostly lies in the range 1.2×10^{-28} m^2 for ^{123}Sb to 8×10^{-28} m^2 for ^{176}Lu while deutron has a value $Q = 2.74 \times 10^{-31}$ m^2. (See **article 15.18**)

15.5 WAVE MECHANICAL PROPERTIES

The nucleus has the following wave-mechanical properties (1) *Statistics* (2) *Parity* and (3) *Isospin*.

1. Statistics. The quantum mechanical description of a system with a number of particles like the nucleus is given by either Bose-Einstein or Fermi-Dirac statistics.

Bose-Einstein statistics. *All particles with integral spin* (in units of $\hbar$) *or zero obey Bose-Einstein (B.E.) statistics* and are called bosons *e.g.*, photon, π-meson and deutron. All nuclei with *even* mass number *A* obey *B.E.* statistics. The wave function of a system obeying Bose-Einstein statistics is *symmetric*. This means that the wave function for such a system remains unaltered by the interchange of all the co-ordinates for any pair of identical particles. thus, if $\psi\,(x_1 ...., x_i ....., x_j .... x_n)$ is the wave function of a system of n identical particles obeying Bose-Einstein statistics and x_i stands for all the co-ordinates of the particle i, the new wave function resulting from interchange of i and j will be given by $\psi\,(x_1 ... x_j ... x_i ... x_n)$. In other words

$$\psi\,(x_1 ...., x_i ...., x_j ... x_n) = \psi\,(x_1 .... x_j .... x_i .... x_n).$$

Fermi-Dirac statistics. All particles with half integral spin ($\hbar/2$) obey *Fermi-Dirac (F.D.) statistics* and are called *fermions e.g.*, electrons, protons and neutrons. All nuclei with *odd* mass number *A* obey *F.D.* statistics. All fermions obey Pauli's exclusion principle.

The wave function of a system obeying Fermi-Dirac statistics is *anti-symmetric*. This means that if all the co-ordinates of any pair of identical particles are interchanged in the wave function, the new system will be identical with the original except for a change of sign in the wave function. In other words

$$\psi\,(x_1 ... x_i ... x_j ... x_n) = -\,\psi\,(x_1 .... x_j ... x_i ... x_n).$$

2. Parity. The parity of a system refers to the behaviour of the wave function ψ under inversion of co-ordinates through the origin *i.e.*, when x is replaced $-x$, y by $-y$ and z by $-z$.

If the change of sign of x, y, z does not change the sign of the wave function *i.e.*,

$$\psi\,(x, y, z) = \psi\,(-x, -y, -z)$$

then the wave function (or the particle with which it is associated) is said to have *even* or *positive* parity. If the change of sign of x, y, z changes the sign of the wave function *i.e.*,

$$\psi\,(x, y, z) = -\,\psi\,(-x, -y, -z)$$

then the wave function (or the particle) is said to have *odd* or *negative* parity.

In general $\psi\,(x, y, z) = P\psi\,(-x, -y, -z)$

where $P = \pm 1$. P can be taken as a *quantum number* and the property defined by it is called the *parity of the system*.

$$P = +1 \text{ means positive or even parity}$$
$$P = -1 \text{ means negative or odd parity.}$$

In the case of hydrogen like atoms, it is found that parity is related to the orbital quantum number l and is given by

$$P = (-1)^l$$

Hence if l is *even* ψ does not change sign, the parity is *positive* or *even*. If l is *odd* ψ does change sign, the parity is *negative* or *odd*. The intrinsic parity of the proton, neutron, neutrino and μ-meson is even whereas the intrinsic parity of π meson is odd. The parity of a whole system is the product of the parities of individual particles.

A system having an even number of odd parity particles and any number of even parity particles will have even parity. A system with an odd number of odd parity particles and any number of even parity particles will have odd parity.

Parity is a purely quantum mechanical concept having no classical analogue. Nuclear states are characterised by a definite parity which may be different for different states of the same nucleus. In a nuclear reaction parity remains conserved.

3. Isospin. Neutrons and protons are similar in all respects except charge. On this basis these particles are considered to be just different manifestations of the same inherent particle, the *nucleon*. To describe this quantum state of the nucleon the quantum number used is known as '*Isotopic spin quantum number*' or *Isospin* or *T-spin*.

Isospin quantum number is conserved in nuclear transformations. A nucleon is assigned isospin $\frac{1}{2}$ and in an electromagnetic field two charge states with isospin components of $+\frac{1}{2}$ and $-\frac{1}{2}$ can be distinguished as the *proton* and *neutron*.

The isospin is a vector in three dimensional space called *iso-spin space*. The isospin space has no relation with physical space. The component of isospin that is measured is termed as T_3, the component of T in the direction of *third* axis in isospin space.

15.6 RELATION BETWEEN MASS NUMBER AND RADIUS OF NUCLEUS

If A is the mass number and r the radius of the nucleus, then

$$r = r_0 A^{1/3}$$

where r_0 is a constant equal to 1.3×10^{-15} m = 1.3 Fermi.

For example, the mass number and radius of $_{13}Al^{27}$.

The mass of a nucleus of mass number $A = M_n = Am_n$ where m_n is the mass of a nucleon $= 1.67 \times 10^{-27}$ kg, supposing that the proton as well as the neutron have the same mass $= 1.67 \times 10^{-27}$ kg

$$\therefore \quad M_n = 27 \times 1.67 \times 10^{-27} = 45.09 \times 10^{-27} \text{ kg.}$$

Radius of $_{13}Al^{27}$ nucleus $r = r_0 (27)^{1/3} = 3\, r_0$ where $r_0 = 1.3 \times 10^{-15}$ m

$$\therefore \quad r = 3 \times 1.3 \times 10^{-15} \text{ m} = 3.9 \times 10^{-15} \text{ m.}$$

SOLVED EXAMPLES

Example 15.1. *Calculate the mass number of the nucleus whose radius is*

(i) 4.8×10^{-15} *m*

(ii) 3.66×10^{-15} *m* **(P.U. 2002)**

given $r_0 = 1.3$ *fm*

Solution. Radius, $\quad r = r_0 A^{1/3}$

or $\quad r^3 = r_0^3 A$

Hence $\quad A = \left(\frac{r}{r_0}\right)^3$

(*i*) $\quad A = \left(\frac{4.8 \times 10^{-15}}{1.3 \times 10^{-15}}\right)^3 = \left(\frac{4.8}{1.3}\right)^3 = 50.3 \qquad \therefore \; A = 50$

(*ii*) $$A = \left(\frac{3.66 \times 10^{-15}}{1.3 \times 10^{-15}}\right)^3 = \left(\frac{3.66}{1.3}\right)^3 = 22.3 \quad \therefore \quad A = 22.$$

Example 15.2. *Assuming that the protons and neutrons posses equal masses calculate how many times the nuclear matter is denser than water.*

Nuclear radius is = $1.2 \times 10^{-15} A^{1/3}$ m, where A is the mass number. Mass of proton 1.67×10^{-27} Kg. Hence show that nuclei of atoms have constant density or nuclear density is independent of mass number. ***(H.P.U. 2001)***

Solution. Nuclear density = $\dfrac{\text{Nuclear mass}}{\text{Nuclear volume}}$

Nuclear mass = Am_N

where m_N = mass of nucleon = 1.67×10^{-27} kg and A is atomic mass number.

$$\text{Nuclear volume} = \frac{4}{3}\pi r^3 = \frac{4}{3}\pi (r_0 A^{1/3})^3 = \frac{4}{3}\pi \times (1.2 \times 10^{-15})^3 \ A \text{ and}$$

$$\therefore \quad \text{Nuclear density } \rho_N = \frac{Am_N}{\frac{4}{3}\pi r_0^3 A} = \frac{m_N}{\frac{4}{3}\pi r_0^3} = \frac{1.67 \times 10^{-27}}{\frac{4}{3}\pi (1.2 \times 10^{-15})^3} = 2.3 \times 10^{17} \text{ kg m}^{-3}$$

$$\text{Again} \quad \frac{\text{Nuclear density}}{\text{Density of water}} = \frac{2.3 \times 10^{17}}{1 \times 10^3} = 2.3 \times 10^{14}$$

Hence nuclear matter is 2.3×10^{14} times heavier than water. This shows that nuclear matter is in a highly compressed state.

Example 15.3. *Estimate the nuclear density if $m_n = m_p = 1$ u.; $R = 1.2 \times 10^{-15}$ m* ***(Calicut U. 2003; K.U. 2001)***

Solution. Take 1 $u = 1.67 \times 10^{-27}$ kg. Then nuclear density = 2.3×10^{17} kg m^{-3} as calculated above.

Nuclei of atoms have constant density. The density of nucleus $\rho_N = \dfrac{m_N}{\frac{4}{3}\pi r_0^3}$ and is independent of *A the mass number. As the density of nucleus is independent of A the atomic mass, its value is almost the same for all nuclei. In other words nuclei of atoms have constant density.*

Example 15.4. *Find the density of $_{12}C^6$ nucleus.* ***(G.N.D.U. 1997)***

Solution. The radius of a nucleus is given by $R = 1.2\ A^{1/3}$ fm

where fm = Fermi = 10^{-15} metre (It is also called a *fermtometer*).

∴ Radius of carbon nucleus

$$R = 1.2 \times 10^{-15} \times 12^{1/3} = 2.7 \times 10^{-15} \text{ m.}$$

$$\text{Now density of the nucleus } \rho_n = \frac{12 \times 1.67 \times 10^{-27}}{\frac{4}{3}\pi (2.7 \times 10^{-15})^3}$$

$$= 2.43 \times 10^{17} \text{ kg m}^{-3}.$$

Example 15.5. *Comment on the following properties of $^{208}Pb_{82}$ nucleus (i) charge (ii) spin (iii) size.* ***(G.N.D.U. 2004, 1997; H.P.U. 1992)***

Solution. (*i*) **Charge.** As the atomic number $Z = 82$, the charge on the nucleus is

$$+Ze = 82 \times 1.6 \times 10^{-19} \text{ C} = 131 \times 10^{-19} \text{ C.}$$

(*ii*) **Spin.** The mass number $A = 208$. Thus there are 82 protons $208 - 82 = 126$ neutrons. As the nucleus has an even number of particles it has an *integral spin.*

(*iii*) **Size.** The radius of the nuclei is given by

$$r = r_0 A^{1/3}$$

where A is the atomic number. In this case $A = 208$. If $r_0 = 1.4 \times 10^{-15}$ m.

$$\text{Radius of Pb}^{208} \text{ nucleus } r = r_0 A^{1/3} = 1.4 \times 10^{-15} \times (208)^{1/3}$$

$$= 1.4 \times 10^{-15} \times 5.926 = 8.296 \times 10^{-15} \text{ m}$$

$$= 8.296 \text{ fm.}$$

Example 15.6. *Express the size of the atom and size of the nucleus in metres. Assuming them to be spheres, what is the ratio of their volumes ?* ***(Pbi.U. 2000)***

Solution. The radius of the atom is of the order of 0.53×10 m and the radius of the nucleus is of the order of 1.2×10^{-15} m

$$\therefore \quad \frac{\text{Volume of the atom}}{\text{Volume of the nucleus}} = \frac{\frac{4}{3}\pi\left[0.53 \times 10^{-10}\right]^3}{\frac{4}{3}\pi\left[1.2 \times 10^{-15}\right]^3} = \frac{(0.53)^3}{(1.2)^3} \times 10^{15}$$

$$= 0.086 \times 10^{15} = 8.6 \times 10^{13}$$

Example 15.7. *Assuming nuclear radius to be given by $R = r_0 A^{1/3}$ and $r_0 = 1.2 \times 10$ m show that the neutron and proton densities in a nucleus with $N = Z$ are 0.069 nucleon per fm^3 for each.* ***(Cal.U. 2003)***

Solution. If A is the mass number of the nucleus and $N = Z$, then $N = A/2$ and $Z = A/2$; $r_0 = 1.2 \times 10^{-15}$ m $= 1.2$ fm

$$\therefore \text{ Proton density} = \text{neutron density} = \frac{A/2}{\frac{4}{3}\pi r_0^3 A} = \frac{1}{2 \times \frac{4}{3}\pi r_0^3}$$

$$= \frac{3}{8\pi r_0^3} = \frac{3}{8 \times 3.142 \times 1.2 \times 1.2 \times 1.2}$$

$$= 0.069 \text{ nucleon/fm}^3$$

Example 15.8. *Find the nuclear density of U^{235} if $R_0 = 1.1 \times 10^{-15}$ m.* ***(H.P.U. 2002)***

Solution. Radius of U^{235} nucleus, $r = R_0 A^{1/3} = 1.1 \times 10^{-15} A^{1/3}$

$$\text{Density of U}^{235} \text{ nucleus, } \rho = \frac{Am_n}{\frac{4}{3}\pi r^3} = \frac{Am_n}{\frac{4}{3}\pi R_0^3 A} = \frac{m_n}{\frac{4}{3}\pi R_0^3}$$

$$= \frac{1.67 \times 10^{-27}}{\frac{4}{3}\pi (1.1 \times 10^{-15})^3} = 2.99 \times 10^{17} \text{ kg m}^{-3}$$

15.7 EXPERIMENTAL DETERMINATION OF NUCLEAR SIZE

The first experimental evaluation of nuclear size (or radius) was made by Rutherford by his experiment on the scattering of α-particles by nuclei. He measured the distance of closest approach of the α-particle from various heavy nuclei like that of gold, silver, copper etc.

Consider an α-particle at a very lage distance from the nucleus having a kinetic energy E and moving towards the centre of the positively charged nucleus in the direction of head on collision. The α-particle being positively charged goes straight towards the nucleus and at the point A the coulomb repulsive force is so strong that the α-particle comes to rest momentarily and retraces its path.

This distance d between the point A and the centre of the nucleus is called the distance of closest approach. At this point all the kinetic energy of the α-particle is converted into its electrostatic potential energy.

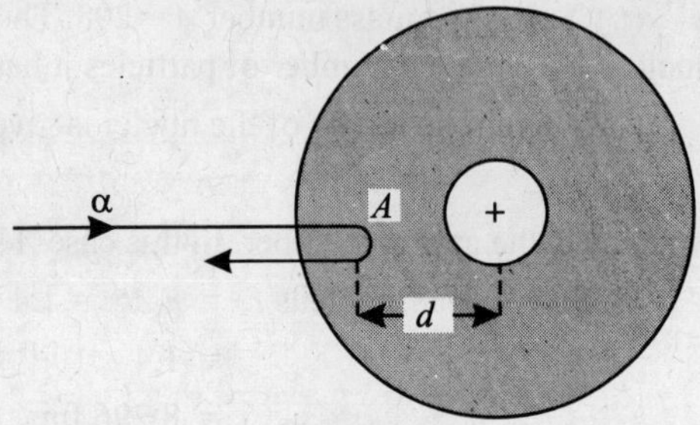

Fig. 15.2

Let Ze be the charge on the nucleus, then the potential at the point $A = \frac{1}{4\pi\varepsilon_0}\frac{Ze}{d}$ where Z is the atomic number of the nucleus.

The potential energy of the α-particle of charge $+2e$ at the point A

$$= \frac{1}{4\pi\varepsilon_0}\frac{Ze}{d}2e = \frac{2Ze^2}{4\pi\varepsilon_0 d}$$

At the point of closest approach kinetic energy E = Potential energy

or $$E = \frac{2Ze^2}{4\pi\varepsilon_0 d}$$

or Distance of closest approach $$d = \frac{2Ze^2}{4\pi\varepsilon_0 E}$$

The distance of closest approach gives the value of the upper limit of nuclear radius. From his experiments Rutherford found the distance of closest approach as under

Gold ($Z = 79$), $r_0 = 3.2 \times 10^{-14}$ m
Silver ($Z = 47$), $r_0 = 2 \times 10^{-14}$ m
Helium ($Z = 2$), $r_0 = 3 \times 10^{-15}$ m

15.8 MAGNETIC MOMENT OF AN ATOM

An atom has a magnetic moment due to the orbital motion of the electron around the nucleus.

Consider an electron of mass (m), charge ($-e$) revolving round the nucleus of an atom in a circular orbit of radius r with a velocity (of magnitude) v. The charge circulating in a loop is equivalent to current

$$\therefore \quad i = -\frac{e}{T} = -\frac{ev}{2\pi r}$$

A current loop has a magnetic moment.

∴ Orbital magnetic moment of the electron

$$\mu_l = \text{current} \times \text{area} = iA = -\frac{ev}{2\pi r}\pi r^2 = -\frac{evr}{2}$$

Multiplying and dividing by m we have

$$\mu_l = -\frac{emvr}{2m} = -\frac{e}{2m}L$$

where $L = mvr$ is the orbital angular momentum of the electron. In vector notation

$$\vec{\mu}_l = -\frac{e}{2m}\vec{L} \quad \ldots(i)$$

The negative sign shows that the direction of the orbital magnetic moment is opposite to that of the orbital angular momentum. Equation (*i*) is generally written as

$$\vec{\mu}_l = -g_l \frac{e}{2m} \vec{L}$$

where $g_l = 1$ and is known as '*orbital g factor*' of the electron.

According to quantum mechanics

$$\vec{L} = \sqrt{l(l+1)}\,\hbar$$

where l is the orbital quanum number having values 0, 1, 2 ...*etc.*

$$\therefore \quad \vec{\mu}_l = -g_l \frac{e}{2m}\sqrt{l(l+1)}\,\hbar = -\frac{e\hbar}{2m}\sqrt{l(l+1)} \qquad ..(ii)$$

For $l = 0$, $\vec{\mu}_l = 0$, For $l = 1$, $\vec{\mu}_l = \sqrt{2}\dfrac{e\hbar}{2m}$;

For $l = 2$, $\vec{\mu}_l = \sqrt{6}\dfrac{e\hbar}{2m}$.

Bohr magneton. *A Bohr magneton is the unit of (atomic) magnetic moment.* It has the value

$$\mu_B = \frac{e\hbar}{2m}$$

Substituting the value of $e = 1.6 \times 10^{-19}$ Coulomb, $m = 9.1 \times 10^{-31}$ Kg and $\hbar = \dfrac{h}{2\pi} = 1.055 \times 10^{-34}\ J$, we have

$$\mu_B = \frac{1.6 \times 10^{-19} \times 1.055 \times 10^{-34}}{2 \times 9.1 \times 10^{-31}} = 9.27 \times 10^{-24}\ \text{Am}^2 \text{ or J/Tesla}$$

$$= \frac{9.27 \times 10^{-24}}{1.6 \times 10^{-19}} = 5.79 \times 10^{-5}\ \text{eV/T}$$

15.9 NUCLEAR MAGNETIC DIPOLE MOMENT (μ)

The spin angular momentum and orbital angular momentum of protons within the nuclei produce magnetic fields which can be decribed in terms of a resultant magnetic dipole moment located at the centre of the nucleus. A spinning positive charge produces a magnetic field whose *N*-pole direction is parallel to the direction of spin. The magnetic moment in such a case is said to be *positive*.

If a particle, say a proton having a charge *e* circulates about a force centre with a frequency v, then the equivalent current

$$i = ev$$

A current loop posseses a magnetic moment called '*orbital magnetic dipole moment* $\vec{\mu}_l$, given by

$$\vec{\mu}_l = i\vec{A} = evA = \frac{e}{T}\pi r^2 \hat{r}$$

where r is the radius of the loop, πr^2, the area of the loop and $\hat{r}$ a unit vector in the direction of area vector *i.e.* perpendicular to the plane of the loop.

As $$T = \frac{2\pi r}{v}, \quad \vec{\mu}_l = \frac{ev}{2\pi r}\pi r^2 \hat{r} = \frac{evr}{2}\hat{r}$$

or $$\vec{\mu}_l = \frac{e}{2m} mvr\,\hat{r}$$ where m is the mass of the particle.

But $mvr = l$ angular momentum of the proton.

$$\therefore \quad \vec{\mu}_l = \frac{e}{2m} l\hat{r} = \frac{e}{2m}\vec{l} = g_e \frac{e}{2m}\vec{l} \qquad ...(i)$$

where $g_e = 1$ is called '*orbital gyromagnetic ratio*' for the proton and $\vec{l}$ is the *orbital angular momentum vector.*

In addition to orbital angular momentum, spin also produces a magnetic dipole moment given by

$$\vec{\mu}_s = g_s \frac{e}{2m}\vec{s} \qquad ...(ii)$$

where g_s is known as *spin gyromagnetic ratio* and $\vec{s}$ is the *spin angular momentum vector*.

A proton has both '*orbital magnetic dipole moment*' and '*spin magnetic dipole moment*' but a neutron being an uncharged particle only has a spin magnetic dipole moment. The total magnetic dipole moment of the proton is, therefore, given by

$$\vec{\mu}_p = \vec{\mu}_s + \vec{\mu}_l = \frac{e}{2m}\left[g_s\,\vec{s} + g_e\,\vec{l}\right] \qquad ...(iii)$$

The *spin magnetic dipole moment* of the neutron is also given by the relation

$$\vec{\mu}_s = g_s \frac{e}{2m}\vec{s} \qquad ...(iv)$$

where g_s is the spin gyromagnetic ratio for the neutron. The factor g_s is different for electron, proton and neutron.

For a nucleus of mass number A and atomic number Z, the number of protons = Z and the number of protons + neutrons = A.

$\therefore$ Total magnetic moment of the nucleus

$$\vec{\mu}_N = \frac{e}{2m}\left[\sum_{k=1}^{Z} g_l\,\vec{l}_k + \sum_{k=1}^{A} g_s\,\vec{s}_k\right]$$

$$= \frac{e}{2m} g\,\vec{I} = g\left(\frac{e\hbar}{2m}\right)\frac{\vec{I}}{\hbar} \qquad ...(v)$$

where $\vec{I}$ is the total angular momentum of the nucleus (or nuclear spin) given by

$$\vec{I} = \sum_{k=1}^{Z}\vec{l}_k + \sum_{k=1}^{A}\vec{s}_k$$

and $$g\,\vec{I} = \left[\sum_{k=1}^{Z} g_l\,\vec{l}_k + \sum_{k=1}^{A} g_s\,\vec{s}_k\right]$$

Here g is the *gyromagnetic ratio* (or *g-factor*) of the nucleus. Thus g is a dimensionless quantity giving the ratio of magnetic dipole moment $\vec{\mu}_N$ of the nucleus in terms of $\frac{e\hbar}{2m}$ (the nuclear magneton) to the total nuclear angular momentum in terms of $\hbar$.

Using quantum mechanics, $\vec{l} = \sqrt{l(l+1)}\,\hbar$ and $\vec{s} = \sqrt{s(s+1)}\,\hbar$ where l is the *orbital quantum number*, we have

$$g = \frac{1}{2}(g_l + g_s) + \frac{1}{2}(g_l - g_s)\frac{l(l+1) - s(s+1)}{I(I+1)}$$

For a nucleus, if *A* is *even*, $\vec{I}$ is an *integral multiple* of $\hbar\left(\frac{h}{2\pi}\right)$. If *A* is *odd*, $\vec{I}$ is an *odd half integral multiple of* $\frac{1}{2}\hbar$ *i.e. integral multiple* of $\frac{1}{2}\hbar$. If both *Z* and *A* are *even*, the value of $\vec{I}$ in the *ground state* is always *zero* and hence according to Eq. (*v*) the *magnetic dipole moment* of the nucleus is always *zero.* All odd-odd nuclei *i.e.* nuclei with *odd* number of *protons* and *odd* number of *neutrons* and hence even *A* have *integral spin I.*

On the other hand all *odd-even* nuclei *i.e.* nuclei with *odd* number of *protons* and *even* number of *neutrons* or *odd* number of *neutrons* and *even* number of *protons* and hence *odd A* have *half integral* spin which in the *ground state* is $\vec{I} = \frac{1}{2}\hbar$. Thus, the magnetic moment of a nucleus in the *ground state* of *even-odd* or *odd-even* nuclei is due only to the single unpaired nucleon.

If the single unpaired nucleon is a proton, then according to Eq. (*v*), the magnetic moment is given by

$$\vec{\mu}_p = g_p\left(\frac{e\hbar}{2m_p}\right)\frac{\vec{I}}{\hbar} = \frac{1}{2}g_p\left(\frac{e\hbar}{2m_p}\right) = \frac{1}{2}g_p\beta_N = 2.792\ \beta_N$$

because $\frac{\vec{I}}{\hbar} = \frac{\frac{1}{2}\hbar}{\hbar} = \frac{1}{2}, \left(\frac{e\hbar}{2m_p}\right) = \beta_N$ the nuclear magneton defined in [**article 15.4** (*vii*)] and g_p = 5.584 or $\frac{1}{2}g_p = 2.792$.

If the unpaired nucleon is a neutron, then the magnetic dipole moment of the nucleus is only due to the single neutron given by

$$\vec{\mu_n} = \frac{1}{2}g_n\beta_N = -1.913\ \beta_N \text{ where } g_n = -3.826$$

The negative sign shows that $\vec{\mu_n}$ is *anti parallel* to nuclear spin.

Example 15.9. *Compare the atomic magnetic moment and nuclear magnetic moment.*

(P.U. 2002)

Solution. Atomic magnetic is measured in units of *Bohr magneton* μ_B = 5.79×10^{-5} eV/T whereas nuclear magnetic moment is measured in units of *nuclear magneton*

$$\beta_N = 3.15 \times 10^{-8} \text{ eV/T}$$

Bohr magneton $\mu_B = \frac{e\hbar}{2m_e}$ and Nuclear magneton $\beta_N = \frac{e\hbar}{2m_p}$

$$\therefore \quad \frac{\beta_N}{\mu_B} = \frac{3.15 \times 10^{-8}}{5.79 \times 10^{-5}} = \frac{m_e}{m_p} = \frac{1}{1836}$$

15.10 NUCLEAR ELECTRIC QUADRUPOLE MOMENT (Q)

Electric quadrupole moment of a nucleus is a measure of the deviation of nuclear charge distribution from spherical symmetry.

Expression for electrical quadrupole moment. Consider a nucleus having charge density ρ represented as ρ (*x*, *y*, z) with its charge centre at the origin. The nucleus is surrounded by its orbital electrons which produce an electrostatic potential ϕ represented as ϕ (*x*, *y*, *z*) at the origin. The interaction between potential ϕ and charge density ρ gives rise to an electrostatic interaction energy given by

$$U = \int_v \rho(x, y, z)\ \phi\ (x, y, z)\, dv \qquad ...(i)$$

According to Taylor's series, the expression of $\phi(x, y, z)$ about the origin gives

$$\phi(x,y,z) = \phi_0 + \left[\left(\frac{\partial \phi}{\partial x}\right)_0 x + \left(\frac{\partial \phi}{\partial y}\right)_0 y + \left(\frac{\partial \phi}{\partial z}\right)_0 z\right] +$$

$$\left[\frac{1}{2!}\left(\frac{\partial^2 \phi}{\partial x^2}\right)_0 x^2 + \frac{1}{2!}\left(\frac{\partial^2 \phi}{\partial y^2}\right)_0 y^2 + \frac{1}{2!}\left(\frac{\partial^2 \phi}{\partial z^2}\right)_0 z^2\right] +$$

$$\left[\left(\frac{\partial^2 \phi}{\partial x \partial y}\right)_0 xy + \left(\frac{\partial^2 \phi}{\partial x \partial z}\right)_0 xz + \left(\frac{\partial^2 \phi}{\partial y \partial z}\right)_0 yz\right] + \ldots$$

The subscript (0) means that the quantity is evaluated at the origin.

Substituting the above values of $\phi(x, y, z)$ in Eq (i), we get

$$U = \phi_0 \int \rho\, dv + \left[\left(\frac{\partial \phi}{\partial x}\right)_0 \int x\rho\, dv + \left(\frac{\partial \phi}{\partial y}\right)_0 \int y\rho\, dv + \left(\frac{\partial \phi}{\partial z}\right)_0 \int z\rho\, dv\right]$$

$$+\left\{\left[\frac{1}{2}\left(\frac{\partial^2 \phi}{\partial x^2}\right)_0 \int x^2\rho\, dv + \frac{1}{2}\left(\frac{\partial^2 \phi}{\partial y^2}\right)_0 \int y^2\rho\, dv + \frac{1}{2}\left(\frac{\partial^2 \phi}{\partial z^2}\right)_0 \int z^2\rho\, dv\right]\right.$$

$$\left.+\left[\left(\frac{\partial^2 \phi}{\partial x \partial y}\right)_0 \int xy\rho\, dv + \left(\frac{\partial^2 \phi}{\partial x \partial z}\right)_0 \int xz\rho\, dv + \left(\frac{\partial^2 \phi}{\partial y \partial z}\right)_0 \int yz\rho\, dv\right]\right\} + \ldots \text{ higher terms}$$

The first term gives simply the interaction energy of a point charge (monopole).

The three terms in the first bracket give the energy of a *dipole*.

The six terms in the second bracket are the *quadrupole* energy terms. For an ellipsoid of revolution, because of symmetry the three integrals involving xy, yz and zx vanish. When the Z-axis is the axis of symmetry, the integral over y^2 gives the same result as integral over x^2. Therefore, the quadrupole interaction energy is given by

$$\Delta U_2 = \frac{1}{2}\left(\frac{\partial^2 \phi}{\partial x^2} + \frac{\partial^2 \phi}{\partial y^2}\right)\int \frac{x^2 + y^2}{2}\rho\, dv + \frac{1}{2}\left(\frac{\partial^2 \phi}{\partial z^2}\right)\int z^2 \rho\, dv$$

Also, according to Laplace's equation

$$\frac{\partial^2 \phi}{\partial x^2} + \frac{\partial^2 \phi}{\partial y^2} + \frac{\partial^2 \phi}{\partial z^2} = 0$$

$$\therefore \quad \frac{\partial^2 \phi}{\partial x^2} + \frac{\partial^2 \phi}{\partial y^2} = -\frac{\partial^2 \phi}{\partial z^2}$$

and

$$\Delta U_2 = \frac{1}{2}\left(\frac{\partial^2 \phi}{\partial z^2}\right)_0 \int \left[z^2 - \frac{(x^2 + y^2)}{2}\right]\rho\, dv \qquad \ldots(ii)$$

Now

$$x^2 + y^2 + z^2 = r^2$$

$$x^2 + y^2 = r^2 - z^2$$

Substituting in Eq. (ii), we have

$$\Delta U_2 = \frac{1}{2}\left(\frac{\partial^2 \phi}{\partial z^2}\right)_0 \int \left[z^2 - \frac{(r^2 - z^2)}{2}\right]\rho\, dv$$

$$= \frac{1}{4}\left(\frac{\partial^2 \phi}{\partial z^2}\right)_0 \int_v (3z^2 - r^2)\rho\, dv$$

$$= \frac{1}{4} eQ \left(\frac{\partial^2 \phi}{\partial z^2}\right)_0$$

where $$Q = \frac{1}{e}\int_v (3z^2 - r^2)\rho\, dv \qquad ...(iii)$$

and is called *nuclear electric quadrupole moment.*

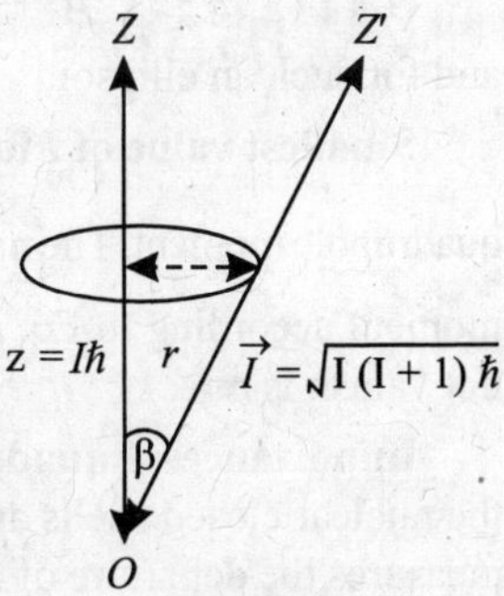

Fig. 15.3

The nuclear symmetry axis is not the *space fixed Z-axis* but it is the I (or Z') axis which can be regarded as symmetry axis. According to quantum mechanics, the total angular momentum vector $\vec{I}$ can never remain fixed in any direction but always keeps on precessing about the direction of the external field. Nuclear quadrupole moments are, therefore, not defined along the body axis of the charge distribution *i.e.*, Z' direction but along the precession axis *i.e.*, Z axis, as shown in Fig. 15.3.

The magnitude of

$$\vec{I} = \sqrt{I(I+1)}\,\hbar$$

where I is the total angular momentum quantum number. The maximum projected vlaue of $\vec{I}$ along Z-axis is $I\hbar = z$. Thus the smallest possible value of β the angle between Z and Z' is given by

$$\cos\beta = \frac{z}{r} = \frac{I\hbar}{\sqrt{I(I+1)}\,\hbar}$$

$$= \frac{I}{\sqrt{I(I+1)}}$$

$$\therefore \qquad (3z^2 - r^2) = r^2\left(3\frac{z^2}{r^2} - 1\right) = r^2\left(3\frac{I^2}{I(I+1)} - 1\right)$$

$$= r^2\left[\frac{3I^2 - I^2 - I)}{I(I+1)}\right] = \left[\frac{I(2I-1)}{I(I+1)}\right]r^2 = r^2\,\frac{2I-1}{I+1}$$

Substituting the above value of $(3z^2 - r^2)$ in Eq (*iii*) and transferring to Z-axis, we get

$$Q = \frac{1}{2}\,\frac{2I-1}{I+1}\int (3z'^2 - r^2)\rho\, dv \qquad ...(iv)$$

If Z is the atomic number of the nucleus, then charge on the nucleus $= Ze$ and $\rho = \frac{\text{Charge}}{\text{Volume}} = \frac{Ze}{\int dv}$

Hence from relation (*iii*),

$$Q = \frac{1}{e}\,\frac{Ze}{\int dv}\int (3z^2 - r^2)\,dv = Z\int\frac{(3z^2 - r^2)\,dv}{\int dv}$$

or $$Q = Z\,[< 3z^2 - r^2 >_{av}] \qquad ...(v)$$

Spherically symmetrical charge distribution. For spherically symmetrical charge distribution *i.e.* when the nucleus has symmetrically situated protons

$$\langle x^2\rangle = \langle y^2\rangle = \langle z^2\rangle = \tfrac{1}{3}r^2$$

$$\therefore \quad \langle 3z^2 - r^2\rangle = 0$$

Hence from relation (v), Quadrupole moment $Q = 0$.

Nucleus in the form ellipsoid of revolution. For a nucleus in the form of ellipsoid of revolution whose diameter is $2b$ along the symmetrical axis and $2a$ along the direction at right angles to it, the ellipsoid is defined by the equation

$$\frac{x^2 + y^2}{a^2} + \frac{z^2}{b^2} = 1$$

and for such an ellipsoid $\quad Q = \tfrac{2}{5}Z(b^2 - a^2)$

Smallest value of I for $Q \neq 0$. The nuclei which have total nuclear spin $I = 0$ have zero electric quadrupole moment. The nuclei which have total nuclear spin $I = \frac{1}{2}$ also have zero electric quadrupole moment according to Eq. (iv). Hence the smallest value of angular momentum I for which Q does not vanish is **one.**

Importance of quadrupole moment. The study of quadrupole moment gives an idea about the nuclear closed shells and helps in the study of nuclear models. The electric quadrupole moment measures the departure of the nucleus from spherical symmetry. The electric quadrupole moment of nuclei with protons and neutrons equal to magic number being zero, it predicts that these nuclei are spherically symmetric. On the other hand the electric quadrupole moment of nuclei in the neighbour hood of magic numbers has a finite value thereby predicting that these nuclei have a distorted shape.

The fact that deutron possesses a small quadrupole moment is important for the basic nature of nuclear forces.

15.11 ATOMIC MASS UNIT (a.m.u.)

One atomic mass unit a.m.u. (or 1 u) is equal to 1/12 of the mass of neutral carbon atom taken as 12 (${}_6C^{12}$).

$$1 \text{ a.m.u.} = 1.6604 \times 10^{-27} \text{ kg}$$

Taking velocity of light $\quad c = 2.998 \times 10^8 \text{ ms}^{-1}$ and 1 Joule = 1.602×10^{13} MeV

we have,

$$1 \text{ a.m.u.} = 1.6604 \times 10^{-27} \times 2.998 \times 10^8 \times 2.998 \times 10^{16}$$

$$= 14.92 \times 10^{-11} \text{ Joules}$$

$$= \frac{14.92 \times 10^{-11}}{1.602 \times 10^{13}}$$

$$= 931.5 \text{ MeV}$$

15.12 MASS DEFECT

The nucleus is formed by bringing protons and neutrons together. The mass of the nucleus so formed is less than the sum of the masses of the constituent protons and neutrons. This mass difference is called *mass defect* and is denoted by Δm.

If Z is the number of protons in the nucleus (Atomic number Z), then the number of neutrons in the nucleus is $(A - Z)$ (A is atomic mass). If m_p is the mass of the proton and m_n that of the neutron, then

Sum of the masses of the protons and neutrons

$$= Zm_p + (A - Z)\, m_n$$

If M_N is the actual mass of the nucleus, then

Mass defect $\Delta m = Zm_p + (A - Z)\, m_n - m_N$

Mass defect is therefore, defined as the *difference between the sum of the rest masses of the nucleons forming the nucleus and the actual rest mass of the nucleus.*

For example in the case of *deutron* which contains one proton and one neutron the combined mass is [1.0073 + 1.0087] = 2.1060 a.m.u., whereas the actual mass of deutron nucleus is 2.0136 a.m.u.

$\therefore$ Mass defect $\Delta m = 2.1060 - 2.0136 = 0.0024$ a.m.u.

15.13 PACKING FRACTION

It has been found that atomic masses, though very close to whole number values invariably differ from the integral value by a small amount. For example the atomic masses of H^1, He^4 and Li^6 are H^1 = 1.007825, He^4 = 4.002603, Li^6 = 6.015126.

The deviation of atomic mass from whole number value is expressed in the form of a quantity known as *packing fraction.* Packing fraction is defined as *the ratio of the difference between the atomic mass of the atom M and its mass number A i.e., (M – A) to its mass number A.*

$$\therefore \quad \text{Packing fraction } f = \frac{M - A}{A}$$

Variation of packing fraction with mass number. A graph between packing fraction f (multiplied by a factor 10^4) and mass number A is shown in Fig. 15.4. It is known as *packing fraction curve.*

From the graph we find that :-

(*i*) For very light nuclei the packing fraction f is maximum indicating thereby that these nuclei are unstable.

(*ii*) As the value of A increases packing fraction goes on decreasing (with the exception of ${}_2He^4$ and ${}_6C^{12}$) till it becomes zero for $A = 16$. Thus the packing fraction for ${}_8O^{16}$ is zero.

(*iii*) As the mass number A increases beyond 16 the value of f goes on decreasing further and becomes negative. This continues approximately upto $A = 180$ (Ta, Tantalum) beyond which f again becomes *positive.*

(*iv*) The value of f is minimum for iron (${}_{26}Fe^{56}$) which has a significance as the most stable nucleus.

(*v*) Greater part of the curve is below zero line with negative values of packing fraction indicating that there is loss of mass and hence liberation of energy due to packing.

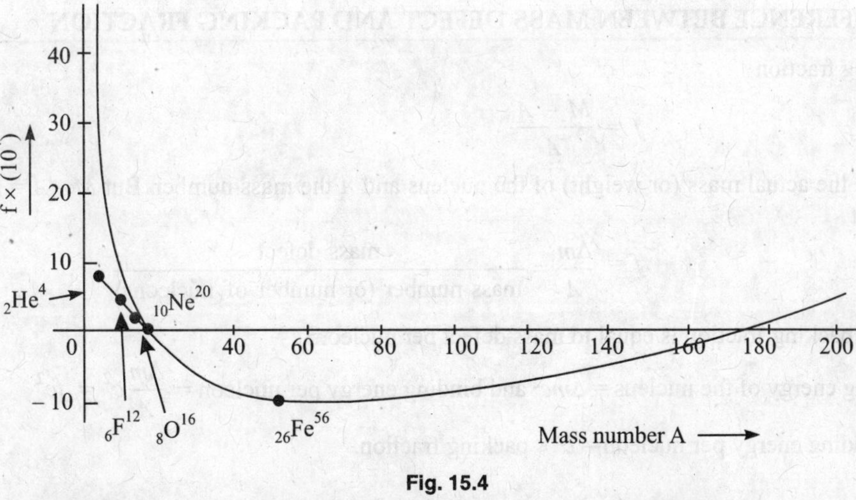

Fig. 15.4

(*vi*) For heavy nuclei beyond $A = 180$ packing fraction is *positive*. Thus we conclude that nuclei of the middle range are stable while light (below $A = 20$) and heavy (beyond $A = 200$) are unstable. All nuclei have a tendency to move from a region of higher packing fraction to lower packing fraction to attain higher stability. This is why *fusion* of light nuclei and *fission* of heavy nuclei takes place with a release of energy.

15.14 BINDING ENERGY

The mass of any permanently stable nucleus is found to be less than the sum of the masses of the neutrons and protons which it contains. The fact is accounted for by the conversion of a part of the mass energy of the particles into energy of binding, the relation between the change in mass and binding energy being given by Einstein's equation

$$\text{Binding energy} = E_B = \Delta\, mc^2$$

the total binding energy for all atoms except the lighter atoms is given approximately by the empirical relation

$$E_B = 15.6\, Z\, \text{MeV}$$

The binding energy of the nucleus with charge number Z and mass number A is given by

$$E_B = c^2\, [Zm_p + (A - Z)\, m_n - M_N] \quad ...(i)$$

where M_N is the mass of the nucleus, m_p the mass of the proton and m_n the mass of the neutron. The nuclear mass is, therefore, given by

$$M_N = Zm_p + (A - Z)\, m_n - E_B/c^2$$

and the atomic mass

$$M = M_N + Zm_e = Zm_e + Zm_p + (A - Z)\, m_n - E_B/c^2$$
$$= ZM_H + (A - Z)\, m_n - E_B/c^2$$

where m_e is the mass of the electron and M_H the mass of hydrogen atom. The term E_B/c^2 represents the mass equivalent of the total binding energy *i.e.*, the *energy which must be added to the nucleus* in order to break it up into Z protons and $(A - Z)$ neutrons.

Binding energy of a nucleus is, therefore, defined as the energy which must be supplied to the nucleus to break it into its constituent nucleons.

Since the mass of a nucleus is less than the sum of the masses of the total number of protons and neutrons in it, the binding energy is *a positive* quantity. The experimental value of nulcear binding energy varies from 2.23 MeV for deutron, the lightest stable atom containing more than one nucleon to 1640 MeV for the heaviest stable nucleus $^{209}\text{Bi}_{83}$.

15.15 DIFFERENCE BETWEEN MASS DEFECT AND PACKING FRACTION

Packing fraction

$$f = \frac{M - A}{A}$$

where M is the actual mass (or weight) of the nucleus and A the mass number. But $M - A = \Delta m$ = mass defect

$$\therefore \quad f = \frac{\Delta m}{A} = \frac{\text{mass defect}}{\text{mass number (or number of nucleons)}}$$

Hence packing fraction is equal to mass defect per nucleon.

Binding energy of the nucleus = Δmc^2 and binding energy per nucleon = $\dfrac{\Delta m}{A} c^2 = fc^2$

$\therefore$ Binding energy per nucleon = $c^2 \times$ packing fraction.

15.16 BINDING ENERGY PER NUCLEON OR AVERAGE BINDING ENERGY

The average binding energy is defined as binding energy per nucleon. It is the energy required to release a nucleon from the nucleus. From relation (*i*) **article 15.14,** It is given by $\frac{E_B}{A}$

$$\therefore \quad \frac{E_B}{A} = \frac{c^2}{A}[Zm_p + (A - Z)\, m_n - M_N] = \frac{\Delta mc^2}{A}$$

$$= c^2 \left[m_n - \frac{Z}{A}(m_n - m_p) - M_N/A\right]$$

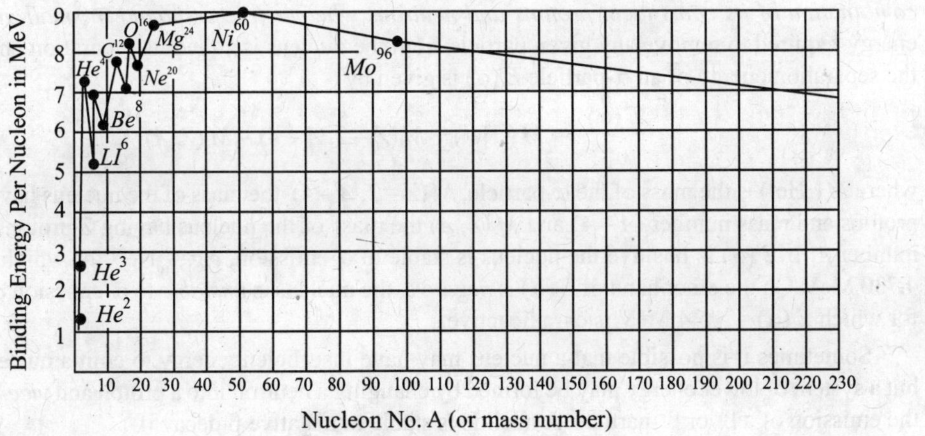

Fig. 15.5

Separation energy. *The separation energy is the minimum energy that must be supplied to remove the least tightly bound nucleon from the nucleus i.e. it is the binding energy of the least tightly bound nucleon.*

Graph between binding energy per nucleon and mass number. For a deutron the average binding energy per nucelon $= \frac{2.23}{2} = 1.11$ MeV. Thus it is a very loosely bound system. If we plot a graph between E_B/A *i.e.*, binding energy per nucleon and the nucleon number A, the curve obtained is as shown in Fig. 15.5. We shall now discuss the variation of average binding energy per nucleon with mass number and stability of the nucleus.

15.17 BINDING ENERGY AND NUCLEAR STABILITY

The binding energy of a nucleus gives a quantitative measure of its stability. Consider the nuclide $_2He^4$ which is made up of two protons and two neutrons for which the binding energy comes out to be 28.299 MeV. In other words the energy required to break up a $_2He^4$ nucleus (α-particle) in its constituent nucleons we must spend about 28 MeV of energy. Thus the binding energy per nucleon for $_2He^4$ nucleus is about 7 MeV. *Greater the binding energy, more stable is the nucleus.* We also find that for nuclides like $_2He^4$, $_4Be^8$, $_6C^{12}$, $_8O^{16}$, $_{10}Ne^{20}$ and $_{12}Mg^{24}$ there is a sharp increase in binding energy. For these nuclides $A = 4n$ where $n = 1, 2, 3, 4, \ldots$ etc. All these nuclides can be built up from α-particles, which as proved above, is a very highly bound system.

There is a broad maximum in the mass number range 40 to 120 at an average value of 8.5 MeV. The maximum value of binding energy per nucleon = 8.8 MeV for mass number A very nearly equal to 60. Thereafter, the binding energy per nucleon falls slowly to 7.4 MeV.

It is this dimunition binding energy per nucleon which is mainly responsible for the release of energy in the *fission* of nuclei. The energy released in fission is several million times that released in chemical changes, since the binding energies of nucleus involved in the process of fission are much higher than the binding energy of electrons involved in the chemical process. Similarly, if two very light nuclei are made to fuse into a single nucleus, a large amount of energy will be released. This is known as *nuclear fusion.*

Nuclides beyond $A = 238$ will have even smaller binding energy per nucleon and hence are less stable.

The fact that the total binding energy of a nucleus is positive is not a sufficient condition for its stability. *For a nucleus to be stable its mass should be less than the sum of the masses of any other combination of its constituent protons and neutrons whether free or bound in small groups.* The energy required to remove any given particle from the nucleus is called its separation energy. Thus the separation energy of an α-particle $E(\alpha)$ is given by

$$\frac{E(\alpha)}{c^2} = M({}_2\text{He}^4) + M(Z-2, A-4) - M(Z, A)$$

where $M({}_2\text{He}^4)$ is the mass of the α-particle, $M(Z-2, A-4)$ the mass of the nucleus having $(Z-2)$ protons and mass number $(A-4)$ and $M(Z, A)$ the mass of the nucleus having Z protons and mass number A. If $E(\alpha)$ is positive the nucleus is stable to α-emission, *e.g.*, ${}_{10}\text{Ne}^{20}$ for which $E(\alpha) = +$ 4.730 MeV. On the other hand, if $E(\alpha)$ is *negative*, the nucleus is *unstable* to α-emission *e.g.*, ${}_{84}\text{Po}^{210}$ for which $E(\alpha) = -5.4$ MeV is α-radioactive.

Sometimes it is possible that a nucleus may have insufficient energy to emit a nuclear particle but a system of lower energy may be formed by changing a neutron into a proton and *vice-versa* with the emission of a β^- or β^+ particle. A nuclide is stable to negative β decay if

$$M(Z, A) \leq M(Z+1, A) + m(\beta^-)$$

and for stability to positive β decay, the condition is

$$M(Z, A) \leq M(Z-1, A) + m(\beta^+)$$

A third type of β process is also possible and is known as decay be *electron capture.* In this process an orbital electron, usually from the K-shell, is absorbed into the nucleus and a neutron is emitted which may have zero energy. The corresponding condition for stability is

$$M(Z, A) \leq M(Z-1, A)$$

Release of nuclear energy in fission and fusion. From the graph between binding energy per nucleon and mass number we find that *the value of binding energy per nucleon is small both for very light and very heavy nuclei so that these substances are unstable.* The nuclei of intermediate masses (mass number 40 to 120) are the most stable and very high amount of energy has to be supplied to liberate each of their nucleons.

It, therefore, follows that when a heavy nucleus breaks up into lighter ones (nuclear fission) the end products formed have higher value of binding energy thereby resulting in the liberation of energy. Similarly when two very light nuclei combine to form a heavy nucleus (nuclear fusion) the higher binding energy per nucleon of the latter again results in the liberation of energy.

Nuclear Binding Energy is More than Atomic Binding Energy

Nuclear binding energy is much more than atomic binding energy. For example, the binding energy per nucleon of α-particle is of the order of 7 MeV whereas the binding energy of the orbital electron in hydrogen atom is only 13.6 eV.

$$\therefore \quad \frac{\text{Nuclear binding energy}}{\text{Atomic binding energy}} = \frac{7 \times 10^6}{13.6} = 0.51 \times 10^6$$

$$= 5.1 \times 10^5$$

Thus, nuclear binding energy is much more than atomic binding energy.

Maxima and minima peaks in B.E curve. The five peaks in B.E curve correspond to $_2He^4$, $_4Be^8$, $_6C^{12}$, $_8O^{16}$, and $_{10}Ne^{20}$ nuclei. These nuclei have even number of protons and even number of neutrons, *i.e.* these do not have any unpaired nucleon. These nuclei are, therefore, more stable than their immediate neighbours and have a higher value of binding energy giving rise to a *maxima.*

The binding energy on either side of these nuclei falls sharply due to the unpaired nucleon and hence give rise to *minima* at $_1H^1$, $_3Li^7$, $_5B^9$ etc.

Significance of binding energy curve. The B.E curve is highly significant. The fact that binding energy exists at all means that nuclei more complex than single proton of hydrogen can be stable. Such stability, in turn, accounts for the existence of various elements and hence explains the reason for the existence of different forms of matter. The B.E curve rises slowly as mass number A increases, has a peak value in the middle at $_{26}Fe^{56}$ and then slowly slopes downward.

As the binding energy per nucleon increases with mass number A, it explains the cause of release of energy in the fusion of light nuclei into heavier ones. The energy which comes from the fusion of protons and neutrons to form heavier nuclei explains how sun and stars get their energy *i.e.* indirectly the curve explains th evolution of the universe.

On the other hand breaking of heavier nuclei into ligher ones (fission) also releases energy which is being used for production of electric energy in nuclear reactors for peaceful purposes and also in weapons of war thereby bringing about a change in modern civilisation.

15.18 DEUTRON BINDING ENERGY

A nucleus of atomic number Z and mass number A contains Z protons and $A - Z$ neutrons. If M_N is the mass of the nucleus, then

Mass defect $\Delta m = Zm_p + (A - Z)\, m_n - M_N$

and binding energy $E_B = \Delta mc^2 = [Zm_p + (A - Z)\, m_n - M_N]$

For a **deutron** $Z = 1$ (proton) and $A - Z = 2 - 1 = 1$ (neutron)

$\therefore$ Mass defect $\Delta m = m_p + m_n - M_D$

and binding energy $E_B = c^2 (m_p + m_n - M_D)$

$$\therefore \text{Binding energy per nucleon} = \frac{E_B}{A} = \frac{c^2 (m_p + m_n - M_D)}{2}$$

Ground state properties of deutron. (*i*) The ground state is only bound state of the deutron.

(*ii*) The deutron consists of two particles of roughly equal mass so that the reduced mass of the system $\mu = \frac{1}{2} M$.

(*iii*) The binding energy of the deutron is small. Its value is 2.23 MeV. As the energy required to eject a nucleon from a nucleus of medium mass number is about 8 MeV, the deutron is comparatively *loosely bound.*

(*iv*) The total angular momentum of the deutron $I = 1$. The total angular momentum is given by

$$\vec{I} = \vec{L} + \vec{S}$$

Since the spin of both particles, proton and neutron is $\frac{1}{2}$, we have $S = \frac{1}{2} + \frac{1}{2} = 1$ or 0.

The two possibilities can be considered as parallel (↑↑) or anti-parallel (↑↓) spin. The ground state is 1*s* *i.e.*, $l = 0$. Therefore, $\vec{L} = 0$ and as $\vec{I} = \vec{L} + \vec{S}$, $\vec{S} = 1$. This means that the ground level of the deutron is a triplet state $^3S_1$ since the multiplicity is $(2S + 1)$.

(*v*) The *r.m.s.* distance between the proton and neutron is 4.2 Fermi.

(*vi*) The parity is even.

(*vii*) As the deutron consists of a proton and a neutron its magnetic moment is the sum of their magnetic moments. Now $\mu_p = 2.792\ \beta_N$ and $\mu_n = -1.913\ \beta_N$

∴ Magnetic moment of the deutron $\mu_d = \mu_p + \mu_n = 2.792 - 1.913$

$$= 0.879\ \beta_N$$

Where β_N is a nuclear magneton.

Precision measurements, however, give $\mu_d = 0.857\ \beta_n$.

The difference between the observed and calculated value is explained in part (*b*).

(*viii*) The electric quadruple moment of the deutron is

$$Q_d = 2.82 \times 10^{-27}\ \text{cm}^2$$

If the deutron nucleus is spherically symmetric its quadruple moments should be zero. The quadruple moment of deutron though small is not zero.

15.19 NON-EXISTENCE OF ELECTRONS IN NUCLEUS

Electrons do not exist in the nucleus. This fact can be proved from wave mechanical, angular momentum, statistical and other considerations too.

(*a*) Wave mechanical consideration. If a particular type of particle is to exist within the nucleus, then the uncertainty in its *position* in the nucleus should not exceed the radius of the nucleus. The radius of a typical atomic nucleus is of the order of 10^{-14} m. Hence if the electron is to exist within the nucleus the uncertainty in its position is given by

$$\Delta x \lesseqqgtr 10^{-14}\ \text{m}.$$

According to Heisenberg's uncertainty principle, $\Delta x \,.\, \Delta p_x = \dfrac{\hbar}{2}$

where Δp_x is the uncertainty in the *x*-component of momentum or the momentum itself and $\hbar = \dfrac{h}{2\pi}$.

∴ Uncertainty in the momentum of the electron if it is to exist in the nucleus

$$\Delta p_x \gtreqqless \frac{\hbar}{2\Delta x} = \frac{6.6 \times 10^{-36}}{2 \times 2\pi \times 10^{-14}} = 5.3 \times 10^{-21}\ \text{kg. ms}^{-1}$$

This means that the total momentum of the electron in the nucleus must be at least of the order of 5.3×10^{-21} kg. ms^{-1}.

The rest mass of the electron $= 9.1 \times 10^{-31}$ kg.

Hence the above value of the momentum is relativistic as a non-relativistic relation for momentum gives a value of $v > c$.

According to the theory of relativity, the total relativistic energy of the electron

$$E = [m_0^2c^4 + p^2c^2]^{1/2}$$

$$= [(9.1 \times 10^{-31})^2 \times (3 \times 10^8)^4 + (5.3 \times 10^{-21})^2\,(3 \times 10^8)^2]^{1/2}$$

$$= 3 \times 10^8\,[9.1 \times 9.1 \times 9 \times 10^{-46} + 5.3 \times 5.3 \times 10^{-42}]^{1/2}$$

$$= 3 \times 10^{-13}\,[9.1 \times 9.1 \times 9 \times 10^{-4} \times 5.3 \times 5.3]^{1/2}$$

$$= 16 \times 10^{-13} \text{ Joule} = \frac{16 \times 10^{-13}}{1.6 \times 10^{-19}} = 10 \text{ MeV}$$

Thus the kinetic energy of a free electron confined within a nucleus should have minimum value of 10 MeV. The rest mass energy of an electron is only about 0.5 MeV and experimentally, the highest value of the kinetic energy of a β-particle emitted spontaneously by a nucleus is found to be about 4 MeV. Since this is much less than the calculated value of the minimum kinetic energy which the electron, if it exists inside the nucleus, should possess, we come to the conclusion that electrons cannot be constituents of the atomic nucleus.

(*b*) Angular momentum consideration. The spin angular momentum of the proton as well as the electron is $\frac{1}{2}\hbar$ as both these are Fermi particles. Thus nucleus with *even* number of protons + electrons will have an *integral spin* and the nuclei with *odd* number of protons + electrons will have *half integral* spin. This rule is not obeyed by nuclei. For example deutron which is the heavier isotope of hydrogen with atomic number (charge) $Z = 1$ and mass number $A = 2$, if composed of protons and electrons must have two protons and one electron, giving a nuclear spin of $\frac{1}{2}$ or $\frac{3}{2}$ depending upon the orientation of the particles. The observed spin of the deutron is, however, 1 from which it is again seen that the electrons are not the constituents of the nucleus.

(*c*) Statistical considerations. If electrons exist within the nucleus, then $_7N^{14}$ nucleus should contain 14 protons and 7 electrons each having spin 1/2 and should, therefore, have a resultant spin 1/2 *i.e.*, it should behave as a Fermion and obey *Fermi-Dirac statistics.* But from a study of band spectra and Raman effect $_7N^{14}$ nuclei were found to obey *Bose-Einstein statistics* showing thereby that electrons do not exist in the nucleus.

(*d*) Other considerations. (*i*) If nuclei consist of protons and electrons, the existence of isotopes cannot be explained. Consider for instance the two isotopes of uranium having masses 238 and 235. If electrons exist within the nucleus, then U^{235} should consist of 235 protons and 143 electrons with 92 extra nuclear electrons. Similar U^{238} should consist of 238 protons and 146 electrons with 92 extra nuclear electrons. This is an ambiguous situation which requires that these two isotopes should be placed in different groups in the periodic table.

(*ii*) The size of an electron calculated on the basis of Coulomb law is too large to be contained in the small space of the nucleus.

(*iii*) The spin magnetic moment of the proton is about 0.15% of that of the electron and is almost negligible. If the electrons exist within the nucleus the spin magnetic moment of the nucleus should be nearly the same as that of the electron. But the observed magnetic moment of the nuclei is comparable to that of proton and is thus much smaller than the spin magnetic moment of the electron. This again shows that electrons are not the constituents of the nucleus.

(*iv*) On the basis of forces between nuclear particles, the binding energy per nuclear particle comes out to be of the order about 8 MeV. If the electrons exist within the nucleus, these nuclear electrons should interact with protons very strongly.

The orbital electrons, on the other hand, only interact electrically with the proton which is comparatively a weaker interaction. Even when fast electrons are scattered by nuclei, these behave in such a way as if these were acted upon only by electric forces. On the contrary when protons are scattered by nuclei, these show departure from electrical forces, which is supposed to be due to strong nuclear forces. It is not possible for the electron to behave in two different ways *i.e.* for one half to have a strong interaction and for the other half to have only weak interaction. This again indicates that electrons are not the the constituents of the nucleus.

(*v*) Several theoretical considerations suggest that a bound fundamental particle cannot be localised in the region smaller than Compton wavelength $\frac{h}{m_0 c}$. The Compton wavelength for electron is 0.02425×10^{-10} m = 242.5×10^{-14} m which is about 1000 times the nuclear radius the value of which is about 1.2×10^{-15} m.

(*vi*) **β^+-decay.** If electrons exist within the nucleus then it is not possible to explain β^+ decay *i.e.*, the emission of positrons from the nucleus, because if both the electron and positron exist within the nucleus annihilation of electron-positron pair should take place.

Example 15.10. *The mass of deutron which is the nucleus of heavy hydrogen is not equal to the sum of the masses of its constituents which are the proton and the neutron. Calculate the mass difference and determine energy equivalent. Given m_p = 1.00728 u ; m_n = 1.00866 u, and M_D = 2.013355 u.* **(Calicut. U. 2002)**

Solution. Mass defect

$$\Delta m = m_p + m_n - M_D$$
$$= (1.00728 + 1.00866 - 2.01355)\, u$$
$$= 0.00239\, u$$

Energy equivalent (binding energy) = 0.00239×931.48

$$= 2.226 \text{ MeV} \qquad [\because 1u = 931.48 \text{ MeV}]$$

Example 15.11. *Calculate binding energy per nucleon for the deutron. Given $m_n = 1.675 \times 10^{-27}$ kg; $m_p = 1.672 \times 10^{-27}$ kg; $M_D = 3.343 \times 10^{-27}$ kg; $c = 3 \times 10^8$ ms^{-1}.* **(Meerut. U., 2003; Agra.U. 2002)**

Solution. A deutron nucleus consists of a proton and a neutron.

$\therefore$ Mass of proton + mass of neutron = $(m_p + m_n)$

$$= 1.672 \times 10^{-27} + 1.675 \times 10^{-27} = 3.347 \times 10^{-27} \text{ kg}$$

Mass of deutron = 3.343×10^{-27}

Mass defect $\Delta m = (3.347 - 3.343) \times 10^{-27}$ kg $= 0.004 \times 10^{-27}$ kg

$\therefore$ Binding energy B.E. = $\Delta mc^2 = 0.004 \times 10^{-27} \times 9 \times 10^{16}$ Joules

$$= \frac{0.004 \times 10^{-27} \times 9 \times 10^{16}}{1.6 \times 10^{-13}} = 2.25 \text{ MeV.}$$

$\therefore$ Binding energy per nucleon = $\frac{2.25}{2} = 1.125$ MeV.

Example 5.12. *Calculate the binding energy of an α-particle in Joules and MeV from the following data*

Mass of helium nucleus = 4.002870 a.m.u.

Mass of proton = 1.007825 a.m.u.

Mass of neutron = 1.008665 a.m.u.

1 a.m.u. = 14.92×10^{-11} J = 931.48 MeV

(Calicut.U. 2000, H.P.U. 2000; G.N.D.U. 2000; P.U. 2003)

Solution. Mass of α-particle = $_2He^4$ nucleus = 4.002870 a.m.u.

As it contains two protons and two neutrons, the binding energy

$$\text{B.E.} = 2m_p + 2m_n - M_{He}$$
$$= 2 \times 1.007825 + 2 \times 1.008665 - 4.002870 = 0.03011 \text{ a.m.u.}$$
$$= 0.03011 \times 14.92 \times 10^{-11} \text{ J}$$
$$= 44.9 \times 10^{-13} \text{ J}$$
$$= 0.03011 \times 931.48 = 28.047 \text{ MeV.}$$

15.20 NUCLEAR FORCES

The radius r_n of the nucleus of mass number A is given by

$$r_n = r_0 A^{1/3}$$

where r_0 is a constant for all nuclei and has a value 1.4×10^{-15} m. Hence the nuclear mass and nuclear volume are both proportional to A. Therefore, the nuclear density is nearly constant for all nuclides.

The nucleus consists of protons and neutrons. As the protons carry a positive charge, the electrostatic force of repulsion between them should cause a disruption of the nucleus. The gravitational force of attraction between the neutrons is too weak to account for the observed binding energy of the nuclei. *Thus, there must be some other forces which bind the nucleons.* As there are two types of nucleons, the proton and the neutron, there are three types of attractive forces acting in the nucleus. These are:

(*i*) Force between two protons ($p-p$ force)

(*ii*) Force between a proton and a neutron ($p-n$ force)

(*iii*) Force between two neutrons ($n-n$ force)

***p* – *p* force.** The $(p-p)$ force has been examined by proton-proton scattering experiment. These experiments show that the force can be represented by the potential curve *i.e.*, a graph between potential binding energy of the nucleon versus the distance from the centre of the nucleus shown in Fig. 15.6.

At large distances of separation the protons repel one another by the Coulomb electrostatic force. At a distance of approximately 3×10^{-15} m a fairly sharp break in the potential curve occurs. At smaller distances the protons strongly attract each other and this strong attractive force is the nuclear force between a pair of protons.

***p* – *n* force.** The $(p-n)$ force has been examined by neutron-proton scattering experiment. These experiments show that the neutron-proton force can be represented by potential curve shown in Fig. 15.7. At large distances of separation there is no force between the two particles but at distances of about 2×10^{-15} m the neutron and proton attract one another by a strong nuclear force.

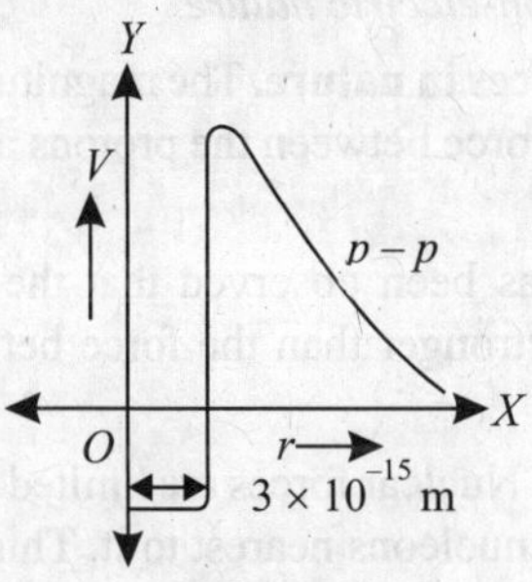

Fig. 15.6

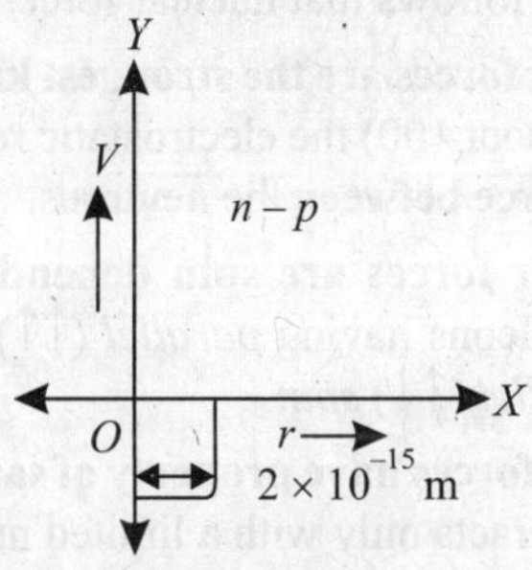

Fig. 15.7

The existance of $(p-n)$ force of attraction is proved by the stability of deutron nucleus consisting of 1 proton and 1 neutron which has a binding energy of 2.2 MeV.

We know that the atomic mass number A is approximately equal to twice the atomic number Z for the light and intermediate nuclei. It shows that nuclei prefer to add nucleons in pairs, *i.e.*, there is a strong interaction between the neutrons and protons.

***n* – *n* force.** The neutron excess in heavy nuclei confirms that $(n-n)$ *forces are attractive* but not sufficiently large to lead to a stable *dineutron* just as a stable *diproton* does not exist. The scattering length of $(n-n)$ interaction appears to have the same (negative) sign as the $(p-p)$ interaction.

Properties of nuclear forces. (*i*) **Nuclear forces are short range forces** so that each nucleon comes under the influence of only those nucleons which are in its immediate close vicinity. Nuclear forces are appreciable only when the distance between the nucleons is of the order of 2.2×10^{-15} m = 2.2 fm. These forces vanish for all practical purposes at distances greater than about 4.2×10^{-15} m.

If these are long range forces, then each nucleon will come under the influence of all other nucleons in the nucleus. In such a case, the binding energy will be proportional to (*i*) the number of nucleons exerting the force and (*ii*) the number of nucleons on which the force is being exerted. Hence the binding energy will be proportional to the square of the number of nucleons in the nucleus *i.e.*, $\propto A^2$. But the binding energy per nucleon is almost a constant and, therefore, the total binding energy of the nucleus is proportional to the number of nucleons *i.e.*, $\propto A$. On the other hand, if these are short range forces, then each nucleon will come under the influence of only those nucleons which are in its immediate vicinity in the nucleus and these alone will contribute to the binding energy of the nucleons making it almost proportional to A.

(*ii*) **Nuclear forces are charge independent.** From a study of *mirror nuclides* we find that the three types of forces *i.e.*, $(p-p)$, $(p-n)$ and $(n-n)$ are almost of equal magnitude. *Mirror nuclides are those nuclides which contain the same number of nucleons but in which the proton and the neutron numbers are interchanged.* For example, the nuclide ${}_1H^3$ which contains 1 proton and 2 neutrons and the nuclide ${}_2He^3$ which contains 2 protons and 1 neutron are mirror nuclides. The binding energy of 8.5 MeV of ${}_1H^3$ nucleus is due to 2 $(n-p)$ forces 1 $(n-n)$ force and that of ${}_2H^3$ nucleus equal to 7.7 MeV is due to 2 $(n-p)$ forces and 1 $(p-p)$ force. The three types of forces being equal in magnitude, the difference of $8.5 - 7.7 = 0.8$ MeV in the binding energies of ${}_1H^3$ and ${}_2He^3$ nuclides is due to the fact that the $(n-n)$ force in ${}_1H^3$ is greater than $(p-p)$ force in ${}_2He^3$ due to the *electrostatic force of repulsion* between two protons. If allowance is made for this force, then $(n-n)$, $(p-p)$ and $(n-p)$ forces are equal.

The equality of $(n-n)$, $(p-p)$ and $(n-p)$ attractive forces is also established from a study of other mirror nuclei after making allowance for the electrostatic force of repulsion between 2 protons and it is found that total binding energy depends only on the total number of nucleons and not on their nature. In other words, *nuclear forces are charge independent*. This conclusion is also confirmed by other experiments like scattering of protons by hydrogen nuclei and neutrons by neutrons.

It, therefore, follows that nuclear forces are of *non-electric nature.*

(*iii*) **Nuclear forces are the strongest known forces in nature.** The magnitude of nuclear forces is many times (about 100) the electrostatic repulsive force between the protons and about 10^{38} times the gravitional force between the neutrons.

(*iv*) **Nuclear forces are spin dependent.** It has been observed that the force of attraction between two nucleons having *parallel* (↑↑) *spin* is stronger than the force between two nucleons having *anti-parallel* (↑↓) *spin.*

(*v*) **Nuclear forces have property of saturation.** Nuclear forces are limited in range. As a result each nucleon interacts only with a limited number of nucleons nearest to it. This effect is known as *saturation of nuclear forces*.

The saturation of nuclear forces arises from the dependence of total binding energy on the mass number A. If there were no saturation and each nucleon could interact with the remaining $(A-1)$ nucleons, then the total binding energy would be propotional to the number of nucleon pairs in the nucleus *i.e.*, $\frac{A(A-1)}{2}$. In other words binding energy will be proportional to A^2, but the binding energy is proportional to A. Hence *nuclear forces are saturated forces.*

(*vi*) **Nuclear forces are non-central.** The force existing between the two nucleons has a non-central component that does not point along the line joining the two nucleons. This non-central component depends upon how the nuclear spins are oriented relative to the line joining the nucleons.

(*vii*) **Nuclear forces are exchange forces.** Nuclear forces have an exchange character as these forces are brought into the existence due to the exchange of π^+, π^- and π^0 mesons (pions) between the nucleons.

(*viii*) **At too close** distances of approach between the nucleons, nuclear forces become *force of repulsion.*

15.21 NUCLEAR FORCES ARE EXCHANGE FORCES

It is supposed that internuclear forces in the nucleus are, at least in part, some type of '*exchange forces*'. To explain the observed energy distribution among the β-particles emitted by radio-active nuclei, it was assumed that a proton and a neutron were two different states of the same particle-the nucleon and a β-particle, positive or negative, was continuously being exchanged between them. But, the corresponding force due to the exchange of electrons was, however, too small to account for the observed nuclear energies.

Meson theory. On theoretical grounds Yukawa predicted a new particle, the *meson* which may have a positive or a negative charge or may be neutral. He assumed that this new particle, (instead of the electron), is exchanged between the nucleons and the corresponding '*exchange force*' is responsible for the binding energy of the nucleons. Just as a photon is a quantum of electromagnetic field, a meson is a quantum of nuclear field.

Nuclear forces are thus produced by a *meson field* similar to the electromagnetic field but *of a much shorter range.* This is because the exchange of mesons between the nucleons can take place only when the nucleons are *close together*, otherwise the principle of conservation of energy will be violated because a meson has a finite rest mass and energy will have to be supplied to remove it away from the nucleon.

The prediction of Yukawa was confirmed by the experimental discovery of a μ-menson in 1936 and of a π-meson in 1947. It was the π-meson or *pion* which was found to explain the observed nuclear binding energies. The pion may be neutral, denoted as π^0 or it may have a positive or a negative charge equal to that of the electron and denoted as π^+ or π^-. The intrinsic spin of a pion is zero.

Mass of π-meson. The mass of the π-meson was roughly estimated theoretically from Heisenberg's uncertainty principle. If ΔE is the rest mass energy of the π-meson, then this energy must be exchanged between the nucleons (due to the exchange of π-meson) in a time Δt so that $\Delta E \cdot \Delta t \gtreqless \hbar$.

The distance that a meson is to travel in time Δt should not exceed the distance between two nucleons *i.e.*, 1.4 Fermi = 1.4×10^{-15} m. Taking the velocity of light 3×10^8 ms^{-1} to be the velocity of meson exchange.

$$\Delta t = \frac{1.4 \times 10^{-15}}{3 \times 10^8} \text{ sec}$$

$$\therefore \quad \Delta E = \frac{\hbar}{\Delta t} = \frac{6.62 \times 10^{-34} \times 3 \times 10^8}{2\pi \times 1.4 \times 10^{-15}} = 2.25 \times 10^{-11} \text{ J}$$

$$= \frac{2.25 \times 10^{-11}}{1.6 \times 10^{-19}} = 1.4 \times 10^8 \text{ eV}$$

$$= 140 \text{ MeV}$$

The mass of the electron = 0.5 MeV (approximately)

∴ Mass of the meson = 280 × mass of the electron.

The experimental value comes out to be 270 times the mass of the electron which is in close agreement.

The attraction between a nuclear proton and neutron is represented as

$$\text{proton} \underset{+ve\ \pi\ \text{meson}\ \pi^+}{\overset{-ve\ \pi\ \text{meson}\ \pi^-}{\rightleftharpoons}} \text{neutron}$$

The emission of a β-particle by a nucleus and the consequent conversion of a neutron into a proton in the nucleus takes place when a π^- meson is ejected. The π^- meson decays almost instantly and appears as a β-particle outside the nucleus.

$$n \longrightarrow p + \underset{\substack{\downarrow \\ e^-}}{\pi^-} + \bar{\nu}$$

where n represents a neutron, p a proton, $\bar{\nu}$ an anti-neutrino and e^- an electron. The emission of a positron is represented as

$$p \longrightarrow n + \underset{\substack{\downarrow \\ e^+}}{\pi^+} + \nu$$

where ν represents a neutrino and e^+ a positron.

Process of π-meson exchange. To understand the process of π-meson exchange forces, the proton and the neutron are each supposed to be surrounded by a *virtual meson cloud*. The transfer of a π^+ meson from the proton cloud to the neutron cloud only results in a change of identity of the particle *i.e.*, in the final state we again have a proton and a neutron with the same energy as in the initial state. This is represented as under

$$p + n = n' + \pi^+ + n \longrightarrow n' + p'$$

where n' indicates the neutron formed from the proton and p' a proton formed from the neutron *i.e.*, the 'primes' indicate change in identity of the nucleons. This change in identity of the nucleons without change of energy can also take place through the following processes:

$$n + p \longrightarrow p' + \pi^- + p \longrightarrow p' + n'$$

$$p + p \longrightarrow p' + \pi^0 + p \longrightarrow p' + p'$$

$$n + n \longrightarrow n' + \pi^0 + n \longrightarrow n' + n'$$

Thus, the force field between two protons or between two neutrons is carried by π^0 meson. But the force field between a proton and a neutron can be carried by a charged π^+ or π^- meson. In this way, the charge is carried from one nucleon to other and a proton is changed into a neutron and *vice-versa*.

Facts in support. The experimental values of magnetic moment of a free proton and a free neutron further support '*Yukawa's meson field theory*'. A free proton is for a part of its life time a neutron with a closely bound (π^+) meson. Thus the net magnetic moment of a free proton will exceed that given by simple theory. Similarly a neutron is, for a fraction of its life time dissociated into a proton and a (π^-) meson. This combination will have a *negative* magnetic moment. It, therefore, follows that a neutron though uncharged will have a negative magnetic moment.

Saturation nature of nuclear forces. A purely attractive force, however, does not account for the *saturation nature* of nuclear forces. Therefore, it is necessary to introduce the possibility of forces of repulsion for *too close* distances of approach betwen the nucleons. For this purpose we require nuclear forces of attraction at inter-nuclear distances of the order of nuclear radii and forces of repulsion at shorter distances. Thus, at distance much smaller than 1 *fm* (Fermi) the nucleon forces change character from being attractive to being repulsive. This is known as '*repulsive nuclear core*'. According to quantum mechanics when a system can be represented by two equi-energy states, two actual combined states of the system are possible; one of these being more stable than the other. The more stable state implies a force of attraction between the nucleons and the less stable state a force of repulsion between them. The actual situation, therefore, is an appopriate combination of these two states.

Example 15.13. *What range of nuclear force is mediated by η^0 having a mass corresponding to 550 MeV ?* **(Luck. U. 2001)**

Solution. Mass (energy) of the particle $\Delta E = 550$ MeV $= 550 \times 1.6 \times 10^{-13}$ J

According to Heisenberg's uncertainty principle

$$\Delta E\, \Delta t = \frac{h}{2\pi}$$

$$\therefore \qquad \Delta t = \frac{h}{2\pi} \cdot \frac{1}{\Delta E}$$

If we assume that the emitted particle moves (nearly) with the velocity of light, then the range of nuclear force

$$r_0 = c\,\Delta t = c \frac{h}{2\pi} \frac{1}{\Delta E}$$

$$= \frac{3 \times 10^8 \times 6.62 \times 10^{-34}}{2\pi \times 550 \times 1.6 \times 10^{-13}}$$

$$= 0.36 \times 10^{-15} \text{ m} = 0.36 \text{ Fermi}$$

15.22 NUCLEAR, ELECTROSTATIC AND GRAVITATIONAL FORCES

Nuclear force is classified as a *strong interaction*. The electrostatic (or elecromagnetic) force is weaker as compared to nuclear force and the gravitational force is still weaker. The order of their strength is as under

Nuclear force = 10^{40}

Electrostatic (or electromagnetic) force = 10^{38}

Gravitational force = 1

Difference between nuclear, electrostatic and gravitational forces. Nuclear force differs from electrostatic and gravitational forces in some of the following respects.

(*i*) Nuclear forces are short range forces, the range being of the order of 3 *fm* only. On the other hand electrostatic (or electromagnetic) as well as gravitational forces are long range forces.

(*ii*) Nuclear forces are charge independent. On the other hand electrostatic forces are charge dependent. There is a force of repulsion between like charges and a force of attraction between unlike charges. However, gravitational force is also charge independent.

(*iii*) Nuclear forces are strong forces. The strength of nuclear force is about 100 times that of electrostatic force and about 10^{38} times that of gravitational force.

(*iv*) Nuclear forces are non-central whereas both electrostatic as well as gravitational forces are central forces.

(*v*) Nuclear forces are exchange forces. These forces arise due to π-meson exchange. On the other hand electrostatic forces are due to photon exchange and gravitational forces are supposed to be due to the exchange of a hypothetical particle called *graviton* (still not discovered).

Why protons do not fly apart ? The nucleus consists of protons and neutrons. The protons carry a positive charge. Thus there is an electrostatic force of repulsion between them due to which the protons should fly apart. But this does not happen, because there is a strong nuclear force of attraction between the protons known as *p - p* force, which extends upto a distance of 3×10^{-15} m (3*fm*) only. It is this force which keeps the protons bound. Thus because within the nucleus the nuclear attractive forces predominate, these hold the protons together and do not allow them to fly apart.

15.23 NUCLEAR STABILITY

All nuclei with $Z > 83$ spontaneously disintegrate with emission of α and β-particles. This means that the emission of α and β-particles provides stability to heavier nuclei. At present more than 900 different isotopic nuclei are known, out of which about 280 are found to be stable. These are classified according to even and odd number of protons and neutrons.

Number of protons	*Number of neutrons*	*Number of stable nuclei*
even	even	164
even	odd	57
odd	even	53
odd	odd	6
	Total	280

It is seen from above that more than 160 nuclides have an *even* number of *protons* and an *even* number of *neutrons.* Thus the combination of an even number of protons and an even number of neutrons composing the nucleus is evidently preferred by nature for stable nuclides.

The number of *even-odd* or *odd-even* combination is almost the same – about 50 each. Only six nuclides have a combination of odd number of protons and an odd number of neutrons. This combination is found only in light elements and except for $_7N^{14}$ all others are very rare. Six even-even nuclei $_8O^{16}$, $_{12}Mg^{24}$, $_{14}Si^{28}$, $_{20}Ca^{40}$, $_{22}Ti^{48}$ and $_{26}Fe^{56}$ form about 80% of earth's crust. Taking the number of protons along the *X*-axis and the corresponding number of neutrons in stable nuclides along the *Y*-axis, the graph obtained is shown in Fig. 15.8.

Why *N* exceeds *Z* in stable nuclei. We find from the graph that for $Z < 20$ the *stability line* is a straight line with $Z = N$. For heavier nuclides $Z > 20$, $N > 20$ the number of neutrons starts increasing *i.e.*, the stability curve bends in the direction of $N > Z$. For example $_{20}Ca^{48}$ has $N = 28$, $Z = 20$. For larger values of *Z*, there is still greater proportion of neutrons and in the case of $_{92}U^{238}$, $N = 146$ and $N = 92$. It is because for large values of *Z*, the coulomb force of repulsion between the protons becomes dominant and there has to be a greater number of neutrons to compensate for this repulsive effect. So the stability curve bends more and more towards the neutron axis and deviates more and more from the $N = Z$ line as *Z* increases.

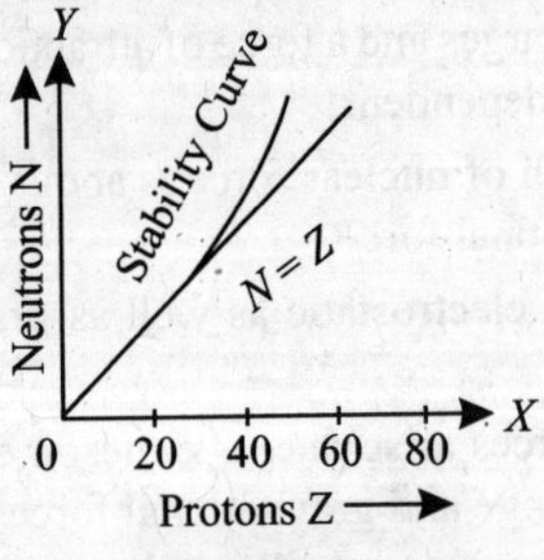

Fig. 15.8

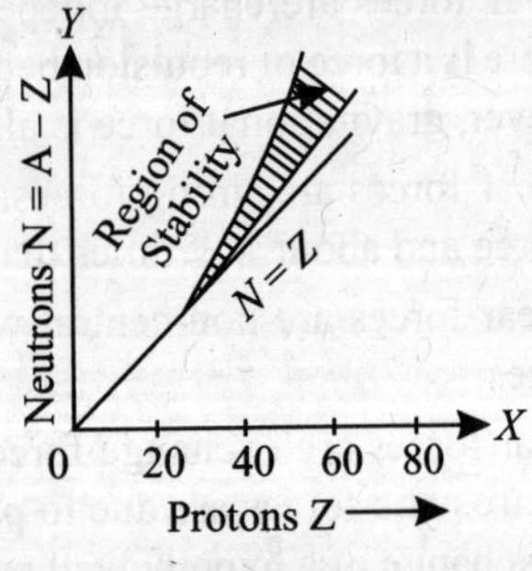

Fig. 15.9

For the stability to be maximum there is a *optimum* value of the ratio of neutrons to protons. The number of neutrons $N (= A - Z)$ required for maximum stability plotted as a function of proton number *Z* is shown in Fig. 15.9. All stable nuclei fall within the shaded region. Nuclei above and below the shaded zone are unstable. Nuclei lying at the fringe of the region of stability are artificially radioactive.

Thus, stable nucleus has-*even*-number of protons and-*even*-number of neutrons.

15.24 α, β AND γ RADIATIONS

The radiations emitted by radioactive substances are of three types (*i*) α-rays, (*ii*) β-rays and (*iii*) γ-rays.

Nature. (*i*) Alpha rays consist of nuclei of helium atoms $_2He^4$ moving with velocities of the order of 10^7 ms^{-1}. They have a positive charge twice in magnitude of the charge on an electron and mass equal to 4 atomic mass units.

(*ii*) β-rays consist of electrons moving with a velocity of the order of 1% to 99% of the velocity of light. As these rays consist of electrons they have a negative charge equal to 1.6×10^{-19} coulomb and the same mass as that of electron.

(*iii*) γ-rays are electromagnetic waves of very short wavelength ranging from 0.005 to 0.5 Å.

Properties of α-rays. (*i*) The α-rays are shot out from the radioactive material with large velocities ranging from 1.4×10^7 to 1.7×10^7 metres per second.

The velocity of α-rays depends upon the radioactive substance from which they are ejected and from a given substance it is always the same.

(*ii*) *They produce intense ionisation* in the gas through which they pass. Their ionising power is 100 times greater than that of β-rays and 10,000 times greater than that of γ-rays. A thickness of 0.0005 cm. of aluminium foil reduces the ionising power to half. The tracks of α-particles observed by means of a *Wilson cloud chamber* are continuous, thick, straight lines with a slight bend in the end in certain cases.

(*iii*) They affect a photographic plate. The effect is, of course, very feeble.

(*iv*) *They produce fluorescence* in substance like zinc sulphide or barium platinocyanide. On observing the fluorescence through a low power microscope, *i.e.*, *spinthariscope* it is found to consist of successive *scintillations*, produced by the impact of an individual α-particle. The impact of a single α particle can thus be observed and the number of such particles given out per second by a radioactive substance can be found out. The fact that these scintillations are due to α-particles and not due to β or γ-rays can be proved by interposing a thin sheet of mica between the radium salt and the fluorescent screen. Since α-rays are absorbed by mica sheet the scintillations disappear.

(*v*) *Range of α particles*. Scintillations from α-particle cease abruptly after it has traversed a certain distance though matter. *The distance that an α-particle can travel in air, at atmospheric pressure is called its range in air*. It varies from 2.70 cms for α-particles given by uranium to 8.62 cms. for those of thorium C. The range of an α-particle depends upon

(*a*) The radioactive substance from which the rays are given out.

(*b*) The nature of the medium through which the rays travel.

(*c*) Velocity of emission. The range is proportional to V^3.

The number of α-particle in a beam remains constant up to the end of the range, but their energy goes on decreasing. Ultimately the α-particles do not have sufficient energy to produce either scintillations or ionisation.

The ionisation first increases slightly, reaches a maximum and then suddenly drops almost to zero value. The distance at which the ionisation drops almost to zero gives the range of α-particles.

(*vi*) The α-rays are scattered when they pass through thin sheets of mica, gold foil, etc. The divergence of α-particle from its straight line path is 2 to 3 degrees. Geiger and Marsden found that a few particles, some

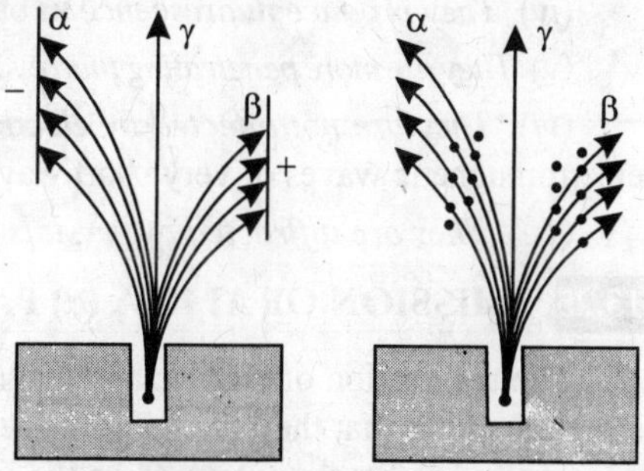

Fig. 15.10

times, were deflected by more than 90°. This was explained by Rutherford to be due to the repulsion between the α-particle and the nucleus of the atom scattering it.

(*vii*) *The rays are deflected by electric and magnetic fields showing that they are charged particles.*

If a narrow beam of radioactive radiations proceeding from a deep cavity in a block of lead is subjected to an electric field between parallel plates, the α-particles are deflected towards the negative plate as shown in Fig. 15.10 (*a*). This shows that α-rays consist of positively charged particles.

If a magnetic field is applied perpendicular to the plane of the paper and directed from top to bottom the α-rays are deflected towards the left as shown in Fig. 15.10 (*b*). Application of Fleming's left-hand rule indicates that α-rays consist of *positively charged particles*. The value of *e*/*m* is determined on the same lines as for positive rays and indicates that α-particles are nuclei of helium atoms.

(*viii*) *They produce a heating effect*. A quantity of radium always maintains itself at a temperature higher than that of the surroundings. The evolution of heat is due to the stoppage of α, β and γ-rays by the radioactive substance.

(*ix*) *When exposed to α-rays the body suffers incurable burns.*

Properties of β-rays. (*i*) β-rays are shot out from radioactive substances with very high velocities ranging from 1% to 99% of the velocity of light. The velocity of all the β-particles given out by a substance is not the same.

(*ii*) *They produce ionisation in air* but the number of ions produced is hardly $\frac{1}{100}$th of those produced by α-rays. Although their velocity is very large they posses a comparatively small mass ($\frac{1}{7000}$ of an α-particle) and hence have a small kinetic energy. As β-particles are slowed down by collision with the atoms of the gas and change their path, their tracks in a Wilson cloud chamber are scattered and not continuous as are for the α-particles.

(*iii*) *They affect a photographic plate* and their effect is greater than those of α-rays.

(*iv*) *They produce fluorescence* in barium platinocyanide, calcium tungstate, willemite, etc.

(*v*) *They can penetrate through large thickness of matter*, *e.g.*, they can easily pass through 1 cm thickness of aluminium sheet.

(*vi*) *They are more readily scattered when they pass through matter, because their mass is very small as compared to the mass of the atomic nuclei.*

(*vii*) *They are affected by electric and magnetic fields*. Their direction of deflection indicates that they are *negatively charged particles*. The *e*/*m* value has been found on the same lines as for cathode rays and indicates that β-rays are fast-moving electrons.

Properties of γ-rays. (*i*) They posses the same velocity as that of light *i.e.*, 3×10^8 metres/sec.

(*ii*) *They ionise the gas through which they pass* but the ionisation produced is very small.

(*iii*) *They affect a photographic plate* and their effect is greater than those for β-rays.

(*iv*) *They produce fluorescence* in barium platinocyanide, etc.

(*v*) *They are more penetrating than even* β-rays and can pass easily through 30 cms, thickness of iron.

(*vi*) *They are not affected by electric and magnetic fields*. This property shows that they are electromagnetic waves of very short wavelengths.

(*vii*) *They are diffracted by crystals* in the same way as *X*-rays.

15.25 EMISSION OF ALPHA (α) PARTICLES

The application of electric and magnetic fields to a beam of α-rays given out by a radioactive material shows that they consist of *positively* charged particles. The value of *e*/*m* and velocity for α-particles has been determined on the line of positive rays and it has been found that α-rays from all substances are of a *similar character*.

The value of the charge on an α-particle in 3.2×10^{-19} Coulomb. This shows that the charge is nearly double that on an electron. The value of e/m is 4.8×10^7 C/kg which gives the mass of an α-particle to be 6.60×10^{-27} kg. The mass of a helium atoms is also 6.60×10^{-27} kg. Hence it is simplest to assume that α-particle is an atom of helium which has lost two electrons, that is, *it is the nucleus of the helium atom.*

The nucleus consists of protons and neutrons, collectively known as nucleons. The number of protons is equal to the atomic number Z and the number of neutrons is equal to $(A - Z)$ where A is the mass number. The protons carry a positive charge and there is an electrostatic force of repulsion between them. The total repulsive (or disruptive) force in the nucleus is approximately proportional to Z^2. The gravitational force of attraction between the nucleons is very weak. The nuclear attractive force between the nucleons is of a very short range and the total binding energy of a nucleus is, therefore, approximately proportional to its mass number A.

High energy of α-particles favours emission. The α-particle which consists of two protons and two neutrons has a very high binding energy of more than 28 MeV because the α-particle mass is sufficiently smaller than that of its constituent nucleons. Thus the formation of an α-particle within the nucleus makes available the kinetic energy the particle must have to escape from the nucleus. The kinetic energy Q known as disintegration energy released in the process of emission of a particle from a heavy nucleus is given by

$$Q = [M_p - (M_d + m_x)]\, c^2$$

where M_p is the mass of the *parent nucleus*, M_d the mass of the *daughter* nucleus and m_x the mass of the *particle* emitted. Nuclei which contain 210 or more nucleons are so large that the short range nuclear forces that hold them together are hardly able to counter balance the mutual repulsive force between the protons. The α-decay occurs in such nuclei as a means of increasing their stability by reducing their size and is energetically possible. The α-decay from ${}_{92}U^{238}$ nucleus is accompanied by a release of 5.4 MeV of energy. This value agrees with the values predicted from the nuclear masses involved in the process.

Thus radioactive nuclei whose atomic number Z is greater than 82 (mass number 208) spontaneously disintegrate with *parent* nucleus decaying into a *daughter* nucleus and an α-particle (helium nucleus). The α-particle and the daughter nucleus have a kinetic energy even if the parent nucleus is at rest; the condition necessary for emission of an α-particle is that the sum of the masses of the daughter nucleus and the α-particle must be *less* than the mass of the parent nucleus. The decrease in mass appears as the kinetic energy of the α-particle and the daughter nucleus.

15.26 RANGE OF α-PARTICLES

The α-particles have the property of ionising a gas. As the α-particles pass through a gas, these lose energy by ionising the gas particles and are slowed down. The process continues till the energy of the α-particle falls below the ionisation potential of the gas. After this the α-particle captures two electrons and becomes a neutral helium atom.

The distance the α-particle travels in the gas before its energy falls below the ionisation potential of the gas is called its range in that gas.

The range is usually expressed in cm. in air at 76 cm of mercury pressure and 15° C. The range depends upon

(*i*) The initial energy of the α-particle

(*ii*) The ionisiation potential of the gas.

(*iii*) Nature and pressure of the gas.

Determination of range of α-particles. The range of α-particles in air can be measured with the help of the apparatus shown in Fig. 15.11. The radioactive α-emitting source S is placed on a movable block, the distance of which from an ionisation chamber I.C. can be adjusted and measured on a graduated scale.

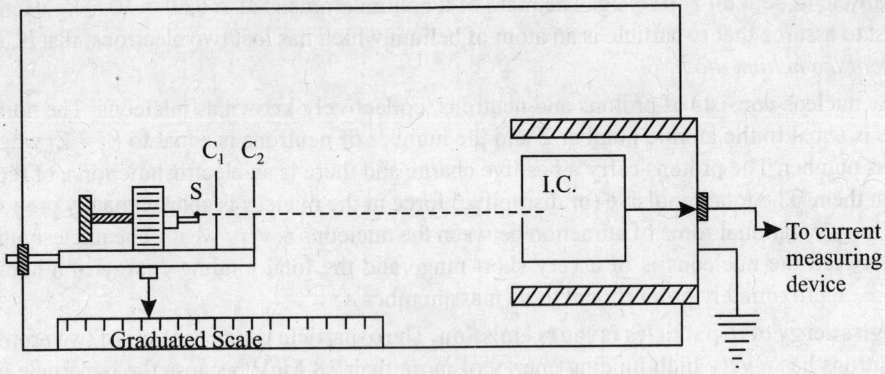

Fig. 15.11

A narrow beam of α-particles coming out of collimating slits C_1, C_2 after passing through a known thickness of air enters the ionisation chamber I.C. where ionisation takes place and an ionisation current is set up. This current is measured by a sensitive current measuring device.

It is observed that as the distance between the source and the ionisation chamber is increased, the ionisation current slowly rises to a maximum up to a certain distance and then drops suddenly to zero as shown in Fig. 15.12. The distance corresponding to maximum ionisation current measures the range of the α-particle.

Straggling and its causes. A graph between *specific ionisation i.e., the number of ions produced by α-particles per unit length of the path* and distance in cm in air at 76 cm of mercury pressure and 15° C is shown in Fig. 15.12. This curve, known as Bragg curve shows that specific ionisation during the path of the α-particle rises slowly to a maximum value and then falls rapidly to zero making a slight *ankle* or *tail* in the curve before the zero line is reached. This tail represents the *straggling effect*.

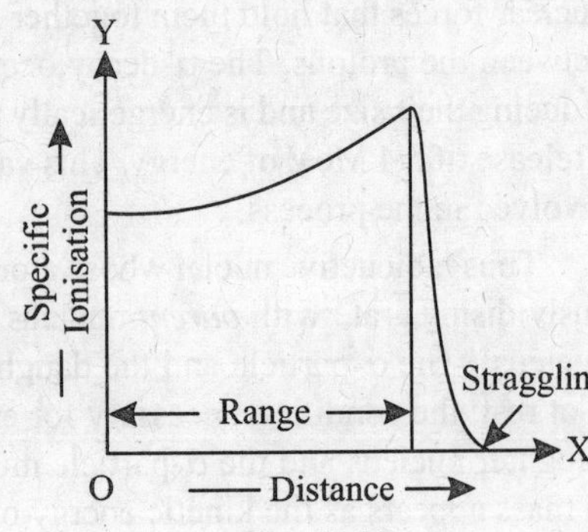

Fig. 15.12

The tail or ankle of the curve showing the straggling effect arises due to the following reasons.

(*i*) All the α-particles are not emitted from the source with the same initial energy (or velocity).

(*ii*) All the α-particles do not lose exactly the same amount of energy in their encounters with the molecules in their path.

All the α-particles, therefore, do not reach the end of the range simultaneously giving rise to straggling.

15.27 GEIGER NUTTAL LAW

In 1911, Geiger and Nuttal made an experimental study of relation between *range* (or *energy*) of α-particle in air and *half-life* (or *decay constant*) of a number of α-emitters. They observed that *longest lived* nuclides (having *smallest value of decay constant*) emit the *least energetic* α-particle, while the *shortest lived* nuclides (having the highest *value of decay constant*) emit the most energetic α-particle and gave following empirical law

$$\log_{10} \lambda = A + B \log_{10} R_\alpha$$

where λ is the *decay constant*, R_α the range of α-particle in air and *A* and *B* are constants. If, therefore, $\log_{10} \lambda$ are plotted against $\log_{10} R_\alpha$ for the different α-emitters in three series, three nearly parallel straight lines are obtained, one for each series as shown in Fig. 15.13.

In the relation $\log_{10} \lambda = A + B \log_{10} R_\alpha$ the constant B has the same value for all the series while the constant A has different values for each series.

According to Geiger's law $R_\alpha \propto v^3$ where v is the velocity of emission of the α-particle. If E is the energy of the α-particle, then $E = 1/2\, mv^2$ or $E \propto v^2$

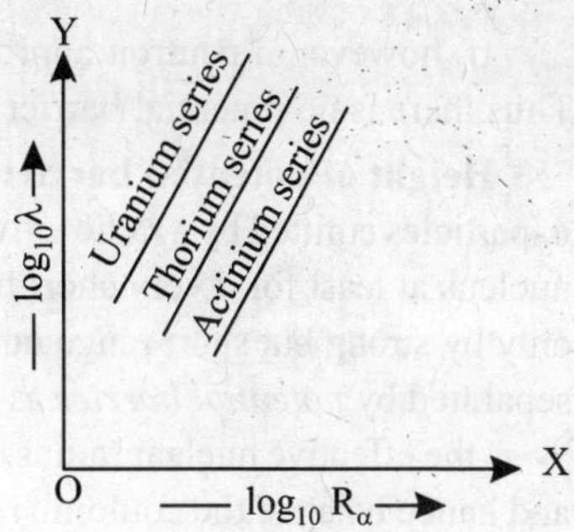

Fig. 15.13

$$\therefore \quad R_\alpha \propto E^{3/2}$$

$$\text{or} \quad \log_{10} R_\alpha \propto \frac{3}{2} \log_{10} E$$

$$\therefore \quad \log_{10} R_\alpha = B' \log_{10} E$$

Hence in terms of energy Geiger-Nuttal law may be put as

$$\log_{10} \lambda = A' + B' \log_{10} E$$

Importance. The decay constants of many radioactive substances can be determined using this empirical law. The ranges of α-particle corresponding to the members of the three series are found experimentally and from Geiger-Nuttal relation, the decay constants are calculated.

15.28 POTENTIAL BARRIER OF A NUCLEUS

Potential Barrier. Consider a proton approaching an atomic nucleus. As the distance from the nucleus decreases, the electrostatic force of repulsion experienced by the proton, increases. As the proton moves against this opposing force, its potential energy goes on increasing. However, when the proton comes closer to the nucleus than about 4×10^{-15} m the *strong nuclear attractive force* begins to act on it. At about 2×10^{-15} m, the nuclear attractive force becomes highly prominent. If we plot the value of potential energy stored between the proton and the nucleus along the Y-axis and the distance of the proton from the nucleus along the X-axis, we get a curve as shown in Fig. 15.14. The curve represents the *potential barrier* and if we rotate it about the Y-axis, this takes the shape of a well, known as *potential well*

A proton, approaching the nucleus, behaves similar to a ball moving towards a sloping hill. If the ball has sufficient energy to reach the top of the slope, it will get into the well, otherwise it will be pulled back by the force of gravity. Similarly, if the proton (or any other positively charged particle) has sufficient energy to surmount the potential barrier of the nucleus, it may interact with the same, otherwise it will be pushed back by the electrostatic force of repulsion.

If the proton succeeds in overcoming the barrier, it would be accelerated towards the bottom of the potential well *i.e.*, the centre of the nucleus and will thus acquire a very high kinetic energy before colliding with one of the nucleons.

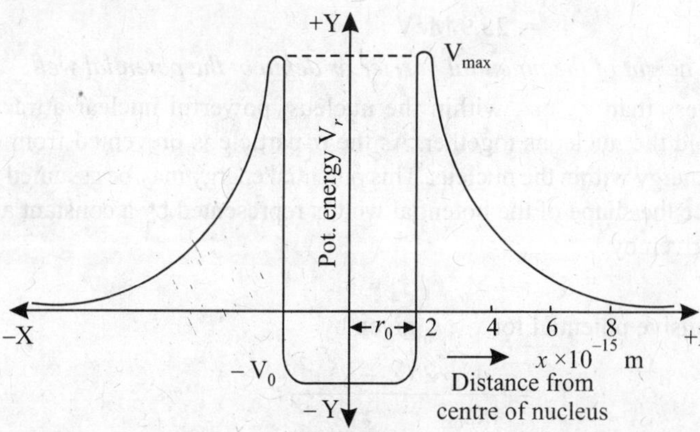

Fig. 15.14

If, however, a neutron approaches the nucleus, no electrostatic repulsion is experienced by it. Thus there is no potential barrier for a neutron.

Height of potential barrier or depth of potential well for an α-particle. If we consider the α-particles emitted by a radioactive nucleus, then those α-particles can be supposed to exist inside the nucleus at least for a very short time before emission. The α-particle can be held inside the nucleus only by strong but short range nuclear forces. Thus the interior of the nucleus can be supposed to be separated by *potential barrier* as explained above. At the boundaries of the barrier, say at a distance $r > r_n$ the effective nuclear radius, the nuclear forces become very weak due to the increase in distance and hence balance the coulomb repulsive force. Therefore, the height of the potential barrier will be equal to the potential energy of the α-particle at this position. The coulomb repulsive electric force at a distance x is given by

$$F = \frac{(Z-2)\, e \cdot 2e}{4\pi\varepsilon_0 x^2}$$

where $2e$ is the charge on the α-particle, $(Z-2)$ the atomic number of the daughter nucleus *i.e.*, after the separation of the α-particle and $x > r_0$ where r_0 is the sum of the radii of the nucleus and α-particle. If the potential energy of α-particle is taken to be zero when it is at an infinite distance from the nucleus, its potential energy will be positive when it is at a distance x, since work has to be done on it to bring it from infinity to x against the forces of repulsion.

∴ Potential energy at a distance x

$$= \int_{\infty}^{x} -F dx = \int_{\infty}^{x} -\frac{(Z-2)\, e \cdot 2e}{4\pi\varepsilon_0 x^2} dx$$

or
$$V_x = \frac{2\,(Z-2)\, e^2}{4\pi\varepsilon_0 x}$$

The maximum positive value of potential energy occurs when $x = r_0$. For ${}_{92}U^{238}$ the value comes out to be approximately 26 MeV calculated as under taking $r_0 = 10^{-14}$ m

$$\therefore \quad V_{max} = \frac{1}{4\pi\varepsilon_0} \frac{2\,(Z-2)\, e^2}{r_0}$$

$$= \frac{9\times 10^9 \times 2 \times 90 \times (1.6\times 10^{-19})^2}{10^{-14}} = 41.47\times 10^{-13}\text{ J}$$

$$= 25.9 \text{ MeV}$$

This given the *height of the potential barrier or depth of the potential well.*

At distances less than r_0, *i.e.*, within the nucleus, powerful nuclear attractive forces come into play which hold the nucleons together. As the α-particle is prevented from escaping, it has a *negative* potential energy within the nucleus. This potential energy may be assumed to be constant and equal to $-V_0$. Hence the shape of the potential well is represented by a constant attractive potential region upto $x = r_0$ given by

$$V_x = -V_0$$

and coulomb's repulsive potential for $x > r_0$ given by

$$V_x = \frac{1}{4\pi\varepsilon_0} \frac{2\,(Z-2)\, e^2}{x}$$

15.29 TUNNEL EFFECT

Tunnel effect can be very well understood from the basic theory of α-decay as given below:

According to the mass-energy condition of α-decay

$$M_p c^2 = M_d c^2 + m_\alpha c^2 + E_d + E_\alpha$$

where M_p is the mass of the parent nucleus, M_d that of the daughter nucleus, m_α of the α-particle, E_α, the kinetic energy with which the α-particle is ejected and E_d the kinetic energy imparted to the daughter nucleus.

But even when the disintegration of a nucleus is energetically possible, all the nuclei in a radioactive substance do not disintegrate immediately. This is because for a radioactive nucleus there exists a *potential barrier* which is a space in which the potential is so high, that according to classical conceptions, an α-particle inside that barrier cannot escape. The potential energy of the α-particle at a distance x from the centre of the nucleus is given by

$$V_x = \frac{2(Z_p - 2)e^2}{4\pi\varepsilon_0 x} = \frac{2Z_d e^2}{4\pi\varepsilon_0 x}$$

where Z_p is the atomic number of the parent nucleus and Z_d that of the daughter nucleus.

Note. *For future discussion we shall denote Z_d by Z only so that*

$$V_x = \frac{2Ze^2}{4\pi\varepsilon_0 x}$$

This gives the height of the potential barrier at a distance x from the centre of the nucleus.

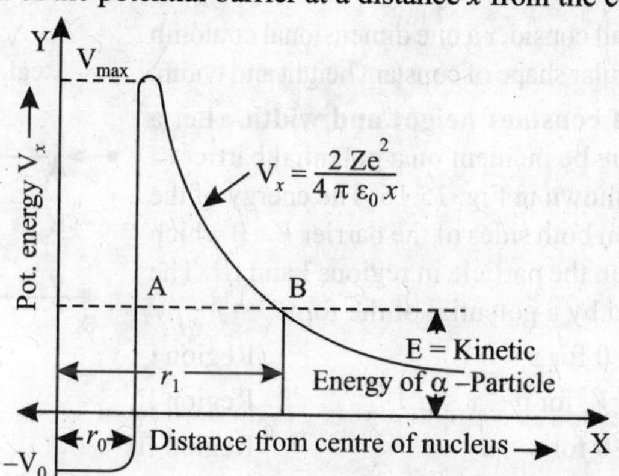

Fig. 15.15

The *maximum height* of the potential barrier

$$V_{max} = \frac{2Ze^2}{4\pi\varepsilon_0 r_0},$$

where r_0 is the sum of the radii of the nucleus and the α-particle. The shape of the potential barrier is shown in Fig. 15.15.

The value of V_{max} for $_{92}U^{238}$ comes out to be 26 MeV. Hence according to classical theory the energy required by the α-particle to escape from the nucleus must be more than 26 MeV, but it has been observed that α-particle emitted by $_{92}U^{238}$ have an energy of only 4.18 MeV. How is it possible for the α-particle of energy 4.18 MeV to cross the barrier of 26 MeV which should require an energy 7 times as much ?

The escape of an α-particle from a radioactive nucleus can be explained on the basis of the quantum theory by an application of wave mechanics. Wave mechanical analysis shows that there is often a small but finite probability for the occurrence of events, which are normally absolutely forbidden by classical mechanics. In quantum mechanics, the probability of locating the α-particle is given by its wave function. The wave function for the potential is oscillatory within the potential well *i.e.*, at distances less than r_0, and is highly attenuated through the potential barrier *i.e.*, for distances greater than r_0 but less than r_1 where r_1 is the distance from the centre of the nucleus where the potential energy is equal to the kinetic energy actually possessed by the α-particle. The value of r_1 for $_{92}U^{238}$ is 5×10^{-14} m. whereas r_0 is nearly 1×10^{-14} m. The value of $(r_1 - r_0)$ gives the thickness of the potential barrier, which the α-particle has to cross. Outside the potential barrier the wave function is again oscillatory and has a small but finite amplitude. In other words, it means that there is a small but finite probability that the α-particle originally within the nucleus may be found outside the nucleus. This unique quantum phenomenon is known as **Tunnel effect.**

15.30 GAMOW'S THEORY OF α-EMISSION

It can be proved that the height of the potential barrier for a U^{238} nucleus is about 26 MeV. According to classical theory, the minimum energy which an α-particle within the U^{238} nucleus, must posses to be able to escape from the nucleus is 26 MeV. But the energies carried by the α-particles emitted by radioactive nuclei are much less than the height of the potential barrier of the respective nuclei. Thus it is very difficult to understand how the α-particle lying within the nucleus can jump over a potential barrier which is more than twice as high as the total energy of the α-particle. However, the escape of an α-particle from a radioactive nucleus can be explained on the basis of quantum theory by the application of wave-mechanics.

For simplicity, we shall consider a one dimensional coulomb potential barrier of rectangular shape of constant height and width.

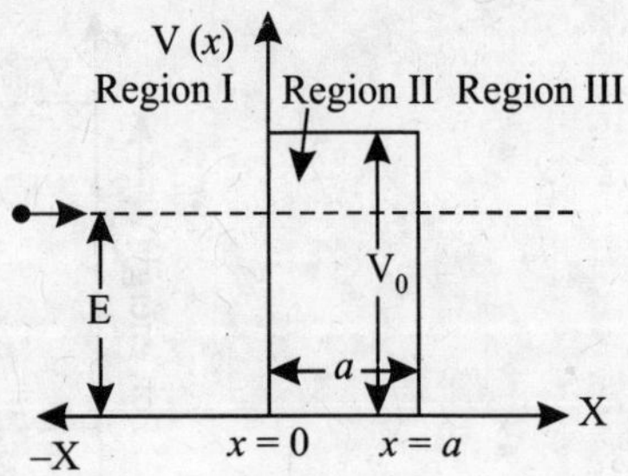

Fig. 15.16

Potential barrier of constant height and width. Let a particle of *reduced* mass *m* be incident on a potential barrier of height V_0 and width *a* as shown in Fig. 15.16. The energy of the particle is *E* and $E < V_0$. On both sides of the barrier $V = 0$ which means that no forces act on the particle in regions I and III. The barrier is thus represented by a potential of the form

$V(x) = 0$ for $x < 0$ Region I

$V(x) = V_0$ for $0 \le x \le a$ Region II

$V(x) = 0$ for $x > a$ Region III

According to Quantum mechanics the motion of the particle is governed by Schrodinger's wave equation for regions I, II and III. As the potential V_0 is independent of time, we have to solve Schrodinger's time independent wave equation to analyse the behaviour of the particle.

It is given by

$$\frac{d^2\psi}{dx^2} + \frac{2m}{\hbar^2}(E - V)\psi = 0 \qquad ...(i)$$

For region I. $(x < 0)$, $V(x) = 0$. Let in this region ψ be denoted by ψ_1.

Substituting $V(x) = 0$ in equation (*i*), we have

$$\frac{d^2\psi_1}{dx^2} + \frac{2m}{\hbar^2}E\psi_1 = 0$$

or $$\frac{d^2\psi_1}{dx^2} + k_1^2\psi_1 = 0 \qquad ...(ii)$$

where $$k_1^2 = \frac{2mE}{\hbar^2}$$

The general solution of equation (*ii*) is given by

$$\psi_1 = Ae^{ik_1x} + Be^{-ik_1x} \qquad ...(iii)$$

where A and B are arbitrary constants.

The first term in equation (*iii*) represents the incident wave ψ_{in} and the second term the reflected wave ψ_{re}

$\therefore$ $$\psi_{in} = Ae^{ik_1x} \qquad ...(iv)$$

and $$\psi_{re} = Be^{-ik_1x} \qquad ...(v)$$

For region II. $(0 \le x \le a)$, $V(x) = V_0$. Let in this region ψ be denoted by ψ_{II}. Substituting the value of $V = V_0$ in equation (*i*), we have

$$\frac{d^2\psi_{II}}{dx^2} + \frac{2m}{\hbar^2}(E - V_0)\,\psi_{II} = 0$$

but $E < V_0$ and hence $(E - V_0)$ is a negative quantity. Putting $(E - V_0) = -(V_0 - E)$, we have

$$\frac{d^2\psi_{II}}{dx^2} - \frac{2m}{\hbar^2}(V_0 - E)\,\psi_{II} = 0$$

or $$\frac{d^2\psi_{II}}{dx^2} - k_2^2\,\psi_{II} = 0 \qquad ...(vi)$$

where $$k_2^2 = \frac{2m}{\hbar^2}(V_0 - E)$$

The general solution of equation (*vi*) is given by

$$\psi_{II} = Ce^{-k_2x} + De^{k_2x} \qquad ...(viii)$$

where C and D are arbitrary constants.

For region III. $(x > a)$, $V(x) = 0$. Let in this region ψ be denoted by ψ_{III}. Substituting $V = 0$ in equation (*i*), we have

$$\frac{d^2\psi_{III}}{dx^2} + \frac{2mE}{\hbar^2}\psi_{III} = 0$$

or $$\frac{d^2\psi_{III}}{dx^2} + k_1^2\,\psi_{III} = 0 \qquad ...(vii)$$

where $$k_1^2 = \frac{2mE}{\hbar^2} \quad \text{(as for region I)}$$

The general solution of equation (*vii*) is given by

$$\psi_{III} = Fe^{ik_1x} + Ge^{-ik_1x} \qquad ...(ix)$$

where F and G are arbitrary constants.

The function Ge^{-ik_1x} represents a wave moving to the left in $-X$ direction. But there is neither a source of particles in region III nor a further potential step to produce a reflected wave. This means, there is no reflected wave and, therefore, $G = 0$.

Substituting $G = 0$ in equation (*ix*), we get

$$\psi_{III} = Fe^{i k_1 x} \quad \text{...}(x)$$

This is the transmitted wave in region III and may be represented as

$$\psi_{tr} = Fe^{i k_1 x} \quad \text{...}(xi)$$

Determination of constants. To evaluate the constant A, B, C and D in terms of F, we apply the boundary conditions which are :

1. The wave function ψ must be continuous at the boundary $x = 0$

$$\therefore \quad \psi_1(0) = \psi_{II}(0)$$

$$\text{or} \quad Ae^{ik_1x_0} + Be^{ik_1x_0} = Ce^{-k_2x_0} + De^{k_2x_0}$$

$$\text{or} \quad A + B = C + D \qquad [\because x_0 = 0] \quad \text{...}(xii)$$

2. The wave function ψ must be continuous at the boundary $x = a$.

$$\psi_{II}(a) = \psi_{III}(a)$$

$$\text{or} \quad Ce^{-k_2a} + De^{k_2a} = Fe^{ik_1a} \quad \text{...}(xiii)$$

3. The first order derivatives of the wave function ψ *i.e.*, $\frac{d\psi}{dx}$ must be continuous at $x = 0$.

$$\therefore \quad \left|\frac{d\psi_I}{dx}\right|_{x=0} = \left|\frac{d\psi_{II}}{dx}\right|_{x=0}$$

$$\text{or} \quad [ik_1Ae^{ik_1x} - ik_1Be^{-ik_1x}]_{x=0} = [-k_2Ce^{-2k_2x} + k_2De^{k_2x}]_{x=0}$$

$$\text{or} \quad ik_1(A - B) = -k_2(C - D)$$

$$\text{or} \quad A - B = \frac{-k_2}{ik_1}(C - D) = \frac{ik_2}{k_1}(C - D) \quad \text{...}(xiv)$$

4. The first order derivatives of the wave function ψ *i.e.*, $\frac{d\psi}{dx}$ must be continuous at $x = a$

$$\left|\frac{d\psi_{II}}{dx}\right|_{x=a} = \left|\frac{d\psi_{III}}{dx}\right|_{x=c}$$

$$\text{or} \quad [-k_2Ce^{-k_2x} + k_2De^{k_2x}]_{x=a} = \left|ik_1Fe^{ik_1x}\right|_{x=a}$$

$$\text{or} \quad -k_2Ce^{-k_2a} + k_2De^{k_2a} = ik_1Fe^{ik_1a}$$

$$\text{or} \quad -Ce^{-k_2a} + De^{k_2a} = i\frac{k_1}{k_2}Fe^{ik_1a} \quad \text{...}(xv)$$

Adding equations (*xiii*) and (*xv*), we have

$$2De^{k_2a} = \left(1 + i\frac{k_1}{k_2}\right)Fe^{ik_1a}$$

$$\text{or} \quad D = \frac{F}{2}e^{ik_1a}\left(1 + \frac{ik_1}{k_2}\right)e^{-k_2a} \quad \text{...}(xvi)$$

Subtracting equation (*xv*) from (*xiii*), we get

$$2Ce^{-k_2a} = \left(1 - \frac{ik_1}{k_2}\right) Fe^{ik_1a}$$

or
$$C = \frac{F}{2}\, e^{ik_1a}\left(1 - \frac{ik_1}{k_2}\right) e^{k_2a} \qquad ...(xvii)$$

Again adding equation (*xvi*) and (*xvii*), we have

$$C + D = \frac{F}{2}\, e^{ik_1a}\left[(e^{k_2a} + e^{-k_2a}) - i\,\frac{k_1}{k_2}\,(e^{k_2a} - e^{-k_2a})\right]$$

$$= Fe^{ik_1a}\left[\left(\frac{e^{k_2a} + e^{-k_2a}}{2}\right) - i\,\frac{k_1}{k_2}\left(\frac{e^{k_2a} - e^{-k_2a}}{2}\right)\right]$$

$$= Fe^{ik_1a}\left[\cosh\,(k_2a) - i\,\frac{k_1}{k_2}\,\sinh\,(k_2a)\right] \qquad ...(xviii)$$

Also subtracting equation (*xvi*) from (*xvii*), we have

$$C - D = \frac{F}{2}\, e^{ik_1a}\left[(e^{k_2a} - e^{-k_2a}) - i\,\frac{k_1}{k_2}\,(e^{k_2a} + e^{-k_2a})\right]$$

$$= Fe^{ik_1a}\left[\sinh\,(k_2a) - i\,\frac{k_1}{k_2}\,\cosh\,(k_2a)\right] \qquad ...(xix)$$

Substituting the value of $(C + D)$ from equation (*xviii*) in (*xii*), we have

$$A + B = Fe^{ik_1a}\left[\cosh\,(k_2a) - i\,\frac{k_1}{k_2}\,\sinh\,(k_2a)\right] \qquad ...(xx)$$

Also substituting the value of $(C - D)$ from equation (*xix*) in (*xiv*), we have

$$A - B = i\,\frac{k_2}{k_1}\left[Fe^{ik_1a}\left\{\sinh\,(k_2a) - i\,\frac{k_1}{k_2}\,\cosh\,(k_2a)\right\}\right]$$

$$= Fe^{ik_1a}\left[\cosh\,(k_2a) + i\,\frac{k_2}{k_1}\,\sinh\,(k_2a)\right] \qquad ...(xxi)$$

Adding (*xx*) and (*xxi*), we get

$$A = Fe^{ik_1a}\left[\cosh\,(k_2a) + \frac{i}{2}\left(\frac{k_2}{k_1} - \frac{k_1}{k_2}\right)\sinh\,(k_2a)\right] \qquad ...(xxii)$$

Subtracting (*xxi*) from (*xx*)

$$B = -\frac{i}{2}\, Fe^{ik_1a}\left[\left(\frac{k_1}{k_2} + \frac{k_2}{k_1}\right)\sinh\,(k_2a)\right] \qquad ...(xxiii)$$

Transmission Co-efficient. The transmission co-efficient T is defined as

$$T = \frac{\text{Transmitted probability density current}}{\text{Incident probability density current}} = \frac{S_{tr}}{S_{in}}$$

Now $\quad S_{tr} = v_{\text{III}}\ \psi_{tr}\ \psi_{tr}{}^{*} = v_1\ \psi_{tr}\ \psi_{tr}{}^{*} \qquad [\because v_{\text{III}} = v_{\text{I}}]$

$\therefore \quad S_{tr} = v_1\ Fe^{ik_1x}\ F^{*} e^{-ik_1x} = v_1\ FF^{*}$

and $\quad S_{in} = v_1\ A_e^{ik_1x}\ A^{*} e^{-ik_1x} = v_1\ AA^{*}$

$$\therefore \quad T = \frac{FF^*}{AA^*} = \frac{1}{\cosh^2 (k_2a) + \frac{1}{4}\left(\frac{k_2}{k_1} - \frac{k_1}{k_2}\right)^2 \sinh^2 (k_2a)}$$

$$= \frac{1}{1 + \sinh^2 (k_2a) + \left(\frac{k_2^2 - k_1^2}{2k_1k_2}\right)^2 \sinh^2 k_2a}$$

$$= \frac{1}{1 + \sinh^2 (k_2a)\left[1 + \left(\frac{k_2^2 - k_1^2}{2k_1k_2}\right)^2\right]} \quad ...(xxiv)$$

$$= \frac{1}{1 + \left(\frac{k_2^2 + k_1^2}{2k_1k_2}\right)^2 \sinh^2 (k_2a)}$$

We know that $\sinh (k_2a) = \frac{e^{k_2a} - e^{-k_2a}}{2} = \frac{e^{-k_2a}}{2}[e^{2k_2a} - 1]$

When $k_2a >> 1$, 1 can be neglected as compared to e^{2k_2a}

and then $\sinh (k_2a) = \frac{e^{-k_2a}.e^{2k_2a}}{2} = \frac{e^{k_2a}}{2}$

and $\sinh^2 (k_2a) = \frac{e^{2k_2a}}{4}$

Substituting this value of $\sinh^2 (k_2a)$ in the expression for T and neglecting 1 as compared to $\sinh^2 (k_2a)$ term, we get

$$T = \frac{1}{\left(\frac{k_2^2 + k_1^2}{2k_1k_2}\right)^2 \frac{e^{2k_2a}}{4}} = 4\left(\frac{2k_1k_2}{k_2^2 + k_2^1}\right)^2 e^{-2k_2a}$$

$$= \left(\frac{4k_1k_2}{k_2^2 + k_1^2}\right)^2 e^{-2k_2a} \quad ...(xxv)$$

The variation in the co-efficient of the exponential term of Eq. (*xxv*) with E and V_0 is negligible compared with the variation in the exponential term itself. Further the co-efficient is never far from unity. Therefore to a very fair approximation, the transmission probability

$$T = e^{-2k_2a} \quad ...(xxvi)$$

Gamow's theory of α-decay. This theory was developed in 1928 by Gamow in collaboration with Gurney and Condon. According to this theory.

1. An α-particle may exist as an entity within the heavy nucleus.

2. The α-particle is in constant motion and is contained in the nucleus by the surrounding potential barrier. It bounces back and forth from the barrier walls. In each collision with the '*wall*' there is a definite probability given by equation (*xxvi*) that the particle will leak through the barrier. In fact, the α-particle within the nucleus must present itself again and again at the barrier surface until conditions are ripe for penetration or leakage. In $_{92}U^{238}$ the α-particle must make 10^{38} tries *i.e.*, 10^{22} tries per second for 10^{16} seconds or 10^9 years before it escapes.

Let ν be the frequency with which the α-particle collides with the walls in order to escape from the nucleus and T the transmission probability in each collision, then

Radioactive constant λ = decay probability per unit time

$$= \nu T$$

Suppose at any time only one α-particle exists as such in the nucleus and that it moves back and forth along the nuclear diameter then

$$\nu = \frac{v}{2r_0}$$

where v is the velocity of the α-particle when it eventually leaves the nucleus and r_0 is the nuclear radius.

Hence $$\lambda = \frac{v}{2r_0} T \qquad ...(xxvii)$$

The value of transmission probability for a barrier of constant height V_0 and width a as given by equation (*xxvi*) is

$$T = e^{-2k_2 a}$$

where $$k_2 = \frac{\sqrt{2m\,(V_0 - E)}}{\hbar}$$

But an α-particle within the nucleus has to overcome a *barrier of varying height* as shown in Fig. 15.14.

As the potential is not constant in the region $x = r_0$ to $x = r$, we can divide it in a series of small steps each of thickness dx, then the probability of transmission through a small thickness dx is given by

$$T = e^{-2k_2(x)\,dx}$$

The total probability is the product of individual probabilities of all the small steps from $x = r_0$ to $x = r_1$

Hence $$T = e^{-2\int_{r_0}^{r_1} k_2(x)\,dx}$$

or $$\log_e T = -2\int_{r_0}^{r_1} k_2(x)\,dx$$

r_1 is the distance from the centre of the nucleus where $V_x = E$. The kinetic energy E is greater than the potential energy V_x for values of $x > r_1$. So, if the α-particle can get past r_1, it will permanently escape from the nucleus.

Now $$V_x = \frac{2Ze^2}{4\pi\varepsilon_0 x}$$ where Z is the atomic number of the daughter nucleus.

$$\therefore \quad k_2(x) = \frac{\sqrt{2m\,(V_x - E)}}{\hbar} = \left(\frac{2m}{\hbar^2}\right)^{1/2} \left[\frac{2Ze^2}{4\pi\varepsilon_0 x} - E\right]^{1/2} \qquad ...(xxviii)$$

As $V_x = E$ when $x = r_1$

$$E = \frac{2Ze^2}{4\pi\varepsilon_0 r_1} \qquad ...(xxix)$$

$$\therefore \quad \frac{2Ze^2}{4\pi\varepsilon_0 x} = \frac{2Ze^2}{4\pi\varepsilon_0 r_1}\,\frac{r_1}{x} = E\,\frac{r_1}{x}$$

Substituting in equation (*xxviii*), we get

$$k_2(x) = \left(\frac{2mE}{\hbar^2}\right)^{1/2}\left(\frac{r_1}{x}-1\right)^{1/2}$$

and
$$\log_e T = -2\int_{r_0}^{r_1} k_2(x)\,dx = -2\left(\frac{2mE}{\hbar^2}\right)^{1/2}\int_{r_0}^{r_1}\left(\frac{r_1}{x}-1\right)^{1/2} dx \quad \text{...(xxx)}$$

To evaluate the integral put $\dfrac{x}{r_1} = \cos^2\phi$

or
$$x = r_1\cos^2\phi$$

$\therefore$
$$dx = -2r_1\cos\phi\sin\phi\,d\phi$$

Hence
$$\left(\frac{r_1}{x}-1\right)^{1/2} dx = -\left(\frac{1}{\cos^2\phi}-1\right)^{1/2} 2r_1\cos\phi\sin\phi\,d\phi$$

$$= -2r_1\sin^2\phi\,d\phi = -r_1(1-\cos 2\phi)\,d\phi$$

$\therefore$
$$\int_{r_0}^{r_1}\left(\frac{r_1}{x}-1\right)^{1/2} dx = -r_1\int_{r_0}^{r_1}(1-\cos 2\phi)\,d\phi$$

$$= -r_1\left[\phi - \frac{\sin 2\phi}{2}\right]_{r_0}^{r_1} = -r_1\left[\phi - \sin\phi\cos\phi\right]_{r_0}^{r_1}$$

$$= r_1\left[\cos^{-1}\left(\frac{r_0}{r_1}\right)^{1/2} - \left(\frac{r_0}{r_1}\right)^{1/2}\left(1-\frac{r_0}{r_1}\right)^{1/2}\right]$$

$\therefore$
$$\log_e T = -2\left(\frac{2mE}{\hbar^2}\right)^{1/2} r_1\left[\cos^{-1}\left(\frac{r_0}{r_1}\right)^{1/2} - \left(\frac{r_0}{r_1}\right)^{1/2}\left(1-\frac{r_0}{r_1}\right)^{1/2}\right] \quad \text{...(xxxi)}$$

since the potential barrier is relatively wide $r_1 >> r_0$

$\therefore$
$$\cos^{-1}\left(\frac{r_0}{r_1}\right)^{1/2} = \frac{\pi}{2} - \left(\frac{r_0}{r_1}\right)^{1/2}$$

and
$$\left(1-\frac{r_0}{r_1}\right)^{1/2} = 1$$

Substituting in equation (*xxxi*), we get

$$\log_e T = -2\left(\frac{2mE}{\hbar^2}\right)^{1/2} r_1\left[\frac{\pi}{2} - 2\left(\frac{r_0}{r_1}\right)^{1/2}\right]$$

Hence
$$\log_e T = -2\left(\frac{2mE}{\hbar^2}\right)^{1/2}\left[r_1\frac{\pi}{2} - 2(r_0 r_1)^{1/2}\right]$$

$\therefore$ From equation (*xxix*)
$$r_1 = \frac{2Ze^2}{4\pi\varepsilon_0 E}$$

$$\therefore \quad \log_e T = -2\left(\frac{2mE}{\hbar^2}\right)^{1/2}\left[\frac{Ze^2}{4\varepsilon_0 E} - 2\left(\frac{r_0 Ze^2}{2\pi\varepsilon_0 E}\right)^{1/2}\right]$$

$$= \frac{4e}{\hbar}\left(\frac{m}{\pi\varepsilon_0}\right)^{1/2} Z^{1/2} r_0^{1/2} - \frac{e^2}{\hbar\varepsilon_0}\left(\frac{m}{2}\right)^{1/2} ZE^{-1/2}$$

Now $\frac{4e}{\hbar}\left(\frac{m}{\pi\varepsilon_0}\right)^{1/2} = 2.97$ and $\frac{e^2}{\hbar\,\varepsilon_0}\left(\frac{m}{2}\right)^{1/2} = 3.95$

$$\therefore \quad \log_e T = 2.97\, Z^{1/2} r_0^{1/2} - 3.95\, ZE^{-1/2} \quad \text{... (xxxii)}$$

where r_0 is in fermi (10^{-15} m) and E is in MeV. The radioactive constant

$$\lambda = \frac{v}{2r_0} T$$

$$\therefore \quad \log_e \lambda = \log_e\left(\frac{v}{2r_0}\right) + \log_e T$$

$$= \log_e\left(\frac{v}{2r_0}\right) + 2.97\, Z^{1/2} r_0^{1/2} - 3.95\, ZE^{-1/2}$$

or

$$\log_{10}\lambda = \log_{10}\left(\frac{v}{2r_0}\right) + 1.29\, Z^{1/2} r_0^{1/2} - 1.72\, ZE^{-1/2} \quad \text{...(xxxiii)}$$

Verification of results. (*i*) A graph between $\log_{10}\lambda$ and $ZE^{-1/2}$ for various α-active nuclei is shown in Fig. 15.17. The slope of straight line = – 1.72 as calculated theoretically.

(*ii*) The intercept on the *Y*-axis gives the value of $\log_{10}\left(\frac{v}{2r_0}\right) + 1.29\, Z_{1/2}\, r_0^{1/2}$.

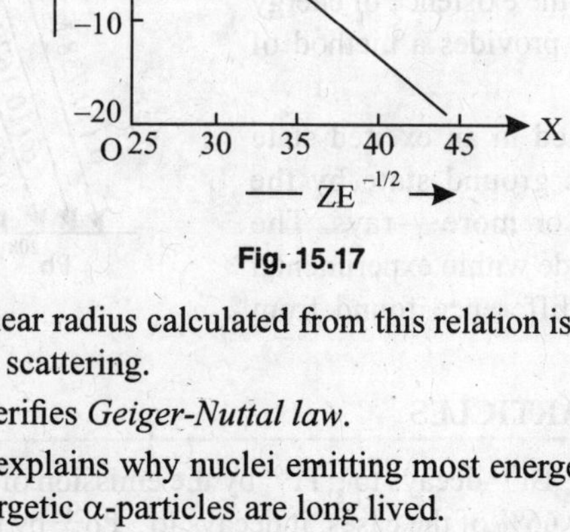

Fig. 15.17

The value of r_0 the nuclear radius calculated from this relation is of the same order as obtained from other experiments like scattering.

(*iii*) Relation (*xxxiii*) verifies *Geiger-Nuttal law*.

(*iv*) This relation also explains why nuclei emitting most energetic α-particles are short lived and those emitting least energetic α-particles are long lived.

15.31 FINE STRUCTURE IN α-RAY SPECTRUM

Accurate energy measurements of the α-particle emitted by radioactive substances indicate that a number of α-emitters show a group of discrete α-particle energies rather than a single energy. This is known as the *fine structure in α-ray energy spectrum*.

As an example consider the decay of ${}_{83}Bi^{212}$ by α-emission to ${}_{81}Tl^{208}$. The most energetic α-ray corresponds to a disintegration energy $Q = 6.201$ MeV. In addition to this, there are α-rays corresponding to disintegration energies of 6.161, 5.873, 5.728, 5.708 and 5.583 MeV giving rise to a fine structure.

The fine structure in the α-ray spectrum can easily be explained in terms of nuclear energy levels. The nuclear energy level diagram for ${}_{83}Bi^{212}$ decaying to ${}_{81}Tl^{208}$ is shown in Fig. 15.18.

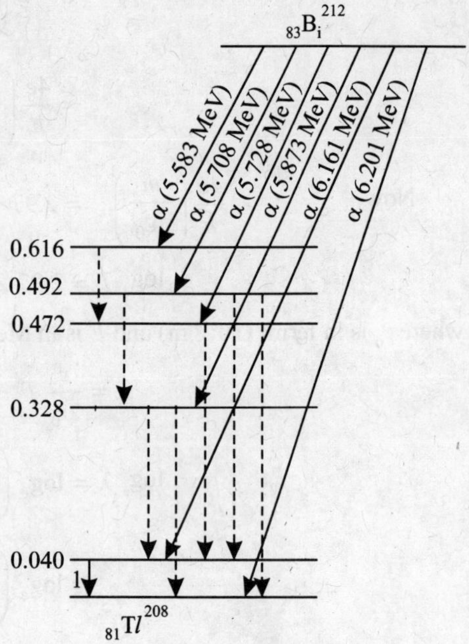

Fig. 15.18

According to the explanation, when a parent element decays by α-emission, the daughter element may go over to any one of the number of allowed energy states. The state corresponding to the lowest energy is known as the *ground state* and those corresponding to higher energies are called *excited states*. The most energetic α-particle corresponds to decay to the *ground state* of the daughter nucleus and other α-particles correspond to decay to the excited state of the daughter nucleus. The difference between pairs of disintegration energies gives directly the energy difference between various levels of the daughter nucleus. Thus, the first excited level of ${}_{81}T_l^{208}$ has an energy

$$6.201 - 6.161 = 0.040 \text{ MeV} = 40 \text{ KeV}$$

and the second excited level has an energy

$$6.201 - 5.873 = 0.328 \text{ MeV} = 328 \text{ KeV}$$

and so on. Thus, the fine structure in the α-ray spectrum not only gives an indication of the existence of energy states in the nucleus but also provides a method of measuring their energies.

Fig. 15.19

A daughter nucleus formed in an excited state sooner or later passes to its ground state by the successive emission of one or more γ–rays. The energies of these γ–rays coincide within experimental error with the energy level difference found from α-particle fine structure.

15.32 LONG RANGE α-PARTICLES

As stated in article 15.31, ${}_{83}Bi^{212}$ decays to ${}_{81}Tl^{208}$ by the emission of an α-particle. This happens in 34% cases. In the remaining 66% of the cases, it decays to ${}_{84}Po^{212}$ by the emission of a β-particle. In has been observed that ${}_{84}Po^{212}$ emits α-particles a few of which have energies much greater than that of the main group.

These are known as **long range α-particles**. For example, the the main group of α-particles from $_{84}Po^{212}$ corresponds to a disintegration energy of 8.945 MeV whereas the long range α-particles have energies of 9.671 MeV, 10.167 MeV and 10.738 MeV. Let us see why these few α-particles have higher energies. The explanation is again based on energy levels. When $_{83}Bi^{212}$ decays by the emission of a β-particle, $_{84}Po^{212}$ which is formed, may exist in the ground state or in one of the excited states. As $_{84}Po^{212}$ is very short lived in the ground state, the excited states are even more short lived. Normally these excited nuclei pass to the lower level within 10^{-12} sec. or so by γ-emission. But this γ-emission is a relatively slow process on a nuclear time scale and sometimes the excited nucleus emits an α-particle in less than 10^{-12} sec. *i.e.*, before the nucleus can de-excite itself by γ-emission. This α-particle carries more energy than the normal *i.e.*, the energy from the ground state of the parent to the ground state of the daughter plus the energy from the excited state of the parent to the ground state of the parent. Thus, the study of the long range α-particles gives us information about the energy levels of the parent α-emitter itself, whereas the study of fine structure in α-ray spectrum gives us information about the energy levels of daughter nucleus.

15.33 RADIOACTIVE EMISSION OF β-RAYS: β-DECAY

The β-particles emitted in radioactive decay of heavy nuclei are ordinary electrons having same charge and mass values. However, the expression β-decay now covers all such phenomenon in which a nucleus makes an isobaric transition due to natural or induced radioactivity by the emission of an electron (β^-) or a positron (β^+) or electron capture *i.e.*, capture of an orbital electron. An electron or a positron does not exist in the nucleus. An electron is produced by the spontaneous conversion of a neutron into a proton whereas a positron is produced due to the spontaneous conversion of a proton into a neutron. Thus in β-decay the mass number remains the same but the charge number changes by one *i.e.*, from Z to $(Z + 1)$ in the case of β^- emission and form Z to $(Z - 1)$ in the case of β^+ emission or electron capture. As the mass of the electron is nearly 1/7000 times the mass of an α-particle, the energy of the β-particle is very small, usually less than 4 MeV though their velocities are as high as 0.9995 c.

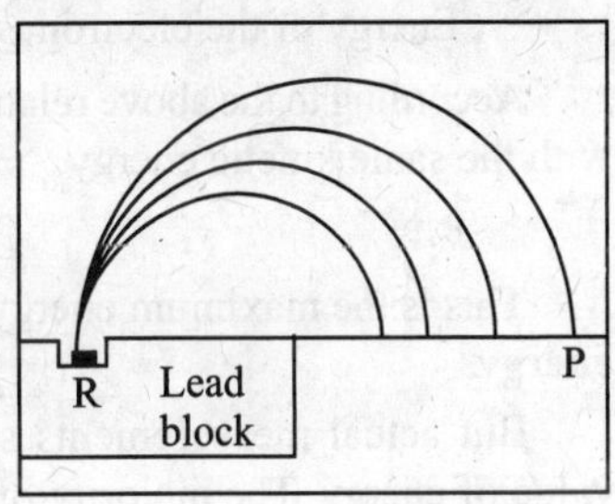

Fig. 15.20

β-ray energy spectrum. The energy spectrum of β-particles has been studied with the help of a magnetic spectrometer shown in Fig. 15.20. The β-particle emitted by the radioactive source R placed in a cavity in a block of lead inside a highly evacuated box are deflected into a circular path by a magnetic field perpendicular to the plane of the paper in accordance with the equation.

$$Bev = \frac{mv^2}{R}$$

where B is the intensity of the magnetic field, e the charge, m the mass, v the velocity of the β-particle and R the radius of the path. Thus particles of different energies (or velocities) are focussed at different points. These are recorded on a photographic plate P and from the photographic impression, the relative number of β-particles and their energies can be determined. A graph is plotted between relative number of β-particles and energy in MeV as shown in Fig. 15.21.

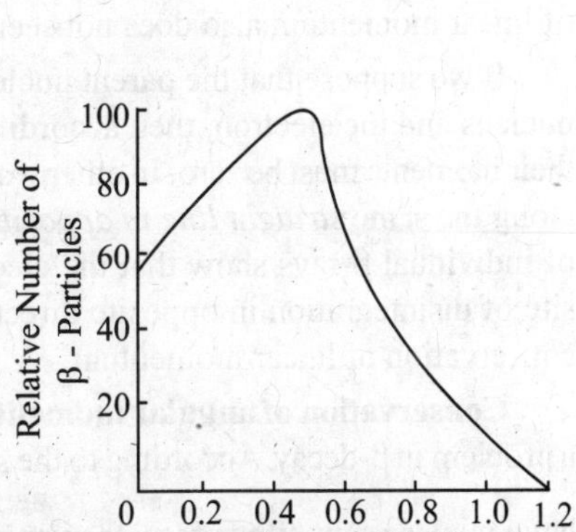

Fig. 15.21

End point energy. The β-spectrum is continuous, having energies ranging from zero to a certain well defined limit known as *End point energy*, which is characteristic of the β-emitter. The energy distribution has a well defined maximum the height and position of which also depends upon the nature of the radioactive substance emitting the β-particles.

Thus end point energy is the maximum energy with which a β-particle is emitted from a radioactive nuclide.

β-spectrum discrete or continuous. We have stated above that β-spectrum is continuous having energies ranging from zero to a certain well defined maximum limit known as *end point energy*.

The β-ray energy spectrum produced by naturally radioactive elements as source is also found to exhibit a series of lines superimposed on the continuous spectrum as background. These sharp lines of the β-spectrum with discrete energies are due to the electrons that have been ejected from *K, L, M* and *N* shells of the atom due to the phenomenon of *internal conversion*.

Laws of conservation of energy and momentum not observed. The explanation of the existence of the *continuous energy spectrum* for the β-particles with *end point energy*, however, presents some difficulties and seems to *contradict the laws of conservation of energy and momentum*.

Law of conservation of energy. When a nucleus emits β-particle a neutron in the nucleus changes to a proton, thereby increasing the atomic number by unity, keeping the mass number the same. If M_p is the mass of the **neutral parent** atoms of atomic number Z and M_d the mass of **neutral daughter atom** of atomic number $(Z + 1)$ and m_e the mass and e the charge on the β-particles, then according to the principle of conservation of mass energy.

Rest mass of the parent nucleus = rest mass of daughter nucleus + rest mass of electron + energy of the electron

or $$(M_p - Z m_e) c^2 = [M_d - (Z + 1) m_e] c^2 + m_e c^2 + Q$$

or $$M_p c^2 = M_d c^2 + Q$$

$$\therefore \text{ Energy of the electron } Q = (M_p - M_d) c^2 \qquad ...(i)$$

According to the above relation all β-particles from a given radioactive substance must be emitted with the same kinetic energy

$$Q = (M_p - M_d) c^2$$

This is the maximum energy with which a β-particle can be emitted and is known as **end point energy.**

But actual measurements show that only a few β-particles are emitted with this maximum value of energy. The majority of β-particle are emitted with smaller energies. What happens to the remaining energy ?

Conservation of linear momentum. In addition to energy conservation, the law of conservation of linear momentum also does not seem to hold good for single particle β^+ or β^- emission.

If we suppose that the parent nucleus is at rest and it decays into just *two* particles, the daughter nucleus and the electron, then according to the principle of conservation of momentum, the sum of their momenta must be zero. In other words, the daughter nucleus and the emitted β-particle must move along the same *straight line in opposite* directions with equal momenta. Experimental observations of individual β-rays show that the electrons and the daughter nucleus do not necessarily leave the site of disintegration in opposite directions. Thus, there is an apparent violation of the principle of conservation of linear momentum.

Conservation of angular momentum. The conservation of angular momentum (spin) also poses a problem in β-decay. According to the shell model of the nucleus, nuclides having even mass numbers have nuclear spin either zero or even multiple of $\frac{1}{2} \cdot \frac{h}{2\pi}$ *i.e.*, these nuclei have integral spin. The nuclides having odd mass numbers have a nuclear spin which is an odd multiple of $\frac{1}{2} \cdot \frac{h}{2\pi}$ *i.e.*, these nuclei have half integral spin. As the number of nucleons in a β-decay remains the same, the mass number remains unchanged. The electron has a spin equal to $\frac{1}{2} \cdot \frac{h}{2\pi}$. How is it possible for a nucleus of even mass number and therefore, an integral spin, to give rise to a daughter nucleus of the same mass number and also an integral spin and yet emit an electron of spin $\frac{1}{2} \cdot \frac{h}{2\pi}$? The same is the difficulty in the emission of an electron from a nuclide of odd mass number. Thus there in an apparent violation of the principle of conservation of angular momentum.

Neutrino hypothesis. All these difficulties have been overcome by supposing the existence of another particle called the **neutrino** and its anti-particle the **anti-neutrino** to be emitted simultaneously along with the β-particle. Its existence was first predicated by **Pauli** on theoretical grounds in 1930 and has been confirmed experimentally in 1956. The neutrino as well as the anti-neutrino has zero rest mass (massless) carries no charge (chargeless) and has an angular momentum or spin

$$= \frac{1}{2} \frac{h}{2\pi} = \frac{1}{2} \hbar$$

Since the neutrino (or anti-neutrino) has zero rest mass, it always travels with the speed of light. Hence its linear momentum

$$p = \frac{E}{c}$$

where E is the total relativistic energy of the neutrino (or anti-neutrino) and c the velocity of light.

The *neutrino* (ν) is defined as a *particle* with its spin vector parallel to its momentum vector and the *anti-neutrino* ($\bar{\nu}$) as a particle with spin opposite in direction or *anti-parallel* to its momentum [Fig. 15.22].

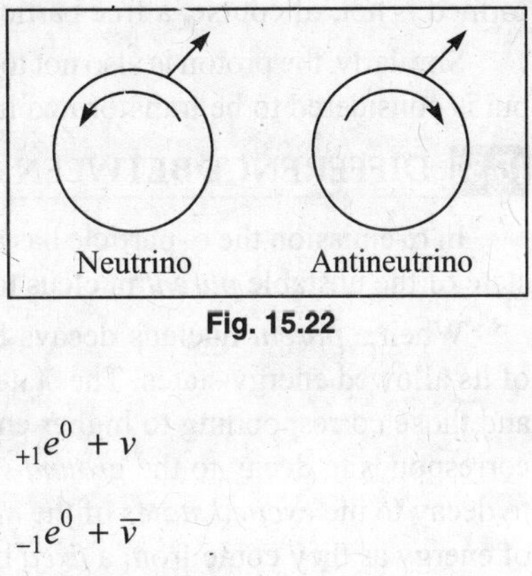

Fig. 15.22

Neutrino theory of β-decay. In accordance with this theory the β-decay of radioactive nuclei involves the following processes

for β^+ emission $\quad {}_1p^1 \longrightarrow {}_0n^1 + {}_{+1}e^0 + \nu$

and for β^- emission $\quad {}_0n^1 \longrightarrow {}_1p^1 + {}_{-1}e^0 + \bar{\nu}$

with spontaneous conversion of a proton into a neutron and the emission of both the positron and neutrino or conversion of a neutron into a proton and emission of both the electron and anti-neutrino. Thus in the above equation ${}_1p^1$ represents the proton, ${}_0n^1$ the neutron, ${}_{+1}e^0$ the positron, ${}_{-1}e^0$ the electron, ν the neutrino and $\bar{\nu}$ the anti-neutrino.

Explanation of continuous β-ray spectrum and conservation of energy. Let us now consider the principle of conservation of energy on the assumption that an electron as well as an anti-neutrino are created in β^- decay process. The total disintegration energy Q is now carried by the electron and the anti-neutrino. The electron will carry the maximum energy when the energy of the anti-neutrino is zero. In all other cases, the electron will carry an energy less than the maximum. The sum of the energies carried by the electron and the anti-neutrino will always be the same and equal to Q. The energy may be shared by the two particles in any proportion. *Hence it explains the continuous β-ray spectrum.*

Conservation of linear momentum. Considering the conservation of linear momentum we find that there are now three particles and not two. The vector sum of the linear momenta of these particles must be zero. In other words, the linear momentum of the daughter nucleus, the electron and the anti-neutrino must be capable of representation by the sides of a triangle taken in order. It is **not** necessary now that the electron and the daughter nucleus should leave the site of decay in opposite directions along the same straight line, because there are a variety of ways in which separate momentum vectors can be arranged to give zero resultant.

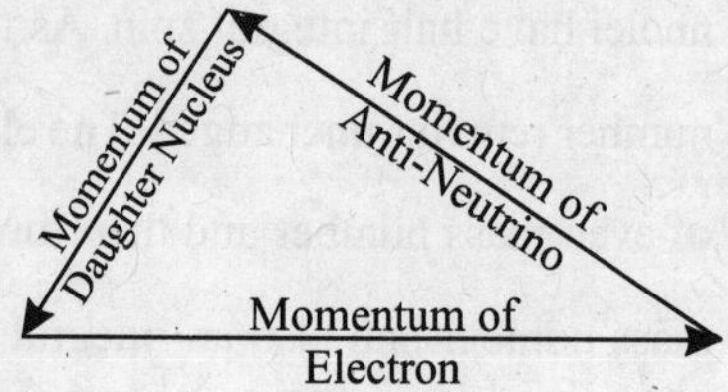

Fig. 15.23

Conservation of angular momentum. The angular momentum or the spin of the anti-neutrino is $\frac{1}{2}\frac{h}{2\pi}$, therefore, when an electron and an anti-neutrino is ejected the total loss of angular momentum is $\frac{h}{2\pi}$. The resultant angular momentum of all particles left after the decay, product nucleus, the electron and the anti-neutrino, is then integral or half-integral according to whether that of the parent nucleus is integral or half integral and thus the total angular momentum can be conserved. This is actually the case with β-emission.

In the same manner it can be shown that the laws of conservation of energy, linear and angular momentum hold good in the case of emission of β^+ particle and the neutrino.

The neutron is not to be regarded as composed of a proton, an electron and an anti-neutrino but is considered to be transformed into these three particles at the instant of β-emission. The proton thus formed is not, of course, a free particle but is bound in the nucleus by the nuclear forces.

Similarly, the proton is also not to be regarded as composed of a neutron, a positron and a neutrino but is considered to be transformed into these three particles at the instant of β^+ emission.

15.34 DIFFERENCE BETWEEN α-SPECTRA AND β-SPECTRA

In α-emission the α-particle has a *discrete* energy corresponding to a well defined higher energy state of the unstable *parent* nucleus to a well defined lower energy state of the *daughter* nucleus.

When a *parent* nucleus decays by α-emission the *daughter* element may go over to any one of its allowed energy states. The state corresponding to the lowest energy is known as *ground state* and those corresponding to higher energies are called *excited states*. The most energetic α-particle corresponds to decay to the *ground state* of the *daughter* nucleus and other α-particles correspond to decay to the *excited states* of the *daughter* nucleus. But all these α-particles have *discrete* values of energy as they come from a *fixed* higher energy state to a *fixed* lower energy state.

On the other hand β-spectrum is *continuous* having energies ranging from zero to a certain well defined maximum limit known as *end point energy*. In β-emission, an electron and an anti–neutrino are created in the β-decay process by the conversion of a neutron into a proton according to the reaction

$$_0n^1 \longrightarrow {}_1p^1 + {}_{-1}e^0 + \overline{\nu}$$

The total energy created in the process is given by

$$Q = (M_p - M_d)\, c^2$$

where M_p is the mass of the *parent* nucleus and M_d that of the *daughter* nucleus. The total disintegration energy Q is carried by the electron and the antineutrino. The electron will carry the maximum energy when the energy of the anti-neutrino is zero. In all other cases the electron will carry an energy less than the maximum. The sum of the energies carried by the electron and the anti–neutrino will always be equal to Q but this energy may be shared by the two particles in any proportion. *This is why the β-ray spectra is continuous.*

15.35 EVIDENCE FOR EXISTENCE OF NEUTRINOS

The inverse β-decay process provides a method of establishing the actual existence of neturinos. It was experimentally proved by Reines and Cowan who made use of the heavy flux of anti-neutrinos produced by β-decay in a nuclear reactor. As soon as an anti-neutrino is absorbed by a proton, a neutron and a positron are produced according to the relation.

$$_1p^1 + \bar{\nu} = {_0n^1} + {_{+1}e^0} - 1.8 \text{ MeV} \quad ...(i)$$

As the reaction is *endothermic* only anti-neutrinos with an energy not less than 1.8 MeV can make it proceed. The reaction cross-section is extremely small being of the order of 10^{-20} **barn**. This is why large proton densities of the order of 10^{28} protons per m^3 have to be employed to get about 10 events per hour. This was achieved by using two large tanks B_1 and B_2 each containing 40 kg of cadmium chloride dissolved in 200 litres of water.

Three γ-ray scintillation detectors A_1, A_2 and A_3 surround the tanks B_1 and B_2 as shown in Fig. 15.24.

As the beam of anti-neutrinos from the reactor is made to bombard the hydrogenous material in the tanks a reaction of the type given in equation (*i*) may occur. A typical event is shown in Fig. 15.24. The emitted positron (β^+) immediately encounters an electron and the pair is annihilated producing two 0.51 MeV (γ-ray) photons in 10^{-9} s. In the mean time the newly created neutron ($_0n^1$) travels through the solution, undergoes elastic collisions with the protons and loses energy. After a few micro-seconds when the neutron is slowed down to thermal energy level it is captured by cadmium in an (*n*γ) reaction as

$$Cd^{113} + {_0n^1} = Cd^{114} + \gamma \text{ (8 MeV)}$$

The heavier cadmium nucleus releases about 8 MeV of excitation energy in the form of 3/4 photons. The *gamma ray scintillation detectors* surrounding the tanks first detect the 0.51 MeV photons from electron-positron annihilation (prompt pulse) and a few micro-seconds later these detectors capture the photons from the de-excitation of cadmium (*delayed pulse*). Thus the arrival of photons at the detector in the above order is a proof of reaction (*i*) having taken place.

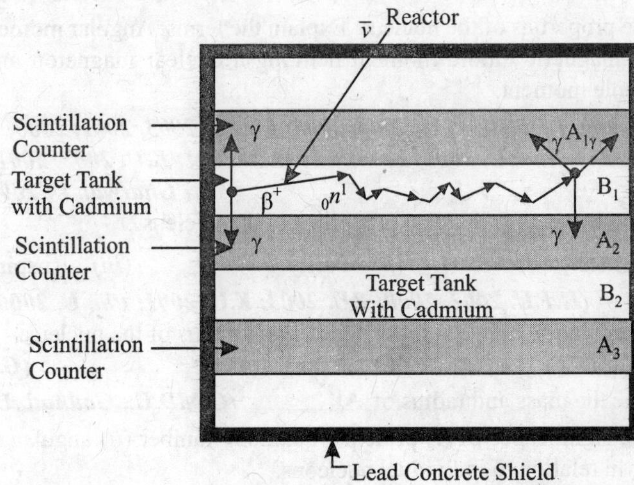

Fig. 15.24

On shutting off the reactor *i.e.*, stopping the neutrino flux a change in delayed pulse detection was noted. The experiment was repeated with alternately 'on' and 'off' position and the difference in the counting rate was considered to represent the anti-neutrino absorption effect. This experiment is by far the best evidence for the existence of anti-neutrino and therefore the neutrino.

Why the existence of neutrino was postulated. The existence of neutrino was postulated to explain the contradictions in the laws of conservation of energy and momentum in the continuous energy spectrum of the β-particles and their end point energy.

15.36 SIGNIFICANCE OF ANTINEUTRINO IN β-DECAY

A nucleus is positively charged and it contains Z protons and $A - Z$ neutrons, where Z is the atomic number and A the mass number of the nucleus. Electrons do not exist within the nucleus. The emission of a β-particle is caused by the spontaneous conversion of a neutron into a proton according to the reaction

$$_0n^1 \longrightarrow {}_1p^1 + {}_{-1}e^0 + \bar{\nu}$$

where $\bar{\nu}$ is an antineutrino. The emission of an antineutrino is necessary in accordance with the law of conservation of energy and linear as well as angular momentum.

Similarly, the emission of a positron β^+ $({}_{+1}e^0)$ is caused by the spontaneous conversion of a proton into a neutron according to the reaction ${}_1p^1 = {}_0n^1 + {}_{+1}e^0 + \nu$ where ν is the neutrino.

This gives an explaination of the emission of β-rays i.e. β-decay even though the electrons do not exist in the nucleus.

EXERCISE

1. (*a*) Qualitatively explain the important constituents and facts which lead to the development of structure of nucleus. ***(G.N.D.U. 2001)***

   (*b*) Discuss proton-electron theory of nuclear composition. Why was it thought? Give at least five causes for the failure of proton-electron hypothesis. ***(P.U. 2001; H.P.U. 2001)***

2. (*a*) What is proton-neutron hypothesis ? Give reasons for acceptance of this hypothesis for the constitution of the nucleus. How does this hypothesis avoid the failures of proton-electron hypothesis? ***(P.U. 2003; K.U. 2002)***

   (*b*) Give any two evidences of neutron-proton model of the nucleus. ***(P.U. 2004)***

3. (*a*) Give the properties of the nucleus. Explain the terms: Angular momentum of the nucleus, nuclear magnetic dipole moment defining a nuclear magneton and nuclear electrical quadrupole moment.

   (Bang. U. 2004; P. U. 2004, 2003; Pbi. U. 2003, 2001, 2000; M.D.U. 2002, 2000; It. P.U. 2002, 2000; G.N.D.U. 2001; K.U. 2002, 2001; Gauhati. U. 2000; Gharwal. U. 2000; Calicut. U. 2002)

   (*b*) Explain the wave mechanical properties of the nucleus

   (*i*) Statistics (*ii*) Parity (*iii*) Isospin

   (H.P.U. 2002, 2000; P.U. 2003; K.U. 2001; Pbi. U. 2000; Gauhati. U. 2000)

4. (*a*) Write the relation between mass number and radius of the nucleus. ***(G.N.D.U. 2001, 2000)***

   (*b*) Calculate the mass and radius of $\mathbf{Al}_{13}^{27}$. ***(G.N.D.U., Gauhati. U. 2000, 2003, 2000)***

5. (*a*) Give the significance of (*i*) principal quantum number (*ii*) angular momentum quantum number in relation to individual nucleons. ***(P.U. 2002)***

   (*b*) What do you understand by nuclear spin ? ***(G.N.D.U. 2004, 2000)***

6. Discuss one method that determines the size of the nucleus. ***(G.N.D.U. 2004; 2003)***

7. Using classical picture of an atom, explain the origin of atomic magnetism. What is a Bohr magneton ? ***(Gauhati.U. 2000)***

8. Explain the concept of nuclear magnetic dipole moment. What is a nuclear magneton ? ***(G.N.D.U. 2004, 2003, 2001; H.P.U. 2002, 2000; M.D.U. 2002, 2000; Gharwal.U. 2000)***

9. Compare the atomic magnetic moment and nuclear magnetic moment. ***(P.U. 2002)***

10. (*a*) Explain the meaning of electric quadrupole moment of a nucleus.

(P.U. 2004, 2003, 2001; Pbi. U. 2003, 2001; H.P.U. 2002, 2000)

(*b*) Derive an expression for the electric quadrupole moment of a nucleus. Show that for a nucleus having symmetrically situated protons, it is zero. ***(Gharwal. U. 2002)***

(*c*) What is the importance (significance) of quadrupole moment in nuclear physics ?

(Pbi. U. 2004, K.U. 2000)

11. (*a*) What do you understand by nuclear spin ? ***(G.N.D.U. 2001, 2000)***

(*b*) What statistics is followed by a nucleus ? Explain the connection between statistics and spin of a nucleus. ***(Pbi.U. 2004; G.N.D.U. 2003; M.D.U. 2001)***

(*c*) What is parity ? What do you mean by even and odd parity ?

(Pbi.U. 2001, G.N.D.U. 2001; M.D.U. 2001)

12. Explain the term atomic mass unit, compute the energy of 1 a.m.u. in MeV.

(Pbi. 2003, H.P.U. 2002; P.U. 2001; Gauhati U. 2000)

13. Explain the terms, mass defect, packing fraction, binding energy of a nucleus and binding energy per nucleon. Sketch and explain the variation of packing fraction with mass number. What is the difference between mass defect and packing fraction ?

(Bang. U. 2004; Pbi. U. 2004, 2004; Cal.U. 2003; Meerut.U. 2003, 2002; G.N.D.U. 2000; P.U. 2004, 2002, 2000; Calicut.U. 2002; Luck.U. 2001; K.U. 2000; Gauhati.U. 2000)

14. (*a*) Explain the term, binding energy per nucleon and separation energy.

(*b*) Discuss graphically the variation of average binding energy per nucleon with mass number and stability of the nucleus. Using the binding energy curve, explain the release of energy in fusion of light nuclei and fission of heavy nuclei.

(P.U. 2004, 2002, 2001, 2000; Pbi.U. 2003, 2001; G.N.D.U. 2001, 2001; Calicut.U. 2003, 2002, 2001; Meerut U. 2002, 2000; K.U. 2000; Gharwal.U. 2000; H.P.U.)

15. Which is more, atomic binding energy or nuclear binding energy ? ***(P.U. 2001)***

16. (*a*) Why we get maxima and minima peaks in B.E curve ? ***(H.P.U. 2002)***

(*b*) Why is binding energy curve so significant ? ***(Pbi. 2001)***

(*c*) What will happen if B/A versus A curve has only positive slope ? Here B refers to binding energy and A to mass number. ***(G.N.D.U. 2000)***

[**Hint : If *B*/*A* versus *A* curve has only positive slope.** If the *B*/*A* (binding energy per nucleon) versus *A* (mass number) curve has only a positive slope it would mean that binding energy per nucleon goes on increasing with mass number. In such a case energy would be released only by *fusion i.e.* combining of lighter nuclei into heavier one and not by fission *i.e.* breking of heavier nuclei into lighter ones.]

17. Prove from wave mechanical, angular momentum, statistical and other considerations that electrons cannot exist in the nucleus.

(P.U. 2004, 2003, 2002, 2001; G.N.D.U. 2003, 2002; Pbi.U. 2002, 2001, 2000; H.P.U. 2002, 2001; Gauhati.U. 2000; K.U. 2000)

18. Calculate the expression for the Deutron binding energy per particle. ***(Merut U. 2002)***

19. What are the constituents of α-particle ? ***(Meerut U. 2002)***

20. Explain why the sum of the masses of 2 neutrons and 2 protons is not equal to the mass of the α-particle.

21. Calculate the energy released when 3 α-particles combine to form a ^{12}C nucleus. Atomic mass of $_2He^4 = 4.002603$ a.m.u. **(Ans. 7.274 MeV) (Calicut U. 2002)**

22. Calculate the pocking fraction and average binding energy per nucleon for $_8O^{16}$ of nuclear mass 15.994916 a.m.u. (Give m_p = 1.007825 a.m.u; m_n = 1.008665 a.m.u.)

(Ans. 8.56275 × 10^{-3} u; 7.976 MeV) (P.U. 2000)

23. Mass of deutron nucleus is 2.014103 a.m.u. If the masses of proton and neutron are 1.007825 a.m.u. and 1.008665 a.m.u. respectively, calculate the mass detect and packing fraction.

(Ans. Δm = 0.002387 a.m.u.; f = 0.0011935 a.m.u) ***(H.P.U. 2003)***

24. (*a*) Explain the nuclear forces responsible for holding the nucleons together. Discuss the nature of these forces giving a qualitative account.

(P.U. 2004, 2001, 2000; G.N.D.U. 2004, 2003, 2000; Meerut.U. 2002; Pbi.U. 2000; Gauhati.U. 2000)

(*b*) What do you mean by charge independence of nuclear forces ?

25. Discuss the meson theory of nuclear forces. Nuclear forces are exchange forces, substantiate.

(Pbi.U. 2004, 2001; Meerut.U. 2002; Calicut.U. 2002)

26. (*a*) Compare the order of strength of nuclear forces with that of electrostatic and gravitational forces. ***(P.U. 2000)***

(*b*) How do the nuclear forces differ from gravitational and electrostatic forces ?

(G.N.D.U. 2003, 2001)

(*c*) Why do the protons in the nucleus not fly apart ? ***(Meerut. U. 2002)***

27. (*a*) Write a short note on nuclear stability with reference to neutron-proton ratio. Why *N* tends to exceed *Z* in stable nuclei ? Why stable nuclei have more neutrons than protons ?

(P.U. 2000; Calicut.U. 2002)

(*b*) Fill in the blanks

A stable nucleus has ... number of protons and ... number of neutrons (even/odd).

28. What radiation do we get from radioactive substances. Give an account of their nature and properties. ***(G.N.D.U. 2004; Calicut. U. 2002)***

29. (*a*) What is meant by range of an α-particle ? On what factors does it depend ? Explain briefly how the range of α-particle can be determined.

(P.U. 2004, 2003, 2002, 2000; M.D.U. 2001, 2000; Calicut. U. 2001, 2000; H.P. U.)

(*b*) Explain straggling and its causes. ***(Pbi. U. 2004, 2002; H.P. U. 2001, 2000)***

30. Write a short note on Geiger-Nuttal law and discuss its importance.

(P.U. 2004, 2003, 2002; Cal. U. 2003; Pbi U. 2002, 2001, 2000; Calicut U. 2002, 2001, 2000, K.U. 2001, 2000; H.P.U. 2001, Gauhati U. 2000; G.N.D.U. 2000)

31. What is potential barrier ? What do you mean by potential well of the nucleus. Find an expression for its depth.

32. Give theoretical basis of α-emission and discuss Gamow's theory of emission of α-particle from a radioactive substance. How far the theoretical results agree with experiments?

(Calicut. U. 2002, 2001)

33. (*a*) What is meant by fine structure in α-ray spectrum ? How does it lead to the existence of energy levels in nuclei ?

(*b*) What are long range α-particles ?

34. (*a*) What is β-decay ? Discuss the energy spectrum curve from β-decay of a radioactive nuclide. What is end point energy ? Is β-spectrum discrete or continuous ?

(*b*) Show that the law of conservation of energy and momentum are not obeyed in β-decay. Show how neutrino hypothesis explains this discrepency and accounts for continuous β-ray spectrum. ***(P.U. 2004, 2003, 2001; Bang. U. 2004, 2000; Cal. U. 2003; G.N.D.U. 2003, 2002, 2000; Pbi. U. 2003; Calicut. U. 2002; H.P. U. 2002, 2001, 2000; K.U. 2000; M.D. U. 2000, Gauhati. U. 2000)***

35. Explain why the α-spectra is discrete whereas β-spectra is continuous. *(Pbi. U. 2001)*

36. Enunciate the difference between α-spectra and β-spectra.

37. Give the evidence of existence of neutrinos. Why was their existence postulated? *(P.U. 2000)*

38. Although nucleus is positively charged, how will you explain β-rays coming out of it? *(H.P.U. 2002)*

39. Electrons as such do not exist in the nucleus. What then is the cause of β-decay? *(H.P.U. 1994)*

40. Calculate the distance of closest approach of 5 MeV α-particle scattered from copper ($Z=29$) target. Given that $\frac{1}{4\pi\varepsilon_0} = 9\times10^9$; $e = 1.6\times10^{-9}$ C *(Gauhati. U. 2000)*

Ans. 1.67×10^{-14} **m.**

MULTIPLE CHOICE QUESTIONS

Choose the correct answer

1. The tunnel effect involves the leakage of the particles
(*a*) α-particles (*b*) electrons
(*c*) protons and neutrons (*d*) γ-rays. *(G.N.D.U. 2000)*

2. A curie is a standard unit of radio-activity. Its value is
(*a*) 10^9 disintegrations/sec (*b*) 10^{10} disintegrations/sec
(*c*) 3.7×10^{10} disintegrations/sec (*d*) 37000 disintegrations per second. *(H.P. U. 2000)*

3. Geiger Nuttal rule gives the range of
(*a*) α-particles (*b*) β^+ particle
(*c*) β^- particle (*d*) γ-rays *(H.P. U. 2000)*

4. Radioactivity is the phenomenon associated with
(*a*) Electron emission from atoms (*b*) Transformation of nuclei
(*c*) Fission of nuclei (*d*) None of the above.

5. The decay constant of the end product of the radio-active series is
(*a*) zero (*b*) infinity
(*c*) uncertain (*d*) May be zero or infinity

6. Parity is not conserved in
(*a*) α-decay (*b*) β-decay
(*c*) γ-decay (*d*) none of these.

7. A long lived excited nucleus is called
(*a*) isomer (*b*) isotone
(*c*) isotope

ANSWERS

1. (*a*)	**2.** (*c*)	**3.** (*a*)	**4.** (*d*)	**5.** (*a*)
6. (*b*)	**7.** (*a*)			

NUCLEAR MODELS

16
Chapter

INTRODUCTION

Nucleus is the core of an atom. It is composed of protons and neutrons. There are a number of features of the nucleus that need to be explained in terms of a nuclear model. Most important of these are :-

(*i*) The binding energy per nucleon is almost constant.

(*ii*) The shape of the binding energy curve needs an explanation.

(*iii*) The magic number nuclei *i.e.* those containing 2, 8, 20, 28, 50, 82 nucleons are most stable. Nuclei with double magic number *i.e.* where both N and Z correspond to magic number are even more stable.

(*iv*) Even-even nuclei *i.e.* nuclei with even number of protons and even number of neutrons are very stable. There is hardly any isotope with odd values of Z and N.

(*v*) Certain nuclei emit α and β particles though these particles do not exist within the nucleus.

(*vi*) Why do nuclei have spins that is measured for them ?

(*vii*) Although a nucleus in the ground state is the most stable, what is the cause of excited state of the nucleus and nuclear energy levels.

Various models of the nucleus which have been suggested are

(*i*) Alpha particle model (*ii*) Degenerate gas model

(*iii*) Liquid drop model (*iv*) Shell model and

(*v*) Collective model.

We shall study the imortant ones only.

16.1 LIQUID-DROP MODEL

The liquid drop model of the nucleus as proposed by N. Bohr in 1937 uses the concept of potential barrier developed by Gamow.

According to this model, the nuclei of all elements behave like a *liquid drop* of electrically charged, incompressible *liquid* of constant density but of varying mass. The 'liquid' in reality consists of neutrons and protons.

As a first approximation, we can think of each nucleon in a nucleus as interacting solely with its nearest neighbours, like the molecules of a liquid which ideally are free to move about but always maintain a fixed intermolecular distance from their nearest molecules.

Assumptions. The main assumptions of liquid drop model are:

(*i*) The nucleus is supposed to be spherical in shape in its stable state just as a liquid drop is spherical due to the symmetrical forces of surface tension.

(*ii*) The volume as well as the mass of the nucleus being proportional to the atomic mass number A, the density of a nucleus is constant, independent of volume. Similarly, the density of a liquid drop is constant, independent of its volume. [One point of difference should, however, be noted. The density of a nucleus is the *same* for nuclei of *different* elements but the density of a liquid is different for different liquids.]

(*iii*) The nucleons are supposed to move about within a spherical enclosure represented by the nuclear potential barrier just as the molecules move about within a spherical drop of the liquid, the shape of which is determined by properties of surface tension.

(*iv*) The two main properties of nuclear forces are their short range and tendency to saturate. These can be deduced from the fact that (*a*) The binding energy per nucleon is almost a constant quantity (nearly 8 MeV) for all but the lighter nuclei and, therefore, the total binding energy of the nucleus is proportional to the total number of nucleons *i.e.*, its mass number *A* and (*b*) the volume of the nucleus is also proportional to the atomic mass.

(*v*) Just as the latent heat of vaporisation is a constant for a liquid, the binding energy per nucleon is also a constant.

(*vi*) The molecules evaporate from a liquid drop on raising the temperature of the liquid due to their increased energy of thermal agitation. Similarly when energy is given to a nucleus by bombarding it with nuclear projectiles, a compound nucleus is fomed which emits nuclear radiations almost immediately.

(*vii*) When a small drop of a liquid is allowed to oscillate, it breaks up into two smaller drops of equal size. The process of nuclear fission is similar and the nucleus breaks up into two smaller nuclei.

Justification. From the above assumptions we find that the properties of a nucleus are very similar to the properties of a liquid drop. Hence it is justified to call this model of the nucleus as *liquid drop model.*

Expression for binding energy. The *mass M* of a neutral atom whose nucleus contains Z protons and $N = (A - Z)$ neutrons is given by $M(Z, A) = Zm_p + (A - Z)\, m_n - E_b$

or
$$M = Zm_p + Nm_n - E_b$$

where E_b is the binding energy. The total binding energy of a nucleus is contributed by the following factors.

1. Volume energy. Suppose the energy associated with each nucleon-nucleon bond has a value U. The energy is really negative because attractive nuclear forces are involved but is usually taken as positive for convenience. Because each bond energy U is shared by two nucleons, each has a binding energy $\frac{1}{2}U$. When an assembly of spheres of the same size is packed together into the smallest volume each interior sphere has 12 other spheres in contact with it. This is the situation within the nucleus where each interior nucleon has 12 other nucleons in contact with it and hence has a binding energy $12 \times \frac{1}{2}U = 6U$. If all the A nucleons in the nucleus are supposed to be in the interior, the total binding energy of the nucleus would be

$$E_v = 6UA = a_v A$$

where a_v is a constant having a value 14 MeV. This energy is known as *volume energy.*

2. Surface energy. The nucleons on the surface of the drop interact only with nucleons on one side. The above expression which has been derived on the assumption that all nucleons are being equally attracted on all sides, has to be reduced by a factor proportional to the number of nucleons on the surface. The radius of the nucleus is given by

$$R = r_0 A^{1/3}$$

∴ Surface area $= 4\pi R^2 = 4\pi r_0^2 A^{2/3}$

$\therefore$ Surface energy $\qquad E_s = -k\,4\pi r_0^2 A^{2/3} = -a_s A^{2/3}$

where $\qquad a_s$ is a constant = 13 MeV

The negative sign indicates that this energy contribution is in a direction opposite to the volume energy, becasue the nucleons on the surface are not as strongly bound as the nucleons inside. Since the relative number of surface nucleons is pretty high for light nuclei, the surface energy factor is quite large in their case.

3. Coulomb energy. A nucleus contains Z protons which can form $\frac{Z(Z-1)}{2}$ pairs. The electronic repulsion between each pair of protons in the nucleus contributes towards decreasing of its binding energy. The energy due to repulsion in protons is known as Coulomb energy. Thus, the Coulomb energy E_C of a nucleus is the work that must be done to bring together Z protons from infinity into a spherical aggregate the size of the nucelus. The potential energy for a pair of protons at a distance r apart.

$$U = \frac{e^2}{4\pi\varepsilon_0 r}$$

As there are $\frac{Z(Z-1)}{2}$ proton pairs,

$$E_C = \frac{Z(Z-1)}{2}\,\frac{e^2}{4\pi\varepsilon_0 r} = \frac{Z(Z-1)e^2}{8\pi\varepsilon_0}\left(\frac{1}{r}\right)_{av}$$

where $\left(\frac{1}{r}\right)_{av}$ is the value of $\frac{1}{r}$ averaged over all the pairs. If the pairs are uniformly distributed throughout the nucleus of radius R, then $\left(\frac{1}{r}\right)_{av} \propto \frac{1}{R}$ and hence proportional to $\frac{1}{A^{1/3}}$

$\therefore$ Coulomb energy
$$E_C \propto \frac{Z(Z-1)e^2}{8\pi\varepsilon_0}\,\frac{1}{A^{1/3}} = -a_c\,\frac{Z(Z-1)}{A^{1/3}}$$

where a_c is an unkown constant = 0.60 MeV. The Coulomb energy is negative because it arises from an effect which opposes nuclear stability.

4. Asymmetry energy. In general, in heavy nuclei, the number of neutrons $N > Z$ the number of protons. In light nuclei, $N = Z$ and these nuclei are highly stable. As the number of neutrons increases, the nucleus acquires an asymmetrical character due to which a force comes into play which reduces the volume energy. This reduction in energy is directly proportional to the square of the excess of neutrons over protons and inversely proportional to the total number of nucleons.

$$\therefore \qquad E_a \propto \frac{(N-Z)^2}{A} = -a_a\frac{(N-Z)^2}{A} = -a_a\frac{(A-2Z)^2}{A}$$

where $\qquad a_a$ is a constant $= k(a_v - a_s A^{1/3})$ and $k = 1.7826$

The value of a_a can be taken to be 19 MeV for most of the cases.

5. Pairing and shell energy. Nucleons have a tendency to exist in pairs. Hence the nuclei with even number Z of protons and even number N of neutrons are highly stable, even Z-odd N, or odd Z-even N, are less stable. The nucelons in a nucleus exist in the form of shells which may be filled by a certain number given by relevant selection rules. Taking both the effects together the term contributing towards volume energy is given by

$$E_p = (\pm, 0)\, a_p A^{-3/4}$$

where a_p is a constant and has a value – 34 MeV for even Z-even A nuclei, *zero for odd A-odd N* nuclei and + 34 MeV for even A and odd Z nuclei.

Weizacker's semi-empirial formula. Combining all the terms which contribute towards the binding energy of the nucleus, we get Weizacker's semi-empirical binding energy formula which can be written as

Total binding energy $\quad E_b = E_v + E_s + E_c + E_a + E_p$

$$= a_v A - a_s A^{2/3} - a_c \frac{Z(Z-1)}{A^{1/3}} - a_a \frac{(A-2Z)^2}{A} + (\pm, 0)\, a_p A^{-3/4}$$

The *binding energy per nucleon* is given by

$$\frac{E_b}{A} = a_v - \frac{a_s}{A^{1/3}} - a_c \frac{Z(Z-1)}{A^{4/3}} - a_a \frac{(A-2Z)^2}{A^2} (\pm, 0)\, a_p \frac{A^{-3/4}}{A}$$

Semi-empirical mass formula. The complete semi-empirical mass formula on the liquid drop model is

$$M(Z, A) = Zm_p + (A - Z)\, m_n - E_b$$

$$= Zm_p + (A - Z)\, m_n - a_v A + a_s A^{2/3} + a_c \frac{Z(Z-1)}{A^{1/3}} + a_a \frac{(A-2Z)^2}{A} (\pm, 0)\, a_p A^{-3/4}$$

Graph. The contribution of various energies (except pairing and shell energy E_p) per nucleon towards the binding energy of the nucleus is shown in Fig. 16.1.

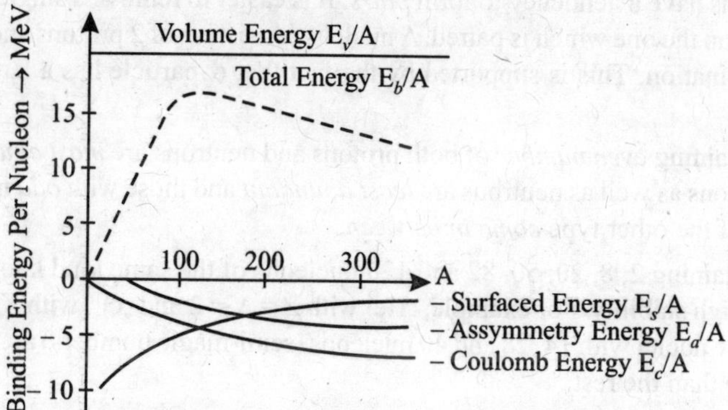

Fig. 16.1.

16.2 MERITS OF LIQUID DROP MODEL

(i) **Stable nucleus.** We can explain the stability of a nucleus on the liquid drop model. The stability of a liquid drop is due to the forces of cohesion between the molecules. Similarly, the stability of a nucleus is due to the binding energy of each nucleon. Just as to remove a molecule from a liquid drop energy has to be supplied to it in the form of heat, energy will have to be supplied to the nucleon equal to or greater than its binding energy to remove it. Hence, the stability of the nucleus.

(ii) **Radioactive nucleus.** In a liquid, with the increase of temperture, the thermal agitation of its molecules becomes more and more rapid and at a particular stage evaporation may take place. A molecule, in a liquid drop, evaporates by gaining energy from its neighbouring molecules during the process of collisions. Similarly, a nucleon or a group of nucleons may leave the nucleus by gaining energy from the neighbouring nucleons during the process of collision, thus exhibiting the phenomenon of radioactivity.

***(iii)* Artificial radioactivity.** The liquid drop model explains the phenomenon of *artificial radioactivity*. It is supposed that when a nucleus is bombarded by fast moving particles, an incoming particle enters the target nucleus forming a *compound nucleus*. It quickly shares its energy with the nucleons already present so that no single particle has sufficient energy for escape. The decay or disintegration of the compound nucleus occurs when the energy is again accidentally concentrated on someone particle which escapes giving rise to the phenomenon of artificial radioactivity, or the energy may by lost by emission of a γ-ray.

***(iv)* Fission.** The liquid drop model also explains the phenomenon of *nuclear fission*.

(v) The liquid drop model helps us to calculate the atomic masses and binding energies of the nuclei quite accurately.

16.3 FAILURES (LIMITATIONS) OF LIQUID DROP MODEL

(*i*) The liquid drop model fails to explain the high stability of nuclei with magic numbers.

(*ii*) This model does not explain the measured spins and magnetic moments of the nuclei.

16.4 SHELL MODEL

In the *liquid drop model* of the nucleus it is assumed that the nucleons exert an influence on one another only at close range *i.e.*, they interact strongly only with their immediate neighbours.

The *shell model* treats the nucleons individually and it is assumed that the nucleons do not, at all interact with each other.

Evidence. The various facts in favour of the shell model are:

1. The nucleons have a tendency to form *pairs*. It is easier to remove a single unpaired nucleon from the nucleus than the one which is paired. A nucleus consisting of 2 protons and 2 neutrons forms a very stable combination. This is supported by the fact that α-particle has a large binding energy of about 28.3 MeV.

2. Nuclei containing *even number* of both protons and neutrons are *most abundant*; nuclei with *odd number* of protons as well as neutrons are *least abundant* and those with *odd* number of one type and *even* number of the other type *come in between*.

3. Nuclei containing 2, 8, 20, 50, 82 and 126 nucleons of the same kind known as *magic numbers*, have a very high stability. For example, ${}_2He^4$ with $Z = N = 2$ and ${}_8O^{16}$ with $Z = N = 8$ are highly stable. Similarly, the nuclei with 14, 28 and 40 nucleons (semi-magic numbers) are slightly less stable but are more stable than the rest.

The nuclides whose proton number Z corresponds to a magic number have the highest number of *isotopes*. Some of the examples are; Ca ($Z = 20$) has six isotopes and Sn ($Z = 50$) has 10 isotopes.

The nuclides whose neutron number N corresponds to magic number has the highest number of *isotones* (Nuclei of different atoms having the same number of neutrons). Thus, there are *five* naturally occuring isotones with $N = 50$.

Isotopes of elements having isotopic abundance of more than 60% fall in the category of magic numbers.

The stable end product of the natural radioactive series, Lead ${}_{82}Pb^{208}$ has $Z = 82$ and $N = 126$ both magic numbers.

4. The *first excitation energy* of magic number nuclei is very *high*. When the number of protons or neutrons exceeds the magic number by one, the binding energy of the extra nucleon is much less than the average value.

Assumption. The various assumptions of the shell model are :-

(*i*) Nucleons form closed sub-shells within the nucleus in a similar manner as the electrons do in the case of atoms. Hence it is named '*shell model*' (*ii*) The nucleons which constitute the nucleus are arranged in some type of shell structure. The shells get closed with a suitable number of protons and neutrons.

(*iii*) The extra nuclear electrons revolve in the Coulomb electrostatic field of the nucleus which is supposed to be heavy and at a very large distance. The electrons revolve only in specific permitted orbits.

(*iv*) Each nucleon moves independently inside the nucleus in a fixed orbit under the effect of central potential produced by the average interaction between remaining $(A - 1)$ nucleons in it.

(*v*) Each nucleon is assumed to posses a spin angular momentum $= \sqrt{s(s+1)}\,\hbar$ where s is the spin quantum number $= \frac{1}{2}$ and an orbital angular momentum $= \sqrt{l(l+1}\,\hbar$ where l is the orbital angular momentum quantum number having values 0, 1, 2, 3, ... etc.

Basis of Shell Model

A nuclear shell model has been developed on the basis of following facts.The model is similar to the Bohr model for electrons in the extra nuclear space. By analogy with the closed sub-shells and shells in the case of atoms it is assumed that nucleons also form similar closed sub-shells and shells within the nucleus. The neutrons and protons constituting the nucleus are supposed to be arranged in some type of shell structure and these shells get closed with a suitable number of protons and neutrons. The extra nuclear electrons are supposed to revolve in the Coulomb electrostatic field of the relatively distant heavy nucleus in specific permitted orbits. From the compelling similarity of stability between the magic nuclei and the inert gases, *it is assumed in the shell model that each nucleon moves independently inside the nucleus in a fixed orbit under the influence of a central field of force or a central potential V (r) produced by the average interaction between all the remaining (A – 1) nucleons in it.*

It is assumed that the nucleons move in an average harmonic oscillator potential given by

$$V = \frac{1}{2}kr^2 = \frac{1}{2}m\omega^2 r^2$$

where m is the mass of the nucleon and ω the oscillator frequency. The Schrodinger's wave equation

$$\left[-\frac{\hbar}{2m}\nabla^2 + V\right]\psi = E\psi$$

then becomes

$$\left[-\frac{\hbar}{2m}\nabla^2 + \tfrac{1}{2}kr^2\right]\psi = E\psi$$

Solving the equation we find that various energy levels are given by

$$E_n = \left(N + \frac{3}{2}\right)\hbar\,\omega$$

Where $N = 0, 1, 2, 3, 4, ...$ and is known as *oscillator quantum number*. This equation shows that all the energy states in the harmonic oscillator model are equally spaced. The wave function ψ contains both the angular (orbital) and the radial part.

Each nucleon is supposed to have an orbital angular momentum $|\vec{l}| = \sqrt{l(l+1)}\,\hbar$ where l is the nuclear orbital quantum number having values 0, 1, 2, 3, ... etc.

Another quantum number similar to the principal quantum number of the electronic orbit characterises the radial part of the nuclear wave function and is denoted by n. The value of $n = 1, 2, 3, ...$ etc.

It can be shown that angular part of ψ requires that the oscillator quantum number N is related to the orbital quantum number l and quantum number n by the relation.

$$N = 2(n-1) + l = 2n + l - 2.$$

The energy levels corresponding to each value of l for the nucleon are represented by a spectroscopic notation similar to the electron as shown

$$l = 0, 1, 2, 3, 4, 5$$

Spectroscopic notation $= s, p, d, f, g, h$

The nucleons are designated by marking their n values followed by spectroscopic notation giving the l value.

The energy levels predicted for a harmonic oscillator potential for $N = 0$ to 4 together with the maximum number of nuclei have been calculated below

N	*n*	*L*	*State*	*Number of nucleons* $2(2l+1)$	
0	1	0	1*s*	2	
0	2	From the relation $N = 2n + l - 2$, l for $N = 0$ and $n = 2$ is negative which is not permissible			
1	1	1	1*p*	6	
1	2	From the relation $N = 2n + l - 2$, l for $N = 1$ and $n = 2$ is negative which is not permissible			
2	1	2	1*d*	10	12
2	2	0	2*s*	2	
2	3	From the relation $N = 2n + l - 2$, l for $N = 2$ and $n = 3$ is negative which is not permissible			
3	1	3	1*f*	14	20
3	2	1	2*p*	6	
3	3	From the relation $N = 2n + l - 2$, l for $N = 3$ and $n = 3$ is negative which is not permissible			
4	1	4	1*g*	18	30
4	2	2	2*d*	10	
4	3	0	3*s*	2	

The series of energy levels calculated above do not agree with the observed sequence of magic numbers. The problem was solved by incorporating spin-orbit interaction.

Spin orbit coupling. Each nucleon is supposed to have a spin angular momentum $|\vec{s}| = \sqrt{s(s+1)}\,\hbar$, where $s = \frac{1}{2}$ is the nuclear spin quantum number and an orbital angular momentum $|\vec{l}| = \sqrt{l(l+1)}\,\hbar$, where l is the nuclear orbital quantum number having values, 0, 1, 2, 3 etc. This *independent particle model* further assumes a strong spin-orbit coupling so that each individual nucleon has a *total angular momentum quantum number* $j = l \pm \frac{1}{2}$ except in case $l = 0$ when j has only one value $= l + \frac{1}{2} = \frac{1}{2}$. Thus each energy level splits up into two sub-levels with $j\uparrow\uparrow = l + \frac{1}{2}$ ($\vec{s}$ and $\vec{l}$ parallel) and $j\uparrow\downarrow = l - \frac{1}{2}$ ($\vec{s}$ and $\vec{l}$ anti-parallel). The sub-level $j + \frac{1}{2}$ has a lower energy than the sub-level $j - \frac{1}{2}$. Thus, the $j + \frac{1}{2}$ state represents a more lightly bound nucleon. The separation between two sub-levels increases rapidly with l and for $l \geq 4$, the two sub-levels lie in different shells. Each sub-level can have a maximum of $(2j + 1)$ nucleons of the same kind *i.e.*, it can accommodate $(2j + 1)$ protons and $(2j + 1)$ neutrons.

We have already introduced a quantum number similar to the principal quantum number for the electronic orbit which characterises the radial part of the nuclear wave function and is denoted by n. The value of n = 1, 2, 3, ...etc. Thus, a nuclear quantum state is represented by three quantum numbers n, l and j. In accordance with Pauli's exclusion principle no two identical nucleons can be in the same quantum state *i.e.*, have the same value of all the three quantum numbers. Pauli's principle is thus, applied to the protons and neutrons separately and two protons or two neutrons with spins $+\frac{1}{2}$ and $-\frac{1}{2}$ are allowed in each level. The neutron levels are equally spaced whereas the spacing between the proton levels increases gradually as we move from the ground level to the higher levels. This is due to the Coulomb force of repulsion between the protons.

For example, for $l = 0$, $j = l + \frac{1}{2} = \frac{1}{2}$ and the number of nucleons in the level = $(2j + 1) = 2$.

For $l = 1, j = 3/2$ and $\frac{1}{2}$ and the number of nucleons in the two sub-levels are $\left(2 \times \frac{3}{2} + 1\right) = 4$ and $\left(2 \times \frac{1}{2} + 1\right) = 2$ repectively. The total number of nucleons in this level = 4 + 2 = 6 giving the progressive total of the number of nucleons in the nucleus = 8. In this way, the completed energy levels or shells and the corresponding total number of nucleons can be predicted.

As already stated the energy levels corresponding to each value of l for the nucleon are represented by a spectroscopic notation similar to the electrons as shown,

$$l = 0, 1, 2, 3, 4, 5, ...$$

$$\text{Spectroscopic notation} = s, p, d, f, g, h, ...$$

The nucleons are designated by writing their n values followed by the spectroscopic notation giving their l values with their j values written as subscript and the number of nucleons required to complete a sub-level written as superscript.

Under this scheme 1s state has $l = 0$. Taking $s = \frac{1}{2}$; $j = l + \frac{1}{2} = \frac{1}{2}$ ($j = l - \frac{1}{2}$ not allowed).

This state is now represented as $(1s_{1/2})^2$ as the number of nucleons in this state = $(2j + 1) = 2$. 1p state has $l = 1$ and therefore $j = l + \frac{1}{2} = \frac{3}{2}$ and $j = l - \frac{1}{2} = \frac{1}{2}$. This state is therefore split into two states denoted as $(1p_{3/2})^4$ [as it has $\left(2 \times \frac{3}{2} + 1\right) = 4$ nucleons] and $(1p_{1/2})^2$ [as it has $\left(2 \times \frac{1}{2} + 1\right) = 2$ nucleons].

The quantum numbers n, l and j, and designation of energy levels according to above notation in order of the energy of the levels upto magic number 82 is shown in **Table 16.1.**

It should be noted from the table that level $(2s_{1/2})^2$ lies between $(1d_{3/2})^4$ and $(1d_{5/2})^6$. Similarly level $(2p_{3/2})^4$ lies between $(1f_{5/2})^6$ and $(1f_{7/2})^8$ wheres level $(2p_{1/2})^2$ lies between $(1g_{9/2})^{10}$ and $(1f_{5/2})^6$ and so on.

16.5 ACHIEVEMENTS (PREDICTIONS)

1. Magic numbers. The shell model explains the existence of *magic numbers*. By making a refrence to the table 16.1 and by adding the nucleons from below the shell closure occurs at neutron or proton number 2, 8, 20, 50, 82 and 126 which correspond to magic numbers.

n	*l*	*j*	*Designation*	*Level*	*2j + 1*	*Total no. of protons or neutrons in the closed shell*	*Progressive Total*
1	5	$\frac{11}{2}$	$(1h_{11/2})^{12}$ ...		12		
3	0	$\frac{1}{2}$	$(3s_{1/2})^{2}$ ...	...	2		
2	2	$\frac{3}{2}$	$(2d_{3/2})^{4}$ ...	...	4	32	
2	2	$\frac{5}{2}$	$(2d_{5/2})^{6}$ ...	...	6		
1	4	$\frac{7}{2}$	$(1g_{7/2})^{8}$ ...	...	8		82
1	4	$\frac{9}{2}$	$(1g_{9/2})^{10}$ ...	...	10		
2	1	$\frac{1}{2}$	$(2p_{1/2})^{2}$ ...	...	2		
1	3	$\frac{5}{2}$	$(1f_{5/2})^{6}$ ...	...	6	30	
2	1	$\frac{3}{2}$	$(2p_{3/2})^{4}$ ...	...	4		50
1	3	$\frac{7}{2}$	$(1f_{7/2})^{8}$ ...	...	8		
1	2	$\frac{3}{2}$	$(1d_{3/2})^{4}$ ...	...	4		
2	0	$\frac{1}{2}$	$(2s_{1/2})^{2}$ ...	...	2	12	
1	2	$\frac{5}{2}$	$(1d_{5/2})^{6}$ ...	...	6		20
1	1	$\frac{1}{2}$	$(1p_{1/2})^{2}$ ...	...	2		
1	1	$\frac{3}{2}$	$(1p_{3/2})^{4}$ ...	...	4	6	8
1	0	$\frac{1}{2}$	$(1s_{1/2})^{2}$...		2	2	

Table 16.1

2. Spin. The shell model successfully explains the *ground state spins* and *magnetic moments* of the nuclei. The neutrons and protons with opposing spin *pair off* so that their mechanical and magnetic moments cancel. It is, therefore, the odd or left proton or neutron which could not be paired that contributes to the spin and the magnetic moment of the nucleus as a whole.

The total angular momentum quantum number *I*, generally referred to as the spin quantum number, is obtained by adding the *j*-values of individual nucleons in it, subject to the usual quantum conditions.

For example, $_1H^2$ has $Z = 1$ *i.e.*, one proton and $N = A - Z = 1$ neutron. Each nucleus will, therefore, be in the ground state $l = 0$ and will have a j-value $\frac{1}{2}$ for either. The j's of the proton and the neutron are oriented parallel to each other so that $I = 1$ which agrees with the experimentally observed value of spin. For $_1H^3$ and $_2He^3$ nuclei we have two nucleons of the same kind which are paired off and one nucleon of the other kind with j value $\frac{1}{2}$. Thus the I value of each of these two nuclei is $\frac{1}{2}$.

Similarly for $_6C^{13}$, the nuclear configuration is $(1s_{1/2})^{2+2}\ (1p_{3/2})^{4+4}\ (1p_{1/2})^1$ and hence $I = \frac{1}{2}$ the j-value for the last unpaired neutron. For $_8O^{17}$ we have $(1s_{1/2})^{2+2}\ (1p_{3/2})^{4+4}\ (1p_{1/2})^{2+2}\ (1d_{5/2})^1$ and $I = 5/2$ the j-value for the last unpaired neutron.

3. Magnetic moment. The observed nuclear magnetic moments are also explained in terms of j-values of the odd nucleon in the nucleus.

The magnetic moment of a nucleus depends upon the configuration whether it is an *even-even* nucleus, *even-odd* or *odd-even* nucleus. For an *even-even* nucleus the nuclear magnetic moment is zero. For *odd-even* (or *even-odd*) nucleus, the nuclear magnetic moment depends upon whether the last unpaired nucleon is a proton or a neutron.

The magnetic moment associated with orbital angular momentum vector $\vec{l}$ is given by

$$\mu_l = g_l\, \vec{l} = g_l \sqrt{l(l+1)}$$

Where g_l is the g-factor for orbital motion and l is the orbital quantum number.

The magnetic moment associated with spin angular momentum vector $\vec{s}$ is given by

$$\mu_s = g_s\, \vec{s} = g_s \sqrt{s(s+1)}$$

Where g_s is the g-factor for spin motion and s is the spin quantum number.

Therefore, total magnetic moment μ = sum of the components of vectors $g_l\ \vec{l}$ and $g_s\ \vec{s}$ along -vector. Hence in units of $\hbar$.

$$\mu = g_l \sqrt{l(l+1)}\, \cos(\vec{l}\,.\,\vec{j}) + g_s \sqrt{s(s+1)}\, \cos(\vec{s}\,.\,\vec{j})$$

Now $$\cos(\vec{l}\,.\,\vec{j}) = \frac{l^2 + j^2 - s^2}{2\,\vec{l}\,.\,\vec{j}} = \frac{\vec{l}.\vec{l} + \vec{j}.\vec{j} - \vec{s}.\vec{s}}{2\,\vec{l}\,.\vec{j}}$$

$$= \frac{l(l+1) + j(j+1) - s(s+1)}{2\sqrt{l(l+1)}\sqrt{j(j+1)}}$$

Similarly, $$\cos(\vec{s}\,.\,\vec{j}) = \frac{s^2 + j^2 - l^2}{2\,\vec{s}\,.\,\vec{j}} = \frac{s(s+1) + j(j+1) - l(l+1)}{2\sqrt{s(s+1)}\sqrt{j(j+1)}}$$

$\therefore$ $$\mu = g_l \sqrt{l(l+1)}\, \frac{l(l+1) + j(j+1) - s(s+1)}{2\sqrt{l(l+1)}\sqrt{j(j+1)}} +$$

$$g_s \sqrt{s(s+1)}\, \frac{s(s+1) + j(j+1) - l(l+1)}{2\sqrt{s(s+1)}\sqrt{j(j+1)}}$$

or $$\mu = g_l \frac{l(l+1)+j(j+1)-s(s+1)}{2\sqrt{j(j+1)}} + g_s \frac{s(s+1)+j(j+1)-l(l+1)}{2\sqrt{j(j+1)}}$$

Magnetic moment of a single nucleon. For a single particle, spin $s = \frac{1}{2}$ in units of $\hbar$ and there are two possible cases

(*i*) $\vec{l}$ is parallel to $\vec{s}$. This is known as *stretch case.*

For such a case $$j = l + s = l + \frac{1}{2}$$

$\therefore$ $$\mu = g_l\left(j - \tfrac{1}{2}\right) + \frac{1}{2} g_s$$

(*ii*) $\vec{l}$ is anti-parallel to $\vec{s}$. This is known as *Jack-knife case.*

For such a case $$j = l - s = l - \tfrac{1}{2}$$

$\therefore$ $$\mu = g_l\left(\frac{j}{j+1}\right)\left[\left(j + \tfrac{3}{2}\right) - \tfrac{1}{2} g_s\right]$$

For single nucleons

$g_l = 1$ and $g_s = 5.58$ for proton.

$g_l = 0$ and $g_s = -3.82$ for neutron.

Thus we find that the odd proton has a positive magnetic moment due to spin whereas the odd neutron has a negative magnetic moment. Thus, the spins and magnetic moments of isotopic nuclei with odd *A* differing in their *A*-values by 2 are the *same* since the two added neutrons are paired off and do not have any effect on the single odd nucleon which contributes to the spin and magnetic moment of the nucleus. For example, $_{47}Ag^{107,\,109}$ and $_{55}Cs^{133,\,135,\,137}$ isotopes have the same value of nuclear spin and magnetic moment.

4. Stability. The very great stability and high binding energy is explained on the basis of *closed shells.* For example, for $_2He^4$ the first shell is completely filled up by protons as well as neutrons (2 each) so that $I = 0$. Further the closure of the shell gives it a very high binding energy of 28.3 MeV. Thus the α-particle $_2He^4$ nucleus is a very stable configuration. The case of $_8O^{16}$ is also similar with first two shells for neutrons and protons completely filled up with 2 and 6 nucleons *i.e.* a total of 8 nucleons respectively.

In adding nucleons to the nucleus the proton and the neutron levels should be equally filled up for maximum stability. Hence for small values of mass number *A*, the number of neutrons is very nearly equal to the number of protons. As the number of nucleons becomes large, the most stable nuclides will again be formed when the lowest energy levels available to protons and neutrons are filled up. As the spacing between proton levels goes on increasing, the number of neutrons will be greater than the number of protons.

5. Nuclear isomerism. The phenomenon of *nuclear isomerism i.e.*, existense of isobaric, isotopic nuclei in different energy states observed in nuclei with odd *A* values lying between 39 to 49, 69 to 81, 111 to 125 has been explained on the shell model due to the large difference in the nuclear spins of the isomeric states of these nuclei because they have their *A* values close to the magic numbers required to complete the various nuclear shells.

Failures and limitations. 1. The shell model does not predict correct values of *I* in case of certain nuclei. For example, it predicts $I = 5/2$ for $_{11}Na^{23}$ whereas the correct value is $\frac{3}{2}$.

2. The four stable nuclei $_1H^2$, $_3Li^6$, $_5B^{10}$, $_7N^{14}$ do not fit in the scheme of this model.

3. This model cannot explain the observed first excited states in even-even nuclei at energies much lower than those expected from single particle excitation.

4. This model could not explain the observed large 'quadruple moment' of odd A nuclei, particularly those which have their nucleon number far away from the magic numbers. The shell model assumes the nucleus to be spherical and a new model known as *collective model* has been developed which assumes that the last nucleon of an odd-A nucleus interacts with the remaining even number of nucleons, deviating it from spherical symmetry. Thus a nucleus with all filled shells like $_8O^{16}$ has a spherical shape but a nucleus with odd A like $_8O^{17}$ acquires a non-spherical deformity and hence has a *quadrupole moment*.

The quadrupole moment for a given I is given by

$$Q = -\frac{2I-1}{2I+2}(r_{r.m.s.})^2$$

where $I = l \pm s$ and $(r_{r.m.s.})^2$ is the mean square radius of the orbit. For uniform charge distribution $(r_{r.m.s.})^2 = \frac{3}{5}R_c^2$ where R_c is the Coulomb radius = 1.2 $A^{1/3}$ Fermi.

16.6 NUCLEAR MAGIC NUMBERS

It has been observed that nuclei having either the number of protons Z or the number of neutrons $N = A - Z$ equal to one of the numbers 2, 8, 20, 50, 82 and 126 have a very high stability as compared to their neighbours. These numbers are called *magic numbers.* Similarly nuclei with 14, 28 and 40 nucleons are slightly less stable but are more stable than the rest. These numbers are called *semi magic numbers.*

Magic numbers on shell model. The shell model explains the existence of magic numbers. The ground state of the nucleus is represented as $(1s_{1/2})^2$. It has radial quantum number $n = 1$ orbital angular momentum quantum number $l = 0$ and total angular momentum quantum number $j = l + s = \frac{1}{2}$ since spin quantum number $s = \frac{1}{2}$. The maximum number of nucleons it can contain is $(2j + 1)$ $= \left(2 \times \frac{1}{2} + 1\right) = 2$. Thus the first shell closure occurs at nucleon (proton or neutron) number 2, which is, therefore, a magic number.

The second shell has two sub-shells represented as $(1p_{3/2})^4$ and $(1p_{1/2})^2$. It has a radial quantum number $n = 1$, orbital angular momentum quantum number $l = 1$ and total angular momentum quantum number $j = l \pm \frac{1}{2}$ *i.e.,* $\left(1 + \frac{1}{2}\right) = \frac{3}{2}$ and $\left(1 - \frac{1}{2}\right) = \frac{1}{2}$. The sub-shell $1p_{3/2}$ can contain a maximum of $\left(\frac{3}{2} \times 2 + 1\right) = 4$ nucleons and sub-shell $1p_{1/2}$ can contain a maximum of $\left(\frac{1}{2} \times 2 + 1\right) = 2$ nucleons. Thus the second shell closure takes place at total nucleon (proton or neutron) number 8, (2 for the first shell + 6 for the second shell) which is therefore, a magic number.

The third shell has three sub-shells $(1d_{5/2})^6, (1d_{3/2})^4$ and $(2s_{1/2})^2$.

For $1d_{5/2}, n = 1,\ l = 2$ and $j = 2 + \frac{1}{2} = 5/2$. It can contain a maximum of $\left(\frac{5}{2} \times 2 + 1\right) = 6$ nucleons.

For $1d_{3/2}, n = 1,\ l = 2$ and $j = 2 - \frac{1}{2} = 3/2$. It can contain a maximum of $\left(\frac{3}{2} \times 2 + 1\right) = 4$ nucleons.

For $2s_{1/2}, n = 2, l = 0$ and $j = \frac{1}{2}$. It can contain a maximum of $\left(\frac{1}{2} \times 2 + 1\right) = 2$ nucleons.

Thus the third shell contains in all (6 + 4 + 2) = 12 nucleons. Adding to it the number of nucleons in the second shell (6) and in the first shell (2) we find that the third shell closure takes place at total nuclear number 20 (2 + 6 + 12) (proton or neutron) which is therefore, a magic number.

The first subshell of the fourth shell is $1f_{7/2}$. It has $n = 1, l = 3$ and $j = 3 + \frac{1}{2} = \frac{7}{2}$. It can contain a maximum of $\left(\frac{7}{2} \times 2 + 1\right) = 8$ nucleons. Combined with the 20 nucleons of the fires three shells there are a total number of 28 nucleons which is, therefore a semi-magic number.

Designation of various subshells upto magic number 126 is shown in **Table 16.2.** It is clear from the figure that shell closures occur at total nucleon number 2, 8, 20, 28, 50, 82 and 126. (Shown by dotted lines).

Subshell	Nucleons	Magic number (dotted line)
		126
$1i_{13/2}$	14	
$1h_{9/2}$	10	
$2f_{5/2}$	6	
$3p_{1/2}$	2	
$3p_{3/2}$	4	
$2f_{7/2}$	8	
		82
$1h_{11/2}$	12	
$1g_{7/2}$	8	
$2d_{3/2}$	4	
$3s_{1/2}$	2	
$2d_{5/2}$	6	
		50
$1g_{9/2}$	10	
$2p_{1/2}$	2	
$1f_{5/2}$	6	
$2p_{3/2}$	4	
		28
$1f_{7/2}$	8	
		20
$1d_{3/2}$	4	
$2s_{1/2}$	2	
$1d_{5/2}$	6	
		8
$1p_{1/2}$	2	
$1p_{3/2}$	4	
		2
$1s_{1/2}$	2	

Table 16.2

Experimental Evidence. (*i*) **Stability.** A very important experimental evidence in favour of magic number nuclei comes from a study of stable nuclides. It has been experimentally observed that ${}_2He^4$ with $Z = N = 2$ and ${}_8O^{16}$ with $Z = N = 8$ are highly stable. A graph between binding energy per nucleon and mass number shows sudden increase in binding energy per nucleon for

${}_{20}Ca^{40}$ ($Z = 20$ $N = 20$)
${}_{30}Sr^{88}$ ($Z = 38$ $N = 50$)
${}_{50}Sn^{120}$ ($Z = 50$ $N = 70$)
${}_{58}Ce^{140}$ ($Z = 58$ $N = 82$)
${}_{82}Pb^{208}$ ($Z = 82$ $N = 126$)

(*ii*) **Relative abundance.** It has been observed from experimental data that nuclei having number of nuclei as magic number are in abundance as compared to other nuclei in nature. The most abundant nuclei are ${}_8O^{16}$ ($N = Z = 8$), ${}_{20}Ca^{40}$ ($N = Z = 20$), ${}_{50}Sn^{120}$ ($Z = 50$), ${}_{38}Sr^{88}$ ($N = 50$), ${}_{56}Ba^{138}$ ($N = 82$) and ${}_{82}Pb^{208}$ ($Z = 82, N = 126$).

(*iii*) **Number of isotopes.** The nuclides whose proton number Z corresponds to a magic number have the highest number of isotopes. For example, ${}_{20}Ca^{40}$ ($Z = 20$) has six isotopes whereas the average number of isotopes in that region is about two. Similarly ${}_{50}Sn^{120}$ ($Z = 50$) has 10 isotopes whereas the average number in this region is about 4.

(*iv*) **Number of isotones.** The nuclides whose neutron number N corresponds to magic number has the highest number of isotones. There are five naturally occuring isotones with $N = 50$.

(*v*) **End product of radioactive series.** The stable end product of the natural radioactive series is Lead (${}_{82}Pb^{208}$). It has $Z = 82$, $N = 126$ both magic numbers.

(*vi*) **Energy of emitted β-particle.** The energy of the emitted β-particle is very high when the number of protons or neutrons in the daughter nucleus is equal to magic number.

(*vii*) **Neutron capture cross-section.** For nuclei with number of neutrons equal to magic number and number of protons $Z = 50$ and $Z = 82$ the neutron capture cross-section is particularly low as compared to their neighbours.

(*viii*) **Electric quadruple moment.** The electric quadruple moment of nuclei with protons and neutrons equal to magic number is zero. These nuclei are, therefore, spherically symmetric because electric quadruple moment measures the departure of nuclear charge distribution from spherical symmetry. Nuclei in the neighbourhood of magic number nuclei have a certain value of quadruple moment *i.e.*, these have a distorted shape.

(*ix*) **The first excitation energy** of magic number nuclei is *very high*. When the number of neutrons or protons exceeds the magic number by *one*, the bindings energy of the extra nucleon is much less than the average value.

Significance. The significance of magic number lies in the fact that the corresponding number of protons and/or neutrons gives a high stability to the nucleus. The presence of magic numbers leads to the development of the shell model of the nucleus, by alalogy with the closed shells and sub-shells in the case of atoms. There is a compelling similarity of stability between magic number nuclei and the inert gas atoms. Corresponding to magic numbers 2, 8, 20, 50, 82 and 126, the first, second, third etc. shell closures in nuclei occurs at number of protons or neutrons in the nucleus equal to 2, 8, 20, 50, 82 and 126.

EXERCISE

1. (*a*) Give the main assumptions of liquid drop model of the nucleus. Justify the name liquid drop model.

   (*b*) Obtain the expression for the binding energy of a nucleus based on liquid drop model. State the semi-empirical formula of Weizacker. Show the graph of binding energy per nucleon as the sum of volume, surface, coulomb and asymmetry energies.

   (*c*) What are its achievements and failures ?

   (P.U. 2004, 2003, 2001, 2000; G.N.D.U. 2003, 2002, 2001; Pbi.U. 2004; Caclicut.U . 2002: H.P.U. 2001, 2000; Meerut.U. 2001, 2000; Bang.U. 2000)

2. What is the basic point of difference between the liquid drop model and shell model of the nucleus ?

   What is the evidence for the shell structure of the nuclei ? Stating the main assumptions explain the shell model of the nucleus. Discuss its achievements, failures and limitations.

   (G.N.D.U. 2004, 2003, 2002, 2001; P.U. 2003, 2001, 2000; Meerut.U. 2003, 2001; Pbi.U. 2002; H.P.U. 2002, 2001; K.U. 2002; Calicut.U. 2001; Luck.U. 2001)

3. What are magic number nuclei ? How does the shell model explain the existence of magic number, 2, 8, 20 and 28 only? Give experimental evidence of nuclear magic numbers? What is their significance ?

   (Pbi. U. 2004, 2003, 2002, 2001; P.U. 2004; G.N.D.U. 2003; H.P.U. 2002, 2001; K.U. 2002; Meerut. U. 2002; Calicut.U. 2003, 2001)

NUCLEAR REACTIONS

17 Chapter

17.1 NUCLEAR REACTION

It is possible to bring about a change in the structure of nuclei by bombarding them with fast moving particles. Such a collision, in which the identity or the characteristic of the struck particle are changed, is known as a **nuclear reaction.**

The bombarding particles may be α, β or γ-rays given out by natural radioactive substances or these may be artificially accelerated protons, deutrons, α-particles, neutrons or electrons from particle accelerators.

When the bombarding particle is a light one, having moderate bombarding energy (less than 10 MeV), two products generally appear as a result of nuclear reaction; a light particle, usually one of those mentioned above, and a heavy one called the product nucleus. The first nuclear reaction was produced in 1919 by Lord Rutherford. He bombarded Nitrogen nuclei with α-particles from a radioactive source and showed that protons were given out according to the reaction

$$_7N^{14} + {_2He^4} \longrightarrow {_8O^{17}} + {_1H^1}$$

The above reaction is briefly written as $_7N^{14}(\alpha, p)\,{_8O^{17}}$. This indicates that a nitrogen $_7N^{14}$ nucleus called the *target nucleus* interacts with an α-particle (α) referred to as *incident particle* or *projectile*; a proton (*p*) is ejected and an oxygen $_8O^{17}$ nucleus known as *recoil* or *product nucleus* is formed.

17.2 TYPES OF REACTIONS

Elastic Scattering. In this case, the incident particle strikes the target nucleus and leaves without loss of energy and in general with altered direction of motion. A familiar example of elastic scattering is that of α-particles from thin gold foil, first studied by Rutherford. It is expressed in the form

$$_2He^4 + {_{79}Au^{197}} \longrightarrow {_{79}Au^{197}} + {_2He^4}$$

The target nucleus remains unaffected. In general, if a particle '*a*' is incident on a target 'X' then elastic scattering process is written as

$$a + X = X + a$$

(*ii*) **Inelastic scattering.** In this case, the kinetic energy is not conserved but a part of the energy of the incident particle is taken up by the target nucleus which is excited to a higher quantum state and later decays to the ground state radiating the excess energy in the form of γ-ray photon.

As an example

$$_3Li^7 + {_1H^1} \longrightarrow {_3Li^{7*}}\ (\text{excited}) + {_1H^1}$$

$$^*_3Li^7\ (\text{excited}) \longrightarrow {_3Li^7} + h\nu$$

In general, $a + X = X^* + a.$

The outgoing particle remains the same. The {*} is used to indicate that after scattering the target nucleus is left in an excited state which ultimately emits its energy as γ-rays known as *inelastic scattered γ-rays*.

(*iii*) **Simple or radiative capture.** In this case, the incident particle is captured by the target nucleus and a new nucleus is formed. The new nucleus, in general, has a considerable excess of energy and decays with the emission of one or more γ-ray photons. As an example

$$_6C^{12} + {}_1H^1 \longrightarrow {}_7N^{13} + h\nu.$$

(*iv*) **Disintegration.** This is the most general type of nuclear reaction in which the incident particle is absorbed and a different particle is ejected. The various important examples are as under.

(*a*) **Protons.** When the bombarding particle is a proton, it may give rise to (p, α), (p, D) or a (p, n) reaction. Reaction of the type (p, α) and (p, D) are less common and (p, n) are more likely. Familiar examples are

$$_4Be^9 + {}_1H^1 \longrightarrow {}_3Li^6 + {}_2He^4 \; (p, \alpha)$$
$$_4Be^9 + {}_1H^1 \longrightarrow {}_4Be^8 + {}_1H^2 \; (p, D)$$
$$_3Li^7 + {}_1H^1 \longrightarrow {}_4Be^7 + {}_0n^1 \; (p, n)$$

(*b*) **Deutrons.** Accelerated deutrons give rise to a number of nuclear reaction. Various types are (D, α) (D, p) and (D, n). Fimiliar examples are

$$_3Li^6 + {}_1H^2 \longrightarrow {}_2He^4 + {}_2He^4 \; (D, \alpha)$$
$$_6C^{12} + {}_1H^2 \longrightarrow {}_6C^{13} + {}_1H^1 \; (D, p)$$
$$_{11}Na^{23} + {}_1H^2 \longrightarrow {}_3Mg^{24} + {}_0n^1 \; (D, n)$$

(*c*) **α-Particles.** In this case, the α-particle is generally absorbed by the light nuclei by the process of resonance capture with the emission of a γ-ray photon.

$$_3Li^7 + {}_2He^4 \longrightarrow {}_5B^{11} + h\nu \; (\alpha, \gamma)$$

Reaction of (α, p) and (α, n) may also sometimes occur.

$$_7N^{14} + {}_2He^4 \longrightarrow {}_8O^{17} + {}_1H^1 \; (\alpha, p)$$
$$_5B^{10} + {}_2He^4 \longrightarrow {}_7N^{13} + {}_0n^1 \; (\alpha, n)$$

(*d*) **Neutrons.** The interaction of neutrons with nuclei may result in (n, γ), (n, α), (n, p), (n, n) and (n, $2n$) reactions respectively. Familiar examples are

$$_{13}Al^{27} + {}_0n^1 \longrightarrow {}_{13}Al^{28} + h\nu \; (\gamma\text{-radiation}) \qquad (n, \gamma)$$
$$_5B^{10} + {}_0n^1 \longrightarrow {}_3Li^7 + {}_2He^4 \qquad (n, \alpha)$$
$$_{12}Mg^{24} + {}_0n^1 \longrightarrow {}_{11}Na^{24} + {}_1H^1 \qquad (n, p)$$
$$_{47}Ag^{107} + {}_0n^1 \longrightarrow {}_{47}Ag^{107} + {}_0n^1 \qquad (n, n)$$
$$_{19}K^{39} + {}_0n^1 \longrightarrow {}_{19}K^{38} + 2\,{}_0n^1 \qquad (n, 2n)$$

In this case, the product nucleus invariably decays with the emission of a positron.

(*v*) **Photo-disintegration.** Nuclear reactions brought about by high energy radiations are known as *photo-disintegration*. To produce photo-disintegration, the energy of the incident photon must be greater than the binding energy of the particle to be ejected. For example,

$$_1H^2 + \gamma = {}_1H^1 + {}_0n^1$$

This requires photon of energy 2.225 MeV.

If the energy is not sufficient even to remove a neutron, the nucleus may enter one of its excited states and later emit excess energy as radiation. For example,

$$_{36}Kr^{83} + \gamma \longrightarrow ({}_{36}Kr^{83})^* \rightarrow {}_{36}Kr^{83} + \gamma'$$

(*vi*) **Stripping and pick-up reaction.** In *stripping* reaction, one or more nucleons from the projectile are captured by the target while the remainder continues on its way. For example,

$$_{29}Cu^{63} + {}_1D^2 \longrightarrow {}_{29}Cu^{64} + {}_1H^1$$

When deutron is accelerated to a high energy and then subjected to a magnetic field, the proton gets deflected while the neutron continues to pass straight.

In *pick-up* reaction, an incident particle collides with target nucleus and immediately pulls out one of the nucleons from the target nucleus. This happens when the de-Broglie wave length associated with the incident particle is almost equal to the average distance between the nucleons so that interaction can be restricted to one nucleon.

The stripping and pick up reaction do not proceed through the formation of compound nucleus as intermediate stage.

(*vii*) **Fission and spallation reaction.** In *fission* reaction the target nucleus after interacting with the projectile splits mainly into two lighter nuclei of comparable mass and at the same time it emits a few neutrons. The typical example of fission reaction is

$$_{92}U^{235} + {_0n^1} \longrightarrow {_{56}Ba^{141}} + {_{36}Kr^{92}} + 3\ {_0n^1} + \text{energy}$$

When the projectile has sufficiently large energy, the target nucleus splits into 20 or 30 nucleons and some α-particles leaving behind a number of nuclei of significantly smaller mass. This is known as *spallation*.

(*viii*) **Heavy ion and high energy reaction.** When the incident particle is heavier than α-particle, the reaction is known as *heavy ion reaction*. This type of reaction was observed in 1950 at Berkley with accelerated carbon ion and Aluminium target

$$_{13}Al^{27} + {_6C^{12}} \longrightarrow {_0n^1} + {_{19}K^{38}}$$

This type of reaction has the properties of both compound nucleus and of stripping and pickup mechanism.

In high energy reaction above 150 MeV, the new kind of particles like π^+, π^- and π^0 and strange particles are ejected along with nucleons.

(*ix*) **Fusion.** In fusion reaction, lighter elements combine to give a new element along with release of large amount of energy.

17.3 LAWS GOVERNING NUCLEAR REACTIONS

Nuclear reactions are governed by certain conservation laws. These are

1. Conservation of charge. *The total charge before and after the reaction is conserved.* In other words, the sum of the charges on the reactant side is equal to the sum of the charges on the product side. For example,

$$_7N^{14} + {_2He^4} \longrightarrow {_8O^{17}} + {_1H^1} + E\ \text{(energy)}$$

Total charge on the reactant side $= 7 + 2 = 9$

Total charge on the product side $= 8 + 1 = 9$

Thus the total charge is conserved or $\Sigma Z = 0$

2. Conservation of mass number. *The total mass number or the total number of nucleons before and after the reaction remains the same.* This is because the nucleons can neither be created nor destroyed. In the above example,

Total number of nucleons on the reactant side = $14 + 4 = 18$

Total number of nucleons on the product side = $17 + 1 = 18$

Thus the total number of nucleons or the mass number is conserved *i.e.* $\Sigma A = 0$.

3. Conservation of mass-energy. *The total mass energy in a nuclear reaction remains unchanged.* If a fast moving particle '*a*' with kinetic energy E_a is incident on a nucleus *X*, supposed to be at rest and a new light particle '*b*' is given out with kinetic energy E_b and the product nucleus *Y* also has a kinetic energy E_y, then

According to the law of mass-energy conservation, the following relation must hold in a nuclear reaction

$$M_x c^2 + (m_a c^2 + E_a) = (M_y c^2 + E_y) + (m_b c^2 + E_b)$$

where M_x, M_y, m_a and m_b are the rest masses of the target nucleus, product nucleus, incident particle and the emitted particle respectively,

$$\therefore \quad E_b + E_y - E_a = (M_x + m_a)\, c^2 - (M_y + m_b)\, c^2 = Q$$

where Q is known as Q-value of the reaction.

Thus the nuclear energy released in a reaction can be calculated from the masses of the participating particles or if one of the masses is not known, it can be calculated from the Q-value of the nuclear reaction.

4. Conservation of linear momentum. *The total linear momentum of the particles taking part in a nuclear reaction must be the same before and after the reaction.* If v_a, v_b and V_y are the velocities of the incident particle, the emitted particle and the product nucleus respectively, then their linear momenta $m_a v_a$, $m_b v_b$ and $M_y V_y$ must be capable of representation by the sides of a triangle taken in order because their vector sum must be zero.

5. Conservation of angular momentum. The total anguarr momentum $\vec{I}$ is composed of intrinsic spin angular momentum $\vec{s}$ and relative orbital angular momentum $\vec{l}$.

In a nuclear reaction, the vector sum of the total angular momenta of the atoms must be conserved before and after the reaction.

6. Conservation of spin and statistics. The spin and statistical character must remain the same before and after a nuclear reaction. Thus the statistics followed by the products must be the same as that followed by the reactants; either Fermi-Dirac (for even A) or Bose-Einstein (for odd A).

7. Conservation of parity. The total parity of the system is the product of the intrinsic parities of the *target nucleus and the bombarding (nucleus) particle. The net parity before and after the reaction must be equal.*

No violation of parity has been observed in a nuclear reaction for strong nuclear forces. However, parity does not appear to be conserved in weak interactions.

Physical quantities not conserved. The most important physical characteristics which are not conserved in nuclear reaction are (*i*) *the magnetic dipole moment* and (*ii*) *the electrical quadrupole moment* of the reacting nuclei. These moments depend upon internal distribution of mass, charge and current within the nuclei involved and are not subjected to conservation laws.

SOLVED EXAMPLES

Example 17.1. *An atom of mass number 15 and atomic number 7 captures an α-particle and then emits a proton. What will be the mass number and atomic number of the resulting atom ?*

(Meerut. U. 2003)

Solution. The reaction can be represented as

$${}_7X^{15} + {}_2He^4 \longrightarrow {}_1H^1 + {}_8Y^{18}$$

∴ Mass number of the resulting atom is 18 and atomic number 8.

17.4 Q-VALUE OF NUCLEAR REACTION

The Q-value of a nuclear reaction is defined as the total energy released or absorbed in the reaction. If a fast moving particle 'a' with kinetic energy E_a is incident on a nucleus X supposed to be at rest and a new light particle 'b' is given out with kinetic energy E_b and the product nucleus Y also has a kinetic energy E_y, then the reaction is represented by the equation.

$$a + X = Y + b$$

or briefly $X(a, b)Y$

and its Q-value is given by the relation

$$Q = (E_y + E_b) - E_a$$

According to the law of mass-energy conservation the following relation must hold in nuclear reaction

$$M_x c^2 + (m_a c^2 + E_a) = (M_y c^2 + E_y) + (m_b c^2 + E_b)$$

where M_x, M_y, m_a and m_b are the rest masses of the target nucleus, product nucleus, incident particle and the emitted particle respectively.

$$\therefore \quad Q = (E_y + E_b) - E_a = (M_x + m_a)c^2 - (M_y + m_b)c^2 \quad ...(i)$$

Hence Q-value may also be defined as the difference between the kinetic energies of the products and that of the incident particle.

Expression for Q-value. From equation (*i*), Q-value is given by

$$Q = (E_y + E_b) - E_a \quad ...(ii)$$

The term E_y representing the recoil energy of the product nucleus is small and hence difficult to measure experimentally. We, therefore, eliminate it by using the law of conservation of linear momentum.

Applying the law of conservation of linear momentum, we have (from Fig. 17.1)

Along x-direction.

$$m_a v_a = M_y V_y \cos\phi + m_b v_b \cos\theta \quad ...(iii)$$

Along y-direction

$$M_y V \sin\phi = m_b v_b \sin\theta \quad ...(iv)$$

Rearranging equation (*iii*), we get

$$M_y V_y \cos\phi = m_a v_a - m_b v_b \cos\theta \quad ...(v)$$

Squaring (*iv*) and (*v*) and adding, we have

$$M_y^2 V_y^2 \sin^2\phi + M_y^2 V_y^2 \cos^2\phi = m_a^2 v_a^2 - 2m_a v_a m_b v_b \cos\theta + m_b^2 v_b^2 \cos^2\theta + m_b^2 v_b^2 \sin^2\theta$$

or

$$M_y^2 V_y^2 = m_a^2 v_a^2 - 2m_a v_a m_b v_b \cos\theta + m_b^2 v_b^2 \quad ...(vi)$$

Now

$$E_a = \frac{1}{2} m_a v_a^2 \quad \text{or} \quad v_a^2 = \frac{2E_a}{m_a}$$

$$E_b = \frac{1}{2} m_b v_b^2 \quad \text{or} \quad v_b^2 = \frac{2E_b}{m_b}$$

$$E_y = \frac{1}{2} M_y V_y^2 \quad \text{or} \quad V_y^2 = \frac{2E_y}{M_y}$$

Putting the values of v_a^2, v_b^2 and V_y^2 in equation (*vi*), we get

$$M_y^2 \frac{2E_y}{M_y} = m_a^2 \frac{2E_a}{m_a} - 2m_a \left(\frac{2E_a}{m_a}\right)^{\frac{1}{2}} m_b \left(\frac{2E_b}{m_b}\right)^{\frac{1}{2}} \cos\theta + m_b^2 \frac{2E_b}{m_b}$$

or

$$2M_y E_y = 2m_a E_a - 4(m_a m_b E_a E_b)^{1/2} \cos\theta + 2m_b E_b$$

or

$$E_y = \frac{m_a}{M_y} E_a + \frac{m_b}{M_y} E_b - \frac{2}{M_y} (m_a m_b E_a E_b)^{1/2} \cos\theta \quad ...(vii)$$

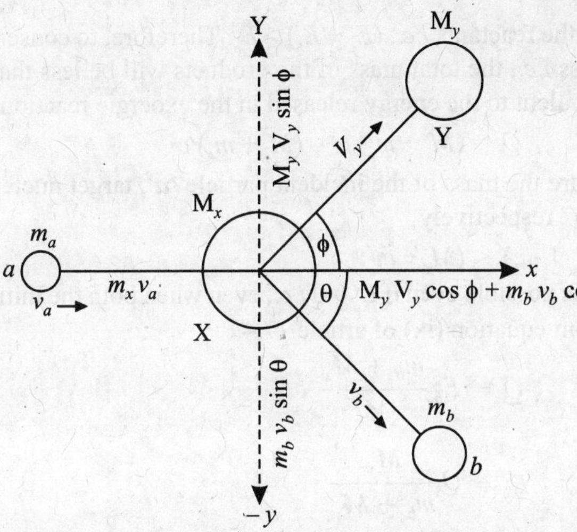

Fig. 17.1.

Substituting the value of E_y from eq. (*vii*) in equation (*ii*), we have

$$Q = \frac{m_a}{M_y}E_a + \frac{m_b}{M_y}E_b - \frac{2}{M_y}(m_a m_b E_a E_b)^{1/2}\cos\theta + E_b - E_a$$

$$= E_a\left(\frac{m_a}{M_y} - 1\right) + E_b\left(\frac{m_b}{M_y} + 1\right) - \frac{2}{M_y}(m_a m_b E_a E_b)^{1/2}\cos\theta$$

$$= E_a\left(\frac{m_a - M_y}{M_y}\right) + E_b\left(\frac{m_b + M_y}{M_y}\right) - \frac{2}{M_y}(m_a m_b E_a E_b)^{1/2}\cos\theta \qquad ...(viii)$$

Equation (*viii*) gives the Q-value of the reaction.

Product particle emerges at right angles to incident direction. For a special case when we observe the ejected particle at an angle $\theta = 90°$ to a collimated beam of incident particle 'a', then $\cos\theta = \cos\pi/2 = 0$

$$\therefore \qquad Q = E_a\left(\frac{m_a - M_y}{M_y}\right) + E_b\left(\frac{m_b + M_y}{M_y}\right)$$

or

$$Q = E_b\left(1 + \frac{m_b}{M_y}\right) - E_a\left(1 - \frac{m_a}{M_y}\right) \qquad ...(ix)$$

It is possible to measure E_a, E_b, m_a, m_b and M_y quite accurately as also the value of Q and hence the above relation can be verified.

17.5 EXO-ERGIC AND ENDO-ERGIC REACTIONS

Exo-ergic reaction. The value of Q is given by

$$Q = (E_y + E_b) - E_a \qquad ...(i)$$

where E_a is the kinetic energy of the (projectile) particle incident on a target nucleus X supposed to be at rest, E_b the kinetic energy of the new light particle 'b' given out and E_y the kinetic energy of the product nucleus Y.

If Q is positive i.e., the process is accompanied by the liberation of energy, it is known as an exothermic or exoergic reaction. In this case, the kinetic energy of the products of the reaction is greater than the kinetic energy of the reactants *i.e.*, $(E_y + E_b) > E_a$. Therefore, to conserve mass energy there must be a decrease of mass *i.e.*, the total mass of the products will be less than that of the reacting nuclei by an amount equivalent to the energy released in the exoergic reaction.

As $$Q = (M_x + m_a)\,c^2 - (M_y + m_b)\,c^2$$

where m_a, M_x, m_b and M_y are the mass of the incident particle 'a', target nucleus X, ejected particle 'b' and product nucleus 'Y' respectively

Hence $$(M_y + m_b) < (M_x + m_a)$$

An exoergic reaction is possible even if $E_a = 0$ *i.e.*, even when both the initial particles are at rest

For $E_a = 0$, we get from equation (ix) of article 17.4

$$Q = E_b \frac{m_b + M_y}{M_y}$$

or $$E_b = Q\frac{M_y}{m_b + M_y} \qquad ...(iii)$$

Also for $E_a = 0$, from equation (i), we have

$$Q = E_y + E_b$$

Endoergic reaction. *If Q is negative i.e., the process is accompanied by the absorption of energy it is known as an endothermic or endoergic reaction.* This energy is supplied by the kinetic energy of the incoming particle. But all the kinetic energy of the incident particle is not available for the nuclear reaction because part of the energy is used to give kinetic energy to the product nucleus and emitted particle. Hence for a reaction in which $(-Q)$ energy is to be absorbed, the incident particle should supply some energy in addition to $(-Q)$.

The minimum kinetic energy which the projectile must posses so that the nuclear reaction may take place is called threshold energy. The reaction cannot occur unless the bombarding particle has the minimum (threshold) energy.

Threshold energy. To find the value of threshold energy we suppose that before the product nucleus and the product particle are given out there is an *intermediate stage* when a *compound nucleus* is formed by the combination of the incident particle 'a' and the target nucleus 'X'. The mass of the compound nucleus,

$$M_c = M_x + m_a$$

The incident particle on combining with the target nucleus imparts all its momentum to the compound nucleus so that it has a velocity V_c. Then, according to the principle of conservation of linear momentum

$$m_a v_a = M_c V_c$$

or $$V_c = \frac{m_a}{M_c} v_a = \left(\frac{m_a}{M_x + m_a}\right) v_a \qquad ...(iv)$$

The Q-value of the endoergic reaction is given by

$$Q = (E_y + E_b) - E_a$$

As we are considering the intermediate stage

$$(E_y + E_b) = E_c$$

where E_c is the kinetic energy of the compound nucleus given by

$$E_c = \frac{1}{2} M_c V_c^2$$

$$\therefore \qquad Q = \frac{1}{2} M_c V_c^2 - \frac{1}{2} m_a v_a^2$$

or
$$-Q = \frac{1}{2}m_a v_a^2 - \frac{1}{2}M_c V_c^2$$
$$= \frac{1}{2}m_a v_a^2 - \frac{1}{2}(M_x + m_a)\left(\frac{m_a}{M_x + m_a}\right)^2 v_a^2$$
$$= \frac{1}{2}m_a v_a^2 - \frac{1}{2}m_a v_a^2\left(\frac{m_a}{M_x + m_a}\right)$$
$$= \frac{1}{2}m_a v_a^2\left[1 - \frac{m_a}{M_x + m_a}\right]$$
$$= \frac{1}{2}m_a v_a^2\left[\frac{M_x}{M_x + m_a}\right]$$

The threshold energy $E_{th} = \frac{1}{2}m_a v_a^2$ and is given by
$$E_{th} = \frac{1}{2}m_a v_a^2 = -Q\left[\frac{M_x + m_a}{M_x}\right] = -Q\left[1 + \frac{m_a}{M_x}\right] \quad ...(v)$$

If the reaction is induced by γ-rays $m_a = 0$ and
$$E_{th} = -Q$$

As an example consider the endothermic reaction Na^{23} $(n\ \alpha)$ F^{20} having a Q-value = – 5.4 MeV for which threshold energy E_{th} has been calculated in **Example 17.2.**

If $Q = 0$ and if the particles are the same before and after the reaction, it is an *elastic collision.*

Significance. The Q-value of a nuclear reaction denotes the energy released or absorbed in such a reaction. This gives a physical verification of conversion of mass into kinetic energy and vice-versa.

For an exoergic reaction, Q-value is positive *i.e.* the process is accompanied by the liberation of energy.

For an endoergic reaction Q-value is negative *i.e.*, the process is accompanied by the absorption of energy.

Example 17.2. *The Q-value of Na^{23} (n, α) F^{20} reaction is – 5.4 MeV. Determine the threshold energy of the neutrons for this reaction. Given mass of neutron = 1.00866 a.m.u. and mass of Na^{23} = 22.99097 a.m.u.* ***(P.U. 2001)***

Solution. Mass of the incident particle (neutron), m_a = 1.00866 a.m.u., mass of the target nucleus (Na^{23}) = M_x = 22.99097 a.m.u.
$$Q\text{-value} = -5.4 \text{ MeV}$$

Now threshold energy of Na^{23} $(n\ \alpha)$ F^{20} reaction which is as under
$${}_{11}Na^{23} + {}_0n^1 = {}_{10}F^{20} + {}_2He^4$$
is given by
$$E_{th} = -Q\left[1 + \frac{m_a}{M_x}\right] = -Q\left[\frac{M_x + m_a}{M_x}\right]$$
$$= -(-5.4)\frac{22.99097 + 1.00866}{22.99097} = 5.637 \text{ MeV}$$

Example 17.3 *Show by mass-energy calculations whether reactions ^{14}N (α, p) ^{17}O and ^{7}Li (p, α) ^{4}He are exothermic or endothermic. Given atomic masses of ${}_7N^{14}$, ${}_2He^4$, ${}_8O^{17}$ and ${}_1H^1$ are 4.00753, 4.00260, 17.00450, 7.01822, and 1.00814 a.m.u. respectively.*

Solution. The reaction $^{14}N\ (\alpha, p)\ ^{17}O$ is represented as

$$_7N^{14} + {_2He^4} \longrightarrow {_8O^{17}} + {_1H^1}$$

Taking all the particles at rest, we have

$$\begin{aligned} Q &= (14.00753 + 4.00260) - (17.00450 + 1.00814) \\ &= -(0.00251)\text{ a.m.u.} = -(0.00251 \times 931.5)\text{ MeV} \\ &= -2.338\text{ MeV} \end{aligned}$$

As Q is *negative* (< 0) the reaction is *endothermic.*

(*ii*) The reaction $^7Li\ (p,\alpha)\ ^4He$ is represented as

$$_3Li^7 + {_1H^1} \longrightarrow {_2He^4} + {_2He^4}$$

$$\begin{aligned} \therefore \quad Q &= (7.01822 + 1.00814) - (4.00260 + 4.00260) \\ &= +(0.02116)\text{ a.m.u.} = +(0.02116 \times 931.5)\text{ MeV} = +19.7\text{ MeV} \end{aligned}$$

As Q is *positive* (> 0) the reaction is *exothermic.*

17.6 NUCLEAR REACTION CROSS-SECTION

Nuclear reaction cross-section is a convenient way to express the probability that a bombarding particle will interact in a certain way with a target particle. It is supposed that each target particle presents a certain area known as its *cross-section* to the incident particles. Any incident particle that is directed at this area interacts with the target particle. Hence the greater the cross-section the greater is the likelihood of interaction.

The nuclear reaction cross-section may be defined as either

(*i*) *The probability that an event may occur when a single nucleus is exposed to a beam of particles of total flux one particle per unit area.*

OR

(*ii*) *The probability that an event may occur when a single particle is shot perpendicularly at a target consisting of one particle per unit area.*

Determination of cross-section. Suppose we have a slab of some material whose area is A and whose thickness is dx.

Then Volume of the slab $= A\ dx$

Let the material contain n atoms per unit volume, then

Total number of nuclei in the slab $= n\ Adx$

Further, we assume that each nucleus has a cross-section σ for some particular interaction, then

Aggregate cross-section for all the nuclei in the slab $= \sigma\ n\ A\ dx$

Let N be the number of incident particles in a bombarding beam and dN the number of particles that interact with nuclei in the slab, then

$$\frac{\text{Number of interacting particles}}{\text{Number of incident particles}} = \frac{\text{Aggregate cross-section}}{\text{Target area}}$$

or
$$\frac{dN}{N} = \frac{\sigma\, n\, Adx}{A} = \sigma\, n\, dx$$

or
$$\sigma = \frac{dN/N}{n\, dx}$$

Here σ give the cross-section per nucleus or *microscopic cross-section.*

Unit of cross-section. In the above discussion, we have assumed that the target is so thin that there is no over lapping of nuclei and every nucleus presents its entire effective cross-section to the incoming beam.

σ has the dimensions of area. The unit normally used for nuclear cross-section is a *barn*.

$$1 \text{ barn} = 1 \text{ b} = 10^{-28} \text{ m}^2 = 100 \text{ fm}^2$$

Number of particles emerging out of slab of finite thickness. Now consider the same beam of particles incident on a slab of finite thickness *x*. If each particle can interact only once, then dN represents the number of particles removed from the beam in passing through the first dx thickness of the slab Fig. (17.2). Hence

$$\frac{-dN}{N} = \sigma\, n\, dx$$

The *negative* sign indicates that the number *N decreases* by dN in passing through the thickness dx.

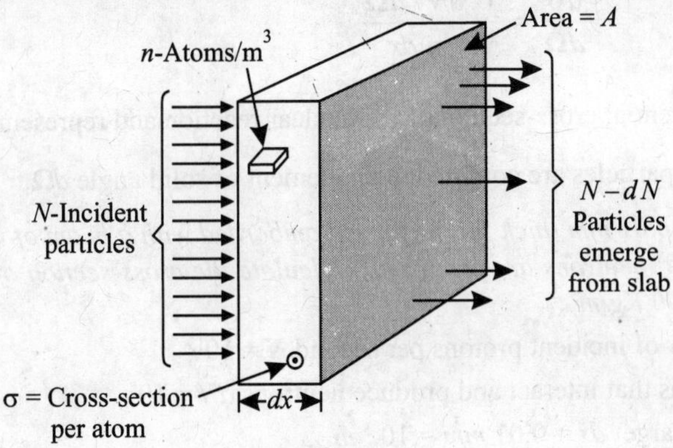

Fig. 17.2

If N_0 is the initial number of incident particles, then

$$\int_{N_0}^{N} \frac{dN}{N} = -\sigma n \int_0^x dx$$

or

$$\ln \frac{N}{N_0} = -\sigma\, nx$$

$\therefore$

$$N = N_0\, e^{-\sigma n x} \quad \text{....}(ii)$$

Thus *N* the number of surviving particles decreases exponentially with increasing slab thickness *x*.

Macroscopic cross-section. The product $\sigma\, n$ *i.e.*, the product of microscopic cross-section σ and *n* the number of nuclei per unit volume is known as *macroscopic cross-section* and is denoted by Σ

$\therefore$

$$\Sigma = \sigma\, n \quad \text{..}(iii)$$

and relation (*ii*) can be put in the form

$$N = N_0 e^{-\Sigma x} \quad \text{..}(iv)$$

Σ has the dimensions of L^{-1}.

From relation (*i*) and (*ii*) we find that σ as well as Σ are both *independent of A the target area.*

Differential cross-section. In many nuclear reactions, the product particles are not produced in an isotropic manner. In such a case, if dN is the number of product nuclei emitted per unit time in a small solid angle $d\Omega$ at some angle Θ with the direction of the incident beam as shown in Fig. 17.3, then

$$\frac{dN}{N} = n\, dx\, d\sigma$$

where $d\sigma$ is the small cross-section corresponding to small solid angle $d\Omega$.

$$\therefore \quad \frac{1}{N}\frac{dN}{d\Omega} = n\,dx\left(\frac{d\sigma}{d\Omega}\right)$$

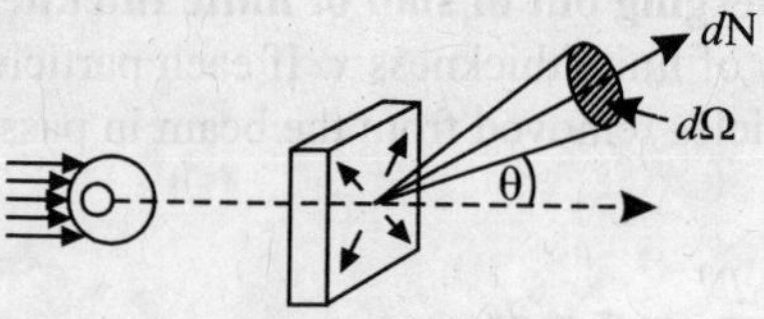

Fig. 17.3

or
$$\frac{d\sigma}{d\Omega} = \frac{1}{N}\frac{dN/d\Omega}{n\,dx}$$

$\left(\frac{d\sigma}{d\Omega}\right)$ is called differential cross-section of the nuclear reaction and represents the probability that the out going product particles are confined to an element of solid angle $d\Omega$.

Example 17.4. *A 0.01 mm thick $_3Li^7$ target is bombarded with a beam of intensity 10^{13} protons per sec. As a result 10^8 neutrons are produced. Calculate the cross-section of the reaction. Given density of lithium = 500 Kg/m³.*

(P.U. 2002, 2001)

Solution. Number of incident protons per second $N = 10^{13}$

Number of protons that interact and produce neutrons $dN = 10^8$

Thickness of the target $dx = 0.01\ mm = 10^{-5}\ m$

Number of atoms of $_3Li^7$ per m^3, $n = \dfrac{500}{7 \times 1.66 \times 10^{-27}} = 4.3 \times 10^{28}$

Now (microscope) cross-section $\sigma = \dfrac{dN/N}{n\,dx} = \dfrac{10^8/10^{13}}{10^{-5} \times 4.3 \times 10^{28}}$

$$= \frac{1}{4.3} \times 10^{-28} = 0.23 \times 10^{-28}\ m^3$$

$$= 0.23 \text{ barn}$$

17.7 DIRECT NUCLEAR REACTION

A nuclear process in which energy introduced into the atomic nucleus is imparted preferentially to one necleon or to a small group of nucleons. A number of types exist. Direct nuclear reaction can be caused by all possible incident particles, over a wide range of energies (from a few million to several billion electron volts). They are characterized by marked angular anisotropy. The nucleus formed as a result of a direct nuclear reaction is generally is either a weakly excited state or the ground state.

Direct nuclear reactions were discovered in early 1950's the first to be detected were deuteron stripping (d, p) and pickup (p, d) reactions involving light nuclei the protons and deuterons emitted in these reactions emerge primarily in the direction of the beam of incident particles. The characteristics of direct nuclear reactions can be explained by assuming that the particles emitted from the nucleus acquire energy and momentum in the process of direct reaction with the incident particle, which the remainder of the target nucleus is involved in the reaction only as on "on-looker" or "spectator". In this respect, direct nuclear reactions are quite the opposite of nuclear reactions that pass through a stage of formation of a compound nucleus, where the energy introduced in to the nucleus is statistically distibuted among all the nucleus because of multiple collisions of the nucleons with each other.

Direct nuclear reactions are used to study the spectrum of nuclear leavels and the structure of the nuclear periphery, especially peripheral correlated groups of nucleons (Clusters).

17.8 NUCLEAR REACTION CHANNELS

A set of coupled integral equations (CIE) for multichannel nuclear reactions is comsidered which appears to have same advantages over the conventional coupledreaction on channel (CRC) equations when transfer channels must be included. Both the CIE and CRC formalism are based on the assumption that the system has a limited number of cluster type intermediate states available to it. The CIE have a connected kernel while those of the CRC do not. The statistical model of Nuclear reaction is devoted to the treatment of beinary reactions of the type

$$a + A \longrightarrow CN^* \longrightarrow b + B$$

wherer a is the projectile, A the target, CN* the intermediate excited compound nucleus, b the ejectile and B the residual nucleus. The Schrödinger equations for such a complicated interaction is impossible to solve by taking into account of all the interacting particles.

Nuclear reactions are described by several models which are linked together to calculate nuclear cross-sections. In this process, the basic model is the *optical model,* whch enables us to separate the total cross-section into various components. One mainly distinguished between two kinds of processes according to their reaction times. On the one hand, one has the fast processes, with reaction times camparable to the time needed by the poojectile to cross the nucleus without interacting with it and on the other hand, rather slow processes which may take much longer time to occur. The statistical (or compound nucleus) models of nuclear reactions belongs to the second kind of processes. In the statistical model, it is assumed that a compound nucleus is formed with an excitation energy high enough so that many states may be excited by the interaction of the incident particle with the target nucleus. Mereover, the incident energy is shared between the individual components of the nucleus which fully equilibrates before decay takes place. Thus picture implies an independance hypothesis, meaning that the compound nucleus decays without remembering the way it has been formed. In other words, in the statistical model picture, there exists no corelation between the formation and decay of the compound nucleus.

17.9 COMPOUND NUCLEUS

The theory of formation of compound nucleus was proposed by N. Bohr in 1936. According to this theory, the nuclear reaction takes place in two steps.

(*i*) formation of the compound nucleus and

(*ii*) decay of the compound nucleus giving rise to product nucleus and emitted particles.

(*i*) Formation of compound nucleus. The incident particle strikes the target nucleus and the two combine to form a new nucleus called a *compound nucleus.* The atomic and mass numbers of the compound nucleus are respectively the sum of the atomic numbers and sum of the mass numbers of the original particles. The compound nucleus has no '*memory*' how it was formed since its nucleons are mixed together regardless of their origin.

If 'X' is the target nucleus and 'a' the incident particle, then the interaction between 'X' and 'a' leads to the formation of compound nucleus C as under

$$X + a = C^*$$

The incident particle captured by a target nucleus makes available to the compound nucleus an energy equal to its kinetic energy plus its binding energy in the compound nucleus. This energy is quickly distributed among all the nucleons of the compound nucleus due to the interaction of these nucleons with all others. The compound nucleus is, therefore, in an *excited state.* If E_a is the kinetic energy of the incident particle of mass m_a, the excitation energy of the compound nucleus is given by

$$E^* = E_a \frac{M_x}{M_x + m_a} + E_B$$

where M_x is the mass of the target nucleus X and E_B the binding energy of the incident particle in the compound nucleus.

Life time of compound nucleus. The compound nucleus has a life time of the order of 10^{-16} s (10^{-14} s to 10^{-18} s). This life time may appear very short even to observe such nuclei directly but it is sufficiently long as compared to the **natural nuclear time** *i.e.*, the time taken by a nuclear particle with an energy of several MeV to pass through a nucleus (10^{-22} s).

***(ii)* Decay scheme of compound nucleus.** During the life time of 10^{-16} s of the nucleus there can always be statistical fluctuations in the energy distribution among various nucleons. The distribution of available energy between nucleons, therefore, takes place in a random manner. At a given instant, the excitation energy may be shared among several nucleons and at a later time it may be shared by some other nucleons. It is also possible that the excitation energy is concentrated on one nucleon or on a group of nucleons. If this excitation energy is large enough the nucleon or the combination of nucleons may escape and the compound nucleus may disintegrate into a product nucleus and an outgoing particle *i.e.*,

Compound nucleus = product nucleus + emitted particle

The emitted particle may be a proton, a neutron, α-particle or a γ-ray. The energy that must be concentrated on the particle or group of particles in order to separate it from the compound nucleus is called the *separation or dissociation energy* and is usually of the order of 8 MeV.

Explanation on liquid drop model. The formation and decay of a compound nucleus can be explained on the basis of '*liquid drop*' model of the nucleus. According to this model an excited nucleus is similar to a drop of hot liquid with binding energy of the emitted particle corresponding to the heat of vaporisation of the liquid molecules. The evaporation process occurs when random fluctuations in energy distribution within the drop cause a particular molecule to have enough energy to escape. Similarly a compound nucleus persists in its excited state until a particular nucleon or a group of nucleons momentarily happens to possess a sufficiently large fraction of the exciting energy to leave the nucleus.

Example 17.5. *What does the following reaction represent ? Also write about the compound nucleus formed.* ***(P. U. 2004)***

Solution. The notation ${}_7N^{14}\ (\alpha, p)_8O^{17}$ represents the following nuclear reaction

$${}_7N^{14} + {}_2He^4 \rightarrow \underbrace{{}_9F^{18}}_{\text{Compound nucleus}} \rightarrow {}_8O^{17} + {}_1H^1$$

${}_9F^{18}$ is the *compound nucleus* formed before the final products are obtained.

17.10 NUCLEAR TRANSMUTATION REACTIONS

Transmutation is defined as conversion of one element into another i.e., one type of atom into another. Natural radioactivity provides examples of transmutation of elements occurring through natural process. Artificial transmutation is the conversion of one element into another by artificial means, say, by using fast moving charged particles or neutrons or high energy photons. The first **successful** attempt to convert one element into another was made by Rutherford.

Transmutation by charged particles. The charged particles generally used are protons, deutrons and α-particles. The first state in majority of such reactions is the formation of a *compound nucleus* and for this purpose the incident particle must overcome or penetrate the coulomb potential barrier.

1. Transmutation by protons. When the bombarding particle is a proton it may give rise to a $(p-\alpha)$, $(p-d)$ $(p-n)$ or a $(p-\gamma)$ reaction.

***(i)* $(p-\alpha)$ reaction.** A general $(p-\alpha)$ reaction may be represented as

$$_ZX^A + {}_1H^1 \longrightarrow ({}_{Z+1}D^{A+1})^* \rightarrow {}_{Z-1}Y^{A-3} + {}_2He^4$$

A few familiar examples are

$$_5B^{11} + {}_1H^1 \longrightarrow ({}_6C^{12})^* \rightarrow {}_4Be^8 + {}_2He^4$$

$$_{11}Na^{23} + {}_1H^1 \longrightarrow ({}_{12}Mg^{24})^* \rightarrow {}_{10}Ne^{24} + {}_2He^4$$

$$_{13}Al^{27} + {}_1H^1 \longrightarrow ({}_{14}Si^{28})^* \rightarrow {}_{12}Mg^{24} + {}_2He^4$$

Note In the first reaction, $_4Be^8$ formed is unstable and splits into two more α-particles

$$_4Be^8 \longrightarrow {}_2He^4 + {}_2He^4$$

***(ii)* $(p-n)$ reaction.** A general $(p-n)$ reaction may be represented as

$$_ZX^A + {}_1H^1 \longrightarrow ({}_{Z+1}D^{A+1})^* \rightarrow {}_{Z+1}D^A + {}_0n^1$$

A few familiar examples are

$$_5B^{11} + {}_1H^1 \longrightarrow ({}_6C^{12})^* \rightarrow {}_6C^{11} + {}_0n^1$$

$$_{11}Na^{23} + {}_1H^1 \longrightarrow ({}_{12}Mg^{24})^* \rightarrow {}_{12}Mg^{23} + {}_0n^1$$

$$_{29}Cu^{65} + {}_1H^1 \longrightarrow ({}_{30}Zn^{66})^* \rightarrow {}_{30}Zn^{65} + {}_0n^1$$

The reaction is generally endo-ergic.

***(iii)* $(p-d)$ reaction.** A general $(p-d)$ reaction may be represented as

$$_ZX^A + {}_1H^1 \longrightarrow {}_ZX^{A-1} + {}_1H^2$$

A few familiar examples are

$$_3Li^7 + {}_1H^1 \longrightarrow {}_3Li^6 + {}_1H^2$$

$$_4Be^9 + {}_1H^1 \longrightarrow {}_4Be^8 + {}_1H^2$$

Such reactions are more common with heavier elements and protons of fairly high energy. These are *pick-up* type reactions and *do not form any compound nucleus.*

***(iv)* $(p-\gamma)$ reaction.** This is also known as *photon-capture* reaction. A general $(p-\gamma)$ reaction may be represented as

$$_ZX^A + {}_1H^1 \longrightarrow ({}_{Z+1}D^{A+1})^* \rightarrow {}_{Z+1}D^{A+1} + \gamma$$

A few familiar examples are

$$_3Li^7 + {}_1H^1 \longrightarrow ({}_4Be^8)^* \rightarrow {}_4Be^8 + \gamma$$

$$_7N^{14} + {}_1H^1 \longrightarrow ({}_8O^{15})^* \rightarrow {}_8O^{15} + \gamma$$

$$_{24}Cr^{50} + {}_1H^1 \longrightarrow ({}_{25}Mn^{51})^* \rightarrow {}_{25}Mn^{51} + \gamma$$

2. Transmutation by deutrons. Accelerated deutrons give rise to a number of reactions. Various types are $(d-\alpha)$, $(d-p)$, $(d-n)$ and $(d-t)$.

***(i)* $(d-\alpha)$ reaction.** A general $(d-\alpha)$ reaction may be represented as

$$_ZX^A + {}_1H^2 \longrightarrow ({}_{Z+1}D^{A+2})^* \rightarrow {}_{Z-1}Y^{A-2}\ {}_2He^4$$

A few familiar examples are

$$_8O^{16} + {}_1H^2 \longrightarrow ({}_9F^{18})^* \rightarrow {}_7N^{14} + {}_2He^4$$

$$_{10}Ne^{20} + {}_1H^2 \longrightarrow ({}_{11}Na^{22})^* \rightarrow {}_9F^{18} + {}_2He^4$$

$$_{20}Ca^{40} + {}_1H^2 \longrightarrow ({}_{21}Sc^{42})^* \rightarrow {}_{19}K^{38} + {}_2He^4$$

***(ii)* $(d-p)$ reaction.** A general $(d-p)$ reaction may be represented as

$$_ZX^A + {}_1H^2 \longrightarrow ({}_{Z+1}D^{A+2})^* \rightarrow {}_ZX^{A+1} + {}_1H^1$$

A few familiar examples are

$$_3Li^7 + {_1H^2} \longrightarrow ({_4Be^9})^* \rightarrow {_3Li^8} + {_1H^1}$$

$$_{15}P^{31} + {_1H^2} \longrightarrow ({_{16}S^{33}})^* \rightarrow {_{15}P^{32}} + {_1H^1}$$

$$_{48}Cd^{114} + {_1H^2} \longrightarrow ({_{49}In^{116}})^* \rightarrow {_{48}Cd^{115}} + {_1H^1}$$

***(iii)* (*d* – *n*) reaction.** A general (*d* – *n*) reaction may be represented as

$$_ZX^A + {_1H^2} \longrightarrow ({_{Z+1}D^{A+2}})^* \rightarrow {_{Z+1}D^{A+1}} + {_0n^1}$$

A few familiar examples are

$$_3Li^7 + {_1H^2} \longrightarrow ({_4Be^9})^* \rightarrow {_4Be^8} + {_0n^1}$$

$$_6C^{12} + {_1H^2} \longrightarrow ({_7N^{14}})^* \rightarrow {_7N^{13}} + {_0n^1}$$

$$_{83}Bi^{209} + {_1H^2} \longrightarrow ({_{84}Po^{211}})^* \rightarrow {_{84}Po^{210}} + {_0n^1}$$

***(iv)* (*d* – *t*) reaction.** The emitted particle in this case is a tritium $_1H^3$. The probability of a (*d* – *t*) reaction is *low.* A general (*d* – *t*) reaction may be represented as

$$_ZX^A + {_1H^2} \longrightarrow ({_{Z+1}D^{A+2}})^* \rightarrow {_ZX^{A-1}} + {_1H^3}$$

A few familiar examples are

$$_3Li^7 + {_1H^2} \longrightarrow ({_4Be^9})^* \rightarrow {_3Li^6} + {_1H^3}$$

$$_{15}P^{31} + {_1H^2} \longrightarrow ({_{16}S^{33}})^* \rightarrow {_{15}P^{30}} + {_1H^3}$$

$$_{47}Ag^{109} + {_1H^2} \longrightarrow ({_{48}Cd^{111}})^* \rightarrow {_{47}Ag^{108}} + {_1H^3}$$

When the energy of the incident deutron is about 20 MeV or more reactions of the type (*d* – 2*n*), (*d* – 2*p*), (*d* – *p*α) are also possible.

3. Transmutation by α-particles. ($_2He^4$). High energy α-particles give rise to a number of reactions. Various types are, (α – *n*), (α, *p*) and (α – γ).

***(i)* (α – *n*) reaction.** A general (α – *n*) reaction may be represented as

$$_ZX^A + {_2He^4} \longrightarrow ({_{Z+2}D^{A+4}})^* \rightarrow {_{Z+2}D^{A+3}} + {_0n^1}$$

A few familiar examples are :–

$$_4Be^9 + {_2He^4} \longrightarrow [{_6C^{13}}]^* \rightarrow {_6C^{12}} + {_0n^1}$$

$$_7N^{14} + {_2He^4} \longrightarrow [{_9F^{18}}]^* \rightarrow {_9F^{17}} + {_0n^1}$$

$$_{13}Al^{27} + {_2He^4} \longrightarrow [{_{15}P^{31}}]^* \rightarrow {_{15}P^{30}} + {_0n^1}$$

***(ii)* (α – *p*) reaction.** A general (α – *p*) reaction may be represented as

$$_ZX^A + {_2He^4} \longrightarrow ({_{Z+2}D^{A+4}})^* \rightarrow {_{Z+1}Y^{A+3}} + {_1p^1}$$

A few familiar examples are :-

$$_7N^{14} + {_2He^4} \longrightarrow [{_9F^{18}}]^* \rightarrow {_8O^{17}} + {_1H^1}$$

$$_9F^{19} + {_2He^4} \longrightarrow [{_{11}Na^{23}}]^* \rightarrow {_{10}Ne^{22}} + {_1H^1}$$

$$_{13}Al^{27} + {_2He^4} \longrightarrow [{_{15}P^{31}}]^* \rightarrow {_{14}Si^{30}} + {_1H^1}$$

***(iii)* (α – γ) reaction.** A general (α – γ) *reaction* may be represented as

$$_ZX^A + {_2He^4} \longrightarrow ({_{Z+2}D^{A+4}})^* \rightarrow {_{Z+2}D^{A+4}} + \gamma$$

A familiar example is

$$_3Li^7 + {_2He^4} \longrightarrow [{_5B^{11}}]^* \rightarrow {_5B^{11}} + \gamma$$

When the energy of the incident α-particle is *very high*, the product may contain two or more nucleons i.e. protons or neutrons and reaction of the type (α–2*n*), (α–4*n*), (α - 3*p*) (α – 4*p*) may occur.

4. Transmutation by triton. Triton ($_1H^3$) induced reactions include (*t* – *p*), (*t* – *d*), (*t* – *n*), (*t* – α) *reactions.*

(*i*) **(*t* – *p*) reaction.** $_{27}Co^{59} + {}_1H^3 \rightarrow ({}_{28}Ni^{62})^* \rightarrow {}_{27}Co^{61} + {}_1H^1$

(*ii*) **(*t* – *d*) reaction.** $_{29}Cu^{63} + {}_1H^3 \rightarrow ({}_{30}Zn^{66})^* \rightarrow {}_{29}Cu^{64} + {}_1H^2$

(*iii*) **(*t* – *n*) reaction.** $_{16}S^{32} + {}_1H^3 \rightarrow ({}_{17}Cl^{35})^* \rightarrow {}_{17}Cl^{34} + {}_0n^1$

(*iv*) **(*t* – α) reaction.** $_3Li^7 + {}_1H^3 \rightarrow ({}_4Be^{10})^* \rightarrow {}_2He^6 + {}_2He^4$

5. Transmutation by neutrons. Please refer article. 17.12.

6. Transmutation by photons. Reactions brought about by high energy photons (γ-rays) are classified as '*photo-disintegration*'. These reactions are endoergic and have threshold energies of the order of 10 MeV. As a photon has zero mass it can supply only its kinetic energy to the nuclear reaction. Photo-disintegration does not occur with γ-rays from natural radioactive substances. However, there are two exceptions

(*i*) deutron - binding energy 2.2 MeV and

(*ii*) $_4Be^9$ nuclide in which one neutron is loosely bound. Reactions of this type are usually (γ – *n*) and (γ – *p*).

(*i*) (γ – *n*) reaction. $_1H^2 + \gamma \longrightarrow ({}_1H^2)^* \rightarrow {}_1H^1 + {}_0n^1$

$$_4Be^9 + \gamma \longrightarrow ({}_4Be^9)^* \rightarrow {}_4Be^8 + {}_0n^1$$

$$_{15}P^{31} + \gamma \longrightarrow ({}_{15}P^{31})^* \rightarrow {}_{15}P^{30} + {}_0n^1$$

(*ii*) (γ – *p*) reaction. $_4Be^9 + \gamma \longrightarrow ({}_4Be^9)^* \rightarrow {}_3Li^8 + {}_1H^1$

$$_{12}Mg^{25} + \gamma \longrightarrow ({}_{12}Mg^{25})^* \rightarrow {}_{11}Na^{24} + {}_1H^1$$

(*iii*) Sometime a deutron ($_1H^2$), a tritium ($_1H^3$) and an α-particle ($_2He^4$), may also be emitted. For example

(γ – *d*) reaction. $_5B^{10} + \gamma \longrightarrow ({}_5B^{10})^* \rightarrow {}_4Be^8 + {}_1H^2$

(γ – *t*) reaction. $_5B^{11} + \gamma \longrightarrow ({}_5B^{11})^* \rightarrow {}_4Be^8 + {}_1H^3$

(γ – α) reaction. $_6C^{12} + \gamma \longrightarrow ({}_6C^{12})^* \rightarrow {}_4Be^8 + {}_2He^4$

(*iv*) With γ-rays of energy more than 10 MeV reactions of the type (γ – *n*, *p*), (γ – 2*n*), (γ – *n*, 2*p*) may occur. For example

$$_8O^{16} + \gamma \longrightarrow ({}_8O^{16})^* \rightarrow {}_6C^{11} + 3\,{}_0n^1 + 2\,{}_1H^1$$

$$_{13}Al^{27} + \gamma \longrightarrow ({}_{13}Al^{27})^* \rightarrow {}_{11}Na^{24} + {}_0n^1 + 2\,{}_1H^1$$

17.11 NEUTRON AND IT'S PROPERTIES

Neutron is one of the fundamental particles constituting the nuclei of the atoms. It is a *neutral* particle having a mass 1.00898 atomic mass units - nearly equal to that of the proton. It is a spin 1/2 particle and obeys Fermi-Dirac statistics.

Discovery of Neutron. In 1930, Bothe and Becker bombarded beryllium with α-particles and found that a very penetrating radiation was produced. As this radiation was found to be uncharged it was thought to be high energy γ-radiation produced according to the nuclear reaction,

$$_4Be^9 + {}_2He^4 \longrightarrow {}_6C^{13} + h\nu.$$

The energy of the γ-ray was found to be 7 MeV which is greater than the energy of any γ-ray known at that time. In 1932, Curie and Joliot found that when the penetrating radiation from this reaction was made to fall on paraffin wax or other materials containing. hydrogen, protons with energies of about 6 MeV were ejected. It was calculated that to produce protons of 6 MeV the original photon must have an energy of 60 MeV according to Compton collision of the high energy photon with the proton. An energy of 6 MeV could not be released in the reaction given according to the decrease in mass that took place. Chadwick showed that the experimental results could be explained if we assume that an uncharged and, therefore, highly penetrating particle, having a mass almost equal to that of a proton was emitted in the above reaction as

$$_4Be^9 + {}_2He^4 \longrightarrow {}_6C^{12} + {}_0n^1 \qquad ...(i)$$

This new particle was given the name neutron ($_0n^1$). By the principle of conservation of mass energy, the neutron would be emitted with an energy of 6 MeV and when it strikes against a proton in a head-on collision, the neutron would come to rest and proton would be released with the energy of 6 MeV.

Chadwick confirmed his hypothesis by the following experiment.

α-particles obtained from the source- Polonium (Po) deposited on a disc *D* were made to bombard a beryllium (Be) target *T* (Fig. 17.4). Both the disc and the target were enclosed in an evacuated chamber *B*. The neutrons released from the reaction given in (*i*) escape through the thin walls of the chamber and enter an ionisation chamber *C* though a thin window *W*. The ionisation chamber is connected to an amplifier and a counter. As the neutrons are uncharged, neutrons themselves cannot produce ionisation in the chamber directly. Therefore, when the radiations from *B* entered *C*, very few pulses per minute were registered. These pulses are due to the ions ejected by the neutrons from the walls of the chamber. On placing a lead sheet about 2 cm thick in the path of the radiation in front of *W*, it was found that the number of pulses was not reduced appreciably. This showed that the radiations coming out of the *Be* target were highly energetic.

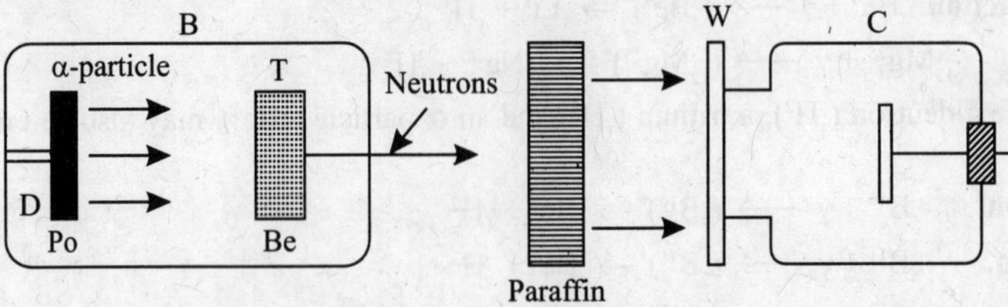

Fig. 17.4

Further, on introducing a thin sheet of paraffin in place of lead the number of counts increased considerably. This is due to the fact that neutrons eject fast moving protons from paraffin which enter the ionisation chamber through the window *W* and are counted.

Production of neutrons. The important methods of producing neutrons are :

(*i*) Radium-α-Beryllium source. This method has been extensively used to provide weak sources of neutrons. In this (α – *n*) reaction, neutrons are produced by the bombardment of beryllium with high energy α-particles according to the reaction

$$_4Be^9 + {}_2He^4 \longrightarrow {}_6C^{12} + {}_0n^1$$

(*ii*) Photo-disintegration source. Neutrons can be knocked out from nuclei by high energy photons such as γ-rays obtained from naturally, or artificially radioactive materials. As for example,

$$_4Be^9 + \gamma \longrightarrow {}_4Be^8 + {}_0n^1.$$

The neutrons thus obtained are mono energetic.

(*iii*) Particle accelerators. Charged particles accelerated by particle accelerators produce neutrons of definite energy, by bombarding suitable targets. As an example, bombardment of Tritium $_1H^3$ with high energy deutrons or lithium with high energy protons produced neutrons according to the reactions.

$$_1H^3 + {}_1D^2 \longrightarrow {}_2He^4 + {}_0n^1$$

$$_3Li^7 + {}_1H^1 \longrightarrow {}_4Be^7 + {}_0n^1$$

***(iv)* From atomic reactors.** The best source of neutrons is an atomic reactor working on the principle of nuclear fission.

***(v)* Stripping action.** Neutrons may also be obtained when a high energy deutron strikes a target and is *stripped* of its proton, leaving the neutron alone.

Detection. Neutrons produce charged particles by their interaction with matter. The ionisation produced by these charged particles is used for detection of neutrons. The kinetic energy of the neutrons can also be measured by these methods after suitable calibration of the experiment.

***(i)* Ionisation chamber method.** The ionisation chamber is filled with a gaseous boron compound such as BF_3 (Boron trifluoride) or lined with a solid boron compound. When a neutron strikes boron, it liberates α-particles according to the reaction.

$$_5B^{10} + {_0n^1} \longrightarrow {_3Li^7} + {_2He^4}$$

The α-particle is detected as usual by the ionisation chamber.

***(ii)* Induced radioactivity method.** Many materials become radioactive under neutron bombardment and can serve as neutron detectors, *e.g.,* when a neutron strikes a silver foil, the new silver isotope formed is radio-active. The nuclear reaction that takes place is

$$_{47}Ag^{107} + {_0n^1} \longrightarrow {_{47}Ag^{108}} + \gamma$$

The product decays according to the reaction

$$_{47}Ag^{108} \longrightarrow {_{48}Cd^{108}} + {_{-1}e^0}$$

The β-particle emitted can be detected and its energy measured.

***(iii)* Elastic collision method.** In this method, the ionisation chamber is filled with hydrogen or is provided with a window made of hydrogenous material. The neutron makes an elastic collision with a proton and is brought to rest. The proton moves forward with the same energy and is detected by the ionisation it produces or by any suitable method.

Fast, slow and thermal neutrons. Neutrons with energies above 1.2 MeV are called fast neutrons, because 1.2 MeV is the minimum energy for fission of $_{92}U^{235}$. These neutrons have energies upto 10 MeV. Many nuclear reactions which are energetically not possible at lower neutron energies take place with fast neutrons.

Neutrons within the energy interval 10-50 MeV are called **very fast neutrons** and those beyond 50 MeV are called **ultra fast** or **ultra high energy neutrons.**

Neutrons with energies below 1eV are called **slow** neutrons.

Neutrons with energy between 1 eV and 1.2 MeV are called *epithermal neutrons. Those neutrons which are in thermal equilibrium with the matter through which they pass and have a Maxwellian distribution of energies with a most probable value depending upon temperature, are called* **thermal neutrons.** The most probable energy of thermal neutron at 27°C is 0.025 eV. When neutrons pass through a material they are slowed down by collisions with the nuclei of the material and lose a part of their energy. For heavier nuclei the loss is very small and a very large number of collisions are required to slow down the neutron. For light nuclei the loss is large and neutrons are slowed down quickly. Materials rich in hydrogen are very efficient in slowing neutrons and are known as *moderators.* Cadmium has the property of absorbing neutrons and is known as *absorber.*

Almost all elements can be transmuted by neutrons. Some of the elements require fast neutrons whereas others require slow ones. Thus a fast neutron ejects an α-particle from a nitrogen nucleus thereby converting it into a boron nucleus.

$$_7N^{14} + {_0n^1} = {_5B^{11}} + {_2He^4}$$

When slow moving neutrons fall on certain substances, they are captured with the emission of a high energy γ-radiation. The new substance formed is radioactive, *e.g.,*

$$_{29}Cu^{65} + _0n^1 \longrightarrow _{29}Cu^{66} + h\nu$$

$$_{29}Cu^{66} \longrightarrow _{30}Zn^{66} + _{-1}e^0 \text{ (electron)}$$

The half-life period of radioactive copper $_{29}Cu^{66}$ is only 5 minutes.

17.12 BASIC PROPERTIES OF A NEUTRON

1. The neutron is one of the fundamental particles constituting the nuclei of the atoms.

2. Charge. A neutron is a neutral particle having no charge. It is, therefore, not deflected by electric and magnetic fields. Being uncharged it has a very low ionising power. Hence it does not lose energy on passing through matter and is readily able to collide with nuclei bringing about various types of nuclear reactions.

3. Mass. The neutron has a mass 1.00898 atomic mass units (1.675×10^{-27} kg = 940 MeV). The mass of the proton is 1.00759 atomic mass units and that of the electron is 0.00055 atomic mass units. The mass of the neutron, therefore, exceeds the sum of the masses of the proton and electron by 0.00084 a.m.u.

4. Spin and magnetic moment. A neutron has a pin 1/2 *i.e.,* its spin angular momentum is $\frac{1}{2}\hbar = \frac{1}{2}\frac{h}{2\pi}$. It is thus a fermion and obeys Fermi-Dirac statistics. The magnetic moment of the neutron = – 1.913 nuclear magnetons. The negative sign for magnetic moment indicates that the magnetic moment and spin vectors are in opposite directions.

5. Penetrating power. The neutrons are very penetrating. Being uncharged a neutron is neither effected by the positive charge on the nucleus nor by the negative charge on the extra nuclear electrons. Their intensity is reduced to one half in passing through about 1.5 cms thickness of lead. They have the property of expelling high speed protons from hydrogenous matter, such as paraffin, water, paper, cellophane, etc. From the range of protons knocked out from paraffin, the energy of neutrons has been calculated. Neutrons resemble X-rays and γ-rays in so far as they do not produce a track in a Wilson cloud chamber but differ in one respect that neutrons produce no ionisation. They are less penetrating than cosmic rays.

6. Biological effect. The neutrons are now being widely used in biological and medical sciences. Because of their intense biological effects they are now being used in the treatment of cancer.

Fast neutrons have been used for disintegration of other nuclei and slow ones in atomic fission.

7. Decay mode. In a free state, the neutron is unstable. A free neutron decays into a proton with the emission of an electron (*and an anti-neutrino*) *to conserve the angular momentum* according to the reaction

$$_0n^1 \longrightarrow _1H^1 + _{-1}e^0 + \bar{\nu}$$

where $_1H^1$ represents a proton, $_{-1}e^0$ an electron and $\bar{\nu}$ an antineutrino. The maximum energy of the electron is 0.783 MeV.

17.13 ROLE OF NEUTRON IN NECLEAR REACTIONS

Neutrons as bombarding agents. Neutrons carry no charge. As they pass through matter, they are neither affected by the positive charge on the nucleus nor by the negative charge on the extra nuclear electrons. It, therefore, does not ordinarily lose energy on its passage through matter and is readily able to collide with nuclei bringing about various types of nuclear reactions.

The interaction of neutrons with nuclei may result in the capture of the neutrons to from temporarily a compound nucleus. This compound nucleus will have the same atomic number but its mass number will be increased by one due to the addition of the neutron. The compound nucleus is generally unstable and emits energy in the form of a γ-ray photon, α-particle, a proton and very rarely a deutron, a single neutron or a pair of neutrons. These reactions are respectively known as (n, γ), (n, α), (n, p), (n, n) and $(n, 2n)$ reactions.

Nuclear reactions. *(i)* n, γ reaction. Low energy neutrons are captured by the nuclei which is raised to an excited state. It then returns to the ground state by the emission of a γ-ray photon. The general reaction is represented by

$$_{Z}X^{A} + {}_{0}n^{1} \longrightarrow ({}_{Z}X^{A+1})^{*} \rightarrow {}_{Z}X^{A+1} + h\nu \text{ (}\gamma\text{-ray)}$$

The final product is the isotope of the target nucleus. An example of this type is

$$_{13}Al^{27} + {}_{0}n^{1} \longrightarrow ({}_{13}Al^{28})^{*} \rightarrow {}_{13}Al^{28} + h\nu \text{ (}\gamma\text{–ray)}$$
$$H^{1} + {}_{0}n^{1} \longrightarrow ({}_{1}H^{2})^{*} \rightarrow {}_{1}H^{2} + h\nu \text{ (}\gamma\text{-ray)}$$
$$_{92}U^{238} + {}_{0}n^{1} \longrightarrow ({}_{92}U^{239})^{*} \rightarrow {}_{92}U^{239} + h\nu \text{ (}\gamma\text{-ray)}$$

***(ii)* (n, α) reaction.** When a nucleus is bombarded with a neutron and the product nucleus decays with the emission of an α-particle, the general reaction is represented by

$$_{Z}X^{A} + {}_{0}n^{1} \longrightarrow ({}_{Z}X^{A+1})^{*} \rightarrow {}_{z-2}Y^{A-3} + {}_{2}He^{4}$$

The final product is a new element whose atomic number is 2 less than the target nucleus and atomic weight is 3 less than that of the target nucleus. An example of this type is

$$_{5}B^{10} + {}_{0}n^{1} \longrightarrow ({}_{5}B^{11})^{*} \rightarrow {}_{3}Li^{7} + {}_{2}He^{4}$$
$$_{13}Al^{27} + {}_{0}n^{1} \longrightarrow ({}_{13}Al^{28})^{*} \rightarrow {}_{11}Na^{24} + {}_{2}He^{4}$$
$$_{11}Na^{23} + {}_{0}n^{1} \longrightarrow ({}_{11}Na^{24})^{*} \rightarrow {}_{9}F^{20} + {}_{2}He^{4}$$

Followed by the emission of an electron $_{9}F^{20} \rightarrow {}_{10}Ne^{20} + {}_{0}e^{-1}$

***(iii)* (n, p) reaction.** When a nucleus is bombarded with a neutron and the product nucleus decays with the emission of proton, the general reaction is represented by

$$_{Z}X^{A} + {}_{0}n^{1} \longrightarrow ({}_{Z}X^{A+1}) \rightarrow {}_{Z-1}Y^{A} + {}_{1}H^{1}$$

The final product is a new element whose atomic number is one less than the target nucleus and atomic weight is the same. As an example

$$_{12}Mg^{24} + {}_{0}n^{1} \longrightarrow ({}_{12}Mg^{25}) \rightarrow {}_{11}Na^{24} + {}_{1}H^{1}$$

The product in all cases of (n, p) reaction is radioactive and decays with the emission of β-particle *e.g.*,

$$_{11}Na^{24} \longrightarrow {}_{13}Mg^{24} + {}_{-1}e^{0}$$

The half life of radio-sodium ($_{11}Na^{24}$) is 15 hours and the maximum energy of the β-particle emitted is 1.39 MeV. The final product is the same as the target nucleus. Thus the overall effect is that the bombarding neutron splits up into a proton and an electron.

Other examples are $_{7}N^{14} + {}_{0}n^{1} \longrightarrow ({}_{7}N^{15})^{*} \rightarrow {}_{6}C^{14} + {}_{1}H^{1}$

followed by $\quad {}_{6}C^{14} \longrightarrow {}_{7}N^{14} + \beta^{-} ({}_{0}e^{-1}) + E$

$$_{13}Al^{27} + {}_{0}n^{1} \longrightarrow ({}_{13}Al^{28})^{*} \rightarrow {}_{12}Mg^{27} + {}_{1}H^{1}$$

***(iv)* (n,n) reaction.** Sometimes the bombarding neutron produces the target nucleus in an excited state which passes on to the ground state with the emission of a neutron accompanied by a γ-ray photon. As an example

$$_{47}Ag^{107} + {}_{0}n^{1} \longrightarrow ({}_{47}Ag^{108}) \rightarrow {}_{47}Ag^{107} + {}_{0}n^{1}$$

The product nucleus $_{47}Ag^{107}$ is in a metastable state and decays with the emission of γ-ray.

***(v)* (n, 2n) reaction.** When the kinetic energy of the bombarding neutron is about 10 MeV, it is able to eject 2 neutrons from the nucleus. For higher energies more than 2 neutrons may be ejected. The general $(n, 2n)$ reaction is

$$_{Z}X^{A} + {}_{0}n^{1} \longrightarrow {}_{Z}X^{A-1} + 2\,{}_{0}n^{1}$$

As for example

$$_{19}K^{39} + _0n^1 \longrightarrow _{19}K^{38} + 2\ _0n^1$$

In these cases, the product nucleus invariably decays with the emission of a positron. Thus

$$_{19}K^{38} \longrightarrow _{18}K^{38} + _{+1}e^0$$

(*vi*) (*n* – 3*n*) reaction. If the energy of the incident neutron approaches 30 MeV, then sufficient energy is available to overcome the potential barrier, three neutrons or even two neutrons and a proton are ejected from the compound nucleus.

Bombardment of 90 MeV neutrons may give rise to (*n* – *d*), and (*n* – *t*) reactions. These reactions do not produce any compound nucleus and are direct reactions of the pick-up type.

With neutrons of more than 100 MeV, nuclei of moderate mass number may undergo *spallation* and those of high mass number may undergo *fission.*

17.14 NUCLEAR FISSION

Artificial transmutation gathered tremendous momentum after the discovery of neutron in 1932. Around 1934, Fermi was carrying on experiments to produce transuranic elements by bombarding uranium (natural uranium is dense black in colour) with neutrons. He could produce number of such elements having atomic number greater than 92 (greatest atomic number of natural element uranium) and hence name given to these elements as transuranic elements. The process of fission was discovered by Otto Hahn and his two associates Meitner and Strassmann is 1939. In the similar experiments fission has proved as boon to man kind because it is going to be an inexhaustible source of energy in the near ficture when all fossil fuels with reach to the lowest bottom of availability.

Nuclear fission. *The process of breaking up the nucleus of a heavy atom into two, more or less equal segments with the release of a large amount of energy is known as fission.*

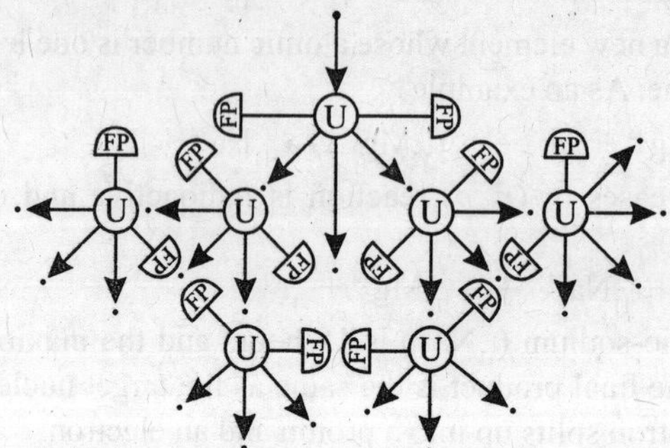

Fig. 17.5

In 1938, Hahn bombarded uranium with thermal neutrons (energy 0.025 eV) and observed that the isotope if uranium of atomic weight 235 showed a queer disintegration. It broke up into two nuclei of atomic numbers 56 and 36 and liberated 3 neutrons in addition to a tremendous amount of energy. The element of atomic number 56 *is barium* and that of 36 is *krypton.* The nuclear reaction is represented as follows:

$$_{92}U^{235}\ _0n^1 \longrightarrow _{56}Ba^{141} + _{36}Kr^{92} + 3\ _0n^1 + E$$

The presence of Barium in the disintegration products was chemically identified.

The U^{235} nuclei do not all split up into those of Ba and Kr. They may divide into the nuclei of several pairs of elements lying in the central region of the Periodic Table with slightly unequal nuclear masses. These are known as *Fission fragments.* Thus another mode of U^{235} fission is

$$_{92}U^{235} + _0n^1 \longrightarrow _{92}U^{236} \rightarrow _{40}Zr^{98} + _{52}Te^{135} + 3\ _0n^1 + E$$

If we plot percentage yield as function of the atomic mass number *A*. we get a curve of the type shown in Fig. 17.6.

The yield is maximum at $A = 95$ and 139 and the distribution is called *asymmetric.* The yield is minimum at $A = 118$ which corresponds to *symmetric* splitting of U_{235} nucleus.

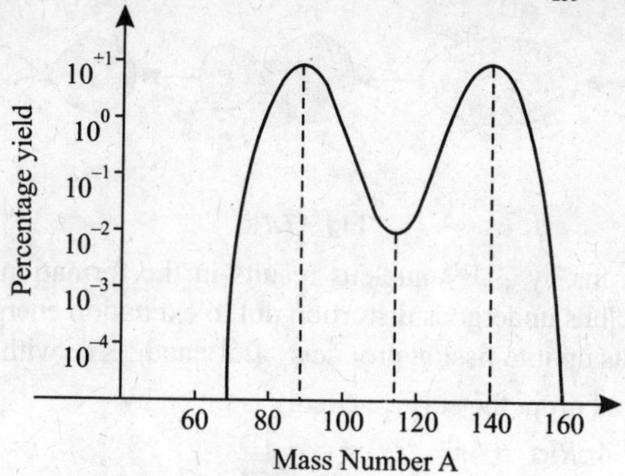

Fig. 17.6

Fission may occur spontaneously or it may be induced. Most of the heavier isotopes of the elements with $Z > 82$ show *spontaneous fission.* In such a case, the number of protons in the nucleus is very large so that the electrostatic force of repulsion between them exceeds the nuclear binding force.

Induced fission takes places when energy is supplied to the heavy nucleus by an impinging particle, thus making it unstable. Positively charged projectiles require sufficient energy to overcome the nuclear potential barrier of the target. The neutrons are the most effective fission producing particles. They carry no charge and are, therefore, neither effected by the presence of orbital electrons nor by the positive charge on the nucleus. The elements of low atomic weight have an atomic number approximately half the atomic weight. Thus the nuclei of the atoms consist of an equal number of protons and neutrons. The proton-neutron pair being the most stable formation in atomic structure, the light elements are highly stable. Heavy nuclei with even Z and odd $(A - Z)$ can undergo fission by *slow or thermal neutrons.* As these nuclei are less tightly held than those with even $(A - Z)$, the neutron binding energy alone is sufficient to cause fission in $_{92}U^{235}$ and $_{94}Pu^{239}$ etc. The nuclei with even $(A - Z)$ can be split by *fast neutrons* having kinetic energy greater than 1 MeV.

About 99% of the neutrons ejected as a result of fission of Uranium emerge within a time interval of 10^{-14} sec and are called *prompt neutrons.* The remaining 1% are ejected later and are called *delayed neutrons.* The delayed neutrons play an important role in nuclear reactors.

It is now known that fission can be induced under proper conditions in practically all the heavy nuclides. The fission activation energy can be supplied to the nucleus by particles other than neutrons, such as protons, deutrons, α-particles, γ-rays, negative π mesons, high energy carbon ion and so on. The fission induced by γ-rays is called *photo fission.* The threshold photon energy for inducing fission in many heavy nuclides is just more than 5 MeV.

Explanation on liquid drop model. Bohr and Wheeler explained the phenomenon of nuclear fission on the liquid drop model of the nucleus. The fissile nucleus is normally maintained in equilibrium under the combined action of short range nuclear forces of attraction among the nucleons in it which try to maintain the shape of the nucleus as such and the coulomb forces of repulsion among the protons in it which try to distort its shape. When some energy is imparted to the drop, say through the capture of a neutron, oscillations are set up in the drop which tend to distort the spherical shape of the nucleus, while the surface tension forces try to restore it. When the excitation energy is sufficiently large, the compound nucleus formed is in an excited state and is sufficiently distorted in shape like that of a dumb-bell. When the coulomb force of repulsion between the two halves of this dumb-bell exceeds the nuclear forces holding the nucleons, the nucleus beaks up into two fragments and fission is said to take place. The various steps from neutron capture to fission of U^{235} nucleus are shown in Fig.17.7.

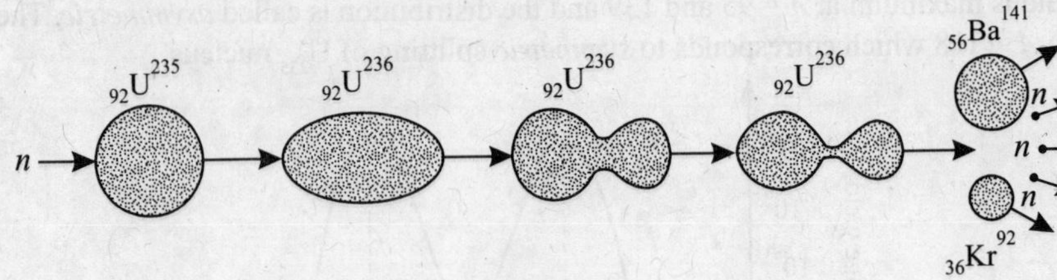

Fig. 17.7

The capture of neutrons by $_{92}U^{235}$ nucleus results in the formation of a compound nucleus $_{92}U^{236}$. The compound nucleus undergoes distortion dut to excitation energy to attain the shape of a dumb-bell and finally splits up into fission products $_{56}Ba^{141}$ and $_{36}Kr^{92}$ with release of three neutrons.

For a spherical charged drop, the surface energy is given by

$$E_s = 4\pi R^2\sigma = 4\pi r_0^2 A^{2/3}\sigma = a_s A^{2/3}$$

and the coulomb energy,

$$E_c = \frac{3}{5}\frac{Z^2e^2}{R} = \frac{3}{5}\frac{Z^2e^2}{r_0 A^{1/3}} = a_c\frac{Z^2}{A^{1/3}}$$

As shown above for the fission process to begin, the shape of the nucleus most be deformed. This means R must become large resulting in an increase of surface energy and decrease in coulomb energy. The net change in energy

$$\Delta E \propto (2E_s - E_c)$$

Hence $\Delta E = 0$ if $2E_s - E_c = 0$ or $E_c = 2E_s$

or $$\frac{a_c Z^2}{A^{1/3}} = 2a_s A^{2/3}$$

or $$Z^2/A = 2a_s/a_c$$

Now $a_s = 18$ MeV and $a_c = 0.8$ MeV approximately

$\therefore$ $$Z^2/A = \frac{2\times 18}{0.8} = 45 \text{ (approx.)}$$

Thus, according to Bohr-Wheeler theory, spontaneous fission should occur in nuclides with Z^2/A values greater than 45. For nuclides with Z^2/A value less than 45, fission is not expected to occur unless some particle is captured by the nuclei which supplies the activation energy required.

Source of energy. Taking the case of fission of U^{235} into barium and krypton, it is found that the mass of barium nucleus + krypton nucleus + 3 neutrons is less than that of the parent $_{92}U^{235}$ + neutron. This decrease in mass results in the corresponding release of energy which can be computed from the following data.

Mass of $_{92}U^{235}$ = 235.1175 a.m.u.

Mass of neutron $_1n^0$ = 1.00898 a.m.u.

$\therefore$ Total initial mass = 236.12648 a.m.u.

Mass of Barium $_{56}Ba^{141}$ = 140.9577 a.m.u.

Mass of Krypton $_{30}Kr^{92}$ = 91.9264 a.m.u.

Mass of 3 neutrons $3\ _1n^0$ = 3.02694 a.m.u.

Total final mass = 235.91104 a.m.u.

Loss in mass = 236.12648 – 235.91104 a.m.u.

= 0.21544 a.m.u.

= 0.21544 × 931.2 MeV = 200.5 MeV

Out of this energy, 170 MeV is carried by the fission fragments as kinetic energy, 5 MeV by fission neutrons, 15 MeV by β and γ-rays which are produced due to the radioactivity of the fission fragments and 10 MeV by the neutrons associated with β-decay.

The fast neutrons produced in the fission process can be thermalised *i.e*, brought down to thermal energies with the help of moderators like heavy water (D_2O) and graphite (C^{12}) and then used to initiate the fission in another U^{235} atom.

17.15 CHAIN REACTION

A chain reaction is a self-propagating process in which the number of neutrons goes on multiplying rapidly almost in geometrical progression during fission till whole of the fissionable material is disintegrated.

A single neutron causing fission in a uranium nucleus produces three *prompt* neutrons. If the three neutrons produced in the reaction disintegrate three more ${}_{92}U^{235}$ nuclei, we shall get 9 neutrons, which in turn can produce fission of nine ${}_{92}U^{235}$ nuclei and so on. Thus a **chain reaction** would start and a tremendous amount of energy will be liberated in a very short time. The chain reaction is diagrammatically illustrated in Fig. 17.5, where it has been supposed that out of the three neutrons produced, one is absorbed and remaining two neutrons keep the chain reaction going. The letters *F.P.* denote fission products and U uranium. For the chain reaction to proceed the ${}_{92}U^{235}$ lump should have a certain minimum size as otherwise the neutrons produced would escape out of the piece and the reaction will not continue.

Nuclear reactor. *A nuclear reactor is a device in which a controlled nuclear chain reaction can be initiated and sustained to produce energy.*

Principle. Each fission of a U^{235} nucleus releases 2 to 3 neutrons (say average 2.5) and an energy of about 200 MeV. The neutrons liberated by the fission of U^{235} nuclei are either lost through the reactor surface or are absorbed without producing further fission. For the reaction to be self-sustained at least **one** neutron should always be available to continue the chain reaction. For this purpose the size of the reactor is increased so that it has a small surface area as compare to its volume. Secondly the reacting fuel is surrounded by a material that scatters neutrons without absorbing them so that most of escaping neutrons are reflected back.

Moderation of neutrons to thermal speeds. Natural uranium contains two isotopes; one of atomic weight 238 (${}_{92}U^{238}$) and the other of atomic weight 235 (${}_{92}U^{235}$), the latter being just 0.7% of the total. The U^{238} isotope readily captures *fast* neutrons and changes to ${}_{92}U^{239}$ in accordance with the reaction

$${}_{92}U^{238} + {}_{0}n^{1} \longrightarrow {}_{92}U^{239}$$

without undergoing fission. It is, therefore, not possible to start the chain reaction in natural uranium. On the other hand, the capture cross-section of U^{238} is very small for *slow* neutrons whereas for U^{235} isotope it is extremely large being about 582 *barns*. Therefore, the fast neutrons liberated in the fission process have to be *slowed* down to prevent their unproductive absorption by U^{238} and thus also increase the fission rate of U^{235} nuclei. To bring about the slowing down of fission neutrons the *uranium reactor* consists of a large number of uranium blocks dispersed in a matrix of a *moderator. This arrangement moderates* neutrons to thermal speeds so that the chain reaction can continue in natural uranium.

Function of Moderator. *A moderator is a substance whose nuclei absorb energy from incident fast neutrons and slow them down to thermal levels with energies less than 1 eV at typical reactor operating temperatures.*

The moderators in common use are ordinary 'light' water (H_2O) 'heavy' water (D_2O or ${}^2H_2O$) and *graphite* a from of pure carbon.

Light water is the best moderator as the energy transfer is maximum when the colliding particles are of equal mass ; which in this case are *two* protons (hydrogen nuclei) and neutrons. However, the protons in '*light*' water have a tendency to capture neutrons to form deutrons. On the other hand, deutrons, are less likely to capture neutrons. Thus '*heavy*' water sometimes makes a better moderator than '*light*' water inspite of the greater deutrium mass.

Similarly, graphite also has certain properties which compensate for its relative inefficiency in slowing down neutrons. These are:

(*i*) Graphite has good mechanical properties and is readily available

(*ii*) It has a very small cross-section for neutron capture

17.16 NUCLEAR REACTORS

Absorbers. In order to control the chain reaction, the reactor makes use of rods made of materials such as *cadmium* or *boron* which readily absorb *slow* neutrons. As these rods are inserted-further and further into the reactor the nuclear reaction is progressively further damped due to larger and larger absorption of neutrons.

Enrichment of U^{235}. Although ordinary '*light*' water is the most efficient moderator its protons tend to capture neutrons to form deutrons in accordance with the reaction

$$_1H^1 + {}_0n^1 \longrightarrow {}_1H^2 + \gamma \text{ (Energy)}$$

'*Light*' water reactors, therefore, cannot use natural uranium for fuel, but require uranium of which the U^{235} content has been enriched about 4 times to 3%. The enrichment of uranium can be performed by many methods but the most important methods are.

(*i*) Magnetic separation. One of the important methods of separation of $_{92}U^{235}$ from $_{92}U^{238}$ is by the application of a magnetic field utilising the principle of a mass spectrograph. The arrangement is known as **calutron** and is shown in Fig. 17.8.

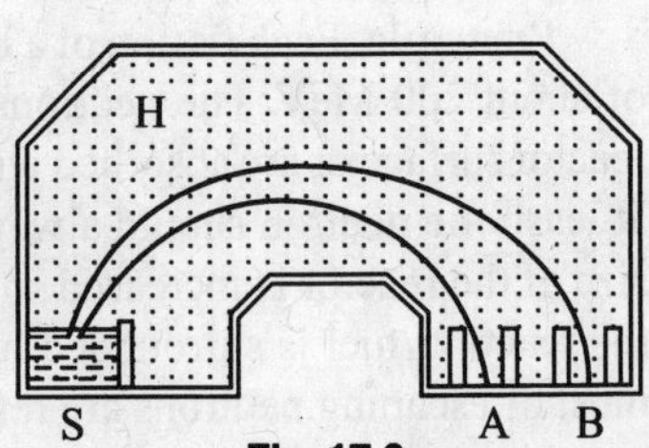

Fig. 17.8

The magnetic field *H* is applied perpendicular to the plane of the paper. *S* is the source of uranium ions. Due to the presence of the magnetic field the lighter $_{92}U^{235}$ ions collect at *A* and heavier $_{92}U^{238}$ ions at *B*, thereby bringing about separation of the two kinds of ions.

(*ii*) Gaseous diffusion. Uranium can also be enriched by *gaseous diffusion* process. In this process, uranium hexaflouride (UF_6) gas is passed through about 2000 successive permeable membranes used as *barriers*. The molecule of $U^{235}F_6$ being lighter diffuse more rapidly through each barrier than those of heavier $U^{238}F_6$ and the uranium hexafluoride gas contains more and more $U^{235}F_6$ after each diffusion till the desired enrichment is obtained.

(*iii*) Laser enrichment. Gaseous diffusion and centrifuge installations are expensive to build and operate. A new and cheaper method of isotope separation of $_{92}U^{238}$ and $_{92}U^{235}$ is by making use of *laser beam.* For this purpose, a laser is used to produce photons of exactly the right energy to ionise U^{235} atom but not U^{238} atoms. This is possible because isotopes of an element have slightly different energy levels and therefore, slightly different ionisation energies. As soon as U^{235} atoms are ionised an applied electric field separates the ionised isotopes from neutral ones.

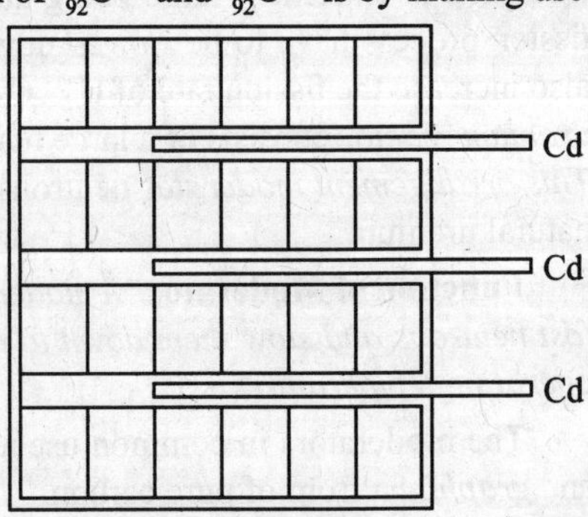

Fig. 17.9

Construction. The nuclear reactor or atomic pile as it is also called, consists of a large number of uranium blocks placed at calculated distances in the form of a lattice in a large pool of '*light*' or '*heavy*' water or a big block of graphite used as *moderator.* To reduce the size of the reactor uranium enriched with U^{235} isotope is used. Rods of cadmium used as *absorber* are placed in the pile as shown in Fig. 17.9. Cadmium absorbs the neutrons and thus does not allow the reaction to proceed.

Working. To work the pile all the cadmium rods except one are removed. The last one is then slowly pulled out. If now, a single neutron from cosmic radiation or from a spontaneous fission strikes a U^{235} nucleus and causes it to split, two or three neutrons are released. These neutrons are slowed down to thermal levels by collision with moderator nuclei and then proceed to produce further fission. The intensity of neutrons produced by fission is allowed to reach the desired value and power is produced at the required rate. The energy may be used to produce electrical power or steam to work a steam engine.

The fuel for '*light*' water reactor is in the form of uranium oxide (UO_2) sealed in long, thin zirconium alloy tubes that are assembled together with movable control rods into a core that is enclosed in a steel pressure vessel. A pressurised water reactor PWR is shown in Fig. 17.10. In it a pressure of about 150 atmospheres is maintained to prevent boiling. Water enters the reactor vessel at about 280°C and leaves at about 320°C passing through the heat exchanger which produces steam and drives the turbine. The turbine in turn operates the electric generator.

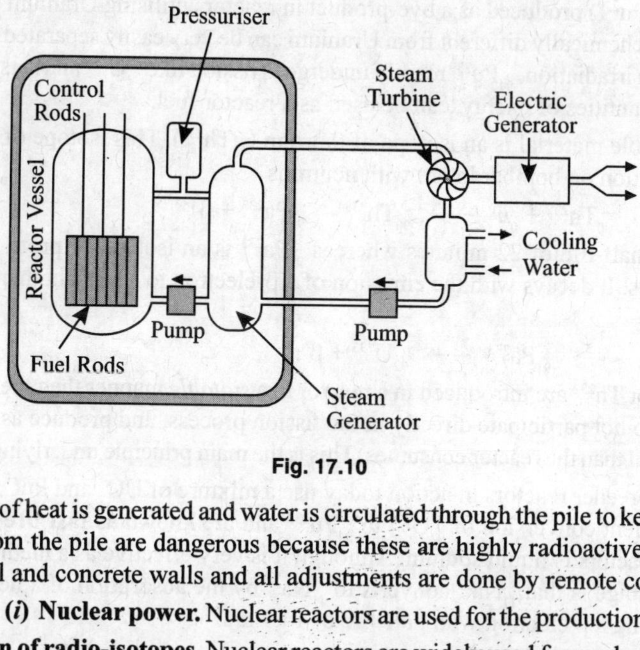

Fig. 17.10

Large amount of heat is generated and water is circulated through the pile to keep the temperature low. Radiations from the pile are dangerous because these are highly radioactive. Hence the pile is constructed in steel and concrete walls and all adjustments are done by remote controls.

Applications. (*i*) Nuclear power. Nuclear reactors are used for the production of electric energy.

(*ii*) Production of radio-isotopes. Nuclear reactors are widely used for producing a large number of radio-isotopes for medical and other purposes *e.g.,* industry and agriculture.

To produce a radio-isotope, a suitable compound is placed in the centre of the reactor core where the flux of neutrons is very high (more than 10^{16} per m^2 per sec).

Radio-sodium is prepared in this manner according to the reaction

$$_{11}Na^{23} + {}_0n^1 \longrightarrow {}_{11}Na^{24*}$$

(*iii*) Scientific research. The radio-isotopes and other radio-active material produced by a nuclear reactor is used in further scientific research. Several nuclear reactions have been studied with the help of high energy neutrons produced in the reactors. Reactors also help us to study '*radiation damage*' and '*biological effect*' of neutrons on human and other animal tissues.

(*iv*) Nuclear reactors can be used for propulsion of ships, submarines and air-crafts.

17.17 BREEDER REACTOR

Certain nuclides which are not fissionable can be transmuted into fissionable nuclei by absorbing neutrons. For example, the bombardment of $_{92}U^{238}$ nuclei with *fast* neutrons brings about the following reaction

$$_{92}U^{238} + {_0n^1} \longrightarrow {_{92}U^{239}} + \beta^- \xrightarrow[24\text{ min}]{} {_{93}Np^{239}} + \gamma\ (\text{energy})$$

i.e., a Uranium 238 nucleus after capturing a fast neutron converts itself into the isotope $_{92}U^{239}$ which decays with the emission of a β^- electron into an isotope of the *transuranic* element Neptunium. *Np* in turn decays to the stable isotope plutonium with the emission of another β electron in accordance with the reaction

$$_{93}Np^{239} \xrightarrow[2.3\text{ days}]{} {_{94}Pu^{239}} + \beta^-$$

The half-life of $_{92}U^{239}$ is 24 minutes whereas that of $_{93}Np^{239}$ is 2.3 days. The Plutonium isotope $_{94}Pu^{239}$ decays with emission of an α-particles and has a half-life of 24000 years.

$$_{94}Pu^{239} \xrightarrow[24000\text{ years}]{} {_{92}U^{235}} + {_2He^4}\ (\alpha-\text{particle})$$

Plutonium ($_{94}Pu^{239}$) produced as a bye-product in reactors utilising Uranium (of which 99.3% is U^{238}) as fuel being chemically different from Uranium can be very easily separated from the remaining $_{92}U^{238}$ after neutron irradiation. $_{94}Pu^{239}$ readily undergoes fission like $_{92}U^{235}$ and sets up a chain reaction liberating large quantities of energy, can be used as a reactor fuel.

Another suitable material is an isotope of thorium ($_{90}Th^{232}$). This isotope of thorium undergoes the following reaction on bombardment with neutrons

$$_{90}Th^{232} + {_0n^1} \longrightarrow {_{90}Th^{233}} \rightarrow {_{91}Pa^{233}} + \beta^-$$

$_{90}Th^{233}$ has a half-life of 22 minutes whereas $_{91}Pa^{233}$ is an isotope of proto actinium and has a half-life of 27 days. It decays with the emission of a β-electron to $_{92}U^{233}$ another fissionable isotope of uranium.

$$_{91}Pa^{233} \longrightarrow {_{92}U^{233}} + \beta^-.$$

Now, if U^{238} or Th^{232} are introduced in a reactor in a *suitable* manner then these will absorb those neutrons which do not participate directly in the fission process, and produce as *much* or *even more fissionable* material than the reactor consumes. This is the main principle underlying **a breeder reactor.**

Most of the breeder reactors in action today use a mixture of UO_2 and PuO_2. These require *fast* neutrons for efficient conversion of $_{92}U^{238}$ to $_{94}Pu^{239}$ and are known as **fast breeder reactors.** The coolant in such reactors is liquid sodium. Although it is very effective as a medium for heat transfer its main disadvantage is that $_{11}Na^{23}$ converts to $_{11}Na^{24}$ by the absorption of a neutron ($_0n^1$) which is radioactive emitting a β electron with half-life of 14.9 hrs.

Due to the abundant availability of thorium, thorium breeder reactors are also being developed. The thorium breeding cycle is most efficient with *slow* neutrons and is therefore, known as a **slow breeder reactor.**

With the widespread use of breeder reactors it is possible that the known reserves of nuclear fuel may be available for a much longer period.

17.18 ATOMIC ENERGY

It has been calculated that one pound of uranium 235, can produce as much energy by fission as is produced by an explosion of 10,000 tons of gun powder. The principle of *fission* is used in the construction of the **atomic bomb.**

An *atomic bomb* consists essentially of two pieces of $_{92}U^{235}$ (or $_{94}Pu^{239}$) (plutonium)] each smaller than the *critical* size and a source of neutrons. These pieces are brought together automatically by a mechanical device when the bombardment of neurons starts a chain reaction and a tremendous amount of energy is released in an incredibly small time.

In the hydrogen bomb, the synthesis of a heavier atom like helium, from hydrogen, results in the release of energy. The process of manufacture and the actual details of atomic bomb and hydrogen bomb are closely guarded secrets.

Atomic energy is now being harnessed for industrial and other peaceful uses by having a controlled release. The ordinary chain reaction is so vigorous and of so short a duration that the tremendous amount of energy released can cause nothing but destruction. Fermi suggested a device of starting a chain reaction in natural uranium containing $_{92}U^{238}$ with the help of **moderators.** A moderator is an element of low atomic number like heavy hydrogen (in the form of heavy water D_2O) or carbon in the form of graphite. These elements reduce considerably the speed of neutrons produced due to fission of $_{92}U^{235}$ nucleus and corresponding fission is thus very much increased. It is, therefore not necessary to separate U^{235} from U^{238}.

It is hoped that the atomic energy will open up fresh avenues and in time will become much cheaper than the present price of electricity, coal or petrol.

17.19 NUCLEAR FUSION

The process of formation of a heavy nucleus from two lighter nuclei is known as nuclear fusion. If two very light nuclei like $_1H^1$ or $_1D^2$ are made to fuse, the heavier nucleus formed has a higher binding energy per nucleon than that of the reactants. Thus there is a decrease in nculear mass and the corresponding difference in energy is released in the process. Fusion can take place at very high pressures and temperature of the order of 10^6 K because it is only under these conditions that nuclei are able to overcome their mutual coulomb repulsion. This energy is known as *thermonuclear energy* and it is possible to produce it under controlled conditions.

Thermonuclear Energy

The simplest process of fusion known as *proton-proton cycle* consists in fusion of two hydrogen nuclei which finally give a $_2He^4$ nucleus. in accordance with the following reaction

$$_1H^1 + {_1H^1} = {_1D^2} + e^+ + \nu + 0.42\text{MeV}$$

$$\downarrow$$

$$_1D^2 + {_1H^1} \longrightarrow {_2He^3} + \gamma + 5.49 \text{ MeV}$$

$$_2He^3 + {_2He^3} \longrightarrow {_2He^4} + 2\,{_1H^1} + 12.86 \text{ MeV.}$$

In the first stage, two hydrogen nuclei combine to form a Deutron with release of 0.42 MeV of energy, a positron and a neutrino. In the second stage $_1D^2$ nucleus combines with another $_1H^1$ nucleus to from a $_2He^3$ nucleus with release of 5.49 MeV of energy and a γ-radiation. The above cycle is repeated to produce another $_2He^3$ nucleus. Finally the two $_2He^3$ nuclei combine to give a $_2He^4$ nucleus and two $_1H^1$ nuclei with release of 12.86 MeV of energy. The two hydrogen nuclei $_1H^1$ allow the process to repeat and make it self-sustaining. The total energy released = 2 (0.42 + 5.49) + 12.86 = 24.68 MeV. If the annihilation energy of two positrons (2.04 MeV) is added, the total energy released = 26.72 MeV.

Another self-sustaining process is known as *carbon-nitrogen cycle.* In this process, carbon only acts as a catalyst and four hydrogen nuclei fuse to yield a helium nucleus, positrons, neutrons and as stated above 26.72 MeV of energy. The stepwise process is as under.

$$_1H^1 + {_6C^{12}} \rightarrow {_7N^{13}} \rightarrow {_6C^{13}} + e^+ + \nu$$

$$\downarrow$$

$$_1H^1 + {_6C^{13}} \rightarrow {_7N^{14}}$$

$$\downarrow$$

$$_1H^1 + {_7N^{14}} \rightarrow {_8O^{15}} \rightarrow {_7N^{15}} + e^+ + \nu$$

$$\downarrow$$

$$_1H^1 + {_7N^{15}} \rightarrow {_6C^{12}} + {_2He^4}$$

It is supposed that the above cycle takes place in the sun and the stars and is the source of their energy.

Importance of thermonuclear energy in the universe. The thermonuclear energy liberated by the fusion of hydrogen nuclei into helium nuclei by the two processes explained above is probably the source of nearly all energy in the universe.

17.20 SOURCE OF ENERGY OF SUN AND STARS

Self-sustaining fusion reactions can occur only under conditions of very high temperature and pressure so that the participating nuclei have sufficient energy to react inspite of their mutual electric repulsion. The reaction under these conditions occurs very frequently and produces much more energy than that lost to the surroundings and thus becomes a continuous source of energy. The interior of the sun and stars have conditions favourable for thermo-nuclear reactions. The *proton-proton cycle* predominates in stars having internal temperature below 1.6×10^7 K and *carbon* cycle in stars having internal temperature higher than that. The temperature of the sun is about 1.5×10^7 K. Therefore, the proton-proton cycle is responsible for about 90% of the energy produced by the sun.

The thermo-nuclear energy makes use of cheap and easily available materials like deuterium (available in sea water) and Tritium (obtained by neutron bombardment of the two isotopes of Lithium *i.e,* ${}_3Li^6$ and ${}_3Li^7$). The greatest difficulty in the way of producing thermonuclear energy on the earth is the production of very high temperatures of the order of 10^8 K and high pressures. At such high temperatures and pressures matter exists as fully ionised gas known as '*plasma*' If n is the particle density of plasma and t the time for which the density is maintained then the product nt must be at least 10^{28} m^{-3} for the reaction to take place. Such high temperture is difficult to maintain for a sufficiently long time. The second difficulty is that the plasma cannot be contained within ordinary walls since every material will melt at such high temperatures.

It has been proposed that a magnetic field in the form of a *torus* (dough nut) can contain the plasma. The high temperature and pressure required for the purpose is obtained by the explosion of a '*fission*' bomb. Alternatively it has been suggested that energetic '*Laser*' beams or beams of high energy electrons, protons may be used to heat and compress tiny deuterium, tritium pallets resulting in miniature hydrogen bombs, a succesion of which could supply a steady stream of energy. The fusion energy can thus be obtained from readily available fuel, has almost no radioactive waste and has no possibility of being used for war purposes. Herein lies the greatest importance of thermo-nuclear energy.

17.21 RADIOACTIVE HAZARDS

Like *X*-rays, the various radiations from radioactive nuclides are able to ionize the matter through which they pass. All ionising radiations are harmful to the living tissues. If the damage is slight the living tissue can repair itself and there is no permanent damage done. However, a strong dose of radiation may prove permanently harmful. It is not very easy to know the radiation hazard immediately because usually there is a delay of many years between the exposure and some possible harmful effects. In some cases radiation is an unwanted by-product particularly in the operation of nuclear reactors and the disposal of nuclear waste.

The harmful effects of radiation include cancer, lukemia and genetic changes that may causes the children to be born handicapped.

17.22 RADIOACTIVE TRACES

The artificially prepared radio-isotopes of certain elements are useful in biological, medical and agricultural research as **tracers.**

The procedure is to incorporate the radioactive nuclide in a chemical compound and follow what happens to the compound in a living organism by monitoring the radiation from the nuclide.

Radio-sodium has been lergely used in medical research to study the action of various medicines. A certain amount of NaCl having radioactive sodium in it is mixed with the medicine and administered to the patient. With the help of radiation from radio-sodium the circulation of the medicine in the body can be detected and hence its effects can be studied.

Brain and thyroid tumours can be detected by using *radio-iodine. Radio oxygen* ${}_8O^{15}$ mixed with normal air breathed by a patient enables radio-detection of the path of inhaled oxygen in the lungs.

Radioactive lead has been used as *tracer* in lead movements of plants. Radio-phosphorus is used to study the phosphorus uptake by plants from the soil.

17.23 DISTINCTION BETWEEN FISSION AND FUSION

Similarities. In fission as well as fusion there is a decrease in mass in the nuclear reaction i.e. the mass of the product particles is less than the mass of the parent particles. This decrease in mass is converted into energy according to Einstein's mass energy relation $E = mc^2$.

Dissimilarities. In fission, a heavy nucleus splits into two or more smaller ones. For example, a slow neutron brings about the fission of ${}_{92}U^{235}$ nucleus according to the reaction.

$$ {}_{92}U^{235} + {}_0n^1 \longrightarrow {}_{56}Ba^{141} + {}_{36}Kr^{92} + 3\,{}_0n^1 + E $$

In fusion, two or more light nuclei combine together to form a single heavy nucleus.

For example,

$$ {}_1H^1 + {}_1H^1 \longrightarrow {}_1D^2 + e^+ + \nu + 0.42\text{ MeV} $$

Also
$$ {}_2He^3 + {}_2He^3 \longrightarrow {}_2He^4 + 2\,{}_1H^1 + 12.86\text{ MeV}. $$

17.24 CARBON CYCLE

Carbon-nitrogen cycle needs 4 protons, to complete carbon cycle as under:

$$ {}_6C^{12} + {}_1H^1 \longrightarrow {}_7N^{13} + \lambda $$
$$ {}_7N^{13} \longrightarrow {}_6C^{13} + e^+ + \lambda $$
$$ {}_6C^{13} + {}_1H^1 \longrightarrow {}_7N^{14} + \lambda $$
$$ {}_7N^{14} + {}_1H^1 \longrightarrow {}_8O^{15} + \lambda $$
$$ {}_8O^{15} \longrightarrow {}_7N^{15} + e^+ + \nu $$
$$ {}_7N^{15} + {}_1H^1 \longrightarrow {}_6C^{12} + {}_2He^4 $$

Thus, carbon nucleus is obtained back at the end of this reaction acting like a catalyst for fusion of 4 protons into helium nucleus.

EXERCISE

1. What is a nuclear reaction ? Explain briefly the various types of nuclear reactions which may occur when a high energy particle approaches a nucleus. Make special mention of radiative capture, photo-disintegration, stripping and pick-up and spallation.
   (K.U. 2000; Pbi. U. 2004; P.U. 2001; G.N.D.U. 2004, 2003)
2. (*a*) Discuss various conservation laws in nuclear reactions with illustrative examples.
   (P.U. 2001; M.D.U.2000; Bang. U. 2000)

   (*b*) Name the physical quantities which are not conserved in nuclear reactions.
   (Pbi. U. 2000; G.N.D.U. 2002)
3. Define Q-value of a nuclear reaction. Derive an expression for the Q-value of the reaction X (a, b) Y in terms of kinetic energies of the incident and product particles and masses of the various particles and nuclei. Assume the target nucleus to be at rest in the laboratory. Discuss the case when the product particle emerges at right angles to the incident direction.
   (Calicut. U. 2002, 2001; Pbi. U. 2004, 2003; G.N.D.U. 2004, 2003; Cal. U. 2003; P.U. 2004, 2001, 2000)

4. What do you mean by *Q*-value of a nuclear reaction? Discuss the energetics of endo-ergic and exoergic reactions. Define threshold energy and derive an expression for the threshold energy of an endoergic reaction. What is the significance of positive and negative sign of *Q*?
(Pbi. U. 2000; Calicut. U. 2001; Bang. U. 2004; G.N.D.U. 2002, 2001, 2000; P.U. 2003, 2001)

5. (*a*) Define and explain the term nuclear reaction cross-section. What are its units? If a beam of N_0 particles is incident on a slab of thickness *x* of the material, how many particles will emerge out of the slab. Given that the slab contains *n* atoms per unit volume and σ is the cross-section of the reaction.

   (*b*) Define and explain the term differential cross-section.
(Pbi.U. 2004; Calicut. U. 2003; P.U. 2004, 2003, 2002, 2001, 2000; G.N.D.U. 2004, 2001)

6. Discuss the compound nucleus theory of nuclear reactions ? Comment on the life time and decay scheme of compound nucleus. How the liquid drop model of the nucleus can be used as the basis of compound nucleus theory ? ***(P.U. 2004; Calicut. U. 2002, G.N.D.U. 2001)***

7. Write notes on :

   (*i*) Direct nuclear reaction.

   (*ii*) Nuclear reaction chennels

8. (*a*) What do you mean by transmutation ? Give examples of nuclear transmutation reactions using charged particles, neutrons and photons. ***(P. U. 2002, 2000; Pbi. U. 2003, 2002)***

   (*b*) What does Si^{29} (*d,n*) P^{30} denote? Atomic number of Si = 14, P = 15.

9. (*a*) What is a neutron ? How was it discovered? Outline various methods of producing and detecting neutrons.

   (*b*) What are fast, slow and thermal neutrons ? ***(P.U. 2003, 2000; H.P.U. 2000)***

10. (*a*) State the basic properties of a neutron. Discuss the decay mode and half-life period of a neutron. ***(K.U. 2001)***

    (*b*) Under what conditions is a neutron stable ? ***(Pbi.U. 2000)***

11. Why are neutrons useful as bombarding agents for nuclear reactions ? Briefly outline various types of nuclear reactions produced by neutrons. ***(P.U. 2001)***

12. (*a*) Define chain reaction. What is a nuclear reactor ? State its principle. Explain the function of moderator in uranium fed nuclear reactors.

    (*b*) Explaining the use of absorbers and methods of enrichment of U^{235}, give the construction, working and applications of a nuclear reactor.
(K.U. 2001; M.D.U. 2001, 2000; Meerut. U. 2002; Gauhati. U. 2000; G.N.D.U.)

13. Write short notes on (*i*) breeder reactors and (*ii*) atomic energy.
(Calicut.U. 2003, 2002, 2001; Meerut.U. 2002)

14. (*a*) Explain the term thermonuclear energy or nuclear fusion. Discuss its importance in universe. Where do the sun and other stars get their energy from?
(Calicut.U. 2003, 2000, Gauhati.U. 2000)

    (*b*) What is nuclear fission and nuclear fusion ? Discuss.
(K.U. 2001, Meerut.U. 2003, 2001; Calicut.U. 2001)

15. What do you mean by redioactive tracers? Give its two affliсation.
(H.P.U. 2002; Pbi. U. 2004, GNDU 2001)

16. Explain carbon-nitrogen cycle.

SECTION VI

SOLID STATE PHYSICS

18. CRYSTAL STRUCTURE

19. X-RAY DIFFRACTION

20. MAGNETIC PROPERTIES OF SOLIDS

21. THERMAL PROPERTIES OF SOLIDS (Lattice Vibration)

22. FREE ELECTRON THEORY OF METALS

23. BAND THEORY OF SOLIDS

SOLID STATE PHYSICS

Solid-state physics, *a revolutionary branch of Modern Technology related particularly to 'material science'* and *'Electronics'* is the study of rigid matter, or solids, through methods such as quantum mechanics, crystallography, electromagnetism, and metallurgy. It is the largest branch of condensed matter physics. Solid-state physics studies how the large-scale properties of solid materials result from their atomic-scale properties. Thus, solid-state physics forms the theoretical basis of materials science. It also has direct applications, for example in the technology of transistors and semiconductors.

Solid materials are formed from densely-packed atoms, which interact intensely. These interactions produce the mechanical (e.g. hardness and elasticity), thermal, electrical, magnetic and optical properties of solids. Depending on the material involved and the conditions in which it was formed, the atoms may be arranged in a regular, geometric pattern (crystalline solids, which include metals and ordinary water ice) or irregularly (an amorphous solid such as common window glass).

The bulk of solid-state physics theory and research is focused on crystals. Primarily, this is because the periodicity of atoms in a crystal — its defining characteristic — facilitates mathematical modeling. Likewise, crystalline materials often have electrical, magnetic, optical, or mehanical properties that can be exploited for engineering purposes.

The forces between the atoms in a crystal can take a variety of forms. For example, in a crystal of sodium chloride (common salt), the crystal is made up of ionic sodium and chlorine, and held together with ionic bonds. In others, the atoms share electrons and form covalent bonds. In metals, electrons are shared amongst the whole crystal in metallic bonding. Finally, the noble gases do not undergo any of these types of bonding. In solid form, the noble gases are held together with van der Waals forces resulting from the polarization of the electronic charge cloud on each atom. The differences between the types of solid result from the differences between their bonding. Let us have a dip look into crystal structure, some important properties of solids and applications in succeeding chapters in this book.

18 Chapter

CRYSTAL STRUCTURE

INTRODUCTION

The sizes of the individual crystals in a crystalline solid material vary depending on the material involved and the conditions when it was formed. Most crystalline materials encountered in everyday life are polycrystalline, with the individual crystals being microscopic in scale, but macroscopic single crystals can be produced either naturally (e.g. diamonds) or artificially. Many properties of materials are affected by their crystal structure. This structure can be investigated using a range of crystallographic techniques, including X-ray crystallography, neutron diffraction and electron diffraction. Real crystals feature defects or irregularities in the ideal arrangements, and it is these defects that critically determine many of the electrical and mechanical properties of real materials.

The crystal lattice can vibrate. Theses vibrations are found to be quantised, the quantised vibrational modes being known as phonons. Phonons play a major role in many of the physical properties of solids, such as the transmission of sound. In insulating solids, phonons are also the primary mechanism by which heat conduction takes place. Properties of electrical conduction and heat capacity are investigated by solid state physics. By assuming that the material contains immobile positive ions and an "electron gas" of classical, non-interacting electrons, the Drude model was able to explain electrical and thermal conductivity and the Hall effect in metals, although it greatly overestimated the electronic heat capacity. By introducing the idea of electronic bands, the theory explains the existence of conductors, semiconductors and insulators.

18.1 CRYSTAL

A solid of definite shape with its atoms, ions or molecules arranged in some regular repetitious three dimensional pattern is termed as a crystal.

Crystalline solid. The crystalline state of a solid is characterised by regular and periodic arrangement of atoms or molecules.

When a crystal grows in a constant environment, the shape remains unchanged during growth. It appears as if identical building blocks are continuously added. The building blocks are atoms or group of atoms.

A periodic and regular arrangement of points in space is known as a *lattice* arrangement. In a crystalline structure, the various atoms and molecules are attached to the lattice in an identical manner. The distances between the various particles constituting a crystal are of the order of the size of these particles. The mean potential energy of every particle is more than its mean kinetic energy and their vibrations about the position of equilibrium are extremely limited. There is always internal regularity of particles, which is responsible for the symmetry of external shapes. A crystal, therefore, presents a sort of *spatial regularity* extending over a large volume *i.e.*, a long range *order* in its structure. The various properties like thermal conductivity, compressibility, tensile strength etc., depend on the structure of the crystal or the arrangement of particles in it. The crystals have generally sharp melting points and get converted into liquid state very soon.

Amorphous solid. There are certain substances like pitch, plastics, silicate glass, in which the molecules do not form a regular spatial lattice and are in a more or less chaotic and random distribution throughout the solid, even though the distance between nearest molecules is of the same order as that of the molecules in crystalline substances. Such substances are known as **'amorphous'** *i.e., substances without any regular form or shape*. They are uncrystallised solids. Like liquids they are isotropic in nature *i.e.*, the value of conductivity, elasticity, tensile strength etc., remains the same, no matter in which direction it is measured within the substance. Also these substances do not have long range spatial order. They have only short range order. They can be regarded to be highly viscous, super-cooled liquids without any sharp melting points and no periodicity and regularity of the arrangement of atoms.

In a crystalline solid, the changing of long range order into short-range order takes place after breaking such bonds, whose strength is nearly the same; while in case of amorphous substances, the strength of bonds vary. At low temperature, only the weakest bond gets ruptured. This accounts for the sharp melting points in crystals, while there is gradual softening in case of the amorphous solids.

Most solids are crystalline in nature due to the reason that crystalline state is a low energy state and is, therefore, preferred by most solids. The energy released during formation of an ordered structure is more than that released during the formation of disordered structure thereby making the crystalline state a low energy state.

Difference between crystalline and amorphous solids.

(*i*) A crystalline solid has *spatial regularity* extending over a large volume *i.e.*, there is a *long range order* in its structure whereas an amorphous solid has only a *short range order.*

(*ii*) A crystalline solid has different properties along different directions or crystal axes, like conductivity, elasticity or tensile strength *i.e.*, it is *anisotropic* in character whereas an amorphous solid is *isotropic* in character *i.e.*, it has the same value of these properties no matter in which direction it is measured.

(*iii*) Crystalline solids have sharp melting points whereas in amorphous solids there is a gradual softening of the material.

18.2 SOME DEFINITIONS

Lattice OR Crystal lattice. In a crystalline substance, the atoms are regularly arranged. *The periodic arrangement of atoms in a crystal is called the lattice.*

Plane lattice. If the arrangement of atoms is represented in a plane, it will be somewhat as shown in Fig. 18.1. Thus in two dimensions, any atom has got the same environment as any other atom. Such an array of atoms is called a *plane lattice*. In case of three dimensional picture, any atom will similarly have the same environment as any other atom and this array of atoms is called *space lattice.*

Space lattice. *A space lattice is defined as an infinite array of points (atoms or molecules) in three dimensions in which every point has identical surroundings.*

Basis. The structure of all crystals can be described in terms of a lattice with a group of atoms attached to each *lattice point*. The group of atoms is called the *basis* (*pattern or motif*). It is repeated in space to form the structure of the crystal.

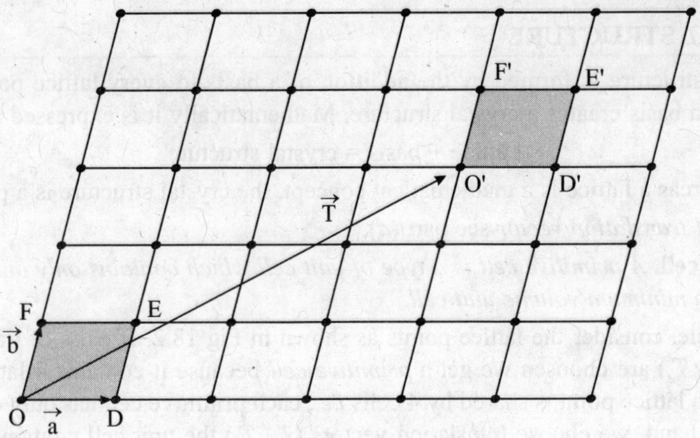

Fig. 18.1

A **basis** *is defined as an assembly of atoms, ions or molecules identical in composition, arrangement and orientation, which when repeated in three dimensions generates a crystal structure.*

The number of atoms in a basis may be *one* as in many metals and inert gases, but could be as large as 1000 in many structures.

Since there are many different ways of placing atoms in a volume, the distance and orientation between atoms can take many forms.

Unit-cell. Consider a plane lattice which is a part of a two dimensional crystal, the atoms of which are arranged in a pattern as shown in Fig. 18.1. It is seen that when a parallelogram like *ODEF* is repeatedly translated by the vectors $\vec{a}$ and $\vec{b}$ corresponding to $\vec{OD}$ and $\vec{OF}$, the pattern may be reproduced. For example, if the vector $\vec{a}$ is translated by 4 $\vec{a}$ and vector $\vec{b}$ by 3 $\vec{b}$ the parallelogram *ODEF* is translated and repeated as parallelogram *O'D'E'F'*.

The region ODEF is known as a **unit cell** *and the vectors* $\vec{a}$ and $\vec{b}$ as **basis vectors.**

Translation vectors. In two dimensions $\vec{a}$ and $\vec{b}$ are also the *translation vectors* of the lattice and the area of the unit cell is $\vec{a} \times \vec{b}$.

If we consider a space lattice, which is a part of a three dimensional crystal, then a **unit cell** *is defined as that volume of the solid from which the entire crystal can be constructed by the translational repetition in three dimensions.* For a three dimensional space lattice we have correspondingly **three basis vectors,** say $\vec{a}$, $\vec{b}$ and $\vec{c}$.

In three dimensions, $\vec{a}$, $\vec{b}$ and $\vec{c}$ are the **translation vectors** of the lattice and the volume of the unit cell is $\left|\left(\vec{a} \times \vec{b}\right) . \vec{c}\right|$.

A crystal may not possess a unique unit cell. Many unit cells can exist for the same structure. A cell of size 2*a*, 2*b*, 2*c* is as much correct as the cell of size *a*, *b*, *c*. For certain structures, unit cells of completely different geometry are possible. Most of the structures have smallest unit cells containing only a few atoms, but some elements like manganese have even hundreds of atoms in a unit cell.

18.3 CRYSTAL STRUCTURE

A crystal structure is formed by the addition of a basis to every lattice point *i.e.*, a lattice combined with a basis creates a crystal structure. Mathematically it is expressed as

Lattice + basis = crystal structure

Thus, whereas a lattice is a mathematical concept, the crystal structure is a physical concept. For *crystal translation vector* see part (*d*).

Primitive cell. *A primitive cell is a type of unit cell which contains only one lattice point at corners and is a minimum volume unit cell.*

For example, consider the lattice points as shown in Fig 18.2. If pairs of translation vectors (T_1, T_2) and (T_3, T_4) are choosen we get a *primitive cell* because it contains 4 lattice points at its corners and each lattice point is shared by 4 cells *i.e.*, each primitive cell has only one lattice point. If, on the other hand, we choose translation vectors (T_5, T_6) the unit cell contains more than one lattice point and is called a *non-primitive or multiple cell.*

Primitive basis vectors. If the vectors $\vec{a}$, $\vec{b}$ and $\vec{c}$ define a primitive cell of space lattice in three dimensions then $\vec{a}$, $\vec{b}$, $\vec{c}$ are known as *primitive basis vectors.*

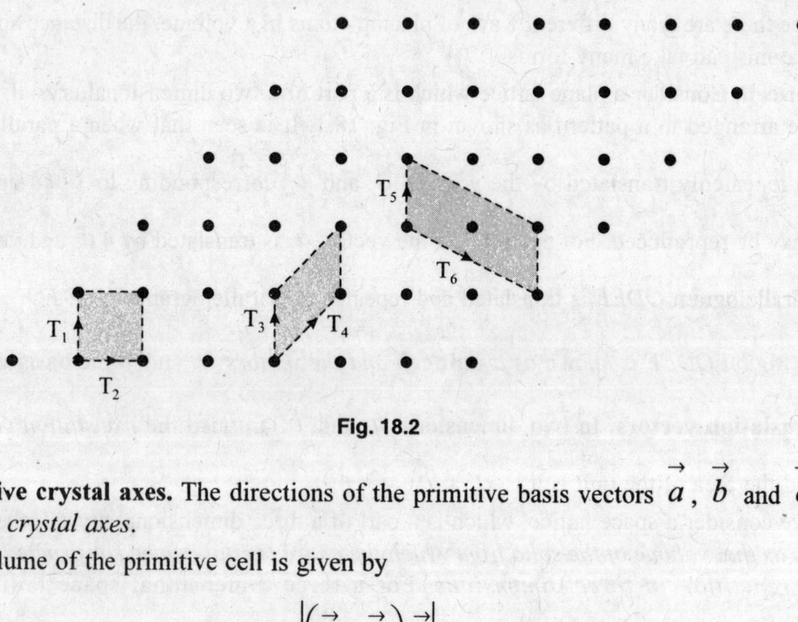

Fig. 18.2

Primitive crystal axes. The directions of the primitive basis vectors $\vec{a}$, $\vec{b}$ and $\vec{c}$ define the *primitive crystal axes.*

The volume of the primitive cell is given by

$$V = \left| \left(\vec{a} \times \vec{b} \right) . \vec{c} \right| \qquad \ldots(i)$$

where $\vec{a}$, $\vec{b}$ and $\vec{c}$ are primitive axes vectors.

Can a unit cell be primitive? A primitive cell is a type of unit cell which contains lattice points at corners only and is a *minimum volume cell.* A unit cell is a conventional cell which may be equal to a primitive cell or a multiple of it. Thus *a unit cell can be primitive.*

Crystal translation vectors. An ideal crystal is composed of atoms arranged on a lattice defined by three fundamental or lattice translation vectors $\vec{a}$, $\vec{b}$, $\vec{c}$ such that the atomic arrangement looks the same in every respect when viewed from any point O as when viewed from the point O' as shown in Fig. 18.1. The vector $OO' = \vec{T}$ is given by

$$\vec{T} = n_1 \vec{a} + n_2 \vec{b} + n_3 \vec{c}$$

where n_1, n_2 and n_3 are arbitrary integers and T is called *Lattice or crystal translation vector.*

Translation operation. In a crystal, in addition to the geometrical arrangement of atoms a function $f(r)$ which describes any property of a crystal must also be invariant or periodic *i.e.*,

$$f(r) = f(r + T)$$

where T is the *translation operation* given by

$$\vec{T} = n_1 \vec{a} + n_2 \vec{b} + n_3 \vec{c}.$$

Hence a translation operation is a kind of symmetry operation which when performed leaves the crystal invariant.

The displacement of a crystal parallel to itself by a crystal translation vector is thus a *crystal translation operation*. Translation repeats the *motif* infinite number of times in a given direction whereas all other operations repeat the *motif* only a finite number of times in a given direction.

Wigner Sietz primitive cell. This is another method of choosing a cell of volume V given by relation (*i*).

1. First draw lines to connect a given lattice point to all nearby lattice points.
2. At the midpoints and normal to these lines draw new lines or planes. The smallest volume enclosed in this way is a *new primitive cell.*

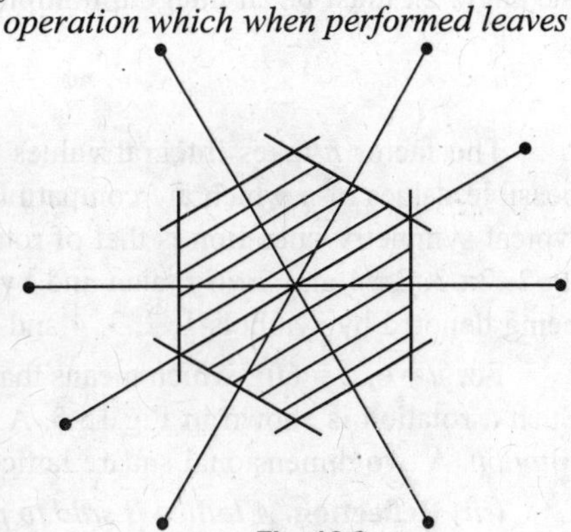

Fig. 18.3

It is known as Wigner Sietz primitive cell.

18.4 SYMMETRY OPERATION

Crystal lattices can be carried into themselves by lattice translation vector $\vec{T}$ and by various other symmetry operations.

A symmetry operation is that which transforms the crystal to itself i.e., a crystal remains invariant under a symmetry operation.

Types of symmetry operations. We shall consider a group of atoms or molecules called *motif* which can be repeated periodically in a number of permissible ways to form the crystal. For simplicity we shall represent the *motif* in a crystal by a figure of seven (7). There are four major methods of repeating the *motif* in space which are known as symmetry operations.

(*i*) **Translation operation.** The *motif* represented by a figure of seven (7) can be repeated by a translation operation as shown in Fig 18.4.

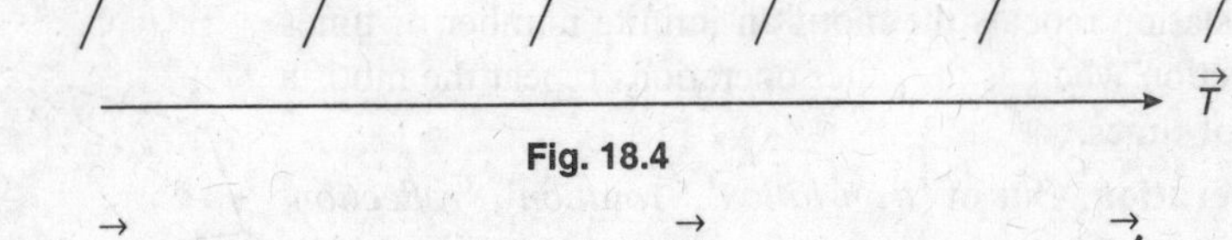

Fig. 18.4

A lattice point $\vec{r}$ under translation operation $\vec{T}$ gives another point $\vec{r}'$, which is exactly identical to $\vec{r}$; *i.e.*,

$$\vec{r'} = \vec{r} + \vec{T}$$

Translation repeats the *motif* an infinite number of times.

(*ii*) **Rotation operation.** In this operation the *motif* can be rotated through an angle ϕ about an axis of rotation as shown in Fig 18.5. This operation repeats the *motif* a finite number of times.

A lattice is said to possess rotational symmetry if its rotation by an angle ϕ about an axis transforms the lattice to itself.

As the lattice remains invariant under a rotation of 2π, the angle 2π must be an integral multiple of ϕ *i.e.*,

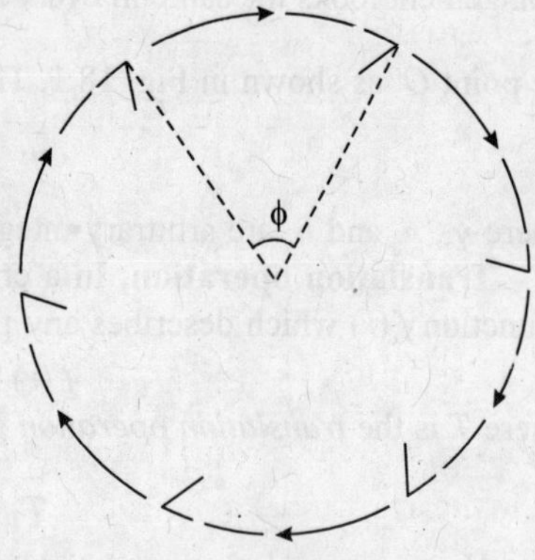

Fig. 18.5

$$n\phi = 2\pi \text{ or } \phi = \frac{2\pi}{n}.$$

The factor n takes integral values only and is known as *Multiplicity of rotation axis*. The possible values of n which are compatible with symmetry requirements are 1, 2, 3, 4 and 6. Thus, typical symmetry operation is that of rotation about an axis passing through a lattice point by 2π, $2\pi/2$, $2\pi/3$, $2\pi/4$ and $2\pi/6$ radian and by their integral multiples, the corresponding rotation axis being denoted by symbols 1, 2, 3, 4 and 6.

For $n = 6$, $\phi = 60°$, which means that the lattice repeats itself with a minimum rotation of 60°. Such a rotation is shown in Fig 18.5. A rotation corresponding to the value of n is called *n-fold rotation*. A two dimensional square lattice has 4-fold rotational symmetry.

(*iii*) **Reflection.** *A lattice is said to possess reflection symmetry if there exists a plane (or line in two dimensions) in the lattice which divides it into two identical halves which are mirror images of each other.*(fig. 18.6).

Reflection across a line or plane changes the character of the lattice from *left handed* to *right handed* and *vice-versa.*

All the above symmetry operations are applicable to a *two dimensional lattice.*

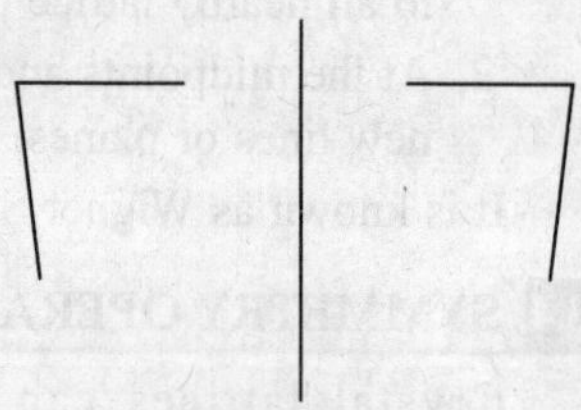

Fig. 18.6

(*iv*) **Inversion.** Inversion is a point operation which is applicable to *three dimensional lattice* only. The symmetry element implies that each point located at $\vec{r}$ relative to a lattice point has an identical point located at $-\vec{r}$ relative to the same lattice point. Inversion through a point is shown in Fig 18.7. It also changes the character of the *motif* from right handed to left handed.

To sum up:-

Symmetry operation like translation and rotation leave the motif unchanged, whereas reflection or inversion changes the character of the motif from right handed to left handed and vice-versa. Further, a translation repeats the motif an infinite number of times in a given direction whereas the other operations repeat the motif a finite number of times.

Point operation. Out of '*translation*', '*rotation*', '*reflection*' and '*inversion*' operations, the *translation operation* applies to lattices only while the remaining operations and their combinations apply to all objects and are collectively known as *point symmetry operations.*

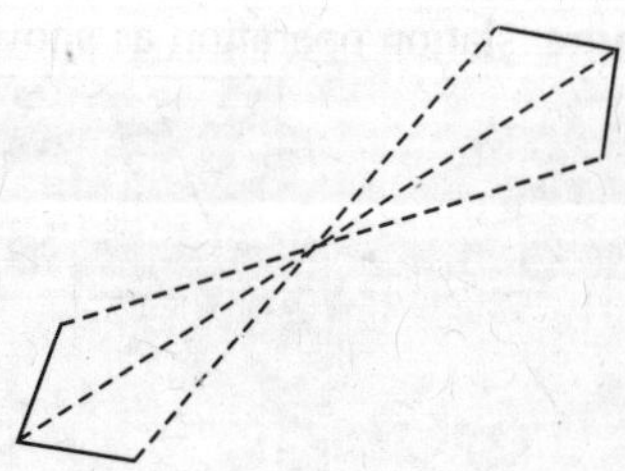

Fig. 18.7

Crystal lattice cannot have five fold symmetry. It is not possible to find a lattice that goes into itself under rotations like $2\pi/5$ or $2\pi/7$ or by any other symmetry operation like mirror reflection or inversion. In other words, a *crystal cannot have five fold symmetry*.

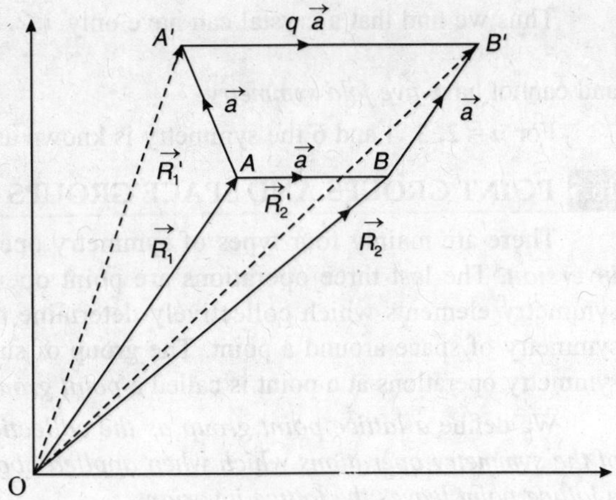

Fig. 18.8

To prove it mathematically let $AB = \vec{a}$ be a primitive translation vector and $OA = \vec{R}_1$, $OB = \vec{R}_2$ the two lattice vectors connecting the origin to A and B - the two closest lattice points of the crystal, then

$$|\vec{a}| = |\vec{R}_1 - \vec{R}_2|$$

Now consider a line perpendicular to the plane of the paper and passing through the lattice point A. Let the vector $\vec{a}$ be rotated about this line as axis through an angle $\theta = \frac{2\pi}{n}$ in the anti-clockwise direction, so that it produces a new lattice vector $OA' = \vec{R}_1'$. Similarly, let the vector $\vec{a}$ be rotated about an axis passing through the lattice point B and perpendicular to the plane of the paper through the same angle $\theta = \frac{2\pi}{n}$ in the clockwise direction so that it produces a new lattice vector $OB' = \vec{R}_2'$.

$$\therefore \qquad AA' = BB' = |\vec{a}|$$

and $$\angle BAA' = \angle ABB' = \theta.$$

The quadrilateral is a trapezium with $A'B'$ parallel to AB. Further $A'B'$ must also be an integral multiple of the primitive translation vector $\vec{a}$

or $$A'B' = q|\vec{a}|$$

where q is an integer.

Now $$A'B' = q|\vec{a}| = |\vec{R}_1 - \vec{R}_2|$$

$$= |\vec{a}| + |\vec{a}| \sin\left(\theta - \frac{\pi}{2}\right) + |\vec{a}| \sin\left(\theta - \frac{\pi}{2}\right)$$

$$= |\vec{a}| + 2|\vec{a}| \sin\left(\theta - \frac{\pi}{2}\right)$$

$$= |\vec{a}| [1 - 2\cos\theta]$$

$$\therefore \qquad q = 1 - 2\cos\theta \text{ or } |\cos\theta| = \left|\frac{1-q}{2}\right|$$

As $\cos\theta < 1$, the above relation can be satisfied only when $q = 3, 2, 1, 0, -1$.

or $$\theta = \frac{2\pi}{2}, \frac{2\pi}{3}, \frac{2\pi}{4}, \frac{2\pi}{6}, \frac{2\pi}{1} \text{ or } n = 1, 2, 3, 4, 6 \text{ only.}$$

Thus we find that a crystal can have only $\bar{1}, \bar{2}, \bar{3}, \bar{4}$ and $\bar{6}$ fold axis of rotation or *symmetry* and cannot have *five fold symmetry.*

For n = 2, 3, 4 and 6 the symmetry is known as *diad, triad, tetrad* and *hexad* respectively.

18.5 POINT GROUPS AND SPACE GROUPS

There are mainly four types of symmetry operations; *translation, rotation, reflection* and *inversion*. The last three operations are point operations and their combinations give certain symmetry elements which collectively determine the symmetry of space around a point. The group of such symmetry operations at a point is called a *point group*.

We define *a lattice point group as the collection of the symmetry operations which when applied about a lattice point leaves the lattice invariant.*

In other words, in point groups all the possible symmetry elements must pass through a point.

Location of equivalent points are shown for four point groups in Fig 18.9.

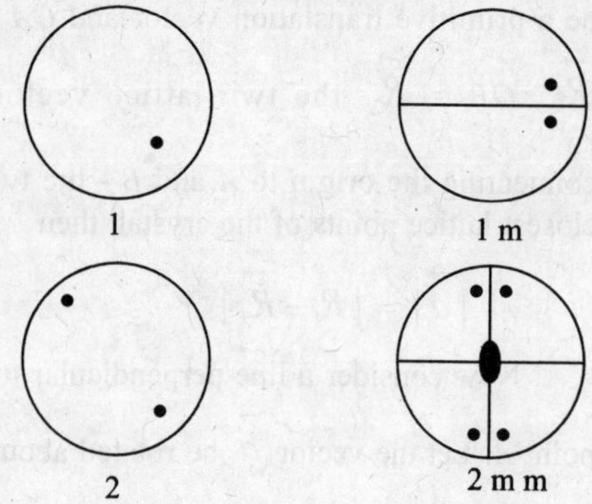

Fig. 18.9

The dots denote equivalent points. The point group 1 has no symmetry element, so this point has no other point equivalent to it. The point group 1 m has a *mirror plane*, so this point (dot) on reflection across the mirror plane becomes the second point (dot). The point group 2 has a two fold axis, so this point (dot) on rotation through π goes into the other point (dot). With a two-fold axis and one mirror plane there is a second mirror plane normal to the first and we have 4 equivalent points (2 m m). The point symmetry of a crystal strucutre as a whole is determined by the point symmetry of the lattice as well as the basis.

The group of all the symmetry elements of a crystal structure is called **space group.**

It determines the symmetry of the crystal structure as a whole.

Number of point groups and space groups. In *two dimensional space* there are 10 different *point groups* designated as 1, 1 m, 2, 2 mm, 3, 3 m, 4, 4 mm, 6 and 6 mm.

In *three dimensional space* there are a total of 32 point groups in a three dimensional lattice.

The point symmetry of crystal structure as a whole is determined by the point symmetry of the lattice as well as of the basis. In order to determine the point symmetry of a crystal it should be noted that

(*i*) a unit cell might show point symmetry at more than one locations inside it and (*ii*) the symmetry elements comprising combined point and translation operations might be existing at these locations.

There are 17 distinct *space groups* possible in 2 dimensions and 230 in three dimensions.

Characteristics of a space group. (*a*) The symmetry of a crystal structure is specified completely when the space group is known.

(*b*) The space group is characterised by Bravais lattice and by the location of the point group and other symmetry elements in a unit cell.

(*c*) The space group through its symmetry elements determines the position of equivalent points within the unit cell.

(*d*) For a crystal structure specified by a particular space group, if only one point in a particular cell is occupied it is necessary that all equivalent points be occupied by identical atoms or molecules.

18.6 BRAVAIS LATTICE

The group of four types of symmetry operations at a point is called '*point group*', there being 10 point groups in two dimensions and 32 in three dimensions respectively. These point groups form the basis for the construction of different types of lattices known as '*Bravais lattices*'

Two dimensional lattice types. In a two dimensional lattice there is no restriction regarding the lengths of the primitives $\vec{a}$ and $\vec{b}$ and the angle ϕ between them. The only condition is that the lattice should be invariant under symmetry operations.

In a two dimensional case there are in all *five* types of lattices depending upon the relative values of $|\vec{a}|$ and $|\vec{b}|$ and ϕ. These are known as **Bravais lattices** in two dimensions. These are:

(*i*) **Oblique lattice.** A general lattice is known as *oblique lattice* and is invariant under rotation of 2π and $2\pi/2$ about any lattice point. An oblique lattice is shown in Fig. 18.10 (*i*) where $|\vec{a}| \neq |\vec{b}|$ and $\phi \neq 90°$. The conventional unit cell is a parallelogram. *Special lattices* of oblique type can be invariant under rotation of $2\pi/3$, $2\pi/4$ and $2\pi/6$.

(*ii*) **Square lattice.** A square lattice is shown in Fig. 18.10 (*ii*). Here $|\vec{a}| = |\vec{b}|$ and $\phi = 90°$. It is invariant under a rotation of $2\pi/4$. The conventional unit cell is a square.

(*iii*) **Hexagonal lattice.** A hexagonal lattice is shown in Fig. 18.10 (*iii*). Here $|\vec{a}| = |\vec{b}|$ and $\phi = 120°$. This is invariant under a rotation of $2\pi/6$. The conventional unit cell is a 60° rhombus.

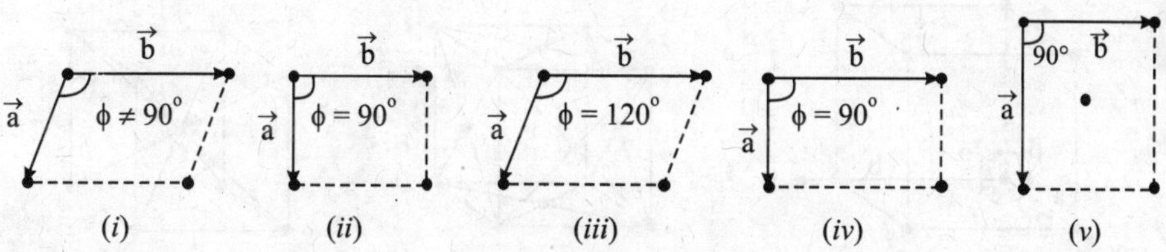

Fig. 18.10

(*iv*) **Primitive Rectangular lattice.** A primitive rectangular lattice is shown in Fig. 18.10 (*iv*). Here $|\vec{a}| \neq |\vec{b}|$ and $\phi = 90°$. This is invariant under mirror reflection. The conventional unit cell is a rectangle.

(*v*) **Centred rectangular lattice.** A centred rectangular lattice is shown in Fig. 18.10 (*v*). Here $|\vec{a}| \neq |\vec{b}|$ and $\phi = 90°$ but $|\vec{a}| > |\vec{b}|$. This is invariant under inversion operation. The conventional unit cell is again a rectangle.

The two dimensional Bravais lattices are given below:

	Lattice	*Conventional unit cell*	*Crystal axes*
1.	Oblique	Parallelogram	$a \neq b;\ \phi \neq 90°$
2.	Square	Square	$a = b;\ \phi = 90$
3.	Hexagonal	60° Rhombus	$a = b;\ \phi = 120°$
4.	Primitive rectangular	Rectangle	$a \neq b;\ \phi = 90°$
5.	Centred rectangular	Rectangle	$a \neq b;\ \phi = 90°$

Three dimensional lattice types. In three dimensional space lattices, there can be in all fourteen types (one general and 13 special). The general type is the *triclinic lattice*. All these are known as **Bravais space lattices.**

With these fourteen types of lattices and on the basis of primitive cell, the crystals have been classified into seven systems. Fig. 18.11 gives the convention used for the angles between primitives *a*, *b*, *c*, also called *lattice parameters*. Thus the angles between (*a*, *b*), (*b*, *c*) and (*c*, *a*) are denoted by γ, α, β respectively. Different specifications for *a*, *b* and *c* as well as α, β and γ generate different crystal systems. There are seven crystal systems in all (*i*) Cubic (*ii*) Tetragonal (*iii*) Hexagonal (*iv*) Rhombohedral (*v*) Orthorhombic (*vi*) Monoclinic and (*vii*) Triclinic. The seven systems of crystals with their types of lattices, angles between the axes and relation between the length of primitives is given below.

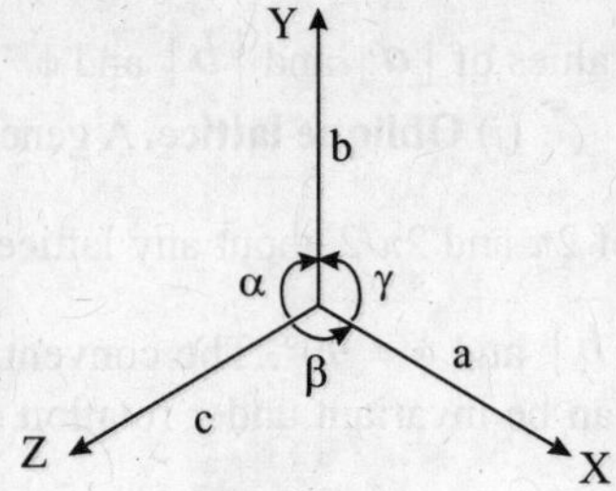

Fig. 18.11

(*i*) **Cubic.** The crystal axes are perpendicular to one another *i.e.*, α = β = γ = 90°. The repetitive interval or the length of the primitives is the same along the three axes *i.e.*, *a* = *b* = *c*. Cubic lattices may be *simple* (*sc*) or *primitive* (*P*), *face centred* (*fcc*) or (*F*) and *body centred* (*bcc*) or (*I*) as shown in Fig. 18.12 (*i*), (*ii*) and (*iii*).

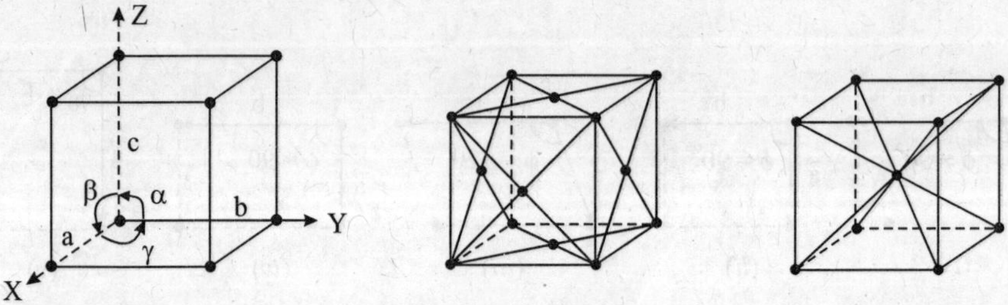

Simple (*sc*) (*i*) Face-Centred (*fcc*) (*ii*) Body Centred (*bcc*) (*iii*)

Fig. 18.12

(*ii*) **Tetragonal.** The crystal axes are perpendicular to one another *i.e.*, α = β = γ = 90°. The repetitive interval or the lengths of the primitives along two axes are the same but the interval along the third axis is different *i.e.*, *a* = *b* ≠ *c*. Tetragonal lattices may be *simple* (*P*) or *body centred* (*I*) as shown in Fig. 18.13 (*i*) and (*ii*).

(*iii*) **Hexagonal.** Two of the crystal axes are 120° apart *i.e.*, γ = 120° while the third is perpendicular to both of them *i.e.*, α = β = 90°. The repetitive intervals are the same along the axes that are 120° (or 60°) apart but the interval along the third axis is different *i.e.*, *a* = *b* ≠ *c* as shown in Fig. 18.14 (*a*).

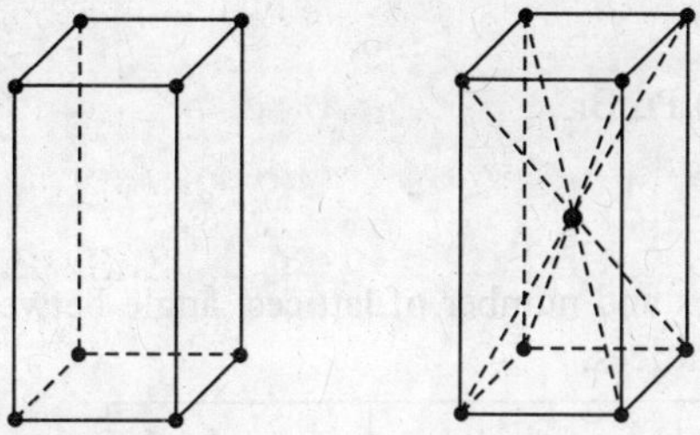

Simple (*i*) Body Centred (*ii*)

Fig. 18.13

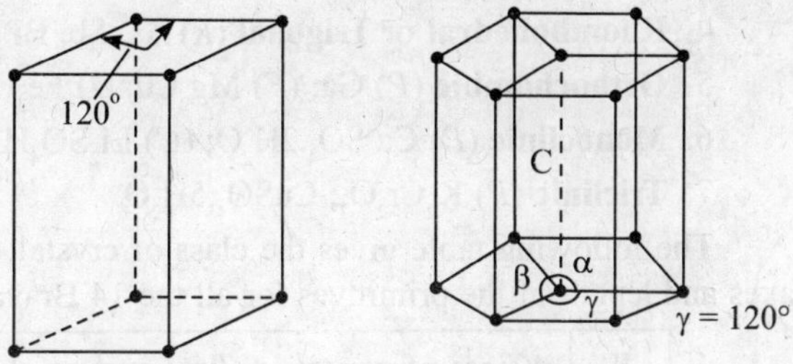

Hexagonal (*a*) Hexagonal (*b*)

Fig. 18.14

The *hexagonal* system is so called because a simple hexagonal structure can be formed from such unit cells as shown in Fig. 18.14 (*b*).

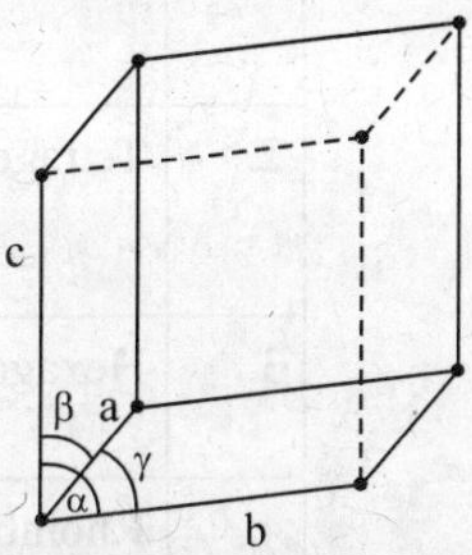

Trigonal

Fig. 18.15

(*iv*) **Trigonal.** (Also known as *rhombohedral*). The angle between each pair of crystal axes are the same but are not 90° *i.e.*, $\alpha = \beta = \gamma \neq 90°$. The repetitive interval (or length of primitives is the same along all the three axes as shown in Fig. 18.15 *i.e.*, $a = b = c$.

(*v*) **Orthorhombic.** The crystal axes are perpendicular to one another but the repetitive interval (or length of primitives) are different along all the three axes *i.e.*, $\alpha = \beta = \gamma = 90°$ and $a \neq b \neq c$. The orthorhombic lattices may be *simple, base centred* (or *end centred*), *body centred or face centred* as shown in Fig. 18.16 (*a*), (*b*), (*c*) and (*d*) respectively.

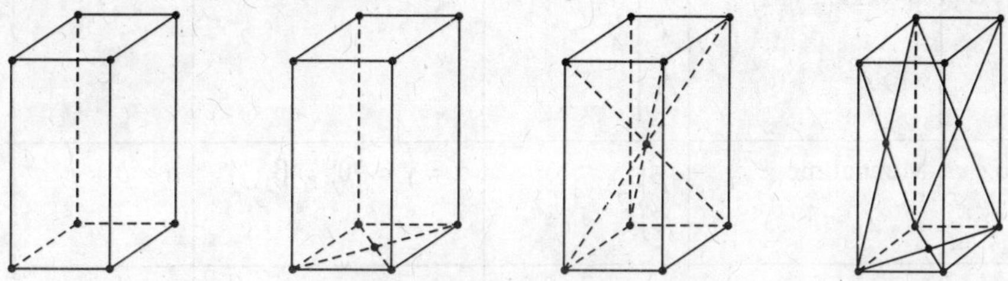

Simple (*a*) Base Centered (*b*) Body-Centered (*c*) Face-Centered (*d*)

Fig. 18.16

(*vi*) **Monoclinic.** Out of the three crystal axes, two are not perpendicular to each other but the third is perpendicular to both of them *i.e.*, $\alpha = \gamma = 90° \neq \beta$. Monoclinic lattices may be *simple* or *base-centred* as shown in Fig. 18.17 (*a*) and (*b*).

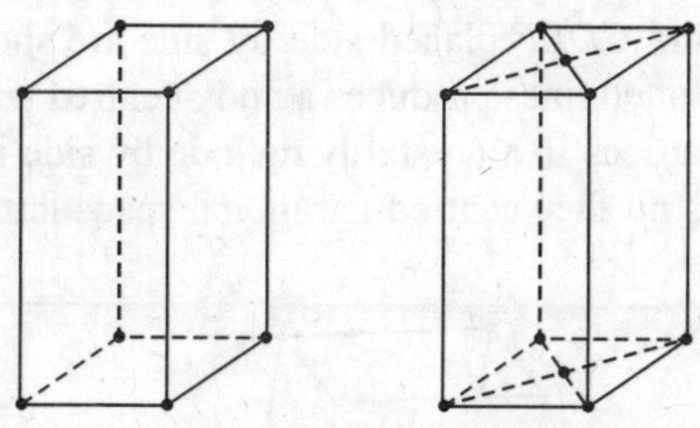

Simple (*a*) Base Centered (*b*)

Fig. 18.17

(*vii*) **Triclinic.** None of the crystal axes is perpendicular to any of the others $\alpha \neq \beta \neq \gamma$. The repetitive intervals are also different ($a \neq b \neq c$) as shown in Fig. 18.18 (*a*).

Example of various classes of crystals. Some examples of various classes of crystals are:—

1. **Cubic** (*P*) Cu, Ag, Fe; (*F*) Cs Cl; (*I*) Na, NaCl.
2. **Tetragonal** (*P*) β – Sn; (*I*) TiO_2.
3. **Hexagonal** (*P*) Mg, Zn, Cd.

4. **Rhombohedral** or **Trigonal** (*R*) As, Sb, Bi.
5. **Orthorhombic** (*P*) Ga; (*F*) Mg_2Cu; (*I*) Fe_3C (*C*) Pu Br_3
6. **Monoclinic** (*P*) Ca $SO_4.2H_2O$; (*C*) $Li_2SO_4H_2O$
7. **Triclinic** (*P*) $K_2Cr_2O_7$; $CuSO_4.5H_2O$.

The following table gives the class of crystal, types and number of lattices, angle between axes and length of the primitives for all the 14 Bravais lattices.

S. No.	*Class of crystal*	*Type and no. of Lattices*	*Angle between axes*	*Length of primitives*
1.	Cubic	$\left.\begin{matrix} sc(P) \\ fcc(F) \\ bcc(I) \end{matrix}\right\} 3$	$\alpha = \beta = \gamma = 90°$	$a = b = c$
2.	Tetragonal	$\left.\begin{matrix} P \\ I \end{matrix}\right\} 2$	$\alpha = \beta = \gamma = 90°$	$a = b \neq c$
3.	Hexagonal	$P\} 1$	$\alpha = \beta = 90°$ $\gamma = 120°$	$a = b \neq c$
4.	Rhombohedral (Trigonal)	$R\} 1$	$\alpha = \beta = \gamma \neq 90°$ $< 120°$	$a = b = c$
5.	Orthorhombic	$\left.\begin{matrix} P \\ F \\ I \\ C \end{matrix}\right\} 4$	$\alpha = \beta = \gamma = 90°$	$a \neq b \neq c$
6.	Monoclinic	$\left.\begin{matrix} P \\ C \end{matrix}\right\} 2$	$\alpha = \gamma = 90° \neq \beta$	$a \neq b \neq c$
7.	Triclinic	$P\} 1$	$\alpha \neq \beta \neq \gamma$	$a \neq b \neq c$

P = Primitive; *sc* (simple cubic); *F* = *fcc* (Face centred); *I* = *bcc* (body centred); *C* = base centred; *R* = Rhombohedral.

Why no face centred tetragonal space lattice. Two face centred tetragonal lattices *ABCD* and *CDEF* placed side by side are shown in Fig. 18.18(*b*). The arrangement of points shown by dotted lines produces a body centred tetragonal space lattice. Thus the face centred tetragonal space lattices in a crystal lying side by side merge into a body centred space lattice and, therefore there is no face centred tetragonal space lattice.

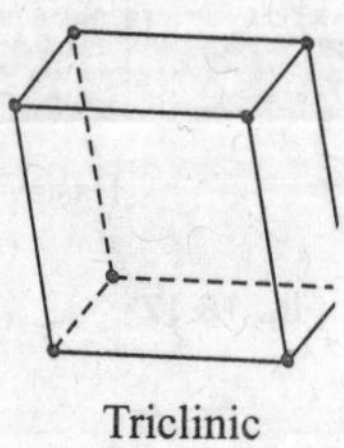

Fig. 18.18(a)

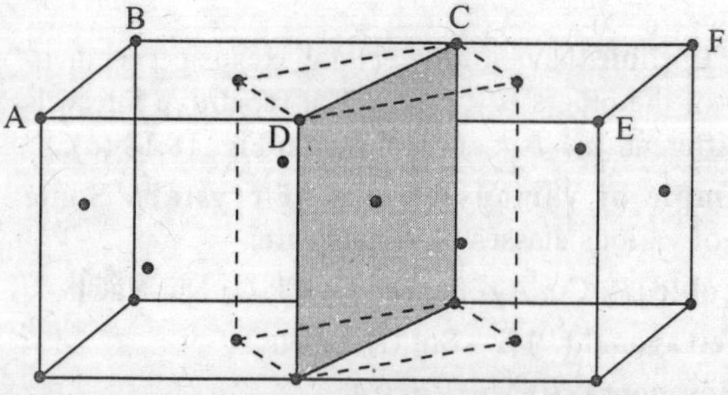

Fig. 18.18(b)

18.7 CHARACTERISTICS OF A UNIT CELL

The characteristics of a unit cell are:

(*i*) Volume, (*ii*) Atoms per unit cell, (*iii*) Co-ordination number, (*iv*) Atomic radius and (*v*) Atomic packing fraction.

1. Volume of a unit cell. For cubic crystals $a = b = c$.

Hence volume of a unit cell $V = a \times b \times c = a^3$.

2. Atoms per unit cell. A study of extended cubic crystal lattice shows that every corner atom in each unit cell participates in the corner position with seven other unit cells. Therefore, each of the eight corner atoms in a cubic unit cell contributes only 1/8th of an atom to the unit cell. Thus:

(*i*) An atom lying at the corner of a cubic unit cell is shared equally by eight unit cells and, therefore, counts for one eighth of an atom. Similarly

(*ii*) An atom lying completely within a unit cell belongs wholly to that cell.

(*iii*) An atom lying in the face of a unit cell belongs equally to two unit cells and, therefore, counts as one half for that particular cell.

(*iv*) An atom lying on one edge of a unit cell is shared equally by four unit cells and, therefore, counts as one quarter of an atom per unit cell.

Simple cubic cell (*sc*). As stated above, each corner atom in a cubic unit cell is shared by a total number of eight unit cells so that each corner atom contributes only 1/8 of its effective part to a unit cell. Since there are in all 8 corner atoms their total contribution is equal to 8/8 =1.

Thus the number of atoms per unit simple cubic cell is *one*.

Face centred cubic (*fcc*). Each face centred atom is shared by two unit cells. There are six faces of a cube and six face centred atoms. These are in addition to eight corner atoms.

∴ Number of atoms per unit face centred cubic cell

$$= \frac{6}{2} + \frac{8}{8} = 4.$$

Body centred cubic cell (*bcc*). The contribution of a body centred atom is full *i.e.*, *one*. This is in addition to eight corner atoms.

∴ Number of atoms per unit body centred cubic cell

$$= 1 + \frac{8}{8} = 2.$$

Hence we conclude that

$$\text{Total number of atoms per unit cell } N = \frac{N_c}{8} + \frac{N_f}{2} + N_i$$

where N_c = number of *corner* atoms, N_f = the number of *face centred* atoms and N_i = The number of *body centred* atoms (which are completely inside the cell). Thus

For *sc* $\quad N = \frac{8}{8} + 0 + 0 = 1$

For *fcc* $\quad N = \frac{8}{8} + \frac{6}{2} + 0 = 4$

For *bcc* $\quad N = \frac{8}{8} + 0 + 1 = 2$

18.8 CO-ORDINATION NUMBER

In a crystal the number of nearest neighbours to an atom in a given structure is called co-ordination number. It gives an idea of the closeness of the packing of the atoms.

Consider any corner atom in a *simple cubic cell*. It has one close neighbour along + X axis and another along – X axis as shown in Fig. 18.12 (*i*). The same is true for Y and Z axes. Thus the coordination number for each corner atom is 6.

For a body centred cell the atom inside the cell has all the corner atoms as its close neighbours as shown in Fig. 18.12 (*iii*). Thus the coordination number for the interior atom is 8.

For a *face centred cell* each corner atom is having four close neighbours in each of the three XY, YZ and ZX planes as shown in Fig. 18.19. Thus the coordination number in this case is $4 \times 3 = 12$.

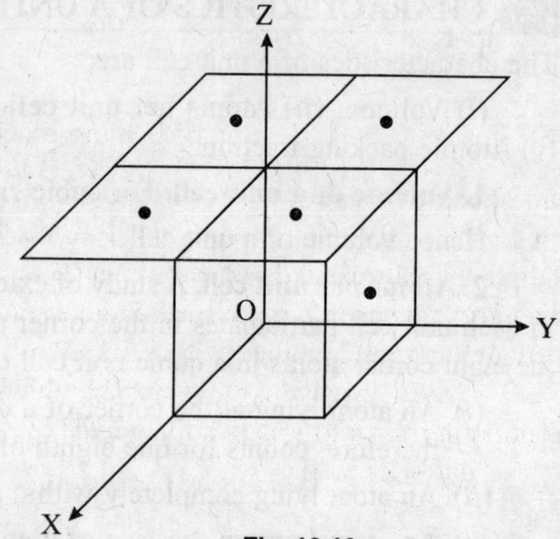

Fig. 18.19

18.9 ATOMIC RADIUS

It is supposed that all the atoms in a crystal have the *same* size and are touching each other. *Thus the atomic radius is the distance between the centres of two neighbouring atoms.*

In the case of a *simple cube*, if r is the atomic radius and lattice parameter is a, then

$$a = 2r$$

or $$r = a/2$$

In the case of a *body centred cube*, the atoms touch each other along the diagonal of the cube as shown in Fig. 18.20 and we have

$$(4r)^2 = (\sqrt{2}a)^2 + a^2$$

or $$r = \frac{\sqrt{3}}{4}a.$$

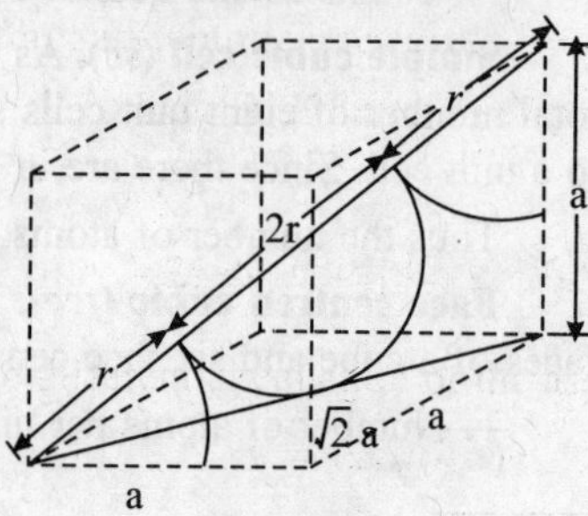

Fig. 18.20

In the case of a face-centred cube as shown in Fig. 18.21, the atoms are in contact along the diagonal of the faces.

$$\therefore \qquad 4r^2 = 2a^2$$

or $$r = \frac{\sqrt{2}}{4}a = \frac{a}{2\sqrt{2}}.$$

18.10 ATOMIC PACKING FRACTION (OR FACTOR)

Also known as *relative packing density is defined as the ratio of the volume of the atoms occupying the unit cell to the volume of the unit cell relating to the structure of the crystal.*

In a **simple cubic structure** the number of atoms per unit cell is *one*. The atomic radius is given by half the primitive *i.e.*, $r = a/2$.

$\therefore$ Volume occupied by the atom in the unit cell

$$= \frac{4}{3}\pi r^3 = \frac{4}{3}\pi\left(\frac{a}{2}\right)^3$$

Volume of the unit cell $= a^3$

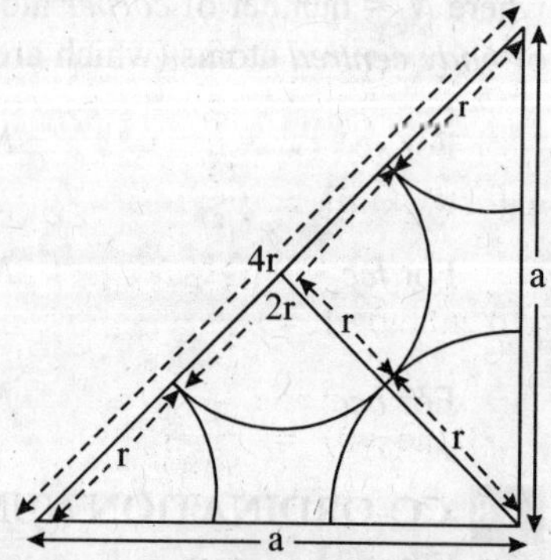

Fig. 18.21

$$\text{Packing fraction } f = \frac{\frac{4}{3}\pi\left(\frac{a}{2}\right)^3}{a^3} = \frac{\pi}{6}$$

$$= 52\%.$$

Thus the atoms are loosely packed. Only polonium at a certain temperature is known to exhibit such a structure. Potassium chloride which has alternate ions of K and Cl also behaves like a simple lattice as regards scattering of X-rays because the two ions are almost identical.

In a **body centred cubic lattice** there are *two* atoms per unit cell. The atomic radius $r = \frac{\sqrt{3}}{4}a$

$\therefore$ Volume of the unit cell $= a^3$

$$\therefore \text{ Packing fraction } f = \frac{\frac{\sqrt{3}}{8}\pi a^3}{a^3} = \frac{\sqrt{3}}{8}\pi = 68\%.$$

Common substances with *bcc* lattice are barium, chromium, sodium, iron and caesium chloride. In caesium chloride, for example, each caesium ion has eight nearest neighbours of chlorine ions. The caesium ion is at the centre of the cube while chlorine ions are at the corners of the cube.

In a **face centred cubic lattice** there are *four* atoms per unit cell. The atomic radius $r = \frac{\sqrt{2}}{4}a$

$\therefore$ Volume occupied by the atoms in the unit cell $= 4\times\frac{4}{3}\pi r^3 = \frac{16}{3}\pi\frac{2^{3/2}}{4^3}a^3 = \frac{\pi}{3\sqrt{2}}a^3$

Volume of the unit cell $= a^3$

$$\therefore \text{ Packing fraction } f = \frac{\frac{\pi}{3\sqrt{2}}a^3}{a^3} = \frac{\pi}{3\sqrt{2}} = 74\%$$

Common examples of this type of structure are Ni, Cu, Au, Al, Ag, Li, K etc.

18.11 NUMBER OF LATTICE POINTS

We have calculated the number of atoms per unit cell in a *simple cubic* (*sc*), *face centred cubic* (*fcc*) and *body centred cubic* (*bcc*) cell. The same discussion is equally valid, if we use the word 'lattic point' instead of 'atom'.

Thus, a *lattice point* lying at the corner of a cubic unit cell is shared equally by *eight* such unit cells. Therefore, the contribution of a lattice point lying at the corner towards a particular cell is $\frac{1}{8}$.

A *lattice point* lying completely within a unit cell belongs wholly to the cell.

A *lattice point* lying on the face of a unit cell belongs equally to *two* unit cells and, therefore, counts as *one half* of a particular cell.

The following equation is used to calculate the effective number of lattice points N belonging to a particular cell

$$N = N_i + N_f/2 + N_c/8$$

where N_i represents the number of lattice points present completely inside the cell, N_f and N_c represent the lattice points occupying the face centre and corner positions of the cell respectively.

Using the relation the effective number of lattice points in a cubic unit cell is $\frac{8}{8} = 1$.

[*Note.* By the same relation the effective number of lattice points in a body centred cubic unit cell is 2 *i.e.*, $8 \times \frac{1}{8} + 1 = 2$ and face centred cubic cell is 4 *i.e.*, $8 \times \frac{1}{8} + 6 \times \frac{1}{2} = 4$]

SOLVED EXAMPLES

Example 18.1 (*a*) *Prove that the packing fraction for a simple cubic (sc) structure is 0.52; for body centred cubic (bcc) structure is 0.68 and for a face centred cubic (fcc) structure is 0.74.*

(Pbi.U. 2004, 1997; G.N.D.U 2003, 1997, 1995; Calicut U. 2003; P.U. 1992, Nagpur U. 2012)

(*b*) *If the lattice parameter for a fcc crystal is 2.14Å find the radius of the particle.*

(Nag. U. 2002)

Solution. (*a*) For proof See article 18.10 Atomic packing fraction.

(*b*) For an *fcc* crystal the radius of the atom (particle)

$$r = \frac{a}{2\sqrt{2}} \quad \text{where } a \text{ is the lattice parameter.}$$

Here, $a = 2.14\text{Å}$

$$\therefore \quad r = \frac{2.14}{2\sqrt{2}}\text{Å} = 0.76\text{Å}.$$

Example 18.2. *For a cubic (fcc) crystal, lattice constant*

$$\boldsymbol{a} = \left[\frac{4M}{\rho N}\right]^{1/3}$$

where M is the gm molecular weight of molecules at lattice points, ρ is the density of crystal and N is Avogadro's number.

Solution. Mass per molecule $= \frac{M}{N}$

Let n be the number of molecules in a unit cell.

$$\therefore \quad \text{Mass of a unit cell} = \frac{nM}{N}$$

$$\text{Density of the crystal} = \rho$$

$$\therefore \text{ Volume of a unit cell} \quad V = \frac{nM}{\rho N}$$

For a (*fcc*) crystal the number of lattice points (molecules) per unit cell = 4

$$\therefore \text{ Volume of a unit cell} \quad V = \frac{4M}{\rho N}$$

If a is the lattice constant, $V = a^3$

$$\therefore \quad a = (V)^{1/3} = \left(\frac{4M}{\rho N}\right)^{1/3}.$$

Example 18.3. *A substance has fcc lattice, molecular weight 60.2 and density 6250 Kg/m³. Calculate its lattice constant 'a'.*

(P.U. 2000)

Solution. For (*fcc*) lattice $a = \left(\frac{4M}{\rho N}\right)^{1/3}$

Here $M = 60.2$; $\rho = 6250$ Kg/m³; $N = 6.02 \times 10^{26}$ per Kg mole.

$$\therefore \qquad a = \left[\frac{4 \times 60.2}{6250 \times 6.02 \times 10^{26}}\right]^{1/3} = \left[\frac{8 \times 10^{-27}}{125}\right]^{1/3}$$

$$\text{or} \qquad a = \frac{2}{5} \times 10^{-9}\text{m} = 4\text{Å}.$$

18.12 HEXAGONAL CLOSE PACKED STRUCTURE

The most efficient packing of atoms–considered to be spheres–in a plane is obtained as shown in Fig. 18.22 with each atom touching six atoms in the plane. A second similar layer can be added by placing atoms in contact with three atoms of the bottom layer in positions marked as *B* or equivalently *C*. The third layer may have atoms placed in two ways:

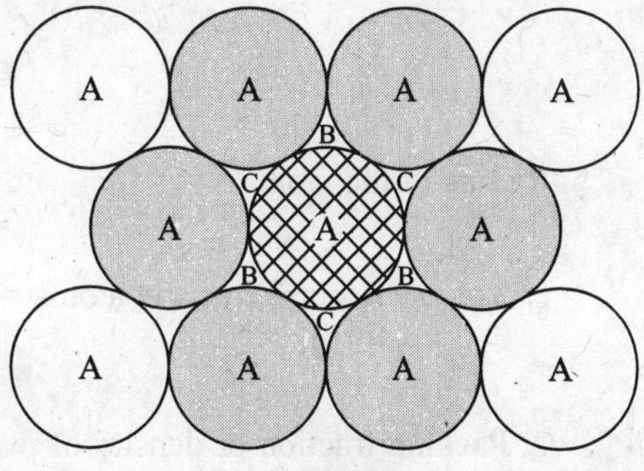

Fig. 18.22

(*i*) The atoms are placed over the holes in the first layer not occupied by the second layer *i.e.*, if the atoms in the second layer are placed in positions marked *B*, those in the third layer are placed in positions marked *C* or *vice-versa*. The positions of atoms in successive layers is thus indicated as *ABC*, *ABC*. This is the *face-centred cubic* (*fcc*) structure. The *close-packed* layers of the *fcc* structure are in the body diagonal planes as shown in Fig. 18.23 or in (111) plane. Familiar examples of elements that have face *centred cubic structures* are Cu, Ag, Au, Al, Pd and Pt.

(*ii*) The atoms are placed directly over the atoms placed in the first layer. The position of atoms in successive layers of this structure may be indicated as *AB*, *AB*, *AB* ..... . This leads to *hexagonal close-packed* (*hcp*) structure as shown in Fig. 18.24.

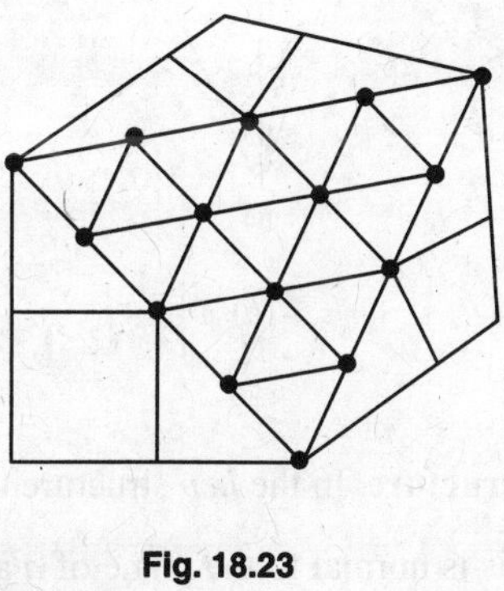

Fig. 18.23

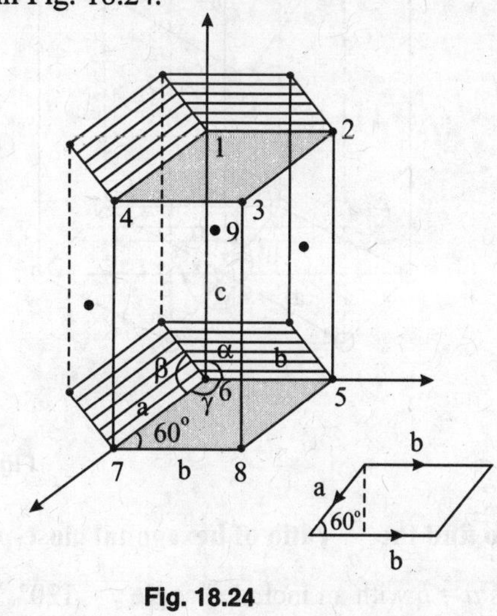

Fig. 18.24

Co-ordination number. According to the above discussion each atom touches *six* atoms in its plane, *three* atoms in the layer immediately above it and similarly *three* atoms in the layer immediately below it. Hence co-ordination number of *hcp* structure is 6 + 3 + 3 = 12.

Atomic Packing fraction. A unit cell of *hcp* structure is a hexagonal cell and three such cells form one *hcp* unit. The number of atoms present in a hexagonal cell is *eight* at the corners and one inside so that the number of atoms per unit cell $= 1 + \frac{8}{8} = 2$.

The base is a parallelogram with each side = $a[b = a]$ and containing angle 60° as shown.

$$\therefore \quad \text{Area of the base} = a \times a \sin 60 = \frac{a^2}{2}\sqrt{3}$$

$$\text{Volume of the hexagonal cell} = \frac{a^2}{2}\sqrt{3}\, c$$

$$= \frac{a^2\sqrt{3}}{2}\frac{2a\sqrt{2}}{\sqrt{3}} \qquad \left[\because c = \frac{2a\sqrt{2}}{\sqrt{3}}\right]$$

$$= a^3\sqrt{2}$$

Radius of the atom $\quad r = \dfrac{a}{2}$

$$\therefore \text{Volume occupied by the atoms} = 2 \times \frac{4}{3}\pi r^3 = 2 \times \frac{4}{3}\pi\frac{a^3}{8} = \frac{\pi a^3}{3}$$

$$\therefore \text{Packing fraction or density of packing} = \frac{\frac{\pi a^3}{3}}{a^3\sqrt{2}} = \frac{\pi}{3\sqrt{2}} = 0.74 = 74\%$$

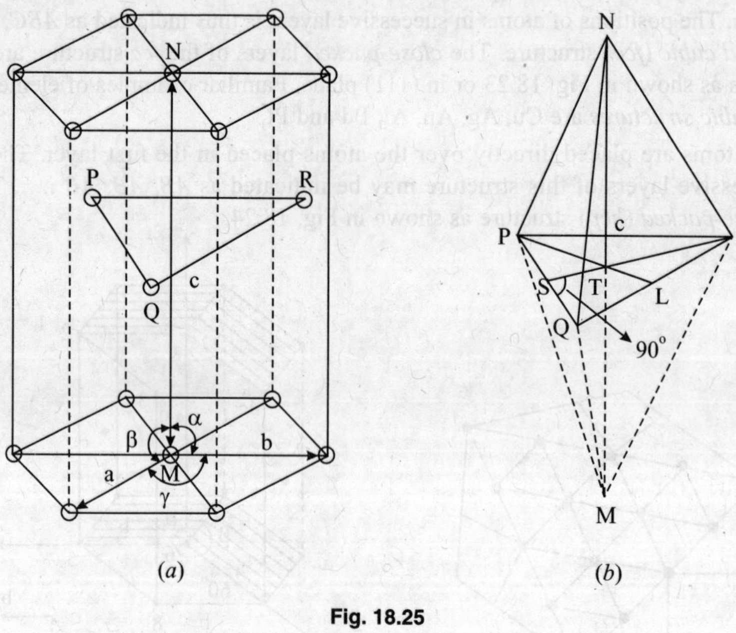

Fig. 18.25

To find the $\frac{c}{a}$ ratio of hexagonal close-packed structure. In the *hcp* structure the primitive cell has $a = b$ with an included angle $\gamma = 120°$. The c axis is normal to the plane of a and b so that α (angle between b and c axes) = β, (angle between a and c axes) = 90°. Let P, Q and R be the centres of the adjacent atoms in the middle plane and N and M the centres of the adjacent atoms in the planes immediately above and below the plane PQR as shown in Fig. 18.25 (*a*). This is also shown separately in Fig. 18.25 (*b*).

Join MN. Also join M and N to P, Q and R. Thus we get two tetrahedrons $NPQR$ and $MPQR$ with the common base PQR which forms an equilateral triangle. The line MN passes through the plane of the triangle PQR at the point T which is the point of intersection of the three medians.

So if we join R to the mid point S of the side PQ, then RS is a median and the point T lies on it such that

$$TR = 2TS$$

As N and M are the two adjacent lattice points on the c-axis, the distance between them is the unit distance c along this axis. [See Fig. 18.25 (a)]

Now $$PQ = QR = a$$

As the median of an equilateral triangle is also perpendicular to the opposite side, the angle $QSR = 90°$.

$\therefore$ $$RS = \sqrt{a^2 - (a/2)^2} = \frac{\sqrt{3}}{2}a$$

Also $$RT = 2TS = \frac{2}{3}(TS + TR) = \frac{2}{3}RS$$

$$= \frac{2}{3}\cdot\frac{\sqrt{3}}{2}a = \frac{a}{\sqrt{3}}$$

Now $$NT = \sqrt{NR^2 - RT^2} = \sqrt{a^2 - \frac{a^2}{3}} = \sqrt{\frac{2}{3}}a$$

because $$NR = NQ = NP = a$$

Again $$c = 2NT = 2a\sqrt{\frac{2}{3}} = \sqrt{\frac{8}{3}}\,a = 1.633a$$

or $$\frac{c}{a} = \sqrt{\frac{8}{3}} = 1.633$$

18.13 SOME CRYSTAL STRUCTURES

(1) Sodium chloride structure. In sodium chloride (NaCl), Na atom loses its outer electron and acquires an excess *positive* charge. On the other hand a Cl atom acquires the electron lost by the Na atom and thus acquires an excess negative charge. In this way, we get Na^+ and Cl^- ions. These ions attract each other due to the electric force of attraction between unlike charges. The two ions, however, cannot approach very close to each other because of the strong forces of repulsion which are called into play when their outer electron shells come into close proximity. When the centres of the two ions come within a certain distance of one another, attraction and repulsion balance and they cannot approach closer. It is in this sense that we think of them as possessing a definite radius.

Co-ordinates of the atoms. Sodium chloride NaCl is an example of face centred cubic (*fcc*) Bravais lattice. The basis consists of one Na atom and one Cl atom separated by one half the body diagonal of a unit cube. There are four NaCl molecules in a unit cube in the position coordinates given below.

Na	$0\ 0\ 0$	$\frac{1}{2}\ \frac{1}{2}\ 0$	$\frac{1}{2}\ 0\ \frac{1}{2}$	$0\ \frac{1}{2}\ \frac{1}{2}$
Cl	$\frac{1}{2}\ \frac{1}{2}\ \frac{1}{2}$	$0\ 0\ \frac{1}{2}$	$0\ \frac{1}{2}\ 0$	$\frac{1}{2}\ 0\ 0$

Number of Na ions in a unit cell. As there are four NaCl molecules in a unit cell, the number of sodium (Na) ions is four per unit cell. A sketch of sodium chloride lattice is shown in Fig. 18.26.

A sodium chloride cubic lattice can be considered to be made up of two face centred cubic sub-lattices, one of Na ions having origin at the point (0, 0, 0) and the other of Cl ions having its origin mid way along the cube edge *i.e.*, at point $\left[\frac{a}{2}, 0, 0\right]$. Each atom has as nearest neighbours six atoms of the opposite kind so that the coordination number is 6. Other crystal having a similar structure are LiH, KBr, KCl, PbS.

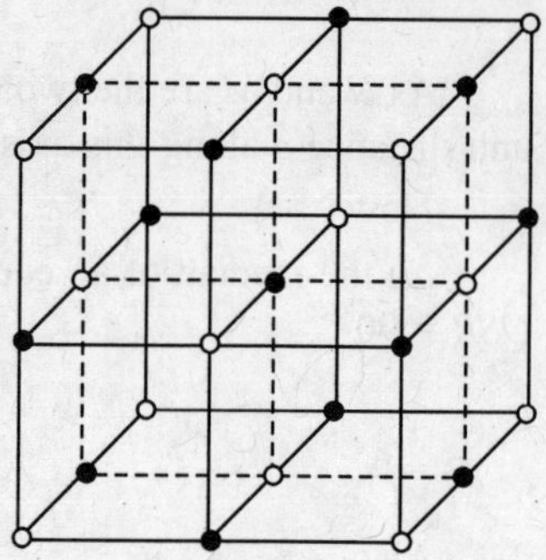

Fig. 18.26

(2) Diamond structure. A typical example of *fcc* structure is diamond. The diamond structure is of considerable importance as the basic lattice structure for many important semiconductors like Si, Ge are diamond type. In many compound semiconductors, atoms are arranged in basic diamond structure but are different on alternating sites. This is called *zinc blende lattice.*

The diamond lattice can be supposed to be an *fcc* structure with an extra atom placed at $\frac{\vec{a}}{4}+\frac{\vec{b}}{4}+\frac{\vec{c}}{4}$ from each of the *fcc* atoms. In other words, the diamond lattice is a face centred cubic (space) lattice with primitive basis of two identical atoms at (0, 0, 0), and $\left(\frac{1}{4}, \frac{1}{4}, \frac{1}{4}\right)$ associated with each lattice point. A two dimensional view of the cell is shown in Fig. 18.27. Fractions denote height above the base in fractions of cube edge. The points 0 and 1/2 are on the *fcc* lattice. Those at 1/4 and 3/4 are on a similar lattice displaced along the body diagonal by 1/4 of its length.

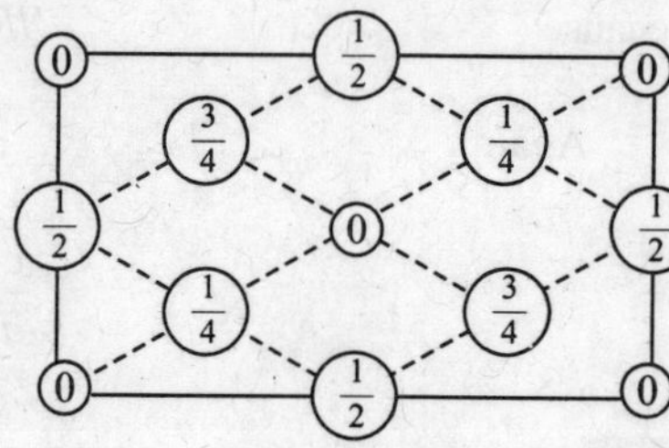

Fig. 18.27

The three dimensional picture is obtained by inserting one *fcc* lattice into another *fcc* lattice displaced along the space diagonal by a quarter of its length. In the way a corner atom of one face centred cube comes in contact with a corner atom of the second cube and not with an atom in a face. Each atom of one cube may, therefore, be regarded to be at the centre of a tetrahedron formed by the four nearest neighbours belonging to the other cube. Thus each atom in a diamond structure forms *four* covalent bonds with its four nearest neighbours. The bond length is 1.544 Å and energy per bond is 3.5×10^5 Joule/mole. The crystal structure of diamond showing the tetrahedral bond arrangement is given in Fig. 18.28. Each atom has four nearest neighbours and twelve next nearest neighbours.

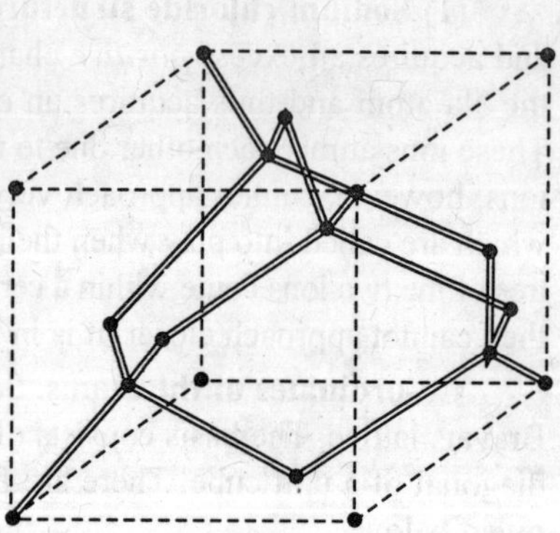

Fig. 18.28

Number of nearest neighbours. The diamond structure is obtained by inserting one *fcc* lattice into another *fcc* lattice displaced along the space diagonal by a quarter of its length. Therefore, each atom has 4 *nearest neighbours* and 12 *next nearest neighbours.*

Therefore, the *co-ordination number* of each carbon atom is 4 and the *nearest neighbour distance is* $\frac{\sqrt{3}a}{4}$.

Number of atoms per unit cell. There are eight atoms per unit cell — 4 inside, 3 in the faces and one at the corners.

Packing fraction. As stated above, each atom of diamond crystal has four nearest neighbours *i.e.*, the coordination number is 4. There are eight atoms per unit cell — 4 inside, 3 in the faces and one at the corners. The body diagonal is equal to 8 r as there are 3 full spheres and 2 half spheres in it. If a is the side of the cube, then

$$(8r)^2 = 3a^2$$

or $$r = \frac{\sqrt{3}}{8}a$$

∴ Volume occupied by the atoms in the unit cell

$$= 8 \times \frac{4}{3}\pi r^3 = 8 \times \frac{4}{3}\pi \frac{3\sqrt{3}}{8^3}a^3 = \frac{\sqrt{3}}{16}\pi a^3$$

Volume of unit cell $= a^3$

∴ Packing fraction $$f = \frac{\frac{\sqrt{3}}{16}\pi a^3}{a^3} = \frac{\sqrt{3}}{16}\pi = 34\%$$

Thus we find that packing in diamond structure is rather loose.

Example 18.4. *Calculate the distance between two atoms of a basis of the diamond structure if the lattice constant of the structure is 5Å.* **(P.U. 2001)**

Solution. Distance between the two atoms = nearest neighbour distance = $\frac{\sqrt{3}}{4}a$

Here, $a = 5$Å

∴ Distance between the two atoms $= \frac{\sqrt{3}}{4} \times 5\text{Å} = \frac{1.732 \times 5}{4} = 2.16\text{Å}$

Example 18.5. *How many atoms are there in the primitive cell of diamond? What is the length of a primitive translation vector if the cube edge* $\vec{a} = 3.56$ *Å.*

Solution. Diamond has *fcc* Bravais lattice with two carbon atoms in a primitive cell. The number of atoms in a conventional cube is 8.

If $\vec{a}$ = cube edge, then

Primitive translation vector $= \frac{a}{\sqrt{2}} = \frac{3.56}{1.41} = 2.52$ Å

18.14 MILLER INDICES

The position and orientation of a crystal plane are determined by any three points in the plane provided the points are not collinear. If each point lies on a crystal axis, the plane can be specified by giving the position of the points along the axes in terms of the lattice constants. For example, if the points have the co-ordinates (2, 0, 0) (0, 3, 0) (0, 0, 1) relative to axis vectors from some origin as shown in Fig. 18.29, the plane may be specified by three numbers 2, 3, 1.

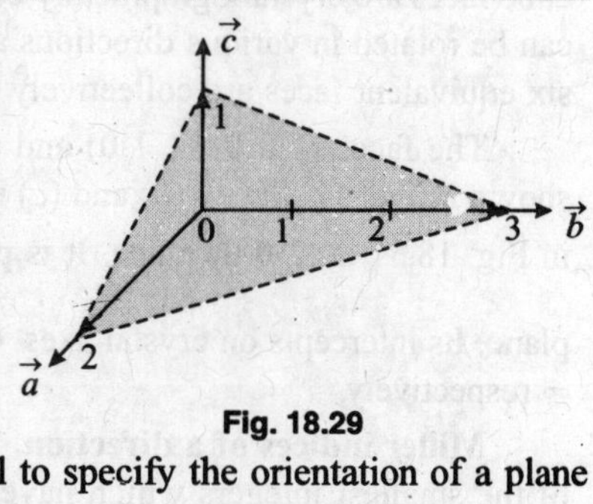

Fig. 18.29

However, in crystal structure analysis it is more useful to specify the orientation of a plane in accordance with the following rules.

(*i*) Find the intercepts of the plane on the crystal axes $\vec{a}$, $\vec{b}$ and $\vec{c}$ in terms of the lattice constant. The axes may be primitive or non-primitive.

(*ii*) Take the reciprocals of these numbers and then reduce to the *smallest* three integers having the same ratio. Denote the result as $(h\ k\ l)$ enclosed in parentheses.

The three integers h, k and l are called **Miller indices.**

Orientation of a plane by Miller indices. The Miller indices $(h\ k\ l)$ may denote a single plane or a set of parallel planes. If a plane cuts an axis on the **negative** side of the origin, the index is *negative* and denoted as $(h\ \bar{k}\ l)$ by placing a minus (–) sign above the corresponding index (in this case $\bar{k}$). For an intercept at infinity the corresponding index is **zero**. For the crystal plane having intercepts $2a$, $3b$ and c on the crystal axes, $\vec{a}$, $\vec{b}$ and $\vec{c}$, the reciprocals of the numbers 2, 3 and 1 are $\frac{1}{2}$, $\frac{1}{3}$ and 1. The smallest three integers having the same ratio are 3, 2 and 6. The plane is, therefore, referred to as (3 2 6) plane.

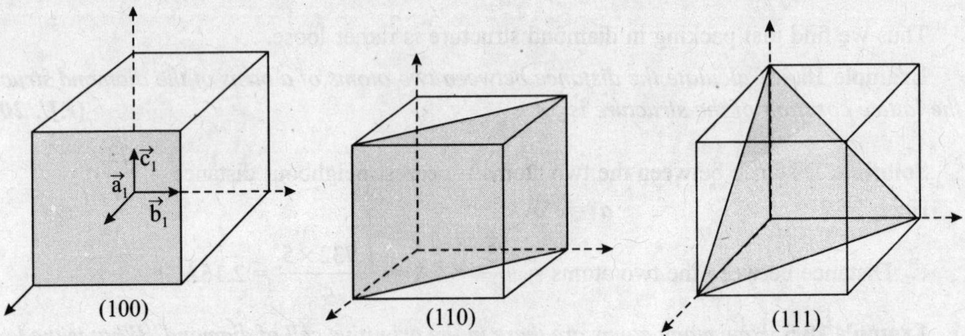

(100) (110) (111)

(*a*) (*b*) (*c*)

Fig. 18.30

The six faces of a cubical crystal are (1 0 0) (0 1 0) (0 0 1) $(\bar{1}\ 0\ 0)$ $(0\ \bar{1}\ 0)$ $(0\ 0\ \bar{1})$. From a crystallographic point of view many planes in a lattice are equivalent *i.e.*, a plane with given Miller indices can be shifted about in the lattice simply by a choice of the position and orientation of the unit cell. The indices of such equivalent planes are enclosed in braces {} instead of in parenthesis ( ). For example, all the six cube faces are crystallographically equivalent because the unit cell can be rotated in various directions and still appear the same. The six equivalent faces are collectively designated as {1 0 0}.

The faces (1 0 0) (1 1 0) and (1 1 1) of a cubic crystal are shown in Fig. 18.30 (*a*), (*b*) and (*c*) respectively. The plane shown in Fig. 18.31 is (2 0 0) plane. It is parallel to (1 0 0) and $[\bar{1}\ 0\ 0]$ plane. Its intercepts on crystal axes $\vec{a}$, $\vec{b}$ and $\vec{c}$, are $\frac{1}{2}a$, ∞ and ∞ respectively.

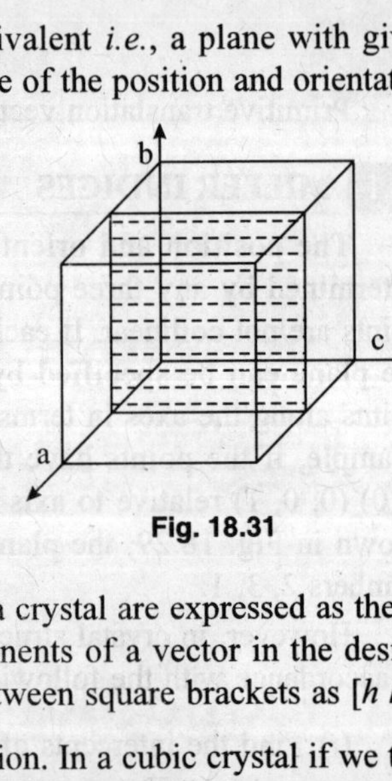

Fig. 18.31

Miller indices of a direction. The *indices of a direction* in a crystal are expressed as the set of the smallest integers which have the same ratio as the components of a vector in the desired direction referred to the axes vectors. The integers are written between square brackets as $[h\ k\ l]$. The $\vec{a}$ axis is the [1 0 0] direction, the $-\vec{b}$ axis is $[0\ \bar{1}\ 0]$ direction. In a cubic crystal if we take the origin at one of the corners and the axes parallel to the edges, the body diagonal would be represented as [1 1 1]. Similarly the $\vec{a}$ axis is [1 0 0] direction, the $\vec{b}$ axis [0 1 0] direction and the $\vec{c}$ axis [0 0 1] direction. In a cubic crystal the direction [*h k l*] is always perpendicular to a plane (*h k l*) having the same indices.

Important features. (*i*) If a plane is parallel to a co-ordinate axis, then the intercept of this plane is *infinity* and hence its Miller index is *zero.*

(*ii*) A plane passing through the origin is defined in terms of a parallel plane having non-zero intercept.

(*iii*) Parallel planes equally spaced have the same index numbers.

(*iv*) It is the ratio of the indices which is important and determines a plane. For example, (2 4 6) and (1 2 3) represent the same plane.

(*v*) The direction [*h k l*] is perpendicular to the plane (*h k l*). [For proof see **example 18.8**].

Purpose of taking reciprocals. The purpose of taking reciprocals to find Miller indices is to bring all the planes inside a single unit cell so that we can discuss all crystal planes in terms of the planes passing through the single unit cell. As the entire crystal can be generated from a single unit cell, the unit cell is studied as a representative of the whole crystal. The unit cell is equivalent to the molecule. Just as we write the chemical equation for a single molecule and this equation applies to the whole sample, similarly the study of the single unit cell gives the characteristics of the whole crystal.

18.15 SOME IMPORTANT PLANES

Let us discuss some important planes in a simple cubic unit cell viz. (100), (001), (110), (101), (011), (111), (200), (020), (002), (201), (112) planes.

Taking the three perpendicular directions of the simple cubic cell along the co-ordinate axes $\vec{x}$, $\vec{y}$ and $\vec{z}$ respectively, we have

(*i*) **(100) plane.** The (100) plane lies in the *y-z* plane and has an intercept 1 along the *x*-axis. Fig. 18.32 (*i*).

(*ii*) **(010) plane.** The (010) plane lies in the *x-z* plane and has an intercept 1 along the *y*-axis. Fig. 18.32 (*ii*).

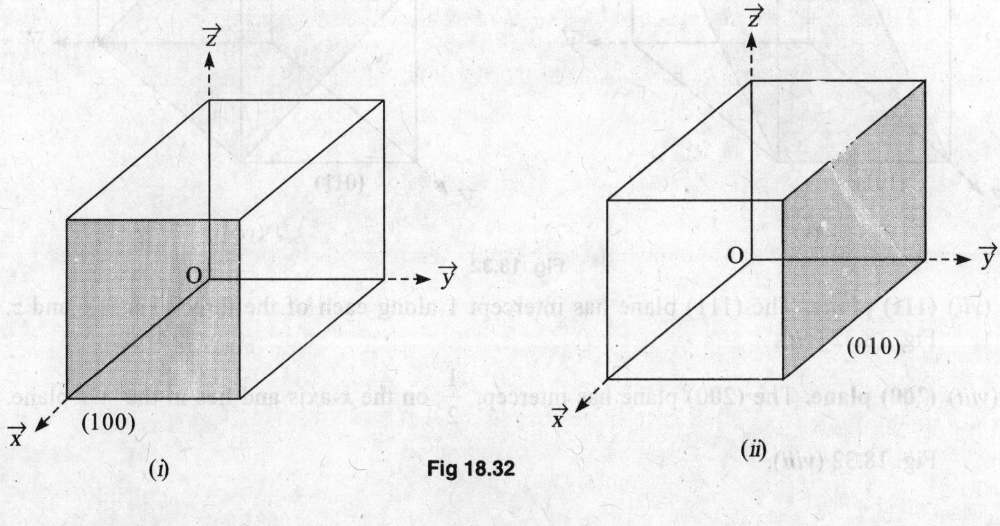

Fig 18.32

(*iii*) **(001) plane.** The (001) plane lies in the $x - y$ plane and has an intercept 1 along the z-axis. Fig. 18.32 (*iii*).

(*iv*) **(110) plane.** The (110) plane is parallel to z-axis and has an intercept 1 on x-axis as well as y-axis. Fig. 18.32 (*iv*).

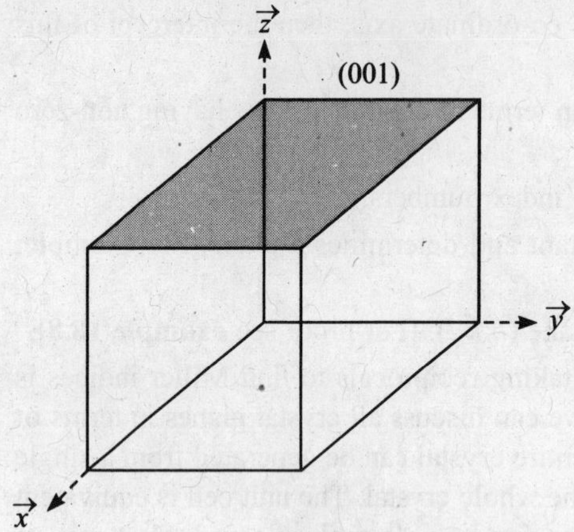

(*iii*)

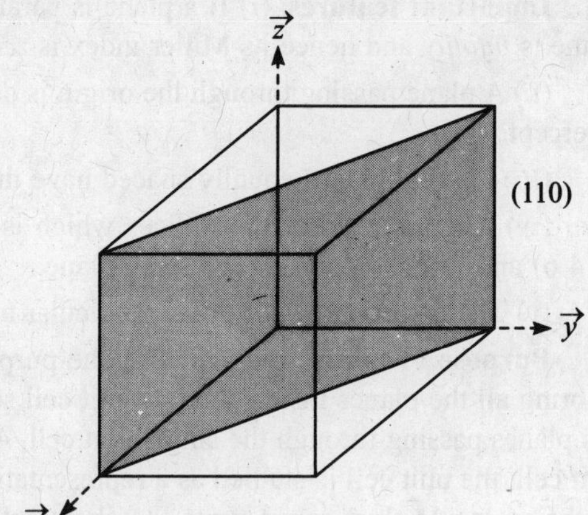

(*iv*)

Fig. 18.32

(*v*) **(101) plane.** The (101) plane is parallel to y-axis and has an intercept 1 on x-axis as well as on z-axis. Fig. 18.32 (*v*).

(*vi*) **(011) plane.** The (011) plane is parallel to x-axis and has an intercept 1 on y-axis as well as on z-axis. Fig. 18.32 (*vi*).

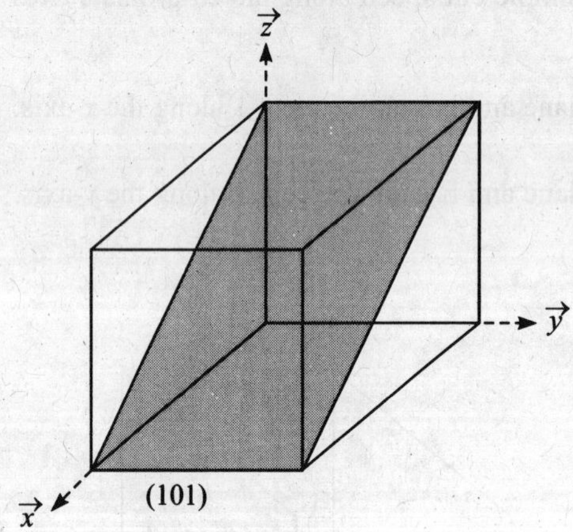

(*v*)

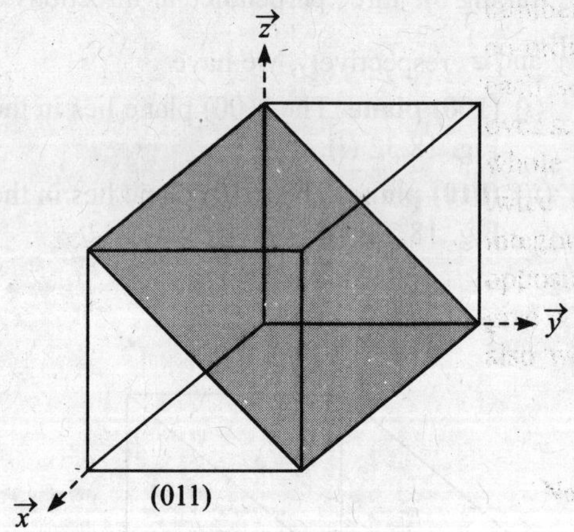

(*vi*)

Fig. 18.32

(*vii*) **(111) plane.** The (111) plane has intercept 1 along each of the three axis x, y and z. Fig. 18.32 (*vii*).

(*viii*) **(200) plane.** The (200) plane has intercept $\frac{1}{2}$ on the x-axis and lies in the y-z plane.

Fig. 18.32 (*viii*).

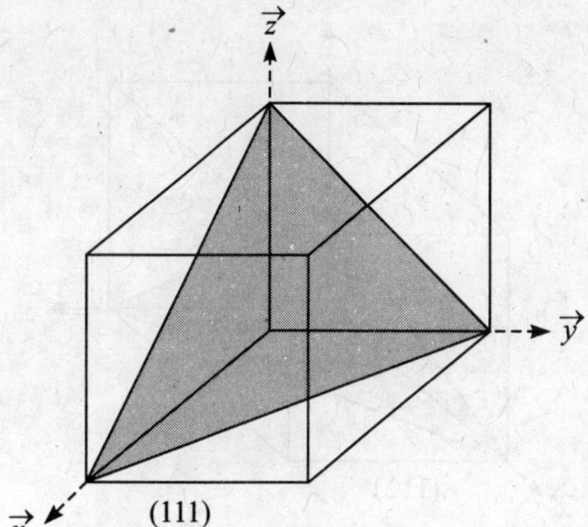

(*vii*)

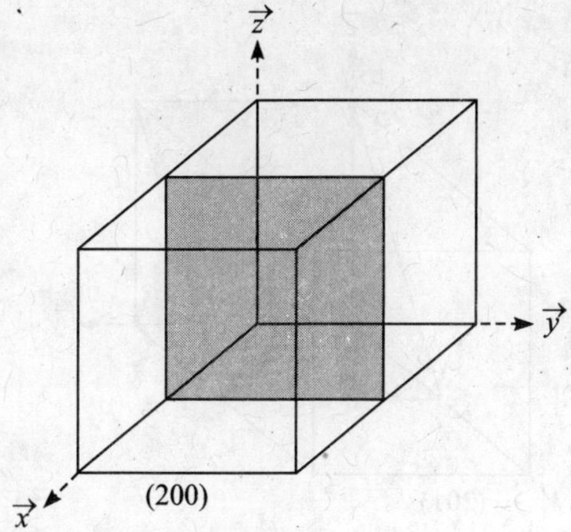

(*viii*)

Fig. 18.32

(*ix*) **(020) plane.** The (020) plane has intercept $\frac{1}{2}$ on the *y*-axis and lies in the *x-z* plane. Fig. 18.32 (*ix*).

(*x*) **(002) plane.** The (002) plane has intercept $\frac{1}{2}$ on the *z*-axis and lies in the *x-y* plane. Fig. 18.32 (*x*).

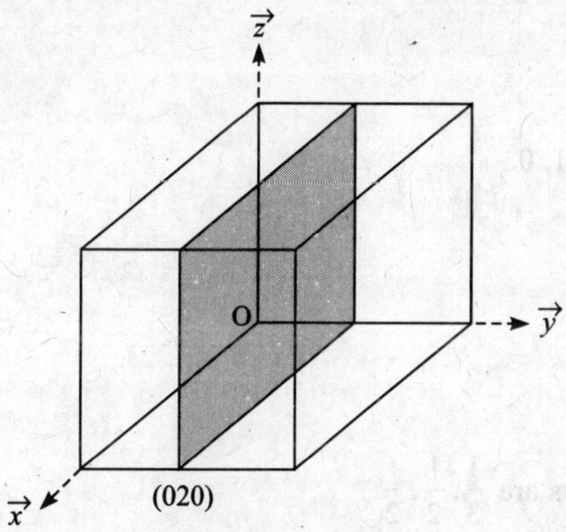

(*ix*)

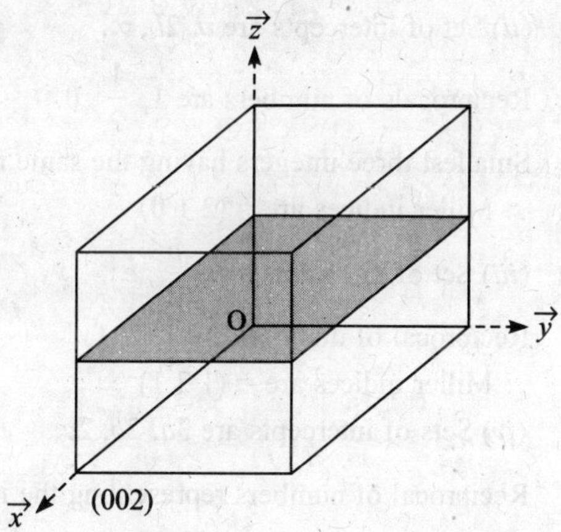

(*x*)

Fig 18.32

(*xi*) **(201) plane.** The (201) plane has intercept $\frac{1}{2}$ on the *x*-axis, is parallel to *y*-axis and has an intercept 1 on the *z*-axis. Fig. 18.32 (*xi*)

(*xii*) **(112) plane.** The (112) plane has intercept 1 each on *x*-axis and *y*-axis and intercept $\frac{1}{2}$ on the *z*-axis. Fig. 18.32 (*xii*)

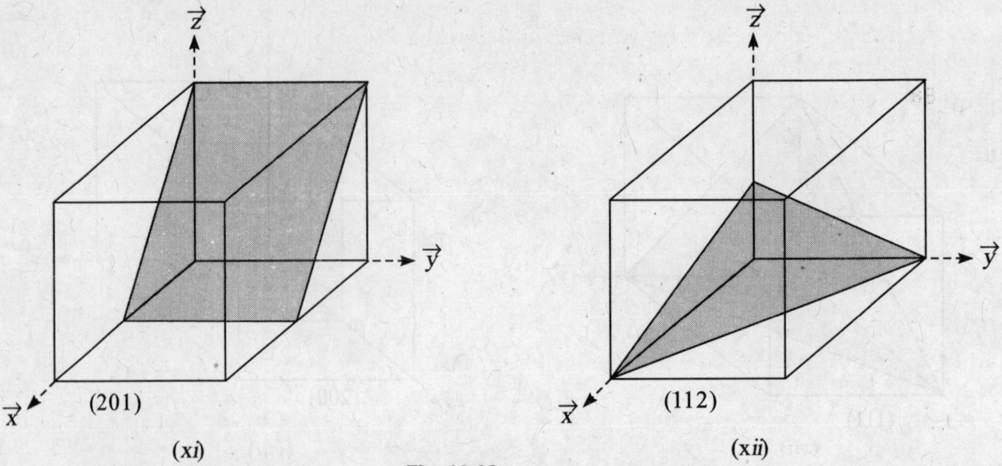

Fig. 18.32

Example 18.6. *Find the Miller indices for planes in each of the following sets which intercept* $\vec{a}$, $\vec{b}$ *and* $\vec{c}$ *at (i) 3a, 3b, 2c; (ii) a, 2b, ∞; (iii) a, b/2, c; (iv) 3a, 2b, 2c.* ***(P.U. 2003)***

Solution. (*i*) Set of intercepts = $3a$, $3b$, $2c$

Reciprocal of numbers representing the intercepts are $\frac{1}{3}, \frac{1}{3}, \frac{1}{2}$.

Smallest three integers having the same ratio are 2, 2, 3.

∴ Miller indices are = (2 2 3)

(*ii*) Set of intercepts are a, $2b$, ∞.

Reciprocals of numbers are 1, $\frac{1}{2}$, 0.

Smallest three integers having the same ratio 2, 1, 0.

∴ Miller indices are = (2 1 0)

(*iii*) Set of intercepts = a, $\frac{b}{2}$, c

Reciprocal of numbers are 1, 2, 1.

∴ Miller indices are = (1 2 1)

(*iv*) Sets of intercepts are $3a$, $2b$, $2c$.

Reciprocal of numbers representing the intercepts are $\frac{1}{3}, \frac{1}{2}, \frac{1}{2}$.

Smallest three integers having the same ratio are 2,3,3.

∴ Miller indices are (2 3 3).

18.16 SPACING OF PLANES IN CRYSTAL LATTICES

Consider a simple unit cell in which the crystal axes $\vec{a}$, $\vec{b}$ $\vec{c}$ are *orthogonal*. Suppose these coincide with co-ordinate axes *X*, *Y*, *Z* respectively with *O* as origin taken at a lattice point[Fig. 18.33].

Now consider a set of planes defined by Miller indices (*h k l*). One such plane *ABC* will have intercepts $\frac{a}{h}$, $\frac{b}{k}$ and $\frac{c}{l}$ on the $X(\vec{a})$, $Y(\vec{b})$ and $Z(\vec{c})$ axes respectively.

Draw ON perpendicular from O on the plane ABC and let $ON = d$, then $d = ON$ is the length of the normal from the origin to the plane and represents the distance between adjacent planes or *interplanar spacing* between parallel planes.

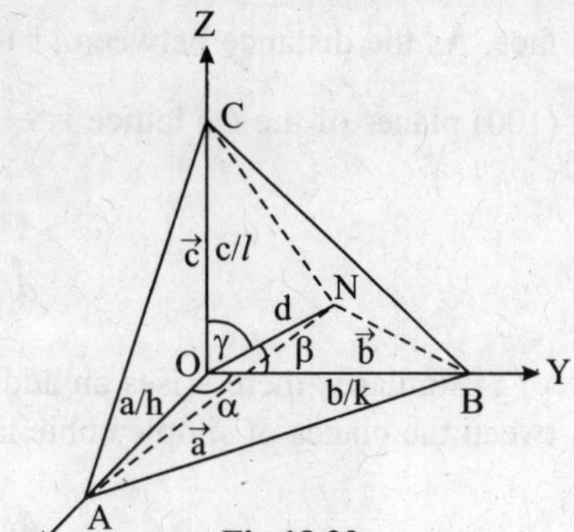

Fig 18.33

Let ON make angles α, β and γ with $X(\vec{a}), Y(\vec{b})$ and $Z(\vec{c})$ axes respectively, then according to the law of direction cosines,

$$\cos^2\alpha + \cos^2\beta + \cos^2\gamma = 1 \quad ...(i)$$

Also $\angle ONA = \angle ONB = \angle ONC = 1rt\angle$

Now ΔAON is a right angled triangle with $\angle AON = \alpha$ and $\angle ONA = 1rt$ angle

$$\therefore \quad \frac{ON}{OA} = \cos\alpha \text{ or } ON = OA\cos\alpha$$

or
$$d = \frac{a}{h}\cos\alpha \text{ or } \cos\alpha = d\left(\frac{h}{a}\right)$$

Similarly
$$\cos\beta = d\left(\frac{k}{b}\right) \text{ and } \cos\gamma = d\left(\frac{l}{c}\right)$$

Substituting the values of cos α, cos β and cos γ in (*i*), we have

$$d^2\left[\frac{h^2}{a^2} + \frac{k^2}{b^2} + \frac{l^2}{c^2}\right] = 1$$

or
$$d = \frac{1}{\left[\frac{h^2}{a^2} + \frac{k^2}{b^2} + \frac{l^2}{c^2}\right]^{1/2}}$$

For a cubical crystal $a = b = c$.

$$\therefore \quad d = \frac{a}{(h^2 + k^2 + l^2)^{1/2}}$$

Case I: Simple cubic lattice (*sc*). A simple cubic lattice has (1 0 0), (1 1 0) and (1 1 1) planes. The (1 0 0) planes cut the *X*-axis and are parallel to *Y* and *Z*-axis.

$$\therefore \quad d_{100} = \frac{a}{[1^2 + 0 + 0]^{1/2}} = a$$

The (1 1 0) planes cut obliquely across *X* and *Y*-axis but are parallel to *Z*-axis.

$$\therefore \quad d_{110} = \frac{a}{[1^2 + 1^2 + 0]^{1/2}} = \frac{a}{\sqrt{2}}$$

The (1 1 1) planes cut obliquely across the three axes.

$$\therefore \quad d_{111} = \frac{a}{(1^2 + 1^2 + 1^2)^{1/2}} = \frac{a}{\sqrt{3}}$$

Case II: Face centred cubic lattice. In *fcc* lattice there is one unit situated at each corner and one unit at the centre of each of the six faces. Thus an additional (1 0 0) plane arises half way between the (1 0 0) planes. These additional planes pass through the centre of the front and back face. As the distance between (1 0 0) planes of a simple cubic lattice is a, the distance between (100) planes of the *fcc* lattice is $\frac{a}{2}$,

$$\therefore \quad d_{100} = \frac{1}{2}a$$

Similarly, there arises an additional set of (1 1 0) planes parallel to the first set midway between the planes of simple cubic lattice.

$$\therefore \quad d_{110} = \frac{1}{2}\frac{a}{\sqrt{2}}$$

The (1 1 1) planes of simple cubic lattice pass through the centre of all the faces of *fcc* lattice and hence no new planes arise in this case.

$$\therefore \quad d_{111} = \frac{a}{\sqrt{3}}$$

Case III: Body centred cubic lattice. In *bcc* lattice one unit is situated at each corner and one at the centre of the cube. An additional (1 0 0) plane arises half way between (1 0 0) planes of a simple cubic lattice.

$$\therefore \quad d_{100} = \frac{1}{2}a$$

The (1 1 0) plane is the same as in a simple cubic lattice.

$$\therefore \quad d_{110} = \frac{a}{\sqrt{2}}$$

The (1 1 1) plane of a *bcc* lattice lies midway between the (1 1 1) planes of a simple cubic lattice.

$$\therefore \quad d_{111} = \frac{1}{2}\frac{a}{\sqrt{3}}.$$

Example 18.7. *Lattice constant of a cubic lattice is a. Calculate the spacing between (2 1 1), (1 1 2), (0 0 1), (1 1 0), (0 1 1), (1 0 0), (1 1 1) and (1 0 1) planes.* ***(Meerut U. 2003)***

Solution. The spacing d of plane ($h\ k\ l$) in a simple cubic lattice of side a is

$$d = \frac{a}{(h^2 + k^2 + l^2)^{1/2}}$$

For (2 1 1) plane

$$d_{211} = \frac{a}{(2^2 + 1^2 + 1^2)^{1/2}} = \frac{a}{\sqrt{6}}$$

For (1 1 2) plane

$$d_{112} = \frac{a}{\sqrt{1^2 + 1^2 + 2^2}} = \frac{a}{\sqrt{6}}$$

For (0 0 1) plane

$$d_{001} = \frac{a}{(0 + 0 + 1^2)^{1/2}} = a$$

For (1 1 0) plane

$$d_{110} = \frac{a}{(1^2 + 1^2 + 0)^{1/2}} = \frac{a}{\sqrt{2}}$$

For (0 1 1) plane

$$d_{011} = \frac{a}{\sqrt{0+1^2+1^2}} = \frac{a}{\sqrt{2}}$$

For (1 0 0) plane

$$d_{100} = \frac{a}{(1^2+0+0)^{1/2}} = a$$

For (1 1 1) plane

$$d_{(111)} = \frac{a}{(1^2+1^2+1^2)^{1/2}} = \frac{a}{\sqrt{3}}$$

For (1 0 1) plane

$$d_{101} = \frac{a}{(1^2+0+1^2)^{1/2}} = \frac{a}{\sqrt{2}}$$

Example 18.8. *Show that for a crystal of cubic symmetry the direction [h, k, l] is perpendicular to the plane* ***(h, k, l). (P.U. 2004; G.N.D.U. 2000; Pbi. U. 2002, 2001)***

Solution. If $\vec{a}$, $\vec{b}$ and $\vec{c}$ are the cubic axes, then the intercepts of the plane (*h, k, l*) on these axes are

$$\vec{r_1} = \frac{\vec{a}}{h}, \vec{r_2} = \frac{\vec{b}}{k} \text{ and } \vec{r_3} = \frac{\vec{c}}{l}$$

Therefore, the vectors $\vec{r_1} - \vec{r_2}$ and $\vec{r_1} - \vec{r_2}$ lie in the (*h, k, l*) plane.

$$\text{Now} \quad \vec{r_1} - \vec{r_2} = \frac{\vec{a}}{h} - \frac{\vec{b}}{k} \text{ and } \vec{r_1} - \vec{r_3} = \frac{\vec{a}}{h} - \frac{\vec{c}}{l}$$

The vector $h\vec{a} + k\vec{c} + l\vec{c}$ represents [*h, k, l*]. It is perpendicular to $\vec{r_1} - \vec{r_2}$ as vell as $\vec{r_1} - \vec{r_3}$ because the scalar product

$$(\vec{r_1} - \vec{r_2}).\left(h\vec{a} + k\vec{b} + l\vec{c}\right)$$

$$= \left(\frac{\vec{a}}{h} - \frac{\vec{b}}{k}\right) \cdot (h\vec{a} + k\vec{b} + l\vec{c})$$

$$= \vec{a}\cdot\vec{a} - \vec{b}\cdot\vec{b} = 0$$

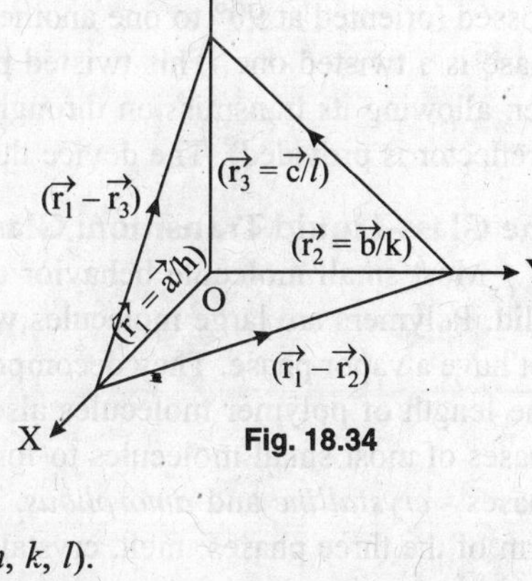

Fig. 18.34

Since in a cubic crystal $a = b = c$.

Similarly $(\vec{r_1} - \vec{r_3}).\left(h\vec{a} + k\vec{b} + l\vec{c}\right) = 0.$

Thus, direction [*h, k, l*] is perpendicular to the plane (*h, k, l*).

18.17 LIQUID CRYSTALS

Liquid crystals are a state of matter that have properties between those of a conventional liquid and those of a solid crystal. For instance, an liquid crystals may flow like a liquid, but its molecules may be oriented in a crystal-like way as shown in Fig. 18.35. There are many different types of liquid crystal phases, which can be distinguished by their different optical properties (such as birefringence). When viewed under a microscope using a polarized light source, different liquid crystal phases will appear to have distinct textures. The contrasting areas in the textures correspond to domains where the liquid crystal molecules are oriented in different directions. Within a domain, however, the molecules are well ordered. Liquid crystal materials may not always be in an liquid crystals phase (just as water may turn into ice or steam).

Liquid crystals can be divided into thermotropic, lyotropic and metallotropic phases. Thermotropic and lyotropic crystals consist of organic molecules. Thermotropic liquid crystals exhibit a phase transition into the liquid crystal phase as temperature in changed. Lyotropic liquid crystals exhibit phase transitions as a function of both temperature and concentration of the liquid crystal molecules in a solvent (typically water). Metallotropic liquid crystals are composed of both organic and inorganic molecules; their liquid crystal transition depends not only on temperature and concentration, but also on the inorganic-organic composition ratio.

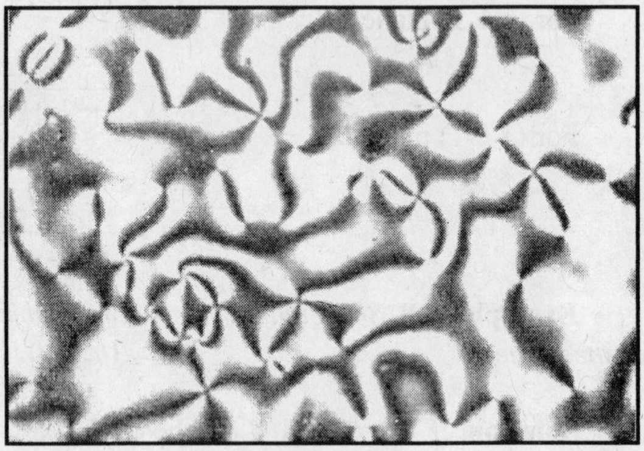

Fig. 18.35.

Examples of liquid crystals can be found both in the natural world and in technological applications. Most electronic displays are currently based on liquid crystals. Lyotropic liquid crystalline phases are abundant in living systems. For example, many proteins and cell membranes are liquid crystals. Other well-known liquid crystal examples are solutions of soap and various related detergents, as well as the tobacco mosaic virus.

Applications of Liquid Crystals

Liquid crystals find wide use in liquid crystal displays (LCDs), which rely on the optical properties of certain liquid crystalline substances in the presence or absence of an electric field. In a typical device, a liquid crystal layer (typically 10 μm thick) sits between two polarizers that crossed (oriented at 90° to one another). The liquid crystal alignment is chosen so that its relaxed phase is a twisted one. This twisted phase reorients light that has passed through the first polarizer, allowing its transmission through the second polarizer (and reflected back to the observer if a reflector is provided). The device thus appears transparent.

The Glass-Liquid Transition: Glassy State

Most small molecule behavior can be understood in terms of three states: gas, liquid, and solid. Polymers are large molecules with strong intermolecular forces and tangled chains, and do not have a vapor phase. They decompose before the temperature gets high enough to form a vapor. The length of polymer molecules also makes it difficult for the large crystals found in the solid phases of most small molecules to form. Instead solid polymers can be modeled in terms of two phases - *crystalline* and *amorphous*. Thus the behavior of polymers can better be understood in term of the three phases: melt, crystalline, and amorphous.

At low temperatures molecular motion in an amorphous region is restricted to molecular vibrations, but the chains cannot rotate or move in space. This form is the **glassy state** of the amorphous region. The glass-liquid transition is the reversible transition in amorphous materials from a hard and relatively brittle state into a molten or rubber-like state. The glassy state can be thought of as being a supper-cooled liquid where the molecular motions have been frozen in. In other words, the glassy state is hard, rigid, and brittle like a crystalline solid, but retains the molecular disorder of a liquid. Super-cooling of a viscous liquid into the glass state is called vetrification. The glass transition temperature T_g is always lower than the melting temperature T_m of the crystalline state of the material.

EXERCISE

1. (*a*) What is a crystal? Distinguish between a crystal and an amorphous solid.

(Nagpur University 2012; Pbi.U. 2004; Meerut U. 2002; Gauhati U. 2000)

(*b*) Why are most solids crystalline in nature?

(P.U. 2003)

(*c*) Give three main differences between crystalline and amorphous solids. ***(H.P.U. 2002)***

2. (*a*) Define crystal lattice, plane lattice, space lattice, basis, unit cell and translation vectors.

(Pbi.U. 2004, 2003, 2002, 2000; Osm. U. 2004; G.N.D.U. 2002, 2001; P.U. 2000; K.U. 2001; A.U. 1995)

(*b*) What is crystal structure? State the relation between crystal structure, lattice and basis.

(Pbi. U. 2002, 2000)

(*c*) Define a primitive cell. Distinguish between a primitive unit cell and non-primitive unit cell with the help of two dimensional diagram. Can a unit cell be primitive?

(Pbi. U. 2004, 2000; Luck. U. 2001; K.U. 2001; P.U. 2001, 2000)

(*d*) Define and explain translation operation.

(*e*) How a Wigner Sietz unit cell is drawn? ***(G.N.D.U. 2001; Luck. U. 2001)***

3. (*a*) What do you mean by a symmetry operation? Explain briefly the various symmetry operations applicable to two dimensional and three dimensional lattices. What is a point operation?

(*b*) Prove that a crystal cannot have five fold symmetry. ***(P.U. 2003, 2001)***

4. What are point groups and space groups? Give their numbers for two and three dimensional lattices. Also give the characteristics of space group.

(G.N.D.U. 2004; P.U. 2000; Pbi. U. 2000)

5. (*a*) What are Bravais lattices? Discuss Bravais lattices in two dimensions. Write down symmetry operations in which each is invariant.

(Pbi. U. 2004; K.U. 2002, 2001; M.D.U. 2001; P.U. 2000)

(*b*) Discuss seven crystal systems by giving one example of each and describe the various types of Bravais lattices in case of three dimensions with the help of neat and clear diagrams.

(Patna U. 2004; Calicut U. 2003; Meerut U. 2003, 2000; Pbi. U. 2001)

(*c*) Why there is no face centred tetragonal space lattice?

6. (*a*) Discuss the characteristics of a unit cell of a cubic system and explain the terms 'volume of a unit cell' "atoms per unit cell", "co-ordination number", "atomic radius" and "atomic packing fraction" as applied to crystal structure. Calculate their values for a simple cubic, face centred cubic and body centred cubic cell.

(*b*) How many lattice points are there in the cubic unit cell of *fcc* structure?

(P.U. 2004, 2001, 2000; H.P.U. 2003, 2001; Meerut. U. 2000; Nagpur University 2012; G.N.D.U. 2004; Pbi. U. 2003, 2000; Nag. U. 2002)

7. (*a*) Describe a hexagonal close-packed structure (*hcp*). What is co-ordination number in (*hcp*)? Calculate its atomic packing fraction.

(*b*) Show that $\frac{c}{a}$ ratio for an ideal (hcp) structure is $\frac{\sqrt{8}}{3} = 1.633.$

(G.N.D.U. 2003)

(G.N.D.U., 2003, 1999; Luck. U., 1995, 1992; H.P.U. 1991, 1992)

8. (*a*) Explain the crystal structure of sodium chloride (NaCl). Draw a sketch of sodium chloride lattice and write down the co-ordinates of the atoms in the cell. What is the number of sodium ions in a unit cell of NaCl? ***(Meerut U. 2003)***

(*b*) Choose the correct answer.

NaCl crystal has (*a*) *fcc* structure (*b*) *sc* structure (*c*) *bcc* structure (*d*) *hpc* structure.

(H.P.U. 2000)

(*c*) Say true or false.

The Bravais lattice of NaCl crystal is *bcc*. ***(Meerut U. 2002)***

9. Explain the crystal structure of diamond. In diamond crystal what is the number of nearest neighbours, the number of atoms per unit cell and packing fraction? Show that it has comparatively loose packing. ***(P.U. 2003, 2001; Luck. U. 2001)***

10. (*a*) Explain the concept of Miller indices. How are they calculated? How the orientation of a plane is specified by Miller indices? Define Miller indices of a direction. State their important features.

(G.N.D.U. 2003, 2001, 2000, 1996; H.P.U. 2003, 1994; P.U. 2003, 2001, 2000, 1995; Pbi. U. 1999; Meerut U. 2002, 2000; Osmania U. 2004; Nag. U. 2002; Cal. U. 2003, 1991; A.U. 1993)

(*b*) Why the reciprocals of intercepts of the plane are taken to find Miller indices?

(H.P.U. 2001, 1999; P.U. 2000; M.D.U. 2001)

11. Draw the planes (100), (010), (001), (110), (101), (011), (111), (200), (020), (002), (201) and (112) in a simple cubic unit cell.

(H.P.U. 2003, Meerut U. 2002; Pbi. U. 2001; K.U. 2002; P.U. 2004; Nag. U. 2009, 2002; Luck. U. 2001, 1996)

12. Derive the expression for the interplanar spacing between two parallel planes with Miller indices ($h\,k\,l$) and show that for a simple cubic lattice of lattice constant *a*.

$$d\,(h\,k\,l) = \frac{a}{\sqrt{h^2 + k^2 + l^2}}.$$

(P.U. 2004, 2000; G.N.D.U. 2003, 2001; H.P.U. 2003; Meerut U. 2002; Nag. U. 2002; Luck. U. 2001, Bang. U. 2001)

13. (*a*) Calculate the separation between lattice planes in a simple cubic, face centred cubic (*fcc*) and body centred cubic (*bcc*) lattice.

(Meerut U. 2002, 2001)

(*b*) For a cubic lattice calculate the distance of (1 2 3) and (2 3 4) planes from a plane passing through the origin.

14. (*a*) What are liquid crystls? Give applications of liquid crystals.

(*b*) Explain in brief the glass-liquid transition.

15. In three dimensional case, how manyu rembers of crystal systems and Brovais lattices are found? ***(Nagpur Uni. S/2012)***

MULTIPLE CHOICE QUESTIONS

Choose the correct answer

1. The atomic radius for (*fce*) lattice is

(*a*) $\frac{a}{2}$ (*b*) $\frac{\sqrt{3}}{4}a$ (*c*) $\frac{\sqrt{2}}{4}a$ (*d*) none of these

2. Packing fraction has maximum value for

(*a*) *fcc* strucure (*b*) *hcc* structure (*c*) *Sc* structure (*d*) *bcc* structure

3. Diamand has.... Bravais lattice

(*a*) *fcc* (*b*) *bcc* (*c*) *hcp* (*d*) *sc*

4. The atomic radius for (fcc) lattic is

(*a*) $\frac{a}{2}$ (*b*) $\frac{\sqrt{3}}{4}a$ (*c*) $\frac{\sqrt{2}}{4}a$ (*d*) $\frac{\sqrt{2}}{3}a$

ANSWERS

1. (*c*)

2. (*a*) and (*b*). The packing fraction for both (*fcc*) and (*hcp*) structure is 0.74 as compared to 0.52 for (*sc*) and 0.68 for (*bcc*) structure.

3. (*a*)

4. (*c*)

X-RAY DIFFRACTION

19

Chapter

INTRODUCTION

X-ray crystallography is a method of determining the arrangement of atoms within a crystal in which a beam of X-rays strikes a crystal and causes the beam of light to spread into many specific directions. From the angles and intensities of these diffracted beams, a crystallographer can produce a three-dimensional picture of the density of electrons, within the crystal. From this electron density, the mean positions of the atoms in the crystal can be determined, as well as their chemical bonds, their disorder and various other information.

Since many materials can form crystals—such as salts, metals, minerals, semiconductors, as well as various inorganic, organic and biological molecules—X-ray crystallography has been fundamental in the development of many scientific fields. The method also reveals the structure of many biological molecules, including vitamins, drugs, proteins and nucleic acids such as DNA.

19.1 X-RAYS FOR CRYSTAL STRUCTURE

X-rays are used for producing diffraction effect in crystals due to the fact that wavelength of X-rays is comparable to the interatomic distances in actual crystals.

For example, X-rays produced by acceleration of electrons through a potential difference V have a wavelength given by

$$eV = h\nu = \frac{hc}{\lambda}$$

or

$$\lambda = \frac{hc}{eV}$$

for $V = 10$ kV, we have

$$\lambda = \frac{6.62\times10^{-34}\times3\times10^{8}}{1.6\times10^{-19}\times10^{4}} = 1.24\times10^{-10}\text{ m} = 1.24\text{ Å}$$

The wavelength is comparable to the inter atomic spacing of an actual crystal. For example, in NaCl crystal atomic spacing is 2.8×10^{-10} m = 2.8 Å.

Visible or ultra violet light has a larger wavelength. Hence these waves give rise to familiar effects of optical reflection and refraction. These rays, therefore, cannot be used for exploring the crystal at an atomic scale.

On the other hand, radiations of wavelength shorter than that of X-rays *i.e.*, γ-rays are diffracted through very small angles which cannot be conveniently measured.

It is only X-rays which are scattered elastically without change of wavelength by the charged particles of the atom.

Why γ-rays are not used? Gamma rays are not used to study crystal structure as the wave length of γ-rays is much smaller than that of X-rays and, therefore, the γ-rays are diffracted through very small angles which cannot be conveniently measured.

Why a crystal diffracts X-rays? A crystal diffracts X-rays because X-rays are scattered elastically without change of wavelength by the charged particles of the atoms in the crystal as explained below:

When an atom is exposed to a monochromatic beam of X-rays, the atomic electrons start vibrating with a frequency equal to that of the incident beam under the effect of the electric field vector of the radiation. These electrons, therefore, undergo acceleration and the accelerated electrons emit secondary radiations of the same frequency as that of the incident X-rays. The secondary X-rays spread out in all directions. If the wavelength of original X-rays is quite large as compared to the atomic dimensions, all the radiations emitted by the electrons shall be in phase with one another.

On the other hand, if the incident X-rays have the same order of wave length as that of the atomic dimensions, then the radiations emitted by the electrons are, in general, out of phase with one another. These radiations may, therefore, undergo constructive or destructive interference producing maxima or minima in certain directions *i.e.*, undergo diffraction phenomenon.

19.2 BRAGG'S LAW

W.L. Bragg discovered that X-rays can be regularly reflected by the *cleavage planes* of the crystals when the rays are incident on their surface nearly at glancing angle. The cleavage planes are *successive atomic planes* in the crystal. X-rays that penetrate more deeply into the crystal are reflected from the lower planes and thus several beams of X-rays reflected from various planes are obtained. The intensity of the reflected beam at certain angles will be maximum, where the two reflected waves from two different planes have a path-difference equal to an integral multiple of the wavelength of X-rays, while at some other angles, the intensity will be minimum.

Consider parallel lattice planes equidistant from one another in a crystal structure, separated by a distance d. Suppose a beam of X-rays AB of wavelength λ is incident on a layer of atoms in one cleavage plane P of a crystal at a glancing angle θ at B and after reflection goes along BC. The ray DE incident on cleavage plane Q after reflection goes along EF. Draw BG perpendicular on DE and BH perpendicular on EF, then BG is the incident wave front and BH the reflected wave front.

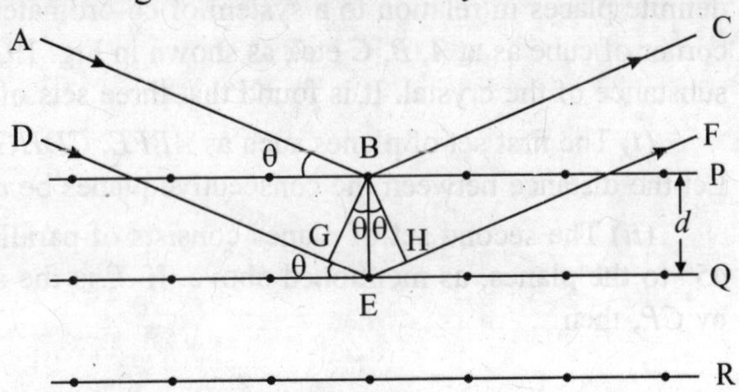

Fig. 19.1

Path difference between the two wave fronts

$$= GE + EH$$

Now in ΔBGE

$$GE = BE \sin\theta = d\sin\theta$$

and in ΔBEH,

$$EH = BE \sin\theta = d\sin\theta$$

$\therefore$ Path-difference

$$GE + EH = 2d\sin\theta$$

If this path-difference is an integral multiple of wavelength λ, constructive interference will take place between the reflected beams. Thus the intensity will be maximum if

$$2d\sin\theta = n\lambda \qquad \text{where } n = 1, 2, 3...$$

For $n = 1$, we get the first order spectrum and for $n = 2$, we get the second order spectrum and so on. This phenomenon is similar to that of the optical diffraction with diffraction grating with only this difference that here various parallel planes of the crystal act as lines in the grating.

The above equation $2d\sin\theta = n\lambda$ is Bragg's equation and represents *Bragg's law*.

Significance. The interpretation of Bragg's law gives that the diffraction intensities can be stronger only at certain values of θ for specific values of λ and *d*. Therefore, a knowledge of experimentally observed diffraction angles and wavelength of X-rays can provide the information about the inter planar spacing of the crystal. Thus, the space, shape and orientation of the unit cell can be determined from any diffraction experiment using monochromatic X-rays. This will be clear from the example of crystal structure determination given below.

Determination of crystal structure. To find the structure of a crystal, the *lattice constant* *d* is determined by using different planes of the crystal as reflecting surface for the same known wavelength of X-rays. Expressing the lattice constants as ratios with one of them taken as unity, it is seen that the ratios depend upon the way in which the atoms are arranged in the crystal. Thus the structure of the crystal is analysed.

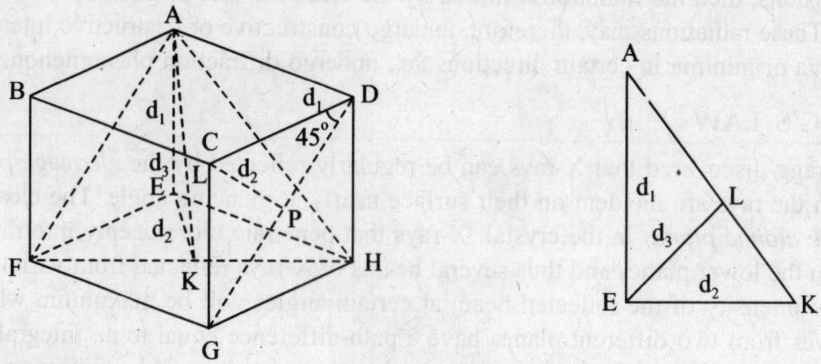

Fig. 19.2

For example, consider a crystal of cubic system. In such a crystal, the atoms are situated at definite places in relation to a system of co-ordinates or **lattice**. Suppose there is an atom at each corner of cube as at *A*, *B*, *C* etc., as shown in Fig. 19.2, the structure being repeated throughout the substance of the crystal. It is found that three sets of planes rich in atoms exist.

(*i*) The first set of planes such as *ABFE*, *CDHG*, *ADHE*, *BCGF* and *ABCD*, *EFGH* are alike. Let the distance between the consecutive planes be d_1. These are known as (100) planes.

(*ii*) The second set of planes consists of parallel planes like *ADGF*, inclined at an angle of 45° to the planes, as mentioned above. If d_2 is the spacing between these planes, as represented by *CP*, then

$$\frac{d_2}{d_1} = \sin 45 = \frac{1}{\sqrt{2}}$$

$$\therefore \quad d_2 = \frac{d_1}{\sqrt{2}}$$

These are known as (110) planes.

(*iii*) The third set of planes consists of planes like *AFH*. To find the spacing between these planes, draw *EK* perpendicular to *FH* and join *AK* to complete the right-angled triangle *AEK* shown separately. From *E* draw *EL* perpendicular to *AK*; then $EL = d_3$ measures the spacing between these planes. Now

In ΔAEK, $\quad AE = d_1,\ EK = d_2$ and $AK = \sqrt{d_1^2 + d_2^2}$

Also $\quad \sin A = \frac{d_3}{d_1} = \frac{d_2}{\sqrt{d_1^2 + d_2^2}}$

or
$$d_3 = \frac{d_1 d_2'}{\sqrt{d_1^2 + d_2^2}} = \frac{d_1 d_2}{\sqrt{d_1^2 + \frac{1}{2}d_1^2}} = \frac{d_1}{\sqrt{3}}$$

These are known as (111) planes.

Hence
$$\frac{1}{d_1}:\frac{1}{d_2}:\frac{1}{d_3} = 1:\sqrt{2}:\sqrt{3}$$

Using a crystal of KCl, Bragg found the **following values** of θ for reflection from the faces *ABCD*, *ADGF* and *AFH* respectively.

$$\theta_1 = 5.22°, \theta_2 = 7.30°, \text{ and } \theta_3 = 9.05°$$

For the first order spectrum

$$\frac{1}{d} = \frac{2\sin\theta}{\lambda}$$

$$\therefore \quad \frac{1}{d_1}:\frac{1}{d_2}:\frac{1}{d_3} = \sin 5.22 : \sin 7.30 : \sin 9.05$$
$$= 0.0910 : 0.1272 : 0.1570$$
$$= 1 : 1.40 : 1.73$$

These ratios, within the limits of experimental error, are

$$1:\sqrt{2}:\sqrt{3}$$

Hence we conclude that KCl has a cubic crystal structure. Similarly other forms can be investigated.

Difference between Bragg reflection and ordinary reflection. In ordinary reflection all the reflected waves are in phase and there are no maxima and minima. The phenomenon of Bragg reflection is similar to that of diffraction of light from a diffraction grating with only this difference that in Bragg reflection various parallel planes of the crystal serve the same purpose as the ruled lines on the grating which produce opacities and transparencies.

19.3 VERIFICATION OF BRAGG'S LAW FOR NaCl CRYSTAL

The NaCl crystals are cubic and every elementary cube has *eight* ions at its corners, *four* of sodium and *four* of chlorine, which form four molecules of sodium chloride. Every elementary cube is surrounded on all sides by other such cubes, as shown in Fig. 19.3. Thus each ion of sodium chloride is shared between the two cubes and each elementary cube contains half a molecule of sodium chloride.

$\therefore$ The mass of elementary cube $= \frac{M}{2N}$

where M is the molecular weight of NaCl and *N is the Avogadro's number.*

$\therefore$ Mass of elementary cube $= \frac{23 + 35.5}{2 \times 6.06 \times 10^{26}}$ kg.

As each side of the elementary cube is equal to d, the volume of the elementary cube

$$d^3 = \frac{\text{Mass}}{\text{Density}}.$$

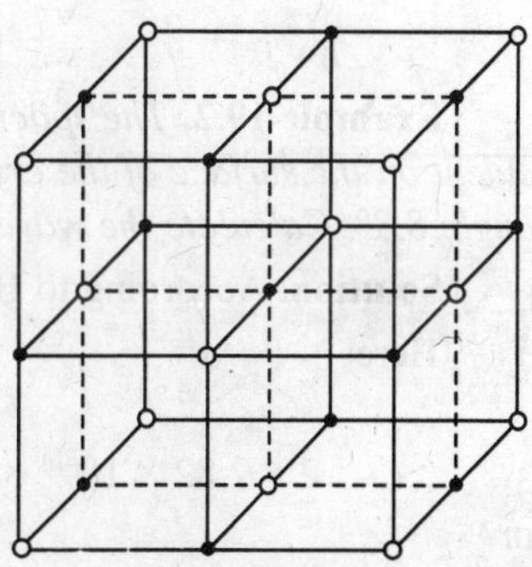

Fig. 19.3

Now density of NaCl is 2.17×10^3 kg. m^{-3}.

$$\therefore \qquad \text{Volume } d^3 = \left[\frac{58.5}{2\times 6.06\times 10^{26}\times 2.17\times 10^3}\right]$$

$$\text{or} \qquad d = \left[\frac{58.5}{2\times 6.06\times 10^{26}\times 2.17\times 10^3}\right]^{1/3}$$

$$= 2.814 \times 10^{-10} \text{ m}$$

According to Bragg's law $2d \sin\theta = n\lambda$

$\therefore$ For the first order diffraction $n = 1$,

$$\therefore \qquad \lambda = 2d \sin\theta$$

$$\text{or} \qquad \lambda = 2 \times 2.814 \times 10^{-10} \times \sin 11.8°$$

$$= 5.628 \times 0.2 \times 10^{-10}$$

$$= 1.12 \times 10^{-10} \text{ m} = 1.12 \text{ Å}$$

For the second order diffraction $n = 2$,

$$\therefore \qquad 2\lambda = 2d \sin\theta$$

$$\text{or} \qquad \lambda = d \sin\theta = 2.814 \times 10^{-10} \times \sin 23.5°$$

$$= 1.12 \times 10^{-10} \text{ m} = 1.12 \text{ Å}$$

For the third order diffraction $n = 3$, $3\lambda = 2d \sin\theta$

$$\text{or} \qquad \lambda = \frac{2}{3} d \sin\theta = \frac{2}{3} \times 2.814 \times 10^{-10} \times \sin 36°$$

$$= 1.12 \times 10^{-10} \text{ m} = 1.12 \text{ Å}$$

Thus we find that for values of $\theta = 11.8°$, $23.5°$ and $36°$ for first, second and third order spectrum $2d \sin\theta = n\lambda$.

Hence diffraction from a NaCl crystal verifies Bragg's law.

SOLVED EXAMPLES

Example 19.1. *Calculate the lattice constant a for NaCl crystal with density $\rho = 2189$ kg/m³. Avogadro's number $N = 6.02 \times 10^{26}$/kg mole.* **(Meerut U. 2003)**

Solution. Molecular weight of NaCl = 23 + 35.5 = 58.5

$$\therefore \text{ Lattice constant} \qquad a = \left[\frac{58.5}{2\times 6.02\times 10^{26}\times 2189}\right]^{1/3}$$

$$= 2.81 \times 10^{-10} \text{ m} = 2.81\text{Å}$$

Example 19.2. *The spacing between successive (1 0 0) planes in NaCl is 2.82 Å. X-rays incident on the surface of the crystal is found to give rise to first order Bragg reflection at glancing angle 8.8°. Calculate the wavelength of X-rays.*

Solution. According to Bragg's law $2d \sin\theta = n\lambda$.

Here $\qquad d = 2.82 \text{ Å} = 2.82 \times 10^{-10}$ m, $\theta = 8.8°$

$$\therefore \qquad \sin\theta = 0.153, \; n = 1$$

$$\therefore \qquad 2 \times 2.82 \times 10^{-10} \times 0.153 = \lambda$$

$$\text{or} \qquad \lambda = 0.863 \times 10^{-10} \text{ m}$$

$$= 0.863 \text{ Å}.$$

Example 19.3. *Calculate the glancing angle on the cube face (100) of a rock salt crystal (a = 2.814 Å) corresponding to second order reflection of X-rays of wavelength 0.710Å.*

(M.D.U. 2001)

Solution. $2d \sin \theta = n\lambda$; $d = a = 2.814 \times 10^{-10}$ m for cube face (100); $n = 2$; $\lambda = 0.710 \times 10^{-10}$m

$$\therefore \quad \sin \theta = \frac{2 \times 0.710 \times 10^{-10}}{2 \times 2.814 \times 10^{-10}} = 0.2523$$

or $$\theta = 14.6°.$$

Example 19.4. *X-rays of wavelength 1.5Å make a glancing angle of 16° in the first order when diffracted from NaCl crystal. Find the lattice constant of NaCl.* *(Osmania U. 2004)*

Solution. $2d \sin \theta = n\lambda$; $\lambda = 1.5 \times 10^{-10}$ m; $n = 1$; $\theta = 16°$

$$\therefore \quad \sin \theta = 0.2756$$

$$\therefore \quad d = \frac{n\lambda}{2 \sin \theta} = \frac{1 \times 1.5 \times 10^{-10}}{2 \times 0.2756} = 2.72 \times 10^{-10} \text{ m} = 2.72 \text{ Å}.$$

Example 19.5. *Calculate the glancing angle on the plane (110) of a cubic rock salt crystal (a = 2.81 Å) corresponding to second order diffraction maxima of wavelength 0.71Å.*

(Bang. U. 2001)

Solution. $a = 2.81$Å

$$\therefore \quad \text{Spacing for (110) plane, } d = \frac{a}{\sqrt{1^2 + 1^2 + 0}} = \frac{a}{\sqrt{2}} \text{ Å} = \frac{2.81}{\sqrt{2}} = 1.993\text{Å}$$

Now, $2d \sin \theta = n\lambda$; $\lambda = 0.71 \times 10^{-10}$ m; $n = 2$

$$\therefore \quad \sin \theta = \frac{n\lambda}{2d} = \frac{2 \times 0.71 \times 10^{-10}}{2 \times 1.993 \times 10^{-10}} = 0.3562 \text{ or } \theta = 20.87°.$$

Example 19.6. *The molecular weight of rock salt (a simple cubic) crystal is 58.5 kg per kilo-mole and its density is 2.16 × 10³ kg/m³. Taking Avogadro's number = 6.02 × 10²⁶, calculate the grating space d_{100} of rock salt.*

Using this value of lattice constant, calculate the wavelength of X-rays in second order, if angle of diffraction θ = 26°. **(Similar Nagpur University 2011)**

Solution. Mass of one kilo-mole of NaCl = 58.5 kg

Number of molecules in one kilo-mole = 6.02×10^{26}

$$\therefore \text{ Mass of one NaCl molecule} = \frac{58.5}{6.02 \times 10^{26}} = 9.72 \times 10^{-26} \text{ kg}$$

Density of NaCl crystal = 2.16×10^3 kg/m³

$$\text{Volume of one NaCl molecule} = \frac{9.72 \times 10^{-26}}{2.16 \times 10^3} = 4.5 \times 10^{-29} \text{ m}^3$$

As there are two atoms per NaCl molecule, the volume per atom

$$= \frac{4.5 \times 10^{-29}}{2} = 2.25 \times 10^{-29} \text{ m}^3$$

If d is the atomic spacing for NaCl crystal, then volume per atom = d^3

$$\therefore \quad d^3 = 2.25 \times 10^{-29}$$

or $$d = [2.25 \times 10^{-29}]^{1/3} = 2.82 \times 10^{-10} \text{ m} = 2.82 \text{ Å}$$

$$\therefore \quad d_{100} = 2.82 \text{ Å}$$

According to Bragg's law $2d \sin \theta = n\lambda$.

Here $d = 2.82 \times 10^{-10}\text{m}, \theta = 26°$ or $\sin\theta = 0.4383$

and $n = 2$

$\therefore$ $2\lambda = 2 \times 2.82 \times 10^{-10} \times 0.4383$

or $\lambda = 2.82 \times 0.4383 \times 10^{-10}$

$= 1.24 \times 10^{-10}\text{ m} = 1.24 \text{ Å}.$

Example 19.7. *If X-rays of wavelength 0.5 Å are diffracted at an angle of 5° in the first order, what is the spacing between the adjacent planes of the crystal? At what angle will second maximum occur?*

Solution. According to Bragg's law $2d \sin = n\lambda$

Here $\lambda = 0.5 \text{ Å} = 0.5 \times 10^{-10}\text{ m}; \theta = 5°$

$\therefore$ $\sin\theta = 0.0871$

$n = 1$

$\therefore$ $2d \times 0.871 = 0.5 \times 10^{-10}$

or $$d = \frac{0.5\times10^{-10}}{0.1742}$$

$= 2.87 \times 10^{-10}\text{ m} = 2.87 \text{ Å}$

For second order maximum, let θ' be the angle of Bragg reflection, then

$$2d\sin\theta' = 2\lambda$$

or $$\sin\theta' = \frac{\lambda}{d} = \frac{0.5\times10^{-10}}{2.87\times10^{-10}} = 0.1742$$

$\therefore$ $\theta' = 10.03°$

Example 19.8. *The Bragg angle for first order reflection from (111) plane in a crystal is 60°. Calculate the interatomic spacing if X-rays of wavelength 1.8 × 10^{-10}m are used.*

(P.U. 2004; M.D.U. 2003; K.U. 2001)

Solution. According to Bragg's law $2d \sin\theta = n\lambda$

Here $\theta = 60° \therefore \sin\theta = \sin 60 = 0.866, \lambda = 1.8 \times 10^{-10}\text{ m}; n = 1$

$\therefore$ $2d \times 0.866 = 1.8 \times 10^{-10}$

or $$d = \frac{1.8\times10^{-10}}{1.732} = 1.039 \times 10^{-10}\text{ m}$$

This gives us d_{111}

or $d_{111} = 1.039 \times 10^{-10}\text{ m}$

If the interatomic spacing for a simple cubic lattice

$$d_{100} = a, \text{ then } d_{111} = \frac{a}{\sqrt{3}}$$

$\therefore$ $a = d_{100} = d_{111} \times \sqrt{3}$

$= 1.039 \times 1.732 \times 10^{-10}\text{ m}$

$= 1.8 \times 10^{-10}\text{ m}.$

Example 19.9. *A monochromatic beam of X-rays of wavelength 1.24 Å is reflected by cubic crystal of KCl. Determine the inter planar distances for (100), (110) and (111) planes. Given density of KCl = 1980 kg/m³ and molecular weight M = 74.5 kg. Avogadro's number N = 6.023 × 10^{26} kg mole^{-1}.*

Solution. For a cubic crystal, we have

$$a^3 = \frac{nM}{N\rho}$$

where a is the lattice constant, n the number of molecule in a unit cell, M the molecular weight, N Avogadro's number and ρ the density.

For KCl cubic crystal $n = 4$

$$\therefore \quad a^3 = \frac{4 \times 74.5}{6.023 \times 10^{26} \times 1.98 \times 10^3} = 25 \times 10^{-29}$$

or $$a = [250 \times 10^{-30}]^{1/3} = 6.3 \times 10^{-10} \text{ m}$$

Now $$d_{100} = \frac{a}{\sqrt{1^2 + 0^2 + 0^2}} = a = 6.3 \times 10^{-10} \text{ m}$$

$$d_{110} = \frac{a}{\sqrt{1^2 + 1^2 + 0^2}} = \frac{a}{\sqrt{2}} = \frac{6.3}{\sqrt{2}} \times 10^{-10} = 4.45 \times 10^{-10} \text{ m}$$

$$d_{111} = \frac{a}{\sqrt{1^2 + 1^2 + 1^2}} = \frac{a}{\sqrt{3}} = \frac{6.3}{\sqrt{3}} \times 10^{-10} = 3.63 \times 10^{-10} \text{ m.}$$

Example 19.10. *From the following data calculate the wavelength of neutron beam and its speed. Spacing between successive (100) planes = 3.84 Å; grazing angle 30°; order of Bragg reflection = 1.*

Solution. According to Bragg's relation

$$2d \sin\theta = n\lambda$$

Here $$d = 3.84 \text{ Å} = 3.84 \times 10^{-10} \text{ m}, \theta = 30°; \sin 30 = \frac{1}{2}; n = 1$$

$$\therefore \quad 2 \times 3.84 \times 10^{-10} \times \frac{1}{2} = 1 \times \lambda$$

or $$\lambda = 3.84 \times 10^{-10} \text{ m} = 3.84 \text{ Å.}$$

According to de-Broglie relation

$\lambda = \frac{h}{p} = \frac{h}{mv}$ where h is Planck's constant having a value 6.62×10^{-34} Js and neutron mass $m = 1.67 \times 10^{-27}$ kg

$$\therefore \quad v = \frac{h}{m\lambda} = \frac{6.62 \times 10^{-34}}{1.67 \times 10^{-27} \times 3.84 \times 10^{-10}}$$

$$= 1.03 \times 10^3 \text{ ms}^{-1}$$

19.4 LAUE'S EQUATIONS

Laue's diffraction equations give the conditions under which X-rays scattered from different atoms of a crystal combine to form a diffracted beam. For this purpose we study the X-ray diffraction pattern produced by identical scattering centres located at the *lattice points* of a three dimensional crystal lattice. Let P_1 and P_2 be two such lattice points separated by a vector $\vec{r}$. Suppose a parallel beam of X-rays is incident on P_1P_2 along the direction of unit vector $\hat{i}$. The scattered beam is also parallel and let it be in any arbitrary direction given by unit vector $\hat{s}$. Draw

P_1A perpendicular to the incident wave direction and P_2B perpendicular to the scattered wave direction, then P_1A is incident and P_2B the scattered wave front. The path difference between the two scattered waves – one scattered from P_2, and the other scattered from P_1 is given by

$$P_2A - P_1B = \vec{r}\cdot\hat{i} - \vec{r}\cdot\hat{s} = \vec{r}\cdot(\hat{i}-\hat{s}) = \vec{r}\cdot\vec{S}$$

where $\vec{S} = (\hat{i}-\hat{s})$ is a vector in the direction of the normal to the reflecting plane.

If θ is the angle that the incident beam makes with the reflecting plane, then the angle that the scattered (reflected) beam makes with the reflecting plane is also equal to θ and the angle between the unit vectors $\hat{i}$ and $\hat{s}$ = 2θ as shown in Fig. 19.4.

$$\therefore \qquad |\vec{S}| = 2\sin\theta \qquad [\because \hat{i} \text{ and } \hat{s} \text{ are unit vectors}]$$

The phase difference between the waves scattered at the lattice points P_1 and P_2 is given by

$$\phi = \frac{2\pi}{\lambda}(\vec{r}\cdot\vec{S})$$

The intensity of the diffracted beam is a *maximum* in the direction in which φ is an integral multiple of 2π.

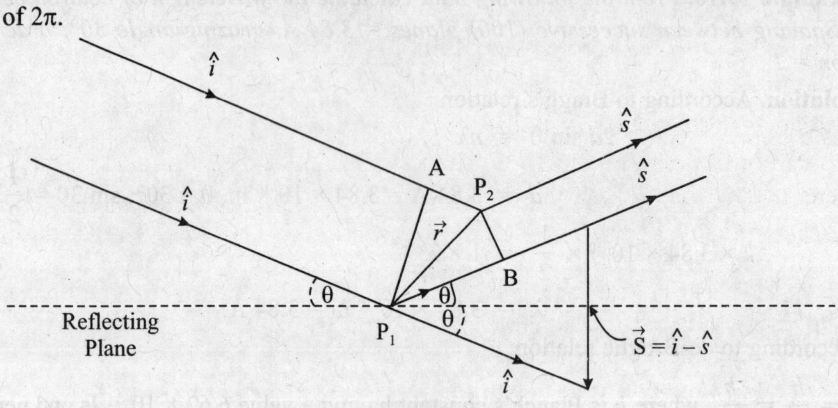

Fig. 19.4

If $\vec{a}$, $\vec{b}$ and $\vec{c}$ are the primitive lattice vectors of the point P_2 with respect to the point P_1, then

$$\vec{r} = \vec{a} + \vec{b} + \vec{c}$$

and the above condition can be put in the form of three separate conditions *i.e.*,

$$\phi_a = \frac{2\pi}{\lambda}(\vec{a}\cdot\vec{S}) = 2\pi h \qquad ...(i)$$

$$\phi_b = \frac{2\pi}{\lambda}(\vec{b}\cdot\vec{S}) = 2\pi k \qquad ...(ii)$$

and

$$\phi_c = \frac{2\pi}{\lambda}(\vec{c}\cdot\vec{S}) = 2\pi l \qquad ...(iii)$$

where *h, k, l* are integers and ϕ_a, ϕ_b and ϕ_c are phase differences between the waves scattered from the two ends of the primitive lattice vectors $\vec{a}$, $\vec{b}$ and $\vec{c}$ respectively.

If α, β, γ are the angles between the normal to the reflecting plane and $\vec{a}$, $\vec{b}$ and $\vec{c}$ axes respectively, then

$$\vec{a}\cdot\vec{S} = |\vec{a}||\vec{S}| \cos\alpha = 2a \sin\theta \cos\alpha$$

$$\vec{b}\cdot\vec{S} = |\vec{b}||\vec{S}| \cos\beta = 2b \sin\theta \cos\beta$$

$$\vec{c}\cdot\vec{S} = |\vec{c}||\vec{S}| \cos\gamma = 2c \sin\theta \cos\gamma$$

From (*i*), (*ii*) and (*iii*), we have

$$\vec{a}\cdot\vec{S} = h\lambda, \quad \vec{b}\cdot\vec{S} = k\lambda \quad \text{and} \quad \vec{c}\cdot\vec{S} = l\lambda$$

$$\therefore \quad 2a \sin\theta \cos\alpha = h\lambda \qquad ...(iv)$$

$$2b \sin\theta \cos\beta = k\lambda \qquad ...(v)$$

$$2c \sin\theta \cos\gamma = l\lambda \qquad ...(vi)$$

These are the Laue's equations for X-ray diffraction.

Derivation of Bragg's relation from Laue's equations. If θ gives the direction of Bragg diffraction for a wavelength λ, then direction cosines of $\vec{S}$ the scattering normal from equations (*iv*), (*v*) and (*vi*) are given by

$$\cos\alpha = \frac{\lambda}{2\sin\theta}\frac{h}{a}$$

$$\cos\beta = \frac{\lambda}{2\sin\theta}\frac{k}{b}$$

$$\cos\gamma = \frac{\lambda}{2\sin\theta}\frac{l}{c}$$

i.e., the direction cosines are respectively proportional to $\frac{h}{a}$, $\frac{k}{b}$ and $\frac{l}{c}$.

Now, the adjacent lattice planes whose Miller indices are (*h*, *k*, *l*) intersect the $\vec{a}$, $\vec{b}$, $\vec{c}$ axes at intervals *a*/*h*, *b*/*k* and *c*/*l*.

Thus we find that the scattering normal $\vec{S}$ is identical to the normal to (*hkl*) planes. The (*hkl*) planes may also be regarded as the reflecting planes of Bragg's diffraction.

If *d*(*hkl*) is the spacing between two adjacent planes of a set (*hkl*), then

$$d(hkl) = \frac{a}{h}\cos\alpha = \frac{b}{k}\cos\beta = \frac{c}{l}\cos\gamma$$

According to Laue's equations

$$\frac{a}{h}\cos\alpha = \frac{b}{k}\cos\beta = \frac{c}{l}\cos\gamma = \frac{\lambda}{2\sin\theta}$$

∴ Laue's equations can be put in the form

$$d(hkl) = \frac{\lambda}{2\sin\theta}$$

or $$2d(hkl)\sin\theta = \lambda$$

The integers *h*, *k*, *l* of Laue's equation are not necessarily identical with Miller indices of an actual crystal plane but these integers may contain a common factor say *n*, while in the indices the common factor has been eliminated.

Introducing the common factor n, we have

$$2d \sin \theta = n\lambda$$

where d is the spacing between adjacent planes with Miller indices $\frac{h}{n}$, $\frac{k}{n}$ and $\frac{l}{n}$. The equation $2d \sin \theta = n\lambda$ is Bragg's relation for an angle θ between the incident beam and the reflecting plane.

19.5 EXPERIMENTAL X-RAY DIFFRACTION METHODS

Diffraction of X-rays by the periodic lattice of a crystal provides a method of studying the crystal structure. Bragg's formula gives a simple relation between wavelength of X-rays λ, and angle of diffraction θ. It states

$$2d \sin \theta = n\lambda$$

where d is the spacing between parallel atomic planes and n the order of diffraction maxima. For an experimental study the values of λ and θ must be matched in accordance with this relation either by varying λ or θ.

The standard experimental methods are:

(*i*) *Laue's method.* In this method a *single* crystal is *held stationary* in a beam of X-rays having continuous wavelength. The crystal selects and produces diffraction corresponding to discrete values of λ for which planes of spacing d exist and angle of incidence θ satisfies Bragg's relation.

(*ii*) *Rotating crystal method.* In this method a *single* crystal is *rotated* about a fixed axis in a beam of monochromatic *X*-rays. The variation of θ brings different atomic planes into position for Bragg reflection.

(*iii*) *Powder method.* In this method a *powdered* sample of crystalline material is placed in a fixed position in a monochromatic beam. Among the distribution of various crystalline orientations there will be some for which the angle of incidence satisfies Bragg's law. We shall describe these method in detail.

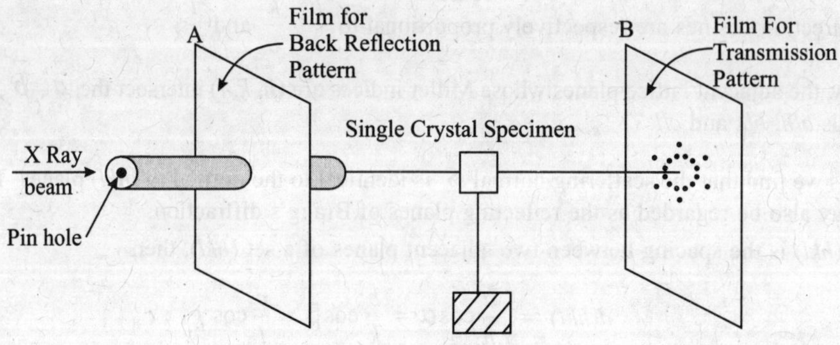

Fig. 19.5

(*i*) **Laue's method.** Laue's *X*-ray camera consists of a source producing *X*-rays of continuous wavelength over a wide range (perhaps from 0.2Å to 2 Å), a pin hole collimator which produces a very fine narrow beam, a single crystal specimen not more than one mm in size held stationary in a stand and two flat films on which the diffraction patterns are recorded as shown in Fig. 19.5. The film *A* records the back reflection pattern and the film *B* records the transmission pattern.

The crystal selects and diffracts the discrete values of λ for which planes exist of spacing d and of incidence angle θ satisfying Bragg's relation. The diffraction pattern consists of a series of spots known as *Laue spots* showing the symmetry of the crystal. If a crystal with four-fold axis of symmetry is oriented with the axis parallel to the beam, the Laue pattern will show four-fold symmetry. This feature makes the Laue method specially useful for checking the orientation of crystals in solid state experiments.

***(ii)* Rotating crystal method.** In the rotating crystal method, a single crystal (the dimensions of which are generally not greater than 1 mm) is rotated about a fixed axis in a beam of monochromatic X-rays. (The incident beam is made monochromatic by using a monochro-matic filter or by reflection from an earlier crystal). The variation in the angle brings different atomic planes into position for reflection. The film is mounted in a cylindrical holder concentric with a rotating spindle on which the single crystal specimen is mounted, as shown in Fig. 19.6.

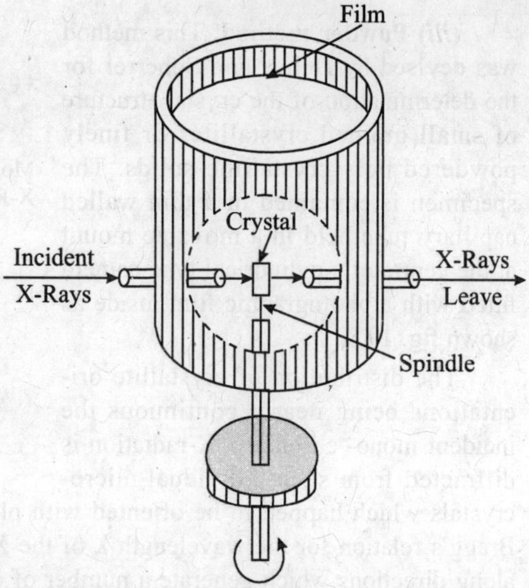

Fig. 19.6

As the crystal is rotated, the X-ray beam undergoes diffraction from a given crystal plane when the value of θ satisfies Bragg's relation for the wavelength of X-rays used. Beams from all planes parallel to the vertical rotation axis will lie in the horizontal plane. Planes with other orientations will reflect in layers above and below the horizontal plane. The reflected spots form parallel lines called '*layer lines*'. As the wavelength is known the spacing *d* can be calculated using Bragg's relation $2d \sin \theta = n\lambda$.

It may be noted that plane which always contains the incident beam during the whole rotation cannot reflect. Similarly a plane with spacing so small that $\frac{\lambda}{2d} > 1$ also cannot reflect. Reflections from planes in different orientations may overlap and make the identification of spots difficult. For this purpose the crystal is oscillated through a limited angular range instead of being rotated through full 360°. The limited range reduces the possibility of overlapping. Modern diffractometers use scintillation counters or proportional counters to detect the diffracted X-radiation and allow automatic recording and collection of data.

Formation of layer lines. If the crystal is placed such that its $\vec{c}$ axis coincides with the axis of rotation, all the planes with Miller indices of the type (*h k* 0) will produce the *central layer line.* Similarly the planes having Miller indices of the type (*h k* 1) and $(h\,k\,\bar{1})$ will produce the layer lines above and below the central line as shown in Fig. 19.7. The vertical spacing between the layer lines dependes upon the distance between the lattice points along the $\vec{c}$ axis. Thus the distance $\vec{c}$ or translation vector $\vec{c}$ can be measured from the photographic film. Similarly, by mounting the crystal along $\vec{a}$ and $\vec{b}$ axes respectively the translation vectors $\vec{a}$ and $\vec{b}$ can be measured and thus the dimensions of the unit cell can be determined.

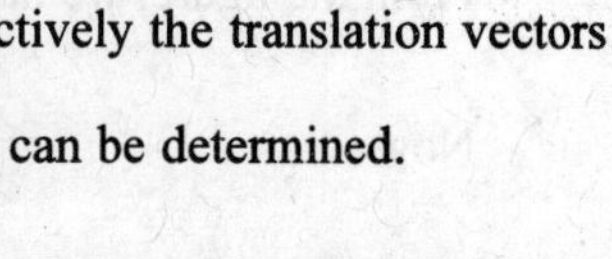

Fig. 19.7

(iii) **Powder method.** This method was devised by Debye and Scherrer for the determination of the crystal structure of small grained crystallites or finely powdered poly-crystalline solids. The specimen is contained in a thin walled capillary tube held in a movable mount at the centre of a cylindrical type camera fitted with a photographic film inside as shown fig. 19.8.

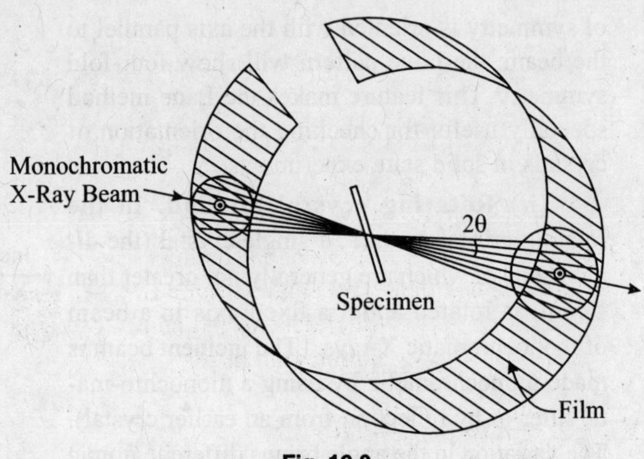

Fig. 19.8

The distribution of crystallite orientations being nearly continuous the incident mono- chromatic X-radiation is diffracted from such individual micro-crystals which happen to be oriented with planes making an angle θ with the beam satisfying Bragg's relation for the wavelength λ of the X-radiation. The diffracted rays leave the specimen along directions which generate a number of cones concentric with the directions of the original beam and make an angle 2θ with it, where θ is Bragg angle. The cones intercept the film in a series of concentric rings. The specimen is rotated slowly during exposure to ensure that all sets of possible planes face the incident X-ray beam.

Diameter of the powder camera. To determine the crystal structure from the diffraction pattern, let R be the radius of the camera and S be the distance on the film between the diffraction arcs corresponding to a particular plane, then the full opening angle of the corresponding cone as shown in Fig. 19.9 is 4θ.

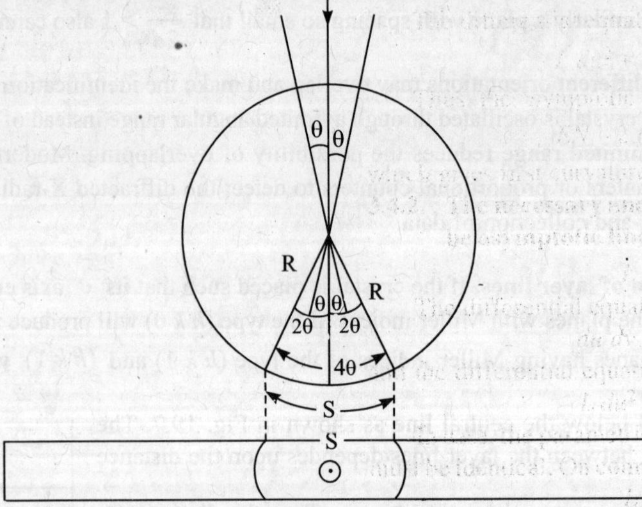

Fig. 19.9

From the figure, we have $4\theta = \frac{S}{R}$ radian $= \frac{180}{\pi R} S$ degrees.

Now $\frac{180}{\pi} = 57.3$

$$\therefore \qquad 4\theta \text{ (degrees)} = 57.3\frac{S}{R}$$

The calculation can be made simple by taking $R = 57.3$ mm or a multiple of that because then the glancing angle can be directly calculated. If we take $R = 57.3$ mm, then

$$\theta \text{ (degrees)} = \frac{S}{4} \text{ mm}$$

Thus, one fourth of the distance between the corresponding arcs of a particular pair in mm is a measure of angle θ in degrees.

This is why the diameter of the powder camera is 57.3 mm or a multiple of that.

Usefulness of Laue, rotating crystal and powder methods. The *Laue method* is suitable for the rapid determination of crystal orientation and symmetry. For example, if the crystal has four fold axial symmetry and is so oriented that its axis is parallel to the *X*-ray beam, the resulting Laue pattern will also show four fold symmetry. The symmetry of the pattern helps to determine the shape of the unit cell.

An imperfect or strained cyrstal has atomic planes which are not exactly plane but are slightly curved. Therefore, instead of sharp diffraction spots we get streaks in Laue's pattern. The formation of streaks on Laue's photographs is known as *asterism.*

This method is, therefore, used for studying the extent of crystalline imperfection under mechanical and thermal treatment.

This method is not suitable for determining crystal structure. It is because out of the continuous range of wavelengths, several wavelengths are reflected in different orders from a single plane and hence different order reflections may overlap on a single spot.

The *rotating crystal method* is the best suited method for structure determination when single crystal specimen is available. If rotation photographs are taken separately about three axes the translation vector $\vec{a}$, $\vec{b}$, $\vec{c}$ and hence the dimensions of the *unit cell* of the crystal is determined. Therefore, the rotating crystal method is a very powerful device for determining the size of a unit cell.

The *powder method* can be applied to any type of matter with crystalline arrangement as it does not require single crystal. It is , therefore, used in the investigation of metals and alloys and in applied work.

The sample used is in the form of a fine powder containing a large number of tiny crystallites with random orientations. It is prepared by crushing a small amount of a poly crystalline material. Thus it eliminates the tedious process of growing single crystals.

19.6 ATOMIC SCATTERING FACTOR

The atomic scattering factor f is defined as the ratio of the radiation amplitude scattered by the actual electron distribution in an atom to that scattered by a single electron localised at a point.

or
$$f = \frac{\text{Amplitude of radiation scattered from an atom}}{\text{Amplitude of radiation scattered from an electron}}$$

The value of f depends upon the *number* and *distribution* of electrons in an atom and *wave-length* and *angle of scattering* of the radiation wave.

To calculate the atomic scattering (or form) factor arrange the co-ordinate system in such a way that the incident and reflected wave normals make equal angles with the normal $\vec{S}$ to the reflecting plane at *A* the centre of the atom as shown in fig. 19.10.

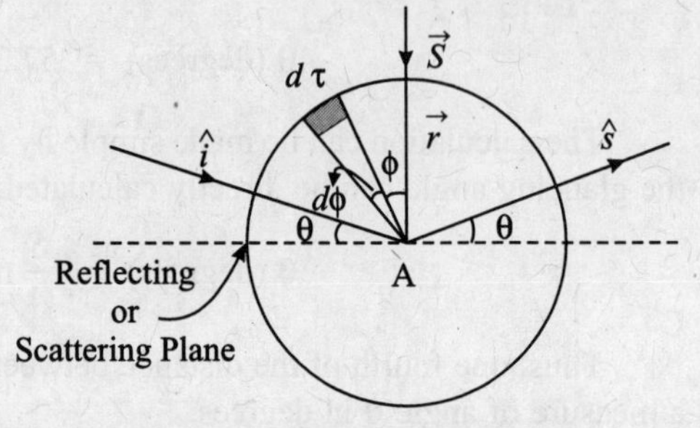

Fig. 19.10

Let the incident wave normal and the reflected (or scattered) wave normal make an angle θ with the reflecting or scattering plane corresponding to Bragg reflection.

Now consider an element of charge $\rho(r)\, d\tau$ in a small volume $d\tau$ at the position (r, ϕ). $\rho(r)$ defines the contribution of the atom to the electron *concentration* at $\vec{r}$. $\rho(r)\, d\tau$ represents the probability of finding the electron in the element of volume $d\tau$ at $\vec{r}$.

The phase difference between the radiation scattered by the element of charge $\rho(r)\, d\tau$ at (r, ϕ) and the radiation that would be scattered by the same amount of charge located at the centre of the atom is given by

$$\phi_r = \frac{2\pi}{\lambda}(\vec{r} \cdot \vec{S}) \qquad ...(i)$$

where

$$|\vec{S}| = 2 \sin\theta$$

Let the amplitude of the wave scattered from the point electron at A in the direction of unit vector $\hat{s}$ be represented by $Ae^{i(kx - \omega t)}$ where x is the distance co-ordinate along $\hat{s}$, k is the magnitude of the wave vector *i.e.*, $|\vec{k}| = \frac{2\pi}{\lambda}$ and ω is the angular frequency of the wave.

The amplitude of the scattered wave from the volume element $d\tau$ in the same direction will be proportional to the amount of charge $\rho(r)\, d\tau$ in the volume element $d\tau$ and have a phase difference ϕ_r stated in (*i*). It is given by

$$Ae^{i[(kx - \omega t) + \phi_r]}\,\rho(r) d\tau$$

The ratio of the amplitude of the radiation scattered by the element of charge $\rho(r)\, d\tau$ at $\vec{r}$ to that scattered by a point electron at the centre of the atom (or origin)

$$df = \frac{Ae^{i[(kx - \omega t) + \phi_r]}\,\rho(r) d\tau}{Ae^{i(kx - \omega t)}}$$

$$= e^{i\phi_r}\rho(r) d\tau$$

$$= e^{i\left(\frac{2\pi}{\lambda}\right)(\vec{r} \cdot \vec{S})}\rho(r) d\tau$$

Now $\quad \vec{r} \cdot \vec{S} = rS\cos\phi = 2r\sin\theta\cos\phi \qquad$ [See **article 19.4**]

$$\therefore \quad \left(\frac{2\pi}{\lambda}\right)(\vec{r} \cdot \vec{S}) = \frac{4\pi}{\lambda} r \sin\theta\cos\phi$$

$$= \mu r \cos\phi \qquad \left(\text{where } \frac{4\pi}{\lambda}\sin\theta = \mu\right)$$

Also $\quad d\tau = 2\pi r^2 \sin\phi\, d\phi\, dr$

as it is the volume of an annular ring of radius $r \sin\phi$, width $r d\phi$ and thickness dr

$$\therefore \quad f = \int df = \int_0^{\infty}\int_0^{\pi} \rho(r) e^{i\mu r\cos\phi}\, 2\pi r^2 \sin\phi\, d\phi\, dr$$

Now $\quad \displaystyle\int_0^{\pi} e^{i\mu r\cos\phi}\sin\phi\, d\phi = \frac{2\sin\mu r}{\mu r}$

$$\therefore \qquad f = \int_0^\infty 4\pi r^2 \rho(r) \frac{\sin \mu r}{\mu r}\, dr$$

As $\theta \to 0$ $\mu \to 0$ and $\dfrac{\sin \mu r}{\mu r} \to 1$

$$\therefore \qquad \text{Lt. } \mu \to 0 \quad f = \int_0^\infty 4\pi r^2 \rho(r)\, dr$$

$$= \text{Total number of electrons in the atom} = Z$$

because $4\pi r^2 \rho(r) dr$ gives the probability that an electron lies between radii r and $r + dr$.

Significance. The atomic scattering factor f is equal to the total number of electrons Z in the atom for $\theta = 0$. For $\theta > 0$ the atomic scattering factor is $< Z$. A knowledge of f gives the charge distribution in the atom.

In other words, the integrand $\int_0^\infty 4\pi r^2 \rho(r)\, dr$ represents the charge inside a spherical shell of radius r and thickness dr and hence the integral gives the total electronic charge Z inside the atom (sphere).

19.7 RECIPROCAL LATTICE

Every crystal has two lattices associated with it, the crystal lattice (or direct space lattice) and the reciprocal lattice.

The concept of reciprocal lattice was devised for the purpose of tabulating two important properties of crystal planes; their *slope* and their *interplanar spacing*.

Each set of parallel planes in a direct lattice can be represented by a *normal* to these planes having length equal to the reciprocal of the interplanar spacing. The normals are drawn with reference to any arbitrary origin and points are marked at their ends. *These points form a regular arrangement which is called reciprocal lattice.* Thus each point in a reciprocal lattice is a representative point of a particular parallel set of planes. It is easier to deal with such points than to deal with set of planes. This is why we use reciprocal lattice.

Thus reciprocal lattice vector is a vector whose magnitude is equal to the reciprocal of the inter planar spacing i.e., $\dfrac{1}{d_{hkl}}$ *and direction is parallel to the normal to (hkl) plane.*

Why reciprocal lattice is so named? The reciprocal lattice is so named because the length assigned to each normal representing the reciprocal lattice is proportional to the reciprocal of the interplanar spacing of that plane in ordinary space.

Primitive translation vectors of reciprocal lattice. Let $\vec{a}$, $\vec{b}$, $\vec{c}$ be the primitive translation vectors of a direct space lattice for a crystal forming a primitive unit cell. The volume of this unit cell is given by

$$V = \vec{a} \cdot (\vec{b} \times \vec{c}) = \vec{b} \cdot (\vec{c} \times \vec{a}) = \vec{c} \cdot (\vec{a} \times \vec{b})$$

The vectors given by
$$\vec{A} = 2\pi \frac{\vec{b} \times \vec{c}}{\vec{a} \cdot (\vec{b} \times \vec{c})}$$

$$\vec{B} = 2\pi \frac{\vec{c} \times \vec{a}}{\vec{b} \cdot (\vec{c} \times \vec{a})}$$

and $$\vec{C} = 2\pi \frac{\vec{a} \times \vec{b}}{\vec{c} \cdot (\vec{a} \times \vec{b})}$$

are known as *reciprocal lattice vectors* or *primitive translation vectors of reciprocal lattice.*

Reciprocal lattice vectors are orthogonal to two axis vectors. Each of the reciprocal lattice vectors is *orthogonal* to two of the axis vectors of a direct space lattice of the crystal as proved below.

$$\vec{A} \cdot \vec{a} = 2\pi, \; \vec{A} \cdot \vec{b} = 0 \text{ and } \vec{A} \cdot \vec{c} = 0$$

Thus $\vec{A}$ is normal to $\vec{a}$ and $\vec{c}$.

Similarly $$\vec{B} \cdot \vec{a} = 0, \; \vec{B} \cdot \vec{b} = 2\pi \text{ and } \vec{B} \cdot \vec{c} = 0$$

$\therefore$ $\vec{B}$ is normal to $\vec{b}$ and $\vec{c}$.

Also $$\vec{C} \cdot \vec{a} = 0, \; \vec{C} \cdot \vec{b} = 0 \text{ and } \vec{C} \cdot \vec{c} = 2\pi$$

$\therefore$ $\vec{C}$ is normal to $\vec{a}$ and $\vec{b}$.

Choosing $\vec{A}$, $\vec{B}$ and $\vec{C}$ as the primitive translation vectors of the reciprocal lattice (also known as reciprocal axes) we can construct the reciprocal lattice points or reciprocal lattice vectors given by

$$\vec{G} = h\vec{A} + k\vec{B} + l\vec{C}$$

where *hkl* are integers and define the co-ordinates of the point in the reciprocal space.

The crystal lattice is a lattice in *real or ordinary space* whereas reciprocal lattice is a lattice in the *reciprocal space or associated* $\vec{k}$ *space or Fourier space.* The wave vector $\vec{k}$ is always drawn in Fourier space. Vectors in the crystal lattice have dimensions of length $[L^1]$ and vectors in the reciprocal lattice have dimensions of 1/Length $[L^{-1}]$.

Construction of reciprocal lattice. A reciprocal lattice from direct lattice is constructed by the following method:

(*i*) Take origin at some arbitrary point.

(*ii*) From the origin draw normals to every set of parallel planes of direct lattice. The direction of the normal specifies the orientation of the plane and thus the two dimensional plane is represented by the normal which has only one dimension.

Fig. 19.11

(*iii*) Take length of each normal equal to or 2π times the reciprocal of the interplanar spacing for the corresponding set of planes.

(*iv*) The terminal points of these normals form the reciprocal lattice because the distances in these lattices are the reciprocal to those in the direct crystal lattice.

As an example, consider a unit cell of a monoclinic crystal in which $a \neq b \neq c$; $\alpha = \gamma = 90°$ and $\beta > 90°$. Orient the unit cell in such a way that the $\vec{b}$ axis is perpendicular to the plane of the paper *i.e.*, the $\vec{a}$ and $\vec{c}$ axes lie in the plane of the paper (Fig. 19.11).

Consider planes of the type ($h0l$) which are parallel to $\vec{b}$ axis (Fig. 19.12) and hence perpendicular to the plane of the paper. Therefore, the normal to these planes lies in the plane of the paper. The planes ($h0l$) being perpendicular to the plane of the paper are represented by lines. Thus the line (101) in fact means the plane (101) and so on. (Fig. 19.12)

Taking the point of intersection of the three axes as the origin draw normals to the planes ($h\ 0\ l$) and take their lengths equal to $\dfrac{1}{d_{h0l}}$ where d_{hol} is the interplanar spacing for the planes.

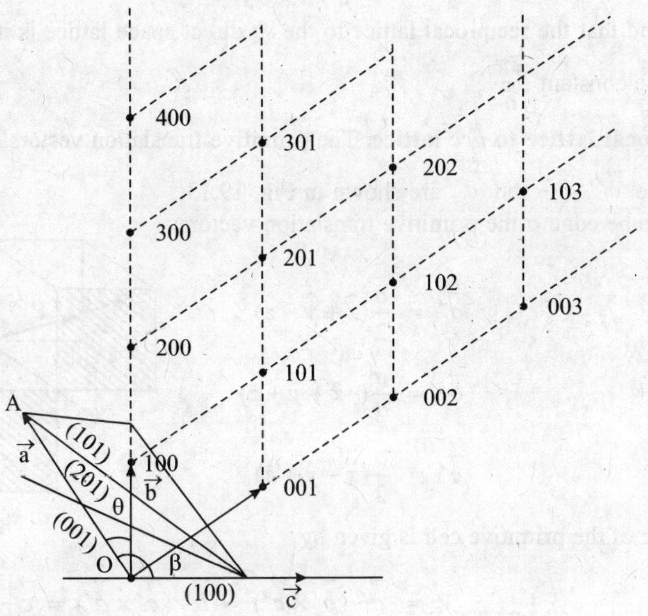

Fig. 19.12

We know that (200) planes have half the interplanar spacing as compared to (100) planes, the reciprocal lattice point (200) is, therefore, twice as far away from the origin as the point (100). In this manner if we draw normals to the ($h\ k\ l$) planes, we get a three dimensional reciprocal lattice.

Properties of reciprocal lattice. (*i*) The direct lattice is the reciprocal lattice to its own reciprocal lattice. Simple cubic lattice is *self reciprocal* whereas *bcc* and *fcc* lattices are reciprocal of each other.

(*ii*) Each point in a reciprocal lattice corresponds to a particular set of parallel planes of the direct lattice.

(*iii*) The distance of a reciprocal lattice point from an arbitrarily fixed origin is inversely proportional to the interplanar spacing of the corresponding parallel planes of the direct lattice.

(*iv*) The unit cell of the reciprocal lattice is not necessarily a parallelopiped. We deal with the Wigner Seitz cell of the reciprocal lattice which constitutes the *Brillouin zone*.

(*v*) The volume of a unit cell of the reciprocal lattice is inversely proportional to the volume of the corresponding unit cell of the direct lattice.

The volume of the cubic cell

$$V = \vec{a}.(\vec{b} \times \vec{c}) = \vec{b}.(\vec{c} \times \vec{a}) = \vec{c}.(\vec{a} \times \vec{b}) = a^3$$

The primitive translation vectors of the reciprocal *sc* lattice are given by

$$\vec{A} = 2\pi\frac{\vec{b} \times \vec{c}}{\vec{a}\cdot(\vec{b}\times\vec{c})} = 2\pi\frac{a\hat{y}\times a\hat{z}}{a^3} = \frac{2\pi}{a}\hat{x}$$

$$\vec{B} = 2\pi\frac{\vec{c} \times \vec{a}}{\vec{b}\cdot(\vec{c}\times\vec{a})} = 2\pi\frac{a\hat{z}\times a\hat{x}}{a^3} = \frac{2\pi}{a}\hat{y}$$

and

$$\vec{C} = 2\pi\frac{\vec{a} \times \vec{b}}{\vec{c}\cdot(\vec{a}\times\vec{b})} = 2\pi\frac{a\hat{x}\times a\hat{y}}{a^3} = \frac{2\pi}{a}\hat{z}$$

Thus we find that the reciprocal lattice to the *sc* direct space lattice is itself a simple cubic lattice with lattice constant $\frac{2\pi}{a}$.

(*ii*) **Reciprocal lattice to *bcc* lattice.** The primitive translation vectors of the body centred cubic (*bcc*) lattice $\vec{a}'$, $\vec{b}'$ and $\vec{c}'$ are shown in Fig. 19.13. In terms of the cube edge *a* the primitive translation vectors are

$$\vec{a}' = \frac{a}{2}(\hat{x}+\hat{y}-\hat{z})$$

$$\vec{b}' = \frac{a}{2}(-\hat{x}+\hat{y}+\hat{z})$$

$$\vec{c}' = \frac{a}{2}(\hat{x}-\hat{y}+\hat{z})$$

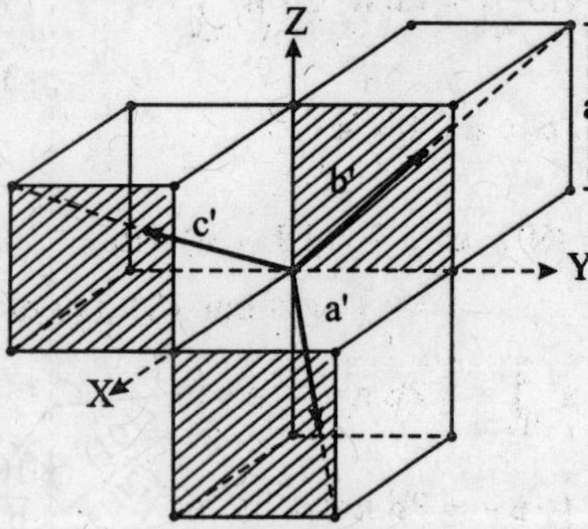

Fig. 19.13

The volume of the primitive cell is given by

$$V = \vec{a}'\cdot(\vec{b}'\times\vec{c}') = \vec{b}'\cdot(\vec{c}'\times\vec{a}') = \vec{c}'\cdot(\vec{a}'\times\vec{b}')$$

Now

$$(\vec{b}'\times\vec{c}') = \frac{a^2}{2}(\hat{x}+\hat{y})$$

$\therefore$

$$V = \frac{a}{2}(x+y-z)\cdot\frac{a^2}{2}(x+y) = \frac{a^3}{2}$$

The primitive translation vectors of the reciprocal *bcc* lattice are given by

$$\vec{A} = 2\pi\frac{\vec{b}'\times\vec{c}'}{\vec{a}'\cdot(\vec{b}'\times\vec{c}')} = 2\pi\frac{\frac{a^2}{2}(\hat{x}+\hat{y})}{a^3/2} = \frac{2\pi}{a}(\hat{x}+\hat{y})$$

Similarly

$$\vec{B} = 2\pi\frac{\vec{c}'\times\vec{a}'}{\vec{b}'\cdot(\vec{c}'\times\vec{a}')} = 2\pi\frac{\frac{a^2}{2}(\hat{y}+\hat{z})}{a^3/2} = \frac{2\pi}{a}(\hat{y}+\hat{z})$$

$$\vec{C} = 2\pi\frac{\vec{a}'\times\vec{b}'}{\vec{c}'\cdot(\vec{a}'\times\vec{b}')} = 2\pi\frac{\frac{a^2}{2}(\hat{z}+\hat{x})}{a^3/2} = \frac{2\pi}{a}(\hat{z}+\hat{x})$$

Thus we find that the reciprocal *bcc* lattice vectors are just the primitive vectors of *fcc* lattice given in (*iii*) with lattice constant $\frac{2\pi}{a}$.

(*iii*) **Reciprocal lattice to *fcc* lattice.** The primitive translation vectors of the face centred cubic (*fcc*) lattice $\vec{a}', \vec{b}', \vec{c}'$ are shown in Fig. 19.14. In terms of the cube edge *a* the primitive translation vectors are

$$\vec{a}' = \frac{a}{2}(\hat{x}+\hat{y})$$

$$\vec{b}' = \frac{a}{2}(\hat{y}+\hat{z})$$

$$\vec{c}' = \frac{a}{2}(\hat{z}+\hat{x})$$

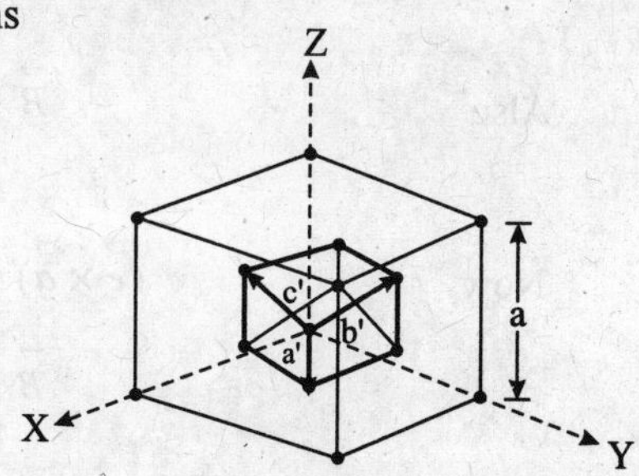

Fig. 19.14

The volume of the primitive cell is given by

$$V = \vec{a}'\cdot(\vec{b}'\times\vec{c}') = \vec{b}'\cdot(\vec{c}'\times\vec{a}') = \vec{c}'\cdot(\vec{a}'\times\vec{b}')$$

Now $$\vec{b}'\times\vec{c}' = \frac{a}{2}(\hat{y}+\hat{z})\times\frac{a}{2}(\hat{z}+\hat{x})$$

$$= \frac{a^2}{4}(\hat{x}+\hat{y}-\hat{z})$$

$$\therefore \quad V = \vec{a}'\cdot(\vec{b}'\times\vec{c}') = \frac{a}{2}(\hat{x}+\hat{y})\cdot\frac{a^2}{4}(\hat{x}+\hat{y}-\hat{z}) = \frac{a^3}{4}$$

The primitive translation vectors of the reciprocal *fcc* lattice are given by

$$\vec{A} = 2\pi\frac{\vec{b}'\times\vec{c}'}{\vec{a}'\cdot(\vec{b}'\times\vec{c}')} = \frac{2\pi\frac{a^2}{4}(\hat{x}+\hat{y}-\hat{z})}{a^3/4} = \frac{2\pi}{a}(\hat{x}+\hat{y}-\hat{z})$$

Similarly $$\vec{B} = \frac{2\pi}{a}(-\hat{x}+\hat{y}+\hat{z}) \text{ and } \vec{C} = \frac{2\pi}{a}(\hat{x}-\hat{y}+\hat{z})$$

Thus we find that reciprocal *fcc* lattice are just the primitive vectors of *bcc* lattice given in (*ii*) with lattice constant $\frac{2\pi}{a}$.

Example 19.11. *A three dimensional lattice has the basis vectors*

$$\vec{a} = \hat{i}-\hat{j},\ \vec{b} = 2\hat{i}+\hat{j} \text{ and } \vec{c} = \hat{k}.$$

Find the basis vectors of reciprocal lattice. **(H.P.U. 2002)**

Solution. Let the basis vectors of reciprocal lattice be $\vec{A}$, $\vec{B}$ and $\vec{C}$, then

$$\vec{A} = 2\pi\frac{(\vec{b}\times\vec{c})}{\vec{a}\cdot(\vec{b}\times\vec{c})}$$

Now $$(\vec{b}\times\vec{c}) = (2\hat{i}+\hat{j})\times\hat{k} = -2\hat{j}+\hat{i} = \hat{i}-2\hat{j}$$

and $$\vec{a}\cdot(\vec{b}\times\vec{c}) = (\hat{i}-\hat{j})\cdot(\hat{i}-2\hat{j}) = 1+2 = 3$$

$\therefore$ $$\vec{a}\cdot(\vec{b}\times\vec{c}) = \vec{b}\cdot(\vec{c}\times\vec{a}) = \vec{c}\cdot(\vec{a}\times\vec{b}) = V = 3$$

and $$\vec{A} = 2\pi\frac{\hat{i}-2\hat{j}}{3} = \frac{2\pi}{3}(\hat{i}-2\hat{j})$$

Also $$\vec{B} = \frac{2\pi(\vec{c}\times\vec{a})}{\vec{b}\cdot(\vec{c}\times\vec{a})}$$

Now $$(\vec{c}\times\vec{a}) = \hat{k}\times(\hat{i}-\hat{j}) = \hat{j}+\hat{i} = \hat{i}+\hat{j}$$

$\therefore$ $$\vec{B} = \frac{2\pi}{3}(\hat{i}+\hat{j})$$

Again $$\vec{C} = 2\pi\frac{\vec{a}\times\vec{b}}{\vec{c}\cdot(\vec{a}\times\vec{b})}$$

Now $$\vec{a}\times\vec{b} = (\hat{i}-\hat{j})\times(2\hat{i}+\hat{j}) = \hat{k}+2\hat{k} = 3\hat{k}$$

$\therefore$ $$\vec{C} = \frac{2\pi}{3}3\hat{k} = 2\pi\hat{k}$$

Example 19.12. *The primitive translation vectors of space lattice are*

$$\vec{a} = 2\hat{i}+\hat{j},\ \vec{b} = 2\hat{j} \text{ and } \vec{c} = \hat{k}.$$

Find the volume of the primitive cell and primitive translation vectors of the reciprocal lattice.

(H.P.U. 2001)

Solution. The volume of the primitive cell $V = \vec{a}\cdot(\vec{b}\times\vec{c}) = (2\hat{i}+\hat{j})\cdot(2\hat{j}\times\hat{k}) = (2\hat{i}+\hat{j})\cdot(2\hat{i}) = 4$

$\therefore$ $$\vec{a}\cdot(\vec{b}\times\vec{c}) = \vec{b}\cdot(\vec{c}\times\vec{a}) = \vec{c}\cdot(\vec{b}\times\vec{c}) = V = 4$$

Let the primitive translation vectors of the reciprocal lattice be $\vec{A}$, $\vec{B}$ and $\vec{C}$.

$\therefore$ $$\vec{A} = 2\pi\frac{(\vec{b}\times\vec{c})}{\vec{a}\cdot(\vec{b}\times\vec{c})} \quad \text{and } (\vec{b}\times\vec{c}) = 2\hat{j}\times\hat{k} = 2\hat{i}$$

Also $$\vec{a}\cdot(\vec{b}\times\vec{c}) = 4$$

$\therefore$ $$\vec{A} = 2\pi\frac{2\hat{i}}{4} = \pi\hat{i}$$

$$\vec{B} = 2\pi\frac{(\vec{c}\times\vec{a})}{\vec{b}\cdot(\vec{c}\times\vec{a})} \quad \text{and } (\vec{c}\times\vec{a}) = \hat{k}\times(2\hat{i}+\hat{j}) = 2\hat{j}-\hat{i}$$

Also $\quad \vec{b}\cdot(\vec{c}\times\vec{a}) = 4$

$$\therefore \quad \vec{B} = 2\pi\frac{(2\hat{j}-\hat{i})}{4} = \frac{\pi}{2}(2\hat{j}-\hat{i})$$

And $\quad \vec{C} = 2\pi\dfrac{(\vec{a}\times\vec{b})}{\vec{c}\cdot(\vec{a}\times\vec{b})}$ and $(\vec{a}\times\vec{b}) = (2\hat{i}+\hat{j})\times 2\hat{j} = 4\hat{k}$

Also $\quad \vec{c}\cdot(\vec{a}\times\vec{b}) = 4$

$$\therefore \quad \vec{C} = 2\pi\frac{4\hat{k}}{4} = 2\pi\hat{k}.$$

Important note. In this book we have denoted *primitive translation vectors of a direct space lattice* for a crystal forming primitive unit cell by $\vec{a}$, $\vec{b}$ and $\vec{c}$ and the *reciprocal lattice vectors* or *primitive translation vectors of reciprocal lattice* by $\vec{A}$, $\vec{B}$ and $\vec{C}$.

Some books, however, represent the primitive translation vectors of reciprocal lattice by $\vec{a}^*$, $\vec{b}^*$ and $\vec{c}^*$ *i.e.,*

$$\vec{A} = \vec{a}^*;\ \vec{B} = \vec{b}^* \text{ and } \vec{C} = \vec{c}^*$$

In this notation the *reciprocal of the reciprocal lattice* vector is represented by $(\vec{a}^*)^*$, $(\vec{b}^*)^*$ and $(\vec{c}^*)^*$. According to the notation used in this book, we shall represent the *reciprocal of the reciprocal lattice* by $\vec{A}^*$, $\vec{B}^*$ and $\vec{C}^*$ *i.e.,*

$$\vec{A}^* = (\vec{a}^*)^*,\ \vec{B}^* = (\vec{b}^*)^* \text{ and } \vec{C}^* = (\vec{c}^*)^*$$

We shall now prove that

$$\vec{A}^* = (\vec{a}^*)^* = a;\ \vec{B}^* = (\vec{b}^*)^* = b \text{ and } \vec{C}^* = (\vec{c}^*)^* = c$$

Now $\quad \vec{A} = \vec{a}^* = 2\pi\dfrac{(\vec{a}\times\vec{b})}{\vec{c}\cdot(\vec{a}\times\vec{b})}$

$$\therefore \quad \vec{A}^* = (\vec{a}^*)^* = 2\pi\frac{\vec{B}\times\vec{C}}{\vec{A}\cdot(\vec{B}\times\vec{C})} = 2\pi\frac{\vec{b}^*\times\vec{c}^*}{\vec{a}^*\cdot(\vec{b}^*\times\vec{c}^*)} \quad \ldots(i)$$

We have already proved that $\quad \vec{A}\cdot\vec{a} = 2\pi$ *i.e.,* $\vec{a}^*\cdot\vec{a} = 2\pi$

$$\vec{B}\cdot\vec{b} = 2\pi \textit{ i.e., } \vec{b}^*\cdot\vec{b} = 2\pi$$

$$\text{and } \vec{b}^*\cdot\vec{b} = 2\pi \textit{ i.e., } \vec{c}^*\cdot\vec{c} = 2\pi$$

$$\therefore \qquad \frac{\vec{a}^* \cdot \vec{a}}{2\pi} = 1; \quad \frac{\vec{b}^* \cdot \vec{b}}{2\pi} = 1 \text{ and } \frac{\vec{c}^* \cdot \vec{c}}{2\pi} = 1$$

Multiplying the right hand side of Eq. (*i*) by $\frac{\vec{a}^* \cdot \vec{a}}{2\pi}$ *i.e.,* 1, we get

$$\vec{A}^* = (\vec{a}^*)^* = \frac{\vec{a}^* \cdot \vec{a}}{2\pi} \times \frac{2\pi(\vec{b}^* \times \vec{c}^*)}{\vec{a}^* \cdot (\vec{b}^* \times \vec{c}^*)}$$

$$= \vec{a}\,\frac{\vec{a}^* \cdot (\vec{b}^* \times \vec{c}^*)}{\vec{a}^* \cdot (\vec{b}^* \times \vec{c}^*)} = \vec{a} \qquad \ldots(ii)$$

Similarly $\qquad \vec{B}^* = (\vec{b}^*)^* = \vec{b}$

and $\qquad \vec{C}^* = (\vec{c}^*)^* = \vec{c}$

We shall also prove that

$$\vec{a} = 2\pi \frac{\vec{b}^* \times \vec{c}^*}{\vec{a}^* \cdot (\vec{b}^* \times \vec{c}^*)} \text{ or } 2\pi \frac{\vec{B} \times \vec{C}}{\vec{A} \cdot (\vec{B} \times \vec{C})}$$

$$\vec{b} = 2\pi \frac{\vec{c}^* \times \vec{a}^*}{\vec{b}^* \cdot (\vec{c}^* \times \vec{a}^*)} \text{ or } 2\pi \frac{\vec{C} \times \vec{A}}{\vec{B} \cdot (\vec{C} \times \vec{A})}$$

and
$$\vec{c} = 2\pi \frac{\vec{a}^* \times \vec{b}^*}{\vec{c}^* \cdot (\vec{a}^* \times \vec{b}^*)} \text{ or } 2\pi \frac{\vec{A} \times \vec{B}}{\vec{C} \cdot (\vec{A} \times \vec{B})}$$

We know that
$$\vec{A} = \vec{a}^* = 2\pi \frac{\vec{b} \times \vec{c}}{\vec{a} \cdot (\vec{b} \times \vec{c})}$$

$$\therefore \qquad \vec{A}^* = (\vec{a}^*)^* = 2\pi \frac{\vec{b}^* \times \vec{c}^*}{\vec{a}^* \cdot (\vec{b}^* \times \vec{c}^*)}$$

but
$$\vec{A}^* = (\vec{a}^*)^* = \vec{a} \qquad \text{[As proved in } (ii)\text{]}$$

$$\therefore \qquad \vec{a} = 2\pi \frac{\vec{b}^* \times \vec{c}^*}{\vec{a}^* \cdot (\vec{b}^* \times \vec{c}^*)} = 2\pi \frac{\vec{B} \times \vec{C}}{\vec{A} \cdot (\vec{B} \times \vec{C})}$$

Similarly
$$\frac{1}{2} = 2\pi \frac{\vec{c}^* \times \vec{a}^*}{\vec{b}^* \cdot (\vec{c}^* \times \vec{a}^*)} = 2\pi \frac{\vec{C} \times \vec{A}}{\vec{B} \cdot (\vec{C} \times \vec{A})}$$

and
$$\vec{c} = 2\pi \frac{\vec{a}^* \times \vec{b}^*}{\vec{c}^* \cdot (\vec{a}^* \times \vec{b}^*)} = 2\pi \frac{\vec{A} \times \vec{B}}{\vec{C} \cdot (\vec{A} \times \vec{B})}.$$

19.8 EWALD'S CONSTRUCTION

Ewald's construction is a geometrical construction to get a relation between wave vector $\vec{k}$ and the direction of incident X-ray beam with the help of reciprocal lattice and thus derive Bragg's law in vector form.

Draw a vector $\vec{OA} = \vec{k} = \frac{2\pi}{\lambda}$ in the direction of incident X-ray beam originating at O and terminating at any reciprocal lattice point A. Draw a sphere of radius $k = \frac{2\pi}{\lambda}$ with origin O as centre. This sphere is known as *reflex sphere* (The section of this sphere in a plane is a circle.). Suppose this sphere intersects some point B of the reciprocal lattice whose indices are $(h'k'l')$, then $\vec{AB}$ represents the reciprocal lattice vector $\vec{G}$ and is normal to some set of direct lattice planes *e.g.*, ON.

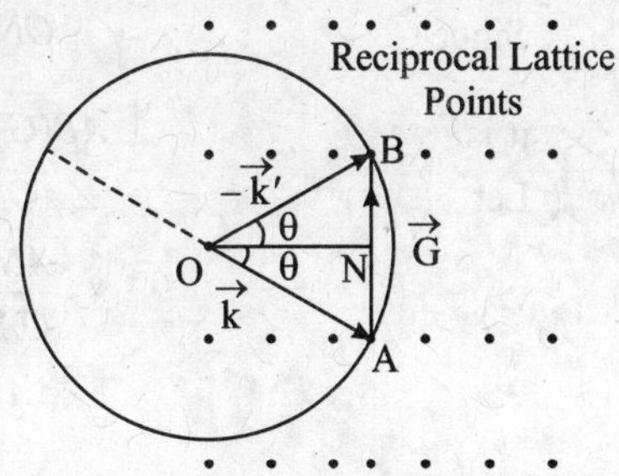

Fig. 19.15

Now
$$AB = |\vec{G}| = \frac{2\pi n}{d}$$
where n is the largest integral factor common to numbers $(h'k'l')$ and d is the inter-planer spacing for the set.

Bragg's diffraction condition in terms of reciprocal lattice. As the wave vector $\vec{OA} = \vec{k}$ is the incident wave vector, $\vec{OB} = \vec{k'}$ is the diffracted (or reflected) wave vector and is given by

$$\vec{k'} = \vec{k} + \vec{G}$$

This shows that scattering changes only the direction of the wave vector $\vec{k}$ and the scattered wave vector differs from the incident wave vector by a reciprocal lattice vector $\vec{G}$.

For diffraction, the magnitude of $(\vec{k} + \vec{G})$ must be equal to the wave vector $\vec{k'}$.

i.e.
$$\vec{k'}^2 = (\vec{k} + \vec{G})^2 = (\vec{k} + \vec{G}) \cdot (\vec{k} + \vec{G})$$
$$= k^2 + 2\vec{k} \cdot \vec{G} + \vec{G} \cdot \vec{G} \qquad ...(i)$$

But the magnitude of $\vec{k'}$ = magnitude of $\vec{k}$

$$\therefore \qquad \vec{k'}^2 = k^2$$

Hence relation (*i*), becomes

$$2\vec{k} \cdot \vec{G} + \vec{G} \cdot \vec{G} = 0 \text{ or } 2\vec{k} \cdot \vec{G} + G^2 = 0$$

or
$$\left(\vec{k} + \frac{\vec{G}}{2}\right) \cdot \vec{G} = 0$$

This is Bragg's law in vector form and gives the Bragg's diffraction condition in terms of reciprocal lattice vector $\vec{G}$.

Bragg's law in scalar form. To convert the relation $\left(\vec{k} + \frac{\vec{G}}{2}\right) \cdot \vec{G} = 0$ into the usual scalar form of Bragg's law, we have

$$\begin{aligned}
&\text{Since} \quad & \overrightarrow{ON} &= \overrightarrow{OA} + \overrightarrow{AN} = \vec{k} + \frac{\vec{G}}{2} \\
&\text{it is} \quad & &\perp AB(\vec{G}) \\
&\text{Let} \quad & \angle AON &= \theta \\
&\therefore \quad & \frac{AN}{OA} &= \sin\theta \text{ or } AN = OA \sin\theta \\
&\therefore \quad & AB &= 2AN = 2OA \sin\theta = 2k \sin\theta = 2\frac{2\pi}{\lambda}\sin\theta \\
&\text{Now} \quad & AB &= |\vec{G}| = \frac{2\pi n}{d} \\
&\therefore \quad & \frac{2\pi n}{d} &= 2\frac{2\pi}{\lambda}\sin\theta \\
&\text{or} \quad & 2d\sin\theta &= n\lambda
\end{aligned}$$

which is Bragg's law or diffraction condition.

Thus the concept of reciprocal lattice and Ewald's construction helps in the determination of crystal structure through the vector form of Bragg's law.

19.9 COHESIVE ENERGY

From the very existence of solids, two general conclusions can be drawn.

(*i*) There must be an *attractive force* between atoms or molecules in a solid which keeps them together.

(*ii*) There must also be a *repulsive force* acting between the atoms. This is why large external pressure is required to compress a solid to any appreciable extent.

When two identical or non-identical atoms, which form a molecule, are brought from infinity to a close proximity, they first attract each other and then, if they are brought closer than a certain distance, repel each other. In other words, a kind of spring effect takes place. As the potential V is related to the force by the relation $F = \frac{-dV}{dr}$, the resulting potential is a function of distance of separation r. The variation of corresponding potential energy V in eV with distance r in Å is shown in Fig. 19.16.

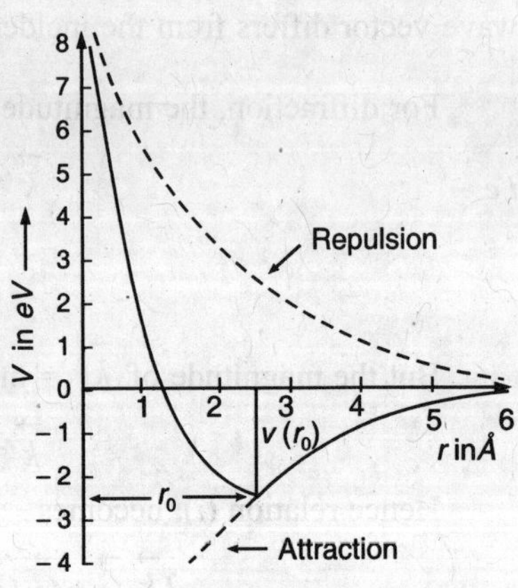

Fig. 19.16

The attractive force gives rise to a negative potential and the repulsive force gives rise to a positive potential shown by dotted curves. The resultant potential energy is shown by full line curve. It has a minimum at a distance r_0 and represents the equilibrium position, where the attractive and the repulsive forces balance each other and potential energy has a maximum negative value. The potential energy $V(r_0)$ at a distance r_0 is thus responsible for keeping the atoms bound in a solid and gives the *cohesive* or *binding energy* of the solid.

The potential energy $V(r_0)$ is a negative quantity. The positive quantity $-V(r_0) = D$ is called the *dissociation energy* of the molecule.

The cohesive energy of a solid is defined as the energy which will be given out in the process of the formation of a crystal by bringing neutral atoms from infinity to the position of equilibrium separation.

If the potential energy due to attraction V_a is supposed to vary as the *m*th power of the distance and that due to repulsion V_r as *n*th power of *r*, then

$$V_a \propto \frac{1}{r^m} = -\frac{A}{r^m} \text{ and } V_r \propto \frac{1}{r^n} = \frac{B}{r^n}$$

$\therefore$ Resultant potential or cohesive energy

$$V = V_a + V_r = -\frac{A}{r^m} + \frac{B}{r^n}$$

where A and B are constants of proportionality. The cohesive energy is generally expressed in eV per atom.

The force between the two atoms is given by

$$F = \frac{-dV}{dr} = \frac{mA}{r^{m+1}} - \frac{nB}{r^{n+1}}$$

At $r = r_0$ the attractive and repulsive forces balance each other and $F(r_0) = 0$. In other words at $r = r_0$

$$\frac{mA}{r_0^{m+1}} - \frac{nB}{r_0^{n+1}} = 0$$

or

$$\frac{m}{n} = \frac{B}{A} r_0^{m-n}$$

or

$$r_0^{m-n} = \frac{A}{B}\frac{m}{n} \qquad \text{... } (i)$$

Hence, the potential energy in the equilibrium state

$$V(r_0) = \frac{-A}{r_0^m} + \frac{B}{r_0^n} = \frac{-A}{r_0^m}\left[1 - \frac{B}{A} r_0^{m-n}\right] = \frac{-A}{r_0^m}\left[1 - \frac{m}{n}\right]$$

Although the attractive and repulsive forces are in equilibrium at $r = r_0$, the potential energy is not zero as $m \neq n$. If $n >> m$ the total energy is essentially the energy of attraction given by $\frac{-A}{r_0^m}$.

For the potential energy to have a minimum at $r = r_0$ we have $\left|\frac{d^2V}{dr^2}\right|_{r=r_0}^{0} > 0$ *i.e.*, it must be positive.

Now,

$$\frac{d^2V}{dr^2} = \frac{-m(m+1)A}{r^{m+2}} + \frac{n(n+1)B}{r^{n+2}} > 0$$

Substituting the value of $r = r_0$ from (*i*), we have

$$-m(m+1)A \times \frac{B}{A}\frac{n}{m}\left(\frac{1}{r_0^{n+2}}\right) + \frac{n(n+1)B}{r_0^{n+2}} > 0$$

or

$$-m - 1 + n + 1 > 0$$

or

$$n - m > 0 \text{ or } n > m$$

In other words, a minimum in the energy curve is possible only if $n > m$. *Thus the formation of a chemical bond requires that the repulsive forces be of shorter range than the attractive forces.*

The energy $V(r_0)$ at the equilibrium distance r_0 is called the *binding energy, the energy of cohesion* or *dissociation energy* of the molecule. This much energy is required to separate the atoms of a diatomic molecule to an infinite distance apart.

The cohesive energy may also be defined as the energy released when two atoms are brought close to each other at the equilibrium distance r_0.

Larger the energy released, more stable the bond formed and hence more stable is the crystal structure.

Separation for a Stable compound

The energy of two particles in the field of each other at a separation r is given by

$$U = \frac{\alpha}{r} + \frac{\beta}{r^8}$$

They will form a stable compound at a separation r_0 such that at that separation the energy U is a minimum, *i.e.*,

$$\left|\frac{dU}{dr}\right|_{r=r_0} = 0$$

Now $$\frac{dU}{dr} = -\frac{\alpha}{r^2} - \frac{8\beta}{r^9}$$

$\therefore$ $$\left|\frac{dU}{dr}\right|_{r=r_0} = -\frac{\alpha}{r_0^2} - \frac{8\beta}{r_0^9} = 0$$

or $$\frac{\alpha}{r_0^2} + \frac{8\beta}{r_0^9} = 0 \quad \text{or} \quad \alpha = \frac{8\beta}{r_0^7}$$

$\therefore$ $$r_0^7 = \frac{8\beta}{\alpha} \quad \text{Hence } r_0 = \left[\frac{8\beta}{\alpha}\right]^{1/7}$$

19.10 COHESIVE ENERGY OF A CRYSTAL

The energy of the two particles in the field of each other at a separation r is given by

$$U = -\frac{\alpha}{r^2} + \frac{\beta}{r^{10}} \quad \text{... (i)}$$

They will form a stable compound at a separation r_0 such that at that separation the energy U is minimum *i.e.*,

$$\left|\frac{dU}{dr}\right|_{r=r_0} = 0$$

Now, $$\frac{dU}{dr} = \frac{2\alpha}{r^3} - \frac{10\beta}{r^{11}} = 0$$

or $$r^8 = \frac{5\beta}{\alpha}$$

$\therefore$ Separation for a stable compound $r_0 = \left[\frac{5\beta}{\alpha}\right]^{1/8}$

Putting the value of $r = r_0$ in Eq. (*i*), we have

$$U_{min} = -\alpha\left(\frac{\alpha}{5\beta}\right)^{1/4} + \beta\left(\frac{\alpha}{5\beta}\right)^{5/4}$$

$$= \frac{\alpha^{5/4}}{(5\beta)^{1/4}}\left[\frac{\beta}{5\beta}-1\right] = -\frac{4}{5}\left[\frac{\alpha^{5/4}}{(5\beta)^{1/4}}\right]$$

$$= -\frac{4\alpha}{5}\left[\frac{\alpha}{5\beta}\right]^{1/4} = -\frac{4\alpha}{5r_0^2} = -\frac{4}{5}\left(\frac{\alpha}{r_0^2}\right)$$

$\therefore$ Cohesive energy of the crystal $U_{min} = -\frac{4}{5}\left(\frac{\alpha}{r_0^2}\right)$.

19.11 BONDING

There are three states of matter solid, liquid and gas. The atoms in a solid are closely packed but are in a state of vibration about their fixed position. The nucleus of an atom can be considered at rest due to its large mass. It is, therefore, the orbital electrons which are responsible for most of the characteristics of the solid. *The ability to hold the atoms or ions together is called bonding.* The basic requirements of all types of bonding is that all bound systems should have minimum energy in their stable form. When some atoms combine to form a stable molecule, it is implied that in so doing they have decreased the energy of the system to a minimum.

The forces acting between the atoms are electrostatic in nature. These forces are determined essentially by the way in which the outer electrons of the composing atoms are distributed in space. The physical properties of the solids are determined, to a large extent by the electron distribution and it is thus possible on an empirical basis to divide solids into different groups corresponding to different types of electron distribution.

Types of bonding: The individual atoms of a crystalline solid are held together as one system with different types of bonds existing between them. These bonds are classified into *five* idealised types.

(*i*) **Ionic bonding.** Examples:- *NaCl*, *LiF* (Due to transfer of valence electrons).

(*ii*) **Co-valent bonding.** Examples:- diamond, *Si C* (Sharing of valence electrons).

(*iii*) **Van der Waal bonding.** Example:- solid Argon (electrons remain associated with original molecules).

(*iv*) **Hydrogen bonding.** Example:- ice

(*v*) **Metallic bonding.** *Cu*, *Ag*, *Fe* are examples. (valence electrons are essentially free)

These bonds are fundamentally electric in origin, but except for ionic bonds which arise through essentially quantum mechanical considerations.

It may be added that the above classification is arbitrary. Many solids fall into intermediate categories. Let us study them one by one in detail.

19.12 IONIC CRYSTALS

The crystals, the molecules of which have ionic bonding are known as *ionic crystals*. Familiar examples are *NaCl*, *KCl*, *ZnS* etc.

Ionic bonding. The ionic bonding arises from the transfer of one or more electrons from an electropositive element to an electronegative element, thereby, creating a positive and a negative ion. The two types of atoms involved in ionic bonding are different from each other. The electro-positive atoms have a *low ionising energy* (the energy required to remove an electron from the atom). The electro-negative elements are of high *electron affinity* and are ready to take up an additional electron to complete their outermost valence orbit and thus acquire stable configuration. The two ions thus formed, being oppositely charged, attract each other with an electrostatic coulomb force which gives rise to an attractive potential V_a the value of which in the S.I. units is given by

$$V_a = -\frac{Z_1 Z_2 e^2}{4\pi\varepsilon_0 r} = -\alpha\frac{e^2}{4\pi\varepsilon_0 r}$$

where $Z_1 e$ is the charge, say on the positive ion, $Z_2 e$ the charge on the negative ion separated by a distance r and α is a constant known as *Madelung constant*, which has the same value for all crystals of the same structure (The value of α for *NaCl* crystal = 1.748, for *CsCl* = 1.763 and *ZnS* = 1.638). This energy keeps the atoms bound inside the solids.

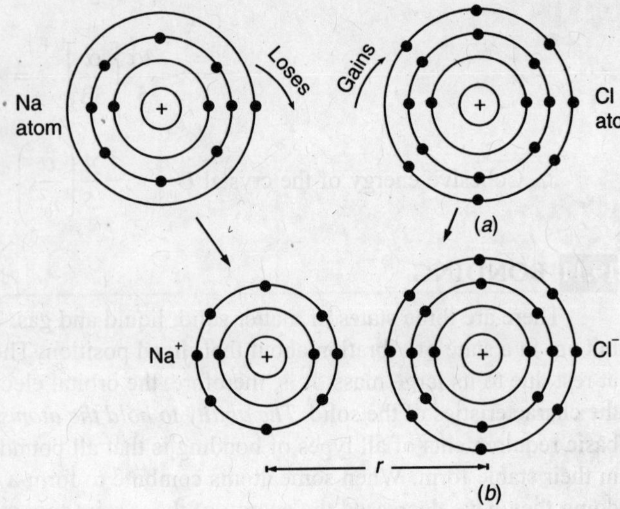

Fig. 19.17

As an example for ionic bonding, let us consider the case of *NaCl* crystal. The *Na* atom loses the only valence electron in its outermost valence orbit and becomes singly charged positive Na^+ ion. The *Cl* atom is short of one electron in its outermost valence orbit. It, therefore, takes up the electron from *N* and becomes a singly charged negative Cl^- ion with its outermost electron orbit now completely filled. The Na^+ and Cl^- ions are bound strongly due to electrostatic force of attraction to form a stable *NaCl* crystal as sown in Fig. 19.17.

Hence an ionic bond can exist between two atoms when one of them has as many valence electrons as are required by the other to complete its outermost valence orbit.

It is clear from the above discussion that after the transfer of electrons, the ions attain electronic configuration similar to inert gas atoms. The charge distribution in the two ions is, therefore, spherically symmetric and as a result *the bond is non-directional*.

As permitted by radius ratio rule, each Na^+ ion is surrounded by six Cl^- ions and in turn each Cl^- ion is surrounded by six Na^+ ions to maintain the charge neutrality.

Thus the binding together of atoms by virtue of the net electrostatic forces of attraction between ions formed by the shifting of electrons from one atom to the other is called *ionic bonding*. These are usually very strong. *NaCl*, *KCl*, *LiF*, *NaBr* and *ZnS* are some examples of this types of bonding.

Properties of ionic crystals. Following are the properties of crystals formed by ionic bonds.

(*i*) The ionic crystals are *strong, hard* and *brittle*. The cohesive energy of ionic crystals is very high, *i.e.*, the ionic bonds are very strong. Thus the ionic crystals have very high melting point and possess high latent heat of fusion. For example, binding energy of *NaCl* = 7.8 *eV* and of *LiI* = 10.4 *eV*.

Strong electrostatic forces between positive and negative ions make ionic crystals hard. However, when a shearing force is applied to an ionic crystal the ions tend to slip past one another with relatively more ease, as the ionic bonding is non-directional. A stage is reached when strong repulsive forces cause a fracture of the crystal, *i.e.*, the crystal is brittle.

(*ii*) The ionic crystals are generally insulators. Their electrical conductivity is very low at ordinary temperatures but increases with increase of temperature.

The ionic crystals consist of ions having closed shell structure, *i.e.*, they do not have free electrons. Hence they are electric insulators. The small conductivity present in the ionic crystals is due to the diffusive motion of ions and hence the conductivity of ionic crystals increases with increase in temperature.

(*iii*) Ionic crystals are normally transparent to visible light while they exhibit a single characteristic optical reflection peak in the infra-red region.

(*iv*) Their dielectric constant changes with the frequency of A.C. applied.

(*v*) They are generally brittle in nature and cannot be drawn into sheets or wires.

(*vi*) Polar liquids like water are able to dissolve many ionic crystals but they are not soluble in non-polar liquids like ether due to low value of di-electric constant.

19.13 POTENTIAL ENERGY OF IONIC CRYSTAL

To find the value of total binding energy and contribution to the potential of an ionic crystal we shall take up the typical case of sodium chloride. The force of attraction between the Na^+ and Cl^- ions is given by

$$F = \frac{Z_1 Z_2 e^2}{4\pi\varepsilon_0 r^2}$$

This force gives rise to an attractive potential energy term V_a the value of which in S.I units is given by

$$V_a = -\frac{Z_1 Z_2 e^2}{4\pi\varepsilon_0 r^2} = -\alpha \frac{e^2}{4\pi\varepsilon_0 r}$$

where α is a constant known as *Madelung constant*. The force of attraction between the Na^+ and Cl^- ions increases as the distance r between the two ions decreases. When they come *very* close to each other, their electron shells interact, as their core begin to overlap and Pauli's exclusion principle leads to a repulsive force which increases rapidly with decreasing inter-nuclear distance r. The repulsive energy term arising due to this force of repulsion is given by

$$V_r = \frac{B}{r^n}$$

The sign of the repulsive energy term is positive.

$\therefore$ Resultant potential energy $V = V_a + V_r = -\dfrac{\alpha e^2}{4\pi\varepsilon_0 r} + \dfrac{B}{r^n}$ ... (*i*)

At the equilibrium separation r_0, V has a minimum value

$$\therefore \quad \left(\frac{dV}{dr}\right)_{r=r_0} = \frac{\alpha e^2}{4\pi\varepsilon_0 r_0^2} - \frac{nB}{r_0^{n+1}} = 0$$

or
$$\frac{\alpha e^2}{4\pi\varepsilon_0 r_0^2} = \frac{nB}{r_0^{n+1}}$$

$$\therefore \quad B = \frac{\alpha e^2 r_0^{n-1}}{4\pi\varepsilon_0 n}$$

Hence
$$V = -\frac{\alpha e^2}{4\pi\varepsilon_0 r_0}\left(1 - \frac{1}{n}\right) \quad \text{... (ii)}$$

The total binding energy of a crystal having N positive and N negative ions is given by

$$V = \frac{N\alpha e^2}{4\pi\varepsilon_0 r_0}\left[1 - \frac{1}{n}\right] \quad \text{... (iii)}$$

In addition to the total potential energy V, as calculated above we must take into account the energy needed to transfer an electron from Na atom to a Cl atom to yield Na^+ and Cl^- ion. If we denote this energy by V_i, then the total potential energy, known as cohesive energy, is given by

$$V_c = V_a + V_r + V_i$$

in which V_a is negative and V_r and V_i are both positive.

From a knowledge of the compressibility of ionic crystals the value of n in relation (i) has been found to be $\simeq 9$.

Example 19.13. *If the total energy of an ionic crystals expressed as*

$$U = -\frac{Ae^2}{r} + \frac{B}{r^n}$$

Show that the equilibrium energy U_0 at $r = r_0$ is given by

$$U_0 = -\frac{Ae^2}{r_0}\left[1 - \frac{1}{n}\right].$$ **(H.P.U. 2003)**

Solution. Total binding energy of the ionic crystals

$$U = -\frac{Ae^2}{r} + \frac{B}{r^n}$$

This relation is the same as given by Eq. (i), where $A = \frac{\alpha}{4\pi\varepsilon_0}$ and $V = U$. Hence, the equilibrium energy at $r = r_0$ will be the same as given by Eq. (ii) by replacing $\frac{\alpha}{4\pi\varepsilon_0} = A$ and $V = U_0$.

$$\therefore \quad U_0 = -\frac{Ae^2}{r_0}\left[1 - \frac{1}{n}\right]$$

Example 19.14. *Calculate the cohesive energy per atom for NaCl crystal from the following data:*

Equilibrium distance between ions	$r_0 = 2.81$ Å
Madelung constant	$\alpha = 1.748$
Ionisation energy of	$Na = +\,5.1$ eV
Electron affinity for	$Cl = -\,3.6$ eV
	$n = 9$

Solution. The potential energy for ionic bonding which binds the $NaCl$ crystal is given by

$$V = V_a + V_r = -\frac{\alpha e^2}{4\pi\varepsilon_0 r_0}\left[1 - \frac{1}{n}\right]$$

where V_a is the potential due to attractive force, V_r potential due to repulsive forces, e the electronic charge and $\frac{1}{4\pi\varepsilon_0} = 9 \times 10^9$ N – m²/C²

$$\therefore \quad V = \frac{1.748\,(1.60\times10^{-19})^2\,(9\times10^9)\,(1-1/9)}{2.81\times10^{-10}}$$

$$= -\,1.27 \times 10^{-18} \text{ J} = -\,7.97 \text{ eV}$$

One-half of the potential energy represents the contribution *per ion.*

$$\therefore \quad V \text{ per ion} = -\,3.99 \text{ eV}$$

In order to remove the outermost electron from the *Na* atom and leaving it as Na^+ ion 5.1 *eV* of work must be done. This is known as *ionisation energy* of *Na*.

When this electron is brought to the *Cl* atom it absorbs it, to complete the outermost electron sub-shell and becomes Cl^- ion. The formation of Cl^- ion from *Cl* atom gives out 3.6 *eV* of energy. This is known as *electron affinity* of chlorine. The net result is that to form Na^+ and Cl^- ions from *Na* and *Cl* atoms, we have to spend 5.1 – 3.6 = 1.5 *eV* of energy. Thus $V_i = 1.5\ eV$.

$$\therefore \quad \text{Contribution of } V_i \text{ per ion} = \frac{1}{2}\ (+\ 1.5\ \text{eV}) = +\ 0.75\ \text{eV}$$

$$\text{Cohesive energy per ion} \quad V_c = V_a + V_r + V_i = V + V_i$$

$$= -\ 3.99 + 0.75 = -\ 3.24\ eV/\text{atom}$$

An empirical value of cohesive energy can also be obtained from measurements on heat of vaporisation, dissociation energy and electron exchange energy. The result thus obtained is 3.28 *eV*/atom for *NaCl* which is in close agreement with the above result.

19.14 LATTICE ENERGY OF IONIC CRYSTALS

Lattice energy is the energy evolved when a crystal is formed from individual ions (not atoms).

The calculation of lattice energy and cohesive energy of ionic crystals was made by Born and Madelung. According to this theory, let U_{ij} be the interaction energy between ions *i* and *j*, and U_i the interaction energy on the ion *i* due to all other ions, then

$$U_i = \sum_j U_{ij}$$

where the summation extends over all ions except $j = i$. The energy U_i consists of two parts.

(*i*) A central field *repulsive* potential supposed to be of the form $\lambda e^{-r/\rho}$ where λ and ρ are empirical parameters. ρ is a measure of the range of the repulsive interaction. When $r = \rho$ the repulsive interaction is reduced to $\frac{1}{e}$ of the value at $r = 0$.

(*ii*) A coulomb potential $\pm \frac{q^2}{r}$ (in C.G.S. units) depending upon the sign of the charges on the two ions under consideration.

$$\therefore \quad U_{ij} = \lambda e^{-(r_{ij}/\rho)} \pm q^2 / r_{ij}$$

The total lattice energy U_{tot} of a crystal composed of *N* molecules (or 2*N* ions)

$$U_{tot} = NU_i$$

We do not use 2*N* because interaction of *each pair* is to be considered only *once*.

For the sake of convenience we put

$$r_{ij} = p_{ij} R$$

where p_{ij} is a dimensionless quantity and *R* the nearest neighbour separation in the crystal. Considering the repulsive interaction only among nearest neighbours, we have $r_{ij} = R$ and

$$U_{ij} = \lambda e^{(-R/\rho)} - \frac{q^2}{R}$$

$$\therefore \quad U_{tot} = NU_i = N \sum U_{ij}$$

$$= N\left(z\lambda e^{(-R/\rho)} - \sum \frac{\pm\ q^2}{p_{ij}\ R} \right)$$

$$= N\left(z\lambda e^{-R/\rho} - \alpha.\frac{q^2}{R}\right) \quad \text{... (i)}$$

where z = the number of nearest neighbours for any ion and $\alpha = \sum \frac{\pm}{p_{ij}}$ = *Madelung constant* also known as coulomb energy constant. If we take the reference ion as a negative charge the plus sign will be used for positive ions and minus sign for negative ions.

At the equilibrium separation $R = r_0$ the crystal energy U_{tot} is a minimum.

$$\therefore \quad \frac{dU_{tot}}{dR} = 0$$

or
$$N\frac{dU_i}{dR} = -\frac{Nz\lambda}{\rho} e^{\left(-\frac{R}{\rho}\right)} + \frac{N\alpha q^2}{R^2} = 0$$

Putting $R = r_0$, we have

$$r_0^2 e^{(-r_0/\rho)} = \frac{\rho\alpha q^2}{z\lambda} \quad \text{... (ii)}$$

If the values of ρ and λ are known, the equilibrium separation r_0 can be calculated. Thus U_{tot}, the total lattice energy of the crystal of $2N$ ions at their equilibrium separation r_0 is given by putting $R = r_0$ in (*i*) and, we get

$$U_{tot} = N\left(z\lambda e^{-r_0/\rho} - \alpha\frac{q^2}{r_0}\right)$$

Now from (*ii*),
$$e^{(-r_0/\rho)} = \frac{\rho\alpha q^2}{r_0^2 z\lambda}$$

$$\therefore \quad U_{tot} = N\left(\frac{\rho\alpha q^2}{r_0^2} - \frac{\alpha q^2}{r_0}\right)$$

$$= -\frac{N\alpha q^2}{r_0}\left(1 - \frac{\rho}{r_0}\right)$$

The term $\frac{-N\alpha q^2}{r_0}$ is the *Madelung energy*.

Hence Potential energy per ion pair $U_{tot} = .\frac{-\alpha q^2}{r_0}\left(1 - \frac{\rho}{r_0}\right)$

In S.I. units U_{tot} per ion pair $= \frac{-\alpha q^2}{4\pi\varepsilon_0 r_0}\left(1 - \frac{\rho}{r_0}\right)$ and Madelung energy $= \frac{-N\alpha q^2}{4\pi\varepsilon_0 r_0}$

The value of for *NaCl* is of the order of 1/9, $q = e$ the electronic charge, $\alpha = 1.748$ and $r_0 = 2.81$ Å. Thus the calculated value of lattice energy per ion pair comes out to be – 7.97 *eV*. This is in close agreement with experimental value of – 7.96 *eV*/ion pair.

19.15 MADELUNG CONSTANT

The principal contribution to the cohesive energy of an ionic crystal is the electric potential energy V_a due to the electrostatic force of coulomb attraction given by

$$V_a = -\frac{Z_1 Z_2 e^2}{4\pi\varepsilon_0 r} = -\frac{\alpha e^2}{4\pi\varepsilon_0 r}$$

in S.I. units where $Z_1 e$ is the charge, say on the positive ion, $Z_2 e$ the charge on the negative ior separated by a distance r and α is a dimensionless constant known as *Madelung's constant*.

To find the value of Madelung constant α, we use the relation for potential energy per ion pair. This relation is given by

$$U_{tot} = \frac{-\alpha q^2}{r_0}\left(1-\frac{\rho}{r_0}\right)$$

where α is a constant known as Madelung constant, q is the charge on each ion, ρ is a constant giving a measure of the range of repulsive interaction and r_0 the equilibrium distance between the ions.

The Madelung constant $\alpha = \sum \frac{\pm 1}{p_{ij}}$ where p_{ij} is a dimensionless quantity given by the relation

$$r_j = p_{ij} R$$

R being the nearest neighbour separation in the crystal. If we take the reference ion as a negative charge, the plus (+) sign will be used for positive ion and minus (–) sign will be used for negative ion.

Since $r_j = p_{ij} R$, the value of α can be put in the form

$$\frac{\alpha}{R} = \sum \frac{\pm 1}{p_{ij} R} = \sum \frac{\pm 1}{r_j}$$

where r_j is the distance of the jth ion from the reference ion and is always taken as positive.

One dimensional lattice. Consider a one dimensional lattice of oppositely charged ions as shown in Fig 19.18. Take a negatively charged ion as the reference ion and let R be the distance between adjacent ions, then

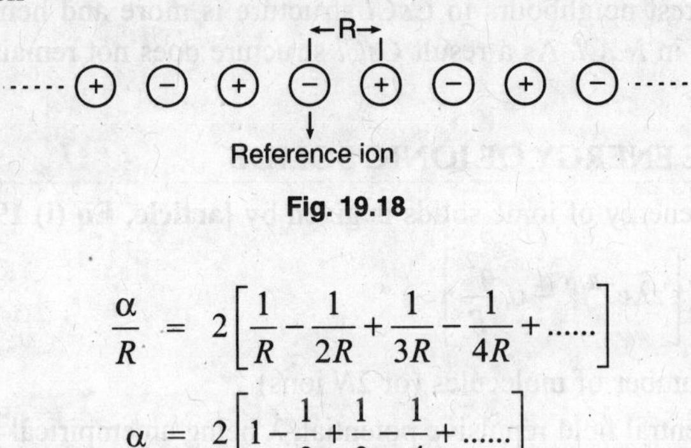

Fig. 19.18

$$\frac{\alpha}{R} = 2\left[\frac{1}{R}-\frac{1}{2R}+\frac{1}{3R}-\frac{1}{4R}+......\right]$$

or $$\alpha = 2\left[1-\frac{1}{2}+\frac{1}{3}-\frac{1}{4}+......\right] \quad ...(i)$$

The factor 2 has been taken due to the fact that there are two ions, one to the right and one to the left of the reference ion and each at a distance $r_j = n R$, where $n = 1, 2, 3...$. The series given in (i), can be summed by the series expression of

$$\log_e(1+x) = x-\frac{x^2}{2}+\frac{x^3}{3}-\frac{x^4}{4}+......$$

Putting $x = 1$, we get

$$\log_e 2 = 1-\frac{1}{2}+\frac{1}{3}-\frac{1}{4}+......$$

$$\therefore \quad \alpha = 2\left[1-\frac{1}{2}+\frac{1}{3}-\frac{1}{4}+......\right] = 2\log_e 2 \quad ...(ii)$$

$$= 2 \times 0.69 = 1.38$$

Hence for a one dimensional array of ions of alternate signs the value of Madelung constant $\alpha = 1.38$.

Madelung constant for NaCl crystal. In *NaCl* structure the nearest neighbours to the negative reference ion are six positive ions at $p_{ij} = p = 1$ giving a positive contribution to α of $\frac{6}{1}$. There are twelve negative ions at $p = \sqrt{2}$ giving a negative contribution $-\frac{12}{\sqrt{2}}$. Again there are eight positive ions at $p = \sqrt{3}$ giving a positive contribution of $\frac{8}{\sqrt{3}}$. Further there are six negative ions at $p = \sqrt{4}$ giving a negative contribution of $\frac{6}{\sqrt{3}}$ and so on.

$$\therefore \quad \text{Madelung constant } \alpha = \left[\frac{6}{\sqrt{1}} - \frac{12}{\sqrt{2}} + \frac{8}{\sqrt{3}} - \frac{6}{\sqrt{4}} + ...\right]$$

$$= -1.748$$

The *Madelung's constant* α has the same value for all crystals of the same structure. For Caesium chloride (*CsCl*), α = 1.763 and for *ZnS*, α = 1.638.

It is a dimensionless quantity, *i.e.*, it is a pure number and has no units. It is

$$= \left[\frac{6}{\sqrt{1}} - \frac{12}{\sqrt{2}} + \frac{8}{\sqrt{3}} - \frac{6}{\sqrt{4}} + ...\right]$$

Why is *NaCl* type structure more stable than *CsCl* type? The value of *Madelung constant* α as calculated as above is 1.748 for *NaCl*. The value of α for *CsCl* crystal is 1.763. Thus, we find that α for *CsCl* has slightly higher value than that of *NaCl* type structure. It may, therefore, appear that *CsCl* type structure should be preferred by the crystal but this is not so. It is because the number of nearest neighbours in *CsCl* structure is more and hence the repulsive energy is also more than that in *NaCl*. As a result *CsCl* structure does not remain more stable structures as compared to *NaCl*.

19.16 COHESIVE ENERGY OF IONIC SOLIDS

The cohesive energy of ionic solids is given by **[article, Eq (i) 19.14]**

$$U_{tot} = N\left[z\lambda e^{-R/\rho} - \alpha\frac{q^2}{R}\right] \quad ...(i)$$

where N = Number of molecules (or $2N$ ions)

$\lambda e^{-R/\rho}$ = Central field repulsive potential, λ being an empirical parameter and ρ a measure of the range of repulsive interaction

z = The number of nearest neighbours of an ion

$\frac{q^2}{R}$ = The coulomb potential (in C.G.S. units) due to the charge on the two ions, and

α = Madelung constant (or coulomb energy constant)

Bulk Modulus. The bulk modulus

$$k = -\frac{dp}{dV/V} \quad ...(ii)$$

where dp is a small change in pressure which produces a volumetric strain dV/V.

According to the first law of thermodynamics,

$$dQ = dU + pdV$$

For an *adiabatic compression, i.e.*, when no heat is gained or lost to the surroundings, $dQ = 0$ and

$$dU = -pdV$$

or $$p = -\frac{dU}{dV}$$

$$\therefore \quad \frac{dp}{dV} = -\frac{d^2U}{dV^2}$$

Substituting in Eq. (*ii*), we have

$$k = V\frac{d^2U}{dV^2}$$

Now, $$\frac{dU}{dV} = \frac{dU}{dR}\cdot\frac{dR}{dV}$$

$$\therefore \quad \frac{d^2U}{dV^2} = \frac{d^2U}{dR^2}\left(\frac{dR}{dV}\right)^2 + \frac{dU}{dR}\frac{d^2R}{dV^2} \qquad \text{... }(iii)$$

At the equilibrium distance $R = R_0$ and $\frac{dU}{dR} = 0$

$$\therefore \quad \frac{d^2U}{dV^2} = \frac{d^2U}{dR^2}\left(\frac{dR}{dV}\right)^2$$

Hence, $$k = V\frac{d^2U}{dR^2}\left(\frac{dR}{dV}\right)^2 \qquad [R = R_0] \qquad \text{... }(iv)$$

***NaCl* type structure.** For *NaCl* structure $R = \frac{a}{2}$ or $a = 2R$, where a is the lattice parameter.

The volume of the unit cell containing 4 molecules of *NaCl*, $a^3 = 8R^3$.

$\therefore$ Volume occupied by N molecules $V = \frac{N}{4}\cdot 8R^3 = 2NR^3$

Hence, $$\frac{dV}{dR} = 6NR^2$$

Substituting in Eq. (*iv*), we get

$$k = 2NR_0^3\frac{d^2U}{dR^2}\left(\frac{1}{36N^2R_0^4}\right)$$

$$= \frac{1}{18NR_0}\left[\frac{d^2U}{dR^2}\right]_{R=R_0} \qquad \text{... }(v)$$

From Eq. (*i*), we have

$$\frac{dU_{tot}}{dR} = \frac{-N\lambda z}{\rho}e^{-R/\rho} + \frac{N\alpha q^2}{R^2}$$

and $$\frac{d^2U_{tot}}{dR^2} = \frac{N\lambda z}{\rho^2}e^{-R/\rho} + \frac{2N\alpha q^2}{R^3}$$

Substituting in Eq. (*v*), we get

$$k = \frac{1}{18NR_0}\left[\frac{N\lambda z}{\rho^2}e^{-R_0/\rho} + \frac{2N\alpha q^2}{R_0^3}\right]$$

Now $$e^{-R_0/\rho} = \frac{\rho\alpha q^2}{R_0^2 z\lambda}$$ [For proof, **Eq. (*ii*) of article 1914**]

$$\therefore \quad k = \frac{1}{18NR_0}\left[\frac{N\lambda z}{\rho^2}\cdot\frac{\rho\alpha q^2}{R_0^2 z\lambda}+\frac{2N\alpha q^2}{R_0^3}\right]$$

or $$\text{Bulk modulus } k = \frac{\alpha q^2}{18R_0^4}\left[\frac{R_0}{\rho}-2\right]$$

This equation can also be used to find the value of ρ the range of repulsive interaction if R_0, α, q and k are known.

Example 19.15. *Given for NaCl crystal, r_0 = 2.81 Å, α = 1.75, ρ = 0.298 Å, e = 4.8 × 10^{-10} e.s.u. Calculate the binding energy per molecule.*

Solution. Potential energy (or binding energy) per ion pair (or molecule) in C.G.S. units is given by

$$U_{tot} = \frac{-\alpha q^2}{r_0}\left(1-\frac{\rho}{r_0}\right)$$

Here α = 1.75; $q = e = 4.8 \times 10^{-10}$ e.s.u, $r_0 = 2.81$ Å $= 2.81 \times 10^{-8}$ cm

$$\rho = 0.298 \text{ Å} = 0.298 \times 10^{-8} \text{ m}$$

$$\therefore \quad U_{tot} = \frac{-1.75\times4.8\times4.8\times10^{-20}}{2.81\times10^{-8}}\left[1-\frac{0.298}{2.81}\right]$$

$$= \frac{-1.75\times4.8\times4.8\times0.89\times10^{-12}}{2.81}$$

$$= 12.77 \times 10^{-12} \text{ ergs} = \frac{12.77\times10^{-12}}{1.6\times10^{-12}}$$

$$= 7.98 \ eV.$$

19.17 COVALENT BONDING

Atoms can also attain stable electron configuration by sharing the unpaired electrons amongst them.

The *type of bonding resulting from sharing of one or more electrons by the atoms is know as co-valent bonding*. Each atom participating in a covalent bond contributes an electron to the bond and these electrons are shared by both the atoms, rather than being virtually the exclusive property of one of them, as in ionic bond. The sharing takes place in such a way that an electron with *spin up* pairs with another electron with *spin down*, if that electron can occupy the state in accordance with Pauli's exclusion principle. In other words, we can say that *attractive potential* V_a arises due to the pairing of electrons with *anti-parallel spins* and the *repulsive* potential V_r arises due to electrons with *parallel spin*. The covalent bonding is also known as *shared electron pair bonding*.

The simplest example of electron sharing is the hydrogen molecule. When two hydrogen atoms come close to each other, each of the two electrons is attracted by both nuclei fig. 19.19. If the electrons have anti-parallel spins, they are capable of moving about both the nuclei and this movement results in the increase of density of electron cloud in the space between the nuclei and the electron clouds of the two overlap and coalesce to form a

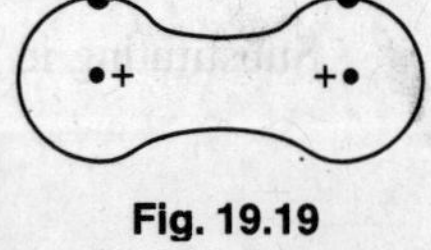

Fig. 19.19

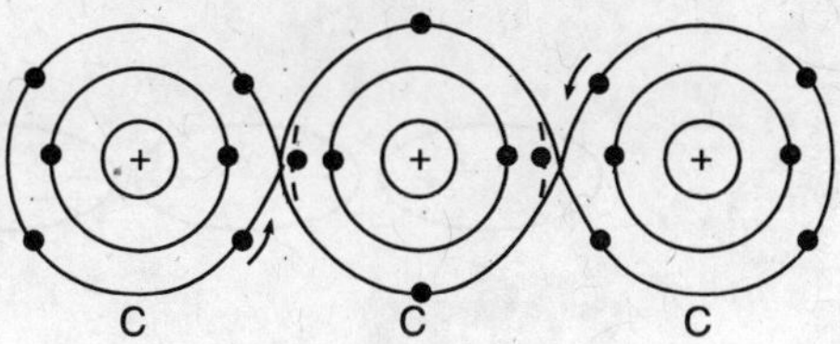

Fig. 19.20

sort of dumb-bell shaped molecular orbit with the two nuclei at a certain distance and two electrons with anti-parallel spins moving about in this orbit. The negatively charged region of high electron density, in between the positively charged nuclei, draws them closer and the potential energy of the system is lowered to its minimum, resulting into a stable structure. The separation of the two protons in the stable position, is of the order of 0.74 Å and the binding energy is of the order of 4.5 eV.

Examples. Purely covalent crystals are relatively few in number. Familiar examples are *diamond* (*carbon*), *silicon*, *germanium*, *SiC*. The covalent bond in carbon, formed by sharing of electrons is shown in Fig. 19.20.

Properties of covalent bonding. The main properties of covalent crystals are:

(*i*) These substances are made up of individual covalent molecules with weak intermolecular forces.

(*ii*) These substances are generally insoluble in ordinary liquids but soluble in ether and benzene etc.

(*iii*) Covalent crystals are very hard since the bond is very strong, *e.g.*, diamond.

(*iv*) Their dielectric constant is independent of applied field.

(*v*) They generally exhibit structural and space isomerism due to their directional nature.

(*vi*) The conductivity of covalent crystal varies over a wide range. Diamond is a good insulator, whereas germanium and silicon are semi-conductors and grey-tin is almost a conductor. The conductivity also increases with temperature.

Directional nature of covalent bond. For the formation of co-valent bond, there should be a net decrease in potential energy of the system as a result of mutual sharing of electrons. This happens when the participating orbits overlap effectively and also when the vacant electronic states are available in the outermost orbitals of each atom. The overlap is more effective when the participating orbitals are *directionally oriented* rather than having spherical symmetry.

In co-valent bonding the two sharing electrons have *anti-parallel spin* and the electron pair is concentrated along the line joining the two atoms. Hence, *a covalent bond is always directional in character*. Since *p*-orbitals are directional in nature, they readily participate in the formation of co-valent bonds.

19.18 VAN DER WAAL BONDING

The atoms of inert gases like those of *He* and *Argon* are spherically symmetrical and have no valence electron, as their outermost electron orbits are completely filled. They are, therefore, incapable of forming any bond. Also certain organic molecules like CH_4 and molecules containing atoms of halogens, oxygen, nitrogen, sulphur, phosphorous have the bonding capacity of all constituent atoms satisfied and the outer orbits of all these atoms in the molecules attain inert gas configuration. However, these molecules condense to liquid and solid state with a decrease of energy. To explain this, Van der Waal suggested some type of attractive force between these molecules and atoms of inert gases. These are weak attractive forces which become relatively stronger in the liquid or solid states. The Van der Waal attraction between two molecules, a distance r apart, is proportional to r^{-7} so that it is significant only for molecules very close together. Thus Van der Waal force is a short range force falling rapidly as the atoms separate.

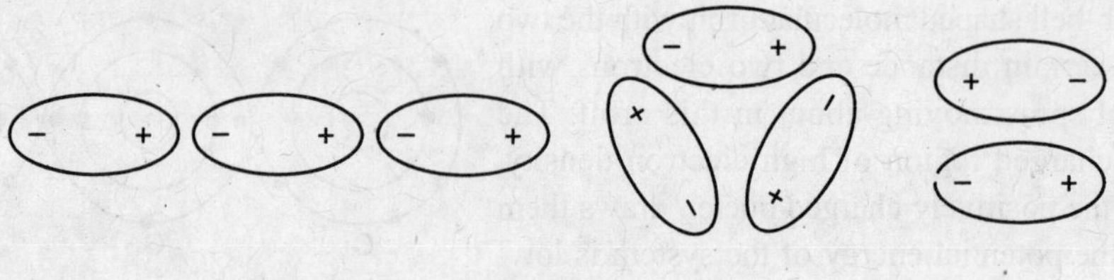

Fig. 19.21

To explain the mechanism of these weak attractive forces, let us first consider *polar molecules, i.e.*, those molecules which have a permanent electric dipole moment of which H_2O molecule is an example. In H_2O, the concentration of electrons around the oxygen atom makes that end of the molecule more negative than the end where the hydrogen atoms are. Such molecules tend to align themselves with ends having opposite kind of charge adjacent to each other because in this orientation the molecules strongly attract each other as shown in Fig. 19.21.

Let us next consider a *polar* and a *non-polar* molecule as shown in Fig. 19.22.. The electric field of the polar molecule causes a separation of charge in the other (non-polar) molecule so that the positive charge lies nearer to the negative end, thus resulting in an attractive force.

Even two *non-polar* molecules can attract each other by the above process. Suppose two non-polar molecules are extremely close to each other. The nuclei and the electrons of these atoms are in ceaseless motion. It may be imagined that the electron cloud may be concentrated at one end of the atom or molecule at a certain moment and a fraction of a second later, it may be located at the other end of the atom or molecule. This instantaneous concentration of the electron cloud sets up a temporary dipole called the *instantaneous dipole*. The energy of the system is decreased by an attractive interaction between two such instantaneous dipoles. The dipole in atom *A* induces a similar dipole in the adjacent atom *B* as shown in Fig. 19.23 and the electron cloud shifts to the other side, pulling back the electron cloud in *B*. These temporary dipoles, both oriented in the same direction, lead to an attractive force between the atoms or molecules. These forces arise from the fact that the electrons in adjacent atoms or molecules are oscillating to produce fluctuating dipoles which produce immediate attraction with the decrease of energy of the system. Thus Van der Waal forces are caused by dipole-dipole and dipole-induced dipole interaction. These forces depend upon the following factors:

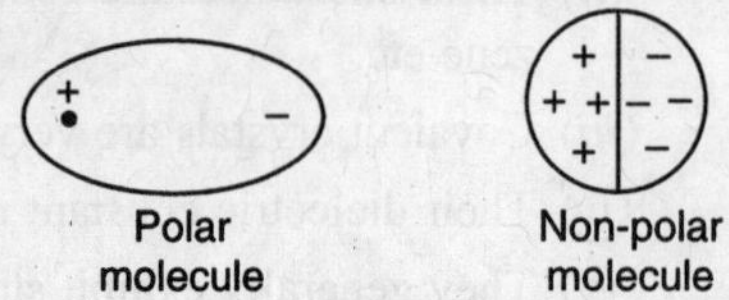

Fig. 19.22

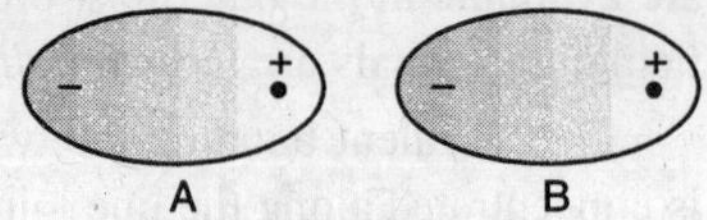

Fig. 19.23

(*i*) *Number of electrons present in the molecule*. The forces increase with the increase in number of electrons in a molecule.

(*ii*) *Molecular size*. The larger the molecular size, greater will be the surface area and stronger is the Van der Waal interaction.

(*iii*) *Effect of pressure*. If the pressure is increased, the molecules are pressed together, resulting in a stronger attraction.

(*iv*) *Condition of temperature*. Lower the temperature lesser is the movement of the molecules and stronger the molecular interactions and hence greater is Van der Waal force.

Van der Waal bonds are the weakest. Van der Waal bond arises due to some type of attractive force between the molecules or the atoms of inert gases. This is a very weak force as Van der Waal attraction between two molecules a distance r apart is proportional to r^{-7}. As such it is significant only for molecules very close together. This is a *short range* force falling rapidly as the atoms separate. Because of this *short range, weak force*, Van der Waal bond is very weak.

Physical characteristics of inert gas crystals. Inert gas crystals have Van der Waal bonding. Their characterisation are:

(*i*) As Van der Waal force is weak, the inert gas crystals have low melting and low boiling points.

(*ii*) The inert gas crystals change their state easily with a small amount of thermal energy.

(*iii*) Van der Waal forces in inert gas crystals give rise to properties like friction, surface tension, viscosity, cohesion, adhesion etc.

(*iv*) Due to Van der Waal forces, there is always a difference between the pressure exerted by a real gas and that of an ideal gas.

(*v*) The inert gas crystals usually adopt cubic close packed *fcc* structure except He^3 and He^4. These crystals are insulators and are transparent to electromagnetic radiation down to ultraviolet range.

(*vi*) The inert gas crystals are brittle and lack strength.

(*vii*) The inert gas crystals have very high ionisation energy.

Inert gas Crystals Exist Only at Very low Temperature

The inert gas crystals are formed due to Van der Waal's bonding. The nuclei and the electrons of these atoms are in ceaseless motion. It is possible that the electron cloud of these atoms may be concentrated at one end of the atom (or molecule) at a certain moment and a fraction of a second later, it may be located at the other end of the atom (or molecule). This instantaneous concentration of the electron cloud sets up a temporary dipole. The electrons in adjacent atoms or molecules are oscillating to produce fluctuating dipoles which produce immediate attraction with decrease of energy of the system. This type of dipole induced dipole interaction is called Van der Waal bond. It is non-directional in character.

The Van der Waal's forces are very weak forces being of the order of 0. 1 *eV* per bond only. At room temperature the thermal energy acquired by these atoms (or molecules) is sufficient to make these forces ineffective. Thus inert gas molecular solids exist in gaseous state at room temperature. At very low temperatures Van der Waal's forces dominate over the thermal forces and play a significant role in the transformation of gaseous state to liquid and solid state.

This is why the inert gas crystals exist only at very low temperatures and not at room temperature.

19.19 INTERACTION IN INERT GAS CRYSTALS

The interaction responsible for the binding among inert gas atoms in inert gas crystals consists of two parts:

(*i*) *Attractive interaction*. It is known as Van der Waals-London interaction, and

(*ii*) *Repulsive interaction*.

(*i*) **Attractive interaction.** Consider two identical inert gas atoms 1 and 2 separated by a distance R large as compared to the radii of the atoms. The atoms of inert gas induce instantaneous dipole moments which gives rise to an attractive interaction between the atoms.

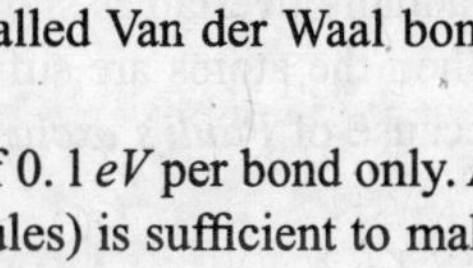

Fig. 19.24

Let p_1 be the magnitude of the instantaneous electric dipole moment induced in atom 1 due to the presence of atom 2 at any instant. This dipole moment produces an electric field E at the centre of atom 2, the magnitude of which in C.G.S. units is given by

$$E = \frac{2p_1}{R^3} \qquad ...(i)$$

as the centre of atom 2 lies on the axial line of atom 1. This electric field will induce an instantaneous dipole moment of magnitude p_2 in C.G.S. units given by

$$p_2 = \alpha E$$

where α is *electronic polarisability* defined as the dipole moment induced in an atom by a unit electric field. Substituting the value of E from (*i*), we have

$$p_2 = \frac{2\alpha p_1}{R^3}$$

The potential energy of the dipole moments is given by

$$-\frac{2p_1p_2}{R^3} = -\frac{2p_1}{R^3}\cdot\frac{2\alpha p_1}{R^3} = -\frac{4\alpha p_1^{\,2}}{R^6} = -\frac{C}{R^6}$$

where $$C = 4\alpha p_1^{\,2} = \text{a constant.}$$

The minus sign indicates that the interaction is *attractive* and varies inversely as the sixth power of the separation between atomic oscillators.

(In S.I. units the potential energy of dipole moments is given by $-\dfrac{1}{4\pi\varepsilon_0}\dfrac{4\alpha p_1^{\,2}}{R^6}$.)

It is known as *Van der Waal interaction, the London interaction* or *the induced dipole interaction.* It is the principal attractive interaction in crystals of inert gases as also in crystals of many organic molecules.

(*ii*) **Repulsive interaction.** As the two atoms are brought closer, their charge distributions gradually overlap as shown in Fig. 19.25, thereby changing the electrostatic energy of the system. When the atoms are sufficiently close to each other the overlap energy is repulsive in large part because of *Pauli's exclusion principle.*

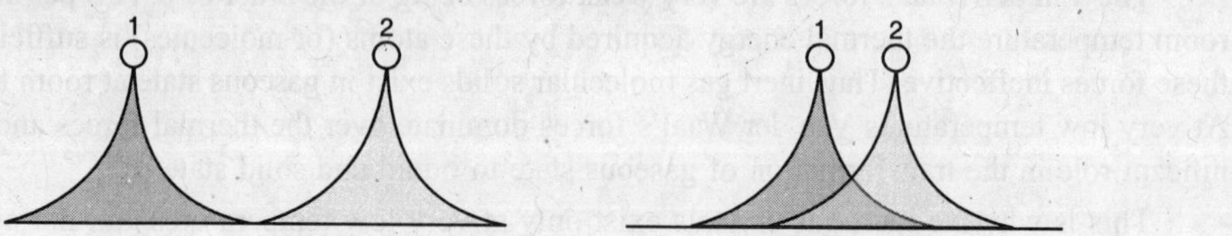

Fig. 19.25

According to this principle, no two electrons can have all their quantum numbers equal. When the charge distributions of two atoms overlap, there is a tendency for electrons from atom 2 to occupy in part states of atom 1, already occupied by electrons of atom 1 and *vice-versa.* As more than one electron cannot occupy a single state, the electron distribution of atoms with closed shells can overlap only if accompanied by the shifting of some electrons to the unoccupied high energy states of the atom, thereby increasing the total energy of the system. This gives a *repulsive contribution* to the interaction between the two atoms. Experimental data on the inert gases can be fitted well if we assume an empirical repulsive potential of the form $\dfrac{B}{R^{12}}$ where B is a positive constant.

19.20 LENARD-JONES POTENTIAL ENERGY

The total potential energy of separation of the two atoms at a distance R is given by

$$U(R) = 4\varepsilon\left[\left(\frac{\sigma}{R}\right)^{12} - \left(\frac{\sigma}{R}\right)^{6}\right] \qquad \text{... (ii)}$$

where ε and σ are new parameters with $4\varepsilon\sigma^{12} = B$ and $4\varepsilon\sigma^6 = C$. The potential given by relation (*ii*) is known as *Lenard-Jones potential*. The plot of $U(R)/4\varepsilon$ against R/σ is shown in Fig. 19.26. The minimum occurs at = 1.12. Before the minimum is reached the curve is very steep but after the minimum has been reached the curve is flat. The value of U at the minimum is $-\varepsilon$ and $U = 0$ at $R/\sigma = 1$ or at $R = \sigma$.

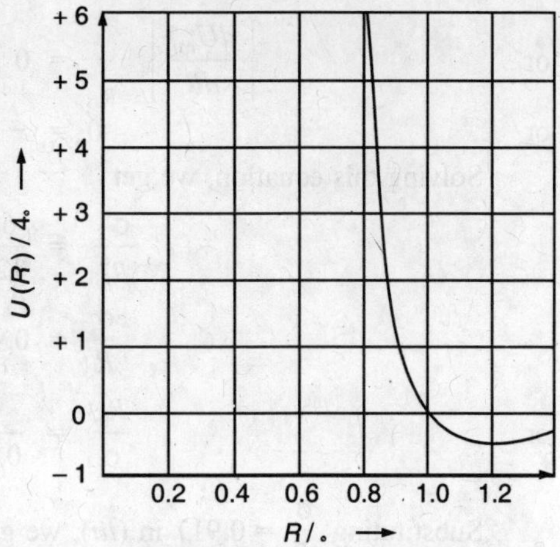

Fig. 19.26

19.21 COHESIVE ENERGY OF AN INERT GAS CRYSTAL

A crystal can be stable only when its total energy is less than the total energy of the atoms (or molecules) when free. The difference

Free atom energy – crystal energy

is known as the cohesive energy of the crystal.

Lenard Jones potential. On the basis of attractive Van der Waal's or London interaction and repulsive interaction at short range due to Pauli's exclusion principle, it is found that resultant Lenard Jones potential is given by

$$U(R) = 4\varepsilon\left[\left(\frac{\sigma}{R}\right)^{12} - \left(\frac{\sigma}{R}\right)^{6}\right]$$

where ε and σ are the parameters, the values of which can be obtained from measurements made in the gas phase of the crystal and R is the nearest neighbour distance.

Expression for cohesive energy. If the kinetic energy of the inert gas atoms is neglected, then the cohesive energy of an inert gas crystal is given by summing the Lenard Jones potential over all pairs of atoms in the crystal. If there are N atoms in the crystal, then the total energy is given by

$$U_{tot} = \frac{1}{2} N \sum U(R)$$

or

$$U_{tot} = \frac{1}{2} N (4\varepsilon)\left[\sum_j \left(\frac{\sigma}{p_{ij} R}\right)^{12} - \sum_j \left(\frac{\sigma}{p_{ij} R}\right)^{6}\right] \quad \text{... (i)}$$

where p_{ij} is a dimensionless quantity and $p_{ij} R$ is the distance between reference atoms i and any other atom j, R being the nearest neighbour distance. The factor 1/2 appears to avoid counting each pair of atoms twice.

For *fcc* structure $\sum_j p_{ij}^{-12} = 12.132$ and $\sum_j p_{ij}^{-6} = 14.454$ ... (*ii*)

Substituting these values of $\sum p_{ij}^{-12}$ and $\sum_j p_{ij}^{-6}$ in equation (*i*), we get

$$U_{tot} = 2N\varepsilon\left[12.131\left(\frac{\sigma}{R}\right)^{12} - 14.454\left(\frac{\sigma}{R}\right)^{6}\right] \quad \text{... (iii)}$$

There are 12 nearest neighbour sites in *fcc* structure. These nearest neighbours contribute most of the interaction energy of the inert gas crystals.

The equilibrium value of $R = R_0$ is given by imposing the condition that U_{tot} be minimum with respect to variation in the nearest neighbour distance R

or $$\left|\frac{dU_{tot}}{dR}\right|_{R=R_0} = 0$$

or $$0 = -2N\varepsilon\left[12\times 12.131\frac{\sigma^{12}}{R^{13}} - 6\times 14.454\frac{\sigma^6}{R^7}\right]$$

Solving this equation, we get

$$\frac{\sigma^6}{R^6} = \frac{6\times 14.454}{12\times 12.131} = 0.5957$$

$$\therefore \quad \frac{\sigma}{R} = 0.917$$

or $$\frac{R_0}{\sigma} = \frac{1}{0.917} = 1.09$$

Substituting $\frac{\sigma}{R} = 0.917$ in (*iii*), we get the cohesive energy of inert gas crystal at absolute zero and zero pressure at $R = R_0$

$$U_{tot}(R_0) = -(2.15)\, 4N\varepsilon$$

This is the calculated value of binding energy, if the kinetic energy of the atom is zero. This relation is true for all inert gases as we have not assumed any particular gas in our discussion.

Example 19.16. *If the potential energy function is expressed as*

$$U = \frac{M}{R^{12}} - \frac{N}{R^6}$$

Show that (i) The intermolecular distance for which potential energy is minimum is given by $\left(2\frac{M}{N}\right)^{1/6}$ *and (ii) The minimum potential energy* $U_{min} = \frac{-N^2}{4M}$ **(H.P.U. 2002)**

Solution. (*i*) For minimum potential energy

$$\left|\frac{dU}{dR}\right|_{R=R_0} = 0$$

Now, $$\frac{d}{dR}\left[\frac{M}{R^{12}} - \frac{N}{R^6}\right] = -12R^{-13}M + 6R^{-7}N$$

At $R = R_0$, $$-12R_0^{-13}M + 6R_0^{-7}N = 0$$

or $$\frac{-2M}{R_0^{\ 6}} + N = 0$$

$$\therefore \quad R_0 = \left(2\frac{M}{N}\right)^{1/6}$$

(*ii*) The potential energy is minimum for $R = R_0$.

$$\therefore \quad U_{min} = \frac{M}{R_0^{\ 12}} - \frac{N}{R_0^{\ 6}} = \frac{1}{R_0^{\ 6}}\left[\frac{M}{R_0^{\ 6}} - N\right]$$

$$= \frac{N}{2M}\left[M\times\frac{N}{2M} - N\right]$$

$$= -\frac{N}{2M}\times\frac{N}{2} = -\frac{N^2}{4}M$$

19.22 HYDROGEN BONDING

A specially strong type of Van der Waal bond occurs between certain molecules containing hydrogen atoms. The hydrogen atom, having a single electron, loses the same to the other constituent atom (or atoms) leaving behind a poorly shielded proton, which holds together two electro-negative atoms with a binding energy of the order of 0.1 eV. The H^+ ion (proton) can have only two near neighbour negative ions because its radius is 10^{-15} m which is thousands of times smaller than the radius of other ions. The Van der Waal forces are, therefore, very strong and vary in accordance with inverse square law or

$$V \text{ (Hydrogen bond)} \propto -\frac{1}{r^2}$$

Examples. The hydrogen bond is important in ice (solid H_2O), in hydrogen fluoride and in many organic molecules particularly proteins and genesis (like *DNA* molecules). If we study the case of $NaHF_2$ we find that hydrogen loses its only electron to either of F atoms with equal probability. [Fig. 19.27]. As explained above, the small size of the proton allows the bond only with F atoms. The Na atom in $NaHF_2$ loses its only valence electron to the other F atom and does not stay bound with H atom. In a typical ice crystal, each oxygen atom is bound by connecting links between four hydrogen atoms with each hydrogen atom linked between two oxygen atoms.

19.23 METALLIC BONDING

In the case of metals, the ionisation energy is low. They have vacant valency orbitals and have a few valency electrons as compared to the number of valency orbitals. Thus in the atoms of metals, the electrons in the outermost orbits are loosely bound and are called *free electrons* or *conduction electrons*. One or two such electrons can be detached from the parent atom due to the attraction of adjacent atomic cores. The remainder portion of atom, which is a positively charged sphere is called a *kernel*. The free electrons are mobile in nature and move from one kernel to another throughout the metal lattice. Thus the metal crystal may be pictured *as an arrangement of positive ions immersed in a sea of mobile electrons*. The motion of these electrons is such that they are simultaneously near to two or more kernels and bind them together. The bond thus formed between the metal atoms is known as **metallic bond** and is shown in Fig. 19.28. *Hence a metallic bond may be defined as a result of simultaneous attraction of an electron by two or more than two positive ions of the metal*. Thus a metallic bond is also electrostatic in nature though partially. Because of proximity of the ions (positive), each free electron is on an average close to one nucleus or another, than it would have been if it belonged to the isolated atom. This results in the decrease of potential energy of the electron forming the crystal. It is this decrease in the energy which is responsible for the metallic bond.

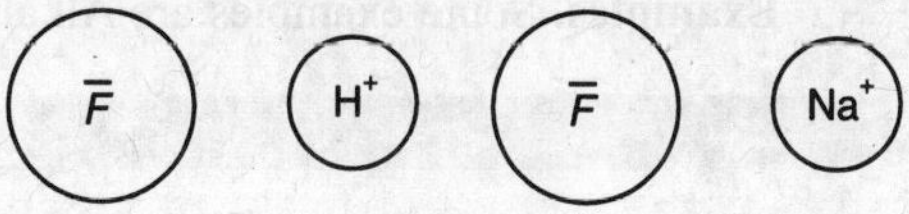

Fig. 19.27

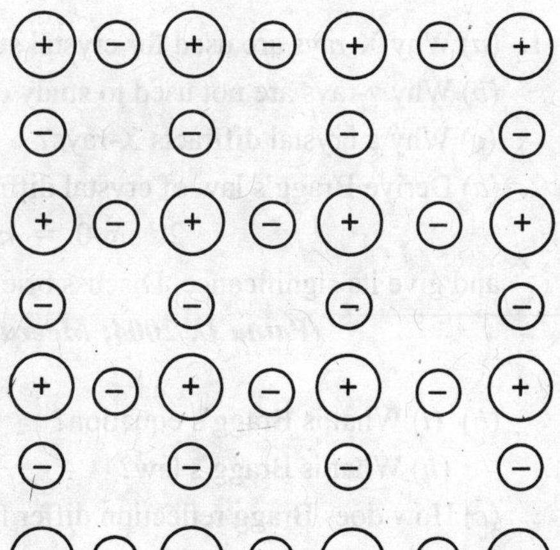

Fig. 19.28

The covalent bond is of directional character because the electrons are localised in a way that fixes the positions of the atoms rigidly and these electrons tend to remain concentrated in the region between the two atoms. But in case of metallic bonds the electrons are spread uniformly throughout the crystal and do not exert directional influence.

The metallic bond is weaker than the covalent bond due to the fewer electrons bonding the nuclei. However, it becomes stronger for those metals in which the number of valence electrons is greater. Thus strength of metallic bond tends to increase as we move from left to right along a row in the periodic table.

Properties of metallic bonding: Following are the properties of metallic crystals formed by the metallic bonding:

(*i*) They are good conductors, as free electrons are accelerated under a potential difference.

(*ii*) Heat is produced in the metals when an electric current is passed. This is due to the collision of free electrons with atomic cores under the influence of current. The increase in the average K.E. of the cores results in heating.

(*iii*) The electrical resistance increases with the heating of metals, because the heating increases the probability of collisions between free electrons and atomic cores.

(*iv*) Metal crystals are of unlimited size due to the unsaturated nature of the bonds.

(*v*) As the metals have uniformly positive ions placed in uniform electron gas, the displacement of ions produces same environments. Hence metals can be deformed and they are malleable and ductile.

(*vi*) Metals have high tensile strength because of stronger attraction between positive ions and electrons.

(*vii*) They have high thermal and electrical conductivity because of the presence of free electrons.

(*viii*) The properties of an alloy formed by a mixture of metals do not depend upon the relative proportion of the metals because atoms interact through the medium of common electron gas.

(*ix*) The opacity and lustre of metals is also due to the presence of free electrons in bonding. These electrons first absorb energy from the electromagnetic radiations and then radiate the same producing lustrous appearance.

Examples. Some examples are Alkalie metals like *Na* and *K*.

EXERCISE

LONG QUESTIONS

1. (*a*) Why X-rays are used for crystal structure analysis?

(*b*) Why γ-rays are not used to study crystal structure?

(*c*) Why a crystal diffracts X-rays? ***(M.D.U. 2001)***

2. (*a*) Derive Bragg's law of crystal diffraction

$$2d \sin \theta = n\lambda$$

and give its significance. Discuss briefly the method of crystal structure determination.

(Patna U. 2004; Meerut U. 2003; Pbi. U. 2003, 2000; K.U. 2001, 2000, 1996; Nagpur U. 2009, 2011; Luck. U. 2001; Osm. U. 2004)

(*b*) (*i*) What is Bragg's equation? ***(Pbi. U. 2004)***

(*ii*) What is Bragg's law? ***(P.U. 2004; Meerut U. 2003, 2002; Bang. U. 2000)***

(*c*) How does Bragg reflection differ from ordinary reflection? ***(Pbi. U. 2000)***

3. State and prove Bragg's law of X-ray diffraction and verify it for NaCl crystal from the following data

Bragg angles 11.8°; 23.5°; 36°

4. Derive Laue's equations of diffraction of X-rays and obtain Bragg's diffraction condition from them.

(G.N.D.U. 2004, 2003, 2002, 2001; H.P.U. 2003, 2001; P.U. 2003, 2001; Pbi. U. 2002; Meerut U. 2001; Cal. U. 2001)

5. (*a*) Name the standard experimental methods of X-ray diffraction.

(M.D.U. 2000; Pbi. U. 2000; G.N.D.U. 2000)

(*b*) Describe the various experimental X-ray diffraction methods for study of crystal structure.

(H.P.U. 2002; G.N.D.U. 2001)

6. Describe in detail (*i*) Laue's method. **(P.U. 2002, H.P.U. 1996)**

(*ii*) Rotating crystal method. How do layer lines form?

(Pbi. U. 2004, 2003, 2002, 2000; Meerut U. 2002)

(*iii*) Powder method. The diameter of a powder camera is 57.3 mm or a multiple of that. Explain. **(Cal. U. 2002; Bang. U. 2001; Meerut U. 2002)**

7. Describe the usefulness of Laue's powder and rotating crystal methods in the study of crystal structure. **(P.U. 2000; G.N.D.U. 2000; H.P.U. 2000)**

8. What is atomic scattering factor? Derive an expression for it. Explain its significance.

(P.U. 2004, 2000; H.P.U. 2000; G.N.D.U. 2000;)

9. What is reciprocal lattice? Why it is so named? Derive the expression for the primitive translation vectors of the reciprocal lattice. How is reciprocal lattice constructed? List its important properties.

(G.N.D.U. 2004, 2002; Nagpur U. 2012; P.U. 2001, 2000; H.P.U. 2000; Meerut U. 2001)

10. (*a*) What is reciprocal lattice? Discuss how this concept helps in the Ewald's construction and determination of crystal structure.

(*b*) What is Bragg diffraction condition? Explain it in terms of reciprocal lattice.

(P.U. 2003)

(*c*) Starting from $2\vec{k}\cdot\vec{G} = G^2$ obtain the diffraction condition $2d \sin \theta = n\lambda$. **(P.U. 2000)**

11. (*a*) Explain the term bonding. Giving suitable examples, describe how crystals have been classified according to their bonding. What are different kinds of bonding?

(Patna Univ. 2004)

(*b*) Define cohesion energy as applied to solids. Explain the formation of a stable bond using potential energy versus interatomic distance curve. Show that the formation of a chemical bond requires that the repulsive forces be of a shorter range than the attractive one.

(Pbi., Univ. 2003, 2002, 2001; G.N.D.U. 2003)

12. The energy of two particles in the field of each other at a separation r is given by

$$U = \frac{\alpha}{r} + \frac{\beta}{r^8}$$

where α and β are constants. At what separation they will form a stable compound?

13. The energy of two particles in the field of each other is given by

$$U = \frac{\alpha}{r^2} + \frac{\beta}{r^{10}}$$

where α and β are constants and r, the separation between the particles. Determine the separation between the particles for a stable compound and also determine the cohesive energy of the crystal. **(G.N.D.U. 2000)**

14. What are ionic crystals and ionic bonding? State the properties of ionic crystals and explain:

(*i*) Ionic crystals are hard but brittle.

(*ii*) Ionic solids are poor conductors.

(P.U. 2004; Pbi. U 2004, 2001; H.P.U. 2002; Bang. U. 2002, 2001; G.N.D.U. 2000)

15. (*a*) What are the different contributions to the potential of an ionic crystal? Obtain an expression for total binding energy of ionic crystals.

(*b*) If the total energy of an ionic crystal is expressed as

$$U = \frac{Ae^3}{r} + \frac{B}{r^n}$$

show that the equilibrium energy U_0 at $r = r_0$ is given by

$$U_0 = -\frac{Ae^2}{r_0}\left[1 - \frac{1}{n}\right]$$

(H.P.U. 2003)

16. What are ionic crystals? Derive an expression for lattice energy (cohesive energy) of ionic crystals from Born's theory. Discuss the result by comparing calculated and experimental values in a typical case. What is Madelung energy? ***(P.U. 2004, 2000; G.N.D.U. 2004, 2002; H.P.U. 2002; Luck. U. 2001)***

17. (*a*) What is Madelung constant? What are its units? Calculate the value of Madelung constant for a linear ionic crystal and show that for a one dimensional array of ions of alternate signs its value is 1.38.

(*b*) How do you determine the Madelung constant for a *NaCl* crystal?

(P.U. 2004, 2000; H.P.U. 2002; G.N.D.U. 2003)

(*c*) Why is *NaCl* type structure more stable than *CsCl* type structure? ***(Luck. U. 2002)***

18. Find the equilibrium cohesive energy of ionic solids and determine expression for bulk modulus. Calculate the value of bulk modulus for *NaCl* type structure.

(Luck. U. 2002)

19. Explain the process of Van der Waal bonding in crystals. Explain Van der Waal bonds are the weakest. Point out the physical characteristics of inert gas crystals.

20. Explain why the inert gas crystals exist only at very low temperature and not at room temperature.

(H.P.U. 2001)

21. What type of interaction is responsible for the binding among inert gas atoms in the inert gas crystals. Explain the origin of both the attractive and repulsive part of the interaction. Discuss the Lenard Jones potential energy between two inert gas atoms.

(G.N.D.U. 2000)

22. Define cohesive energy for inert gas crystal. State Lenard Jones two body potential and use it to obtain expression for the cohesive energy of an inert gas crystal.

(G.N.D.U. 2001, 2000; H.P.U. 2003, 2001, 2000; P.U. 2001, 2000)

23. Explain how crystals have been classified according to their bonding. State the properties of crystals of each type. Give examples. ***(Luck. U. 2001)***

24. What do you understand by ionic, co-valent, Van der Waal, hydrogen and metallic bonds? Explain.

(Pbi. U. 2004; G.N.D.U. 2002; Bang. U. 2000; Meerut U. 2002, 2001)

SHORT QUESTIONS

1. Give the characteristics of co-valent bond. What is cause of directional nature of co-valent bond? *(G.N.D.U. 2004; P.U. 2001)*
2. All metals are opaque to light of all wavelengths and have high lustre. Explain. *(P.U. 2003, 2001, 2000; G.N.D.U. 2000)*

   [**Hints:** The opacity and lustre of metals is due to the presence of free electrons in metallic bonding. The electrons first absorb energy from light and other electromagnetic radiation and then radiate the same producing lustrous appearance.]

   As the light is completely reflected metals are opaque to light of all wavelengths.
3. **Metal crystals shear easily. Explain.**

   [**Hint:** Metal crystals shear easily as they are held together by a cloud of free electrons and not by covalent bonds. The energy of these electrons is affected remarkably by change in volume so that there is, in general, good *cleavage strength.* The energy is affected very little by the movement of atoms which does not change the volume making the metals malleable and ductile.

   Again metals have uniformly positive ions placed in uniform electron gas. Therefore, the displacement of ions produces the same environment. Hence metals can be easily deformed, *i.e.,* they are malleable and ductile.]

MULTIPLE CHOICE QUESTIONS

Mark the correct answer:

1. *bcc* lattice is reciprocal of
   (*a*) *bcc* lattice itself (*b*) *sc* lattice
   (*c*) *fcc* lattice (*d*) none of these.
2. The primitive translation vectors in the reciprocal lattice have dimensions of
   (*a*) length (*b*) $(\text{length})^2$
   (*c*) $(\text{length})^{-1}$ (*d*) $(\text{length})^{-2}$.
3. Transfer of electrons takes place in
   (*a*) ionic bonding (*b*) covalent bonding
   (*c*) metallic bonding
4. An ionic molecule is a
   (*a*) polar molecule (*b*) non-polar molecule
   (*c*) none of these

ANSWERS

1. (*c*) 2. (*c*) 3. (*a*) 4. (*a*)

MAGNETIC PROPERTIES OF SOLIDS

20 Chapter

INTRODUCTION

In our book 'physics for Degree students, B. Sc. First Year', in chapter no. 22 (article 22.16-22.21), we have discussed magnetisation, magnetic susceptibility, (χ_m), and permeability (μ). We have also studied the classification of magnetic materials on the basis of their susceptibility and permeability. Let us start with their brief introduction and then we shall study dia-, para-, and ferromagnetism, ferromagnetic domains and hysteresis.

20.1 INTENSITY OF MAGNETISATION

Intensity of magnetisation also called *magnetisation* is defined as *magnetic moment per unit volume developed inside a solid*. It is denoted by $\vec{M}$.

If $\vec{H}$ is the strength of the applied magnetic field also called *magnetic field intensity which induces a magnetisation $\vec{M}$ in the material, then the magnetisation produced per unit applied magnetic field is called magnetic susceptibility* χ_m.

$$\therefore \quad \text{Magnetic susceptibility } \chi_m = \frac{\vec{M}}{\vec{H}} \quad \text{...(i)}$$

The magnetic induction or magnetic flux density $\vec{B}$ produced inside the material as a result of applied magnetic field $\vec{H}$ is given by

$$\vec{B} = m_0\left(\vec{H}+\vec{M}\right) \quad \text{... (ii)}$$

where μ_0 is the permeability of free space or vacuum and has a value $4\pi \times 10^{-7}$ Henry per metre (Hm^{-1}). The quantity $\vec{B}$ is measured in Weber per metre square ($Wb\ m^{-2}$) or Tesla T.

Putting $\vec{M} = \chi_m \vec{H}$ in Eq. (*ii*), we get

$$\vec{B} = \mu_0\left(\vec{H}+\chi_m \vec{H}\right)$$

or

$$\vec{B} = \mu_0 (1+X_m) \vec{H} \quad \text{...(iii)}$$

or

$$\vec{B} = \mu \vec{H} \quad \text{...(iv)}$$

where μ is the absolute permeability of the medium. The relative permeability of a medium μ_r is the ratio of absolute permeability of the medium to the permeability μ_o of the vacuum or free space

$$\therefore \quad \mu_r = \frac{\mu}{\mu_0}$$

or $\mu = \mu_0 \mu_r$

Hence we can put $\vec{B} = \mu_0 \mu_r \vec{H}$

From Equation (iii), we get

$$\mu_r = 1 + \chi_m \quad ...(v)$$

For free space $\vec{B} = \mu_0 \vec{H}$ ...(vi)

For paramagnetic materials $\chi_m = \chi_p$ is a positive quantity and for diamagnetic materials $\chi_m = \chi_d$ is a negative quantity. For ferromagnetic materials $\mu >> 1$, for paramagnetic $\mu > 1$ and for diamagnetic $\mu < 1$ (but not negative).

[In C. G. S units $\mu = 1 + 4\pi\, \chi_m$ and $\mu_0 = 1$ for vacuum or free space. Also $\vec{B} = \vec{H} + 4\pi \vec{M}$]

20.2 CLASSIFICATION OF MAGNETIC MATERIALS

Magnetic materials have been classified into *diamagnetic, paramagnetic* and *ferromagnetic* on the basis of their *permeability* and *susceptibility*.

(*i*) The susceptibility of diamagnetic materials is negative and has a value (between 0 and –1) For example, for Bismuth $\chi_d = -1.4 \times 10^{-8}$.

(*ii*) The susceptibility of *paramagnetic materials* is positive and has a value (between 0 and +1). For example, for platinum $\chi_p = +1.71 \times 10^{-6}$.

(*iii*) The susceptibility of *ferromagnetic materials* is positive and has a very large value.

The permeability of diamagnetic materials is less than 1 (but > 0) i.e. it can not be negative.

The permeability of paramagnetic material is slightly greater than 1 (one) and that of ferromagnetic materials is very large.

20.3 REPULSION OF A DIAMAGNETIC MATERIAL

Let *SN* be a strong bar magnet which produces a magnetic field *B* along the *X*-axis. The field is non-uniform and goes on decreasing as we move away from the magnet in + *X* direction.

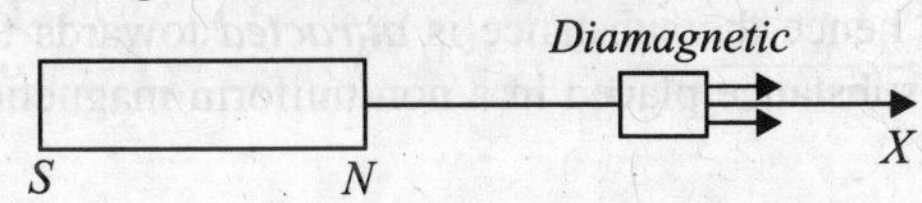

Fig. 20.1

$\therefore$ Rate of change of magnetic field along *X*-axis $= -\dfrac{\partial B}{\partial x}$ *i.e.,* $\dfrac{\partial B}{\partial x}$ acting along – *X* direction.

If we place a slab of diamagnetic material of unit volume at a position where the average magnetic field is *B*, then the magnetic dipole moment per unit volume produced in the slab,

$$\vec{P}_m = -\frac{Ne^2}{6m} < r^2 > \vec{B} = \chi_d \vec{B}$$

where χ_d is the diamagnetic susceptibility and has a negative value.

The magnitude of the force acting on a dipole of magnetic moment $\vec{P}_m$ when placed in a non-uniform magnetic field varying at the rate $\dfrac{\partial B}{\partial x} = P_m \dfrac{\partial B}{\partial x}$.

But $\dfrac{\partial B}{\partial x}$ as well as p_m are both negative. Therefore the force acting on the material is $+ \chi_d \vec{B} \dfrac{\partial B}{\partial x}$.

i.e., it acts along the $+X$ direction and hence the substance is *repelled* away from the magnet.

In other words, a diamagnetic material placed in a non-uniform magnetic field moves from strong to a weaker part of the field.

20.4 ATTRACTION OF A PARAMAGNETIC MATERIAL

Let SN be a strong magnet which produces a magnetic field B along the X-axis. The field is non-uniform and goes on decreasing as we move away from the magnet in the $+X$ direction.

$\therefore$ The rate of change of magnetic field along X-axis $= -\dfrac{\partial B}{\partial x}$

i.e., $\dfrac{\partial B}{\partial x}$ acts along $-X$ direction.

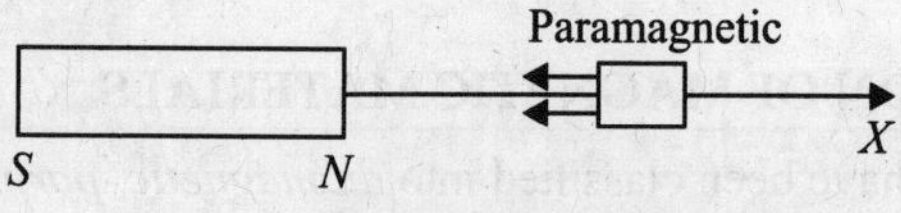

Fig. 20.2

Place a slab of paramagnetic material at a position where the average magnetic field is B, then the magnetic dipole moment per unit volume produced in the slab $\vec{P}_m = \chi_p B$,

where χ_p is the paramagnetic susceptibility of the material. If V is the volume of the slab, then

The magnetic dipole moment $= \chi_p\ VB$

The magnitude of the force acting on a dipole of magnetic moment $\vec{P}_m$ when placed in a non-uniform magnetic field varying at the rate $\dfrac{\partial B}{\partial x} = \vec{P}_m \dfrac{\partial B}{\partial x}$.

$\therefore$ Force on the paramagnetic slab $F = \chi_p\ V\ B \dfrac{\partial B}{\partial x}$.

But $\dfrac{\partial B}{\partial x}$ acts along the $-X$ direction. It means that the force on the paramagnetic substance acts in the $-X$ direction and hence the substance is *attracted* towards the magnet. In other words, a sample of paramagnetic substance placed in a non-uniform magnetic field moves from weaker to stronger parts of the field.

20.5 DIAMAGNETIC, PARAMAGNETIC AND FERROMAGNETIC SUBSTANCES

20.5.1 Diamagnetic substances

Diamagnetic substances are those which are repelled by magnets and when placed in a non-uniform magnetic field move from stronger to weaker part of the field. Familiar examples of these are bismuth, phosphorous, antimony, copper, water, alcohol and hydrogen.

Properties. (*i*) When a diamagnetic substance is placed in a magnetic field it sets itself at right angles to the direction of the lines of force as shown in Fig. 20.3. This indicates that the lines of force tend to have minimum path through the diamagnetic substances.

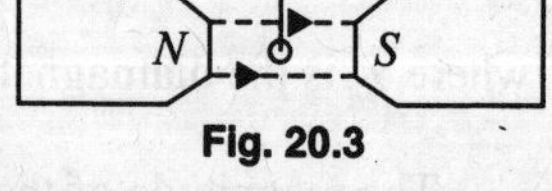

Fig. 20.3

(*ii*) When a diamagnetic material is placed within a magnetic field the lines of force tend to go away from the material as shown in Fig. 20.4.

(*iii*) When a diamagnetic material is placed in a non-uniform magnetic field it moves from stronger to weaker parts of the field.

When a diamagnetic liquid is placed in a watch glass on the pole pieces of a magnet the liquid accumulates on the sides causing a depression at the centre which is the strongest part of the field as shown in Fig. 20.5. When the distance between the pole pieces is large, the effect is reversed.

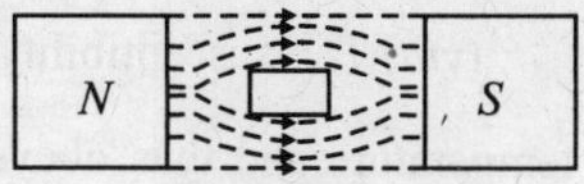

Fig. 20.4

Fig. 20.5

(*iv*) A diamagnetic liquid in a *U*-tube placed in a magnetic field shows a depression.

(*v*) When a diamagnetic gas is allowed to ascend between the pole pieces of an electromagnet it spreads across the field.

(*vi*) The susceptibility of diamagnetic material is negative and has a small value. For example, for bismuth $\chi_d = -1.4 \times 10^{-8}$. The permeability is less than one but can never be negative.

(*vii*) The susceptibility neither changes with the strength of the field nor with temperature.

20.5.2 Paramagnetic Substances

Paramagnetic substances are those which are attracted by magnets and when placed in a non-uniform magnetic field move from weaker to stronger parts of the field. Familiar examples are aluminium, manganese, platinum, rare earth elements like *Gd*, crown glass solutions of salts of iron and oxygen.

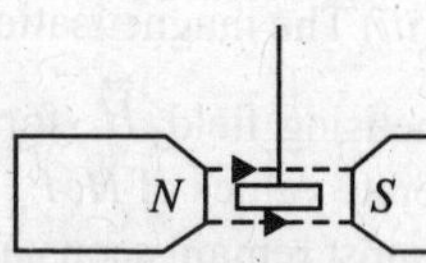

Fig. 20.6

Properties. (*i*) If a bar of paramagnetic material is suspended in between the pole pieces of an electromagnet, it sets itself parallel to the lines of forces as shown in Fig. 20.6 showing thereby that the lines of force try to have their longest path through the paramagnetic materials.

Fig. 20.7

(*ii*) When a bar of paramagnetic material is placed in a magnetic field, the lines of force tend to accumulate in it as shown in Fig. 20.7.

(*iii*) When a paramagnetic material is placed in a non-uniform magnetic field it moves from weaker to stronger parts of the field.

Fig. 20.8

If a paramagnetic liquid is placed in a watch glass on the pole pieces (not more than 2 to 3 mm apart) of an electromagnet then it accumulates in the middle as shown in Fig. 20.8. It is because in the central region the field is the strongest. If the pole pieces are not close together the field is *strongest near the poles* and the liquid moves away from the centre giving an almost opposite effect.

(*iv*) If one end of a narrow *U*-tube containing a paramagnetic liquid is placed within the pole pieces of an electromagnet in such a manner that the level of the liquid is in line with the field, then on applying the field, the level of the liquid rises. The rise is proportional to the susceptibility of the liquid. This is shown in Fig. 20.9.

(*v*) When a paramagnetic gas rendered visible by traces of ammonia and hydrogen chloride is allowed to ascend between the pole pieces of an electromagnet it spreads along the direction of the field.

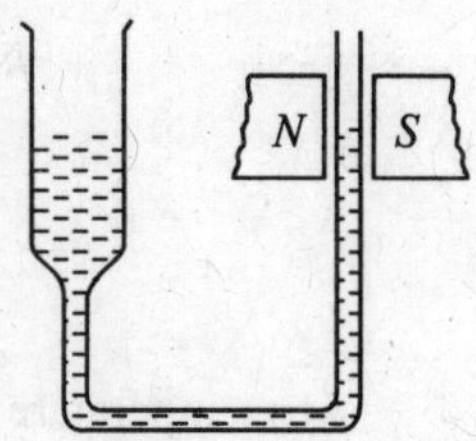

Fig. 20.9

(*vi*) The susceptibility of the paramagnetic material is positive though it has a small value (< 1). The permeability is slightly greater than one. For example, for platinum $\chi_p = 1.71 \times 10^{-6}$ and permeability $\mu = 1.00002$.

(*vii*) When paramagnetic substances are subjected to a magnetic field these are feebly magnetised in the direction of the field.

The magnetisation (or magnetic dipole moment per unit volume) P_m is proportional to the magnetising force. Therefore, susceptibility and hence permeability remains constant for varying values of the field.

(*viii*) The susceptibility for a given magnetising force is inversely proportional to the absolute temperature and thus obeys *Curie's law* $\chi_p \propto \frac{1}{T}$. If, therefore, the temperature is increased the susceptibility decreases and at some high temperature it even becomes negative *i.e.* the substance becomes diamagnetic.

20.5.3 Ferromagnetic Substances

Ferromagnetic substances are those which are attracted by the magnets and can also be magnetised. Familiar examples are iron, nickel, cobalt and their *alloys*.

Properties. (*i*) Ferromagnetic substances show all the properties of a paramagnetic substance to a much greater degree.

(*ii*) The susceptibility has a positive value and the permeability is also very large.

(*iii*) The magnetisation (or magnetic dipole moment per unit volume) $\vec{P}_m$ is proportional to the magnetising field $\vec{H}$ for small values as shown by part *OA* of the curve in Fig. 20.10 (*a*). For moderate values of H, P_m increases rapidly as shown by the part *AB* and for very large values of H, P_m almost remains constant due to the approach of saturation stage as shown by the part *BC* of the same curve.

(*iv*) The susceptibility χ_m remains constant for very small values of the magnetising field, increases with the medium values of the field and begins to decreases for very large values of the field as shown by the parts *PQ*, *QR* and *RS* respectively in the curve between χ_m and H in Fig. 20.10 (*a*).

(*v*) The susceptibility decreases with temperature and obeys **Curie's law**, which is expressed as

$$\chi_m \propto \frac{1}{T} \quad \text{or} \quad \chi_m\, T = \text{constant}$$

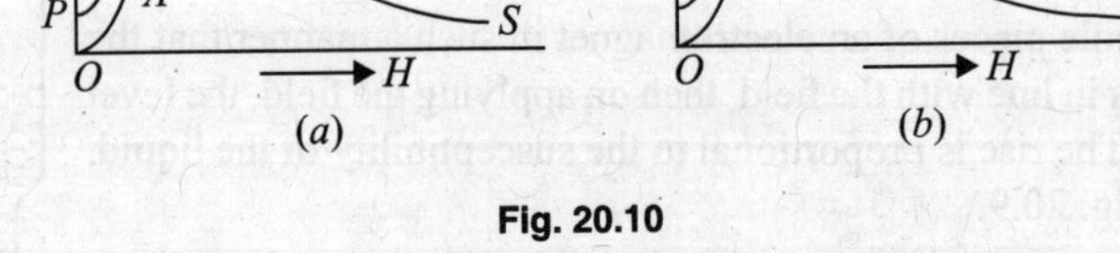

Fig. 20.10

At a temperature called the critical temperature or **Curie point** the value of susceptibility drops suddenly and the body becomes paramagnetic.

The Curie point for steel is 770°C, for cobalt 1100°C and for nickel 400°C. The susceptibility of a ferromagnetic substance above its Curie point is proportional to the difference of temperature of the material and Curie point. This is called **Curie-Wiess law.**

The magnetic induction B varies nearly in the same manner with H as P_m does except that the value of B does not become constant for large values of H. This is shown in Fig. 20.10 (*b*). The permeability μ varies with H in a similar manner as susceptibility does.

20.6 ATOMIC MAGNETIC MOMENT

All substances are composed of atoms and an atom consists of a central positively charged nucleus where the whole mass of the atom is supposed to be concentrated, with suitable number of electrons revolving round it in more or less circular orbits. An electron carries a *negative* charge of 1.6×10^{-19} coulomb. The revolution of an electron in a clockwise direction is equivalent to a conventional current in the anticlockwise direction and the electronic orbit behaves like a *magnetic dipole.*

Magnetic moment of an atomic dipole. An atomic dipole consists of an electron of charge $-e$ and mass m revolving around an atomic nucleus with a linear velocity v in an orbit of radius r. The revolving electron is equivalent to a current $I = -\frac{e}{2\pi r/v} = -\frac{ev}{2\pi r}$

The magnitude of the dipole magnetic moment of a current loop of area A carrying a current I is given by

$$p_m = IA$$

$\therefore$ Magnetic moment of the atomic dipole $= -\frac{ev}{2\pi r} \times \pi r^2$

$$= -\frac{1}{2} evr = -\frac{1}{2}\frac{e}{m} mvr = -\frac{1}{2}\frac{e}{m} L$$

where $L = mvr$ = the magnitude of the angular momentum of the electron revolving around the nucleus.

Consider the simplest atom of hydrogen. It consists of one proton in the nucleus and one electron revolving round it. If m is the mass of the electron, $-e$ the charge on it, v its vleocity of revolution and r the radius of the orbit, then the electron is maintained in the orbit due to the electrostatic force of attraction between the electron and the nucleus which provides the necessary centripetal force. Taking magnitudes only

$$\therefore \quad \frac{mv^2}{r} = \frac{1}{4\pi\varepsilon_0}\frac{e.e}{r^2} = \frac{1}{4\pi\varepsilon_0}\frac{e^2}{r^2}$$

$$\text{or} \quad v = \left[\frac{1}{4\pi\varepsilon_0}\frac{e^2}{mr}\right]^{1/2}$$

Now $m = 9 \times 10^{-31}$ kg; $e = 1.6 \times 10^{-19}\ C$

For hydrogen $r = 0.5 \times 10^{-10}$ m; $\frac{1}{4\pi\varepsilon_0} = 9 \times 10^9$ Nm^2 C^{-2}

$$\therefore \quad v = \left[\frac{9\times10^9 \times 1.6 \times 1.6 \times 10^{-38}}{9 \times 10^{-31} \times 0.5 \times 10^{-10}}\right]^{1/2} = 2.262 \times 10^6 \text{ m/sec}$$

and time taken by the electron to complete one revolution

$$= \frac{2\pi r}{v} = \frac{2\pi \times 0.5 \times 10^{-10}}{2.262 \times 10^6} = 1.39 \times 10^{-16} \text{ sec}$$

Current in the orbit. From the above calculations we find that the time taken by the electron to complete one orbit is *very very small* and hence the electron will go round the nucleus millions of times during the time of observation. In other words, the electron going round the orbit will not behave as an isolated particle but will appear to be a current loop carrying a current I given by

$$I = \frac{\text{Charge}}{\text{Time taken to complete the orbit}} = \frac{e}{2\pi r / v} = \frac{ev}{2\pi r}$$

Orbital magnetic moment. The magnitude of the dipole magnetic moment of the current loop is given by

$$p_m = IA = \frac{ev}{2\pi r}\pi r^2 = \frac{1}{2}evr$$

The magnitude of the angular momentum of the electron revolving round the nucleus

$$L = mvr$$

$$\therefore \quad p_m = \frac{1}{2}\frac{e}{m}L \quad ...(i)$$

As $\vec{p}_m$ and $\vec{L}$ are both vector quantities and electron charge e is negative, relation (*i*) can be put as

$$\vec{p}_m = -\frac{e}{2m}\vec{L} \quad ...(i)\ (a)$$

Orbital magnetic moment integral multiple of $\frac{e\hbar}{2m}$. According to Bohr's theory, an electron can only revolve in an orbit in which its total angular momentum is an integral multiple of $\frac{h}{2\pi}$ where h is Planck's constant.

$$\therefore \quad L = n\frac{h}{2\pi}$$

where n is an integer.

Substituting the value of L in (*i*), we have

$$p_m = \frac{1}{2}\frac{e}{m}\frac{nh}{2\pi} = \frac{neh}{4\pi m}$$

$$= \frac{ne\hbar}{2m} \quad ...(ii) \quad \left[\text{where } \hbar = \frac{h}{2\pi}\right]$$

From relation (*ii*) we find that the magnetic moment due to orbital motion of an electron must be an integral multiple of $\frac{e\hbar}{2m}$.

Bohr magneton. The smallest value of orbital magnetic moment of the electron for $n = 1$ is given by

$$p_m = \frac{e\hbar}{2m} = \frac{eh}{4\pi m} \quad ...(iii)$$

This is the unit of magnetic moment and is known as a *Bohr Magneton*. It is denoted by μ_B.

Numerical value of Bohr magneton. Now $e = 1.6 \times 10^{-19}$ coul.

$$h = 6.6 \times 10^{-24} \text{ Joule sec; } m = 9 \times 10^{-31} \text{ kg}$$

Substituting these values in Eq. (*iii*), we get

$$\mu_B = \frac{eh}{4\pi m} = \frac{1.6 \times 10^{-19} \text{ coul} \times 6.6 \times 10^{-34} \text{ Joule sec}}{4\pi \times 9 \times 10^{-31} \text{ kg}}$$

$$= 9.27 \times 10^{-24} \text{ Joule } \frac{\text{coul met sec}^{-1}}{\text{kg met sec}^{-2}}$$

$$= 9.27 \times 10^{-24} \text{ Joule } \frac{\text{coul met sec}^{-1}}{\text{Newton}}$$

$$= 9.27 \times 10^{-24} \frac{\text{Joule}}{\text{Tesla}} = 9.27 \times 10^{-24} \text{ Joule/Tesla}$$

[The unit of Bohr magneton is also the same as that of p_m. As $p_m = IA$, the unit of Bohr magneton is also Am^2]

Bohr magneton is used as a unit of measurement of atomic magnetic moments.

Orbital gyromagnetic ratio. It is defined as *the ratio of magnetic moment of the atomic dipole to its angular momentum*.

From relation (*i*) we have,

$$\text{Orbital gyromagentic ratio} = \frac{p_m}{L} = \frac{e}{2m}.$$

20.7 LANGEVIN'S THEORY OF DIAMAGNETISM

An electron revolving in a circular orbit around the nucleus in an atom is equivalent to a current loop. If $-e$ is the charge on the electron and ω its angular velocity in the orbit, then the equivalent current

$$I = \frac{\text{Charge}}{\text{Time taken to complete the orbit}} = \frac{-e}{2\pi/\omega} = \frac{-e\omega}{2\pi}$$

Electron revolving in counter clockwise direction. Suppose the electron revolves in a circular orbit of radius r in a counter clockwise direction in the $X-Y$ plane, with the nucleus at the origin, then the magnetic moment of the electron

$$\vec{p}_m = I\vec{a} = \frac{-e\omega}{2\pi}\pi r^2\,\hat{k} \qquad [\because \text{ Area vector } \vec{a} \text{ is along } +Z \text{ axis}]$$

$$= \frac{-e\omega r^2\,\hat{k}}{2}$$

The direction of dipole magnetic moment $\vec{p}_m$ is along the $-Z$ axis.

As the electron is moving in a circular orbit, the centripetal force acting on it

$$F = m\,\omega^2 r \qquad ...(i)$$

where m is the mass of the electron. The direction of this force is *radially* inward and it is provided by the *electrostatic force of attraction* between the nucleus and the electron.

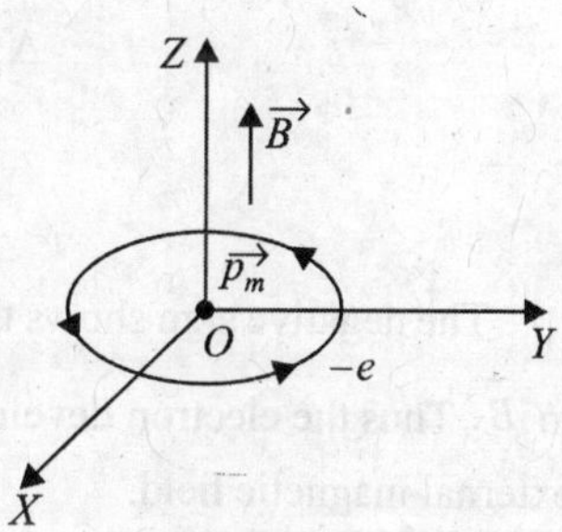

Fig. 20.11

If an external magnetic field $\vec{B}$ is now applied perpendicular to the plane of the orbit of this single electron *i.e.*, in the $+Z$ direction, then an additional magnetic force ΔF will act on it, given by

$$\Delta\vec{F} = e\,\vec{v} \times \vec{B}$$

Since $\vec{B}$ is perpendicular to $\vec{v}$, the magnitude of the force

$$\Delta F = evB = er\omega B = Ber\omega \qquad ...(ii)$$

This force will also act *radially inward* on the electron at right angles to both the orbit and the magnetic field. Thus the force on the electron due to the external magnetic field and the electrostatic force acting on it due to the nucleus both act in the same direction.

∴ Net force acting on the electron in the magentic field

$$= -(m\omega^2 r + Ber\omega)$$

Under the effect of this increased force the electron can either move in an orbit of smaller radius or have a greater velocity in the same orbit or both. But according to the laws of quantum mechanics which are applicable to the electronic orbits, the electron can move only in specific orbits of fixed radii governed by quantum laws. Hence the only alternative for the electron is to increase its angular velocity in the original orbit. If $\omega + \Delta\omega$ is the angular velocity of the electron after the application of the magnetic field, the new magnetic moment of the electron

$$\vec{p}_m + \Delta\vec{p}_m = -\frac{er^2}{2}(\omega + \Delta\omega)\,\hat{k}$$

∴ Change in the magnetic moment

$$\Delta\vec{p}_m = -\left[\frac{er^2}{2}(\omega + \Delta\omega) - er^2\,\omega\right]\hat{k}$$

$$= -\frac{er^2}{2}\Delta\omega\,\hat{k} \qquad ...(iii)$$

According to relation (*i*) $F = m\omega^2 r$

$$\therefore \quad \Delta F = 2mr\omega\,\Delta\omega \qquad ...(iv)$$

From relation (*ii*) $\Delta F = Ber\omega$

$$\therefore \quad 2mr\omega\Delta\omega = Ber\omega$$

or

$$\Delta\omega = \frac{e}{2m}B$$

Substituting the value of $\Delta\omega$ in (*iii*), we have

$$\Delta\vec{p}_m = \frac{-e^2r^2}{4m}B\,\hat{k}$$

$$= \frac{-e^2r^2}{4m}\vec{B} \qquad ...(v)$$

The negative sign shows that the change in magnetic moment takes place in a direction opposite to $\vec{B}$. Thus the electron develops an induced magnetic moment $\Delta\vec{p}_m$ in a direction opposite to the external magnetic field.

The change in magnetic moment of the orbiting electron is $\frac{-e^2r^2}{4m}\vec{B}$.

Electron revolving in clockwise direction. If we consider the electron as moving in the *clockwise* direction, the area vector of the current loop

$$\vec{a} = -\pi r^2\,\hat{k}$$

and the magnetic moment of the electron before the application of the magnetic field is

$$\vec{p}_m = \left(-\frac{e\omega}{2\pi}\right)\left(-\pi r^2\,\hat{k}\right) = \frac{e\omega r^2}{2}\,\hat{k}$$

The directon of $\vec{p}_m$ is, therefore, along + Z axis.

As in this case the direction of the velocity vector has also changed, the force due to the external magnetic field will now act on the electron in the *radially outward* direction *i.e.*, opposite to that of the electrostatic force. The net force on the electron will decrease and the electron will move in the same orbit with a smaller angular velocity $(\omega - \Delta\omega)$. The change in the magnetic moment will, therefore be given by

$$\Delta\,\vec{p}_m = \left[\frac{er^2}{2}(\omega - \Delta\omega) - \frac{er^2}{2}\,\omega\right]\hat{k}$$

$$= -\frac{er^2}{2}\,\Delta\,\omega\,\hat{k} \qquad \text{...}(v)\,(a)$$

which on substituting the value of $\Delta\omega = \frac{e}{2m}\,B$ gives

$$\Delta\,\vec{p}_m = \frac{-e^2\,r^2}{4m}\,\vec{B} \qquad \text{...}(vi)$$

Thus the value of $\Delta\,\vec{p}_m$ is the same in direction as well as in magnitude whether the electron is moving in the clockwise or anticlockwise direction.

If N is the number of electrons per unit volume and Z the atomic number, then

$$\vec{P}_m = \frac{-NZe^2\,r^2}{4m}\,\vec{B} = \frac{-NZe^2\,r^2\,\mu_0}{4m}\,\vec{H}$$

and susceptibility $\qquad \chi_m = \chi_m = \frac{\vec{P}_m}{\vec{H}} = \frac{-NZe^2\,r^2\,\mu_0}{4m}$

Langevin's explaination of diamagnetism. In the expression, (*vi*) if the electronic orbit is not perpendicular to the applied field, then r is replaced by r_1, the projection of the radius r of the orbit on the plane perpendicular to the magnetic field *i.e.*, in the $X-Y$ plane.

If x, y, z are the co-ordinates of the radius r of the orbit, then

$$r^2 = x^2 + y^2 + z^2$$

and $\qquad r_1^{\,2} = x^2 + y^2$

If the radii are arranged in all possible directions, then the average value of

$$r^2 = x^2 + y^2 + z^2 = 3x^2$$

and $\qquad r_1^{\,2} = x^2 + y^2 = 2x^2 = \frac{2}{3}\,r^2$

If we denote the average value of r^2 as $< r^2 >$, we have relation (*v*) as

$$\Delta\,\vec{p}_m = \frac{-e^2}{4m}\cdot\frac{2}{3} < r^2 > \vec{B}$$

$$= \frac{-e^2}{6m} < r^2 > \vec{B}$$

If N is the number of electrons per unit volume and Z the atomic number, then the diamagnetic dipole moment per unit volume

$$\vec{P}_m = \frac{-NZe^2}{6m} < r^2 > \vec{B} = \frac{-NZe^2}{6m} < r^2 > \mu_0 \vec{H}$$

$$= \chi_d \vec{H} \qquad ...(vii)$$

where χ_d is the diamagnetic susceptibility of the material.

$$\therefore \qquad \chi_d = -\frac{NZe^2\mu_0}{6m} < r^2 > \qquad ...(viii)$$

According to the laws of atomic structure, the electrons in an atom have a tendency to exist in pairs with their angular momentum vectors pointing in opposite directions. Thus for most of the atoms which have an even number of electrons, the total angular momentum is zero. Such atoms have zero magnetic dipole moment and are *diamagnetic*.

If we consider an atom having two electrons moving in the same orbit but in opposite directions, the net magnetic dipole moment of such an atom will be zero as the dipole moment due to one electron will be equal and opposite to that of the other. When an external magnetic field is applied; as explained in Eq (*vi*) above both the electrons will acquire a magnetic moment

$$\Delta \vec{p}_m = \frac{-e^2 r^2}{4m} \vec{B}$$

irrespective of the fact whether the motion of the electron is clockwise or anticlockwise. The induced dipole magnetic moment will act in a direction opposite to that of the applied field and hence the atom will experience a force in the direction of decreasing B and the material is *repelled*.

From the above discussion we note the following points :

(*i*) The diamagnetic dipole moment per unit volume $\vec{p}_m$ is proportional to the magnetic field $\vec{B}$ (or $\vec{H}$).

(*ii*) The direction of $\vec{p}_m$ is opposite to that of $\vec{B}$ (or $\vec{H}$).

(*iii*) The diamagnetic susceptibility χ_d is always negative and almost independent of temperature.

20.8 SPIN ANGULAR MOMENTUM

An electron in an atom revolves round the nucleus in a more or less circular orbit and possessses an orbital angular momentum which is an integral miultiple of $\frac{h}{2\pi}$ where h is Planck's constant. In addtion, the electron possesses an angular momentum which according to classical theory is supposed to arise due to the rotation of the electron about an axis passing through its centre. This property is called *spin* and the electron is said to possess a spin angular momentum equal to $\frac{1}{2}\frac{h}{2\pi} = \frac{1}{2}\hbar$.

Spin magnetic moment. Electron spin is a quantum phenomenon and in accordance with quantum laws, there is an *intrinsic magnetic moment* attached to the spin angular momentum of *constant magnitude* equal to one Bohr magneton. Its value $\frac{eh}{4\pi m} = \frac{e\hbar}{2m} = 9.27 \times 10^{-24}$ Joule/Tesla

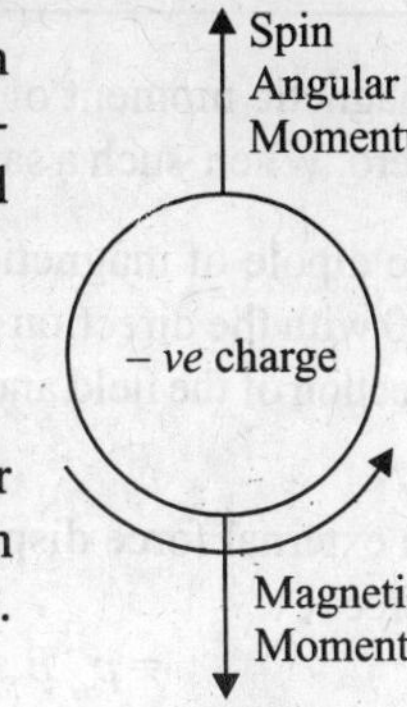

Fig. 20.12

The magnetic moment vector points *antiparallel* to the spin angular momentum vector. This is in accordance with the fact that the electron can be considered as a small sphere of negative charge spinning around its axis.

[**Note.** The orbital magnetic moment of the electron

$$\vec{p}_m = -\frac{e}{2m}\vec{L}_0 \text{ where } \vec{L}_0 \text{ is the orbital angular momentum.}$$

Similarly, according to the classical theory the spin magnetic moment of the electron should be given by $(\vec{p}_m)_{spin} = -\frac{e}{2m}\vec{L}_s$

where $\vec{L}_s$ is the spin angular momentum $= \frac{1}{2}\frac{h}{2\pi}$

$$\therefore \quad (\vec{p}_m)_{spin} = \frac{e}{2m}\frac{1}{2}\frac{h}{2\pi} = -\frac{1}{2}\frac{eh}{4\pi m}$$

This value of spin magnetic moment of the electron is half the value given above. But this derivation is based on laws of classical Physics whereas the spin motion of the electron obeys quantum laws according to which the spin magnetic moment of the electron is given by

$$(\vec{p}_m)_{spin} = 2\left(-\frac{e}{2m}\right)\vec{L}_s$$

$$= 2\left(-\frac{e}{2m}\right) \times \frac{1}{2}\frac{h}{2\pi} = -\frac{eh}{4\pi m} = -\frac{e\hbar}{2m}$$

20.9 LANGEVIN'S THEORY OF PARAMAGNETISM

According to the laws of atomic structure the electrons in an atom have a tendency to exist in pairs with their angular momentum vectors pointing in opposite directions. Thus for most of the atoms which have an *even* number of electrons the total angular momentum is zero. Such atoms have zero net magnetic dipole moment and are dimagnetic. Atoms with an *odd* number of electrons have at least one electron which is *unpaired*. Such atoms tend to have a net magnetic moment and are *paramagnetic*. In a molecule consisting of two such atoms, the two electrons which are originally unpaired in the atom will pair off with their angular momenta pointing in opposite directions. The molecule thus has zero magnetic moment and behaves as a diamagnetic molecule. However, there are some exceptions like O_2 whose electronic structure is such as not to favour complete pair formation and cancellation of spin angular momentum. Such substances also have a net magnetic dipole moment and are paramagnetic. In certain groups of elements, notably the elements around gadolinium and iron in the periodic table the atoms contain unpaired electron spin which are relatively free to orient in the magnetic field.

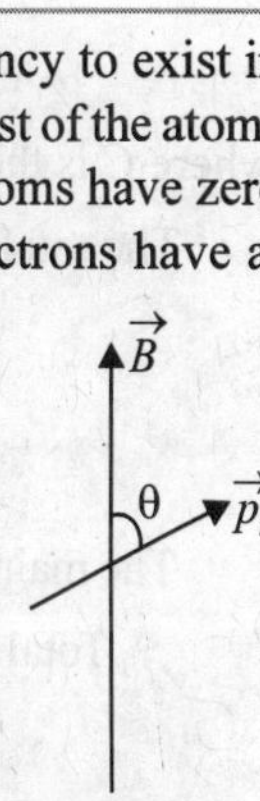

Fig. 20.13

Potential energy of an atomic dipole. If we have a bulk sample containing a very large number of paramagnetic atoms, their dipole moments will normally be oriented in random direction so that the net magnetic moment of the sample as a whole on the macroscopic scale is zero. When such a sample is placed in an external magnetic field, an atomic dipole of magnetic moment $\vec{p}_m$ lying in a direction making an angle θ with the direction of the magnetic field will tend to align itself in the direction of the field and the *magnitude* of the corresponding torque.

$$= p_m B \sin\theta$$

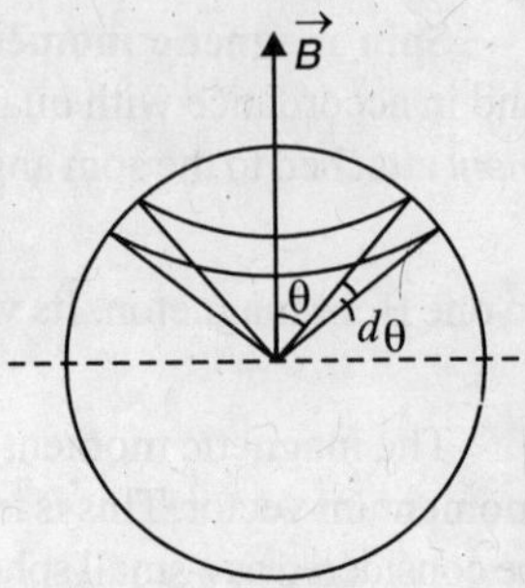

Fig. 20.14

If an external force displaces it through an angle $d\theta$, the work done by the force

$$= p_m B \sin\theta \, d\theta$$

Hence the total work done in displacing the atomic dipole to the position making an angle θ with the direction of $\vec{B}$

$$= \int_0^\theta p_m B \sin\theta \, d\theta = -p_m B \cos\theta$$

This work is stored in the atomic dipole as its potential energy.

∴ Potential energy of atomic dipole $U = -p_m B \cos\theta$

Average dipole moment induced in the atom. If we consider a spherical volume having a very large number of atomic dipoles, the number of those possessing an energy $U = p_m B \cos\theta$ will be proportional to the solid angle enclosed between two cones with semi-vertical angle θ and $\theta + d\theta$ *i.e.*, within a solid angle

$$2\pi \sin\theta \, d\theta$$

Further, the tendency of the magnetic dipoles to align themselves parallel to the applied magnetic field is opposed by their thermal agitation. Hence the probability that an atomic dipole will lie in the position making an angle θ with the direction of the magnetic field $\vec{B}$, according to the laws of probability is $e^{-U/kT}$ where k is Boltzmann constant and T the absolute temperature *i.e.*, the number of such atoms will be proportional to $e^{-U/kt}$.

If dN is the number of such atoms, then

$$dN \propto e^{-U/kT} 2\pi \sin\theta \, d\theta$$

$$= Ce^{-U/kT} 2\pi \sin\theta \, d\theta$$

where C is the constant of proportionality.

The total number of atoms

$$= C \int_0^\pi e^{-U/kT} 2\pi \sin\theta \, d\theta \qquad ...(i)$$

The magnetic moment contributed by each atomic dipole in the direction of field $\vec{B} = p_m \cos\theta$.

∴ Total magnetic moment contributed by the atomic dipoles

$$= C \int_0^\pi p_m \cos\theta \, e^{-U/kT} 2\pi \sin\theta \, d\theta \qquad ...(ii)$$

But $\qquad U = -p_m B \cos\theta$

∴ $\qquad e^{-U/kT} = e^{p_m B \cos\theta/kT}$

Now $\dfrac{p_m B}{kT}$ is of the order of 10^{-2} and << 1

$$\therefore \qquad e^{p_m B \cos\theta / kT} = 1 + \frac{p_m \; B \cos\theta}{kT}$$

Substituting the value of

$$e^{-U/kT} = 1 + \frac{p_m \; B \cos\theta}{kT}$$

in (*i*) and (*ii*), we have

Average magnetic moment per atom in the direction of the field

$$< p_m > = \frac{C \int_0^{\pi} p_m \cos\theta \left(1 + \dfrac{p_m \; B\cos\theta}{kT}\right) 2\pi \sin\theta \, d\theta}{C \int_0^{\pi} \left(1 + \dfrac{p_m \; B\cos\theta}{kT}\right) 2\pi \sin\theta \, d\theta}$$

Now $\int_0^{\pi} \cos\theta \sin\theta \, d\theta = 0$; $\int_0^{\pi} \cos^2\theta \sin\theta \, d\theta = \left[\dfrac{-\cos^3\theta}{3}\right]_0^{\pi} = \dfrac{2}{3}$ and $\int_0^{\pi} \sin\theta \, d\theta = 2$

$$\therefore \qquad < p_m > = \frac{\dfrac{2}{3} \dfrac{2\pi p_m^2 B}{kT}}{4\pi} = \frac{p_m^2 \; B}{3kT} \qquad ...(iii)$$

This gives the average magnetic moment induced in each atom by the external magnetic field.

Magnetic dipole moment per unit volume. If N is the total number of magnetic dipoles (atomic or molecular), per unit volume, then the magnetic dipole moment per unit volume or intensity of magnetisation is given by

$$\vec{P}_m = \frac{N \; p_m^2 \; \vec{B}}{3kT} = \chi_p \; \vec{B} \qquad ...(iv)$$

where χ_p is the paramagnetic susceptibility.

Now $\qquad \vec{B} = \mu_0 \; (\vec{H} + \vec{P}_m)$

where $\vec{H}$ is the *magnetic field vector.* But for paramagnetic materials $\vec{P}_m$ is very small.

$$\therefore \qquad \vec{B} = \mu_0 \; \vec{H}$$

Hence $\qquad \vec{P}_m = \dfrac{N \; p_m^2 \; \mu_0 \; \vec{H}}{3kT} = \chi_p \; \vec{H} \qquad ...(v)$

where $\qquad \chi_p = \dfrac{\vec{P}_m}{\vec{H}} = \dfrac{\mu_0 \; N \; p_m^2}{3kT} \qquad ...(vi)$

is the magnetic susceptibility for a paramagentic material.

Equation (*v*) shows that the *paramagnetic sample acquires magnetisation in the direction of the magnetic field.*

Thus (*i*) The magnetisation $\vec{P}_m$ induced in the paramagnetic material is proportional to $\vec{H}$.

(*ii*) The direction of $\vec{P}_m$ is the same as the direction of magnetic field.

(*iii*) The susceptibility of a paramagnetic substance χ_p is always positive.

Curie law. The susceptibility of a paramagnetic substance is given by

$$\chi_p = \frac{\mu_0 N p_m^2}{3kT} \quad \text{or} \quad \chi_p \propto \frac{1}{T}$$

as $\frac{\mu_0 N p_m^2}{3k}$ is a constant. The relation $\chi_p \propto \frac{1}{T}$ is known as *Curie's law*. Hence Curie's law states that

'*The susceptibility of a paramagnetic substance varies inversely as the absolute temperature*'.

20.10 FERROMAGNETISM

Ferromagnetic substances are those which can be magnetised and retain magnetism even when the external magnetising field has been removed. Familiar examples are *Fe, Co, Ni, Gd, Dy* and their alloys. The magnetic dipole moment per unit volume or the intensity of magnetisation $\vec{P}_m$ produced in such substances is very high. It is also not proportional to the magnetising force $\vec{H}$ *i.e.*, The susceptibility χ_m is not a constant but a function of $\vec{H}$ as well as of absolute temperature *T*. The susceptibility decreases with temperature and above a certain temperature known as the *Curie point* the substance loses its ferromagnetic character and behaves as a paramagnetic material.

Domain theory. Ferromagnetism like paramagnetism is caused by the spin magnetic moment of the electron. Thus individually each atom of a ferromagnetic substance is paramagnetic and has a permanent magnetic dipole moment due to *unpaired electron spins*. In a paramagnetic substance, in the unmagnetised state the atomic or molecular dipoles are themselves arranged at random so that the net resultant magnetic dipole moment of the substance is zero. But in ferromagnetic substances adjacent atomic dipoles are locked in rigid parallelism as each atomic or molecular dipole is very strongly coupled to the neighbouring atom by a typical quantum mechanical interaction known as *exchange interaction* with the result that a large number of atomic dipoles are aligned in the same direction. This group of atomic dipoles is known as a *ferromagnetic domain.* A domain contains about 10^{16} or more atoms and has a volume between 10^{-8} to 10^{-12} cubic metre. All the atoms in a domain are lined up in the same direction and hence it is magnetised to the saturation stage. *Thus domains are local regions within which there is essentially perfect alignment of atomic dipoles.*

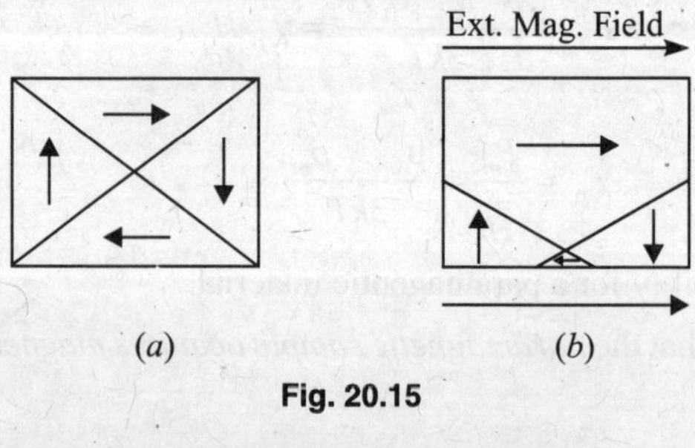

Fig. 20.15

The direction of magnetisation of each domain in a ferromagnetic material depends upon the crystalline state of the substance. In the case of single crystal domains magnetisation is along the *axes of symmetry or easy directions.* For examples, a single crystal of iron has three axes of symmetry perpendicular to one another and if we denote them as X, Y, Z, the domain magnetisation will be along the $\pm X$, $\pm Y$, and $\pm Z$ direction. There is not a sudden transition from one domain to the other. The domain wall, knwon as '*Block wall*' has a finite thickness extending over a number of atoms whose spins change gradually in direction as we proceed through the wall and in an extreme case may change from 0° to 180°.

Thus we find that ferromagnetic substances are intrinsically magnetised even in the absence of an external magnetic field due to the presence of these domains.

Why a piece of iron ordinarily does not behave as a magnet. Every piece of iron does not ordinarily behave as a magnet because the magnetic domains within the material are, so arranged as to form closed chains and cancel the magnetic effect of one another with the result that the material is on the whole neutral (unmagnetised). A possible arrangement of domains forming closed chains is shown in Fig. 20.15 (*a*). Here all the domains are of the same size and their direction of magnetisation is shown by arrow heads. These are so arranged that the net magnetic moment of the sample is zero as shown in Fig. 20.15 (*a*).

Magnetisation. When a *small* external magnetic field is applied in a particular direction two effects take place.

(*i*) The size of the domains within the material which have their magnetic orientation in the direction of the applied magnetic field increases at the expense of those that are not favourably oriented, and

(*ii*) More and more atomic dipoles within a domain become parallel to the magnetising field.

As a result of both these effect the size of domains oriented in unfavoured directions decreases as shown in Fig. 20.15 (*b*). The material, therefore, acquires a net magnetic dipole moment.

Saturation stage. As the strength of the magnetising field is increased further a stage is reached when all the dipoles are oriented parallel to the magnetising field in a single domain. On further increasing the magnetising field, there is no increase in magnetism. This stage is known as *saturation stage.*

Hysteresis. When an external magnetic field is applied, the domain walls of the favourably oriented domains expand and start moving out. If the field is extremely smali the process is reversible *i.e.*, as soon as the field is removed the material gets demagnetised and domains regain their original size and orientation. It is observed that when the magnetic field is increased the imperfections in the crystalline structure of the material offer obstruction to the outward movement of the domain walls and the impurities resist the favourable orientation of the atomic dipoles. However, when the magnetising field is sufficiently increased the domain walls snap past these imperfections with a *sudden jerk.* Thereafter, the movement of the domain wall is again quite smooth and rapid till the next imperfection is reached.

If the field is now *decreased* the domain walls start moving inward and there is an obstruction again in moving past an imperfection. The field has to be decreased to a much lower value in order to overcome the obstruction and some energy is lost in the process. Even when the magnetising field is reduced to zero, some magnetism may still be present in the material. This explains *residual magnetism.* In order to reduce the residual magnetism to zero a field in the reverse direction has to be applied. This measures the *coercive force.* As the magnetisation always lags behind the magnetising force the phenomenon is known as *hysteresis* and the energy lost in the process as stated above is the loss of energy due to hysteresis.

Let us study hysteresis in detail.

20.11 VARIATION OF $\vec{H}$ WITH P_m (HYSTERESIS LOOP)

A speciment of iron can be magnetised by placing it in a solenoid and passing a current through it. If the value of current is gradually increased the magnitude of the magnetising field H increases and the magnitude of the intensity of magnetisation P_m developed in the specimen also increases. A graph between P_m and H is shown in Fig. 20.16. At O when H is zero P_m is also zero. When H increases P_m also increases but not very uniformly till we reach A. Beyond A if H is further increased P_m remains constant. The substance in this condition is said to be *saturated*.

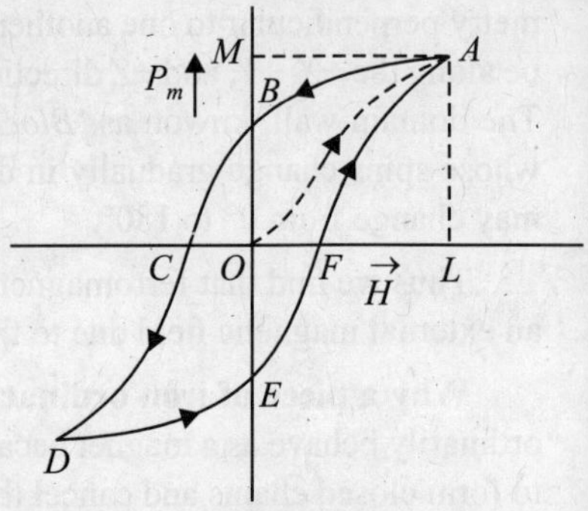

Fig. 20.16

(*ii*) If H is now decreased the value of P_m decreases but at a much lower rate and when H is zero the value of P_m is not zero but has finite value represented by OB.

(*iii*) If the value of H is now increased in the opposite direction the value of P_m further decreases and becomes zero when the value of the magnetising field H is equal to OC.

(*iv*) If the value of H is further increased again the saturation stage is reached in the opposite direction and the curve CD is obtained.

(*v*) On decreasing the value of H to zero, we get the portion DE of the curve. It is clear from the curve that P_m has a value given by OE even when H is zero.

(*vi*) On increasing the value of H in the positive direction the part EFA of the cycle is repeated till we again reach the saturation stage at A. This completes the cycle of magnetisation.

The graph between magnetic induction B and the magnetising field H as shown in Fig. 20.17 is almost similar in shape as that between P_m and H but is bigger in size. Since at the saturation stage, increase in B does not stop but is equal to the increase in H, this part of the B–H curve slopes very slightly upwards and unlike the $P_m – H$ curve is not parallel to the H-axis.

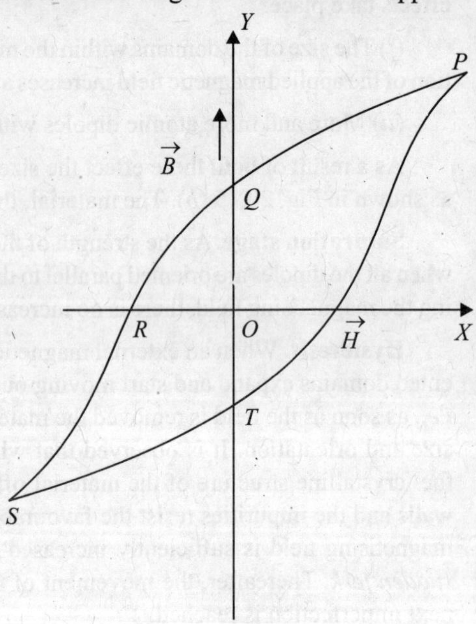

Fig. 20.17

In the above experiment it is seen that the intensity of magnetisation or magnetic induction does not become zero although the magnetising field is brought down to zero, *i.e.*, the value of P_m or B always lags behind H. This shows that the magnetic conditions of iron tend to persist or oppose any change.

This lagging of magnetic induction or intensity of magnetisation behind the magnetising force producing it when a specimen of iron is taken through a cycle of magnetisation is called **hysteresis.**

Hysteresis loop. *The graph showing how P_m or B increases with H from zero to a maximum in one direction and then taken through zero to a maximum in the opposite direction and finally back again through zero to the first maximum is called a* **hysteresis loop.**

Remanence (or retentivity). *The value of magnetic induction B which remains after the material has been magnetised and the magnetising force i.e., H has been reduced to zero is known as* ***remanent induction.***

The value of remanent induction depends upon the degree of magnetisation of the specimen before making H zero. *If the magnetic induction B reaches the saturation value before H is reduced to zero, the limiting value of remanent induction is known as* **remanence** *or* **retentivity**. Thus in Fig. 20.17 the magnetic induction left behind when H is reduced to zero after the specimen has reached the saturation state is represented by OQ.

$\therefore$ Remanence = OQ.

Coercivity. *The coercive force is the magnitude of the reverse field required to reduce the remanent induction to zero.*

The limiting value of coercive force required to reduce remanence to zero is known as **coercivity**. In Fig. 20.17 the reverse field required to make remanence zero is OR.

$\therefore$ Coercivity = OR

Magnetic materials having a **small** coercive force are known as **soft** whereas those possessing a **large** coercive force are known as **hard**.

20.12 DISSIPATION OF ENERGY DUE TO HYSTERESIS

Consider a unit volume $a\ b\ c\ d$ of a material placed in a magnetising field of strength $\vec{H}$. Let $\vec{P}_m$ be the magnetic moment of one of the molecular or atomic magnets of the specimen and let θ be the angle between the direction of magnetic dipole moment of this molecule and the direction of the magnetising field $\vec{H}$.

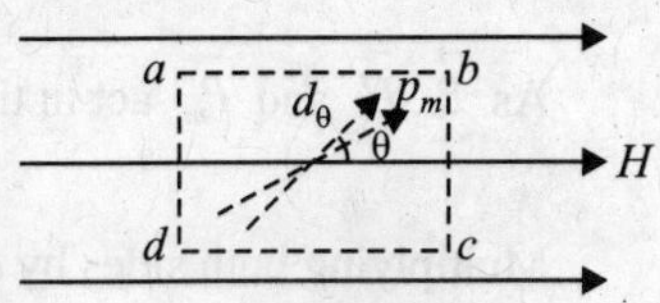

Fig. 20.18

The component of $\vec{p}_m$ parallel to the field $\vec{H} = p_m \cos\theta$ and component of $\vec{p}_m$ perpendicular to the field $\vec{H} = p_m \sin\theta$

$\therefore$ The total magnetic moment parallel to the field $= \Sigma p_m \cos\theta$ and total magnetic moment perpendicular to the field $= \Sigma p_m \sin\theta$ where Σ denotes the sum of the components of the magnetic moment due to all the molecular magnets in a particular direction.

Now $\Sigma p_m \cos\theta = P_m$

where P_m denotes the density of magnetisation, *i.e.*, magnetic moment per unit volume.

Since there is no magnetisation perpendicular to $\vec{H}$, the sum of all the magnetic moments in a direction perpendicular to the field = 0 or

$$\Sigma p_m \sin\theta = 0$$

When P_m is increased by a very small amount dP_m, it is given by

$$dP_m = d\,\Sigma p_m \cos\theta = -\Sigma p_m \sin\theta\, d\theta \qquad ...(i)$$

The negative sign simply indicates that P_m decreases as θ increases.

The torque acting on a dipole of magnetic moment $\vec{p}_m$ in a magnetic induction field $\vec{B}$

$$= \vec{\tau} = \vec{p}_m \times \vec{B} = p_m\, B \sin\theta$$

If now the angle is *decreased* by a small amount $d\theta$, then

Work done in rotating the dipole through $(-d\theta)$

$$= \tau(-d\theta) = -B\,p_m \sin\theta\, d\theta$$

$\therefore$ Work done in rotating all the molecular magnets in a unit volume

$$dW = -\sum B\, p_m \sin\theta\, d\theta = -B\sum p_m \sin\theta\, d\theta \qquad ...(ii)$$

Substituting $-\sum p_m \sin\theta\, d\theta = dP_m$ from Eq. (*i*) in Eq. (*ii*), we have

$$dW = B\, dP_m$$

or the work done in changing the value of P_m from P_{m1}, to P_{m2}

$$= \int_{P_{m_1}}^{P_{m_2}} B\, dP_m$$

The total work done in taking the sample through a complete cycle of magnetisation

$$W = \oint B\, dP_m$$

But $\qquad \vec{B} = \mu_0 (\vec{H} + \vec{P}_m)$

As $\vec{B}, \vec{H}$ and $\vec{P}_m$ act in the same direction

$$B = \mu_0 (H + P_m) = \mu_0 H + \mu_0 P_m$$

Multiplying both sides by dP_m and integrating over a complete cycle, we have

$$\oint B\, dP_m = \mu_0 \oint H dP_m + \mu_0 \oint P_m\, dP_m$$

$$= \mu_0 \oint H\, dP_m$$

as $\oint P_m\, dP_m = 0$ for if we plot a curve for P_m against P_m, it is a straight line and $\oint P_m\, d P_m$ which represents the area of this closed curve for a complete cycle is zero.

$$\therefore \qquad W = \oint B\, d P_m = \mu_0 \oint H\, d P_m \qquad ...(iii)$$

If we plot a graph between $\vec{H}$ and $\vec{P}_m$ for a complete cycle we get a closed curve. $\oint H\, dP_m$ represents the area of this curves as proved below.

***(i)* Loss of energy per cycle.** Consider a complete cycle of magnetisation *ABCDEFA*. Let *pq* be the small step in the process of magnetisation in which $\vec{P}_m$ increases by a small amount $d\vec{P}_m$.

$\therefore$ Work done per unit volume

$$= \mu_0 H dP_m = \mu_0 \times \text{area } pqrs$$

where H is the magnitude of the magnetising field $\vec{H}$ and dP_m that of change in intensity of magnetisation $\vec{P}_m$.

(*i*) When the magnetising force H is increased from 0 to *OL*, the intensity of magnetisation changes from *OE* (in the negative direction) to *OM* (in the positive direction).

$\therefore$ Work done **on** the specimen per unit volume in increasing $\vec{P}_m$ from the value $-OE$ to $+OM$

$$= \mu_0 \int_{-}^{M} H\, dP_m = \mu_0 \times \text{area } EFAMBOE$$

(*ii*) When H is decreased from OL to zero P_m decreases from OM to OB.

Work done **by** the specimen per unit volume

$$= \mu_0 \int_M^B H\, dP_m = \mu_0 \times \text{area } AMBA$$

(*iii*) When H is increased in the opposite direction from zero to OL', P_m changes from $+ OB$ to $- ON$

$\therefore$ Work done **on** the specimen per unit volume

$$= \mu_0 \int_B^N H\, dP_m = \mu_0 \times \text{area } BCDNEOB$$

(*iv*) When H is again decreased to zero and the cycle is completed P_m decreases from $- ON$ to $-OE$.

$\therefore$ Work done **by** the specimen per unit volume

$$= \mu_0 \int_N^E H\, dP_m = \mu_0 \times \text{area } DNED$$

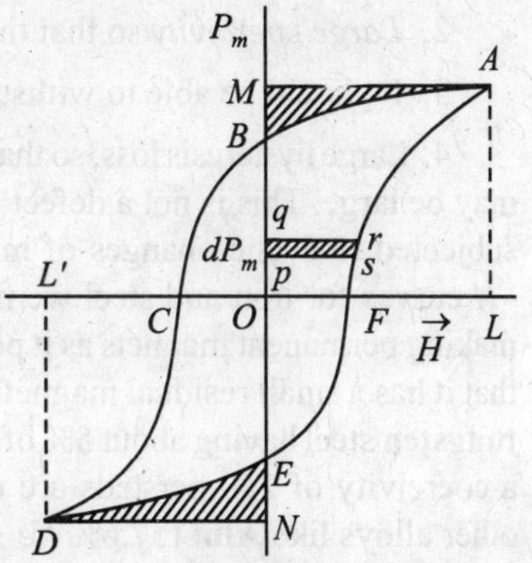

Fig. 20.19

It is seen that when the specimen is taken through a complete cycle of magnetisation more work is done **on** the material than is recovered **from** it.

Thus there is a net loss of energy in each cycle of hysteresis which is given by

Loss of energy per cycle per unit volume

$$= \mu_0 \text{ (area } EFAMBOE + \text{area } BCDNEOB) - \mu_0 \text{ (area } AMBA + \text{area } DNED)$$

$$= \mu_0 \times \text{ area of the hysteresis loop } ABCDEFA$$

Hence the loss of energy per unit volume per cycle due to hysteresis is equal to μ_0 × *area of* P_mH *hysteresis loop.*

(*ii*) Hysteresis loss from *B-H* cycle. We know that

$$B = \mu_0 (H + P_m)$$

$$\therefore \quad dB = \mu_0 (dH + dP_m) = \mu_0\, dH + \mu_0\, dP_m$$

Multiplying both sides by H and integrating over a complete cycle we have

$$\oint HdB = \mu_0 \oint HdH + \mu_0 \oint Hd\, P_m = \mu_0 \oint Hd\, P_m$$

as $\oint HdH = 0$ for if we plot a curve of H against H, it is a straight line and the area included by it for a complete cycle is zero.

But $\mu_0 \oint HdP_m$ gives the loss of energy per unit volume per cycle of hysteresis and $\oint HdB$ the area of B-H hysteresis loop.

Hence loss of energy per unit volume per cycle of hysteresis is equal to $\oint HdB$ = *Area of B-H loop.*

20.13 PERMANENT MAGNETS

The material for a permanent magnet should have the following properties :

1. *High residual magnetism* even in the presence of the mechanical disturbances.
2. *Large coercivity* so that the magnetisation is not wiped out by stray external fields.
3. It should be able to withstand mechanical ill-treatment and temperature changes.
4. Large hysteresis loss, so that the energy stored in the material may be large. This is not a defect as permanent magnets are never subjected to cyclic changes of magnetisation. Looking at the P_m $-H$ curves for iron and steel we find that steel is better suited for making permanent magnets as it possesses all the properties except that it has a small residual magnetism. In actual practice, however, **tungsten** steel having about 5% of tungsten and **cobalt steel** having a coercivity of 240 oersteds are used for the purpose. There are other alloys like **Alni** (57.6% *Fe* + 29% *Ni* 13.4% *Al*) and **Alnico** (18% *Ni*, 13% *Al*, 12% *Co*, 3% *Cu* and 54% *Fe*) which have very high coercive force.

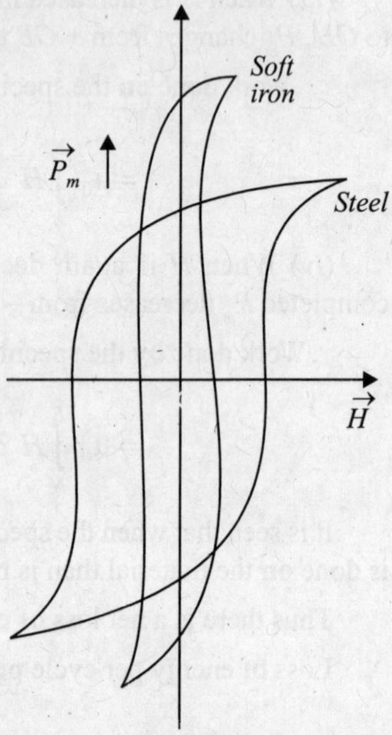

Fig. 20.20

Very recently a new magnetic alloy **vicalloy** has been developed. It contains iron, cobalt and vanadium and has the highest value of coercive force so far obtained. It can be made into the form of a wire or ribbon (tape) and is used in magnetic recording of sound.

Electromagnets (or cores of transformers and chokes). The material for the electromagnet should have the following properties.

1. Large value of P_m with comparatively small magnetising field.
2. High susceptibility for low fields and high permeability.
3. Low hysteresis loss because the material is subjected to cyclic changes of magnetisation.

All these properties are possessed by soft iron which is thus the best suited material for the purpose.

SOLVED EXAMPLES

Example 20.1. *The volume of the core of a transformer is 1000 cm³. It is fed with A.C of 50 Hz. If the loss of energy due to hysteresis per hour is 36 Joules calculate the area of B-H loop.*

(P.U. 2000; H.P.U. 2003; Pbi. U. 2007, 2002)

Solution. Loss of energy due to hysteresis per second

$$= \frac{36}{60 \times 60} = 10^{-2} \text{ J/s}$$

$$\therefore \text{ Loss of energy per cycle } = \frac{10^{-2}}{50} = 2 \times 10^{-4} \text{ J/s/cycle}$$

$$\text{Volume of transformer core } = 1000 \text{ cm}^3 = \frac{1000}{10^6} = 10^{-3} \text{ m}^3$$

$$\therefore \text{ Loss of energy per cycle per unit volume } = \frac{2 \times 10^{-4}}{10^{-3}}$$

$$= 2 \times 10^{-1} = 0.2 \text{ Jm}^{-3}$$

$$= \text{area of } B\text{-}H \text{ loop}$$

Example 20.2. *Hysteresis loss per cubic metre for soft iron is 10^3 Joules. Taking the density 7.5 gm per c.c. and specific heat 100 cal k $gm^{-1} {}^0C^{-1}$ calculate the rise of temperature per minute when the specimen is subjected to an alternating magnetic field of frequency 50 cycles/second.*

Solution. (*a*) Loss of energy per unit volume per cycle = 10^3 Joules

∴ Energy lost per minute per cubic metre = $10^3 \times 50 \times 60$ Joule

$$\text{Heat generated per minute} = \frac{10^3 \times 50 \times 60}{4.2} \text{ cals}$$

$$\text{Mass of iron per cc} = 7.5 \text{ gm}$$

$$\text{Mass of iron per cubic metre in kg} = \frac{7.5 \times 10^6}{10^3} = 7.5 \times 10^3$$

If θ is the rise in temperature then

$$7.5 \times 10^3 \times 100 \times \theta = \frac{10^3 \times 50 \times 60}{4.2}$$

$$\therefore \quad \theta = 0.95°C$$

Example 20.3. *The hysteresis loop of a transformer has area 2500 ergs cm^{-3}. Calculate the loss of energy per hour at 50 Hz frequency. Density of iron is 7.5 gm cm^{-3} and weight is 10 kg.*

(Kan. U., 1995, 1991)

Solution. Loss of energy per unit volume per cycle = 2500 ergs cm^{-3}

∴ Loss of energy per unit volume per hour = $50 \times 60 \times 60 \times 2500$ ergs cm^{-2}

$$\text{Volume of iron} = \frac{10 \times 1000}{7.5} \text{ c.c.}$$

$$\therefore \text{Total loss of energy per hour} = \frac{50 \times 60 \times 60 \times 2500 \times 10 \times 1000}{7.5} = 6 \times 10^{11} \text{ ergs}$$

$$= 6 \times 10^4 \text{ Joules}$$

20.14 USE OF HYSTERESIS CURVE

The loss of energy per unit volume per cycle of hysteresis is equal to the area of a *B-H* loop. Therefore, in electrical and magnetic applications where alternating currents are used particularly of high frequency the hysteresis loss has to be kept a minimum. Transformer cores, telephone diaphragms and chokes are usually subjected to cyclic changes. Therefore the magnetic material suitable for these must have the following essential properties.

(*i*) *High initial permeability* so that the material acquires a high flux density even for low magnetising fields.

(*ii*) *Low hysteresis loss* so that there is a small dissipation of energy. This will avoid heating and reduce the possibility of breakdown of the insulation of the windings.

(*iii*) *High specific resistance* to reduce the eddy current losses.

Taking into consideration these requirements, *soft iron is better than steel* for these appliances.

Transformer steel is an alloy of soft iron containing 4% silicon. **Mumetal** is an alloy having high initial permeability. It contains 76% Ni, 5% Cu, 17% Fe and 1.5% Cr.

20.15 MAGNETISM IN MATTER

The Langevin theory of diamagnetism and the theory of paramagnetism has been discussed in article 20.7 and 20.9. We shall now discuss

(*i*) Quantum theory of paramagnetism

(*ii*) Weiss theory of ferromagnetism

(*iii*) Anti-ferro magnetism and

(*iv*) Ferri magnetism.

Quantum Theory of Paramagnetism. The magnetic moment of an atom or ion in free space is given by

$$\vec{\mu}_j = \gamma\hbar\vec{J} = -g_j\mu_B\vec{J} \qquad ...(i)$$

where $|\vec{J}| = \sqrt{J(J+1)}$, J being the *total angular momentum* quantum number.

In equation (*i*) $\hbar\vec{J}$ is the total angular momentum *i.e.*, the sum of the orbital angular momentum $\hbar\vec{L}$ and spin angular momentum $\hbar\vec{S}$. The constant γ is the ratio of the magnetic moment $\vec{\mu}$ to the total angular momentum $\hbar\vec{J}$. It is called **gyromagnetic ratio** or **magnetogyric ratio.**

The quantity g_j is called the *g-factor* or *Lande's splitting factor*. For a free atom it is given by

$$g_j = 1 + \frac{J(J+1) + S(S+1) - L(L+1)}{2J(J+1)}$$

μ_B is the *Bohr magneton* and has a value $\frac{e\hbar}{2m}$ (in S.I. units) = 9.27×10^{-24} Joule/Tesla.

The energy levels of the system in a magnetic field are

$$E = -\vec{\mu}_j \cdot \vec{B} = m_j g_j \mu_B B = \mu B \qquad ...(ii)$$

where $\mu = m_j g_j \mu_B$ and m_j is the *azimuthal* or *total magnetic quantum number* given by $m_j = m_l + m_s$. It has $(2J + 1)$ values from $-J$ through zero to $+J$ *i.e.*, $(-J)$, $(-J+1)$ ... 0, 1 ... $(J-1)$ (J) *i.e.*, there are $(2J+1)$ ways in which the atom can align itself in a magnetic field.

Consider a simple case where $L = l = 0$ and single spin *i.e.*,

$$S = s = \frac{1}{2}, \text{ then}$$

$$m_l = 0 \quad m_s = \pm\frac{1}{2} \text{ and } m_j = m_l + m_s = \pm\frac{1}{2}. \text{ Also } g_j = 2$$

$$\therefore \quad \mu = m_j g_j \mu_B = \pm\mu_B$$

and

$$E = \pm\mu_B B$$

Thus the system has two energy levels

$$E_1 = +\mu_B B \text{ and } E_2 = -\mu_B B$$

with energy difference

$$E_1 - E_2 = 2\mu_B B$$

The energy level splitting for one electron in a magnetic field B directed along the positive Z-axis is shown in Fig. 20.21.

For an electron the magnetic moment $\vec{\mu}$ *is opposite* in sign to the spin $\vec{S}$ so that $\vec{\mu} = -g_j\mu_B\vec{S}$. In the *lower* energy state the magnetic moment is *parallel* to the magnetic field and in the *upper* energy state it is *anti-parallel*.

Let us suppose that the system contains N atoms per unit volume out of which the spins of N_1 are parallel to the field (lower energy state) and that of N_2 is anti-parallel (upper energy state), then

$$N = N_1 + N_2$$

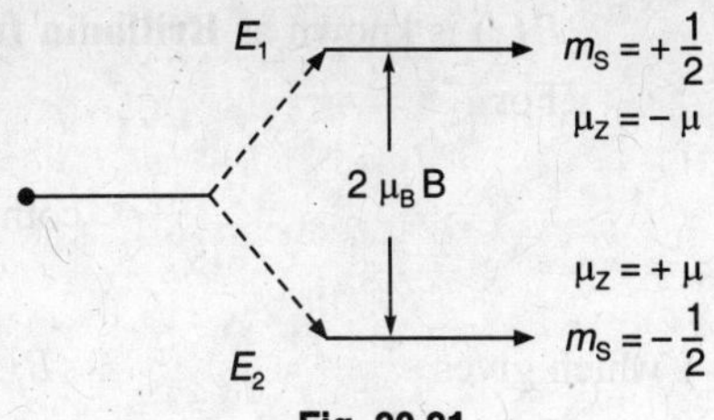

Fig. 20.21

The value of N_1 and N_2 is given by Maxwell-Boltzmann statistics. In thermal equilibrium at a temperature T and in a magnetic field B

$$\frac{N_1}{N} = \frac{e^{+\mu B/kT}}{e^{+\mu B/kT} + e^{-\mu B/kT}}$$

and

$$\frac{N_2}{N} = \frac{e^{-\mu B/kT}}{e^{+\mu B/kT} + e^{-\mu B/kT}}$$

The component of the magnetic moment of the upper state in the direction of the magnetic field is $-\mu$ and of the lower state is $+\mu$.

∴ Resultant magnetic moment per unit volume

$$M = \mu(N_1 - N_2) = N\mu \frac{e^{+\mu B/kT} - e^{-\mu B/kT}}{e^{+\mu B/kT} + e^{-\mu B/kT}}$$

$$= N\mu \tanh x \quad \text{...(iii)}$$

where $x = \mu B/kT$

For $x \ll 1$, $\tanh x \cong x$

and we have

$$M = N\mu x = \frac{N\mu^2 B}{kT} \quad \text{...(iv)}$$

$$= \frac{N m_j^2 g_j^2 \mu_B^2 B}{kT} \quad \text{...(v)}$$

For $m_j = \frac{1}{2}$, $g_j = 2$, this gives the value of

$$M = \frac{N\mu_B^2 B}{kT} \quad \text{...(vi)}$$

Comparing with relation $P_m = \frac{N p_m^2 B}{3kT}$ (and substituting $P_m = M$ and $p_m = \mu_B$) we have

$$M = \frac{N\mu_B^2 B}{3kT}$$

Thus the result derived in Eq. (*vi*) appears to differ from classical result by a factor 1/3. But relation (*vi*) is only a special case where we have taken $L = l = 0$ and $S = s = \frac{1}{2}$ so that $J = j = \frac{1}{2}$.

As a general case, in a magnetic field an atom with total angular momentum quantum number J has $(2J + 1)$ equally spaced energy levels. The magnetisation M is then given by

$$M = N g_j J \mu_B B_j(x) \quad \text{...(vii)}$$

where

$$x = \frac{\mu_j B}{kT} = \frac{g_j J \mu_B B}{kT}$$

and

$$B_j(x) = \frac{2J+1}{2J} \coth\left(\frac{2J+1}{2J} x\right) - \frac{1}{2J} \coth\left(\frac{x}{2J}\right)$$

$B_j(x)$ is known as **Brillouin function.**

For $x << 1$, we have

$$\coth x = \frac{1}{x} + \frac{x}{3} - \frac{x^3}{45} + \ldots$$

which gives $$B_j(x) = \frac{(J+1)}{J}\frac{x}{3} = \frac{(J+1)g_j\mu_B B}{3kT} \qquad \ldots (viii)$$

Substituting in (*vii*), we get

$$M = \frac{NJ(J+1)g_j^2\mu_B^2 B}{3kT} \qquad \ldots (viii)\ (a)$$

Paramagnetic susceptibility. Paramagnetic susceptibility in S.I units

$$= \frac{\mu_0 M}{B}$$

or $$\chi_{para} = \frac{N\mu_0\mu_B^2 g_j^2}{3kT} J(J+1)$$

$$= \frac{\mu_0 N p^2 \mu_B^2}{3kT} = \frac{\mu_0 N \mu_{eff}^2}{3kT} = \frac{C}{T} \qquad \ldots (ix)$$

where $$p^2 = g_j^2 J(J+1),\ \mu_{eff}^2 = p^2\mu_B^2 = g_j^2 J(J+1)\mu_B^2 \qquad \ldots (ix)\ (a)$$

As $\mu_{eff} = p\mu_B$, p gives the effective number of Bohr magnetons. C is a constant given by

$$C = \frac{\mu_0 N \mu_B^2 g_j^2 J(J+1)}{3k}$$

$$= N\mu_0 \left(\frac{e\hbar g_j}{2m}\right)^2 \frac{J(J+1)}{3k}$$

and is known as **Curie's constant.** The result

$$\chi_{para} = \frac{C}{T}$$ is known as **Curie's law.**

This result is identical with classical result.

From relation (*ix*) (*a*) we find that

$$\mu_{eff}^2 = \mu_j^2 = g_j^2 J(J+1)\mu_B^2$$

or $$\mu_j = g_j\mu_B\sqrt{J(J+1)}$$

Paramagnetism of ionic crystals. The property of paramagnetism is due to the existence of partly filled electronic shells. The paramagnetic compounds are those which contain the *transition group elements i.e.*, rare earth group (incomplete 4*f* shell) and the iron group (incomplete 3*d* shell).

The prediction of equation (*ix*) for χ_{para} with J values given by Hund's rule *i.e.*,

$J = L - S$ for a shell less than half full

$J = 0$ for a shell just half full

and $J = L + S$ for a shell more than half full

are generally in good agreement with experimental results for many rare earth ions.

The predictions of equation (*ix*) are not in good agreement with experimental results for the ions of iron group. The reason for this is that the partially filled 3*d* shell for these ions is the outermost shell and is exposed to the strong field due to the neighbouring ions in the crystal. This field is known as **Crystal field.** It breaks the rotational symmetry and the total angular momentum quantum number is no longer the proper quantum number for the purpose.

Again the average values of L_z may be reduced to zero, the effect being known as *quenching of the orbital angular momentum.* It further means that the relation $E = m_j g_j \mu_B B$ given in equation (*ii*) should be replaced by

$$E = 2m_s \mu_B B \qquad [\because g_j = g_s = 2 \text{ for spin}]$$

$$\therefore \quad E = 2\frac{e\hbar}{2m} m_s B = \frac{e\hbar}{m} m_s B$$

Using this value of E and $m_s = \pm \frac{1}{2}$, we get

$$\chi_{para} = N\mu_0 \left(\frac{e\hbar}{m}\right)^2 \frac{S(S+1)}{3kT} \qquad ...(x)$$

The results given by equation (*x*) are in good agreement with experimental results for iron group. For example the value of χ_{para} for Mn^{3+} is zero according to Eq. (*ix*) but according to Eq. (*x*) with $S = 2$ it is in good agreement with experimental results.

According to Eq. (*ix*) $p = g_j \sqrt{J(J+1)}$ and according to Eq. (*x*) $p = 2\sqrt{S(S+1)}$. Considering the basic levei $5D_o$ for Mn^{3+}, $J = 0$, $L = 2$ and $S = 1$ which gives the calculated value of p from equation (*ix*) equal to zero and from equation (*x*) $p = 4.90$. The experimental value of p for Mn^{3+} = 4.9 in perfect agreement with the result of Eq. (*x*).

20.16 WEISS THEORY OF FERROMAGNETISM

Ferromagnetism. Ferromagnetic substances are those which can be magnetised and retain magnetism even when the external magnetic field has been removed. Familiar examples are *Fe, Co, Ni, Cd, Dy* and their alloys.

Main features. (*i*) The magnetic dipole moment per unit volume $\vec{M}$ produced in such substances is very high even in weak magnetic fields.

(*ii*) The value of $\vec{M}$ is not linearly proportional to the applied magnetic field $\vec{B}a$. In other words, their magnetic susceptibility $\chi_m = \frac{\mu_0 M}{B_a}$ (in S.I. units) is not constant but varies with the applied field.

(*iii*) Above a certain critical temperature known as *Curie temperature* (or *Curie point*) T_c, they behave as paramagnetic substances. Their susceptibility at any temperature $T (T > T_c)$ is given by *Curie Weiss law i.e.*,

$$\chi_m = \frac{C}{T - T_c}$$

The susceptibility has a *singularity* at $T = T_c$. At this temperature and below, there exists a *spontaneous magnetisation* because if χ_m is infinite, we have a finite $\vec{M}$ for zero value of $\vec{B}a$. This is due to the interaction between the magnetic ions which is strong enough to align their magnetic moments against the disorder produced by thermal effects. The Curie temperature is the temperature above which the spontaneous magnetisation vanishes. It separates the disordered paramagnetic phase at $T > T_c$ from the ordered ferromagnetic phase at $T < T_c$.

Weiss field theory of ferromagnetism. A ferromagnetic specimen of macroscopic dimensions contains a number of small regions called **domains** which are spontaneously magnetised. Within each domain the spontaneous magnetisation is due to the alignment of the magnetic moments of adjacent atoms – all pointing in the same direction. The interaction which aligns the magnetic moments is quantum mechanical in origin and is due to the exchange properties of the electron wave functions. The *exchange interaction* also known as *Weiss field* arises when the wave functions of the two atoms overlap. The electrons being indistinguishable belong to both the atoms. In such cases the symmetry or the antisymmetry of the wave functions strongly influences the energy of the system. It is the *exchange symmetry* between the spins and the *extent of overlap* of the wave functions which determines the nature and extent of the exchange interaction.

We can treat the *exchange field* as equivalent to an internal magnetic field $\vec{B_e}$ which is proportional to the magnetisation $\vec{M}$ within a domain. Thus in the mean *field approximation* each atom experiences a magnetic field proportional to magnetisation *i.e.*,

$$\vec{B_e} = \lambda \vec{M}$$

where λ is *Weiss constant* independent of temperature. If the external applied field is $\vec{B_a}$, then the effective field $\vec{B}$ acting on an ion or an atom is given by

$$\vec{B} = \vec{B_a} + \vec{B_e} = \vec{B_a} + \lambda \vec{M}$$

As $\vec{B}$, $\vec{B_a}$, $\vec{B_e}$ and $\vec{M}$ are all vectors and act along the same direction, we have

$$B = B_a + B_e = B_a + \lambda M \qquad ...(i)$$

Suppose a unit volume of the ferromagnetic material contains N atoms, then as in a paramagnetic case [Eq. (*vii*), of **article 20.15**]

$$M = Ng_j J\mu_B B_j(x) \qquad ...(ii)$$

where
$$B_j(x) = \frac{2J+1}{2J}\coth\left(\frac{2J+1}{2J}x\right) - \frac{1}{2J}\coth\left(\frac{x}{2J}\right)$$

$B_j(x)$ being the Brillouin function and for ferromagnetic substances

$$x = \frac{\mu_j B}{kT} = \frac{g_j J\mu_B B}{kT}$$

$$= \frac{g_j J\mu_B (B_a + \lambda M)}{kT}$$

For spontaneous magnetisation when there is no external applied magnetic field $B_a = 0$; $B = \lambda M$ [From Eq. (*i*)], then

$$x = \frac{g_j J\mu_B \lambda M}{kT} \qquad ...(iii)$$

Suppose when all the domains are aligned along the magnetic field the magnetic moment per unit volume known as *saturation magnetic moment* is given by

$$M_s = Ng_j J\mu_B \qquad ...(iv)$$

Comparing (*ii*) and (*iv*), we have

$$\frac{M}{M_s} = B_j(x) \quad \text{...(v)}$$

From (*iii*)

$$M = \frac{xkT}{g_j J \mu_B \lambda} \quad \text{...(vi)}$$

Comparing (*vi*) and (*iv*), we get

$$\frac{M}{M_s} = \frac{kT}{Ng_j^2 J^2 \mu_B^2 \lambda} x \quad \text{... (vii)}$$

The simultaneous solutions of equations (*v*) and (*vii*) can be obtained by plotting $\frac{M}{M_s}$ against x. The plot of $\frac{M}{M_s}$ against x in accordance with Eq. (*vii*) for $T > T_c$, $T = T_c$ and $T < T_c$ and the plot of $\frac{M}{M_s}$ against x in accordance with Eq. (*v*) for $T = T_c$ is shown in Fig. 20.22. The plot of Eq. (*vii*) gives a straight line with different values of T and that of Eq. (*v*) gives a curve.

(*i*) **At critical temperature or (Curie point).** At the ferromagnetic Curie temperature T_c the straight line given by Eq. (*vii*) is tangent to the curve given by Eq. (*v*) at the origin.

The slope of the straight line given by

$$\frac{M}{M_s} = \frac{kT}{Ng_j^2 J^2 \mu_B^2 \lambda} x$$

at

$$T = T_c$$

$$\frac{M / M_s}{x} = \frac{kT_c}{Ng_j^2 J^2 \mu_B^2 \lambda} \quad \text{...(viii)}$$

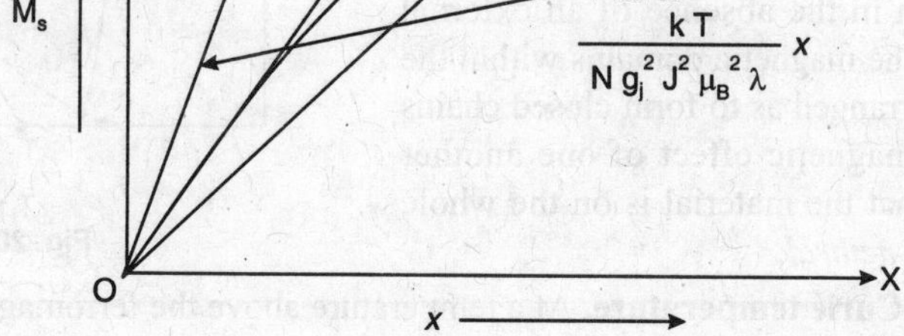

Fig. 20.22

For $x \ll 1$, the Brillouin function

$$B_j(x) \cong \frac{J+1}{3J} x \quad \text{[See Eq. (viii) article 20.15]}$$

∴ Tangent to the curve at the origin has a slope given by

$$\frac{B_j(x)}{x} = \frac{J+1}{3J} \quad \text{...(ix)}$$

Equating (*viii*) and (*ix*), we get

$$\frac{kT_c}{Ng_j^2 J^2 \mu_B^2 \lambda} = \frac{J+1}{3J}$$

or

$$T_c = \frac{Ng_j^2 \mu_B^2 \lambda}{3k} J(J+1) \quad ...(x)$$

$$= \frac{N\mu_j^2 \lambda}{3k} \quad ... [x(a)]$$

where μ_j is the total magnetic moment per atom given by

$$\mu_j = g_j \mu_B \sqrt{J(J+1)} \quad \text{[See Eq. (i) article 20.15]}$$

This shows that T_c is proportional to the *Weiss field constant* λ.

(*ii*) **Below Curie temperature.** Below Curie temperature the straight line given by Eq. (*vii*) intersects the curve given by Eq. (*v*) at two points $x = 0$ and P. The intersection at $x = 0$ is classically unstable condition. The intersection at P corresponds to *spontaneous magnetisation*. Thus below the critical temperature spontaneous magnetisation results in ferromagnetic materials even in the absence of magnetic field. Substituting

$$\frac{k}{Ng_j^2 J^2 \mu_B^2 \lambda} = \frac{1}{T_c}\frac{J+1}{3J} \quad \text{from [Eq. (x)] in Eq. (vii)}$$

we get

$$\frac{M}{M_s} = \frac{T}{T_c}\left(\frac{J+1}{3J}\right)x \quad ...(xi)$$

In Eq. (*xi*) M_s is the maximum value of spontaneous magnetisation and it gives the dependence of spontaneous magnetism on temperature.

However, a macroscopic sample of a ferromagnetic material at a temperature $T < T_c$ is generally not magnetised. It only becomes a magnet under the influence of an external magnetic field. This behaviour is due to the fact that the ferromagnetic material is usually subdivided into small regions known as *domains*. Each of these domains is spontaneously magnetised even in the absence of an external magnetic field. The magnetic domains within the material are so arranged as to form closed chains and cancel the magnetic effect of one another with the result that the material is on the whole unmagnetised or *neutral*.

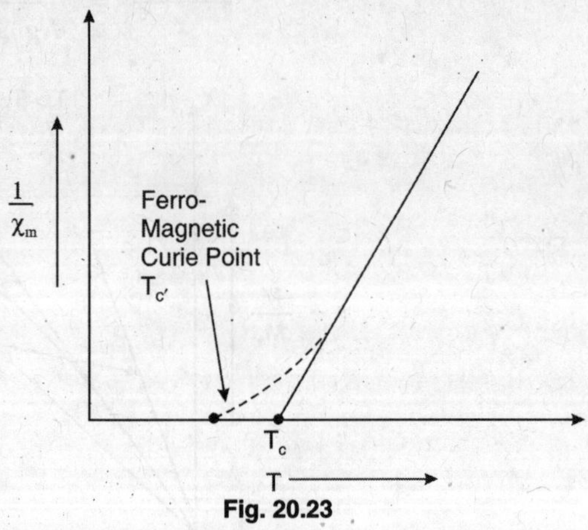

Fig. 20.23

(*iii*) **Above Curie temperature.** At a temperature above the ferromagnetic Curie temperature, the magnetisation occurs only when the external magnetic field $\vec{B_a}$ is applied. There is no spontaneous magnetisation. This region is known as *paramagnetic region*. Therefore, as in a paramagnetic case

$$M = \frac{NJ(J+1)g_j^2\mu_B^2 B}{3kT} \quad \text{[Eq. (viii) (a) article 20.15]}$$

$$= \frac{NJ(J+1)g_j^2\mu_B^2 (B_a + \lambda M)}{3kT}$$

or $$M\left[1-\frac{NJ(J+1)g_j^2\mu_B^2\lambda}{3kT}\right] = \frac{NJ(J+1)g_j^2\mu_B^2 B_a}{3kT}$$

$$\therefore \qquad M\left(1-\frac{T_c}{T}\right) = \frac{C}{\mu_0 T}B_a$$

where $$C = \frac{\mu_0 NJ(J+1)g_j^2\mu_B^2}{3k}.$$
is the Curie constant

or $$\frac{\mu_0 M}{B_a}\left(\frac{T-T_c}{T}\right) = \frac{C}{T}$$ [Eq. (*viii*) (*a*) **article 20.15**]

$$\therefore \qquad \chi_m = \frac{\mu_0 M}{B_a} = \frac{C}{T-T_c} \qquad \text{...}(xii)$$

This is *Curie Weiss law*. According to this law

$$\frac{1}{\chi_m} = \frac{T}{C}-\frac{T_c}{C}$$

Thus the graph between $\frac{1}{\chi_m}$ and T is a straight line. For $\frac{1}{\chi_m} = 0 \;\; T = T_c$.

This law agrees quite accurately with experimental results. There is a slight curvature near the Curie point T_c. This slight curvature leads to the distinction between ferromagnetic and paramagnetic Curie points. In case of *Fe*, *Co* and *Ni* their values in degrees Kelvin are

	Fe	*Co*	*Ni*
T_c (Ferro)	1043	1393	631
T_c (Para)	1093	1428	650

Thus T_c (ferro) is slightly lower than T_c (para).

20.17 ANTI-FERROMAGNETISM

It is seen that ferromagnetism is due to the exchange interaction or exchange field which arises when the wave functions of the two atoms overlap. This gives rise to an *exchange energy*. The exchange energy is *positive* when neighbouring spins are parallel. This leads to a ground state in which the neighbouring spins are anti-parallel and hence there is a co-operative alternating alignment below a certain temperature T_N known as the *Neel temperature*.

In these crystals the alternate atoms have their spins parallel to each other but not the adjacent atoms. Such crystals can be considered to be composed up of two inter-penetrating sub-lattices *A* and *B* — one with atoms having their spin moments oriented in one direction and the other with atoms having their spin moments oriented in the opposite direction fig. 20.24 (*b*). Thus one sub-lattice gets aligned in one direction, whereas the second gets aligned in the other. Such substances are known as *anti-ferromagnetics*.

In the absence of the external magnetic field and below the Neel temperature their net magnetic moment is zero.

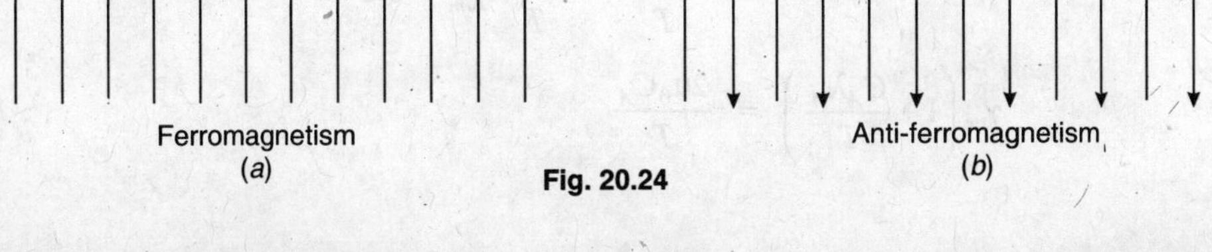

Fig. 20.24

The phenomenon of anti ferromagnetism can be analysed by assuming that alternate lattice points are occupied by atoms A and B with anti-parallel spins. Let the average magnetic moment of A and B sets of atoms be $\vec{M}_A$ and $\vec{M}_B$ respectively. For anti-parallel alignment the total magnetic field at the sites of atoms A and B respectively will be

$$\vec{B}_A = \vec{B}_0 - \lambda_A \vec{M}_B \quad ...(i)$$

and

$$\vec{B}_B = \vec{B}_0 - \lambda_B \vec{M}_A \quad ...(ii)$$

where $\vec{B}_0$ is the external applied magnetic field, λ_A, λ_B constants independent of temperature and $\vec{B}_A$ and $\vec{B}_B$ the total magnetic field for the A and B sets of atoms respectively. For temperatures above T_N, the Neel temperature, according to Curie law, we have

$$\frac{C_A}{T} = \frac{\vec{M}_A}{\vec{B}_0 - \lambda_A \vec{M}_B}$$

or

$$\vec{M}_A = \frac{C_A}{T}(\vec{B}_0 - \lambda_A \vec{M}_B) \quad ...(iii)$$

Similarly

$$\vec{M}_B = \frac{C_B}{T}(\vec{B}_0 - \lambda_B \vec{M}_A) \quad ...(iv)$$

where

$$C_A = N_A \left(\frac{e\hbar g_A}{2m}\right)^2 \frac{J_A(J_A+1)}{3k}$$

and

$$C_B = N_B \left(\frac{e\hbar g_B}{2m}\right)^2 \frac{J_B(J_B+1)}{3k}$$

If atoms A and B are magnetically equivalent, then susceptibility

$$\chi_m = \frac{\mu_0(\vec{M}_A + \vec{M}_B)}{\vec{B}_0}$$

$$= \frac{\mu_0 C_A}{T}\left(1 - \lambda_A \frac{\vec{M}_B}{\vec{B}_0}\right) + \frac{\mu_0 C_B}{T}\left(1 - \lambda_B \frac{\vec{M}_A}{\vec{B}_0}\right)$$

$$= \frac{2\mu_0 C_A}{T} - \frac{C_A \lambda_A}{T}\left(\frac{\mu_0(\vec{M}_A + \vec{M}_B)}{\vec{B}_0}\right) \quad \left[\because \begin{array}{l} C_A = C_B \\ \lambda_A = \lambda_B \end{array}\right]$$

$$= \frac{2\mu_0 C_A}{T} - \frac{C_A \lambda_A}{T}\chi_m \quad ...(v)$$

or

$$\chi_m\left(1 + \frac{C_A \lambda_A}{T}\right) = \frac{2\mu_0 C_A}{T}$$

Substituting $C_A\lambda_A = T_c$, we have

$$\chi_m\left(1+\frac{T_c}{T}\right) = \frac{2\mu_0 C_A}{T}$$

or
$$\chi_m = \frac{2\mu_0 C_A}{T+T_c} = \frac{C}{T+T_N}$$

where $T_c = T_N$ = Neel temperature and $C = 2\,\mu_0 C_A$.

Thus in the case of anti-ferromagnets χ_m *is inversely proportional to* $T + T_c$.

When a magnetic field is applied, small magnetisation in the direction of the field is developed. This magnetisation increases as the temperature of the specimen is increased. With further increase of temperature, the interaction between sub-lattices becomes less effective and the susceptibility increases. For a particular temperature the susceptibility becomes maximum and above this temperature it starts decreasing. This critical temperature at which the susceptibility is maximum is known as *Neel temperature*. The variation of susceptibility χ_m with temperature T is shown in Fig. 20.25.

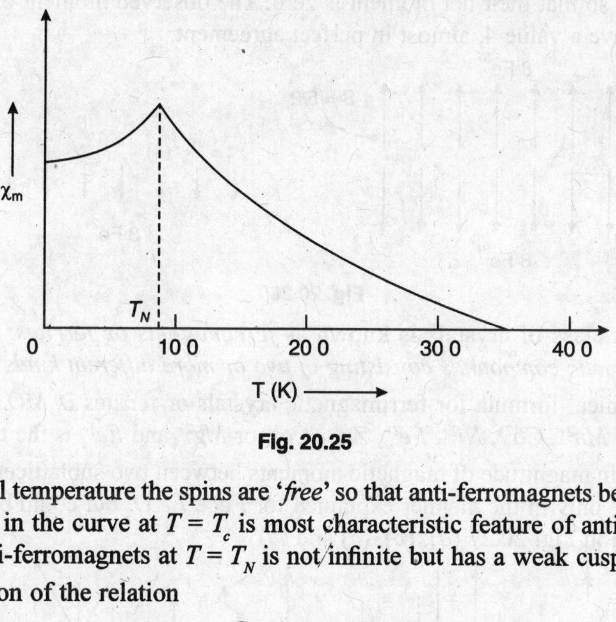

Fig. 20.25

Above the Neel temperature the spins are '*free*' so that anti-ferromagnets become paramagnets. The peak occurring in the curve at $T = T_c$ is most characteristic feature of anti-ferromagnets. The susceptibility of anti-ferromagnets at $T = T_N$ is not infinite but has a weak cusp as shown.

Another solution of the relation

$$\vec{M}_A = \frac{C_A}{T}(\vec{B}_0 - \lambda_A \vec{M}_B)$$

can be found for $T = T_c$ and $\vec{B}_0 = 0$. Then, we have

$$\vec{M}_A = -\frac{C_A\lambda_A}{T}\vec{M}_B$$

But
$$C_A\lambda_A = T_c \text{ and } \frac{T_c}{T} = 1$$

$$\therefore \quad \vec{M}_A = -\vec{M}_B$$

It means that at temperature below $T_c = T_N$ (Neel temperature) the sets of atoms A and B are spontaneously magnetised in the opposite directions and the net magnetisation is zero. This is why such materials are known as *antiferromagnets*. In antiferromagnets the crystal field aligns the magnetic moments along a preferred direction below T_C. A field applied perpendicular to this direction is associated with a susceptibility $\chi_\perp$ which is independent of $T(T < T_c)$ while a field applied parallel to this direction is associated with a susceptibility $\chi_\parallel$ which is equal to $\chi_\perp$ at $T = T_c$ but decreases to zero as $T \to 0$ K.

20.18 FERRIMAGNETISM

Ferrimagnetism is a special case of anti-ferromagnetism in which opposing magnetic moments of the two sets of atoms in the sub-lattices are of different magnitude giving rise to a large net magnetisation. A familiar example is that of *magnetite* Fe_3O_4 or FeO,Fe_2O_3. The ferric Fe^{3+} ions are in a state with spin $S = 5/2$ and zero orbital moment. Thus each ion should contribute $5\mu_B$ to the saturation moment. The ferrous Fe^{2+} ions have a spin $S = 2$ and should contribute $4\mu_B$ in addition to any residual orbital magnetic moment contribution. Thus the effective number of Bohr magnetons per Fe_3O_4 unit should be $2 \times 5 + 4 = 14\ \mu_B$ if all spins were parallel. The observed value is 4.1. The discrepancy is explained if the moments of the two Fe^{3+} ions are anti-parallel to each other as shown in Fig. 20.26, so that their net moment is zero. The observed moment only arises from the Fe^{2+} ions and will have a value 4, almost in perfect agreement.

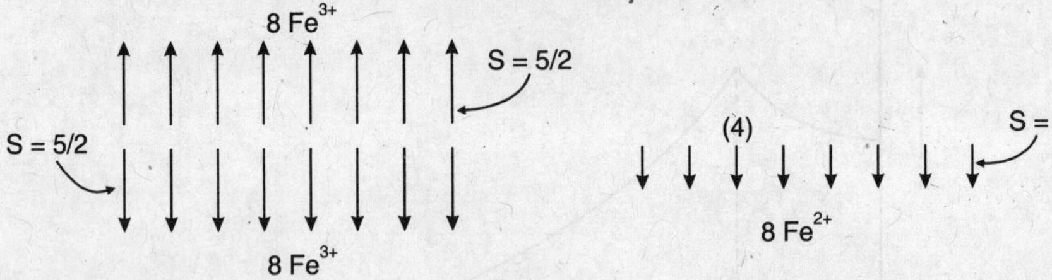

Fig. 20.26

This important class of crystals is known as *ferrimagnets or ferrites. The ferrimagnetic crystals are the magnetic compounds consisting of two or more different kinds of atoms.*

The usual chemical formula for ferrimagnetic crystals or ferrites is $MO.\ Fe_2O_3$ where M is a divalent metal like Mn^{2+}, Co^{2+}, Ni^{2+}, Fe^{2+}, Zn^{2+}, Cd^{2+} or Mg^{2+} and Fe^{3+} is the trivalent ferric ion.

The difference in magnitude of magnetic moments between two sublattices of a paramagnetic crystal does not arise only in the manner explained for FeO, Fe_2O_3 but could be possible in many other ways as shown in Fig. 20.27 (*a*), (*b*), (*c*) and (*d*).

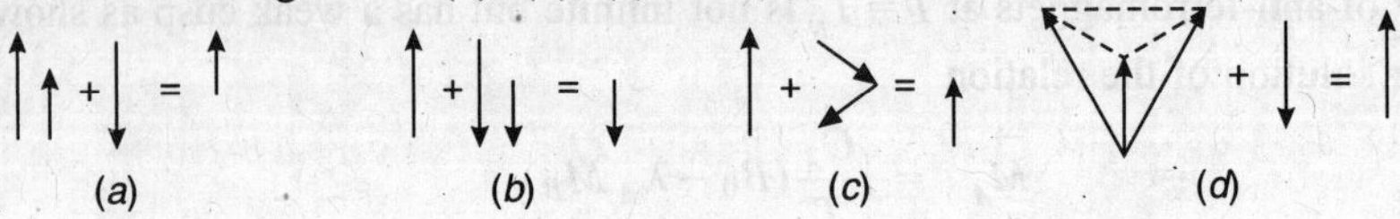

Fig. 20.27

Thus if the sets of atoms A and B forming the two sub-lattices are magnetically not equivalent, then from the relations

$$M_A = \frac{C_A}{T}(B_0 - \lambda_A M_B) \qquad ...(i)$$

and

$$M_B = \frac{C_B}{T}(B_0 - \lambda_B M_A) \qquad ...(ii)$$ [See **article 20.17**]

Substituting the value of M_B from (*ii*), in (*i*), we have

$$M_A = \frac{C_A}{T}B_0 - \frac{C_A\lambda_A}{T}\left[\frac{C_B}{T}(B_0 - \lambda_B M_A)\right]$$

$$= \frac{(TC_A - C_A C_B \lambda_A)}{T^2} B_0 + \frac{C_A C_B \lambda_A \lambda_B}{T^2} M_A$$

Substituting $C_A C_B \lambda_A \lambda_B = T_c^2$ and re-arranging, we have

$$M_A \left(1 - \frac{T_c^2}{T^2}\right) = \frac{TC_A - C_A C_B \lambda_A}{T^2} B_0 \quad \text{...(iii)}$$

Similarly, we get

$$M_B \left(1 - \frac{T_c^2}{T^2}\right) = \frac{TC_B - C_A C_B \lambda_B}{T^2} B_0 \quad \text{...(iv)}$$

Adding (*iii*) and (*iv*), we have

$$(M_A + M_B)(T^2 - T_c^2) = [T(C_A + C_B) - C_A C_B(\lambda_A + \lambda_B)]B_0$$

$$\therefore \text{ Susceptibility} \qquad \chi = \frac{\mu_0 (M_A + M_B)}{B_0}$$

$$= \frac{\mu_0 [T(C_A + C_B) - C_A C_B (\lambda_A + \lambda_B)]}{(T + T_c)(T - T_c)} \quad \text{... (v)}$$

Thus χ tends to infinity as $T \to T_c$, which implies that there is a net spontaneous magnetisation for $T < T_c$ (M_A and M_B are opposite in sign but $M_A + M_B \neq 0$).

20.19 FERRITE

An important class of ferrimagnets is *ferrite*. Ferrites are of great technical importance since they have large magnetisation at room temperature and high resistivity. They are, therefore, more suitable than ferromagnets for use at high frequencies where eddy current losses are a serious problem. They are also used for *memory storage* in computers.

The mixed ferrites are known under the trade name *Feroxcube*. The most important of these are *MnZn* ferrites. The *iron garnets* are cubic ferri-magnetic insulators of which a familiar example is *Yttrium iron garnet* $Y_3 Fe_5 O_{12}$ known as *YIG*.

In addition to the diamagnetic, paramagnetic and ferromagnetic materials, there are some other material having extra ordinary magnetic properties. Some of these are '*Antiferromagnetic*' and '*Ferrimagnetic*'.

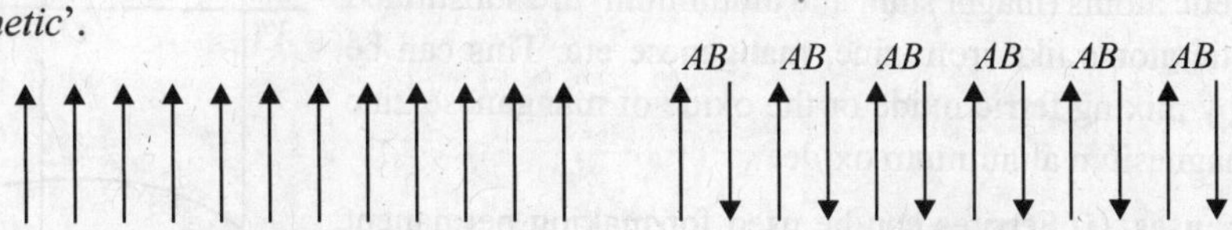

Fig. 20.28

According to the domain theory ferromagnetism is due to the exchange interaction or exchange field which arises when the wave functions of two atoms overlap. This gives rise to an **exchange energy**. The **exchange energy** is positive when neighbouring spins are parallel [Fig. 20.28 (*a*)] and the material is ferromagnetic.

When the neighbouring spins are antiparallel [(Fig. 20.28 (*b*)] and hence there is co-operative alternating alignment, the situation leads to a *ground state*.

In these crystals the alternate atoms have their spins parallel but not the adjacent atoms. Such crystals can be considered to be composed of two interpenetrating sub lattices *A* and *B* -one with atoms having their spin moments oriented in one direction and the other with atoms having their spin moments oriented in the other direction. Thus one sub-lattice gets aligned in one direction whereas the second gets aligned in the other. Such substances are known as *antiferromagnetics*. In the absence of an external magnetic field the net magnetic moment is zero.

Ferrimagnetism is a special case of antiferromagnetism in which opposing magnetic moments of the two sets of atoms in the sub-lattices are of different magnitudes giving rise to a large net magnetisation (Fig. 20.29). This important class of crystals is known as *ferrimagnets* or *ferrites.*

The ferrimagnetic crystals are magnetic compounds consisting of two or more different kinds of atoms. The usual chemical formula for ferrimagnetic crystals or ferrites is MO Fe_2O_3 where M is a divalent metal like Mn^{2+}, Co^{2+}, Ni^{2+}, Fe^{2+}, Zn^{2+}, Cd^{2+}, or Mg^{2+} and Fe^{3+} is the trivalent ferric ion.

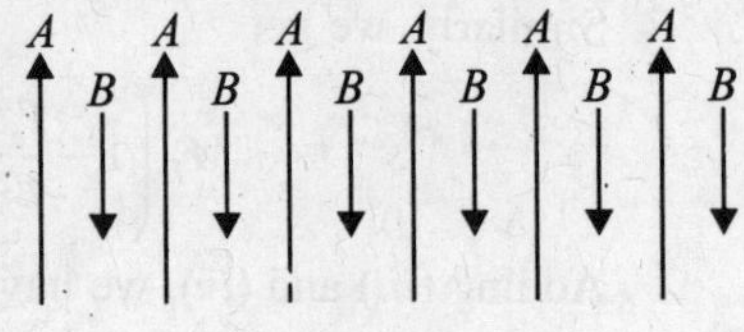

Ferrimagnetism
Fig. 20.29

Ferrites are ferrimagnetic materials having negligible conductivity (resistivity 10^9 times that of ferromagnetics).

20.20 DIFFERENCE BETWEEN FERRITES AND FERROMAGNETIC MATERIALS

Ferrites differ from ferromagnetic materials in following respects :

(*i*) Ferrites have a negligible electrical conductivity.

(*ii*) Magnetic induction of ferrites is much less than that of ferromagnetics.

(*iii*) Ferrites have a higher permeability and lower hysteresis loss as compared to ferromagnetics.

20.21 USE OF FERRITE MATERIAL

Ferrites have a very low electrical conductivity. They are, therefore, used as the cores of *a.c.* inductors and transformers in radios and communication devices operating with high frequency currents. Because of poor conductivity, negligible eddy currents are set up in them and loss of energy due to Joule heating and hysteresis is very small.

Ferrites can be obtained by replacing the magnesium and aluminium atoms from $MgAl_2O_4$ (Magnesium Aluminium oxide or spinel) by iron, zinc, nickel, cobalt, manganese etc. *i.e.*, some of non-magnetic atoms (magnesium and aluminium) are substituted by magnetic atoms like iron, zinc, manganese etc. This can be achieved by mixing ferric oxide or the oxide of manganese zinc etc. wih magnesium aluminium oxide.

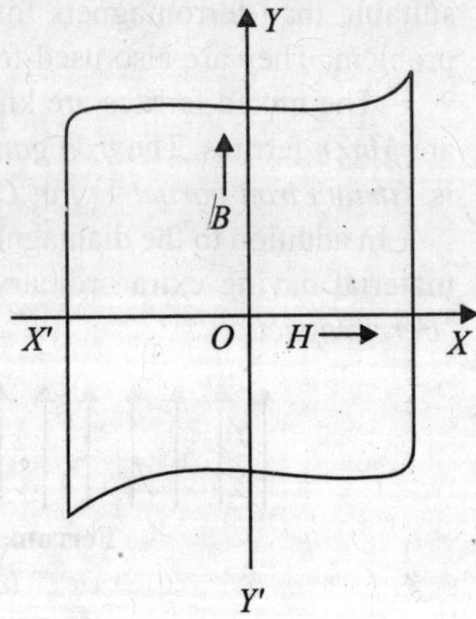

Fig. 20.30

Other uses. (*i*) Ferrites can be used for making permanent magnets. Barium ferrite is as good as alnico magnet. The properties of a ferrite depend upon its composition. For example, the coercivity of Nickel zinc ferrite is 1 A/m and that of cobalt ferrite is 2×10^5 Am^{-1}.

(*ii*) The hysteresis loop of magnesium manganese ferrite is nearly rectangular in shape (Fig. 20.30). Such materials can be magnetised only in one direction or the other. The reversal of the magnetic field can remagnetise the material in the reverse direction almost instantaneously and to the same strength as before. This makes such material very useful in computers and in memory circuits.

EXERCISE

1. Explain why a diamagnetic material is repelled by a bar magnet. ***(Nagpur U. 2002, 2011)***
2. Explain why a diagmagnetic material placed in a non-uniform magnetic field moves from strong to weaker part of the field. ***(Nagpur U. 2011)***
3. Explain why a paramagnetic material is attracted by a bar magnet.
4. Why does a sample of paramagnetic substance placed in a non-uniform magnetic field move from weaker to stronger part of the field? ***(G.N.D.U. 2004)***
5. What are the characteristics of diamagnetic, paramagnetic and ferro-magnetic substance? Illustrate by simple experiments.

   (Pbi. U. 2003, 1995; G.N.D.U. 2004, 2003, 2002; M.D.U. 2001; Nagpur University, 2011, 200; Meerut U. 2003, 2002, 2001; Gauhati.U. 2000)
6. Calculate the magnetic moment of an atomic dipole. ***(P.U. 2003)***
7. (*a*) Prove that the magnetic moment of an electron is given by $\vec{p}_m = -\frac{e}{2m}\vec{L}$ where e = charge on the electron, m = electron mass and $\vec{L}$ = orbital angular momentum due to orbital motion.

   Hence show that magnetic moment due to orbital motion of an electron must be an integral multiple of $\frac{e\hbar}{2m}$. How do you define a Bohr magneton ? Give its numerical value.

   (P.U. 2006, 2000; G. N. D.U. 2000; Gauhati U. 2000; H.P.U. 2001, 2000; Pbi. U. 2003, 2002, 2001; M.D.U. 2002)

   (*b*) What is orbital gyromagnetic ratio? ***(Pbi.U. 2006, 2001; P.U. 2000)***
8. (*a*) Show that when a magnetic field is applied on a diamagnetic substance the change in magnetic moment of an orbiting electron is given by $\frac{-Be^2r^2}{4m}$ where the symbols have their usual meanings.

   Prove that the change is the same whether the electron is orbiting around the nucleus in clockwise or anticlockwise direction. Hence prove that magnetic dipole moment per unit volume is given by $P_m = \frac{-NZ^2e^2r^2\mu_0 H}{4m}$ and diamagnetic susceptibility $\chi_m = \frac{-NZ^2e^2r^2\mu_0}{4m}$

   (*b*) Discuss Langevin's theory of diamagnetism. ***(G.N.D.U. 2009, 2004; Nagpur.U. 2002; K.U. 2002, 2001; P.U., 2009; Gharwal. U. 2000; M.D.U. 2008, 2003)***
9. Explain the term spin angular momentum and spin magnetic moment of an electron.
10. (*a*) Give Langevin's electronic theory of paramagnetism and hence prove that susceptibility χ_p of paramagnetic substances is inversely proportional to absolute temperature. What is Curie's law? ***(H.P.U. 2003, 2001; Pbi. U. 2006, 2001, 2000; G.N.D.U. 2003, Nagpur U. 2012, 2009)***

    (*b*) Why are some substances diamagnetic while others are paramagnetic? Explain.
11. What is ferromagnetism? Explain ferromagnetism on the basis of domain theory. Why does a piece of iron ordinarily not behave as a magnet?

    (Meerut U. 2002; H.P.U. 2007, 2002; P.U. 2001, 2000; M.D.U. 2002)

12. Explain what you understand by Hysteresis, remanence (retentivity) and coercivity. What is hysteresis loop? How will you determine the value of remanence and coercivity from a loop? ***(Meerut. U. 2008, 2003, 2002, 2001; P.U. 2002, 2001)***

13. Show that the loss of energy due to hysteresis per unit volume of the material per cycle of magnetisation is given by (*i*) μ_0 × area of P_m – H loop and (*ii*) area of B–H loop. What happens to this energy? ***(M.D.U. 2007, 2001; Gauhati.U. 2000; K.U. 2000; G.N.D.U. 2006, 2001; P.U. 2001, 2000; Pbi.U. 2002, 2001)***

14. What type of material should be used for making (*a*) permanent magnets (*b*) electromagnets? (or Cores of transformers and chokes) ***(P.U. 2007, 2001, 2000; H.P.U. 2002, 2000; M.D.U. 2002, 2000)***

15. Explain the use of Hysteresis curve. What type of magnetic material is suitable for transformer cores, telephone diaphragm and chokes? ***(K.U. 2001; Pbi.U. 2007, 2006, 2000; H.P.U. 2003, 2001)***

16. Give an account of quantum theory of paramagnetism and derive an expression for susceptibility. How does this result agree with experimental results for rare earth and iron group elements?

17. What is ferromagnetism? Discuss the Weiss field theory of ferromagnetism and explain how magnetic susceptibility varies with temperature.

18. Discuss the phenomenon of anti-ferromagnetism. Show that the susceptibility of anti-ferromagnetic materials is inversely proportional to $(T + T_N)$ where T_N is the Neel temperature.

19. Explain Ferrimagnetism. What are ferrites? Discuss their applications.

20. What are ferrites? How do ferrites differ from ferromagnetic substances? Why do we use ferrite material in radios and other communication equipment? Discuss their other applications. ***(G.N.D.U. 2000; H.P.U. 2002, 2000; P.U. 2002, 2001; Pbi.U. 2003, 2002)***

MULTIPLE CHOICE QUESTIONS

Mark the correct answer:

1. Dipole moment of a current loop does not depend upon
(*a*) Current in the loop (*b*) Area of the loop
(*c*) Number of turns in the loop (*d*) Shape of the loop

2. Diamagnetism is explained in terms of
(*a*) spin motion of the electron (*b*) orbital motion of electron
(*c*) both orbtial and spin motion (*d*) spin motion of the nucleus

3. The magnetic moment of an atom is due to
(*a*) orbital motion of electron only (*b*) spin motion of electron
(*c*) both orbtial and spin motion (*d*) none of these

4. The unit of Bohr magneton is
(*a*) JT (*b*) J^{-1} T (*c*) JT^{-1} (*d*) JT^{-2}

5. Susceptibility of a diamagnetic material is
(*a*) very large (*b*) small but positive
(*c*) small but negative (*d*) zero

6. Which of the following statements are true about magnetic susceptibility χ_m?

(*a*) χ_m may be positive or negative

(*b*) χ_m for paramagnetic material has values close to 1

(*c*) At a given temperature the values of χ_m increases with increasing magnetic field.

(*d*) For paramagnetic substances χ_m is inversely proportional to absolute temperature of the same.

7. The relative permeability of a material is 0.99. It will essentially be

(*a*) paramagnetic substance (*b*) Diamagnetic substance

(*c*) ferromagnetic substance (*d*) none of these.

8. A current loop of magnetic moment $\vec{p}_m$ when placed in a non-uniform magnetic field experiences a force

(*a*) $\vec{\nabla}\,(\vec{p}_m \cdot \vec{B})$ (*b*) $(\vec{\nabla} \cdot \vec{B})$ (*c*) $\vec{\nabla} \cdot \vec{p}_m\, \vec{B}$ (*d*) $(\vec{p}_m \cdot \vec{\nabla})\, \vec{B}$

9. The units of magnetic susceptibility are

(*a*) A/m (*b*) A/m^2 (*c*) Am^2 (*d*) no units

10. Ferrites are

(*a*) Ferrimagnetic having large electrical conductivity

(*b*) Ferrimagnetic having negligible electrical conductivity

(*c*) Diamagnetic having large electrical conductivity

(*d*) Ferromagnetic having negligible electrical conductivity

11. Which of the following substance can have positive permeability and negative susceptibility ?

(*a*) Ferromagnetic (*b*) paramagnetic (*c*) diamagnetic (*d*) none

ANSWERS

1. (*d*)	**2.** (*b*)	**3.** (*c*)	**4.** (*c*)	**5.** (*c*)
6. (All)	**7.** (*b*)	**8.** (*d*)	**9.** (*d*)	**10.** (*b*)
11. (*c*)				

THERMAL PROPERTIES OF SOLIDS (Lattice Vibrations)

21 Chapter

21.1 LATTICE VIBRATIONS

A solid is composed of atoms or molecules which are closely packed. This is because atoms and molecules attract each other. As the distance between the atoms or molecules is made smaller and smaller a stage reaches when a repulsive force comes into play. At the equilibrium distance between the atoms or molecules the attractive and repulsive forces balance each other and the atom or molecule is fixed to this equilibrium position. The interatomic interaction, however, causes the atoms or molecules to vibrate about their mean position. As these atoms or molecules are present at the lattice points, these vibrations are known as *lattice vibrations*.

Quantizatiion of Lattice Vibrations

According to quantum theory, the energy of an elastic mode of vibration of angular frequency ω, when the mode is excited to a quantum number n is given by $E = \left(n + \frac{1}{2}\right)\hbar\,\omega$. The term $\frac{1}{2}\hbar\,\omega$ is the zero point energy of the mode. It occurs both for electromagnetic wave quanta-photons and elastic wave quanta-phonons as a consequence of their equivalence to a harmonic oscillator of angular frequency ω for which the energy eigenvalues are $\left(n + \frac{1}{2}\right)\hbar\,\omega$. The energy of a quantum of elastic wave of lattice vibration—the phonon—is, therefore quantised and each phonon is taken to have an energy $\hbar\,\omega$.

21.2 PHONON

The energy in a lattice vibration or an elastic wave is quantised. Just as the quantum of energy in an electro-magnetic wave is a photon, the quantum of energy in an elastic wave is known as a *phonon*. A phonon wave travels in a medium with the velocity of sound v. The energy of a phonon is given by $h\nu = \hbar\,\omega = \hbar\, v\, K$, where h is Planck's constant ($\hbar = h/2\pi$), ν *the frequency*, ω *the angular frequency and* $\vec{K}$ *the wave vector* $\left(|\vec{K}| = \frac{2\pi}{\lambda}\right)$ *corresponding to the phonon wave.*

The vibration spectrum of the phonon wave occupies a wide range of frequencies from $10^4 - 10^{12}$ Hz. The low frequency part lies in the acoustic region and consists of sound waves in a crystal latice. The high frequency part lies in the infra-red region and is due to thermal vibrations in a crystal analogous to thermally excited photons in black body radiation.

Statistics obeyed by phonons. Phonons like photons have zero spin. Hence phonons obey *Bose-Einstein statistics* and Bose-Einstein distribution law.

$$n = \frac{1}{e^{h\nu/k_B T} - 1}$$

Phonon representation. The pictorial representation of a phonon is shown in Fig. 21.1 (*a*). The pictorial representation of a photon is shown in Fig. 21.1 (*b*) for comparison.

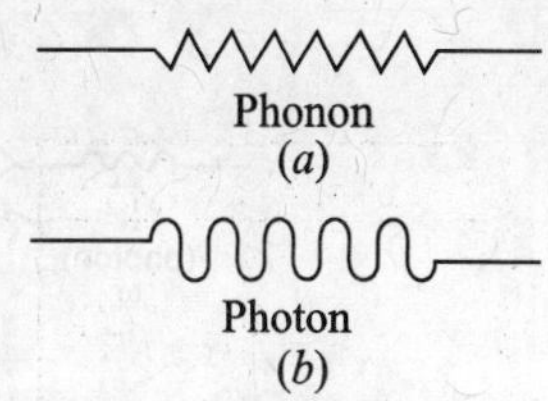

Fig. 21.1

Evidence for existence of phonons. The experimental evidence that the energy of an elastic wave is quantised or for the existence of phonons "the quanta of lattice vibration energy" are:

(*i*) The lattice contribution to the heat capacity of solids approaches zero at absolute zero of temperature. This can be explained only if the lattice vibration is quantised.

(*ii*) Another experimental evidence for phonons is found in the inelastic scattering of *X*-ray photons (or neutrons) by crystals with momentum and energy changes according to the *creation* or *absorption* of one or more *phonons*.

Energy of phonon. The energy of a phonon of frequency ν is given by $E = h\nu = \hbar\,\omega$ $= h v \vec{K}$ where v is the velocity of phonon wave and $\vec{K}$ the phonon wave vector.

Momentum of a phonon. A phonon on a lattice really has no momentum in the physical sense. On the other hand, a phonon of wave vector $\vec{K}$ interacts with other particles such as photons, electrons and neutrons and fields as if it has a momentum $\hbar\vec{K}\left(\frac{h}{2\pi}\cdot\frac{2\pi}{\lambda}=\frac{h}{\lambda}\right)$, the same as for a photon of equivalent wavelength λ. The momentum $\hbar\vec{K}$ is called *crystal momentum*.

Normal process. In an elastic scattering of *X*-ray photon by a crystal, *i.e.*, Bragg's diffraction the wave vector conservation law gives

$$\vec{k'} = \vec{k} + \vec{G} \qquad \text{... } (i)$$

where $\vec{k}$ and $\vec{k'}$ are the wave vectors of incident and scattered photons respectively and $\vec{G}$ is the reciprocal lattice vector of the crystal. Equation (*i*) may be put in the form

$$\hbar\vec{k'} = \hbar\vec{k} + \hbar\vec{G}$$

In this process the crystal as a whole recoils with a momentum $\hbar\vec{G}$ and the frequency of the incident photon remains unchanged, *i.e.*, $|\vec{k'}| = |\vec{k}|$

Such a process in which the frequency of the incident photon is the same as that of the scattered photon is called normal or N-process. In this process no phonon is created or absorbed.

21.3 UMKLAPP PROCESS

In an inelastic scattering of a photon by a crystal, the frequency of the incident photon changes and a new photon is emitted alongwith a phonon which may be emitted or absorbed.

If a phonon of wave vector $\vec{K}$ is emitted or created during the scattering, then according to wave vector conservation law

$$\vec{k'}\text{ (scattered)} + \vec{K} = \vec{k}\text{ (incident)} + \vec{G}$$

or

$$\hbar\vec{k'} + \hbar\vec{K} = \hbar\vec{k} + \hbar\vec{G}$$

If a phonon of wave vector $\vec{K}$ is absorbed during the scattering, then according to the wave vector conservation law

$$\vec{k'}\text{ (scattered)} = \vec{k}\text{ (incident)} + \vec{K} + \vec{G}$$

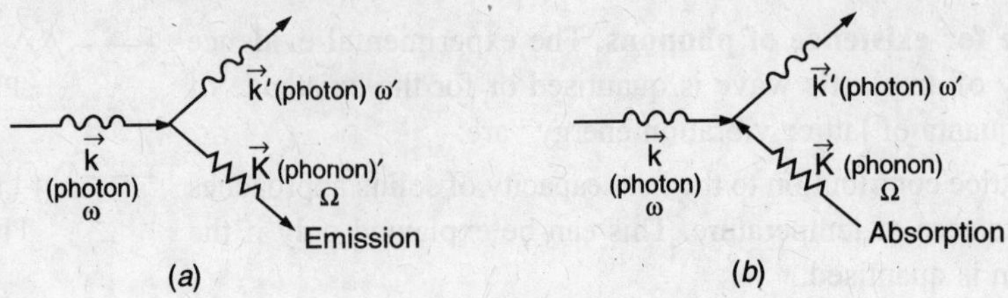

Fig. 21.2

The emission and absorption of phonon in inelastic scattering is shown in Fig. 21.2 (*a*) and (*b*) respectively.

Such a process in which the frequency of the photon changes is called umklapp (after the German for flipping over) or *U-process*. In this process, a phonon is created or absorbed in addition to the Bragg reflection and the momentum is transferred to the crystal as a whole.

21.4 INELASTIC SCATTERING OF PHOTONS BY PHONONS

In an elastic scattering of *X*-ray photons by a crystal there is only a change in direction of the wave vector $\vec{k}$ of the incident photon but the magnitude of the wave vector (or the frequency) remains unchanged.

In inelastic scattering of photon by a crystal the frequency of the incident photon also changes and a *phonon* is emitted or absorbed. The frequency of the emitted phonon is such that:

Energy of incident photon = Energy of scattered photon + energy of emitted phonon

and Momentum of incident photon = momentum of scattered photon + momentum of emitted phonon.

Consider a photon of frequency $\nu = \omega/2\pi$ which is propagating in a crystal considered to be a continuum of refractive index μ. The velocity of photon in the medium is c/μ where c is the velocity of light in vacuum. The wave vector of the photon is given by

$$k = \frac{2\pi}{\lambda} = \frac{2\pi}{c/\mu} \cdot \frac{\omega}{2\pi} = \frac{\omega\mu}{c} \qquad \text{... (i)}$$

Suppose the photon interacts with a phonon beam (sound wave) in the crystal and is scattered. This is because the refractive index of the crystal changes due to change in concentration of atoms brought about by the elastic strain field of the sound waves. In the inelastic scattering of the photon by the crystal a phonon is either created or absorbed. Let the wave vector of the photon change from $\vec{k}$ to $\vec{k'}$ and the frequency from ω to ω′ due to inelastic scattering and a phonon of wave vector $\vec{K}$ and frequency Ω be created as shown in Fig. 21.3.

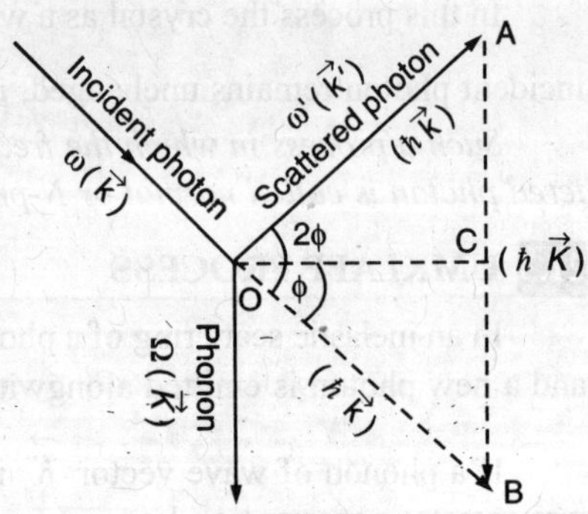

Fig. 21.3

Then, according to the **law of conservation of energy**

$$\hbar\,\omega = \hbar\,\omega' + \hbar\,\Omega \qquad \text{... (ii)}$$

and according to the **principle of conservation of momentum**

$$\hbar\,\vec{k} = \hbar\,\vec{k'} + \hbar\,\vec{K}$$

If the velocity of sound v is a constant, then (angular) frequency Ω of the phonon of wave vector of magnitude K is given by

$$v = \frac{\Omega}{2\pi} \cdot \frac{2\pi}{K} = \frac{\Omega}{K}$$

or $$\Omega = vK \qquad ...(iii)$$

The emitted phonon carries only a very small fraction of energy of the incident photon. The velocity of sound $v <<$ the velocity of light c. If the magnitude of $\vec{K}$, *i.e.*, K is comparable with magnitude of $\vec{k}$, *i.e.*, k, then

$$vK << ck$$

Also $\omega = ck$ and $\Omega = vK$

$$\therefore \qquad \omega >> \Omega$$

Hence from (*ii*), we have $\omega' \simeq \omega$ and, therefore, $k' \simeq k$ ... (*iv*)

The momentum of the photon as well as the phonon is proportional to and in the direction of corresponding wave vector. Therefore, OAB is the momentum vector diagram. According to relation (*iv*), the magnitude of k' is very nearly equal to that of k. Taking $k' = k$

we have $\quad OA = OB$ and the $\Delta\, OAB$ is isosceles.

$\therefore \qquad AB =$ magnitude of $\vec{K} = 2k \sin\phi$

where $\angle AOB = 2\phi$ and OC is perpendicular to AB.

But from relation (*i*) $k = \dfrac{\omega\mu}{c}$

$$\therefore \qquad K = \frac{2\omega\mu}{c} \sin\phi$$

and $$\Omega = vK = \frac{2v\,\omega\,\mu}{c} \sin\phi \qquad ...(v)$$

Relation (v) gives the (angular) frequency of the emitted phonon.

A very close agreement of the calculated and experimental values of Ω is a very strong evidence in favour of existence of phonons and their behaviour.

21.5 COMPARISON BETWEEN PHOTONS AND PHONONS

Concepts common to photons and phonons. (*i*) A photon and a phonon both exhibit wave particle duality.

(*ii*) The energy of a photon of frequency ν is $h\nu$. The same is true of a phonon as its energy is also $h\nu = \hbar\omega$ where ω is the frequency of vibration.

(*iii*) The momentum of a photon $p = \dfrac{h}{\lambda} = \hbar k$ where k is the propagation constant.

Although a phonon in a lattice really has no momentum in the physical sense, a phonon of wave vector $\vec{K}$ interacts with other particles such as photon, electron etc. as if it has a momentum $\hbar\vec{K} = \dfrac{h}{\lambda}$, the same as for a photon of equivalent wavelength λ. The momentum $\hbar\vec{K}$ is called *crystal momentum*.

(*iv*) Phonons are indistinguishable particles like photons and also obey Base Einstein statistics.

Difference between photons and phonons. (*i*) A photon is a quantum of energy in electromagnetic waves, whereas a phonon is a quantum of energy in an elastic wave.

(*ii*) A photon travels with the velocity of light whereas phonon travels with the velocity of sound in the solid.

(*iii*) Photons do not require any material medium for their propagation, *i.e.*, photon can travel through vacuum whereas phonons do require a material medium to travel, *i.e.*, they cannot travel through vacuum.

(*iv*) A phonon is emitted or absorbed in the inelastic scattering of a photon by a crystal. The reverse process, *i.e.*, emission or absorption of a photon by scattering of a phonon is not possible.

21.6 LATTICE WAVES IN A MONATOMIC LINEAR LATTICE

To study the vibrations of one dimensional monatomic lattice, consider the vibrational motion of a non-homogeneous medium, *e.g.*, a linear chain of identical atoms of mass *m* spaced at a distance *a* – the lattice constant, connected by ideal springs. In practice this is met with, in the case of elastic vibrations of a crystal with one atom in the primitive cell. When a wave propagates in such a cubic crystal along the directions [100], [110], [111], the entire planes of atoms move in phase with displacements either parallel or perpendicular to the direction of the wave vector. In such a case the displacement of a plane from its equilibrium position can be described by a *single* coordinate, *i.e.*, the problem becomes *one dimensional*. However, for each wave vector there are *three* modes–one of longitudinal polarisation and two of transverse polarisations.

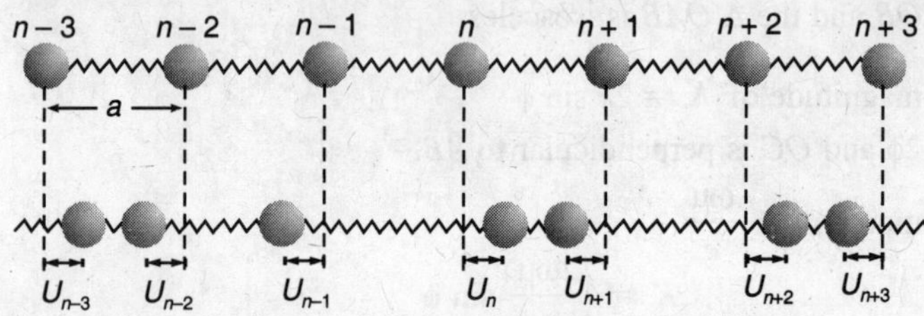

Fig. 21.4

Let the single coordinate be *x*. The co-ordinate of the *n*th atom is represented by $x = na$. Thus at equilibrium the atoms are situated at *equally spaced sites* represented by ..... $n - 3, n - 2, n - 1, n, n + 1, n + 2, n + 3$, ... with *x* coordinates as $(n - 3)\,a$, $(n - 2)\,a$, $(n - 1)\,a$, na, $(n + 1)\,a$, $(n + 2)\,a$, $(n + 3)\,a$, and so on.

As the atoms in the crystal start vibrating, let U_n denote the displacement of the *n*th atom, then U_{n-1}, U_{n+1}, U_{n-2}, U_{n+2}, ... denote the displacement of $n - 1, n + 1, n - 2, n + 2$th atom respectively.

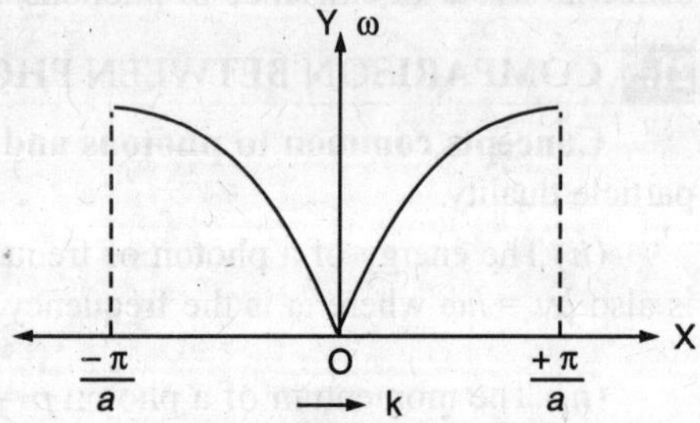

Fig. 21.5

Let β be the *spring constant* or *stiffness* or nearest neighbour interaction given by the relation

$$F = \beta U$$

where *F* is the applied force and *U* the displacement of the corresponding atom. This means that the force required to produce an *atomic displacement* is proportional to the *displacement* itself. In other words, the vibrational motion of each atom is *simple harmonic*. Therefore, the net force acting on the *n*th atom in terms of the extensions of the two springs which bind it to the $(n - 1)$th and $(n + 1)$th atom is given by

$$F_n = \beta\,(U_{n+1} - U_n) - \beta\,(U_n - U_{n-1})$$

$$= \beta\,(U_{n+1} + U_{n-1} - 2U_n)$$

neglecting the direct effect of other atoms like $(n + 2)$, $(n - 2)$ ... etc which are supposed to act via atoms $(n + 1)$ and $(n - 1)$.

If $\frac{d^2U_n}{dt^2}$ denotes the acceleration of the *n*th atom, then $$m \frac{d^2U_n}{dt^2} = \beta (U_{n+1} + U_{n-1} - 2U_n) \quad \text{... } (i)$$

This is a differential equation in displacements and has a periodic travelling wave solution of the type

$$U_n = U_0 \, e^{i(\omega t + kx)}$$

where ω and k are constants, ω being the angular frequency and k the wave vector.

Here $x = na$ where a is the spacing between the two nearest atoms of the linear chain. Thus

$$U_n = U_0 \, e^{i[\omega t + kna]}$$

$$U_{n+1} = U_0 \, e^{i[\omega t + k(n+1)a]} = U_0 \, e^{i[\omega t + kna]} \, e^{ika} = U_n \, e^{+ika}$$

$$U_{n-1} = U_0 \, e^{i[\omega t + k(n-1)a]} = U_0 \, e^{i[\omega t + kna]} \, e^{-ika} = U_n \, e^{-ika}$$

$$\frac{dU_n}{dt} = i\omega \, U_0 \, e^{i(\omega t + kna)}$$

and $$\frac{d^2U_n}{dt^2} = -\omega^2 \, U_0 \, e^{i(\omega t + kna)} = -\omega^2 \, U_n$$

Substituting in (*i*), we have

$$-m\omega^2 \, U_n = U_n \, \beta \, [e^{+ika} + e^{-ika} - 2]$$

or $$-m\omega^2 = \beta \, [e^{+ika/2} - e^{-ika/2}]^2$$

Now, $$\frac{e^{i\theta} - e^{-i\theta}}{2i} = \sin\theta$$

$\therefore$ $$\frac{e^{+ika/2} - e^{-ika/2}}{2i} = \sin\frac{ka}{2}$$

or $$\sin^2\frac{ka}{2} = -\frac{1}{4}\left[e^{+ika/2} - e^{-ika/2}\right]^2$$

Hence $$-m\omega^2 = -4\beta \sin^2\frac{ka}{2}$$

or $$\omega^2 = 4\frac{\beta}{m}\sin^2\frac{ka}{2} \quad \text{or} \quad \omega = \pm 2\sqrt{\frac{\beta}{m}}\sin\frac{1}{2}(ka) \quad \text{... } (i)\,(a)$$

Dispersion relation. If we denote the nearest neighbour interaction by C instead of β, we get

$$\omega = \pm 2\sqrt{\frac{C}{m}}\sin\frac{1}{2}ka \quad \text{... } (ii)$$

The above relation is known as *dispersion relation* because ω depends upon the value of $\sin\frac{1}{2}ka$ which varies from $+1$ to zero taking only *positive* values.

$\therefore$ Relation (*ii*) can be put as

$$\omega = \sqrt{\frac{4C}{m}}\left|\sin\frac{1}{2}ka\right| \quad \text{... } (iii)$$

Allowed values of phonon wave vector. The energy in a lattice vibration or an elastic wave is quantised. By analogy with *photon* – the quantum of energy in an electromagnetic wave, the quantum of energy in an elastic wave is known as a *phonon*. A phonon wave travels in a medium with the velocity of sound v. The energy of a phonon is given by $h\nu = \hbar\omega = \hbar nk$ where h is Planck's constant ($\hbar = h/2\pi$), ν the frequency, ω the angular frequency, and $\vec{k}$ the wave vector $\left(k = |\vec{k}| = \dfrac{2\pi}{\lambda}\right)$ corresponding to the phonon wave.

A graph between ω taken along the Y-axis and values of k taken along the X-axis from $-\dfrac{\pi}{a}$ to $+\dfrac{\pi}{a}$ is shown in Fig. 21.5.

For elastic waves only those values of *phonon wave vector* are *allowed* (significant) for which the *phase ka* lies between $-\pi$ and $+\pi$ or k lies between $-\dfrac{\pi}{a}$ and $+\dfrac{\pi}{a}$. We take both positive and negative values of k because the wave can propagate to the right ($+x$ direction) as well as to the left ($-x$ direction).

For any other value of relative phase ka say $ka = 1.2\pi$ $\left(\text{or } k = 1.2\dfrac{\pi}{2}\right)$, it is equivalent to a relative phase $ka = (1.2\pi - 2\pi) = -0.8\pi$ lying between $-\pi$ and $+\pi$. Similarly a relative phase of 4.2π is equivalent to $(4.2\pi - 2 \times 2\pi) = 0.2\pi$ which again lies between $-\pi$ to $+\pi$.

Brillouin zones. Thus it is clear that only allowed values of phonon wave vector k are those lying between $-\dfrac{\pi}{a}$ to $+\dfrac{\pi}{a}$. This range of phonon wave vector k between $-\dfrac{\pi}{a}$ and $+\dfrac{\pi}{a}$ gives the *first* Brillouin zone of the linear lattice. The values of k outside the first Brillouin zone simply reproduce lattice motions described by values with the limits $\pm\dfrac{\pi}{a}$.

In general, let the value of k lying between $-\dfrac{\pi}{a}$ and $+\dfrac{\pi}{a}$ corresponding to a vibrational motion of the lattice of frequency ω_0 be k_0. If k is another value given by

$$k = k_0 - \frac{2\pi}{a} n, \quad \text{where } n \text{ is an integer, then}$$

$$\sin\frac{ka}{2} = \sin\left(k_0 - \frac{2\pi}{a} n\right)\frac{a}{2} = \sin\left(k_0 \frac{a}{2} - n\pi\right)$$

$$= \left| \sin k_0 \frac{a}{2} \right|$$

Thus, the vibrational motion of the lattice corresponding to k is also ω_0. In other words, associated with a given frequency ω_0 there are many possible values of wave vector k, *i.e.*, the lattice can propagate several wavelengths at the same frequency ω_0.

21.7 DISPERSION RELATION IN TERMS OF WAVE VELOCITY

The dispersion relation for the lattice waves in a monatomic linear lattice of mass m spacing a and β the stiffness or the nearest neighbour interaction is given by (according to relation [(*i*) (*a*) **article 21.6**]

$$\omega = \pm 2\sqrt{\frac{\beta}{m}} \sin\frac{1}{2} ka \qquad \text{... (i)}$$

If instead of individual atoms we consider the entire crystal plane, then the *linear density* along the single *co-ordinate* (or line)

$$\rho = \frac{m}{a} \quad \text{or } m = a\rho$$

$$\therefore \qquad \omega = \pm 2 \sqrt{\frac{\beta}{a\rho}} \sin \frac{1}{2} ka$$

If E is the longitudinal stiffness (or elasticity), then $E = \beta a$ or $\beta = \dfrac{E}{a}$

$$\therefore \qquad \omega = \frac{2}{a} \sqrt{\frac{E}{\rho}} \sin \frac{1}{2} ka$$

But $\sqrt{\dfrac{E}{\rho}} = v_0$ the velocity of the wave along the line

$$\therefore \qquad \omega = \frac{2}{a} v_0 \left| \sin \frac{ka}{2} \right| \text{ taking only positive values.}$$

(*i*) **At low frequencies.** For values of $k \to 0$, $\sin \dfrac{ka}{2} = \dfrac{ka}{2}$ because for small values of θ, $\sin \theta = \theta$

$$\therefore \qquad \omega = \frac{2}{a} v_0 \frac{ka}{2} = v_0 k$$

In this case phase velocity $\quad v_p = \dfrac{\omega}{k} = v_0$ ... (*ii*)

and group velocity $\quad v_g = \dfrac{d\omega}{dk} = v_0$ ... (*iii*)

i.e., phase velocity = group velocity = v_0

The relation $v_0 = \dfrac{\omega}{k}$ is the same as that for a homogeneous continuous line. Thus for long wavelength the atomic nature of linear chain has no importance, so for as the dynamical behaviour of the system is concerned, *i.e.*, the discrete nature of the structure is not sensitive to long wavelengths. In other words, a large number of atoms together participate in all displacements as on a homogeneous line.

(*ii*) **At high frequencies.** At high frequencies

Phase velocity $\quad v_p = \dfrac{\omega}{k} = \dfrac{2}{ka} v_0 \sin \dfrac{ka}{2}$ ... (*iv*)

and group velocity $\quad v_g = \dfrac{d\omega}{dk} = v_0 \cos \dfrac{ka}{2}$ ... (*v*)

Thus both the velocities are a function of frequency (as $k = \dfrac{2\pi}{\lambda}$). This is similar to the optical case in which as light passes through a medium the refractive index being a function of frequency is responsible for the phenomenon of dispersion of light. By analogy the present medium is also said to be *dispersive*.

(*iii*) ***At frequency*** $\omega = \sqrt{4\dfrac{C}{m}} \left(\text{or } \sqrt{\dfrac{4\beta}{m}} \right)$ (*maximum value*). At frequency $\omega = \sqrt{4\dfrac{C}{m}} \left(\text{or } \sqrt{\dfrac{4\beta}{m}} \right)$ which represents the maximum angular frequency of vibration, relation (*i*) gives

$$\sin \frac{1}{2} ka = 1 \quad \text{or} \quad \frac{ka}{2} = \frac{\pi}{2}$$

$$\therefore \qquad k = \frac{\pi}{a} \quad \text{or} \quad \frac{2\pi}{\lambda} = \frac{\pi}{a}$$

$$\lambda = 2a$$

Substituting, $k = \dfrac{\pi}{a}$ and $\dfrac{ka}{2} = \dfrac{\pi}{2}$ in Eq. (*iv*), we get

$$v_p = \frac{2}{\frac{\pi}{a}a} v_0 \sin\frac{\pi}{2} = \frac{2}{\pi} v_0 \qquad \text{... (vi)}$$

Substituting $k = \dfrac{\pi}{a}$ and $\dfrac{ka}{2} = \dfrac{\pi}{2}$ in Eq. (*v*), we get

$$v_g = v_0 \cot\frac{\pi}{2} = 0 \qquad \text{... (vii)}$$

From Eqs. (*vi*) and (*vii*) we find that the group velocity is zero and phase velocity has a finite value $\dfrac{2v_0}{\pi}$. The group velocity represents the velocity of energy transmission or signal. Therefore, for values of $k = \dfrac{\pi}{a}$ there is no propagation of signal or energy, *i.e.*, the corresponding wave is a *standing wave*.

This situation is similar to Bragg condition of reflection of *X*-rays from successive planes of atoms in a crystal given by

$$2d \sin\theta = n\lambda$$

In the first order reflection ($n = 1$) and for normal incidence ($\theta = \dfrac{\pi}{2}$ or $\sin\theta = 1$), $\lambda = 2d$. Compare with $\lambda = 2a$ given above. Thus at the edges of the Brillouin Zone $k = -\pi/a$ and $k = +\dfrac{\pi}{a}$, $v_g = 0$ – a condition necessary for standing waves.

Further such a linear chain can propagate waves (or vibrations) only of angular frequency

$$\omega \le \frac{2}{a} v_0$$

Thus a monoatomic linear lattice acts as a low pass filter which allows only the frequencies in the range zero to $\dfrac{2v_0}{a}$. In other words, $\omega = \dfrac{2}{a} v_0$ is the maximum or *cut off* frequency for a linear monoatomic lattice.

SOLVED EXAMPLES

Example 21.1 *Compute the cut off frequency for a linear monatomic lattice if the velocity of sound and interatomic spacing in the lattice are 3×10^3 ms^{-1} and 3×10^{-10} m respectively.*

(*Pbi.U. 2000*)

Solution. Velocity of sound $v_0 = 3 \times 10^3$ ms^{-1}; Interatomic spacing in the lattice $a = 3 \times 10^{-10}$ m.

$$\therefore \quad \text{Cut off frequency } \omega = \frac{2}{a} v_0 = \frac{2 \times 3 \times 10^3}{3 \times 10^{-10}} = 2 \times 10^{13} \text{ Hz}$$

21.8 VIBRATIONS OF ONE DIMENSIONAL LINEAR DIATOMIC LATTICE

Consider a diatomic crystal such as NaCl or diamond in which there are two atoms per unit primitive cell. Let the distance between nearest neighbours be '*a*' so that the '*repeat distance*' is $2a$. Let the mass of one atom be m and that of the other M ($m < M$). These atoms are placed alternately along the *X*-axis and in a state of equilibrium let these atoms be at places..... $(2n - 3)$, $(2n - 2)$,

$(2n-1)$, $2n$, $(2n+1)$, $(2n+2)$ $(2n+3)$ ...... the atoms of mass m are at even numbered points and those of mass M at odd numbered points. Let U_{2n} be the displacement of the atom corresponding to the $2n$th site at any time during the vibratory motion of the atoms.

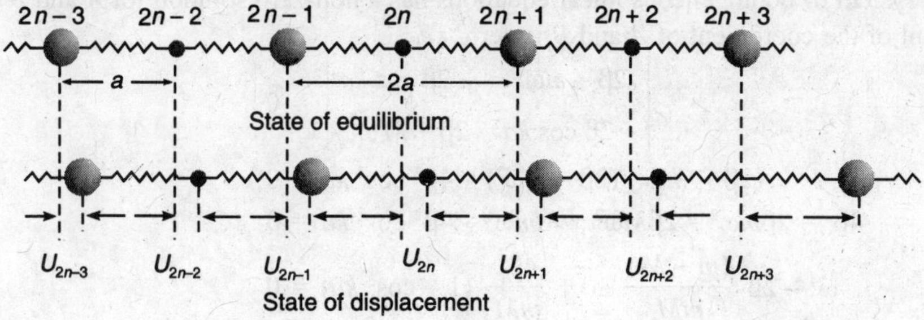

Fig. 21.6

By analogy with the monatomic case and using similar assumptions we obtain the following two equations of motion, one for the lighter atom (mass m) and the other for the heavier atom (mass M).

For atoms of mass m at *even* points

$$F_{2n} = m\frac{d^2U_{2n}}{dt^2} = \beta\,(U_{2n+1} + U_{2n-1} - 2U_{2n}) \quad \text{... (i)}$$

For atoms of mass M at *odd* points

$$F_{2n+1} = M\frac{d^2U_{2n+1}}{dt^2} = \beta\,(U_{2n+2} + U_{2n} - 2U_{2n+1}) \quad \text{... (ii)}$$

The solution of these equations in the form of travelling waves with different amplitudes is given by

$$U_{2n} = A \exp i\,[\omega t - 2n\,ka]$$

$$U_{2n+1} = B \exp i\,[\omega t - (2n+1)\,ka]$$

Similarly, $$U_{2n+2} = A \exp i\,[\omega t - (2n+2)\,ka]$$

and $$U_{2n-1} = B \exp i\,[\omega t - (2n-1)\,ka]$$

where k is the wave vector of a particular mode of vibration. The vibrational frequency of both the types of atoms have been taken to be the same because both types of atoms participate in the same wave motion. The amplitudes A and B are different because of different masses of the atoms

Now, $$\frac{d^2U_{2n}}{dt^2} = -A\omega^2 \exp i\,[\omega t - (2n)\,ka]$$

Hence, $$m\frac{d^2U_{2n}}{dt^2} = -Am\omega^2 \exp i\,[\omega t - (2n)\,ka]$$

$$\therefore\ -Am\omega^2 \exp i\,[\omega t - (2n)\,ka] = \beta B\,\{\exp i\,[\omega t - (2n+1)\,ka] + \exp i\,[\omega t\,(2n-1)\,ka]\} - 2\beta A \exp i\,[\omega t - (2n)\,ka]$$

or $$-m\omega^2 A = \beta B\,(e^{ika} + e^{-ika}) - 2\beta A \quad \text{... (iii)}$$

Similarly, $$\frac{Md^2U_{2n+1}}{dt^2} = -BM\omega^2 \exp i\,[\omega t - (2n+1)\,ka]$$

$$\therefore\ -M\omega^2 B \exp i\,[\omega t - (2n+1)\,ka] = \beta A\,\{\exp i\,[\omega t - (2n+2)\,ka] + \exp i\,[\omega t - (2n)\,ka]\} - 2\beta B \exp i\,[\omega t - (2n+1)\,ka]$$

or $$-M\omega^2 B = \beta A\,(e^{ika} + e^{-ika}) + 2\beta B \quad \text{... (iv)}$$

Now, $$e^{ika} + e^{-ika} = 2\cos ka$$

$\therefore$ Eq. (*iii*) becomes $(2\beta - m\omega^2)\,A - (2\beta \cos ka)\,B = 0$ ... (*v*)

and Eq. (*iv*) becomes $(-\,2\beta \cos ka)\,A + (2\beta - \omega^2 M)\,B = 0$ ... (*vi*)

This system of homogeneous linear equations has a non-zero solution for A and B only if the determinant of the coefficient of A and B is zero, *i.e.*,

$$\begin{vmatrix} 2\beta - m\omega^2 & -\,2\beta \cos ka \\ -\,2\beta \cos ka & 2\beta - M\omega^2 \end{vmatrix} = 0$$

or $$(2\beta - M\omega^2)(2\beta - m\omega^2) - 4\beta^2 \cos^2 ka = 0$$

or $$4\beta^2 - 2\beta m\omega^2 - 2\beta M\omega^2 + Mm\omega^4 - 4\beta^2 \cos^2 ka = 0$$

or $$\omega^4 - 2\beta\,\frac{(m+M)}{mM}\,\omega^2 + \frac{4\beta^2}{mM}\,(1 - \cos^2 ka) = 0$$

or $$\omega^4 - 2\beta\,\frac{(m+M)}{mM}\,\omega^2 + \frac{4\beta^2}{mM}\,\sin^2 ka = 0$$

This is a quadratic equation in ω^2 and gives the value of ω^2 as

$$\omega^2 = \beta\left[\frac{1}{m} + \frac{1}{M}\right] \pm \beta\sqrt{\left(\frac{1}{m} + \frac{1}{M}\right)^2 - \frac{4\sin^2 ka}{mM}} \quad \text{... (vii)}$$

This is the dispersion relation for a linear diatomic lattice. The corresponding dispersion relation for a monatomic lattice is

$$\omega^2 = \frac{4\beta}{m}\sin^2\frac{ka}{2}$$

Since ω should be positive each value of ω^2 leads to a single value of ω. Thus we find that in the monatomic case there is only one value of ω for a single value of k whereas in the diatomic case there are two values of ω. These values are denoted as ω_+ and ω_-

$$\therefore \quad \omega_+^2 = \beta\left[\frac{1}{m} + \frac{1}{M}\right] + \beta\sqrt{\left(\frac{1}{m} + \frac{1}{M}\right)^2 - \frac{4\sin^2 ka}{mM}} \quad \text{... (viii)}$$

and $$\omega_-^2 = \beta\left[\frac{1}{m} + \frac{1}{M}\right] - \beta\sqrt{\left(\frac{1}{m} + \frac{1}{M}\right)^2 - \frac{4\sin^2 ka}{mM}} \quad \text{... (ix)}$$

Different branches of the dispersion relation curve. The graph between ω and k leads to two branches: one corresponding to ω_+ and is called *optical branch* and the second corresponding to ω_- and is called the *acoustical branch* as shown in Fig. 21.7.

(*i*) **Optical branch (case of ω_+).** In the case of ω_+ for $k \to 0$, $\sin ka \to 0$, *i.e.*, it is negligible. Therefore, putting $\sin ka = 0$ in Eq. (*viii*), we get

$$\omega_+ = \sqrt{2\beta\left(\frac{1}{m} + \frac{1}{M}\right)} \quad \text{... (x)}$$

For $k \to \dfrac{\pi}{2a}$, $\sin ka \to 1$, and we have

$$\omega_+^2 = \beta\left(\frac{1}{m} + \frac{1}{M}\right) + \beta\sqrt{\left(\frac{1}{m} + \frac{1}{M}\right)^2 - \frac{4}{mM}}$$

$$= \beta\left(\frac{1}{m} + \frac{1}{M}\right) + \beta\left(\frac{1}{m} - \frac{1}{M}\right)$$

or $$\omega_+ = \sqrt{\frac{2\beta}{m}} \quad \text{... (xi)}$$

(*ii*) **Acoustical branch (case of ω_-).** In the case of ω_-, for $k \to 0$, we write $\sin ka = ka$. We do not put $\sin ka = 0$ because if we do so the value of ω_- will be zero, which is not acceptable.

$$\therefore \quad \omega_-^2 = \beta\left(\frac{1}{m}+\frac{1}{M}\right)-\beta\left(\frac{1}{m}+\frac{1}{M}\right)\sqrt{1-\frac{4k^2a^2}{mM}\left(\frac{mM}{m+M}\right)^2}$$

$$= \beta\left(\frac{1}{m}+\frac{1}{M}\right)\left[1-\sqrt{1-\frac{mM}{(m+M)^2}4k^2a^2}\right]$$

$$= \beta\left(\frac{1}{m}+\frac{1}{M}\right)\left[1-1+\frac{mM}{(m+M)^2}2k^2a^2+......\right]$$

[Using Binomial theorem and neglecting higher powers]

$$= \frac{2\beta}{m+M}k^2a^2$$

$$\therefore \quad \omega_- = ka\sqrt{\frac{2\beta}{(m+M)}} \qquad \text{... } (xii)$$

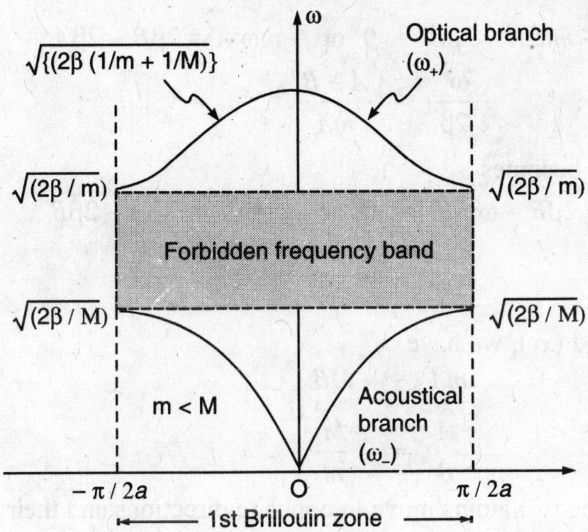

Fig. 21.7

For $k = \dfrac{\pi}{2a}$, we get a solution similar to result obtained in Eq. (*xi*) proved as under.

For $k = +\dfrac{\pi}{2a}$, the sine term becomes $\dfrac{-4}{mM}$, and we get

$$\omega_-^2 = \beta\left(\frac{1}{m}+\frac{1}{M}\right)-\beta\sqrt{\left(\frac{1}{m}+\frac{1}{M}\right)^2-\frac{4}{mM}}$$

$$= \beta\left(\frac{1}{m}+\frac{1}{M}\right)-\beta\left(\frac{1}{m}-\frac{1}{M}\right)$$

or
$$\omega_- = \sqrt{\frac{2\beta}{M}} \qquad \text{... } (xiii)$$

The complete curves for ω_+ and ω_- versus k are shown in Fig. 21.7.

Forbidden band. It is clear from Fig. 21.7, that wave like solutions do not exist for frequency lying between $\sqrt{\frac{2\beta}{m}}$ and $\sqrt{\frac{2\beta}{M}}$. In other words, there is a band of frequencies between the two branches which cannot propagate. This band is known as *Forbidden Frequency band.* The width of the forbidden band depends upon the difference of masses. The larger the mass ratio $\frac{M}{m}$ the wider the frequency gap between the two branches. If $M = m$, the optical and the acoustical frequency coincide at $k = \pm\frac{\pi}{2a}$ and the forbidden band vanishes. The first Brillouin zone is defined by k-values which lie in the range

$$-\frac{\pi}{2a} < k < \frac{\pi}{2a}$$

Thus, the smallest possible wavelength of this zone is $4a$ which corresponds to $k = \frac{\pi}{2a}$ at the zone boundary.

Difference between two branches. For the optical branch as $k \to 0$, $\cos ka \to 1$ and Eq. (*v*) becomes.

$$2\beta A - m\omega^2 A - 2\beta B = 0 \text{ or } -m\omega^2 A = 2\beta B - 2\beta A$$

or
$$\frac{\omega^2}{2\beta} = \frac{A-B}{mA} \quad \text{... (xiv)}$$

Similarly Eq. (*vi*) becomes

$$-2\beta A + 2\beta B - \omega^2 MB = 0 \text{ or } -M\omega^2 B = 2\beta A - 2\beta B$$

or
$$\frac{\omega^2}{2\beta} = -\frac{A-B}{MB} \quad \text{... (xv)}$$

From Eq. (*xiv*) and (*xv*), we have

$$mA = -MB$$

or
$$\frac{A}{B} = -\frac{M}{m} \quad \text{... (xvi)}$$

This means that the two atoms move in opposite directions and their amplitudes are inversely proportional to their masses so that the centre of mass of the unit cell remains unchanged. Such a mode of vibration is shown in Fig. 21.8 (*a*).

For $\qquad M = m, \ \frac{A}{B} = -1$

Thus, even when their masses are equal, the atoms always move in opposite directions.

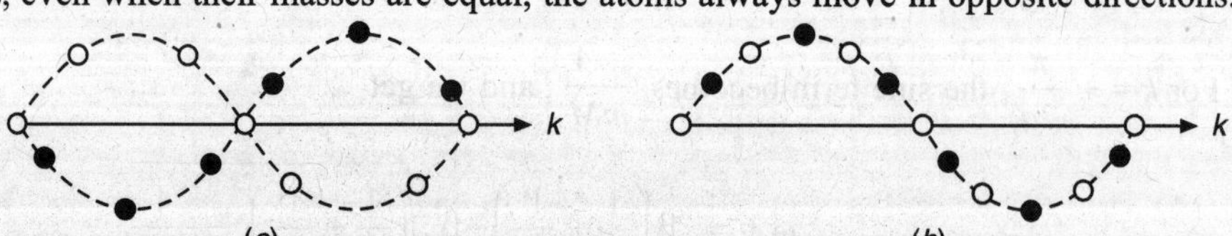

Fig. 21.8

For the acoustical branch as $k \to 0$

$$\cos ka = 1 - \frac{k^2a^2}{2} + ... \qquad \left[\because \cos\theta = 1 - \frac{\theta^2}{2} + ...\right]$$

In this case, the second term becomes significant and hence cannot be neglected. Eq. (*v*) now becomes

$$(2\beta - m\omega^2) A - 2\beta B \left[1 - \frac{k^2a^2}{2}\right] = 0$$

or $$-m\omega^2 A = 2\beta B \left[1 - \frac{k^2a^2}{2}\right] - 2\beta A \quad \text{... (xvii)}$$

Similarly, Eq. (*vi*) now becomes

$$-2\beta A \left[1 - \frac{k^2a^2}{2}\right] + (2\beta - \omega^2 M) B = 0$$

or $$-M\omega^2 B = 2\beta A \left[1 - \frac{k^2a^2}{2}\right] - 2\beta B \quad \text{... (xviii)}$$

Adding Eqs. (*xvii*) and (*xviii*), we get

$$-\omega^2 (mA + MB) = -k^2a^2\beta (A + B)$$

But $$\omega = ka\sqrt{\frac{2\beta}{m+M}} \quad \text{according to Eq. (xii)}$$

$$\therefore \quad \omega^2 = k^2a^2 \frac{2\beta}{m+M}$$

$$\therefore \quad -k^2a^2 \cdot \frac{2\beta}{m+M}(mA + MB) = -k^2a^2\beta (A + B)$$

or $$2(mA + MB) = (m + M)(A + B)$$

or $$mA + MB = MA + mB$$

or $$(m - M) A = (m - M) B$$

$$\therefore \quad \frac{A}{B} = +1$$

This means that in the acoustical branch the two atoms of different masses move in the same direction with the same amplitude as shown in Fig. 21.8 (*b*).

In this case, there is a movement of their centres of masses as well. We, therefore, conclude that, in the optical case, the neighbouring atoms move in opposite directions whereas, in the acoustical case they move in the same direction.

The vibrations of the acoustical branch can be excited by a force which makes the atoms in a crystal move in the same direction such as sound waves. This is the reason why it is called acoustical branch. Monatomic crystals respond to such vibration.

The vibration of the optical branch can be excited by a force which makes the two neighbouring atoms move in opposite directions. The motion of this type can be excited by means of the electric field of the light waves which tends to move the ions in opposite directions.

The diamond and *hcp* lattices having two atoms per primitive cell exhibit optical branches.

To sum up:

(*i*) In diatomic lattice, the allowed frequencies constitute two frequency bands, one corresponding to the acoustical mode and the other to the optical mode of vibration.

(*ii*) If the masses m and M are equal, the frequency range in both the monatomic and diatomic lattice is the same. In this case, there is no forbidden band.

(*iii*) At the Brillouin zone edge $\left(k = \frac{\pi}{2a}\right)$ only one of the sub-lattice vibrates. The lighter atom (m) vibrates in the optical branch and the heavier atom (M) vibrates in the acoustical branch.

(*iv*) As the heavier mass M increases the optical branch flattens and the acoustical branch yields downward. For $M \to \infty$, the acoustical branch disappears and the optical branch becomes flat (parallel to k-axis).

(*v*) As the lighter mass m decreases, the optical branch moves upward while the acoustical branch is not effected. As $m \to 0$ the optical branch disappears altogether.

(*vi*) The optical mode is, in fact, the natural frequency of the lighter atoms (m) perturbed by the heavier sub-lattice. At $k = \frac{\pi}{2a}$ the heavy atom is stationary and the natural frequency of the vibrating lighter atom is $\sqrt{\frac{2\beta}{m}}$.

21.9 SPECIFIC HEAT

In general, whenever heat is imparted to a body a part of it is used to increase its internal energy, *i.e.*, to raise its temperature and the rest in doing external work. If dQ is the heat energy absorbed by a system, dU the increase in internal energy, and dW the external work done by it, then if all quantities are measured in units of work, we have

$$dQ = dU + dW$$

If pressure remains constant at P and a change of volume dV takes place, then $dW = PdV$ and

$$dQ = dU + PdV \qquad \text{... } (i)$$

If dT is the corresponding increase in temperature, then dQ/dT is known as specific heat of the substance.

There are in general two specific heats:

(*i*) Specific heat at constant pressure, and

(*ii*) Specific heat at constant volume.

Molar specific heat. If dQ is the amount of heat energy required to raise the temperature of one kilo-mole of a substance through dT °C keeping the pressure constant, then

Molar specific heat at constant pressure

$$C_p = \left(\frac{dQ}{dT}\right)_p$$

Similarly if dQ is the amount of heat energy required to raise the temperature of one kilo-mole of a substance through dT °C, keeping the volume constant, then

Molar specific heat at constant volume

$$C_v = \left(\frac{dQ}{dT}\right)_v$$

In case of solids the two specific heats do not differ much particularly at low temperatures as the volume expansion in solids is small. By the term specific heat of a solid we shall mean specific heat at constant volume.

Now, at constant volume $\quad dV = 0$

$\therefore$ Relation (*i*) becomes

$$dQ = dU$$

Hence,

$$C_v = \left(\frac{dQ}{dT}\right)_v = \left(\frac{dU}{dT}\right)_v$$

Lattice heat capacity. In solids the increase in internal energy takes place in two ways:

(*i*) Due to vigorous vibrations of atoms of the solid (assumed to be free) about their equilibrium positions as in crystals.

(*ii*) Due to thermal excitation of electrons to higher energy states as in metals and semiconductors.

If we denote the energy contribution due to atomic vibrations in crystal lattice as $U_{Lattice}$ and thermal excitation of electrons as U_{Elect}, then

$$U = U_{Lattice} + U_{Elect}$$

and correspondingly, we have

$$C_v = \left(\frac{dU}{dT}\right)_v = \frac{d}{dt}(U_{Lattice}) + \frac{d}{dt}(U_{Elect})$$

$$= C_{Lattice} + C_{Elect}$$

The quantity $C_{Lattice}$ is known as lattice specific heat or *Lattice heat capacity*. In discussing the specific heat of solids like crystals, we assume that no free electrons are present and the entire specific heat of a crystal is due to the excitations of thermal vibrations of the lattice. Therefore, for our discussion

$$C_v = C_{Lattice}$$

21.10 DULONG AND PETIT'S LAW

According to classical theory, an atom is taken to be a point particle having no structure. It, therefore, has *three degrees of freedom*. According to kinetic theory, the average energy of a classical system is $\frac{1}{2}$ kT per degree of freedom where k is Boltzmann's constant. This result can be extended to particles interacting with harmonic forces. In such a case the *average kinetic energy* remains the same, *i.e.*, $\frac{1}{2}$ kT per degree of freedom and the *average potential energy* also has the same value, *i.e.*, $\frac{1}{2}$ kT. Hence for particles which interact with harmonic forces the average *total energy* per degree of freedom $= \frac{1}{2}kT + \frac{1}{2}kT = kT$. As a point particle has three degrees of freedom the total energy per atom is 3 kT.

Therefore, the total vibrational energy of a crystal composed of N identical atoms can be written as

$$U = 3NkT$$

as a crystal is being taken equivalent to $3N$ harmonic oscillators.

It is supposed that atoms in a solid behave as free classical harmonic oscillators vibrating about their mean position which gives the lattice contribution to the molar heat capacity at constant volume.

As
$$U = 3NkT$$

$$C_v = \left(\frac{dU}{dT}\right)_v = 3Nk = 3R$$

If N is equal to Avogadro's number, R is the universal gas constant having a value $R = Nk = 8.31 \times 10^3$ Joules/kilo-mole K = 1.99 kilo cal/kilo-mole K

$$\therefore \quad C_v = 3R = 5.97 \text{ kilo-cals/kilo-mole}$$

This result $C_v = 5.97$ = a constant is known as Dulong and Petit's law.

21.11 AGREEMENT OF RESULTS WITH EXPERIMENTS

According to Dulong and Petit's law and classical theory, specific heat is independent of temperature. This result is in quite good agreement with observed heat capacity of many solids including metals at high temperature but the classical formula breaks down at low temperatures. The specific heat of all solids drops sharply at low temperatures and approaches zero at temperatures approaching 0 K. This is against the Dulong and Petit's law.

A graph showing the variation of heat capacity with temperature for a typical case is shown in Fig. 21.9.

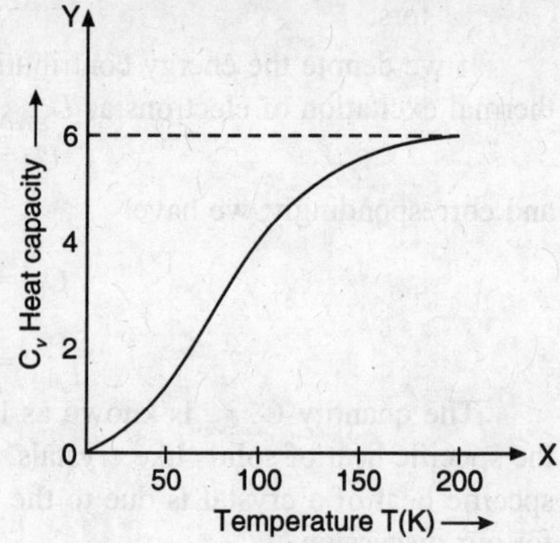

Fig. 21.9

(*i*) At high temperature $C_v = 3R$ per mole. This is in accordance with Dulong and Petit's law. This is true for several compounds and heavy elements.

(*ii*) For light elements (such as gases) the contribution at room temperature is less than $3R$.

(*iii*) At low temperature C_v decreases exponentially as the temperature is reduced and tends to zero as $T \rightarrow 0$.

In the temperature range 0 – 20 K

For non-metals $C_v = \alpha T^2$

For metals $C_v = \alpha T^3 + \gamma T$ (γ is volume expansion coefficient)

For layered compounds $C_v = \alpha T^2$

For linear chained compounds $C_v = \alpha T$

Discrepency removed by Einstein. In the derivation of Dulong and Petit's law on the basis of *classical theory*, it is assumed that each atom is a classical three dimensional harmonic oscillator which vibrates independently of all other atoms in the crystal and has any value of energy ranging *continuously* from zero to infinity.

The results obtained were not, however, in agreement with experimentally observed values and gave rise to a number of discrepencies stated above.

The discrepency in classical theory was removed by Einstein by the application of *quantum* theory. According to quantum theory, the energy spectrum of atomic oscillators is not continuous but has discrete values of energy given by $E_n = \left(n + \frac{1}{2}\right) h\nu$ where ν is the frequency of the oscillator, *h* the Planck's constant and *n* = 0, 1, 2, ... etc. (integral values) called quantum númber. According to this theory, the ground state energy of a harmonic oscillator is not zero (as assumed in classical theory) but has a value $\frac{1}{2} h\nu$. Thus, the atom has vibrational energy even at $T = 0$ K.

21.12 EINSTEIN'S THEORY OF LATTICE HEAT CAPACITY (*SPECIFIC HEAT*)

The results of the classical theory of specific heat were not in agreement with experimental results for many solids particularly at low temperatures. The discrepancies in the classical theory of specific heat were removed by Einstein to some extent by the application of quantum theory. For this purpose Einstein made the following assumptions:

Assumptions. (*i*) The atoms are identical independent simple harmonic oscillators.

(*ii*) All atoms vibrate in identical environments and have the same natural frequency.

(*iii*) The energy spectrum of atomic oscillators is not continuous as assumed in classical theory but has discrete values as given by quantum theory. According to this theory an oscillator of frequency ν has only *discrete* values of energy given by $E_n = nh\nu$ where *n* = 0, 1, 2 ... etc., called quantum number, *h* the Planck's constant = 6.63 × 10^{-34} Joule-sec.

(*iv*) The vibration of each atom in the solid may be resolved into components along three perpendicular axes so that each atom is represented by *three* harmonic oscillators. Thus a solid containing N atoms is considered to be equivalent to $3N$ harmonic oscillators.

(*v*) The oscillators form an assembly of systems which are distinguishable (or identifiable) due to their location at separate and distinct lattice sites in the crystal. These, therefore, obey Maxwell Boltzmann statistics and have corresponding distribution of energy.

Lattice heat capacity. According to above assumptions the *relative probability* that the (atomic) oscillator has the energy $E_n = nh\nu$ at the temperature T is given by Boltzmann's distribution of energy function $e^{-E_n/kT}$.

To find $\bar{E}$ the *average* energy of the oscillator find the sum of $E_n e^{-E_n/kT}$ over all n from $n = 0$ to $n = \infty$, *i.e.*,

$$\sum_{n=0}^{n=\infty} E_n\, e^{-E_n/kT}$$

and divide the result by the sum of $e^{-E_n/kT}$ over all n from $n = 0$ to $n = \infty$, *i.e.*,

$$\sum_{n=0}^{n=\infty} e^{-E_n/kT}$$

$$\therefore \qquad \bar{E} = \frac{\sum_{n=0}^{n=\infty} E_n\, e^{-E_n/kT}}{\sum_{n=0}^{n=\infty} e^{-E_n/kT}} = \frac{\sum_{n=0}^{n=\infty} nh\nu\, e^{-nh\nu/kT}}{\sum_{n=0}^{n=\infty} e^{-nh\nu/kT}}$$

$$= \frac{h\nu\,[0 + e^{-h\nu/kT} + 2e^{-2h\nu/kT} + \ldots\ldots \text{ to } \infty]}{[1 + e^{-h\nu/kT} + e^{-2h\nu/kT} + \ldots\ldots \text{ to } \infty]}$$

Substituting $-h\nu/kT = x$, we have

$$\bar{E} = \frac{h\nu\,[0 + e^{x} + 2e^{2x} + \ldots\ldots \text{ to } \infty]}{[1 + e^{x} + e^{2x} + \ldots\ldots \text{ to } \infty]}$$

Each term in the numerator is the differential coefficient of the corresponding term in the denominator.

$$\therefore \qquad \bar{E} = h\nu \frac{d}{dx} \log_e [1 + e^x + e^{2x} + \ldots\ldots \text{ to } \infty]$$

Now, $1 + e^x + e^{2x} + \ldots\ldots$ to ∞ is a geometric series having first term 1 and common ratio e^x.

The sum of the series to $\infty = \dfrac{1}{1 - e^x}$.

$$\text{and} \qquad \log_e \frac{1}{1 - e^x} = -\log_e (1 - e^x)$$

$$\therefore \qquad \bar{E} = -h\nu \frac{d}{dx} \log_e (1 - e^x)$$

$$= \frac{h\nu\, e^x}{1 - e^x} = h\nu \frac{1}{e^{-x} - 1} = \frac{h\nu}{e^{h\nu/kT} - 1}$$

According to wave mechanics the ground state energy of a harmonic oscillator is not zero but $\frac{1}{2} h\nu$ and thus the energy of subsequent states is given by

$$E_n = \left(n + \frac{1}{2}\right) h\nu$$

If we start with this value of E_n, we get

$$\bar{E} = h\nu \left(\frac{1}{2} + \frac{1}{e^{h\nu/kT} - 1}\right)$$

The term $\frac{1}{2}\ h\nu$ is temperature independent and gives the value of $\bar{E}$ at $T = 0$. In other words, according to quantum mechanics, the atoms have vibrational energy even at absolute zero. However, this term does not make any contribution to the specific heat (or lattice heat capacity).

If N_0 is the number of atoms in a kilo-mole, it is equivalent to $3N_0$ oscillators (See assumption *iv*) and the total energy

$$U = 3N_0\,\bar{E} = \frac{3N_0 h\nu}{e^{h\nu/kT} - 1}$$

and molar lattice heat capacity

$$C_v = \left(\frac{dU}{dT}\right)_v = 3N_0 k \left(\frac{h\nu}{kT}\right)^2 \frac{e^{h\nu/kT}}{(e^{h\nu/kT} - 1)^2} \quad \text{... (i)}$$

This result can be put in a simple form by substituting $h\nu/k = \theta_E$ known as Einstein temperature in the above relation and, get

$$C_v = 3N_0 k \left(\frac{\theta_E}{T}\right)^2 \frac{e^{\theta_E/T}}{(e^{\theta_E/T} - 1)^2} \quad \text{... (ii)}$$

The frequency ν of the atomic oscillator corresponding to Einstein temperature θ_E is known as Einstein frequency or frequency of Einstein oscillator and is given by

$$\nu = \frac{k\theta_E}{h}$$

Predictions and comparison with experimental results

(*i*) At *high temperatures* the relation

$$\bar{E} = \frac{h\nu}{e^{h\nu/kT} - 1}$$

reduces to the relation $\bar{E} = kT$ because for large values of T, $h\nu << kT$ and

$$e^{h\nu/kT} = 1 + \frac{h\nu}{kT} + \ldots\ldots \qquad \left[\because\ e^x = 1 + x + \frac{x^2}{2!} + \ldots\ldots\right]$$

neglecting higher powers.

$$\therefore \qquad \bar{E} = \frac{h\nu}{1 + \frac{h\nu}{kT} - 1} = kT$$

and

$$C_v = \frac{dU}{dT} = \frac{d}{dT}(3N_0\bar{E}) = 3N_0 k = 3R$$

i.e., the lattice heat capacity at high temperatures approaches the classical value $3R$. This is in fair agreement with experimental results.

(*ii*) *At low temperatures*. As T decreases $\bar{E}$ falls off rapidly and at very low temperatures $h\nu >> kT$ so that

$$e^{h\nu/kT} >> 1 \text{ and } e^{h\nu/kT} - 1 = e^{h\nu/kT}$$

Making this approximation relation (*i*) becomes

$$C_v = 3N_0 k \left(\frac{h\nu}{kT}\right)^2 \frac{e^{h\nu/kT}}{(e^{h\nu/kT})^2}$$

$$= 3N_0 k \left(\frac{h\nu}{kT}\right)^2 e^{-h\nu/kT} = 3R\left(\frac{h\nu}{kT}\right)^2 e^{-h\nu/kT}$$

$$= 3R\left(\frac{\theta_E}{T}\right)^2 e^{-\theta_E/T}$$

i.e., the lattice heat capacity is proportional to $e^{-h\nu/kT}$ $(e^{-\theta_E/T})$ which is the dominant factor and in the limit approaches zero as $T \to 0$.

This is not in agreement with experimental observations because it has been observed that for most of the solids the heat capacity at very low temperatures varies as T^3, *i.e.*, it obeys the T^3 law.

(*iii*) If we plot a graph between C_v and T/θ_E, we find that Einstein curve fits the experimental results fairly well at all temperatures except very low temperatures in the range of liquid helium. The Einstein temperature θ_E distinguishes between the high temperature and low temperature behaviour.

(*iv*) Einstein assumed the atoms in the crystal to be independent oscillators but this is not in conformity with actual facts. As each atom is under the influence of the field of force of a number of other vibrating atoms the situation is a very complex one.

21.13 EINSTEIN TEMPERATURE

According to Einstein theory of lattice heat capacity, the atoms of a solid are identical independent simple harmonic oscillators having a frequency ν and energy $E = nh\nu$ where n is an integral number 0, 1, 2, ... etc. known as quantum number. According to this theory, the molar lattice heat capacity at a temperature T is given by

$$C_v = 3N_0 k \left(\frac{h\nu}{kT}\right)^2 \frac{e^{h\nu/kT}}{[e^{h\nu/kT} - 1]^2}$$

The quantity $\frac{h\nu}{k} = \theta_E$ is known as **Einstein temperature** and the corresponding value of ν is known as **Einstein frequency** or frequency of Einstein oscillator.

Example 21.2. *Calculate Einstein temperature given Einstein frequency as 9 × 10¹¹ Hz.* **(*H.P.U. 2001*)**

Solution. Einstein frequency $\nu = 9 \times 10^{11}$ Hz

∴ Einstein temperature $\theta_E = \frac{h\nu}{k} = \frac{6.6\times10^{-34}\times9\times10^{11}}{1.38\times10^{-23}}$

$= 43$K

Example 21.3. *Calculate the frequency of Einstein oscillator for $\theta_E = 236$ K. Given $k_B = 1.38 \times 10^{-23}$ JK^{-1} and $h = 6.6 \times 10^{-34}$ J sec.* **(*M.D.U. 2000*)**

Solution. Einstein temperature $\theta_E = 236$K

Now, $\theta_E = \frac{h\nu}{k}$

∴ Frequency of Einstein oscillator $\nu = \frac{k\theta_E}{h} = \frac{1.38\times10^{-23}\times236}{6.6\times10^{-34}}$

$= 49.34 \times 10^{11}$ Hz

21.14 FAILURE OF DULONG AND PETIT'S LAW AT LOW TEMPERATURE

According to Einstein's theory of specific heat the total energy is given by

$$U = 3N_0\bar{E} = 3N_0 \frac{h\nu}{e^{h\nu/kT} - 1} \quad ...(i)$$

and

$$C_v = \frac{dU}{dT} = 3N_0 k \left(\frac{h\nu}{kT}\right)^2 \frac{e^{h\nu/kT}}{(e^{h\nu/kT} - 1)^2} \quad ...(ii)$$

The photon energy $h\nu$ indicates the difference between the allowed energies of the oscillator. At very high temperatures $h\nu/kT$ is a very small quantity and hence

$$e^{h\nu/kT} = 1 + \frac{h\nu}{kT} + ... \text{ higher powers}$$

$$\therefore \quad e^{h\nu/kT} - 1 = 1 + \frac{h\nu}{kT} - 1 = \frac{h\nu}{kT}$$

Substituting in (*i*), we get

$$U = 3N_0 h\nu \frac{kT}{h\nu} = 3N_0 kT$$

$$\therefore \quad C_v = \frac{dU}{dT} = 3N_0 k = 3R$$

i.e., Dulong and Petit's law is obeyed.

At low temperatures $\frac{h\nu}{kT}$ is a large quantity and unity can be neglected as compared to $e^{h\nu/kT}$

$$\therefore \quad e^{h\nu/kT} - 1 = e^{h\nu/kT}$$

Substituting in equation (*ii*), we have

$$C_v = 3N_0 k \left(\frac{h\nu}{kT}\right)^2 \frac{e^{h\nu/kT}}{(e^{h\nu/kT})^2} = 3N_0 k \left(\frac{h\nu}{kT}\right)^2 e^{-h\nu/kT}$$

Hence C_v is proportional to $e^{-h\nu/kT}$, *i.e.*, Dulong and Petit's law fails at low temperatures.

21.15 DEBYE MODEL FOR LATTICE SPECIFIC HEAT

In Einstein model of lattice specific heat of solids each atom is supposed to vibrate as a single unit independently of its neighbours. The vibrational motion of the crystal as a whole was considered to be the same as the vibrational motion of a single atom and, therefore, all the atomic vibrations of the crystal were assigned a common frequency ν which is the natural frequency of vibration of a single atom. This assumption is, however, over simplified because the atoms in a crystal are bound together to form a system of *coupled oscillators* which cannot vibrate independently. Each atom is under the influence of the field of force of a number of other vibrating atoms. This is why Einstein's theory of lattice heat capacity was not capable of giving correct behaviour particularly at low temperatures.

Assumptions of Debye model of lattice specific heat

1. In Einstein's theory of specific heat of solids each atom is supposed to vibrate as a single unit independently of its neighbours. But the situation is not so simple as each atom is under the influence of the field of force of a number of other vibrating atoms.

Debye took into account the effect of *coupling* between adjacent atoms and considered the *solid as a continuous elastic body*.

2. According to Debye the internal energy of a solid instead of residing in the vibrations of individual atoms was present in the *elastic standing waves* produced due to the vibrations of individual atoms.

3. Debye assumed that the *energy* of these elastic standing waves was *quantised.* A quantum of vibrational energy is called a *phonon.* It has an energy $h\nu$ similar to that of a *photon* where ν is the frequency of vibration and h Planck's constant. A *phonon* travels with the speed of sound as sound waves are elastic waves in nature.

4. The number of possible vibrational modes of any kind in the frequency range ν and $(\nu + d\nu)$ is given by $Z(\nu)\, d\nu = \left(\frac{4\pi V}{v^3}\right) \nu^2\, d\nu$ where $Z(\nu)$ is the number of vibrational modes per unit frequency range and is known as *density of modes,* $V = L^3$ is the volume of the solid taken in the form of a cube of each side L and v the velocity of propagation of the wave through the medium.

5. We shall first calculate the *density of vibrational modes* of crystal treating it as a continuous medium.

Density of modes. For simplicity we shall consider one dimensional continuous string of length L. Suppose the string is fixed at both the ends and is vibrating. If the displacement of the string at any point x at any instant of time t is $u(x, t)$, then the wave equation is given by

$$\frac{\partial^2 u}{\partial x^2} = \frac{l}{v^2}\frac{\partial^2 u}{\partial t^2} \qquad \text{... (i)}$$

where v is the velocity of propagation of the wave in the string.

Since both the ends of the string are fixed, the solution of Eq. (*i*) corresponds to standing waves.

$$u(x, t) = A \sin\left(\frac{n\pi}{L}x\right) \cos 2\pi \nu_n t \qquad \text{... (ii)}$$

where n is a positive integer ≥ 1. The length of the string L is an integral multiple of half wave-lengths

$$\therefore \quad L = n\frac{\lambda_n}{2}$$

or

$$\lambda_n = \frac{2L}{n}$$

$$\therefore \quad \text{Frequency } \nu_n = \frac{v}{\lambda_n} = \frac{nv}{2L} \qquad \text{... (iii)}$$

This shows that the frequency of the string can have discrete values only which are integral multiples of $\frac{v}{2L}$.

For example, $\nu_1 = \frac{v}{2L}$, $\nu_2 = \frac{v}{L}$;

$$\nu_3 = \frac{3v}{2L} \text{ etc.,}$$

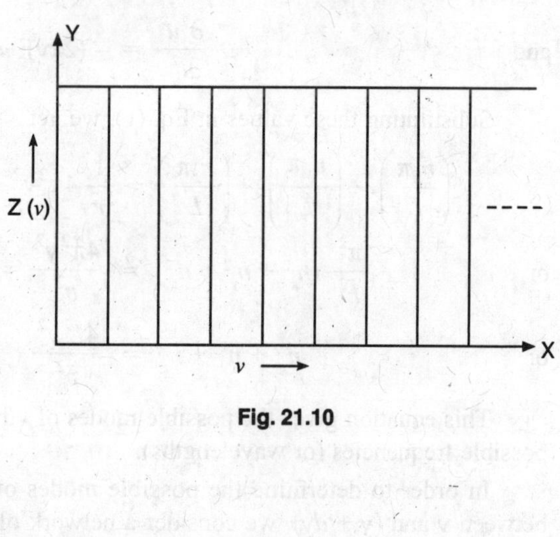

Fig. 21.10

i.e., the frequency spectrum of the continuous string is discrete and contains an infinite number of parallel lines equidistant from each other as shown in Fig. 21.10.

From Eq. (*iii*) $$n = \frac{2L}{v} \nu_n$$

$$\therefore \quad dn = \frac{2L}{v} d\nu_n \qquad \text{... (iv)}$$

Eq. (*iv*) gives the number of possible modes of vibration in a frequency interval $d\nu$.

Three dimensional case. If we consider a three dimensional case the wave equation (*i*) takes the three dimensional form

$$\frac{\partial^2 u}{\partial x^2} + \frac{\partial^2 u}{\partial y^2} + \frac{\partial^2 u}{\partial z^2} = \frac{1}{v^2}\frac{\partial^2 u}{\partial t^2} \qquad \text{... (v)}$$

Taking the three dimensional continuous medium as a cube of side L, with fixed faces, the standing wave solution of the standing wave equation (v) becomes

$$u(x, y, z, t) = A \sin\left(\frac{n_x \pi}{L} x\right) \sin\left(\frac{n_y \pi}{L} y\right) \sin\left(\frac{n_z \pi}{L} z\right) \cos 2\pi\nu t \qquad \text{... (vi)}$$

where n_x, n_y and n_z are positive integers ≥ 1.

Differentiating Eq. (*vi*) twice with respect to x, we get

$$\frac{\partial^2 u}{\partial x^2} = -\left(\frac{n_x \pi}{L}\right)^2 A \sin\left(\frac{n_x \pi}{L} x\right) \sin\left(\frac{n_y \pi}{L} y\right) \sin\left(\frac{n_z \pi}{L} z\right) \cos 2\pi\nu t$$

$$= -\left(\frac{n_x \pi}{L}\right)^2 u(x, y, z, t)$$

Similarly differentiating Eq. (*vi*) twice with respect to y and z respectively, we get

$$\frac{\partial^2 u}{\partial y^2} = -\left(\frac{n_y \pi}{L}\right)^2 u(x, y, z, t)$$

$$\frac{\partial^2 u}{\partial z^2} = -\left(\frac{n_z \pi}{L}\right)^2 u(x, y, z, t)$$

and $$\frac{\partial^2 u}{\partial t^2} = -(2\pi\nu)^2 u(x, y, z, t)$$

Substituting these values in Eq. (*v*), we get

$$\left(\frac{n_x \pi}{L}\right)^2 + \left(\frac{n_y \pi}{L}\right)^2 + \left(\frac{n_z \pi}{L}\right)^2 = \frac{1}{v^2} 4\pi^2\nu^2 = \frac{4\pi^2}{\lambda^2}$$

or $$\frac{\pi^2}{L^2}(n_x^2 + n_y^2 + n_z^2) = \frac{4\pi^2\nu^2}{v^2} = \frac{4\pi^2}{\lambda^2}$$

or $$n_x^2 + n_y^2 + n_z^2 = \frac{4L^2\nu^2}{v^2} \qquad \text{... (vii)}$$

This equation gives the possible modes of vibration. The integers n_x, n_y and n_z determine the possible frequencies (or wavelengths).

In order to determine the possible modes of vibration $Z(\nu)\, d\nu$ in the frequency interval between ν and $(\nu + d\nu)$, we consider a network of points, the position of each point being given by three Cartesian positive integer coordinates n_x, n_y and n_z. Thus the radius vector R from the origin to any point is given by

$$R^2 = n_x^2 + n_y^2 + n_z^2 \qquad \text{... (viii)}$$

[This is the equation of a sphere in integral space]

From Eq. (*vii*) and (*viii*), we get

$$R^2 = \frac{4L^2\nu^2}{v^2} \qquad \text{... (ix)}$$

$$\therefore \quad R = \frac{2L}{v}\nu$$

and

$$dR = \frac{2L}{v}\, d\nu \qquad \text{... (x)}$$

As each point occupies, on the average, a unit volume in the integer space, the number of points in the spherical shell between radii R and $(R + dR)$ is equal to the volume of an octant of a spherical shell in integer space, *i.e.*, $\frac{1}{8}$ × volume of spherical shell.

$\therefore$ Volume of octant of spherical shell = $\frac{1}{8} \times 4\pi R^2\, dR$

Therefore, the number of possible modes of vibration, *i.e.*, frequencies being between ν and (ν + *d*ν) are given by

$$Z(\nu)\, d\nu = \frac{4\pi}{8} R^2\, dR \qquad \text{... (xi)}$$

Substituting the value of R^2 from Eq. (*ix*) and dR from Eq. (*x*) in Eq. (*xi*), we get,

$$Z(\nu)\, d\nu = \frac{4\pi}{8} \times \frac{4L^2\nu^2}{v^2} \cdot \frac{2L}{v}\, d\nu$$

$$= 4\pi \frac{L^3}{v^3}\, \nu^2\, d\nu$$

But $L^3 = V$ the volume of the cube. Therefore, the number of modes of vibration between ν and (ν + *d*ν) are

$$Z(\nu)\, d\nu = \left(\frac{4\pi V}{v^3}\right) \nu^2\, d\nu \qquad \text{... (xii)}$$

In general there are two kinds of elastic waves that can occur in a solid, *viz.*, *longitudinal waves* having a velocity v_l and *transverse* waves having a velocity v_t. The velocity of propagation v_t of transverse waves is generally different from the velocity of propagation v_l of longitudinal waves. Again transverse waves have two perpendicular directions of polarisation or two vibrational modes perpendicular to the direction of propagation, whereas the longitudinal waves have only one mode which lies along the direction of propagation.

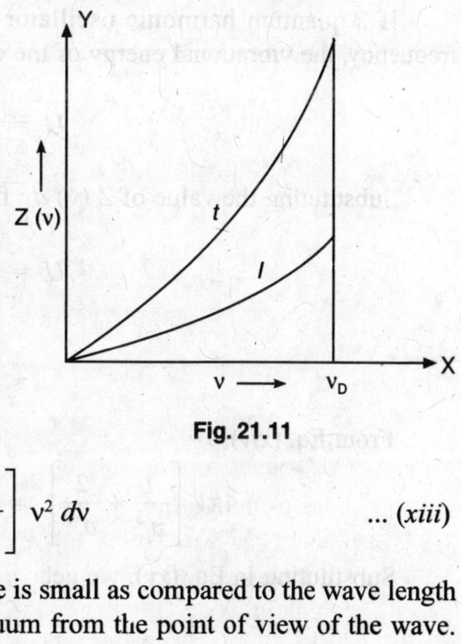

Fig. 21.11

In such a case the total number of vibrational modes is given by

$$Z(\nu)\, d\nu = 4\pi V \left[\frac{1}{v_l^3} + \frac{2}{v_t^3}\right] \nu^2\, d\nu \qquad \text{... (xiii)}$$

6. **Debye approximation.** If the interatomic distance is small as compared to the wave length of elastic waves, the crystal can be regarded as a continuum from the point of view of the wave.

Debye, therefore, assumed that the continuum model was applicable to all possible vibrational modes of the crystal. As the crystal consists of a finite number of atoms (N) the total number of vibrational modes is taken to be $3N$. This fact, puts an upper limit to the frequency of the elastic waves which can propagate through the crystal. This highest frequency propagating through a crystal is known as *Debye frequency*. It is common to longitudinal and transverse modes of vibration. Thus the frequency spectrum corresponding to the perfect continuum is cut off at Debye frequency ν_D so as to comply with a total of $3N$ modes as shown in Fig. 21.11. Thus Debye frequency can be determined by equating the total number of possible vibrations to $3N$.

Hence $$\int_0^{\nu_D} Z(\nu)\,d\nu = 3N$$

But according to Eq. (*xiii*),

$$Z(\nu)\,d\nu = 4\pi V\left[\frac{1}{v_l^3}+\frac{2}{v_t^3}\right]\nu^2\,d\nu$$

$$\therefore \int_0^{\nu_D} 4\pi V\left[\frac{1}{v_l^3}+\frac{2}{v_t^3}\right]\nu^2\,d\nu = 3N$$

or $$4\pi V\left[\frac{1}{v_l^3}+\frac{2}{v_t^3}\right]\frac{\nu_D^3}{3} = 3N$$

or $$\nu_D^3 = \frac{9N}{4\pi V}\left[\frac{1}{v_l^3}+\frac{2}{v_t^3}\right]^{-1} \quad \text{... }(xiv)$$

7. Debye further assumed that phonon gas behaves like a photon gas and average energy E per standing wave at a temperature T is given by

$$E = \frac{h\nu}{e^{h\nu/kT}-1}$$

If a quantum harmonic oscillator is associated with each vibrational mode of the same frequency, the vibrational energy of the crystal is given by

$$U = \int_0^{\nu_D} EZ(\nu)\,d\nu = \int_0^{\nu_D} Z(\nu)\frac{h\nu}{e^{h\nu/kT}-1}\,d\nu$$

Substituting the value of $Z(\nu)\,d\nu$ from Eq. (*xiii*), we get

$$U = \int_0^{\nu_D} 4\pi V\left[\frac{1}{v_l^3}+\frac{2}{v_t^3}\right]\frac{h\nu^3}{e^{h\nu/kT}-1}\,d\nu$$

$$= 4\pi V\left[\frac{1}{v_l^3}+\frac{2}{v_t^3}\right]\int_0^{\nu_D}\frac{h\nu^3}{e^{h\nu/kT}-1}\,d\nu \quad \text{... }(xv)$$

From Eq. (*xiv*),

$$4\pi V\left[\frac{1}{v_l^3}+\frac{2}{v_t^3}\right] = \frac{9N}{\nu_D^3}$$

Substituting in Eq. (*xv*), we get

$$U = \frac{9N}{\nu_D^3}\int_0^{\nu_D}\frac{h\nu^3}{e^{h\nu/kT}-1}\,d\nu \quad \text{... }(xvi)$$

Substituting $\frac{h\nu}{kT} = x$ and $\frac{h\nu_D}{kT} = x_m$, we get

$$\nu = \frac{x\,kT}{h} \qquad \therefore\ d\nu = \frac{kT}{h}\,dx$$

Eq. (*xvi*), therefore, becomes,

$$U = \frac{9Nh}{\nu_D^3}\left(\frac{kT}{h}\right)^4 \int_0^{x_m} \frac{x^3}{e^x - 1}\,dx$$

$$= 9N\left(\frac{kT}{h\nu_D}\right)^3 kT \int_0^{x_m} \frac{x^3}{e^x - 1}\,dx$$

We now introduce a characteristic temperature θ_D, known as *characteristic Debye temperature* given by $\theta_D = \frac{h\nu_D}{k}$

$$\therefore \qquad x_m = \frac{\theta_D}{T}$$

and

$$U = 9NkT\left(\frac{T}{\theta_D}\right)^3 \int_0^{\theta_D/T} \frac{x^3}{e^x - 1}\,dx \qquad \text{... (xvii)}$$

$$= 9RT\left(\frac{T}{\theta_D}\right)^3 \int_0^{\theta_D/T} \frac{x^3}{e^x - 1}\,dx \qquad \text{... (xvii)(a)}$$

when N is Avogadro's number.

Specific heat. From Eq. (*xvii*), we get

$$C_v = \left(\frac{\partial U}{\partial T}\right)_v = 9NK\left(\frac{T}{\theta_D}\right)^3 \int_0^{\theta_D/T} \frac{e^x\,x^4}{(e^x - 1)^2}\,dx$$

$$= 3R\left(\frac{\theta_D}{T}\right) F_D$$

where

$$F_D = 3\left(\frac{T}{\theta_D}\right)^4 \int_0^{\theta_D/T} \frac{e^x\,x^4}{(e^x - 1)^2}\,dx \qquad \text{... (xviii)}$$

The function is called *Debye function*.

21.16 SUCCESS OF DEBYE MODEL

On the basis of above assumptions Debye predicted the variation of specific heat with temperature, particularly at very high and very low temperatures.

1. **For very high temperatures.** At very high temperatures ($T >> \theta_D$), $x = \frac{h\nu}{kT}$ has a very small value so that $e^x = 1 + x$ and $\frac{x^3}{e^x - 1} = x^2$ and $\int_0^{\theta_D/T} x^2\,dx = \left[\frac{x^3}{3}\right]_0^{\theta_D/T} = \frac{1}{3}\left(\frac{\theta_D}{T}\right)^3$

Substituting these values in Eq. (*xvii*), we get

$$U = 9NkT\left(\frac{T}{\theta_D}\right)^3 \frac{1}{3}\left(\frac{\theta_D}{T}\right)^3$$

$$= 3NkT \quad \text{... (xix)}$$

∴ Specific heat at constant volume

$$C_v = \frac{dU}{dT} = 3Nk = 3R$$

This is Dulong and Petit's law. Thus at very high temperatures Debye's theory agrees with both the classical and Einstein's quantum theory results. In other words, quantum considerations are of no importance at high temperatures.

2. **At very low temperatures.** At very low temperatures ($T << \theta_D$)

$$x_m = \frac{\theta_D}{T} \to \infty \text{ and } \int_0^{\theta_D/T} \to \int_0^{\infty}$$

$$\therefore \int_0^{\theta_D/T} \frac{x^3}{e^x - 1} dx = \int_0^{\infty} \frac{x^3}{e^x - 1} dx = \frac{\pi^4}{15}$$

which on substituting in Eq. (*xvii*) (*a*) gives

$$U = 9RT \left(\frac{T}{\theta_D}\right)^3 \frac{\pi^4}{15} = \frac{3}{5}\pi^4 R \frac{T^4}{\theta_D^{\ 3}}$$

∴ Specific heat at constant volume

$$C_v = \frac{dU}{dT} = \frac{12}{5}\pi^4 \frac{R}{\theta_D^{\ 3}} T^3 = 234R\left(\frac{T}{\theta_D}\right)^3 \quad \text{... (xx)}$$

Debye's T^3 law. From relation (*xx*) we find that the specific heat is proportional to T^3 at very low temperatures as $\theta_D = \frac{h\nu_m}{k}$ = a constant for the substance. This is *Debye's T^3 law*. This law has been verified for a number of substances and experimentally observed values agree remarkably with theoretical values. This law holds for $T \leq \frac{\theta_D}{10}$.

21.17 LIMITATIONS (DRAWBACKS OR FAILURES) OF DEBYE MODEL

1. The Debye's continuum model is valid for long wavelengths only, *i.e.*, only low frequencies are active in the solid. Thus this theory neglects the dispersion of waves.

2. The total number of vibrational modes are assumed to be $3N$. As the solid is taken to be an elastic continuum, it should have an infinite number of frequencies.

3. The cut off frequency ν_D is assumed to be the same for both longitudinal and transverse waves. But as the velocities v_l and v_t are unequal, there should be two different values of ν_D for transverse and longitudinal waves.

4. The experimental results agree with the results from Debye's theory of specific heat fairly well at high temperatures.

At low temperatures Debye's T^3 law does not hold in the temperature range $\leq 0.1\ \theta_D$.

According to Debye's theory $\theta_D = \frac{h\nu_D}{k}$ is independent of temperature but actually θ_D is found to vary up to 10% with temperature.

5. This theory assumes that the solid consists of identical atoms. Hence the theory is not applicable to crystals where the atoms are not identical, *e.g.*, *NaCl*.

6. Debye's theory completely neglects the inter-atomic interaction and electronic contribution to lattice specific heat.

21.18 COMPARISON WITH EINSTEIN MODEL

1. At high temperatures both Debye and Einstein models give the same result $C_v = 3R$ which agrees with classical result of Dulong and Petit.

2. At low temperatures Einstein's theory predicts $C_v \propto e^{-h\nu/kT}$ whereas Debye's theory predicts $C_v \propto T^3$ which is more closely in agreement with experimental results. Thus, Einstein's theory fails at low temperatures. For example, for silver at $14K$ the value calculated from Einstein's formula is almost 28 times lower than the experimental value. Einstein assumed that atoms vibrate independently in the presence of other atoms with a constant frequency corresponding to optical branch whereas Debye proposed that waves could be propagated through solids and their frequencies ranged from a lower value corresponding to sound waves to a higher value corresponding to infra-red part of electromagnetic spectrum. He considered the vibrational mode of a crystal as a whole.

21.19 DEBYE TEMPERATURE

According to Debye's model, the lattice heat capacity is given by

$$C_v = 3R\left(\frac{\theta_D}{T}\right) F_D$$

where θ_D is the Debye temperature and F_D the Debye function.

The Debye temperature is given by

$$\theta_D = \frac{h\nu_D}{k}$$

where ν_D is Debye frequency, h Planck's constant and k Boltzmann constant.

Significance. The significance of Debye temperature θ_D is that at temperatures $T >> \theta_D$ the molar specific heat at constant volume (C_v) has the same value as given by classical theory (Dulong and Petit's law) or Einstein's quantum theory and its value is

$$C_v = 3Nk = 3R = 5.97 \text{ kilo-cals/kilo-mole}$$

At very low temperatures $T << \theta_D$, the specific heat at constant volume obeys the T^3 law as

$$C_v = \frac{12}{5}\pi^4 \frac{R}{\theta_D{}^3} T^3 = 234R\left(\frac{T}{\theta_D}\right)^3$$

or

$$C_v \propto T^3$$

i.e., C_v is proportional to T^3 at temperatures much below the Debye temperature.

Example 21.4. *If the frequencies of vibration of a solid are all equal and* $\frac{h\nu}{k_B} = 100$ *K, calculate molar specific heat of the solid at 500 K.* **(*M.D.U. 2001*)**

Solution. It is given that $\frac{h\nu}{k_B} = \theta_D$ = Debye's temperature = 100 K.

As $T = 500 \text{ K} > \theta_D$, the molar specific heat of the solid is given by Dulong and Petit's law and $3R = 5.97$ kilo-cals per kilo-mole.

21.20 LATTICE HEAT CAPACITY AT LOW TEMPERATURE

At low temperature $T \to 0$ K, the heat capacity decreases sharply and follows T^3 law for insulators and T – law for metals.

According to classical theory — Dulong and Petit's law — the molar heat capacity of all solids is constant and independent of temperature being equal to $3R = 5.97$ cals.

According to Einstein's theory of specific heat for low temperature $T << \theta_E$ where θ_E is the Einstein temperature, the lattice heat capacity is proportional to $e^{-\theta_E/kT}$. This does not, however, agree with experimental results.

According to Debye's theory, at low temperatures, the specific heat is proportional to T^3. This is known as *Debye's T^3 law* and holds for $T \leq \frac{\theta_D}{10}$, where θ_D is the Debye temperature. In actual practice, however, this is true only upto a temperature of $\theta_D/50$.

Example 21.5 *What can we say about the room temperature specific heat if Debye temperature of the solid is 3000 K?*

Solution. The lattice specific heat at constant volume, at low temperatures is given by

$$C_v = \frac{12}{5}\pi^4 \frac{R}{\theta_D^3} T^3 = 234R\left(\frac{T}{\theta_D}\right)^3$$

The law is true for temperatures $T \leq 0.1\ \theta_D$.

Here $\theta_D = 3000$ K. Therefore, Debye's law is true for temperatures $\leq \frac{3000}{10} \leq 300$ K.

The average room temperature can be assumed to be 27°C or less. Taking room temperature to be 27°C = 27 + 273 = 300 K, we have

$$C_v = 234R\left(\frac{300}{3000}\right)^3 = 0.234R = \frac{0.234}{3} \times 5.97$$

$$= 0.466 \text{ cal/mole K}$$

EXERCISE

1. (*a*) Explain briefly quantisation of lattice vibrations.
   (*b*) Explain the concept of phonon. Which statistics does it obey?
   (*c*) Give an evidence for the existence of phonons.
   (*d*) Derive an expression for the energy and momentum of phonons and explain 'normal' process and Umklapp process.
   (*e*) Discuss the inelastic scattering of photons by phonons and obtain an expression for the frequency of phonon emitted in the process.
   (G.N.D.U. 2000; Pbi.U. 2000; P.U. 2004, 2000; H.P.U. 2002)
2. (*a*) State the concepts which are common to both photons and phonons.
   (*b*) What is the difference between photons and phonons?
   (Pbi.U. 2004, 2003; G.N.D.U. 2003; P.U. 2003)
3. Show that the dispersion relation for the lattice waves in a monatomic linear lattice of mass *m*, spacing *a* and nearest neighbour interaction *C* is $\omega = 2\sqrt{\frac{C}{M}}\sin\left|\frac{1}{2}\vec{k}a\right|$ where ω is the angular frequency and $\vec{k}$ the wave vector. What are the allowed values of phonon wave vector?
   (Meerut U. 2003; Pbi.U. 2002, 2001; G.N.D.U. 2000)

4. (*a*) State the dispersion relation for the lattice waves in a monatomic linear lattice in terms of wave velocity. Discuss the dispersion behaviours (*i*) at low frequencies (*ii*) at high frequencies and (*iii*) at maximum frequency of vibration.
   (*b*) Compute the cut off frequency for a linear monatomic lattice if the velocity of sound and inter atomic spacing in the lattice are 3×10^3 ms^{-1} and 3×10^{-10} m respectively.
   (Pbi.U. 2000)
5. What is a phonon? Derive the vibrational modes of a diatomic linear lattice. Name the different branches of the dispersion relation curve. What is the difference between the two branches?
   (Meerut U. 2002; Luck.U. 2001)
6. (*a*) Define 'Molar specific heat' (*i*) at constant pressure (*ii*) at constant volume and lattice heat capacity.
   (*b*) Explain Dulong and Petit's law for the specific heat of a solid. Discuss the agreement of the result with that of the experiment. What discrepency was removed by Einstein?
   (G.N.D.U. 2002; Gharwal U. 2000)
7. What are the assumptions of Einstein's theory of specific heat of solids? Derive relation for lattice heat capacity following Einstein model. Discuss the relation at very high and very low temperatures. Give the predictions of the model and compare with experimental observations. ***(G.N.D.U. 2004, 2003, 2002, 2001, 2000; Meerut U. 2003; Gauhati U. 2000; H.P.U. 2002; K.U. 2001; Gharwal U. 2000)***
8. (*a*) What is Einstein temperature? ***(P.U. 2001)***
   (*b*) Calculate Einstein temperature given Einstein frequency as 9×10^{11} Hz. ***(H.P.U. 2001)***
   (*c*) Calculate the frequency of Einstein oscillator for $\theta_E = 236$ K. Given $k_B = 1.38 \times 10^{-23}$ JK^{-1} and $h = 6.6 \times 10^{-34}$ J sec. ***(M.D.U. 2000)***
   (*d*) Discuss as why Einstein's theory of lattice heat capacity is not capable of giving correct behaviour at low temperatures. ***(H.P.U. 2002)***
9. Why does Dulong and Petit's law fail at low temperature? ***(G.N.D.U. 2004; Pbi. U. 2001)***
10. What are the basic drawbacks of Einstein model of specific heat? Explaining the assumptions clearly, derive an expression for the lattice specific heat according to Debye model. Discuss the high and low temperature limits and define Debye's T^3 law. Give limitations (drawbacks) of Debye model. Compare Debye model with Einstein model.
    (P.U. 2004, 2003, 2001; Pbi.U. 2004, 2000; G.N.D.U. 2004; H.P.U. 2003, 2002, 2001, 2000; M.D.U. 2003, 2001; Calicut U. 2003; Kolkata U. 2002; K.U. 2000)
11. (*a*) Describe Debye's T^3 law of specific heat of solids.
    (G.N.D.U. 2002, 2000; Pbi.U. 2000; P.U. 2000)
    (*b*) (*i*) What is Debye frequency?
    (*ii*) What is Debye temperature? What is its significance?
    (Pbi.U. 2002, 2000; P.U. 2000)
    (*c*) If the frequencies of vibration of a solid are all equal and $\frac{h\nu}{k_B} = 100$ K, calculate molar specific heat of the solid at 500 K. ***(M.D.U. 2001)***
12. Write the temperature dependence of lattice heat capacity at low temperatures.

MULTIPLE CHOICE QUESTIONS

1. As the temperature approaches zero, the lattice contribution to heat capactiy of solids approaches

(*a*) infinity (*b*) zero

(*c*) any value between zero and infintiy

2. In Einstein theory of specific heat of solids the atoms are assumed as

(*a*) independent oscillator (*b*) coupled oscillator

(*c*) none of these

3. A phonon is emitted or asborbed in

(*a*) elastic scattering of photon by a crystal

(*b*) inelastic scattering of a photon by a crystal

(*c*) both the elastic and inelastic scattering of a photon by a crystal

(*d*) None of these.

4. Quantum of elastic vibration is

(*a*) photon (*b*) phonon

(*c*) graviton (*d*) meson

(H.P.U. 2001)

ANSWERS

1. (*b*)	**2.** (*a*)	**3.** (*a*)	**4.** (*b*)

FREE ELECTRON THEORY OF METALS

22
Chapter

INTRODUCTION

A number of important physical properties of metals can be explained in terms of '*free electron model*'. According to this model the *most weakly bound* electrons of the constituent atoms move about freely through the volume of the metal. Thus the *valence electrons* of the metal become the *conduction electrons*. The forces between the conduction electrons and ion cores are neglected. The total energy is all kinetic, the potential energy is also neglected. Hence the conduction electrons move about within the metal freely without any collisions similar to the molecules of an ideal gas.

It should be, however, noted that the '*electron gas*' is charged while ordinary gas is generally neutral. Secondly, the electron concentration in metals is much higher than that in ordinary gas.

The simplest metals are alkali metals. For example, in *Na* atom ($1s^2\ 2s^2\ 2p^6\ 3s^1$) the valence electron is in a $3s$ state. In the metal this electron becomes a conduction electron moving throughout the metal crystal. The remaining 10 electrons of Na^+ ion core fill the $1s$, $2s$ and $2p$ states having 2, 2 and 6 electrons respectively in the free ion, the distribution of core electrons being *essentially* the *same* in the free ion as in the metal.

The *metal crystal* is thus supposed to consist of positive metal ions with free electron gas of valence electrons of the atom *permeating* the space between the positive ions. These free electrons behave like the atoms or molecules in a perfect gas and are, therefore, sometimes known as '*free electron gas*' or '*fermi gas*' or '*Electron cloud*'.

22.1 FREE ELECTRON FERMI GAS

By a *free electron Fermi gas* we mean a gas of free and non-interacting electrons subject to Pauli's exclusion principle.

Interpretation of metallic properties. The interpretation of metallic properties in terms of free electron gas model was developed by Drude and Lorentz on the basis of classical theory applying Maxwell Boltzmann statistics to the electron gas. The mutual repulsion between the negatively charged electrons was neglected and the potential field due to the positive ion core was assumed to be constant everywhere. Thus this theory could explain:

(*i*) *Ohm's law i.e.*, the current i is proportional to potential difference E

$$i \propto E \text{ or } \frac{E}{i} = R \text{ (a constant)}$$

(*ii*) For *pure metals* electrical conductivity σ is inversely proportional to absolute temperature

$$\sigma \propto \frac{1}{T}$$

but for *impure metals* electrical conductivity is inversely proportional to the square root of absolute temperature

$$\sigma \propto \frac{1}{\sqrt{T}}$$

(*iii*) For most of the metals the resistivity ρ is inversely proportional to pressure *P*

$$\rho \propto \frac{1}{P}$$

(*iv*) The ratio of thermal conductivity *K* to electrical conductivity σ is proportional to absolute temperature

$$\frac{K}{\sigma} \propto T$$

This is known as *Wiedemann-Franz* law.

(*v*) A very characteristic property of metals is their *complete opacity* and *high lustre.*

A metal behaves like an opaque body because when light energy falls on a metal crystal, the free electrons are set into forced oscillation having the same frequency as the incident light. In this way the whole of the light is absorbed and hence the metal behaves as an opaque body.

Metals have high lustre because the incident light makes the free electrons go to the excited states. When these electrons come back to the initial state they emit light of the same energy as was absorbed initially. Thus the metal appears to reflect virtually all the light incident on it which gives it the characteristic metallic lustre.

(*vi*) The specific heat and paramagnetism of metals can be understood on the basis of interaction between free electrons and external energy source, whether thermal or magnetic in nature. The results predicted on the basis of classical theory, using Maxwell Boltzmann statistics were much larger than the actually observed values. However, making use of quantum mechanics and Fermi-Dirac statistics, it was found that resulting values of specific heat and paramagnetic susceptibility were in much better agreement with experimental results.

22.2 STATIC AND TRANSPORT PROPERTIES

The electronic properties of metals are broadly studied under two heads.

(*i*) Static properties, (*ii*) Transport properties.

(*i*) **Static properties.** The static properties are those properties which can be treated effectively by considering the energy levels only or the distribution of energy levels to which the electrons belong. These properties include, various electron emission properties, magnetic properties and properties like heat capacity and contact potential. These properties arise as a result of the excitation of electrons by light, thermal energy, strong electric and magnetic fields.

(*ii*) **Transport properties.** The transport properties are those which can be treated by considering the detailed response of the electrons to an external field *i.e.*, by taking into consideration the properties due to the acceleration of electrons such as electrical and thermal conductivity, Hall effect, thermoelectric effect etc.

22.3 MEAN FREE PATH (λ)

At a given temperature, the electrons in a metal suffer continuous collisions with one another Between any two collisions, the electron travels freely a certain distance in a straight line. This distance is known as *free path. Thus, the path covered by an electron between any two consecutive collisions in a straight line is called the free path.* The direction of the electron is changed after every collision. After a number of collisions, the total path appears zig-zag and free path is not constant. Therefore, a term 'mean free path' is used and is defined as the average distance travelled by a gas molecule between two successive collisions. It is denoted by λ. If the total distance travelled after N collisions is S, then mean free path (λ) is given by

$$\lambda = \frac{S}{N}$$

22.4 ELECTRICAL CONDUCTIVITY

Electrical conductivity is defined as the quantity of electricity (charge q) that flows in a unit time through a unit area of cross-section of the conductor per unit potential gradient (or electric field).

If a charge q flows through a conductor of area of cross-section A in a time t under an electric field E, then

$$\text{Electrical conductivity } \sigma = \frac{q}{AtE}$$

According to free electron model, the electrons in a solid move freely. If e is the charge, m the mass of the electron and E the applied electric field, then acceleration produced in the electron

$$a = \frac{d^2x}{dt^2} = \frac{Ee}{m}$$

Integrating, we have

$$v = \frac{dx}{dt} = \frac{Eet}{m} + C$$

At $t = 0$, $\frac{dx}{dt} = 0$ $\qquad \therefore \quad C = 0$

$$\therefore \qquad v = \frac{Eet}{m} \qquad \ldots(i)$$

and Energy of the electron $= \frac{1}{2}mv^2$.

If T is the absolute temperature and k Boltzmann's constant, then

$$\text{Energy of the electron} = \frac{3}{2}kT$$

$$\therefore \qquad \frac{1}{2}mv^2 = \frac{3}{2}kT$$

$$\text{or} \qquad mv = \frac{3kT}{v} \qquad \ldots(ii)$$

If λ is the mean free path of the electron, then relaxation time between two successive collisions

$$T = \frac{\lambda}{v}$$

From Eq. (i) $v = \frac{Eet}{m}$

$\therefore$ Average velocity between two successive collisions

$$\overline{v} = \frac{1}{T}\int_0^T \frac{Ee}{m} t\, dt = \frac{Ee}{mT}\int_0^T t\, dt = \frac{Ee}{mT}\left[\frac{t^2}{2}\right]_0^T = \frac{1}{2}\frac{Ee}{mT}\cdot T^2$$

$$\text{or} \qquad \overline{v} = \frac{1}{2}\frac{EeT}{m}.$$

Substituting $T = \frac{\lambda}{v}$, we get

$$\bar{v} = \frac{1}{2}\frac{Ee\lambda}{mv} \qquad \text{... (iii)}$$

Substituting the value of $mv = \frac{3kT}{v}$ from Eq. (*ii*) in Eq. (*iii*), we have

$$\bar{v} = \frac{1}{2}\frac{Ee\lambda v}{3kT} = \frac{Ee\lambda v}{6kT}$$

If n is the number density of electrons in the conductor, then current density

$$J = ne\bar{v} = \frac{ne^2 E\lambda v}{6kT} \qquad \text{...(iv)}$$

If q is the quantity of charge flowing through the conductor of cross-sectional area A in time t, then

$$q = \sigma AEt$$

or
$$J = \frac{q}{At} = \sigma E \qquad \text{...(v)}$$

Comparing Eqs. (*iv*) and (*v*), we get

$$\sigma = \frac{ne^2\lambda v}{6kT} \qquad \text{...(vi)}$$

Ohm's law. From Eq. (*iv*), we get

$$J = \frac{ne^2 E\lambda v}{6kT}$$

or
$$\vec{J} = \sigma\vec{E} \text{ where } \sigma = \frac{ne^2\lambda v}{6kT}$$

which is vector form of Ohm's law.

22.5 THERMAL CONDUCTIVITY

The coefficient of thermal conductivity of a material is defined as the quantity of heat energy that flows in a unit time through a unit area of cross-section of the material of unit thickness when its opposite faces are maintained at a unit difference of temperature.

The quantity of heat energy Q that flows in a time t through two opposite faces each of area A at a distance dx apart maintained at a temperature difference dT is given by

$$Q = -KA\frac{dT}{dx}t$$

When $A = 1$, and $t = 1$, $Q = -K\frac{dT}{dx}$

or
$$Q = K\frac{dT}{dx} \qquad \text{...(i) [Ignoring negative sign]}$$

Thermal conductivity of a metal. To discuss the thermal conductivity of metals on the basis of electronic theory we suppose that there exists an *electron gas* in the metal and instead of a voltage gradient as in the case of electrical conductivity there now exists a temperature gradient $\frac{dT}{dx}$ across a metallic rod. The electrons at the higher temperature end lose energy and electrons on the lower temperature end gain energy. The amount of heat energy Q passing through a cross-section of the rod per unit area, per second according to kinetic theory of gases is given by

$$Q = \frac{1}{3} nv\lambda \frac{dE}{dx} \qquad \text{...}(ii)$$

where $\frac{dE}{dx}$ is the *energy gradient*, n the number density of electrons and λ the mean free path. Also on the basis of Kinetic theory of gases

$$E = \frac{3}{2} kT$$

where k is Boltzmann's constant and T the absolute temperature.

$$\therefore \qquad \frac{dE}{dx} = \frac{3}{2} k \frac{dT}{dx}$$

Substituting in Eq. (*ii*), we have

$$Q = \frac{1}{3} \cdot \frac{3}{2} knv\lambda \frac{dT}{dx} = \frac{1}{2} nv\lambda k \frac{dT}{dx} \qquad \text{...}(iii)$$

Comparing (*i*) and (*iii*), we have

$$K \frac{dT}{dx} = \frac{1}{2} nv\lambda k \frac{dT}{dx}$$

or Thermal conductivity $K = \frac{1}{2} nv\lambda k$ ...(*iv*)

22.6 WIEDEMANN FRANZ LAW

All good conductors of electricity are also good conductors of heat. This fact is given in the form of Wiedemann Franz law. The law states that '*The ratio of thermal conductivity to electrical conductivity is proportional to the absolute temperature*'.

or $$\frac{K}{\sigma} \propto T$$

Now, thermal conductivity $K = \frac{1}{2} nv\lambda k$ [Refer article 22.5 Eq. (*iv*)]

and Electrical conductivity $\sigma = \frac{ne^2\lambda v}{6kT}$ [Refer article 22.4 Eq. (*vi*)]

$$\therefore \qquad \frac{K}{\sigma} = \frac{\frac{1}{2} nv\lambda k \cdot 6kT}{ne^2\lambda v} = 3\left(\frac{k}{e}\right)^2 T$$

or $\frac{K}{\sigma} \propto T$ as $\left(\frac{k}{e}\right)^2$ is a constant

This is Wiedemann Franz law. This law is, however not true at very low temperatures.

22.7 SOMMERFELD THEORY OF METALS

The classical theory of free electron gas considered electrons as distinguishable particles and applied Maxwell Boltzmann statistics to electron gas. Sommerfeld investigated the behaviour of free electron gas by taking into account quantum theory and Pauli's exclusion principle. He used Fermi Dirac statistics and succeeded in explaining many experimental facts.

The electron ion interaction cannot be ignored in metals because the conduction electrons remain confined within the volume of the metal. Since no emission of electrons is observed at room temperature, the potential energy of an electron at rest inside the metal must be lower than that of an electron at rest outside. This level is called the *vacuum level.* As the electrons move in the field of periodic potential of positive ions, the potential energy of electrons should be periodic in the crystal lattice. Sommerfeld in his model assumed that electrons find themselves in a potential which is constant everywhere inside the metal and arrived at a physical model in which the interior of the metal is represented by a potential energy box with an infinite potential barrier at the surface taking the potential energy as a constant within the box. The box may be considered one dimensional or three dimensional and may be discussed separately.

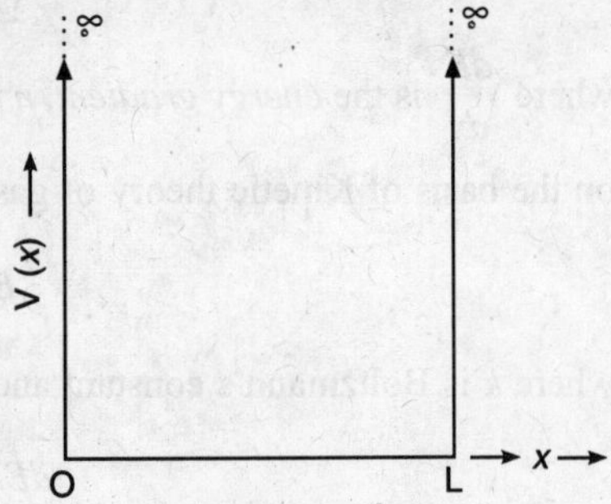

Fig. 22.1

In Sommerfeld's model, the free electrons are assumed to be the valence electrons composing the atoms. Thus the alkali metals are assumed to contain one free electron per atom.

22.7.1 Free Electron Gas in One Dimensional Box

Consider an electron of mass m confined to remain within a one dimensional line in a crystal of length L bounded by infinite barriers at the ends of the length as shown in Fig. 22.1. This is equivalent to the case of a free electron in a one dimensional box. It is supposed that as the electron is free the potential energy everywhere in the crystal is constant and equal to zero. At the ends of the crystal the electron is prevented from leaving the crystal by an infinite potential energy barrier. This means

$$V(x) = 0;\ 0 \leq x \leq L$$

$$= \infty;\ x \leq 0 \text{ and } x \geq L.$$

Wave function. As the potential energy is independent of time, we shall use Schrodinger's time independent wave equation *i.e.*,

$$\frac{\partial^2 \psi(x)}{\partial x^2} + \frac{2m}{\hbar^2}(E - V)\,\psi(x) = 0$$

Since the potential energy inside the crystal is zero the above equation becomes

$$\frac{\partial^2 \psi(x)}{\partial x^2} + \frac{2mE}{\hbar^2}\psi(x) = 0 \qquad \text{... } (i)$$

where E is the total energy of the electron.

Now
$$E = \frac{1}{2}mv^2 = \frac{m^2v^2}{2m} = \frac{p_x^2}{2m}$$

where p_x is the momentum of the electron

$$\therefore \qquad \frac{2m}{\hbar^2}E = \frac{2m}{\hbar^2}\frac{p_x^2}{2m} = \frac{p_x^2}{\hbar^2} = k^2$$

where k is the *propagation constant* given by

$$k = \frac{\sqrt{2mE}}{\hbar}$$

Eq. (*i*) may now be written as

$$\frac{\partial^2\psi(x)}{\partial x^2} + k^2\psi(x) = 0 \quad \text{...}(ii)$$

The general solution of Eq. (*ii*) is given by

$$\psi(x) = A \sin kx + B \cos kx \quad \text{...} (iii)$$

where A and B are arbitrary constants, the values of which can be determined from boundary condition.

As the probability of finding the electron across the walls (of infinite potential energy) is zero, we get the following boundary condition.

$$\psi(x) = 0 \quad \text{when } x = 0$$
$$\psi(x) = 0 \quad \text{when } x = L.$$

Substituting $\psi(x) = 0$, when $x = 0$ in Eq (*iii*), we get

$$0 = B$$

$\therefore$ Eq. (*iii*) now becomes

$$\psi(x) = A \sin kx \quad \text{...} (iv)$$

Substituting $\psi(x) = 0$, when $x = L$ in Eq. (*iv*), we have $0 = A \sin kL$

or
$$kL = n\pi \text{ or } k = \frac{n\pi}{L} \text{ where } n = 1, 2, 3.....$$

We do not take $n = 0$, because if $n = 0$, $k = 0$ and $k = \frac{p_x}{\hbar}$, which means $p_x = 0$ or momentum = 0 *i.e.*, the particle is at rest for all times which is not physically acceptable.

Hence equation (*iv*) becomes

$$\psi_n = A \sin\left(\frac{n\pi}{L}\right)x \quad \text{...} (v)$$

Evaluation of A. To find the value of A, we *normalise* the wave function $\psi(x)$. The probability of finding the electron anywhere on the line between $x = 0$ to $x = L$ is given by $\psi(x)\psi^*(x)dx = |\psi(x)|^2dx$. As there is no chance of the electron being located in the region $x < 0$ or $x > L$, we have

$$\int_0^L |\psi(x)|^2\, dx = 1$$

or
$$A^2 \int_0^L \sin^2 \frac{n\pi}{L} x\, dx = 1$$

or
$$\frac{A^2}{2} \int_0^L 2\sin^2 \frac{n\pi}{L} x\, dx = 1$$

But
$$\int_0^L 2\sin^2 \frac{n\pi}{L} x dx = \int_0^L \left(1 - \frac{\cos 2\pi\, nx}{L}\right) dx = L$$

$$\therefore \quad \frac{A^2}{2} L = 1 \text{ or } A = \left(\frac{2}{L}\right)^{1/2}$$

Substituting in Eq. (v), we get

$$\psi_n(x) = \left(\frac{2}{L}\right)^{1/2} \sin \frac{n\pi}{L}x \qquad \text{... } (vi)$$

This is the expression for the wave function.

Eigenvalues of energy. According to Eq. (v),

$$\psi_n = A \sin\left(\frac{n\pi}{L}\right)x$$

$$\therefore \qquad \frac{d\psi_n}{dx} = A\left(\frac{n\pi}{L}\right)\cos\left(\frac{2\pi}{L}\right)x$$

and
$$\frac{d^2\psi_n}{dx^2} = -A\left(\frac{n\pi}{L}\right)^2 \sin\left(\frac{n\pi}{L}\right)x$$

Substituting in Eq. (i), we get

$$-A\left(\frac{n\pi}{L}\right)^2 \sin\left(\frac{n\pi}{L}\right)x + \frac{2mE_n}{\hbar^2} A\sin\left(\frac{n\pi}{L}\right)x = 0$$

or
$$\left(\frac{n\pi}{L}\right)^2 = \frac{2mE_n}{\hbar^2}$$

$$\therefore \qquad E_n = \frac{\hbar^2}{2m}\left(\frac{n\pi}{L}\right)^2 \qquad \text{... } (vii)$$

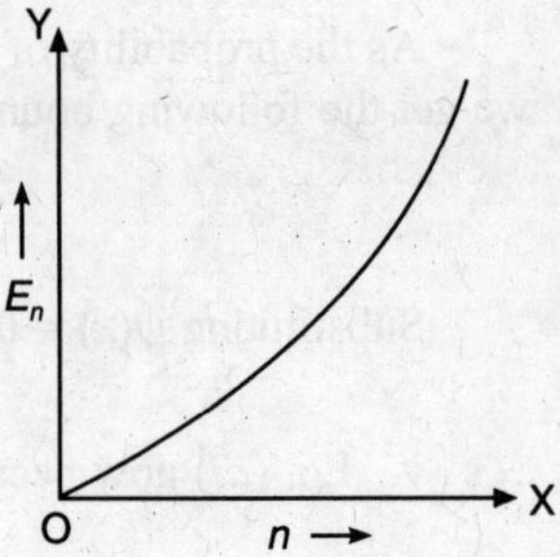

Fig. 22.2

i.e., the eigenvalues of energy E of the electron are

$$E_1 = \frac{\hbar^2}{2m}\frac{\pi^2}{L^2}\ ;\ E_2 = \frac{\hbar^2}{2m}\cdot\frac{2^2\pi^2}{L^2}\ ;\ E_3 = \frac{\hbar^2}{2m}\frac{3^2\pi^2}{L^2}$$

Hence $E_n \propto n^2$

Thus energy is proportional to n^2. A graph between energy E_n and n is shown in Fig. 22.2.

Fermi level. Let us suppose we want to accommodate N electrons on the line. According to Pauli's exclusion principle, no two electrons can have all their quantum numbers identical *i.e.*, each quantum state can be occupied by at the most one electron. In a solid an electron in a conduction electronic state has quantum numbers n and m_s, where n is the principal quantum number and m_s, the magnetic spin quantum number. Each set of values of n and m_s, defines a quantum state. For each value of n, m_s, can have two possible values $+\frac{1}{2}$ and $-\frac{1}{2}$. This means that each energy level defined by the quantum number n can have two quantum states and hence can accommodate a maximum of two electrons, one with *spin up* (↑) and the other with *spin down* (↓).

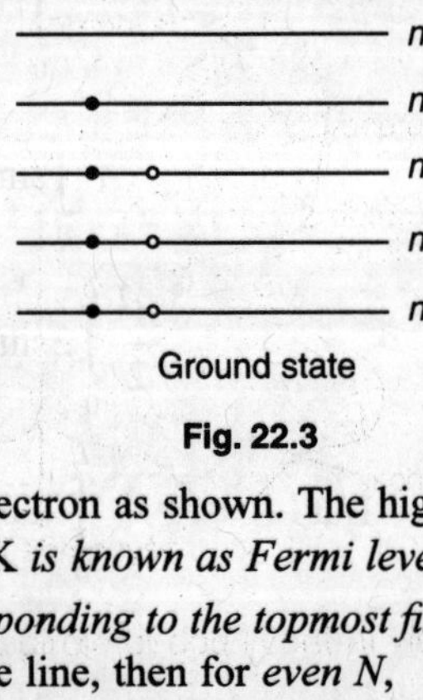

Ground state

Fig. 22.3

As an example, let us consider a case when the total number of electron to be accommodated is seven, then the ground state (n = 1) of the system has two electrons, next state (n = 2) has two, the third state (n = 3) also has two but fourth last state (n = 4) has only one electron as shown. The higher states n = 5 or above are empty. *The topmost filled energy level at* 0 K *is known as Fermi level.*

Fermi energy. *The Fermi energy is defined as the energy corresponding to the topmost filled level.* If N is the total number of electrons to be accommodated on the line, then for *even N*,

$$2n_F = N$$

where n_F represents the principal quantum number of the Fermi level *i.e.*, n_F represents the number of the topmost filled level.

Substituting $n = n_F$ in Eq. (*vii*), we get

$$E_F = \frac{\hbar^2}{2m}\left(\frac{n_F\pi}{L}\right)^2 \quad \text{... (viii)}$$

But $\quad n_F = \frac{N}{2}$, substituting in Eq. (*viii*), we get

$$E_F = \frac{\hbar^2}{2m}\left(\frac{N}{2}\frac{\pi}{L}\right)^2 = \frac{1}{4\pi^2}\cdot\frac{h^2}{2m}\left(\frac{N}{2L}\right)^2\pi^2 = \frac{h^2}{2m}\left(\frac{N}{4L}\right)^2 \quad \text{... (ix)}$$

Thus the value of Fermi energy depends upon $\left(\frac{N}{L}\right)^2$ where $\frac{N}{L}$ is the number of electrons per unit length.

Density of states. *It is defined as the number of electronic states (or quantum states) per unit energy range.* The density of states for energy E is denoted as $D(E)$.

Now $$E = \frac{\hbar^2}{2m}\left(\frac{n\pi}{L}\right)^2 = \frac{\hbar^2}{2m}\left(\frac{\pi}{L}\right)^2 n^2$$

$$\therefore \quad \frac{dE}{dn} = \frac{\hbar^2}{2m}\left(\frac{\pi}{L}\right)^2 2n = \frac{h^2}{4\pi^2}\cdot\frac{1}{2m}\frac{\pi^2}{L^2}\cdot 2n = \frac{h^2}{m}\frac{n}{4L^2}$$

$$\therefore \quad \frac{dn}{dE} = \frac{4mL^2}{h^2 n}$$

where $\frac{dn}{dE}$ gives the number of energy levels in the energy range E and $E + dE$.

Since there are two quantum states corresponding to each energy level, one with *spin up* (↑) and the other with *spin down* (↓) so that the density of states for a free electron gas in one dimension is

$$D(E) = 2\frac{dn}{dE} = \frac{8mL^2}{h^2}\cdot\frac{1}{n} \quad \text{... (x)}$$

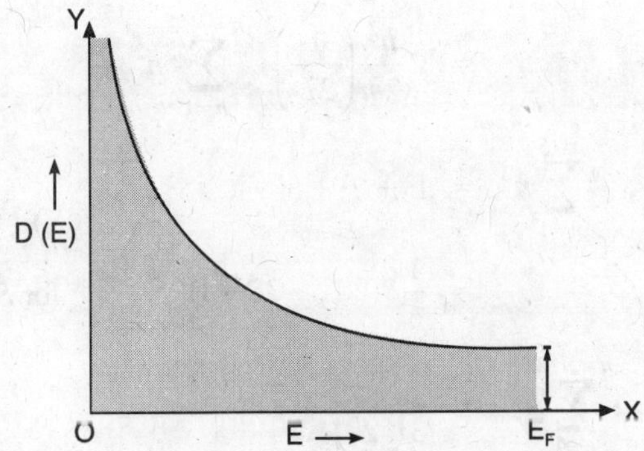

Fig. 22.4

According to Eq. (*vii*)

$$E = \frac{\hbar^2}{2m}\left(\frac{n\pi}{L}\right)^2 = \frac{h^2}{4\pi^2}\frac{1}{2m}\frac{n^2\pi^2}{L^2} = \frac{h^2}{8m}\frac{n^2}{L^2}$$

$$\therefore \quad n^2 = \frac{8mL^2 E}{h^2} \text{ or } n = \sqrt{8mE}\,\frac{L}{n}$$

$$\therefore \quad \frac{1}{n} = \frac{h}{L\sqrt{8mE}}$$

Substituting in Eq. (*x*), we get

$$D(E) = \frac{8mL^2}{h^2}\cdot\frac{h}{L\sqrt{8mE}} = \frac{4L}{h}\left(\frac{m}{2E}\right)^{1/2} = \frac{L}{\hbar^2\pi}\left(\frac{2m}{E}\right)^{1/2} \quad \text{... }(xi)$$

The variation of density of states $D(E)$ with energy E is shown in Fig. 22.4. It shows that levels are filled from $E = 0$ to $E = E_F$ where E_F denotes the Fermi energy. Beyond E_F all the states are empty.

Total Kinetic Energy

The energy of an electron in the *n*th energy state on a one dimensional line of length L in a crystal is given by

$$E_n = \frac{\hbar^2}{2m}\left(\frac{n\pi}{L}\right)^2 \quad \text{... }(xii)$$

The total energy E_0 of all the N electrons in the *ground state* is found by summing up the energies of individual electron from $n = 1$ to $n = n_f$ where n_f is the quantum number corresponding to Fermi level. In the ground state all the energy levels for $n > n_f$ are empty.

For N electrons the number of filled energy states is $\frac{N}{2}$ or $n_f = \frac{N}{2}$ as each energy state can accommodate a maximum of two electrons.

$$\therefore \text{ Total energy} \quad E_0 = 2\sum_{n=1}^{n=N/2} E_n$$

$$= 2\sum_{n=1}^{n=N/2}\frac{\hbar^2}{2m}\left(\frac{n\pi}{L}\right)^2 = 2\frac{h^2}{4\pi^2}\frac{\pi^2}{2m}\frac{1}{L^2}\sum_{n=1}^{n=N/2} n^2$$

$$= 2\frac{h^2}{2m}\left(\frac{1}{2L}\right)^2\sum_{n=1}^{n=N/2} n^2 \quad \text{... }(xiii)$$

$$\text{Now} \quad \sum_{n=1}^{n=S} n^2 = 1^2 + 2^2 + 3^2 + \dots S^2$$

$$= \frac{1}{6}S(2S^2 + 3S + 1) \cong \frac{1}{3}S^3 \quad \text{for } S >> 1$$

$$\text{Hence} \quad \sum_{n=1}^{n=N/2} n^2 \cong \frac{1}{3}\left(\frac{N}{2}\right)^3$$

Substituting in Eq. (*xiii*), we have

$$E_0 = 2\frac{h^2}{2m}\left(\frac{1}{2L}\right)^2\frac{1}{3}\left(\frac{N}{2}\right)^3 = \frac{1}{3}\frac{h^2}{2m}\left(\frac{N}{4L}\right)^2 N$$

But Fermi energy $$E_F = \frac{h^2}{2m}\left(\frac{N}{4L}\right)^2$$

∴ Total energy $$E_0 = \frac{1}{3}NE_F$$

Thus the total kinetic energy of N electrons is $\frac{1}{3}NE_F$, where E_F is the Fermi energy.

Average kinetic energy. The average kinetic energy $\overline{E_0}$ is the total energy divided by the number of electrons present.

∴ $$\overline{E_0} = \frac{E_0}{N}$$

or Average Kinetic energy $$\overline{E_0} = \frac{1}{3}E_F$$

In other words, the average kinetic energy per electron in the ground state is $\frac{1}{3}$rd of Fermi energy.

SOLVED EXAMPLES

Example 22.1. *Given the number of electrons per unit length of a crystal is 0.5 electrons per Å, calculate the Fermi energy.*

$$h = 6.62 \times 10^{-34} \text{ Js}, \; m = 9 \times 10^{-31} \text{ kg}$$

Solution. Fermi energy $$E_F = \frac{\hbar^2}{2m}\left(\frac{N\pi}{2L}\right)^2 = \frac{h^2}{2m}\left(\frac{N}{4L}\right)^2$$

Number of electrons per unit length $\frac{N}{L}$ = 0.5 electrons/Å

$$= 5 \times 10^9 \text{ electrons m}^{-1}$$

∴ Fermi energy $$E_F = \frac{6.62\times6.62\times10^{-68}\times5\times5\times10^{18}}{2\times9\times10^{-31}\times16} = 3.8 \times 10^{-19} \text{ J}$$

$$= \frac{3.8\times10^{-19}}{1.6\times10^{-19}} = \mathbf{2.37\ eV.}$$

22.7.2 Free Electron Gas in Three Dimensional Box

A metal can be described by a free electron gas which is confined to a cube of each side L. The potential inside the crystal is zero while outside the crystal the potential is infinitely large.

Wave function. For a free electron in one dimension say along the direction of X-axis, the wave function ψ_x is given by Schrodinger's wave equation

$$\frac{\partial^2\psi_x}{\partial x^2} + \frac{p_x^2}{\hbar^2}\psi_x = 0 \qquad ...(i)$$

where p_x is the x-component of the momentum of the electron and $\hbar = \frac{h}{2\pi}$. Substituting $\frac{p_x^2}{\hbar^2} = k_x^2$ where k_x is the x-component of the wave vector $\vec{k} = \frac{2\pi}{\lambda}$, λ being the wavelength associated with the moving electron, we have

$$\frac{\partial^2\psi_x}{\partial x^2} + k_x^2\psi_x = 0 \qquad ...(ii)$$

A solution of equation (*ii*) is

$$\psi_x = e^{ik_x x}$$

Similarly for the *y* and *z* direction respectively

$$\psi_y = e^{ik_y y} \text{ and } \psi_z = e^{ik_z z}$$

where k_y and k_z are the *y* and *z* components of the wave vector $\vec{k}$.

Considering a three dimensional case the wave function

$$\psi(r) = \psi(x, y, z) = \psi_x\psi_y\psi_z = Ae^{i(k_x x + k_y y + k_z z)}$$

$$= Ae^{i(\vec{k}\cdot\vec{r})} \qquad ...(iii)$$

where $|\vec{k}| = \sqrt{k_x^2 + k_y^2 + k_z^2}$ and the magnitude of position vector

$$|\vec{r}| = \sqrt{x^2 + y^2 + z^2}$$

The free electron gas is confined to a cube of each side *L*, and the potential inside the cube is zero while it has a large value outside it. To describe the electron within such a lattice, we apply the *periodic boundary condition* in which we quantise the electron energies in the cube (of each side *L*) *i.e.*, we require the wave function to be periodic in *x*, *y*, *z* with period *L*. Mathematically, it means

$$\psi(r) = \psi(x, y, z) = \psi(x + L, y, z) + \psi(x, y + L, z) + \psi(x, y, z + L)$$

As $\psi(r) = Ae^{i(k_x x + k_y y + k_z z)}$

The above condition can be satisfied only if

$$k_x = 0 : \pm\frac{2\pi}{L}; \pm\frac{4\pi}{L} : ....$$

and similarly for k_y and k_z *i.e.*, any component of $\vec{k}$ is of the form $n\frac{2\pi}{L}$ where *n* is a positive or negative integer.

To find the value of *A*. The constant *A* in Eq. (*iii*) can be determined by using the normalisation condition *i.e.*,

$$\int_0^v \psi^*(r)\psi(r)dV = 1$$

where *V* is the volume of the cube of each side *L*.

$$\therefore \quad \int_0^L\int_0^L\int_0^L A^2 e^{-i(\vec{k}\cdot\vec{r})}e^{i(\vec{k}\cdot\vec{r})}dx\,dy\,dz = 1$$

or $$A^2L^3 = 1$$

$$\therefore \quad A = \left(\frac{1}{L^3}\right)^{1/2} = \left(\frac{1}{V}\right)^{1/2}$$

Hence the normalised wave function for a free electron gas in three dimensions is

$$\psi_k(r) = \left(\frac{1}{V}\right)^{1/2} e^{i(\vec{k}\cdot\vec{r})}$$

Fermi energy. The wave vector $\vec{k}$ is given by

$$|\vec{k}| = \sqrt{k_x^2 + k_y^2 + k_z^2}$$

$$\therefore \quad k^2 = k_x^2 + k_y^2 + k_z^2 = \frac{p_x^2 + p_y^2 + p_z^2}{\hbar^2} = \frac{2mE}{\hbar^2}$$

or $$E = E_k = \frac{\hbar^2}{2m}k^2$$

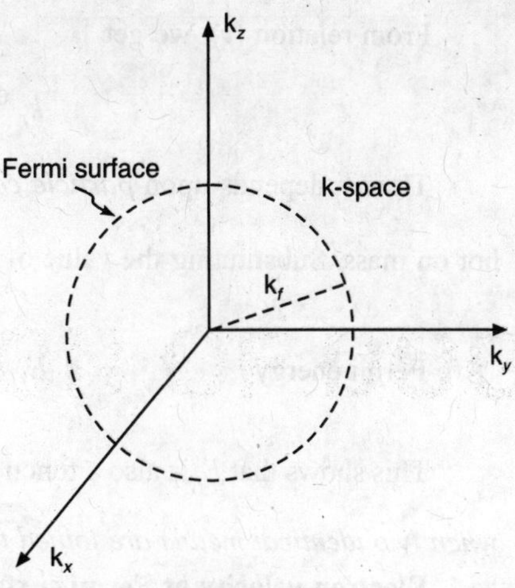

Fig. 22.5

We denote it by E_k as the components of $\vec{k}$ give the quantum numbers in this case in addition to the spin quantum numbers $m_s \pm \frac{1}{2}$.

In the ground state of a system of N free electrons the *occupied orbitals* may be represented as points inside a sphere in $\vec{k}$ space. The energy at the surface of the sphere is Fermi energy E_f and the wave vectors at the *Fermi surface* have a magnitude k_f such that

$$E_f = \frac{\hbar^2}{2m}k_f^2 \quad \text{...}(iv)$$

For the volume element $\left(\frac{2\pi}{L}\right)^3$ in $\vec{k}$ space *i.e.*, for $n = 1$ and $k_x = k_y = k_z = \frac{2\pi}{L}$ there is one allowed wave vector $\vec{k}$ with one distinct triplet of quantum numbers k_x, k_y and k_z.

$\therefore$ Number of allowed values of wave vector $\vec{k}$ upto the value $\vec{k} = \vec{k}_f$ is given by

$$\frac{\frac{4}{3}\pi k_f^3}{\left(\frac{2\pi}{L}\right)^3}$$

and total number of orbitals or energy states

$$N = 2 \times \frac{\frac{4}{3}\pi k_f^3}{\left(\frac{2\pi}{L}\right)^3} = \frac{L^3}{3\pi^2}k_f^3 = \frac{V}{3\pi^2}k_f^3 \quad \text{...}(v)$$

where $$V = L^3$$

The multiplied factor 2 appears on the right side as there are two allowed values of spin quantum number $m_s = +\frac{1}{2}$ (spin up ↑) and $m_s = -\frac{1}{2}$ (spin down ↓) for each allowed value of $\vec{k}$.

From relation (*v*), we get

$$k_f = \left(3\pi^2 \frac{N}{V}\right)^{1/3} \qquad ...(vi)$$

Thus k_f depends upon *particle concentration i.e.*, number of electrons per unit volume and not on mass. Substituting the value of $k_f = \left(3\pi^2 \frac{N}{V}\right)^{1/3}$ in (*iv*), we have

Fermi energy $$E_f = \frac{\hbar^2}{2m}\left(3\pi^2 \frac{N}{V}\right)^{2/3} \qquad ...(vii)$$

This shows that E_f is also a function of $\frac{N}{V}$ only. *As a result the Fermi energy does not change when two identical metals are joined together.*

Electron velocity at Fermi surface. If v_f is the velocity of the electron at the Fermi surface, then

$$mv_f = \frac{\hbar^2}{k_f}$$

$$\therefore \quad v_f = \frac{\hbar}{m} k_f = \frac{\hbar}{m}\left(3\pi^2 \frac{N}{V}\right)^{1/3}$$

Density of States. *The total number of available electronic energy states (or orbitals) per unit energy range is known as density of states*. It is denoted as $D(E)$.

$$\therefore \quad D(E) = \frac{dN}{dE}$$

The Fermi energy E_f is given by

$$E_f = \frac{\hbar^2}{2m}\left[3\pi^2 \frac{N}{V}\right]^{2/3}$$ [For proof See **Eq.** (*vii*) **article 22.7.1**]

Hence $$N^{2/3} = \left(\frac{V}{3\pi^2}\right)^{2/3} \frac{2m}{\hbar^2} E_f$$

or $$N = \frac{V}{3\pi^2}\left(\frac{2m}{\hbar^2}\right)^{3/2} E_f^{3/2}$$

As all the orbitals below the Fermi level are filled, as a general case we have for a value of energy $E < E_f$

$$N = \frac{V}{3\pi^2}\left(\frac{2m}{\hbar^2}\right)^{3/2} E^{3/2} \qquad ...(viii)$$

where E refers to the highest filled level energy.

$$\therefore \quad D(E) = \frac{dN}{dE} = \frac{V}{3\pi^2}\left(\frac{2m}{\hbar^2}\right)^{3/2} \frac{3}{2} E^{1/2}$$

$$= \frac{V}{2\pi^2}\left(\frac{2m}{\hbar^2}\right)^{3/2} E^{1/2} \qquad ...(ix)$$

This relation is true at 0 K. The density of filled electronic states [denoted as $N(E)$] at a particular temperature T K is given by

$$N(E) = D(E)f(E)$$

where $f(E)$ is the *Fermi function* which gives the probability that a particular quantum state at energy E is filled in accordance with Fermi-Dirac distribution law and has a value

$$f(E) = \frac{1}{e^{(E-E_f)/kT}+1}$$

where E_f is the Fermi energy and k Boltzmann's constant.

At absolute zero $f(E) = 1$ for $E < E_f$ *i.e.*, all energy states for $E < E_f$ are occupied.

And $f(E) = 0$ for $E > E_f$ *i.e.*, all energy states for $E > E_f$ are vacant (empty).

$\therefore$ Number of filled states between energy E and $E + dE$ at a temperature T is given by

$$N(E)dE = D(E)f(E)dE$$

$$= \frac{V}{2\pi^2}\left(\frac{2m}{\hbar^2}\right)^{3/2} \frac{E^{1/2}}{e^{(E-E_f)/kT}+1} dE$$

For energies above E_f such that $(E - E_f) > kT$, the term 1 in the denominator may be neglected as compared to the other term

$$\therefore \qquad N(E)dE = \frac{V}{2\pi^2}\left(\frac{2m}{\hbar^2}\right)^{3/2} \frac{E^{1/2}}{e^{(E-E_f)/kT}}$$

Putting $\hbar = \frac{h}{2\pi}$, we get

$$N(E)dE = \frac{\pi V}{2}\left(\frac{8m}{h^2}\right)^{3/2} \frac{E^{1/2}}{e^{(E-E_f)/kT}}.$$

22.8 AVERAGE KINETIC ENERGY AT 0 K

At 0 K all the electrons have energy less than E_f the Fermi energy.

$$\therefore \qquad \text{Fermi function } F(E) = \frac{1}{e^{(E-E_f)/kT}+1} = 1$$

As proved in Eq. (*ix*) of article 22.7.2, the number of energy states per unit energy range at energy E,

$$D(E) = \frac{V}{2\pi^2}\left(\frac{2m}{\hbar^2}\right)^{3/2} E^{1/2}$$

Now average kinetic energy per electron

$$\langle E_0 \rangle = \frac{1}{N}\int_0^{E_f} ED(E)F(E)dE = \frac{1}{N}\int_0^{E_f} ED(E)dE \quad [\because F(E) = 1]$$

$$= \frac{1}{N}\frac{V}{2\pi^2}\left(\frac{2m}{\hbar^2}\right)^{3/2}\int_0^{E_f} E^{3/2}dE$$

$$= \frac{1}{N}\frac{V}{2\pi^2}\left(\frac{2m}{\hbar^2}\right)^{3/2}\left[\frac{2}{5}E^{5/2}\right]_0^{E_f}$$

$$= \frac{1}{N}\frac{V}{2\pi^2}\left(\frac{2m}{\hbar^2}\right)^{3/2}\frac{2}{5}E_f^{5/2}$$

Now $$N = \frac{V}{3\pi^2}\left(\frac{2m}{\hbar^2}\right)^{3/2} E_f^{3/2}$$ **[Eq. (*viii*) article 22.7.2]**

$$\therefore \quad \langle E_0\rangle = \frac{3}{5}E_f$$

Thus average kinetic energy of a free electron at 0 K is $\frac{3}{5}E_f$.

22.9 FERMI VELOCITY

The magnitude of the wave vector $\vec{k}$ at the Fermi surface is given by

$$k_f = \left(3\pi^2\frac{N}{V}\right)^{1/3}$$ [For proof see **article 22.7.2., Eq.** (*vi*)]

If v_f is the velocity of the electron at the Fermi surface (level), then

$$mv_f = \hbar k_f$$

$$\therefore \quad v_f = \frac{\hbar}{m}k_f = \frac{\hbar}{m}\left(3\pi^2\frac{N}{V}\right)^{1/3}$$

or $$v_f^3 = \left(\frac{\hbar}{m}\right)^3 3\pi^2\frac{N}{V}$$

or $$v_f = \left(\frac{\hbar}{m}\right)\pi^{2/3}\left(\frac{3N}{V}\right)^{\frac{1}{3}}$$

This gives fermi velocity of the electron at the Fermi surface (level).

$\therefore$ The value of $$N = \frac{V}{3\pi^2}\left(\frac{m}{\hbar}\right)^3 v_f^3$$

As all the orbitals (energy states) below Fermi level are filled, we have as a general case

$$N = \frac{V}{3\pi^2}\left(\frac{m}{\hbar}\right)^3 v^3$$

Hence the number of states having velocity between v and $v + dv$

$$dN = \frac{V}{3\pi^2}\left(\frac{m}{\hbar}\right)^3 3v^2 dv$$

Now average speed at 0 K

$$\langle v\rangle = \frac{1}{N}\int_0^{v_f} v\, dN$$

$$= \frac{1}{N}\frac{V}{3\pi^2}\left(\frac{m}{\hbar}\right)^3\int_0^{v_f} 3v^2 v\, dv$$

$$= \frac{1}{N}\frac{V}{3\pi^2}\left(\frac{m}{\hbar}\right)^3\left[\frac{3}{4}v^4\right]_0^{v_f}$$

$$= \frac{1}{N}\frac{V}{3\pi^2}\left(\frac{m}{\hbar}\right)^3\frac{3}{4}v_f^4 = \frac{3}{4}v_f. \quad ...(i)$$

22.10 SIGNIFICANCE OF FERMI ENERGY LEVEL

The value of energy upto which all the energy states in a Fermi gas are filled at 0 K and above which all the energy states are empty is known as Fermi energy and corresponding energy level as Fermi energy level.

Units. As Fermi level is an energy level, it is measured in units of energy *i.e.*, Joule or electron Volt (*eV*).

Importance and physical meaning. Fermi energy level plays a very important role in semiconductors and other electronic grade material. The free electrons in a conductor or semiconductor behave like an electron gas. The distribution of electrons over a range of allowed energy levels at thermal equilibrium is given by

$$f(E) = \frac{1}{e^{(E-E_f)/k_BT}+1}$$

where k_B is Boltzmann's constant = 8.62×10^{-5} eV/K = 1.38×10^{-23} J/K where $f(E)$ known as *Fermi- Dirac distribution function* gives the probability that an available energy state at E will be occupied by an electron at absolute temperature T. The energy state E_f is called Fermi level. For an energy E = Fermi level energy E_f

$$f(E) = \frac{1}{1+1} = \frac{1}{2}$$

In other words an energy state at the Fermi level has a probability of $\frac{1}{2}$ of being occupied by an electron.

At 0 K *i.e.*, with $T = 0$ and $E < E_f$

$$f(E) = 1 \text{ and for } E > E_f \;\; f(E) = 0$$

which implies that at $T = 0$ every energy state *upto* E_f is *filled* with electrons and all energy states *above* E_f are *empty*. However, at temperatures higher than 0 K there is some probability that even states above Fermi level E_f will be filled and there is a corresponding probability that states below Fermi level be empty. The Fermi function is symmetrical about E_f and this property makes the Fermi level a natural reference point in calculating the electron and hole concentration in semiconductors.

Position of Fermi level. For *intrinsic semiconductor* material the concentration of electrons in the conduction band is equal to the concentration of holes in the valence band and hence Fermi level E_f lies at the middle of the band gap. [fig. 22.6 (a)]

In *n-type material* there is a higher concentration of electrons in the conduction band as compared to hole concentration in the valence band. The Fermi level, therefore, lies near the conduction band below the minimum value of conduction band energy level E_c. [Fig. 22.6b)]. On the other hand, for a *p-type material* there is a higher concentration of holes in the valence band as compared to the electron concentration in the conduction band. The Fermi level, therefore, lies near the valence band above the maximum value of valency band energy level E_v. [Fig. 22.6 (c)].

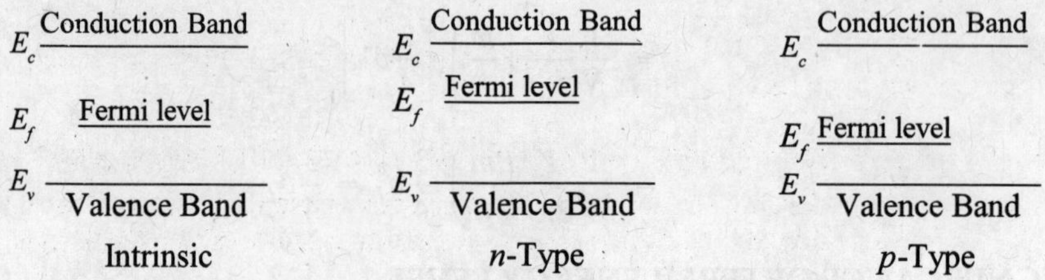

Fig. 22.6

Whenever a *p-n* junction diode is formed, the barrier potential V_B is given by

$$eV_B = E_{fn} - E_{fp}$$

where E_{fn} is the Fermi level energy in *n*-type and E_{fp} in the *p*-type material.

Thus Fermi level regulates the flow of electrons and holes in intrinsic, *n*-type and *p*-type semiconductor material and in junction diode and transistors.

22.11 FERMI TEMPERATURE

Fermi temperature T_f is defined as the ratio of Fermi energy to Boltzmann's constant k_B and is given by

$$T_f = \frac{E_f}{k_B}$$

As there is only one value of E_f at 0 K there is only one Fermi temperature. Thus a metal cannot be associated with two Fermi temperatures.

22.12 FERMI-DIRAC STATISTICS

A system of particles, identical and indistinguishable, which obey Pauli's exclusion principle and have half integral spin angular momentum in units of $\frac{h}{2\pi} = \hbar$ *i.e.*, $\frac{1}{2}\hbar$ *is known as Fermi gas*. These particles are known as *Fermions* and obey laws of Fermi-Dirac statistics. Free electrons inside a metallic conductor behave like an electron gas and obey *Fermi Dirac statistics.*

22.13 FERMI - DIRAC ENERGY DISTRIBUTION

The equation $$n_i = \frac{g_i}{e^{\alpha+\beta E_i}+1} \qquad ...(i)$$

represents the most probable distribution of the particles among various energy levels for a system obeying Fermi - Dirac statistics and is therefore known as Fermi - Dirac distribution law, for an assembly of fermions*.

The parameter $\beta = \frac{1}{kT}$, where k is Boltzman's constant, has the same role as in the case of M.B. distribution law. Substituting for β is Eq (*i*), we get

$$n_i = \frac{g_i}{e^{\alpha+E_i/kT}+1} = \frac{g_i}{e^{\alpha}e^{E_i/kT}+1} \qquad ...(ii)$$

* For details, refer article 12.8 of Author's book: Physics for Degree Students, B.Sc., II Year.

The quantity

$$f(E_i) = \frac{1}{e^{\alpha} e^{E_i/kT} + 1} \qquad ...(iii)$$

is known as Fermi - Dirac distribution function. The value of $f(E_i)$ for the Fermi - Dirac distribution at a particular energy E_i is the probability that under equilibrium a quantum state of that eneryg is occupied by a particle.

For particles like electrons of spin angular momentum $\pm \frac{1}{2} \hbar$, there are two possible spin orientations. For a system of such particles $g(E)\, dE$ in the energy lange E to $(E + dE)$ is given by

$$n(E)\, dE = 2 \times 2\pi V \left(\frac{2m}{h^2}\right)^{3/2} \cdot \frac{E^{\frac{1}{2}} dE}{e^{\alpha} . e^{E/kT} + 1} \qquad ...(iv)$$

This is the Fermi - Dirac energy distribution law for continuous variation of energy among free particles like electrons protons, neutrons, (all the fermions) with spin $\frac{1}{2}$.

22.14 FAILURE OF FREE ELECTRON THEORY

The free electron theory was successful in explaining various electronic and thermal properties of metals like heat capacity, thermal and electrical conductivity, magnetic susceptibility etc. However, there are some other properties which could not be explained by this theory. For example:–

(*i*) The theory could not explain why certain solids have a large number of free electrons and behave as good conductors, while certain others have hardly any free electrons and behave as insulators.

(*ii*) The theory could not explain the properties of semi-conductors.

(*iii*) It is found that divalent metals like *Be*, *Cd* etc. and trivalent metals like *Al*, *In* etc. are not good conductors. On the other hand, some monovalent metals are good conductors although these metals have a much smaller concentration of electrons as compared to divalent and trivalent metals.

(*iv*) According to this theory, the shape of the Fermi surface should be spherical but it is found to be non-spherical.

(*v*) Some of the metals like *Be* and *Zn* exhibit positive Hall co-efficient while the free electron theory predicts negative Hall effect for all metals.

The failure of the free electron theory is due to the over simplified assumption that the conduction electrons in a metal move in a region of zero or constant potential due to ion cores and are free to move about in the metal crystal – the motion being retarded only by the surface of the crystal. In fact, the potential due to ion cores is not constant and changes with the position of the electron in the crystal in a complicated manner. Some contributions to potential also arises due to the presence of other electrons in the metal crystal which has not been considered.

To a reasonable approximation the ion cores may be considered at rest and the potential experienced by the electron in a metal crystal assumed to be periodic with a period equal to lattice constant. This periodic potential is supposed to extend upto infinity in all directions and is interrupted only at the surface of the crystal.

Thus, the failure of the free electron theory of metals was due to the fact that regular arrangement of atoms in a crystal was not taken into account.

EXERCISE

1. (a) What is meant by free electron gas model of metals? Define free electron Fermi gas. Which properties of solids are explained by free electron gas theory?
(G.N.D.U. 2004; Pbi. U. 2004, 2002, 2001; P.U. 2003, 2001; H.P.U. 2000)

   (b) How far the free electron theory is inferior or superior to band theory? Discuss the successes and failures of these theories. ***(Pbi. U. 2002, 2000)***

2. (a) What are static and transport properties in the case of free electron model?
(Pbi. U. 2004; H.P.U. 2002)

   (b) Obtain an expression for the electrical conductivity of a metal on the basis of free electron theory. Hence prove Ohm's law.
(Bang. U. 2004, 2000; Nagpur U. 2012; Luck. U. 2002, Gharwal U. 2000)

3. (a) Obtain an expression for thermal conductivity of a metal on the basis of free electron theory. ***(Luck. U. 2002; Gharwal U. 2000)***

   (b) State Wiedemann Franz law. ***(Gauhati U. 2000)***

4. (a) Give an account of Sommerfeld theory of metals.

5. What is Sommerfeld model? ***(Luck. U. 2002; Gharwal U. 2000; P.U. 2000)***

   (b) Derive expressions for wave function and energy eigenvalues for electrons confined to a line of length L. Derive expressions for Fermi energy and density of states of the system.
(G.N.D.U. 2004, 2002, 2001; P.U. 2004, 2000; Poorvanchal U. 2001)

6. (a) Show that for a one dimensional crystal the average kinetic energy in the ground state is $\frac{1}{3}$rd of Fermi energy. ***(Poorvanchal U. 2001; G.N.D.U. 2000)***

7. (a) Obtain expression for the wave function of a free electron gas in three dimensions.

   (b) Derive the expression for Fermi energy E_f of a free electron gas in the three dimensions.
(Meerut U. 2000; Luck. U. 2002; H.P.U. 2001)

8. (a) Explain the concept of density of states for a free electron gas.
(Poorvanchal U. 2001; Luck. U. 2001)

   (b) Briefly explain Fermi function. ***(G.N.D.U. 2001)***

9. Show that average kinetic energy of a free electron at 0 K is $\frac{3}{5}E_f$ where E_f is the Fermi energy and average speed is $\frac{3}{4}v_f$ where v_f is the velocity at Fermi surface.
(P.U. 2003, 2001; Meerut U. 2003; Bang. U. 2001)

10. (a) Define Fermi energy level. What are its units? What is its physical meaning and importance in electronic grade material? ***(Nagpur U. 2012)***

    (b) Fermi level of an intrinsic semiconductor lies near the middle of the forbidden gap but for an n-type semiconductor it lies near the conduction band. Explain.
(G.N D.U. 2002, 1992; Pbi.U. 2002, 2001; Bang. U. 2001)

11. (a) What do you understand by Fermi gas? Can a metal be associated with two Fermi temperatures? ***(Pbi. U. 2003)***

    (b) Does the Fermi energy of a metal depend upon temperature and total number of electrons in the sample? ***(Pbi. U. 2003)***

12. What is the cause of failure of free electron theory? ***(P.U. 2000)***

MULTIPLE CHOICE QUESTIONS

Mark the correct answer:

1. Fermi level is defined as
   (*a*) Lowest filled level at 0K, (*b*) Highest filled level at 0K,
   (*c*) Lowest filled level at 300K, (*d*) Highest filled level at 300 K.
   (Porvanchal U. 2003)

2. The average kinetic energy of the electron in the ground state in one dimension is
   (*a*) $\frac{1}{3}$ of Fermi energy (*b*) $\frac{1}{2}$ of Fermi energy
   (*c*) $\frac{1}{4}$ of Fermi energy (*d*) Equal to Fermi energy
   (H.P.U. 1997)

3. Average kinetic energy of an electron at 0K in three dimension is
   (*a*) $\frac{1}{3}$ of Fermi energy (b) $\frac{1}{2}$ of Fermi energy
   (*c*) $\frac{2}{5}$ of Fermi energy (d) $\frac{3}{5}$ of Fermi energy

4. Fermi temperature (T_f) is given by
   (*a*) E_f / k_β (*b*) k_β / E_f
   (*c*) $E_f \cdot k_\beta$ (*d*) none of these.

ANSWERS

1. (*b*) **2.** (*a*) **3.** (*d*) **4.** (*a*)

BAND THEORY OF SOLIDS

23

Chapter

INTRODUCTION

The free electron theory is very useful in explaining experimental facts that essentially depend upon the kinetic properties of conduction electrons and in this respect is *superior* to band theory.

For example, the free electron theory conveniently explains the properties of metals like specific heat, thermal conductivity, electrical conductivity, Ohm's law, magnetic susceptibility, Wiedemann-Franz law, complete opacity of metals and their high lustre etc.

However, the free electron theory is unable to explain the properties which depend upon the interaction of electrons with neighbouring atoms or ions and in this respect is *inferior* to band theory.

For example, the model fails to explain the distinction between metals, semi-metals, semiconductors and insulators and why some crystals are metallic. It does not explain the occurrence of positive values of Hall effect, the relation between conduction electrons in the metals to the valence electrons in the free atom.

The theory also fails to correctly predict the temperature dependence of resistivity. According to this theory, resistivity varies as $\sqrt{T}$ whereas actually it is found to vary linearly with temperature.

The theory could not correctly explain the heat capacity and paramagnetic susceptibility of conduction electrons.

It does not explain the occurrence of long mean free paths of the order of $10^8 - 10^9$ interatomic space or more than one cm at low temperatures.

All these properties can be explained on the basis of band theory according to which electrons in crystals are arranged in energy bands separated by energy gaps or band gaps. This band gap arises from the interaction of conduction electron waves with ion cores of the crystal.

23.1 SPLITTING OF ATOMIC ENERGY LEVELS INTO BANDS

The atoms in a solid are closely packed, but are in a state of vibration about their mean positions. The nucleus of an atom can be considered at rest due to its large mass and it is the orbital electrons which are responsible for most of the characteristic properties of the solids.

When the atoms are brought together in a solid, the interaction between them perturbs the initial atomic levels. Most of the symmetry of the electronic states in the isolated atom is deformed and the level splits into bands of levels. A band constitutes a sort of energy continuum, in which separate levels due to individual atoms cannot be identified. In the process of inter atomic interaction, the inner shell electron states are the least effected, whereas the valence electrons, which are closest to the neighbouring ions, are the most effected. The effect of bringing one atom close to the other is to split a single sharp energy level into a pair of levels.

Let ψ_1 and ψ_2 be the electronic wave functions of the two atoms respectively when they are infinite distance apart. As the two atoms are brought together the wave function ψ_1 and ψ_2 overlap. The resultant wave function is either $\psi_1 + \psi_2$ or $\psi_1 - \psi_2$ as shown in Fig. 23.1 (*a*) and (*b*).

The electron has finite probability of existing midway between the two nuclei for wave function $\psi_1 + \psi_2$. In this region the binding energy increases because the electron experiences the force of both the nuclei. For the wave function $\psi_1 - \psi_2$ the probability of finding the electron between the two nuclei is zero. Thus the extra contribution of energy disappears. In other words, there is a difference between the energy states $\psi_1 + \psi_2$ and $\psi_1 - \psi_2$. Thus when two isolated atoms are brought together each energy state splits up into two distinct energy states.

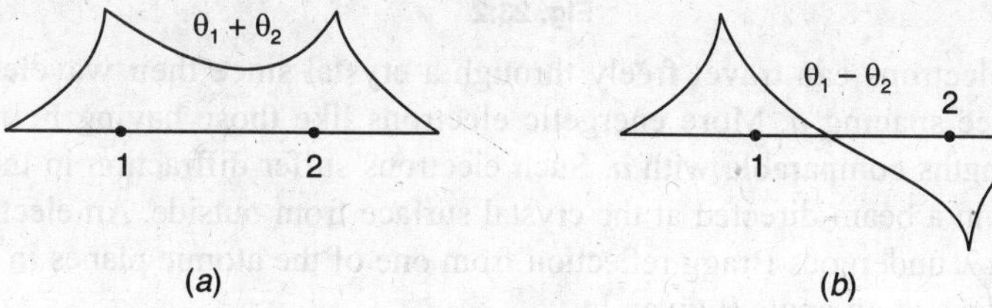

Fig. 23.1

If there are many interacting atoms say N and N is very large, each energy state will split into N number of energy states. The separation between these energy states will be so small that each level will appear to have spread into a continuous band of energy levels. This is why energy bands rather than specific energy levels exist in solids. The width of a band depends upon the strength of interaction.

This explains how the atomic energy levels are split into bands when a number of atoms are brought close together to form a crystal.

23.2 ORIGIN OF BAND STRUCTURE

The band structure of a solid crystal can be described by the *nearly free electron model* in which the band electrons in a crystal are supposed to move in a region of *periodically varying potential due to the ion cores of the crystal.*

The energy of a free electron in terms of the wave vector $\vec{k}$ is given by

$$E_k = \frac{\hbar^2}{2m}k^2 = \frac{\hbar^2}{2m}(k_x^2 + k_y^2 + k_z^2)$$

where k_x, k_y and k_z are respectively the x, y and z components of $\vec{k}$. The corresponding free electron wave function is of the form

$$\psi_k(r) = e^{i(\vec{k}\cdot\vec{r})} = e^{i(k_x x + k_y y + k_z z)}$$

where x, y, z are the components of position vector $\vec{r}$.

If we consider one dimensional case

$$\psi_k(x) = e^{ikx} \text{ or } \psi = e^{ikx}$$

We shall also consider the interaction between electrons and ion cores in *one dimensional lattice* with ion spacing a as shown in Fig. 23.2.

The ion cores bear a positive charge and the electrostatic potential energy of a electron in the field of a positive ion is *attractive, i.e.*, negative. This is why the potential energy is shown along the −V axis. (Fig. 23.2)

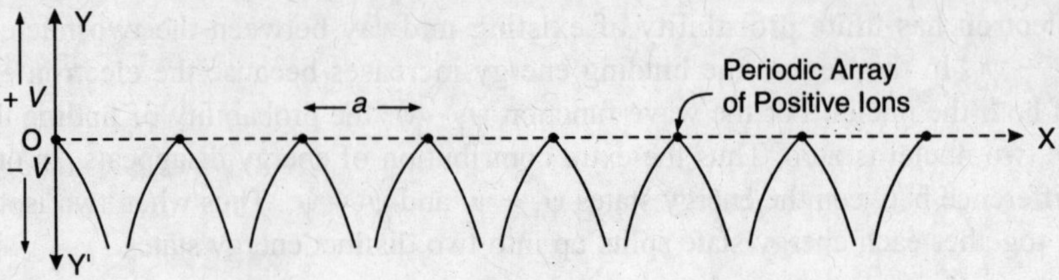

Fig. 23.2

Low energy electrons can travel freely through a crystal since their wavelengths are long relative to the lattice spacing a. More energetic electrons like those having Fermi energy in a metal have wavelengths comparable with a. Such electrons suffer diffraction in the same way as X-rays or electron in a beam directed at the crystal surface from outside. An electron having de-Broglie wavelength λ undergoes Bragg reflection from one of the atomic planes in a crystal when it approaches the plane at an angle θ given by

$$(a + b) \sin \theta = n\lambda$$

As lattice spacing $a = b$, $\quad n\lambda = 2a \sin \theta$

In terms of wave vector $\vec{k} = \dfrac{2\pi}{\lambda}$, the above relation becomes

$$k = \pm \frac{n\pi}{a \sin \theta} \qquad \text{... (i)}$$

In a one dimensional linear lattice $\theta = \pi/2$ and $k = \pm n\pi / a$. Thus a line representing k values is divided by energy discontinuities into segments of length $\pm\pi / a$. These line segments are known as *Brillouin zones*. (Fig. 23.3)

Note. The same result $k = \pm \dfrac{n\pi}{a}$ can be obtained from Bragg's reflection condition in reciprocal lattice which states $\left(\vec{k} + \dfrac{\vec{G}}{2}\right) \cdot \vec{G} = 0$ [For proof See article 19.8]

or $\quad k = \pm \dfrac{G}{2} = \pm \dfrac{n\pi}{a} \qquad \left(\because G = \pm \dfrac{2n\pi}{a}\right)$

A free electron travelling wave in one dimension, say along X-axis is given by

$$\psi = e^{ikx}$$

At the boundaries $k = \pm \dfrac{\pi}{a}$ the wave functions are not travelling waves $e^{i\frac{\pi}{a}x}$ and $e^{-i\frac{\pi}{a}x}$ but these are standing waves as explained below.

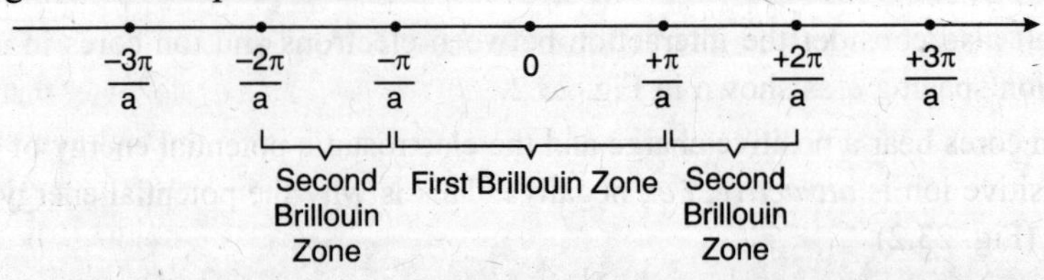

Fig. 23.3

For $k = \pm\frac{\pi}{a}$ Bragg condition is satisfied so that the wave travelling in one direction is Bragg reflected at the boundary and then travels in the opposite direction. Each subsequent Bragg reflection reverses the direction of travel again. The only time independent situation is formed by standing waves. There can be two types of standing waves which can be formed from the two travelling waves $e^{i\frac{\pi}{a}x}$ and $e^{-i\frac{\pi}{a}x}$

$$\psi(+) = e^{i\frac{\pi}{a}x} + e^{-i\frac{\pi}{a}x} = 2\cos\left(\frac{\pi}{a}x\right)$$

and

$$\psi(-) = e^{i\frac{\pi}{a}x} - e^{-i\frac{\pi}{a}x} = 2i\sin\left(\frac{\pi}{a}x\right)$$

The standing waves ar labelled (+) if they do not change sign when $(-x)$ is substituted for $(+x)$ and (–) if they change sign.

In quantum mechanics the probability density of a particle $\rho = |\psi|^2$. For a pure travelling wave given by $\psi = e^{ikx}$, the probability density $\rho = e^{ikx}e^{-ikx} = 1$, so that the charge density $e|\psi|^2 =$ a constant where e is the charge on the electron. However, the charge density is not constant for standing waves. For ψ (+) charge density $e|\psi_+|^2 \propto \cos^2\left(\frac{\pi}{a}x\right)$.

For $x = 0, a, 2a$ etc. $\cos^2\frac{\pi}{a}x = 1$ and has a maximum value so that negative electron charge density is concentrated at the positive ion cores, thereby lowering the potential energy. Similarly for ψ (–) charge density $e|\psi_-|^2 \propto \sin^2\frac{\pi}{a}x$. For $x = \frac{a}{2}, \frac{3}{2}a$ etc., $\sin^2\frac{\pi}{a}x = 1$ and has a maximum value so that negative electron charge is concentrated midway between the positive ion cores, thereby raising the potential energy in comparison with that of the travelling wave. The distribution of electronic charge density corresponding to a travelling wave, and that for the standing wave ψ (+) and ψ (–) is shown in Fig. 23.4. If we calculate the average values of potential energy over

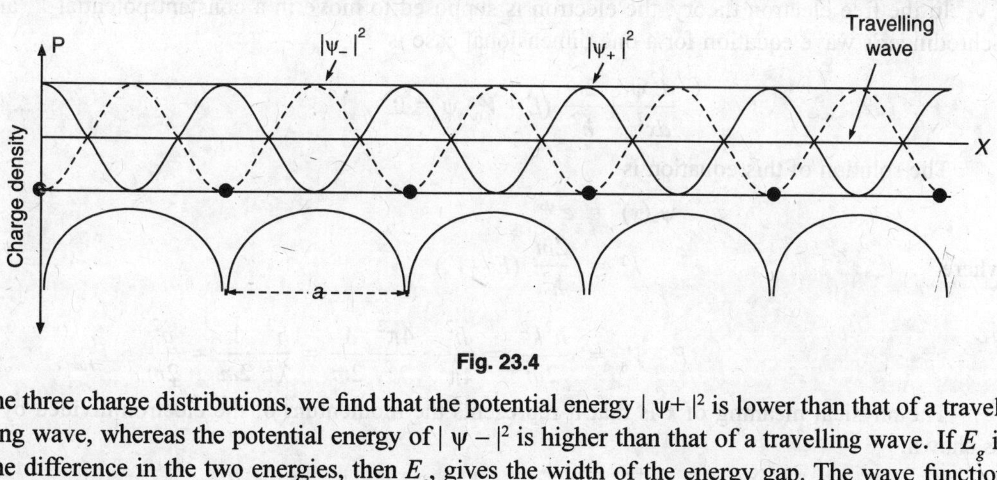

Fig. 23.4

the three charge distributions, we find that the potential energy $|\psi+|^2$ is lower than that of a travelling wave, whereas the potential energy of $|\psi-|^2$ is higher than that of a travelling wave. If E_g is the difference in the two energies, then E_g, gives the width of the energy gap. The wave function just below the energy gap is ψ (+) at points A and that just above at points B is ψ (–), BA giving the value of energy gap E_g as shown in Fig. 23.10.

23.3 PERIODIC POTENTIAL IN A CRYSTAL

The potenial energy of an electron at a distance r from an atomic nucleus of charge +Ze is given by

$$V = -\frac{1}{4\pi\varepsilon_0}\frac{Ze^2}{r}$$

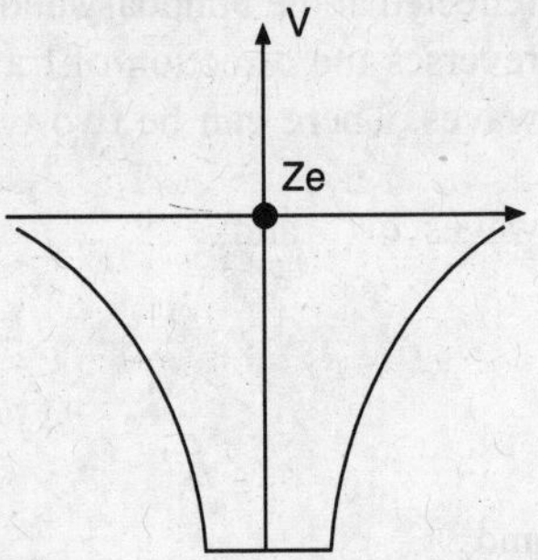

Fig. 23.5

The variation of potential energy V with distance r for an isolated system is shown in Fig. 23.5.

When a crystal is formed a number of such atomic nuclei are brought close to each other. The potential of an electron in such a case is the sum of the potential energies due to individual nuclei. The variation of potential energy with distance in a *one dimensional* case is shown in Fig. 23.6.

The atomic nuclei have a positive charge and the electrostatic potential energy of an electron in the field of positive charge is *attractive, i.e., negative*. This is why the potential energy is shown along the – V axis. To a reasonable approximation the nuclei (or ion cores) have been considered

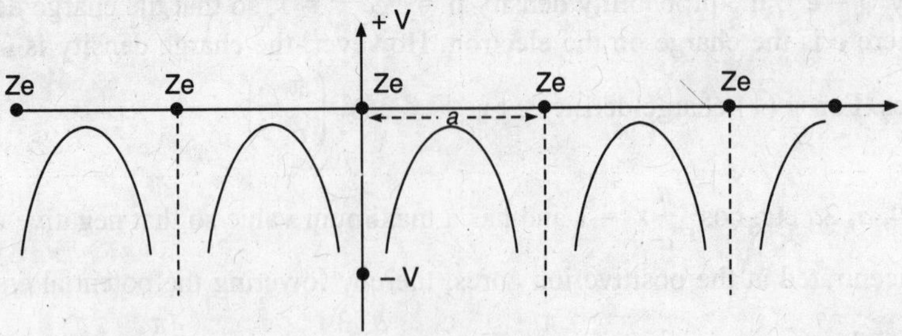

Fig. 23.6.

at rest and the potential experienced by an electron in the crystal is assumed to be periodic with the period equal to the lattice constant a. The quantity 'a' is also known as lattice parameter.

23.4 BLOCH THEOREM

In the free electron theory, the electron is supposed to move in a constant potential V_0 and Schrodinger's wave equation for a one dimensional case is

$$\frac{d^2\psi}{dx^2} + \frac{2m}{\hbar^2}(E - V_0)\,\psi = 0 \qquad \ldots (i)$$

The solution of this equation is

$$\psi(x) = e^{\pm ikx}$$

where

$$k^2 = \frac{2m}{\hbar^2}(E - V_0)$$

$$\therefore \quad E - V_0 = \frac{\hbar^2 k^2}{2m} = \frac{h^2}{4\pi^2}\cdot\frac{4\pi^2}{\lambda^2}\frac{1}{2m} = \frac{h^2}{\lambda^2}\frac{1}{2m} = \frac{p^2}{2m} = E_{kinetic}$$

The physical meaning of k is that it represents the momentum of the electron divided by $\hbar$ as shown:

The magnitude of $\quad \vec{k} = |\vec{k}| = \frac{2\pi}{\lambda} = 2\pi\frac{mv}{h} = \frac{mv}{\hbar} = \frac{p}{\hbar}$

Each electron will have its own $\vec{k}$ and its own energy.

For an electron moving in a one dimensional periodic potential the potential energy is given by the relation

$$V(x) = V(x+a)$$

where a is the period equal to lattice constant. In this case Schrödinger's equation becomes

$$\frac{d^2\psi}{dx^2} + \frac{2m}{\hbar^2}[E - V(x)]\,\psi = 0 \quad \text{... (ii)}$$

With reference to the solution of this equation there is an important theorem known as *Bloch theorem* or *Floquet's theorem* which states that there exist solutions of the form

$$\psi(x) = e^{\pm ikx}\,\mu_k(x) \quad \text{... (iii)}$$

where

$$\mu_k(x) = \mu_k(x+a) \quad \text{... (iv)}$$

Thus the solutions are plane waves of the type $e^{\pm ikx}$ modulated by the function $\mu_k(x)$ which has the same periodicity as the lattice constant.

The wave function of the type $\psi(x) = e^{\pm ikx}\,\mu_k(x)$ is called *Bloch function.* The wave vector $\vec{k}$ gives the direction of Bloch wave.

Proof. Suppose $f(x)$ and $g(x)$ are two real and independent solutions of equation (*ii*). Then the general solution is

$$\psi(x) = A\,f(x) + B\,g(x) \quad \text{... (v)}$$

where A and B are arbitrary constants. As the potential is periodic, *i.e.*, $V(x) = V(x+a)$, therefore, the functions $f(x+a)$ and $g(x+a)$ must also be the solutions of equation (*ii*). Since a differential equation of the second order can have only two independent solutions, the functions $f(x+a)$ and $g(x+a)$ must be able to be expressed in terms of functions $f(x)$ and $g(x)$, *i.e.*,

$$f(x+a) = \alpha_1 f(x) + \alpha_2\,g(x) \quad \text{... (vi)}$$

and

$$g(x+a) = \beta_1 f(x) + \beta_2\,g(x) \quad \text{... (vii)}$$

where α_1, α_2, β_1 and β_2 are real functions of energy E.

According to Eq. (*v*), the general solution of Schrodinger's equation is

$$\psi(x) = A\,f(x) + B\,g(x)$$

and

$$\psi(x+a) = A\,f(x+a) + B\,g(x+a)$$

According to Eq. (*vi*) and (*vii*), we have

$$\begin{aligned}\psi(x+a) &= A\alpha_1 f(x) + A\alpha_2\,g(x) + B\beta_1 f(x) + B\beta_2\,g(x)\\ &= (A\alpha_1 + B\beta_1)\,f(x) + (A\alpha_2 + B\beta_2)\,g(x) \quad \text{... (viii)}\end{aligned}$$

Now select A and B such that

$$A\alpha_1 + B\beta_1 = \lambda A \quad \text{... (ix)}$$

and

$$A\alpha_2 + B\beta_2 = \lambda B \quad \text{... (x)}$$

where λ is a constant. Substituting in Eq. (*viii*), we get

$$\begin{aligned}\psi(x+a) &= \lambda A\,f(x) + \lambda B\,g(x)\\ &= \lambda\,(A\,f(x) + B\,g(x)) = \lambda\psi(x) \quad \text{... (xi)}\end{aligned}$$

Equations (*ix*) and (*x*) will give non-zero values of A and B if the determinant of their coefficient vanishes, *i.e.*,

$$\begin{vmatrix} \alpha_1 - \lambda & \beta_1 \\ \alpha_2 & \beta_2 - \lambda \end{vmatrix} = 0$$

or

$$\lambda^2 - (\alpha_1 + \beta_2)\,\lambda + \alpha_1\beta_2 - \alpha_2\beta_1 = 0 \quad \text{... (xii)}$$

We shall first prove that $\alpha_1\beta_2 - \alpha_2\beta_1 = 1$ as given below.

As $f(x)$ and $g(x)$ are two real and independent solutions of Eq. (*ii*), we have

$$\frac{d^2 f(x)}{dx^2} + \frac{2m}{\hbar^2}[E - V(x)]f(x) = 0 \quad \text{... (xiii)}$$

and

$$\frac{d^2 g(x)}{dx^2} + \frac{2m}{\hbar^2}[E - V(x)]g(x) = 0 \quad \text{... (xiv)}$$

Multiplying Eq. (*xiii*) by $g(x)$ and Eq. (*xiv*) by $f(x)$ and subtracting former from the latter, we get

$$f(x)\frac{d^2 g(x)}{dx^2} - g(x)\frac{d^2 f(x)}{dx^2} = 0$$

or

$$f(x)\frac{dg(x)}{dx} - g(x)\frac{df(x)}{dx} = \text{a constant}$$

The left hand side is called Wronskian, $W(x)$ of the solution and is constant in this case.

$$\therefore \quad W(x) = f(x)\frac{dg(x)}{dx} - g(x)\frac{df(x)}{dx}$$

Now from Eq. (*vi*) and (*vii*), we get

$$W(x+a) = f(x+a)\frac{dg(x+a)}{dx} - g(x+a)\frac{df(x+a)}{dx}$$

$$= f(x+a)\left[\beta_1\frac{df(x)}{dx} + \beta_2\frac{dg(x)}{dx}\right] - g(x+a)\left[\alpha_1\frac{df(x)}{dx} + \alpha_2\frac{dg(x)}{dx}\right]$$

$$= [\alpha_1 f(x) + \alpha_2 g(x)]\left[\beta_1\frac{df(x)}{dx} + \beta_2\frac{dg(x)}{dx}\right]$$

$$- [\beta_1 f(x) + \beta_2 g(x)]\left[\alpha_1\frac{df(x)}{dx} + \alpha_2\frac{dg(x)}{dx}\right]$$

$$= (\alpha_1\beta_2 - \alpha_2\beta_1)\left[f(x)\frac{dg(x)}{dx} - g(x)\frac{df(x)}{dx}\right]$$

$$= (\alpha_1\beta_2 - \alpha_2\beta_1)W(x) \quad \text{... (xv)}$$

Thus, $\quad W(x+a) = (\alpha_1\beta_2 - \alpha_2\beta_1)W(x)$

But $\quad W(x+a) = W(x) = \text{a constant}$

$\therefore \quad \alpha_1\beta_2 - \alpha_2\beta_1 = 1$

Hence, Eq. (*xii*) becomes

$$\lambda^2 - (\alpha_1 + \beta_2)\lambda + 1 = 0 \quad \text{... (xvi)}$$

The quantity $(\alpha_1 + \beta_2)$ is a real function of energy E. There are two roots λ_1 and λ_2 of this quadratic equation, so there will be two functions $\psi_1(x)$ and $\psi_2(x)$ which have the property $\psi_1(x+a) = \lambda\psi(x)$. The product of the two roots $\lambda_1\lambda_2 = 1$.

We consider the following cases:

(*i*) *For energy ranges such that* $(\alpha_1 + \beta_2)^2 < 4$. In such a case Eq. (*xvi*), will have complex roots. Since $\lambda_1\lambda_2 = 1$ these roots will be *complex* conjugate of each other. Therefore, we write

$$\lambda_1 = e^{ika} \quad \text{and} \quad \lambda_2 = e^{-ika}$$

where k is real.

The corresponding functions $\psi_1(x)$ and $\psi_2(x)$ will then have the property

$$\psi_1 (x + a) = e^{ika} \psi_1 (x) \qquad ... (xvii)$$

and
$$\psi_2 (x + a) = e^{-ika} \psi_2 (x) \qquad ... (xviii)$$

Thus in general

$$\psi (x + a) = e^{\pm ika} \psi (x) \qquad ... (xix)$$

It can be seen that a function having the property given by Eq. (*xix*) is the Bloch function of the type given in Eq. (*iii*)

i.e.,
$$\psi (x) = e^{\pm ikx} \mu_k (x)$$

by replacing x by $(x + a)$, then we get

$$\psi (x + a) = e^{\pm ik (x + a)} \mu_k (x + a)$$
$$= e^{\pm ika} e^{\pm ikx} \mu_k (x)$$

because
$$\mu_k (x + a) = \mu_k (x) \quad \text{Also } e^{\pm ikx} \mu_k (x) = \psi (x)$$

$\therefore$
$$\psi (x + a) = e^{\pm ika} \psi (x) = \lambda\psi (x) \qquad ... (xx)$$

where
$$\lambda = e^{\pm ika}$$

We, therefore, find that Eq. (*xix*) and Eq. (*xx*) are the same. This proves Bloch theorem.

(*ii*) *For the energy range such that* $(\alpha_1 + \beta_2)^2 > 4$. In such a case Eq. (*xvi*) will have *real* roots λ_1 and λ_2, which may be taken as

$$\lambda_1 = e^{\mu a} \quad \text{and} \quad \lambda_2 = e^{-\mu a}$$

where μ is real. The corresponding solutions to the Schrödinger's equation are

$$\psi_1 (x) = e^{\mu x} \mu (x) \text{ and } \psi_2 (x) = e^{-\mu x} \mu (x) \qquad ... (xxi)$$

Though mathematically valid, these wave functions are not allowed because these functions become infinite at $+\infty$ or $-\infty$, *i.e.*, they are not bounded functions.

The roots $\lambda_1 = e^{ika}$ and $\lambda_2 = e^{-ika}$ (*allowed roots*) as well as the roots $\lambda_1 = e^{\mu a}$ and $\lambda_2 = e^{-\mu a}$ (forbidden roots) are functions of $(\alpha_1 + \beta_2)$, hence of energy E. The allowed roots correspond to the *allowed energy regions* and the forbidden (or disallowed) roots are associated with *forbidden energy regions*. This means that the energy spectrum of an electron moving in a periodic potential consists of allowed and forbidden energy regions or bands.

The theory given by Bloch leading to the existance of allowed and forbidden energy regions (called bands) is supported by model suggested by kronig and Penney. Let us study this Kronig-Penney model in detail.

23.5 KRONIG PENNEY MODEL

In the free electron theory it is assumed that the potential to which the valence electrons are subjected is constant and, therefore, it can be taken to be equal to zero, while solving Schrödinger's equation. But this is not true for the valence electron in ionic and co-valent crystalline solids because in such solids the electrons are localised near the parent nuclei. This gives rise to a periodically varying potential. Thus, in an actual case the constant potential should be replaced by a periodically varying potential.

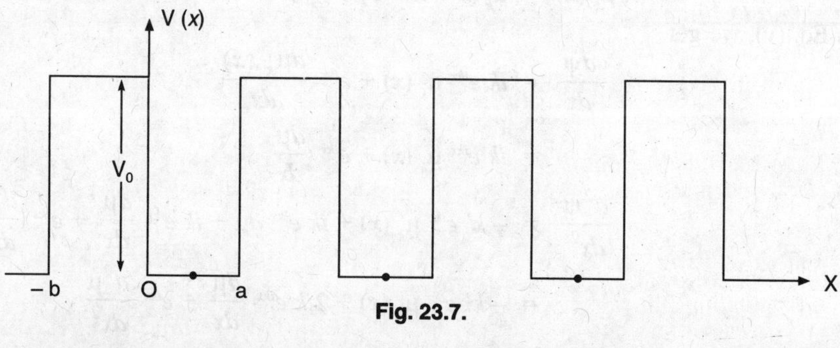

Fig. 23.7.

Kronig and Penney model illustrates the behaviour of electrons in a periodic potential by assuming a relatively simple one dimensional model of periodic potential as shown in Fig. 23.7, which shows the periodic potential energy of a valence electron in the *X*-direction.

Basic Assumptions

In Kronig Penney model it is assumed that the potential energy of an electron in a linear array of positive nuclei has the form of a periodic array of square wells with period $(a + b)$.

At the bottom of the well, *i.e.*, for $0 < x < a$ the electron is assumed to be in the vicinity of a nucleus and the potential energy is taken as zero, whereas outside a well, *i.e.*, for $-b < x < 0$ the potential energy is assumed to be V_0.

Analytical Treatment

Although this model employs a very crude approximation to the type of periodic potential existing inside a lattice, yet it is very useful to illustrate various important features of the quantum behaviour of the electron in the periodic lattice.

The Schrodinger's equations for the two regions are

$$\frac{d^2\psi}{dx^2} + \frac{2m}{\hbar^2} E\psi = 0 \quad \text{for} \quad 0 < x < a \qquad \text{... (i)}$$

and

$$\frac{d^2\psi}{dx^2} + \frac{2m}{\hbar^2} (E - V_0)\psi = 0 \quad \text{for} \quad -b < x < 0 \qquad \text{... (ii)}$$

We shall assume that the energy E of the electron under consideration is smaller than V_0 and define two real quantities α and β given by

$$\alpha^2 = \frac{2mE}{\hbar^2} \text{ and } \beta^2 = \frac{2m(V_0 - E)}{\hbar^2} \qquad \text{... (iii)}$$

Eqs. (*i*) and (*ii*) can now be written as

$$\frac{d^2\psi}{dx^2} + \alpha^2 \psi = 0 \quad \text{for} \quad 0 < x < a \qquad \text{... (iv)(a)}$$

and

$$\frac{d^2\psi}{dx^2} - \beta^2 \psi = 0 \quad \text{for} \quad -b < x < 0 \qquad \text{... (iv)(b)}$$

The potential is periodic, and is given by

$$V(x) = V(x + a + b)$$

where $(a + b)$ is the period.

Using Bloch theorem, the solution of wave equation for a periodic potential will be of the form of a plane wave modulated with the periodicity of the lattice and will be given by

$$\psi(x) = e^{ikx} \mu_k(x) \qquad \text{... (v)}$$

where $\mu_k(x)$ is the periodic function in x with period $(a + b)$, *i.e.*,

$$\mu_k(x) = \mu_k(x + a + b)$$

From Eq. (*v*), we get

$$\frac{d\psi}{dx} = ik\, e^{ikx} \mu_k(x) + e^{ikx} \frac{d\mu_k(x)}{dx}$$

$$= ik\, e^{ikx} \mu_k(x) + e^{ikx} \frac{d\mu}{dx} \qquad \text{... (vi)}$$

and

$$\frac{d^2\psi}{dx^2} = -k^2 e^{ikx} \mu_k(x) + ik\, e^{ikx}\, d\mu + ik\, e^{ikx} \frac{d\mu}{dx} + e^{ikx} \frac{d^2\mu}{dx^2}$$

$$= -k^2 e^{ikx} \mu_k(x) + 2ik\, e^{ikx} \frac{d\mu}{dx} + e^{ikx} \frac{d^2\mu}{dx^2} \qquad \text{... (vii)}$$

Substituting the values of $\psi(x)$ from Eq. (*v*) and $\frac{d^2\psi}{dx^2}$ from Eq. (*vii*) in Eqs. (*iv*) (*a*) and (*iv*) (*b*), we get

$$-k^2 e^{ikx}\mu + 2ik\, e^{ikx}\frac{d\mu}{dx} + e^{ikx}\frac{d^2\mu}{dx^2} + \alpha^2 e^{ikx}\mu = 0$$

(In the above Equation instead of $\mu_k(x)$, we have used μ only)

or
$$\frac{d^2\mu}{dx^2} + 2ik\frac{d\mu}{dx} + (\alpha^2 - k^2)\mu = 0 \qquad \text{... (viii)}$$

for
$$0 < x < a$$

and
$$\frac{d^2\mu}{dx^2} + 2ik\frac{d\mu}{dx} - (\beta^2 + k^2)\mu = 0 \qquad \text{... (ix)}$$

for
$$-b < x < 0$$

The solutions of these equations are

$$\mu_1 = A\, e^{i(\alpha-k)x} + B\, e^{-i(\alpha+k)x} = 0 \qquad \text{... (x)}$$

for $0 < x < a$

and
$$\mu_2 = C\, e^{(\beta-ik)x} + D\, e^{-(\beta+ik)x} = 0 \qquad \text{... (xi)}$$

for $-b < x < 0$

Where *A*, *B*, *C* and *D* are constants.

These constants must be chosen in such a manner that the following boundary conditions are satisfied, *i.e.*, the wave function ψ and its derivative $\frac{d\psi}{dx}$ are single valued and continuous

$$\therefore \quad (\mu_1)_{x=0} = (\mu_2)_{x=0};\ \left(\frac{d\mu_1}{dx}\right)_{x=0} = \left(\frac{d\mu_2}{dx}\right)_{x=0} \qquad \text{... (xii)}$$

and
$$(\mu_1)_{x=a} = (\mu_2)_{x=a};\ \left(\frac{d\mu_1}{dx}\right)_{x=a} = \left(\frac{d\mu_2}{dx}\right)_{x=b} \qquad \text{... (xiii)}$$

The first two conditions are imposed because of the requirement of continuity and the other two conditions are required because of periodicity of $\mu_k(x)$.

The application of these boundary condition to Eqs. (*x*) and (*xi*), gives

$$A + B = C + D$$
$$A\, i(\alpha - k) - B\, i(\alpha + k) = C(\beta - ik) - D(\beta + ik)$$
$$A\, e^{i(\alpha-k)a} + B\, e^{-i(\alpha+k)a} = C\, e^{-(\beta-ik)b} + D\, e^{(\beta+ik)b}$$
$$Ai(\alpha - k)\, e^{i(\alpha-k)a} - B\, i(\alpha+k)\, e^{-i(\alpha+k)a} = C(\beta - ik)\, e^{-(\beta-ik)b} - D(\beta+ik)\, e^{(\beta+ik)b}$$

These four equations have a non-zero solution only if the determinant of the coefficients *A*, *B*, *C* and *D* vanishes, *i.e.*,

$$\begin{vmatrix} 1 & 1 & 1 & 1 \\ i(\alpha-k) & -i(\alpha+k) & (\beta-ik) & -(\beta+ik) \\ e^{i(\alpha-k)a} & e^{-i(\alpha+k)a} & e^{-(\beta-ik)b} & e^{(\beta+ik)b} \\ i(\alpha-k)\,e^{i(\alpha-k)a} & -i(\alpha+k)\,e^{-i(\alpha+k)a} & (\beta-ik)\,e^{-(\beta-ik)b} & -(\beta+ik)\,e^{(\beta+ik)b} \end{vmatrix} = 0$$

On solving the determinant and after simplification, we obtain

$$\frac{\beta^2+\alpha^2}{2\beta\alpha}\ \sinh \beta b \sin \alpha a + \cosh \beta b \cos \alpha a = \cos k\,(a+b) \qquad \text{... }(xv)$$

To simplify this equation Kronig and Penney considered the case when $V_0 \to \infty$ and $b \to 0$, but the product $V_0 b$ has a finite value, *i.e.*, the potential barriers become *delta* functions. Under these conditions the model is modified in such a way that it represents a series of wells separated by infinitely thin potential barriers of infinitely large potential. The limiting value of $V_0 b$ for $V_0 \to \infty$ and $b \to 0$ is known as *barrier strength.*

As $\quad b \to 0$, $\sinh \beta b \to \beta b$ and $\cosh \beta b \to 1$.

Also from Eq. (*iii*),

$$\beta^2+\alpha^2 = \frac{2mV_0}{\hbar^2}$$

or

$$\frac{\beta^2+\alpha^2}{2\alpha\beta} = \frac{mV_0}{\alpha\beta\hbar^2}$$

$\therefore$ Eq. (*xv*) becomes

$$\frac{mV_0}{\alpha\beta\hbar^2}\ \beta b \sin \alpha a + \cos \alpha a = \cos ka$$

$$\left[\frac{mV_0 b}{\alpha\hbar^2}\right] \sin \alpha a + \cos \alpha a = \cos ka \qquad \text{... }(xvi)$$

Let us now define a quantity $P = \dfrac{mV_0 ba}{\hbar^2}$, which is a measure of the area $V_0 b$ of the potential barrier.

$$\therefore \qquad P\,\frac{\sin \alpha a}{\alpha a} + \cos \alpha a = \cos ka \qquad \text{... }(xvii)$$

The physical significance of the quantity P is that if P is increased the area of the potential barrier is increased and the given electron is bound more strongly to a particular potential well. When $P \to 0$, the potential barrier becomes very weak which mean that electrons are *free electrons.* In this case we obtain from Eq. (*xvii*)

$$\alpha a = ka \quad \text{or} \quad \alpha = k$$

Now,
$$\alpha^2 = \frac{2mE}{\hbar^2} = k^2 \qquad \text{... }(xviii)$$

Hence,
$$E = \frac{k^2\hbar^2}{2m} = \frac{h^2k^2}{8\pi^2 m} \qquad \text{... }(xviii)(a)$$

This result is the same as obtained by the free electron model. Equation (*xvii*) also gives the condition which must be satisfied so that solutions of the wave equation may exist. Since $\cos k\alpha$ lies between + 1 and – 1, the left hand side [of Eq. (*xvii*)] should take up only those values of αa for which its values lie between + 1 and – 1. Such values of αa, therefore, represent wave like solutions of the form

$$\psi(x) = e^{ikx}\,\mu_k(x)$$

and are *allowed values*. The other values of αa are *not allowed.*

If we plot a graph between $\left(\dfrac{P \sin \alpha a}{\alpha a} + \cos \alpha a\right)$ and αa for the value of $P = \dfrac{3\pi}{2}$, we get the curve shown in Fig. 23.8.

From Eq. (*xviii*) $\alpha = \sqrt{\dfrac{2mE}{\hbar^2}}$. Therefore, the abscissa is a measure of energy and by finding the value of αa at any point the energy represented by the function at that point is calculated. The values of αa satisfying the equation

$$P\frac{\sin \alpha a}{\alpha a} + \cos \alpha a = \cos ka$$

are obtained by drawing lines parallel to αa –axis at a distance cos *ka* from it and if *ka* is continuously varied from 0 to π, *i.e.*, cos *ka* from + 1 to – 1, we obtain all possible values of αa and hence that of energy can be calculated. These possible values of αa are shown by thick lines.

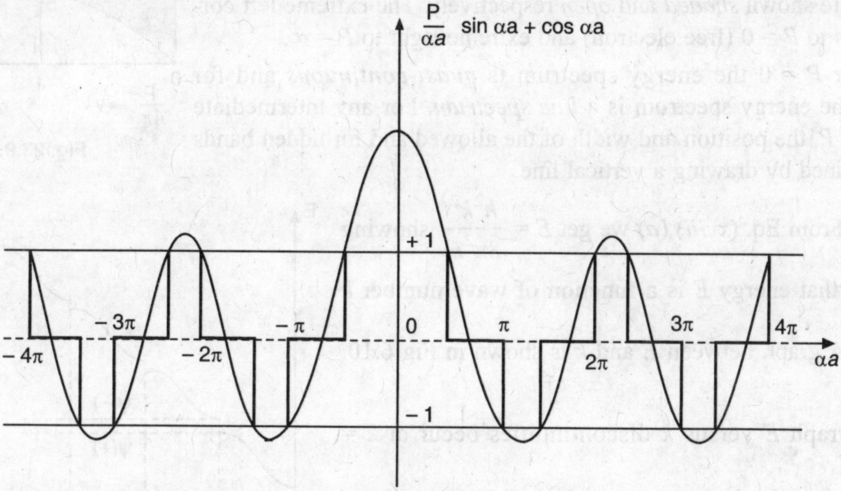

Fig. 23.8

From Fig. 23.8, the following conclusions are drawn:

Conclusions. 1. *Allowed and forbidden energy bands.* The energy spectrum consists of an infinite number of *allowed energy bands* (shown thick) separated by intervals in which there are no energy levels (shown dotted). These are known as *forbidden regions* or *bands.*

The boundaries of the allowed energy levels correspond to the values of cos *ka* = ± 1

or $$ka = n\pi \quad \text{or} \quad k = \frac{n\pi}{a}$$

2. As αa increases the term $P\dfrac{\sin \alpha a}{\alpha a}$ on the left hand side of Eq. (*xvii*) decreases so that width of allowed energy bands increases and hence forbidden energy regions become narrower.

3. The width of allowed energy bands decreases with the increasing value of *P* (*i.e.*, with the increasing binding energy of the electron). When $P \to \infty$, the allowed energy bands become infinitely narrow and are independent of *k*, *i.e.*, energy spectrum becomes a line spectrum.

When $P \to \infty$ the allowed energy ranges of αa reduces to points given by

$$\alpha a = \pm n\pi$$

or $$\alpha^2 = \frac{n^2\pi^2}{a^2} = \frac{2mE}{\hbar^2}$$

∴ $$E = \frac{n^2\pi^2\hbar^2}{2ma^2} \qquad \text{... (xix)}$$

From Eq. (*xix*), we find that E is independent of k. The energy levels in this case are discrete and the electron is completely bound, *i.e.*, the electrons remain within the potential wells and move only in one cell of width a. This case applies to crystals where the electrons are tightly bound to their nuclei.

If P is small, the barrier strength is small and in the limit $P \to 0$ (no barrier) the electron can be considered to be moving freely through the potential well. This case applies to crystals where the electrons are almost free of their nuclei. The effect of varying P on the band structure is shown in Fig. 23.9. Allowed and forbidden energy ranges are shown *shaded* and *open* respectively. The extreme left corresponds to $P = 0$ (free electron) and extreme right to $P = \infty$.

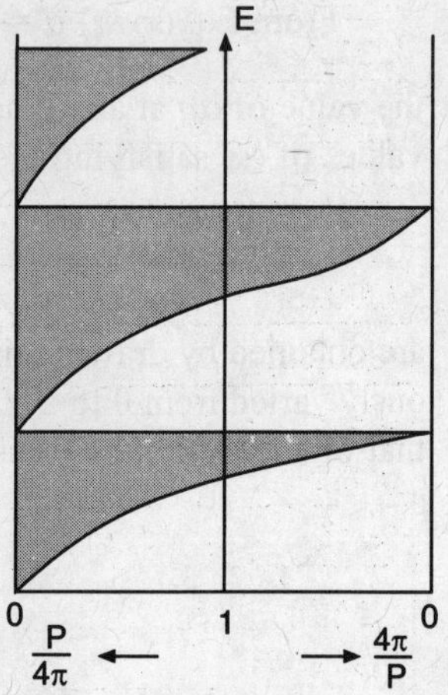

Fig. 23.9.

For $P = 0$ the energy spectrum is *quasi-continuous* and for $P = \infty$ the energy spectrum is a *line spectrum*. For any intermediate value of P, the position and width of the allowed and forbidden bands are obtained by drawing a vertical line.

4. From Eq. (*xviii*) (*a*) we get $E = \dfrac{h^2k^2}{8\pi^2 m}$ showing thereby that energy E is a function of wave number k $\left(\dfrac{2\pi}{\lambda}\right)$. A graph between E and k is shown in Fig 6.10.

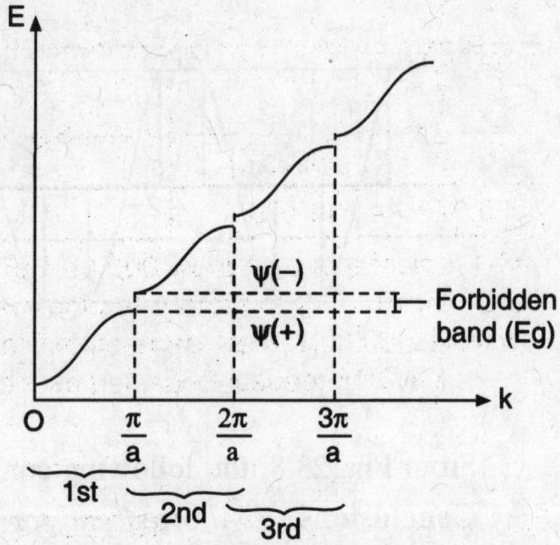

Fig. 23.10.

In the graph E versus k discontinuities occur at $k = \dfrac{\pi}{a}, \dfrac{2\pi}{a}, \dfrac{3\pi}{a}, \ldots\ldots, \dfrac{n\pi}{a}$ where $n = 1, 2, 3, \ldots\ldots$

The k values define the first, second, third etc., *Brillouin Zones*. For $k = +\dfrac{\pi}{a}$ to $-\dfrac{\pi}{a}$, there exists the first Brillouin Zone. The second Brillouin zone consists of two parts, from $+\dfrac{\pi}{a}$ to $+\dfrac{2\pi}{a}$ and the second from $-\dfrac{\pi}{a}$ to $-\dfrac{2\pi}{a}$. Each portion of the curve may be called a *band*. The curves are

(*i*) horizontal at the top and the bottom

(*ii*) parabolic near the top and the bottom with curvature in opposite directions

(*iii*) $\dfrac{d^2E}{dk^2}$ is positive in the lower portion of the band and negative in the upper portion.

(*iv*) The zone boundaries represents the maximum energies that the electron can have without any discontinuity.

(*v*) Within a given energy band, the energy is a periodic function of k. For, example, if we replace k by $k + \dfrac{2\pi n}{a}$, where n is an integer in the right hand side of equation

$$P\frac{\sin \alpha a}{\alpha a} + \cos \alpha a = \cos ka$$

the equation remains the same because $\cos\left(k + \frac{2\pi n}{a}\right)a = \cos(ka + 2\pi n) = \cos ka.$

Energy of lowest energy band

According to Kronig Penney model the solution of Schrödinger's equation for one dimensional periodic lattice is given by

$$\frac{P \sin \alpha a}{\alpha a} + \cos \alpha a = \cos ka$$

where $\alpha = \left(\frac{2mE}{\hbar^2}\right)^{1/2}$, $P = \frac{mV_0ba}{\hbar^2}$ and $k = \pm\frac{n\pi}{a}$

Thus, the energy of the lowest band corresponds to $n = 1$ or $k = \pm\frac{\pi}{a}$

or $$\cos ka = \pm \cos \pi = \pm 1$$

Considering only the magnitude, we have

$$\frac{P \sin \alpha a}{\alpha a} + \cos \alpha a = 1$$

or $$\frac{P}{\alpha a} = \frac{1 - \cos \alpha a}{\sin \alpha a} \quad \text{... (i)}$$

Now, $$\cos 2\theta = 1 - 2\sin^2\theta \quad \text{or} \quad 1 - \cos 2\theta = 2\sin^2\theta$$

$$\therefore \quad 1 - \cos \alpha a = 2\sin^2\left(\frac{\alpha a}{2}\right)$$

and $$\sin 2\theta = 2 \sin\theta \cos\theta$$

$$\therefore \quad \sin \alpha a = 2\sin\left(\frac{\alpha a}{2}\right)\cos\left(\frac{\alpha a}{2}\right)$$

Substituting in (*i*), we have

$$\frac{P}{\alpha a} = \frac{2\sin^2\left(\frac{\alpha a}{2}\right)}{2\sin\left(\frac{\alpha a}{2}\right)\cos\left(\frac{\alpha a}{2}\right)} = \tan\left(\frac{\alpha a}{2}\right)$$

Since $P << 1$, $$\frac{P}{\alpha a} = \tan\left(\frac{P}{\alpha a}\right), \text{ for lowest energy band}$$

$$\therefore \quad \tan\left(\frac{\alpha a}{2}\right) = \tan\left(\frac{P}{\alpha a}\right)$$

or $$\frac{\alpha a}{2} = \frac{P}{\alpha a}$$

or $$\alpha^2 = \frac{2P}{a^2}$$

But $\alpha^2 = \frac{2mE}{\hbar^2}$

$$\therefore \quad \frac{2P}{a^2} = \frac{2mE}{\hbar^2}$$

or $$E = \frac{P}{ma^2}\hbar^2$$

This gives the energy of lowest energy band.

23.6 ORIGIN OF ENERGY GAP

A free electron travelling wave in one dimension (say along X-axis) is given by

$$\psi = e^{ikx}$$

For values of $k = \pm \frac{\pi}{a}$ (where a is the lattice spacing), the wave functions are not travelling waves $e^{i\frac{\pi}{a}x}$ and $e^{-i\frac{\pi}{a}x}$ but these are *standing waves*. For value of $k = \pm \frac{\pi}{a}$ Bragg condition is satisfied and the wave travelling in one direction is Bragg reflected at the boundary and then travels in the opposite direction. The incident and the reflected wave form two types of standing waves

$$\psi(+) = e^{i\frac{\pi}{a}x} + e^{-i\frac{\pi}{a}x} = 2\cos\left(\frac{\pi}{a}x\right)$$

or $$\psi(-) = e^{i\frac{\pi}{a}x} - e^{-i\frac{\pi}{a}x} = 2i\sin\left(\frac{\pi}{a}x\right)$$

For the ψ (+) standing wave the charge density $e(\psi_+)^2 \propto \cos^2\left(\frac{\pi}{a}x\right)$. Thus, for $x = 0, a, 2a$ etc. $\cos^2 \frac{\pi}{a} = 1$ and has a maximum value so that negative electron charge density is concentrated at the positive ion cores thereby lowering the potential energy. Similarly, for the ψ (–) standing wave the charge density $e(\psi_-)^2 \propto \sin^2\left(\frac{\pi}{a}x\right)$. For $x = \frac{a}{2}, \frac{3a}{2}$, ... etc. $\sin^2 \frac{\pi}{a} x = 1$ and has a maximum value so that negative charge density is concentrated midway between the positive ion cores thereby raising the potential energy as compared to that of the travelling wave. Thus, potential energy for $(\psi_+)^2$ is lower and for $|\psi_-|^2$ is higher than that of a travelling wave thereby producting an *energy gap*. If E_g is the difference in the two energies then E_g gives the width of the energy gap.

Value of energy gap. The approximate value of energy gap in a semi-conductor is about 1 eV. For Silicon (Si) the energy gap is 1.1 eV and for Germanium (Ge) it is 0.7 eV.

23.7 VALENCE BAND, CONDUCTION BAND AND FORBIDDEN BAND

Valence band. The valence electrons, including electrons *transferred* in the case of solids having *ionic bonds* or *shared in the case of solids having covalent* bonds, are not free to gain energy from an applied electric field and are, therefore, unable to migrate through the crystal (solid). As explained earlier, the sharp atomic levels are spread out in solids and all the allowed energy states upto a *forbidden energy region* are occupied (at least at 0° K) and there are no more electrons to be accounted for.

The highest filled energy band, which includes electrons shared in covalent bonds or electrons transferred in ionic bonds, is known as valence band.

The energy level corresponding to the top of the valence band is denoted as E_v.

Conduction band. When the number of valence electrons in one atom is less than the number of electrons required to fill the outer orbit of the other atom in a solid, the valence electrons, which are ordinarily weakly held by the atom, are *free* to move through the solid and form a sort of '*free electron-gas*'. These electrons can move under the influence of an applied electric field. Due to the mutual interaction between these electrons, there is a change in their energy, the value of which varies from zero to a maximum value E_f called *Fermi energy*. In other words, a band of energy of valence orbits consisting of closely spaced energy levels of these *free* electrons is formed. *This is known as conduction band.* The energy level corresponding to the bottom of the conduction band is denoted as E_c.

Forbidden band. The conduction band has a higher energy level than the valence band. Even in excellent crystalline insulators, there are conduction bands above the valence bands to which electrons can be excited if these are given sufficient energy. But, there is a *forbidden energy region* between the two energy bands known as the **forbidden band**. The energy gap between the conduction band and valency band is denoted as E_g. When the forbidden energy gap is very small, thermal excitation may excite the electrons from the valence band to the conduction band so that a material, although a good insulator at 0 K, may become a conductor at a higher temperature. The bands may be completely filled, partially filled, or empty in accordance with Pauli's exclusion principle. The energy gap between the valence band and the conduction band may be small or large or even the two bands may overlap. The band structure obtained at the equilibrium distance r_e determines most of the properties of the solid.

23.8 BEHAVIOUR OF CONDUCTORS, INSULATORS AND SEMICONDUCTORS (ON THE BASIS OF BAND THEORY)

The behaviour of a solid as a good conductor, semiconductor or insulator can be explained on the basis of band theory.

The energy levels of individual atoms which are sharply defined when the atoms have large separation, results in the formation of bands as the interatomic distance r decreases. Depending upon the nature of the solid, the broadening of a level into a band begins to take place at different values of r the interatomic distance. In some cases, the broadening of various levels at equilibrium distance is so large that energy bands overlap, whereas in others the broadening may be so small that a large gap may exist between them. When the two bands overlap, it means there is a continuous distribution of allowed energies. On the other hand, the presence of a gap indicates the forbidden energies that the electrons are not allowed to have. It may be noted that the presence of forbidden band is due to the quantum nature of the energies of the electron.

The behaviour of metal insulator and semiconductor can be explained in terms of rate of change of electron energy with wave vector K. We define a factor f_k

$$f_k = \frac{m}{m^*} = \frac{m}{\hbar^2}\frac{d^2E}{dk^2}$$

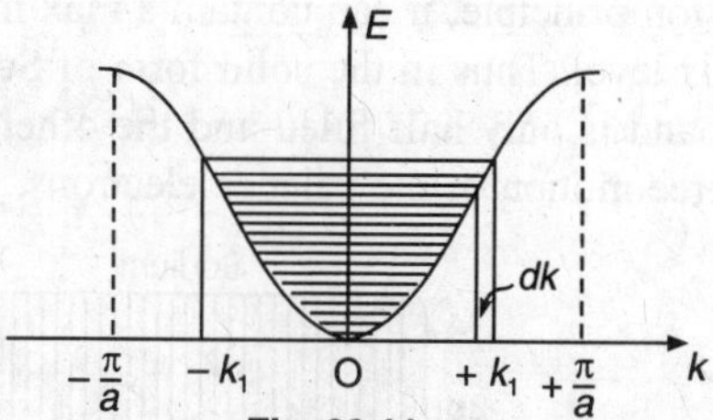

Fig. 23.11

which is a measure for the extent to which an electron in a state k is free to take part in electrical conduction. Where m* is effective mass of electron in the crystal. The total number of free electrons in a band effective for conduction is given by $N_{eff} = \sum f_k$ where the summation extends over all the occupied energy states. Now consider a particular energy band which is filled with electrons up to a certain value k_1 which is less than $\frac{\pi}{a}$.

In one dimensional lattice of length L, the number of states dn in the interval dk is given by

$$dn = \frac{L}{2\pi} dk$$

Since two electrons occupy each state, the effective number of free electrons in the shaded region of Fig. 23.11 is given by

$$N_{eff} = 2\int_{-k_1}^{+k_1} f_k dn = 2\int_{-k_1}^{+k_1} \frac{m}{\hbar^2}\frac{d^2E}{dk^2}\frac{L}{2\pi}dk$$

$$= \frac{mL}{\pi\hbar^2}\int_{-k_1}^{+k_1}\frac{d^2E}{dk^2}dx = \frac{2mL}{\pi\hbar^2}\int_{0}^{+k_1}\frac{d^2E}{dk^2}dk$$

$$= \frac{2mL}{\pi\hbar^2}\left[\frac{dE}{dk}\right]_0^{k_1} = \frac{2mL}{\pi\hbar^2}\left[\frac{dE}{dk}\right]_{k=k_1} \quad \text{... (i)}$$

as $\frac{dE}{dk}$ at $k = 0$ is zero.

It is clear that the effective number of electrons N_{eff} in a band depends upon the value of $\frac{dE}{dk}$.

When the band is completely full, *i.e.*, at the top of the band $\frac{dE}{dk} = 0$ and hence $N_{eff} = 0$.

Thus a completely filled band does not contribute any thing towards conductivity and *cannot be a metal* because one of the characteristic properties of a metal is its good conductivity.

The value of $\left(\frac{dE}{dk}\right)$ is the maximum at the points of inflexion of $E-k$ curve [Fig. 23.11 and hence the effective number of electrons is also a maximum. Thus in a material with incompletely filled valence band, there are quite a large number of electrons available for conduction, *i.e.*, they behave as free electrons and hence as charge carriers. Such materials are good conductors of electricity.

In case of conductors, *e.g.*, metal crystals the last occupied valence band is only partially filled. Let us consider the case of sodium atom. It has 2 electrons in $1p$, 2 in $2s$, 6 in $2p$ and only one electron in $3s$ level. According to Pauli's exclusion principle, it can contain a maximum of 2 electrons in the $3s$ level. Thus in the solid form of Sodium, the last, (valence) band is only half filled and the other half is available for the free motion of the valence electrons.

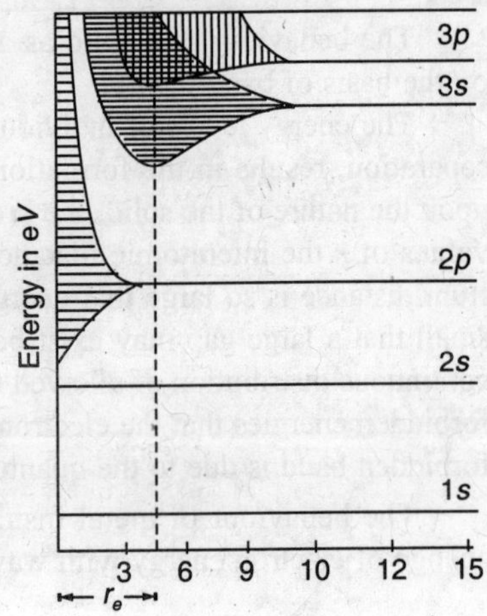

Fig. 23.12

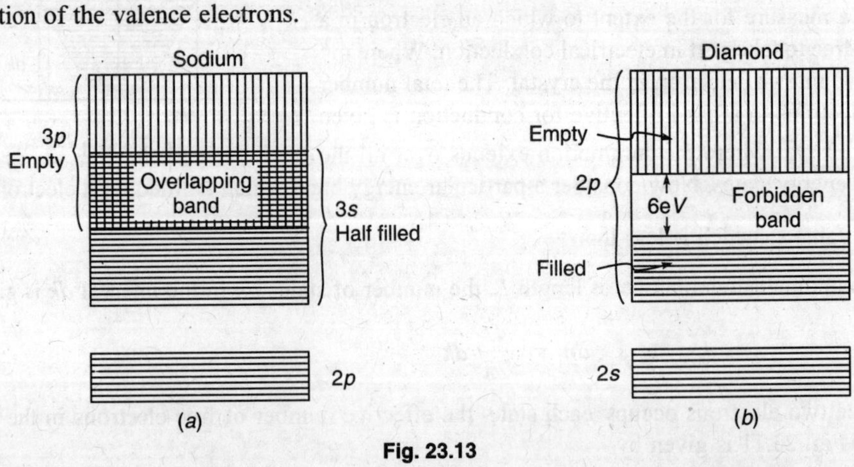

Fig. 23.13

At equilibrium separation, the interaction energy is minimum as the situation corresponds to maximum stability. As the atoms are brought closer, the value of Δx decreases and that of Δp and hence the energy, increases in accordance with uncertainty principle.

The energy bands of solid sodium, as a function of inter-atomic distance r, are shown in Fig. 23.12. The equilibrium separation $r_e = 0.67$ Å. The energy of the bands rises upward from a minimum value at r_e as explained above. At distance r_e, the $2p$ level does not spread into a band (it does at a smaller distance), whereas the $3s$ and $3p$ levels spread into bands and overlap. The overlapping of $3p$ (conduction) band with $3s$ (valence) band as shown separately in Fig. 23.13 (*a*) provides a continuous set of unfilled levels for the valence electrons to move readily under the effect of applied electric field and thus gives rise to an electric current. Hence sodium like crystals are very good conductors of electricity. There are about 70 metal crystals which behave as good conductors.

Insulators. As stated above $N_{eff} = \frac{2mL}{\hbar^2}\frac{dE}{dk}$. At the top (and bottom) of the band, *i.e.*, for the $E - k$ curve for $k = 0, \frac{\pi}{a}$; $\frac{dE}{dk} = 0$ and hence the effective number of electrons is also zero. Thus when the valence band is completely filled (and the conduction band completely empty) the solid is a perfect insulator.

Diamond is a very good example of such an insulator. Chemically it is the purest form of carbon. The simplified diagram of energy bands of diamond is shown in Fig. 23.13 (*b*). The outer most $2p$ energy band is completely filled with electrons and is separated from the empty band by a gap of 6 eV above it. This means an additional energy of 6 eV must be imparted to an electron in the diamond crystal to jump across the forbidden band. But an electron moving through a diamond crystal has a mean free path of about 10^{-6} cm, *i.e.*, it has a collision thereafter, in which it loses most of the energy gained by it from the electric field. Hence an electric field of the order of 6×10^6 volts/cm is required to impart to the electron an energy of 6 eV in a distance of 10^{-6} cm, *i.e.*, before suffering a collision. This value is about 10^{10} times greater than the electric field required to cause a current to flow in sodium. Diamond is, therefore, a very poor conductor of electricity and hence a good insulator. In fact, all solids like diamond with large gap (forbidden energy band) between the *filled* (valence) and *empty* (conduction) energy bands are good electrical insulators.

Semiconductors. Silicon, Germanium etc., are familiar examples of semiconductors. Silicon has a crystal structure similar to that of diamond, but the forbidden band between the filled (valence) energy band and the empty (conduction) band above it, has a gap of only 1.1 eV. It is, therefore, possible for some electrons at room temperature to have sufficient *thermal kinetic energy* to jump across the forbidden band into the empty (conduction) band above it. When an electric field is applied, these electrons constitute a limited amount of current. Thus silicon has an electrical conductivity, intermediate between a good conductor and an insulator. The silicon like crystals, which show partial conductivity, are known as *semiconductors*.

Semiconductor is insulator at 0 K. At 0 K the electrons in the valence band do not have sufficient energy to jump from the valence band to the conduction band across the forbidden band as the band gap is about 1 eV. Thus the conduction band being empty at 0 K, the semiconductor behaves as an insulator.

Effect of temperature. An important distinction between metals (good conductors) and semiconductors is the temperature dependence of their conductivities. The conductivity of metals decreases with increase of temperature. This is because, as the temperature is raised more and more phonons are excited which scatter electrons and reduce their mobility. In semiconductors the decrease in mobility is more than compensated by the increase in the number of charge carriers because as the temperature increases more and more electrons in the valence band have sufficient energy to jump into the conduction band. Hence the conductivity of a semiconductor increases with temperature.

23.9 EFFECT OF IMPURITY ON CONDUCTIVITY OF SEMICONDUCTOR

The conductivity of a semiconductor can be considerably improved by adding a small amount of *impurity*. Such semiconductors are known as *impurity semiconductors* or *extrinsic* or *doped* semiconductors. As the intrinsic semiconductors are generally elements of group IV, *i.e.*, they have 4 electrons in the outermost electronic orbit, the impurity element chosen has 5 electrons in the outermost orbit (*i.e.*, one more than that required to completely fill the outermost shell of the intrinsic semiconductor) or it may have 3 electrons (*i.e.*, one less than the required number). The column V *donor* level is approximately 0.01 eV below the conduction band in *Ge* and about 0.03 to 0.06 eV in Silicon. Similarly, the column III *acceptor* level is approximately 0.01 eV above the valence band in *Ge* and 0.03 – 0.06 eV in silicon. The band gap for *Ge* is about 0.7 eV and for silicon 1.1 eV.

***n*-type semiconductor.** If we incorporate a few *arsenic* atoms (or any other group V element) in a silicon crystal, then 4 out of the 5 electrons in its outermost shell form covalent bonds with the 4 electrons in the outermost shell of silicon atom. The *fifth* electron is free to move about in the crystal. Thus the presence of arsenic as an impurity provides energy levels *just below* the conduction band as shown in Fig. 23.14 (*a*).

This level is filled with electrons at 0 K and very little thermal energy is required to excite these electrons to the conduction band. At about 50 – 100 K almost all the electrons in the impurity level are '*donated*' to the conduction band. Such levels are, therefore, termed as *donor levels* and the impurity a *donor impurity*. The semiconductor *doped* with '*donor*' atoms is known as an *n-type semiconductor* because electric current in it is carried by negative charges.

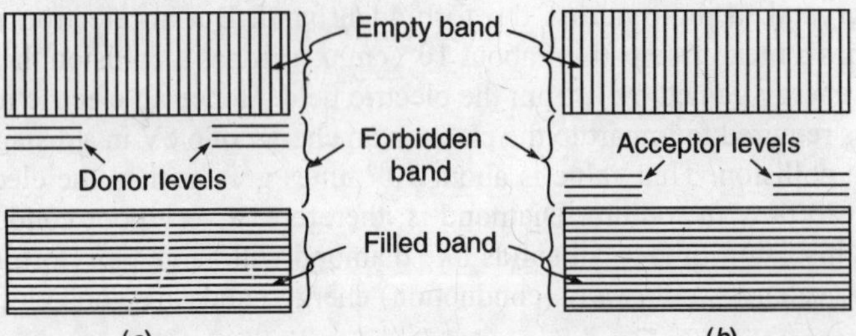

Fig. 23.14

***p*-type semiconductor.** On the other hand if we incorporate a few *gallium* atoms (or any other group III element) in silicon crystal, only three electrons out of four in the outermost shell of silicon form covalent bonds with the three electrons in the outermost shell of gallium atom, thereby leaving a vacancy or a *hole* in the electron structure of the crystal. An electron readily enters this *hole* and in the process leaves a new *hole* in its former position. The presence of gallium, therefore, provides energy levels *just* above the highest filled band known as *acceptor level*, as shown in Fig 23.14 (*b*). The levels contain no electrons at 0 K but even at very low temperature (50–100 K) the thermal energy is sufficient to excite electrons from the valence band into this impurity or *acceptor level* (because it *accepts* electrons from the valence band). An electron that occupies these levels leaves behind a '*hole*' in the filled valency band. If an electric field is applied, the *holes* behave like a *positive charge* and move from positive to negative electrode, thus producing an electric current.

The semiconductor *doped* with '*acceptor*' atoms is known as *p-type semiconductor* because electric current in it is carried by positive charges (*holes*).

Difference. In an intrinsic or pure semiconductor, the charge carriers are electrons, whereas in an impurity semiconductor the charge carriers may be electrons as in *n*-type or 'holes' as in *p*-type.

The conductivity of a pure semiconductor is low. The addition of impurity increases the conductivity.

SOLVED EXAMPLES

Example. 23.1. *An insulator has an optical absorption which occurs for all wavelengths shorter than 1800 Å. Find the width of the forbidden energy band for this insulator.*

Solution. As the insulator can absorb only wavelengths shorter than 1800 Å, the minimum energy required to move the electron from the valence band to the conduction band across the forbidden energy band of the insulator is the energy equal to the light of wavelength 1800 Å.

Now, $$\lambda = 1800 \text{ Å} = 1800 \times 10^{-10} \text{ m}$$

Corresponding frequency, $$\nu = \frac{3\times10^{8}}{1800\times10^{-10}} = \frac{10^{16}}{6}$$

$\therefore$ Energy gap of the forbidden band

$$E_g = h\nu$$

$$= \frac{6.626\times10^{-34}\times10^{16}}{6} \text{ Joules}$$

$$= \frac{6.626\times10^{-34}\times10^{16}}{6\times1.6\times10^{-19}}\ eV = 6.9\ eV$$

23.10 RESISTIVITY RANGE OF GOOD CONDUCTOR AND INSULATOR

The electrical resistivity of a good conductor is as low as 10^{-8} ohm metre at 0°C, and that of a good insulator can be as high as 10^{15} ohm metre. The resistivity of semi-conductors lies somewhere in the middle of these two extreme values.

For example, the electrical resistivity of good conductor like silver (Ag) 1.6×10^{-8} ohm metre and copper (Cu) 1.7×10^{-8} ohm metre. For insulators like glass resistivity is $10^{10} - 10^{14}$ ohm metre and mica $10^{11} - 10^{15}$ ohm metre. For semi-conductor like Germanium resistivity is 0.46 and silicon 2.5×10^{3} ohm metre.

EXERCISE

1. Explain how the atomic energy levels split into bands when a number of atoms are brought close together to form a crystal. ***(Poorvanchal U. 2002; G.N.D.U. 2002, 2001; Pbi.U. 2001; Bang.U. 2000; H.P.U. 2000)***
2. Prove that the motion of an electron through the periodic potential of solids gives rise to band structure. ***(Meerut U. 2001; P.U. 2000)***
3. Discuss the periodicity character of potential in a crystal. State and prove Bloch theorem in this reference. ***(LUCK. U. 2001, G.N.D.U. 2000; Meerut U. 2001, HPU 2000)***
4. Discuss Kronig Penney model. Using the model show that the energy spectrum of electron consists of a number of allowed energy bands separated by forbidden regions. ***(Pat. U. 2004; Poorvanchal U. 2003, 2001; Cal. U. 2002; G.N.D.U. 2003; H.P.U. 2002; P.U. 2001, 2000)***

5. (*a*) Starting with the solution of Schrödinger's equation for one dimensional periodic lattice

$$\frac{P \sin \alpha a}{\alpha a} + \cos \alpha a = \cos ka \text{ with } \alpha = \left(\frac{2mE}{\hbar^2}\right)^{1/2}$$ discuss the formation of energy bands in solids. **(*P.U. 2003*)**

(*b*) Using Kronig Penney model show that for P << 1, the energy of the lowest energy band is

$$E = \frac{\hbar^2 P}{ma^2}$$ **(*P.U. 2001*)**

6. (*a*) What is the basic assumption in Kronig Penney model?

(*H.P.U. 2003, 2001, 2000; P.U. 2003*)

(*b*) What are the conclusions of Kronig Penney model for energy band structure of solids? **(*H.P.U. 2001*)**

(*c*) What is the effect of changing the binding energy of an electron on the energy bands? **(*Pbi. U. 2000*)**

7. (*a*) What is the origin of energy gap? **(*Pbi. U. 2004; P.U. 2001*)**

(*b*) Write approximate value of energy gap of a semi-conductor.

8. (*a*) What are valence and conduction bands? What do you mean by forbidden band?

(*P.U. 2001; Meerut U. 2001; Gharwal U. 2000*)

(*b*) Explain why a solid whose energy bands are full cannot be a metal. **(*P.U. 2004*)**

9. (*a*) Distinguish between metals (good conductors), insulators and semi-conductors on the basis of band theory of solids. Why a semiconductor acts as an insulator at 0 K?

(*b*) The conductivity of a metal decreases with rise of temperature whereas the conductivity of a semiconductor increases with increase of temperature. Explain.

(*Osm. U. 2004; Poorvanchal U. 2003, 2002, 2001; Luck. U. 2002; Meerut U. 2002; Pbi. U. 2004, 2003; H.P.U. 2003, 2002, 2001, 2000; G.N.D.U. 2004, Gharwal U. 2000; Gauhati U. 2000*)

10. Discuss the effect of impurities on the conduction properties of Ge or Si crystal.

11. Differentiate a good conductor and a good insulator taking resistivity range. **(*Pbi. U. 2001*)**

MULTIPLE CHOICE QUESTIONS

Mark the correct answer

1. A pure semi-conductor behaves as insulator at

(*a*) 273 K (*b*) – 273°C (*c*) 373 K

2. Separation between valence band and conduction band is measured in

(*a*) metres (*b*) electron volt (*c*) metre^{-1} **(*H.P.U. 2001*)**

3. The energy region separating the highest occupied band and the empty band is called

(*a*) conduction band (*b*) valence band (*c*) forbidden band

(*H.P.U. 2000*)

4. In Kronig Penney model of electrons in a linear lattice if the strength of the periodic potential increases, the width of the allowed energy band

(*a*) increases (*b*) decreases (*c*) remains constant (*d*) none of these

(*Poorvanchal U. 2003*)

5. Quantum of lattice vibrational energy is:

(*a*) photon (*b*) phonon (*c*) electron (*d*) proton

6. The energy gap n pure Germanium is:

(*a*) 1.1 eV (*b*) 1.7 eV (*c*) 0.7 eV (*d*) 0.1 eV

7. The motion of electron in the periodic crystal lattice gives rise to:

(*a*) energy bands (*b*) continuous energy levels (*c*) none of these

ANSWERS

1. (*b*)	**2.** (*b*)	**3.** (*c*)	**4.** (*b*)	**5.** (*b*)
6. (*c*)	**7.** (*a*)			

[Hint for Q.4 The width of allowed energy band decreases with the increasing value of P. But $P = \dfrac{mV_0ba}{\hbar^2}$ where V_0 is the strength of the periodic potential, *i.e.*, $P \propto V_0$. Therefore, the width of allowed energy band decreases with increasing value of V_0 – the periodic potential.]

SECTION VII

SOLID STATE DEVICES

24. SEMICONDUCTORS

25. SEMICONDUCTOR DEVICES

SEMICONDUCTORS

24

Chapter

INTRODUCTION

Electronic instrumentation has a deep impact on todays advanced life style and modernization. Its contribution to efficient running of power stations, steel plants, textile mills, chemical plants, aviation, transport, computer for mass communication and satellite linkage, banking, railway reservations, radio, T.V., telephone, mobile etc. is really amazing. The invention of transistor in 1948 by Shockly, Brattain and Bardeen, a team of scientists of BELL Laboratories (U.S.A.) was the beginning of this new era of electronics. The inception of semiconducting devices has made a tremendous revolution in electronics industry. The main features of semiconducting devices are the miniaturization, light weight, lower power consumption, no heater requirement, low cost, no ageing problem hence long life, almost zero warm-up time hence instant start, high efficiency and many more as compared to vacuum tube devices.

24.1 SEMICONDUCTING MATERIALS

A class of materials which is neither a good conductor of electricity nor an insulator and whose conductivity lies between a good conductor and an insulator, is known as semiconductor.

The resistivity (ρ) of material is defined as

$$\rho = \frac{RA}{l}$$

where R is its resitance, A its cross-section and l its length. The reciprocal of resistivity is known as conductivity (σ) i.e. $\sigma = \frac{1}{\rho}$. The resistivity ρ for a good conductor (say copper) is $\cong 10^{-6}$ ohm.cm and for an insulator (say, mica) it is 10^{12} ohm.cm. For germanium (Ge), it is 50 ohm.cm. and for silicon (Si), it is 50×10^3 ohm.cm. These values lie in-between a good conductor and an insulator; hence Ge and Si are classified as semiconductors. Moreover, band gap E_g for semiconductors vary from 0.2 to 2.5 eV which is quite small as compared to that of insulators. For example, the band gap of diamond (a typical insulator) is 6 eV.

Why Only Ge and Si are Preferred as Semiconducting Materials

In fact, there are large number of elements in periodic table whose resistivity lies in-between a good conductor and an insulator. However, Ge and Si have received much favour as semiconducting materials for a number of reasons. Some of them are:

(*i*) a very high purity level

The essential criteria for selecting a semiconducting material is its purity level. The purity level must be very high of the order of 10^{10} : 1. The advanced techniques have reduced impurity levels in the pure material to 1 part in 10 billion (1:10,000,000,000). Ge and Si can be manufactured to such a very high purity level.

(*ii*) ability to change its conductivity significantly by doping:

By addition of 1 part impurity (of proper type) per millian i.e. $1:10^6$, in a wafer of material, can change that material from a relatively poor conductur to a good conductor of electricity. Ge and Si has this ability to change its conductivity characteristics significantly if doped by proper type of impurity.

(*iii*) ability to alter its characteristics significantly when exposed to heat and light:

The characteristics of Ge and Si can be altered significantly through the application of heat or light. This led Ge and Si an important consideration in the development of heat and light sensitive devices. For example, thermistors, LED, phototransistors etc.

24.2 INTRINSIC SEMICONDUCTOR

If a semiconductor crystal contains no impurities, i.e. crystal has highest purity level by carefully refining using modern techniques, the only charge carriers present are those produced by the thermal break-down of the covalent bonds. The conducting properties are thus characteristic of pure semiconductor. Such a crystal is termed as intrinsic semiconductor.

Ge (32) and Si (14) crystal belong to Group IV of periodic table and have the three dimensional diamond (i.e. tetrahedral) structure, each having 4 valence electrons. Consider the case of Si. The atomic member is 14. The electronic configuration is (2,8,4) i.e. it contains 2 electrons in K-shell, 8 in L-shell, and 4 in outermost M-shell. Since Si has a tetrahedral structure, each atom of silicon shares one electron each, with four neighbouring atoms by convalent bonds as shown in Fig. 24.1 At 0K or at absolute zero, all the four valence electrons are engaged in covalent bonds, no free electrons exist in the solid. Consequently, the semiconductor at 0K can not conduct and acts as a pure insulator.

As the temperature rises above absolute zero (T > 0K), the thermal energy causes each atom in the crystal to vibrate about its position. When the vibrations become violent, some of the electrons acquire sufficient energy and break away from the covalent bonds. The electrons liberated from bonds become free electrons which move randomly in the empty spaces that exist between the fixed positions of the atoms. This increased number of carriers will increase the conductivity, as a result its resistance decreases. Semiconducting materials such as Ge and Si that show a reduction in resistance with increase in temperature are said to have a negative resistance coefficient.

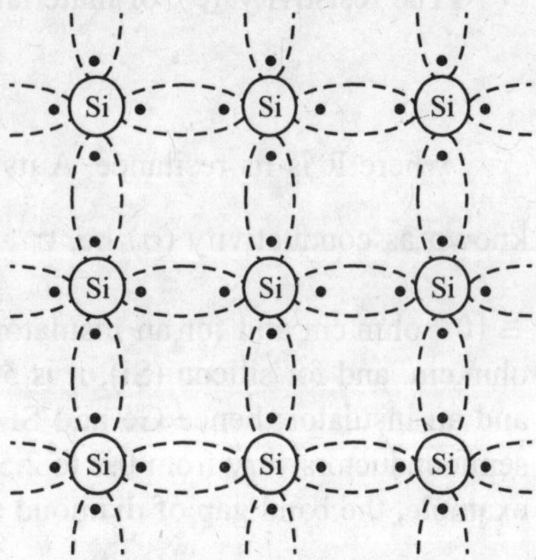

Fig. 24.1. Silicon Crystal at 0K

Intrinsic Carriers : Electron and Hole pair

In a silicon crystal, when a covalent bond is broken (Fig. 24.2) and an electron is set free, (T > 0K), there arises a **quantum vacancy** at the site of the broken bond. The removal of negative charge leaves an effective positive charge (+e). This quantum vacancy created is known as **hole.** Thus, breaking of one covalent bond, an 'electronhole pair' is generated. It means 'the conduction band electron' and 'the valence band hole' are created simultaneously. An electron from the covalent bond of a neighbouring atom shifts to fill this hole, thereby creating a new hole in the atom from which it shifts and so on. Thus, the hole shifts atom to atom and behaves like a free positive electric charge.

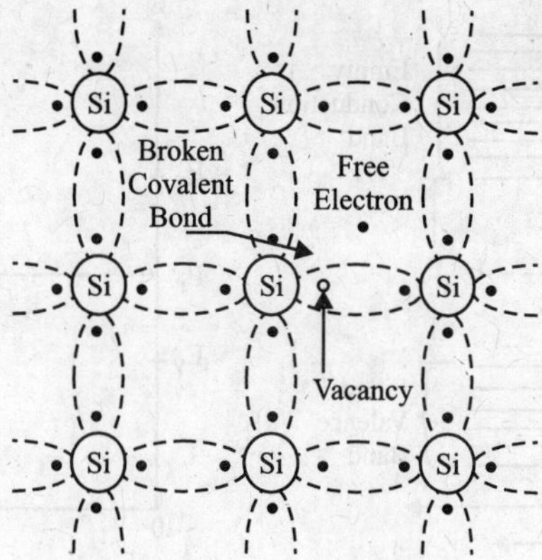

Fig. 24.2. Silicon Crystal at Room Temperature (T > 0k)

24.3 RECOMBINATION

At temperature T > 0K, in an intrinsic semiconductor, along with the generation of free electron-hole pairs, another process called 'recombination' also take place. A free electron moving at random in a semiconductor crystal, may encounter a hole and combine with it to construct a broken bond. thus, an electron-hole pair disappears and the free electron is converted in to bound electron. This recom-bination process is equivalent to the electron jumping from the conduction band to the valence band with release of energy ($E = n\nu$) in the form of electromagnetic waves equal to band gap energy (Eg). In an intrinsic semiconductor, when an equilibrium state is reached, the rate of generation of electron-hole pairs (g) is equal to their recombination rate (r). i.e. g = r, as shown in fig. 24.2 (A)

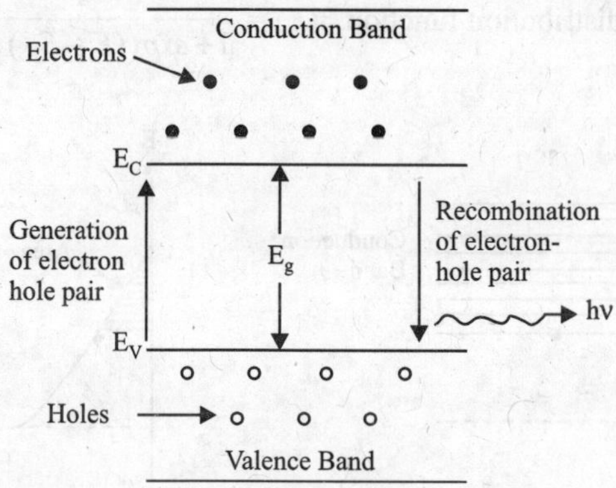

Fig. 24.2 (A)

24.4 ENERGY BAND STRUCTURE OF AN INTRINSIC SEMICONDUCTOR

The main feature of semiconductor is that it is characterized by a conduction band and valence band seperated by a smaller energy gap (E_g). The band gap for germanium is $E_g = 0.7$ eV and for silicon is Eg = 1.1 eV.

1. At absolute zero i.e. T = 0 K

All energy levels below E_V are completely occupied and above E_C are completely empty at 0K. Further E_V to E_C represents a forbidden gap. Its width for semiconductor is about 1 eV. At 0K, no electron in valence band can jump to conduction band, hence semiconductor at 0 K behaves as an insulator Fig. 24.3 (a). Fig. 24.3 (b) shows the Fermi-Dirac distribution at 0 K.

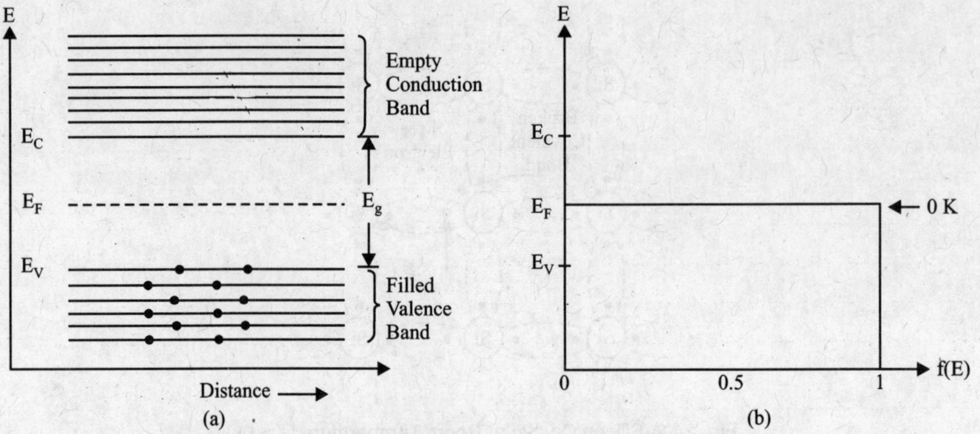

Fig. 24.3. (a) Energy band diagram for intrinsic semiconductor at 0 K
(b) The Fermi-Dirac distribution at 0 K.

2. At room temperature (T > 0 K)

As the temperature of intrinsic semiconductor increases, a significant number of electrons are thermally excited from valence band to conduction band Fig. 24.4 (*a*). At 300 K, in silicon crystal $1.5 \times 10^{16}/\text{m}^3$ covalent bonds are broken. There will be 1.5×10^{16} electrons/m^3 in the conduction band which has energy levels that can accomodate 2×10^{29} electrons/m^3. Therefore, the conduction band is partly occupied. An equal number of holes are produced in the valence band.

Fermi level

The Fermi-Dirac distribution function, $f(E) = \dfrac{1}{1 + exp\left((E - E_F)/kT\right)}$ is applicable to a semi-conductor also.

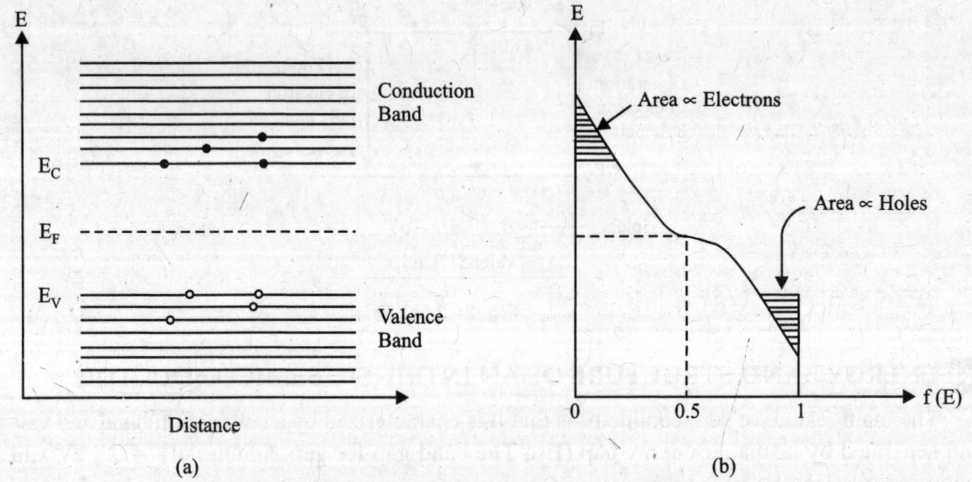

Fig. 24.4 (a) Energy band structure of semiconductor at TK
(b) Fermi-Dirac distribution function

From Fig. 24.4 (b), it is seen that the probability function *f*(*E*) rapidly approaches zero with increasing E. It suggests that the electrons in the conduction band are clustered very close to the bottom edge of the band. In a similar way, the holes are grouped very close to the top edge of the valence band. Therefore, electron may be assumed to be located at the bottom edge of the conduction bond whereas the holes at the top edge of the valence band. The Fermi level lies at the middle of the energy gap.

24.5 ELECTRON AND HOLE CONCENTRATION

Since a semiconductor becomes conducting at $T > 0$ K, we anticipate that a large number of electrons can be found in conduction band and also large number of holes in valence band. Let us calculate the number of electrons n in conduction band and, the number of holes p in valence band.

Each energy band in crystal accomodates a large number of electron energy levels. Each energy level consists of two states and each state accomodates only one electron.

Thus, an energy level can be occupied by two electron only, having opposite spin.

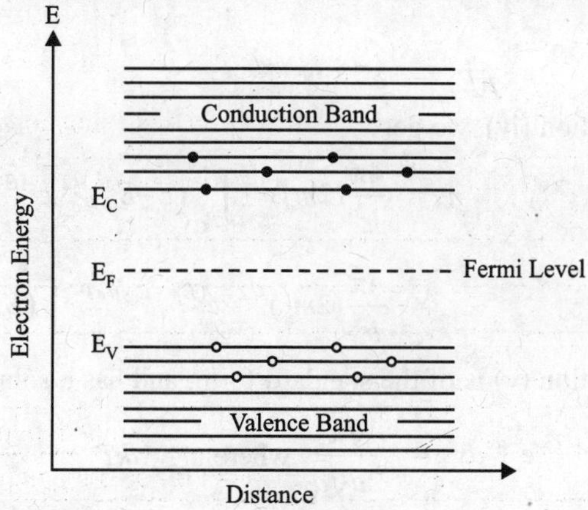

Fig. 24.5

The actual number of electrons in the conduction band is given by

$$n = \int_{E_C}^{\text{top of band}} \underbrace{f(E)}_{\text{Probability of occupancy}} \quad \underbrace{g_c(E)}_{\text{Available energy state per unit volume in the range E to (E+dE)}} \quad dE \qquad \text{... (i)}$$

As the bottom edge of the conduction band E_C in Fig. 24.5 corresponds to the potential energy of an electron at rest in the conduction band, the quantity $(E–E_C)$ represents the kinetic energy of conduction electron at higher energy levels. Here E represents the total energy of an electron. Similarly, the top level in the valence band E_V is the potential energy of a hole at rest. Then $(E_V–E)$ represents the kinetic energy of a hole at higher energy levels. Note that the levels below the level E_V in the diagram are higher energy levels for holes.

The Fermi-Dirac function $f(E)$ is given by

$$f(E) = \frac{1}{1+\exp\left((E - E_F / kT)\right)}$$

rapidly approaches zero for higher energies, the integral in equation (*i*) can be rewritten as

$$n = \int_{E_C}^{\infty} f(E) g_c(E) dE \qquad \text{... (ii)}$$

The density of states g_c (*E*) in the conduction band, which gives the number of available states per unit volume per unit energy interval centred around *E*, is given by

$$g_c(E) = \frac{4\pi}{h^3}(2m_e)^{3/2}(E–E_C)^{1/2} dE \quad \text{for } E > E_C \qquad \text{... (iii)}$$

Substituting, we get

$$n = \int_{E_C}^{\infty} \frac{4\pi}{h^3}(2m_e)^{3/2} \frac{(E-E_C)^{1/2}}{1+exp((E-E_F/kT))} dE \quad ...(iv)$$

Here m_e is effective mass of an electron. [m_e highly depends on its location in the conduction band. Near the bottom of conduction band, m_e is nearly equal to mass of free electron, but when an electron is located at the top of conduction band, it has negative effective mass.]

When the number of particles is very small compared to the available energy levels, the probability of an energy level being occupied by more than one electron is very small. Under this condition $(E_C–E_F) >> 3kT$ and the number of available states in the conduction band are far larger than the number of electrons present in the band. Then Fermi-Dirac function can be approximated to Boltzmann function.

$$f(E) = e^{-(E-E_F)/kT}$$

Using this into equation (iv), we get

$$n = \frac{4\pi}{h^3}(2m_e)^{3/2}\int_{E_C}^{\infty} (E-E_C)^{1/2}.e^{-(E-E_F)/kT} dE$$

$$= \frac{4\pi}{h^3}(2m_e)^{3/2} \cdot e^{(E_F-E_C)/kT} \int_{E_C}^{\infty} (E-E_C)^{1/2} \cdot e^{-(E-E_C)/kT} dE$$

The integral in equation (v) is of the standard form, and has a solution as

$$\int_0^{\infty} x^{1/2} \cdot e^{-ax} dx = \frac{\sqrt{\pi}}{2a\sqrt{a}} \quad \text{where } a = 1/kT$$

$$\therefore \quad n = \frac{4\pi}{h^3}(2m_e)^{3/2} \cdot e^{(E_F-E_C)/kT} \cdot \left[\frac{\sqrt{\pi}}{2}(kT)^{3/2}\right]$$

$$= 2\left[\frac{2\pi m_e kT}{h^2}\right]^{3/2} \cdot e^{-(E_C-E_F)/kT}$$

or

$$n = N_C \cdot e^{-(E_C-E_F)/kT} \quad ... (vi)$$

where $N_C = 2\left[\frac{2\pi m_e kT}{h^2}\right]^{3/2}$. It is a temperature dependent material constant, and is known as 'the effective density of states' in the conduction band. For silicon, $N_C = 2.8\times10^{25}/m^3$ at 300 K. Equation (vi) gives the electron concentration in conduction band.

To Evaluate hole Concentration

Let $f(E)$ be the probability for occupancy of an energy state at E by an electron. Therefore the probability that the energy state is vacant is given by $[1-f(E)]$. Since a hole represents the probability of absence of electron at that level. Thus, the hole concentration in valence band will be

$$p = \int_{\text{Bottom of band}}^{E_V} [1-f(E)]g_v(E)dE \quad ... (vii)$$

or

$$p = \int_{-\infty}^{E_V} [1-f(E)]g_v(E)dE \quad ... (viii)$$

Therefore,

$$[1-f(E)] = 1-\frac{1}{1+exp[(E-E_F/kT)]}$$

$$= 1-\left[1+exp\left[(E-E_F / kT)\right]\right]^{-1}$$

$$\cong e^{+(E-E_F)/kT} \cong e^{-(E_F-E)/kT} \quad \text{... (ix)}$$

$$\therefore \quad p = \int_{-\infty}^{E_V} \frac{4\pi}{h^3} \cdot (2m_h)^{3/2} (E_V - E)^{1/2} \cdot e^{-(E_F - E)/kT} dE$$

$$= \frac{4\pi}{h^3}(2m_h)^{3/2} \cdot e^{-(E_F - E_V)/kT} \int_{-\infty}^{E_V} (E_V - E)^{1/2} \cdot e^{-(E_V - E)/kT} dE$$

$$= 2\left[\frac{2\pi m_h kT}{h^2}\right]^{3/2} \cdot e^{-(E_F - E_V)/kT}$$

or

$$p = N_V \cdot e^{-(E_F - E_V)/kT} \quad \text{... (x)}$$

where $N_V = 2\left[\frac{2\pi m_h kT}{h^2}\right]^{3/2}$. Here N_V is also a temperature dependent material constant and is known is 'the effective density of states' in the valence band. For silicon, $N_V = 10^{25}/m^3$ at 300 K. Thus we observed that the ratio N_C/N_V is 2.8.

24.6 FERMI LEVEL IN INTRINSIC SEMICONDUCTOR

We know that in an intrinsic semiconductor, the free electron and hole concentrations are equal

$$\therefore \quad n = p = n_i$$

Substituting for n and p from equation (*vi*) and (*x*), we get

$$N_C \cdot e^{-(E_C - E_F)/kT} = N_V \cdot e^{-(E_F - E_V)/kT}$$

Taking logarithm of both sides,

$$\log_e N_C + [-(E_C - E_F)/kT] = \log_e N_V + [-(E_F - E_V)/kT]$$

or

$$-\frac{E_C - E_F}{kT} = \log_e \frac{N_V}{N_C} - \frac{E_F - E_V}{kT}$$

$$-E_C + E_F = kT \log_e \frac{N_L}{N_C} - E_F + E_V$$

$$2E_F = kT \log_e \frac{N_V}{N_C} + (E_C + E_V)$$

or

$$E_F = \frac{E_C + E_V}{2} + \frac{1}{2} kT \log_e \left(\frac{N_V}{N_C}\right) \quad \text{... (i)}$$

Now, substituting the values of N_C and N_V and after simplification, we get

$$E_F = \frac{E_C + E_V}{2} + \frac{1}{2} kT \log_e \left(\frac{m_h}{m_e}\right)^{3/2}$$

or

$$E_F = \frac{E_C + E_V}{2} + \frac{3}{4} kT \log_e \left(\frac{m_h}{m_e}\right) \quad \text{... (ii)}$$

Since kT is small and the effective masses m_e and m_h do not differ much, the second term of equation (ii) may be neglected i.e. if $m_e \cong m_h$, then

$$E_F = \frac{E_C + E_V}{2} \qquad \text{... (iii)}$$

Thus, we conclude that the Fermi-level in an intrinsic semiconductor lies exactly at the centre of the energy gap.

Alternatively, $$E_F = \frac{E_C + E_V + E_V - E_V}{2} = \frac{E_V - E_V}{2} + E_V$$

As $(E_C - E_V) = E_g$, $\therefore$ $$E_F = \frac{E_g}{2} + E_V$$

$$E_F = \frac{E_g}{2} + \frac{3}{4} kT \log_e \left(\frac{m_h}{m_e} \right)$$

If the top of the valence band E_V is taken as zero level,

$$E_F = \frac{E_g}{2} \qquad \text{... (iv)}$$

In other words, the Fermi level E_F is located in the middle of the forbidden band.

24.7 EFFECT OF TEMPERATURE ON FERMI LEVEL

Variation of Fermi level with temperature in an intrinsic semiconductor is shown in (Fig. 24.6). The variation is mainly due to the changes of effective masses of electron (m_e) and hole (m_h) with temperature.

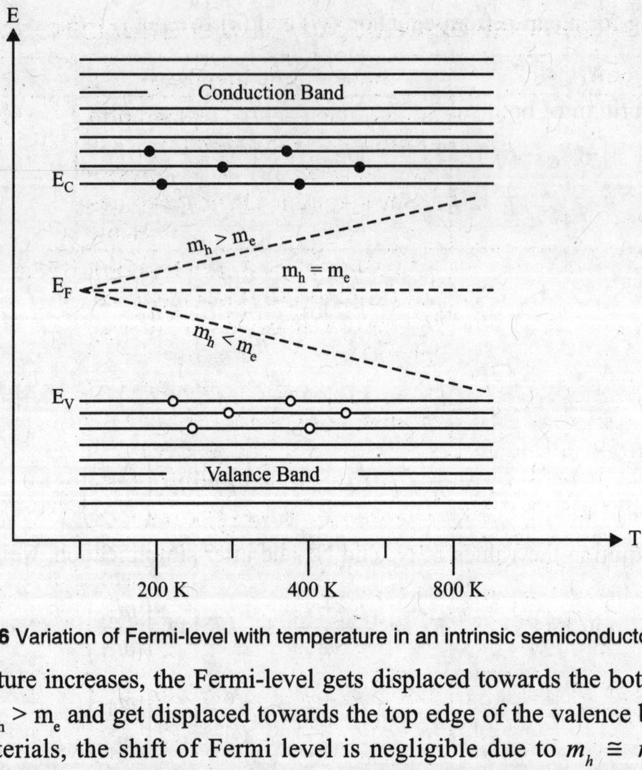

Fig. 24.6 Variation of Fermi-level with temperature in an intrinsic semiconductor.

As the temperature increases, the Fermi-level gets displaced towards the bottom edge of the conduction band if$m_h > m_e$ and get displaced towards the top edge of the valence bond if $m_h < m_e$. For most of the materials, the shift of Fermi level is negligible due to $m_h \cong m_e$. However in practice, the Fermi level in an intrinsic semiconductor is almost independent of temperature and remains in the middle of forbidden gap.

24.8 INTRINSIC DENSITY (n_i)

As we know that in an intrinsic semiconductor at temperature T, the electron concerntration in the conduction band is equal to the hole concentration in the valence band.

$$n = p = n_i$$

$$\therefore \quad n_i^2 = np$$

$$= \left[N_C \cdot e^{-(E_C - E_F)/kT}\right] \times \left[N_V \cdot e^{-(E_F - E_V)/kT}\right]$$

$$= N_C N_V \cdot e^{-(E_C - E_V)/kT}$$

But $(E_C - E_V) = E_g$

$$\therefore \quad n_i^2 = N_C N_V \cdot e^{-E_g/kT} \quad \text{or } n_i = (N_C N_V)^{1/2} \cdot e^{-E_g/2kT}$$

Substituting the values of N_C and N_V we get

$$n_i^2 = 2\left[\frac{2\pi m_e kT}{h^2}\right]^{3/2} \times 2\left[\frac{2\pi m_h kT}{h^2}\right]^{3/2} \times e^{-E_g/kT}$$

$$= 4\left[\frac{2\pi k}{h^2}\right]^3 \times (m_e \cdot m_h)^{3/2} \times e^{-E_g/kT}$$

$$\therefore \quad n_i = 2\left[\frac{2\pi k}{h^2}\right]^{3/2} (m_e \cdot m_h)^{3/4} \cdot T^{3/2} \cdot e^{-E_g/2kT}$$

Thus, (*i*) the intrinsic density is independent of Fermi level,

(*ii*) it strongly depends on the temperature T,

(*iii*) It is a function of band gap E_g, which represents the energy needed to break a covalent bond.

(*iv*) the factor 2 in the exponent shows that for one covalent bond broken, two charge carriers are produced.

SOLVED EXAMPLES

Example 24.1 : *Determine the position of the Fermi-level in an intrinsic semiconductor from the centre of forbidden gap at room temperature, if the effective mass of an electron is equal to twice the effective mass of hole.*

Solution : Given : $m_e = 2m_h$; $T = 300\ K$

$$\therefore \quad kT = 0.026 \text{ eV}$$

$$E_F = \frac{E_g}{2} + \frac{3}{4}kT \log_e\left(\frac{m_h}{m_e}\right)$$

$$= \frac{E_g}{2} + \frac{3}{4} \times 0.026 \times 2.303 \log_{10}\left(\frac{1}{2}\right)$$

$$= \frac{E_g}{2} + 0.75 \times 0.26 \times 2.303 \times 0.69$$

$$= \frac{E_g}{2} + 0.0135 \text{ eV}$$

$$\therefore \quad \left(\frac{E_g}{2} - E_F\right) = -0.0135 \text{ eV}$$

Thus, the Fermi-level is below the centre of the forbidden gap by 0.0135 eV.

24.9 INTRINSIC CONDUCTIVITY (σ)

Fig. 24.7 shows an intrinsic semiconductor sample connected to a D.C. source. The electric field E due to the potential difference V causes a current I to flow through the sample. The current in semiconductor is bipolar. i.e. it is composed of two currents, one due to electrons drifling in conduction band and other due to holes drifring in the valence band. Thus, total current

$$I = I_e + I_h$$

The total current density

$$J = I/A$$

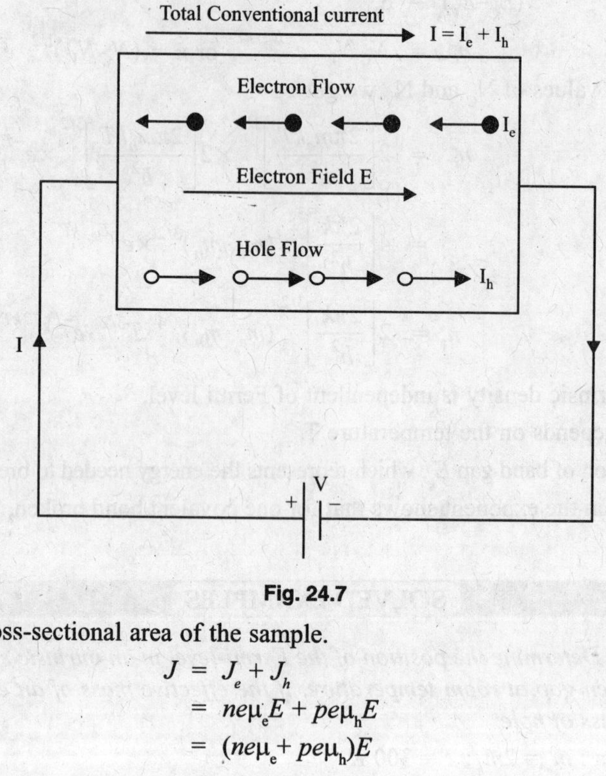

Fig. 24.7

where A is cross-sectional area of the sample.

$$\therefore \quad J = J_e + J_h$$

$$= ne\mu_e E + pe\mu_h E$$

$$= (ne\mu_e + pe\mu_h)E$$

Using $\frac{J}{E} = \sigma$, we get the total conductivity as

$$\sigma = ne\mu_e + pe\mu_h \quad \text{... (i)}$$

or $\sigma = \sigma_e + \sigma_h$, where $\sigma_e = ne\mu_e$ and $\sigma_h = pe\mu_h$

In an intrinsic semiconductor, $n = p = n_i$. Hence

$$\sigma = en_i(\mu_e + \mu_h) \quad \text{... (ii)}$$

Substituting for n_i from equation (*i*) of article 24.8 in equation (*ii*) we have

$$\sigma = 2\left[\frac{2\pi k}{h^2}\right]^{3/2} (m_e \cdot m_h)^{3/4} \cdot T^{3/2} \cdot e(\mu_e + \mu_h) \cdot e^{-E_g/2kT}$$

$$\sigma = \sigma_0 \cdot e^{-E_g/2kT} \quad \text{... (iii)}$$

where $\sigma_0 = 2\left[\frac{2\pi k}{h^2}\right]^{3/2} (m_e \cdot m_h)^{3/4} \cdot T^{3/2} \cdot e(\mu_e + \mu_h)$

Equation (iii) is of practical importance in determining E_g, the band gap. Taking logarithm of both sides of Eqn. (iii), we get

$$\log_e \sigma = \log_e \sigma_0 - E_g/2kT$$

A plot of $\log_e \sigma$ verses $\frac{1}{T}$ represents a straight line having slope $E_g/2k$. (using $y = mx + c$).

24.10 DOPING

The electrical conductivity of an intrinsic semiconductor crystal is very poor. However, a judicious introduction of impurity atoms in a semiconductor crystal produces useful modifications of its electrical conductivity. An intentional introduction of controlled quantity of suitable impurity into to an intrinsic semiconductor is called doping. When a crystal is doped with impurity atoms is called an extrinsic semiconductor.

Extrinsic Semiconductor : The intrinsic semiconductors are not of much use due to their small and fixed conductivity. The conductivity of a semiconductor can be considerably improved by adding a small amount of impurity. Such semiconductors are known as extrinsic semiconductors or doped semiconductors. The impurity added is called **dopant**. One of the important methods of doping is to add precisely determined quantities of impurity to the melt from which the silicon crystal is grown. Diffusion method can also be employed for doping. Typical doping levels range from 10^{20} to 10^{27} impurity atoms / m^3 or in the impurity ratio of about 1 in 10^6 to 10^8 atoms of the semiconductor. An intrinsic semiconductors Ge or Si belongs to Group IV of periodic table. Pentavalent elements from Group V or trivalent elements from Group III are used as dopants. The atoms belonging to these two groups are nearly of the same size as silicon or germanium atoms and can easily substitute themselves in place of some of the host atoms in the semiconductor crystal.

The impurity atoms of Group V are know as 'donar' or n-type and gives rise to n-type semiconductor. The impurity atoms of Group III are known as 'acceptor' or p-type and gives rise to p-type semiconductor.

24.11 DONAR OR *n*-TYPE SEMICONDUCTOR

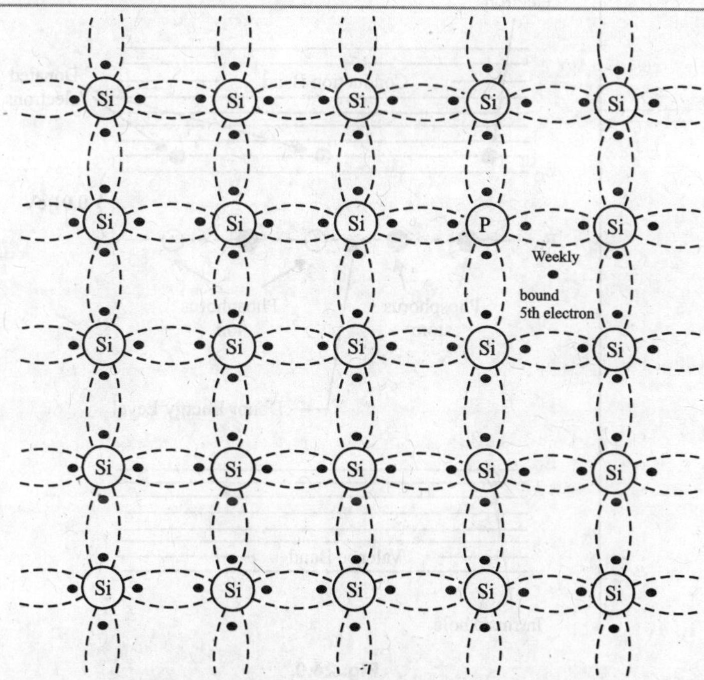

Fig. 24.8. Effect of doping pure silicon with pentavalent impurity.

When a pure semiconductor Ge or Si is doped with a pentavalent impurity like arsenic (As), antimony (Sb) or phosphorus (P), an n-type semiconductor is produced. Let us examine the effect of these pentavalent impurities on the carrier concentration and electrical conductivity. Suppose a phosphorus atom (P) has replaced a silicon atom in the crystal. Phosphorus atom has five valence electrons. Out of the five electrons, only four electrons are used up in forming covalent bonds with neighbouring four Si atoms. The fifth electron remains excess and being loosely bound, it becomes free to move from one atom to another as shown in Fig. 24.8. Each electron liberated leaves behind a positive phosphorus ion that is fixed in the crystal. These ions are immobile. Therefore, they do not contribute to current. The impuruty atoms which contribute electrons to the conduction band are called **donar atoms**. They produce electrons without producing holes in the valence band.

Fig. 24.9 shows the effect of donar atoms on the energy band structure of the semiconductor. Since the impurity density is low, the donar atoms are distantly spaced and hance can not interact with each other and is shown by a single dotted line in figure.

This donar energy level is slightly below ($\cong 0.01eV$) the bottom edge of conduction band. This is because, the ionization energy of donor is about $\frac{1}{25}$ th of band gap energy Eg and secondly, an electron detached from donar atom becomes a conduction electron to breaking of covalent bonds, jump into the conduction band, leaving holes in valence band. The conduction band, thus contains electrons from two different sources. The holes in the valence band are smaller at ordinary temperatures. The electrons which are out-numbering the holes are called majority carriers whereas the holes are called minority carriers. As the current is mainly carried by electrons, the semiconductor so obtained is called donor or n-type extrinsic semiconductor.

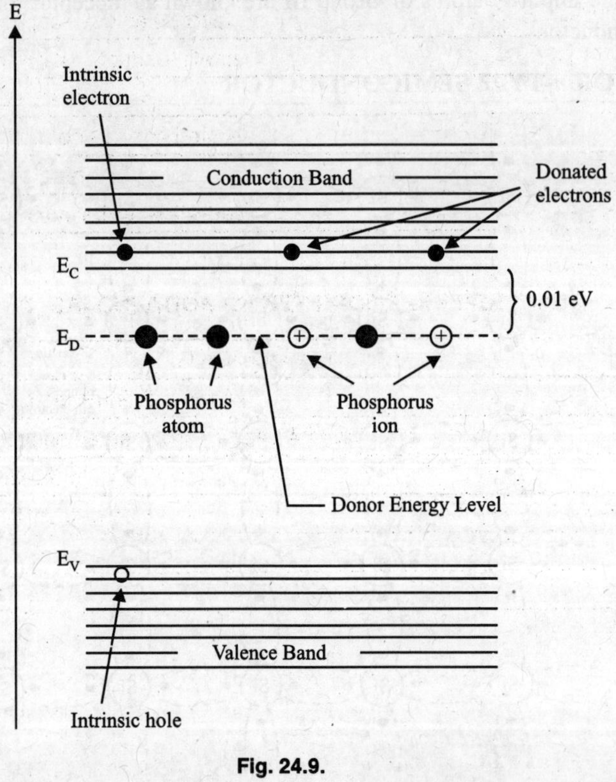

Fig. 24.9.

24.12 ACCEPTOR OR p-TYPE SEMICONDUCTOR

When an intrinsic semiconductor Ge or Si is doped with trivalent impurity like boron (B). Indium (In) or aluminium (Al), p-type semiconductor is produced. When a trivalent impurity atom (say Boron B) substitutes a host silicon atom, it falls short of one electron for completing the four covalent bonds with its neighbours. The non-formation of covalent bond is equivalent to a hole. The environment in the crystal is initially neutral. The substitution of a host Si atom with B does not disturb the neutral environment around the boron atom. However, when an electron from a neighbouring atom acquires energy and jumps to form a bond, it leaves behind a hole. The boron atom having acquired an additional electron becomes a negative ion. The hole can move freely in the valence band whereas impurity ion (B^-) is locked in position by the covalent bonds. Consequently, it is **immobile**. The impurity atoms which accept electrons from the valence band are known as **acceptor** atoms. The acceptor impurity atoms produce holes without the simultaneous generation of electrons, in the conduction band.

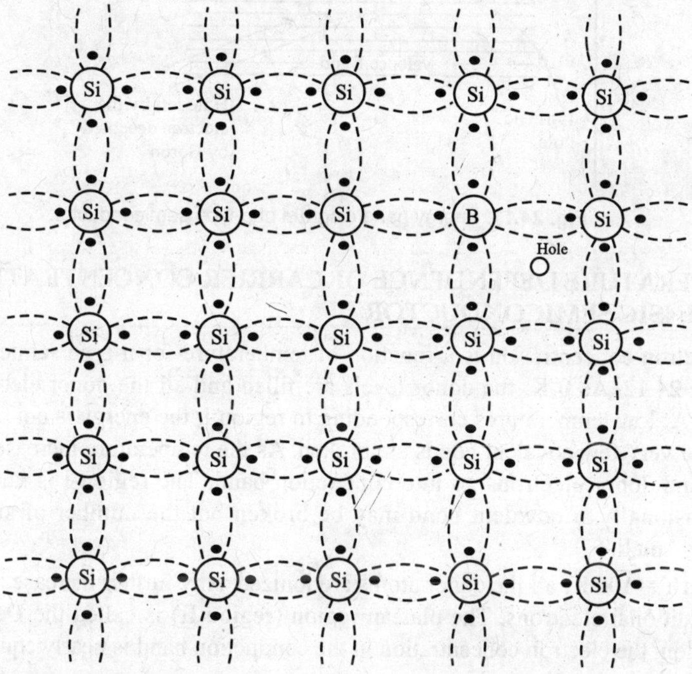

Fig. 24.10. Effect of doping pure silicon with trivalent impurity.

At room temperatures (T > 0 K), holes are produced due to intrinsic process also. Some bonds are broken, electrons jumps to the conduction band leaving holes in valence band. The net result is that holes outnumber electrons in the semiconductor. Therefore, holes constitute the **majority carriers** and electrons are **minority carriers** in this type of semiconductor. As the positively charged carriers (i.e. holes) are mainly contributing to the conduction process; this type of semiconductor is known as **p-type semiconductor**.

Fig. 24.11 shows the effect of doping of an acceptor impurity on the energy band structure of silicon. As valence electrons readily jump from the valence band to boron atoms after acquiring a small amount of energy, the acceptor level, shown by a dotted line, situated closer to and above the top edge ($\cong 0.01 eV$) of the valence band.

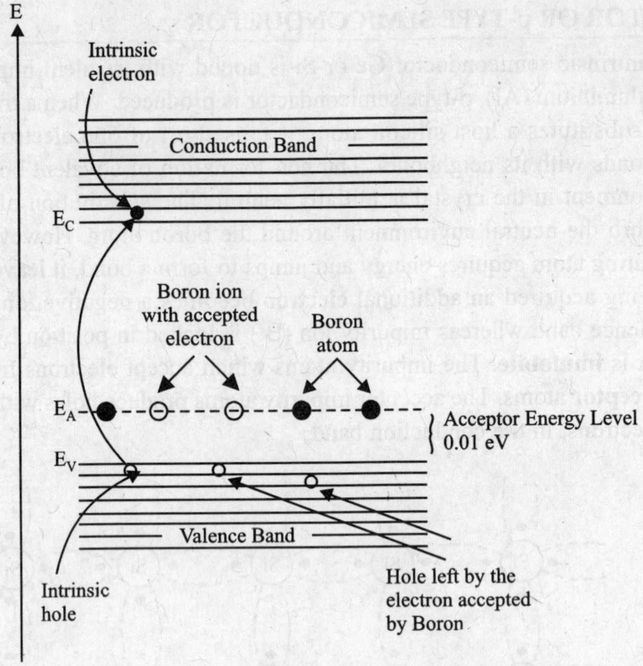

Fig. 24.11. Energy band diagram of p-type semiconductor.

24.13 TEMPERATURE DEPENDENCE OF CARRIER CONCENTRATION IN EXTRINSIC SEMICONDUCTOR

The electron concentration is a function of temperature for n-type semiconductor and is as shown in Fig 24.12. As 0 K, the donor levels are filled and all the donor electrons are bound to donor atoms. At low temperatures corresponding to region I, the energy is not enough to ionize all the donors but very few covalent bonds are broken. As the temperature increases, the donor atoms get ionized and donor electrons go into conduction band. The region-I is known as **Ionization region.** Occasionally, a covalent bond may be broken but the number of such events will be insignificantly small.

At about T = 100 K, all the donor atoms are ionized. Any further increase in temperature does not create additional electrons. The plateau region (region II) is called the **Depletion region**. In depletion region, the electron concentration in the conduction band is nearly equal to the concentration of the dopant atoms. ($n_n = N_D$).

As the temperature increases further, electron transitions from valence band to conduction band enhances. At high temperature (region III), the number of electron transitions become so large that the intrinsic electron concentration exceeds the electron concentration due to donors. This regions is known as **Intrinsic region.**

The dependence of hole concentration on temperature in a p-type semiconductor is very much similar to that shown in Fig. 24.12

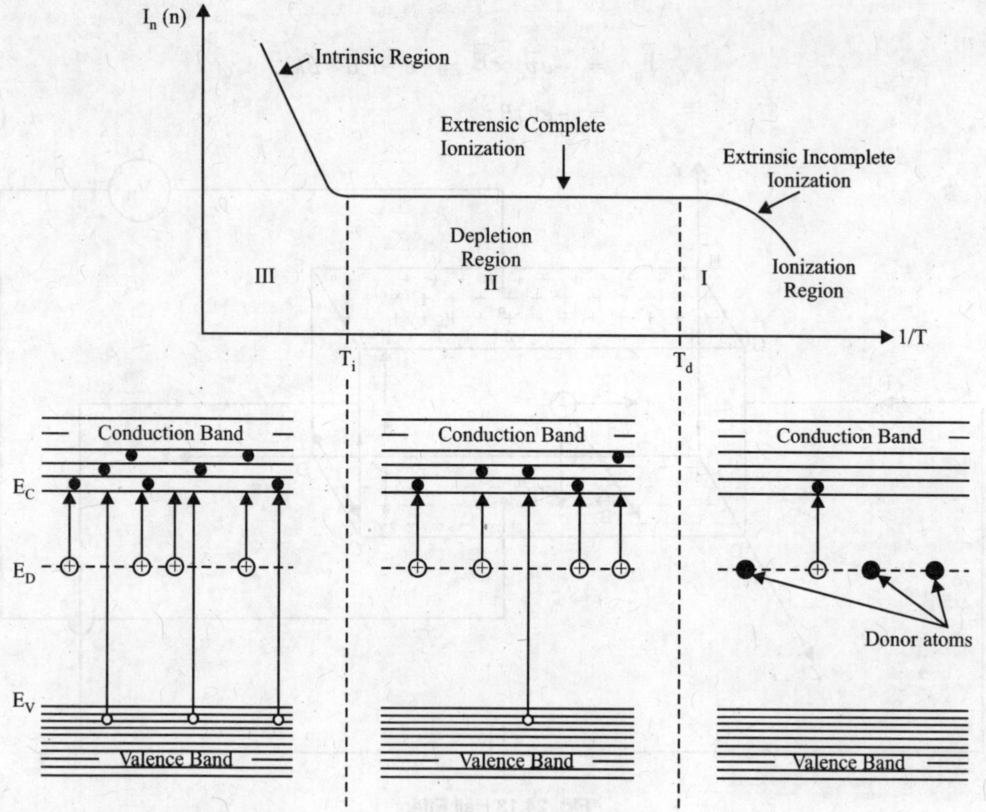

Fig. 24.12

24.14 HALL EFFECT

If a semiconductor carrying a current is placed in magnetic field acting perpendicular to the direction of current, then a potential difference is developed in a direction perpendicular to both, the current as well as the magnetic field. This phenomenon is known as **Hall effect** in honour of E.H. Hall, the physicist who discovered it in 1879.

The importance of Hall effect is that it helps

(i) to determine the sign of charge carriers,

(ii) to determine the charge carrier concentration and

(iii) to determine the mobility of charge carriers if conductivity of the material is known.

(*i*) Negative charge carriers (n-type semiconductor)

Let us consider n-type semiconductor ABCDEFGH carrying a current I in the + X direction. A magnetic field $\vec{B}$ is applied to the plate in the + Z direction, then $\vec{B} = B\hat{K}$

Since material is n-type semiconductor, the charge carriers are electrons. As the current is flowing in + X direction, the free electrons are difting in the – X direction. If $\vec{v}_d$ is the drift velocity of electrons. then

Magnetic force $\vec{F}_m$ on each electron $\vec{F}_m$ is given by

$$\vec{F}_m = -e\vec{v}_d \times \vec{B} = -e(-v_d\hat{i} \times B\hat{k})$$

$$= -ev_d B\hat{j} \quad \text{... (i)}$$

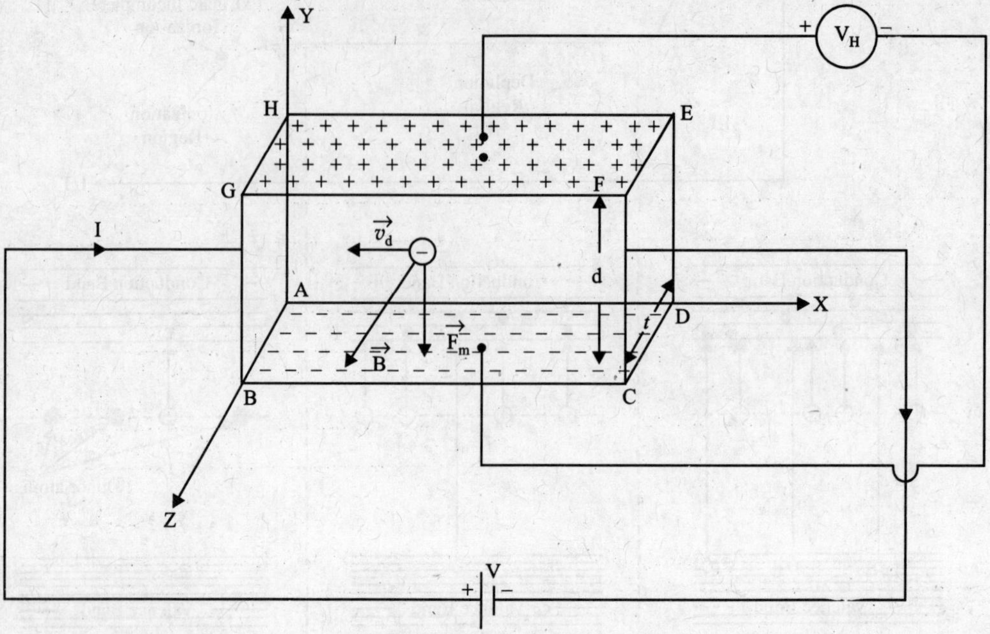

Fig. 24.13 Hall Effect

The force $\vec{F}_m$ acts in the – Y direction. This results the electrons to bend downward towards the face ABCD. The electrons accumulate on the lower surface producing a net negative charge on face ABCD. As a result, a net positive charge is developed on the upper surface EFGH. This sets up an electric field between these two surfaces acting in the downward – Y direction. This electric field is known as **Hall electric field** E_H, given by

$$\therefore \quad \vec{E}_H = -E_H\hat{j}$$

The electron flow towards the lower face will stop when the force on the electron due to the Hall field $-\vec{F}_e$ and the force due to the magnetic field $\vec{F}_m$ balance each other.

$$\text{or} \quad \vec{F}_m = -\vec{F}_e$$

$$\text{or} \quad \vec{E}_m + \vec{F}_e = 0 \quad \text{... (ii)}$$

Now $\vec{F}_e = -e\vec{F}_H$ ($\because$ –ve sign indicates the charge of an electron)

$$= -e(-E_H\hat{j}) = eE_H\hat{j} \quad \text{... (iii)}$$

Subsituting from Eqn (*i*) and (*iii*) in Eqn. (*ii*),

$$\therefore \quad eE_H\hat{j} - ev_d B\hat{j} = 0$$

$$\text{or} \quad E_H = v_d B \quad \text{... (iv)}$$

(ii) Positive charge carriers (p-type semiconductor)

Now consider a p-type semiconductor in which the charge carriers are 'holes'. The direction of positively charged particles in along the direction of the current. Hence

$$\vec{v}_d = v_d\hat{i}$$

If $+q$ is the charge on each positive charge carrier, magnetic force $\vec{F}_m$ on each positive charge carrier will be

$$\vec{F}_m = q\vec{v}_d \times \vec{B} = qv_d\hat{i}\times B\hat{k}$$

$$= -qv_dB\hat{j} \qquad \text{... (v)}$$

This equation shows that positive charge carriers will also experience a force in the – Y direction. The positive charge carriers are deflected dowards, towards the lower face ABCD. This face positively charged As a result, the upper face EFGH is negatively charged. This gives rise to an electric field in + Y direction. This gives Hall electric field $\vec{F}_H$, given by

$$\vec{E}_H = E_H\hat{j}$$

Thus, the direction of Hall electric field for p-type semiconductor is opposite to that for *n*-type semiconductor.

Now $$\vec{E}_q = +q\vec{E}_H = qE_H\hat{j} \qquad \text{... (vi)}$$

In equilibrium, further flow of holes towards lower force stops, when

$$\vec{F}_q + \vec{F}_m = 0$$

$$qE_H\hat{j} - qv_dB\hat{j} = 0$$

or $$E_H = v_dB$$

Hall Voltage

Let V_H be the potential difference developed between the two faces ABCD and EFGH and d is thickness or the distance between these two faces, then

$$E_H = \frac{V_H}{d} = v_dB$$

or $$V_H = v_d d \cdot B \qquad \text{... (vii)}$$

This equation gives Hall-voltage (V_H)

24.15 HALL COEFFICIENT

The current density $\vec{J}$ is given by

$$J = nev_d$$

where n is the number of charge carriers per unit volume. Considering the magnitudes,

$$v_d = \frac{J}{ne} \qquad \text{... (viii)}$$

From equation (*iv*), $$v_d = \frac{E_H}{B} \qquad \text{... (ix)}$$

Equating (*viii*) and (*ix*), we get

$$\frac{J}{ne} = \frac{E_H}{B}$$

$$\frac{E_H}{JB} = \frac{1}{ne}$$

The Hall electric field per unit current density per unit magnetic field i.e. (E_H/JB) is called Hall coefficient and is denoted by R_H.

$$R_H = \frac{1}{ne} \quad \text{... } (x)$$

Thus, knowing the Hall coefficient R_H, the value of n, the density of charge carriers (either electrons or holes) can be calculated.

However, the experimentally observed quantities are the Hall voltage V_H and the current I through the sample. If d is the width and t its thickness, then

$$V_H = E_H \cdot d$$

and I = J × Area of cross-section

Hence, the Hall coefficient,

$$R_H = \frac{E_H}{JB} = \frac{V_H/d}{(I/d \cdot t)B}$$

$$R_H = \frac{V_H \cdot t}{IB} \quad \text{... } (xi)$$

Knowing t, I and B and measuring V_H, the value of Hall coefficient R_H can be calculated.

24.16 MOBILITY

The conductivity of a semiconducting material is given by

$\sigma = ne\mu$

where μ is the mobility of the charge carrier.

But $ne = \frac{1}{R_H}$, from equation (x). On substitution, we get

$$\sigma = \frac{\mu}{R_H} \quad \text{... } (xii)$$

$\therefore$ Mobility, $\mu = \sigma R_H$

Knowing, the Hall coefficient R_H and the conductivity (σ) of the sample, the mobility (μ) of charge carrier can be evaluated.

24.17 EXTRINSIC CONDUCTIVITY

The depletion region i.e. region-II of Fig. 24.12 is of practical importance, since the carrier concentration is independent of temperature and the concentration can be controlled by adding impurities from outside. The conductivity of an n-type semiconductor is given by

$$\sigma_n = n_n e\mu_e$$

or $\sigma_n = N_D e\mu_e$ Depletion Region ... (i)

The conductivity of p-type semiconductor is given by

$$\sigma_p = p_p e\mu_h$$

or $\sigma_p = N_A e\mu_h$ Saturation Region ... (ii)

where n_n is electron concentration in n-type material,

p_p is hole concentration in p-type material,

N_D is impurity concentration in depletion region,

and N_A is impurity concentration in saturation region.

In the intrinsic region, the conductivity of both types of extrinsic semiconductor is given by

$$\sigma_i = \sigma_e + \sigma_h = n_e e\mu_e + n_h e\mu_h = e(n_e\mu_e + n_h\mu_h) \quad \text{... (iii)}$$

$$\sigma_i = n_i e\,(\mu_e\,\mu_h) \quad \text{if } n_e = n_h = n_i$$

Example 24.2 : *The intrinsic carrier density of Ge at 27°C is 2.4×10¹⁷ m⁻². Calculate its intrinsic resistivity, if the electron and hole mobilities are 0.35 m² and 0.18 m² V⁻¹ S⁻¹.*

(Nagpur U. 2011)

Solution : Conductivity of an intrinsic semiconductor is given by

$$\sigma_i = \sigma_e + \sigma_h = e(n_e\,\mu_e + n_h\,\mu_h)$$

but $n_e = n_h = n_i$

$$\therefore \quad \sigma = en_i(\mu_e + \mu_h)$$

$$= 1.6\times10^{-19}\times2.4\times10^{17}\times(0.35\times0.18) = 1.6\times2.4\times0.53\times10^{-2}$$

$$\therefore \text{ Intrinsic resistivity, } \rho = \frac{1}{\sigma} = \frac{10^2}{1.6\times2.4\times0.53} = 49\times10^{-2} \text{ ohm metre.}$$

EXERCISE

1. What is an intrinsic semiconductor ? Why are Si and Ge preferred as semiconducting material? ***(Nagpur U. 2009)***
2. What is meant by a hole ? Explain energy band structure of an intrinsic semiconductor at absolute zero and at room temperature. What do you understand by recombination ?
3. Derive an expression for the electron contentration in conduction band of an intrinsic semiconductor. ***(Nagpur U. 2011)***
4. Derive an expression for the hole concentration in valence band of an intrinsic semiconductor. ***(Nagpur U. 2009)***
5. Show that for an intrinsic semiconductor, the Fermi level is given by

$$E_F = \frac{E_C + E_V}{2} + \frac{3}{4}kT\log_e\left(\frac{m_h}{m_e}\right)$$

where the symbols have their usual meaning.

6. Draw a neat energy band diagram of intrinsic semiconductor. Lable the various energy levels. Prove that the Fermi level lies at the middle of the band gap.
7. Explain the variation of Fermi level with temperature in an intrinsic semiconductor. ***(Nagpur U. 2012)***
8. Derive an expression for intrinsic conductivity of semiconductor.
9. Find the resistivity of intrinsic germanium at 300 K. Given that the intrinsic density of carriers is $2.5\times10^{19+}/m^3$. **[Ans : 0.43 ohm metre]**
10. What is doping ? What are extrinsic semiconductor ? Describe n-type and p-type semiconductors clearly explaining the terms 'donor' and 'acceptor'. What are majority and minority charge carriers in each type of semiconductor ? Draw energy level diagrams for the concept of n-type (donor) and p-type (acceptor) material. ***(Nagpur U. 2011, 2012)***
11. How does the carrier concentration change with temperature in extrinsic semiconductor. Sketch the diagram.

12. Explain the carrier concentration of an extrinsic semiconductor in
    (*i*) Ionization region
    (*ii*) Depletion region
    (*iii*) Intrinsic region
13. What is Hall effect ? Mention its importance in the field of semiconductors. ***(Nagpur U. 2011, 2012)***
14. Obtain an expression for Hall coefficient for an extrinsic semiconductor. Show that the result of Hall effect may be used to deduce carrier concentration.
15. What is Hall effect ? Give expression for each of the following :
    (*i*) Hall coefficient
    (*ii*) Hall voltage
    (*iii*) Hall mobility

    Mention some applications of Hall effect.
16. Why does a semiconductor behave as an insulator at absolute zero ? ***(Nagpur U. 2012)***
17. Show that $E_H = R_H \vec{J}.\vec{B}$

    where symbols have their usual meaning ***(Nagpur U. 2009)***

SEMICONDUCTOR DEVICES

25 Chapter

INTRODUCTION

The extrinsic semiconductors as such are of limited use. However, when a p-type semiconductor is joined together with an n-type, a pn junction is formed. This junction is known as **semiconductor diode**. A pn junction a basic but very important device. Its rectifying action is utilised in coverting an a.c. voltage into d.c. voltage. The different diodes like a varicap that acts as a variable capacitor, a tunnel diode and a Gun diode as oscillator, a zener diode as a voltage stabalizer, a photodiode as a light detector, a solar cell as a voltage source, LED and a laser as light sources are all pn junctions. A junction transitors and a phototransitor are fabricated with two pn junctions in close proximity. Thus, a pn junction diode is a most basic component of solid state devices.

25.1 PN JUNCTION DIODE

When a p-type semiconductor and n-type semiconductor are joined metallurgically, a pn junction is formed. The interface lying between the p-type and n-type regions is called a pn junction. A pn junction can not be produced simply by mechanically joining the two semiconductors. Actual process is a complex one. In practice, such a pn junction diode may be prepared by employing one of the following two methods :

1. GROWN JUNCTION

In a grown junction method, a single crystal is obtained from a melt which first contains impurities of either p or n-type. In the middle of the growth process, impurities of the opposite type are added to the melt so that the remainder of the crystal grows into the opposite type.

For example : An n-type Si crystal containing pentavalent impurity P (Phosphorus) atoms is first grown from a melt. The growth process is stopped in the middle and sufficient trivalent B (Boron) atoms are added so that the remaining crystal grown is p-type.

2. FUSED JUNCTION

A fused junction type is obtained by fusing a metal pallet of the opposite type of a semiconducting material.

For example : A small pallet of trivalent Indium (In) is pressed on a thin wafer of the opposite n-type semiconductor. After a few minutes of heat treatment some of the Indium atoms get fused in to the surface of Germanium (or Silicon). The Indium diffuses into the Germanium crystal and produces p-type semiconductor along with n-type Germanium wafer. Thus, pn junction is formed.

DIFFUSION OF MAJORITY CARRIERS

Suppose each block of *p*-type and *n*-type semiconductors are suddenly joined together, as shown in Fig. 25.1 (a). Initially both the blocks are electrically neutral. The moment two blocks are joined, electrons tends to diffuses from n-type rigion into p-type region while holes tends to diffuse from p-type to n-type region. If J_{hp} represents hole diffusion current density and J_{en}, the electron diffusion current density components, then the net diffusion current will be

$$J_{(diffusion)} = J_{hp} + J_{en} \quad \text{... } (i)$$

25.2 DEPLETION REGION AND POTENTIAL BARRIER

Near the junction, a hole diffusing from p-type to n-type, leaves behind a negative acceptor ion. When it arrives in the n-region, it becomes a minority carrier. There it is surrounded by a large number of electrons leading to a large probability of recombination. Therefore, the hole recombines with an electron and both hole and free electron disappear. For each hole that leaves the p-region to corss the junction into the *n*-region there is left-behind an acceptor atom that now have a net negative charge. Similarly, for each electron that leaves the n-region to cross the junction into the p-region, there is left-behind a donor atom. The donor atom acquires a net positive charge. Thus, the two kinds of majority carriers diffusing across the junction, undergo recombinations, leaving a distribution of fixed negative ions on the p-side i.e. acceptor side and a distribution of fixed positive ions on the n-side i.e. donor side of the junction. This charge distribution on each side is shown in Fig. 25.1 (a) and is known as **space charge**.

Due to recombination of electrons and holes, a narrow region of a small width of the order of 10^{-16} m about the junction becomes depleted of mobile carriers (i.e. electrons and holes) and contains only localised ions. This region is known as the **depletion region**.

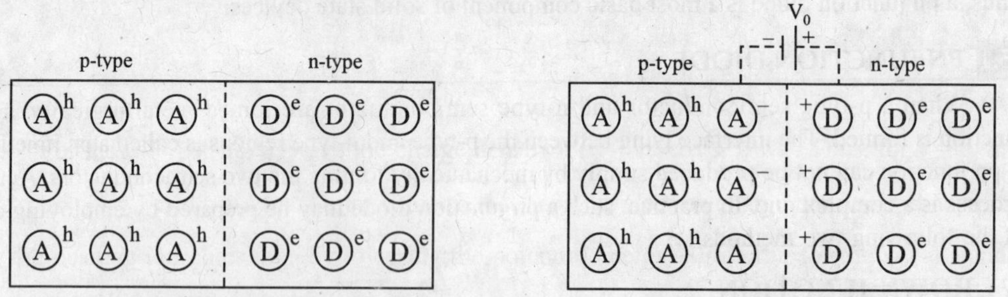

(a) Blocks of p-type and n-type at the instant they joined.

(b) pn junction showing charged ions after hole and electrons diffusion.

Fig. 25.1

Any accumulation of electric charge of opposite polarities due to acceptor ions and donor ions on either side of the junction, produces an electric field across the junction. Electric lines of force start from donor ions on n-type side of the junction and terminates on an equal number of acceptor ions on the p-type side. The process of majority charge carriers crossing the junction, their recombination and building of charge continue till a sufficient electric field across the junction is established. This electric field is known as **built-in-field**. The electric field builds up to such a strength, that it brings majority carriers diffusion to a near halt. i.e. the further diffusion comes to an end. Thus, the electric field acts now as a **barrier** to the diffusion of carriers. Since, potential is the integral of electric field, a potential difference will appear across the junction. This voltage V_0 appearing across the junction is called the **built-in-potential barrier**. The product eV_0 is called the **height of barrier**, where e is electron charge. It should be noted that the barrier potential V_0 is not a true emf. It can be source of energy to provide current, because the mobile charges are not available in the depletion region. The depletion region is more like an insulator having high resistance. The pn-junction resembles closely a charged capacitor. V_0 can not be measured using a volt meter.

The width of the depletion layer depends on the barrier potential and on the carrier concentration.

The barrier potential V_0 is given by

$$V_0 = \frac{kT}{q}\log_e\left(\frac{N_A N_D}{n_i^2}\right)$$

where V_0 = Boltzmann's constant = 1.38×10^{-23} J/K
T = temperature in Kelvin
q = electron charge = 1.6×10^{-19} coulombs
N_A = acceptor doping density in p-type
N_D = donor doping density in n-type
n_i = intrinsic impurity density.

Note that barrier potential is directly proportional to temperature. The quantity $\frac{kT}{q}$ has the units of volts and is called the thermal voltage (V_T).

$$\therefore \quad V_T = \frac{kT}{q}$$

$$\text{or} \quad V_0 = V_T \log_e\left(\frac{N_A N_D}{n_i^2}\right) \quad \text{... (ii)}$$

SOLVED EXAMPLES

Example 25.1. *A silicon pn junction diode is formed from p-material doped with 10^{22} acceptors/m^3 and n-material doped with 1.2×10^{21} donors/m^3. Find the thermal voltage and borrier voltage at 25^{0}C.*

Solution : T = 273 + 25 = 298 K

$$\text{Thermal voltage} \quad V_T = \frac{kT}{q} = \frac{(1.38\times10^{-23})(298)}{1.6\times10^{-19}} = 25.7 \text{ mV}$$

$$n_i^2 = (1.5\times10^{16})^2 = 2.25\times10^{32}$$

Using relation for barrier potential V_0, we get

$$V_0 = V_T \log_e\left(\frac{N_A N_D}{n_i^2}\right) = 25.7\times10^{-3} \log_e\left(\frac{10^{22}\times1.2\times10^{21}}{2.25\times10^{32}}\right)$$

$$= (0.0257)\,(24.6998)$$

$$= 0.635 \text{ V}$$

25.3 DRIFT OF MINORITY CARRIERS

The minority carriers are generated through breaking of covalent bonds and in each region the minority carrier concentration is solely determined by the temperature. It is seen that built-in potential barrier V_0 across the junction prevents the diffusion of majority carriers. However, the built-in-electric field has the right direction to promote the flow of minority carriers across the junction. Electrons arriving at the junction from the bulk of p-region (or electrons produced in the depletion region due to rupture of covalent bonds) are assisted by the electric field to move into n-region. Similarly, holes in the n-region are helped to move into p-region. As a consequence, an electric current flows across the junction. As the current is caused by an electric field, it is a drift-current. The two components, namely electron and hole drift current densities are designated by J_{ep} and J_{hn} respectively. The direction of J_{ep} and J_{hn} are the same. Therefore, the net drift current through the junction is given by

$$J_{(drift)} = J_{ep} + J_{hn}$$

Four important points regarding minority carriers and drift current are to be, noted:

(*i*) Since the minority carrier concentration is solely determined by the temperature, the drift current is constant at a given temperature.

(*ii*) Minority carriers are very small in number and hence the drift current will be very small.

(*iii*) Minority carriers can move across the junction only when the barrier potential exists across the junction. When the barrier potential is small, they move slower and when it is larger, they move faster.

(*iv*) An externally applied voltage can not change the magnitude of the drift current. It can only cause a decrease or an increase in the kinetic energy of the minority carriers.

Equilibrium condition

Until an external voltage source is not connected to the pn junction, electric current cannot flow across it. Also, when the junction is in thermal equilibrium, there can not by any current across it. It means that the drift of minority carriers is counterbalanced by the diffusion of the same member of majority carriers across the junction. The barrier height adjusts itself till the current due to majority carriers is exactly compensated by the current due to minority carriers across the junction. Thus,

$$J_{(diffusion)} = J_{(drift)}$$

$$J_{hp} + J_{en} = J_{hn} + J_{ep} \quad \text{... (i)}$$

As the electrical neutrality of each side is to be retained, the diffusion and drift current components due to each type of carriers should cancel each other. Then, the condition for equilibrium may be written as

$$J_{hp} - J_{hn} = 0 \text{ and } J_{en} - J_{ep} = 0$$

or

$$J_{hp} = J_{hn} \text{ and } J_{en} = J_{ep} \quad \text{... (ii)}$$

Thus, at equilibrium, the drift current due to minority carriers (J_{hn}) and (J_{ep}) is compensated by the diffusion current due to majority carriers (J_{hp}) and (J_{en}) respectively and the net current across the (unbiased) junction is zero.

25.4 ENERGY BAND STRUCTURE OF PN JUNCTION AT EQUILIBRIUM

The energy band diagram of p and n-semiconductors separately are as shown in Fig. 25.2. At the instant of joining, the Fermi levels E_{F_p} and E_{F_n} are at different levels.

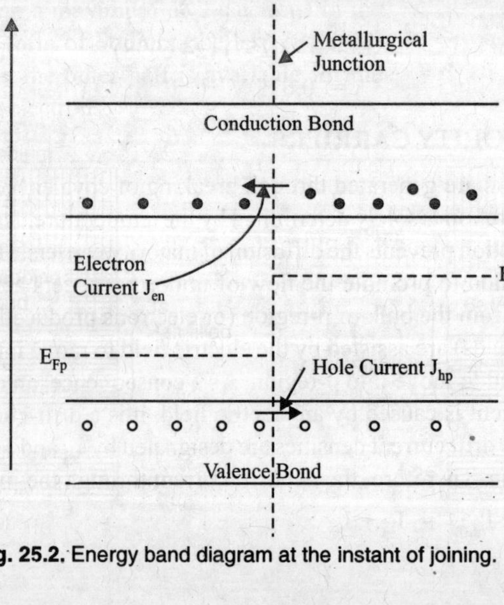

Fig. 25.2. Energy band diagram at the instant of joining.

Joining of two semiconductor pieces together having two different energy levels, electrons and holes will cross the junction and the process migration will continue till an equilibrium is set

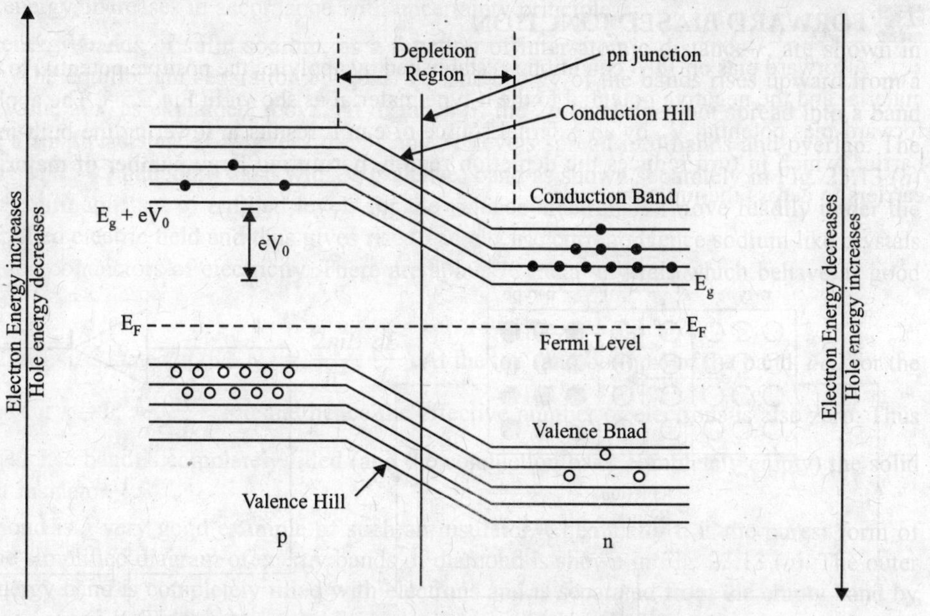

Fig.25.3. Energy Band diagram of pn-junction at equilibrium

up. In other words, both Fermi levels E_{F_p} and E_{F_n} must be equilevel at equilibrium. This phenomenon can be understood with the help of an analogy of two level containers connected by a value. Initially valve is closed and the containers are filled with water to different levels as shown in Fig. 25.4 (a). When the valve is opened. Water flows from container (2) into container (1). The water level in contain (1) raises and water level in container (2) falls down till the level of both containers become equal as shown in Fig. 25.4 (b). Therefore the conduction and valence band positions in p and n-type materials would shift relative to each other. E_{F_p} slowly moves up along with its valence band and conduction band and at the same time E_{F_n} slowly falls down along with its valence band and conduction band. During this process of shifting their Fermi levels, the diffusion of electrons and holes and their recombination continues. Only the immobile ions are left behind,

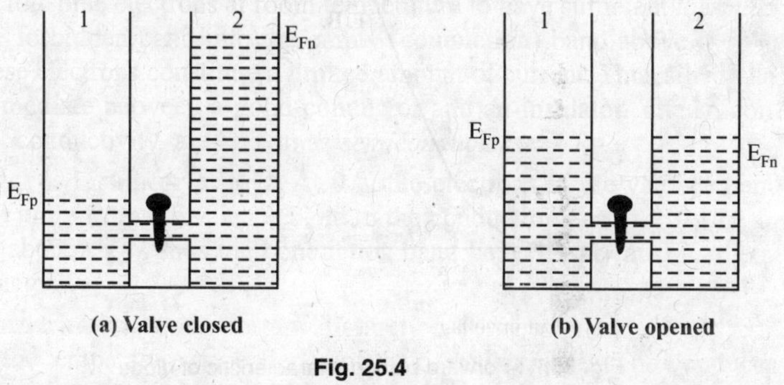

Fig. 25.4

on either side of junction and built-in potential strarts building. The charge flow stops when both E_{F_p} and E_{F_n} become equal and attains a common Fermi level E_F. The mutual displacement of energy levels on both sides causes a bending of the energy bands in the region around the metallurgical junction as shown in Fig. 25.3. It results in a barrier potential V_0 and an energy hill of height eV_0 electrons in the conduction band of n-region face a raising energy hill, called as **a**

conduction hill. Since the direction of increasing energy is downward for holes, the holes in the valence band of p-region faces an energy hill, and is called as **a valence hill.**

25.5 FORWARD BIASED JUNCTION

A forward bias or 'ON' condition is established by applying the positive potential to the p-type material and the negative potential to the n-type material as shown in Fig. 25.5. The application of forward-bias potential V_D by an external source of e.m.f. results in lowering the built-in potential barrier, which in turn reduces the depletion region, permitting large number of majority charge carriers to across the junction.

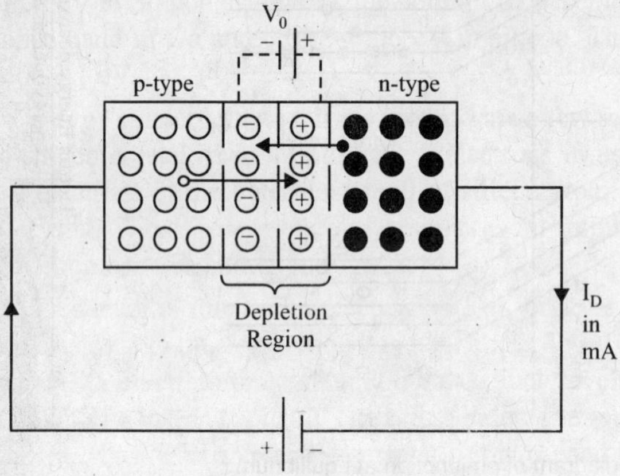

Fig. 25.5. Forward Biased Junction

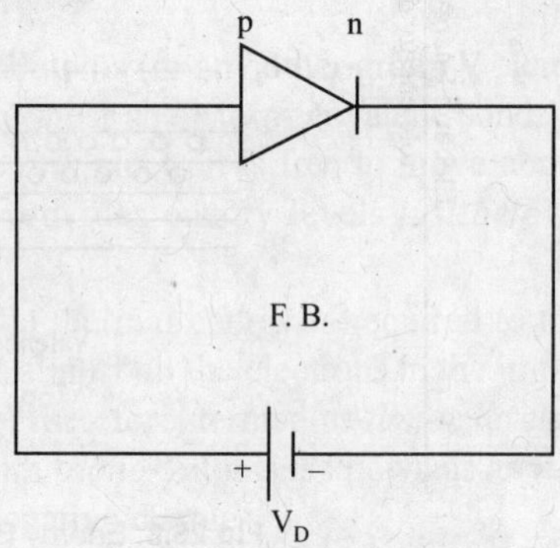

Fig. 25.6. Forward Biased Junction Using symbol of diode.

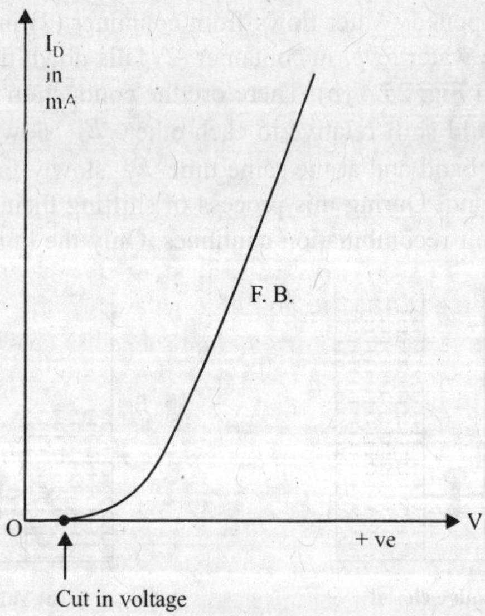

Fig. 25.7 : Forward biased characteristic of diode

If the external applied e.m.f. is increased, the junction potential difference falls, till it becomes minimum (but not zero) and thus allows a very large number majority carriers to cross the junction and hence a large current flows as shown in Fig. 25.7. The forward characteristic curve is initially a non-linear while it is fairly linear afterwards. With sufficient bais, the height of the barrier almost disappears and the flow of current is predominated by only the acceleration of the carriers by the field. Initially, applied e.m.f. has to overcome the build help in potential barrier and hence current flowing through the diode is zero. The minimum external voltage needed for the diode to start conducting is called cut-in voltage. It is 0.3V for germanium and 0.7V for silicon junction. Fig. 25.6 alow shows forward biased junction using a symbol of diode.

REVERSE BIASED JUNCTION

If the negative terminal of an external source of e.m.f. is connected to the p-region and potive terminal to the n-region, as shown in Fig. 25.8 (a), we get a reverse biased junction. The external battery is in parallel with and aiding the potential barrier across the junction. The net effect, therefore, is a widening of the depletion region. No majority charge carrier is able to overcome such a huge potential barrier. This almost stops the flow of majority carriers across the junction.

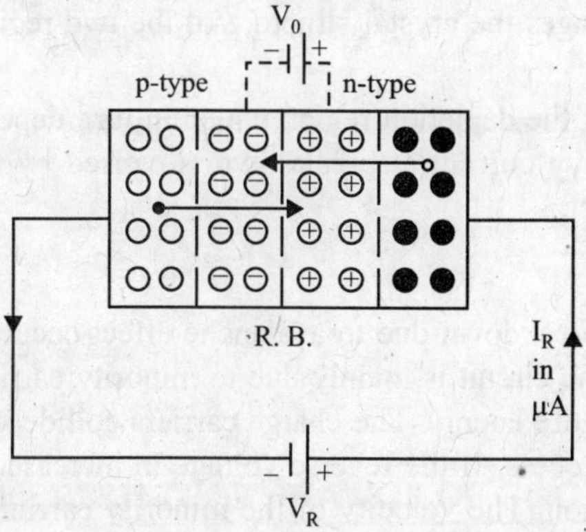

Fig 25.8 (a): Reverse Biased Junction

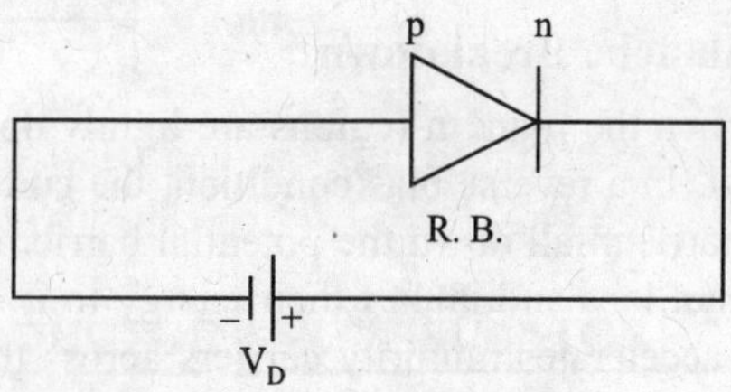

Fig. 25.8. (b) Reverse Biased Junction using symbol of diode

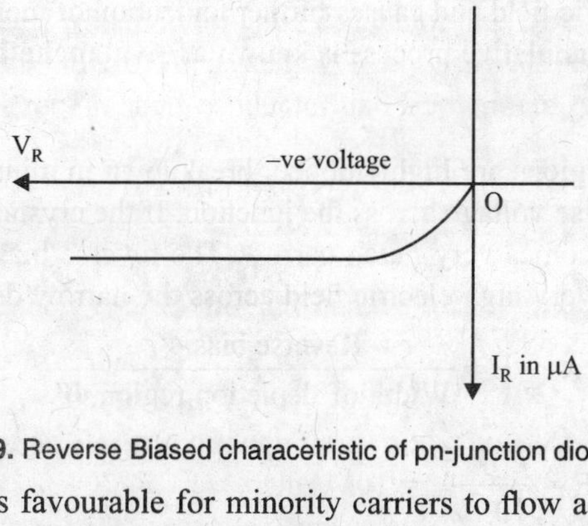

Fig. 25.9. Reverse Biased characetristic of pn-junction diode

However, the situation is favourable for minority carriers to flow across the junction. Thus, the current, across the junction in reverse bias condition, is due to only minority carriers crossing the junction per unit time and naturally is very small (a few micro amps). The minority carriers are due to the breaking of covalent bonds. Hence increase in the temperature of crystal produces more minority carriers and an increase in reverse current. The reverse bias characteristic is shown by Fig. 25.9.

The reverse current is given by

$$I_R = I(\text{drift}) - I(\text{diffusion})$$
$$= (-I_0) - I_0 e^{-eV/kT}$$

The reverse current increases repidly in the initial stage due to an expontial decrease of diffusion current with increasing reverse voltage. At higher V_R values, the diffusion current drops almost to zero and reverse current (I_R) becomes equal to the number of minority due to minority carriers. As the number of minority carriers is constant, at a given temperature and does not change with the bias voltage, the reverse current stays constant.

25.6 REVERSE BREAKDOWN

An ordinary pn junction diode does not conduct when it is reverse biased. If the reverse bias is increased, at a particular point the junction breaks down and starts conducting heavily. This, critical value of the voltage is called the **breakdown voltage** of the junction. Once the breakdown occurs, a very small increase in voltage causes a large change in the reverse current. The resistance offered by the junction will be nearly zero at this point. In normal operation the condition of breakdown should be avoided as it permanently damages the crystal structure of the two regions and makes the junction useless.

The breakdown voltage depends on the width of the depletion region which in turn depends upon the doping levels of p-and n-regions. The heavy current at breakdown is caused by two mechanisms : (i) Avalanche effect and (ii) Zener effect.

(*i*) Avalanche Breakdown

When the p-and n-regions are lightly doped, a breakdown due to avalanche effect occurs in a crystal. In a reverse bias condition, the current in the circuit is mainly due to minority carriers. These carriers fall down the potential barrier and acquire energy. The charge carriers collide with the crystal ions and impart their energy to it in the process. If the reverse voltage in increased, it further accelerates minority carriers across the junction. The velocity of the minority carriers is proportional to the applied bias voltage. The accelerated minority carriers acquire sufficient kinetic energy to knock off electrons by disrupting the covalent bonds during collision in the depletion region. This breakdown creates an electron-hole pair, which acquire sufficient energy under the influence of external, electric field and causes further ionization of more and more atoms, resulting in a chain reaction. This cumulative process is known as **Avalanche breakdown.**

(*ii*) Zener Breakdown

When the p- and n-regions are highly doped, breakdown in a junction diode occurs even by applying a very small reverse voltage across the junction. If the crystal is heavily doped, the width of the depletion region becomes very much narrow. The reverse bias voltage (even small of the order of 1 Volt), sets up a very high electric field across the narrow depletion layer. It is given by

$$E = \frac{\text{Reverse bias, } V_R}{\text{Width of depletion region, } W}$$

If $V_R = 1$ Volt, then $E = \frac{1V}{10^{-7}\,m} = 10^7$ V/m

This shows that, a small reverse bias can produce a strong electric field in the depletion region. This electric field $\left(\cong 10^7 V/m\right)$ is strong enough to cause rupture of covalent bands of atoms. Under the influence of external voltage, the strong electric field E generates large number of electron-hole pairs leading to a sharp increase in the reverse current, such a breakdown is known as ***Zener breakdown***. This breakdown does not involve collisions of carriers with the ions of the crystal, as in the case of Avalanche breakdown.

25.7 ZENER DIODE

Zener diode are more heavily doped than ordinary diodes and they have narrower depletion regions. Zener diodes are designed so as to exhibit breakdown at a very low voltage, say –5V. The breakdown is not due to Avalanche effect, but due to Zener effect, which does not involve ionization by collision.

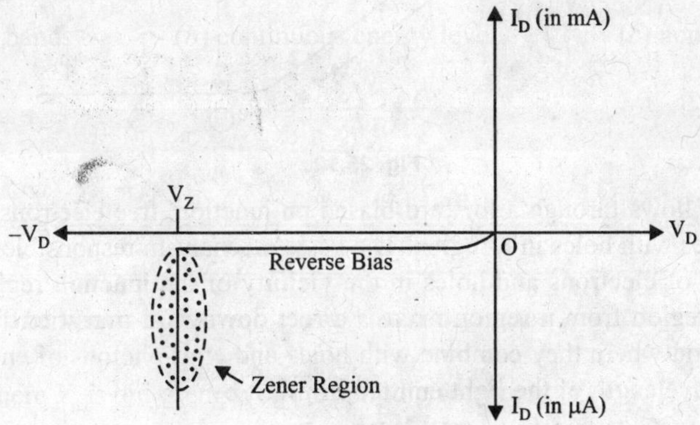

Fig. 25.10

Zener breakdown is due to the rupture of covalent bonds due to strong electric field and spontaneous generation of hole-electron pairs in the depletion region take place. The vertial characteristic in the breakdown region means the voltage (V_Z) across the diode remains constant in that region, independent of the current (reverse) flowing through it. This property is useful in many applications where the zener diode serves as a **voltage reference**, similar to an ideal voltage source. Fig. 25.10 shows the I-V characteristic of a zener diode. The symbols and the required polarity of the applied voltage of semiconductor diode and zener diode is shown in Fig. 25.11

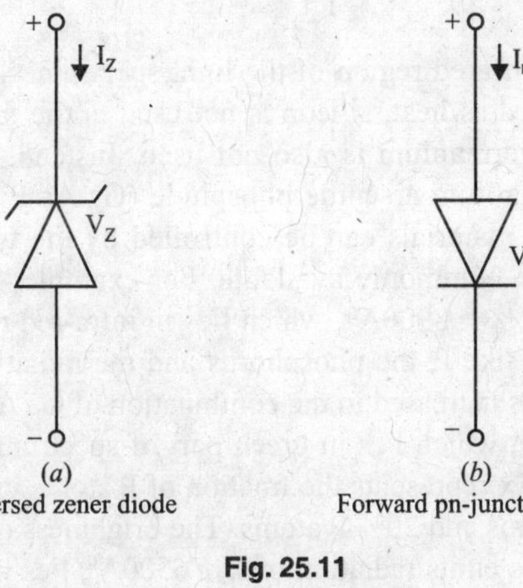

(*a*) Reversed zener diode (*b*) Forward pn-junction diode

Fig. 25.11

For a semiconductor diode, the 'ON' state will support a current in the direction of arrow, and the diode is in forward bias. On the contrary, the same polarity and the same direction of conduction current I_Z, the zener is connected in reverse bias.

25.8 LIGHT EMITTING DIODE (LED)

A semiconductor pn junction device that gives-off light when it is forward biased, is called a light emitting diode or simply LED.

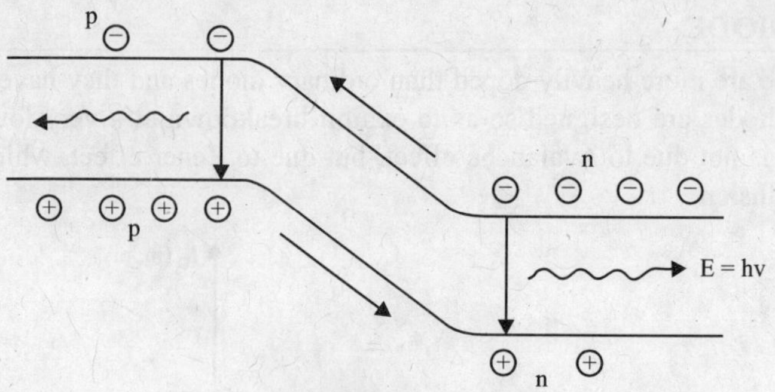

Fig. 25.12

When a current flows through a forward-biased pn junction, free electrons cross from the n-region and recombines with holes in p-region. The basic mechanism responsible for light emission is the recombination of electrons and holes in the vicinity of the junction region. The electrons injected in to the p-region from n-region make a direct downward transition from a conduction band into valence band where they combine with holes and emit photons of energy Eg as shown in Fig. 25.12. The wavelength of the light emitted in given by

$$hv = (E_C - E_V) = E_g$$

$$\frac{hc}{\lambda} = E_g$$

$$\therefore \quad \lambda = \frac{hc}{E_g} = \frac{1.24}{E_g} \mu m$$

This luminescence mechanism is known as injection electroluminiscence. For silicon, $E_g = 1.1$ eV

and
$$\lambda = \frac{1.24 \times 10^{-6}\, m}{1.1} = 1.13 \times 10^{-6} \text{m}$$

This wavelength is in the infrared region of the light spectrum. Since infrared is not visible and most of the energy is released as heat, silicon is not used in the fabrication of light emitting diodes. For the same reasons, germanium is also not used. Instead, gallium arsenide (GaAs), gallium phosphide (Ga P) and gallium arsenide phosphide (Ga As P) are commonly used. The colour of light emitted by these materials can be controlled by the type and degree of doping. Red, green and yellow LEDs are commonly available. For example, for Ga As, the energy gap Eg = 1.4eV which corresponds to $\lambda = 8900 A^0$, which lies in infra-red region. The energy gap can be increased by adding impurity like P, the phosphorus and the radiation in the visible region is obtained. As the percentage of P is increased in the combination of Ga As and Ga P, the energy gap increases from 1.4 eV to 2.26 eV, which lies in green part of spectrum. The resulting material is represented by Ga As_{1-x} P_x where x represents the fraction of P atoms viz. a crystal represented by Ga $As_{0.8}$ $P_{0.2}$ contains 80% As atoms and 20% P atoms. The brightness of such diodes is maximum at x = 0.4, when Eg = 1.9eV. This emits radiation of $\lambda = 6500 A^0$, lies in red part of the spectrum. By changing x, it is possible to obtain orange, yellow and green emission.

USES :

(*i*) LEDs are used in the manufacture of signal lamps and indicators.

(*ii*) The infra-red LED is used in optical fibre communication.

(*iii*) The light emitted by LED can be coherent under certain conditions. Such a LED can be used as a laser.

(*iv*) They may be arranged in the form of a seven-segment display to display decimal numbers 0 to 9 or/and alphabetical character in the form of 5 × 7 matrix to generate alphanumeric display.

25.9 METAL-SEMICONDUCTOR JUNCTION : SCHOTTKY DIODE

The metal-semiconductor junction differs from the conventional semiconductor diode in that the junction consists of a metal and semiconductor rather than usual two different semiconductors. The metal-semiconductor junction provides better performance at higher frequencies for such applications as detectors, mixers, switches, discriminators, limiters, logarithmic converters, gating circuits, sampling circuits and clamping circuits.

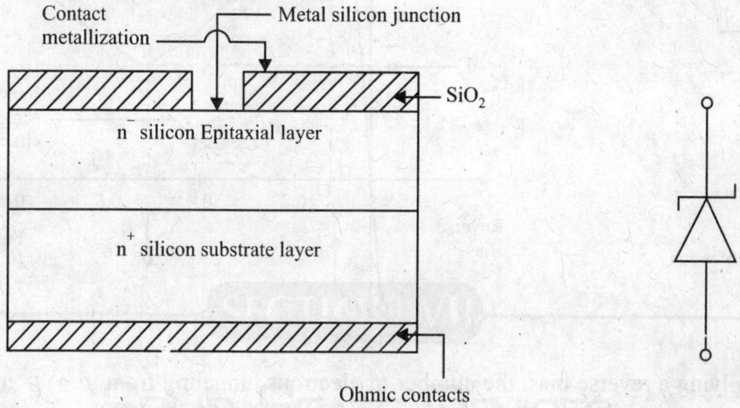

(a) Schottky diode (b) Symbol

Fig. 25.13

When metal-semiconductor junction is forward biased, electrons in the conduction band of the semiconductor gain sufficient energy to overcome junction barrier and are injected into the metal. These electrons have energy substantially above the energy of the metal's free electrons, that is why the injected electrons are labelled 'hot' carriers. In the metal, the hot electrons give up their excess energy very quickly (10^{-15}s) after which they become a part of the sea of free electrons in the metal.

Reverse bias increases the potential barrier, leaving the semiconductor electrons with sufficient energy to be injected into the metal.

25.10 TUNNEL DIODE

The tunnel diode was first introduced by Leo Esaki in 1958 who was awarded Nobel Prize in 1973 for this work.

A tunnel diode is a special type of pn-junction between two different semiconductors operated in a region of negative resistance of its I-V characteristic. Both the semiconductors of pn-junction are highly doped and hence the depletion region is greately reduced of the order of 10^{-6} cm i.e. about 1:100 the width of depletion region for a general purpose smiconductor diode. It is this depletion region that many carriers can **tunnel** through at low forward-bias potentials that accounts for the peak (V_P) in Fig. 25.14 for a comparison purpose, a typical semiconductor diode characteristic has been superimposed on the tunnel diode characteristic of Fig. 25.14.

Due to heavy doping the donor states form a band which may overlap the bottom of the conduction band so that the Fermi-level E_{Fn} lies within the conduction band, instead of lying within the forbidden band, as shown in Fig. 25.15. In the equilibrium condition, the p-side Fermi-level E_{Fp} lies below the valence band edge E_V and on the n-side, Fermi-level E_{Fn} lies above the conduction band edge E_C, although both E_{Fp} and E_{Fn} have the same value.

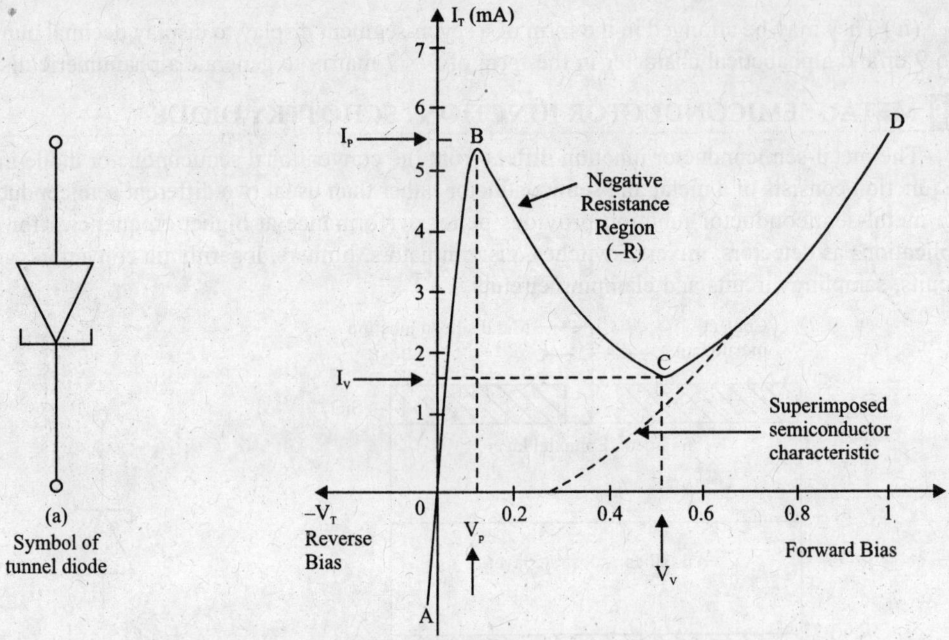

Fig. 25.14

On applying a reverse bias, the number of electrons tunneling from $p \rightarrow n$ increases. This is equivalent to an increasing conventional current from n to p (portion OA).

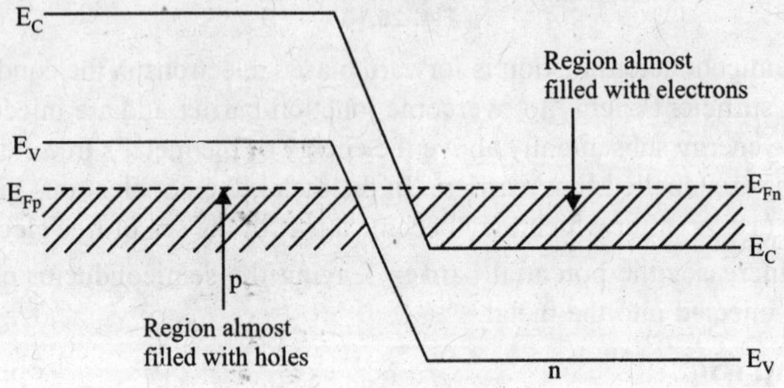

Fig. 25.15. Fermi-level in equilibrium of Tunnel diode.

For small forward bias applied E_{Fn} moves up with respect to E_{Fp}. Thus, electron tunneling takes place now from n to p (portion OB). However, as E_{Fn} continues to move up with respect to E_{Fp}, a point (B) is reached at V_P, at which the bands begin to pass by each other, resulting in a decrease in tunneling and corresponding decrease in current (portion BC) of I-V curve which represent a **negative-resistance region**.

On increasing forward bias further beyond V_V, the current begins to increase, similar to conventional diode.

Applications : The negative resistance of a tunnel diode is used in the design of oscillators, switching network, pulse generators and amplifiers. The tunneling process takes place very fast, so a tunnel diode can work at very high frequencies of the order of 10 GHZ.

25.11 SOLAR CELL

A solar cell is basically a pn junction diode that converts sun light directly into electric current. These are used extensively in satellites, space vehicles etc. as important long duration power supply source.

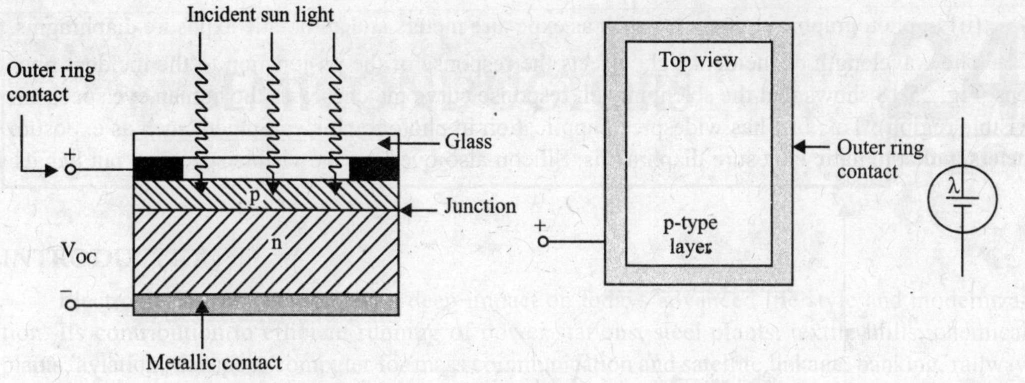

(a) Cross-section (b) Top view (c) Symbols

Fig. 25.16. Solar Cell.

The schematic diagram of silicon pn-junction solar cell is shown in Fig. 25.16. Fig. 25.16 (a) shows the cross-section while Fig. 25.16 (b) shows the top view of solar cell. Every effort is made to ensure that the surface area perpendicular to the sun is a maximum. The metallic conductor connected to the p-type material and thickness of p-type material are such that maximum number of photons of light-energy reaches the junction. A photon of light energy in this region many collide with a valence electron and impart to it sufficient energy to leave the parent atom. The result is a generation of free electrons and holes. This phenomenon will occur on each side of the junction. In the p-type material, the newly generated electrons are minority carriers and will move freely across the pn-junction with no applied bias. A similar discussion is true for the holes generated in the n-type material. This results in an increase of reverse current. Under open-circuit condition ($i_d = 0$), the photon voltage V_{OC} will result. Selenium and silicon are the most widely used materials for solar cells, although gallium arsenide, indium arsenide and cadmium sulfide are also used.

Applications :

(i) In solar power applications, solar cells are connected in series/parallel arrays. Parallel connections increase the total current that can be supplied by the array and series connections increase the voltage as shown in Fig. 25.17.

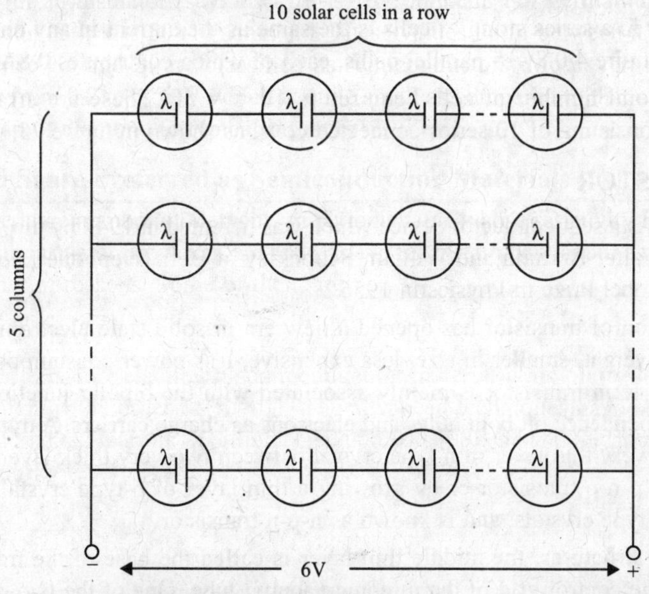

Fig. 25.17. A pannel of solar cells.

(ii) In photographic equipments such as exposure meters and automatic exposure diaphragms.

The wavelength of incident light affects the response of the pn-junction to the incident photons. Fig. 25.18 shows that the selenium cell response curve matches with the human eye spectrum (visible region). This fact has widespread application in photographic equipment such as exposure meters and automatic exposure diaphragms. Silicon also overlaps the visible spectrum but has its

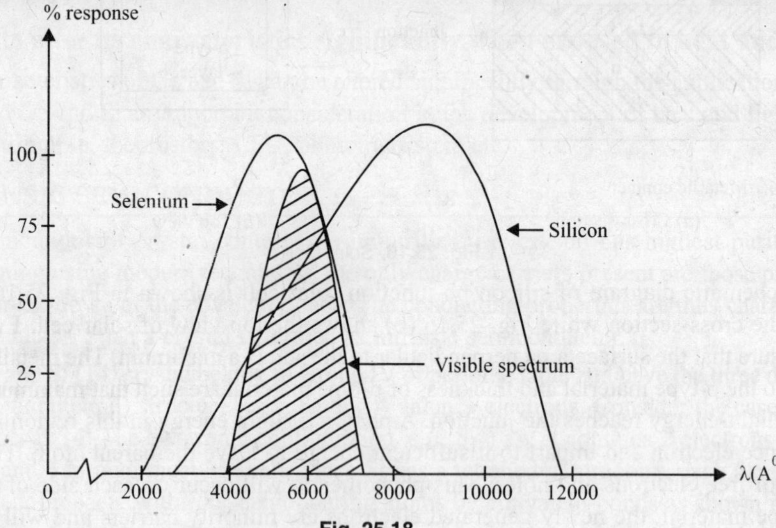

Fig. 25.18

peak at 8000A^0, which is in the infra-red region. Both materials have excellent temperature characteristics. Hence, they can withstand at extreme high or low temperatures without any significant drop-off in efficiency.

Example 25.2 : *An array of solar cells is to be constructed so it will deliver 4A at 6V under a certain light intensity. If each cell delivers 0.8A at 0.6V under the given intensity, how many cells are required and how should they be connected ?*

Solution : To obtain 6V output from cells that produce 0.6V each, we require 6/0.6 = 10 series-connected cells.

The current in a series string of cells is the same as the current in any one cell i.e. 0.8A . To obtain 4A, we require 4/0.8 = 5 parallel paths, each of which contributes 0.8A.

Thus, the total number of cells required is 10×5 = 50. These are arranged in 5 parallel branches, each consisting of 10 series-connected cells, as shown in figure 25.17.

25.12 TRANSISTOR

A transistor is a semiconductor device which was invented in 1947 by the American physicists John Bardeen, Walter Brattain and William Schockley at Bell Telephone Laboratories and were honoured with Nobel Prize in Physics in 1956.

The invention of transistor has opened a new era in solid state electronics with electronics devices of light weight, smaller in size, less expensive, less power consumption to operation and many more. The term transistor is mainly associated with the bipolar junction transistor (BJT), because of its dependence on both holes and electrons as charge carriers. A transistor is formed by growing either a very thin layer of n-type crystal between two very thick layers of p-type crystals and is known as p-n-p transistor or by growing a thin layer of p-type crystal between two very thick layers of n-type crystals, and is known as n-p-n transistor.

In both the structures, the middle thin layer is called the base of the transistor and can be compared with the control grid of therminonic vacuum tube. One of the two outer thick layers is called **emitter**, often compared with the cathode of a vacuum tube. The emitter is highly doped and is the source of majority carriers i.e. holes in case of p-n-p transitor and electrons in case of n-p-n transitor. The another thick layer is known as the **collector**, often compared with the place of a vacuum tube.

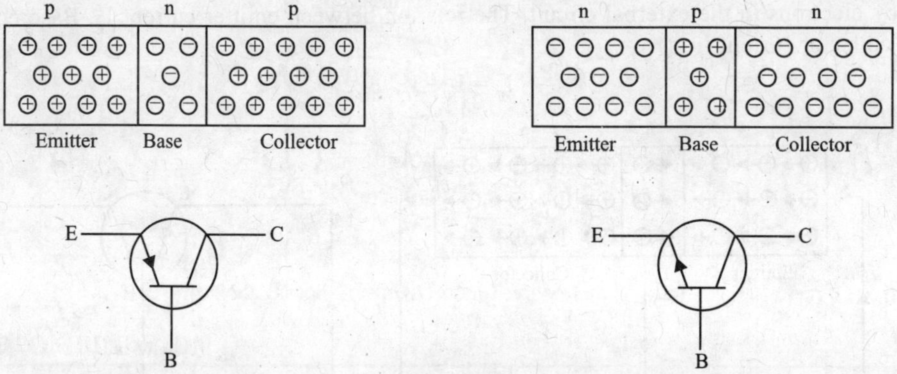

(a) p-n-p transistor (a) n-p-n transistor

Fig. 25.19

The collector is comparatively less doped and its function is to collect the charge carriers. Moreover, the length of the collector is greater than that of the emitter and hence the emitter and collector can not be interchanged. The two types of transistors p-n-p and n-p-n are symbolically shown in Fig. 25.19 (a) and (b) respectively. The junction between the emitter and the base is called the Emitter-Base junction and the junction between the collector and the base is called collector-base junction. The entire system is sealed against moisture inside a metal or plastic case with metal leads comming out for connection to emitter, base and collector respectively. The n-region contains free electrons which are negative charge carriers and p-region contains mobile holes which are positive charge carriers. Thus, two types of charge carriers, namely electrons and holes are involved in current flow through n-p-n or p-n-p transistor. Therefore, the transistor is known as bipolar junction transistor (BJT).

Origin of the Term 'Transistor'

A transistor consists of two pn-junctions. One junction Emitter-Base is always forward biased and the other Collector-Base is reverse biased. The forward biased junction has a low resistance path, whereas, the reverse biased junction has a very high resistance path. A weak signal is introduced in the low resistance circuit as input and the output is taken from the high resistance circuit. Therefore, a transistor transfers a signal from a low resistance to a high resistance. The prefix 'trans' means the signal transfer property of device while 'istor' classifies it as a solid state element in the same general family with resistors. So a transfer of resistor is a transistor.

25.13 WORKING OF P-N-P TRANSISTOR

To study the working of a p-n-p transistor, the emitter-base junction is forward biased with a small voltage V_{EB} while collector-base junction is reverse biased with relatively large voltage V_{CB} as shown in Fig. 25.20.

The emitter-base junction being forward biased, the holes in p-region and electrons in n-region are forced towards the junction between emitter and base. In other words, the holes in the emitter and the electrons in the base are repelled by the positive and negative terminals respectively of the battery V_{CB}. Out of total holes reaching in base region, a small fraction about 5% recombines with the electrons in base region and gets neutralised. As the base region is very thin and collector is given a high negative potential, 95% of the holes are able to reach to the collector, where they are collected. For every hole reaching the collector, and electron is released from the negative of the collector-base battery V_{CB} and neutralises the hole. Also for each hole which is lost in the collector region, a covalent bond near the emitter breaks down and an electron is liberated which enters the positive terminal of V_{EB}. Thus, we see that the current is carried by the holes in side the crystal and by electrons in the external circuit. The relation between emitter current I_E, Base current I_B and collector current I_C is

$$I_E = I_B + I_C.$$

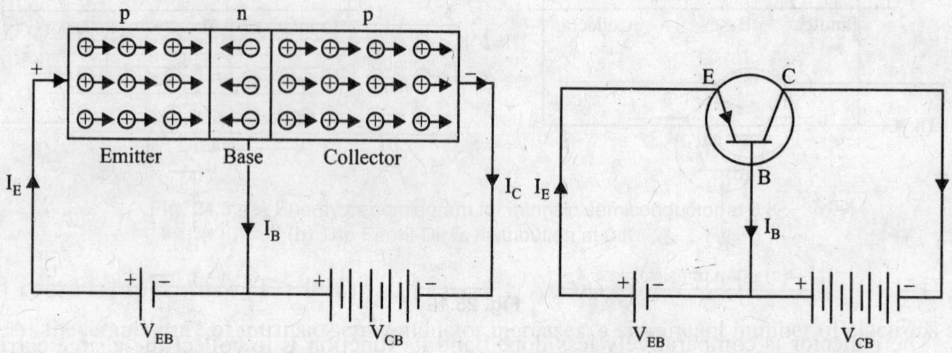

(a) Biasing of p-n-p transistor (b) Biasing using symbol of p-n-p transistor

Fig. 25.20

25.14 WORKING OF N-P-N TRANSISTOR

In this case also, the emitter-base junction is forward biased while the collector-base circuit is reverse biased. The holes in the base being repelled by positive terminal of the forward bias and electrons in the emitter being repelled by the negative terminal move towards the emitter-base junction. On reaching the junction, a small fraction of about 5% of total number of electrons entering p-region, combine with the holes at the base to get neutralised. As the base is very thin and the collector is given a very high positive potential, nearly 95% of electrons will rush towards the collector and these electrons are collected at collector. These electrons collected by the collector flow towards the positive terminal of battery V_{CB}. The defficiency of the electrons in the emitter is made up by the negative of the battery V_{EB}. In the base circuit the current is

$$I_B = I_E - I_C \quad \text{or} \quad I_E = I_B + I_C$$

25.15 DIFFERENT TECHNIQUES FOR FABRICATION OF BJT

The following four techniques are generally employed for the fabrication of bipolar junction transistor (BJT).

***(i)* Grown Technique:** In this technique, a single crystal is grown from a melt of silicon or germanium consisting of say, n-type impurity. During the growth process, the type of impurity of a melt is changed by adding p-type impurity. After this gain n-type impurity is added to form a n-p-n trainsistor. In the same manner, p-n-p transistor can be constructed.

***(ii)* Alloy or Fused Technique:** In this technique, a small dot of acceptor element say indium (In) is placed on each side of a thin wafer of n-type semiconductor. The combination is heated to a high temperature for a short time. During the process, indium dissolves and penetrates into the semiconductor. This structure on cooling gives a p-n-p transistor. Similarly n-p-n transistor can be fabricated.

***(iii)* Diffusion Technique:** In this technique, we employ a batch process and large number of transistors can be fabricated in a single operation. An extrinsic semiconductor, say n-type silicon, serving as collector is heated in a furnance containing p-type impurity. The impurity, being in gaseous form diffuses in to one surface of the semiconductor forming a p-region and work as base of the transistor. This part is covered with a mask containing a large number of apertures and heated again in a furnance containing n-type impurity in gaseous form. This forms a n-type layer over a p-type layer. This layer of n-type works as emitter of the transistor. Thus, n-p-n-type transistor is fabricated. Similarly p-n-p transistor can be fabricated.

***(iv)* Epitaxial Technique:** The word 'epi' means 'on' and 'taxis' means arrangement (both Greek words). Thus, epitaxial means growing a thin crystal layer on a crystalline substrate where the atoms of the growing layer become a copy of the arrangement of the substrate. In this technique, thin n or p-type layers of a semiconductor are developed on a heavily doped semiconducting material known as substrate. The layer so form may serve as emitter, base and collector. This advanced technique enables us to grow very thin layers of the order of 10^{-8} m in thickness.

EXERCISE

1. What is pn-junction diode ? Name different type of diodes as per their utilities.
2. Explain in brief, the different methods employed in the fabrication of pn-junction diode.
3. Explain the terms :
   (*i*) Potential Barrier (*ii*) Depletion region.
4. Explain the mechanism of drifting of the minority charge carriers in a pn-junction diode. How much is the net drift current through the junction, when it is unbiased?
   Explain the equilibrium condition in an unbiased pn-junction.
5. Draw a neat energy band diagram of a pn-junction at equilibrium and explain the formation of a conduction hill and a valence hill.
6. Explain the mechanism of current flow in forward and reverse biased pn-junction. Draw the I-V characteristics of pn-junction diode under both modes.
7. What is reverse breakdown ? Explain the mechanism in details of
   (*i*) Avalanche breakdown (*ii*) Zener breakdown
8. Explain how zener diode is different from a ordinary pn-junction diode. Explain clearly the function of zener diode with the help of its I-V characteristics. ***(Nagpur U. 2009)***
9. Explain the luminiscene mechanism in a light emitting diode (LED). How different colour emitting LEDs are constructed by proper percentage of dopant concentration ? ***(Nagpur U. 2011)***
10. Explain the fabrication of metal-semiconductor junction diode. Give its name and symbol.
11. Explain the working of tunnel diode with its I-V characteristics. What is the importance of negative resistance region ? Draw its energy band diagram showing clearly the Fermi-level in equilibrium. ***(Nagpur U. 2012)***
12. Describe the mechanism of solar cell. Give its applications.
13. Explain the importance of a semiconductor device BJT. Why it is called transistor ?
14. Explain the working of a p-n-p transistor with necessary diagram.
15. Explain the working of a n-p-n transistor with necessary diagram. ***(Nagpur U. 2009, 2012)***
16. Give different techniques employed for the fabrication of BJT ?
17. Explain the effect of forward and reverse biasing on the width of depletion region. ***(Nagpur U. 2012)***
18. Explain why (*i*) the emitter is heavily doped and (*ii*) the base is thin and lightly doped in a transistor. ***(Nagpur U. 2011)***
19. Describe how zener diode can be used as a voltage regulator. ***(Nagpur U. 2009)***

SECTION VIII

ELECTRONICS

26. POWER SUPPLY
27. CHARACTERISTICS OF TRANSISTORS
28. FIELD EFFECT TRANSISTORS
29. SMALL SIGNAL AMPLIFIERS
30. HYBRID EQUIVALENT CIRCUITS AND NOISE

POWER SUPPLY

26
Chapter

INTRODUCTION

Electricity is almost universally generated and transmitted in A.C. form due to the fact that alternating currents and voltages permit the use of transformers. A desired level of D.C. (direct current) voltage for a particular application, is obtained from a.c. by using either a step-up or step-down transformer. All electronic devices need D.C. voltages for their biasing. The D.C. voltages can be obtained from batteries but this is generally incovenient. Thus, a circuit must be designed to convert an alternating voltage (A.C.) to a direct voltage (D.C.) of the desired value. Such a circuit is called the power supply. The A.C. input power is first rectified to provide a unidirectional D.C. and then filtered to produce a smooth D.C. voltage. Finally, the D.C. voltage is regulated to maintain a constant output level inspite of fluctuations in power line voltage, variation of load or change in temperature.

26.1 DIODE AS A CIRCUIT ELEMENT

A pn-junction diode has an unparallel importance as rectifier in the fabrication of power supplies. Its unidirectional current flow property is an unique feature. 1N4001 and BY127 diodes are commonly used as 'rectifiers'. A modified version zener diode is used in power supplies to get an almost constant and fixed level output D.C. voltage. Hence, diode has become an essential element of an electronic circuits used in the fabrication of power supply.

Ideal Power Supply

The ideal power supply has three essential characteristics :

(1) It will supply a pure D.C. voltage, that does not vary with time. That is, there is no alternating component of voltage, called ripple, riding on the D.C. voltage.

(2) The magnitude of pure D.C. voltage does not change, no matter how much current or power is drawn from it and no matter how the 230V line voltage might change.

(3) It does not generate any internal power that dissipates useless energy in itself.

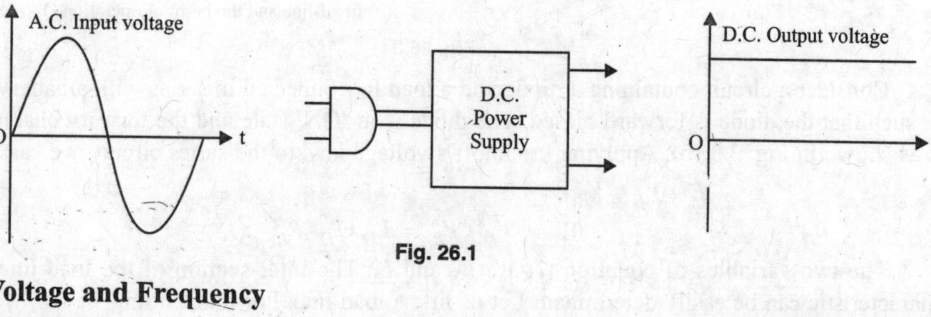

Fig. 26.1

Line Voltage and Frequency

The mains voltage is 230V A.C. rms at a frequency of 50 Hz. This is the value measured with an A.C. voltmeter. The sinusoidal waveform can be observed with an oscilloscope. The peak value of 230V A.C. voltage will be $V_p = \sqrt{2}\, V_{rms} = 1.414 \times 230V = 263V$. The period of A.C. mains = 1/50 sec or 20 ms. The first half cycle from 0 to 10 ms is called the positive half cycle and second half cycle from 10 ms to 20 ms is the negative half cycle.

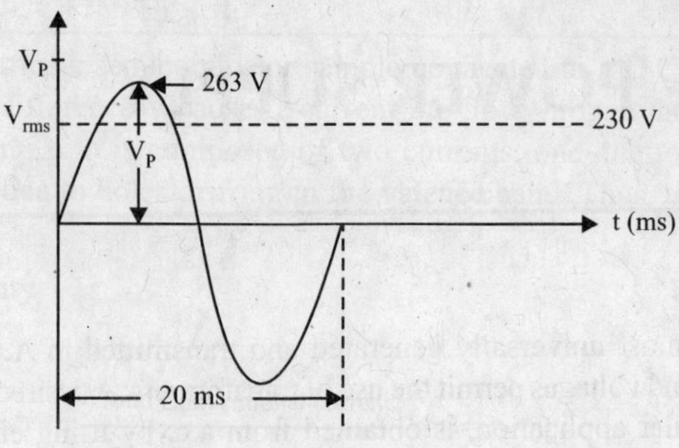

Fig. 26.2

26.2 LOAD-LINE CONCEPT

The applied load R_L has an important impact on the point of operation of a device. If the analysis is performed in a graphical manner, a line can be drawn on the characteristic of the device that represents the applied load. The intersection of the load-line with the characteristic will determine the point of operation of the system. Such an analysis is called load-line analysis.

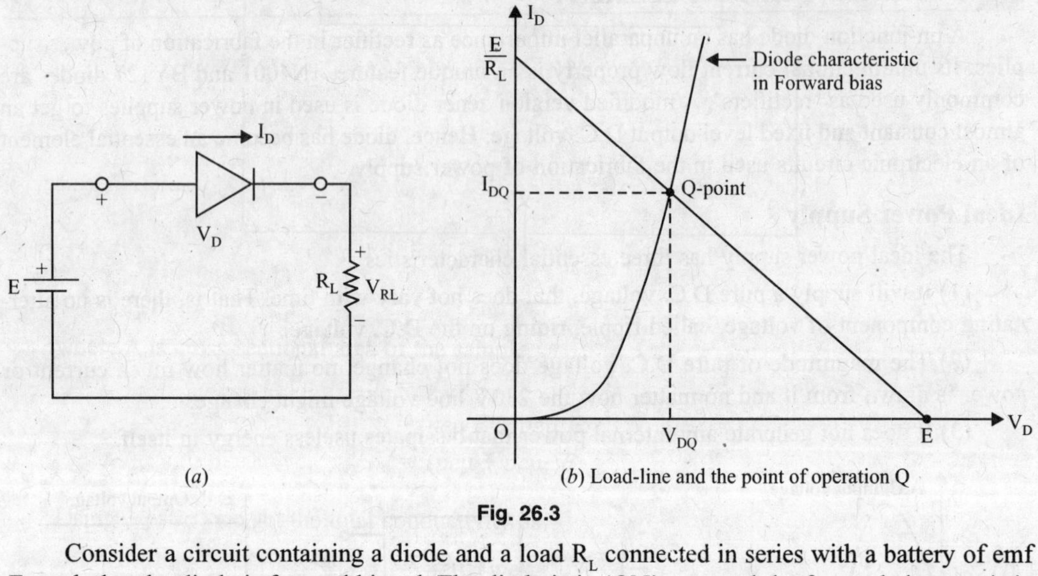

Fig. 26.3

Consider a circuit containing a diode and a load R_L connected in series with a battery of emf E, such that the diode is forward biased. The diode is in 'ON' state and the forward characteristic is as shown in Fig. 3.3 (b). Applying krichhoff's voltage law, to the series circuit, we can write

$$E - V_D - V_{R_L} = 0$$

or $$E = V_D + V_{R_L} = V_D + I_D \cdot R_L \quad \text{... (i)}$$

The two variables of equation (i) are V_D and I_D. The inter-section of the load line on the characteristic can be easily determined. Let us first a load line. For this, putting $V_D = 0$, we get an intercept on I_D axis as

$$E = 0 + I_D R_L$$

or $$I_D = \left.\frac{E}{R_L}\right|_{V_D = 0V} \quad \text{... (ii)}$$

and if we set $I_D = 0$, we get an intercept on V_D axis as

$$E = V_D + I_D R_L$$

$$= V_D + 0 = V_D$$

$$\therefore \quad V_D = E\big|_{I_D = 0A} \qquad \text{... (iii)}$$

Joining these two points given by equations (ii) and (iii) on I_D and V_D axes respectively, we get a straight line, known as load line. By changing the load R_L, the slope of load line can be changed. Hence, the point of intersection Q of load line and forward characteristic of the diode can be changed. The point of intersection Q is called as 'operating point' or 'quiescent point' (abbreviated as 'Q' point).

26.3 HALF-WAVE DIODE RECTIFIER

The simplest way to produce D.C. voltage is to eliminate the negative half-cycle of the sine-wave. Initially, 230 V mains A.C. voltage has been reduced or lowered down to a desired value by

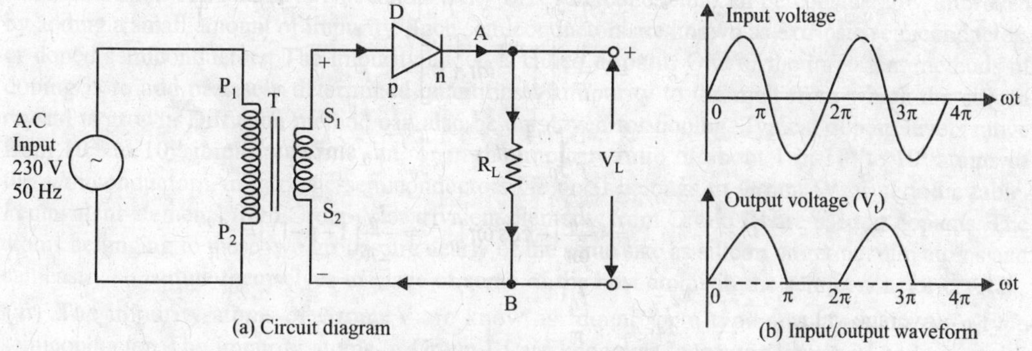

Fig. 26.4. Half-wave diode rectifier

using a step-down transformer T. The secondary A.C. voltage is now converted in D.C. voltage by using the semiconductor diode D, say 1N4001 between the transformer secondary and the load R_L as shown in Fig. 26.4 (a). The D.C. output is drawn across the Load.

Working Principle

During the first half of the A.C. cycle, let us suppose that the end S_1 of secondary is positive with respect to S_2, the diode D is forward biased. A large conventional current flows through load R_L from A to B. During the second half of the A.C. cycle, the polarity is reversed. Now terminal S_1 becomes negative with respect to S_2, the diode D is reverse biased, hence does not conduct. Naturally no current flows through the load resistance R_L. In other words, the output voltage remains zero. This type of rectifier is known as half-wave rectifier as the output is obtained during only half of the input cycle. The waveforms at the input as well as output are shown in Fig. 26.4 (b). Thus, the output voltage obtained across the load R_L is unidirectional, intermittent, consisting of only positive half cycle.

26.4 ANALYSIS OF HALF-WAVE RECTIFIER

Consider a circuit diagram of a half-wave rectifier as shown by Fig. 26.4 (a). If $e = E_0 \sin \omega t$ is the instantaneous value of A.C. voltage applied to diode, which is acting as half-wave rectifier, then output current through load R_L will be

$$i = \frac{e}{r + R_L} = \frac{E_0 \sin\omega t}{r + R_L} = \frac{E_0}{r + R_L} \sin \omega t = t = I_0 \sin \omega t$$

where r is the internal resistance of the diode and $I_0 = \dfrac{E_0}{r + R_L}$ is maximum value of the current.

For half-wave rectifier,

$$i = I_0 \sin \omega t \qquad \text{if } 0 \le \omega t \le \pi$$

or

$$i = 0 \qquad \text{if } \pi \le \omega t \le 2\pi$$

Average Value $I_{d.c.}$

The average value $I_{d.c.}$ of current over one complete cycle is that much direct current which send the same charge through the resistance as is sent by the varying current through the same resistance. This is current measured by D.C. ammeter connected in series with the load.

Total charge passing through load during one complete cycle

$$= \int_0^T i dt = I_{d.c.} T$$

$$\therefore \quad I_{d.c.} = \frac{1}{T}\int_0^T i dt = \frac{1}{T}\left[\int_0^{T/2} i dt + \int_{T/2}^T i dt\right]$$

$$= \frac{1}{T}\left[\int_0^{T/2} i dt + 0\right]$$

$$= \frac{1}{T}\int_0^{T/2} i dt = \frac{1}{T}\int_0^{T/2} I_0 \sin \omega t dt$$

$$= \frac{I_0}{\omega t}\left[-\cos \omega t\right]_0^{T/2} = \frac{I_0}{2\pi}\left[1-(-1)\right] = \frac{I_0}{\pi}$$

$$\therefore \text{ Average value } \quad I_{d.c.} = \frac{I_0}{\pi} \qquad \text{... (iv)}$$

Average D.C. voltage across $R_L = V_{d.c.} = I_{d.c.} \cdot R_L$

$$\therefore \quad V_{d.c.} = \frac{I_0}{\pi} \cdot R_L = \frac{E_0}{\pi(r+R_L)} \cdot R_L \qquad \text{... (v)}$$

Root Mean Square Value ($I_{r.m.s}$)

The square root of the average of the squares of the instantaneous values taken over the whole cycle. This is the current measured by A.C. ammeter connected in series with the load R_L.

$$I_{rms}^2 = \frac{1}{T}\int_0^T i^2 dt = \frac{1}{T}\left[\int_0^{T/2} i^2 dt + \int_{T/2}^T i^2 dt\right]$$

$$= \frac{1}{T}\int_0^{T/2} i^2 dt \quad \left[\because \text{ For half-wave rectifier } \int_{T/2}^T i^2 dt = 0\right]$$

$$= \frac{1}{T}\int_0^{T/2} I_0^2 \sin^2 \omega t dt = \frac{I_0^2}{2T}\int_0^{T/2} (1-\cos 2\omega t) dt$$

$$= \frac{I_0^2}{2T}\left[\int_0^{T/2} dt - \int_0^{T/2} \cos 2\omega t dt\right]$$

$$= \frac{I_0^2}{2T}\left[\{t\}_0^{T/2} - \left\{\frac{\sin 2\omega t}{2\omega}\right\}_0^{T/2}\right]$$

$$= \frac{I_0^2}{2T} \cdot \frac{T}{2} \qquad [\because \sin 2\omega T/2 = \sin \omega T = \sin 2\pi = 0]$$

$$= \frac{I_0^2}{4} \qquad \text{... (vi)}$$

$$\therefore \quad I_{rms} = \frac{I_0}{2}$$

$$\text{R.M.S. voltage across } R_L = I_{rms} \cdot R_L = \frac{I_0}{2} \cdot R_L$$

$$\therefore \quad V_{rms} = \frac{E_0}{2(r+R_L)} \cdot R_L \qquad \text{... (vii)}$$

Efficiency

The efficiency of a rectifier is defined as the ratio of D.C. output power $P_{d.c.}$ to the A.C. input power $P_{a.c.}$ supplied to the rectifier.

$$\text{D.C. output power} = P_{d.c.} = I_{d.c.}^2 R_L$$

$$= \frac{I_0^2}{\pi} R_L \qquad \text{... (viii)}$$

$$\text{A.C. input power} = P_{a.c.} = \frac{1}{T}\int_0^T eidt$$

$$= \frac{1}{T}\int_0^T i(r+R_L)idt \qquad [\because e = i(r+R_L)]$$

$$= \left(\frac{r+R_L}{T}\right)\int_0^T i^2 dt$$

$$= (r+R_L)\left[\frac{1}{T}\int_0^T I_0^2 \sin^2 \omega t dt\right]$$

$$= (r+R_L) I_{rms}^2$$

$$= \frac{I_0^2}{4}(r+R_L) \qquad (\because \text{Eqn. } vi) \qquad \text{... (ix)}$$

$$\text{Hence, Efficiency,} \quad \eta = \frac{P_{d.c.}}{P_{a.c.}} = \frac{I_0^2 R_L}{\pi} \times \frac{4}{I_0^2(r+R_L)}$$

$$= \frac{4}{\pi^2} \cdot \frac{R_L}{(r+R_L)} = \frac{4}{\pi^2} \cdot \frac{1}{(1+r/R_L)}$$

$$= \frac{0.406}{1+r/R_L} \qquad \text{... (x)}$$

$$\text{or} \quad \text{Percentage efficiency} = \frac{40.6}{1+\dfrac{r}{R_L}}$$

Smaller the value of $\frac{r}{R_L}$, the greater is the efficiency. Thus the internal diode resistance must be small and load R_L must be high to produce large efficiency of a half-wave rectifier.

Ripple Factor (γ)

The A.C. component contained in the output of rectifier is known as ripple and is defined as

$$\text{Ripple factor} = \frac{I_{a.c.}}{I_{d.c.}} \qquad \text{... (xi)}$$

The output contains an A.C. component as well as D.C. component. The I_{rms} is given by

$$I_{mns}^2 = I_{d.c.}^2 + I_{a.c.}^2$$

or

$$I_{a.c.} = \sqrt{I_{rms}^2 - I_{d.c.}^2}$$

Hence, Ripple factor,

$$\gamma = \frac{I_{a.c.}}{I_{d.c.}} = \sqrt{\frac{I_{rms}^2 - I_{d.c.}^2}{I_{d.c.}}}$$

$$= \left\{\left(\frac{I_{rms}}{I_{d.c.}}\right)^2 - 1\right\}^{1/2} = \left\{\left(\frac{I_0^2}{4} \times \frac{\pi}{I_0^2}\right) - 1\right\}^{1/2}$$

$$= \sqrt{\frac{\pi^2}{4} - 1} = \sqrt{1.4649}$$

$$= 1.21 \qquad \text{... (xii)}$$

Thus, ripple factor for half-wave rectifier is 1.21. Since A.C. component of output i.e. ripple part exceeds the $I_{d.c.}$, half-wave rectifier is not considered to be suitable for converting A.C. into D.C.

Peak Inverse Voltage

The Peak Inverse Voltage (P.I.V.) is defined as the maximum voltage across the diode when it is non-conducting.

This happens during the negative-half of the applied voltage. At this stage, voltage across the diode is equal to the applied voltage. The P.I.V. in a half-wave rectifier is E_0, the maximum value of applied A.C. voltage. Hence insulation strength of diode must be sufficiently high to withstand at this P.I.V.

26.5 FULL WAVE RECTIFIER

Fig.26.5 (a) shows the circuit diagram of a full wave rectifier. In this case, two pn junction diodes D_1 and D_2 are connected across the secondary S_1 S_2 of transformer. A load resistance R_L is connected between the centre-tap (C) of secondary and a common terminal of n-type of both D_1 and D_2 diodes. The output voltage (V_L) is drawn across the load R_L.

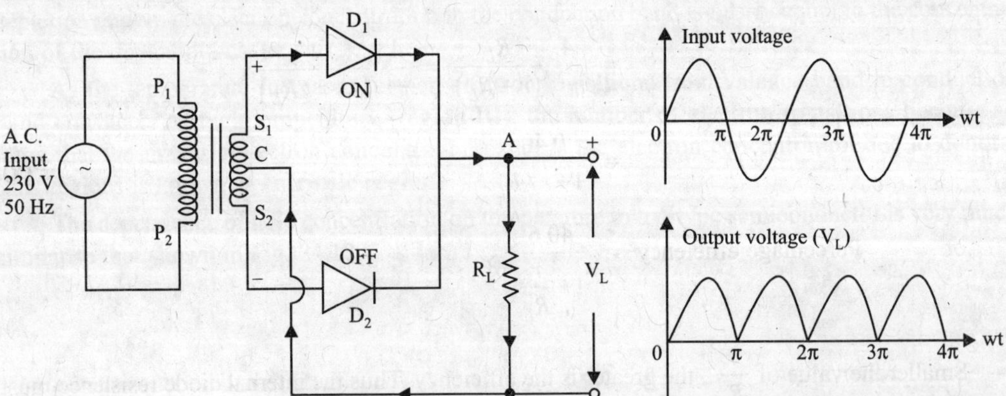

(a) Circuit diagram (b) Input/output waveforms

Fig 26.5. Full-wave rectifier

Working Principle

During the positive half of the input A.C. cycle, let one end, say S_1 is positive with respect to S_2. The junction diode D_1 becomes forward biased and hence it conducts. In other words, diode D_1 is 'ON'. A large conventional current flows along D_1ABCD_1 and produces a voltage drop V_L across load R_L. During this cycle, diode D_2 remains 'off' as it is reverse biased. During the negative half of the input A.C. cycle, now polarities of S_1 and S_2 are reversed. Terminal S_1 becomes negative and S_2 positive. Diode D_2 becomes forward biased and D_1 reverse biased and hence D_1 remains 'off'. During this negative half cycle diode D_2 conducts heavily i.e. D_2 is 'ON'. The direction of conventional current in D_2ABCD_2. During both half cycles, the load R_L receives the current in the same direction. Thus, the output voltage obtained across the load R_L consists of continuous positive pulses shown in Fig. 26.5 (b), thus giving full-wave rectification.

26.6 ANALYSIS OF FULL-WAVE RECTIFIER

Fig. 26.5 (a) shows a full wave rectifier, which uses two diodes D_1 and D_2. If $e_1 = E_0 \sin \omega t$ represents the A.C. supply voltage applied to diode D_1, then A.C. supply voltage applied to diode D_2 will be $e_2 = E_0 \sin(\omega t + \pi)$ i.e. opposite in phase. Therefore, during positive half cycle i_1 is the current due to conduction of D_1 (for D_2, $i_2 = 0$) and during negative half cycle i_2 is the current due to conduction of diode D_2 (for D_1, $i_1 = 0$). Therefore,

At any time, the total current through the load resistance (R_1) will be

$$i = i_1 + i_2 \quad \text{... (xiii)}$$

For the first half-cycle, i.e. $0 \le t \le T/2$ or $0 \le \omega t \le \pi$

$$i_1 = \frac{E_0 \sin \omega t}{r + R_L} \text{ and } i_2 = 0$$

$$i = i_1 + i_2 = \frac{E_0 \sin \omega t}{r + R_L} = I_0 \sin \omega t \quad \text{... (xiv)}$$

For the second half-cyle, i.e. $T/2 \le t \le T$ or $\pi \le \omega t \le 2\pi$

$$i_1 = 0 \text{ and } i_2 = \frac{E_0 \sin(\omega t + \pi)}{r + R_L} = -\frac{E_0 \sin \omega t}{r + R_L}$$

$$\therefore \quad i = i_1 + i_2 = -\frac{E_0 \sin \omega t}{r + R_L} = -I_0 \sin \omega t \quad \text{... (xv)}$$

Average Value $I_{d.c.}$

Total charge through load during one complete cycle

$$= \int_0^T i dt = I_{d.c.} T$$

$$\therefore \quad I_{d.c.} = \frac{1}{T}\int_0^T i dt$$

$$= \frac{1}{T}\left[\int_0^{T/2} i_1 dt + \int_{T/2}^T i_2 dt\right]$$

$$= \frac{I_0}{T}\left[\int_0^{T/2} \sin \omega t dt + \int_{T/2}^T (-\sin \omega t) dt\right]$$

$$= \frac{I_0}{T}\left[\left\{\frac{-\cos \omega t}{\omega}\right\}_0^{T/2} + \left\{\frac{\cos \omega t}{\omega}\right\}_{T/2}^T\right]$$

$$= \frac{I_0}{\omega T}[2+2] = \frac{4I_0}{2\pi} = \frac{2}{\pi}I_0$$

$\therefore$ Average value, $$I_{d.c.} = \frac{2}{\pi}I_0 \qquad \text{... (xvi)}$$

Average D.C. voltage across $R_L = V_{d.c.} = I_{d.c} \cdot R_L$

$$\therefore \quad V_{d.c} = \frac{2I_0}{\pi}R_L = \frac{2E_0}{\pi(r+R_L)}R_L \qquad \text{... (xvii)}$$

Root Mean Square value ($I_{r.m.s.}$)

In practice, this is the value measured by an A.C. ammeter, by connecting it in series with the load R_L. By defination,

$$I^2_{r.m.s.} = \frac{1}{T}\int_0^T i^2 dt$$

$$= \frac{1}{T}\left[\int_0^{T/2} i_1^2 dt + \int_{T/2}^T i_2^2 dt\right]$$

$$= \frac{I_0^2}{T}\left[\int_0^{T/2} \sin^2 \omega t dt + \int_{T/2}^T (-\sin \omega t)^2 dt\right]$$

$$= \frac{I_0^2}{T}\left[\int_0^{T/s} \sin^2 \omega t dt + \int_{T/2}^T \sin^2 \omega t dt\right]$$

$$= \frac{I_0^2}{T}\left[\frac{1}{2}\cdot\frac{T}{2} + \frac{1}{2}\cdot\frac{T}{2}\right] = \frac{I_0^2}{T}\left[\frac{T}{4}+\frac{T}{4}\right]$$

$$= \frac{I_0^2}{2}$$

$$\therefore \quad I_{r.m.s} = \frac{I_0}{\sqrt{2}} \qquad \text{... (xviii)}$$

R.M.S. voltage across $R_L = I_{r.m.s.} R_L$

$$= \frac{I_0}{\sqrt{2}}R_L = \frac{E_0 R_L}{\sqrt{2}(r+R_L)} \qquad \text{... (xix)}$$

Efficiency

The efficiency is given by

$$\eta = \frac{P_{d.c.}}{P_{a.c.}} = \frac{\text{D.C. output power}}{\text{A.C. input power}}$$

D.C. output power is given by

$$P_{d.c.} = I^2_{d.c.} R_L$$

$$= \frac{4I_0^2 R_L}{\pi^2} \qquad \text{... (xx)}$$

A.C. input power $= P_{a.c.} = \frac{1}{T}\int_0^T eidt$

where e and i the instantaneous values of e.m.f. and current.

But $e = i\,(r+R_L)$

where r is internal resistance of the diode.

$$\therefore \quad \text{A.C. input power} = P_{a.c.} = \left(\frac{r+R_L}{T}\right)\int_0^T i^2\,dt$$

$$= (r+R_L)\left[\frac{1}{\mathrm{T}}\int_0^{\mathrm{T}} i^2\,dt\right]$$

$$= (r+R_L)\left[\frac{1}{T}\int_0^T I_0^2 \sin^2 \omega t\right]$$

$$= (r+R_L)I_{r.m.s.}^2$$

$$= \frac{I_0^2}{2}(r+R_L) \qquad (\because \text{Eqn. } xviii) \qquad \ldots (xxi)$$

$$\therefore \quad \text{Efficiency} \qquad \eta = \frac{P_{d.c.}}{P_{a.c.}}$$

$$= \frac{4I_0^2}{\pi^2}\times\frac{2}{I_0^2}\times\frac{R_L}{(r+R_L)} = \frac{8}{\pi^2}\cdot\frac{R_L}{(r+R_L)}$$

$$= \frac{0.812}{1+\dfrac{r}{R_L}} \qquad \ldots (xxii)$$

$$\text{Percentage efficiency} = \frac{81.2}{1+\dfrac{r}{R_L}}$$

When $\frac{r}{R_L}\to 0$, the maximum efficiency is 81.2%, smaller the value of $\frac{r}{R_L}$, greater is its efficiency.

Ripple Factor (γ): It is defined as,

$$\text{Ripple factor,} \qquad \gamma = \frac{I_{a.c.}}{I_{d.c.}}$$

The output current contains both A.C. component and D.C. component. The $I_{r.m.s.}$ is given by

$$I_{r.m.s.}^2 = I_{d.c.}^2 + I_{a.c.}^2$$

$$\text{or} \qquad I_{a.c} = \sqrt{I_{r.m.s.}^2 - I_{d.c.}^2}$$

$$\text{Ripple factor,} \qquad \gamma = \frac{I_{a.c.}}{I_{d.c.}} = \frac{\sqrt{I_{r.m.s.}^2 - I_{d.c.}^2}}{I_{d.c.}}$$

$$= \left\{\left[\frac{I_{r.m.s.}}{I_{d.c.}}\right]^2 - 1\right\}^{1/2} = \left\{\left(\frac{I_0^2}{2}\cdot\frac{\delta^2}{4I_0^2}\right) - 1\right\}$$

$$= \left[\frac{\pi^2}{8} - 1\right]^{1/2}$$

$$= 0.482 \qquad \ldots (xxiii)$$

Thus, in full wave rectifier, the ripple factor (0.482) is much smaller than that for a half-wave rectifier (1.21). The efficiency is also double. Hence, a full-wave rectifier is widely used for converting A.C. into D.C.

Peak Inverse Voltage (P.I.V.)

The Peak Inverse Voltage (P.I.V.) is defined as the maximum voltage across the diode when it is not conducting.

In full-wave rectifier, P.I.V. is $2E_0$. Thus, the diode should have sufficiently high insulation strength so as to withstand the peak inverse voltage.

26.7 FULL-WAVE BRIDGE RECTIFIER

A more efficient way of employing semiconductor diodes as full wave rectifier, is to use them in the form of a bridge shown in Fig. 26.6 (a).

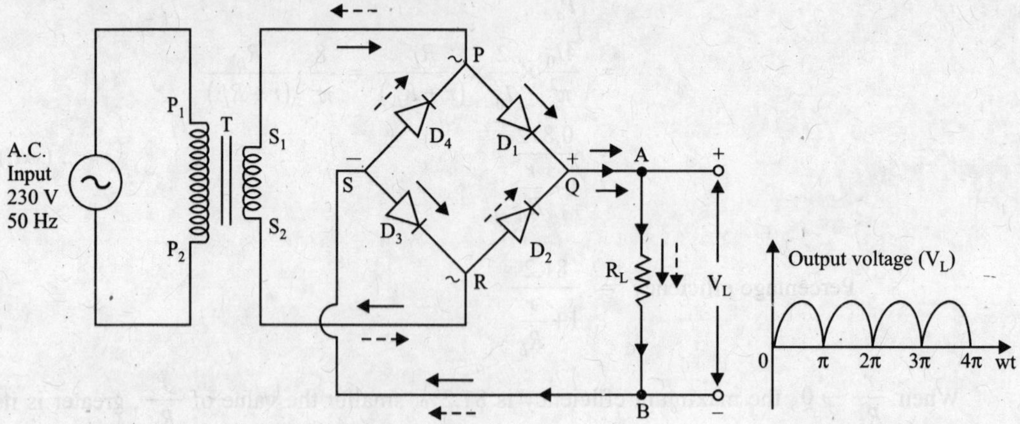

(a) Circuit diagram of bridge rectifier (a) Output waveform

Fig. 26.6

Working Principle

During first half-cycle of input A.C. voltage, suppose terminal S_1 of secondary is positive with respect to S_2, then diode D_1 and D_3 are forward biased and they conduct heavily. During this half cycle, other two diode D_2 and D_4, being reverse biased, remain 'off'. The path of conventional current flow is $S_1\,P\,Q\,A\,B\,S\,R\,S_2$, produces voltage drop V_L across load R_L.

During second of the cycle, i.e. negative half, the polarities of S_1 and S_2 are reversed i.e. terminal S_1 becomes negative and S_2 positive. Now diodes D_2 and D_4 conduct heavily, they being forward biased. In this case, the other two diodes D_1 and D_3 remain 'off' being reverse biased. During this negative half cycle, the path of convential current flow is $S_2\,R\,Q\,A\,B\,S\,P\,S_1$, producing voltage drop V_L across R_L. Thus, the output voltage across the load R_L consists of continuous positive pulses as shown in Fig. 26.6(*b*) are obtained. This gives a full wave rectification.

Advantages : The bridge rectifier is more popular because

(1) There is no necessity of a centre tapped transformer.

(2) For the same secondary voltage, the output is double that of the centre tapped transformer.

(3) The peak inverse voltage is one half that centre tap circuit using two diodes.

Disadvantages: The only disadvantage with a bridge rectifier is that it requires four diodes instead of two as in case of a full-wave rectifier.

26.8 COMPARISON OF A HALF-WAVE AND FULL-WAVE RECTIFIER

Sr. No.	*Quantity*	*Half-wave rectifier*	*Full-wave rectifier*
1.	Peak alternating current	$I_0 = \frac{E_0}{r+R_L}$	$I_0 = \frac{E_0}{r+R_L}$
2.	Average value of current	$I_{d.c.} = \frac{I_0}{\pi}$	$I_{d.c.} = \frac{2I_0}{\pi}$
3.	Effective value of current	$I_{r.m.s.} = \frac{I_0}{2}$	$I_{r.m.s.} = \frac{I_0}{\sqrt{2}}$
4.	Average D.C. output power	$P_{d.c.} = \frac{1}{\pi^2} I_0^2 R_L$	$P_{d.c.} = \frac{4}{\pi^2} I_0^2 R_L$
5.	Average A.C. input power	$P_{a.c.} = \frac{I_0^2}{4}(r+R_L)$	$P_{a.c.} = \frac{I_0^2}{2}(r+R_L)$
6.	Efficiency (η)	$\frac{4}{\pi^2}\left(\frac{R_L}{r+R_L}\right)$	$\frac{8}{\pi^2}\left(\frac{R_L}{r+R_L}\right)$
7.	Max. efficiency	40.60%	81.2%
8.	P.I.V.	E_0	$2E_0$
9.	Ripple factor	1.21	0.482

26.9 FILTERS

The unfiltered output of a rectifier is highly distorted d.c. voltage which contains large a.c. ripple in addition to the desired d.c. voltage. The unwanted a.c. component can be eleminated and smoothened by using filter circuit. Filter circuits generally used are of the following types :

(*a*) a shunt capacitor filter (*b*) series inductor filter (*c*) L-section filter (*d*) π-section filter.

(*a*) A Shunt Capacitor Filter

This is the most commonly used being very sample type of a filter is to connect a capacitor (electrolyte) C of suitable value (generally 1000 μF) across the load R_L as shown in Fig. 26.7.

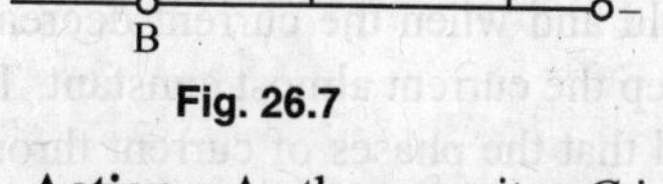

Fig. 26.7

Fig. 26.8

Action : As the capacitor C is connected across the output of the rectifier at points A and B (See Fig. 26.4, 26.5, 26.6), it gets charged when the input positive half cycle is increasing. The condenser C charges to the peak value of the positive cycle and output voltage across R_L reaches to point Q. Now incoming i.e. rectifier voltage starts decreasing towards R but the capacitor C now slowly discharges through load R_L and tries to maintain the output voltage close to the original value. The discharge of condenser is shown by a decay curve QP'. The slope of this decay curve is so chosen, that the time cosntant R_LC is very large and hence C discharges slowly, holding almost the same charge. This discharge continues, till negative half cycle is rectified to give another positive half cycle (action of full-wave rectifier), the discharge curve meets the second positive half (rectified) cycle at point P'. Again condenser C charges to the peak value of the incoming voltage i.e. up to point Q'.

The large value of capacitor is so chosen that it provodes a small reactance $X_C = \dfrac{1}{2\pi f_C}$ to the a.c. harmonics which may be present in the input to the filter and hence bypasses them. Thus, shunt capacitor not only reduces the ripple but also improves the efficiency.

The ripple factor for a half-wave rectifier using shunt capacitor filter is given by $\gamma = \dfrac{\pi}{\sqrt{3}\omega c R_L}$ and for a full-wave rectifier, it is $\gamma = \dfrac{\pi}{2\sqrt{3}\omega c R_L}$, for a line frequency of frequency $f = 50$ Hz, the ripple factors are : for half-wave rectifier, $\gamma = \dfrac{1}{175 R_L C}$ and for full-wave rectifier, $\gamma = \dfrac{1}{350 R_L C}$.

(*b*) A Series Inductor Filter

A series inductor filter is shown in Fig. 26.9 along with the waveform for the load current, for a full-wave rectifier in Fig. 26.10. The dotted line in the figure represents the wave form of the input

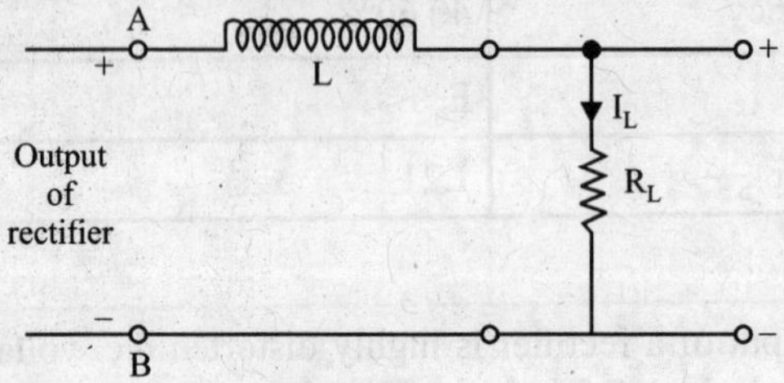

Fig. 26.9

voltage to the series inductor filter obtained as a output of full-wave rectifier, while the continuous line shows the waveform of the current I_L through the load resistor R_L.

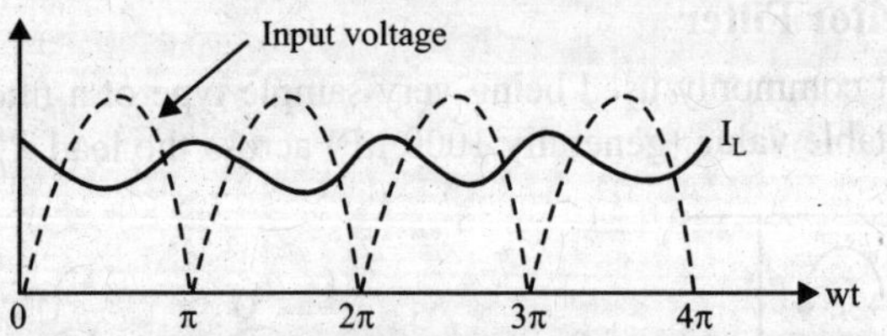

Fig. 26.10

The reactance offered by a choke i.e. inductor of inductance L is given by $X_L = 2\pi f L$. It opposes any change of current through it. When the incoming current increases above its average value, the choke stores energy in the form of its magnetic field and when the current decreases below this value, it releases energy and the field collapses to keep the current almost constant. This provides a smoothening action of the ripple. It should be noted that the phases of current through the load R_L and rectifiers unfiltered output voltage are opposite. The ripple factor γ of the full-wave rectifier with this filter is

$$\gamma = \frac{R_L}{3\sqrt{2}\omega L} = \frac{R_L}{6\sqrt{2}\pi f L}$$

Thus, ripple factor γ reduces for a choke having large value of inductance L.

(c) L-Section Filter

Fig. 26.11 shows a *L*-section filter. The choke of inductance L is connected in series with the output of the rectifier and the capacitor *C* (an electrolytic usually of value 1000 μF) is connected across the load R_L. Since the first component of the filter circuit consists of a choke coil, the circuit is also called as **choke input filter**.

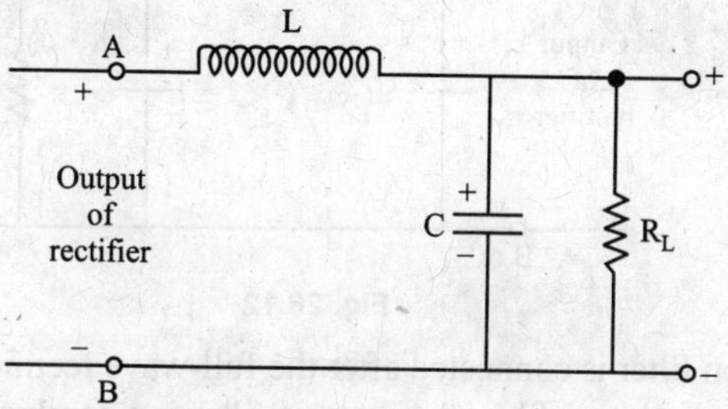

Fig. 26.11

The series inductor L offers a high inductive reactance $X_L = 2\pi f L$ to the A.C. component of the rectifier output and thus blocks the major part of A.C. and allows the D.C. component to pass through it. The capacitance C offers a very low capacitive reactance $X_C = \frac{1}{2\pi f c}$ to the A.C. component as compared to the load R_L and thus offers a low impedance path to the A.C. component which comes out of the inductor. Thus, the combined action of both the inductor L and the capacitor C is to maintain a fairly constant voltage (ripple free almost) across the load resistance R_L.

Ripple Factor (γ)

The ripple factor using a full-wave rectifier and L-section filter is

$$\gamma = \frac{E_{ac}}{E_{dc}} = \frac{\sqrt{2}E_0}{6\pi\omega^2 LC} \Big/ \frac{2E_0}{\pi} = \frac{\sqrt{2}}{3} \cdot \frac{1}{4\omega^2 LC}$$

$$= \frac{\sqrt{2}}{3}\left(\frac{1}{2\omega L}\right)\left(\frac{1}{2\omega C}\right)$$

$$= \frac{\sqrt{2}}{3 \times 4 \times \omega^2 LC} = \frac{1}{6\sqrt{2}\omega^2 LC}$$

At f ≐ 50 Hz, we get $\gamma = \frac{1.19}{LC}$ where C is in microfarad and L in Henry. It is clear from the above expression that ripple factor is independent of load resistance. But in practice, it has been observed that a certain minimum load resistor should be connected. This is known as **bleeder resistor**. The main purpose of bleeder resistance of minimum value is to ensure a minimum current through the inductor for better regulation. Without the bleeder resistor, there would be an initial high voltage surge, when the rectifier is turned on and this might damage the diodes. The bleeder resistor also help to discharge the capacitor *C*, when it has been turned 'off' and helps to prevent from dangerous shocks.

The output voltage for an L-section filter is less than the peak value of the rectifier output pulse because a series inductor does not allow the capacitor to charge to the peak value when the load current is drawn.

(*d*) Capacitor Input Filter or π-section Filter

A π-section filter consists of two capacitor C_1 and C_2 and a choke of inductance L as shown in Fig 26.12. The output is drawn across the load resistor R_L.

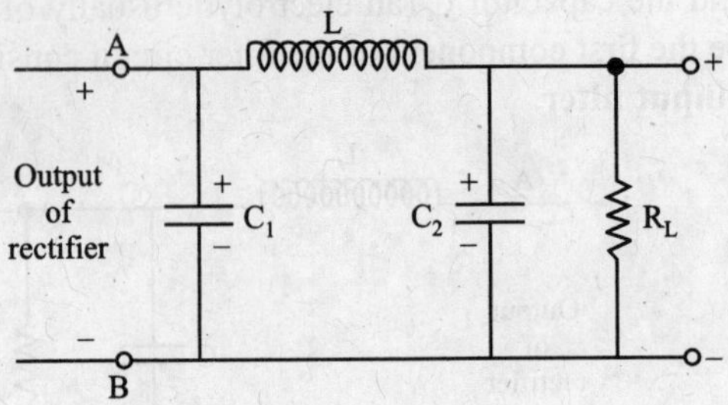

Fig. 26.12

When this π-section filter is connected after the full-wave rectifier, yields a higher D.C. output voltage than that with an L-section filter. A π-section filter is a combination of a shunt capacitor (C_1) and an L-section filter formed by L and C_2. The capacitor C_1 is periodically charged to almost the maximum value of the A.C. input voltage by the positive pulses of the current. Between successive charging and discharging of C_1, some ripple is present. This ripple present is smoothened by the smoothening action of series inductor L, however its amplitude is slightly reduced. Then, the A.C. component of reduced amplitude is bypassed by another condensor C_2. Thus, the π-section filter produces an almost constant level, ripple free output aross the load resistor R_L.

Ripple Factor

The ripple factor is equal to the product of the ripple factor of shunt capacitor and ripple factor of L-section filter.

The ripple factor of a shunt capacitor filter is

$$\gamma_1 = \frac{\pi}{2\sqrt{3}\omega C_1 R_L}$$

and ripple factor of L-section filter is

$$\gamma_2 = \frac{\sqrt{2}}{3} \cdot \frac{1}{4\omega^2 LC_2}$$

$\therefore$ Ripple factor of π-section filter,

$$\gamma = \gamma_1 \times \gamma_2$$

$$= \frac{\pi}{2\sqrt{3}\omega C_1 R_L} \times \frac{\sqrt{2}}{3} \cdot \frac{1}{4\omega^2 LC_2} \qquad \text{... (xxiv)}$$

$$\gamma = \frac{\pi}{12\sqrt{6}\,\omega^3\, C_1 C_2 L R_L}$$

Thus ripple factor $\gamma \propto \frac{1}{R_L}$, the ripple factor reduces as R_L increases. Such a high load resistance connected helps in reducing ripple. The voltage output from π-section filter is, thus highly filtered but not as stable as from a L-section filter, Its regulation is generally poor.

26.10 ELECTRONIC VOLTAGE REGULATORS (STABILIZERS)

For many applications, we need power supplies with a well-regulated output voltage that remains as constant as possible under variations of mains voltage, load current and temperature. The performance of regulated power supply shown in Fig. 26.13 may be characterized by the following.

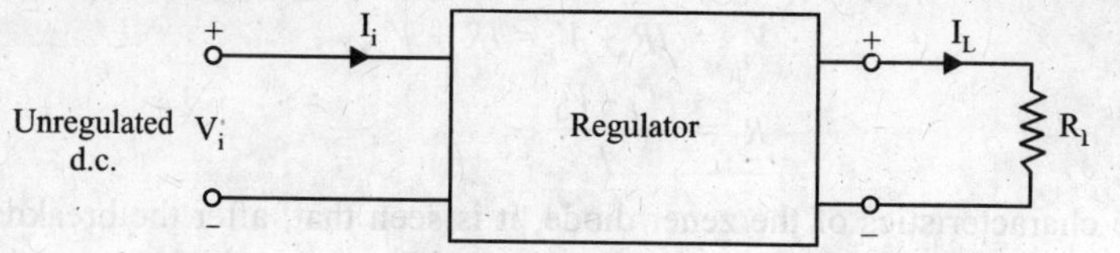

Fig. 26.13

basic relationship relating changes in the output d.c. voltage V_0 to the change in input voltage V_i load current I_L and temperature T as

$$dV_0 = \frac{\partial V_0}{\partial V_i}\cdot dV_i + \frac{\partial V_0}{\partial I_L}\cdot dI_L + \frac{\partial V_0}{\partial T}\cdot dT$$

$$= S_V dV_i - R_0 dI_L + S_T dT \quad \text{... (xxv)}$$

***(i)* Stabilization factor (s_V) :**

$S_V = \dfrac{\partial V_0}{\partial V_i}$ (for I_L and T constant) and is known as 'stabilization factor'.

***(ii)* Output resistance (R_0):**

$R_0 = -\dfrac{\partial V_0}{\partial I_L}$ (for V_i and T constant) and is known as 'output resistance', also known internal resistance of the regulator.

***(iii)* Temperature coefficient (S_T):**

$S_T = \dfrac{\partial V_0}{\partial T}$ (for V_i and I_L constant) and is known as 'temperature coefficient'.

The regulation factor S_r is defined as the relative change in V_0 corresponding to the change in V_i,

$$S_r = \frac{\partial V_0/V_0}{\partial V_i/V_i} = \left(\frac{V_i}{V_0}\right) S_V \text{(for } I_L \text{ and } T \text{ constant)}$$

Hence the power supply would have better regulation if all the three coefficients are kept to a minimum value.

26.11 ZENER DIODE AS A VOLTAGE REGULATOR

A full-wave rectifier followed by a proper filter circuit, can provide almost a ripple free output. But this output may not be stable. Those power supplies in which the output voltage is made

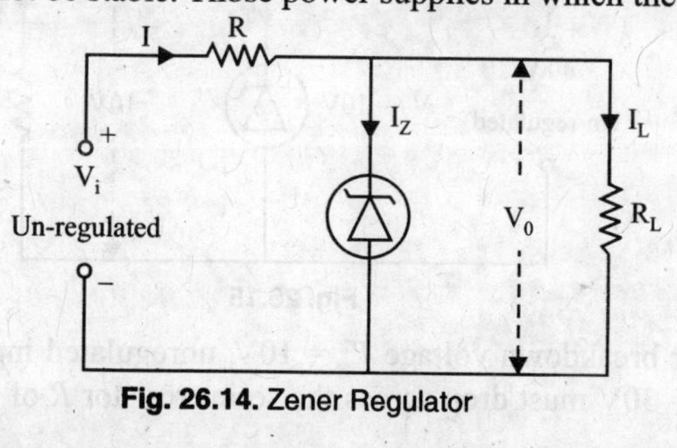

Fig. 26.14. Zener Regulator

stable or is made independent of input voltage (mains supply voltage), the load resistance and temperature are called regulated power supplies. Zener is used for this purpose.

The Fig. 26.14 shows a zener diode used as a shunt voltage regulator. It is necessary, that the unregulated input voltage V_i must be greater than the desired regulated voltage. The breakdown voltage V_Z is chosen to be equal to the desired regulated voltage. We can write

$$V_i = IR + V_Z = IR + V_0$$

Hence,
$$R = \frac{V_i - V_0}{I} \qquad \text{... (xxvi)}$$

From the characteristics of the zener diode, it is seen that, after the breakdown, the voltage across it remains almost constant through the current increases, which means that the resistance of the zener decreases.

Action of zener

The performance of zener as a voltage regulator can be explained under two situations:

(1) When the input voltage or mains voltage increases (Line regulation) :

The increase in line voltage will result in an increase

(*i*) in current I_L, through the load resistance R_L

(*ii*) in current I_Z, through the zener diode.

But an increase in current through zener means decrease in its resistance. This decrease in its resistance, would mean still more increase of current through the diode. Thus the current through the diode increases more. But $I = I_L + I_Z$. Hence, the increase in current I will increase in voltage drop across R, resulting into lesser output voltage. In other words, the output voltage remains closer to its original value. i.e. the output voltage remains unaffected, even through the input voltage changes.

(2) When the load R_L changes (Load regulation) :

Zener offers some resistance R_Z. The load resistance R_L and R_Z, both being parallel, gives an equivalent parallel resistance. This equivalent parallel resistance and series resistance R form a voltage divider. Now, if the load resistance changes, the zener resistance also changes in such a way that the equivalent parallel resistance still remains the same. Hence, the output voltage remains the same. i.e. again the output voltage remains unaffected, even though the load resistance changes.

SOLVED EXAMPLES

Example 26.1: *An unregulated d.c. supply of 40V is applied to a shunt zener regulator to give an regulated output of 10 V. The series resistor is of 3kΩ. Calculate the current through the zener diode and also the current through load resistor of 2kΩ.*

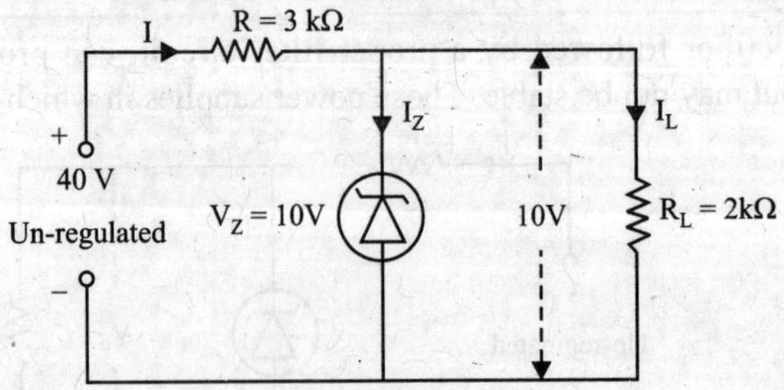

Fig. 26.15

Solution : Zener breakdown voltage V_Z = 10V, unregulated input voltage, V_i = 40V Hence $(V_i - V_0)$ = (40 – 10) = 30V must drop across the series resistor R of 3kΩ.

(i) The current through this resistor R

$$I = \frac{(40-10)V}{3\times10^3\Omega} = \frac{30}{3\times10^3} = 10 \times 10^{-3} \text{ Amp} = 10 \text{ mA}$$

(*ii*) The voltage across load $R_L = 2k\Omega$ is 10V. Hence current through R_L is

$$I_{RL} = \frac{10V}{2\times10^3\Omega} = 5 \times 10^{-3} \text{ A} = 5 \text{ mA}$$

But $I_R = I_Z + I_L$

$\therefore$ $I_Z = I_R - I_L = 10 \text{ mA} - 5 \text{ mA} = 5 \text{ mA}$

Thus, the current through zener = I_Z = 5 mA

and, the current through load resistance = I_L = 5 mA

Example 26.2: *Suppose the source voltage of Example 26.1 changes to 50 V. Calculate I_R, I_Z and I_L.*

Solution :

$$I_R = \frac{V_i - V_0}{R} = \frac{(50-10)V}{3\times10^3\Omega} = \frac{40}{3\times10^3}$$

$$= 13.33 \times 10^{-3} \text{ A} = 13.33 \text{ mA}$$

$$I_L = \frac{10V}{2\times10^3\Omega} = 5\times10^{-3} \text{ A} = 5 \text{ mA}$$

$\therefore$ $I_Z = I_R - I_L = (13.33 - 5) \text{ mA} = 8.33 \text{ mA}$

The increase in input voltage from 40 V to 50 V, to keep the output voltage unaltered, the zener current increases from 5 mA to 8.33 mA. Thus, the increase in source voltage 40 V to 50 V, produces more voltage drop across the series resistor R.

Example 26.3: *Justify the zener shunt regulator action if the load resistance varies from 2kΩ to 10kΩ in the circuit Fig. 26.15, if the zener power rating is 1/2 watt.*

Solution : The current flowing through scrics rcsistor R will be

$$I_R = \frac{V_1 - V_0}{R} = \frac{(40-10)V}{3\times10^3} = 10 \times 10^{-3} \text{ A} = 10 \text{ mA}$$

$$I_L = \frac{V_0}{R_L} = \frac{10V}{10\times10^3} = 1 \times 10^{-3} \text{ A} = 1 \text{ mA}$$

The zener current = $I_Z = I_R - I_L = (10 - 1) \text{ mA} = 9 \text{ mA}$

Since the power through zener = $V_Z \cdot I_Z = 10\text{V} \times 9 \times 10^{-3} \text{ A} = 0.09$ watt

which is must smaller that the rated value.

Example 26.4 : *A full-wave rectifier uses two diodes, the internal resistance of each diode is 20 ohm. The transformer r.m.s. secondary voltage from the centre tap to each end of the secondary is 50 V and load resistance R_L = 980 ohm. Find (i) The average value of current $I_{d.c.}$ (ii) Average d.c. voltage across R_L (iii) r.m.s. value of current $I_{r.m.s.}$ (iv) r.m.s. voltage across R_L (v) Maximum rectification efficiency (vi) Ripple factor (vii) Peak inverse voltage.* **(Nagpur U. 2012)**

Solution : Given $r = 20\ \Omega$, $R_L = 980\ \Omega$

r.m.s. secondary voltage between centre tap and one end of secondary = 50 V

$\therefore$ Maximum a.c. voltage, $E_0 = 50\sqrt{2} = 70.7\text{V}$

$$\text{Maximum load current, } I_0 = \frac{E_0}{r + R_L} \text{V} = \frac{70.7}{20+980} = 70.7 \text{ mA}$$

(*i*) Average value of current, $I_{dc} = \frac{2I_0}{\pi} = \frac{2\times 70.7}{3.14} = 45$ mA

(*ii*) Average d.c. voltage across, $R_L = I_{dc} \times R_L = \frac{45}{1000}\times 980 = 44.13$ V

(*iii*) r.m.s value of current, $I_{rms} = \frac{I_0}{\sqrt{2}} = \frac{70.7}{1.4142} = 50$ mA

(*iv*) r.m.s voltage across, $R_L = I_{rms} \times R_L = \frac{50}{1000} \times 980 = 49$ V

(*v*) Maximum rectification efficiency, $\eta = \frac{81.2}{1+\frac{r}{R_L}} = \frac{81.2}{1+\frac{20}{980}} = \frac{81.2}{1.02} = 79.6\%$

(*vi*) Ripple factor = 0.482

(*vii*) Peak inverse voltage = $2E_0 = 2 \times 70.7 = 141.4$ V

26.12 VOLTAGE REGULATION USING IC317

IC LM317 is popularly used to construct an adjustable voltage regulator. It provides a better line regulation and load regulation.

LM 317 is a three terminal adjustable voltage regulator. This IC chip includes a series pass transistor that can handle more than 1.5 A of load current. Operational amplifier thermal shutdown and current limiting techniques are also included in this chip. The pin connection diagram of LM 317 IC is shown in Fig. 26.16.

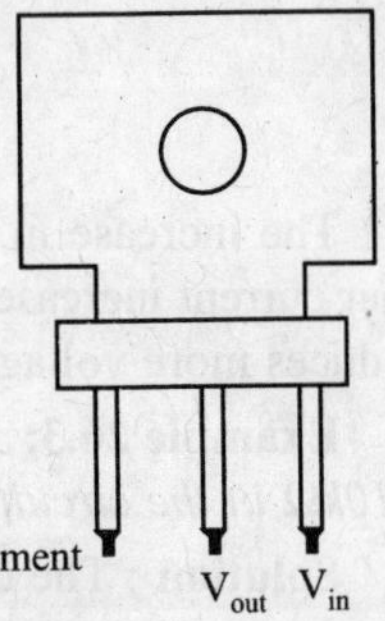

Fig. 26.16. 4 pin diagram of IC 317

The chip will automatically turn itself 'off', if the temperature becomes dangerously high around 175°C. This is called **'Thermal shutdown'**. This provides protection against excessive power dissipation, otherwise the IC chip would have burnt causing permanent damage. A suitable heat sink or cooling arrangement is provided. Due to the use of thermal shutdown and current limiting techniques, LM 317 IC is almost indestructible. The circuit diagram using a full-wave rectifier is shown in Fig. 26.17.

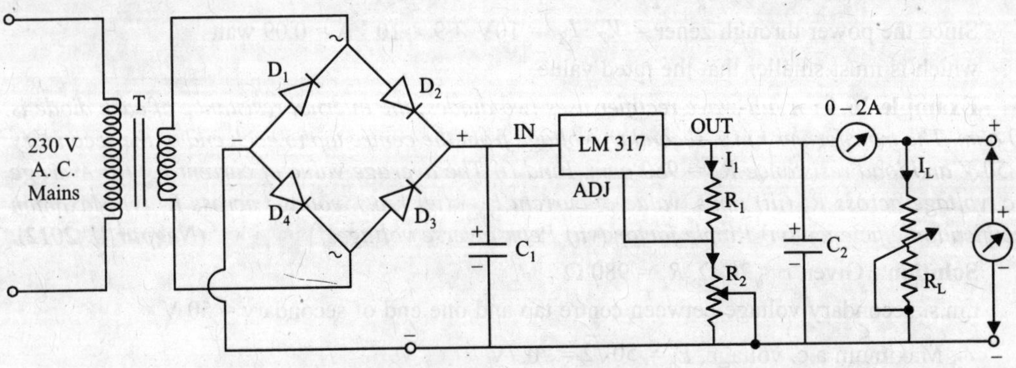

Fig. 26.17. IC Voltage Regulation

Designing :

The circuit can be designed for a wide range of output 1.25 V to 37 V by selecting R_1 and R_2. The output voltage is given by

$$V_{out} = V_{Ref}\left(\frac{R_2}{R_1}+1\right)$$

$$= 1.25\left(\frac{R_2}{R_1}+1\right)$$

For V_{out} = 12.6 V, R_1 is so selected that it insures the requirement of minimum load current is met.

Thus, selecting I_1 = 12.5 mA, we have

$$R_1 = \frac{1.25V}{12.5mA} = 100\ \Omega, \text{ voltage across } R_1 = V_{Ref.} = 1.25\text{ V}$$

Therefore, select R_1 = 100 Ω and R_2 = 910 Ω for fixed voltage regulator. For adjustable voltage regulator, connect R_2 potentiometer of 1kΩ.

This voltage regulator also filters the input ripple. Therefore, there is no need of filters.

Uses : The above circuit shown in Fig. 26.17 can be designed for

(1) an adjustable voltage regulator over a wide range of 1.25 V to 37 V.

(2) 100 mA current regulator.

(3) constant current battery charger etc.

EXERCISE

1. What is the necessity of power supply ? State the essential characteristics of an ideal power supply.
2. What is 'Q' point ? How load line is plotted for a pn-junction diode ?
3. What is rectifier ? Explain the use of a pn-junction diode as a half-wave rectifier.
4. Define average value, r.m.s. value, efficiency, ripple factor, and peak inverse voltage of a half-wave rectifier. Calculate their values for a half-wave rectifier. Show that ripple factor for a half-wave rectifier is 1.21. ***(Nagpur U. 2009, 2012)***
5. Explain the working of a full-wave rectifier using two pn-junction diodes. Draw input-output waveforms. ***(Nagpur U. 2011, 2012)***
6. Calculate average value, r.m.s. value, efficiency, ripple factor and peak inverse voltage of a full-wave rectifier. Show that the ripple factor for a full-wave rectifier is 0.48.
7. What are the advantages of a full-wave rectifier over a half-wave rectifier. Compare the two on the basis of :

   (*i*) Peak alternating current

   (*ii*) Average value of current

   (*iii*) Effective value of current

   (*iv*) Average d.c. output power

   (*v*) Average a.c. output power

   (*vi*) Efficiency, η

   (*vii*) Maximum efficiency

   (*viii*) P.I.V.

(*ix*) Ripple factor

(*x*) Hysterisis loss.

8. What is the necessity of using filter in a rectifier ? Describe the action of

(*a*) a shunt capacitor filter

(*b*) a series inductor filter

(*c*) L-section filter

(*d*) π-section filter

9. What is an electronic voltage regulator ? What distinguishes an unregulated power supply from a regulated one ? ***(Nagpur U. 2011)***

10. Explain the construction and working of a zener diode as a voltage regulator under two situations -

(*a*) line regulation

(*b*) load regulation.

11. Explain main features of IC LM 317. Draw a neat circuit diagram of an electronic voltage regulator using IC chip LM 317 and a full-wave rectifier.

12. The circuit diagram of Fig. 26.18 is required to supply an average voltage of 120 V. Determine the r.m.s. value of secondary voltage. average current delivered to the load, peak current through each diode and PIV of each diode.

[Ans : 266 V, 4.8 mA, current through each diode = 2.4 mA, peak current through each diode = 7.5 mA, 396.8 V]

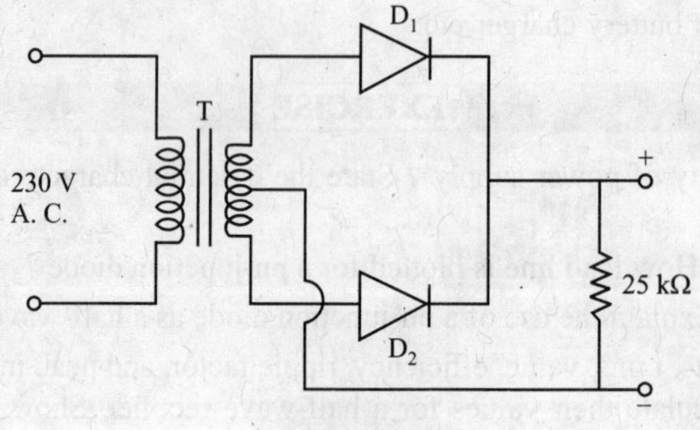

Fig. 26.18

13. A zener diode of 30 V breakdown is connected in series with a resistance of 200 ohm. If a load of 2 k ohm is connected across the zener diode, over what range of input voltage will the circuit operate? (Max. zener current is not to exceed 25 mA.) **[Ans : 33 V and 38 V]**

14. A zener diode 1Z10 along with a series resistance is connected across a 40 V supply. Calculate the minimum value of the resistance required if the max. zener current is 50 mA.

[Ans : 600 Ω] ***(Nagpur U. 2012)***

15. Explain how zener diode is different from ordinary diode. ***(Nagpur U. 2012)***

16. Show that ripple factor of a full wave rectifier is 0.42. ***(Nagpur U. 2011)***

CHARACTERISTICS OF TRANSISTORS

27 Chapter

INTRODUCTION

The use of three terminal element, commonly popular as Bipolar junction transistor (*BJT*) has brought profound changes virtually in every area of our interest. Exploration of outer space and deep sea, advancements in atomic power and arsenal, communications and computers, entertainment and medicine, automatization of production process in huge industries – all these and other fields are enriched due to the use of this small solid state device - the *BJT*.

The term transistor is mainly associated with the bipolar junction transistor (*BJT*), since the working of transistor is due to both, electron and hole charge carriers. There are other types of transistors like *FET*, *MOSFET*, *UJT* which operate on the action of only one type of charge carrier. Transistor basically performs two functions, namely amplification and switching.

27.1 BASIC TRANSISTOR CIRCUITS

A transistor being a three terminal device, it can be connected in the electrical circuit in three different modes, with one terminal common to both, the input and output. The basic three modes are:

(1) Common base (*CB*) mode

(2) Common emitter (*CE*) mode

(3) Common collector (*CC*) mode.

Whenever, a transistor is connected in any mode, it should be remembered that the basic bias conditions are satisfied, i.e. the emitter-base junction is forward-biased, and the collector-base junction is reverse-biased. Fig. 27.1 shows the circuit diagrams using *p-n-p* transistor in three different modes.

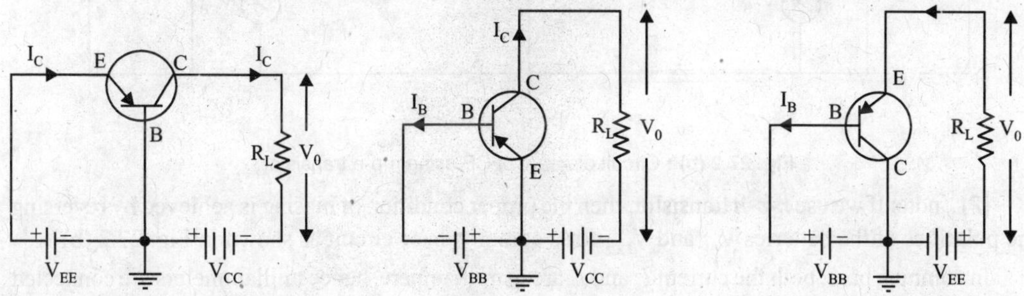

(*a*) Common-base mode (*b*) Common-emitter mode (*c*) Common-collector mode

Fig. 27.1 Three modes using p-n-p transistor.

27.2 CHARACTERISTICS IN COMMON BASE (CB) MODE

The most commonly used common-base characteristics are:

(*i*) **Input Characteristics:** It is a graph between emitter current and emitter-base voltage $(I_E, V_{EB})_{V_{CB}}$ for constant collector-base voltage.

(*ii*) **Output characteristics:** It is a graph between collector current and collector current and collector-base voltage $(I_C, V_{CB})_{IE}$ for constant emitter current.

(*iii*) **Transfer or mutual characteristics:** It is a graph between collector current I_C and emitter current I_E i.e $(I_C, I_E)V_{CB}$ for constant collector-base voltage.

Experimental Circuit Diagram

A circuit diagram for obtaining the static characteristics of a p-n-p transistor is shown in Fig. 27.2 (*a*). In this figure, emitter-base junction is made forward biased by connecting positive terminal of the battery V_{EE} to emitter terminal E and collector-base junction is reverse-biased by connecting negative terminal of another high voltage battery V_{CC} to collector terminal *C*.

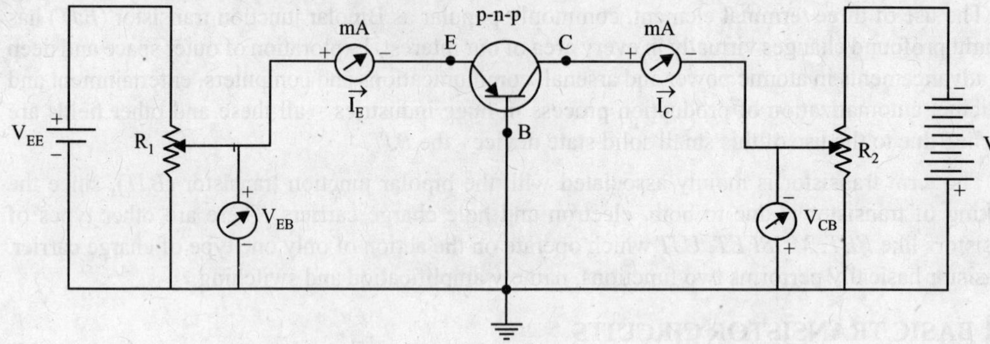

Fig. 27.2. (a) : Circuit diagram of CB using p-n-p transistor.

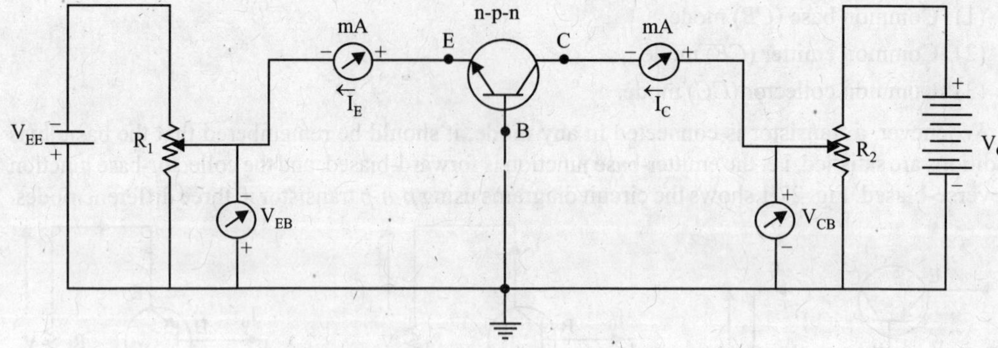

Fig. 27.2 (b) : Circuit diagram of CB using n-p-n transistor.

[PL. note: If we use *n-p-n* transistor, then the proper condition of biasing is achieved by reversing the polarities of the batteries V_{EE} and V_{CC}, for a common base circuit as shown in Fig. 27.2 (b)].

In common base, both the current I_E and I_C are in milliampere, hence milliammeters are connected in input and output sides. The input voltage V_{EB} can be varied by using potentiometer R_1 and V_{CB} voltage can be varied by potentiometer R_2 and are measured by two seperate voltmeters as shown in Fig. 27.2(*a*) and 27.2(*b*).

(*i*) **Input Characteristics i.e. $(I_E, V_{EB})_{V_{CB}}$ Characteristics**

To plot the input characteristics, the collector to base voltage V_{CB} is kept at 0V. The emitter-base voltage V_{EB} is slowly increased and every time emitter current I_E is noted. A graph a plotted between I_E (in mA) taken along y-axis and V_{EB} along *x*-axis.

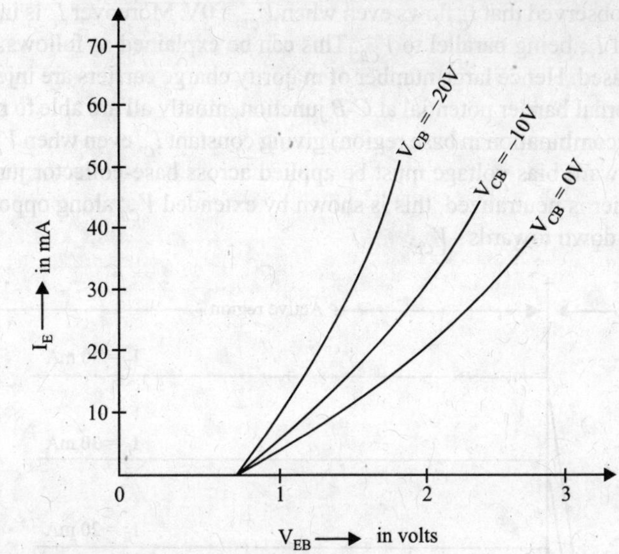

Fig. 27.3

Similar graphs can be ploted by keeping V_{CB} constant at –10V, –20V and so on. The input characteristics ploted are shown in Fig. 27.3

Input Impedance (Z_i)

In a *CB* mode, the input impedance (Z_i) is defined as the ratio of change in emitter base voltage (ΔV_{EB}) to the corresponding change in emitter current (ΔI_E), at constant collector to base (V_{CB}) voltage. Thus

$$Z_i = \left(\frac{\Delta V_{EB}}{\Delta I_E}\right)_{V_{CB}}$$

From the graph, we observe that there is a particular voltage below which the emitter current is almost zero. This voltage is known as the **cut in** or **threshold voltage** of the transistor. Its value for Ge transistor. Its value for Ge transistor is about 0.3 V and for a Si transistor is about 0.7 V at 25°C. The quantity Z_i is an important hybrid parameter denoted by h_{ib}. Thus $h_{ib} = Z_i$.

(*ii*) Output Characteristics i.e. (I_C, V_{CB})$_{I_E}$ Characteristics

The emitter current I_E is kept constant at some value say I_E = 10 mA by adjusting R_1. Keeping I_E constant, the collector voltage V_{CB} is slowly increased from zero onward in equal stapes and the collector current I_C is noted everytime. A graph is plotted between I_C along *Y*-axis and V_{CB} along *X*-axis. Similar graphs can be plotted by keeping the emitter current I_E constant at 20 mA, 40 mA and so on as shown in Fig. 27.4.

The output impedance (Z_0) is given by

$$Z_0 = \left(\frac{\Delta V_{CB}}{\Delta I_C}\right)_{I_E} \quad ...(ii)$$

The collector current I_C is always a little less than the emitter current I_E due to neutralization of few holes and electrons within the base region on account of **recombination**.

For V_{CB} > 0V, the output characteristics are parallel to V_{CB} axis. Therefore, the slope which represents output admittance h_{ob} is negligible and hence the output impedance Z_0 which is reciprocal of h_{oe} is very high for *CB* mode.

From graph, it is observed that I_C flows even when $V_{CB} = 0$V. Moreover I_C is independent of V_{CB}, for a constant value of I_E; being parallel to V_{CB}. This can be explained as follows: The emitter-base junction is forward biased. Hence large number of majority charge carriers are injected into the base region. Due to the internal barrier potential at *C-B* junction, mostly all are able to reach the collector (very few are lost in recombination in base region) giving constant I_C, even when $V_{CB} = 0$V. To reduce I_C to zero, a small forward bias voltage must be applied across base-collector junction, so that the internal potential barrier is neutralized, this is shown by extended V_{CB} along opposite direction. All the curves have slope down towards $-V_{CB}$.

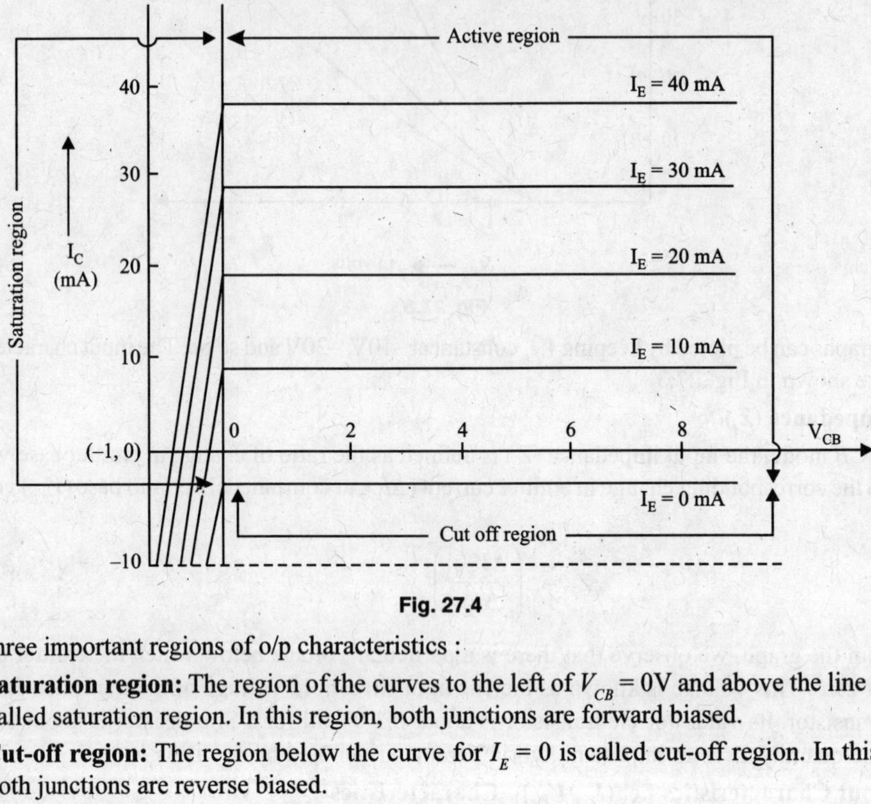

Fig. 27.4

Three important regions of o/p characteristics :

1. **Saturation region:** The region of the curves to the left of $V_{CB} = 0$V and above the line $I_E = 0$ is called saturation region. In this region, both junctions are forward biased.
2. **Cut-off region:** The region below the curve for $I_E = 0$ is called cut-off region. In this region both junctions are reverse biased.
3. **Active region:** The central region where the curves are parallel to V_{CB} axis is called active region. In this region, the emitter-base junction is forward biased and collector-base junction is reverse-biased.

All the above three region are shown in Fig. 27.4

(*iii*) Transfer Characteristics i.e. $(I_C, I_E)_{V_{CB}}$ Characteristics

Keeping collector to base voltage V_{CB} constant, the emitter current (I_E) is slowly increased in steps and the corresponding collector current I_C is noted. A graph is plotted Between I_C along *Y*-axis and I_E along *X*-axis. The graph is a straight line as shown in Fig. 27.5. The straight line graph makes an angle of nearly 45° with the *x*-axis. *i.e.* slope of the graph is 45°. The current gain in common-base mode is denoted by α and is given by

$$\text{Current gain, } \alpha = \left(\frac{\Delta I_C}{\Delta I_E}\right)_{V_{CB}} \cong \tan 45° \cong 1 \qquad \text{...(iii)}$$

In fact α is always less than 1. This is because same charge carries, although very few are lost in base region in recombination. This is also known s forward current transfer ratio h_{fb} or α_{ac}.

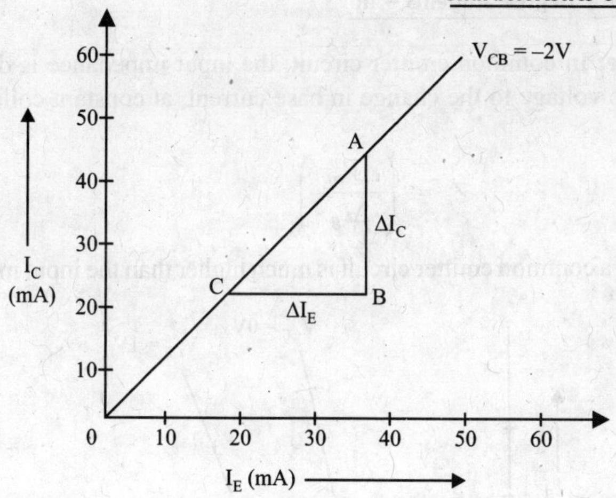

Fig. 27.5

27.3 CHARACTERISTICS IN COMMON EMITTER (CE) MODE

Experimental Arrangement

The circuit diagrams for determining the static characteristic curves of a *p-n-p* and an *n-p-n* transistor in common-emiter (*CE*) mode are shown in Fig. 27.6 (a) and 27.6 (b) respectively. Let us study input –, output – and transfer characteristics of the transistor in CE mode.

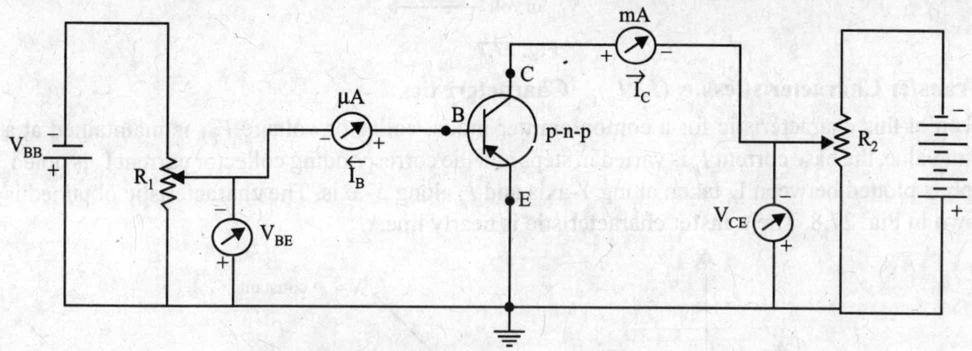

Fig 27.6 (a): Circuit diagram of CE using p-n-p transistor.

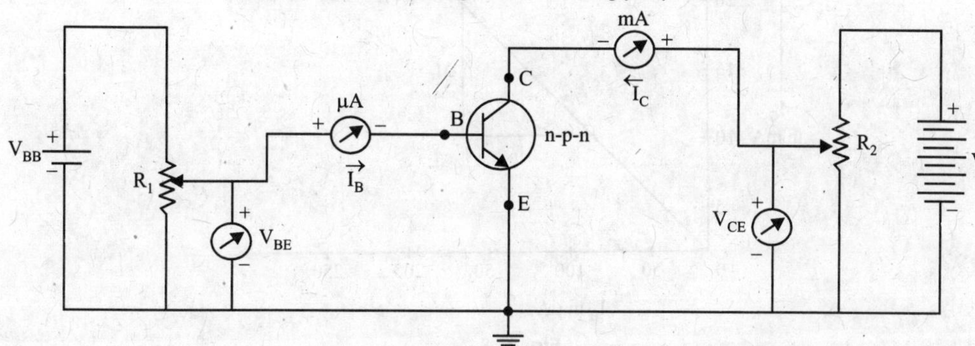

Fig. 27.6 (b): Circuit diagram of CE using n-p-n transistor.

(*i*) Input Characteristics i.e. $(I_B, V_{BE})_{V_{CE}}$ Characteristics

To plot these characteristics, the collector-emiter voltage V_{CE} is kept constant at 0V. The base-emitter voltage V_{BE} is increased in steps from zero and the corresponding base current I_B (in μA) is noted every time. A grath is plotted taking base current I_B along *Y*-axis and V_{BE} along *X*-axis, as shown in Fig 27.7. The same procedure is repeated for higher value of V_{CE}.

Input Impedance: In common emitter circuit, the input impedance is defined as the ratio of change in emitter-base voltage to the change in base current, at constant collector-emitter voltage. Hence

$$Z_i = \left(\frac{\Delta V_{BE}}{\Delta I_B}\right)_{V_{CE}} \quad \text{...(iv)}$$

The value of Z_i for a common emitter circuit is much higher than the input impedance in common-base mode.

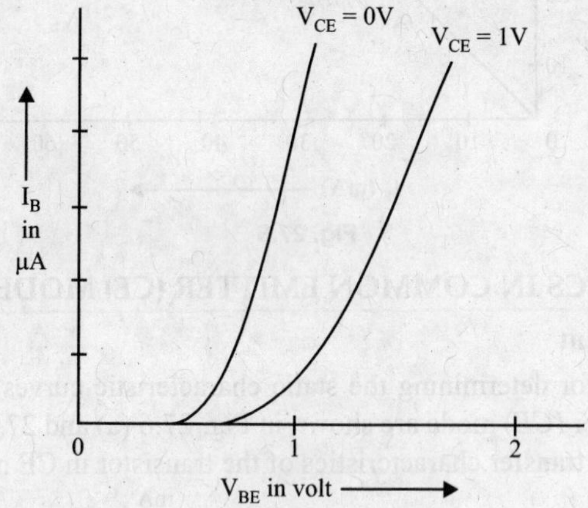

Fig. 27.7

(ii) Transfer Characteristics i.e. $(I_C, I_B)_{V_{CE}}$ Characteristics

To plot this characteristic for a comon-emitter mode, collector voltage V_{CE} is maintained at a constant value, the base current I_B is varied in steps and the corresponding collector current I_C is noted. A graph is plotted between I_C taken along *Y*-axis and I_B along *X*-axis. The characteristic obtained is as shown in Fig. 27.8. The transfer characteristic is nearly linear.

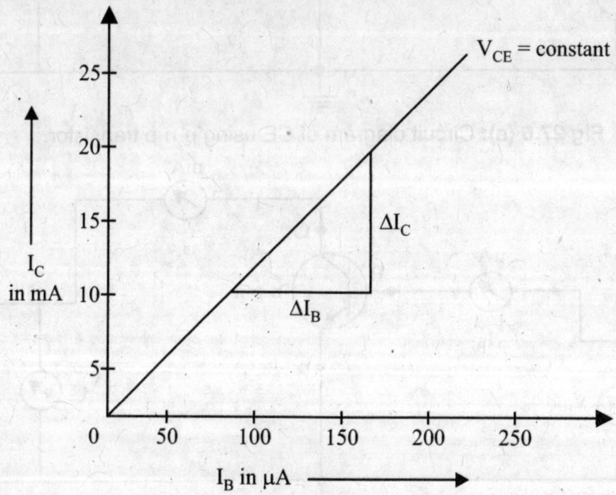

Fig. 27.8

It is clear from the graph that a small change in base current produces a large change in collector current for a constant collector to emitter voltage. The collector current is o predominantly determined by the base current, that it is usually taken as a function of base current. It is because, 95% to 98% of the majority carriers reaches to the collector and only 5% to 2% are absorbed (i.e. neutralized by recombination) in the base region. Thus a small change in base current produces a large change in collector current. In other words, common emitter mode give a large current gain.

Further, even when the base current is zero, there is some collector current flowing. This is known as **leakage current** of about 100 to 500 μA. This is due to minority carriers produced due to braking of covalent bods, which is highly temperature dependent. The leakage current I_{CO} increase with rise in temperature and is due to reverse biasing between the base and collector.

Current gain β

In a common emitter circuit, the current gain is defined as the ratio of the change in collector current (ΔI_C) to change the base current (ΔI_B) at constant collector to emitter voltage. The current gain in common emitter mode is denoted by β.

$$\therefore \text{ Current gain,} \quad \beta = \left(\frac{\Delta I_C}{\Delta I_B}\right)_{V_{CE}} \qquad \ldots(v)$$

The value of β ranges form 80 to 800.

(iii) Output Characteristics i.e. $(I_C, V_{CE})_{IB}$ Characteristics

To plot this characteristic, the base current I_B is maintained at a suitable constant value by adjusting V_{BE}. The collector-emitter voltage V_{CE} is varied in steps from zero and the corresponding collector current I_C is noted. A graph is plotted between collector current I_C taken along *Y*-axis and collector to emitter voltage V_{CF} along the *X*-axis. The procedure is repeated for higher fixed values of I_B. The output characteristics thus obtained are shown in Fig. 27.9

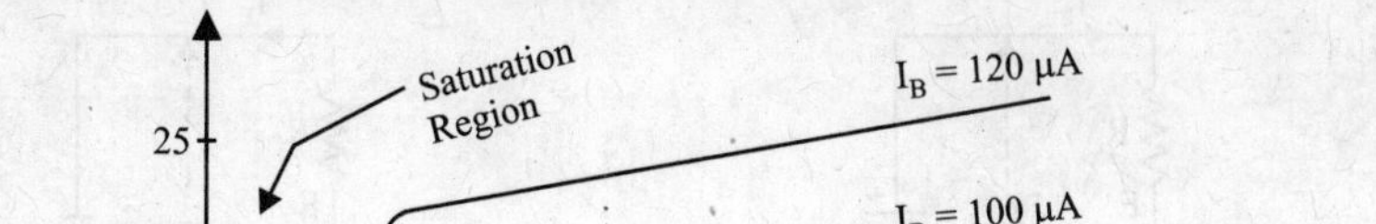

Fig. 27.9

The output admittance h_{oc} of the transistor for a suitable operating point *P* on the output characteristic of I_B = 60 μA in Fig. 27.9 is given by the slope of the characteristic at the point; i.e.

$$h_{oe} = \left(\frac{\Delta I_C}{\Delta V_{CE}}\right)_{I_B}$$

The reciprocal of the admittance is the output resistance or impedance of the transistor.

Hence

$$R_0 = \left(\frac{\Delta V_{CE}}{\Delta I_C}\right)_{I_B} \qquad \ldots(vii)$$

As compared to common base (I_C, V_{CB}) output characteristic, the common emitter (I_C, V_{CE}) characteristic curves have more marked slope. Hence R_0 of a *CE* circuit is smaller than that for *CB* mode. It is of the order of 5kΩ

Three important regions of characteristics:

1. Saturation region: The region of the curves to the left of the line *OA* is called the saturation region and the line *OA* is called saturation line. In this region, both emiter-base and collector-base junctions are foreard biased.

2. Cut-off region: The region below the curve for $I_B = 0$ is called cut-off region. In this region, both the junctions are reverse biased.

3. Active region: The central region where the curves are uniform in spacing and slope, is called the active region. In this region, the emitter-base junction is forward biased and collector base junction is reverse biased.

27.4 COMMON COLLECTOR (CC) CONFIGURATION

Here input is applied between the base and collector, while the output is taken from emitter and collector. In *CC* mode, the base current I_B is the input current and the emitted current I_C is the output current as shown in Fig. 27.10 (a) and (*b*) for *p-n-p* and *n-p-n* transistor respectively.

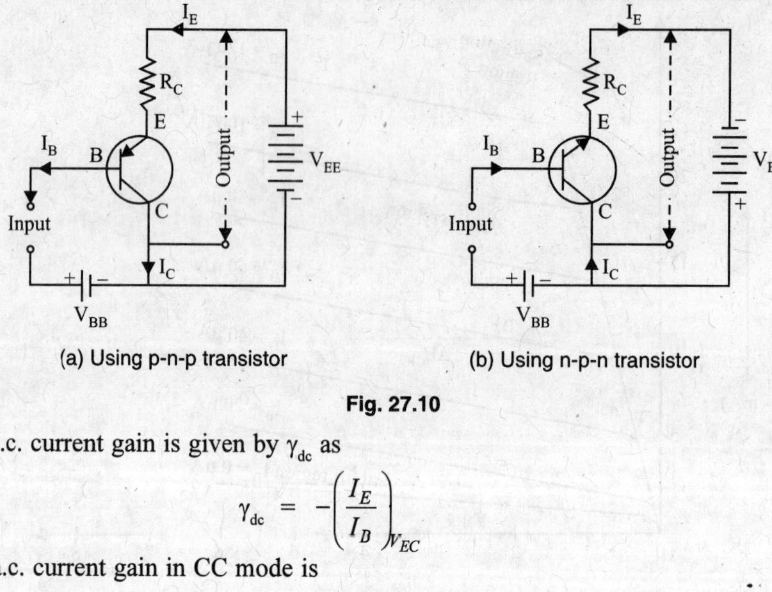

Fig. 27.10

The d.c. current gain is given by γ_{dc} as

$$\gamma_{dc} = -\left(\frac{I_E}{I_B}\right)_{V_{EC}} \qquad \ldots(i)$$

And a.c. current gain in CC mode is

$$\gamma_{ac} = -\left(\frac{\Delta I_E}{\Delta I_B}\right)_{V_{EC}}$$

The common-collector configuration is used primarity for impedance-matching purposes since it has a high input impedance and low output impedance; opposite to that of the common-base and common-emiter configuration.

From a design view-point, there is no need for a set of common-collector characteristics as the curve are not useful and so the curves are not given in manufacturers' data manual.

27.5 RELATION BETWEEN CURRENT GAINS α AND β

We can derive a relation between α and β for their d.c as well as a.c. definitions

(*i*) D.C. current gain

The d.c. current gain for a common-emitter configuration is defined as

$$\beta_{dc} = \frac{I_C}{I_B}$$

and for a common-base configuration is given by

$$\alpha_{dc} = \frac{I_C}{I_E}$$

The emitter current I_E is given by $I_E = I_B + I_C$

or $$I_B = I_E - I_C$$

Hence α_{dc} and β_{dc} are dependent on each other

$$\beta_{dc} = \frac{I_C}{I_E - I_C}$$

Dividing numerator and denominator by I_E, we get

$$\beta_{dc} = \frac{I_C / I_E}{I - I_C / I_E} = \frac{\alpha_{dc}}{1 - \alpha_{dc}} \quad \ldots(i)$$

Similarly, $$\alpha_{dc} = \frac{I_C}{I_B + I_C}$$

Dividing numerator and denominator by I_B, we get

$$\alpha_{dc} = \frac{I_C / I_B}{1 + I_C / I_B} = \frac{\beta_{dc}}{1 + \beta_{dc}} \quad \ldots(ii)$$

(ii) A.C. current gain

By definition, current gain common-emitter and common-base are given by

$$\beta_{ac} = \frac{\Delta I_C}{\Delta I_B}, \ \alpha_{ac} = \frac{\Delta I_C}{\Delta I_E}$$

and $I_E = I_B + I_C$ or $\Delta I_E = \Delta I_B + \Delta I_C$

Therefore, $$\beta_{ac} = \frac{\Delta I_C}{\Delta I_E - \Delta I_C}$$

Dividing numerator and denominator by ΔI_E, we get

$$\beta_{ac} = \frac{\Delta I_C / \Delta I_E}{1 - \Delta I_C / \Delta I_E} = \frac{\alpha_{ac}}{1 - \alpha_{ac}} \quad \ldots(iii)$$

Similarly, $$\alpha_{ac} = \frac{\Delta I_C}{\Delta I_B + \Delta I_C} = \frac{\Delta I_C / \Delta I_B}{1 + \Delta I_C / \Delta I_B} = \frac{\beta_{ac}}{1 + \beta_{ac}} \quad \ldots(iv)$$

SOLVED EXAMPLES

Example 27.1. *A transistor with an $\alpha = 0.99$ is operated in a common emitter configuration. What is the maximum alternating current gain?*

Solution. If β is the maximum current gain in common-emitter mode and α, the alternating current gain in common-base mode, then

$$\beta = \frac{\alpha}{1 - \alpha}$$

$$= \frac{0.99}{1 - 0.99} = 99$$

Example 27.2. *The current gain of transistor in common-emitter (CE) configuration is 49. What will be the current gain of the same transistor in common-base (CB) configuration?*

Solution. The current gain in common-emitter configuration is

$$\beta = \frac{\alpha}{1-\alpha}$$

where α is the current gain in common-base configuration

Now $\beta = 49$

$\therefore$ $$49 = \frac{\alpha}{1-\alpha} \text{ or } 49 - 49\alpha = \alpha$$

or $$50\alpha = 49$$

$$\alpha = \frac{49}{50} = 0.98$$

27.6 ACCURATE COLLECTOR CURRENT IN CB AND CE MODE

Since the collector-base is always connected in reverse biased, some leakage current $I_{leakage}$ flows due to minority carriers. Therefore, the accurate collector current is calculated by taking into account this $I_{leakage}$ current in CB and CE mode.

1. Accurate Collector Current in CB Mode

The current gain in common-base (CB) mode is given by

$$\alpha = \frac{I_C}{I_E} \quad \text{or} \quad I_C = \alpha I_E$$

This current I_C is due to majority charge carriers. In addition to this, there is $I_{leakage}$ current due to minority charge carriers. This is because the collector to base junction is reverse biased.

$\therefore$ Total collector current $= I_C + I_{leakage}$

$$= \alpha I_E + I_{leakage}.$$

It is clear that if $I_E = 0$, i.e. the emitter circuit is open, a small leakage current still flows in the collector circuit. The leakage current $I_{leakage}$ is denoted by I_{CBO}, meaning collector to base current with emitter open. This is also known as **reverse saturation current**. Thus, an accurate expression for collector current is

$$I_C = \alpha I_E + I_{CBO} \quad \quad ...(i)$$

However, I_{CBO} is in μA, being very small, usually it may be neglected.

2. Accurate collector current in CE mode

In case of a common-emitter (*CE*) mode, I_B is the input current and I_C is the output current. According to relation (*i*)

$$I_C = \alpha I_E + I_{CBO}$$

Also $$I_E = I_B + I_C$$

$\therefore$ $$I_C = \alpha(I_B + I_C) + I_{CBO} = \alpha I_C + \alpha I_C + I_{CBO}$$

$$I_C(1 - \alpha) = \alpha I_B + I_{CBO}$$

$\therefore$ $$I_C = \frac{\alpha}{1-\alpha} I_B + \frac{1}{1-\alpha} I_{CBO} \quad \quad ...(ii)$$

But $$\frac{\alpha}{1-\alpha} = \beta \text{ and } \frac{1}{1-\alpha} = \beta + 1$$

$\therefore \quad I_C = \beta I_B + (\beta + 1) I_{CBO}$ ...(*iii*)

Now if $I_B = 0$, the base circuit is open, the collector current will be denoted by I_{CEO} meaning collector to emitter current with base open.

From relation (*ii*), $I_{CEO} = \frac{1}{1-\alpha} I_{CBO}$ and from (iii) $I_{CEO} = (\beta + 1) I_{CBO}$.

Hence, from relation (*ii*)

$$I_C = \frac{\alpha}{1-\alpha} I_B + I_{CEO}$$

and from relation (*iii*),

$$I_C = \beta I_B + I_{CEO} \quad \text{... (iv)}$$

Sometimes, I_{CBO} is also denoted by I_{CO}. Hence using this notation, relation (*i*), (*ii*) and (*iii*) becomes

$$I_C = \alpha I_E + I_{CO}$$

or
$$I_C = \frac{\alpha}{1-\alpha} I_B + \frac{1}{1-\alpha} I_{CO}$$

or
$$I_C = \beta I_B + (\beta + 1) I_{CO} \quad \text{...(v)}$$

Example 27.3. *The current gain a of an n-p-n transistor is 0.98. It is connected in common base mode and gives a reverse saturation current I_{CO} = 12μA. Find the base and collector current for an emitter current of 2mA.*

Solution. If I_C is the collector current, I_E the emitter current and I_{CO} the reverse saturation current in *CB* mode, then

$$I_C = \alpha I_E + I_{CO}$$

Given: $I_E = 2\text{mA} = 2 \times 10^{-3}$ A, $I_{CO} = 12\mu\text{A} = 12 \times 10^{-6}$ A, $\alpha = 0.98$

$\therefore$ Collector current, $I_C = 0.98 \times 2 \times 10^{-3} + 12 \times 10^{-6}$ A

$= (1.96 + 0.012) \times 10^{-3}$ A

$= 1.972$ mA

Base current, $I_B = I_E - I_C = 2\text{ mA} - 1.972\text{ mA} = 0.028\text{ mA}$

27.7 TRANSISTOR AS AN AMPLIFIER

The electrical signals viz. R.F. voltage induced on the antenna of radio receiver, the output of a transducer like microphone, photocell, thermistor etc. are very weak. These weak signals ar amplified to a suitable voltage level so that they can be processed further. Transistor is one of the active devices as an amplifier to boost up their current, voltage or power. Accordingly amplifiers are classified as current amplifiers, voltage amplifiers and power amplifiers respectively.

As we know, a transistor can be employed as an amplifier in any one of the three modes to give (1) Common base amplifier, (2) Common emitter amplifier and (3) Common collector amplifier.

Some important notations:

During the study of working of a transistor as an amplifier, we come across d.c and a.c. quantities. The following notations are recommended:

1. D.C. values are represented by capital letter with capital subscript (I_E, I_C, I_B, V_{BE}). D.C. supply voltage are indicated by repeated subscripts (V_{BE}, V_{EE}, V_{CC}).
2. R.M.S. values of quantities are represented by capital letter with lower-case subscripts (I_e, V_{cb}, I_b, I_c).

3. A.C. components of currents and voltage are represented by lower-case letters with lower-case subscripts (i_e, v_{be})
4. Instantaneous total values are represented by lower-case letters with supper case subscripts (i_E, v_{BE})
5. Maximum or peak values are represented by same letters as for r.m.s. values, but with an additional final subscript *m* (I_{cm}, V_{bcm}).

In a transistor, as emitter current changes, the collector and the base current will also change and will consist of a.c. component in addition to the d.c. component. The instantaneous total values i_E, i_C and i_B of the emitter, collector and base currents respectively are given by

$$i_E = I_E + i_e$$
$$i_C = I_C + i_c$$
$$i_B = I_B + i_b$$

where i_e, i_c and i_b are a.c. components. The instantaneous total values are related by

$$i_E = i_C + i_B$$

27.8 COMMON EMITTER AMPLIFIER

Fig. 27.11 shows a basic circuit of common-emitter amplifier using *n-p-n* transistor.

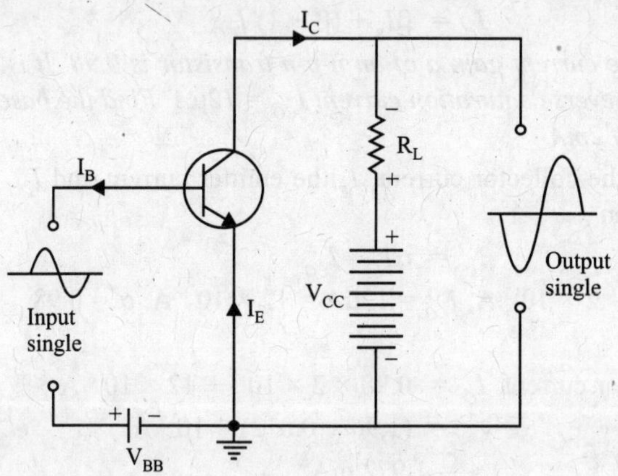

Fig. 27.11

The battery V_{BB} provides the necessary forward bias between emitter-base junction and another battery V_{CC} provides the desired reverse bias between collector-base junction. A load resistance R_L is connected in collector circuit as shown in figure. The signal to be amplified is applied at the input and an amplified output is obtained between collector and the ground.

Working: When no signal is applied, some d.c. collector current flows through the load R_L. Hence voltage drop across R_L is $I_C R_L$ is in opposition to the collector supply voltage V_{CC} and hence the collector voltage is reduced to some value, say V_C and is given by

$$V_C = V_{CC} - I_C R_L$$

Suppose now a signal from signal source, say A.F.O. (Audio Frequency Oscillator) is applied at the input. During positive half-cycle of the first cycle, the forward bias of the junction increases. This increases the emitter current. Since $I_E = I_B + I_C$, the increase in I_E will increase the collector current I_C and there by voltage drop across R_L increases. But the output is given by

$$V_C = V_{CC} - I_C R_L$$

Thus, increase in $I_C R_L$ will result in decrease in V_C, i.e. the collector becomes more negative. In other words, we get a negative half-cycle at the output corresponding the positive half-cycle at the input.

During the negative half-cycle, the forward bias of pn junction (of emitter-base) decreases. This decreases the emitter current. Correspondingly, the collector current decreases, and hence the voltage drop at the load resistance R_L decreases. This, result in increase of collector voltage V_C. Thus, we get a positive going signal at the output for a negative going input signal.

Thus, we conclude that, in a common emitter circuit, the input and output voltages are out of phase i.e. 180°.

Voltage gain of CE Amplifier

Let v_i = a.c. input voltage

R_i = input a.c. resistance of the transistor in CE mode.

Then, the a.c. base current, $i_b = \dfrac{v_i}{R_i}$

The current gain in CE mode is given by $\beta = \dfrac{i_c}{i_b}$

Therefore, the collector a.c. current, $i_c = \beta i_b = \beta \dfrac{v_i}{R_i}$

The magnitude of output voltage $= |v_o| = i_c R_L = \dfrac{\beta v_i R_L}{R_i}$

Hence, the magnitude of voltage gain in CE mode is

$$|a_v| = \frac{|v_o|}{|v_i|} = \frac{\beta v_i R_L / R_i}{v_i} = \beta R_L / R_i \qquad ...(i)$$

Since β is very large for most of the transistors (80 to 800), R_L is taken large and R_i is moderate, the voltage gain of *CE* amplifier is large, Hence, *CE* mode is popular as it has large voltage gain, and thereby large power gain.

Advantages of CE Amplifier

The common emitter amplifier is widely used as it has the following advantages:

1. The current gain (β) is very large.
2. The voltage gain is also very high.
3. The power gain is very high.
4. Since the input impedance and output impedance do not differ much from each other, large number of stages can be cascaded to achieve the desired gain. Moreover, as there is no impedance matching problem, the multistage cascading does not require transformers.
5. Since the input impedance is moderately high, it can be connected to sources of different input impedances, as it does not load them.

Disadvantages of CE Amplifier

1. The frequency response of common emitter amplifier is fairly poor, particularly at high frequency as compared to CB and CC mode, because of its large effective input capacitance.
2. The CE amplifier is highly temperature dependent. Therefore, it needs special bias stabilization.

27.9 COMPARISON OF CB, CE AND CC CONFIGURATIONS

The comparison of various parameter in CB, CE and CC mode is given in a tabular form as under:

No.	*Parameter*	*Common base CB*	*Common emitter CE*	*Common collector CC*
1.	Input resistance (Ri)	Low (about 100 Ω)	Low (about 750 Ω)	Very high (about 750 kΩ)
2.	Output resistance (Ro)	Very high (about 450 kΩ)	High (about 45 kΩ)	Low (about 50Ω)
3.	Current gain	Slightly less than 1 denoted by α	High (80 to 800) denoted by β	Moderate (about 35) denoted by γ
4.	Voltage gain	100 - 1000	250 - 300	Less than 1
5.	Power gain (db)	20 - 30	40	15 - 30
6.	Applications	For high frequency (RF and VHF)	For audio frequency (20 Hz to 20 KHz)	For impedance matching.

27.10 LOAD LINE AND Q POINT

Consider a common-emitter amplifier using *n-p-n* transistor as shown in Fig. 27.11 for which the output characteristics are as shown in Fig. 27.12. When no signal is applied,

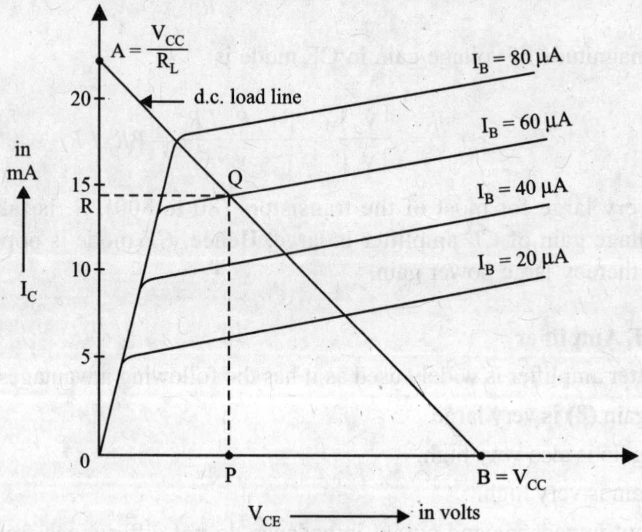

Fig. 27.12

the d.c. conditions prevails in the circuit. As seen from the figure, the collector to emitter voltage V_{CE} is given by

$$V_{CE} = V_{CC} - I_C R_L$$

Hence,

$$I_C = -\frac{V_{CE}}{R_L} + \frac{V_{CC}}{R_L} \quad \ldots(i)$$

Now the quantity $\frac{V_{CC}}{R_L}$ is a constant. $-\frac{1}{R_L}$ is also a constant, V_{CE} and I_C being variables. Hence equation (*i*) is an equation of the form $y = mx + c$, which is a straight line, the slope of the line being

$-\frac{1}{R_L}$ and the intercepts on Y-axis and X-axis are obtained as: putting $V_{CE} = 0$, $I_C = \frac{V_{CC}}{R_L}$, the intercept on Y-axis (point A) and putting $I_C = 0$, $V_{CE} = V_{CC}$, the intercept on X-axis (point B). Joining these two points A and B, we get a straight line, known as **d.c. load line**.

Q-POINT

Zero signal values of I_C and V_{CE} are known as the operating point or Quiescent point or Q point. When a signal is applied, variations of I_C and V_{CE} take place about this point. Hence it is called the operating point. In the absence of a signal, the transistor is said to be silent or quiet. So the point is called as Quiescent (i.e. Silent) or Q-point. The operating point can be determined from the point where d.c. load line intersects $I_C - V_{CE}$ characteristic for proper value of I_B.

27.11 GRAPHICAL ANALYSIS OF CE VOLTAGE AMPLIFIER

A graphical method is used to analyse the working of a transistor as an amplifier, the exact relationship between input and output voltages in terms of their magnitude and phases. (There is another method using h-parameters and will be studied in article 27.13).

Consider a load line AB drawn on output characteristic curves of a common emitter amplifier as shown in Fig. 27.13. The point Q is chosen as operating point. When a.c. signal is not applied, then a constant base current $I_B = 60$ μA flows producing a constant output voltage across R_L.

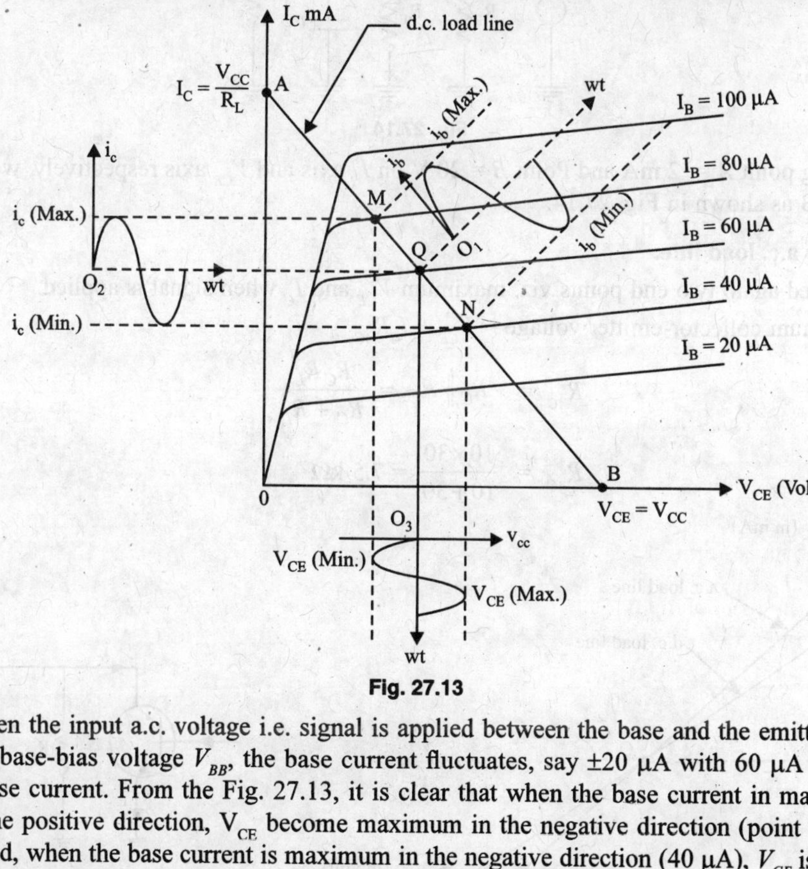

Fig. 27.13

When the input a.c. voltage i.e. signal is applied between the base and the emitter in series with the base-bias voltage V_{BB}, the base current fluctuates, say ±20 μA with 60 μA as the zero signal base current. From the Fig. 27.13, it is clear that when the base current in maximum (80 μA) in the positive direction, V_{CE} become maximum in the negative direction (point M). On the other hand, when the base current is maximum in the negative direction (40 μA), V_{CE} is maximum in the positive sense (point N). Thus, the input and output voltages are in phase opposition or the transistor is said to produce a 180° phase reversal with respect to signal voltage at the input.

27.12 A.C. LOAD LINE

This is the line on the output characteristics of a transistor which gives the value of I_C and V_{CE} when signal is applied. To draw a.c. load line, we require we require two end points, i.e. maximum collector-emitter voltage point and maximum collector current point,when signal is applied.

Example 27.4. *In a CE transistor amplifier shown in Fig. 27.14 (a) $R_C = 10\ k\Omega$, $R_L = 30\ k\Omega$ and $V_{CC} = 20\ V$. The vales of R_1 and R_2 are so as to fix the operating point at 10 V and 1mA. Draw d.c. and a.c. load lies. Assume R_E is negligible.*

Solution: (*a*) To draw d.c. load line: We need two end points viz. Maximum $V_{CE} = V_{CC} =$ 20 V and maximum collector current,

$$I_C = \frac{V_{CC}}{R_C + R_E} = \frac{20V}{10\,k\Omega} = 2\text{mA}.$$

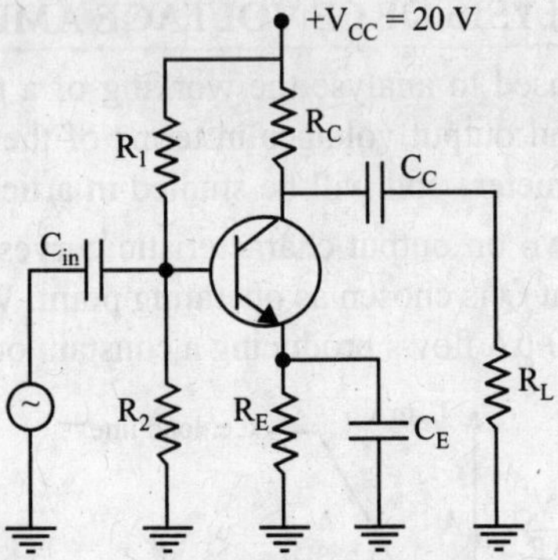

Fig. 27.14

Joining point A = 2 mA and Point B = 20 V on I_C axis and V_{CE} axis respectively, we get d.c. load line AB as shown in Fig. 27.14.

(*b*) To draw a.c. load line:

We need again two end points viz. maximum V_{CE} and I_C when signal is applied.

Maximum collector-emitter voltage = $V_{CE} + I_C R_{AC}$

where $$R_{AC} = R_C \| R_L = \frac{R_C R_L}{R_C + R_L}$$

$$\therefore \quad R_{AC} = \frac{10 \times 30}{10 + 30} = 7.5\ k\Omega$$

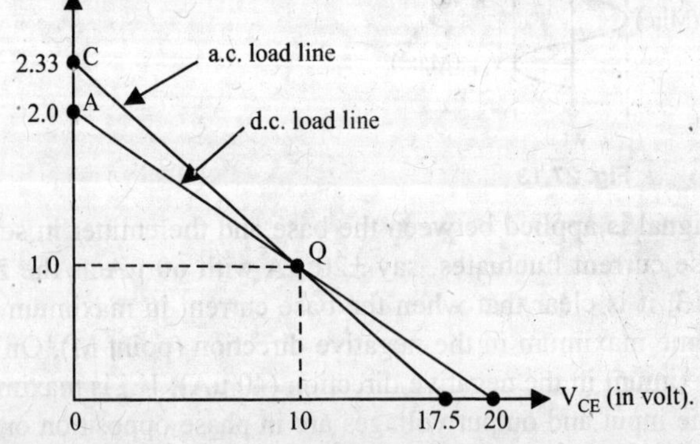

Fig. 27.14

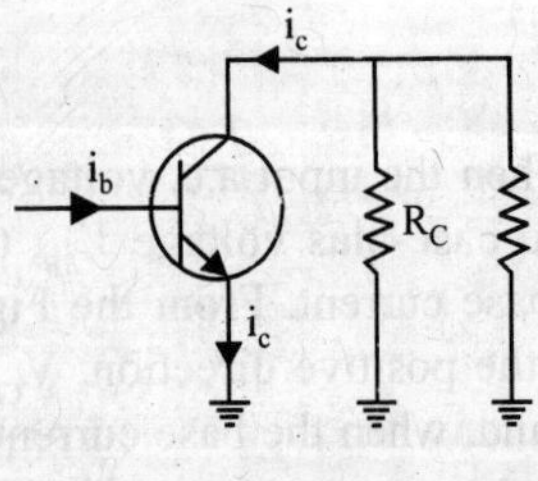

Fig. 27.15

Maximum collector-emitter voltage = $V_{CE} + I_C R_{AC}$

$$= 10 + (1 \text{ mA} \times 7.5 \text{ k}\Omega) = 10 + 7.5 = 17.5 \text{ V}$$

This locates the point D on the V_{CE} axis.

$$\text{Maximum collector current} = I_C + \frac{V_{CE}}{R_{AC}} = 1\text{mA} + \frac{10\text{V}}{7.5 \text{ k}\Omega}$$

$$= 1 \text{ mA} + 1.33 \text{ mA} = 2.33 \text{ mA}$$

This locates the point C on I_C axis. Line CD represents the a.c. load line.

ADVANTAGES OF GRAPHICAL METHOD

A graphical method is useful in evaluating the a.c. output voltage using the static characteristics of a transistor. It reveals the exact relationship between input and output voltages in terms of their magnitude and phases.

Drawbacks: The graphical method has the following short comings:

(1) The method is tedious

(2) It can not evaluate input, output resistances and

(3) It does not give any idea about frequency response of the amplifier. These drawbacks are overcome by using h-parameters.

27.13 HYBRID OR *h* PARAMETERS

There are two types of parameters:

(*a*) hybrid, or *h* parameters and

(*b*) admittance, or *y* parameters.

The standard method for analysing the performance of a small signal amplifier is to represent the circuit by its equivalent circuit. The *h* parameter equivalent circuit is widely used for analysing low frequency devices like transistors while y parameter equivalent circuit is used for analysing high frequency devices and circuits using FET.

A port is a pair of terminals in an electrical network. A transistor is a two-port network having an input port and output port as shown in Fig. 27.16. The currents are assumed positive when they enter the transistor and negative when they leave it. Here i_1 and i_2, both are positive. The voltages are assumed positive from upper to lower terminals and vice-versa. These are standard conventions used all over.

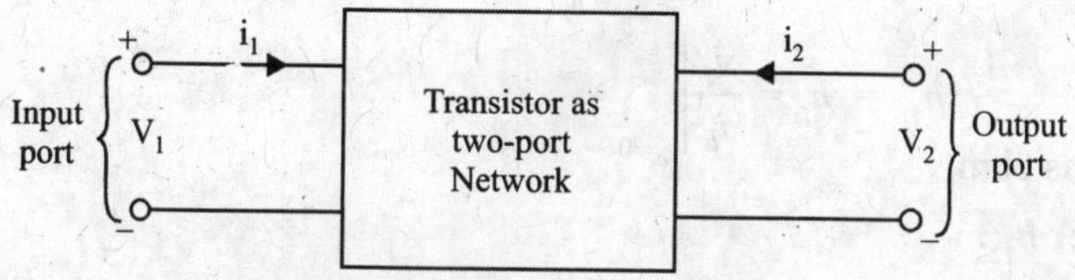

Fig. 27.16

The *h*-parameters of two-port network are designated as h_{11}, h_{12}, h_{21} and h_{22}. Their values for any network are such that the following two network equations are always satisfied:

$$v_1 = h_{11} i_1 + h_{12} v_2 \quad \ldots (i)$$

$$i_2 = h_{21} i_1 + h_{22} v_2 \quad \ldots (ii)$$

h-PARAMETERS OF A TRANSISTOR

Fig. 27.17 (*a*) shows the *n-p-n* transistor in common-emitter mode. When the transistor is property biased and a.c. source *i.e.* input signal a applied at the input port, then a.c. components flowing are as under:

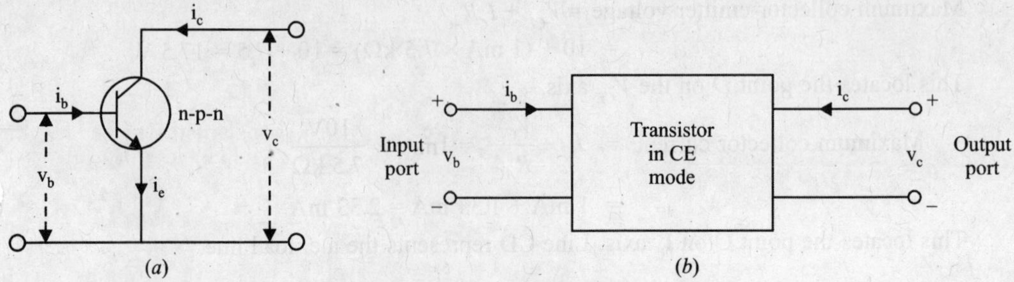

Fig. 27.17

Let $i_1 = i_b$ = instantaneous alternating base current,

$v_1 = v_b$ = instantaneous alternating base voltage,

$i_2 = i_c$ = instantaneous alternating collector current,

and $v_2 = v_c$ = instantaneous alternating collector voltage.

Figure 27.17 (b) shows the equivalents four terminal network of a transistor showing the instantaneous values of alternating base voltage v_b, base current i_b, collector current i_c and collector voltage v_c. Using two-port network equations in term of *h*-parameters given by Eqn. (*i*) and (*ii*) becomes

$$v_b = h_{11}i_b + h_{12}v_c \qquad \ldots (iii)$$

and

$$i_c = h_{21}i_b + h_{22}v_c \qquad \ldots (iv)$$

h_{11}, h_{12}, h_{21} and h_{22} are known as hybrid parameters or *h*-parameters.

1. The parameter h_{11}:

If $v_c = 0$, i.e, the collector voltage does not change, or remains constant at V_C a.c. component is zero, Eqn. (iii) gives

$$v_b = h_{11}i_b \quad \text{or} \quad h_{11} = \left.\frac{v_b}{i_b}\right|_{v_c=0} \text{ at constant } V_C$$

The quantity h_{11} is the ratio of input a.c. voltage to input a.c. current and hence represents *input impedance*, when the output is short circuited. It is also called a short-circuit input impedance'. It is denoted by h_{ie} where the suffix '*i*' relates the input and '*e*' for the common emitter configuration. There are other two similar quantities h_{ib} and h_{ic} related to CB and CC modes.

Thus, $$h_{11} = h_{ie} = \left.\frac{v_b}{i_b}\right|_{v_c=0} \qquad \ldots (v)$$

The unit of h_{ie} is ohm.

2. The parameter h_{12}:

Now if $i_b = 0$ i.e. base current does not change or remains constant at I_B, from equation (*iii*), we get

$$v_b = h_{12}v_c \quad \text{or} \quad h_{12} = \left.\frac{v_b}{v_c}\right|_{i_b=0} \text{ i.e. at constant } I_B$$

The quantity h_{12} is the ratio of a.c. base (input) voltage to a.c. collector (output) voltage and is called as '*reverse voltage transfer ratio*' with input terminals open. It is denoted by h_{re}. Here the sufix '*r*' is for the reverse voltage transfer ratio and '*e*' stands for common emitter configuration.

Thus, $$h_{12} = h_{re} = \left.\frac{v_b}{v_c}\right|_{i_b=0} \qquad \ldots (vi)$$

There are other two similar quantities h_{rb} and h_{rc} related to CB and CC configurations. h_{re} is a pure (dimensionless) number, being ratio of two voltages i.e. no unit.

3. The Parameter h_{21}:

Suppose $v_c = 0$ i.e collector voltage does not change and remains constant at V_c, then equation (iv) gives

$$i_c = h_{21} i_b \text{ or } h_{21} = \left.\frac{i_c}{i_b}\right|_{v_c=0} \quad \text{i.e. at constant } V_C$$

The quantity h_{21} is a ratio of a.c. collector (output) current to a.c. base (input) current and is called as '*forward current transfer ratio*' or current gain with output shorted to a.c. signal.

It is denoted by h_{fe}, suffix '*f*' denotes forward transfer ratio and '*e*' denotes CE configuration. h_{fe} is the same as β_{ac} of the transistor. Thus

$$h_{21} = h_{fe} = \beta_{ac} = \left.\frac{i_c}{i_b}\right|_{v_c=0} \quad \text{... }(vii)$$

The quantity h_{fe} is a pure (dimensionless) number i.e. hence no unit. There are other two similar quantities h_{fb} and h_{fc} related to CB and CC configurations.

4. The Parameter h_{22}:

Let $i_b = 0$ i.e base current does not change or base in open circuited for a.c., the equation (*iv*) gives

$$i_c = h_{22} v_c \text{ or } h_{22} = \left.\frac{i_c}{v_c}\right|_{i_b=0} \quad \text{i.e. at constant } I_B$$

The quantity h_{22} is the ratio of a.c. collector (output) current to a.c. collector (output) voltage and is called as '*open circuit output admittance*' (or *conductance*) for $i_b = 0$ i.e. with input terminals open. It is dented by h_{oe}. Suffix 'o' is for the 'output conductance'. The value of h_{oe} is mho.

Thus, $$h_{22} = h_{oe} = \left.\frac{i_c}{v_c}\right|_{ib=0}$$

There are other two similar parameters h_{ob} and h_{oc} related to CB and CC configuration.

Equation (*iii*) and (*iv*) can be written as

$$v_b = h_{ie} i_b + h_{re} v_c$$

and $$i_c = h_{fe} i_b + h_{oe} v_c$$

CONVERSION FORMULAE:

Transistor manufacturers normally specify the *h*-parameters only in CE mode, When the transistor is to be used in CB or CC mode, it becomes necessary to have the *h*-parameters in that mode. The following table gives the conversion in tabular form.

Common Emitter	*Common Base*	*Common Collector*
h_{ie}	$h_{ib} = \frac{h_{ic}}{1+h_{fe}}$	$h_{ic} = h_{ie}$
h_{re}	$h_{rb} = \frac{h_{ie} h_{oe}}{1+h_{fe}} - h_{re}$	$h_{rc} = 1$

h_{fe}	$h_{fb} = \dfrac{-h_{fe}}{1+h_{fe}}$	$h_{fc} = -(1+h_{fe})$
h_{oe}	$h_{ob} = \dfrac{h_{oe}}{1+h_{fe}}$	$h_{oc} = h_{oe}$

27.14 LOW FREQUENCY *h*-PARAMETER EQUIVALENT CIRCUIT

Fig. 27.18 shows a low frequency *h*-parameter equivalent circuit for a linear four terminal network of a transistor satisfying the two network equations

$$v_b = h_{ie} i_b + h_{re} v_c \qquad ...(ix)$$

and

$$i_c = h_{fe} i_b + h_{oc} v_c \qquad ...(x)$$

The Eqn. (*x*) shows the instantaneous a.c. base voltage made up of (*i*) $h_{ie} i_b$, the voltage drop across the line resistance (h_{ie}) produced by the instantaneous a.c. base current (i_b) and (*ii*) $h_{re} v_c$, an ideal voltage source. The polarity of voltage source is such that the top end is positive w.r. to ground at the instant when the base current enters the base. This is shown by left half portion of Fig. 27.18.

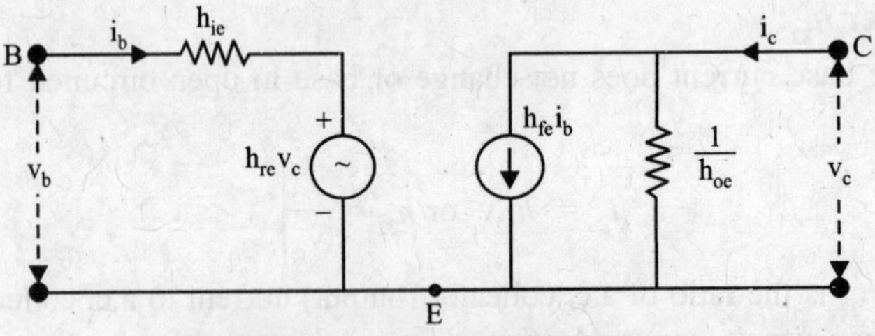

Fig. 27.18 h-parameter equivalent circuit

The equation (x) shows the instantaneous a.c. collector current made up of (*i*) $h_{fe} i_b$, an ideal current source and (ii) $h_{oe} v_c$, a parallel or shunt current. This is achieved by connecting a shunt resistance $\dfrac{1}{h_{oe}}$. Now due to the a.c. collector voltage v_c, a current of $\dfrac{V_c}{1/h_{oe}} = h_{oe} v_c$ flows through the shunt $\dfrac{1}{h_{oe}}$. The current source being ideal, possesses infinite resistance and hence will not draw any current, hence its loading effect is zero. Thus, the total a.c. collector current i_c is the sum of current through $\dfrac{1}{h_{oe}}$ and the current supplied by an ideal current source. The phase of an ideal current source is such that it sends current along the arrow pointing downwards when the collector current i_c enters the combination. This is shown by right half portion of Fig. 27.18.

27.15 FAITHFUL AMPLIFICATION

The basic function of a transistor is to do amplification. Out of three configurations, the CE mode is best suited, even for cascading, to achieve the desired (large) gain at the output. The weak signal is given to the base of the transistor and the amplified output is obtained in the collector circuit. Most important requirements during amplification are: (i) the magnitude of the signal should increase and (ii) there should be no change in the signal shape. Thus, the process of raising the strength of a weak signal without any change in its original shape is known as **Faithful amplification**.

A transistor will function properly if its input circuit (base-emitter junction) remain forward biased and output circuit (collector-base junction) remain reverse biased at all times. For getting faithful amplification, the following basic condition must be satisfied.

(1) Proper zero signal collector current.

(2) Minimum proper base-emitter voltage at any instant.

(3) Minimum proper collector-emitter voltage (V_{CE}) at any instant.

The selection of operating point Q must be proper such that the base-emitter junction shall remain forward biased during all parts of the signal i.e. during positive half & negative half of the cycle. The conditions (1) and (2) are then satisfied. On the other hand, the condition (3) ensures that the base-collector junction shall remain properly reverse-biased at all times.

1. Proper zero signal collector current

This means when no signal is applied, the d.c. current I_C flowing in the collector circuit must be proper due to V_{BB} so as to accommodated both, the positive half and negative half of the cycle, when signal is applied. In other words, during the positive half-cycle of the signal, the circuit is more forward biased ad hence collector current increases. However, during the negative half-cycle

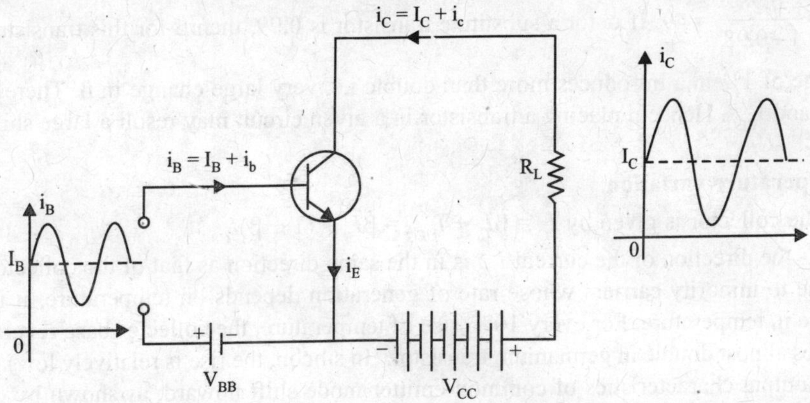

Fig. 27.19 Faithful Amplification

of the signal, input circuit in less forward biased and collector current decreases. In this way, the negative half-cycle of the signal also appears in the output. The shape of input signal and output amplified signal has exactly the same i.e. no distortion. Hence faithful amplification results.

2. Proper minimum base-emitter voltage (V_{BE})

This condition is necessary to achieve faithful amplification because the cut in voltage for Ge being 0.3 V and for Si 0.7 V, the base-emitter voltage (V_{BE}) should never fall below these values otherwise distortion will occur.

3. Proper minimum collector-emitter voltage (V_{CE})

For faithful amplification, the collector-emiter voltage V_{CE} should not fall below the knee voltage V_{knee} which is 0.5 V for Ge and 1.0 V for Si transistor. When V_{CE} is too low than V_{knee} voltage, the collector-base junction is not properly reverse biased. However, when V_{CE} is greater than V_{knee}, the collector-base junction is properly reverse-biased and the value of β remains at constant value, resulting faithful amplification.

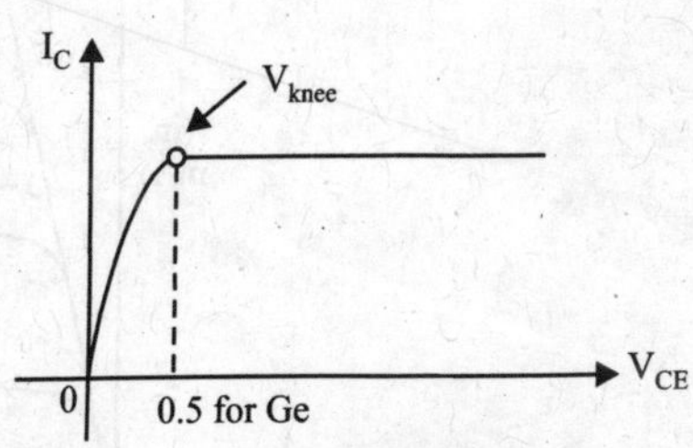

Fig. 27.19 (a)

27.16 STABILISATION

For faithful amplification, it is essential that operating or queiscent point (*i.e.* zero signal I_C nd V_{CE}) must remain fixed. The factors which cause shift in the position of the operating point are:

(1) The change in temperature and

(2) Replacement of the transistor by another transistor.

When the temperature changes or the transistor is replaced by another transistor, the operating point shifts. This is undesired. Therefore, it is of prime importance to make the operating point independent of these variations. This is known as stabilisation. It is defined as: "The process of making operating point independent of temperature changes or variation in transistor parameter is known as stabilisation."

1. Variation in transistor parameter

Any two transistors even of the same type (say *BC* 149 or so) and produced by the same manufacturer, do not have the transistor parameters α, β, V_{BE} exactly the same, i.e. there is usually a considerable spread in characteristics among two transistors of the same type. For example: Suppose the value of α for a transistor is 0.98. The value of β for this transistor will be $\beta = \frac{\alpha}{1-\alpha} = \frac{0.98}{1-0.98} = 49$. If α for a substitute transistor is 0.99, then β for this transistor is 99. Thus, a change of 1% in a introduces more than double i.e. very large change in β. There is also spread in I_{CBO} and I_{CEO}. Hence replacing a transistor in a given circuit may result a large shift in operating point Q.

2. Temperature variation

The collector is given by $I_C = \beta I_B + I_{CEO} = \beta I_B + (1 + \beta)I_{CBO}$

As the direction of the current I_{CO} is in the same direction as that of the collector current, and it is due to minority carriers whose rate of generation depends on temperature, it increases with increase in temperature. For every 10°C rise of temperature, the collector base reverse current I_{CBO} becomes almost double in germanium transistor. (In silicon, the rise is relatively low). Consequently all the output characteristics of common-emitter mode shift upward, as shown by dotted lines in Fig. 27.20. Hence the operating point Q shifts from the middle of the active region to the point Q′ near the saturation line. This results in distortion in output signal and may cause clipping.

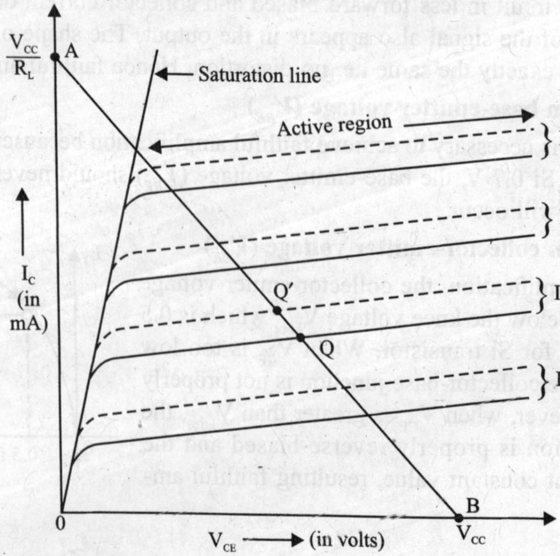

Fig. 27.20

27.17 THERMAL RUNAWAY

In addition to effect of shifting of operating point Q upward (i.e. towards the saturation line) which causes distortion due to rise of temperature, there is another serious effect called as **thermal runaway**. As the temperature increases, there is an increase in collector leakage current I_{CBO}. This increases the power dissipated at the collector junction. This produces heat within the transistor due to which again more and more covalent bonds are broken i.e. further increases the leakage current I_{CBO}. Thus, the sequence is repetitive and hence I_C goes on increasing continuously. This cumulative process increases I_C to a very high value i.e. may exceed the maximum rated value and the transistor will be permanently damaged. Thus, the self destruction of an unstabilised transistor due to the junction temperature exceeding its rated value is known as **thermal runaway.**

Principle: In order to avoid thermal runaway and consequent destruction of the transistor, it is essential to stop the shifting of operating point i.e. stabilisation of operating point Q. In other words, I_C must remain constant. For this purpose, the circuit is so modified that I_B decreases automatically *i.e.* emitter-base junction is made less forward biased automatically with increase in temperature, so that any decrease of βI_B will compensate the increase in $(\beta + 1)\, I_{CBO}$, so that I_C almost remain constant. This is the principle of designing of a stabilised *CE* transistor amplifier.

Heat sink: In case of power amplifiers, which handles large currents and always get heated up, during their operation. Since, the transistor being temperature dependent device, a **heat sink** is provided. Heat sink is nothing but an aluminium or some other metal sheet in which the transistor is fitted in, so that additional heat is dissipated to the surrounding so that the temperature of the transistor remains within permissible limit.

27.18 STABILITY FACTOR

The ability of any circuit in keeping its collector current I_C constant is expressed in terms of the **stability factor**. The stability factor is defined as the ratio of a small change in collector current to the corresponding change in the reverse saturation current.

or Stability factor, $S = \dfrac{\partial I_C}{\partial I_{CBO}} = \dfrac{\partial I_C}{\partial I_{CO}}$ ($\because$ I_{CBO} is also denoted as I_{CO})

If the changes are infinitesimally small, then

$$S = \frac{dI_C}{dI_{CO}} = \frac{\Delta I_C}{\Delta I_{CO}} \qquad ...(i)$$

The collector current I_C is given by Eqn. (v) of article 27.6

$$I_C = \beta I_B + (\beta + 1)\, I_{CO}$$

Differentiating w.r. to I_C, we get

$$1 = 0 + (\beta + 1)\frac{dI_{CO}}{dI_C} \qquad (\because \beta \text{ and } I_B \text{ assuming constant})$$

$$\therefore \quad S = \frac{dI_C}{dI_{CO}} = (\beta + 1) \qquad ...(ii)$$

This shows that the change in collector current is equal to $(\beta + 1)$ times dI_{CO}, the change in the reverse saturation current. For a transistor of $\beta = 400$, the collector current changes 401 times the change in I_{CO}. This is undesired. Hence a common-emitter amplifier, which is most commonly used in practice, should be provided with the bias stabilising circuits.

27.19 BIAS STABILISING CIRCUITS

Fig. 27.21 shows a bias stablising circuit with an arrangement of self-bias or emitter bias with base resistance. The emitter resistance R_E plays an important role in bias stabilisation. Here only one battery E_{CC} is used for biasing. The resistances R_B and R_E are selected such that emitter base diode is forward biased and base collector is reverse biased. The base current flows through R_B, emitter current through R_E and collector current through R_L. The emitter current through R_E, maintains emitter at higher potential than the ground. The base potential in still higher than emitter potential, to make the emitter-base junction forward biased.

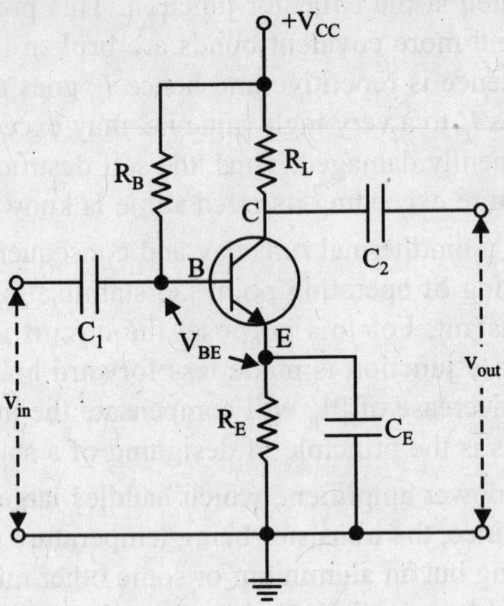

Fig. 27.21

Working: Suppose the temperature rises. This will result in increase in I_C and hence I_E also increases. This increases the voltage drop across R_E. The potential of emitter increases causing the base-emitter junction less forward bias. The reduction in forward bias, reduces base current I_B and hence emitter current decreases. So the tendency of the increase in the collector current is opposed. I_C current tends to remain constant, even through there is a rise in temperature. Thus, the desired condition of bias stability seems to be achieved. But this is not perfectly true, because we have not considered the voltage drop I_BR_B. In fact I_BR_B drop decreases with increase in temperature and hence do not help to reduce the forward bias of emitter-base junction. The potential of the base should be maintained constant and independent of the base-current. This situation is achieved by considering a potential divider arrangement as shown in Fig. 27.22.

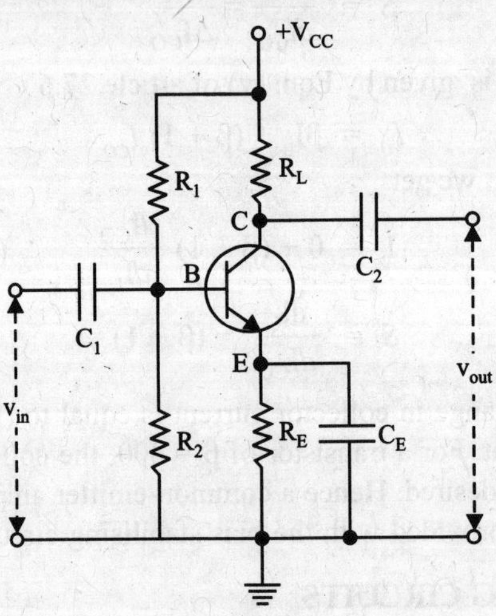

Fig. 27.22

Now, the values of R_1 and R_2 are so selected that the voltage drop across R_2 maintains emitter-base junction forward biased and a suitable operating point Q. The current flowing through the potential divider of R_1 and R_2 is much higher than base current (nearly 10 times or above). The current flowing through R_1 is the sum of the original divider current and the base current I_B. Since the base current is very small (of the order of μA) as compared to the divider current, any change (reduction) in base current due to increase in temperature will have almost no effect on the total current through the resistance R_1. Thus, the potential of base of the transistor is maintained constant and thus kept independent of the base current. This makes the circuit more stable.

An coupling capacitor C_1 blocks d.c. current to pass through the a.c. source i.e. signal source (v_i) but allows the a.c. signal through it. Similarly, another blocking capacitor C_2 is used to block again d.c. current due to $+V_{CC}$ through the next stage but allows the amplified a.c. signal very easily. The purpose of using C_E to bypass the a.c. component across R_E, when an a.c. source is applied. Thus capacitor C_E avoids fluctuations in the self-bias by bypassing the a.c. components. The capacitor C_E, therefore called as bypass capacitor.

EXPRESSION FOR STABILITY FACTOR

Consider the bias stabilising circuit shown by Fig. 27.21.

Applying Kirchhoff's law, we can write

$$V_{CC} = I_B R_B + V_{BE} + I_E R_E \qquad \text{...(i)}$$

Putting $\quad I_E = I_B + I_C$

$$V_{CC} = I_B R_B + V_{BE} + R_E(I_B + I_C)$$

$$= I_B(R_B + R_E) + V_{BE} + I_C R_E$$

But $\quad I_C = \beta I_B + (\beta + 1) I_{CO} \qquad (\because \text{Eqn. } (v) \text{ of article 27.6})$

$$\therefore \quad I_B = \frac{1}{\beta} I_C - \frac{(\beta+1)}{\beta} I_{CO}$$

Substituting, we get

$$V_{CC} = \frac{R_B + R_E}{\beta} I_C - \frac{\beta+1}{\beta}(R_B + R_E) I_{CO} + V_{BE} + I_C R_E$$

Differentiating both sides w.r. to I_{CO} and assuming V_{BE} to be constant, ($V_{BE} = 0.7$ V for Si, and $V_{BE} = 0.3$ V for Ge), we get

$$O = \frac{R_B + R_E}{\beta} \cdot \frac{dI_C}{dI_{CO}} - \frac{\beta+1}{\beta}(R_B + R_E) + O + R_E \frac{dI_C}{dI_{CO}}$$

or $$\frac{dI_C}{dI_{CO}}[R_B + R_E(1+\beta)] = (\beta + 1)(R_B + R_E)$$

$$\therefore \quad \text{Stability factor,} \quad S = \frac{dI_C}{dI_{CO}} = \frac{(\beta+1)(R_B + R_E)}{R_B + R_E(1+\beta)}$$

$$= (\beta+1) \times \frac{R_B + R_E}{R_B + R_E(1+\beta)}$$

$$S = (\beta+1) \times \frac{1 + \dfrac{R_B}{R_E}}{\beta + 1 + \dfrac{R_B}{R_E}} \qquad \text{...(ii)}$$

The stability factor S for the circuit using potential divider i.e. for circuit diagram shown by Fig. 27.22 comes out to be same as given by Eqn. (*ii*).

Thus, in Eqn. (*ii*) R_B stands for base bias resistance or is equivalent resistance of the parallel combination of R_1 and R_2 in case of potential divider arrangement. To examine the circuit performance:

(*i*) if R_E is zero, $\frac{R_B}{R_E} >> 1$, then S become equal to $(\beta + 1)$. This is the case of unstablised circuit.

(*ii*) if R_E is large, $\frac{R_B}{R_E} << 1$, then S = 1, the circuit is most stable, i.e. the stability is maximum.

EXERCISE

1. Explain the three transistor configurations? Draw the circuit diagram using *p-n-p* and *n-p-n* transistor.
2. Explain the input and output characteristics of a common base transistor circuit. How will you study them experimentally. Discuss the three important regions of output characteristics.
3. What do you mean by transfer characteristics of a common base transistor configuration? The slope of the graph is nearly 45°, what does it indicates?
4. With the help of an experimental arrangement, explain how the input, output and transfer characteristics of common emitter mode by studied. Sketch the characteristics is each mode. Explain the condition to use the transistor in the active and cut-off region of output characterstic.
5. Define α and β. Establish the relation between α and β for d.c. as well as a.c.
6. Derive the accurate expression for collector current in *CB* and *CE* mode of a bipolar junction transistor.
7. Why transistor is called an active device? Explain the mechanism of amplification in a transistor. Show that the output of a common emitter amplifier is 180° out of phase with the input signal. Calculate the voltage gain of *CE* amplifier.
8. Discuss the merits and demerits of a common emitter amplifier.
9. What is d.c. load line? How will you draw d.c. load line on the output characteristics of a transistor? What is its importance?
10. With the help of a neat figure, explain the graphical method of analysing the working of a transistor as an amplifier in *CE* mode.
11. Discuss the advantages and drawbacks of graphical method.
12. What is a.c. load line? For a transistor as *CE* amplifier with $R_1 = 10$ kΩ, $R_2 = 5$kΩ, $R_E = 2$kΩ, $R_C = 1$kΩ and $V_{CC} = +15$V Draw
    (*i*) d.c. load line
    (*ii*) Determine the operating point
    (*iii*) Draw a.c. load line. Assume $V_{BE} = 0.7$ V.
    Ans. (i) A(0, 5mA), B(15V), (ii) Q (8.55 V, 2.15 mA), (iii) C (0, 19.25 mA) and D (9.62 V, 0)
13. Considering transistor as a two-port network, define hybrid parameters for a transistor. Draw a low frequency h-parameter equivalent circuit for a transistor. ***(Nagpur U. 2012)***
14. What do you mean by faithful amplification? What are the necessary conditions?
15. What do you mean by stabilisation as applied to a transistor amplifier? Why is stabilisation necessary? Explain thermal runaway. ***(Nagpur U. 2012)***
16. Explain the principle involved in avoiding the thermal runaway. Derive stability factor. ***(Nagpur U. 2009)***
17. Explain with a neat circuit diagram that a potential divider arrangement is much suitable than using base resistance. Derive an expression for stability factor. What is the role of a bypass capacitor in CE? ***(Nagpur U. 2009)***

FIELD EFFECT TRANSISTORS

Chapter 28

INTRODUCTION

The field effect transistor (FET), like the bipolar junction transistor (BJT) is a three terminal semiconductor device. A field effect transistor is called a **unipolar** device because the current through it is carried by only one of the two kinds of majority charge carriers: electrons (in n-channel FET) or holes (in p-channel FET). The name field effect is due to the fact that the current flow is controlled by a electric field set-up in the device by an externally applied voltage. There are mainly two types of FETs: the junction field-effect transistor (JFET) and the metal-oxide-semiconductor FET (MOSFET). The FET has high input impedance and low noise level. FET's can be used as a linear device in amplifier or as a digital device in logic circuits. The MOSFET is the most important component in modern digital integrated circuits, such as microprocessor ad computer memories.

28.1 JUNCTION FIELD EFFECT TRANSISTOR (JFET)

JFET is of two types (a) n-channel JFET and (b) p-channel JFET.

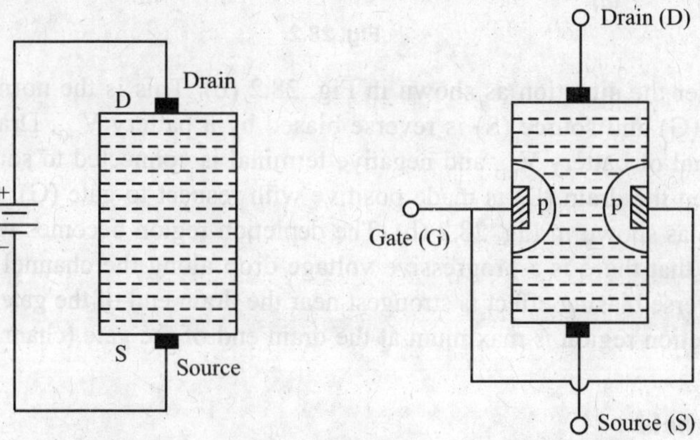

Fig. 28.1

Basic Construction of JFET

(a) n-channel FET: It consists of a lightly doped n-type silicon bar. To the two ends of the bar are connected two ohmic contacts, which are non-rectifying as shown in Fig. 28.1 (a) One of the contacts is called a source and the other one is called a drain.

Through the source, majority charge carriers, which are electrons in this case, enter the n-type material, ad they come out of the drain. The current flowing in this case, depends on the voltage applied and the amount of doping. From S to D, there is a uniform voltage drop due to the external voltage source connected between S and D as shown in Fig. 28.1 (*a*).

Now imagine a situation, that would arise by embedding two heavily doped p-regions, on the sides of the n-channel as shown in Fig. 28.1 (b). This situation is similar to a pn junction. Because of heavy doping of p, at the junction the depletion region is formed and will extend in the n-channel. Ultimately a dynamic equilibrium is set up, and the region near the junction gets depleted of any charge carriers, and hence called the depletion region.

The two heavily doped p-junctions are known as gates, which are internally connected together and only one external lead in taken out as gate (G). When the gate is reverse biased by battery V_{GG}, the depletion region extends into the n-channel and hence the channel width decreases. The width of the n-channel, thus depends on the amount of reverse bias across the gate and source. Larger the reverse bias, smaller becomes the channel width as shown in Fig. 28.2 (a).

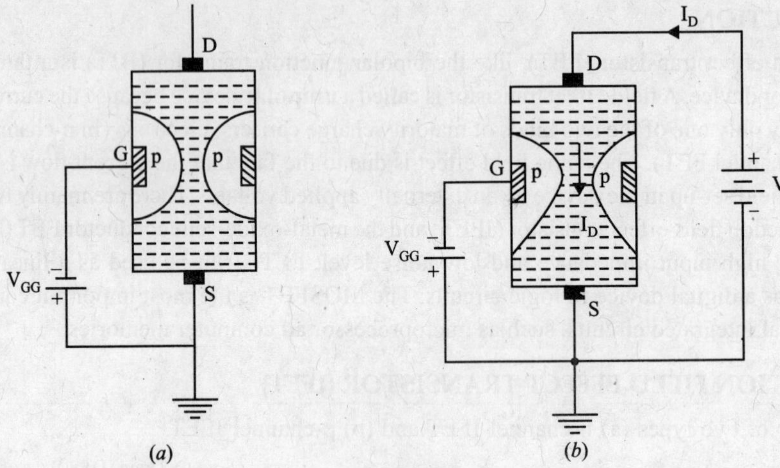

Fig. 28.2

Now consider the situation as shown in Fig. 28.2 (b). This is the normal mode of biasing a FET. The gate (G) and source (S) is reverse biased by a battery V_{GG}. Drain (D) is connected to positive terminal of battery V_{DD} and negative terminal is connected to source (S). The source is grounded. When the drain (D) is made positive with respect to gate (G), the depletion region changes its shape as shown in Fig. 28.2 (b). The depletion region becomes wedge shaped, on account of the fact that there is a progressive voltage drop along the channel (the drain positive) and hence the reverse biasing effect is strongest near the drain end of the gate. In other words, the width of the depletion region is maximum at the drain end of the gate (channel width naturally is minimum here).

Schematic Symbols

The schematic symbols of n-channel and p-channel JFET are shown in Fig. 28.3 (a)

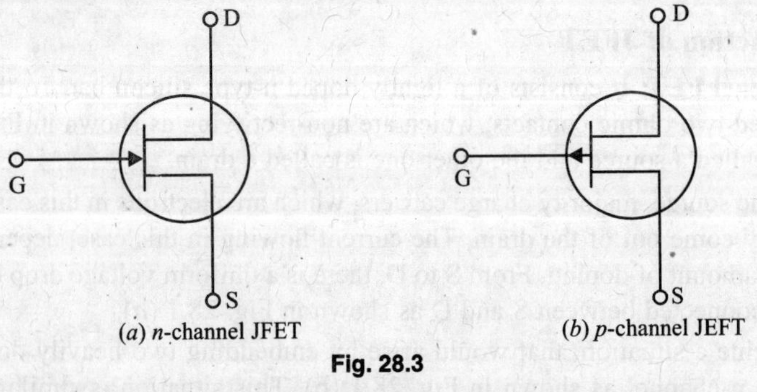

(*a*) *n*-channel JFET (*b*) *p*-channel JEFT

Fig. 28.3

The arrow on the gate terminal signifies the direction of the conventional gate current when the gate-source junction is forward biased. However, in normal operation, the gate-source junction is reverse biased.

Working Principle: The width of the depletion region in Fig. 28.2 (b) on the magnitude of the reverse-biasing voltage V_{GS}. It has also been said that as the rever crease, the depletion region expands, thereby decreasing the channel width. Hence, the a cross-sectional area for the current carriers decreases, which means resistance increases. Th urrent flowing in the channel, is a function of reverse-biasing voltage V_{GS} and the channel s as a variable resistance.

28.2 STATIC CHARACTERISTICS OF JFET

Static characteristics of JFET are the graphical plots of the drain current (I_D) against th ain to source voltage (V_{DS}) keeping gate to source voltage (V_{GS}) constant. It is also known as ut characteristic or common source characteristic.

Experimental Arrangement

To study the characteristics of a n-channel JFET, the experimental arrangement is shown in Fig. 28.4 Keeping gate-source voltage V_{GS} fixed at some value say – 1V, the drain-source voltage (V_{DS}) is slowly increased in steps and the corresponding drain current I_D is noted. A graph between V_{DS} and I_D gives the static characteristic corresponding to V_{GS} = –1V. Similar curves are plotted for V_{GS} = 0V, –2V, -3V and so on.

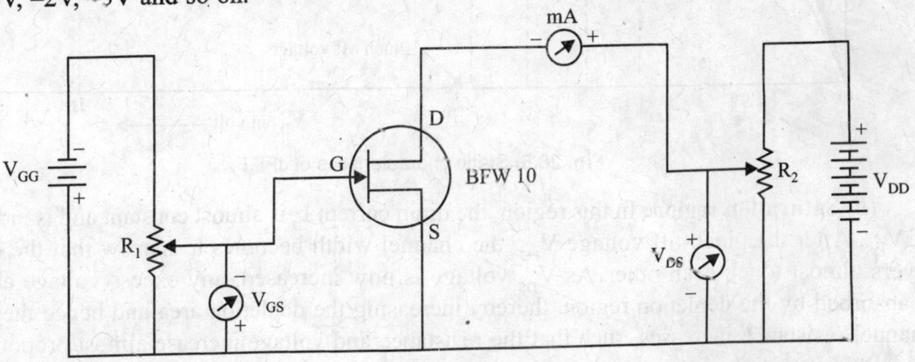

Fig. 28.4

A family of static (output) characteristics is shown in Fig. 28.5 These characteristics can be divided into three regions:

***(i)* Linear region:** In this region, the drain current I_D is very nearly proportional to the drain source voltage V_{DS}. Consider the characteristic for V_{GS} = 0V. When the voltage V_{DS} is increased from 0V by a small amount, the n-channel acts as a simple resistance. Therefore, the current I_D increases linearly with V_{DS} in this region.

This ohmic relationship between V_{DS} and I_D continue till V_{DS} reaches a certain critical value called *pinch-off* voltage (V_{PO}), when the current becomes constant as its maximum value called I_{DSS}. (i.e. I_D saturation). Under pinch-off condition, separation between the depletion regions near the drain i.e. width reaches a minimum value. At V_{GS} = 0V, pinch-off does occur because of the fact that the gate (G) and channel are still reverse-biased because of the potential drop in the channel. Naturally, as the gate becomes progressively more and more negative, the pinch-off occurs at lesser and lesser V_{DS}. The input impedance of the device is also very high of the order of 10^6 to 10^8 ohm, in contrast to a very low input resistance of the order of few ohms in case of BJT.

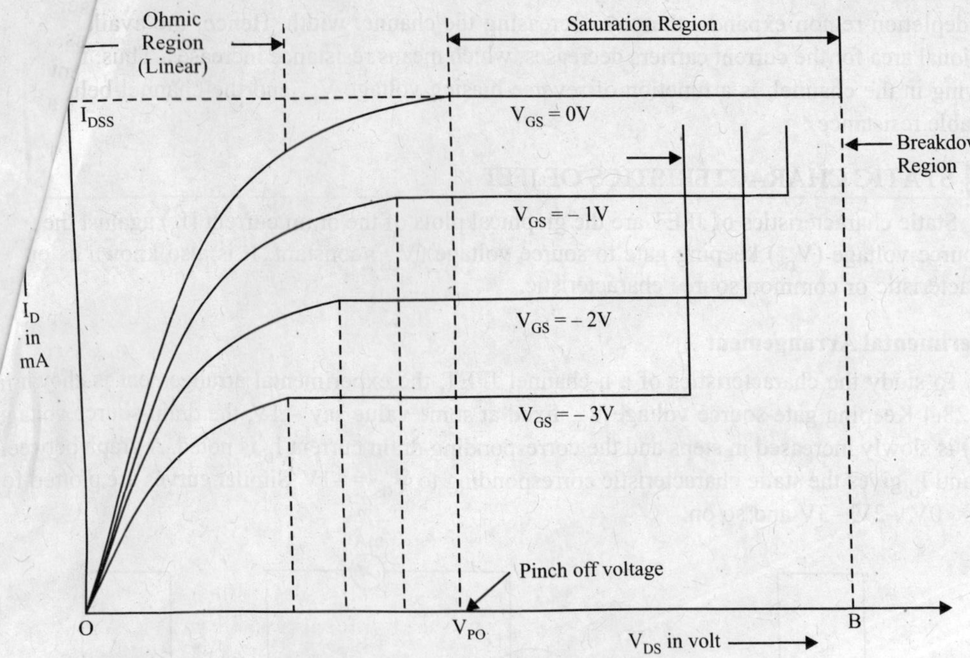

Fig. 28.5. Static characteristics of JFET

***(ii)* Saturation region:** In this region, the drain current I_D is almost constant and is independent of V_{DS}. After the pinch-off voltage V_{PO}, the channel width becomes to narrow that the depletion layers almost touch each other. As V_{DS} voltage is now increased, any excess voltage above V_{PO} is absorbed by the depletion region, thereby increasing the depletion area and hence the effective channel resistance increases, such that the resistance and voltage increase almost proportionately, the drain current I_D remain very nearly constant.

***(iii)* Breakdown region:** If V_{DS} become too large, the drain current I_D rises abruptly, as expected, because this would be the avalanche breakdown voltage. The avalanche multiplication of electrons caused by braking covalent bonds of Si atoms lying in the depletion region between the gate and drain. As the gate to source reverse bias is increased, the breakdown takes place at a lower value of V_{DS} i.e. breakdown voltage decreases.

28.3 TRANSFER CHARACTERISTIC

To plot the transfer characteristic, keep some constant value of V_{DS} at the middle point of the saturation region of the static characteristic and note down the drain current I_D for various values of V_{GS}. A graph is plotted between drain current I_D against V_{GS} as shown in Fig. 28.6. The point A where this curve meets the I_D axis ($V_{GS} = 0$) gives the value of I_{DSS}, drain to soure current with shorted gate. I_{DSS} is the maximum current the JFET can produce. The point B where this curve meets the V_{GS} axis on the negative side give the magnitude of gate to source cut-off voltage V_{GS} (cut-off) i.e. at V_{GS} (cutoff), we get $I_D = 0$.

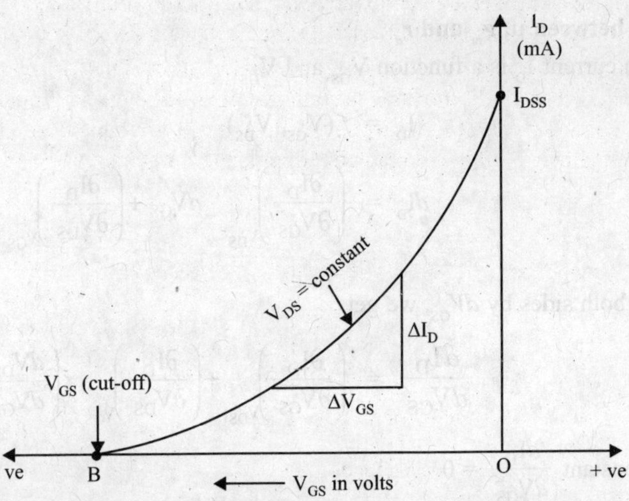

Fig. 28.6

28.4 PARAMETERS OF JFET

One of the ways of selecting a JFET to be used in a particular circuit, is from the knowledge of its parameters or constants. The main parameters are:

1. Drain resistance (r_d): It is defined as the ratio of a small change in drain to source voltage to the corresponding small change in drain current, when gate to source voltage is kept constant. Thus,

$$\text{Drain resistance,} \quad r_d = \left.\frac{\Delta V_{DS}}{\Delta I_D}\right|_{V_{GS}\ \text{constant}} \quad \ldots(i)$$

It is also called a.c. or dynamic drain resistance. Its unit is ohm. It is given by the slope of the drain characteristic in the 'pinch-off region'. The reciprocal of drain resistance is known as output admittance y_{os} . i.e. $y_{os} = \frac{1}{r_d}$. BFW 10 has a very high value of r_d, say $1.17 \times 10^4\Omega$.

2. Trans-conductance (g_m): It is defined as the ratio of a small change in drain current to the corresponding small change in g_{te} to source voltage, keeping drain to source voltage constant. Thus,

$$\text{Trans-conductance,} \quad g_m = -\left.\frac{\Delta I_D}{\Delta V_{GS}}\right|_{V_{DS}\ \text{constant}} \quad \ldots(ii)$$

The quantity g_m is also known as mutual conductance. Its unit is mhos or Amp/volt. This can be calculated from Fig. 28.6 The –ve sign indicates that as V_{GS} is increased (i.e. made more negative), I_D decreases.

3. Amplification factor (μ): It is defined as the ratio of small change in the drain to source voltage to the corresponding change in gate to source voltage, keeping the drain current constant. Thus,

$$\text{Amplification factor,} \quad \mu = \left.\frac{\Delta V_{DS}}{\Delta V_{GS}}\right|_{I_D\ \text{constant}} \quad \ldots(iii)$$

This parameter μ of a JFET indicates how much more control the gate voltage has over drain current than has the drain voltage. It has no unit. The –ve sign indicates that when V_{GS} is increased, V_{DS} decreases to keep I_D constant.

Relation between μ, g_m and r_d

The drain current I_D is a function V_{GS} and V_{DS}

or $$I_D = f(V_{GS}, V_{DS})$$

$$\therefore \quad dI_D = \left(\frac{\partial I_D}{\partial V_{GS}}\right)_{V_{DS}} dV_{GS} + \left(\frac{\partial I_D}{\partial V_{DS}}\right)_{V_{GS}} dV_{DS}$$

Dividing both sides by dV_{GS}, we get

$$\frac{dI_D}{dV_{GS}} = \left(\frac{\partial I_D}{\partial V_{GS}}\right)_{V_{DS}} + \left(\frac{\partial I_D}{\partial V_{DS}}\right)_{V_{GS}} \times \left(\frac{dV_{DS}}{dV_{GS}}\right)$$

If I_D is constant $\frac{dI_D}{dV_{GS}} = 0$

$$\therefore \quad 0 = \left(\frac{\partial I_D}{\partial V_{GS}}\right)_{V_{DS}} + \left(\frac{\partial I_D}{\partial V_{DS}}\right)_{V_{GS}} \times \left(\frac{dV_{DS}}{dV_{GS}}\right)_{I_D}$$

or $$0 = \left(\frac{\Delta I_D}{\Delta V_{GS}}\right)_{V_{DS}} + \left(\frac{\Delta I_D}{\Delta V_{DS}}\right)_{V_{GS}}$$

or $$0 = -g_m + \left(\frac{1}{r_d}\right)(\mu)$$

$$\mu = -g_m \cdot r_d \quad \ldots(iv)$$

This is the required relation between the parameters μ, g_m and r_d.

28.5 D.C. BIASING OF JFET

A JFET may be biased by using either

(i) Two separate power supplies V_{GG} and V_{DD}, as shown in Fig. 28.7 (a) or

(ii) Self biasing arrangement (D.C.) using only one power supply V_{DD}. The circuit as shown in Fig. 28.7 (b) and is called as self-biasing arrangement. The bias is obtained from JFET's own source current (i.e. drain current I_D) through R_S.

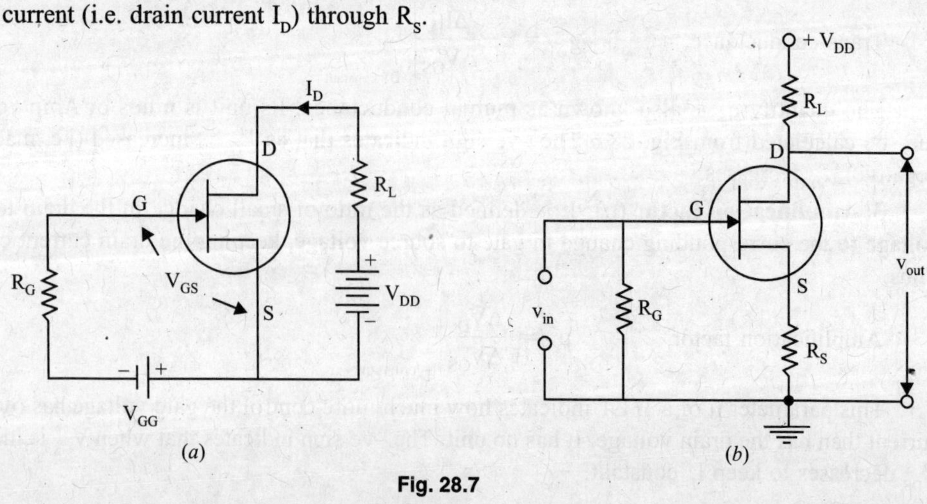

Fig. 28.7

The gate is kept at this much negative potential with respect to the ground. The addition of R_G in Fig. 28.7 (b) does not upset this d.c. bias for the simple reason that no gate current flows through it (the gate leakage current is almost zero). Hence the gate is essentially at ground potential. Without R_G, the gate would be 'floating' which could collect charge and ultimately the JFET undergoes 'cut-off'.

The resistance R_G has been put in the circuit for avoiding short circuting of the a.c. input voltage V_{in}. Moreover, if any leakage current would build up static charge, (voltage) at the gate which could change the bias or even destroy the JFET.

D.C. load line

A d.c. load line for JFET can easily be drawn by knowing the following two points:

(*i*) At $I_D = 0, V_{DS} = V_{DD}$

and (*ii*) At $V_{DS} = 0,$ $\quad I_D = \dfrac{V_{DD}}{R_L}$

The Q-point is generally situated at the middle point of the load line (for Class-A operation), so that

$$V_{DSQ} = \frac{1}{2} V_{DD}$$

Also $$I_{DSQ} = \frac{\frac{V_{DD}}{2}}{R_S + R_L}$$

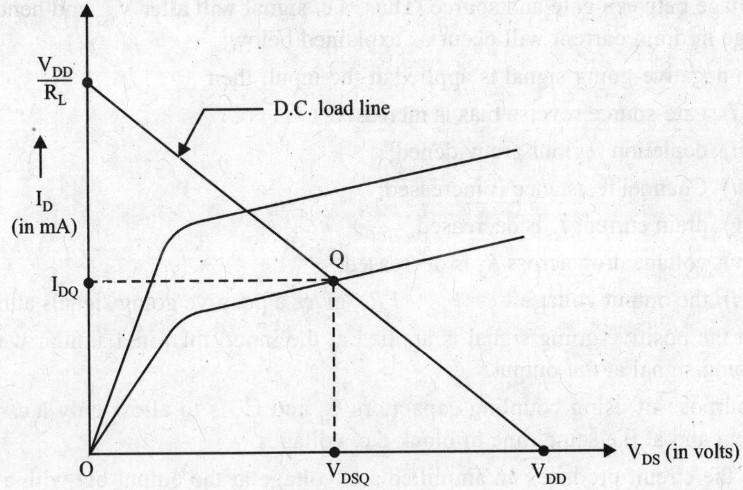

Fig. 28.8

28.6 JFET AS AN AMPLIFIER

Fig. 28.9 shows the circuit diagram using JFET as an amplifier. It is easier to use JFET in the amplifier circuit than using bipolar junction transistor (BJT), because the former has high input impedance. Here R_G serves the purpose of providing leakage path to the gate current, R_S develops self-bias and C_S acts as a bypass capacitor for any a.c. component across R_S.

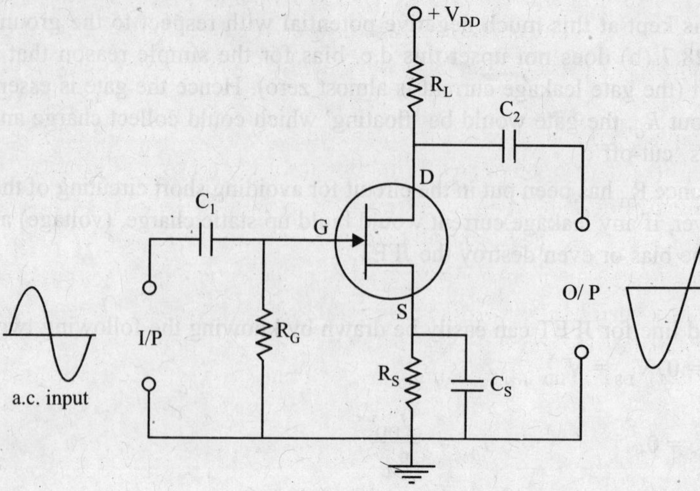

Fig. 28.9

Working:

When input signal is zero, the gate current is practically zero and hence there is no voltage drop across the resistance R_G. The drain current I_D flowing through the resistance R_S, produces a voltage drop $I_D R_S$ across it, which is responsible for maintaining the reverse biasing between the gate and the source. The value of R_S is properly selected to have desired biasing i.e. desired operating point Q.

When the input signal is applied, the a.c. signal voltage is superimposed on the d.c. reverse biasing voltage between gate and source. Thus, a.c. signal will alter V_{GS} and hence corresponding large change in drain current will occur as explained below:

When negative-going signal is applied at the input, then

(*i*) gate-source reverse bias is increased,

(*ii*) depletion regions are widened,

(*iii*) Channel resistance is increased,

(*iv*) drain current I_D is decreased,

(*v*) voltage drop across R_L is decreased,

(*vi*) the output voltage $V_d = V_{DD} - I_d R_L$, gives a positive going signals at the output.

When the positive-going signal is applied at the input, then in a similar way, we will get a negative going signal at the output.

The purpose of using coupling capacitors C_1 and C_2 is to allow only a.c. voltages to pass through them and at the same time to block d.c. voltages.

Thus, the circuit produces an amplified a.c. voltage at the output but with a phase inversion of 180°. Thus, the circuit acts as an amplifier.

Let us evaluate the input impedance (Z_i), output impedance (Z_o) and voltage gain (A_v) of a common-source JFET amplifier.

(1) Input Impedance (Z_i)

The input impedance of JFET is ideally an infinite (i.e. open circuit).

Hence, $Z_i = \infty$.

In practice,l $Zi = R_G \parallel R_{GS} = \dfrac{R_G \cdot R_{GS}}{R_G + R_{GS}}$

In an ideal JFET, R_{GS} is infinite because $I_G = 0$. In practice, however, R_{GS} is not actually infinite but extremely high as compared to R_G. ...(v)

$$\therefore \quad Z_i \cong R_G$$

The practical value of Z_i for JFET is of the order of $10^9\ \Omega$.

(2) Output Impedance (Z_o)

$$Z_o = r_d \parallel R_L = \frac{r_d \cdot R_L}{r_d + R_L}$$

$$\cong R_L \text{ if } r_d >> R_L \quad \text{...(vi)}$$

(3) Voltage Gain (A_v)

Considering the a.c. components, the output voltage is

$$v_o = i_d \times (r_d \parallel R_L)$$

But $g_m = \dfrac{i_d}{v_i}$ or $i_d = g_m \cdot v_i$

Substituting, we get

$$v_o = g_m \cdot v_i \times (r_d \parallel R_L)$$

The voltage gain,

$$A_v = \frac{v_o}{v_i} = \frac{g_m \cdot v_i \times (r_d \parallel R_L)}{v_i}$$

$$= g_m \times (r_d \parallel R_L)$$

where $(r_d \parallel R_L) = \dfrac{r_d \cdot R_L}{r_d + R_L}$

$$\therefore \quad A_v = \frac{g_m \cdot r_d \cdot R_L}{r_d + R_L}$$

$$A_v = \frac{-\mu R_L}{r_d + R_L} \quad (\because \mu = -g_m \cdot r_d) \quad \text{...(vii)}$$

Since $r_d >> R_L$, $A_v = -g_m \cdot R_L$ ... (viii)

28.7 SOURCE-FOLLOWER JFET: A COMMON-DRAIN CIRCUIT

A common-drain or source-follower circuit with an a.c. circuit configuration is shown in Fig. 28.10. The circuit is similar to a bipolar emitter follower configuration.

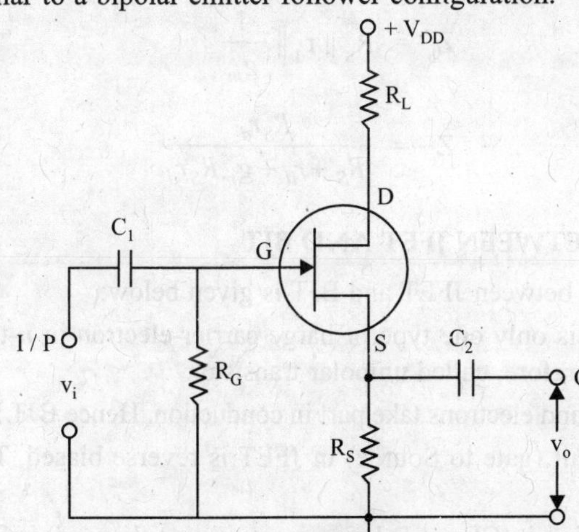

Fig. 28.10. JFET as source-follower

Like BJT, the JFET source-follower provides voltage gain less than 1 with no polarity inversion. In addition, the JFET circuit provides very high input impedance (Zi) and low output impedance (Z_o).

To evaluate voltage gain (A_v)

The a.c. signal output is taken from the source terminal as shown in Fig. 28.10. The output voltage is

$$v_o = i_d \times (R_S \parallel r_d)$$

Using
$$g_m = \frac{i_d}{v_{gs}} \text{ or } i_d = g_m v_{gs}$$

Substituting, we get

$$v_o = (g_m v_{gs})(R_S \parallel r_d)$$

Since, the voltage v_{gs} in source-follower is

$$v_{gs} = v_g - v_s = v_i - v_o$$

$$\therefore \quad v_o = g_m (v_i - v_o)(R_S \parallel r_d)$$

$$= g_m v_i (R_S \parallel r_d) - g_m v_o (R_S \parallel r_d)$$

or
$$v_o [1 + g_m (R_S \parallel r_d)] = g_m v_i (R_S \parallel r_d)$$

$$\therefore \quad \text{Voltage gain, } A_v = \frac{v_o}{v_i} = \frac{g_m (RS \parallel r_d)}{1 + g_m (R_S \parallel r_d)} \qquad \ldots(ix)$$

From equation (*ix*) we can see that the voltage gain is +ve non-inverting and has a value less than unity. The gain approaches a value of 1, as the value of g_m $(R_S \parallel r_d)$ becomes much larger than 1.

Input Impedance

The input impedance of a JFET source-follower circuit is the value of gate resistance

$$Z_i = R_G$$

Output Impedance

The output impedance of a JFET source-follower is the impedance seen on looking back into the source. The source resistance R_S in parallel with device drain resistance r_d, in parallel with the current source, $g_m V_{gs}$. If we replace the current source by a voltage source in series with resistance $1/g_m$, then

$$Z_o = R_S \parallel r_d \parallel \frac{1}{g_m}$$

or
$$Z_o = \frac{R_S r_d}{R_S + r_d + g_m R_S r_d} \qquad \ldots(xi)$$

28.8 DIFFERENCE BETWEEN JFET AND BJT

The the difference between JFET and BJT is given below:

(1) In JFET, there is only one type of harge carrier-electron in *n*-type channel and holes in p-type channel. It is, therefore, called unipolar transistor.

In BJT, both holes and electrons take part in conduction. Hence BJT is called bipolar transistor.

(2) The input circuit (gate to Source) in JFET is reverse biased. Therefore, JFET has very high input impedance.

In BJT, the input circuit (CE mode being most commonly used) is forward biased. Therefore, BJT has low input impedance.

(3) JFET is essentially a 'voltage-driven device'. The input gate to source voltage controls the output current.

While BJT is a 'current-operated device'. The input base-emitter current controls the output collector current.

(4) In JFET, there are no junctions as in BJT. The electric conduction takes place through the n-type channel or p-type channel of semiconducting material. For this reason, noise level in JFET is very low.

Additional features of JFET

Additional distinguishing features of JFET are:

(1) In JFET, the charge carriers move through the channel by the 'drift' process. Whereas in BJT the charge carriers move the base by the process of 'diffusion'. A drift is faster than diffusion. Hence, JFETs can be used at higher frequency of the order of GHz. moreover, high speed switching performance of JFET are better than that of BJT.

(2) At audio frequencies the power gain of JFETs is much greater than that of BJTs.

(3) The fabrication of JFETs in much simpler. These occupy less space in IC (Integrated circuit) configuration and thermally more stable.

(4) JFETs do not show any offset voltage at zero drain current.

(5) JFETs are linear devices as they have square law characteristics.

(6) JFETs have negative temperature coefficient of resistance. Hence, they posses better thermal stability and nearly avoids thermal runaway, which causes burning in the worst case.

Preference of JFET over BJT

JFET is prefered to BJT. A JFET is a voltage controlled device. Input voltage controlled the output current. While BJT is a current controlled device. Here input current controlled device. Here input current controlls the output current. Therefore, JFET a semiconductor device acts like a vacuum tube. A gate, source and drain of JFET corresponds to grid, cathode and anode of a triode, BJT was unable to replace many functions of triode i.e. vacuum tubes, but JFET could replace successfully a vacuum tube. JFET have replaced not only vacuum tubes but BJT too because of its high input impedance, less noise and greater resistance to nuclear radiation.

28.9 APPLICATIONS OF JFET

JFETs can be used in almost every application in which bipolar transistors are used. However, they have certain applications which are exclusive to them.

(1) As input amplifiers in oscilloscopes, electronic voltmeters, and the other measuring and testing equipment because of the very high input impedance, which reduces the loading effect to the minimum.

(2) For mixer operation of FM and TV receivers.

(3) As voltage variable resistor (VVR) in operational amplifiers, tone controls and AGC circuit is radio receivers.

(4) Large-scale integration (LSI) and computer memories because of very small size.

28.10 APPLICATION AS A VOLTAGE VARIABLE RESISTOR

In most of the applications of JFET, the device is operated in the 'pinch-off' region i.e. a constant current portion of its output characteristics. However, before the 'pinch-off' when V_{DS} is small, the JFET is very useful as a 'voltage controlled resistor'. This property makes it a **Voltage Variable Resistor** or VVR. Therefore, *JFET is used to obtain AGC, the automatic gain control in radio receiver*.

The above property of VVR, along with its inherent high input impedance is utilised in constructing a very high input impedance voltmeter, which can replace much more costly VTVM. Fig.

28.11 shows a circuit diagram of JFET as solid-state d.c. voltmeter.

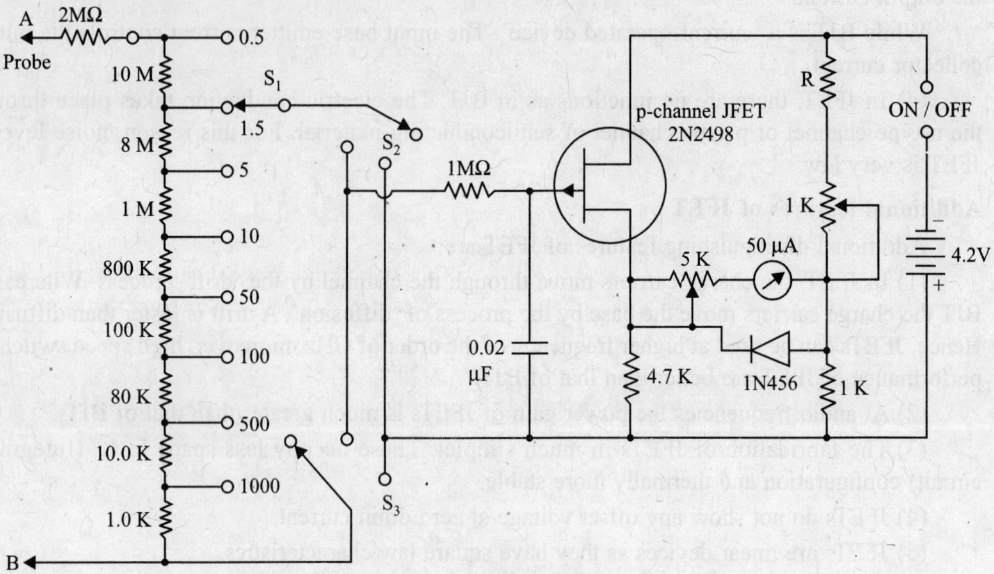

Fig. 28.11. JFET as Voltmeter.

Working

Here the JFET serves as a high input resistance device and also forms one arm of the wheatstone bridge. The other three arms are formed by 4.7 K, 1K and 1K pot. (along with R). The cells of 4.2 V and a micrometer are also placed at their proper places.

With no input signal voltage, the bridge is balanced with the help of 1K potentiometer. When the external voltage is applied (range is selected by switch S_1), the resistance of the JFET changes and unbalances the bridge due to which the micrometer shows deflection. This deflection is proportional to the input voltage due to the VVR (voltage variable resistor) property of the JFET. The scale of micrometer is calibrated to read the desired range of input voltage, hence act as a voltmeter. Diode and 5kΩ potentiometer are provided for protection of the meter from the excess current. A ladder of resistors between A and B is used for a desired voltage range, which can be selected by switch S_1.

If the same voltmeter is to be used for measuring a.c. voltages, then some rectification circuit may be provided in the circuit (not shown).

Disadvantages of JFET

The only disadvantages with JFET are:

(1) Small gain-bandwidth product, and

(2) They are very dilicate and hence they are highly susceptible to damage in handling.

28.11 MOSFET

The working principle and properties of JFET, which is one of the family member of FET, is based on the fact that due to the reverse-bias at pn-junction, a depletion region is created, which alters the channel resistance and ultimately the current flowing through the JFET. The other equally important member of the family, though appeared late on the horizon of the world of electronics, has no doubt gone much ahead of its senior member (JFET), in terms of its applications. This important member is the insulated gate type FET (IGFET) and is widely known as metal-oxide semiconductor FET or MOSFET. If differ from JFET, in that, the gate is not connected to the channel but is insulated from it. The principle of MOSFET was first demonstrated in 1930 and since 1965 it is commercially manufactured.

28.12 TYPES OF MOSFET

There are two basic forms of the MOSFET. They are named according to the way in which they operate, namely,

(1) The Enhancement MOSFET

(2) The Depletion MOSFET.

Principle: MOSFET works on the principle that, "the thickness and hence the resistance of a conducting channel and hence the resistance of a semiconducting material can be controlled by a transverse electric field across an insulating layer deposited on the semiconducting material". This is the fundamental difference between MOSFET and BJT.

28.13 ENHANCEMENT MOSFET

Fig. 28.12 shows the p-channel enhancement MOSFET which is widely used since 1972. Let us describe its construction and operation.

Construction: It consists of a lightly doped substance of n-type silicon. On the n-type substrate, two heavily doped p^+ regions are diffused. One of the two p^+ regions, one is known as drain (D) and the other as source (S). These regions are separated by 10^{-5} m to 2×10^{-5} m. A thin layer of insulating silicon dioxide (SiO_2) (marked with shaded area in figure) is grown over the surface. Two small holes are drilled through the SiO_2 layer so that metallic contacts can be made with the source S and the drain D. Then a thin layer of metal aluminium (Al) is formed over the layer of SiO_2. The metal layer covers the entire channel region and it forms the gate G.

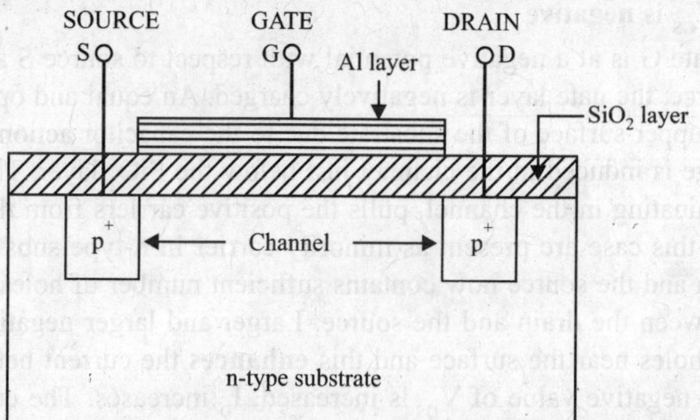

Fig. 28.12. p-channel enhancement MOSFET

The layer of SiO_2 acts as a very good insulator between the surface of substrate and metallic layer. Thus, they form a capacitor, with SiO_2 as a dielectric medium. The length of the gate is about 7×10^{-6} m and the gate-channel capacitance is about 1 to 4 pF. The input resistance of the MOSFET is very high and is of the order of 10^{10} to 10^{15} ohm.

Note that the gate is not in direct contact with the channel but is insulated from it and hence the name 'IGFET'.

Working

The circuit diagram of an p-channel enhancement MOSFET is shown in Fig. 28.13. The substrate is generally kept grounded.

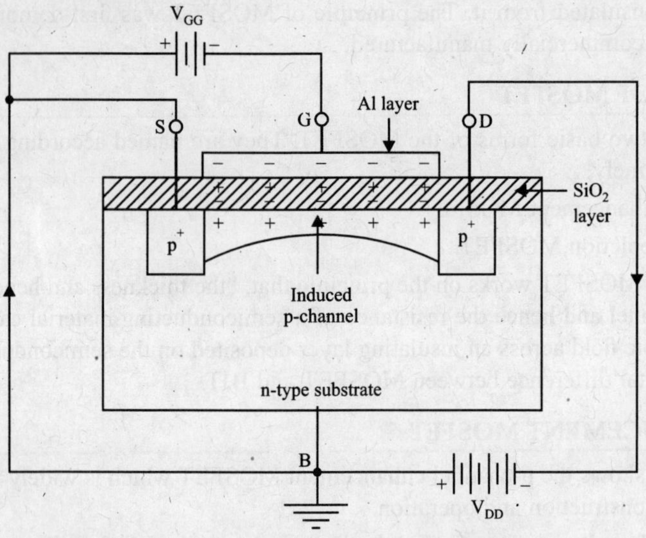

Fig. 28.13. p-channel enhancement MOSFET (Circuit diagram)

(*i*) When V_{GS} = 0 V

Suppose the gate G is also grounded i.e. gate G is short circuited to the source S and the drain D is at a negative potential with respect to the source. It is found that I_D = 0 i.e. no current flows between drain and source. This is because the junction between p-type drain and n-type substrate is reverse biased. Truely speaking very small current I_D (in μA) flows due to minority carriers.

(*ii*) When V_{GS} is negative

When the gate G is at a negative potential with respect to source S and the substrate is connected to the source, the gate layer is negatively charged. An equal and opposite charge (positive) is induced at the upper surface of the substrate due to the capacitor action as shown in Fig. 28.13 i.e. positive charge is induced in the channel just below the SiO_2 layer. This happens because the electric field terminating in the channel, pulls the positive carriers from the substrate close to the surface (holes in this case are present as minority carrier in n-type substrate.). Thus, the region between the drain and the source now contains sufficient number of holes, and an electric current I_D now flows between the drain and the source. Larger and larger negative gate voltage attracts more number of holes near the surface and this **enhances** the current between drain and source. Therefore, as the negative value of V_{GS} is increased, I_D increases. The curve of I_D verses V_{GS} is called the transfer characteristic curve for the MOSFET as shown in Fig. 28.14.

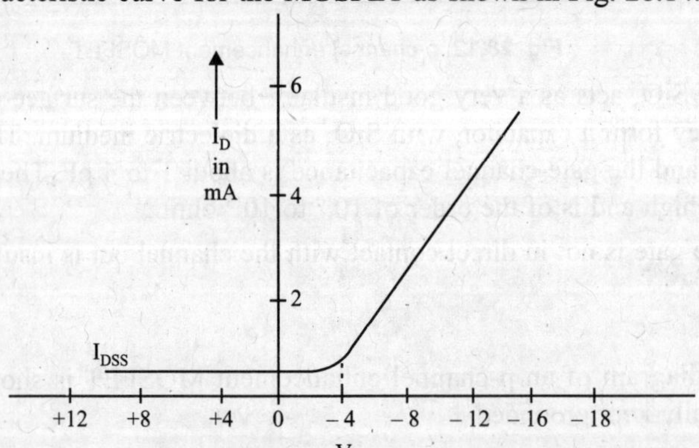

Fig. 28.14 Transfer characteristic

The lowest value of V_{GS} beyond which the drain current enhances is called the **threshold voltage**, and is denoted by V_T which is nearly equal to –4V. When $V_{GS} \geq V_T$, a very small saturation drain current I_{DSS} flow through the channel, while the drain current I_D increases when V_{GS} is made more negative. i.e. $V_{GS} < V_T$.

Thus, the drain current I_D is enhanced by increasing the magnitude of the negative gate voltage, hence the name enhancement MOSFET.

(iii) When | V_{GS} | > | V_T | and | V_{DS} | is increased

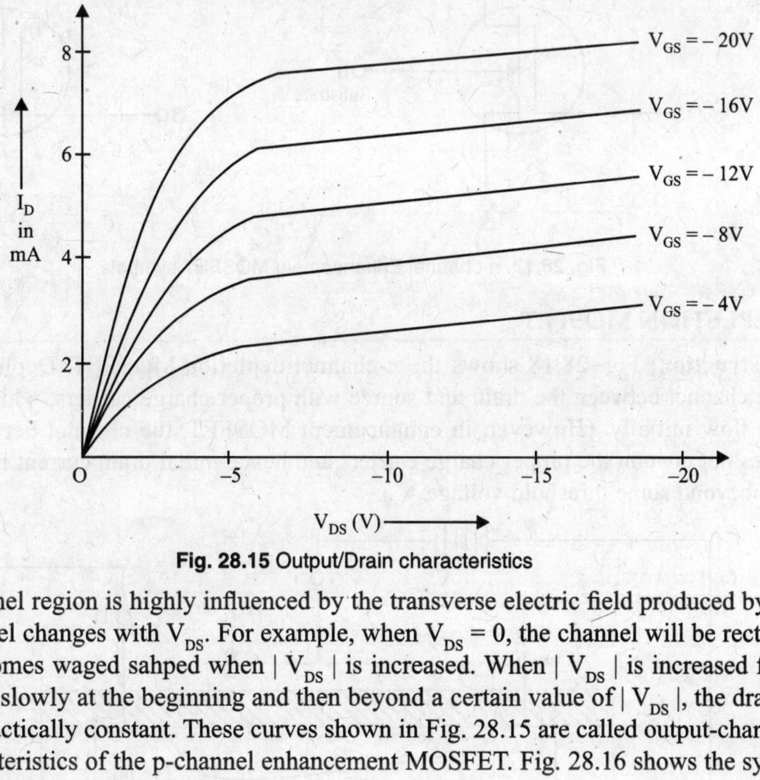

Fig. 28.15 Output/Drain characteristics

The channel region is highly influenced by the transverse electric field produced by V_{GS}. The shape of channel changes with V_{DS}. For example, when V_{DS} = 0, the channel will be rectangular in shape and becomes waged sahped when | V_{DS} | is increased. When | V_{DS} | is increased from zero, | I_D | increases slowly at the beginning and then beyond a certain value of | V_{DS} |, the drain current | I_D | remain practically constant. These curves shown in Fig. 28.15 are called output-characteristics or drain characteristics of the p-channel enhancement MOSFET. Fig. 28.16 shows the symbols for p-channel enhancement MOSFET.

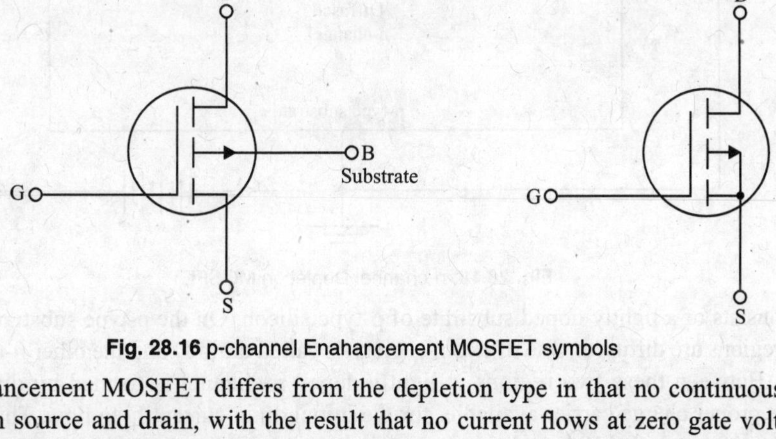

Fig. 28.16 p-channel Enahancement MOSFET symbols

The enhancement MOSFET differs from the depletion type in that no continuous channel exists between source and drain, with the result that no current flows at zero gate voltage. The symbols 'broken channel' represents this condition. At we know, a minimum or threshold gate voltage (V_T) is required before the drain current will flow.

This type of FET is very useful in switching applications, since no gate voltage is required to hold it 'OFF' and when the device is to be switched 'ON', the application, of the same polarity as that of drain will achieve this condition. Also, the threshold voltage (V_T) provides some noise immunity to prevent false-switching.

Fig. 28.17 shows the symbols of n-channel enhancement MOSFET.

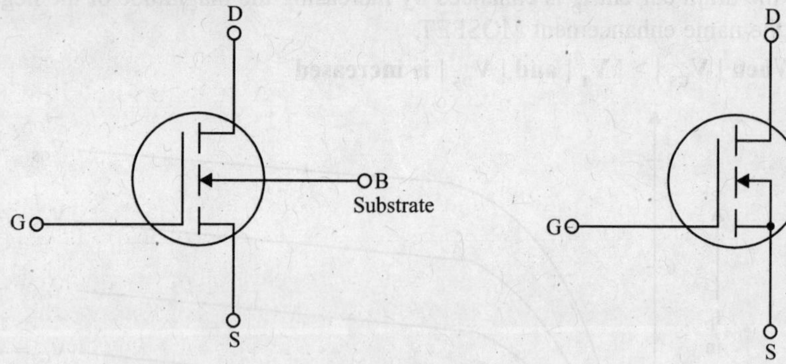

Fig. 28.17. n-channel Enhancement MOSFET symbols.

28.14 DEPLETION MOSFET

Construction: Fig. 28.18 shows the n-channel depletion MOSFET. Depletion MOSFET contains a channel between the drain and source with proper charge carriers, which allows some current to flow initially. (However, in enhancement MOSFET, the channel between drain and source does not contain the proper charge carriers and hence initial drain current is zero. The current starts beyond some threshold voltage V_T).

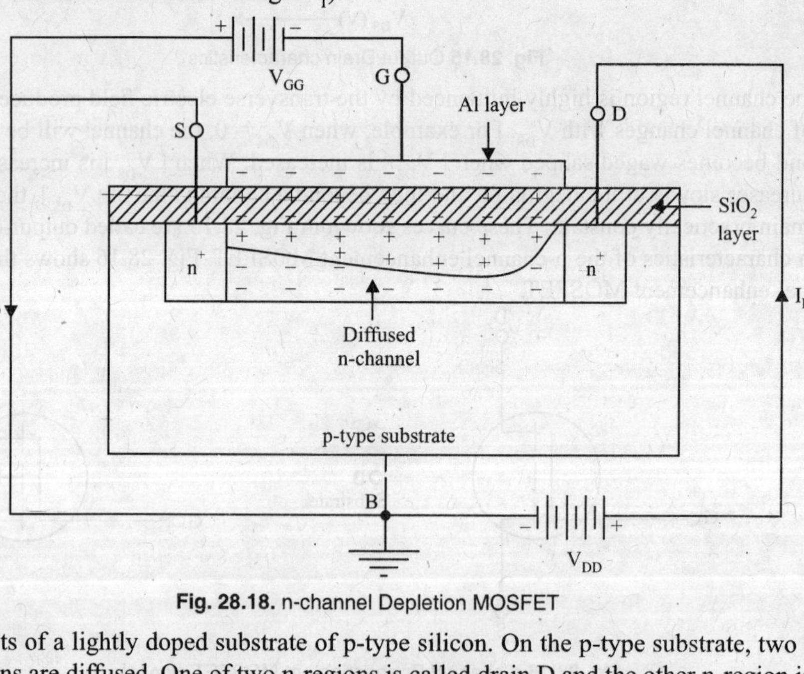

Fig. 28.18. n-channel Depletion MOSFET

It consists of a lightly doped substrate of p-type silicon. On the p-type substrate, two heavily doped n-regions are diffused. One of two n-regions is called drain D and the other n-region is called source S. Between these two regions, a lightly doped n-channel is formed by diffusion, which makes the proper charge carrier available, for the initial drain current I_D to flow. The length of this n-channel is 10^{-5} m to 2×10^{-5} m. A thin layer of insulating silicon-dioxide (SiO_2) is grown over the surface. Two small holes are drilled through the SiO_2 layer, so that metallic contacts can be made with the source S and drain D. Then a thin layer of metal aluminium (Al) is formed over the layer of SiO_2. The metal layer covers the entire n-channel region and it forms the gate G. The layer of SiO_2 works as a dielectric (insulator) between the metallic surface and the surface of the substrate. Thus, a parallel plate capacitor is formed with SiO_2 layer as a dielectric medium between two parallel plates of the capacitor. The length of the gate is about 7×10^{-6} m and the capacitance offered is about 1 to 4 pF. The input impedance of MOSFET is very high, 10^{10} to 10^{15} ohm approximately.

Working: In order to study the static drain characteristics and transfer characteristics of a n-channel depletion MOSFET, the connections are made as shown in Fig. 28.18. The working is as under.

1. Static Drain Characteristics

(*i*) When V_{GS} = 0V

For this, the gate G is shorted to the source S. The drain D is at positive potential with respect to the source. The majority carriers in the n-channel (i.e. electrons) flow from source S to drain D through the n-channel. The conventional current I_D flows through the channel from D to S. Since the gate G is short circuited to the source S, the gate G is at a negative potential with respect to drain D (Due to battery V_{DD}). Hence a positive charge consisting of holes is induced in the channel through SiO_2 of the gate channel capacitor. This results a depletion of mobile electrons in the channel, causing a depeletion region in the channel. The shape of the depletion region depends on V_{GS} and V_{DS}. The channel becomes wedge-shaped as shown in the figure. Thus, when V_{DS} is increased, I_D also increases ad at a certain value of V_{DS}, I_D becomes almost constant as shown in Fig. 28.19 This value of V_{DS} is called the 'pinch-off' voltage. Beyond 'pinch-off' voltage, a saturation region for I_D is obtained. When V_{GS} = 0, VI_{DSS} is appreciable in case of depletion MOSFET, whereas it is very small for enhancement MOSFET.

(ii) When V_{GS} is negative

The ampere-volt characteristics of the depletion MOSFET are similar to those for the JFET. When the gate G is at a negative potential with respect to source, the depletion region further extends into the n-channel region due to capacitor action. This results into increase in thickness of the depletion region.

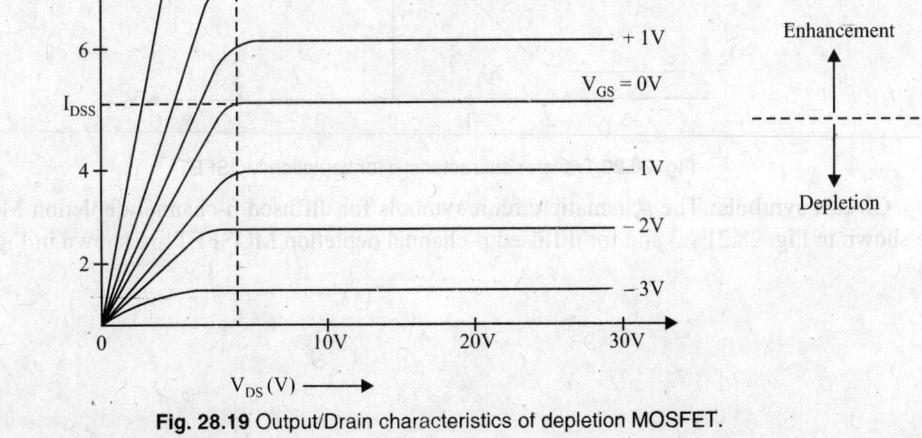

Fig. 28.19 Output/Drain characteristics of depletion MOSFET.

On account of the presence of 'holes' due to induced positive charge, the number of free electrons in the channel region between S and D decreases, making the channel less conducting. Thus, the drain current I_D decreases as the gate-to-source voltage (V_{GS}) is made more negative. The induced positive charges are responsible for the depletion of majority carriers in the channel. This is the reason, that this type of MOSFET is known as depletion MOSFET. The curves for V_{GS} having negative values lie below the curve for V_{GS} = 0 V, as I_D is reduced.

***(iii)* When V_{GS} is positive**

However, unlike the JFET, there is no pn-junction to keep reverse biased, so the gate is allowed to be made positive. Under this condition, the negative charge is induced in the channel, due to capacitor action. The population of charge carriers i.e. electrons in the channel, due to capacitor action. The population of charge carriers i.e. electrons in the channel increases thereby 'enhancing' the current flow. Hence, in the case, the conductivity of the channel increases with V_{GS} becoming more and more positive. Hence these characteristic curves lie above that for V_{GS} = 0V.

Thus, this depletion MOSFET device can be operated in depletion (by maintaining V_{GS} negative) as well as in enhancement (by maintaining V_{GS} positive) mode. Hence, the depletion MOSFET is also called dual mode MOSFET.

2. Transfer Characteristics

A transfer characteristic is a graph between input gate-to-source voltage V_{GS} and the corresponding drain current I_D for a constant value of drain-to-source voltage V_{DS}.

When the drain-to-source voltage V_{DS} is maintained at some positive potential with respect to source and V_{GS} is varied, the variation of drain current I_D with V_{GS} is as shown in Fig. 28.20. The curve I_D against V_{GS} is called transfer characteristic.

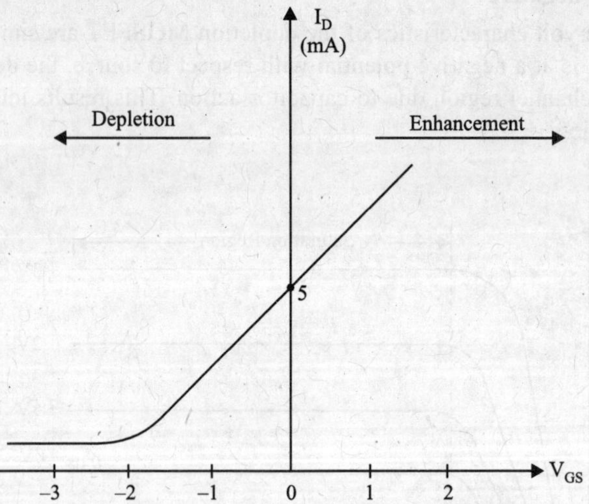

Fig. 28.20 Transfer characteristic for depletion MOSFET

Circuit symbols: The schematic circuit symbols for diffused n-channel depletion MOSFET are shown in Fig. 28.21 (a) and for diffused p-channel depletion MOSFET are shown in Fig. 28.21 (b).

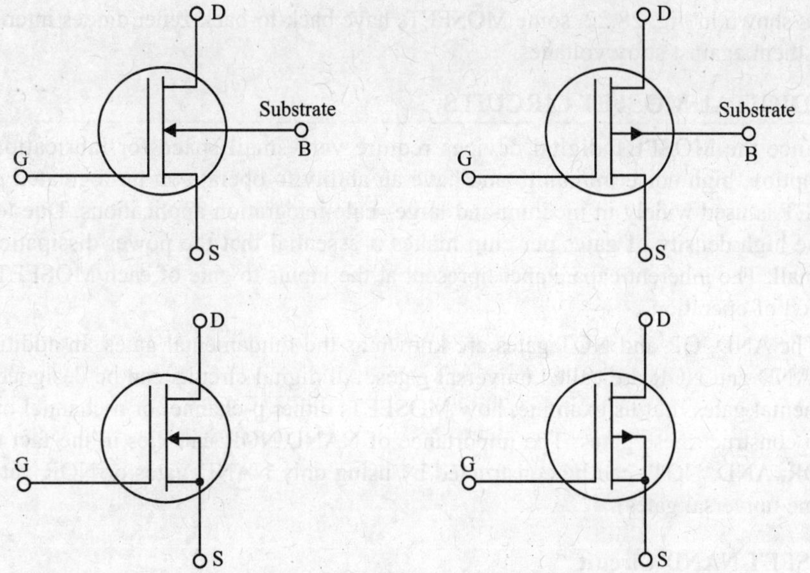

(a) n-channel depletion MOSFET (b) p-channel depletion MOSFET

Fig. 28.21

28.15 SPECIAL FEATURES OF MOSFET

MOSFET possesses almost all the advantages that JFET exhibits, which we have mentioned earlier. In addition to those advantages, the most important is its very high input impedance. As we know, input impedance of JFET ranges from 100 M ohm to 10,000 M ohm. MOSFET offers still higher input impedance ranging from 100 M ohm to 10^{13} ohm.

Another notable special feature of MOSFET is the most simpler technology involved in manufacturing them in large scale. An IC (integrated circuit) chip for large scale integrated circuits, MSO/LSI, MOSFETs are mostly used.

The only few demerits are:

(*i*) MOSFET requires very careful handling particularly when it is out circuit. Even a small jerk or shock may rupture the ultra thin SiO_2 layer between the channel and gate.

(*ii*) MOSFET is highly susceptible to overload voltage picked up from any stray or static charges. Since gate channel junction looks like a capacitor with extremely high resistance and hence, it requires only few electrons to produce a high voltage across it. Even by picking up few stray charges by it leads can destroy it. That is why, grounding ring is provided to short all the leads of a MOSFET for avoiding any voltage build up between them. This ring is removed only after the MOSFET is soldered into its position.

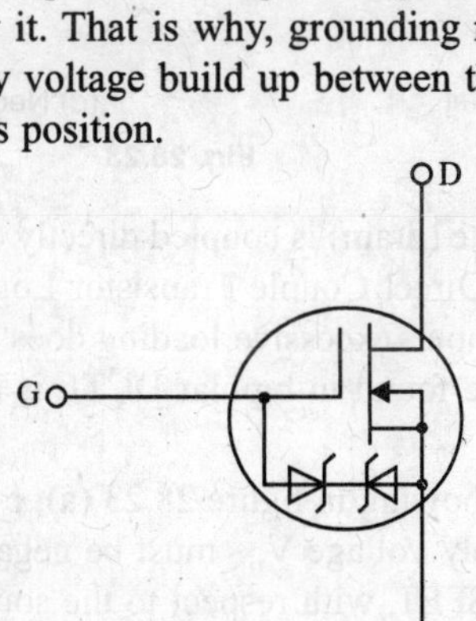

Fig. 28.22

As shown in Fig. 28.22, some MOSFETs have back-to-back zener diodes internally itched to protect them against stray voltages.

28.16 DIGITAL MOSFET CIRCUITS

Since the MOSFET digital devices require very small space for fabrication, low power consumption, high noise immunity and have an ability to operate on un-regulated power supply, MOSFET is used widely in medium-and large-scale-integration applications. Due to compact nature, the high density of gates per chip makes it essential that the power dissipation per gate be very small. The inherent capacitances present at the inputs to gate of each MOSFET, slows down the speed of operation.

The AND, OR and NOT gates are known as the fundamental gates. In addition, two more gate NAND and NOR are called universal gates. All digital circuits can be designed by using the fundamental gates. Let us examine, how MOSFETs either p-channel or n-channel or both are utilised to construct these gates. The importance of NAND/NOR gate lies in the fact that any basic gate (OR, AND, NOT) can be constructed by using only NAND gates or NOR gates (and hence the name universal gates).

1. MOSFET NAND Circuit

Fig. 28.23 shows the schematic circuit of MOSFET NAND gate along with truth table for a NAND gate Fig. 28.23 (b).

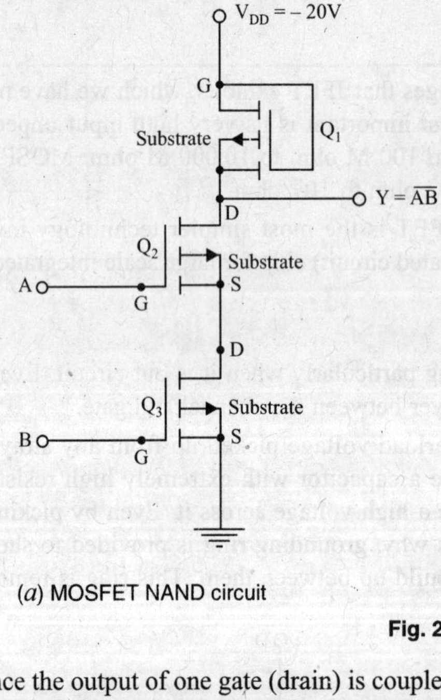

(*a*) MOSFET NAND circuit

Input		*Output*
A	*B*	$A = \overline{AB}$
0	0	1
0	1	1
1	0	1
1	1	0

(*b*) Truth Table of NAND gate

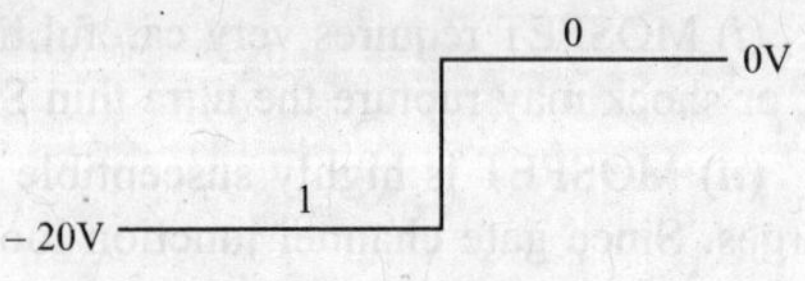

(*c*) Negative-logic showing voltage levels

Fig. 28.23

Since the output of one gate (drain) is coupled directly (without resistors) to the next gate, this give the advantage of DCTL (Direct Couple Transistor Logic). However, due to insulated nature of FET's gate (IGFET) at the input, excessive loading does not take place on the previous devices output, so that the disadvantage found in bipolar DCTL is not present in MOSFET. It has a high fan-out capability.

Principle: In the circuit showing in figure 28.23 (a), p-channel enhancement type MOSFETs are used, so that the drain supply voltage V_{DD} must be negative (V_{DD} = – 20V.) We know that 0V at the gate of a p-channel MOSFET, with respect to the source, holds the transistor "OFF", and a negative voltage-more than the threshold voltage V_T is necessary to turn it "ON".

Logic used: In the gate shown, negative logic is normally used. That is, 0V corresponds to a logical 0 and V_{DD} = –20V corresponds to a logical 1, as shown in Fig. 28.23 (c)

Working: Since the gate (G) of Q_1 is connected to –20 V, Q_1 is always conducting and its drain-to-source resistance acts as the load resistance (of the order of 2kΩ) for the driver MOSFETS Q_3. If either A or B (or both) is 0 (0 V), at least one of two Q_2 or Q_3 (connected in series, totem-pole-arranged) will be 'OFF'. A negligible current flow through Q_1, so that the output voltage is approximately –20V, representing logic 1. Only if both A and B are high (–20V), both Q_2 and Q_3 will be go into saturation and makes output Y = 0V, representing logic 0. This is in accordance with the truth table shown by Fig. 28.23(b).

2. MOSFET NOR Circuit

Fig. 28.24 (a) shows the MOSFET NOR circuit in which MOSFET Q_1 is always Conducting (like NAND) as its gate (G) is maintained at very high negative (VDD = –20 V) potential. The two MOSFETS Q_2 and Q_3 connected parallel, whose common drain drive the source Q1, totem-pole-arranged.

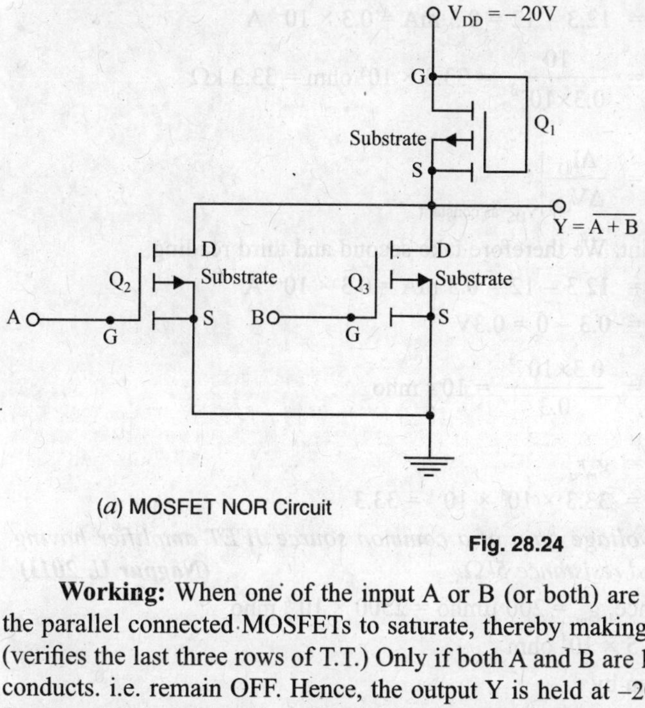

(*a*) MOSFET NOR Circuit

Input		*Output*
A	*B*	$A = \overline{A+B}$
0	0	1
0	1	0
1	0	0
1	1	0

(*b*) Truth Table of NOR gate

Fig. 28.24

Working: When one of the input A or B (or both) are high (–20 V), causes at least one of the parallel connected MOSFETs to saturate, thereby making output Y = 0V, representing logic 0 (verifies the last three rows of T.T.) Only if both A and B are low (0 V), none of the two Q_2 and Q_3 conducts. i.e. remain OFF. Hence, the output Y is held at –20V, representing logic 1, verifies the first row of T.T. Thus, the truth table for NOR is verified as given by Fig. 28.14 (b).

SOLVED EXAMPLES

Example 28.1. *Calculate the dynamic resistance of a JFET with amplification factor 80 and trans-conductance 200 μmho.* ***(Nagpur U. 2011)***

Solution. Given : Amplification factor, μ = 80

Trans-conductiance, g_m = 200 × 10^{-6} mho

Let r_d be the dynamic resistance, then

$$\mu = g_m r_d$$

$$\therefore \quad r_d = \frac{\mu}{g_m} = \frac{80}{200\times10^{-6}} = 400 \times 10^3 \text{ ohm} = 400 \text{ k}\Omega$$

Example 28.2. *The following reading were obtained when a student performs an experiment with JFET*

V_{GS} *(Volts)*	0	0	0.3
V_{DS} *(Volts)*	6	16	16
I_D *(mA)*	12	12.3	12.0

Calculate (i) a.c. drain resistance

(ii) transconductance and

(iii) amplification factor.

Solution. (*i*) a.c. drain resistance, $r_d = \left.\dfrac{\Delta V_{DS}}{\Delta I_D}\right|_{V_{GS} \text{ as constant}}$

Form observations, at V_{GS}, we take first two readings,

$$\Delta V_{DS} = 16 - 6 = 10 \text{ V}$$

$$\Delta I_D = 12.3 - 12 = 0.3 \text{ mA} = 0.3 \times 10^{-3} \text{ A}$$

$$\therefore \quad r_d = \frac{10}{0.3\times10^{-3}} = 33.3 \times 10^3 \text{ ohm} = 33.3 \text{ k}\Omega$$

(*ii*) Transconductance g_m

$$g_m = \left.\frac{\Delta I_D}{\Delta V_{GS}}\right|_{V_{DS} \text{ as constant}}$$

From observations, V_{DS} is constant. We therefore take second and third reading.

$$\Delta I_D = 12.3 - 12 = 0.3 \text{ mA} = 0.3 \times 10^{-3} \text{ A}$$

$$\Delta V_{GS} = 0.3 - 0 = 0.3\text{V}$$

$$\therefore \quad g_m = \frac{0.3\times10^{-3}}{0.3} = 10^{-3} \text{ mho}$$

(*iii*) Amplification factor μ

$$\mu = g_m r_d$$

$$= 33.3 \times 10^3 \times 10^{-3} = 33.3$$

Example 28.3. *Calculate the voltage gain of a common source JFET amplifier having transconductance 2500 μmho and load resistance 5kΩ.* ***(Nagpur U. 2011)***

Solution. Given : Tansconductance, g_m = 200 μmho = 2500 × 10^{-6} mho

Load resistance, $R_L = 5\text{k}\Omega = 5 \times 10^3$ ohm

The voltage gain of JFET is given by

$$A_v = -g_m R_L$$

$$= -2500 \times 10^{-6} \times 5 \times 10^3 = -2.5 \times 10^{-3} \times 5 \times 10^3$$

$$= -12.5$$

Example 28.4. *A common source JFET amplified has a load resistance R_L = 500 kΩ. If the a.c. drain resistance (r_d) and amplification factor (m) of the JFET are 100 kΩ and 24 respectively, calculate the voltage gain of the amplifier.* ***(Nagpur U. 2011, 2012)***

Solution. The voltage gain of a common-source JFET amplifier is

$$A_v = -\frac{\mu R_L}{r_d + R_L}$$

The negative sign indicates that the output is 180° out of phase that of input. Considering the magnitude only

$$|A_v| = \frac{\mu R_L}{r_d + R_L}$$

Substituting $\mu = 24$, $R_L = 500\ k\Omega = 500 \times 10^3\ \Omega$

$$r_d = 100\ k\Omega = 100 \times 10^3\ \Omega$$

$$\therefore \quad |A_v| = \frac{24\times500\times10^3}{100\times10^3+500\times10^3} = \frac{24\times5\times10^5}{1\times10^5+5\times10^5}$$

$$= \frac{120\times10^5}{6\times10^5} = 20$$

EXERCISE

1. What is field effect transistor ? Describe the construction and working of n-channel JFET. Discuss its working principle. ***(Nagpur U. 2011)***
2. Explain the experimental arrangement with a neat circuit diagram necessary to plot the static output characteristics of a JFET. Draw the static characteristic and explain the region:
   (*a*) Linear region (*b*) Saturation region
   (*c*) Breakdown region.
3. Explain the term 'pinch-off voltage.' How much is the separation between the depletion regions near the drain at pinch-off condition? ***(Nagpur U. 2009)***
4. Plot the transfer characteristic of a JFET and explain the cut-off voltage. What dos its slope represent?
5. Define the parameter of JFET
   (*a*) Drain resistance, r_d (*b*) Transconductance, g_m
   (*c*) Amplification factor, μ.
   Derive the relation between μ, g_m and r_d. ***(Nagpur U. 2011)***
6. Explain clearly the terms:
   (*a*) D.C. biasing of a JFET (*b*) D.C. load line
7. Explain with neat circuit diagram the working of JFET as a common-source amplifier. What is the phase relationship between input signal and output voltage?
8. Evaluate the input impedance, output impedance and voltage gain for a common source JFET amplifier.
9. What is source-follower? Explain the working of source-follower JFET with a neat circuit diagram. Evaluate its voltage gain. Also find its input impedance and output impedance.
10. Mention the difference between JFET and BJT transistors. What are the distinguishing features of JFET and BJT? Which one is preferred? and Why?
11. Mention the applications of JFET as a voltage variable resister. Draw a circuit diagram of JFET as solid-state d.c. voltmeter. ***(Nagpur U. 2011)***
12. What are the demerits of JFET?
13. A common source JFET amplifier uses a load resistance $R_L = 200\ k\Omega$. If the drain resistance and transconductance of the JFET are 100 kΩ and 0.15 mA/V respectively calculate the magnitude of the voltage gain of the amplifier.
14. What do you understand by 'pinch-off point' and 'cut-off point'? Are they same? What happens to the pinch-off when V_{GS} is made more and more negative?

15. What is MOSFET? What are its basic forms? State the principle on which MOSFET works. ***(Nagpur U. 2012)***

16. What is the major physical difference between an enhancement type and the depletion type MOSFET?

17. Sketch the structure of a p-channel enhancement MOSFET. Explain its working. Draw a set of static drain characteristics and transfer characteristics and explain. ***(Nagpur U. 2012)***

18. What is meant by the threshold voltage? Is this peculiar to a p-channel enhancement type MOSFET? Draw the circuit symbol of p-channel enhancement MOSFET. ***(Nagpur U,. 2011, 2012)***

19. Explain the working of *n*-channel depletion MOSFET using a neat circuit diagram. Draw a set of static drain characteristics for *n*-channel depletion MOSFET. ***(Nagpur U. 2011)***

20. What is 'pinch-off' voltage for a *n*-channel depletion MOSFET? Explain the saturation region.

21. Explain why a depletion MOSFET is called a 'dual mode MOSFET'.

22. What do you mean by transfer characteristics? Draw a transfer characteristic for a depletion MOSFET. Draw symbols used for (*i*) *n*-channel depletion MOSFET and (*ii*) *p*-channel depletion MOSFET.

23. Point out the special features of MOSFET over BJT and JFET.

24. What accounts for the much smaller value of I_{DSS} in an enhancement MOSFET as compared to depletion MOSFET?

25. Give at least two precautions to observe when handling a MOSFET, that does not have any built-in gate profection.

26. Explain the construction and working of a digital NAND logic gate by using *p*-channel enhancement MOSFET. ***(Nagpur U. 2011, 2012)***

27. Explain the construction and working of a digital NOR logic gate by using p-channel enhancement MOSFET.

27. What makes *p*-channel MOSFET to be used in a medium and large cale integration applications.

SMALL SIGNAL AMPLIFIERS

29

Chapter

INTRODUCTION

The amplitude of the voltage produced by any transducer viz. microphone, thermistor, photocell etc. is extremely small. Such small signals need amplification without any distortion, so that amplified signals can be further processed for useful purposes. In chapter 27, we have seen a CE amplifier with self-biasing arrangement so that operating point remains almost stable. A single stage transistor amplifier does not give sufficient voltage gain. Therefore, large number of amplifier stages are cascaded to produce the output of desired level. Cascading the amplifiers means the output of the first stage becomes the input of the second stage. In general, there are four methods used for coupling one amplifier stage to the another. These are:

(*i*) Resistance capacitance (RC) coupling

(*ii*) Direct coupling

(*iii*) Transformer coupling

(*iv*) Inductive coupling

Due to high voltage gain and stable operating point, the common-emitter amplifier is popularly used for voltage amplification and RC coupling is most commonly used for cascading the stages. In this chapter, we will discuss the classification of amplifiers, distortion and frequency response of an RC coupled amplifier.

29.1 CLASSIFICATION OF AMPLIFIERS

In practice, amplifiers are various types and are classified according to their use, mode of operation, frequency range and method of inter stage coupling, as specified below:

(*i*) According to their Use

According to their use, the amplifiers are classified as

(*a*) Voltage amplifier (*b*) Power amplifier.

The voltage amplifier increases the voltage level of the signal. The power amplifier mainly increases the power level of the signal.

One more class called '*current amplifier*' is also sometimes included. The purpose of this amplifier is almost the same as that of voltage amplifier except that it amplifiers current wave form rather than voltage wave form.

(*ii*) According to Frequency Range

According to frequency range, the amplifiers are classified as:

(a) Audio frequency (A.F.) amplifier.

(*b*) Radio frequency (R.F.) amplifier.

Audio frequency amplifiers are used to amplify audio frequency (20 Hz to 20 kHz) signals. Radio frequency amplifiers are used to amplify high frequency signals. (R.F. includes all ranges viz. H.F. (300 kHz to 30 MHz), V.H.F. (30 MHz – 300 MHz), U.H.F. (300 MHz – 3 GHz) and S.H.F. (3 GHz – 30 GHz).

***(iii)* According to Interstage Coupling**

According to coupling device, they are classified as:

(*a*) RC coupled amplifier (*b*) transformer coupled amplifier (*c*) Direct coupled and (*d*) LC coupled amplifier.

***(iv)* According to Mode of Operation**

According to mode of operation, they are classified as :

(*a*) Class - A (*b*) Class - B (*c*) Class - AB (*d*) Class - C amplifier.

These classifications are according to the positions of operating point (or quiscent point) on the characteristic curve, which decides the duration of conduction of the active device during a period of input cycle.

(a) Class-A amplifier : In Class-A operation, the bias voltage and the amplitude of the externally applied a.c. signal is so adjusted that the output current is never zero during the complete cycle of input a.c. signal. In class A transistor amplifier, operating in CE mode, the collector current

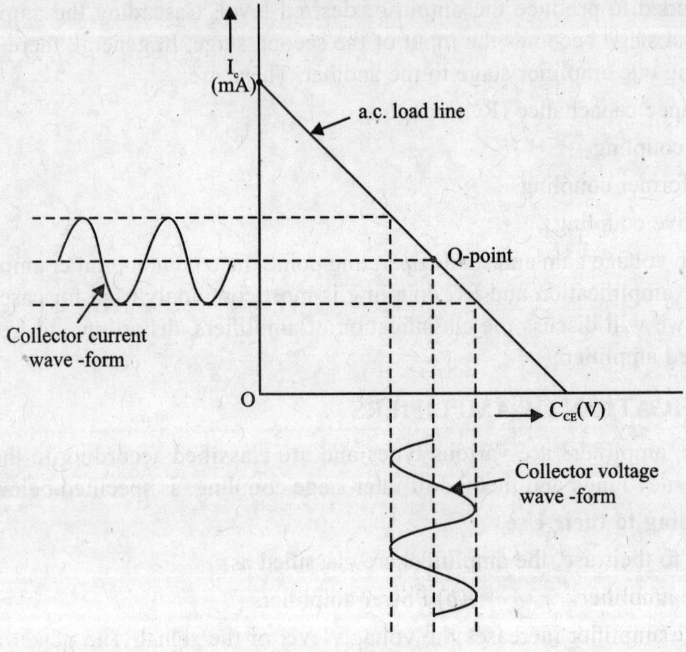

Fig. 29.1. Class-A

flows at all times during the full cycle of the input signal. Fig. 29.1 shows the class-A operation in terms of a.c. load line with output voltage and current wave-forms. The operating point Q is selected almost in the centre of the load line, so that collector current I_C flows at all times throughout the full cycle of the applied signal and no part of the signal is cut-off. The shape of the output wave-form is exactly identical to the input wave-form. Class A amplifiers are used as low-power audio amplifiers. The disadvantage of class A is that the power output and collector efficiency is low.

***(b)* Class-B amplifier:** In class-B operation, the bias voltage and the amplitude of the externally applied a.c. signal is so adjusted that the output current flows for just one half cycle of the input a.c. signal. In class-B, transistor amplifier operating in CE mode, the collector current flows only for one half cycle of the input signal as shown in Fig. 29.2. In this case, the transistor bias is such that the operating point Q lies at the collector cut-off voltage. The output is just amplified half-wave rectification.

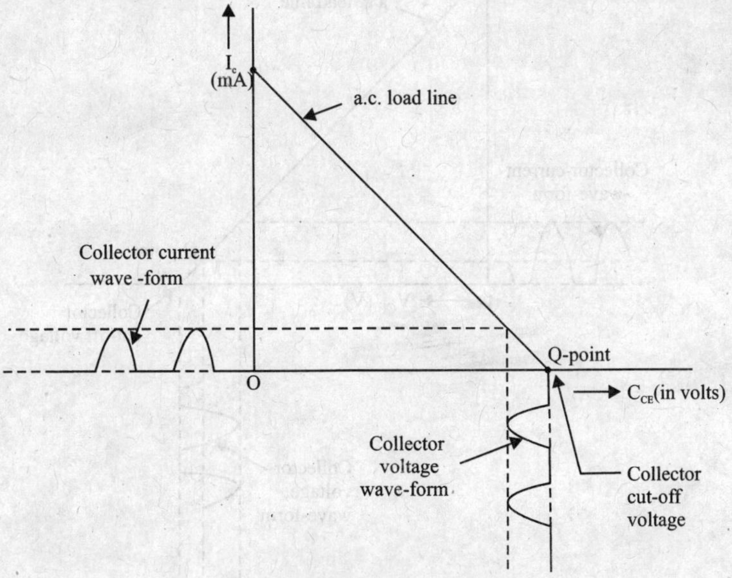

Fig. 29.2. Class-B

As in class B amplifier, negative half of the input a.c. signal is 'cut-off' at the output, a severe distortion takes place. But the amplification given high output. The collector efficiency is also high. It is possible to arrange two class B amplifier in a balanced circuit so that one of them amplifies one half of the input signal and the second half amplifies the other half. The two output are then added together by a transformer. Such an arrangement is known as **push-pull amplifier**. This produces large power output without distortion and are used in radio broadcasting transmitters.

(c) Class-AB amplifier : In class-AB operation, the bias voltage and the amplitude of the a.c. input signal are such that the output (collector) current flows for more than half cycle but less than full cycle of the input a.c. signal. The operating point is located some where between the middle of the a.c. load line and the collector 'cut-off' voltage as shown in Fig.29.3. In class AB amplifiers, distortion is some what less than that in class B amplifiers and power output is greater than in class A amplifiers, but less than that in class B amplifier. Class AB amplifiers are used in low power audio amplifiers, with some distortion.

(d) Class-C amplifiers : In class-C operation, the bias voltage and the amplitude of the a.c. input signal are such that the output (collector) current flows for less than one half cycle of the a.c. input signal, as shown in Fig. 29.4. The output current i.e. collector remains cut-off for more than half the cycle. In other words, the collector current flows in terms of very small pulses. Such amplifiers are used for high power output from which the harmonic components of the current can be eliminated and only the fundamental component is retained (by tuning to fundamental frequency). The power output and efficiency of class-C amplifier are higher than those of a class-B and class AB amplifier. Tuned radio frequency power amplifiers are operated in class-C. Oscillators are usually operated as class-C amplifiers. It is used in Amplitude Modulation (A.M.).

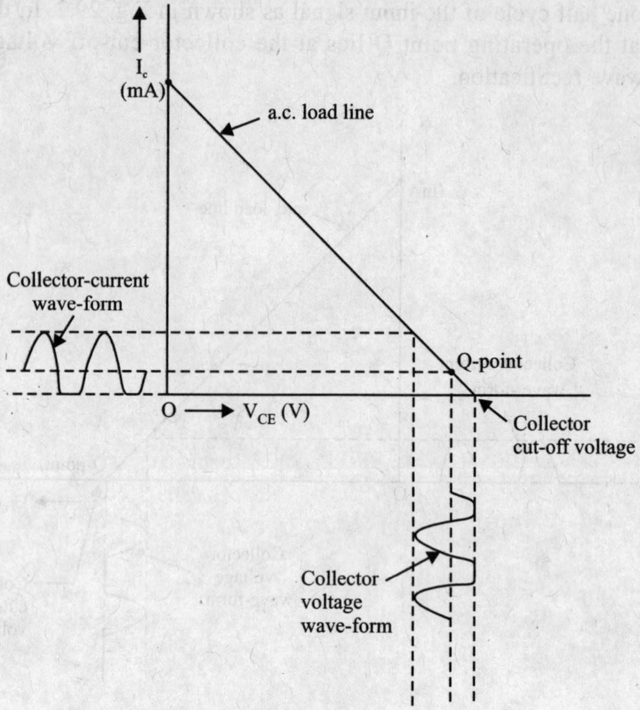

Fig. 29.3: Class-AB

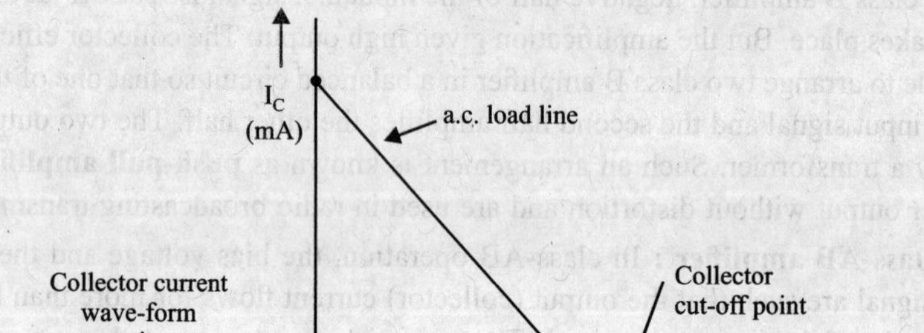

Fig. 29.4: Class-C

29.2 MULTISTAGE AMPLIFIER

The output from a single stage amplifier is usually insufficient to drive an output device. In other words, the gain of a single amplifier is inadequate for practical purposes. Consequently, two or more stages are necessary to get a high amplification. To achieve this, the output of each amplifier is coupled in some way to the input of the next stage. The resulting system is referred to as multistage amplifier.

Multistage Transistor Amplifier

A transistor circuit containing more than one stage of amplification is known as multistage transistor amplifier. Connecting the amplifier stages is known as cascading the stages, i.e. output of first stage is connected to the input of the second stage through a suitable coupling device and so on. The multistage amplifiers may be divided into following two categories:

***(a)* Cascaded amplifiers:** In these amplifiers, each stage is identical as well as the type of interstage coupling used is also identical.

***(b)* Compound amplifiers:** In these amplifiers, each stage may be different from the other ie. one stage may CE and the other may be CB or CC stage and also different types of interstage couplings may be employed.

29.3 CASCADED AMPLIFIER

In cascaded amplifier, the output a.c. voltage of the first stage becomes the input voltage of the second stage and the output of the second stage becomes the input of the third stage and so on.

Fig. 29.5: Block diagram of n-stage amplifier

Gain : The ratio of the output electrical quantity to the input one of the amplifier is called gain.

***(a)* Absolute Gain :** The overall absolute voltage gain (A_V) of the cascaded amplifier is equal to the product (not the sum) of the gain of the individual stages.

$$\therefore \quad A_V = A_{V_1} \times A_{V_2} \times A_{V_3} \times ------ \times A_{V_n}$$

Similarly, the overall current amplification is given by

$$A_i = A_{i_1} \times A_{i_2} \times A_{i_3} \times ------ \times A_{i_n}$$

Hence, the overall power gain will be

$$A_p = A_v \times A_i$$

***(b)* Decibel Gain :** Although, gain, an unitless quantity, ratio being a number, for practical purposes, it is deliberately expressed in some unit. The unit assigned is bel or decibel (db).

The natural logarithm (log to the base 10) of power gain is known as bel power gain i.e.

$$\text{Power gain} = \log_{10} \frac{P_{out}}{P_{in}} \quad \text{bel}$$

$$1 \text{ bel} = 10 \text{ db}$$

$$\therefore \quad \text{Power gain} = G_p = 10 \log_{10} \frac{P_{out}}{P_{in}} \quad \text{db}$$

As $P \propto V^2$ or I^2, the voltage gain and current gain are expressed in db as

$$\text{Voltage gain} = 20 \log_{10} \frac{V_{out}}{V_{in}} \quad \text{db}$$

$$\text{Current gain} = 20 \log_{10} \frac{I_{out}}{I_{in}} \quad \text{db}$$

Advantages

(1) The unit db is a logarithmic unit. Human ear response is also logarithmic, i.e. loudness of sound heard by ear is not according to the intensity of sound but according to the log of intensity of sound. Thus, if the intensity of sound given by a speaker (i.e. power) is increased 100 times, our ear hears only a doubling effect ($\log_{10}100 = 2$).

(2) When gains are expressed in db, the overall gain of a multistage amplifier is the sum of gains of individual stages in db. For instance, if G_{V_1}, G_{V_2} and G_{V_3} are the individual voltage gains in db of a three-stage amplifier, then the total voltage gain G_V is given by

$$G_V = G_{V_1} + G_{V_2} + G_{V_3}$$

However, absolute gain is obtained by multiplying the gains of individual stages. Obviously, it is easier to add than to multiply. This will be very clear from the following example :

Illustration : Suppose in a two-stage cascaded amplifier, the first stage has a voltage amplification of 2000. (The db gain of $20 \log_{10}2000 = 66$ db) and the second stage has voltage gain of 1000 (i.e. db gain = $20 \log_{10}1000 = 66$ db). If the a.c. output of the first stage is fed into the second stage, the overall amplification would become = $1000\times1000 = 2\times10^6$ which corresponds to dB gain of $(66+60) = 126$ i.e. $20 \log_{10} (2\times10^6) = 20\times6.3 = 126$ db.

Loading effect : The above result would be true only when we neglect the loading effect of the first stage by the second stage. Otherwise, the overall gain would be just less.

29.4 COUPLING NETWORK

All amplifiers need some coupling network. Even a single stage amplifier has to be coupled to the input source and the output device. In case of multistage system, there is interstage coupling. The purpose of coupling networks is (*i*) to transfer a.c. output of one stage to the input of the next stage and (*ii*) to isolate the d.c. conditions of one stage from the next. The four basic methods of coupling are :

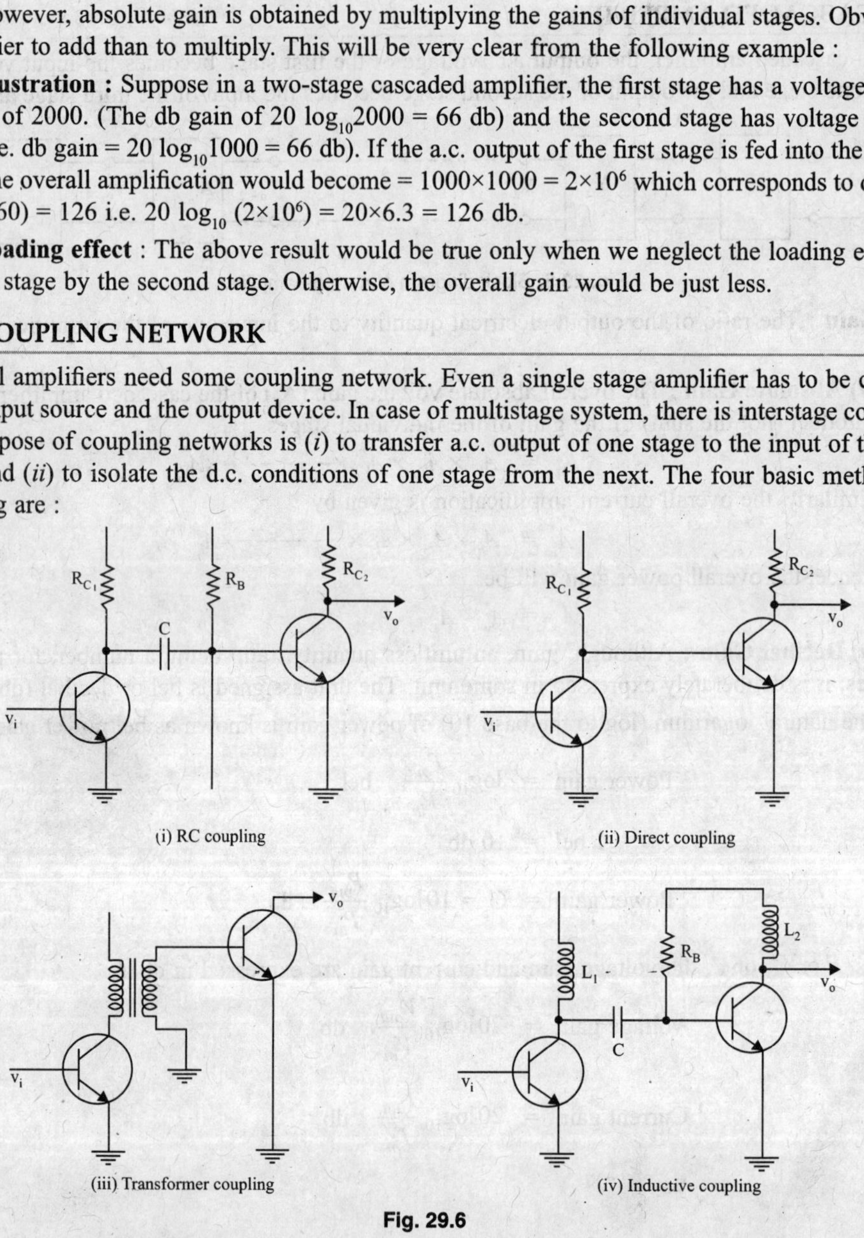

Fig. 29.6

(*i*) Resistance-capacitance (RC) coupling

Fig. 29.6 (*i*) shows RC coupled amplifier in which two stages are coupled by resistor R_{C_1} and capacitor C forming a coupling network. The function or RC-coupling network is two-fold:

(*a*) to allow or to pass signal from one stage to the next.

(*b*) to block the d.c. voltage passing from one stage to the next.

(*ii*) Direct coupling

Fig. 29.6 (ii) shows the direct coupling. This coupling is used where it is desirable to amplify signals of extremely low frequency (less than (10 Hz). The coupling devices such as capacitors and transformers can not be used because reactance offered by these components becomes appreciable and attenuates the voltage at low frequencies. To avoid this, one stage is directly coupled to the next stage without any intervening coupling device, so that the frequency response remains almost flat right from zero frequency (d.c.) to large frequency range. This type of amplifier is used in differential amplifier which is the first stage of an operational amplifier.

(*iii*) Transformer coupling

Transformer coupling is shown in Fig. 29.6 (iii) and is mostly employed for impedance matching. In general. the last stage of multistage amplifier is the power stage. This stage delivers the maximum power to the load e.g. a loudspeaker. For maximum power transfer, the impedance of power source must be equal to that of the load. Since the secondary of the coupling transformer conveys the a.c. component of the signal directly to the base of the second stage, there is no need for coupling capacitor. Moreover, the secondary winding also provides a base return path, and there is no need of base resistance. Amplifiers using this coupling are called transformer-coupled amplifiers.

(*iv*) Inductive coupling

It is also known as choke-capacitance (LC) coupling and is shown in Fig. 29.6 (iv). Amplifiers using this coupling are known as impedance-coupled amplifiers.

We shall now discuss the resistance-capacitance (RC) coupled amplifier.

29.5 R.C. COUPLED AMPLIFIER

This is the most popular type of coupling because it is easily to adjust and provides an excellent audio fidelity over a wide range of frequency. It is usually employed for voltage amplification. A two stage RC coupled transistor amplifier operating in CE mode is shown in Fig. 29.7. Generally the two identical transistors T_1 and T_2 and a common power supply is used. The resistances R_1, R_2 and R_E form the biasing network and establish the quiescent or operating point. The bypass capacitor C_E across R_E is of large value so that its reactance $\frac{1}{\omega C_E}$ is small enough to bypass the a.c. component of the emitter current. Hence only d.c. component of emitter current produces a voltage drop across R_E. Without C_E , the voltage gain of each stage would be lost. The output of the first stage amplifier is fed (or coupled) to the input of the second stage amplifier through the coupling capacitor C_C. This condenser C_C couples only a.c. component from first stage to the second stage and blocks or prevents d.c. component of the output of the first stage from reaching the input of the second stage. Hence C_C is also called a blocking condenser.

The input capacitor C_{in} couples the a.c. signal voltage to the base of the first transistor. If it is not used, the internal resistance of the signal source will be in parallel to R_2 and the bias voltage of the base will change. As a result the quiescent or operating point will shift, which is undesired.

Working : When an a.c. signal is applied (from a.c. source like Audio frequency oscillator) to the base of the first transistor, it appears in the amplified form across its collector load resistance

R_L. The amplified signal developed across R_L is fed to the base of the next stage transistor through the coupling capacitor C_C. The second stage does further amplification of the signal. In this way, the various cascaded stages go on further amplifying the signal and the overall gain is considerably increased.

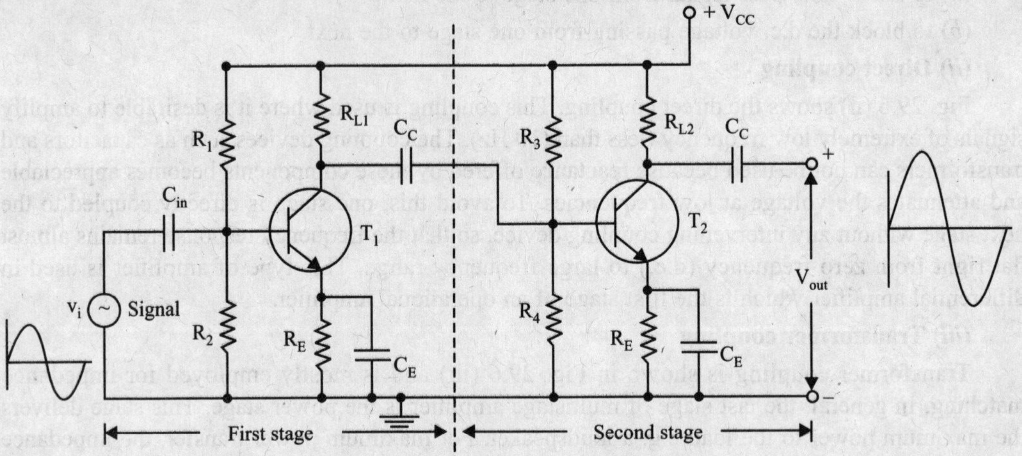

Fig. 29.7. RC coupled CE Amplifier (Two stage)

It is important to note that the total gain, in actual practice, is always less than the product of the gains of the individual stages. It is because, when a second stage is made to follow the first stage, 'the effective load resistance' of the first stage is reduced due to the shunting effect of the input resistance of the second stage. This reduces the gain of the first stage. However, the gain of the last stage which has no loading effect of the subsequent stage remains unchanged.

Frequency Response

Fig. 29.8 shows a graph drawn voltage gain (A_V) against frequency (f) called as the frequency response of a typical RC-coupled CE amplifier. This characteristic can be divided in to three regions:

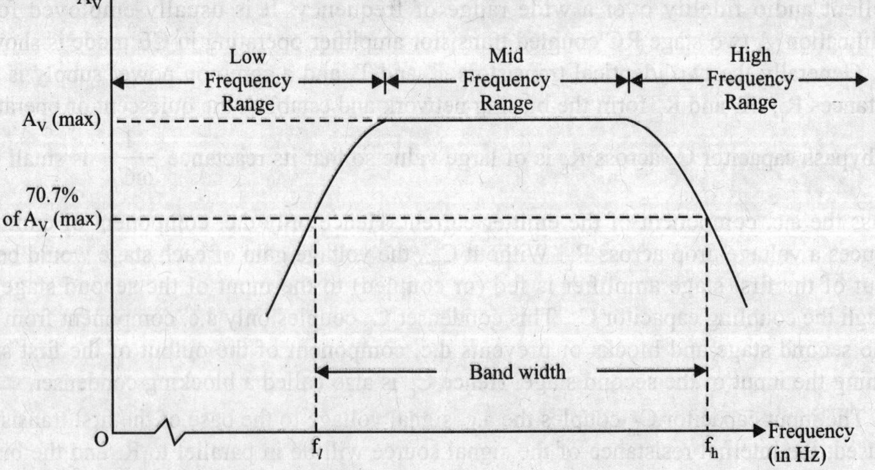

Fig. 29.8. Frequency Response of RC-coupled CE amplifier.

(*i*) The frequency range over which the gain of the amplifier is approximately constant is called **mid frequency range**.

(*ii*) The frequency range below mid-frequency range where the gain decreases with decreasing frequency is called **low frequency range.**

(*iii*) The frequency range above the mid-frequency range where the gain decreases with increasing the frequency is called the **high frequency range** of the amplifier.

(*i*) Mid frequency range: Mid frequency range covers frequencies from 50 Hz to 20 kHz (nearly the audio frequency range). In this range, the voltage gain of the amplifier almost remains constant. The reactance offered by the coupling capacitor C_C for this range $X_C = \frac{1}{2\pi f C_C}$ being very small, is neglected. Moreover, reactance offered by the inherent internal capacitances of the transistor are very large. But these are in parallel with the resistances of the circuit. Hence, the reactance offered by the transistor are also negligible for this range. The results in a uniform mid frequency gain.

(*ii*) Low frequency range: Low frequency range covers frequencies less than 50 Hz. At these frequencies, the reactance of the coupling capacitor C_C is very high so that voltage drop across C_C increases as the frequency decreases. Hence smaller and smaller signal passes from one stage to the next. Moreover, at low frequencies C_E cannot effectively shunt the emitter resistance R_L, because of its large reactance. These two factors combined causes a falling off voltage gain at low frequencies. The internal capacitance of the transistor remains very large.

(*iii*) High frequency range: High frequency range covers frequencies above 20 kHz. At such high frequencies the reactance of C_C is negligible, so it behaves as a short circuit. This increases the loading effect on the first stage, thereby reducing the voltage gain. Stray (wiring) capacitances become effective at these high frequencies. Moreover, capacitative reactance of the base-emitter junction is low, which increases the base current, thereby reducing the emitter current (as $I_E = I_B + I_C$), thereby reducing the current amplification factor. These effects combined together results a drop in voltage gain at high frequencies.

29.6 BAND WIDTH (B.W.)

From the frequency response characteristic shown in Fig. 29.8, we have two frequencies f_l and f_h where the power gain of the amplifier is 'half' its mid-frequency power gain. As a result the voltage gain is $\frac{1}{\sqrt{2}}$ = 0.707 or 70.7 percent of the mid-frequency voltage gain A_V (max), frequency f_l corresponds to the 'lower cut-off frequency' while f_h corresponds to the 'upper cut-off frequency'. The frequency difference $(f_h - f_l)$ gives the measure of band width of the amplifier. Therefore,

Band width (B.W.) = $(f_h - f_l)$

Since the flat portion of the response curve occupies a frequency range of $10f_l$ to $0.1\,f_h$, the mid-frequency gain is nearly 99.5%. Thus, an RC coupled amplifier has excellent frequency response, over AF range. Hence, it is widely used to amplify the audio programmes like speech, music etc. as AF amplifier.

Sometimes f_h and f_l are represented by f_2 and f_1 are represented by f_2 and f_1 respectively. Then

Band width (B.W.) = $(f_2 - f_1)$

29.7 ADVANTAGES OF RC COUPLING

1. It requires no expensive or bulky components and no adjustments. Hence, it is small, light and inexpensive.

2. Its overall amplification is higher than that of the other couplings.

3. It has minimum possible non-linear distortion because it does not use any coil (Inductance) or transformer which might pickup undesirable signals. Hence, there are no magnetic fields to interfere with the signal.

4. The frequency response is very flat. Hence, it gives uniform voltage amplification over a wide range from a few hertz is few megahertz because R, the resistor has no effect on frequency variation.

29.8 DISTORTION IN AMPLIFIER

An ideal amplifier is that which produces an enlarged output wave-form of the input signal. Moreover, a faithful reproduction (measured by Hi-Fi, i.e., high fidelity) at the output of the amplifier is a measure of its quality. Thus, it is expected that—

(a) The relative amplitudes of various output components must be same as that of the input signal.

(b) The output must contain all the frequencies which are present in the input signal, and

(c) The relative phase change in all output components must be equal. In practice the output of all practical amplifiers does not exactly resemble to of the input signal. Any deviation present in the output wave-form from the input signal wave-form is called **distortion** in amplifier. The distortion in amplifiers may either arise due to the imperfection in the transistor or FET (active devices) characteristics or due to circuit elements i.e. reactive components of the circuit. There are mainly two types of distortions present either separately or simultaneously in an amplifier, namely·

(1) Non-linear distortion

(2) Linear distortion

1. Non-Linear Distortion

This occurs, in case of the large-signal inputs as the transistor operates in the non-linear region of it dynamic characteristic.

(i) Amplitude Distortion

When we visualize the signal in time domain, it is called **amplitude distortion** or **wave distortion**. Fig. 29.9 gives the time domain view of this distortion. It is evident from the graph that,

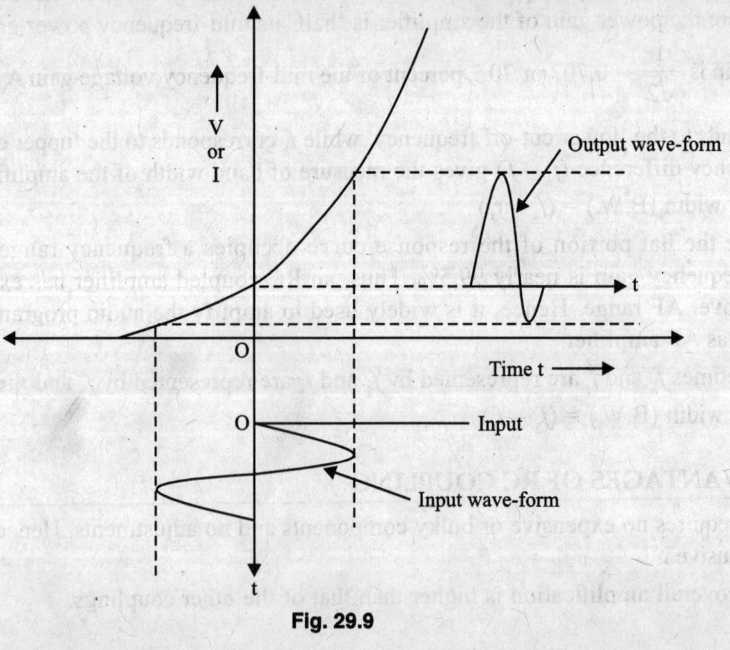

Fig. 29.9

the positive half-cycle of the input signal has been amplified more than its negative half-cycle. Consequently, the wave-shape of the output signal differs from that of the input signal. This is called amplitude distortion. This is due to the non-linearity of the characteristic.

(*ii*) Harmonic Distortion

When we think of the signal in the frequency domain, and the input signal is of only one frequency (i.e. a pure sine wave) the output contains d.c. component and different harmonics i.e. frequencies which are harmonically related with each other. These harmonics are an integral multiple of the input signal frequency. Consequently. the output is distorted.

If suppose an input applied is $V_1 = A \cos\omega$ then due to the non-linear distortion, the output contains the fundamental frequency and harmonics of the fundamental frequency. Thus, the output current may be written as

$$i_C = I_{CQ} + A_O + A_1 \cos \omega t + A_2 \cos 2\omega t + A_3 \cos 3\omega t + \ldots$$

Thus, the current wave form contains the original quiescent current I_{CQ}, which occur with zero input signal, an additional d.c. current A_O due to the non-zero average of the distorted signal, the fundamental component of the distorted a.c. signal A_1 (of the original frequency $\frac{\omega}{2\pi}$); the second harmonic component A_2 at twice the fundamental frequency $\left(\frac{2\omega}{2\pi}\right)$; the third harmonic component A_3 at thrice the fundamental frequency $\left(\frac{3\omega}{2\pi}\right)$; and so on.

The degree of distortion for a particular harmonic is measured by the ratio of the amplitude of that harmonic component to the amplitude of the fundamental component and this is generally expressed in percentage distortion. If the fundamental frequency component has an amplitude A_1 and the *n*th frequency component has an amplitude A_n, the harmonic distortion can be defined as

$$\% \; n^{\text{th}} \text{ harmonic distortion } = \% \; D_n = \frac{|A_n|}{|A_1|} \times 100\%$$

The fundamental component is larger than any other harmonic component.

The total harmonic distortion is given by

$$\% \; \text{THD} = \sqrt{D_2^2 + D_3^2 + D_4^2 + \ldots} \times 100\%$$

(*iii*) Intermodulation (IM) Distortion

It occurs when input signal consists of more than one frequency, like speech or music. Suppose, an input signal contains two frequencies f_1 and f_2. The output signal will contain their harmonics i.e. f_1, $2f_1$, $3f_1$, … etc. and f_2, $2f_2$, $3f_2$, ... etc. In addition there would be components $(f_1 + f_2)$ and $(f_1 - f_2)$ and also the sum and difference of the harmonics. These sum and difference frequencies are called **Intermodulation (IM) frequencies** and are quite undesirable in amplifiers. Since IM frequencies are not harmonically related to the signal, they are easily detected by human ear as 'noise'. Hence, great care is taken to minimise them, particularly in high-fidelity $(H_i - F_i)$ audio amplifiers.

2. Linear Distortion

Even when the active device (transistor, FET, Triode etc.) is working on linear part of its characteristic with **small signal inputs**, linear distortion occurs. It is basically due to frequency-dependent reactances associated with the circuit and active device itself and occurs when input signal is composite i.e. has signals of may different frequencies (say, fundamental and its harmonics). Thus, this may be further subdivided into two type of distortion.

(*i*) Frequency Distortion

It occurs even when the active device is working with small-signal inputs over linear region of its characteristic. It is basically due to change in the amplifier gain with frequency, i.e. the different input signal frequencies are amplified by different amount.

Suppose, the input signal has two frequencies of 50 Hz and 100 Hz. If they are not equally amplified i.e. the first frequency in amplified n times and the second one by $2n$ times, then their frequency distorted output would be as shown in Fig. 29.10.

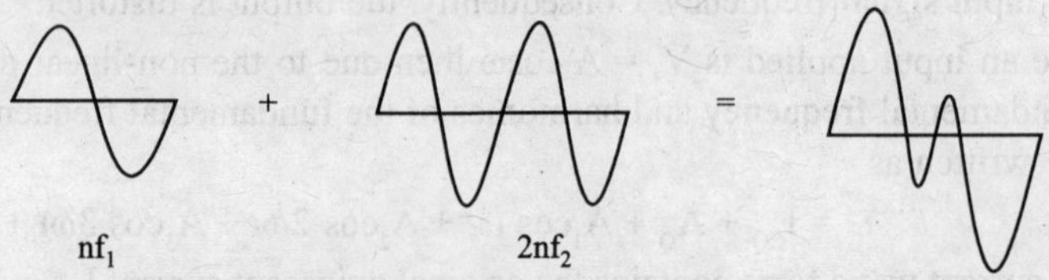

Fig. 29.10

The distortion is due to the various frequency-dependent reactance (both capacitive and inductive) associated either with the circuit or the active device itself. Hence, in the design of untuned or wide-band amplifiers, special care is taken to reduce variation of gain with frequency.

(iii) Phase or Delay Distortion

For faithful reproduction, the amplifier circuit must maintain the exact phase relationship at the output between the various components originally present in the input.

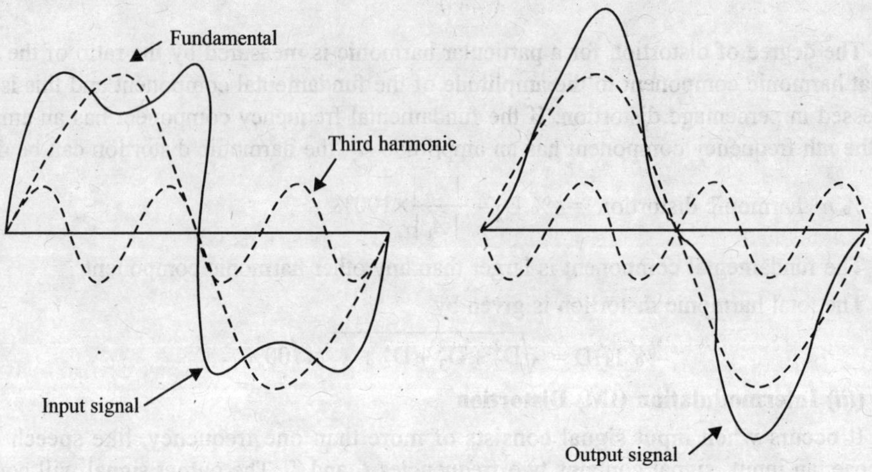

Fig. 29.11

But in practice, the amplifier introduces unequal phase shift for different frequencies of the input signal. This introduces what is known as **phase distortion**. This phase distortion is due to frequency-dependent capacitive and inductive reactances associated with the circuit and the active device of the amplifier. The phase distortion is difficult to eliminate. It should be noted that if all the frequency components in a signal are shifted in phase by an integral multiple of 180°, the resultant output wave-form resembles the input, although the polarity of wave may have altered but changes in polarity does not constitute distortion. Fortunately, the human ear is unable to distinguish phase difference (though eye can) and is thus not sensitive to thus distortion. Consequently, phase distortion is of no practical significance in audio amplifiers. However, for the systems which use wave-shapes for their operation such as Television sets (Video amplifiers), the phase distortion becomes a serious problem. Hence in video amplifiers special phase equalising circuits are employed to eliminate phase distortion.

SOLVED EXAMPLES

Example 29.1. *Calculate the harmonic distortion components for an output signal having fundamental amplitude of 2.5 V, second harmonic amplitude of 0.25 V, third harmonic amplitude of 0.1 V and fourth harmonic amplitude of 0.05 V. Also calculate the total harmonic distortion.*

Solution: We know that percentage n^{th} harmonic distortion is given by

$$\% D_n = \frac{|A_n|}{|A_1|} \times 100\%$$

Therefore,

$$\% D_2 = \frac{|A_2|}{|A_1|} \times 100\% = \frac{0.25\,V}{2.5\,V} \times 100\% = 10\%$$

$$\% D_3 = \frac{|A_3|}{|A_1|} \times 100\% = \frac{0.1\,V}{2.5\,V} \times 100\% = 4\%$$

$$\% D_4 = \frac{|A_4|}{|A_1|} \times 100\% = \frac{0.05\,V}{2.5\,V} \times 100\% = 2\%$$

∴ Total harmonic distortion will be

$$\% \text{THD} = \sqrt{D_2^2 + D_3^2 + D_4^2} \times 100\%$$

$$= \sqrt{(0.10)^2 + (0.04)^2 + (0.02)^2} \times 100\%$$

$$= 0.1095 \times 100\%$$

$$= 10.95$$

Example 29.2. *A three-stage amplifier has a first stage voltage gain 100, second stage voltage gain of 200 and third stage gain of 400. Find the total voltage gain in db.*

(Nagput U., S/2011)

Solution. First stage voltage gain in db

$$= 20 \log_{10} 100$$

$$= 20 \times 2 = 40$$

Second stage voltage gain in db $= 20 \log_{10} 200$

$$= 20 \times 2.36 = 46$$

Third stage voltage gain in db $= 20 \log_{10} 400$

$$= 20 \times 2.6 = 52$$

∴ Total voltage gain $= 40 + 46 + 52 = 138$ db

Example 29.3. *Find the gain in db in the following cases:*

(*i*) *Voltage gain of 30* (*ii*) *Power gain of 10*

Solution: (*i*) Voltage gain $= 20 \log_{10} 30$ db

$$= 20 \times 1.477 \text{ db}$$

$$= 29.54 \text{ db}$$

(*ii*) Power gain $= 10 \log_{10} 100$ db

$$= 10 \times 2 \text{ db}$$

$$= 20 \text{ db}$$

Example 29.4. *Express the following gains as a number:*

(*i*) Power gain of 40 db (*ii*) Power gain of 43 db

Solution. (*i*) Power gain of 40 db = 4 bel

By taking antilog, we get a number.

$$\therefore \quad \text{Power gain} = \text{Antilog } 4 = 10^4 = 10{,}000$$

(*ii*) Power gain of 43 db = 4.3 bel

By taking antilog, we get a number.

$\therefore$ Power gain = Antilog 4.3 = 2×10^4 = 20,000

EXERCISE

1. What are different classes of amplifiers? How they are classified? *(Nagpur U. 2009, 2011, 2012)*
2. Explain in detail the classification mode according to the position of quiescent point. OR Explain Class A, Class B, Class AB and Class C amplifiers. *(Nagpur U. 2009, 2012)*
3. What do you mean by multistage amplifier? What is the voltage gain of n-stage amplifier? Mention its need and explain its two categories.
4. Explain cascaded amplifier. Draw block diagram of *n*-stages amplifier. What is its overall voltage gain in terms of absolute and decibel? *(Nagpur U. 2012)*
5. What is the use of coupling network in amplifiers? Explain different types of coupling networks employed in a multistage amplifier.
6. Explain two stage transistor RC coupled amplifier with reference to frequency response and band width. *(Nagpur U. 2011)*
7. Give advantages of RC coupling.
8. What do you mean by distortion in amplifiers? Give a brief discussion of

   (*i*) Non-linear distortion

   (*ii*) Linear distortion.
9. Explain the followings:

   (*a*) Amplitude distortion

   (*b*) Harmonic distortion

   (*c*) Intermodulation distortion

   (*d*) Frequency distortion *(Nagpur U. 2012)*

   (*e*) Phase distortion. *(Nagpur U. 2012)*
10. Calculate the harmonic distortion components for an output signal having fundamental amplitude of 2.1 V, second harmonic amplitude of 0.3 V, third harmonic component of 0.1 V, and fourth harmonic component of 0.05 V. Also calculate the total harmonic distortion. *(Nagpur U. 2009, 2011)*

    [Ans. % D_2 = 14.3%, %D_3 = 4.8%, % D_4 = 2.4%, %THD = 15.27]
11. Explain the reason for the reduction in the gain of an RC coupled amplifier at high and low frequencies. *(Nagpur U. 2011)*

HYBRID EQUIVALENT CIRCUITS AND NOISE

30
Chapter

INTRODUCTION

We have studied hybrid parameters in Chapter 27 and their interrelations. Any a.c. circuit can be analysed by two methods: the graphical method and the analytical method. Graphical method locates the quiescent or operating point and the nature of input ad output waveform, when large signal voltages are applied. In order to analyse small signal voltages, we use the hybrid equivalent i.e. h-parameter equivalent circuits. Since h-parameters of an active device i.e. transistor can be easily determined, h-parameter equivalent circuit analysis has proved to be an easy tool.

30.1 HYBRID EQUIVALENT CIRCUIT OF CE AMPLIFIER

The h-parameters of a two-port network are designated by h_{11}, h_{12}, h_{21} and h_{22}. Their values for any network are such that the following two network equations are always satisfied.

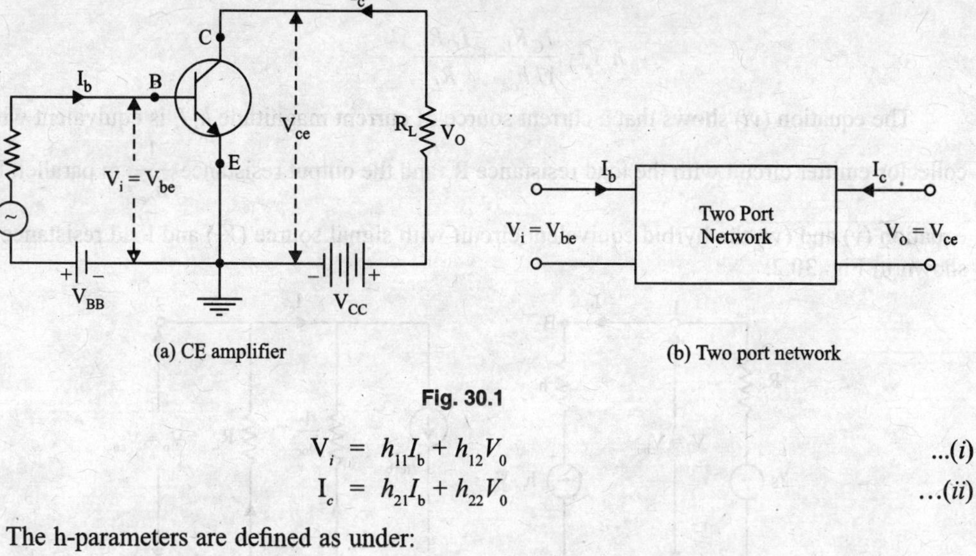

(a) CE amplifier (b) Two port network

Fig. 30.1

$$V_i = h_{11}I_b + h_{12}V_0 \qquad ...(i)$$

$$I_c = h_{21}I_b + h_{22}V_0 \qquad ...(ii)$$

The h-parameters are defined as under:

(1) $$h_{11} = \left.\frac{V_i}{I_b}\right|_{V_0=0} = h_{ie}, \text{ the input impedance}$$

(2) $$h_{12} = \left.\frac{V_i}{V_o}\right|_{I_b=0} = h_{re}, \text{ the reverse voltage ratio}$$

(3) $$h_{21} = \left.\frac{I_c}{I_b}\right|_{V_o=0} = h_{fe}, \text{ the forward current ratio}$$

(4) $$h_{22} = \left.\frac{I_C}{V_0}\right|_{I_b=0} = h_{oe}, \text{ the output admittance.}$$

Substituting these values, equation (*i*) and (*ii*) becomes,

$$V_i = h_{ie}I_b + h_{re}V_o \quad \text{... (iii)}$$

$$I_c = h_{fe}I_b + h_{oe}V_o \quad \text{... (iv)}$$

These are the two simultaneous equations.

From Eqn. (*iii*) we get

$$I_b = \frac{V_i - h_{re}V_o}{h_{ie}} \quad \text{...(v)}$$

The d.c. collector to emitter voltage V_{CE} in given by

$$V_{CE} = V_{CC} - I_C R_L$$

Differentiating, $dV_{CE} = 0 - R_L dI_C$

$$V_{ce} = -R_L i_c$$

or $$V_{ce} = V_o = -R_L I_c$$

Substituting in Eqn. (iv) we get

$$I_c = h_{fe}I_b + h_{oe}(-R_L I_C) = h_{fe}I_b - h_{oe}R_L I_c$$

$$h_{fe}I_b = h_{oe}R_L I_C + I_C$$

or $$h_{fe}I_b = \frac{I_C R_L}{1/h_{oe}} + \frac{I_C R_L}{R_L} \quad \text{...(vi)}$$

The equation (*vi*) shows that a current source of current magnitude $h_{fe}I_b$ is equivalent with the collector-emitter circuit with the load resistance R_L and the output resistance $\frac{1}{h_{oe}}$ in parallel. Using equation (*v*) and (*vi*) the hyrbid equivalent circuit with signal source (V_S) and load resistance R_L is shown in Fig. 30.2.

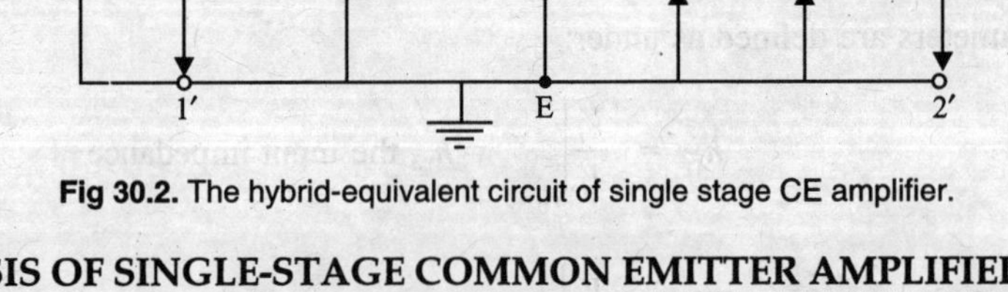

Fig 30.2. The hybrid-equivalent circuit of single stage CE amplifier.

30.2 ANALYSIS OF SINGLE-STAGE COMMON EMITTER AMPLIFIER

(1) Current Gain (A_{ie}): The current gain is defined as the ratio of the load current I_C to the input current I_b and is denoted by A_{ie},

$$A_{ie} = \frac{I_c}{I_b}$$

Let Z be the equivalent impedance of $\frac{1}{h_{oe}}$ and R_L in parallel.

$$\therefore \qquad \frac{1}{Z} = \frac{1}{\frac{1}{h_{oe}}} + \frac{1}{R_L} = h_{oe} + \frac{1}{R_L} = \frac{1 + h_{oe}R_L}{R_L}$$

or
$$Z = \frac{R_L}{1 + h_{oe}R_L}$$

Voltage across R_L = voltage across Z

$$I_C R_L = h_{fe} I_b \left(\frac{R_L}{1 + h_{oe}R_L}\right)$$

or
$$\frac{I_c}{I_b} = \frac{h_{fe}}{1 + h_{oe}R_L}$$

or Current gain,
$$A_{ie} = \frac{h_{fe}}{1 + h_{oe}R_L} \qquad \ldots(i)$$

(2) Input Impedance (Z_{ie}): The impedance between the input terminals 1 and 1′ is called input impedance (Z_{ie}). It is given by

$$Z_{ie} = \frac{V_i}{I_b}$$

But
$$V = h_{ie} I_b + h_{re} V_o$$
$$= h_{ie} I_b + h_{re} (-I_c R_L)$$

Dividing by I_b, we get

Input impedance,
$$Z_{ie} = \frac{V_i}{I_b} = h_{ie} - h_{re} R_L \left(\frac{I_c}{I_b}\right)$$

$$Z_{ie} = h_{ie} - h_{re} R_L A_{ie} \qquad \ldots(ii)$$

This equation indicates that the input impedance depends on the load resistance R_L.

(3) Voltage Gain (A_{ve}): It is defined as the ratio of output voltage V_o to the input voltage V_i.

Thus, Voltage gain,
$$A_{ve} = \frac{V_o}{V_i}$$

$$= \frac{-I_c R_L}{V_i} \qquad (\because V_o = -I_C R_L)$$

$$= -\left(\frac{I_c}{I_b}\right)\left(\frac{I_b}{V_i}\right) R_L$$

$$= -A_{ie}\left(\frac{1}{Z_i}\right) R_L$$

$$= -\frac{A_{ie} R_L}{Z_i}$$

Now substituting for A_{ie} and Z_i from Eqn. (*i*) and (*ii*), we get

$$A_{ve} = \frac{h_{fe} R_L}{1 + h_{oe} R_L} \times \frac{1}{h_{ie} - h_{re} R_L A_{ie}}$$

$$= -\frac{h_{fe}R_L}{(1+h_{oe}R_L)\left(h_{ie} - \frac{h_{re}R_L h_{fe}}{1+h_{oe}R_L}\right)}$$

$$= -\frac{h_{fe}R_L}{h_{ie}(1+h_{oe}R_L) - h_{re}R_L h_{fe}}$$

$$= \frac{h_{fe}\mathrm{R_L}}{h_{ie} + (h_{ie}h_{oe} - h_{fe}h_{re})\mathrm{R_L}}$$

$$= -\frac{h_{fe}\mathrm{R_L}}{h_{ie} + \mathrm{R_L}\Delta h}$$

where $\Delta h = h_{ie}h_{oe} - h_{fe}h_{re}$

In practice Δh is negligible ($\Delta h \to 0$) compared to h_{ie}.

$$\therefore \quad A_{ve} = -\frac{h_{fe}R_L}{h_{ie}} \qquad \ldots(iii)$$

(4) Output Impedance (Z_{oe}): The input voltage source to the amplifier is removed and replaced by its internal resistance R_S. The a.c. voltage source V_O is connected to the output terminals as shown in Fig. 30.3.

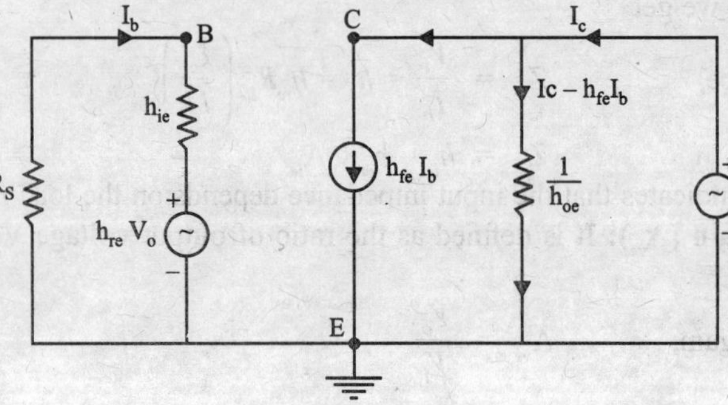

Fig. 30.3

The output impedance $Z_o = \frac{V_o}{I_c}$

Here I_c is the current supplied by the a.c. source V_o.

The base current in given by

$$I_b = -\frac{h_{re}V_o}{h_{fe} + R_s}$$

The current through the output resistance $\frac{1}{h_{oe}}$ of the transistor is $(I_C - h_{fe}\,I_b)$.

Therefore, the output voltage is

$$(I_c - h_{fe}I_b)\cdot\frac{1}{h_{oe}} = V_o$$

$$I_c - h_{fe} I_b = h_{oe} V_o$$

$$I_c = h_{oe} V_o + h_{fe} I_b$$

Substituting for I_b, we get

$$I_c = h_{oe} V_o + h_{fe}\left(-\frac{h_{re} V_o}{h_{fe} + R_s}\right)$$

or

$$I_c = h_{oe} V_o - \frac{h_{fe} h_{re}}{h_{fe} + R_s} \cdot V_o$$

$$= \left(h_{oe} - \frac{h_{fe} h_{re}}{h_{fe} + R_s}\right) V_o$$

or

$$\frac{I_c}{V_o} = h_{oe} - \frac{h_{fe} h_{re}}{h_{fe} + R_s}$$

But $\frac{I_c}{V_o}$ is the output admittance Y_{oe}

$$\therefore \quad Y_{oe} = h_{oe} - \frac{h_{fe} h_{re}}{h_{fe} + R_s}$$

The output impedance, therefore will be

$$Z_{oe} = \frac{1}{Y_{oc}} = \frac{1}{h_{oe} - \frac{h_{fe} h_{re}}{h_{fe} + R_s}}$$

or

$$Z_{oe} = \frac{h_{fe} + R_s}{(h_{fe} + R_s) h_{oe} - h_{fe} h_{re}} \quad \ldots(iv)$$

(5) Power Gain: The power gain is defined as the product of the magnitudes of voltage gain and current gain. Thus.

$$\text{Power Gain} = |A_{ve}|\,|A_{ie}|$$

$$\text{But } |A_{ve}| = \frac{A_{ie} R_L}{Z_{ie}}$$

$$\therefore \quad \text{Power Gain} = \frac{A_{ie}^2 R_I}{Z_{ie}}$$

or

$$\text{Power Gain} = \left(\frac{h_{fe} R_L}{h_{ie} + R_L \Delta h}\right)\left(\frac{h_{fe}}{1 + h_{oe} R_L}\right)$$

$$= \frac{h_{fe}^2 R_L}{(1 + h_{oe} R_L)(h_{ie} + R_L \Delta h)} \quad \ldots(v)$$

where $\Delta h = h_{ie} h_{oe} - h_{fe} h_{re}$.

30.3 A.C. EQUIVALENT CIRCUIT OF TWO STAGE RC COUPLED AMPLIFIER

The general a.c. equivalent circuit of the two-stage RC coupled CE amplifier is shown in Fig. 30.4. This a.c. equivalent circuit corresponds to the circuit diagram shown in Fig. 29.7.

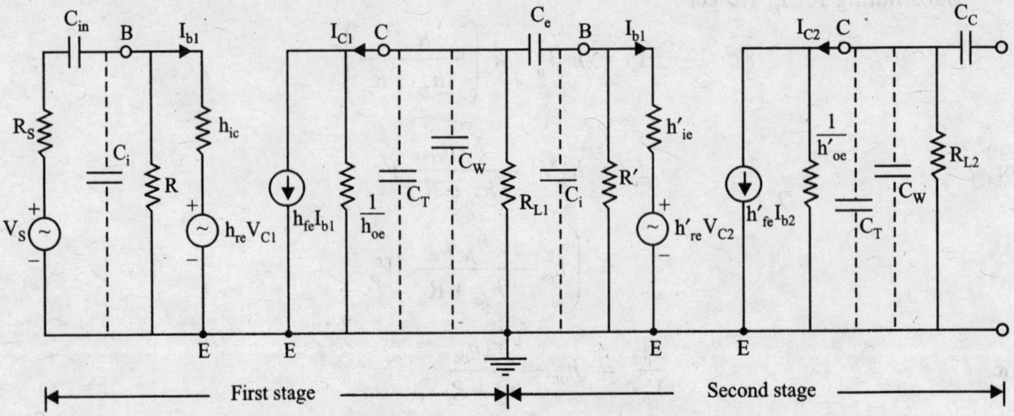

Fig. 30.4. General a.c. equivalent circuit of RC coupled CE amplifier

It should be noted that,

(*i*) The emitter resistances R_E and bypass capacitors C_E are not shown as it is presumed that even at the lowest frequency, the capacitance C_E shorts the resistance R_E. Hence, emitter of each transistor is shown at common ground potential.

(*ii*) Resistance R represents equivalent resistance of R_1 and R_2 in parallel. This is because top of R_1 gets connected to the ground through a negligible internal resistance of the d.c. supply. R′ is equivalent of R_3 and R_4 in parallel.

(*iii*) The diffusion capacitance (C_D) exists in case of a forward biased diode. If the diode is forward biased by sudden application of voltage to the diode, the current does not reach to maximum value instantly. There is always a delay which is identical to the charging of a capacitor through a resistor. This is known as diffusion capacitor C_D, which is of the order of few hundred picofarad.

When a diode is reverse biased, these exists a depletion region which contains strong electric field developed. This is identical to the opposite charges placed on the plate of a condenser. This is denoted by C_T and is known as transition capacitance. In common emiter configuration, C_T appears between the input (base) and output (collector) terminals, while C_D appears between emitter and base. Note that C_D is quite large and C_T is very small. The input capacitance C_i is given by $C_i = C_D + C_T (1 + g_m R_L)$ which is few hundred pico-farad. C_W is the stray capacitance due to wiring. This is also few pico-farad. Let us examine the circuit for mid-frequency, low frequency and high frequency range.

30.4 MID FREQUENCY RESPONSE OF THE AMPLIFIER

To analyse the circuit, lengthy mathematical calculation are involved. This is a tedious job. Hence, in practice, it is sufficient to obtain an approximate expressions. For this purpose, we may some assumptions as under:

(1) The transistors T_1 and T_2 in the two-stage RC coupled CE amplifier are assumed to be identical, so that the input resistance h_{ie} and the forward current transfer ratio h_{fe} of each is the same.

(2) The bias resistors R_1 and R_2 in parallel (due to negligible internal resistance of power supply) give an equivalent resistor R shown in Fig. 30.4 being large compared with h_{ie}, R is omitted.

(3) The transistors are designed such that their h_{re} is of the order of 10^{-4} which can be neglected. (i.e. $h_{re} = 0$). Similarly $\frac{1}{h_{oe}} >> R_L$ and $\frac{1}{h_{oe}}$ being parallel to R_L $\frac{1}{h_{oe}}$ is neglected. Hence, h_{re} as well as $\frac{1}{h_{oe}}$ can be omitted.

(4) The output resistance (R_S) of the input source (v_s) is assumed to be infinitesimally small compared to h_{ie} and the reactance of capacitor C_{in} is also small compared to h_{ie}. Hence R_S and C_{in} can be omitted.

(5) In the mid-frequency range, the reactance $\left(X_{C_E} = \frac{1}{2\pi f C_E}\right)$ offered by bypass capacitor C_E is negligible. Moreover, the reactance $\left(X_{C_C} = \frac{1}{2\pi f C_C}\right)$ offered by C_C are very small. Hence they can be omitted.

(6) At mid-frequency, the capacitance C_i and C_T can also be treated as negligible and hence are absent in the equivalent circuit. The stray or wire capacitance C_W is also ineffective at mid-frequency and hence is omitted in mide-frequency a.c. equivalent circuit.

Under above assumptions, the general a.c. equivalent circuit of RC coupled CE amplifier can be drawn as shown in Fig. 30.5 for mid-frequency range. The first stage of the amplifier extend from its input circuit to the input resistance of the second stage.

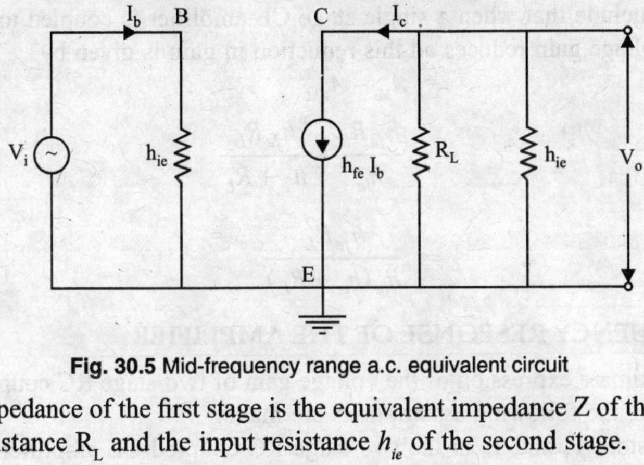

Fig. 30.5 Mid-frequency range a.c. equivalent circuit

The output impedance of the first stage is the equivalent impedance Z of the parallel combination of the load resistance R_L and the input resistance h_{ie} of the second stage.

$$\therefore \quad \frac{1}{Z} = \frac{1}{R_L} + \frac{1}{h_{ie}} \quad ...(i)$$

$$Z = \frac{h_{ie} R_L}{h_{ie} + R_2}$$

The r.m.s value of V_o of the output voltage across Z is given by

$$V_o = -h_{fe} I_b Z \quad .... (ii)$$

The r.m.s value of the input voltage is

$$V_i = I_b h_{ie} \quad ...(iii)$$

Therefore, the voltage gain in the mid-frequency range is given by

$$A_{VM} = \frac{V_o}{V_i} = \frac{-h_{fe} I_b Z}{I_b h_{ie}}$$

$$= -\frac{h_{fe}Z}{h_{ie}}$$

Substituting the value of Z from equation (*i*), we get

$$A_{VM} = -\frac{h_{fe}}{h_{ie}}\left(\frac{h_{ie}R_L}{h_{ie}+R_L}\right)$$

$$= -\frac{h_{fe}R_L}{h_{ie}+R_L} \quad \text{...(iv)}$$

The –ve sign indicates that the phase shift of A_{VM} is 180°. The magnitude of voltage gain in given by

$$|A_{VM}| = \frac{h_{fe}R_L}{h_{ie}+R_L} \quad \text{...(v)}$$

Conclusions: From equation (v), the following conclusions can be drawn:

(1) In the mid-frequency range, the voltage gain of the first stage RC coupled CE amplifier is independent of frequency.

(2) We know that the magnitude of voltage gain of a signal-stage CE amplifier is

$$|A_{Ve}| = \frac{h_{fe}R_L}{h_{ie}}$$

Thus $$|A_{VM}| < |A_{VE}|$$

Hence, we conclude that when a single stage CE amplifier is coupled to another stage with RC coupling, its voltage gain reduces ad this reduction in gain is given by

$$= A_{Ve} - A_{VM}$$

$$= \frac{h_{fe}R_L}{h_{ie}} - \frac{h_{fe}R_L}{h_{ie}+R_L}$$

$$= \frac{h_{fe}R_L^2}{h_{ie}(h_{ie}+R_L)} \quad \text{...(vi)}$$

30.5 LOW FREQUENCY RESPONSE OF THE AMPLIFIER

For the approximate expression of the voltage gain of two-stage RC coupled CE amplifier in low frequency range, the following assumptions are made:

(1) The transistors T_1 and T_2 in the two-stage RC coupled CE amplifier are assumed to be identical so that the input resistance h_{ie} and the forward current transfer ratio h_{fe} of each is the same.

(2) The bias resistors R_1 and R_2 are in parallel (due to negligible internal resistance of power supply) gives an equivalent resistor R shows in Fig. 30.4 being large compared with h_{ie}, R is omitted.

(3) The transistors are designed such that their h_{re} can be neglected (i.e. $h_{re} = 0$). Hence h_{re} omitted, similarly $\frac{1}{h_{oe}} >> R_L$ and $\frac{1}{h_{oe}}$ being parallel to R_L, $\frac{1}{h_{oe}}$ is omitted.

(4) The output resistance (R_S) of the input source (v_s) is assumed to be infinitesimally small compared to h_{ie} and the reactance of capacitor C_{in} is also small compared to h_{ie}. Hence R_S and C_{in} are omitted.

(5) In the low frequency range, the reactance offered $\left(X_{C_E} = \frac{1}{2\pi f C_E}\right)$ by bypass capacitor C_E is negligible. Hence C_E is omitted.

(6) In the low frequency range, the reactance offered by series coupling capacitor $C_C \left(i.e.\ X_{C_C} = \dfrac{1}{2\pi f C_C} \right)$ is appreciable and it is included in the equivalent circuit.

(7) At low frequency, the capacitance C_i and C_T can also be treated as negligible and hence omitted.

Under above assumptions, the a.c. equivalent circuit of first stage of an RC coupled amplifier in low frequency range extending from its input circuit to the input of the second sage is shown in Fig. 30.6.

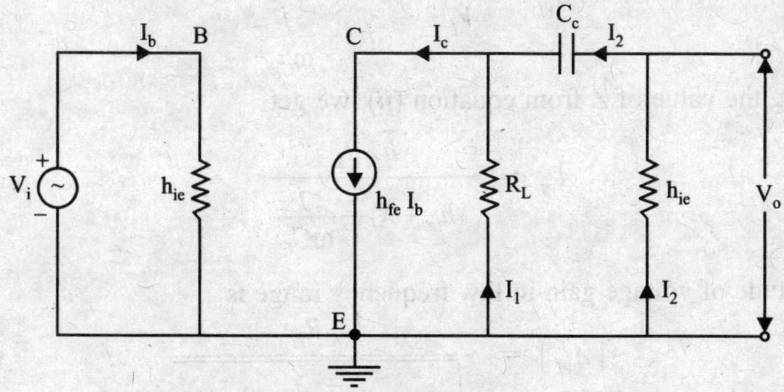

Fig. 30.6: A.C. equivalent circuit in low frequency range.

The r.m.s value of input voltage V_i is given by

$$V_i = I_b h_{ie} \qquad \ldots(i)$$

The output impedance Z of the first stage of RC coupled amplifier is the equivalent impedance of parallel combination of R_L and $\left(h_{ie} + \dfrac{1}{j\omega C_C} \right)$. Thus,

$$\frac{1}{Z} = \frac{1}{R_L} + \frac{1}{h_{ie} + \dfrac{1}{j\omega C_C}}$$

$$= \frac{1}{R_L} + \frac{1}{h_{ie} - \dfrac{j}{\omega C_C}} = \frac{h_{ie} + R_L - \dfrac{j}{\omega C_C}}{R_L \left(h_{ie} - \dfrac{j}{\omega C_C} \right)}$$

or

$$Z = \frac{R_L \left(h_{ie} - \dfrac{j}{\omega C_C} \right)}{h_{ie} + R_L - \dfrac{j}{\omega C_C}} \qquad \ldots(ii)$$

Let I_1 be the current flowing through R_L and I_2 be flowing through h_{ie} and C_C. The potential difference between the emitter and collector along h_{ie} and C_C = the *P.D.* across the load impedance *Z*.

$$\therefore \qquad I_2 \left(h_{ie} - \frac{j}{\omega C_C} \right) = h_{fe} I_b Z$$

or

$$I_2 = \frac{h_{fe} I_b Z}{h_{ie} - \dfrac{j}{\omega C_C}}$$

The output voltage V_o is the P.D. across h_{ie}

$$\therefore \quad V_o = -I_2 h_{ie}$$

Substituting for I_2, we get

$$V_o = -\frac{h_{fe} h_{ie} I_b Z}{h_{ie} - \dfrac{j}{\omega C_C}} \qquad \text{...(iii)}$$

The voltage gain in the low frequency range, is given by

$$A_{VL} = \frac{V_o}{V_i} = -\frac{h_{fe} Z}{h_{ie} - \dfrac{j}{\omega_{C_C}}}$$

Substituting the value of Z from equation (*ii*), we get

$$A_{VL} = -\frac{h_{fe} R_L}{h_{ie} + R_L - \dfrac{j}{\omega C_C}} \qquad \text{...(iv)}$$

The magnitude of voltage gain in low frequency range is

$$|A_{VL}| = \frac{h_{fe} R_L}{\sqrt{(h_{ie} + R_L)^2 + \left(\dfrac{1}{2\pi f C_C}\right)^2}} \qquad \text{...(v)}$$

Thus the magnitude of voltage gain decreases as the frequency of the input voltage decreases.

Lower cut-off Frequency (f_1)

It is defined as the frequency at which the magnitude of the voltage gain in the lower frequency range falls to $\frac{1}{\sqrt{2}}$ or 0.707 of the magnitude of gain in the mid-frequency range. Thus, at $f = f_1$

$$|A_{VL}| = \frac{|A_{VM}|}{\sqrt{2}} \qquad \text{...(vi)}$$

Substituting the values of A_{VL} and A_{VM} from Eqn. (*iv*) and (*iii*) of article 30.4, we get

$$\frac{A_{VL}}{A_{VM}} = \left(\frac{h_{fe} R_L}{h_{ie} + R_L - j/\omega C_C}\right) \times \left(\frac{h_{ie} R_L}{h_{fe} R_L}\right)$$

$$\therefore \quad A_{VL} = \frac{A_{vm}}{1 - \dfrac{j}{2\pi f_1 C_c (h_{ie} + R_L)}}$$

$$\therefore \quad |A_{VL}| = \frac{|A_{VM}|}{\sqrt{1 + \left(\dfrac{1}{2\pi f_1 C_C (h_{ie} + R_L)}\right)^2}} \qquad \text{...(vii)}$$

$$\therefore \quad |A_{VL}| < |A_{VM}|$$

Equating Eqn. (*vi*) and (*vii*), we get

$$f_1 = \frac{1}{2\pi C_C (h_{ie} + R_L)} \qquad \text{...(viii)}$$

This gives the lower cut-off frequency (f_l).

30.6 HIGH FREQUENCY RANGE

In high frequency range, the effect of coupling capacitors C_C is negligible and hence they are omitted.

Miller Effect : In high frequency range, the two internal capacitances in the bipolar transistor become effective. These are: (*i*) the forward-biased capacitor or the capacitance between base and emitter and is known as diffusion capacitor (C_D) and (*ii*) the reverse-biased capacitor between the collector and the base. This is called transition capacitor (C_T). In high frequency range, this capacitor CT provides feedback from the output to the input circuit. This is known as Miller effect. The stray or wiring capacitance (C_W) also becomes effective in high frequency range.

It should be noted that in mid-frequency and low frequency ranges, these capacitances have negligible effect on the frequency response of the amplifier.

30.6.1 High Frequency Equivalent Circuit (Single Stage)

Fig. 30.7 shows the high frequency equivalent circuit for a single stage *CE* amplifier.

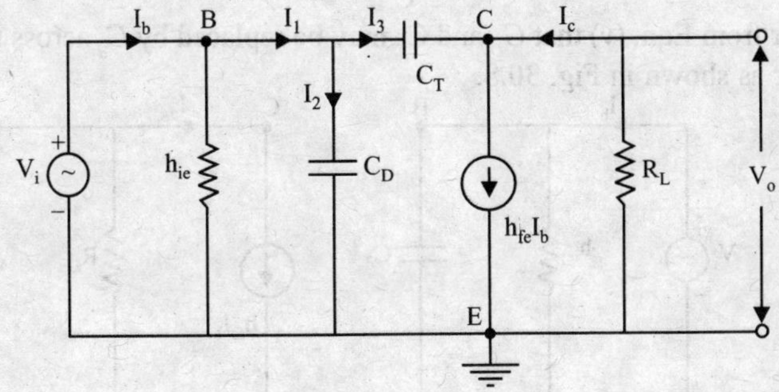

Fig. 30.7

As $\frac{1}{h_{oe}} >> R_L$ and $\frac{1}{h_{oe}}$ being parallel to R_L, $\frac{1}{h_{oe}}$ is neglected as usual. The base-emitter capacitor (C_D) is connected as shunt while capacitor between base and collector (C_T) is connected in series, through which feedback from output to input takes place at high frequency. The current I_1 From Fig. 30.7 is ...(*i*)

$$I_1 = I_2 + I_3$$

$$V_i = \frac{I_2}{j\omega C_D}$$

$$\therefore \quad I_2 = j\omega C_D V_i \quad \text{...(ii)}$$

$$V_o = -I_C R_L \quad \text{...(iii)}$$

Let us apply Kirchhoff's law to the circuit containing C_D, C_T and R_L,

$$I_3 X_{CT} - I_C R_L - I_2 X_{CD} = 0$$

or
$$\frac{I_3}{j\omega C_T} - I_C R_L - \frac{I_2}{j\omega C_D} = 0$$

or
$$\frac{I_3}{j\omega C_T} + V_o - V_i = 0$$

or
$$I_3 = j\omega C_T (V_i - V_o) \quad \text{...(iv)}$$

Substituting the value of I_2 and I_3 in Eqn. (*i*) , we get

$$I_1 = j\omega C_D V_i + j\omega C_T (V_i - V_o)$$

$$= j\omega V_i\left[C_D + \left(1-\frac{V_o}{V_i}\right)C_T\right]$$

Since, $$\frac{V_o}{V_i} = -\frac{h_{fe}R_L}{h_{ie}}$$

$$I_1 = j\omega V_i\left[C_D + \left(1+\frac{h_{fe}R_L}{h_{ie}}\right)C_T\right]$$

$$= j\omega V_i C_d$$

$$= j\omega C_d V_i \quad \text{...(v)}$$

where $$C_d = C_D + \left(1+\frac{h_{fe}R_L}{h_{ie}}\right)C_T \quad \text{...(vi)}$$

Now it is clear from Eqn. (*v*) that C_D and C_T may be replaced by C_d across the input resistance hie of the transistor as shown in Fig. 30.8.

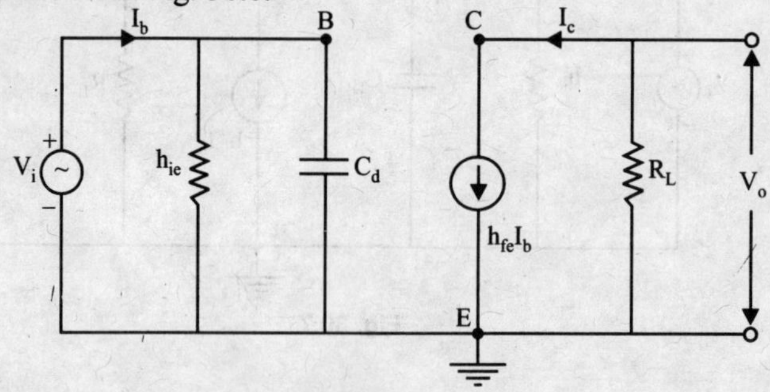

Fig. 30.8

30.6.2 High Frequency Responce (Two-stage)

To evaluate the approximate expression for voltage gain of two-stage RC coupled CE amplifier in high frequency range, we make the following assumptions:

(1) The transistors T_1 and T_2 used in two-stage RC coupled CE amplifier are assumed to be identical so that the input resistance h_{ie} and the forward current transistor ratio h_{fe} of each is the same.

(2) The bias resistors R_1 and R_2 in parallel gives an equivalent resistor R, being large compared with h_{ie}, R is omitted.

(3) The transistors are designed such that their h_{re} can be neglected and hence h_{re} is omitted. Similarly $\frac{1}{h_{oe}} >> R_L$ and $\frac{1}{h_{oe}}$ being parallel to R_L, $\frac{1}{h_{oe}}$ is omitted.

(4) Source of input voltage has negligible internal resistance (R_S) as compared to h_{ie} and hence R_S is ommited.

(5) In high frequency range, the reactance offered by C_E $\left(i.e.\ X_{C_E} = \frac{1}{2\pi f C_E}\right)$ is negligible small and hence C_E is omitted.

(6) In high frequency range, the series reactance due to coupling capacitors C_C $\left(i.e.\ X_{C_C} = \frac{1}{2\pi f C_C}\right)$ is negligible.

(7) In high frequency range, the value of the shunt capacitance C_d in the input circuit of the first stage is small. This is because it depends on the output impedance of the first transistor, which is very small. But in the output circuit of the first stage C_d is enhanced by the stray capacitance (C_W) of the wiring. Therefore, the reactance of C_d in output circuit of the stage will have appreciable shunting effect on R_L and h_{ie}.

Under above assumption, an a.c. equivalent of the first-stage for the high frequency range is shown in Fig. 30.9.

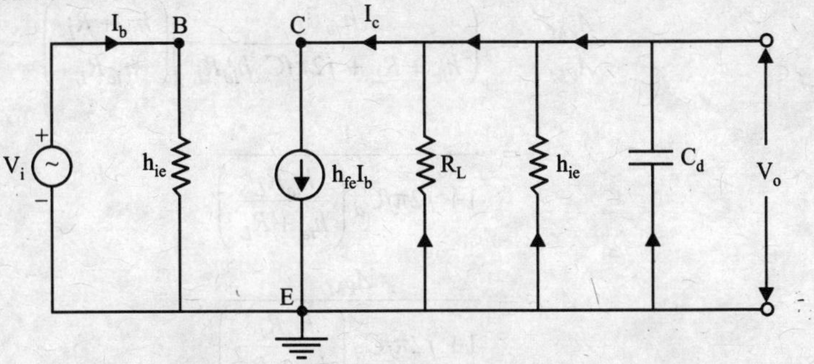

Fig. 30.9: a.c. equivalent circuit of high frequency

The equivalent output impedance Z is a parallel combination of R_L, h_{ie} and C_d and is given by

$$\frac{1}{Z} = \frac{1}{R_L}+\frac{1}{h_{ie}}+\frac{1}{X_{C_d}}$$

$$= \frac{1}{R_L}+\frac{1}{h_{ie}}+\frac{1}{1/j\omega C_d}$$

$$= \frac{1}{R_L}+\frac{1}{h_{ie}}+j\omega C_d$$

$$\therefore \quad Z = \frac{h_{ie}R_L}{h_{ie}+R_L+j\omega C_d h_{ie}R_L} \qquad ...(vii)$$

The r.m.s. value of output voltage V_o across Z is given by

$$V_o = -h_{fe}I_b Z$$

The voltage gain (A_{VH}) in high frequency range is given by

$$A_{VH} = \frac{V_o}{V_i} = -\frac{h_{fe}I_b Z}{h_{ie}I_b} = -\frac{h_{fe}Z}{h_{ie}} \qquad ...(viii)$$

Substituting the value of Z, we get

$$A_{VH} = -\frac{h_{fe}}{h_{ie}}\left(\frac{h_{ie}R_L}{h_{ie}+R_L+j\omega C_d h_{ie}R_L}\right)$$

$$= -\frac{h_{fe}R_L}{h_{ie}+R_L+j\omega C_d h_{ie}R_L}$$

$$= -\frac{h_{fe}R_L}{h_{ie}+R_L+j2\pi f C_d h_{ie}R_L} \qquad ...(ix)$$

The magnitude of voltage gain will be

$$|A_{VH}| = \frac{h_{fe}R_L}{\sqrt{(h_{ie}+R_L)^2+(2\pi f C_d h_{ie}R_L)^2}} \qquad ...(x)$$

Thus as the frequency of input voltage increases, the voltage gain decreases.

Upper cut-off Frequency (f_2):

We know, $$A_{VM} = -\frac{h_{fe}R_L}{h_{ie}+R_L} \quad (\because \text{ Eqn. iv of article 30.4})$$

Dividing Eqn. (*ix*) by A_{VM}, we get

$$\frac{A_{VM}}{A_{VM}} = \left(\frac{h_{fe}R_L}{h_{ie}+R_L+j2\pi fC_d h_{ie}R_L}\right)\left(\frac{h_{ie}+R_L}{h_{fe}R_L}\right)$$

$$= \frac{1}{1+j2\pi fC_d\left(\dfrac{h_{ie}R_L}{h_{ie}+R_L}\right)}$$

$$\therefore \quad A_{VH} = \frac{A_{VM}}{1+j2\pi fC_d\left(\dfrac{h_{ie}R_L}{h_{ie}+R_L}\right)}$$

The magnitudes of gains are related by

$$|A_{VH}| = \frac{|A_{VM}|}{\sqrt{1+2\pi fC_d\left(\dfrac{h_{ie}R_L}{h_{ie}+R_L}\right)^2}} \quad ...(xi)$$

Thus, $$|A_{VH}| < |A_{VM}|$$

Now the upper cut-off frequency (f_2) is defined as the frequency at which the magnitude of the voltage gain in the high frequency range falls to $\frac{1}{\sqrt{2}}$ or 0.707 of the magnitude of the voltage gain in the mid frequency range. Thus, $f = f_2$

$$\therefore \quad |A_{VH}| = \frac{|A_{VM}|}{\sqrt{2}} \quad ...(xii)$$

Comparing equations (*xi*) and (*xii*), we get

$$\frac{|A_{VM}|}{\sqrt{2}} = \frac{|A_{VM}|}{\sqrt{1+2\pi f_2 C_d\left(\dfrac{h_{ie}R_L}{h_{ie}+R_L}\right)^2}}$$

This gives $$f_2 = \frac{1}{2\pi C_d}\left(\frac{h_{ie}+R_L}{h_{ie}R_L}\right)$$

or $$f_2 = \frac{1}{2\pi C_d}\left(\frac{1}{h_{ie}}+\frac{1}{R_L}\right) \quad ...(xiii)$$

This Eqn. (*xiii*) gives the upper cut-off frequency (f_2).

SOLVED EXAMPLES

Example 30.1: *A common emitter RC coupled amplifier with a load resistance R_L = 10 kΩ. The h-parameters of the transistor are h_{fe} = 330 and h_{ie} = 4.5 kΩ. Calculate the overall gain for mid frequency range with three such stages are cascaded. Assume the source resistance (R_S) to be negligible compared to h_{ie}.*

Solution: (1) In the mid-frequency range, the voltage gain of the first stage is given by

$$A_{VM_1} = -\frac{h_{fe}R_L}{h_{ie}+R_L} = -\frac{330\times10\times10^3}{4.5\times10^3+10\times10^3} = -227.59$$

(2) The mid-frequency range voltage gain of the second stage is also the same, i.e.

$$A_{VM_2} = -227.59$$

(3) The output voltage of the last i.e. third stage appears across load R_L only. Therefore, the voltage gain of the third stage will be

$$A_{VM_3} = -\frac{h_{fe}R_L}{h_{ie}} \qquad (\because \text{ Eqn. } (iii) \text{ of article 30.2})$$

$$= -\frac{330\times10\times10^3}{4.5\times10^3} = -733.33$$

Therefore, the overall voltage gain is given by

$$A_{VM} = (-227.59)(-227.59)(-733.33)$$

$$= -3.8\times10^7$$

Example 30.2: *A two stage RC coupled CE amplifier uses transistors of the type BC149B of which h-parameters are h_{fe} = 330 and h_{ie} = 4.5 kΩ. If the load resistance R_L = 5.5 kΩ. Find the required value of the coupling capacitor C_C so that power cut-off frequency is 30 Hz.*

Solution: Given: h_{fe} = 330, h_{ie} = 4.5×10³ Ω, R_L = 5.5×10³ Ω, f_1 = 30 Hz. The lower cut-off frequency (f_1) is given by

$$f_1 = \frac{1}{2\pi C_C(h_{ie}+R_L)} \qquad ((\because \text{ Eqn. (viii) of article 30.5}$$

$$\therefore \quad C_C = \frac{1}{2\pi f_l(h_{ie}+R_L)}$$

$$= \frac{1}{2\times3.14\times30\times(4.5+5.5)\times10^3} = 0.0530\times10^{-5}\text{F}$$

$$= 0.53\times10^{-6}\text{ F}$$

$$= 0.53\ \mu\text{F}.$$

Example 30.3: *A two stage common-emitter RC couple amplifier uses transistors of the type BC 149 C of which the h-parameters and internal resistances are*

$$h_{fe} = 600,\ h_{ie} = 10\text{ k}\Omega,\ C_{be} = C_T = 2.5\text{ pF},\ C_{be} = C_D = 9\text{ pF}$$

If the coupling capacitor C_C is 0.5 mF and the load resistance is 10 kΩ. Find lower and upper cut-off frequencies and Band width.

Solution:

(1) The lower cut-off frequency (f_1) is given by

$$f_1 = \frac{1}{2\pi C_C(h_{ie}+R_L)} = \frac{1}{2\times3.14\times0.5\times10^{-6}\times(10+10)\times10^3}$$

$$= 0.16\times10^2 = 16\text{ Hz}$$

(2) The upper cut-off frequency (f_2) is given by

$$f_2 = \frac{1}{2\pi C_d}\times\left(\frac{1}{h_{ie}}+\frac{1}{R_L}\right) \qquad (\because \text{ Eqn. } (xiii) \text{ of article 30.6.2})$$

where

$$C_d = C_D + \left(1+\frac{h_{fe}}{h_{ie}}R_L\right)C_T \qquad (\because \text{ Eqn. } (vi) \text{ of article 30.6.1})$$

$$= 9\times10^{-12} + \left(1+\frac{600}{10\times10^3}\times10\times10^3\right)\times(2.5\times10^{-12})$$

$$= [9 + (1 + 600)2.5] \times 10^{-12}$$

$$= (9 + 1502.5)\times10^{-12} = 1511 \text{ pF}$$

Substituting, we get

$$f_2 = \frac{1}{2\times3.14\times1511\times10^{-12}}\left(\frac{20\times10^3}{10^8}\right) = \frac{20}{9489}\times10^7$$

$$= 21.07\times10^3 \text{ Hz} = 21.07 \text{ kHz}$$

(3) Band width

$$= \text{B. W.} = (f_2 - f_1)$$

$$= (21.07 \text{ kHz} - 16 \text{ Hz})$$

$$= 21.054 \text{ kHz}$$

$$\cong 21 \text{ kHz.}$$

30.7 EMITTER FOLLOWER

Fig. 30.10 shows the circuit diagram of an emitter follower. It differs from the circuitry of conventional amplifiers by the absence of collector load and emitter bypass capacitor. The emitter resistance R_E it self acts as the load and a.c. output voltage is taken across it, using a coupling capacitor C_C. The biasing is generally provided by base resistor method, using R_b, When the base voltage in the circuit increases, the emitter current will increase. Thus, the emitter potential (i.e. P.D. across R_E) follows the base potential.

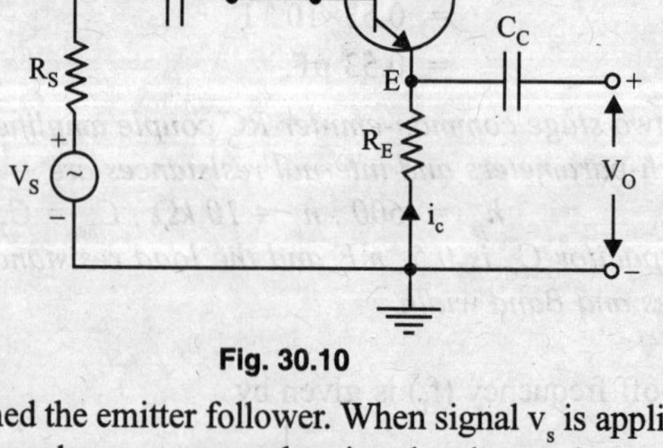

Fig. 30.10

Hence, the circuit is named the emitter follower. When signal v_s is applied, produces an emitter voltage $i_e R_E$ across R_E. This voltage opposes the signal voltage, thus providing negative feed back. Thus, an emitter follower circuit is essentially a voltage series negative feedback amplifier. Basically it is a common collector npn transistor with collector at a.c. ground.

Obviously, it is a current feedback circuit since the voltage feedback is proportional to the emitter current i.e. output current. Its most important characteristic is that it has high input impedance and low output impedance. This make it an ideal circuit for impedance matching.

Circuit Analysis

Current gain (A_{ic}): Considering emitter resistor R_E as load R_L, the general relation for current gain is

$$A_i = -\frac{h_f}{1+h_o R_L} \qquad (\because \text{ Eqn } (i) \text{ of article 30.2})$$

For common collector amplifier, $h_f = h_{fc}$ and h_o and h_{oc}

$$\therefore \quad A_{ic} = -\frac{h_{fe}}{1+h_{oc}R_L}$$

From the conversion table, $h_{fe} = -(1 + h_{fe})$ and $h_{oc} = h_{oe}$

$$\therefore \quad A_{ic} = \frac{1+h_{fe}}{1+h_{oc}R_L} \qquad ...(i)$$

Input Impedance (Z_{ic})

The general relation for input impedance is

$$Z_i = h_i + h_r A_i R_L$$

For a common collector amplifier

$$Z_{ic} = h_{ic} + h_{rc} A_{ic} R_L$$

From the conversion table, $h_{ic} = h_{ie}$ and $h_{rc} = 1$

$$\therefore \quad Z_{ic} = h_{ie} + \frac{(1+h_{fe})R_L}{1+h_{oc}R_L} \qquad ...(ii)$$

Voltage Gain (A_{vc})

The voltage gain is given by

$$A_{vc} = A_{ic}\frac{R_L}{Z_{ic}}$$

The input impedance Z_{ic} is given by

$$Z_{ic} = h_{ie} + \frac{(1+h_{fe})R_L}{1+h_{oc}R_L}$$

Assuming $h_{fe} >> 1$ and $h_{oc} \cong 0$, we get

$$Z_{ic} = h_{ie} + h_{fe}R_L$$

Also

$$A_{ie} = \frac{1+h_{fe}}{1+h_{oe}R_L}$$

Again assuming , $h_{fe} >> 1$ and $h_{oe} \cong 0$, we get

$$A_{ic} = h_{fe}$$

Substituting for A_{ic} and Z_{ic}, we get

$$A_{vc} = \frac{h_{fe}R_L}{h_{ie} + h_{fe}R_L} \qquad ...(iii)$$

Output Impedance (Z_{oc})

The general relation for output impedance is

$$Z_o = \frac{h_f R_S}{h_f h_o + R_S h_o - h_f h_r}$$

For common collector amplifier

$$Z_{oc} = \frac{h_{fc}R_S}{h_{fc}h_{oc} + R_S h_{oc} - h_{fc}h_{rc}}$$

From the conversion table,

$$h_{fc} = -(1 + h_{fe});\ h_{rc} = 1,\ h_{oc} = h_{oe}$$

$$\therefore \qquad Z_{oc} = \frac{R_S - (1+h_{fe})}{h_{oe}R_S - (1+h_{fe})h_{oe} + (1+h_{fe})} \qquad \text{...(iv)}$$

Example 30.4: *A transistor having $h_{ie} = 800\ \Omega$, $h_{fe} = 50$, $h_{oe} = 80\times10^{-6}\ \Omega$ and $h_{re} = 5.5\times10^{-4}$ is used as an emitter follower. If the load resistance is 5 kΩ and the effective source resistance is 500 ohm; calculate the current gain, input impedance, voltage gain, power gain and output impedance.*

Solution:

(1) Current gain,

$$A_{ic} = \frac{1+h_{fe}}{1+h_{oe}R_L}$$

$$= \frac{1+50}{1+(80\times10^{-6}\times5\times10^{3})} = \frac{51}{1+(400\times10^{-3})}$$

$$= \frac{51}{1.4} = 36.42$$

(2) Input impedance,

$$Z_{ic} = h_{ie} + \frac{(1+h_{fe})R_L}{1+h_{oc}R_L}$$

$$= h_{ie} + A_{ic}\cdot R_L$$

$$= 800 + (36.42\times5\times10^3)$$

$$= 800 + 182100 = 182900\ \Omega$$

$$= 182.9\times10^3\Omega = 182.9\ \text{k}\Omega$$

(3) Voltage gain,

$$A_{vc} = \frac{h_{fe}R_L}{h_{ie}+h_{fe}R_L}$$

$$= \frac{50\times5\times10^3}{800+(50\times5\times10^3)} = \frac{250\times10^3}{800+250\times10^3}$$

$$= \frac{250\times10^3}{250.8\times10^3} = 0.99$$

(4) Power gain,

$$A_{pc} = A_{vc}\times A_{ic}$$

$$= (0.99)\times(36.42)$$

$$= 36$$

In decibel,

$$G_p = 10\log_{10} 36 \quad \text{db}$$

$$= 10\times1.5441$$

$$= 15.441 \quad \text{db}$$

(5) Output impedance,

$$Z_{oc} = \frac{R_S - (1+h_{fe})}{h_{oc}R_S - (1+h_{fe})h_{oe} + (1+h_{fe})}$$

$$= \frac{500-(1+50)}{(500\times80\times10^{-6}) - (1+50)80\times10^{-6} + (1+50)}$$

$$= \frac{449}{51.044} = 8.797\ \Omega$$

30.8 JFET AS AMPLIFIER

A JFET can be used as an amplifier in two different configurations :

(1) Common source (CS) : This JFET configuration is similar to common emitter (CE) configuration of a bipolar transistor.

(2) Common drain (CD) : This JFET configuration is similar to common collector (CC) (i.e. emitter follower) configuration of a bipolar transistor.

JFET as a Common Source Amplifier

Fig. 30.11 shows a circuit diagram of JFET as a common source amplifier. The JFET is self-biased by using a biasing network R_S–C_S. The d.c. component of the drain current (I_D) flowing through the source-biasing resistor R_S produces the desired bias voltage.

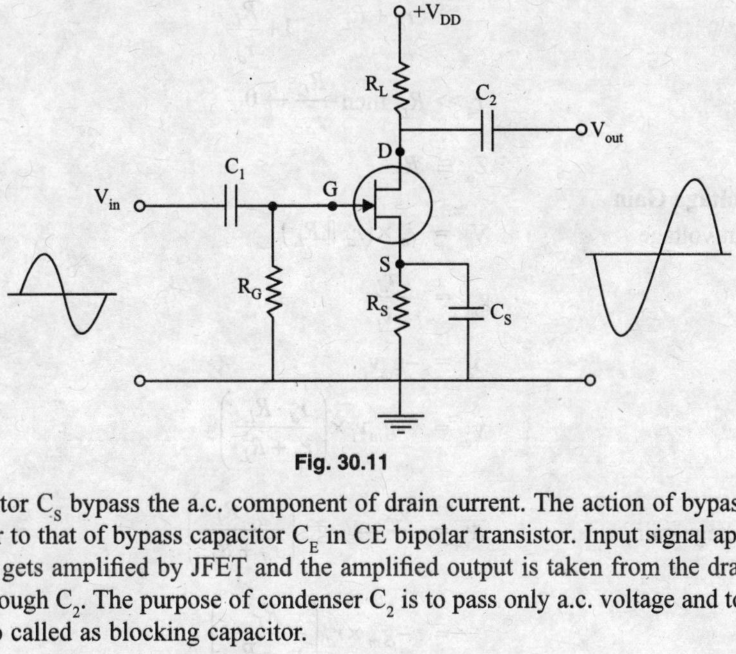

Fig. 30.11

The capacitor C_S bypass the a.c. component of drain current. The action of bypass capacitor C_S is just similar to that of bypass capacitor C_E in CE bipolar transistor. Input signal applied at the input terminals, gets amplified by JFET and the amplified output is taken from the drain of JFET after passing through C_2. The purpose of condenser C_2 is to pass only a.c. voltage and to block d.c. Hence C_2 is also called as blocking capacitor.

Working

When negative going signal is applied to the input then

(*i*) gate bias is increased

(*ii*) depletion regions are widened

(*iii*) channel resistance is increased

(*iv*) drain current I_D is decreased

(*v*) voltage drop across R_L is declared

(*vi*) Consequently, the voltage at drain V_D is increased. In other words, a positive going signal becomes available at the output through the point D in Fig. 30.11

When a positive going signal is applied at the input, then in a similar way, negative going signal is available at the output. Thus, there is a phase inversion of 180° between the input signal applied at the gate and the output signal obtained at the drain.

CIRCUIT ANALYSIS

(1) Input Impedance

$$Z_{in} = R_G \parallel R_{GS}$$

$$= \frac{R_G R_{GS}}{R_G + R_{GS}} = \frac{R_G}{1+\dfrac{R_G}{R_{GS}}}$$

In an ideal JFET, R_{GS} is infinite because $I_G = 0$. In practice, however, R_{GS} in not actually infinite but extremely high as compared to R_G

$\therefore$ $$Z_{in} \cong R_G \qquad ...(i)$$

(2) Output Impedance

$$Z_o = r_d \parallel R_L$$

$$= \frac{r_d \cdot R_L}{r_d + R_L} = \frac{R_L}{1+\dfrac{R_L}{r_d}}$$

If $r_d >> R_L$, then $\dfrac{R_L}{r_d} \to 0$

$\therefore$ $$Z_o \cong R_L \qquad ...(ii)$$

(3) Voltage Gain

Output voltage, $V_o = i_d \times (r_d \parallel R_L)$

But $g_m = -\dfrac{i_d}{v_i}$

$$i_d = -g_m v_i$$

$\therefore$ $$v_o = -g_m v_i \times \left(\frac{r_d \cdot R_L}{r_d + R_L}\right)$$

$\therefore$ Voltage gain, $$A_v = \frac{v_o}{v_i} = -g_m\left(\frac{r_d \cdot R_L}{r_d + R_L}\right)$$

$$= -g_m \times r_d\left(\frac{R_L}{r_d + R_L}\right) \qquad ...(iii)$$

$$A_v = \frac{-\mu R_L}{r_d + R_L} \qquad ...(iv)$$

Rearranging Eqn. (iii), $A_v = -g_m\left(\dfrac{R_L}{1+R_L/r_d}\right)$

Since $r_d >> R_L$, we have

$$A_v = -g_m R_L \qquad ...(v)$$

Example 30.5. *For a common-source JFET voltage amplifier $g_m = 2$ ms and $R_L = 10\ k\Omega$. Find its voltage amplification assuming $r_d > R_L$.*

Solution : Voltage gain, $A_v = -g_m R_L$

$$= 2\times10^{-3}\times10\times10^3$$

$$= 20$$

Example 30.6: *For a JFET amplifier circuit shown in Fig. 30.12, calculate the approximate values of (i) A_v and (ii) output voltage v_o.*

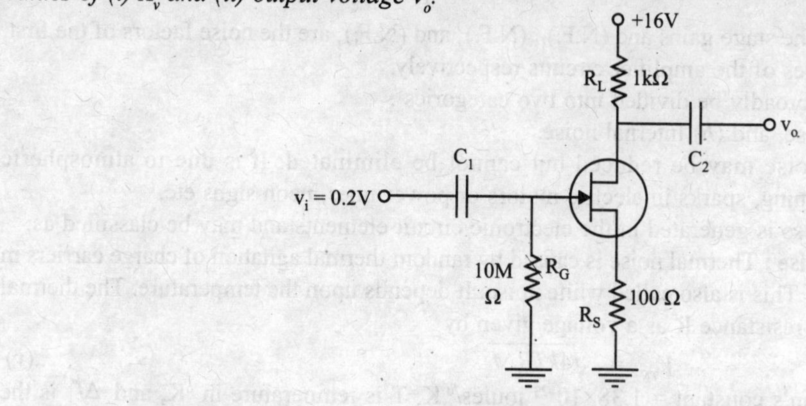

Fig. 30.12

Solution: An approximate expression for voltage gain of an amplifier (assuming class - A operation) is

(*i*)
$$A_v = \frac{1}{2}\left(\frac{R_L}{R_S}\right) = \frac{1}{2} \times \frac{1000}{100} = 5$$

(*ii*)
$$v_o = A_v \times v_i = 5 \times 0.2 = 1.0\text{V}$$

30.9 NOISE IN ELECTRONIC CIRCUITS

Any spurious unwanted signal in an amplifier which is not derived from or related to the input signal is called noise. Just as distortion is the limiting factor in the amplification of large signals, noise is the limiting factor in the case of small signals. When a very weak (radio or TV) signal is applied at the input stage of an amplifier, it is generally not possible to distinguish the output signal from the noise. Hence, noise places a basic lower limit on the signal levels at which a device or an amplifier may operate. In radio reception, noise is noticed as hiss and in television as snow. Noise has no specific frequency and is distributed over broad frequency bands. It is said to be a **white noise** if distributed uniformly over all frequencies.

The signal-to-noise (S/N) ratio should be very high for good signal intelligibility. Since all amplifiers contribute some noise of their own (in addition to external noise picked up) to the signal being processed, their output S/N is bound to be less than their input S/N. High quality amplifiers are designed to have an output S/N as close to input S/N as possible. The amplifier performance is measured by the term "Noise factor" (N.F.) which is defined as

$$\text{Noise Factor N.F.} = \frac{input\ S/N}{output\ S/N}$$

$$= \frac{S_i / N_i}{S_o / N_o} \qquad ...(i)$$

Obviously N.F. is greater than unity.

When expressed in decibels, the noise factor is called "Noise Figure" denoted by F.

$$\therefore \quad F = 10\log_{10} \text{N.F.} \quad \text{db} \qquad ...(ii)$$

Noise in cascaded amplifiers is additive. For the two-stage cascaded amplifier

$$(\text{N.F})_{1,2} = (N.F)_1 + \frac{(N.F)_2 - 1}{A_1} \qquad ...(iii)$$

and for the three-stage cascaded amplifier

$$(\text{N.F.})_{1,2,3} = (N.F.)_1 + \frac{(N.F.)_2 - 1}{A_1} + \frac{(N.F.)_3 - 1}{A_1 A_2} \quad ...(iv)$$

where A_1 and A_2 are the stage gains and $(\text{N.F.})_1$, $(\text{N.F.})_2$ and $(\text{N.F.})_3$ are the noise factors of the first, second and third-stages of the amplifier circuits respectively.

The noise may broadly be divided into two categories :

(*a*) External noise, and (*b*) Internal noise.

The external noise may be reduced but cannot be eliminated. It is due to atmospheric disturbance like lightning, sparks in electric motors or power lines, neon signs etc.

The internal noise is generated in the electronic circuit elements and may be classified as:

(1) Thermal noise : Thermal noise is caused by random thermal agitation of charge carriers in an electric conductor. This is also called white noise. It depends upon the temperature. The thermal noise is generated in resistance R as a voltage given by

$$V_{TN} = \sqrt{4kTR\Delta f} \quad ...(v)$$

where k is Boltzmann's constant = 1.38×10^{-23} joules/°K, T is temperature in °K, and Δf is the band width in Hz over which the power is measured.

To reduce thermal noise, the operating temperature should be lowered and the band width should be made as small as possible.

(2) Shot noise : The short noise arises due to statistical fluctuations of charge carries through the active devices like electrons emitted from cathode in a vacuum tube. This noise may be reduced if the tube operates under 'space charge limited conditions'. Short noise is uniformly distributed throughout the useful frequency spectrum. The r.m.s. shot noise current I_n derived by Schottky for vacuum diode is given by

$$I_n^2 = 2qI\Delta f \quad ...(vi)$$

where q = charge of electron in coulombs

I = d.c. current through diode in amperes

Δf = band width in Hz.

In bipolar junction transistor, shot noise is due to (i) random motion of the charge carriers (either electrons or holes) when they diffuse across the junctions and (ii) recombination of the carriers in base region.

(3) Flicker noise: The flicker noise is present typically in semiconductor devices Its intensity is inversely proportional to the frequency and hence also called as 1/f noise. It has been observed that this noise becomes negligible above 1kHz. Although, the cause is not completely understood, it is assumed that it is due to the irregular distribution and flow of charge carriers on the cystal surface, in the vicinity of the depletion region. This noise is completely absent in metals. Hence, wire-wound resistors are preffered to carbon resistors, in which flicker noise is minimum.

(4) Gas noise: Gas noise is produced by the random current disturbances, when the positive ions produced due to the ionization of the remaining gas in the tube penetrate the space charge. It is not applicable above 10 MHz.

(5) Partition noise: Partition noise is caused due to the irregular division of current between the various electrodes. Thus, the noise is greater in pentode. Moreover, in BJT, the emitter current is devided into the collector and base current ($I_E = I_B + I_C$).

As low frequencies, the mean square value of partition noise current i_c is given by

$$i_c^2 = 2qI_C(1-\alpha)\Delta f$$

where q = charge of an electron in coulomb,

I_C = d.c. collector current,

α = current gain in common base mode,

Δf = band width in Hz.

(6) Contact noise: Contact noise is developed by poor or intermittant electric contact, such as from dirty or damaged switch contacts, terminals, leaky capacitors, faulty resistances etc.

(7) Microphone noise: The noise is produced due to mechanical vibrations of parts of circuit. In case of valves, when the valve is struck, its electrons vibrate, their relative distance change and the anode current varies which produces noise.

(8) Hum: Hum is the extraneous output voltage produced by the associated power circuits. It is produced (i) due to a.c. component present in the power supply voltage. Larger the a.c. component i.e. ripple present, more is the hum at the output (ii) by the pickup due to stray electrostatic and magnetic fields produced by adjacent power leads (iii) through the inter-electrode capacitance between cathode and grid when directly heated filament operates with a.c.

Example 30.7: *The signal input to small signal amplifier consists of 50 μW of signal power and 0.5 μW of noise power. The amplifier generates an internal noise power of 50 μW and has a gain of 20 db. Calculate (a) input S/N (b) output S/N (c) noise factor and (d) noise figure.*

Solution:

(*a*) Input S/N ratio $= \dfrac{S_i}{N_i} = \dfrac{50}{0.5} = 100$

(*b*) Output S/N ratio, $10\log_{10} G = 20$ db

$$\log_{10} G = 2 \text{ or } G = 10^2 = 100$$

$$S_o = GS_i = 100 \times 50 = 5000\ \mu W$$

$$N_o = GN_i + N_A = 100 \times 0.5 + 50 = 100\ \mu W$$

$$\therefore \quad \frac{S_o}{N_o} = \frac{5000}{100} = 50$$

(*c*) Noise factor, $\text{N.F.} = \dfrac{S_i / N_i}{S_0 / N_o} = \dfrac{100}{50} = 2$

(*d*) Noise figure, $F = 10\log_{10} 2$ db $= 10 \times 0.3 = 3$ db

EXERCISE

1. Derive the two equations for the hybrid equivalent circuit of common emitter transistor amplifier.
2. Derive expressions for current gain, input impedance, voltage gain, output impedance, and power gain of a single stage common-emitter transistor amplifier.
3. Prove that the voltage gain for a common-emitter transistor amplifier is

   $$A_{ve} = -\frac{h_{fe} R_L}{h_{ie} + R_L \Delta h}$$

4. Prove that the current gain for a common-emitter transistor amplifier is

   $$A_{ie} = -\frac{h_{fe}}{1 + h_{oe} R_L}$$

5. Show that input impedance of an common-emitter transistor amplifier is

   $Z_{ie} = h_{ie} - h_{re} R_L A_{ie}$

6. Draw and explain the general a.c. equivalent circuit diagram of two-stage RC coupled transistor amplifier. ***(Nagpur U. 2011)***
7. Derive an expression for the voltage amplification (gain) for RC coupled transistor amplifier for

   (*a*) Mid frequency range. ***(Nagpur U. 2011)***

   (*b*) Low frequency range

   (*c*) High frequency range.

8. What is Miller effect ? How does it helps in understanding the high frequency response of an amplifier ?
9. Describe the effect of shunt capacitors in an amplifier circuit at
   (*a*) Mid frequency range (*b*) Low frequency range
   (*c*) High frequency range
10. Describe the a.c. characteristics of an RC coupled amplifier at low frequency range. What factors determine the lower cut-off frequency?
11. Explain the reasons for the reduction in the gain of an RC coupled amplifier at high frequencies. What is upper cut-off frequency ? Derive the relation for f_h.
12. Calculate the lower cut-off frequency of an RC coupled amplifier. Given: input impedance of the amplifier is 2K ohm, load resistance 1K ohm and the capacity of the coupling capacitor is 4.7 μF. **[Ans : f_l = 11.3 Hz]**
13. A two stage common-emitter RC coupled amplifier uses transistor of which the h-parameters and the internal capacitances are : h_{ie} = 1.5 kΩ, h_{fe} = 100 , C_{bc} = 4 pF and C_{be} = 25 pF. If the coupling capacitance is 0.5 μF and the load resistance R_L = 2.5 kΩ; calculate the lower and upper cut-off frequencies. **[Ans : 7936 Hz , 91.4 kHz]**
14. Draw the circuit diagram of an emitter-follower. Why it is called emitter follower? Explain. What are the most important characteristics of an emitter follower ? Can we use it for impedance matching? ***(Nagpur U. 2012)***
15. Derive expressions for current again, voltage gain, input impedance, output impedance, for an emitter follower.
16. Explain JFET as a common source amplifier with a neat circuit diagram. Discuss its working.
17. Derive expressions for input impedance, output impedance, voltage gain, for a common-sources JFET amplifier.
18. What is meant by noise? Define ***(Nagpur U. 2009)***
   (*a*) Noise factor (*b*) Noise figure.
19. What are different sources of noise? Write in brief:
   (*a*) Thermal noise ***(Nagpur U. 2012)***
   (*b*) Shot noise ***(Nagpur U. 2012)***
   (*c*) Flicker noise (*d*) Partition noise
   (*e*) Hum.
20. An RC coupled transistor amplifier has mid-frequency gain A_{VM} = 100. The values of the lower and upper cut-off frequencies are f_1 = 20 Hz and f_2 = 50 kHz. Find the frequencies at which the gain is reduced to 80%. **[Ans : 26.66 Hz]**
21. Find the h-parameters for the circuit as shown in Fig. 30.13 ***(Nagpur U. 2009, 2011)***

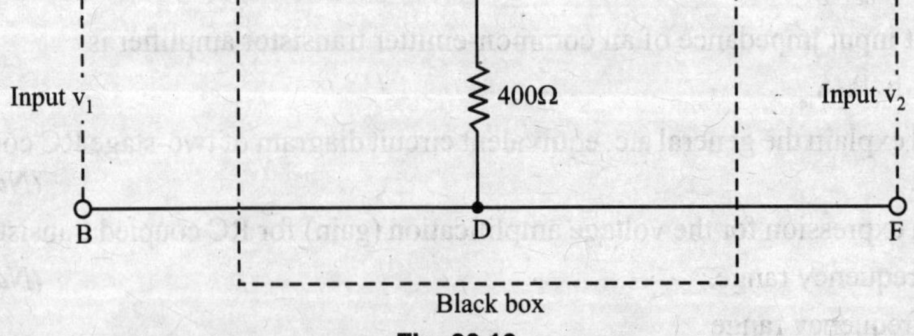

Fig. 30.13

[Ans : h_{11} = 371.4 Ω, h_{12} = 0.57, h_{21} = –0.57, h_{22} = 0.0014 mho]

Index

A

Absorption of X-rays, 410
Anti-neutrino, 548
Atomic energy, 592
Atomic mass unit (a.m.u.), 512
Auger effect, 409
Azimuthal quantum number, 275

B

Band width, 913
Bethe- Bloch formula, 464
Binding energy, 177, 514
BJT, 830
Black body radiation, 92
Black holes, 2
Bloch theorem, 774
Bohr's theory
  Merits, 114
  Limitations, 114
Born's interpretation, 195
Bosons, 362
Breeder reactor, 592
Bremsstrahlung, 395
Bridge rectifier, 844
Bubble chamber, 490

C

Carbon cycle, 595
Cascaded amplifier, 909
Chain reaction, 589
Complementarity principle, 181
Compound nucleus, 577
Compton effect, 118
Conduction band, 784
Conductors, 785
Conservation of angular momentum, 7, 166
Conservation of energy, 9
Conservation of linear momentum, 6, 17
Coupling network, 910
Coupling of LS and JJ, 366
Critical potentials, 307

D

Davisson and Germer's experiment, 149
de-Broglie hypothesis, 132
Debye model, 738, 744
Debye temperature, 745
Degeneracy, 288
Deutron, 517
Diatomic lattice, 726
Direct nuclear reaction, 576
Discovery of spin, 312
Dispersion, 724
Distortion in amplifier, 914
Doping, 805
Dulong and Pettit's law, 733

E

Eigen function, 188
Eigen value, 188
Einstein temperature, 737
Einstein's mass- energy relationship, 69
Einstein's photoelectric equation, 103
Electrical conductivity, 751
Electron microscope, 180
Electronic stabilizer, 849
Emission spectrum of hydrogen, 112
Endo-ergic nuclear reactions, 571
Energy levels, 214
Energy-momentum relation, 78
Event, 4
Excitation potentials, 307
Exo-ergic nuclear reactions, 571
Expectation value, 204
Extrinsic semiconductor, 812

F

Fermi-Dirac statistics, 766
Fermi gas, 749
Fermi temperature, 766
Fermi velocity, 764
Fermions, 362
Filters, 845
Forbidden band, 784
Frank-Hertz experiment, 307

Free and localized particle, 155
Full-wave rectifier, 840

G

G.P. Thomson's experiment, 151
Galilean transformations, 10, 13
Gamow's theory, 534
Geiger-Mueller counter, 476
Geiger-Nuttal law, 530
Group velocity, 145
Gyromagnetic ratio, 314

H

Half-wave rectifier, 837
Hall coefficient, 811
Hall effect, 809
Hamiltonian operator, 188
Harmonic oscillator, 179, 249, 256
Heisenberg's uncertainty principle, 164
Homogeneity of space, 3
Homogeneity of time, 8
Hybrid (h)parameters, 871, 919
Hydrogen atom
  Balmer lines, 338
  Bohr's model, 300
  Bohr's theory, 109
  fine structure, 324
  hyperfine structure, 333

I

Inertial frame of reference, 4
Insulators, 785
Intrinsic conductivity (σ), 804
Intrinsic density (n_i), 803
Intrinsic semiconductor, 796, 801
Invariance of space, 5, 7
Ionization chamber, 470
Ionization potentials, 307
Isotropic space, 3

J

Junction diode, 815
Junction Field Effect Transistor (JFET), 81, 937

K

Kronig Penney model, 777

L

Lande's factor, 329
Larmour precession, 311
Lattice vibrations, 718
Lattice waves, 722
Length contraction, 36
Light emitting diode (LED), 823
Liquid- drop model, 552
Load line, 868
Lorentz transformations, 29
Luminescence, 435

M

Magnetic moment, 310
Magnetic moment of an atom, 506
Mass defect, 512
Matter waves, 149
Mean free path (λ), 750
Michelson -Morley experiment, 20
Minkowski space, 81
Mobility, 812
Molecular spectra, 419
Momentum- kinetic energy relation, 79
Moseley's law, 400
MOSFET, 892
Multistage amplifier, 908

N

Neil- Bohr's formula, 461
Neutrino, 547
Neutrons with matter, 467
Noise, 939
Non-inertial frame of reference, 4
NOR and NAND gates, 900
Normalization factor, 198
n-p-n transistor, 830
Nuclear electric quadrupole moment (Q), 509
Nuclear fission, 586
Nuclear forces, 521
Nuclear fusion, 593
Nuclear magic numbers, 563
Nuclear magnetic dipole moment (μ), 507

Nuclear reaction, 566
Nuclear reactors, 590
Nuclear stability, 526

O

Observer, 4
Old quantum theory, 126
One dimensional box, 210
Operator, 202
Orbital quantum number, 276
Orthogonal wave functions, 223

P

Planck's quantum theory, 94
Planck's radiation law, 96
Photoelectric effect, 98
Particle velocity, 143
Phase velocity, 143
Photon rest mass, 148
Probability density, 197
Particle with zero energy, 222
Parity, 223
Potential step, 232
Potential barrier, 242
Principal quantum number, 279
Precision, 289
Proportional counter, 473
Packing fraction, 513
Phonon, 718
  Inelastic scattering, 720
  Photons and phonons, Comparison, 721

Q

Q point, 868
Quantum mechanics, 206
Quantum model of hydrogen atom, 300

R

Radiations α, β and γ, 527, 543
Radioactive hazards, 594
Radioactive traces, 594
Raman effect, 440
Raman scattering, 448
Raman shift, 444
Raman spectroscopy, 449
Range of a charged particle, 465, 542
Rayleigh scattering, 448
RC coupled amplifier, 911
Relativity of mass, 65
Ritz combination principle, 117
Rotational spectra, 421
Rotational vibrational spectra, 431

S

Schottky diode, 825
Schrödinger's wave equation, 185, 189
Scintillation counter, 482
Semiconductors, 785
Shell model, 556
Single and triplet states, 362
Solar cell, 826
Sommerfeld theory of metals, 754
Sommerfield's theory, 334
Space quantization, 313
Spark chamber, 491
Special theory of relativity, 29
Specific heat, 732
Spectra of alkalie elements, 379
Spectra of alkaline earth elements, 381
Spectrophotometry, 453
Spherical harmonics, 277
Spin angular momentum, 313
Spin-orbit coupling, 316
Spin-orbit interaction, 321
Stabilizing circuits, 877
Stability factor, 877
Stability of atom, 108
Stabilization, 876
Steady state, 197
Stern- Gerlach experiment, 339
Straggling of a charged particle, 465
Structure of the nucleus, 495
Superposition of states, 201

T

Tachyons, 82
Thermal conductivity, 752
Thermal runaway, 877
Time dilation, 43
Transistor
  CB mode, 855
  CC mode, 862
  CE mode, 859
  n-p-n transistor, 830

p-n-p transistor, 829
Transistor as amplifier, 865
Tunnel diode, 825
Tunnel effect, 247, 533
Twin paradox, 53

U

Umklapp process, 719

V

Valence band, 784
Vector model of atom, 319
Vibrational spectra, 427

W

Wave packet, 156
Wave functions: symetric and anti-symmetric, 360
Wiedemann Franz law, 753
Wilson cloud chamber, 488

X

X-rays
Characteristics, 391, 394
Production, 386
Soft and hard, 391
Uses, 388

Z

Zeeman effect, 345, 349
Zener diode, 823, 849
Zero point energy, 177